ELECTRONIC DEVICES

DISCRETE AND INTEGRATED

Stephen R. Fleeman

Associate Professor
Rock Valley College
Rockford, Illinois

PRENTICE HALL,
Englewood Cliffs, New Jersey 07632

Library of Congress Cataloging-in-Publication Data

Fleeman, Stephen.
Electronic devices : discrete and integrated / Stephen Fleeman.
p. cm.
ISBN 0-13-338120-X
1. Transistors. 2. Operational amplifiers. 3. Analog electronic systems. I. Title.
TK7871.9.F58 1990
621.381—dc20 89-28148
CIP

Editorial/production supervision: *Eileen M. O'Sullivan*
Interior design: *Lorraine Mullaney*
Cover design: *Lorraine Mullaney*
Cover photograph: © *Erik Leigh Simmons/The Image Bank*
Manufacturing buyer: *David Dickey*

Printed in the United States of America

10 9 8 7 6 5 4 3 2 1

ISBN 0-13-338120-X

Prentice-Hall International (UK) Limited, *London*
Prentice-Hall of Australia Pty. Limited, *Sydney*
Prentice-Hall of Canada Inc., *Toronto*
Prentice-Hall Hispanoamericana, S.A., *Mexico*
Prentice-Hall of India Private Limited, *New Delhi*
Prentice-Hall of Japan, Inc., *Tokyo*
Prentice-Hall of Southeast Asia Pte. Ltd., *Singapore*
Editora Prentice-Hall do Brasil, Ltda., *Rio de Janeiro*

Contents

4 Bipolar Junction Transistors 116

5 Field-Effect Transistors 164

7 Amplifier Models for Discrete and Integrated Circuits 262

8 Discrete and Integrated Amplifier Configurations 292

9 Cascaded Amplifiers and Frequency Response 360

11 Additional Negative Feedback Concepts and Applications 501

12 Discrete and Integrated Power Amplifiers 529

Appendices

PREFACE

Technological advancement is bringing to reality practices and products that were once considered to be topics of science fiction. Machines that see, speak, and hear are currently available. Artificial intelligence is appearing routinely as one of the featured topics in trade publications, journals, and professional conferences. Truly, "the robots are among us."

The impetus behind much of this advancement rests with development of the 16- and 32-bit microprocessors. However, the realm of analog electronics is an equal partner in these technological achievements. The demands on analog devices, circuits, and systems are growing steadfastly. Improved efficiency, greater speed, and enhanced precision are among the mandates placed on analog electronics.

To meet these evolving technical challenges, the electronics student must clearly have a firm command of analog, digital, microprocessor, and control system fundamentals. *Electronic Devices: Discrete and Integrated* is targeted toward addressing the student's analog electronics needs.

Approach and Philosophy

Electronic Devices: Discrete and Integrated provides an in-depth explanation of essential topics in contrast to the superficial attention offered by many other textbooks. The material takes the reader from an elementary understanding of a topic to the point where the reader can analyze *practical* circuits and possibly *design* his or her own circuits. The reader does not necessarily have to go as deeply into a topic as it is presented; however, the material is there for future reference.

The textbook assumes a prerequisite knowledge of dc circuit theory and that the student has either taken, or is currently enrolled in, an ac circuits class. The textbook also assumes that the student is reasonably proficient in algebra and knows some basic plane trigonometry.

Electronic Devices: Discrete and Integrated is comprehensive in that it gives the student background information to understand how, why, and when various techniques should be used. The intent is to give the student a *working* knowledge rather than cursory exposure. The types of circuits found in manufacturers' application notes have been included routinely. Negative feedback is also used extensively since it is not an isolated topic but an integral component of "real" electronics circuits and systems.

The textbook holds a rather unique class-tested perspective of treating bipolar junction transistors (BJTs), field-effect transistors (FETs), and op amps in a parallel rather than serial fashion. This emphasizes their respective commonalities and differences. "Real" electronics circuits freely mix these three "active devices" to provide an optimal design. *Electronic Devices: Discrete and Integrated* tends to promote this concept.

Chapter Organization and Content

The first four chapters follow the "classical" sequence: solid-state physics, the p-n junction and diodes, applying diode dc and ac models, and the bipolar junction transistor. The hybrid-pi model for small-signal BJT analysis is developed.

The unique sequence begins with the fifth chapter, which introduces the FET. Once again, the hybrid-pi model is described to characterize the small-signal performance. This allows us to use the *same* model for both the BJT and the FET. The cumbersome (and often obscure) two-port parameters are avoided.

Chapter 6 deals with dc biasing. BJT and FET biasing circuits are discussed side by side. This emphasizes to the student that the same problems and basic solutions exist in BJT and FET biasing circuits. Graphical (bias line) techniques are used for FET circuits. Experience dictates that this tends to promote a more intuitive understanding of FET biasing. The biasing of discrete and integrated differential amplifier pairs is also covered along with the fundamental biasing requirements of the integrated-circuit operational amplifier (IC op amp).

Chapter 7 introduces basic amplifier models. The intent is to emphasize that the fundamental parameters such as input and output resistances, voltage and current gains, and loading effects are necessary to characterize *any* amplifier system. The "host device" (BJT, FET, or op amp) does not alter the basic questions. Decibels and cascaded amplifier systems are also introduced.

Discrete and integrated amplifiers are covered in Chapter 8. Voltage amplifiers are grouped into four basic categories: inverting, noninverting, followers, and differential. The discussion parallels the BJT, FET, and IC op amp.

Chapter 9 leads the student through cascaded amplifier systems and frequency response. The hybrid-pi model is now used to describe the BJT, the FET, and even the IC op amp. The Bode plot is used extensively to predict the amplitude and phase responses.

Chapters 10 and 11 deal with negative feedback. Chapter 10 uses voltage-series negative feedback to illustrate the effects of negative feedback on gain, input and output resistances, frequency response, and nonlinear distortion. The nemisis of all feedback systems—the oscillation problem—is also discussed, and its solution—frequency compensation—is considered. These "advanced" topics are usually left for later course work. The inclusion of these topics is made relatively painless by drawing on Bode plots and phase-margin methods. Chapter 11 continues the coverage with cascaded feedback amplifier strategies, *active* negative feedback, and the three other "classical" topologies (current-series, voltage-shunt, and current-shunt).

Power amplifiers are covered in depth in Chapter 12. Placing the negative feedback discussions before the topic of power amplifiers allows us to delve into "real" power amplifier systems. Negative feedback is critical to their performance. Problems such as slew rate, transient intermodulation (TIM), efficiency, drive requirements, device-safe-operating area (SOA), and *reactive* loads are also considered. These topics are usually avoided in fundamental texts, but are essential if a working knowledge is to be achieved.

Chapter 13 introduces dc power supplies, rectification, and filtering. Voltage regulation systems are analyzed in Chapter 14. Working application circuits are discussed and real-world design considerations are also included. Integrated circuits in conjunction with their discrete partners are considered. Ac power control is also included. The SCR and triac are introduced as ac switches. Trigger devices and optocouplers are also explained.

The text concludes with discrete and integrated oscillator circuits (Chapter 15). Each of the circuits analyzed has been lab-tested, and complete component information is included. Harmonic (sinusoidal) and relaxation oscillators are explored thoroughly. Particular attention has been given to crystal (Pierce) oscillators. The crystal oscillator is extremely popular, yet most fundamental texts merely hint at its operation.

Electronic Devices: Discrete and Integrated is comprehensive in its coverage and includes material to form the basis of the first two (semester) introductory analog electronics courses. If the instructor covers the material in complete detail, the textbook can be used in a three-course sequence.

For courses that ''highlight'' the key concepts to form a two-course sequence, Chapters 1 through 9 could appear in the first course and Chapters 10 through 15 could be covered in the second course. A three-course sequence that covers the topics in complete detail could include Chapters 1 through 7 in the first course, Chapters 8 through 11 in the second course, and Chapters 12 through 15 in the third course.

Motivational Features

The learning aides and motivational enhancements features in this book are:

- Each chapter begins with a list of objectives specifying the student's goals.
- A two-color format has been incorporated to emphasize important concepts and illustrations.
- Numerous narrated examples. (Beginning electronics students need all the help they can get to grasp fundamental concepts.)
- Each chapter concludes with a problem set. The problems have been categorized as (1) drill, derivations, and definitions; (2) design; (3) troubleshooting and failure modes; and (4) computer. The computer problems request BASIC programming solutions. (A supplement IBM-compatible disk contains possible solutions.) The computer problems can be included or disregarded at the instructor's discretion.
- Manufacturer data sheets are incorporated directly within the chapters where discussed.

Available Supplements

A coordinated package of supplements is available to support this text:

- A laboratory manual, *Experiments to Acccompany Electronic Devices: Discrete and Integrated*, developed by Linden M. Griesbach and Stephen R. Fleeman, which contains 40 coordinated and class-tested laboratory experiments.
- Coordinated IBM-compatible software.
- An Instructor's Transparency Masters Manual, which contains enlarged versions of over 100 textbook illustrations, is provided free upon adoption.
- A Test Item File, authored by Jack Hall, which contains an abundance of drill problems, is also provided free upon adoption. Also free upon adoption, you may receive this test item file on an IBM disk along with the Prentice Hall Test Generator.
- An Instructor's Resource Manual, authored by Stephen R. Fleeman and Linden M. Griesbach, which contains the detailed solutions to many of the chapter problems, and nominal solutions to the laboratory experiments. Again, this is provided free upon adoption.

Acknowledgments

My deepest gratitude is extended to the many careful reviewers, most notably Professors John Starr of the Milwaukee School of Engineering and Alan Czarapata, Montgomery College, The Department of Applied Technologies. Among the many others who also provided extremely useful critiques are:

Phillip D. Anderson, Muskegon Community College
Lucias B. Day, Metropolitan State College

Kenneth Dunn, Pensacola Junior College
Arch F. Dye, Mountain View College
Theodore E. Fahlsing, Purdue University
Ralph Frank, Hudson Valley Community College
Doug Fuller, Humber College of Applied Arts and Technology
Trevor Glave, British Columbia Institute of Technology
Jack Hall, Illinois Central College
Stephen C. Harsany, Mount San Antonio College
Arnold Jacobs, DeVry Institute of Technology
Bradley E. Jenkins, St. Petersburg Junior College
Vincent Loizza, DeVry Institute of Technology
Maurice J. Nadeau, Central Maine Vocational Institute
Joe M. O'Connell, DeVry Institute of Technology
Jim Pannell, DeVry Institute of Technology
Joseph Phillips, Jr., RETS Electronics Institute
Probhat K. Rastogi, Case Western Reserve University
William R. Srp, previously from DeVry Institute of Technology

I also owe a debt of thanks to my collegues at Sundstrand Advanced Technology Group, particularly Barry Mendeloff and Marvin Proctor, and Rock Valley College, notably John Banaszak, for their constant encouragement. My patient editor, Alice Barr, my production editor, Eileen O'Sullivan, the designer, Lorraine Mullaney, the editorial assistant Bunnie Neuman, and the rest of the staff at Prentice Hall are also to be commended for their help in making this textbook a reality. Last, but not least, my patient wife, Barbara, and my three sons, Kirt, Brennan, and Blake, whose unending sacrifices allowed me to complete this formidable task.

Stephen R. Fleeman

1

SOLID-STATE PHYSICS

After Studying Chapter 1, You Should Be Able to:

- Describe the differences between analog and digital signals.
- Explain the relationship between the orbital radius of an electron and its energy level.
- Define the electron volt (eV).
- Explain the formation of energy bands in a solid and how energy bandgaps can be used to distinguish between conductors, semiconductors, and insulators.
- Describe the difference between hole and electron movement in an intrinsic semiconductor.
- Explain the formation of n- and p-type semiconductors.
- Name the majority and minority carriers in n- and p-type semiconductors.

1-1 Analog versus Digital Electronics

Two broad classifications of electronic circuits and systems are *analog* and *digital*. The distinction between these two terms rests primarily on the voltage waveforms on which they operate (see Fig. 1-1).

An analog waveform (e.g., voltage plotted as a function of time) is a waveform that can have an *infinite* number of values between two finite limits. In general, analog waveforms can be either continuous or discontinuous [see Fig. 1-1(a) and (b)].

Digital waveforms are discontinuous in that they can only assume a finite number of values between two finite limits. For example, a binary digital waveform can have only two discrete voltage levels (HI and LO), and makes abrupt level changes [see Fig. 1-1(c)].

FIGURE 1-1 **(a) Continuous analog waveform; (b) discontinuous analog waveform; (c) binary digital waveform; (d) multiple-level digital signal approximation of an analog waveform.**

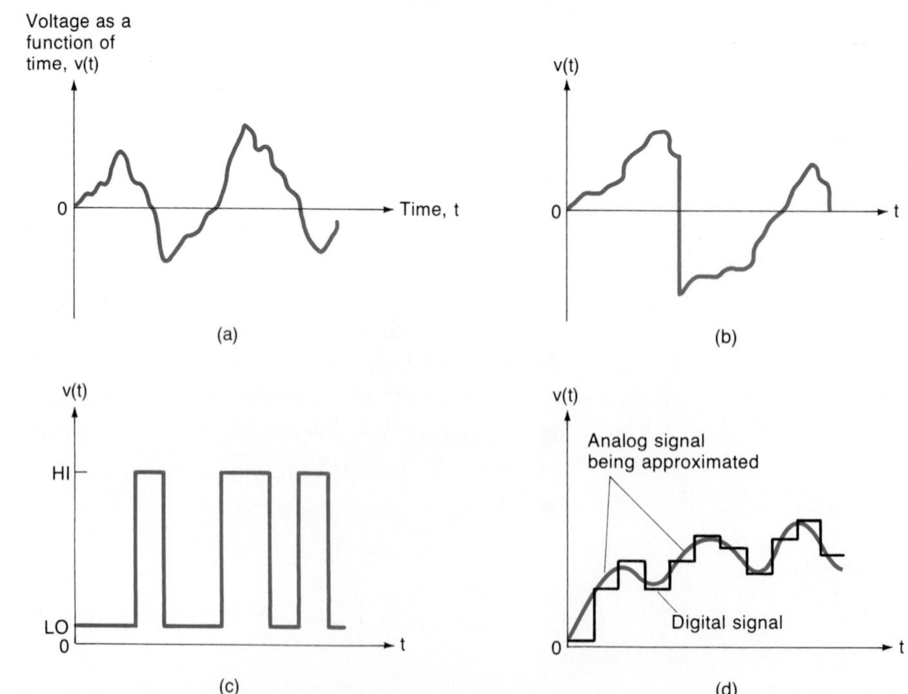

In electronic *signal processing* both analog and/or digital techniques may be used. In this sense an analog waveform (signal) may be an *exact* representation of some physical phenomenon. Specifically, analog signals may be produced by our speech, music, temperature, and virtually all natural phenomena. In digital signal processing, the digital signal is an *approximation* of some physical phenomenon, and the physical signal is said to be quantized [see Fig. 1-1(d)].

Digital waveforms [Fig. 1-1(c)] may be produced by a keyboard, relay contact, or an analog-to-digital converter. The digital waveform shown in Fig. 1-1(d) might represent the output of a digital-to-analog converter. We should note that each level in Fig. 1-1(d) is one of a finite number of distinct values. Therefore, the output is not truly analog.

Analog techniques are utilized in many of the electronic systems in use today. Consequently, our studies will concentrate on the electronic devices, circuits, and systems employed in analog designs. However, many of the same devices and circuit techniques are also incorporated in digital systems.

Exclusively analog systems might include regulated dc power supplies, radio receivers, and high-fidelity stereo amplifiers. Systems which are hybrids (i.e., they contain analog *and* digital subsystems) might include digital voltmeters, function generators, and digital oscilloscopes. Examples of exclusively digital systems are calculators, digital computers, and electronic watches.

1-2 Discrete versus Integrated Designs

Discrete circuit designs center around the use of bipolar and/or field-effect transistors. However, much of the emphasis in new analog designs has been directed toward the use of *linear integrated circuits* (ICs) whenever possible. Integrated circuits may contain an entire amplifier or most of a radio receiver in one tiny package. An IC may contain *hundreds* of transistors and resistors, and occasionally a capacitor or two. The most basic of these linear integrated circuits is the operational amplifier (op amp). (The op amp derives its name from the fact that it can be configured to perform mathematical operations such as adding, subtracting, scaling, integration, and differentiation. It was originally developed for use in *analog computers*.)

However, a great many *discrete designs* (i.e., designs using a great number of individual electronic components) are still in use today. Further, there are some applications that are best served by discrete designs. Consequently, we need to become familiar with both approaches. In addition, to capitalize fully on the advantages offered by linear integrated circuits, we must have at least a rudimentary knowledge of their internal components and design. We shall achieve this end by first considering discrete design approaches.

In discrete designs many components (e.g., resistors and capacitors) are required to allow the active device (a transistor) to amplify (increase the amplitude of) an input signal. In contrast, integrated-circuit amplifier designs require considerably fewer external components to support the "active device," which in this case is often an op amp.

As our studies progress, we shall see that the integrated circuit approach offers not only simplicity, but usually improved performance over discrete designs. We introduce the discrete designs and use them as a vehicle to see exactly how amplifiers and op amps perform their magic. Again, we do not wish to imply that discrete designs are dead. There are still a great many applications that require discrete active components (such as the transistor).

1-3 Importance of Solid-State Physics

The operation of all solid-state devices is based on the properties of semiconductors. Examples of these devices include rectifier, zener, and light-emitting diodes; bipolar and field-effect transistors; and linear integrated circuits such as the op amp.

In general, *semiconductors* are elements whose electrical properties lie between those of electrical *conductors* and electrical *insulators*. Under external influences such as temperature, light, or a magnetic field, conductors and insulators do not substantially change their electrical characteristics. However, semiconductors are very sensitive to these external influences.

For example, semiconductors are used to make *thermistors*, which convert small changes in temperature into relatively large changes in resistance. Thermistors are used in some temperature measurement systems.

In a similar fashion, the light sensitivity of semiconductors is used to advantage in light-measurement devices such as *photoconductive cells*. The sensitivity of semiconductors to magnetic fields is utilized in *Hall effect devices*, which are used to convert changes in magnetic field intensity into resistance changes.

The sensitivity of semiconductors to these various external influences is *not* always desirable. For example, the temperature sensitivity of semiconductors can pose a severe problem in some electronic circuit applications.

1-4 Atomic Structure of Matter

The smallest piece of any element that retains its chemical and physical properties is an *atom*. In our work, the Bohr model will be used. A two-dimensional version of this model is indicated in Fig. 1-2 for the semiconductors silicon and germanium.

FIGURE 1-2 **Bohr atomic model for the semiconductors silicon and germanium.**

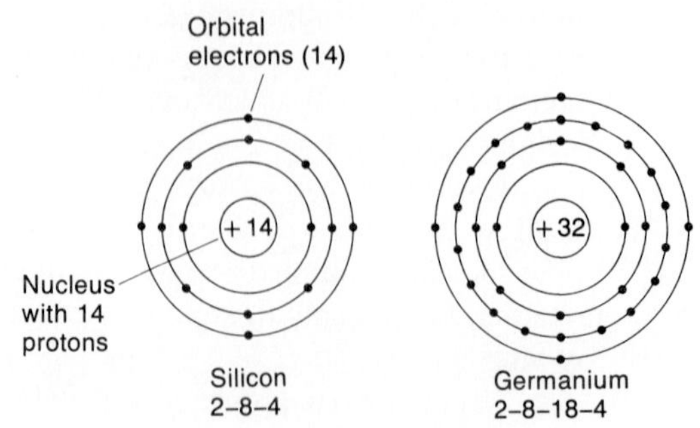

The model depicts electrons orbiting around the nuclei of the atoms. In actuality, there are many particles contained in the nucleus, including protons and neutrons. We shall concern ourselves only with the orbital electrons and the protons in the nucleus.

The electrons possess a negative electrical charge, and the protons possess a positive electrical charge. The magnitudes of the charges are equal. In general, the number of electrons orbiting the nucleus will equal the number of protons contained in the nucleus.

The electrical charge associated with an electron is very small. The reader should recall that it takes the charge of 6.242×10^{18} electrons to equal 1 coulomb (C). Hence

$$6.242 \times 10^{18} \text{ electrons} = -1 \text{ C}$$

and dividing both sides by 6.242×10^{18} gives us

$$1 \text{ electron} = -1.602 \times 10^{-19} \text{ C} \tag{1-1}$$

Since a proton has an equal but opposite charge,

$$1 \text{ proton} = +1.602 \times 10^{-19} \text{ C} \tag{1-2}$$

Consequently, an atom is electrically *neutral* when viewed from a distance.

1-5 Electron Energy Levels

An orbiting electron possesses kinetic energy due to its motion and electrical potential energy. Kinetic energy (KE) is given by

$$\text{KE} = \tfrac{1}{2} \text{ mass} \times \text{velocity}^2 = \tfrac{1}{2} mv^2 \tag{1-3}$$

The electron has electrical potential energy since it is a displaced charge residing in the electric field that exists between it and the nucleus.

According to the field of study referred to as *quantum mechanics*, orbital electrons can possess only certain discrete amounts of energy. Therefore, only particular orbits can exist, and orbits may be referred to as *energy levels*. Between each energy level a *forbidden region* exists. Electrons cannot orbit in forbidden regions. However, they may quickly pass through them [refer to Fig. 1-3(a)]. If electrons in a particular orbit acquire enough energy from some outside source (such as an external electric field or thermal energy), they will jump to a higher energy level by passing quickly through the forbidden region.

Excited electrons are unstable, and they tend to fall back to lower energy levels by giving up their additional energy. These phenomena are illustrated in Fig. 1-3. In Fig. 1-3(c) an electron is shown to emit energy as it falls back to its original orbit. In some cases, the energy it emits can be in the form of light energy (called photons), and the wavelength (or color) of the light is determined by the particular orbits involved. This, in part, is the mechanism that occurs in solid-state light-emitting diodes (LEDs).

The electrons in the highest energy levels are tied so loosely to the nucleus that they can be torn completely away from the atom if they acquire enough additional energy. The electrons that do leave their parent atom are termed *free electrons*. An atom losing an electron (or electrons) is in the *ionized state*. Further, since it has the same number of positive charges in its nucleus and fewer orbiting electrons, it has a net positive charge. Such an atom is termed a *positive ion*.

1-6 The Electron Volt

To quantify our discussion slightly, we introduce a basic unit of *energy* referred to as the *electron volt* (eV). Recall that the *volt* is the basic unit of potential difference. From basic circuit theory:

> **One volt is the potential difference between two points in an electric circuit when the energy required to move 1 coulomb of charge (electrons) from one point to the other is 1 joule.**

Mathematically,

$$V = \frac{W}{Q} \tag{1-4}$$

where V = electrical potential [volts (V)]
W = energy [joules (J)]
Q = charge (coulombs)

FIGURE 1-3 Electron movement: (a) electron receives energy; (b) energized electron has a higher orbit; (c) electron emits energy as it falls back; (d) electron has returned to its original orbit.

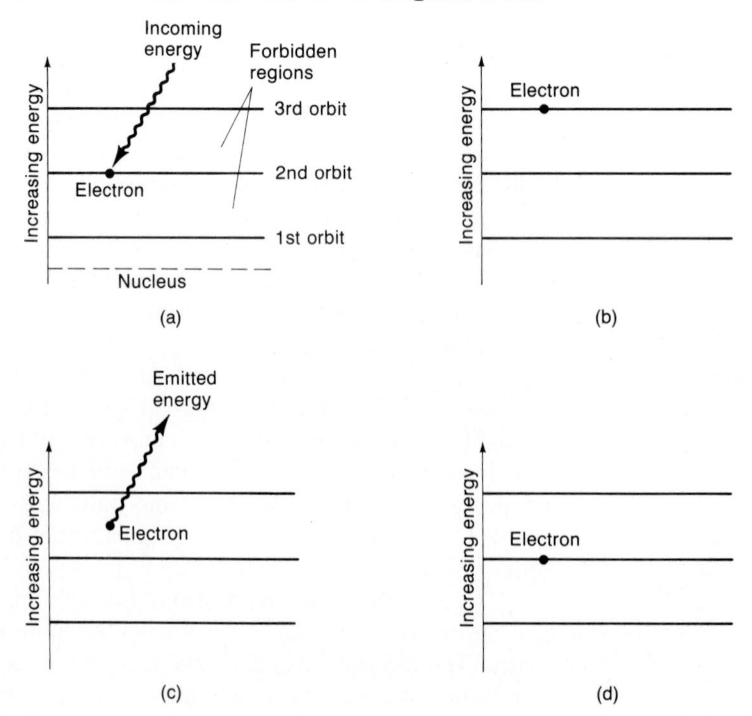

We can apply Eq. 1-4 to an individual electron. Recall from Eq. 1-1 that the charge on one electron has a magnitude of 1.602×10^{-19} C, and if we let the electrical potential be 1 V:

$$V = \frac{W}{Q}$$

$$1 \text{ V} = \frac{W}{1.602 \times 10^{-19} \text{ C}}$$

and solving for the energy W yields

$$W = (1 \text{ V})(1.602 \times 10^{-19} \text{ C}) = 1.602 \times 10^{-19} \text{ J}$$

The result above shows that it takes 1.602×10^{-19} J of potential energy to move one electron through a potential difference of 1 V. For simplicity, we let

$$1 \text{ eV} = 1.602 \times 10^{-19} \text{ J}$$

1-7 Energy Bands

In solids, very large numbers of atoms are squeezed into a small volume and there is considerable interaction between them. As a result, all the electrons traveling in the first orbits will have slightly different energy levels because no two of them will see exactly the same charge environment.

FIGURE 1-4 Development of the energy diagram for atoms in a solid.

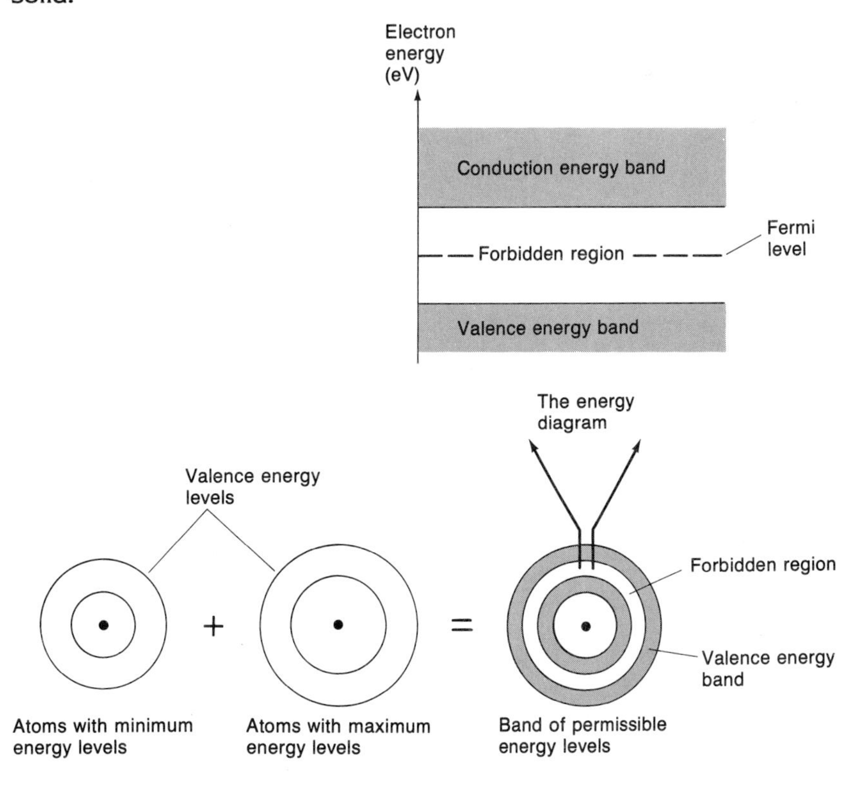

Since there may be billions of first-orbit electrons, the slightly different energy levels will have a *range*, or *band*, of different values. In a similar fashion, all the upper orbits, or energy levels, will be affected. As a result, they too will have a band of possible energy values. Hence, in a single, isolated atom, the electrons are located at definite energy levels, whereas in solids, the electrons are located in energy bands. These energy bands are separated by forbidden regions (see Fig. 1-4).

The energy band of interest is the highest energy band or *valence band*. If a sufficient amount of energy is given to an electron in the valence band, the electron is freed of the atomic structure. Such an electron is said to possess enough energy to be in the *conduction band*, where it can then take part in electric current flow. Again there is a forbidden region that exists between the valence and conduction bands. This is illustrated in the energy diagram also shown in Fig. 1-4.

Free (conduction band) electrons can move readily under the influence of an external electric field. However, electrons with enough energy to place them in the valence band are more or less attached to their parent atoms. As before, electrons in the valence band can be elevated to the conduction band if they receive enough external energy. Normally, electrons do *not* orbit in the forbidden region, which separates the conduction and valence bands as shown in Fig. 1-4.

The last term to be encountered in the study of energy diagrams is the *Fermi level*. The Fermi level is simply a reference energy level. It is the energy level at which the probability of finding an electron n energy units above it in the conduction band is equal to the probability of finding a hole (electron absence) n energy units below it in the valence band. Very simply, it can be considered to be the average energy level of the electrons (see Fig. 1-4).

It is the energy difference between the valence and conduction bands that determines if a solid is a conductor, a semiconductor, or an insulator. The magnitude of the energy difference is expressed in electron volts.

The forbidden region varies in width. For example, in the case of an ''ideal'' insulator, such as a diamond, the forbidden ''gap'' is 7 eV, whereas for a semiconductor, such as germanium, silicon, or gallium arsenide, it is 0.75, 1.12, or 1.4 eV,

FIGURE 1-5 **Energy-band diagrams for solids: (a) insulators (e.g., silicon dioxide has a band gap of 9 eV); (b) semiconductors such as germanium, silicon, and gallium arsenide have band gaps of 0.75, 1.12, and 1.4 eV, respectively; (c) conductors have a band gap of 0 eV.**

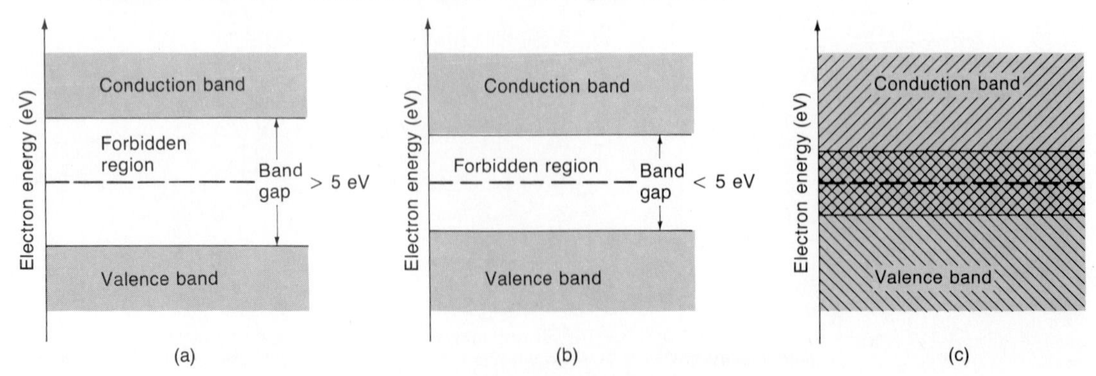

respectively. For an insulator, the valence and conduction bands are so far apart (>5 eV) that at normal ambient temperatures (i.e., 25 °C) an extremely small number of electrons will receive enough energy to jump over (see Fig. 1-5).

However, in a semiconductor material, both bands are so close that the electrons can jump over in considerable numbers, even at normal ambient temperatures. We can see from Fig. 1-5(c) that in conductors, the valence and conduction bands overlap. Therefore, even without any external energy, there are many electrons in the conduction band.

1-8 Semiconductors

The semiconductor group is made up of *elements* and *chemical compounds*. The most interesting semiconductor elements appear in the fourth group of the periodic table of the elements. These elements have *four* valence electrons in their outer shell and are called *tetravalent*. Two of these elements, *germanium* and *silicon*, are the most widely used.

Previously, we illustrated the silicon and germanium atoms in Fig. 1-2. To emphasize that these two elements are tetravalent, they have been redrawn in Fig. 1-6. The "core" of the germanium atom contains its nucleus and 28 of its 32 orbital electrons. The core of the silicon atom contains its nucleus and 10 of its orbital electrons.

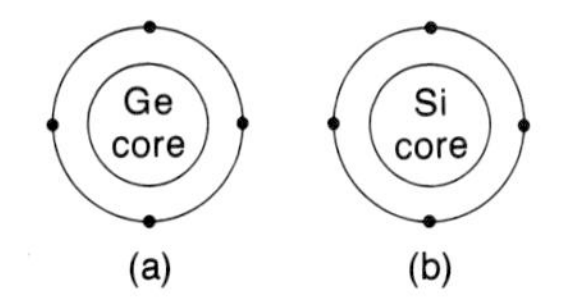

FIGURE 1-6 Tetravalent (a) germanium (Ge) and (b) silicon (Si) atoms.

In Table 1-1 several of the elements of interest in semiconductor physics are given. The *valence* refers to the number of *valence electrons* contained in each atom. With each element appears its chemical symbol and atomic number (which is the total number of electrons within the atom). In the right-hand column, the number of electron shells (collections of energy levels) and their designations are given.

In addition to silicon and germanium, some intermetallic compounds between the *trivalent* (three valence electrons) and the *pentavalent* (five valence electrons) elements behave like tetravalent semiconductors. These are alloys of two or more metals, such as gallium arsenide and indium antimonide.

As we shall see in Chapter 2, gallium arsenide (GaAs), gallium phosphide (GaP), and gallium arsenide phosphide (GaAsP) are the semiconductors that are used

TABLE 1-1
Partial Periodic Table of the Elements

Valence 3	Valence 4	Valence 5	Number of Shells
Boron (B) 5	Carbon (C) 6	Nitrogen (N) 7	2 *K,L*
Aluminum (Al) 13	Silicon (Si) 14	Phosphorus (P) 15	3 *K,L,M*
Gallium (Ga) 31	Germanium (Ge) 32	Arsenic (As) 33	4 *K,L,M,N*
Indium (In) 49	Tin (Sn) 50	Antimony (Sb) 51	5 *K,L,M,N,O*

to make light-emitting diodes (LEDs). The particular semiconductor employed determines the wavelength (or color) of the light it emits. The gallium compounds have also made possible the development of a relatively new ''breed'' of incredibly high-speed analog and digital integrated circuits.

Various *oxides* also behave like tetravalent semiconductors. Examples of these include copper and titanium oxides. These and other oxides are employed in solid-state, temperature-measurement devices called thermistors and in (electrical) transient-suppression devices called *varistors*.

1-9 Semiconductor Crystals

Pure (intrinsic) silicon, which is first melted and then cooled until it solidifies, forms a crystalline structure called a *crystal lattice*. In Fig. 1-7 we see a representation of this structure. Each atom is equidistant from its four neighboring atoms.

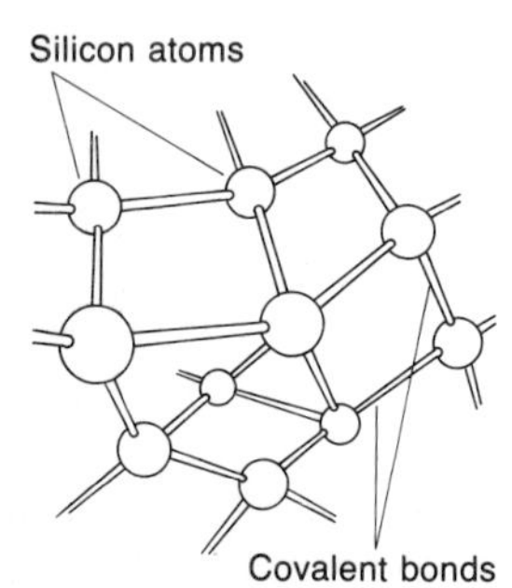

FIGURE 1-7 Crystal lattice structure.

In a crystalline structure each atom *shares* its four valence electrons with its four adjacent atoms. This occurs because each tetravalent semiconductor atom requires a total of eight outer-shell electrons. This sharing of electron pairs is called *covalent bonding*. The covalent bonds are illustrated in Fig. 1-7 by means of the rods connecting the adjacent atoms. To simplify our discussion, the crystal lattice structure has been reduced to two dimensions in Fig. 1-8. It is easy to see that all the valence electrons are bound in the crystalline structure.

The covalent bonds can only be broken by the expenditure of some amount of energy. In silicon the energy required to break a covalent bond is 1.12 eV. In germanium the required energy is only 0.75 eV. (These energy values are the energies required to move the valence electrons into the conduction band.)

The ideally formed crystalline structure rarely occurs in nature. Usually, there are some impurities that disturb the orderly atomic arrangement. Even if the crystal is purified, it is necessary that it be grown, or formed, from *one* center. If the

FIGURE 1-8 Two-dimensional illustration of a semiconductor crystal lattice.

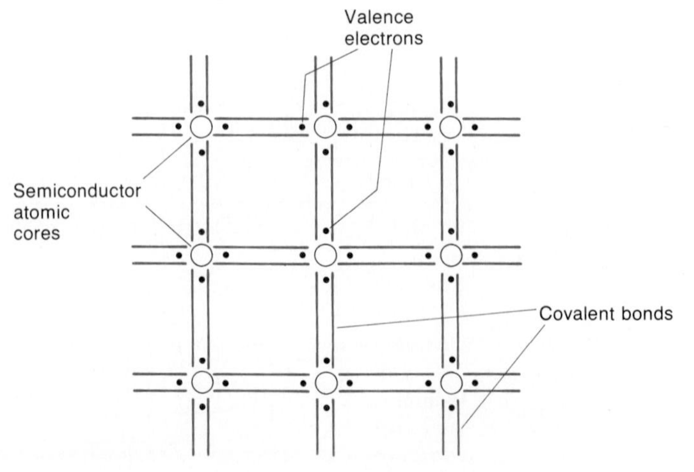

crystallization takes place too rapidly, a multicentered, or *polycrystalline* structure results. When we speak of an ideally built crystal, we understand this to mean a *monocrystal*, which is grown from one center.

1-10 Conductivity of Intrinsic Semiconductors

Silicon, the gallium compounds, and to a lesser degree, germanium are the most widely used semiconductors. Silicon is the most popular semiconductor. The discussion that follows is presented in terms of silicon. However, it applies equally well to the other semiconductor crystals.

In the crystalline structure of a silicon monocrystal, all the valence electrons are tied in covalent bonds. As we know, the free electrons in a solid take part in the

FIGURE 1-9 Formation of a hole-electron pair: (a) silicon crystal lattice; (b) a hole is formed.

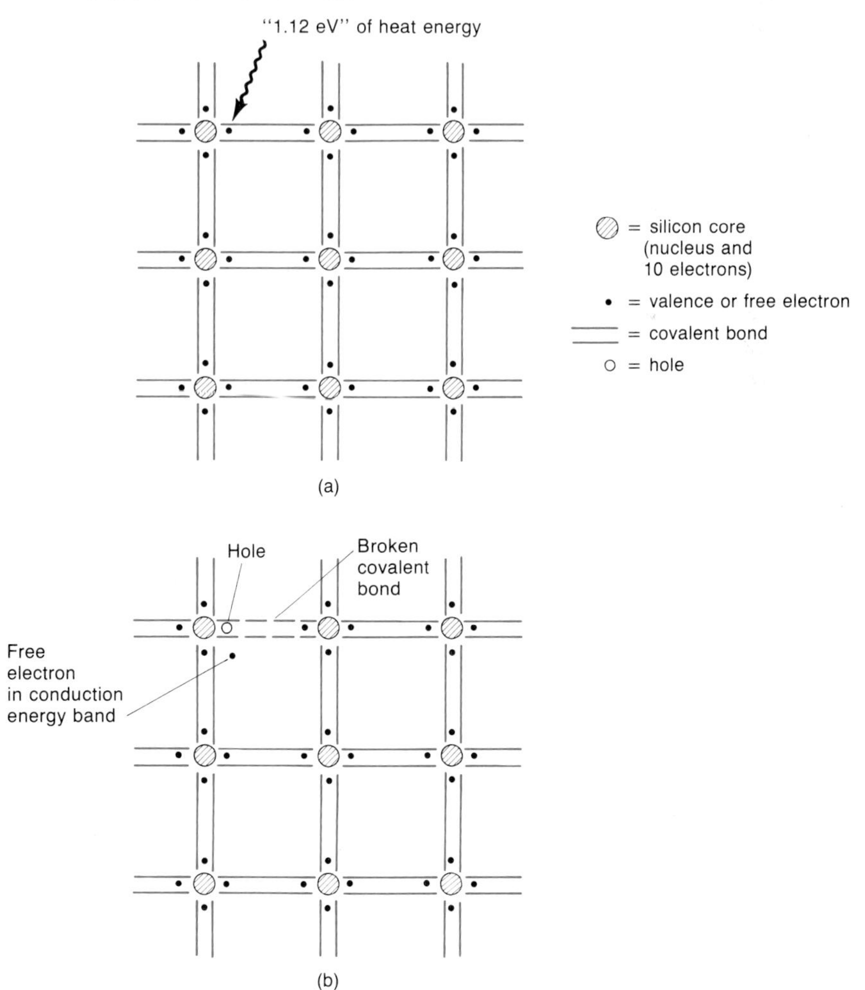

conduction of electric current. Therefore, a silicon monocrystal would appear to be an ideal insulator. However, this is true only when there is *no* thermal energy to agitate the electrons. No thermal energy exists at *absolute zero* when the temperature reaches 0 K (approximately −273°C).

As the temperature is raised from absolute zero, more and more heat energy becomes available, and an occasional covalent bond will be broken. At room temperature (25°C) there is enough heat energy so that many electrons are set free from their covalent bonds. Therefore, the semiconductor crystal will have enough free electrons to transport current through it (refer to Fig. 1-9).

When an electron is set free there remains a broken covalent bond, which is called a *hole*. The hole lacks a negative charge, and will therefore attract electrons. It should also be noted that since the parent atom has become deficient by one electron, it has become a positive ion.

The electrons that are attracted to the hole may come from one of two sources. They may be other free electrons which lose energy and fall into the hole, or they may be electrons from a neighboring atom. The latter case is more likely. This action is depicted in Fig. 1-10. Observe that when a hole captures an electron from a neighboring atom, the neighboring atom will then have a hole.

Note that when an electron fills a hole, it enters into a covalent bond. To make this jump, the electron does *not* have a conduction band energy. The electron has

FIGURE 1-10 Electron-hole movement in a semiconductor crystal.

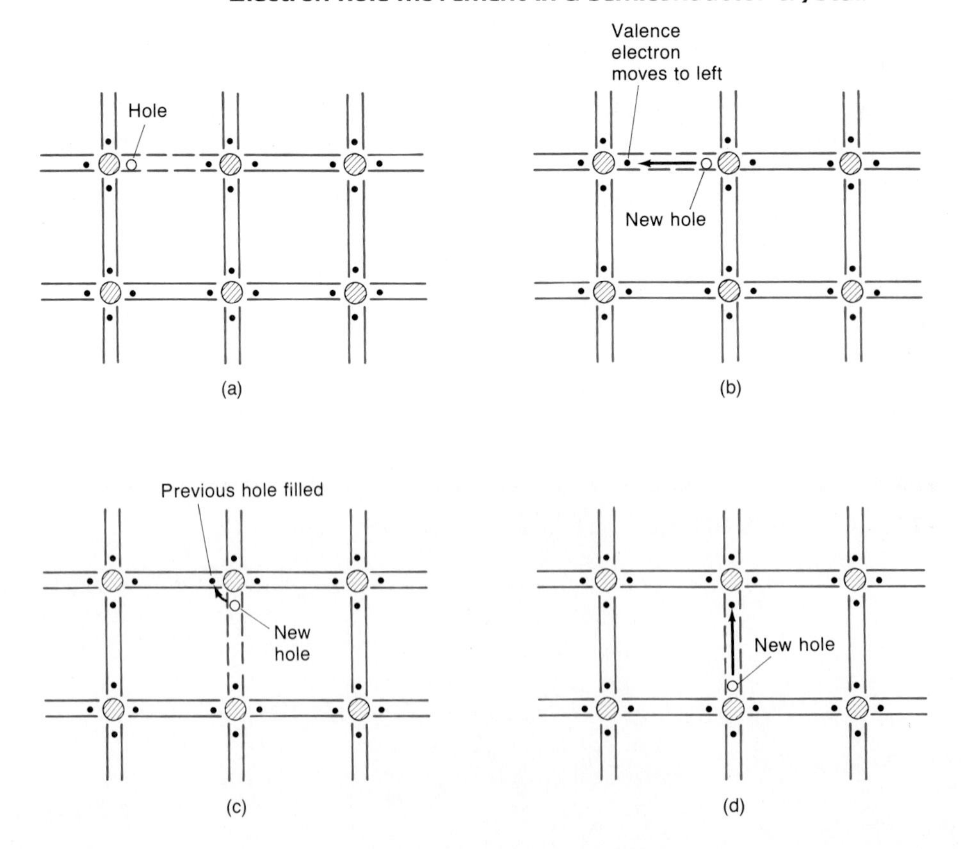

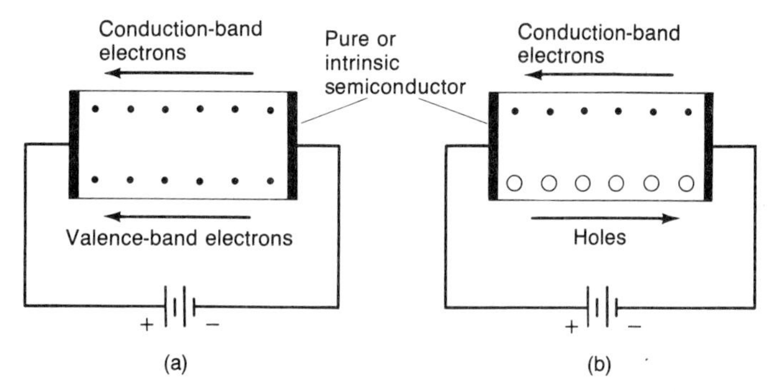

FIGURE 1-11 Electron-hole movement in an intrinsic semiconductor: (a) conduction- and valence-band electron flow; (b) conduction-band electrons and hole flow.

remained in the valence band. Observe in Fig. 1-10 that when an electron moves from a nearby bond to a vacancy, a new vacancy is created at the bond from which the electron left. In this manner, the vacancy, or hole, has moved.

In fact, under the influence of an electric field, the free electrons (and to some extent, the valence electrons) will move toward the plus pole, and the holes will *effectively* move toward the minus pole. Hence it can be said that the holes behave as positive charges. This electron-hole movement is illustrated in Fig. 1-11.

Recalling the energy diagram shown in Fig. 1-4, we can conclude that a free electron jumps from the valance band to the conduction band, whereas a hole remains in the valence band. In other words, electron flow requires electrons with conduction band energies, but hole flow requires electrons with lower (valence band) energies. The electron and the hole from the broken covalent bond form an electron-hole pair. Since each time a free electron is produced, a hole is formed, we have an equal number of holes and electrons in an intrinsic semiconductor.

As mentioned earlier, if a free electron wanders near a hole, it may fall into it. This is referred to as *recombination*. As it turns out, the probability of recombination occurring in regions where the crystal is perfect is small. Consequently, the probability of recombination is much higher in regions where the crystal has imperfections such as impurities. These regions are referred to as *recombination centers*. To promote recombination in silicon crystals, impurity atoms, such as gold, may be built into the crystal to increase the recombination rate.

1-11 Movement of the Charge Carriers

There are two mechanisms by which holes and electrons move through a semiconductor crystal. One of these is termed *diffusion* and the other is *drift*. Diffusion is a process. Specifically, a mixture of easily flowing materials (such as a drop of ink in a glass of water) will ultimately reach a final blend that is uniform in composition. This is precisely what happens in a semiconductor material. In this case concentrations of charge carriers (either holes or electrons) will tend to distribute themselves uniformly throughout a semiconductor crystal. The charge carrier movement that results is called *diffusion current*. Drift current results when available charges move under the influence of an applied electric field. Drift and diffusion occur simultaneously in a semiconductor.

1-12 Doped (or Extrinsic) Semiconductors

In intrinsic semiconductors there are an equal number of holes and electrons. In *extrinsic* semiconductors, impurities are added to increase the number of holes, or to increase the number of free electrons.

Semiconductors that have had impurities added are referred to as *doped semiconductors*. If the semiconductor has been doped to have additional free electrons, it is called *n-type*. If the semiconductor has been doped to have additional holes, it is called *p-type*. These *p*- and *n*-type semiconductors are found in almost all solid-state devices. Understanding the characteristics of these two types of semiconductor materials is instrumental to our future studies.

In an *n*-type semiconductor, the electrons are called the *majority carriers* and the holes are called the *minority carriers*. The reverse is true for *p*-type semiconductors.

Donor impurities are pentavalent elements such as arsenic (as shown in the valence 5 group in Table 1-1). Donor impurities are added to semiconductors to give them an excess of electrons. Since pentavalent elements have five outer-shell electrons, one of them will *not* be taken up in a covalent bond. This is true because in the crystal lattice structure there are only four neighboring atoms. This is illustrated in Fig. 1-12.

As shown in the figure, the unbound electron will remain near its parent atom. This occurs because the impurity atom has a proton in its nucleus which attracts the unbound electron. However, due to the geometry involved, the electron is partially shielded from its parent atom and the attractive force between them will be somewhat reduced.

As can be seen in Fig. 1-13, the energy level of this unbound electron is not far below the semiconductor's conduction energy band. Note that the Fermi level is

FIGURE 1-12 N-type semiconductor.

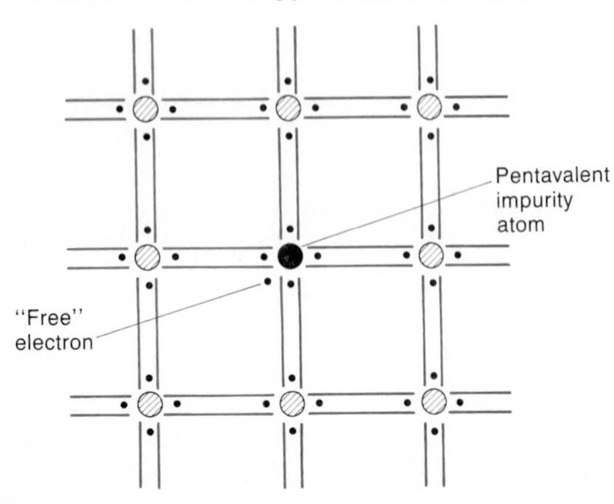

FIGURE 1-13 Energy diagram for a silicon n-type semiconductor.

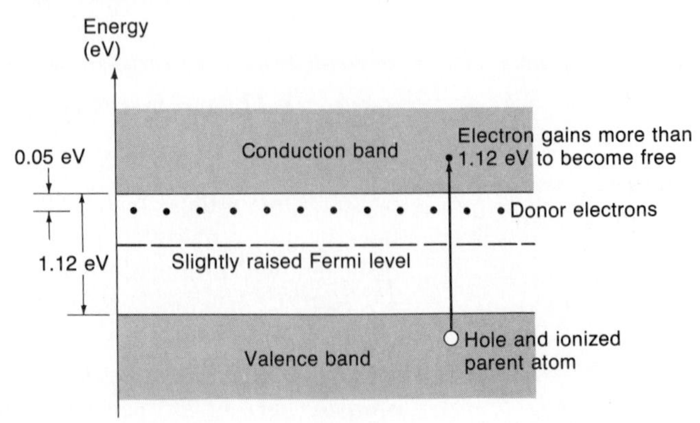

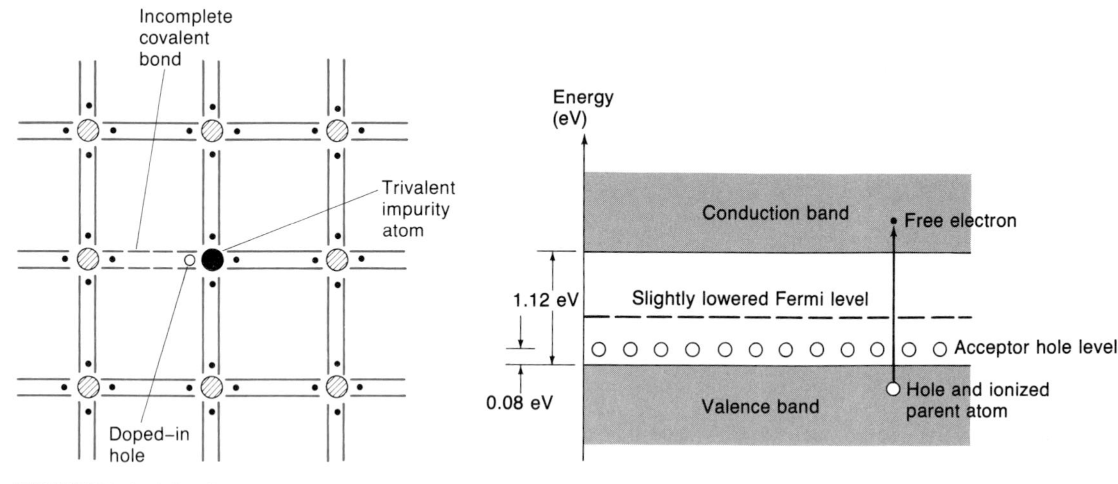

FIGURE 1-14 **P-type semiconductor.**

FIGURE 1-15 **Energy diagram for a p-type silicon semiconductor.**

slightly raised because the *average* energy level of the electrons is higher. Also recall that the hole is a minority carrier. Observe in Fig. 1-13 that the energy required to ionize a donor atom is only 0.05 eV, whereas the energy required to ionize a silicon atom is 1.12 eV (0.7 eV for Ge). As before, when a silicon atom is ionized, a hole is left in the valence energy band.

Acceptor impurities are trivalent elements such as boron (as shown in the valence 3 group in Table 1-1). Acceptor impurities are added to semiconductors to give them an excess of holes. Since trivalent elements have only three outer-shell electrons, one of its four neighboring semiconductor atoms will be lacking a covalent bond (see Fig. 1-14).

As discussed previously, a hole is the absence of an electron required to complete a covalent bond. A hole will capture a nearby electron. However, if the electron is a free electron, it must *lose* energy to fall into a hole. In general, an electron with only about 0.08 eV above the valence band may be captured by a hole. As a final point, when an electron is captured by a hole, the associated acceptor impurity atom will become a *negative ion*. It will have a net negative charge since it will not have enough positive protons in its nucleus to compensate an additional orbital electron.

The energy diagram for a *p*-type semiconductor is shown in Fig. 1-15. Observe that the Fermi level is slightly lowered by adding the acceptor impurity. The free electron is a minority carrier.

PROBLEMS

Drill, Derivations, and Definitions

Section 1-1

1-1. Explain the difference between "analog" and "digital" electronics.

1-2. In the study of electronics, we encounter a wide variety of waveforms. Is the waveform shown in Fig. 3-28 digital, continuous analog, or discontinuous analog?

1-3. Repeat Prob. 1-2 for Fig. 3-30.

1-4. Repeat Prob. 1-2 for Fig. 3-42(b).

1-5. Repeat Prob. 1-2 for Fig. 3-42(d).

Section 1-2

1-6. What is the most basic linear integrated circuit? Explain its name.

Section 1-5

1-7. What is the relationship between an electron's orbit and its energy level?

1-8. From an energy standpoint, discuss how an electron may rise to a higher orbital or fall to a lower one.

Section 1-7

1-9. Define the terms ''energy band,'' ''valence band,'' and ''conduction band.''

1-10. Make a neat sketch of the energy diagrams for an insulator, a semiconductor, and a conductor.

1-11. The energy required to move an electron from the valence band to the conduction band in silicon is 1.12 eV. Express the required energy in joules.

1-12. The energy required to move an electron from the valence band to the conduction band in gallium arsenide is 1.4 eV. Express the required energy in joules.

Section 1-9

1-13. Which semiconductor, silicon or germanium, is more apt to lose a valence electron by thermal ionization? Explain.

Section 1-10

1-14. Define the term ''recombination.''

Section 1-11

1-15. Explain the terms ''diffusion current'' and ''drift current.''

Section 1-12

1-16. The majority carriers in *n*-type semiconductors are ____ (holes, electrons) and the minority carriers are ____ (holes, electrons).

1-17. Pentavalent impurity atoms have ____(5,4,3) valence electrons and are used to form ____ (*p*-, *n*-) type semiconductors.

1-18. The majority carriers in *p*-type semiconductors are _____ (holes, electrons) and the minority carriers are _____ (holes, electrons).

1-19. Trivalent impurity atoms have _____ (5,4,3) valence electrons and are used to form _____ (*p*-, *n*-) type semiconductors.

1-20. Donor impurities are used to make _____ (*p*-, *n*-) type semiconductors and acceptor impurities are used to make _____ (*p*-, *n*-) type semiconductors.

2

THE P-N JUNCTION AND DIODES

After Studying Chapter 2, You Should Be Able to:

- Describe the formation of the p-n junction and its associated depletion region.
- Explain the effects of forward and reverse bias on a p-n junction.
- Name and describe the reverse currents (I_S, I_{SL}, and I_R) through a p-n junction.
- Interpret and develop V-I characteristic curves for two-terminal devices.
- Explain temperature effects on forward- and reverse-biased rectifier diodes.
- Distinguish between zener and avalanche diodes and their respective temperature effects.
- Describe the operation of the light-emitting diode (LED).
- Briefly describe the diode fabrication steps and terminology including floating-zone refining, diffusion, passivation, epitaxial growth, metallization, and planar construction.

2-1 Introduction to the *P-N* Junction

Most semiconductor (solid-state) devices employ one or more *p-n* junctions. The basic *p-n* junction is also fundamental to the various linear and digital integrated circuits. The *p-n* junction is produced by placing a layer of *p*-type semiconductor next to a layer of *n*-type semiconductor. The formation of a *p-n* junction is illustrated in Fig. 2-1(a).

Because of the tendency to diffuse, the negative electrons are effectively attracted to the holes. Each time an electron diffuses across the junction a pair of ions is formed. Negative ions are formed on the *p*-side because the acceptor impurity atoms now have extra electrons. Similarly, positive ions are formed on the *n*-side because the donor impurity atoms are now deficient of electrons. The ions are fixed in the crystal structure because of the covalent bonding.

FIGURE 2-1 Forming a p-n junction.

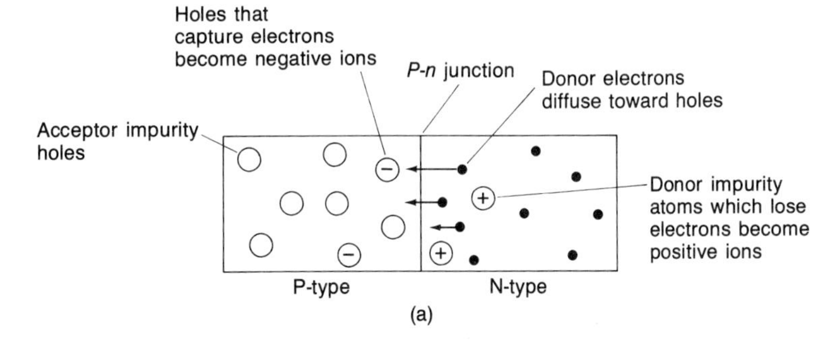

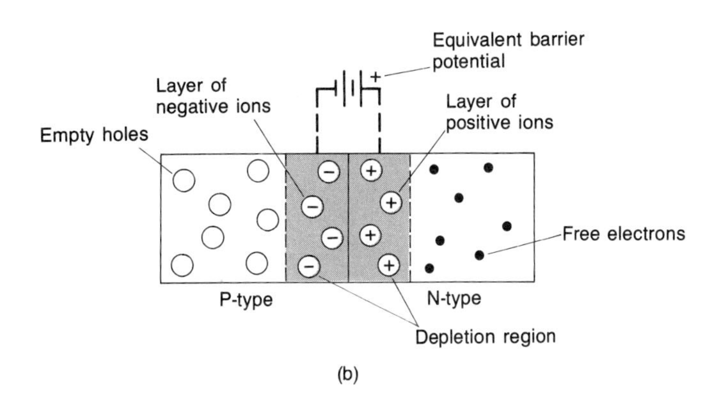

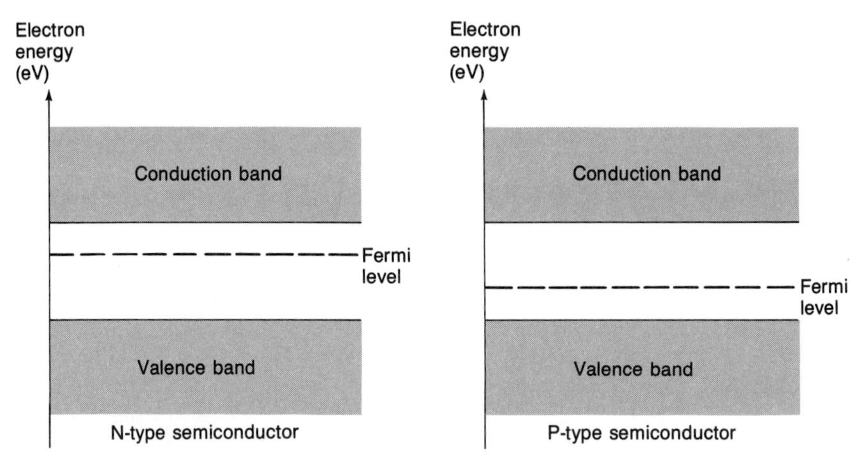

FIGURE 2-2 Separate energy states of n- and p-type semiconductors.

This process continues until the attractive force of the positive ions, and the repulsion produced by the negative ions prevent additional negative electrons from moving across the junction to annihilate more holes. It is the combination of these two effects that stops the action. When this point is reached, it is said that a *barrier potential* has formed.

The term "barrier potential" is derived from the fact that since we have positive and negative ions, we have an electric field between them. From basic circuit theory we recall that with every electric field there is an associated potential difference. The barrier potentials for germanium and silicon are 0.3 V and 0.7 V, respectively. The barrier potentials for the gallium compounds range from 1.2 to 1.8 V. We shall use a nominal value of 1.6 V. These are important numbers to remember since these three semiconductors are very widely used. The barrier potential effect is emphasized in Fig. 2-1(b) by the voltage source shown in phantom.

The region around the junction is completely ionized. As a result, there are no free electrons on the *n*-side, nor are there holes on the *p*-side. Therefore, this region is termed the *depletion region* since it is devoid, or depleted, of majority carriers.

The formation of the barrier potential can also be examined by means of energy diagrams. Consider the illustrations provided in Fig. 2-2. The average energy level (Fermi level) of the electrons is raised for the *n*-type semiconductor and lowered for *p*-type, as discussed in Section 1-12. Electrons seek the lowest energy level. When the *n* and *p* regions are joined, the electrons from the *n*-region will move across the junction to annihilate holes in the *p*-region. As a result, the conduction bands and the valence bands will distort until the *Fermi levels align*. The alignment of the Fermi levels corresponds to the formation of the barrier potential.

The alignment of the Fermi levels is very similar to the situation shown in Fig. 2-3. When the columns of water are considered separately, each may have its own level. However, when the valve is opened, the two water tanks are joined together. Obviously, water will flow from tank *n* to tank *p* until the water levels are both equal.

The Fermi level alignment between the *p* and *n* semiconductors is depicted in

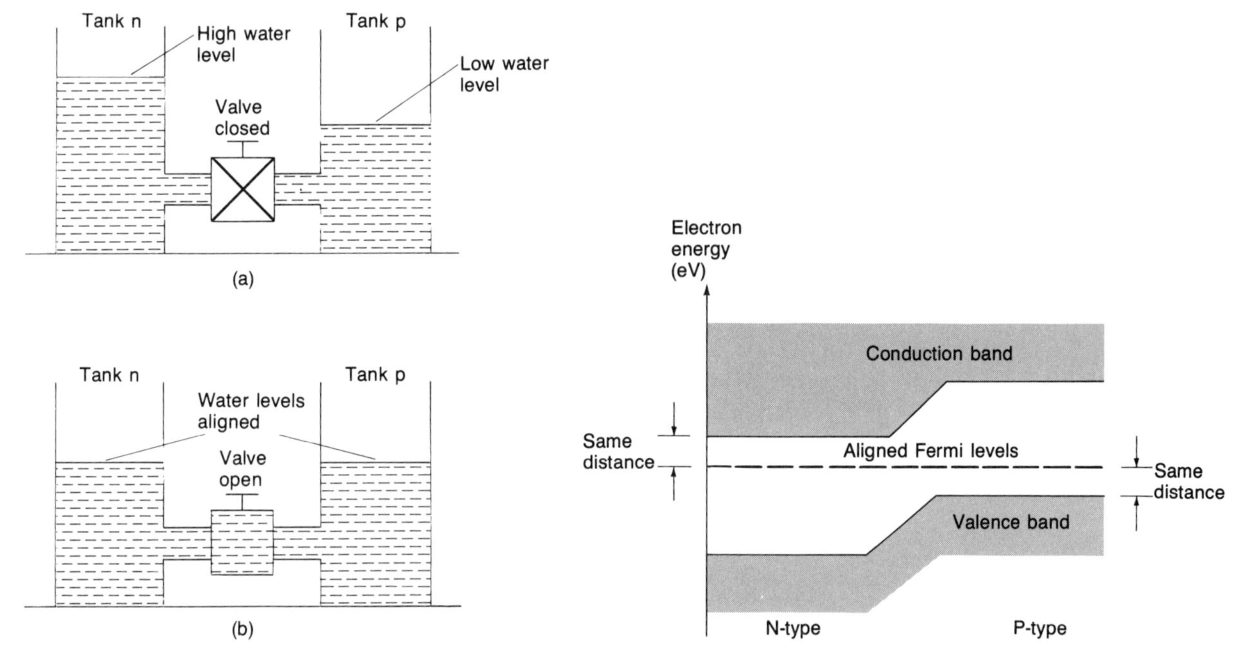

FIGURE 2-3 The alignment of the water levels is analogous to the alignment of the Fermi levels.

FIGURE 2-4 Energy diagram for a p-n junction.

Fig. 2-4. Notice that even though the Fermi level for the *n*- and *p*-type semiconductors has aligned, it has *not* changed its relative position in the two materials. Specifically, its distance from the conduction band in the *n*-type semiconductor is the same, and its distance from the valence band in the *p*-type semiconductor is also unchanged.

From Fig. 2-4 we observe that the *p*-region now has higher energy levels in the conduction band, but some of the levels at the bottom of the valence band have been lost. The reverse is now true for the *n*-region.

As a final comment, it should be mentioned that a *p-n* junction is *not* formed by merging two separate *p*-type and *n*-type materials. If this were done, we would have a polycrystalline structure that would not perform very well.

The *p-n* junction found in the semiconductor devices is actually formed in a single monocrystalline structure by adding carefully controlled amounts of donor and acceptor impurities. We took a simplified approach to illustrate the formation of the depletion region and the resulting barrier potential.

2-2 Forward Bias of a *P-N* Junction

There are *two* ways to apply a voltage (bias) to a *p-n* junction: forward and reverse. First, we shall study the effects of forward bias. Consider Fig. 2-5. Before we discuss the action inside *p*- and *n*-type semiconductors, let us look at the external circuit. The circuit consists of a voltage source, a current-limiting resistor, and the *p-n* junction

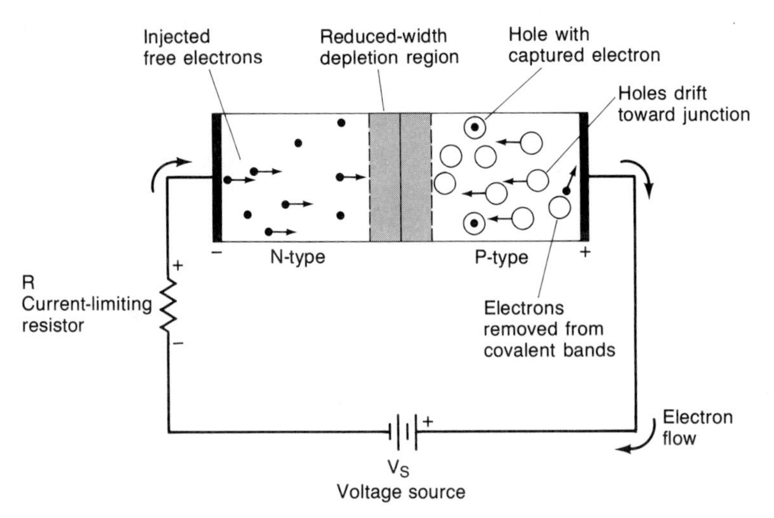

FIGURE 2-5 P-n junction with forward bias.

in series. Also note that in Fig. 2-5 we are using *electron current flow* to describe the physics.

The polarity of the voltage source makes the *n*-type material negative and the *p*-type material positive as shown. The argument to describe forward bias proceeds as follows:

1. Free electrons leave the negative terminal of the voltage source and flow into the *n*-type semiconductor. (Recall that electrons are the majority carriers in *n*-type semiconductor.)
2. The negative potential applied to the *n*-type material causes electrons to drift toward, and into, the depletion region.
3. The conduction band electrons lose energy as they overcome the barrier potential.
4. The electrons fall into the valence energy band on the *p*-side of the junction and are captured by holes.
5. The captured electrons in the valence energy band drift toward the positive terminal. (Alternatively, we can say that the holes move away from the positive terminal.)
6. The positive potential excites the valence-band electrons to raise their energies to the conduction energy band. (The positive potential removes the electrons from the holes.)
7. The free electrons leave the *p*-type semiconductor and flow through the external circuit to the positive supply terminal.
8. The resulting empty holes are repelled by the positive potential, and they drift back to the junction to capture more free electrons.

Now we are in a position to consider the effects of forward bias on the energy diagram for the *p-n* junction. *Any* external bias destroys the equilibrium at the junction

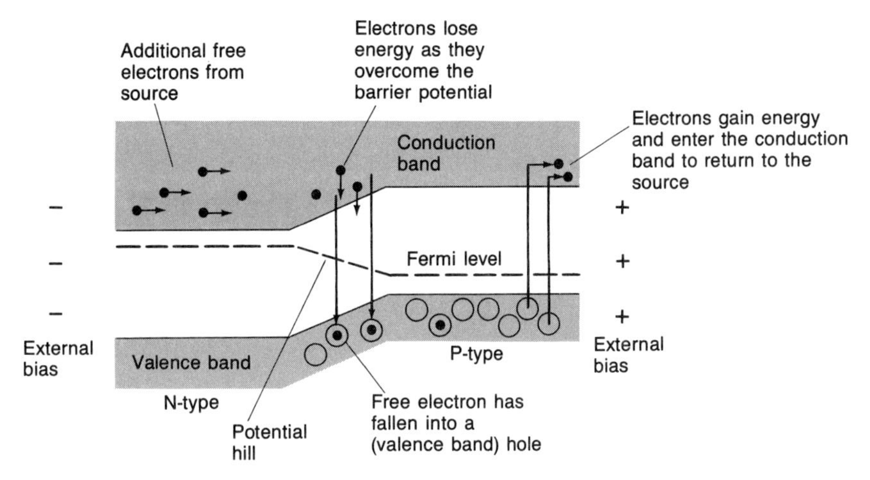

FIGURE 2-6 Electron-hole energy levels in a forward-biased p-n junction.

by bending the Fermi level (see Fig. 2-6). When forward bias is applied, the conduction bands and the valence bands of the *p* and *n* materials at the junction tend to become aligned.

The bend in the Fermi level is often called a *potential hill*. Simply, it shows that the average energy of the electrons decreases as we move from the *n*-type to the *p*-type material. Electrons can move easily from the *n*-type to the *p*-type material because they must move down a potential hill, or lose energy as they propagate.

The conduction-band electrons are repelled by the external bias. They drift to the junction and lose energy (as they overcome the barrier potential). The energy loss places them in the valence band, where they can recombine with holes. At the right-hand side, electrons leave the valence-band holes and go into the conduction band. In this case, they gain energy because of the applied external bias.

2-3 Reverse Bias of a *P-N* Junction

As we saw in Section 2-2, electrons may readily flow through a forward-biased *p-n* junction due to the action of the majority carriers. However, when a *p-n* junction is reverse biased, the *majority* carriers will *not* flow through it. The reverse-bias condition is shown in Fig. 2-7.

Observe that the width of the depletion region is increased by the reverse bias. Since the *n*-type side is now made positive and the *p*-type side is now made negative, the voltage source V_S attracts the majority carriers away from the *p-n* junction. Because the majority carriers are moved away from the junction, the depletion region width will be increased. Since more ions will be ''uncovered'' in the depletion region, the magnitude of the barrier potential will also be increased.

As a result of the external bias and the increased barrier potential, the majority

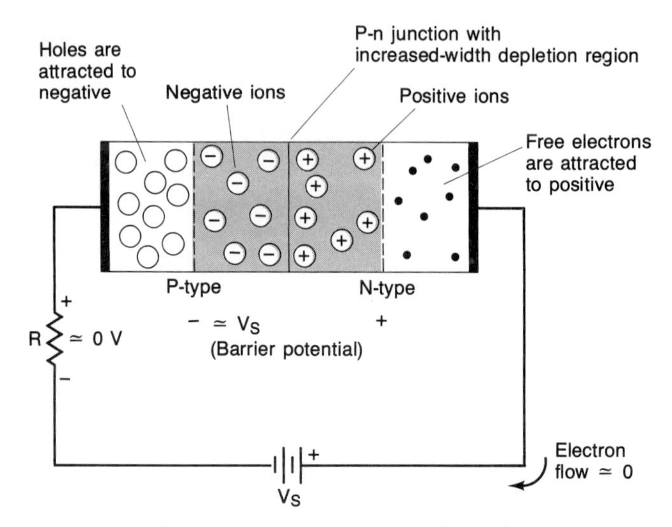

FIGURE 2-7 Reverse-biased p-n junction.

carriers *cannot* easily flow across the junction. Hence very little current will flow through the junction.

The energy diagram for a reverse-biased *p-n* junction is given in Fig. 2-8. Notice that the conduction and valence energy bands tend to become more misaligned with the application of a reverse bias. This is due to the increased width of the depletion region. Also observe that the Fermi level is again bent by the external bias to form a potential hill. However, in the case of reverse bias, electrons must climb, or go up a potential hill. This means that the electrons must *gain* energy in order to travel from the *n* region to the *p* region. Obviously, this is not very likely. The energy

FIGURE 2-8 Energy diagram for a reverse-biased p-n junction.

Electron energy (eV)
Reverse bias increases the misalignment between the valence and conduction bands
Conduction band
Same distance
Fermi level
Valence band
Same distance
Potential hill
N-type
P-type

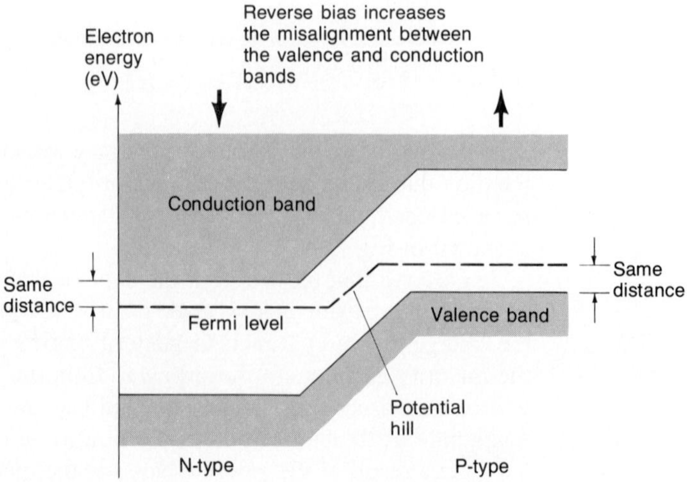

diagram supports the fact that current flow across a reverse-biased *p-n* junction will be approximately zero.

To summarize the effects of the bias conditions of a *p-n* junction we can state:

A *p-n* junction is *forward biased* by an external voltage source which makes its *p-type* end *more positive* than its *n*-type end. A forward-biased junction will allow current flow through it.

A *p-n* junction is *reverse biased* by an external voltage source which makes its *p-type* end *more negative* than its *n*-type end. A reverse-biased *p-n* junction will have a current of approximately zero through it.

2-4 Reverse Leakage Currents

Based on our discussion to this point, one might think that the current through a reverse-biased *p-n* junction must be zero. However, there will be a *small* current flow through a reverse-biased junction. In most cases this small reverse current can be ignored. However, since this reverse current can place a *severe limitation* on the operation of a solid-state device, we must take this opportunity to discuss it.

There are *two* basic mechanisms (at normal ambient temperatures) that permit current flow through a reverse-biased *p-n* junction: (1) reverse current due to *minority carriers*, and (2) leakage current around the *surface* of the semiconductor crystal. First, let us consider the effects of minority-carrier leakage current.

Recall that the minority carriers in *p*-type semiconductors are electrons and the minority carriers in *n*-type semiconductors are holes. These minority carriers are produced by thermal ionization of the semiconductor atoms, as discussed in Section 1-10. Consequently, the reverse current produced by the minority carriers is strongly temperature dependent. (This is explored further in Section 2-7.)

In Fig. 2-9 it is shown that if a minority carrier is produced *in the depletion region*, it will usually be swept across the *p-n* junction. This is called the *reverse saturation current* I_S.

Now let us consider the surface leakage phenomena. Imperfections in a semiconductor crystal lattice promote recombinations. The surface of a semiconductor crystal is an imperfection since there are no atoms outside to form covalent bonds

FIGURE 2-9 Minority-carrier current through a reversed-biased p-n junction.

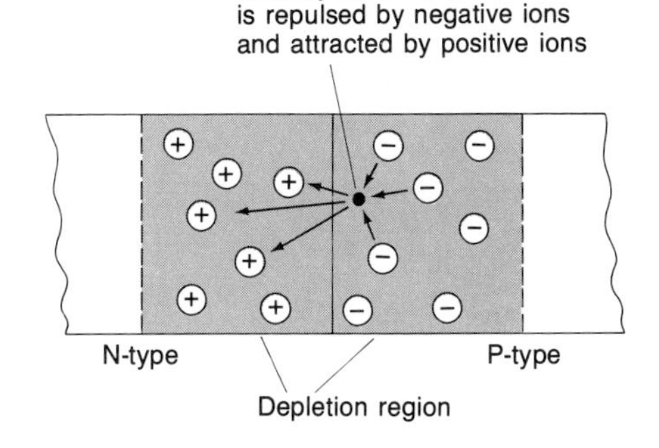

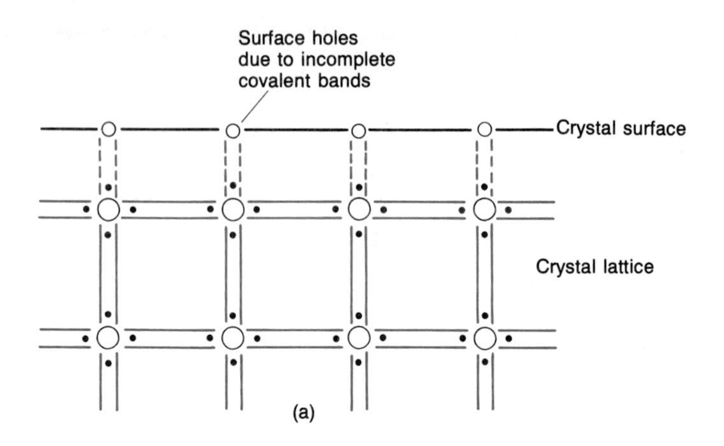

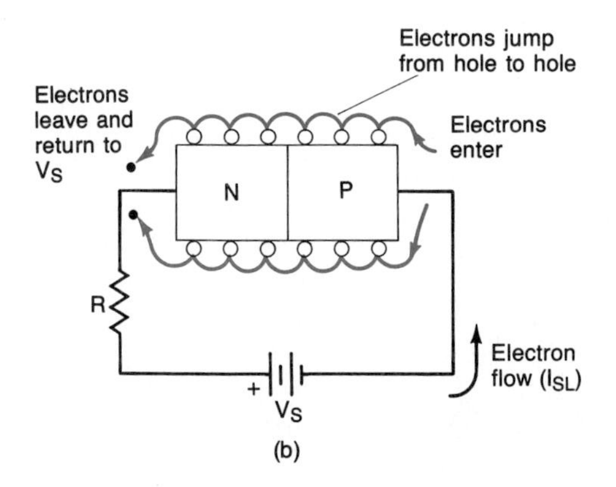

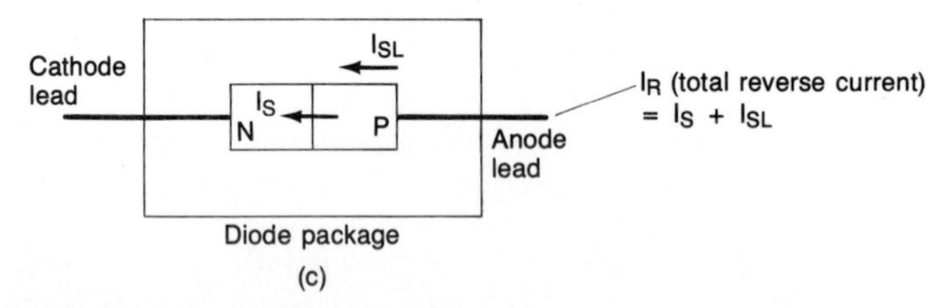

FIGURE 2-10 Surface leakage current: (a) structure; (b) electron movement; (c) I_R.

[see Fig. 2-10(a)]. Figure 2-10(a) shows that holes are formed on the surface of the crystal. In Fig. 2-10(b) we see that the surface leakage current (I_{SL}) results from the electrons moving from hole to hole. Unlike I_S, I_{SL} is *not* a function of temperature. Figure 2-10(c) emphasizes that the total reverse current (I_R) has two components: I_S and I_{SL}.

As a final comment, we should point out that unclean crystal surface conditions due to impurities also serve to promote surface leakage current. Therefore, we regard the surface leakage current as being produced by a combination of these two effects.

2-5 *V–I* Characteristic Curves

The operation of solid-state devices can be described graphically by making a plot of their terminal voltage and current relationships. This is referred to as a voltage–current characteristic curve, or simply a *V–I characteristic curve*. A *V–I* characteristic curve can be generated semiautomatically through the use of an electronic curve tracer, or manually by using the test circuit shown in Fig. 2-11(a).

To grasp completely the information presented by a *V–I* characteristic curve, we must first understand the *reference direction* employed. Consider Fig. 2-11(b). Reference directions for a *positive* voltage and current of a general two-terminal device are defined. If the *actual* voltage and current directions are *reversed*, such as shown in Fig. 2-11(c), we label them as *negative* quantities.

In Fig. 2-11(d) we see the test circuit used to develop the positive portion of the *V–I* curve for a resistor. Note that the *actual* voltages and currents are regarded as *positive* since they *agree* with our defined reference shown in phantom.

FIGURE 2-11 Determining the V-I characteristics of a two-terminal device: (a) test circuit; (b) reference direction for positive voltages and currents; (c) negative voltages and currents; (d) resistor forward (positive) V-I characteristic; (e) resistor reverse (negative) V-I characteristic.

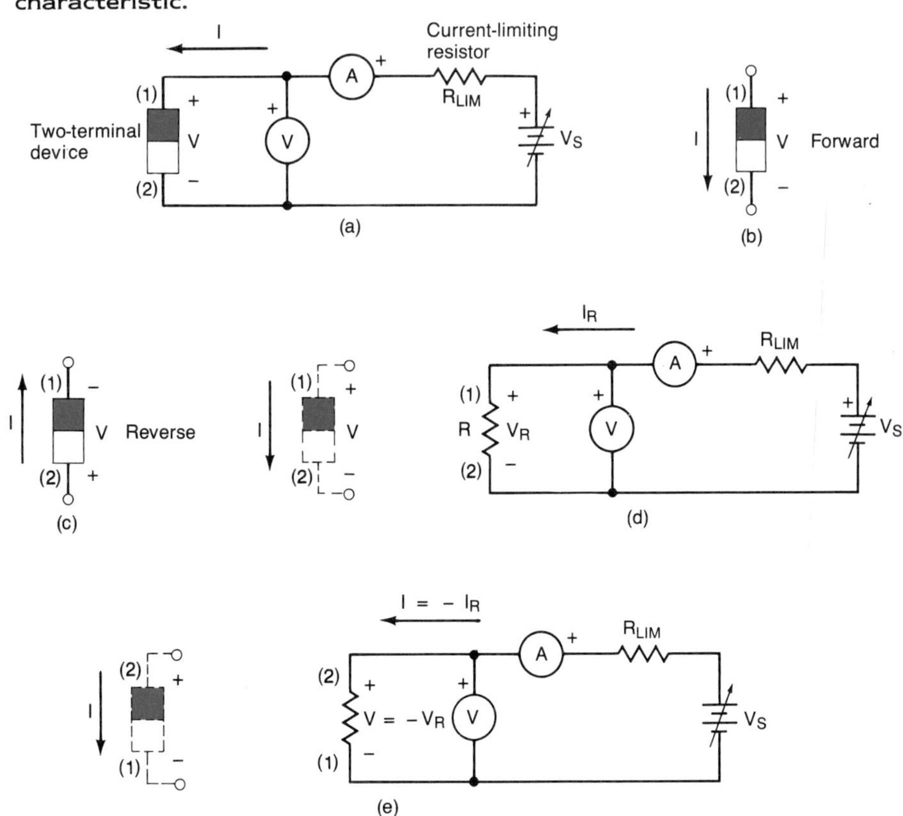

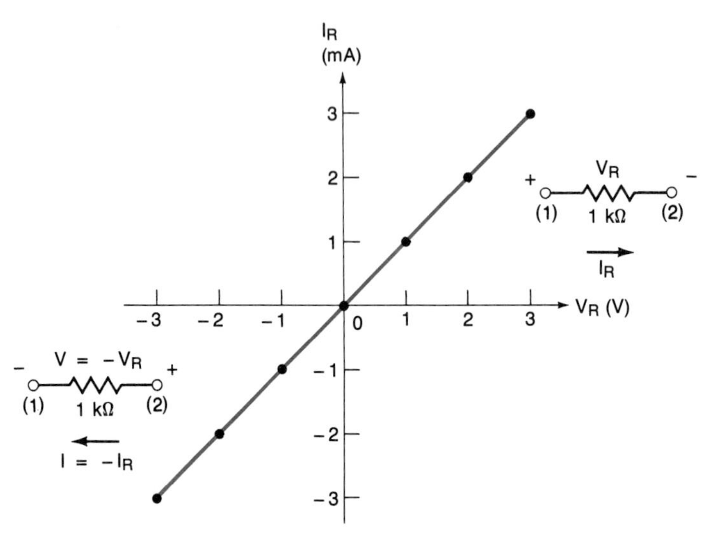

FIGURE 2-12 V-I characteristic of a 1-kΩ resistor.

In Fig. 2-11(e) we have reversed the resistor. Since the *actual* voltages and currents are *opposite* to those of our defined reference, they are regarded as negative.

The reader should recall that a resistor is a *bilateral device*. ''Bilateral'' means that current flows equally well through a resistor independent of its direction. Consequently, the forward and reverse *V–I* characteristic curve of a resistor is linear. A plot of the forward and reverse *V–I* values forms the characteristic curve shown in Fig. 2-12.

2-6 Diode *V–I* Characteristics

The rectifier diode is a *unilateral* device. A ''unilateral'' device tends to pass current better in one direction than the other. The rectifier diode's basic structure, schematic symbol, and a typical package are presented in Fig. 2-13(a). Its bias conditions are summarized in Fig. 2-13(b) and (c).

The *V–I* characteristics of the diode may be obtained by employing the circuits shown in Fig. 2-14. The forward-bias circuit in Fig. 2-14(a) initially has the adjustable voltage source V_S set to 0 V. The diode's voltage V_D and the current I_D through it will also be zero.

If the voltage source is increased such that V_D is 0.1 V, a small current will start to flow. No significant current will flow through the diode until the forward bias is large enough to overcome the barrier potential. This would be about 0.7 V for a silicon diode, about 0.3 V for a germanium diode, and about 1.6 V for an LED. This is called the knee voltage V_K.

With the diode reverse biased as shown in Fig. 2-14(b), only a very small reverse current will flow. The current is essentially constant no matter what reverse bias is applied. This reverse current (I_R) is sometimes approximated as the reverse saturation current I_S. The *V–I* characteristic is shown in Fig. 2-15.

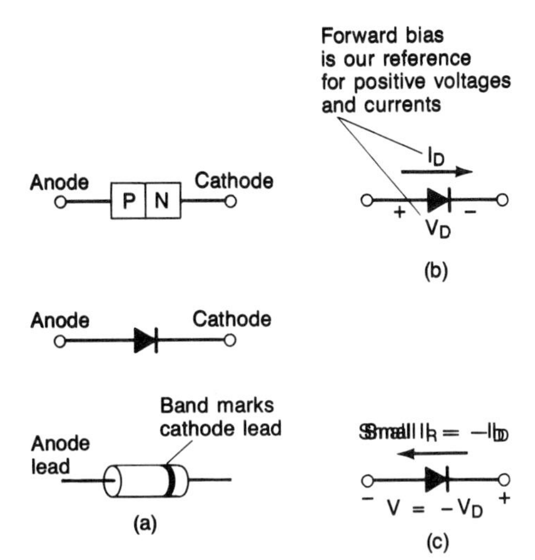

FIGURE 2-13 **Summary of the rectifier diode: (a) p-n junction, symbol, and package; (b) forward-biased diode passes conventional current easily; (c) reverse-biased diode passes only a small reverse current.**

FIGURE 2-14 **Circuit to determine the V-I characteristic of a rectifier diode: (a) forward; (b) reverse.**

If the reverse-bias voltage is increased sufficiently, the diode will begin to pass a large reverse current. This is called *avalanche breakdown* and is discussed in Section 2-8. Avalanche breakdown will occur at a particular reverse-bias voltage. For rectifier diodes this voltage may be called the *reverse breakdown voltage*, the *peak reverse voltage* (PRV), or the *peak inverse voltage* (PIV).

The reverse breakdown of a rectifier may destroy the *p-n* junction if the current

FIGURE 2-15 **Rectifier diode V-I characteristics: (a) silicon; (b) germanium.**

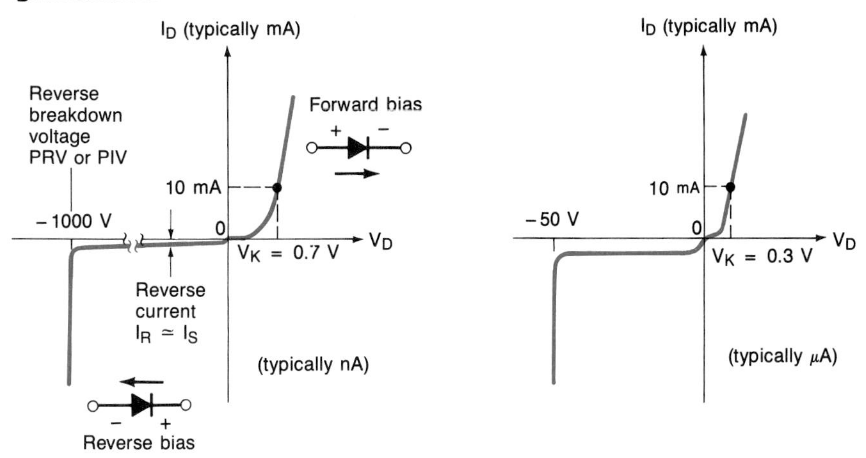

is not limited to a safe value by an external series resistor. In general, reverse breakdown of a rectifier is undesirable. However, zener and avalanche diodes are devices that are used in the breakdown mode. More will be said about these devices in Section 2-8.

As can be seen in Fig. 2-15, the knee voltage of silicon is 0.7 V, while the knee voltage for germanium is only 0.3 V. In general, the reverse saturation current for germanium is greater than that for silicon. The reason for this is simple. Since the silicon atom has 14 electrons, its 4 valence electrons are relatively close to its nucleus. The germanium atom has 32 electrons and its 4 valence electrons are farther away from its nucleus. Therefore, the valence electrons of silicon atoms are more tightly bound to their parent atoms than are those of germanium. Consequently, germanium is more susceptible to thermal ionization, which produces the minority carriers to create reverse current. This is one reason why silicon is the preferred semiconductor for most solid-state devices.

2-7 Temperature Effects on the Diode Curves

The reverse current I_S is dependent on such factors as doping levels, junction area, and of course, the junction temperature. However, with the exception of junction temperature, all the other factors may be regarded as being constant. It has been shown that

I_S approximately doubles for each 10°C increase in temperature.

FIGURE 2-16 Effect of temperature on the reverse saturation current of a diode.

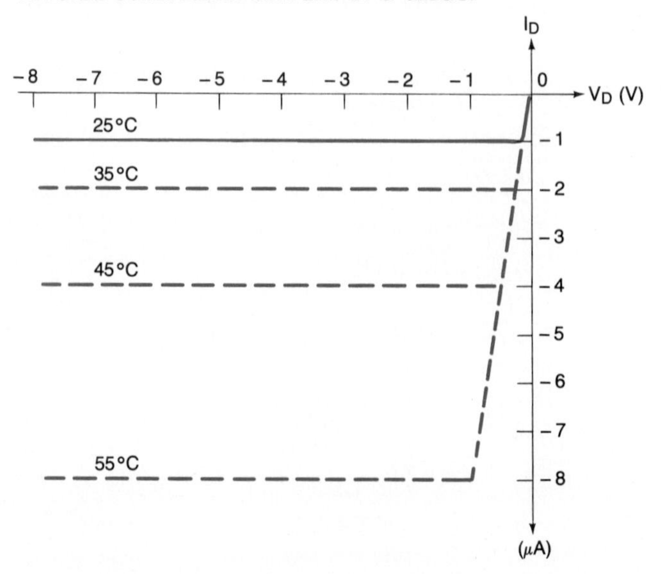

This relationship is described by

$$I_S(T) = [I_S(25°\text{C})][2^{(T-25)/10}] \quad (2\text{-}1)$$

where $I_S(T)$ = reverse saturation current at temperature T (°C)
$I_S(25°\text{C})$ = reverse saturation current at a temperature of 25°C
T = temperature (°C)

In Fig. 2-16 this effect is illustrated. With a little study it is quite easy to predict I_S as long as we have 10°C temperature changes. However, this is not always the case, and we must then draw on Eq. 2-1.

EXAMPLE 2-1

A 1N914B is described as a silicon high-speed switching diode. It has a reverse current of 25 nA ($\simeq I_S$) at 25°C. Find its I_S at 100°C.

SOLUTION The reverse current can easily be found by using Eq. 2-1.

$$\begin{aligned} I_S(T) &= [I_S(25°\text{C})][2^{(T-25)/10}] \\ &= (25\text{nA})[2^{(100-25)/10}] \\ &= (25\text{nA})(181) = 4.525\mu\text{A} \end{aligned}$$

■

The *forward* characteristics of a diode (or any *p-n* junction) are also controlled by temperature. The most notable effect of temperature is its influence on the barrier potential.

As the junction temperature is *increased*, the forward voltage drop required to push a given current through the diode will *decrease*. Since the voltage decreases with increasing temperature, it is described as having a *negative temperature coefficient* (or a negative ''tempco,'' for short). Specifically, it has been found that the forward voltage drop across a silicon *p-n* junction *decreases* about 1.8 mV per degree Celsius, and about 2.02 mV per degree Celsius for germanium. Consider Fig. 2-17. Typically, a tempco used for both silicon and germanium is −2 mV per degree Celsius.

FIGURE 2-17 Effect of temperature on a forward-biased silicon diode.

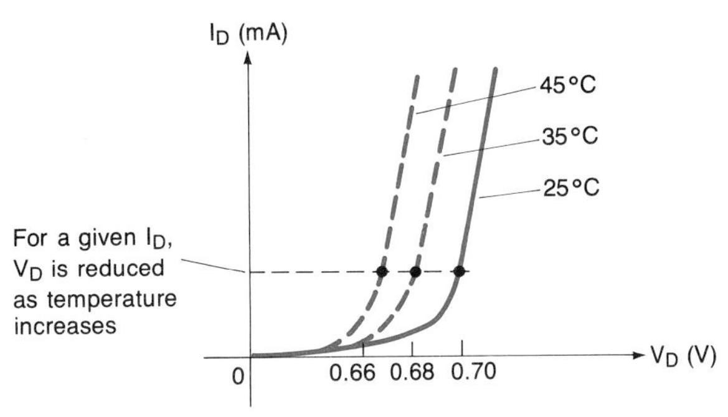

The forward voltage drop tempco ($\Delta V_D/\Delta T$) across either a silicon or germanium diode may be approximated as -2 mV/°C.

The notation ($\Delta V_D/\Delta T$) is read: "the change in the diode voltage drop with respect to a change in temperature." The Greek letter Δ (capital delta) is used to denote a change in a quantity.

2-8 Zener and Avalanche Diodes

The zener and avalanche diodes also employ a single (silicon) *p-n* junction. However, the *doping levels* (amounts of added impurities) are considerably different from those normally found in a rectifier diode. By using exactly the same test circuit shown in Fig. 2-14, we can produce the *V–I* characteristics given in Fig. 2-18. The *V–I* characteristic curve is exactly that for a silicon rectifier diode. However, we have indicated three distinct regions of operation in Fig. 2-18. Also note our indicated reference directions and the zener's schematic symbol.

Operating a rectifier diode in the breakdown region is generally undesirable. However, zener and avalanche diodes are *designed* to be used in the breakdown region. The primary uses for these breakdown diodes are as *voltage references* and *voltage regulators*. (This is covered in Chapter 14.)

There are two mechanisms that cause a reverse-biased *p-n* junction to break down: the *Zener effect* and *avalanche breakdown*. Either of the two may occur independently, or they may both occur simultaneously. Diode junctions that break down *below* 5 V are caused by the Zener effect. Junctions that experience breakdown *above* 5 V are caused by avalanche breakdown. Junctions that break down around 5 V are usually caused by a combination of the two effects.

A zener diode is produced by moderately doping the *p*-type semiconductor and heavily doping the *n*-type material (see Fig. 2-19). Observe that the depletion region extends more deeply into the *p*-type region.

FIGURE 2-18 The V-I characteristic for a zener diode. (The schematic symbol is included.)

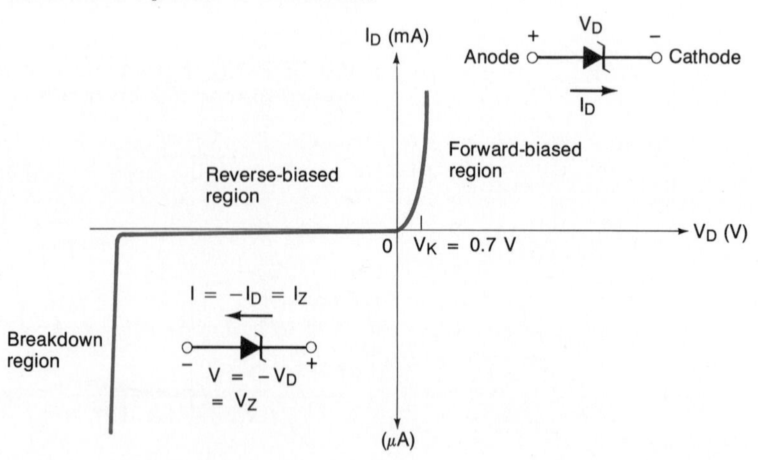

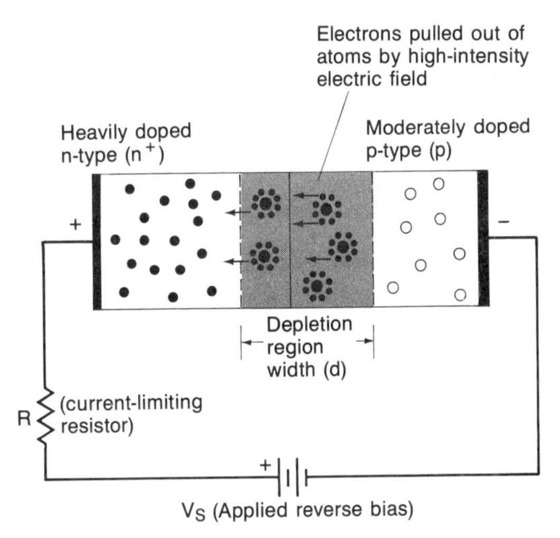

FIGURE 2-19 Reverse-biased p-n junction of a zener diode.

Under the influence of a high-intensity electric field, large numbers of bound electrons within the depletion region will break their covalent bonds to become free. This is ionization by an electric field. When ionization occurs, the increase in the number of free electrons in the depletion region converts it from being practically an insulator, to being a conductor. As a result, a large reverse current may flow through the junction.

The actual electric field intensity required for the Zener effect to occur is approximately 3×10^7 V/m. From basic circuit theory we recall that the electric field intensity E (capital epsilon) is given by

$$\mathrm{E} = \frac{V}{d} \tag{2-2}$$

where E = electric field intensity (volts per meter)
V = potential difference (volts)
d = distance (meters)

In terms of the *p-n* junction depicted in Fig. 2-19, we note that the applied reverse voltage is V and the depletion region width is the distance d. The narrower the depletion region, the smaller the required reverse bias to cause Zener breakdown. A small reverse bias can produce a sufficiently strong electric field in a narrow depletion region.

By controlling the doping levels, manufacturers can control the magnitudes of the reverse biases required for Zener breakdown to occur. Only certain standard zener diode voltages are available. These range from 2.4 to 5.1 V. More will be said about zener diode specifications in Chapter 3.

With lightly doped *p*-type material, the depletion region may be too wide for the electric field intensity to become sufficient for Zener breakdown to occur. In these

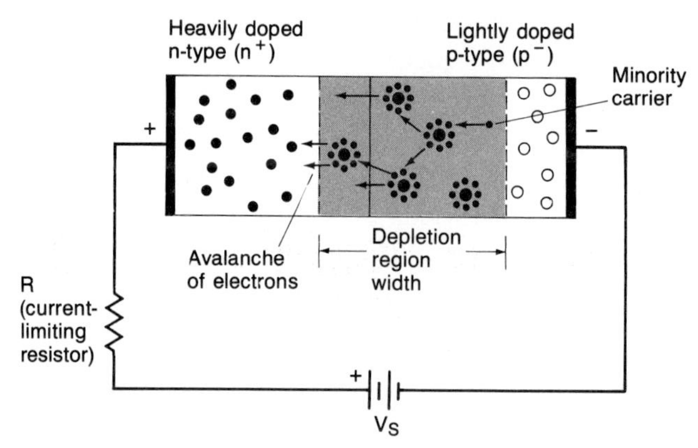

FIGURE 2-20 Reverse-biased p-n junction of an avalanche diode.

cases, the breakdown of the reverse-biased junction is caused by avalanche breakdown (see Fig. 2-20). The depletion region is wider because it extends more deeply into the *p* region.

Recall that the reverse saturation current I_S that flows across a reverse-biased *p-n* junction is due to minority carriers. Even though the electric field strength is not large enough to ionize the atoms in the depletion region, it may accelerate the minority carriers sufficiently to allow them to cause ionization by collision. The specifics may be outlined as follows:

1. The depletion region is too wide to allow an electric field intensity of at least 3×10^7 V/m.
2. The minority carriers are accelerated by the applied electric field.
3. The minority carriers gain kinetic energy.
4. The minority carriers collide with atoms in the depletion region.
5. The valence electrons of the atoms receive enough energy from the collisions to become free (conduction band) electrons.

As a result, the number of free electrons in the depletion region increases to support a large reverse current. This effect is illustrated in Fig. 2-20. This avalanche of carriers is also termed "carrier multiplication" since one minority carrier may ultimately cause many free electrons.

2-9 Temperature Effects on the Zener and Avalanche Curves

Temperature has a pronounced effect on the breakdown voltages of zener and avalanche diodes. When Zener breakdown occurs, a very intense electric field (3×10^5 V/cm, or greater) exists in the reverse-biased *p-n* junction. The intense electric field

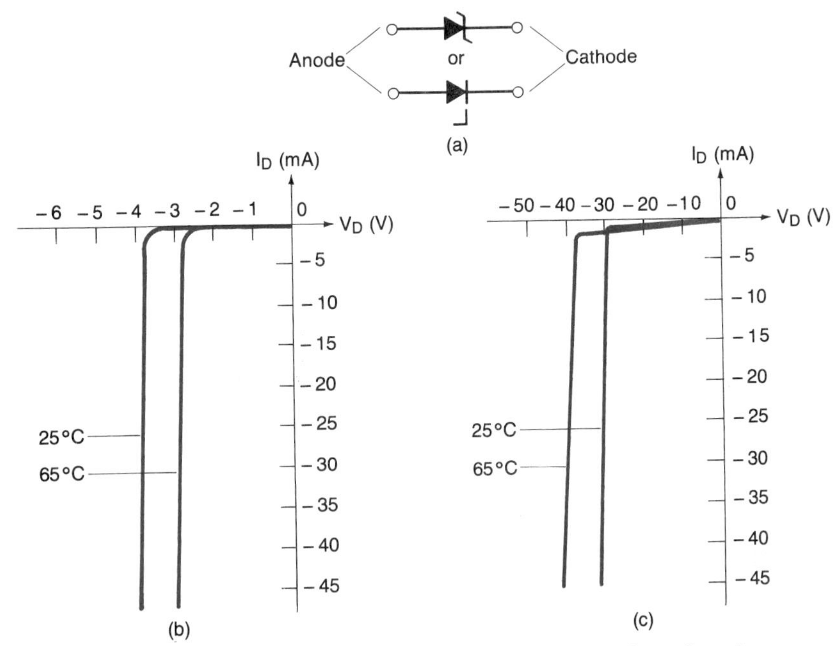

FIGURE 2-21 Temperature effects on zener and avalanche breakdown diodes: (a) schematic symbols; (b) zener breakdown exhibits a negative tempco; (c) avalanche breakdown exhibits a positive tempco.

adds additional energy to the valence electrons, which allows them to transverse the forbidden energy region.

When the junction temperature is increased, the valence electrons gain additional energy, and they orbit at higher energy levels. Consequently, the valence energy band becomes closer to the conduction energy band. Since the energy difference between the valence and conduction energy bands is reduced, the required electric field strength for Zener breakdown to occur becomes less. Hence the required breakdown voltage is also less as temperature is increased [see Fig. 2-21(b)]. Since increasing temperature reduces the Zener breakdown voltage, a zener diode is said to have a *negative temperature coefficient*.

Since avalanche breakdown requires relatively wide depletion regions, the charge carriers must transverse longer distances. Another point to note is that atoms in a solid tend to vibrate. The amplitude of their vibrations is directly proportional to temperature. Hence, as the temperature increases, the charge carriers moving through the *p-n* junction are more likely to collide with the atoms. Since the charge carriers have more collisions, the energy they impart is lowered. As a result, fewer ionizations occur and the multiplication of carriers is reduced. To increase the avalanche breakdown, the reverse breakdown voltage must be *increased*. Therefore, the avalanche breakdown voltage must increase as the temperature increases. The avalanche diodes therefore possess a *positive temperature coefficient*. This is illustrated in Fig. 2-21(c).

By convention, both zener and avalanche breakdown diodes are called zener diodes. Because of this widespread misnomer, it is important that we memorize the following points:

> **Zener diodes have breakdown voltages of less than 5 V and have a negative temperature coefficient.**
>
> **Avalanche diodes have breakdown voltages of greater than 5 V and have a positive temperature coefficient.**
>
> **Diodes that have breakdown voltages around 5 V have a zero temperature coefficient.**

2-10 Light-Emitting Diodes

Light-emitting diodes (LEDs) are used in their forward-bias mode and are employed as light sources. As we mentioned in Section 1-8, alloys between some trivalent elements and pentavalent elements behave as tetravalent semiconductors. Consequently, we mentioned that gallium arsenide (GaAs), gallium arsenide phosphide (GaAsP), and gallium phosphide (GaP) are the semiconductors used to make LEDs.

FIGURE 2-22 (a) Typical LED package; (b) LED construction; (c) LED operations; (d) schematic symbol.

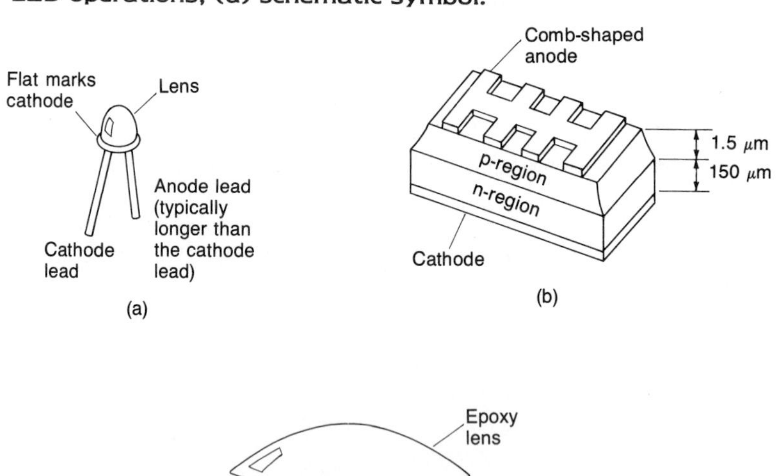

As we saw in Section 2-2 and Fig. 2-6, charge carrier recombination takes place in a forward-biased *p-n* junction as electrons cross from the *n*-side and recombine with holes on the *p*-side. Free electrons are in the conduction band energy levels, while holes are in the valence energy band. Therefore, the electrons are at higher energy levels than the holes. For the electrons to recombine with the holes, they must give up some of their energy. Typically, these electrons give up energy in the form of heat and light.

In silicon and germanium rectifiers, most of the electrons give up their energy in the form of heat. However, with GaAs, GaAsP, and GaP semiconductors, the electrons give up their energy by emitting *photons* (units of light energy). If the semiconductor is translucent, the light will be emitted and the junction becomes a light source.

LEDs made from GaAs emit invisible infrared light. LEDs constructed of GaAsP tend to emit either red or yellow light. The GaP LEDs provide either red or green light.

The typical construction of an LED is depicted in Fig. 2-22. Since recombination occurs in the *p*-type material, the actual light is emitted from the *p*-side of the junction. Therefore, the *p*-region forms the surface of the device. The anode connection to the *p*-type material is patterned to allow most of the emitted light to pass. This is accomplished by depositing a metal film around the outside edges of the *p*-type material. The metal film connection across the center of the *p*-type material has a comb-shaped pattern. A gold film is deposited along the bottom of the *n*-side. This provides the cathode connection and also serves to reflect as much light as possible to the *p*-type surface.

FIGURE 2-23 Forward V-I characteristics for a silicon diode and various gallium compound LEDs.

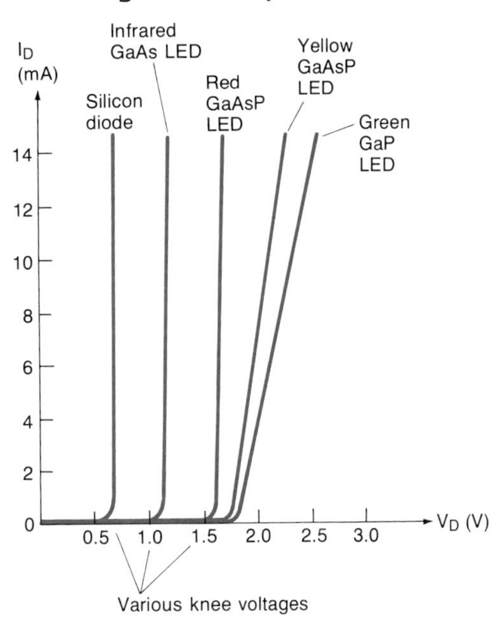

LEDs that emit visible light are widely used as indicators because of their extremely long life and low power consumption compared to incandescent bulbs. For example, highly efficient LEDs are currently available that operate with a forward voltage drop of 1.6 V and a forward current of only *2 mA*! LEDs that emit invisible infrared light find application in remote control schemes, object detectors, and burglar alarm systems.

The typical *V–I* characteristic curves of these LEDs are indicated in Fig. 2-23. Note that the forward knee voltages V_K are higher than those of a silicon or a germanium rectifier diode. Most LEDs also exhibit relatively low breakdown voltages. As a general rule, LEDs should *not* be reverse biased.

2-11 Diode Fabrication

In practice, the *p-n* junction is formed from a single monocrystalline lattice structure by adding carefully controlled amounts of donor and acceptor impurities. In the next few sections we outline some of the techniques employed in the manufacture of diode *p-n* junctions. Our aim is not to become experts in fabrication, but merely to become acquainted with the basic techniques and terminology. Many manufacturers of solid-state components will often indicate the particular fabrication techniques used to construct their devices. For example, the 1N4727 diode is described on one manufacturer's data sheet as a ''silicon planar epitaxial passivated diode.'' We need to be familiar with this terminology. The basic methods described here can also be extended to transistor and integrated circuit fabrication.

2-12 Floating-Zone Technique

The first step in the manufacture of any solid-state device is to purify the semiconductor material. Initially, this may involve subjecting the raw materials to a series of complex chemical reactions. To reduce the number of impurities further, and to ensure the formation of a monocrystalline (one-center) structure, a technique termed *floating zone* is quite often used. This is illustrated in Fig. 2-24.

The key to forming a monocrystalline structure is through the use of a small *seed* of semiconductor (e.g., silicon). The seed itself is a monocrystal that has been very carefully cut along the face of its cubic lattice. Support clamps are used to hold the low-purity polycrystalline rod.

Great care is taken to ensure that the semiconductor is not further contaminated during the floating-zone process. The rod, seed, and support clamps are placed in a quartz cylinder. The process may be carried out in a vacuum or by surrounding the semiconductor with an inert (little or no chemical reaction) gas.

Induction coils encompass the quartz container as shown in Fig. 2-24. The coils are excited by a radio-frequency (RF) voltage. The resulting magnetic field induces circulating (eddy) currents in the semiconductor. The semiconductor will then heat up due to I^2R power dissipation. Consequently, a molten region forms.

By slowly rotating the rod, the semiconductor atoms will tend to align themselves with the atoms in the monocrystalline seed. Therefore, as the induction coils are

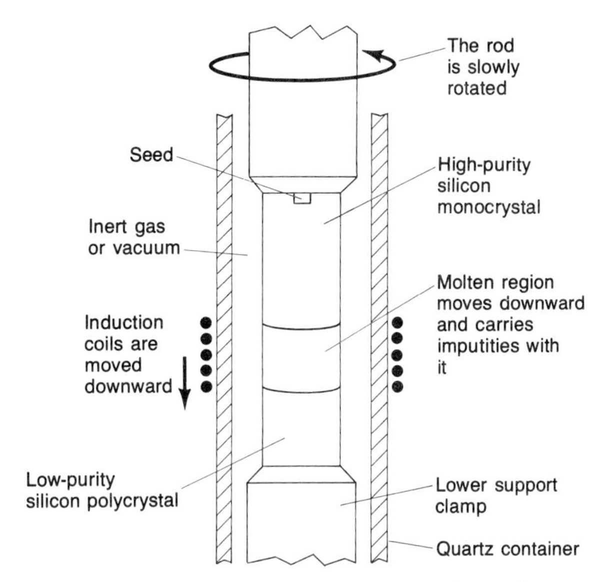

FIGURE 2-24 Floating zone technique for the formation of a high-purity silicon monocrystal.

moved downward, the molten region follows, and a monocrystalline structure continues to grow from the seed.

Purification of the semiconductor rod occurs simultaneously with the formation of the monocrystal. The impurities in the semiconductor rod will tend to become "more liquid" than the semiconductor. Consequently, as the induction coils are moved downward, the impurities will tend to follow the molten region. Once the induction coils have traversed the length of the semiconductor rod, the impurities will be collected in its lower end.

The lower end of the semiconductor rod may then be cut off, and the process repeated until the desired impurity level is reached. Typically, impurity levels of *less* than 1 part in 10^{10} semiconductor atoms are required. (This is roughly equivalent to one grain of salt in 10 buckets of sand.)

Once a pure monocrystalline semiconductor has been produced, we must carefully add impurities to form a *p-n* junction. Specifically, a carefully controlled amount of donor (*n*-type) and acceptor (*p*-type) impurities must be added to the semiconductor without disturbing the orderly monocrystalline structure. Two of the most important methods are *gaseous diffusion* and *epitaxial growth*.

2-13 Gaseous Diffusion

Once a rod of pure silicon monocrystal has been produced, it may be carefully sliced into thin disks as depicted in Fig. 2-25(a). Many diodes can be produced simultaneously on the same disk. To simplify our discussion, we present a side view of a portion of the silicon disk [Fig. 2-25(b)].

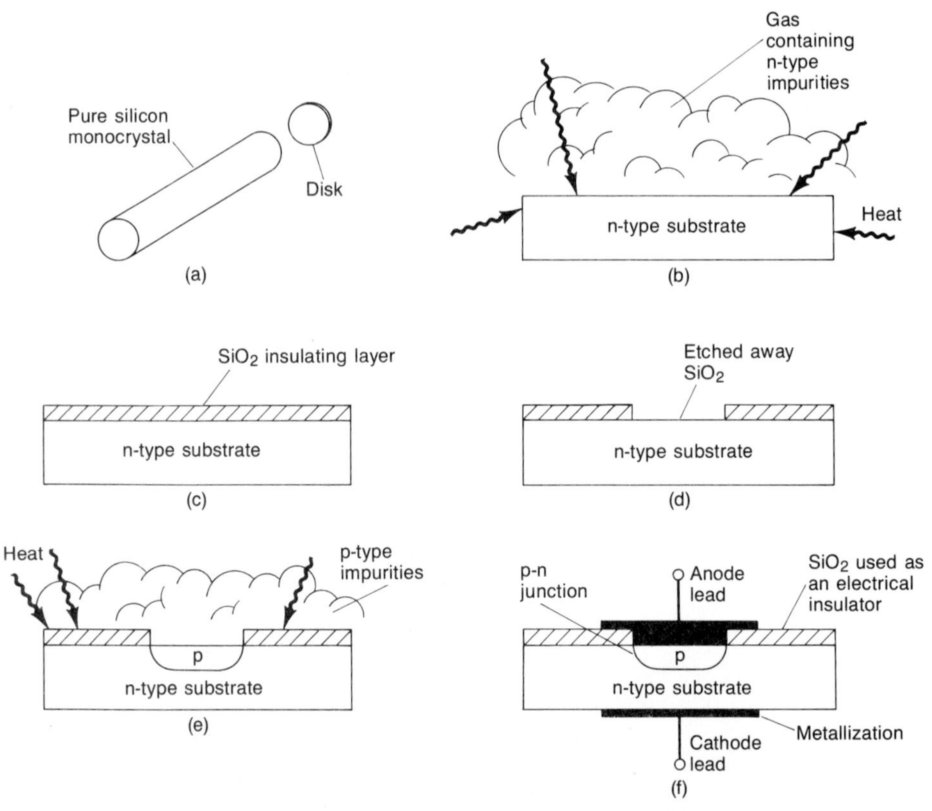

FIGURE 2-25 Gaseous diffusion method: (a) a thin disk is sliced off the rod; (b) forming the n-type substrate; (c) a layer of silicon dioxide results from passivation; (d) an opening is etched into the SiO_2 to permit the forming of the p region; (e) forming the p region via diffusion; (f) metallization and the attachment of leads.

Diffusion is a process in which a heavy concentration of atoms will move into a nearby region of lesser concentration. The diffusion process can be accelerated through the use of heat. This is depicted in Fig. 2-15(b). The silicon is heated and exposed to a gaseous atmosphere containing an *n*-type dopant such as antimony (Sb; refer to Table 1-1). The result is an *n-type substrate*. The substrate serves as the "chassis" on which the diode is built.

The next step is shown in Fig. 2-25(c) and is referred to as *passivation*. If pure oxygen is blown across a heated silicon substrate, the resulting chemical reaction produces a glasslike layer of silicon dioxide (SiO_2). This SiO_2 layer forms a protective coating that seals off the surface of the *n*-type semiconductor. SiO_2 is used to mask the substrate from additionally added impurity dopants. It is also an excellent protective cover for the surface of the device and an outstanding electrical insulator [refer to Fig. 1-5(a)]. Part of the SiO_2 layer is then etched away from the location of the *p*-region to be formed [see Fig. 2-25(d)].

The exposed area of the *n*-type substrate is then subjected to a gaseous atmosphere containing particles of an acceptor impurity (such as indium) and heated. The *p*-type impurity will then diffuse into the *n*-type substrate, counteract the *n*-type impurities, and form a *p*-region. The resultant boundary between the *p*- and *n*-regions is the *p-n* junction depicted in Fig. 2-25(f).

2-14 Epitaxial Growth Method

The epitaxial growth method is described in Fig. 2-26. The term "epitaxial" is derived from the Latin terms *epi*, which means "upon," and *taxis*, which means "arrangement." The fabrication initiates by heavily doping a pure semiconductor with *n*-type impurities. This heavily doped *n*-type material is designated as n^+. The n^+ region has a very low resistance, and it is used as a semiconductor extension of the electrical conductor that forms the cathode connection. This n^+ region serves as the substrate [see Fig. 2-26(a)].

The epitaxial layer is formed by growing additional semiconductor material on the substrate. The epitaxial layer is deposited on the substrate in a special piece of

FIGURE 2-26 Epitaxial growth method: (a) heavily doped substrate; (b) substrate and epitaxial layer; (c) passivation deposits an S_iO_2 layer and a hole is etched in it; (d) the p region is diffused in; (e) metallization and lead connections; (f) silicon planar epitaxial passivated diode.

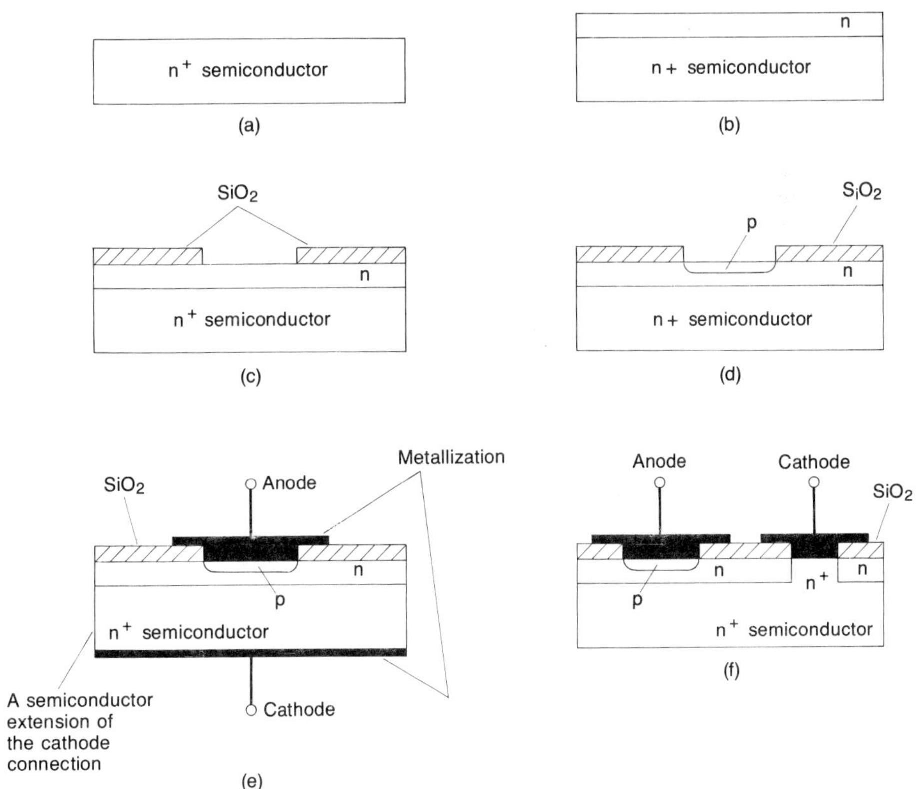

processing equipment called an epi reactor. The epitaxial layer results from the thermal decomposition of a gas that contains silicon, such as silicon tetrachloride ($SiCl_4$) or silane (SiH_4). By including *n*-type impurity dopants in the silicon gas, the resulting *n*-type epitaxial layer shown in Fig. 2-26(b) is formed. The *n*-type epitaxial layer is very uniformly doped. This uniform doping is *not* possible via the diffusion process since the concentration of impurity atoms would tend to be greatest at the surface of the semiconductor.

The next step involves forming an SiO_2 layer over the *n*-type semiconductor, and then etching a hole in it where the *p*-region is to be formed [see Fig. 2-26(c)]. If we then expose the semiconductor to gaseous diffusion, a *p*-region can be formed as shown in Fig. 2-26(d). The last step of the process is *metallization*. This involves depositing a metal film so that we may make our anode and cathode connections.

Observe in Fig. 2-26(f) that it is possible to carry out all of our fabrication steps (including the cathode connection) on one side of the semiconductor. Consequently, the surface is flat and the diode is termed *planar*. At this point, the 1N4727 ''silicon planar epitaxial passivated'' diode, mentioned in Section 2-11, should be a lot less intimidating.

As we pointed out in Chapter 1, tetravalent semiconductors may be formed by alloying trivalent elements (such as gallium) with pentavalent elements (such as arsenic). The resulting gallium compound semiconductor (e.g., gallium arsenide, or GaAs) is incorporated in the manufacture of LEDs and high-frequency semiconductor devices. (In high-speed digital integrated circuits, gallium arsenide clearly performs better than silicon-based devices.)

The gallium compound semiconductors are more difficult to dope than silicon or germanium. This is because they will decompose at elevated temperatures. For example, if GaAs were to be placed in a diffusion furnace, the arsenic (As) would tend to boil off. This would drastically alter the gallium-arsenide compound and would probably render it useless. Therefore, epitaxial techniques are employed since they require much lower temperatures. Hence a GaAs LED may have its *p*-region doped with zinc (valence 2) and its *n*-region doped with sulfur (valence 6) by using epitaxial techniques.

The diffusion and epitaxial methods lend themselves to the simultaneous fabrication of many hundreds of diodes on a single n^+-type substrate. These methods are also used extensively in the production of transistors and integrated circuits.

PROBLEMS

Drill, Derivations, and Definitions

Section 2-1

2-1. When the barrier potential forms at a *p-n* junction, negative ions are formed on the ____ (*p*-, *n*-) side and positive ions are formed on the ____ (*p*-, *n*-) side.

2-2. What are the barrier potentials for silicon, germanium, and the gallium compound semiconductors? What is our nominal value for the gallium compounds?

2-3. Why do electrons move from the *n*-type material to the *p*-type material when a *p-n* junction is formed? What prevents the entire semiconductor crystal from ionizing completely?

Section 2-2

2-4. When a *p-n* junction is forward biased the *p*-type material is more ____ (positive, negative) than the *n*-type material.

2-5. Electrons ____ (lose, gain) energy as they move from the *n*-type material to the *p*-type material.

2-6. The resistance of a forward-biased *p-n* junction is ____ (high, low).

2-7. The depletion region ____ (widens, narrows) as a result of forward bias on a *p-n* junction.

Section 2-3

2-8. When a *p-n* junction is reverse biased the *p*-type material is more ____ (positive, negative) than the *p*-type material.

2-9. In order for the electrons in the *n*-type material to go to the *p*-type material in a reverse-biased *p-n* junction, they must ____ (lose, gain) energy.

2-10. The resistance of a reverse-biased *p-n* junction is ____ (high, low).

2-11. The depletion region ____ (widens, narrows) as a result of reverse bias on a *p-n* junction.

2-12. Ideally, the reverse current through a *p-n* junction is ____ .

Section 2-4

2-13. Explain the reverse saturation current that is produced by the minority charge carriers.

2-14. How is surface leakage current produced?

2-15. Explain the relationship between reverse diode current (I_R), minority carrier current (I_S), and surface leakage current (I_{SL}).

Section 2-5

2-16. Draw the *V–I* characteristic curves for a 50-, a 100-, and a 200-Ω resistor. Place them on the same graph and label them.

2-17. Given that the slope of a line is its rise/run, what are the slopes of the three resistors given in Prob. 2-16? What are the units of the slopes-ohms or siemens?

Section 2-6

2-18. Make a neat sketch of the *V–I* characteristics of a rectifier diode. Indicate V_K, I_S, and PRV.

Section 2-7

2-19. A silicon diode has a reverse saturation current I_S of 1 nA at 25°C. Make a sketch of its reverse characteristics for 25, 30, 40, 50, and 60°C.

2-20. A germanium diode has a knee voltage of 0.3 V at 25°C. Make a sketch of its forward characteristic for 25, 35, 45, and 55°C. Label V_K in each case.

Section 2-8

2-21. Given that a zener diode breaks down at 3.6 V and the depletion region experiences an electric field intensity of 3×10^7 V/m, compute the width of the depletion region in centimeters.

2-22. Repeat Prob. 2-21 for a 4.7-V zener diode.

Section 2-9

2-23. Zener diodes have breakdown voltages _____ (above, below) 5 V and exhibit a _____ (positive, negative) temperature coefficient.

2-24. Avalanche diodes have breakdown voltages _____ (above, below) 5 V and exhibit a _____ (positive, negative) temperature coefficient.

2-25. _____ (Zener, Avalanche) breakdown results from ionization by an electric field.

Section 2-10

2-26. Explain the principle of operation of an LED.

2-27. Make a neat sketch of the structure of an LED, and label it completely.

Section 2-12

2-28. Describe the floating-zone process in your own words.

Section 2-13

2-29. Describe the gaseous diffusion method of forming a *p-n* junction.

Section 2-14

2-30. In your own words, describe the epitaxial growth method of forming a *p-n* junction.

Troubleshooting and Failure Modes

Carbon composition resistors tend to either increase in value or open when they fail. Wire-wound resistors tend to decrease in value or short when they fail. Even though solid-state components are highly reliable, they are more prone to fail than resistors. When rectifier and zener diodes fail, their most common failure mode is to short-circuit. Of course, if large currents continue to flow through them, they will eventually melt open. Diodes may be checked with an ohmmeter. Since diodes are nonlinear, the actual resistance indicated by the ohmmeter is relatively useless. However, the *ratio* of the reverse resistance reading to the forward resistance reading can be used as a quick check. A good diode should possess a ratio on the order of 1000:1.

2-31. Resistors R_1 and R_2 in Fig. 2-27 are carbon composition units. The voltage across R_2 is zero. Select the most likely causes from the following alternatives: (a) resistor R_1 is open, (b) resistor R_2 is shorted, (c) diode D_1 is shorted, (d) resistor R_2 is open, and (e) diode D_1 is open. Explain your reasons for selecting your answers.

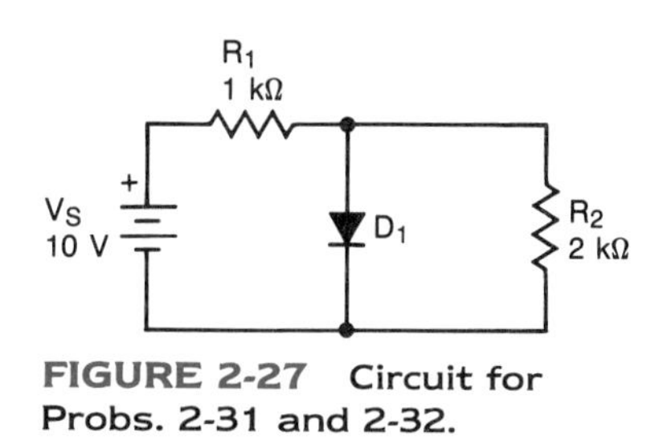

FIGURE 2-27 Circuit for Probs. 2-31 and 2-32.

2-32. The voltage across R_2 in Fig. 2-27 is 6.67 V. Normally, it should be 0.7 V. (This is explained in Chapter 3.) Indicate the most likely failure. Use the alternatives given in Prob. 2-31. Explain your reasons for selecting your answers.

Computer Problems

2-33. Write a computer program in BASIC that computes I_S at any given temperature in degrees Celsius. (Use Eq. 2-1.) It should prompt the user for I_S at 25°C, request the temperature in °C, and then ask the user ''LAST TEMPERATURE? (Y/N).'' If ''Y''es is entered, the program should loop back and request the temperature in °C. (The program should continue to use the previously entered I_S (25°C).) If ''N''o is entered, the program should print ''*** PROGRAM TERMINATED ***'' and end.

2-34. Write a computer program in BASIC that converts temperature in degrees Fahrenheit to degrees Celsius.

2-35. Merge your solutions to Probs. 2-33 and 2-34. Specifically, the program described in Prob. 2-33 should request the temperature in degrees Fahrenheit.

3

Diode Circuit Analysis

After Studying Chapter 3, You Should Be Able to:

- Use the Shockley diode equation to predict diode current.
- Use the Shockley diode equation at temperatures other than 25°C.
- Construct dc load lines for the analysis of rectifier, zener, and light-emitting diode circuits.
- Explain the effects of varying the load resistance or the supply voltage on the Q-point.
- Define and use the ideal, knee voltage, and dynamic resistance forward-bias diode models.
- Explain the difference between static and dynamic resistance.
- Apply the ideal and reverse current source models to analyze reverse-biased diode circuits.
- Apply the ideal and dynamic impedance models to analyze zener diode circuits.
- Employ Thévenin's theorem to analyze diode circuits and to assist in the selection of the proper diode model.

- Explain the effects of a signal on a diode's operating point and define the term ''small signal.''
- Use the diode small-signal model.
- Draw dc and ac equivalent circuits and recognize the role of coupling and decoupling capacitors.
- Describe the limitations in diode switching speed produced by diffusion capacitance, and the meaning of t_{rr}.
- Explain the operation and characteristics of the Schottky diode.
- Describe the nature of junction capacitance and its general relationship to the applied reverse bias.
- Explain the operation of the varactor diode.
- Analyze the operation of the half-wave rectifier and diode clipper circuits.
- Analyze the operation of the diode clamper circuit.
- Recognize diode basic applications and troubleshoot simple diode circuits.

3-1 Approaches to Diode Circuit Analysis

As we saw in Chapter 2, the rectifier, zener, avalanche, and light-emitting diodes are all *nonlinear* devices. Therefore, to analyze circuits containing these electronic components, we must develop some techniques to solve these nonlinear circuit problems. There are essentially *three* basic approaches to the solution of such problems: the use of nonlinear mathematics, graphical techniques, and the use of equivalent circuit models.

Nonlinear mathematics, by its very name, would intimidate all but the most dedicated electronics students. Fortunately, we find that its use is restricted, and we can avoid it for the vast majority of our problems. However, to investigate topics such as the amplitude modulation (AM) circuits found in some transmitters, the diode detector found in AM receivers, or the distortion found in audio power amplifiers, we must draw on some nonlinear concepts.

Graphical techniques can eliminate much of the complex algebra associated with nonlinear mathematics. They can also provide us with considerable insight as to the function of an electronic device or circuit. We employ graphical techniques extensively throughout the balance of the book. Specifically, we use them to perform both dc (bias) and ac (signal) analyses on electronic circuits.

Equivalent-circuit models will be our most powerful analysis and design tools. We shall develop the *models* (which are nothing more than electrical equivalent circuits) for all the solid-state devices that we encounter. We shall create models for both dc and ac analyses.

3-2 Shockley Diode Equation

As an example of a nonlinear mathematical expression, we turn to Eq. 3-1, which was developed by W. Shockley. The Shockley diode equation describes the *ideal* *V–I* characteristics of a *p-n* junction. Our reasons for considering it are threefold. First, it is a quantitative description of the *p-n* junction that serves to reinforce the qualitative descriptions given in Chapter 2. Second, it illustrates a very useful nonlinear mathematical relationship. Third, it provides us with the background to understand additional concepts, such as the equation for dynamic junction resistance r_j that we employ in our ac model of the diode.

$$I_D = I_S(e^{V_D/\eta V_T} - 1) \tag{3-1}$$

where I_D = diode current (amperes)

I_S = reverse saturation current (amperes)

e = natural number, 2.71828 · · ·

V_D = bias voltage across the diode (volts)

η = eta, a constant: 1 for Ge; $\simeq$ 2 for Si

V_T = voltage equivalent of temperature (volts)

The reverse saturation current for a particular diode is determined by its temperature, doping levels, and the geometry of its junction. Similarly, the constant η combines several of the physical parameters of the *p-n* junction together. The voltage equivalent of temperature V_T is a unique term. It is defined by Eq. 3-2. Observe that Boltzmann's constant is used to convert a given absolute temperature (in kelvin) into its equivalent energy in joules. Also recall that 1 volt is equal to 1 joule per coulomb.

$$\boxed{V_T = \frac{kT}{q}} \qquad (3\text{-}2)$$

where V_T = voltage equivalent of temperature (volts)

k = Boltzmann's constant, 1.381×10^{-23} J/K (J/K means joules per degree kelvin)

T = absolute temperature (kelvin)

q = charge on one electron, 1.602×10^{-19} C

Since room temperature is normally taken to be 25°C (about 77°F), we may convert this temperature to its equivalent absolute temperature by applying

$$\boxed{T(\text{K}) = T(°\text{C}) + 273} \qquad (3\text{-}3)$$

where T(K) is the absolute temperature in kelvin and T(°C) is the temperature in degrees Celsius. Therefore, to find the voltage equivalent of temperature at room temperature, 25°C, we must first employ Eq. 3-3. Hence

$$T(\text{K}) = T(°\text{C}) + 273 = 25 + 273 = 298 \text{ K}$$

and the voltage equivalent of temperature is found from Eq. 3-2.

$$V_T = \frac{kT}{q} = \frac{(1.381 \times 10^{-23} \text{ J/K})(298 \text{ K})}{1.602 \times 10^{-19} \text{ C}}$$
$$= 2.569 \times 10^{-2} \text{ J/C} \simeq 26 \text{ mV}$$

The result is an important number that should be committed to memory.

Substituting the 25°C voltage equivalent of temperature into Eq. 3-1, we have

$$I_D = I_S\,(e^{V_D/\eta V_T} - 1) = I_S\,\{e^{V_D/[\eta(26 \text{ mV})]} - 1\}$$
$$= I_S\,(e^{38.46 V_D/\eta} - 1)$$

To illustrate the use of the Shockley diode equation, consider Example 3-1.

EXAMPLE 3-1

The 1N914B is a very commonly used, diffused junction silicon diode. According to one manufacturer, the 1N914B has a maximum reverse current (I_R) of 25 nA at 25°C. Use the Shockley diode equation to obtain enough data points to make a sketch of its *V–I* characteristic curve.

SOLUTION Since the 1N914B is a silicon diode, we shall assume that η is equal to 2. If the diode is forward-biased with 0.7 V and we let $I_S \simeq I_R$,

$$I_D = I_S\,(e^{19.23V_D} - 1) = (25 \times 10^{-9})\,[e^{(19.23)(0.7)} - 1]$$
$$= (25 \times 10^{-9})(701.5 \times 10^{+3} - 1) = 17.5 \text{ mA}$$

and if the diode is reverse biased with -0.1 V, we have

$$I_D = I_S\,(e^{19.23V_D} - 1) = (25 \times 10^{-9})\,[e^{(19.23)(-0.1)} - 1]$$
$$= (25 \times 10^{-9})\,(146.2 \times 10^{-3} - 1) = -21.35 \text{ nA}$$

The rest of the calculations are similar and have been summarized in Table 3-1. ■

Observe that the reverse bias (negative) produces a small negative reverse current. The negative signs indicate that the directions of these quantities are opposite to those of our defined reference conventions.

The *V–I* characteristic curve has been plotted in Fig. 3-1 using the data provided in Table 3-1. The typical *V–I* curve using the parameters provided by the manufacturer has also been plotted. Note that different scaling has been used on the *V–I* axes to exaggerate the diode's characteristics.

As is obvious from Fig. 3-1, the Shockley diode equation does *not* totally predict the *V–I* characteristics for the 1N914B. Let us consider the reasons for the discrepancies. In the forward-biased region, the diode has larger voltage drops than the Shockley diode equation predicts. This occurs because a real diode has forward resistance which results from the collisions between conduction band electrons and the semiconductor atoms. This resistance is referred to as the diode's *bulk resistance* r_B. Further, a real diode also exhibits some small *contact resistance* between its

TABLE 3-1
1N914B Diode *V–I* Values of Example 3-1

V_D	I_D
0.8 V	120 mA
0.6 V	2.56 mA
0.4 V	0.055 mA
0.2 V	0.001 mA
0.0 V	0.000 mA
−0.1 V	−21.35 nA
−1.0 V	−25.00 nA
−10 V	−25.00 nA
−100 V	−25.00 nA
−10 kV	−25.00 nA

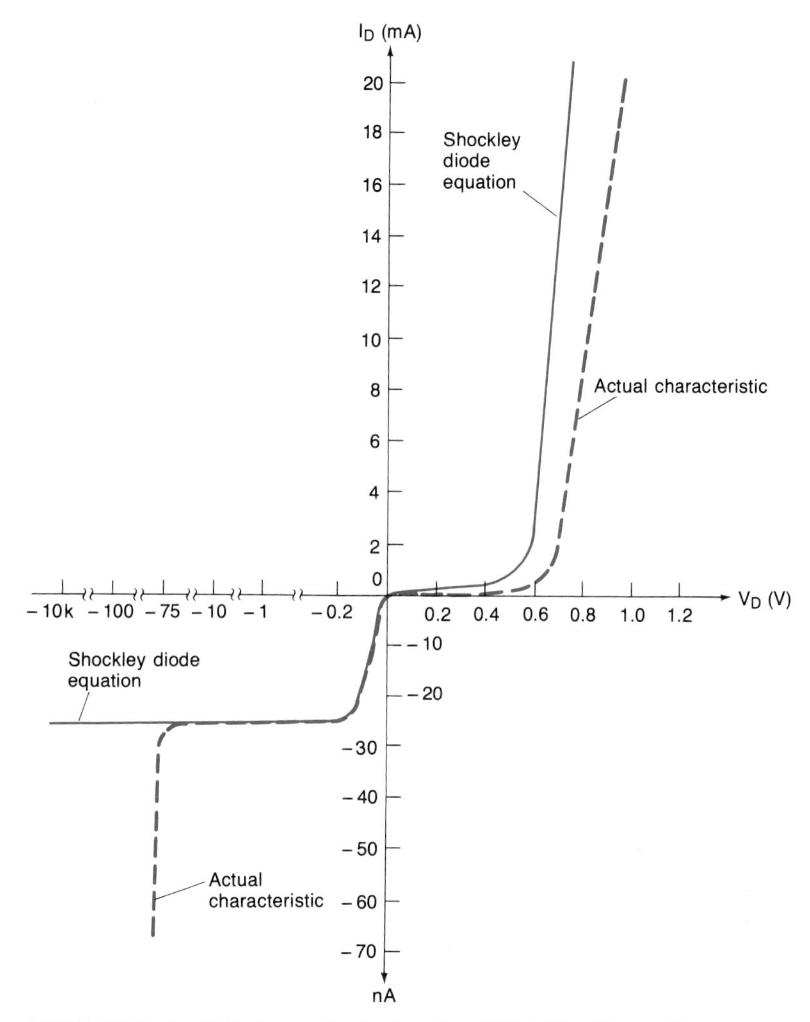

FIGURE 3-1 V-I characteristic of a 1N914B silicon diode.

semiconductor material and the metallic conductors used to form its anode and cathode terminal connections.

In the reverse-bias region, the Shockley diode equation indicates that the reverse current is constant for all reverse voltage greater than a few tenths of a volt. As indicated in Fig. 3-1, this is *not* true for a real diode. For large reverse voltages, the surface leakage current (which was discussed in Section 2-4) becomes significant. Further, if the reverse voltage becomes large enough, the diode will experience avalanche breakdown. (In Fig. 3-1 we see that the 1N914B breaks down at 75 V.) It is possible to improve the Shockley diode equation's fit to a real *V–I* characteristic curve by adjusting η, but we shall not delve into this.

Despite these shortcomings, the Shockley diode equation presents a reasonably accurate mathematical representation. For example, in Section 2-7 we studied the effects of temperature upon the diode *V–I* characteristics. The Shockley diode equation also reflects the effects of temperature. (Equation 2-1 is used to find $I_S(T)$ and Eq. 3-2 is used to determine V_T at any temperature T.)

This concludes our treatment of the nonlinear mathematical approach to finding diode voltages and currents. In the next section we introduce simple graphical techniques to analyze diode circuits. These graphical techniques eliminate the need for nonlinear mathematics.

3-3 DC Load Lines

The graphical approach to the analysis of nonlinear electronic devices is extremely straightforward. It can be applied to both dc (bias) and ac (signal) circuit problems. For dc and ac problems we employ *dc load lines* and *ac load lines*, respectively. In this section we study the dc load line. Consider the general case presented in Fig. 3-2. In Fig. 3-2(a) we have a two-terminal device connected to a *load resistor* R_L in series with a voltage source V_S. Our problem is to find the voltage across the device (V_{DEVICE}) and the current through it (I_{DEVICE}).

The procedure for using the dc load-line approach is as follows:

1. Obtain the V–I characteristic curve for the device.
2. If the current flow and the voltage across the device agree with our reference directions, we use its first quadrant. If the current and voltage disagree with our reference, we use the third quadrant.
3. Find the short-circuit current I_{SH} as shown in Fig. 3-2(b).
4. Find the open-circuit voltage V_{OC} as shown in Fig. 3-2(c).
5. Plot these two points in the appropriate quadrant of the V–I characteristic for the two-terminal device.
6. Connect these two points with a straight line. This is the dc load line.
7. The intersection between the dc load line and the V–I characteristic curve gives us the I_{DEVICE} and V_{DEVICE}. This intersection is called the *dc operating point* or the *quiescent point* (Q-point).

FIGURE 3-2 General circuit to illustrate the dc load line approach: (a) circuit; (b) the device is removed and replaced by a short; (c) the device is removed and replaced by an open.

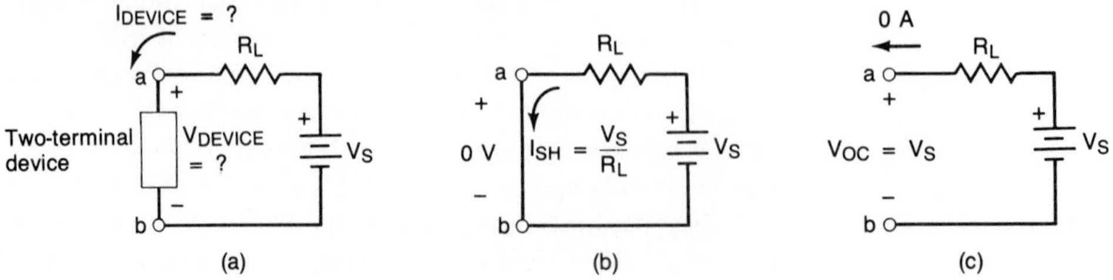

The dc load-line approach is valid for *all* two-terminal devices. It does not matter whether the device is linear or nonlinear. Consider Example 3-2. In this case, the two-terminal device is a 1-kΩ resistor.

EXAMPLE 3-2

Given the circuit shown in Fig. 3-3(a), find the dc operating point using the dc load-line approach.

SOLUTION Following the given procedure:

1. From our work in Section 2-5 we know that the *V–I* characteristic for a resistor (such as R_D) is linear, and it may be drawn as shown in Fig. 3-3(b).
2. Since R_D is bilateral, it really does not matter which direction current flows through it. However, to be consistent with our work in Chapter 2, we shall assume that operation occurs in the first quadrant [see Fig. 3-3(b)].
3. Removing R_D and replacing it with a short circuit as shown in Fig. 3-3(c), we may calculate I_{SH}.
$$I_{SH} = \frac{V_S}{R_L} = \frac{10\text{ V}}{1\text{ k}\Omega} = 10\text{ mA}$$
4. Removing R_D and replacing it with an open circuit as shown in Fig. 3-3(d), we may find V_{OC}.
$$V_{OC} = V_S = 10\text{ V}$$
5. We plot these two points ($V_{RD} = 0$ V, $I_{SH} = 10$ mA and $V_{RD} = V_{OC} = 10$ V, $I_{RD} = 0$ mA) as illustrated in Fig. 3-3(e).
6. Our dc load line is constructed by connecting these two points with a straight line [see Fig. 3-3(e)].
7. The intersection between our dc load line and the *V–I* characteristic for R_D is our *Q*-point. Hence
$$I_{DEVICE} = I_{RD} = 5\text{ mA} \quad \text{and} \quad V_{DEVICE} = V_{RD} = 5\text{ V}$$

■

From basic Ohm's law it should be clear that the dc load-line solution has given us the correct answer. Specifically,

$$R_T = R_D + R_L = 1\text{ k}\Omega + 1\text{ k}\Omega = 2\text{ k}\Omega$$
$$I_{RD} = \frac{V_S}{R_T} = \frac{10\text{ V}}{2\text{ k}\Omega} = 5\text{ mA}$$
$$V_{RD} = I_{RD}R_D = (5\text{ mA})(1\text{ k}\Omega) = 5\text{ V}$$

Obviously, a straight analytical approach using only mathematics is a great deal easier than the dc load-line approach. However, this is only true because our two-terminal device is a *linear* circuit element. When we are dealing with a *nonlinear* device such as a rectifier, zener, or light-emitting diode, a mathematical approach can become extremely involved. Consequently, the dc load-line approach is much more efficient. Consider Example 3-3.

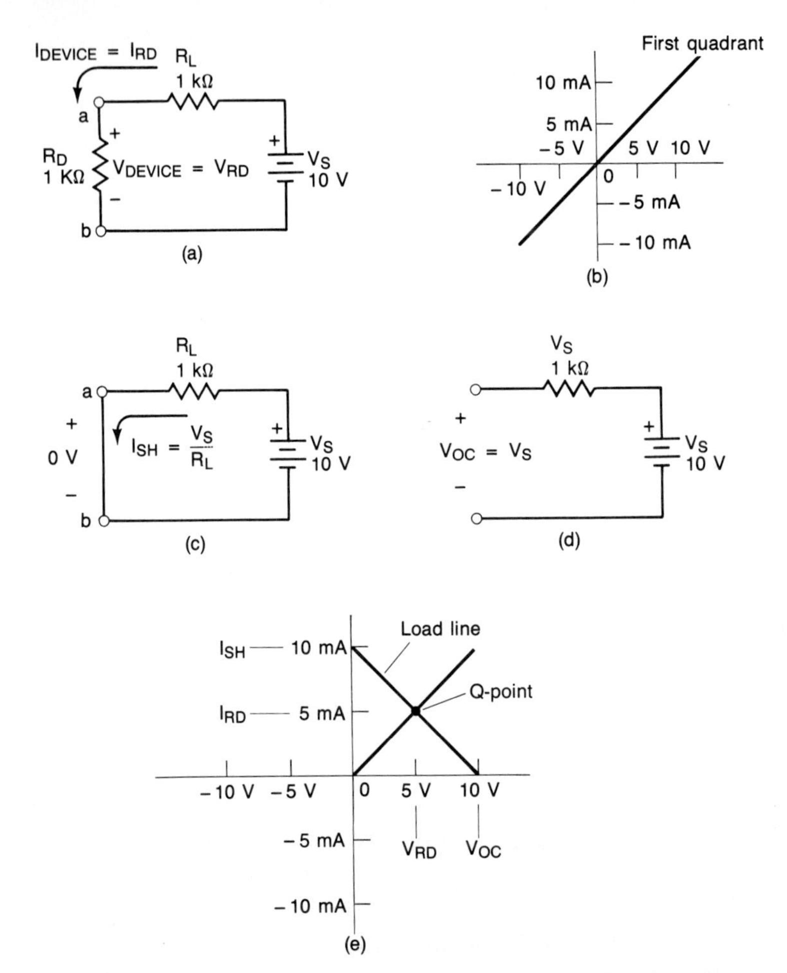

FIGURE 3-3 Dc load line for a resistor: (c) circuit to be analyzed; (b) resistor's V-I characteristic; (c) determining I_{SH}; (d) finding V_{OC}; (e) plotting the load line.

EXAMPLE 3-3

Find the current through the diode and the voltage across it in the circuit shown in Fig. 3-4(a).

SOLUTION Using the same general procedure gives us

$$I_{SH} = \frac{V_S}{R_L} = \frac{3\text{ V}}{200\ \Omega} = 15\text{ mA}$$

$$V_{OC} = V_S = 3\text{ V}$$

These two points are used to draw the dc load line in the first quadrant since the diode is forward biased. By inspection of Fig. 3-4(b), the Q-point gives

$$I_D = 11\text{ mA}$$

$$V_D = 0.8\text{ V}$$

■

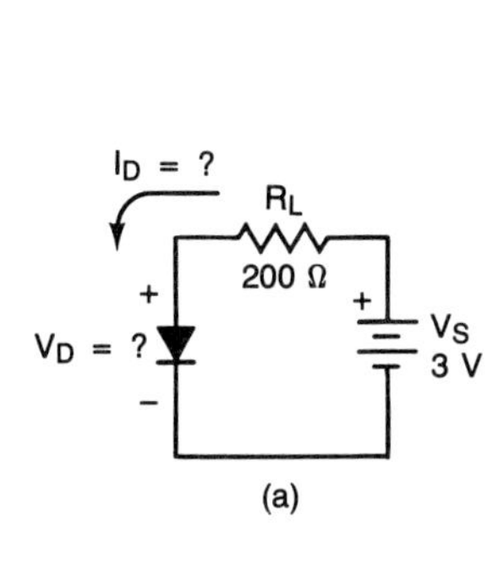

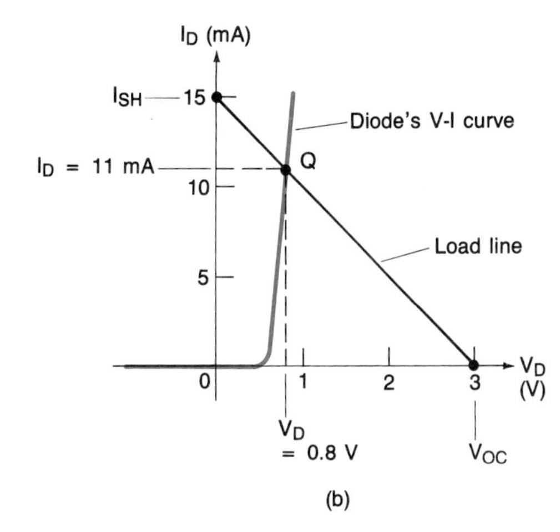

FIGURE 3-4 Dc load-line solution for a rectifier diode.

The dc load line is also valid for other devices, such as the zener diode and the LED. Consider Example 3-4.

EXAMPLE 3-4

The zener diode shown in Fig. 3-5(a) has a nominal breakdown voltage of 5.6 V. Find the current through it and the voltage across it.

SOLUTION We may follow exactly the same basic procedure. However, since the zener diode is being operated with a reverse bias (which is its normal operating mode), its Q-point will be in the third quadrant of its V–I characteristics. We find that

$$I_{\mathrm{SH}} = \frac{V_S}{R_L} = \frac{-10\ \mathrm{V}}{2\ \mathrm{k}\Omega} = -5.0\ \mathrm{mA}$$

$$V_{\mathrm{OC}} = -V_S = -10\ \mathrm{V}$$

and by inspection of Fig. 3-5(b),

$$I_D = -2.2\ \mathrm{mA}$$

$$V_D = -5.6\ \mathrm{V}$$

■

From the examples presented above, it is clear that the dc load-line approach may be used to analyze *any* linear or nonlinear two-terminal circuit element. To conclude our discussion, we consider the effect of changing R_L or V_S on the Q-point. In Fig. 3-6 we see that lowering R_L raises the Q-point, and increasing R_L lowers the Q-point.

In Fig. 3-7 we see that increasing V_S will not only increase I_{SH} but will also increase V_{OC}. Lowering V_S will decrease I_{SH} *and* V_{OC}. Therefore, increasing V_S raises the Q-point, and lowering V_S lowers the Q-point.

Study Fig. 3-7 carefully. The information presented can provide us with the background necessary to understand the relationships between an ac signal and the operating point of a solid-state device. Therefore, it is recommended that the student verify the I_{SH} and V_{OC} values given in Fig. 3-7.

The graphical procedure can provide us with considerable insight as to the

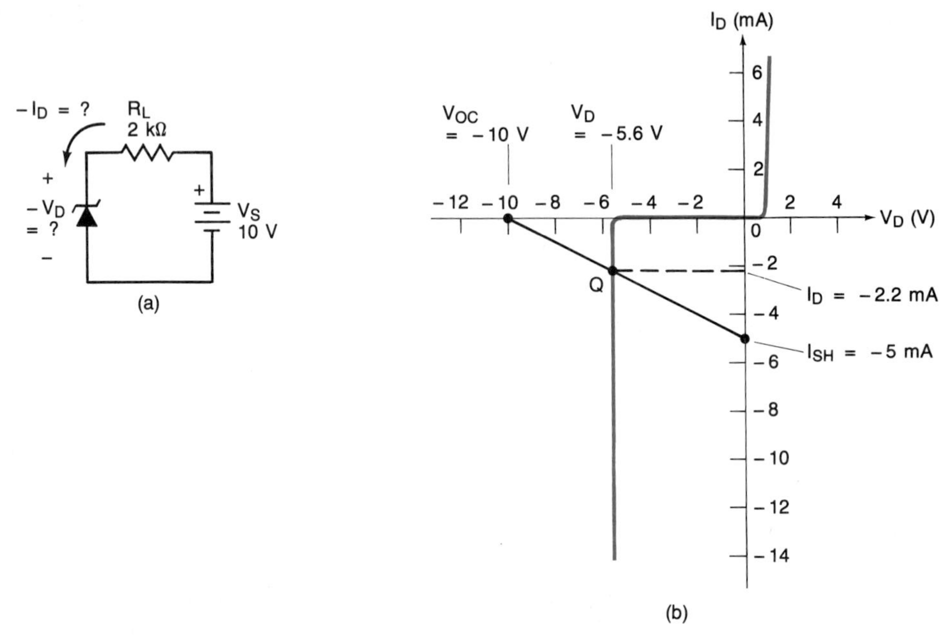

FIGURE 3-5 Dc load-line solution for a zener diode.

operation of a nonlinear circuit element. However, graphical techniques have two inherent disadvantages. First, we must have the *V–I* characteristic curves for the device. (These may be obtained by following the procedure outlined in Section 2-5, using a commercial curve tracer or sketched using a manufacturer's data.) Second, the accuracy of graphical techniques is limited by neatness and how closely we can read the graph. In the next few sections we develop dc models for the diode. This will allow us to analyze diode circuits without the need for nonlinear mathematics and a minimal use of *V–I* characteristic curves.

FIGURE 3-6 Effect of R_L on the Q-point.

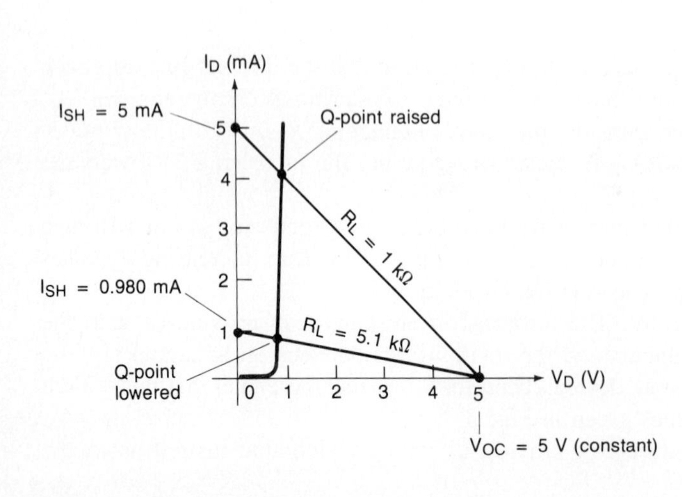

FIGURE 3-7 Effect of V_S on the load line.

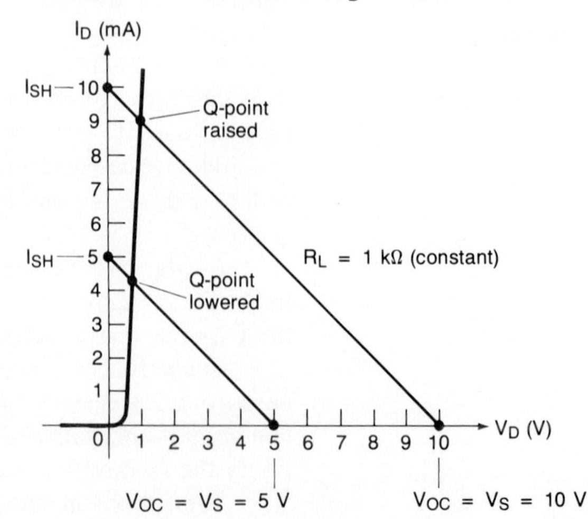

3-4 Diode DC Models: The Ideal Diode

As we mentioned in Section 3-1, the approach we emphasize in the analysis of electronic devices involves the use of *models* or equivalent circuits. For the present we shall be most interested in the *piecewise linear* models.

A piecewise linear model is an electrical equivalent circuit of a nonlinear electronic device. It is composed of linear circuit elements arranged to approximate the characteristics of the electronic device in a particular region of its operation.

In this section we focus our attention on the simplest model we have for the diode—the *ideal diode model*. As we have seen, a forward-biased diode has a very small forward voltage drop across it. A short circuit or a closed switch has zero volts dropped across it, independent of the magnitude or the direction of the current through it [see Fig. 3-8(a)]. An open circuit or an open switch has zero current through it, independent of the magnitude, or the polarity of the voltage across it [see Fig. 3-8(b)]. By combining these concepts, we can model the diode as shown in Fig. 3-8(c) and (d).

To summarize, a forward-biased diode is approximately a closed switch, and a reverse-biased diode is approximately an open switch. This simple model can, in general, be applied to any of the diodes we have studied. The ideal diode model will be designated as shown in Fig. 3-8(e).

3-5 Knee-Voltage Model

The ideal diode model provides a reasonably good first-order approximation in many problems. However, we typically need more accuracy. This can be achieved by recalling that the forward voltage drop across a diode is approximately constant and equal to its knee voltage V_K. (Recall that the knee voltages for germanium, silicon, and the gallium compounds are about 0.3, 0.7, and 1.6 V, respectively.) To incorporate the knee voltage drop into our diode model, we must first recall the definition of an *ideal voltage source*.

Ideal voltage sources are network elements that maintain a constant voltage across their terminals independent of the magnitude or the direction of the current through them.

The *V–I* characteristic of an ideal voltage source is provided in Fig. 3-9(a). As can be seen in Fig. 3-9(b), a forward-biased diode can be replaced by an ideal voltage source in series with the ideal diode model. In Fig. 3-9(c) we should note that no current flows through the diode model until the anode of the ideal diode becomes *more positive* than its cathode. This occurs when the bias voltage across the knee-voltage model attempts to exceed V_K. At that point, the ideal diode switch closes and V_D is clamped to V_K. In the reverse-bias condition, the ideal diode is simply an open switch as shown in Fig. 3-9(c).

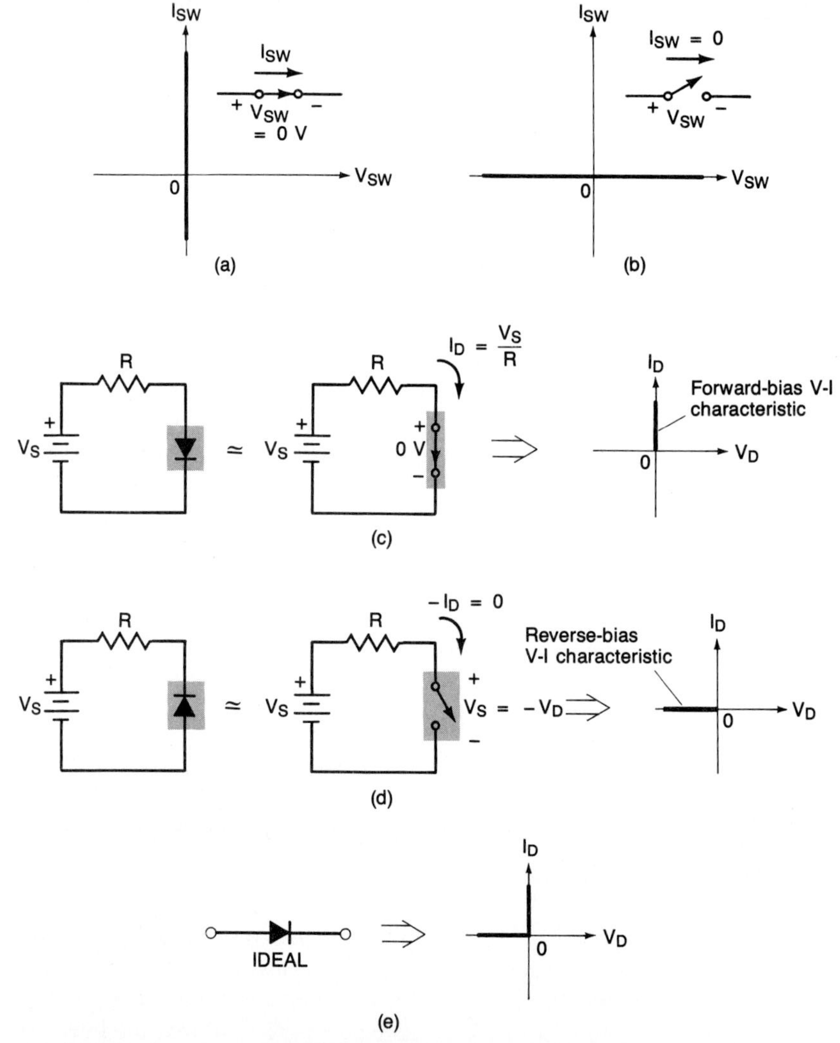

FIGURE 3-8 **Development of the ideal diode model: (a) V-I characteristic of a closed switch; (b) an open switch; (c) the ideal diode model of a forward-biased diode; (d) the ideal diode model of a reverse-biased diode; (e) the ideal diode symbol and V-I characteristic.**

3-6 Static and Dynamic Resistance

The knee-voltage diode model is the one we utilize most often. However, if we require even greater precision, we must recall that the forward voltage drop across a diode is *not* constant. The forward voltage drop tends to increase as the current through the diode increases. This is due to the fact that the diode *V–I* characteristic

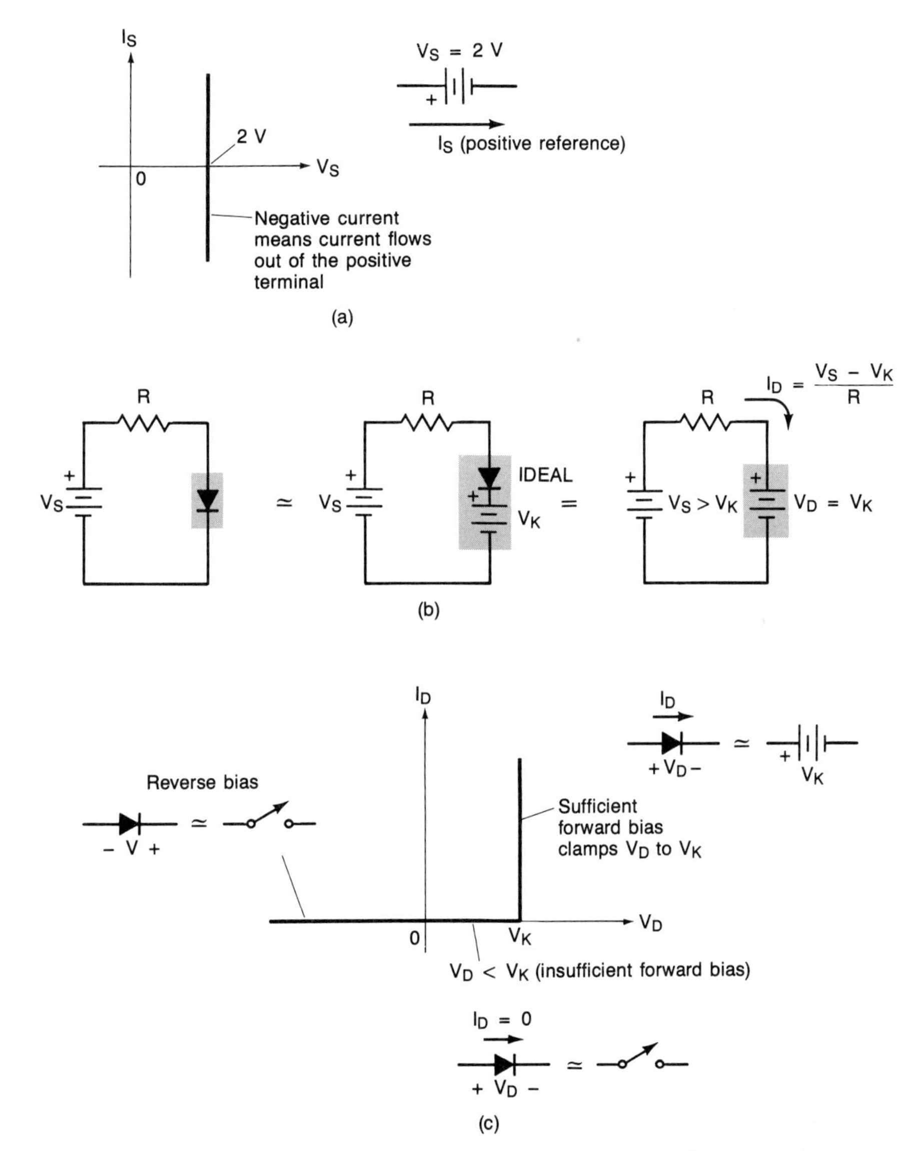

FIGURE 3-9 **Developing the knee-voltage diode model: (a) voltage source V-I characteristic; (b) forward bias; (c) knee voltage model V-I characteristic.**

has a forward slope as predicted by the Shockley diode equation. Diodes also have bulk and contact resistance as explained in Section 3-2.

Before we add the effects of resistance in our diode model, we need to expand on our understanding of resistance. From basic circuit theory we recall the definition of resistance:

> **Resistance is that circuit property which opposes the flow of an electric current. It is given by the ratio of the potential difference between two points to the current flow between them.**

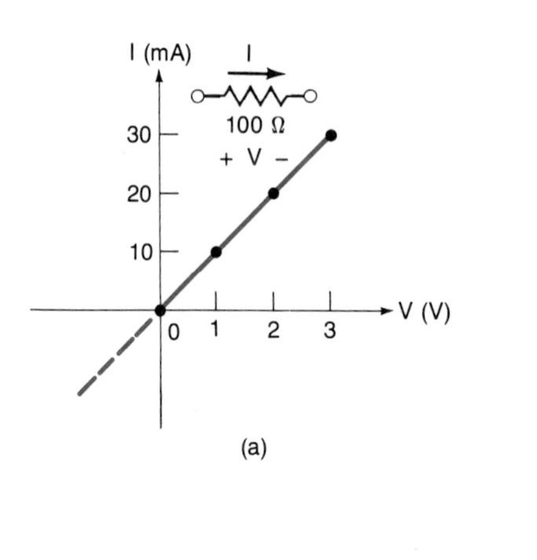

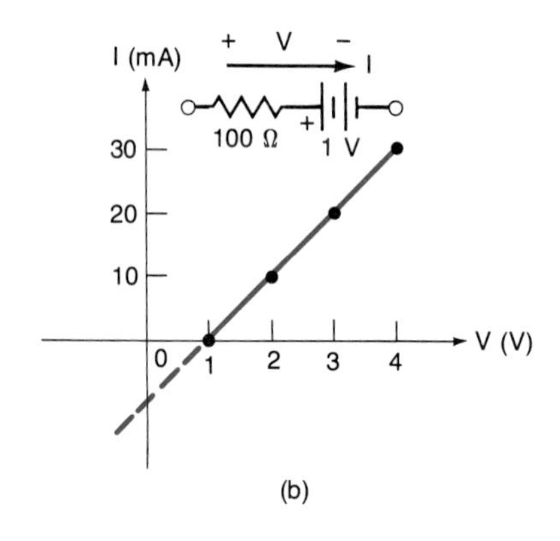

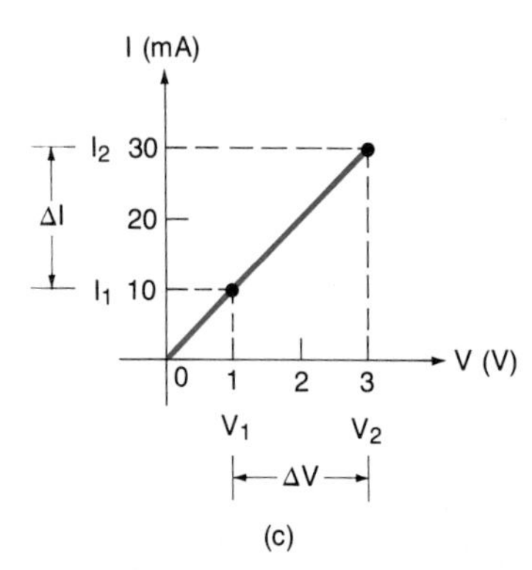

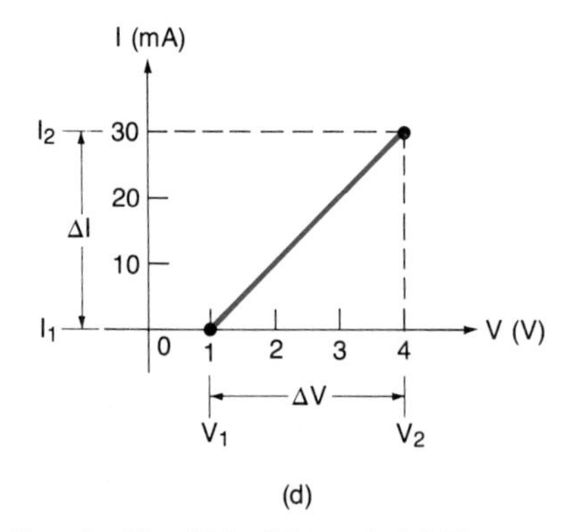

FIGURE 3-10 Resistance computation in the V-I plane: (a) V-I curve of a 100-Ω resistor; (b) V-I curve of the resistor in series with a 1-V source; (c) r = ΔV/ΔI = 100 Ω; (d) r = ΔV/ΔI = 100 Ω.

In the case of a resistor, we simply divide the voltage across it by the current through it [see Fig. 3-10(a)]. We see that at any point on its *V–I* curve, the resistance is a constant (i.e., 100 Ω). Specifically, we see that 1 V/10 mA = 100 Ω, 2 V/20 mA = 100 Ω, and so on. The resistance of a resistor is constant for *two* reasons. First, the *V–I* characteristic is linear, and second, the *V–I* characteristic passes through zero.

Both of these constraints must hold true for the resistance to be a constant. In Fig. 3-10(b) we see a characteristic curve for a 1-V voltage source in series with a 100-Ω resistor. The curve is still linear, but it is *offset* from zero. As can be seen in Fig. 3-10(b), at 2 V we have a current of 10 mA, at 3 V we have a current of 20 mA, and at 4 V the current is 30 mA. If we divide these voltages by their corresponding currents, we obtain 200, 150, and 133 Ω, respectively.

The reason the resistance appears to change is because we are finding the total resistance offered by the resistor and voltage source *combination*. What we have found is the dc (static) resistance at three particular operating points. Ohm's law is still valid, but we must be careful how we interpret our results.

To determine the value of the resistance in series with the voltage source (if it were unknown), we must slightly modify our definition of resistance.

Resistance is that circuit property which opposes the flow of an electric current. It is given by the ratio of the *change* in the potential difference between two points to the corresponding *change* in the current flow between them.

Note that the primary difference in this definition of resistance is the use of the term "change." Stated mathematically, we have

$$R = \frac{\Delta V}{\Delta I} = \frac{V_2 - V_1}{I_2 - I_1} \tag{3-4}$$

where R = resistance (ohms)

ΔV = change in voltage (volts)

ΔI = change in current (amperes)

Recall that the triangular symbol in Eq. 3-4 is called delta. It is used to denote a finitely large change in a quantity.

In Fig. 3-10(c) we have repeated the *V–I* curve for the 100-Ω resistor shown in Fig. 3-10(a). The change in voltage has been taken to be from 1 V to 3 V, with the corresponding change in current being from 10 mA to 30 mA. Applying Eq. 3-4, we have

$$R = \frac{\Delta V}{\Delta I} = \frac{V_2 - V_1}{I_2 - I_1} = \frac{3\text{ V} - 2\text{ V}}{30\text{ mA} - 20\text{ mA}} = \frac{1\text{ V}}{10\text{ mA}} = 100\ \Omega$$

This is exactly the same result as that obtained previously.

If we had taken the change in voltage to be from 0 to 3 V, and the corresponding change in current from 0 to 30 mA, we also arrive at the same result. Specifically,

$$R = \frac{\Delta V}{\Delta I} = \frac{V_2 - V_1}{I_2 - I_1} = \frac{3\text{ V} - 0\text{ V}}{30\text{ mA} - 0\text{ mA}} = \frac{3\text{ V}}{30\text{ mA}} = 100\ \Omega$$

From the above it should be clear that the "Ohm's law" that we were taught in basic circuit theory is merely a special case of our expanded definition of resistance.

Now let us apply Eq. 3-4 to the problem of determining the value of the resistance in series with the voltage source [Fig. 3-10(b)]. We shall let the change in current be from 0 to 30 mA [see Fig. 3-10(d)]. Hence

$$R = \frac{\Delta V}{\Delta I} = \frac{V_2 - V_1}{I_2 - I_1} = \frac{4\text{ V} - 1\text{ V}}{30\text{ mA} - 0\text{ mA}} = \frac{3\text{ V}}{30\text{ mA}} = 100\ \Omega$$

and we find the expected result.

We conclude our discussion by introducing two terms: *static resistance* and *dynamic resistance*.

Static resistance (R) is the ratio of the dc (Q-point) voltage across a device to the dc (Q-point) current through the device.

Dynamic resistance (r) is the ratio of the change in voltage across a device to the change in current through the device.

The dynamic resistance is usually denoted by a lower case letter (r). Dynamic resistance is a very important concept, and we shall continue to draw on it in our endeavors to come.

3-7 Dynamic Resistance Diode Model

To account for the slope in a diode's forward *V–I* characteristic, as depicted in Fig. 3-11(a), we develop the piecewise linear *dynamic resistance diode model* shown in Fig. 3-11(b). The *V–I* curve for the model is also provided in the figure.

To determine the values of the *offset voltage* V_O and the *dynamic resistance* r_O to be used in our model, we employ a graphical approach. The general procedure is outlined below.

1. Obtain the forward-bias *V–I* curve for the diode.
2. Approximate the *Q*-point (Q') by using the knee-voltage model to determine a preliminary value for I_D.
3. Draw a tangent line at Q'.
4. Determine ΔV and ΔI for the tangent line. The ratio gives r_O.
5. The intersection between the tangent line and the voltage axis yields V_O.
6. Recalculate I_D and V_D by using the dynamic resistance diode model.
7. If the *Q*-point (plotted by using *both* V_D and I_D) lies on the diode's *V–I* characteristic, we are done. If it does *not*, we must *repeat* steps 3 through 7, using the *Q*-point determined in step 6.

This procedure may be described as ''iterative'' (in *polite* terms) and can result in a high degree of accuracy.

To obtain the required data to model a given diode, we can quite often draw on the manufacturer's data sheet. An example has been illustrated in Fig. 3-12. Consider the 1N917.

FIGURE 3-11 Real diode versus the dynamic resistance diode model.

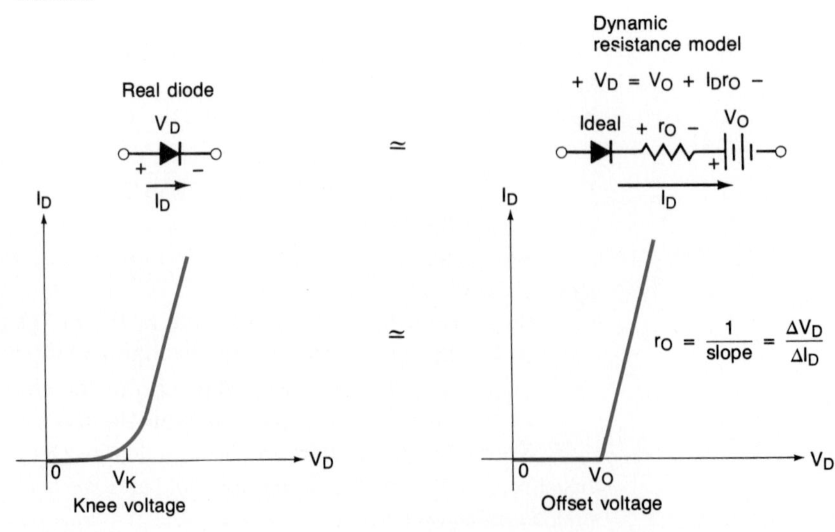

TYPES 1N914, 1N914A, 1N914B, 1N915, 1N916, 1N916A, 1N916B, 1N917 SILICON SWITCHING DIODES

1N914 SERIES AND 1N915

*electrical characteristics at 25°C free-air temperature (unless otherwise noted)

	PARAMETER	TEST CONDITIONS		1N914 MIN	1N914 MAX	1N914A MIN	1N914A MAX	1N914B MIN	1N914B MAX	1N915 MIN	1N915 MAX	UNIT
$V_{(BR)}$	Reverse Breakdown Voltage	I_R = 100 µA		100		100		100		65		V
I_R	Static Reverse Current	V_R = 10 V									25	nA
		V_R = 20 V			25		25		25			
		V_R = 20 V,	T_A = 100°C						3		5	µA
		V_R = 20 V,	T_A = 150°C		50		50		50			
		V_R = 50 V									5	
		V_R = 75 V			5		5		5			
V_F	Static Forward Voltage	I_F = 5 mA						0.62	0.72	0.6	0.73	V
		I_F = 10 mA	See Note 4		1							
		I_F = 20 mA					1					
		I_F = 50 mA									1	
		I_F = 100 mA							1			
C_T	Total Capacitance	V_R = 0,	f = 1 MHz		4		4		4		4	pF

1N916 SERIES AND 1N917

*electrical characteristics at 25°C free-air temperature (unless otherwise noted)

	PARAMETER	TEST CONDITIONS		1N916 MIN	1N916 MAX	1N916A MIN	1N916A MAX	1N916B MIN	1N916B MAX	1N917 MIN	1N917 MAX	UNIT
$V_{(BR)}$	Reverse Breakdown Voltage	I_R = 100 µA		100		100		100		40		V
I_R	Static Reverse Current	V_R = 10 V									50	nA
		V_R = 20 V			25		25		25			
		V_R = 20 V,	T_A = 100°C						3		25	µA
		V_R = 20 V,	T_A = 150°C		50		50		50			
		V_R = 75 V			5		5		5			
V_F	Static Forward Voltage	I_F = 0.25 mA									0.64	V
		I_F = 1.5 mA									0.74	
		I_F = 3.5 mA									0.83	
		I_F = 5 mA						0.63	0.73			
		I_F = 10 mA	See Note 4		1						1	
		I_F = 20 mA					1					
		I_F = 30 mA							1			
C_T	Total Capacitance	V_R = 0,	f = 1 MHz		2		2		2		2.5	pF

NOTE 4: These parameters must be measured using pulse techniques. t_w = 300 µs, duty cycle ≤ 2%.

operating characteristics at 25°C free-air temperature

	PARAMETER	TEST CONDITIONS		1N914, 1N914A, 1N914B, 1N916, 1N916A, 1N916B MIN	MAX	1N915 MIN	1N915 MAX	1N917 MIN	1N917 MAX	UNIT
t_{rr}	Reverse Recovery Time	I_F = 10 mA, R_L = 100 Ω,	I_{RM} = 10 mA, i_{rr} = 1 mA, See Figure 1 (Condition 1)		8		10*		3*	ns
		I_F = 10 mA, R_L = 100 Ω,	V_R = 6 V, i_{rr} = 1 mA, See Figure 1 (Condition 2)		4*					ns
$V_{FM(rec)}$	Forward Recovery Voltage	I_F = 50 mA,	R_L = 50 Ω, See Figure 2		2.5*					V
η_r	Rectification Efficiency	V_r = 2 V, Z_{source} = 50 Ω,	R_L = 5 kΩ, C_L = 20 pF, f = 100 MHz	45*						%

PARAMETER MEASUREMENT INFORMATION

FIGURE 3-12 Diode data sheet. (Reprinted by permission of Texas Instruments.)

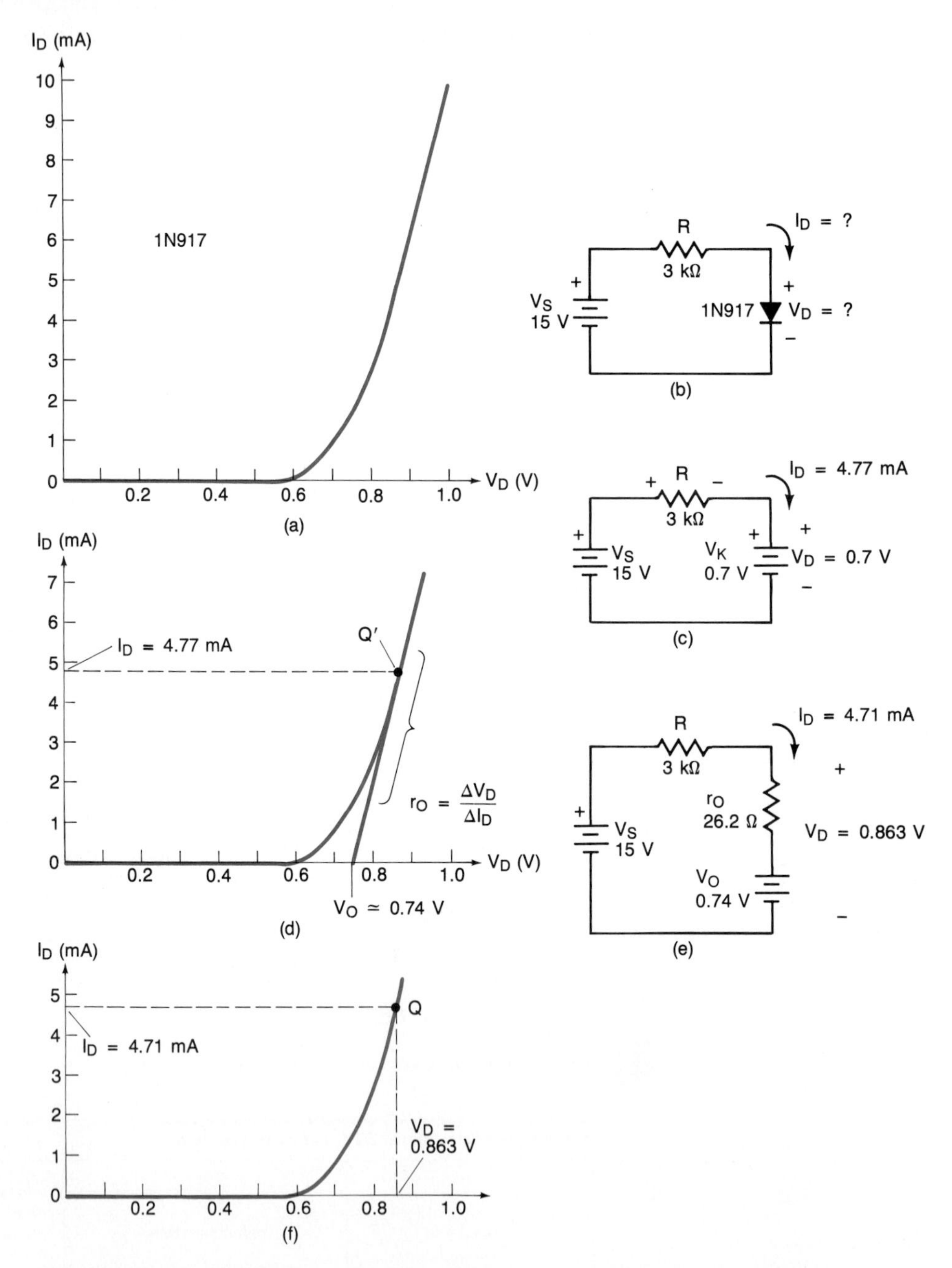

FIGURE 3-13 (a) V-I curve; (b) circuit; (c) using the knee voltage model to find Q′; (d) finding r_O and V_O; (e) reanalyzing using the dynamic resistance model; (f) resulting Q-point.

This silicon diode is designed to serve as a relatively high speed, low-current switch. The general applications of "switching diodes" are widespread and will be demonstrated in our later work. Many of the parameters and ratings associated with switching and rectifier diodes will be discussed later. The dynamic parameters, the capacitance C at a reverse voltage (V_R) of 0 V and the reverse recovery time (t_{rr}), will be introduced later in this chapter. Our immediate concern is with the forward voltage (V_F) and the current (I_F) values. These data will allow us to sketch the diode's V–I characteristic curve without a phone call to the manufacturer or trotting down to the lab [see Fig. 3-13(a)].

The relatively large forward voltage drops are produced by the diode's bulk resistance and make our 0.7-V knee-voltage model relatively inaccurate in this case. Therefore, we shall draw on the dynamic resistance diode model.

EXAMPLE 3-5

The V–I characteristic curve for a 1N917 silicon diode has been sketched in Fig. 3-13(a). Graphically determine the dynamic resistance model for the 1N917, and find the diode's Q-point if it is used in the circuit given in Fig. 3-13(b).

SOLUTION By following the procedure outlined above:

1. We replace the 1N917 diode with its knee-voltage model as shown in Fig. 3-13(c) and by using Kirchhoff's voltage law, we have
$$-V_S + I_D R + V_K = 0$$
and solving for I_D gives us
$$I_D = \frac{V_S - V_K}{R} = \frac{15\text{ V} - 0.7\text{ V}}{3\text{ k}\Omega} = 4.77\text{ mA}$$
2. Plotting $I_D = 4.77$ mA on the diode's V–I characteristic produces Q' [see Fig. 3-13(d)].
3. A straight line is drawn tangent at Q' as shown in Fig. 3-13(d).
4. Determining r_O in Fig. 3-13(d) yields
$$r_O = \frac{\Delta V_D}{\Delta I_D} = \frac{0.9\text{ V} - 0.74\text{ V}}{6.1\text{ mA} - 0\text{ mA}} = \frac{0.16\text{ V}}{6.1\text{ mA}} = 26.2\ \Omega$$
5. By inspection of Fig. 3-13(d), we have
$$V_O \simeq 0.74\text{ V}$$
6. Using the dynamic resistance diode model [Fig. 3-13(e)], we recompute I_D and find V_D.
$$I_D = \frac{V_S - V_O}{R + r_O} = \frac{15\text{ V} - 0.74\text{ V}}{3\text{ K}\Omega + 26.2\ \Omega} = 4.71\text{ mA}$$
and
$$V_D = V_O + I_D r_O = 0.74\text{ V} + (4.71\text{ mA})(26.2\ \Omega)$$
$$= 0.863\text{ V}$$
7. Plotting the resulting Q-point on the V–I curve as shown in Fig. 3-13(f) indicates that our result appears reasonable. ■

The dynamic resistance model is the most accurate piecewise linear model presented thus far. Even though graphical techniques were used to determine the values for r_O and V_O, the results are far more accurate and certainly more convenient to obtain than the dc load-line approach. Even so, the knee-voltage model will suffice in the vast majority of the problems we encounter.

3-8 Reverse Current Source Model

We have considered three different models of a forward-biased diode: the ideal diode model, the knee-voltage model, and the dynamic resistance model. Note that these models may be used for the rectifier, zener, and light-emitting diodes. Up to this

FIGURE 3-14 Development of the reverse current source diode model: (a) the reverse characteristics of a diode show that the reverse current is essentially constant; (b) current source V-I characteristic; (c) using the model.

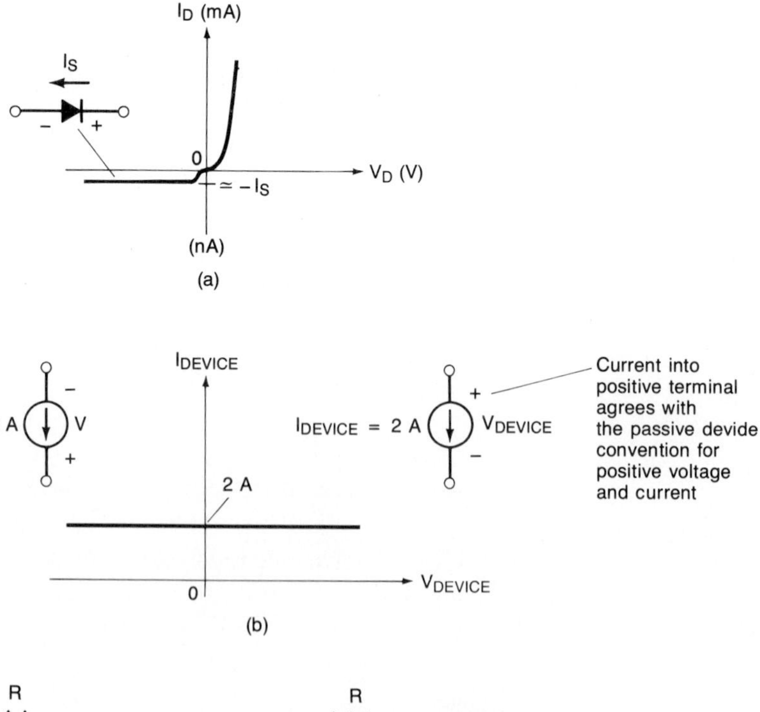

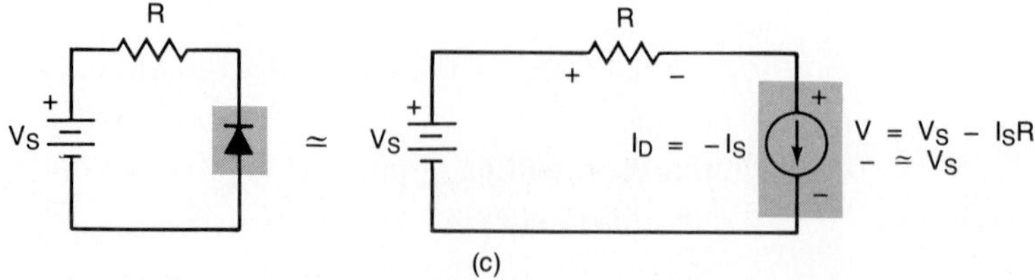

point we have regarded reverse-biased diodes to be equivalent to an open switch. In this section we refine our view of the reverse-biased diode.

Occasionally, a reverse-biased diode is approximated as a large resistance. However, this is not the best approach. By close inspection of Fig. 3-14(a) we see that the reverse current is essentially constant and independent of the magnitude of the reverse voltage across it. To model this reverse current effect, we may utilize an *ideal current source*.

> **Ideal current sources are network elements that maintain a constant current through their terminals independent of the magnitude, or the polarity of the voltage across them.**

The *V–I* characteristic of a current source is shown in Fig. 3-14(b). Notice that the passive device convention has been included for reference. The model of a reverse-biased diode that incorporates a current source is given in Fig. 3-14(c). Quite simply, when a diode is reverse biased, we may regard it as a constant-current source.

3-9 Ideal Zener Diode Model

The reverse-bias-diode models presented above apply only if the *p-n* junction is *not* broken down. Since the zener diode is normally operated in its reverse breakdown mode, we need to develop a model. From Fig. 3-15(a) we recall that a zener diode which has broken down will have essentially a constant voltage across it independent of the reverse current flow through it. Consequently, when it has broken down we may use an ideal diode in series with a voltage source to model it. This is depicted in Fig. 3-15(b).

Note that the ideal diode is an open circuit until the reverse bias across the zener reaches V_Z. Therefore, the current through the zener model is zero until breakdown occurs. To see how this model is used, study Example 3-6.

FIGURE 3-15 Ideal zener diode model: (a) real zener diode characteristic; (b) ideal zener model (similar to the knee voltage model of a forward-biased diode).

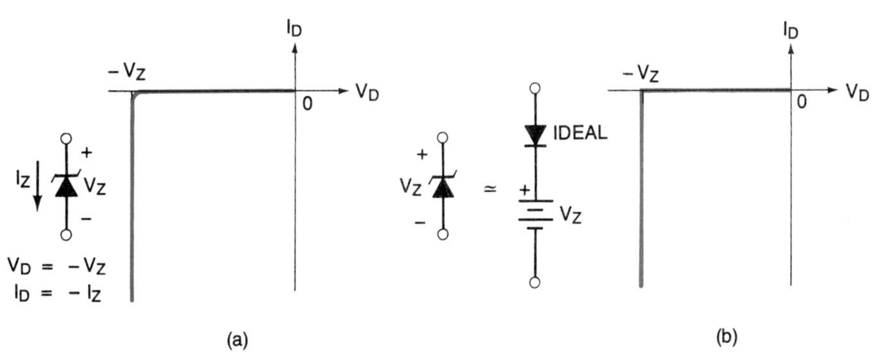

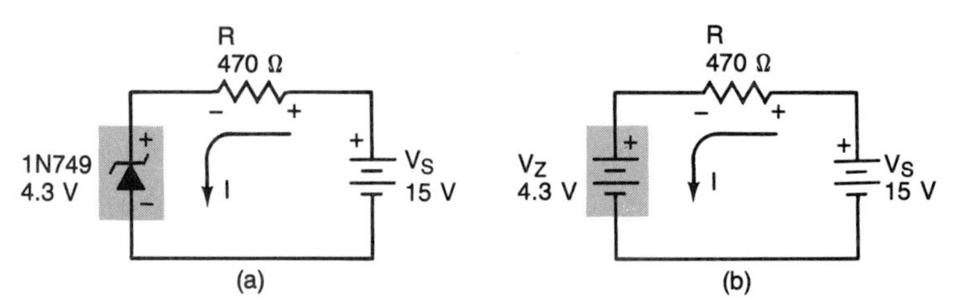

FIGURE 3-16 Circuits for Example 3-6.

EXAMPLE 3-6

Compute the current through the zener diode shown in Fig. 3-16(a).

SOLUTION The 1N749 is a 4.3-V zener diode. Since the voltage source is larger than 4.3 V and the diode is reverse biased, we shall assume that it is broken down. Therefore, we replace it with the model indicated in Fig. 3-16(b). (The ideal diode is a short circuit and has been omitted for clarity.) By applying Kirchhoff's voltage law, we write

$$-V_S + IR + V_Z = 0$$

and solving for I, we have

$$I = \frac{V_S - V_Z}{R} = \frac{15\text{ V} - 4.3\text{ V}}{470\ \Omega} = 22.8\text{ mA}$$ ■

3-10 Zener Dynamic Impedance Model

Careful inspection of the zener diode's characteristic curve indicates that the voltage across the zener will increase slightly as the current through it increases [see Fig. 3-17(a)]. The variation in the zener's terminal voltage occurs because it possesses some small resistance when it is broken down. This resistance is called the zener dynamic impedance Z_Z, and it is usually provided on the manufacturer's specification sheet.

To include the effect of Z_Z in a zener model, we must invoke the dynamic resistance concept presented in Section 3-6. This is more evident if we flip the zener's characteristic over into the first quadrant as illustrated in Fig. 3-17(b). This is accomplished by labeling the vertical axis $I_Z\ (-I_D)$ and the horizontal axis $V_Z\ (-V_D)$.

The zener's dynamic impedance model has been given in Fig. 3-17(c). Note that V_{ZO} is our notation for the zener's offset voltage. Figure 3-17(c) also depicts the model's approximation of the V–I characteristic curve. The computation of Z_Z is relatively straightforward, as illustrated in the figure.

As we mentioned earlier, the zener's Z_Z is often listed on the manufacturer's data sheet. However, Z_Z is often specified under a given set of test conditions. It is typically designated as Z_{ZT}. Consider Fig. 3-18.

The 1N5236 zener diode has a maximum power dissipation of 500 mW and a

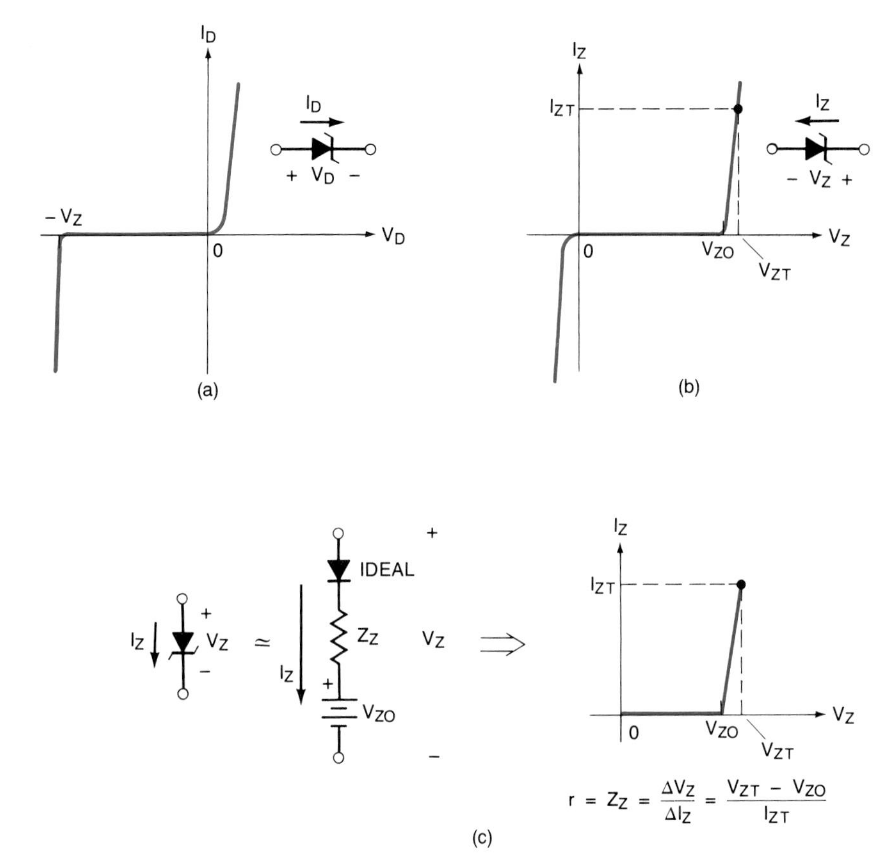

FIGURE 3-17 **Zener dynamic impedance model.**

nominal breakdown voltage (V_{ZT}) of 7.5 V at a reverse test current (I_{ZT}) of 20 mA. These test conditions have been indicated in the actual *V–I* characteristic curve and our approximation of it in Fig. 3-17(b) and (c), respectively.

We also note that the data sheet (Fig. 3-18) cites that the zener's maximum dynamic impedance under the test conditions (Z_{ZT}) is 6.0 Ω. To specify the model for the 1N5236 completely, we must also determine V_{ZO}.

EXAMPLE 3-7

Arrive at the dynamic impedance model for the 1N5236 zener diode using the data sheet parameters given in Fig. 3-18.

SOLUTION From Fig. 3-17(c) we note that

$$V_{ZT} = V_{ZO} + I_{ZT}Z_{ZT}$$

and solving for V_{ZO} yields

$$\begin{aligned} V_{ZO} &= V_{ZT} - I_{ZT}Z_{ZT} = 7.5 \text{ V} - (20 \text{ mA})(6.0 \ \Omega) \\ &= 7.38 \text{ V} \end{aligned}$$

This completely specifies our model for the 1N5236 (see Fig. 3-19). ■

1N5221 thru 1N5281 series (SILICON)

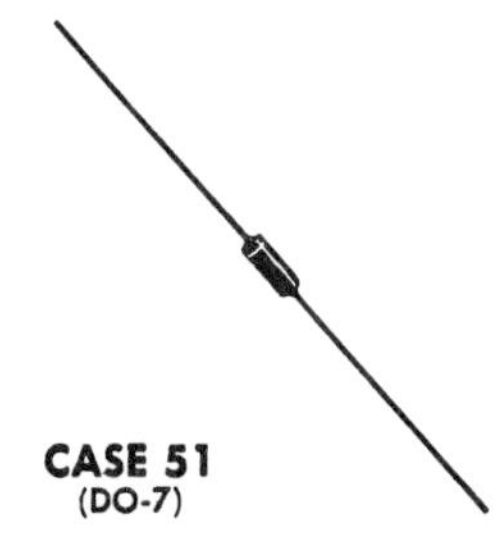

500 Milliwatt surmetic 20 silicon zener diodes-a complete series of Zener Diodes in the popular DO-7 case with higher ratings, tighter limits, better operating characteristics and a full set of designers' curves that reflect the superior capabilities of silicon-oxide-passivated junctions. All this in an axial-lead, transfer-molded plastic package offering protection in all common environmental conditions.

MAXIMUM RATINGS

Junction and Storage Temperature: −65 to +200°C

Lead Temperature not less than 1/16" from the case for 10 seconds: 230°C

DC Power Dissipation: 500 mW @ T_L = 75°C, Lead Length = 3/8"
(Derate 4.0 mW/°C above 75°C)

Surge Power: 10 Watts (Non-recurrent square wave @ PW = 8.3 ms, T_J = 55°C, Figure 16)

MECHANICAL CHARACTERISTICS

CASE: Void free, transfer molded, thermosetting plastic.

FINISH: All external surfaces are corrosion resistant. Leads are readily solderable and weldable.

POLARITY: Cathode indicated by color band. When operated in zener mode, cathode will be positive with respect to anode.

MOUNTING POSITION: Any.

WEIGHT: 0.18 gram (approximately).

FIGURE 1 — POWER-TEMPERATURE DERATING CURVE

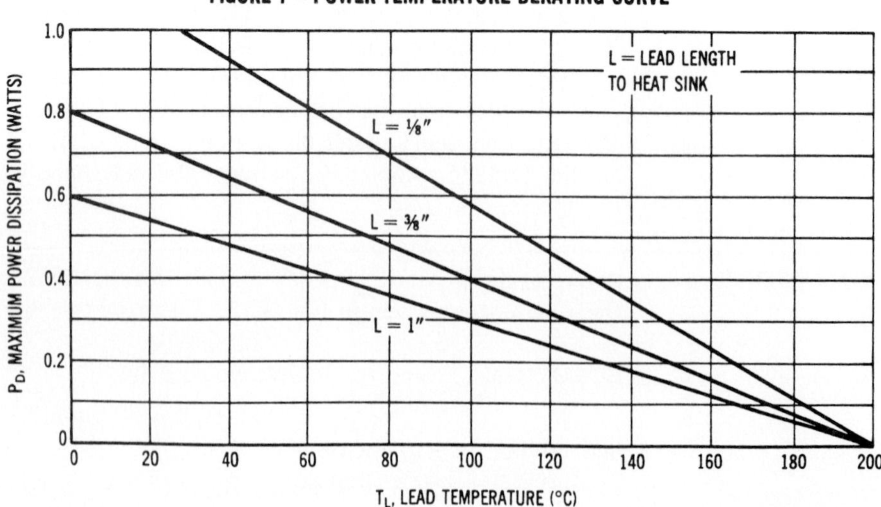

FIGURE 3-18 Zener diode data sheet. (Reprinted with permission of Motorola, Inc.)

ELECTRICAL CHARACTERISTICS (T_A = 25°C unless otherwise noted). Based on dc measurements at thermal equilibrium; lead length = 3/8"; thermal resistance of heat sink = 30°C/W) V_F = 1.1 Max @ I_F = 200 mA for all types.

JEDEC Type No. (Note 1)	Nominal Zener Voltage V_Z @ I_{ZT} Volts (Note 2)	Test Current I_{ZT} mA	Max Zener Impedance A & B Suffix Only		Max Reverse Leakage Current				Max Zener Voltage Temp. Coeff. (A & B Suffix Only) θ_{VZ} (% /°C) (Note 3)
					A & B Suffix Only			Non-Suffix	
			Z_{ZT} @ I_{ZT} Ohms	Z_{ZK} @ I_{ZK} = 0.25 mA Ohms	I_R µA	@ V_R Volts A	B	I_R @ V_R Used For Suffix A µA	
1N5221	2.4	20	30	1200	100	0.95	1.0	200	-0.085
1N5222	2.5	20	30	1250	100	0.95	1.0	200	-0.085
1N5223	2.7	20	30	1300	75	0.95	1.0	150	-0.080
1N5224	2.8	20	30	1400	75	0.95	1.0	150	-0.080
1N5225	3.0	20	29	1600	50	0.95	1.0	100	-0.075
1N5226	3.3	20	28	1600	25	0.95	1.0	100	-0.070
1N5227	3.6	20	24	1700	15	0.95	1.0	100	-0.065
1N5228	3.9	20	23	1900	10	0.95	1.0	75	-0.060
1N5229	4.3	20	22	2000	5.0	0.95	1.0	50	±0.055
1N5230	4.7	20	19	1900	5.0	1.9	2.0	50	±0.030
1N5231	5.1	20	17	1600	5.0	1.9	2.0	50	±0.030
1N5232	5.6	20	11	1600	5.0	2.9	3.0	50	+0.038
1N5233	6.0	20	7.0	1600	5.0	3.3	3.5	50	+0.038
1N5234	6.2	20	7.0	1000	5.0	3.8	4.0	50	+0.045
1N5235	6.8	20	5.0	750	3.0	4.8	5.0	30	+0.050
1N5236	7.5	20	6.0	500	3.0	5.7	6.0	30	+0.058
1N5237	8.2	20	8.0	500	3.0	6.2	6.5	30	+0.062
1N5238	8.7	20	8.0	600	3.0	6.2	6.5	30	+0.065
1N5239	9.1	20	10	600	3.0	6.7	7.0	30	+0.068
1N5240	10	20	17	600	3.0	7.6	8.0	30	+0.075
1N5241	11	20	22	600	2.0	8.0	8.4	30	+0.076
1N5242	12	20	30	600	1.0	8.7	9.1	10	+0.077
1N5243	13	9.5	13	600	0.5	9.4	9.9	10	+0.079
1N5244	14	9.0	15	600	0.1	9.5	10	10	+0.082
1N5245	15	8.5	16	600	0.1	10.5	11	10	+0.082
1N5246	16	7.8	17	600	0.1	11.4	12	10	+0.083
1N5247	17	7.4	19	600	0.1	12.4	13	10	+0.084
1N5248	18	7.0	21	600	0.1	13.3	14	10	+0.085
1N5249	19	6.6	23	600	0.1	13.3	14	10	+0.086
1N5250	20	6.2	25	600	0.1	14.3	15	10	+0.086
1N5251	22	5.6	29	600	0.1	16.2	17	10	+0.087
1N5252	24	5.2	33	600	0.1	17.1	18	10	+0.088
1N5253	25	5.0	35	600	0.1	18.1	19	10	+0.089
1N5254	27	4.6	41	600	0.1	20	21	10	+0.090
1N5255	28	4.5	44	600	0.1	20	21	10	+0.091
1N5256	30	4.2	49	600	0.1	22	23	10	+0.091
1N5257	33	3.8	58	700	0.1	24	25	10	+0.092
1N5258	36	3.4	70	700	0.1	26	27	10	+0.093
1N5259	39	3.2	80	800	0.1	29	30	10	+0.094
1N5260	43	3.0	93	900	0.1	31	33	10	+0.095
1N5261	47	2.7	105	1000	0.1	34	36	10	+0.095
1N5262	51	2.5	125	1100	0.1	37	39	10	+0.096
1N5263	56	2.2	150	1300	0.1	41	43	10	+0.096
1N5264	60	2.1	170	1400	0.1	44	46	10	+0.097
1N5265	62	2.0	185	1400	0.1	45	47	10	+0.097
1N5266	68	1.8	230	1600	0.1	49	52	10	+0.097
1N5267	75	1.7	270	1700	0.1	53	56	10	+0.098
1N5268	82	1.5	330	2000	0.1	59	62	10	+0.098
1N5269	87	1.4	370	2200	0.1	65	68	10	+0.099
1N5270	91	1.4	400	2300	0.1	66	69	10	+0.099
1N5271	100	1.3	500	2600	0.1	72	76	10	+0.110
1N5272	110	1.1	750	3000	0.1	80	84	10	+0.110
1N5273	120	1.0	900	4000	0.1	86	91	10	+0.110
1N5274	130	0.95	1100	4500	0.1	94	99	10	+0.110
1N5275	140	0.90	1300	4500	0.1	101	106	10	+0.110
1N5276	150	0.85	1500	5000	0.1	108	114	10	+0.110
1N5277	160	0.80	1700	5500	0.1	116	122	10	+0.110
1N5278	170	0.74	1900	5500	0.1	123	129	10	+0.110
1N5279	180	0.68	2200	6000	0.1	130	137	10	+0.110
1N5280	190	0.66	2400	6500	0.1	137	144	10	+0.110
1N5281	200	0.65	2500	7000	0.1	144	152	10	+0.110

NOTE 1 — TOLERANCE AND VOLTAGE DESIGNATION

Tolerance designation — The JEDEC type numbers shown indicate a tolerance of ±10% with guaranteed limits on only V_Z, I_R and V_F as shown in the above table. Units with guaranteed limits on all six parameters are indicated by suffix "A" for ±10% tolerance and suffix "B" for ±5.0% units.

Non-Standard voltage designation — To designate units with zener voltages other than those assigned JEDEC numbers, the type number should be used.

EXAMPLE:

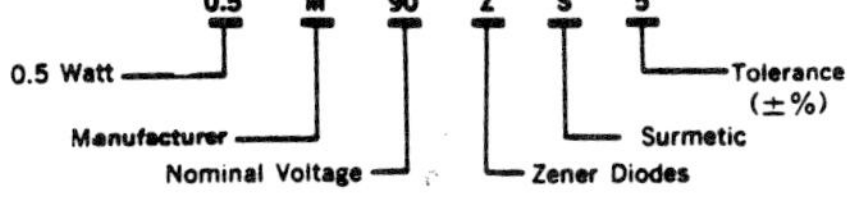

NOTE 2 — SPECIAL SELECTIONS AVAILABLE INCLUDE:

1 — Nominal zener voltages between those shown.

2 — **Matched sets: (Standard Tolerances are ±5.0%, ±3.0%, ±2.0%, ±1.0%) depending on voltage per device.**

a. Two or more units for series connection with specified tolerance on total voltage. Series matched sets make zener voltages in excess of 200 volts possible as well as providing lower temperature coefficients, lower dynamic impedance and greater power handling ability.

b. Two or more units matched to one another with any specified tolerance.

3 — Tight voltage tolerances: 1.0%, 2.0%, 3.0%.

NOTE 3 — TEMPERATURE COEFFICIENT (θ_{VZ})

Test conditions for temperature coefficient are as follows:

a. I_{ZT} = 7.5 mA, T_1 = 25°C,
T_2 = 125°C (1N5221A, B thru 1N5242A, B.)

b. I_{ZT} = Rated I_{ZT}, T_1 = 25°C,
T_2 = 125°C (1N5243A, B thru 1N5281A, B.)

Device to be temperature stabilized with current applied prior to reading breakdown voltage at the specified ambient temperature.

FIGURE 3-18 *(continued)*

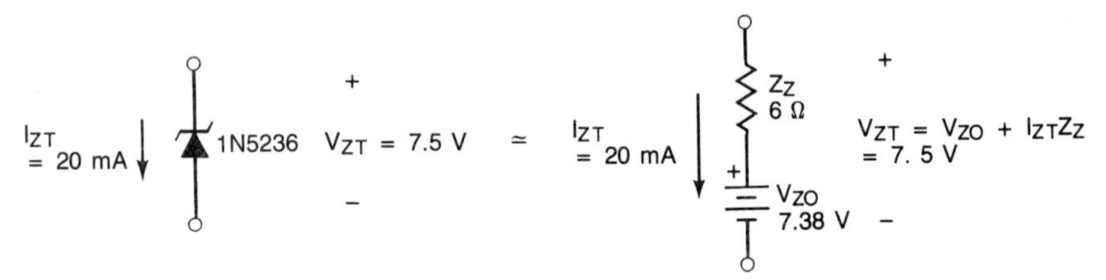

FIGURE 3-19 Zener dynamic impedance model for the 1N5236 (refer to Example 3-7).

As a final comment, the reader is again directed to the data sheet given in Fig. 3-18. Observe that the last column lists the temperature coefficients. The zener diodes (below 5 V) have a negative tempco while the avalanche diodes (above 5 V) have a positive tempco. This supports our discussion in Section 2-9.

3-11 Applying the Ideal Diode Model

Given the general circuit indicated in Fig. 3-20, we immediately recognize that the diode is forward biased. Before we can proceed with an analysis, we must select the appropriate forward-biased diode model.

As a general rule, the diode may be replaced by its ideal model (a closed switch) provided that V_S is at least *10* times larger than the diode's knee voltage.

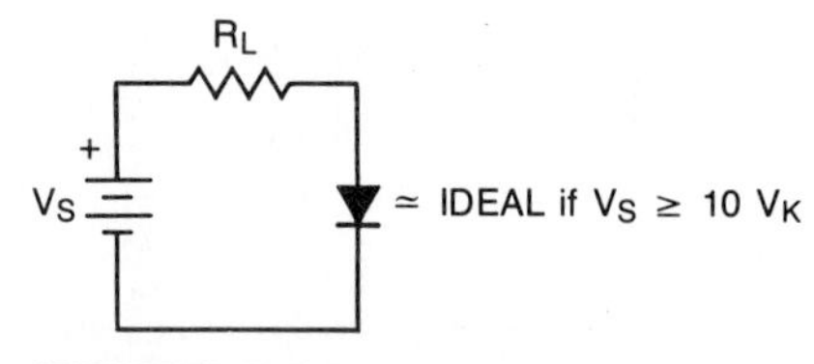

FIGURE 3-20 General diode circuit.

FIGURE 3-21 Circuits for Example 3-8.

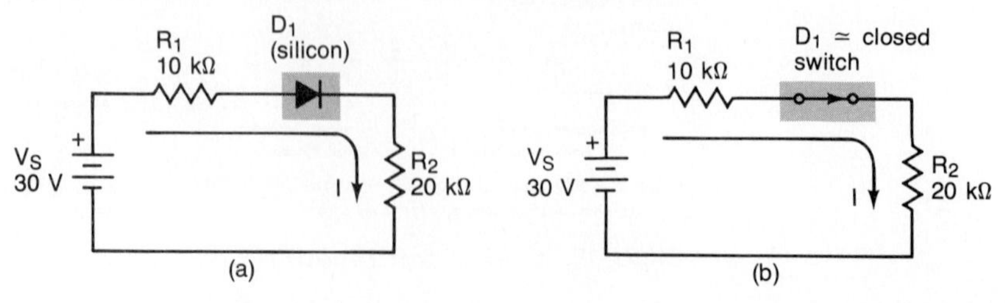

EXAMPLE 3-8

Compute the power dissipated in resistor R_2 in the circuit shown in Fig. 3-21(a).

SOLUTION By inspection of Fig. 3-21(a), we see that D_1 is a forward-biased silicon diode ($V_K = 0.7$ V). Since

$$V_S = 30 \text{ V} \geq 10V_K = 7 \text{ V}$$

we may replace D_1 with a closed switch as shown in Fig. 3-21(b). At this point, the "electronics" is over, and we must apply circuit analysis.

By voltage division,

$$V_{R2} = \frac{R2}{R1 + R2} V_S = \frac{20 \text{ k}\Omega}{10 \text{ k}\Omega + 20 \text{ k}\Omega}(30 \text{ V}) = 20 \text{ V}$$

Since we now know the voltage across R_2, we may now determine P_{R2}. Hence

$$P_{R2} = \frac{V_{R2}^2}{R_2} = \frac{(20 \text{ V})^2}{20 \text{ k}\Omega} = 20 \text{ mW}$$

■

EXAMPLE 3-9

An electronic game requires six standard C-cell 1.5-V batteries. Even though the proper polarities have been marked in the battery compartment, there is always a good chance that the consumer will inadvertently reverse them. *Reverse dc bias can destroy the integrated circuits used in the game.* Design a scheme to protect the product. The maximum current drawn by the electronics is 250 mA.

SOLUTION We can make a sketch of the circuit as shown in Fig. 3-22(a). Even though we do not know the details of the electronic game, we have enough information to model it as a resistor R_L. Specifically, we calculate its *static resistance* at 250 mA.

$$R_L = \frac{V_S}{I} = \frac{9 \text{ V}}{250 \text{ mA}} = 36 \ \Omega$$

Using R_L to model the game, we arrive at the circuit shown in Fig. 3-22(b).

Since a diode will pass current only in one direction, we can protect the integrated circuits in the game by placing a diode in series. This is depicted in Fig. 3-22(c). Now since

$$V_S = 9 \text{ V} \geq 10V_K = 7 \text{ V}$$

we model the use of the diode as shown in Fig. 3-22(d) and (e). As we can see, when the batteries are installed correctly, the diode conducts. However, when the batteries are reversed, the diode acts as an open switch to protect the electronics. ■

As seen in Fig. 3-22, the diode behaves as a switch. However, the primary reason it is so successful in protecting polarity-sensitive electronics is that it is an extremely *fast* switch. To complete the design we must specify the diode's part number. To accomplish this we must go to the manufacturers' data sheets and make a selection. Our specifications include the fact that the diode must have an average forward current rating (I_O) of better than 250 mA. Further, it must have a reverse voltage rating (V_R) of better than 9 V.

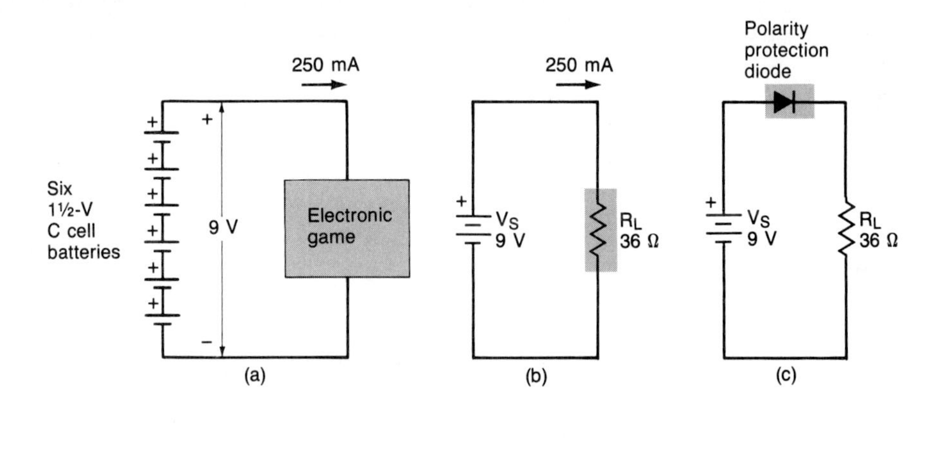

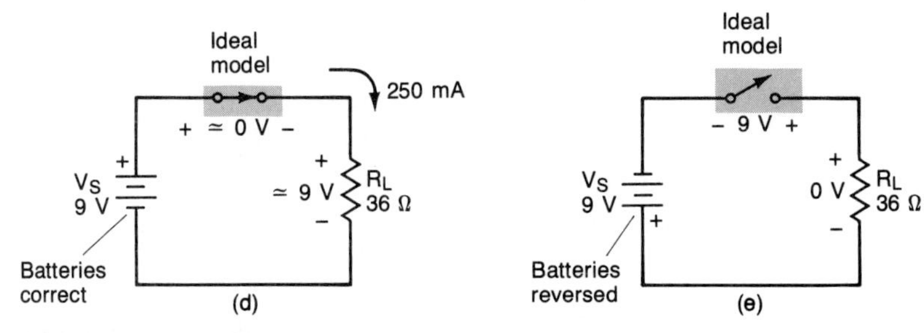

FIGURE 3-22 Solution to Example 3-9.

3-12 Taking the Knee Voltage into Account

It is very important to examine the magnitude of the forward bias across a diode in order to select the best model. In Example 3-10 we *incorrectly* employ the ideal model for the diode.

FIGURE 3-23 Circuits for Examples 3-10 and 3-11: (a) circuit; (b) using the ideal model; (c) using the knee voltage model for greater accuracy.

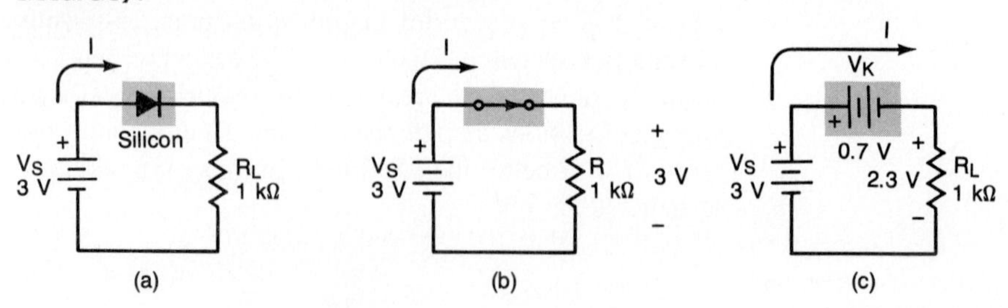

EXAMPLE 3-10

Calculate the current I in the circuit shown in Fig. 3-23(a).

SOLUTION Incorrectly using the ideal diode model, we arrive at the circuit shown in Fig. 3-23(b). Since the diode is a closed switch, all of the voltage is dropped across the resistor and

$$I = \frac{V_S}{R} = \frac{3\text{ V}}{1\text{ k}\Omega} = 3\text{ mA}$$

■

In Example 3-11 we repeat the calculation, but this time we take the knee voltage into account.

EXAMPLE 3-11

Calculate the current I in the circuit shown in Fig. 3-23(a).

SOLUTION This time we employ the knee-voltage model to arrive at the circuit shown in Fig. 3-23(c). By Kirchhoff's voltage law we write

$$-V_S + V_K + IR_L = 0$$

and solving for I, we have

$$I = \frac{V_S - V_K}{R_L} = \frac{3\text{V} - 0.7\text{ V}}{1\text{ k}\Omega} = 2.3\text{ mA}$$

■

By observing the result of Example 3-11, we see that the current is actually much closer to 2.3 mA rather than the 3 mA found in Example 3-10. Although this discrepancy may seem small at first glance, we see that the ideal diode model gave us an answer that is about 30% too large! In some cases this could be a totally unacceptable error.

The importance of taking the knee voltage into account in some problems cannot be understated. Consider Example 3-12.

EXAMPLE 3-12

Compute the currents I_1 and I_2 in the circuit shown in Fig. 3-24(a). All three diodes are silicon.

SOLUTION Initially, we note that V_S is quite large and that all three diodes are forward biased. However, the two diodes in series (D_1 and D_2) require more forward bias (0.7 V + 0.7 V = 1.4 V) to conduct than does D_3 (0.7 V). Consequently, D_3 will turn on first and its forward voltage drop will clamp the forward bias across D_1 and D_2 to 0.7 V. Therefore, D_1 and D_2 will *not* have enough forward bias to turn them on [see Fig. 3-24(b)]. Since $V_S \geq 10V_K$, we use the ideal diode model and

$$I_1 = \frac{V_S}{R} = \frac{30\text{ V}}{1\text{ k}\Omega} = 30\text{ mA}$$

Obviously, the current through diodes D_1 and D_2 is zero:

$$I_2 = 0$$

■

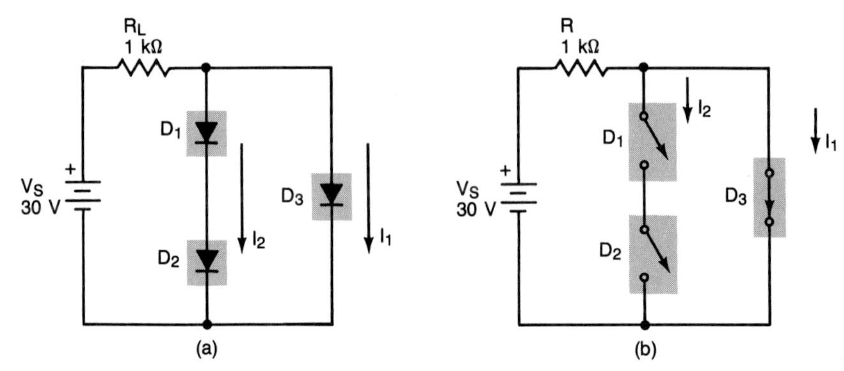

FIGURE 3-24 Circuits for Example 3-12.

In the next section we continue to pursue the problem of determining if sufficient bias is developed across a given diode. The basic tool we shall employ is *Thévenin's theorem*.

3-13 Applying Thévenin's Theorem

One of the most powerful tools we have in the analysis of electronic circuits is *Thévenin's theorem*. Its use can greatly simplify many complex electronic circuits. Thévenin's theorem can be applied to dc (bias) problems and ac (signal) problems. Initially, we restrict our use of it to simple dc problems.

Since Thévenin's theorem will be utilized frequently in our work to come, we shall review it briefly.

> **Thévenin's theorem is based on a property exhibited by all linear networks. In essence it states that any (dc) linear network containing voltage and current sources may be resolved into a single equivalent voltage source in series with a single equivalent resistance.**

In Fig. 3-25 we see that Thévenin's theorem may be employed to simplify complex linear networks. The general approach may be outlined as follows:

1. "Look into" terminals *a–b* [Fig. 3-25(a)] and find the voltage across the terminals with the load disconnected. This is called the open-circuit or Thévenin equivalent voltage, V_{TH}.
2. If the complex network contains constant (independent) sources, replace all voltage sources with short circuits and all current sources with open circuits.

FIGURE 3-25 Thévenin's theorem is used to simplify complex linear networks.

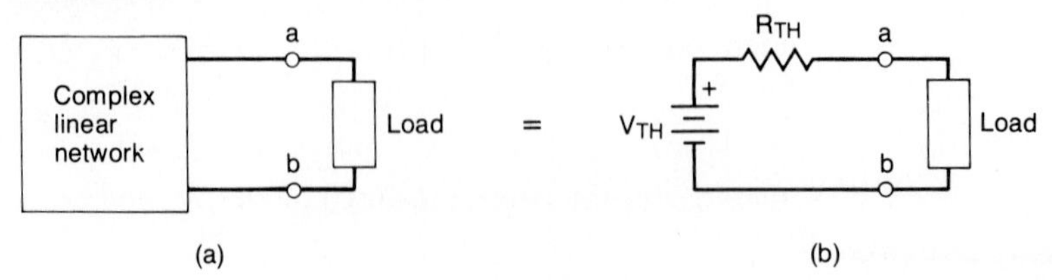

3. Find the equivalent resistance "looking into" terminals a–b with the load disconnected. This is called the Thévenin equivalent resistance R_{TH}.

The resulting Thévenin equivalent circuit is depicted in Fig. 3-25(b).

The only constraint on the application of Thévenin's theorem is that the network to be "Thévenized" must be linear. However, the theorem poses no such constraints on the load connected across the network. In general, the "load" may be a simple resistor, another linear network, or a *nonlinear* circuit element, such as a diode. In Examples 3-13 and 3-14 we illustrate the application of Thévenin's theorem to aid our analysis of circuits containing diodes.

EXAMPLE 3-13

Determine the current through the diode in the circuit shown in Fig. 3-26(a).

SOLUTION By applying our three basic steps, we may arrive at the Thévenin equivalent circuit "seen" by the diode.

1. Disconnecting the load (the silicon diode), we may find V_{TH}. Since no current flows through R_3, it drops no voltage [see Fig. 3-26(b)]. Therefore, V_{TH} is given by simple voltage division between R_1 and R_2.

$$V_{TH} = \frac{R_2}{R_1 + R_2} V_S = \frac{20\ \text{k}\Omega}{10\ \text{k}\Omega + 20\ \text{k}\Omega}(6\ \text{V}) = 4\ \text{V}$$

2. Replacing V_S with a short circuit produces the circuit shown in Fig. 3-26(c).
3. From Fig. 3-26(c),

$$R_{TH} = R_3 + R_1 \parallel R_2 = 10\ \text{k}\Omega + 10\ \text{k}\Omega \parallel 20\ \text{k}\Omega$$
$$= 16.7\ \text{k}\Omega$$

The Thévenin equivalent circuit has been connected to the diode as shown in Fig. 3-26(d). Since the voltage source V_{TH} is small compared to the knee voltage of the diode ($V_{TH} < 10V_K$), we must take the 0.7-V drop across the diode into account, and from Fig. 3-26(e),

$$I_D = \frac{V_{TH} - V_K}{R_{TH}} = \frac{4\ \text{V} - 0.7\ \text{V}}{16.7\ \text{k}\Omega} = 0.198\ \text{mA}$$ ■

EXAMPLE 3-14

Determine the current through the diode shown in Fig. 3-27(a).

SOLUTION Applying Thévenin's theorem, we find that

$$V_{TH} = \frac{R_2}{R_1 + R_2} V_S = \frac{100\ \Omega}{10\ \text{k}\Omega + 100\ \Omega}(15\ \text{V}) = 0.149\ \text{V}$$
$$R_{TH} = R_1 \parallel R_2 = 10\ \text{k}\Omega \parallel 100\ \Omega = 99\ \Omega$$

The Thévenin circuit is illustrated in Fig. 3-27(b). Since

$$V_{TH} = 0.149\ \text{V} < V_K = 0.7\ \text{V}$$

the diode does not have enough forward bias to conduct. Therefore, it is an open switch. Refer to Fig. 3-27(c), and

$$I_D = 0$$ ■

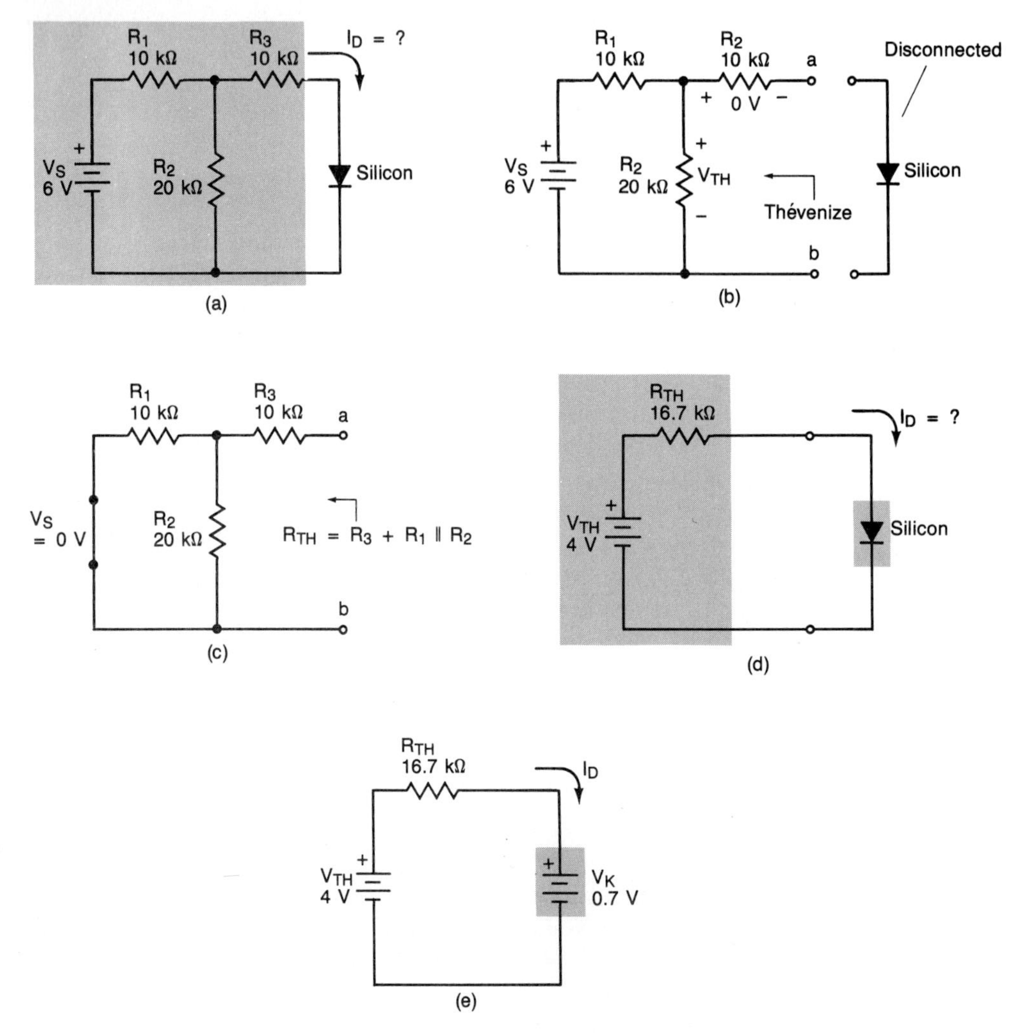

FIGURE 3-26 Applying Thévenin's theorem to analyze the diode circuit of Example 3-13.

FIGURE 3-27 Diode circuits for Example 3-14.

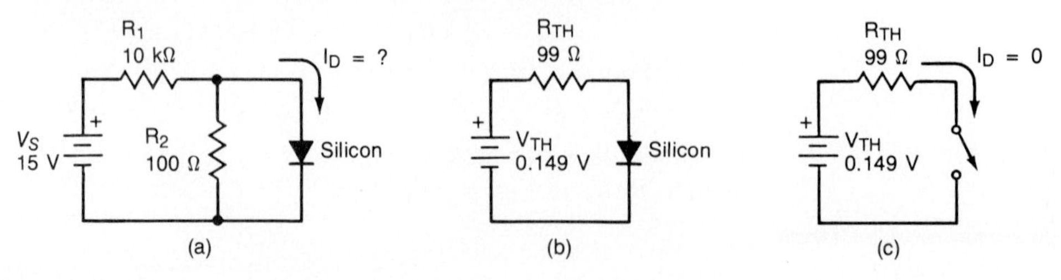

As we saw in Examples 3-12 and 3-14, diodes will not conduct if their applied forward bias does not exceed their barrier potential (their knee voltage). We encounter the same problem in zener diode circuits. If the Thévenin equivalent (reverse) voltage applied to them does not exceed their rated breakdown voltages, they will not break down. This can be a serious problem when they are used as *voltage regulators*. Instead of maintaining a constant terminal voltage, they will act merely as open circuits.

3-14 Sine Waves

In the study of fundamental ac circuit theory, we investigate such topics as capacitive and inductive reactance, complex impedance, phase angles, and a host of other considerations. There are two basic premises upon which this vast body of theory and techniques is based: (1) the networks are *linear*, and (2) the excitation is *sinusoidal* (steady state).

The sine wave is the most fundamental ac waveform. Therefore, we find it used extensively in the analysis of electronic circuits. The basic equation describing the sine wave is given by

$$v(t) = V_m \sin \omega t \tag{3-5}$$

where $v(t)$ = instantaneous voltage
V_m = peak voltage
ω = (omega) angular velocity in radians per second = $2\pi f$
t = time (seconds)

The sine wave and its key attributes are described in Fig. 3-28.

FIGURE 3-28 Sine wave.

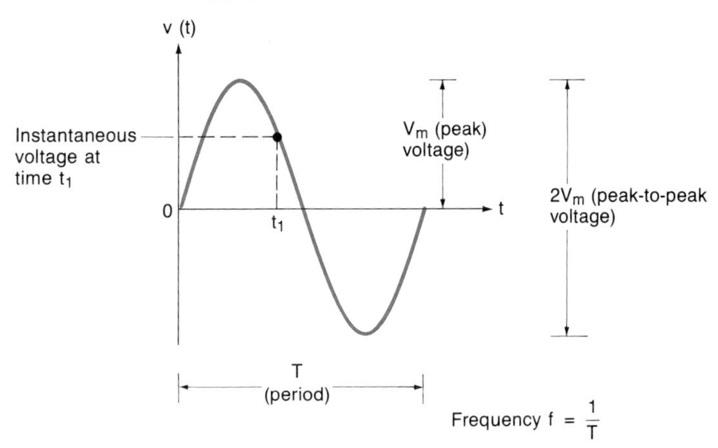

3-15 Signal Effects on the Operating Point

In Fig. 3-7 we examined the effects of varying V_S on the dc load line. Figure 3-7 shows that increasing V_S raises the Q-point, and lowering V_S lowers the Q-point. In this section we pursue this effect further by examining how an ac sinusoidal signal can vary the operating point of a diode. To initiate our study, consider Fig. 3-29.

From basic circuit theory we know that voltage sources in series add. The situation shown in Fig. 3-29 poses no exception to that rule. Paying particular attention to the notation shown in Fig. 3-29, we can write

$$v_S(t) = V_S + v_s(t) \tag{3-6}$$

where $v_S(t)$ = total instantaneous voltage of the ac signal and the dc level

V_S = dc level

$v_s(t)$ = instantaneous voltage of the ac signal

Capital subscripts refer to dc and total instantaneous signals. Lowercase subscripts refer to instantaneous ac quantities.

Equation 3-6 states that the total instantaneous voltage is equal to the sum of the dc level and the instantaneous value of the ac signal riding on it. To illustrate this point, consider Example 3-15.

EXAMPLE 3-15

Given the circuit shown in Fig. 3-29, assume that V_S = 10 V and that $v_s(t)$ is a sine wave with a peak amplitude of 1 V and a frequency of 1 kHz. Write an equation for the total instantaneous voltage $v_S(t)$, and make a sketch of it.

SOLUTION From Eq. 3-5 we may write

$$v_s(t) = V_m \sin 2\pi ft = 1 \sin 2\pi(1000)t = \sin 2000\pi t$$

and from Eq. 3-6 we have

$$\begin{aligned} v_S(t) &= V_S + v_s(t) \\ &= 10 \text{ V} + \sin 2000\pi t \end{aligned}$$

A sketch of this result is provided in Fig. 3-30. The reader should verify it. ■

FIGURE 3-29 Signal source in series with a dc source.

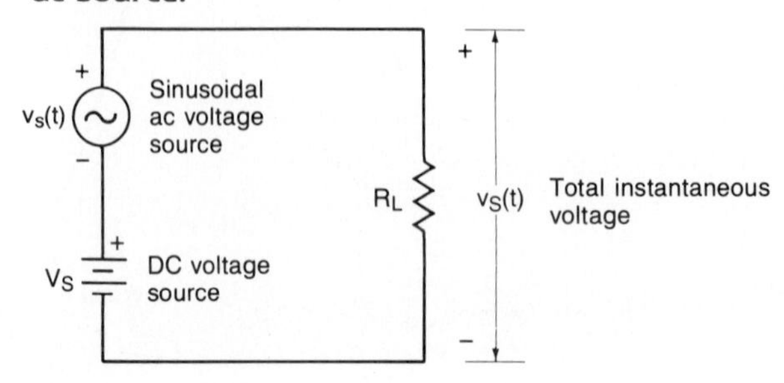

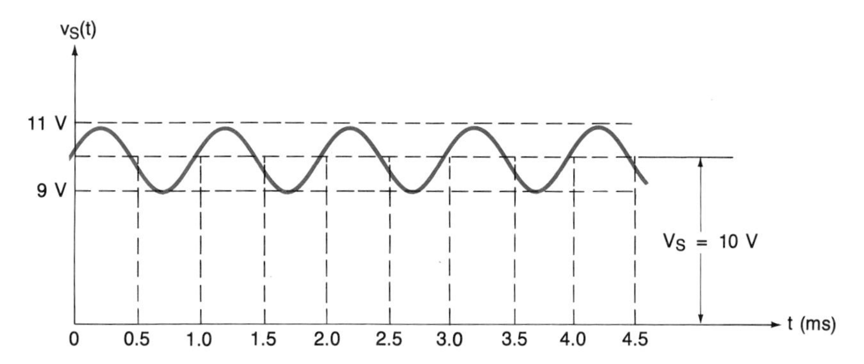

FIGURE 3-30 Solution to Example 3-15.

FIGURE 3-31 Instantaneous operating point for $v_S(t) = 0$ V: (a) diode circuit; (b) equivalent circuit when the ac signal passes through zero; (c) instantaneous load line matches the dc load line.

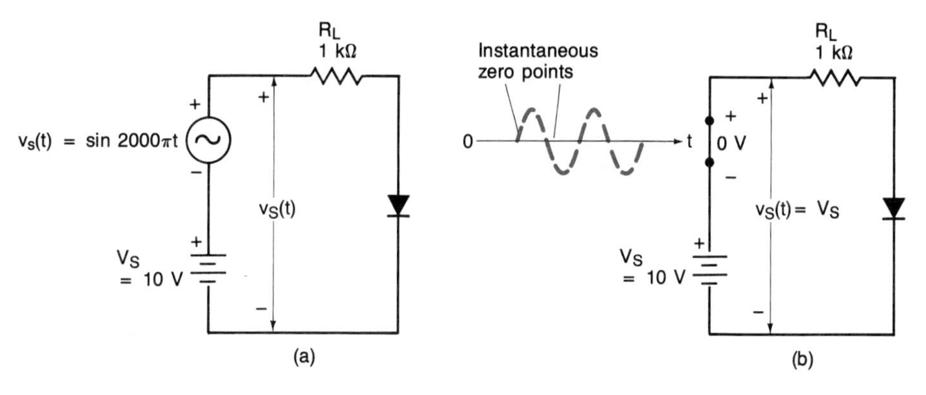

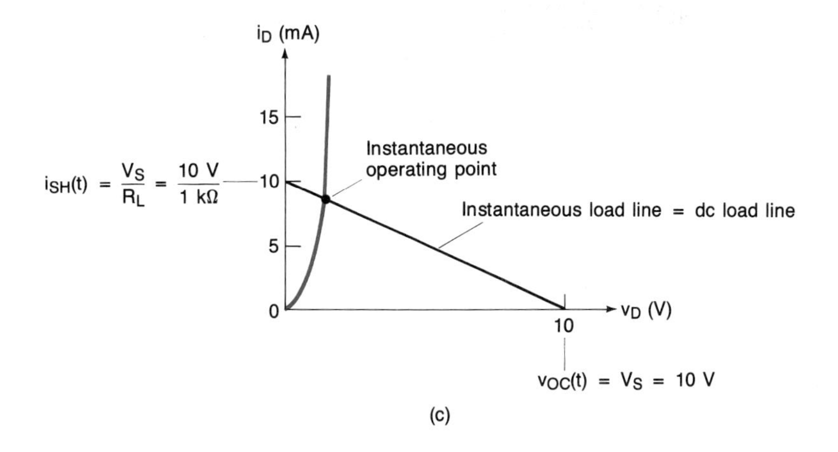

The small sinusoidal signal adds to and subtracts from the dc level. As a final comment, we should note that there are *two* equally valid ways of interpreting Fig. 3-30. First, we have an ac waveform riding on a dc level, and second, we have varying dc. The first interpretation follows directly from Example 3-15. The second interpretation requires that we observe that the waveform is always positive. Consequently, if $v_S(t)$ were applied across a resistive load, the resulting current would vary, but it would always flow in the same direction (dc).

To observe the effects of such a signal on a diode's operating point, consider Fig. 3-31. In this case the diode is being driven by both an ac and a dc source as shown in Fig. 3-31(a). When the instantaneous value of the ac signal is zero, we have the equivalent circuit shown in Fig. 3-31(b). If we were to draw the load line at this instant in time, it would match the dc load line [see Fig. 3-31(c)].

When the sine wave reaches its positive peaks, the total instantaneous voltage is 11 V. See $t = 0.25$, 1.25, and 2.25 ms in Fig. 3-30. Therefore, the instantaneous load line and the operating point will both move up, as shown in Fig. 3-32(a). When the total instantaneous voltage reaches its minimum (9 V) at $t = 0.75$ and 1.75 ms in Fig. 3-30, the instantaneous load line will be lowered as shown in Fig. 3-32(b).

Since the ac signal source is sinusoidal, it follows that the operating-point (the diode's voltage and current) variation will also be *approximately* sinusoidal [see Fig. 3-33(a)]. If the Q-point is lowered (by reducing the dc source V_S), the diode voltage and current waveforms may be *distorted* (or nonsinusoidal), as shown in Fig. 3-33(b). However, even if we maintain the same dc operating point shown in Fig. 3-33(b), the distortion in the diode's voltage and current waveforms may be reduced by *decreasing* the amplitude of the ac signal $v_s(t)$ [see Fig. 3-33(c)].

This concept is an extremely important one that can be generalized to all of the electronic devices that we shall encounter. Specifically, *linear* devices do *not* distort (change the shape of) sinusoidal signals. However, sufficiently small signals in the linear portion of the *V–I* characteristic [Fig. 3-33(a)], or very small signals in the nonlinear portion of the *V–I* characteristic curve [Fig. 3-33(c)] are also essentially undistorted. Hence, by imposing the constraint that we deal only with *small signals*, nonlinear electronic devices may be replaced by their linear ac models.

FIGURE 3-32 **Operating-point variation: (a) load line is moved up until the ac signal reaches its positive peaks; (b) load line is moved down until the ac signal reaches its negative peaks.**

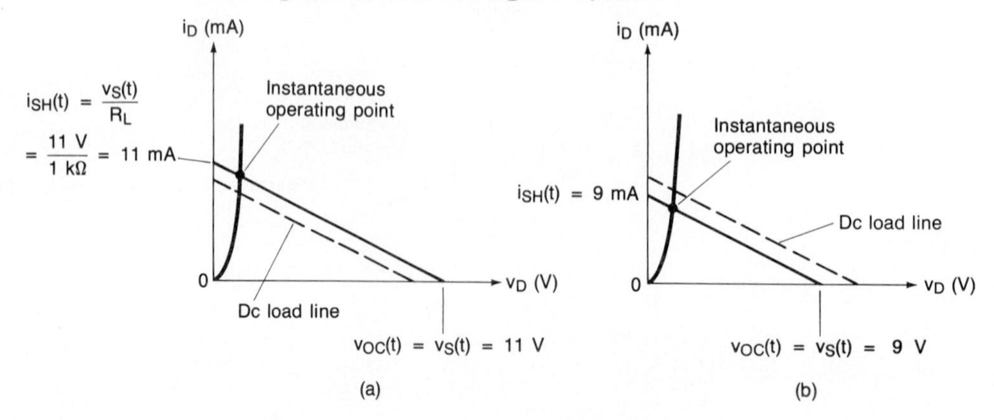

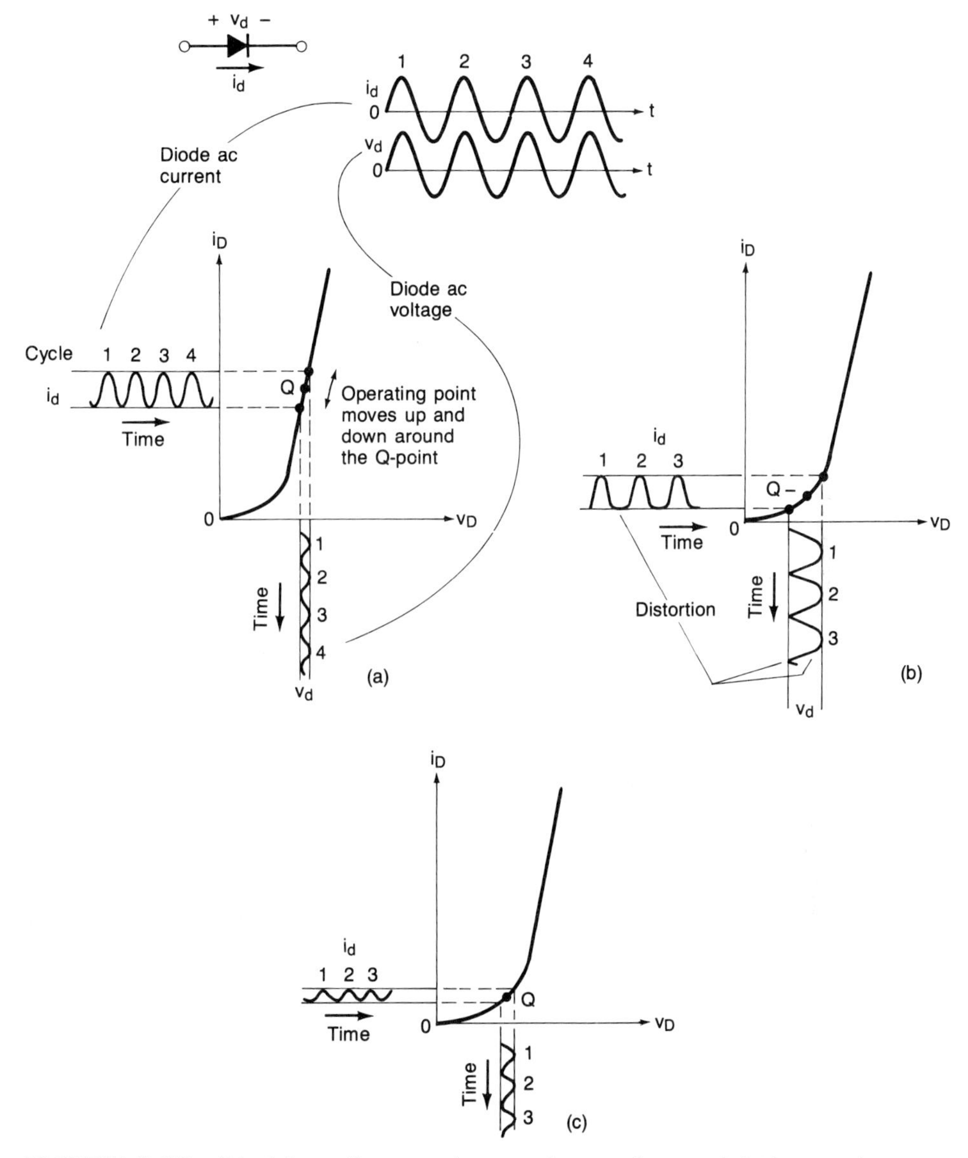

FIGURE 3-33 Diode's voltage and current waveforms: (a) the ac signal produces a sinusoidal response; (b) lowering the Q-point produces distortion; (c) reducing the signal level reduces the distortion.

The next obvious question is: How large can a signal become and still be termed a ''small signal''? To answer this, we set forth two simple rules. *Small signals* are:

1. Sinusoidal ac voltages and/or currents that do not have an amplitude great enough to produce a significantly nonlinear, or distorted, response in the electronic device, or system under consideration

2. Generally, sinusoidal voltage waveforms with a peak-to-peak amplitude of *less* than one-tenth of the knee voltage of a diode

From the second rule, it would seem that most small signals would be less than 70 mV peak to peak when we are dealing with silicon diodes. However, it should be remembered that the first rule *always* supersedes the second. Specifically, in some problems small signals may be as large as 1 V peak to peak, while other problems might restrict our signal size to less than 10 mV peak to peak. We shall pursue the small-signal concept even further when we begin our study of amplifier circuits.

3-16 Superposition Theorem

The *superposition theorem* is another circuit analysis tool that can greatly simplify the analysis and design of electronic circuits. Since we shall draw on the superposition theorem extensively in our work to come, we shall briefly review it and extend its use to our electronics work.

The superposition theorem is used to analyze electric circuits in which there are two or more sources acting simultaneously to produce a net voltage or current. To find this unknown voltage or current by superposition, we find the component produced by each source, in turn, and algebraically add all the individual components to arrive at the net effect. The primary constraint upon the superposition theorem is that the circuit *must be linear*.

From the definition above, it should be clear that the superposition theorem may be applied to linear circuits that contain multiple dc voltage and/or current sources. However, it may also be applied to linear circuits that contain *ac sources*, or *both ac and dc sources*.

It is the latter case that occurs most often in our electronics work. Typically, we first perform a dc analysis and then an ac analysis. The reason a dc analysis must be performed first will become clear in the next section.

3-17 Diode Small-Signal Models

In Section 3-15 we saw that an ac signal causes the load line (and therefore the operating point) to move up and down. In Fig. 3-33 we saw graphically how the operating-point variation produces an ac current through the diode and an ac voltage across it.

Since the diode has an ac voltage across it (v_d) and an ac current (i_d) through it, it follows that the diode must possess some *ac resistance* (r_{ac}). Specifically,

$$r_{ac} = \frac{v_d}{i_d} \tag{3-7}$$

[For simplicity, the functional notation has been dropped. Specifically, v_d is the same as $v_d(t)$ and i_d is equivalent to $i_d(t)$.]

In this section we develop an *ac model* for the forward-biased diode. In Section

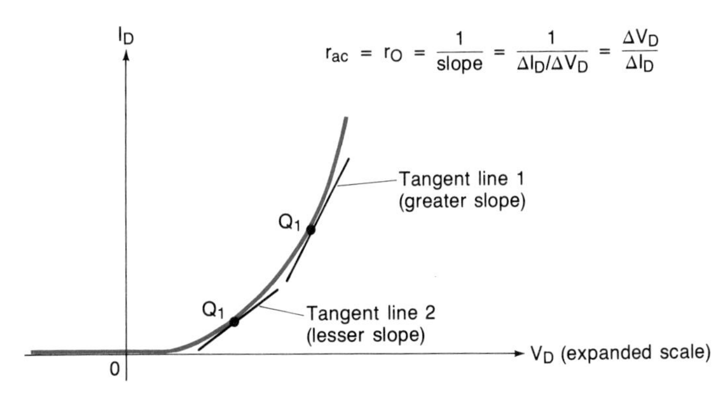

FIGURE 3-34 Changing the dc operating point changes the slope of the tangent line.

3-18 we see exactly how to use this model, together with the superposition theorem and our dc piecewise linear diode models, to make a complete analysis of a diode circuit.

In Section 3-7 we introduced the graphical approach to determine r_O. [Recall that r_O is employed in our dynamic resistance diode (dc) model.] As we shall see, for small ac signals,

$$r_O = r_{ac}$$

In Fig. 3-34 we have indicated a diode's *V–I* curve with two *Q*-points established upon it. To determine the diode's ac resistance, we draw a straight line that is tangent to the *V–I* characteristic curve at the *Q*-point (in exactly the same manner as described in Section 3-7).

The *slope* of a straight line (such as our tangent line) is given by

$$\text{slope} = \frac{\text{rise}}{\text{run}} = \frac{\Delta I}{\Delta V} \tag{3-8}$$

Since the ratio of a current to a voltage is a conductance (g) and the reciprocal of a conductance is a resistance, we arrive at

$$r_{ac} = \frac{1}{\text{slope}} = \frac{1}{g_{ac}} = \frac{\Delta V}{\Delta I} \tag{3-9}$$

By studying Fig. 3-34 and keeping the relationships defined by Eq. 3-9 in mind, we can draw the following conclusions:

1. *Raising* the *dc operating point* (Q_1) increases the slope of the tangent line, and r_{ac} must *decrease*.
2. *Lowering* the *dc operating point* (Q_2) decreases the slope of the tangent line, and r_{ac} must *increase*.

Simply stated, the greater the dc current through the diode, the less its ac resistance will be, and vice versa. This is a very important observation. In more general terms, the ac parameters of a diode are controlled by its dc operating point. This is true not only for the diode, but for *all* solid-state devices.

Just as we developed piecewise linear *dc models* for the diode, now we shall develop a *linear ac model* for the diode. Our objective shall once again be to minimize the cumbersome graphical method. To develop this ac model for the diode, we shall once again call on the Shockley diode equation.

$$I_D = I_S\,(e^{V_D/\eta V_T} - 1) \tag{3-10}$$

The mathematical equivalent of finding the slope of a tangent line involves a technique taught in calculus that is referred to as taking the *derivative*. Since we have already noted that the slope is a conductance, we may write

$$\boxed{g_j = \frac{dI_D}{dV_D}} \tag{3-11}$$

where g_j = conductance to a small signal about a dc operating point (siemens)
dI_D/dV_D = calculus notation for taking a derivative that is read "the derivative of I_D with respect to V_D"

The dI_D/dV_D notation implies a Δ that is infintesimally (approaching zero) small.

For those students who have studied calculus, we shall take the derivative of the Shockley diode equation. All others may skip over the mathematics. Taking the derivative of Eq. 3-10, we write

$$\frac{dI_D}{dV_D} = \frac{d}{dV_D}\,[I_S(e^{V_D/\eta V_T} - 1)]$$

$$= I_S\left(\frac{1}{\eta V_T}\,e^{\,V_D/\eta V_T} - 0\right) = \frac{I_S e^{V_D/\eta V_T}}{\eta V_T}$$

To eliminate the exponential in our result, we may once again call on Eq. 3-10,

$$I_D = I_S\,(e^{V_D/\eta V_T} - 1) = I_S e^{V_D/\eta V_T} - I_S$$

and

$$I_D + I_S = I_S e^{V_D/\eta V_T}$$

Substituting this result into our equation for the derivative yields

$$g_j = \frac{dI_D}{dV_D} = \frac{I_D + I_S}{\eta V_T}$$

Again, since the reciprocal of a conductance is resistance,

$$\boxed{r_j = \frac{1}{g_j} = \frac{\eta V_T}{I_D + I_S}} \tag{3-12}$$

where r_j = small-signal dynamic junction resistance
η = a constant: $\eta = 1$ for Ge; $\eta \simeq 2$ for Si
V_T = voltage equivalent of temperature

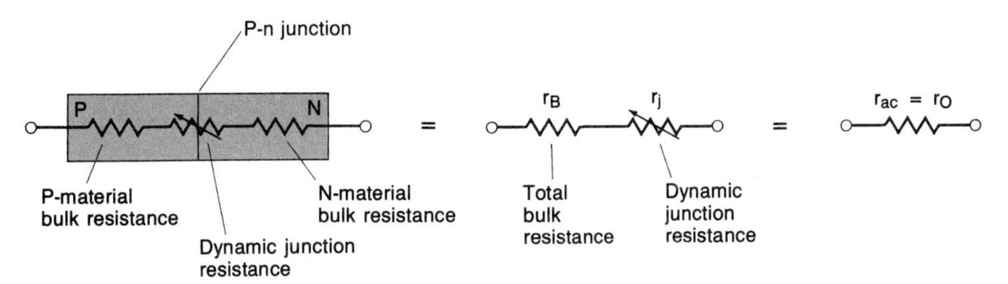

FIGURE 3-35 Components of the ac resistance of a p-n junction diode.

I_D = diode's Q-point current

I_S = diode's reverse saturation current

To simplify Eq. 3-12, we may make some assumptions. First, since we apply it only to forward-biased diodes, it is reasonable to assume that the forward current through the diode is much greater than the diode's reverse saturation current. Hence if

$$I_D >> I_S$$

then

$$r_j \simeq \frac{\eta V_T}{I_D} \tag{3-13}$$

Further, since V_T = 26 mV at 25°C, we may state that the dynamic junction resistance at room temperature is

$$r_j \simeq \eta \frac{26 \text{ mV}}{I_D} \tag{3-14}$$

In Section 3-2 we noted that $\eta = 1$ for germanium diodes and $\eta \simeq 2$ for silicon diodes. This was a slight oversimplification. Whereas η is unity for germanium diodes, it does tend to vary for silicon diodes. Specifically, for diode voltages below about 0.5 V, $\eta \simeq 2$, and for diode voltages above 0.5 V, $\eta \simeq 1$. This variation in η is gradual and does not change abruptly. The point is that for sufficiently forward-biased silicon diodes, η is also approximately 1. Therefore, we arrive at

$$\boxed{r_j \simeq \frac{26 \text{ mV}}{I_D}} \tag{3-15}$$

where r_j is the dynamic junction resistance of forward-biased Ge and sufficiently forward-biased Si diodes at 25°C and I_D is the diode Q-point current.

As we discussed in Section 3-2, the Shockley diode equation provides us with a mathematical description of an idealized *p-n* junction. Consequently, it does not take the bulk resistance (r_B) into account. Therefore, we arrive at the situation shown in Fig. 3-35. Inspection of the figure shows that

$$\boxed{r_{ac} = r_O = r_j + r_B} \tag{3-16}$$

The bulk resistance r_B of diodes can range from 0.1 Ω for high-current rectifiers to as large as 10 Ω or more, for small-signal diodes. The bulk resistance of a diode is reasonably constant over a range of moderate forward currents. To simplify our work we shall make the following assumption:

We shall assume that the bulk resistance (r_B) of rectifier diodes is 1 Ω and the r_B of small-signal diodes is 10 Ω.

Our point is to emphasize the presence of r_B rather than to pursue its precise value in a given diode under a specific set of operating conditions. If a great deal of precision is required, it is recommended that the graphical technique for the determination of r_O be employed.

EXAMPLE 3-16

A silicon rectifier diode is biased at a forward current of 0.1 mA. Compute its approximate ac resistance at this operating point. Assume operation at 25°C. Repeat the calculation for I_D = 1, 10, and 100 mA.

SOLUTION By Eq. 3-15 we arrive at the diode's approximate dynamic junction resistance.

$$r_j \simeq \frac{26\text{mV}}{I_D} = \frac{26\text{mV}}{0.1 \text{ mA}} = 260 \ \Omega$$

The rectifier diode's ac resistance may be approximated by assuming a bulk resistance of 1 Ω and by Eq. 3-16,

$$r_{ac} = r_j + r_B = 260 \ \Omega + 1 \ \Omega = 261 \ \Omega$$

The rest of the calculations are similar and have been summarized in Table 3-2. ■

TABLE 3-2
Effect of the DC Bias Point on r_j and r_{ac}

I_D (mA)	r_j (Ω)	r_{ac} (Ω)
0.1	260.0	261.0
1.0	26.0	27.0
10.0	2.6	3.6
100.0	0.26	1.26

From Table 3-2 it is clear that the small-signal ac resistance *decreases* as the dc forward current through the diode *increases*. This result is in full agreement with the graphical explanation provided in Fig. 3-34.

Again, we emphasize that the ac parameters (such as r_{ac}) of all solid-state devices are controlled by their dc operating points. It is for this reason that we typically analyze electronic circuits by first performing a dc analysis followed by an ac analysis. This point will be demonstrated further in the next section.

3-18 Using the Diode Small-Signal Model

In Fig. 3-36(a) we have the 1N917 diode circuit analyzed in Example 3-5. Note that it is now being driven by *both* the dc voltage source and an ac voltage source. Our problem is to find the total voltage [$v_D(t)$] waveform across the diode. Since the ac signal is small, we may model the diode as a linear resistance r_{ac} to determine its ac signal response. Further, since we shall be using a *linear* model for the ac analysis and a piecewise *linear* model for the diode's dc analysis, the superposition theorem may be used.

FIGURE 3-36 Using the small-signal model: (a) circuit; (b) dc equivalent circuit using the knee-voltage model; (c) ac equivalent circuit; (d) solution based on the knee-voltage model and approximations of r_j and r_B; (e) solution based on the dynamic resistance (r_O) model.

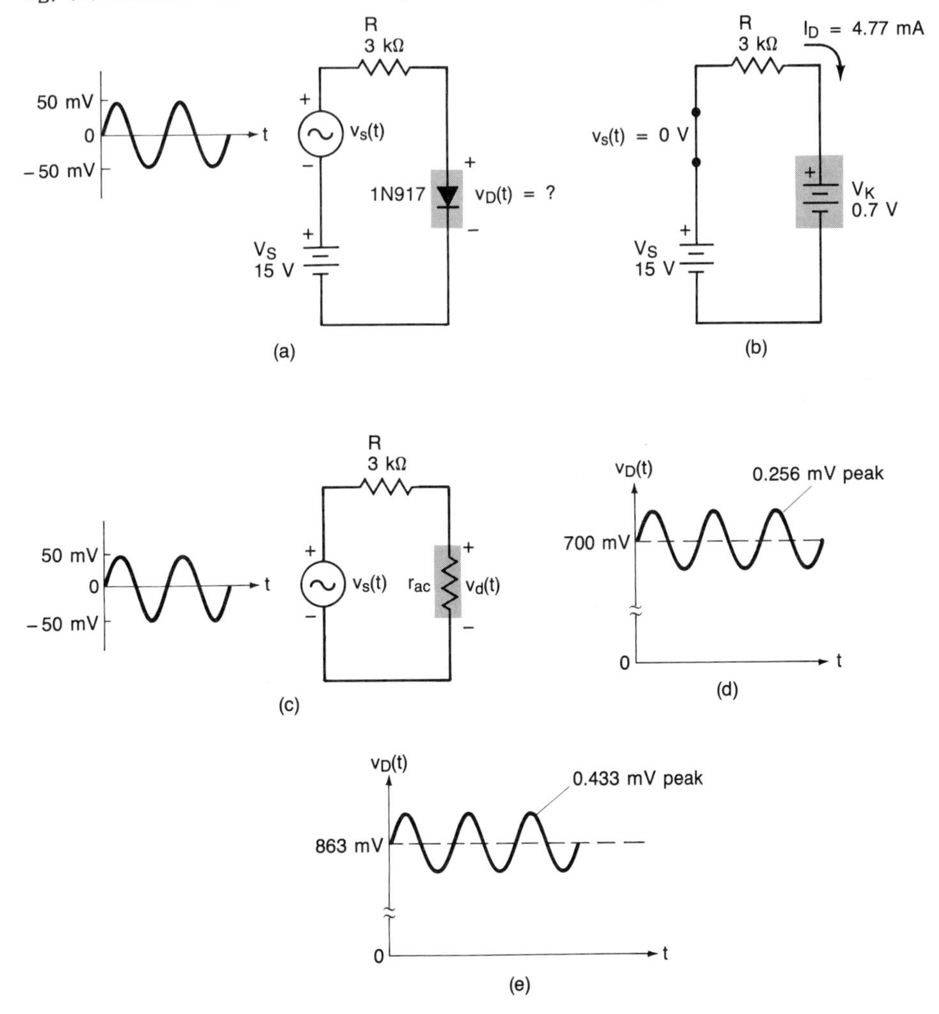

EXAMPLE 3-17

Given the small-signal 1N917 silicon diode employed in the circuit illustrated in Fig. 3-36(a), use the knee-voltage model to approximate the Q-point and Eqs. 3-15 and 3-16 to approximate r_{ac}. Find $v_D(t)$ across the diode.

SOLUTION From Example 3-5, the knee-voltage model resulted in a V_D of 0.7 V and an I_D of 4.77 mA. The dc equivalent circuit has been repeated in Fig. 3-36(b). Now from Eqs. 3-15 and 3-16 we obtain

$$r_j = \frac{26 \text{ mV}}{I_D} = \frac{26 \text{ mV}}{4.77 \text{ mA}} = 5.45\ \Omega$$

$$r_{ac} = r_j + r_B = 5.45\ \Omega + 10\ \Omega = 15.45\ \Omega$$

In Fig. 3-36(c) we have indicated the *ac equivalent* circuit which includes our diode model. We can see that we have a simple voltage-division problem. The peak ac voltage across the diode (V_{dm}) is

$$V_{dm} = \frac{r_{ac}}{r_{ac} + R} V_m = \frac{15.45\ \Omega}{15.45\ \Omega + 3\text{ k}\Omega}(50 \text{ mV}) = 0.256 \text{ mV}$$

Analytically, the total diode voltage waveform is

$$v_D(t) = V_D + v_d(t)$$
$$= 700 \text{ mV} + 0.256 \sin \omega t \text{ mV}$$

A sketch of this result has been provided in Fig. 3-36(d). The voltage axis has been broken so that we can see the very small ac signal riding on the much larger dc level. ■

EXAMPLE 3-18

The 1N917 has considerable bulk resistance which is not accounted for by the knee-voltage model, and the assumption that it is merely 10 Ω. Repeat Example 3-17 using the dynamic resistance diode model to improve our accuracy.

SOLUTION In Example 3-5 we found that $I_D = 4.71$ mA, $V_D = 0.863$ V, and $r_O = 26.2\ \Omega$. Hence

$$r_{ac} = r_O = 26.2\ \Omega$$

Again, by voltage division,

$$V_{dm} = \frac{r_{ac}}{r_{ac} + R} V_m = \frac{26.2\ \Omega}{26.2\ \Omega + 3\text{ k}\Omega}(50 \text{ mV}) = 0.433 \text{ mV}$$

The total instantaneous voltage waveform is

$$v_D(t) = 863 \text{ mV} + 0.433 \sin \omega t \text{ mV}$$

and has been indicated in Fig. 3-36(e). ■

Let us reflect upon the results of Examples 3-17 and 3-18. First, Example 3-18 has provided us with much more accurate results. However, the determination of r_O requires graphical techniques which can only be employed if a V–I curve is available. (Obviously, a V–I curve can be generated if necessary.) The conclusion to be drawn is as follows: If time permits and the situation demands accuracy, use the procedure illustrated in Example 3-18, but if a quick "rough order of magnitude" answer is all that is required, use the procedure given in Example 3-17.

3-19 AC Equivalent Circuits

As we have seen, the dc and ac equivalent circuits are different because the dc model for the diode is different from its ac model. However, in most electronic circuits, the dc equivalent circuits will be *radically* different from the corresponding ac equivalent circuits. This is true because of the use of *coupling* and *bypass capacitors*.

In basic ac circuit theory we see that a capacitor offers an opposition to the flow of ac (sinusoidal) current that is termed *capacitive reactance* X_C. The capacitive reactance of a capacitor is given by

$$X_C = \frac{1}{\omega C} = \frac{1}{2\pi f C} \tag{3-17}$$

where X_C = magnitude of the capacitive reactance (ohms)
f = frequency (hertz)
C = capacitance (farads)

From Eq. 3-17 we can see that X_C is *inversely proportional* to the frequency f. Specifically, as f increases, X_C decreases, and as f decreases, X_C increases. From this fundamental relationship, we can draw two general conclusions.

1. An ideal (coupling or bypass) capacitor is an open circuit to dc (0 Hz).
2. A (coupling or bypass) capacitor may be regarded as a short circuit to the ac signal frequencies.

The second conclusion will be true only if the capacitor has enough capacitance to ensure a low X_C at the lowest signal frequency. The actual sizing of coupling and bypass capacitors will be targeted in Chapter 9.

In Fig. 3-37(a) we have shown a diode that is biased by a dc source and also has an ac signal applied. Observe that the ground symbol has been incorporated in the schematic. (Recall that the ground symbol represents a point of common connection and is defined as zero volts.) Capacitor C_1 is called the *input coupling capacitor*, while capacitor C_2 is referred to as the *output coupling capacitor*. Coupling capacitors are used to pass an ac signal from one point to another in an electronic circuit while simultaneously blocking the dc bias levels. (Consequently, C_1 and C_2 are also called dc blocking capacitors.) This isolation of the dc bias serves to ensure that any dc level produced by the signal source, or the effects of R_L, will not upset the diode's Q-point.

Capacitor C_3 is called a bypass or decoupling capacitor. Its role is to bypass the ac signal around the dc power supply to ground. To analyze a diode circuit such as the one shown in Fig. 3-37(a), we perform a dc analysis followed by an ac analysis,

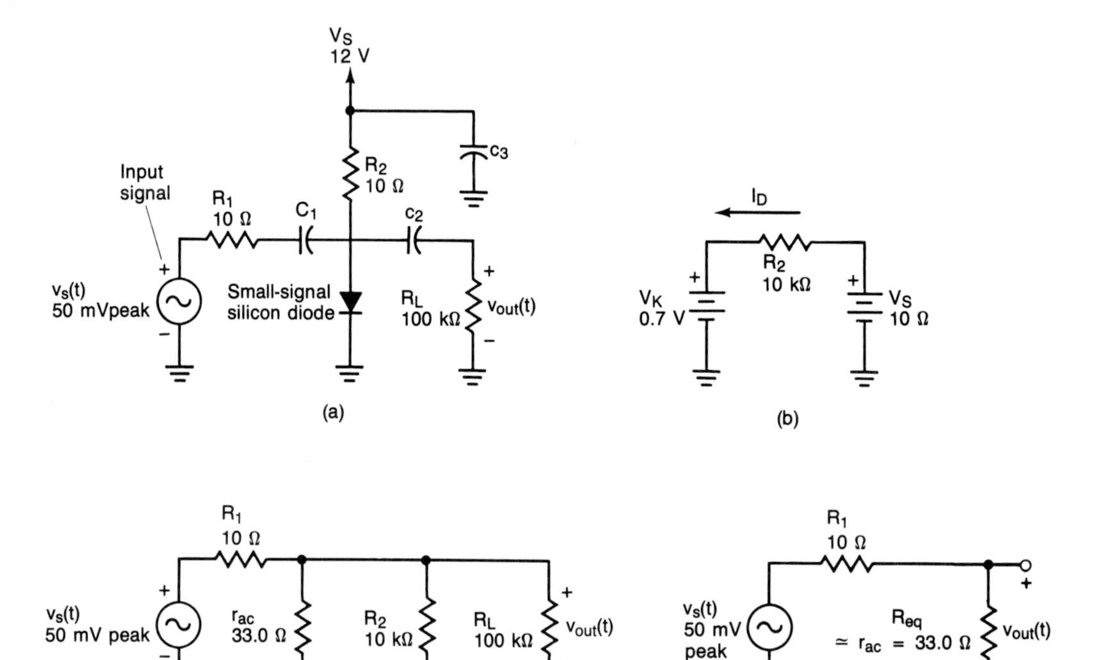

FIGURE 3-37 **Circuits for Example 3-19: (a) complete circuit; (b) dc equivalent circuit; (c) ac equivalent circuit; (d) simplified ac equivalent circuit.**

just as we have done previously. However, now we must also include the effects of the coupling and bypass capacitors.

Hence for the *dc analysis*:

1. Replace all capacitors with open circuits and all ac voltage sources with short circuits.
2. Simplify the resulting circuit as required.
3. Replace the diode (or other solid-state device) with its appropriate dc model.

The dc equivalent circuit of Fig. 3-37(a) is shown in Fig. 3-37(b). The capacitors have disappeared since they are open circuits, and the resulting circuit has been simplified. The knee-voltage diode model has been indicated.

The procedure for the *ac analysis* is:

1. Replace all capacitors and dc voltage sources with short circuits.
2. Simplify the resulting circuit as required.
3. Replace the diode (or other solid-state device) with its appropriate ac model.

The ac equivalent circuit of Fig. 3-37(a) is given in Fig. 3-37(c). The circuit has been simplified in Fig. 3-37(d). Let us now work through the complete analysis.

EXAMPLE 3-19

Determine the peak ac voltage across the load resistance R_L in the circuit given in Fig. 3-37(a). We do not have any data on the diode, but we do know that it is a small-signal silicon unit.

SOLUTION By replacing the capacitors with open circuits, we arrive at the dc equivalent circuit shown in Fig. 3-37(b). Hence

$$I_D = \frac{V_S - V_K}{R_2} = \frac{12V - 0.7V}{10K\Omega} = 1.13 \text{ mA}$$

and

$$r_{ac} = r_j + r_B = \frac{26 \text{ mV}}{1.13 \text{ mA}} + 10\ \Omega = 33.0\ \Omega$$

Replacing the capacitors with short circuits gives us the ac equivalent circuit shown in Fig. 3-37(c). By inspection, we see that v_{out} is across the parallel combination of R_2, R_L, and r_{ac}. Because r_{ac} is so small compared to R_2 and R_L, the equivalent resistance R_{eq} is

$$\begin{aligned} R_{eq} &= r_{ac} \parallel R_2 \parallel R_L \\ &= 33.0\ \Omega \parallel 10\ k\Omega \parallel 100\ k\Omega \simeq 33.0\ \Omega \end{aligned}$$

and we arrive at Fig. 3-37(d). By voltage division,

$$\begin{aligned} v_{out(peak)} &= \frac{r_{ac}}{r_{ac} + R_1} v_{s(peak)} = \frac{33.0\ \Omega}{33.0\ \Omega + 10\ \Omega}(50 \text{ mV}) \\ &= 38.4 \text{ mV} \end{aligned}$$

which completes the problem. ■

3-20 Large-Signal Models and Frequency Effects

In Fig. 3-38(a) we have indicated a diode circuit which is being driven by a *large-signal* sinusoidal ac voltage source.

When we are dealing with large-signal ac voltage sources, the small-signal diode model does *not* apply. However, it *is* appropriate to revert back to our piecewise linear *dc* models.

Consequently, when the sine wave is in its positive half-cycle, we may consider the diode to be a closed switch [see Fig. 3-38(b)]. When the sine wave is in its negative half-cycle, the diode is reverse biased and we may model it as an open switch [see Fig. 3-38(c)]. When the diode is used in this fashion, it is termed a *half-wave rectifier* [refer to Fig. 3-38(d)]. Rectifier circuits are discussed more fully in Chapter 13.

The half-wave rectifier is normally used in dc power supplies to change a bipolar (positive- and negative-going) ac waveform into a unipolar (either positive or negative) pulsating dc. The rectifier diode acts as an extremely fast switch when its switching

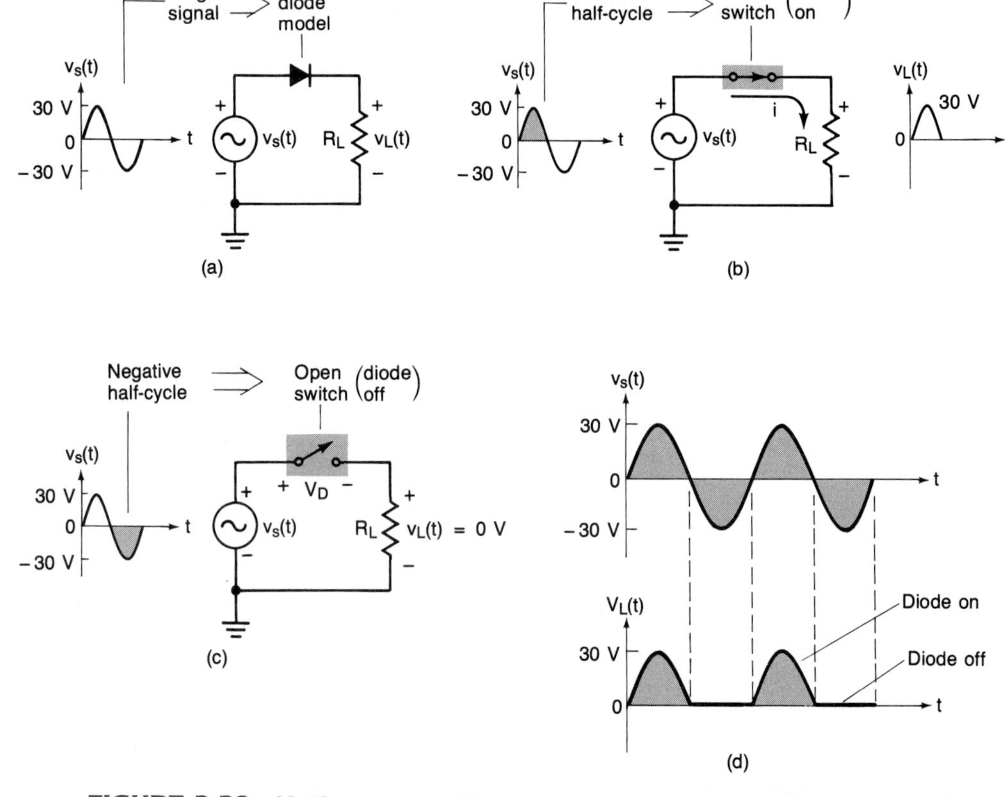

FIGURE 3-38 Half-wave rectifier: (a) circuit; (b) positive half-cycles of input; (c) negative half-cycles of input; (d) waveforms.

speed is compared to a 60-Hz sine wave. (Recall that $T = 1/f$, and in the case of a 60-Hz sine wave the period is 16.7 ms.)

An ordinary rectifier may require from nanoseconds to microseconds to switch from its conducting to its nonconducting state. When this switching time is compared to 16.7 ms, it may be regarded as negligible. However, at higher frequencies the period becomes smaller and the switching time will become significant. This effect may be observed in Fig. 3-39.

As can be seen in Fig. 3-39(b), the diode will conduct significantly in the reverse direction when it is used to rectify higher frequencies. The higher the frequency, the more noticeable the effect.

The reason the diode's switching action deteriorates at higher frequencies is because its *p-n* junction possesses some small capacitance. The diode's junction capacitance limits its switching speed. As we shall see in our later work, junction, and other capacitive effects tend to affect the high-frequency performance of all electronic circuits. Consequently, we need to examine the capacitive effects associated with the *p-n* junction since it is fundamental not only to the diode, but to many other solid-state devices as well.

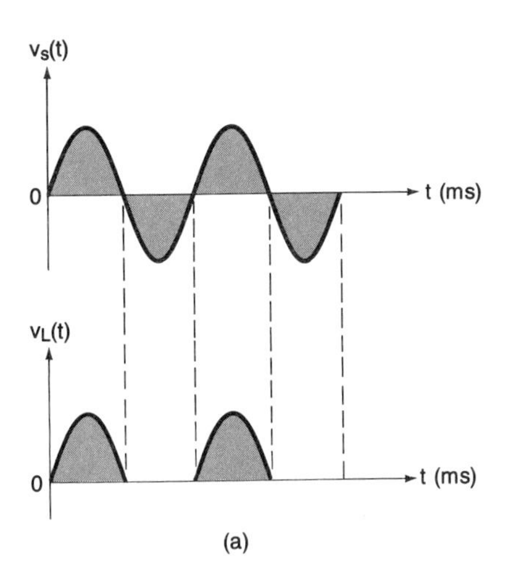

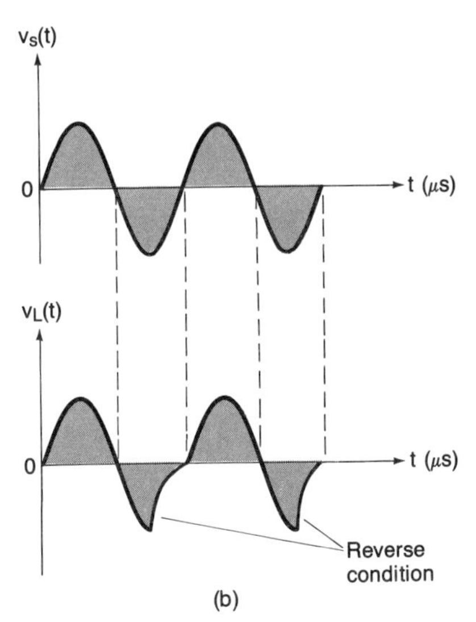

FIGURE 3-39 Frequency effects on diode switching: (a) low-frequency response; (b) high-frequency response.

3-21 Diffusion Capacitance

When a *p-n* junction is forward biased, it exhibits a capacitive effect which is termed *diffusion capacitance*. Recall that forward bias reduces the width of the depletion region at the *p-n* junction, and reverse bias increases the width of the depletion region. It is this readjustment in the depletion region width, as the bias changes from forward to reverse, which produces the capacitive effect.

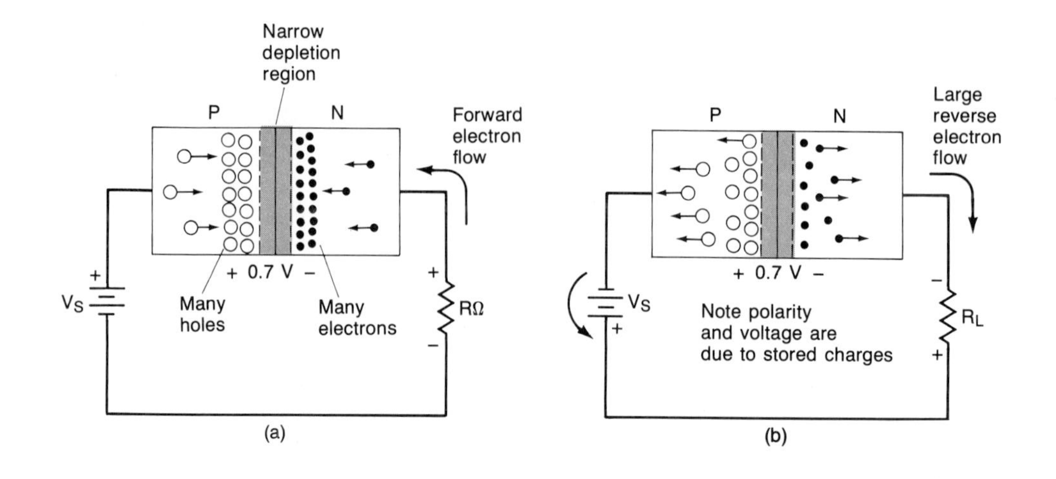

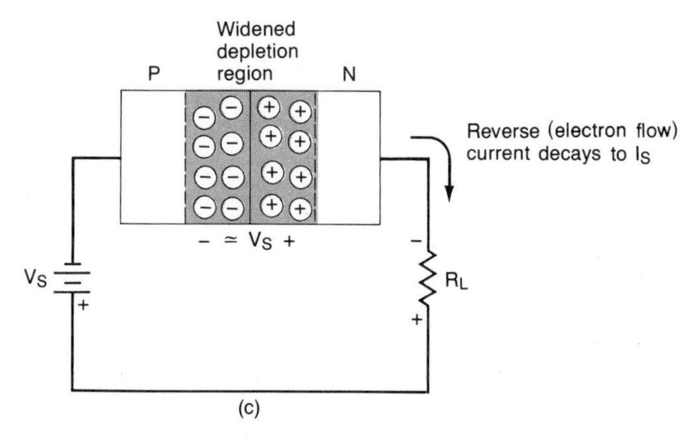

FIGURE 3-40 Diffusion capacitance effects: (a) forward-biased p-n junction; (b) p-n junction is suddenly reverse biased; (c) depletion region has widened fully and the reverse current has decayed to I_S.

In Fig. 3-40(a) we see a forward-biased *p-n* junction. In Fig. 3-40(b) the *p-n* junction is suddenly reverse biased. Observe that the depletion width is still narrow. This is true because there are many holes and electrons at the junction. The stored carriers at the junction prevent the junction voltage from changing instantaneously.

As the majority carriers move away from the junction and the electrons recombine with holes, the depletion region widens and the diode's forward voltage begins to decrease. When enough of the majority carriers have moved out of the junction region, a sufficient number of ions have been "uncovered" to allow the reverse voltage to build up. Ultimately, enough ions exist to produce a reverse voltage across the junction which is approximately equal to V_S. At this point, only the small reverse saturation current I_S flows [see Fig. 3-40(c)].

From Fig. 3-40 it is clear that if a reverse bias is suddenly applied, the forward current ceases immediately, leaving some majority charge carriers in the depletion

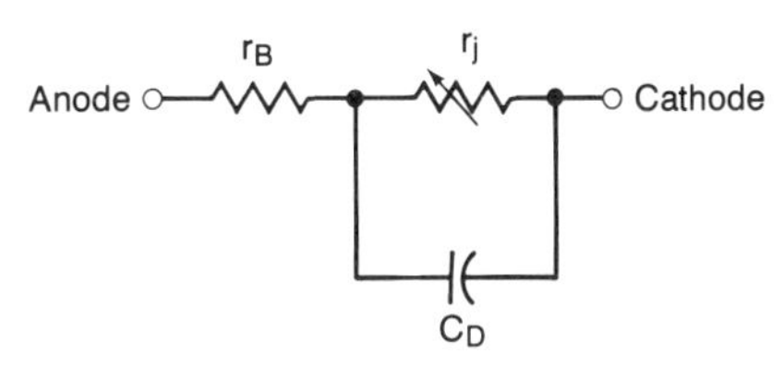

FIGURE 3-41 High-frequency equivalent of a forward-biased diode.

region. These majority carriers must be "swept out" of the depletion region as it is being widened by the reverse bias. As a result, when a forward-biased diode is suddenly reverse biased, a reverse current flows that may be large initially and slowly decays to the level of the reverse saturation current I_S.

The effect is very similar to the discharging of a capacitor. Therefore, it may be modeled as a capacitance. This capacitance is termed a *diffusion capacitance* C_D. C_D is directly proportional to the forward current through the diode. This relationship occurs since the number of majority carriers injected into the depletion region is directly proportional to the diode's forward current.

The diffusion capacitance may be included in the model for a forward-biased diode as shown in Fig. 3-41. The value of C_D can range to values as large as 2000 pF. Since C_D depends on the forward current through the diode, and it is in parallel with the relatively low resistance, r_j, its value is difficult to specify in useful terms for *large signals*. Consequently, manufacturers supply a parameter known as the *reverse recovery time* t_{rr}.

To specify t_{rr}, a diode usually has a step voltage (rectangular waveform) applied to it as shown in Fig. 3-42(b). The diode's voltage waveform is indicated in Fig. 3-42(c), and its corresponding current waveform is given in Fig. 3-42(d).

The typical values for t_{rr} can range from less than 1 ns to approximately 1 μs. The effect of t_{rr} on a diode's rectification is summarized in Fig. 3-43. To minimize the effects of t_{rr} on a diode's rectification of a signal, we provide the condition stated by

$$t_{rr} \leq 0.1T \tag{3-18}$$

This condition is shown in Fig. 3-43(b).

EXAMPLE 3-20

A 1N4148 diode has a t_{rr} of 4 ns. What is the maximum frequency of a sine wave that a 1N4148 can rectify (without significant degradation due to diffusion capacitive effects)?

SOLUTION From Eq. 3-18,

$$T = 10t_{rr} = (10)(4\text{ns}) = 40 \text{ ns}$$

$$f_{\max} = \frac{1}{T} = \frac{1}{40 \text{ ns}} = 25 \text{ MHz}$$

A 25-MHz sine wave would produce a diode current waveform such as that shown in Fig. 3-43(b). ■

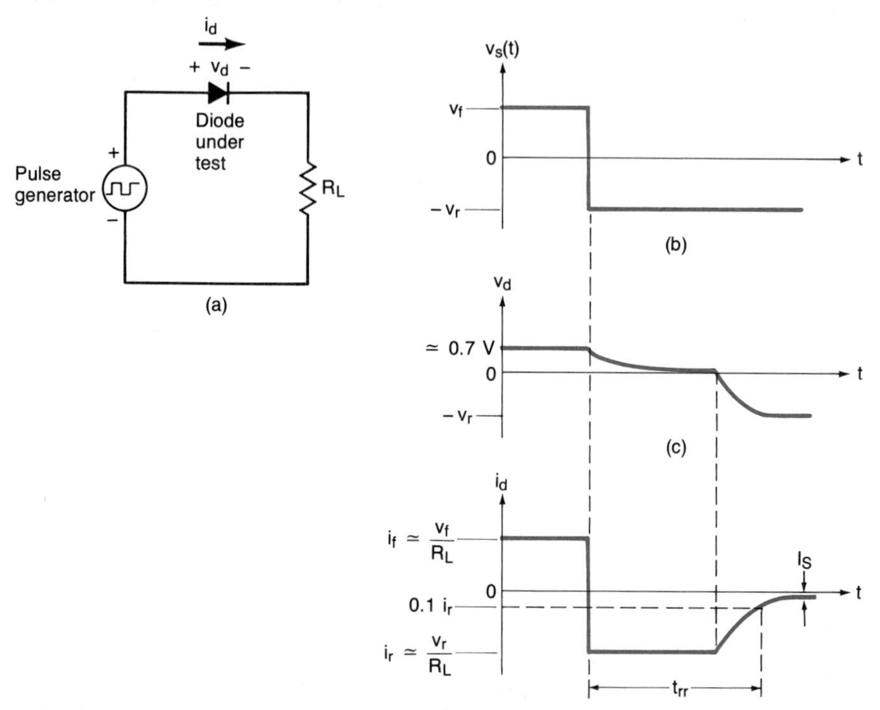

FIGURE 3-42 Reverse recovery time: (a) test circuit (b) generator voltage signal; (c) diode's voltage; (d) diode's current waveform and t_{rr}.

FIGURE 3-43 Relationships between the period T and t_{rr}.

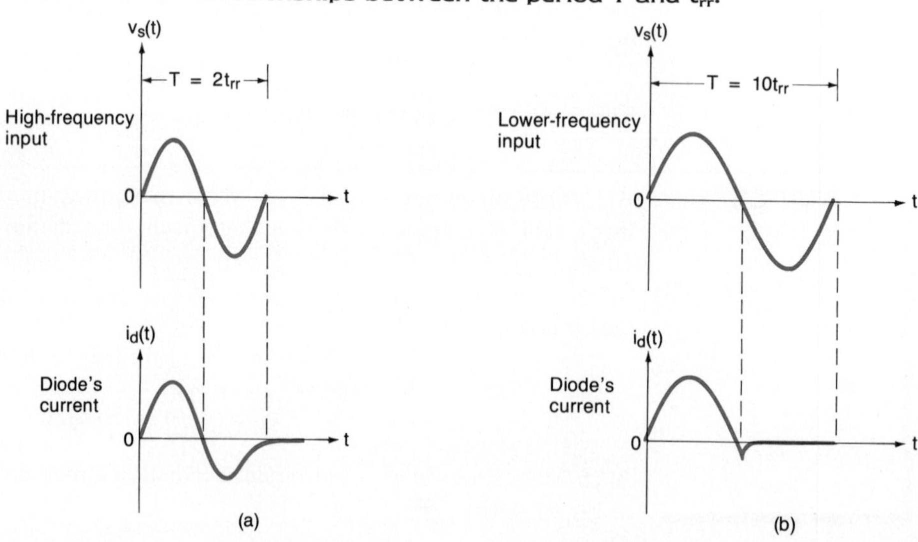

3-22 Schottky Diode

To shorten the reverse recovery time (t_{rr}) of a diode, the carriers stored in the depletion region must be quickly removed. The reverse recovery time can be shortened by providing more recombination centers for the charge carriers. Adding gold atoms to the semiconductor lattice has been proven to be a very effective means for creating additional recombination centers (refer to Section 1-10). A gold-doped silicon *p-n* junction produces very fast power rectifiers. For example, a standard 1-A rectifier diode might have a t_{rr} of 5 μs, and a *fast-recovery* (gold-doped) 1-A rectifier might have a t_{rr} on the order of 350 ns. However, fast-recovery rectifiers tend to have larger reverse leakage currents and slightly larger forward voltage drops when compared with the standard rectifiers.

To provide even faster rectification, the *Schottky diode* was developed. In Section 2-14 we noted that an ohmic contact may be formed by attaching a metal conductor to a heavily doped *n*-type semiconductor (n^+). By "ohmic contact" we mean that current flows equally well in both directions. However, if we attach metal to a more lightly doped *n*-type semiconductor, a rectifying junction will be formed. This is the basis of the Schottky diode. In Fig. 3-44 we illustrate the schematic symbol and the basic structure of the Schottky diode.

To understand the operation of the Schottky metal–semiconductor junction, we shall first parallel its formation to that of the conventional *p-n* junction. Recall that isolated *p*- and *n*-type semiconductors have their majority carriers distributed as shown in Fig. 3-45(a). In Fig. 3-45(b) we see that the *n*-type semiconductor has "free" electrons with higher energy levels than those associated with the free electrons found in the metallic conductor.

In Fig. 3-45(c) we note that the Fermi level of the *n*-type semiconductor is higher than that of the *p*-type semiconductor. Figure 3-45(d) illustrates that the Fermi level of the *n*-type semiconductor is also higher than that for the metallic conductor. Also note in Fig. 3-45(d) that the valence and conduction energy bands of a metallic conductor overlap. Therefore, to simplify our later diagrams, we will omit the details of the valence and conduction bands of the conductor.

When the *p*- and *n*-type semiconductors are merged together, electrons from the *n*-side cross over the junction to annihilate holes on the *p*-side. This results in a layer of positive ions on the *n*-side and negative ions on the *p*-side [see Fig. 3-45(e)]. These positive and negative charges form a barrier potential of 0.7 V (for silicon).

FIGURE 3-44 **Planar Schottky diode and its schematic symbol.**

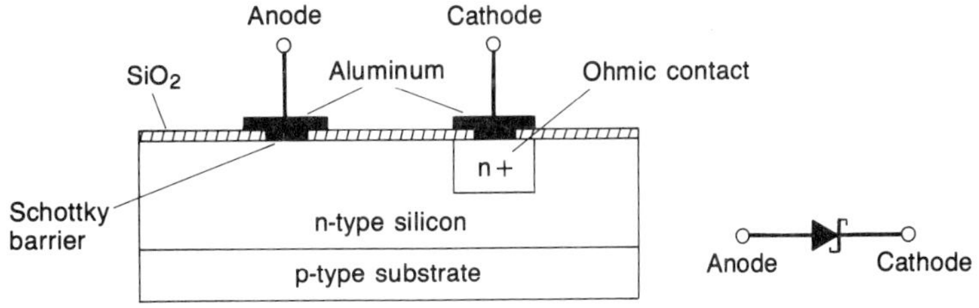

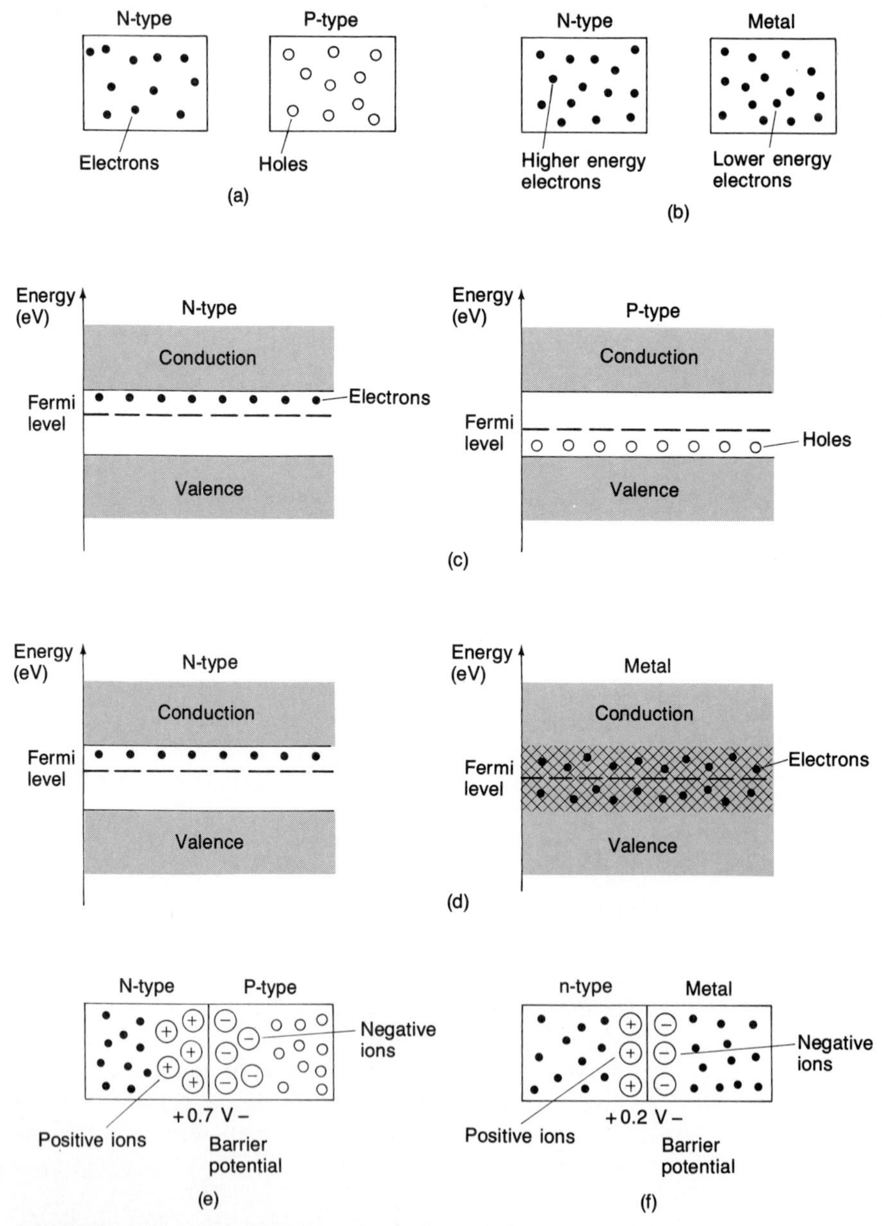

FIGURE 3-45 Analogy between the formation of the p-n junction and the Schottky junction.

Essentially the same phenomena occurs in the Schottky junction. Electrons always seek the lowest energy level. Therefore, electrons diffuse from the *n*-type semiconductor to the metallic conductor. Again, a layer of ions are produced, but the barrier potential is only about 0.2 V. This is illustrated in Fig. 3-45(f).

In both cases the Fermi levels become aligned [see Fig. 3-46(a) and (b)]. The primary difference between a forward-biased *p-n* junction and a Schottky junction

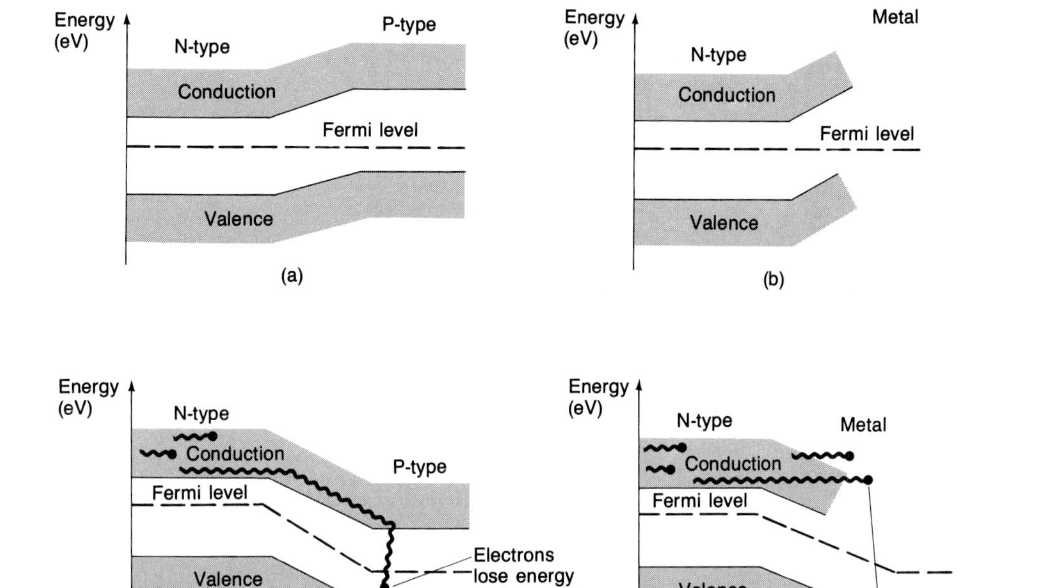

FIGURE 3-46 Forward bias in a p-n junction and a Schottky junction; (a) Fermi level alignment in n- and p-type semiconductors; (b) Fermi level alignment in a Schottky junction; (c) forward bias allows electrons to travel down a potential hill to the p-type semiconductor; (d) forward bias injects high-energy electrons into the metal.

FIGURE 3-47 V-I characteristics.

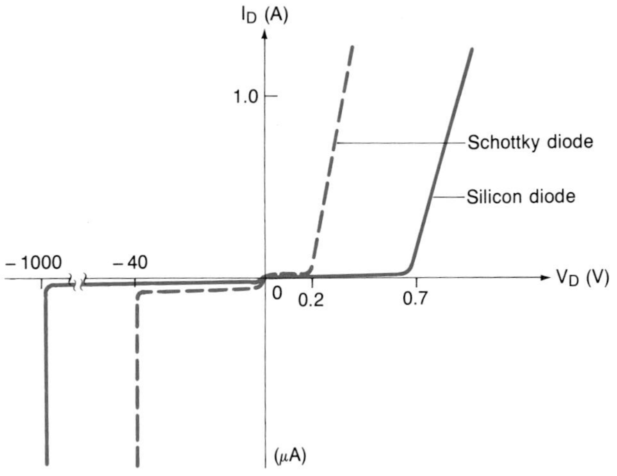

is that electrons maintain high energy levels in the Schottky device [compare Fig. 3-46(c) and (d)]. Because the electrons from the *n*-type semiconductor plunge into the metal with such high energies, they are referred to as "hot" carriers, and the Schottky diode is sometimes called a *hot-carrier diode*. Further, since only electrons are involved in conduction, it is also described as being a *unipolar* device.

The Schottky diode is a much faster switch than the standard silicon diode rectifier. This is true because there is no majority-carrier storage at the *p-n* junction since only electrons are involved in its conduction. Consequently, Schottky diodes demonstrate a t_{rr} of nearly zero. The Schottky diode also has lower forward voltage drops ($V_K = 0.2$ V). This makes it more efficient in high-power applications than silicon rectifiers. However, Schottky diodes do exhibit higher leakage currents and lower reverse breakdown voltages. Consequently, they are only used when t_{rr}, or efficiency are the primary concerns. A comparison of the *V–I* characteristics is given in Fig. 3-47.

3-23 Junction Capacitance and the Varactor Diode

As we have seen, a forward-biased diode exhibits a capacitive effect which is referred to as its diffusion capacitance C_D. A *reverse-biased p-n* junction will also possess a capacitive effect which is called *junction capacitance* (C_j).

One of the basic attributes of any capacitor, or a capacitance, is its ability to store electrical charge. A parallel-plate capacitor contains two conductors separated by a dielectric (or insulator). When the capacitor is charged, electrons are removed from one plate, leaving positive ions, and the other plate contains an excess number of negative electrons. The stored charges are separated by a uniform distance *d*, and the resulting electric field is also very uniform.

As we have seen, the depletion region at a *p-n* junction also contains separated, immobile positive and negative charges. In this sense the *p-n* junction must possess some capacitance. However, the positive and negative ions are distributed randomly within the crystal lattice.

Because of the very uniform electric field associated with a parallel-plate capacitor, we find that its capacitance is given by a very simple formula: $C = \epsilon A/d$. The capacitance (C) is equal to the permittivity (ϵ) times the plate area (A) divided by the distance between the plates (d). When the voltage across a parallel-plate capacitor is changed, its capacitance will remain *essentially* constant, but its charge will change.

However, when the reverse bias across a *p-n* junction is changed, its capacitance will change dramatically. The capacitance associated with a *p-n* junction depends on the *doping levels* on both sides, the *doping profiles* (the changes in the doping density as one moves away from the junction), and the magnitude of the *reverse bias*.

First, let us look at effects of reverse bias on the capacitance. When the reverse bias across a *p-n* junction is increased, we observe the following:

1. The depletion region widens, which increases the net distance between the charges.
2. More positive and negative ions are produced (we gain more charge).

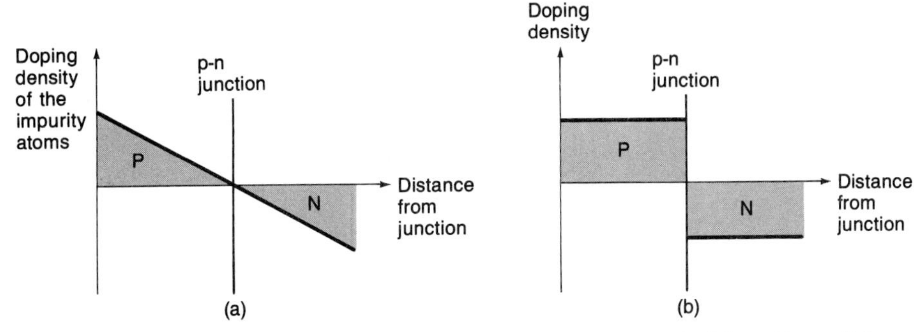

FIGURE 3-48 **Diode doping profiles: (a) linearly-graded junction; (b) abrupt junction.**

Because the net distance (d) between the charges is increased by *increasing* the reverse bias, the capacitance will tend to *decrease*. This is true because in a very general sense, the junction capacitance obeys the parallel-plate capacitor relationship. However, because the geometry of the diode's depletion region is not as simple as that for a parallel-plate capacitor, the variation in the diode's capacitance is a *nonlinear* function of the reverse bias.

The particular doping profile employed in the *p-n* junction fabrication also affects the magnitude of the change in the capacitance that the reverse bias controls. Two doping profiles for a *p-n* junction are shown in Fig. 3-48. The relationship defined by Eq. 3-19 describes how the reverse bias affects the junction capacitance C_j.

$$C_j = \frac{C_{j0}}{(1 + V_R/V_K)^n} \tag{3-19}$$

where C_j = junction capacitance (picofarads)
C_{j0} = reference capacitance measured with V_R = 0V (picofarads)
V_R = reverse bias (volts)
V_K = diode's barrier potential, V (e.g., 0.7 V for silicon)
$n = \frac{1}{3}$ (linearly graded); $\frac{1}{2}$ (abrupt junction)

To illustrate the use of Eq. 3-19 and the effect of the doping profiles, consider Examples 3-21 and 3-22.

EXAMPLE 3-21

A reverse-biased diode with a linearly graded doping profile has a capacitance of 40 pF when V_R = 0 V (C_{j0} = 40 pF). Calculate the C_j of this silicon diode when V_R is 1 V and V_R is 10 V.

SOLUTION Since the diode junction is linearly graded, $n = \frac{1}{3}$, and by Eq. 3–19, if V_R = 1 V,

$$C_j = \frac{C_{j0}}{(1 + V_R/V_K)^n} = \frac{40 \text{ pF}}{(1 + 1 \text{ V}/0.7 \text{ V})^{1/3}}$$

$$= \frac{40 \text{ pF}}{\sqrt[3]{2.4286}} = 29.7 \text{ pF}$$

When $V_R = 10$ V,

$$C_j = \frac{40 \text{ pF}}{\sqrt[3]{1 + 10 \text{ V}/0.7 \text{ V}}} = 16.1 \text{ pF}$$ ■

EXAMPLE 3-22

Repeat Example 3-21 for the diode if its doping profile is abrupt.

SOLUTION Since this diode's junction is abrupt, $n = \frac{1}{2}$, and if $V_R = 1$ V,

$$C_j = \frac{C_{j0}}{(1 + V_R/V_K)^n} = \frac{40 \text{ pF}}{(1 + 1 \text{ V}/0.7 \text{ V})^{1/2}}$$

$$= \frac{40 \text{ pF}}{\sqrt[2]{2.4286}} = 25.7 \text{ pF}$$

When $V_R = 10$ V,

$$C_j = \frac{40 \text{ pF}}{\sqrt[2]{1 + 10 \text{ V}/0.7 \text{ V}}} = 10.2 \text{ pF}$$ ■

If we compare the ratios of the capacitance change for Examples 3-21 and 3-22, we see that for the linearly graded profile,

$$\frac{C_j(\text{max})}{C_j(\text{min})} = \frac{29.7 \text{ pF}}{16.1 \text{ pF}} = 1.84$$

and for the abrupt profile,

$$\frac{C_j(\text{max})}{C_j(\text{min})} = \frac{25.7 \text{ pF}}{10.2 \text{ pF}} = 2.52$$

The *p-n* junction with an abrupt doping profile has a C_j that is affected by V_R much more than a linearly graded *p-n* junction.

A particular diode that is designed to be used in its reverse-biased mode, and serve as a voltage-variable capacitor (VVC) is the *varactor diode*. Its schematic symbol is shown in Fig. 3-49.

A varactor diode is typically constructed from a silicon *p-n* junction with an abrupt doping profile. The typical $C_j(\text{max})/C_j(\text{min})$ ratio is about 4:1. To get a larger $C_j(\text{max})/C_j(\text{min})$ ratio, the doping profile may be hyperabrupt as shown in Fig. 3-50. The hyperabrupt *p-n* junction typically has a $C_j(\text{max})/C_j(\text{min})$ ratio of 10:1.

FIGURE 3-49 Varactor diode schematic symbol.

Cathode —|◄— Anode

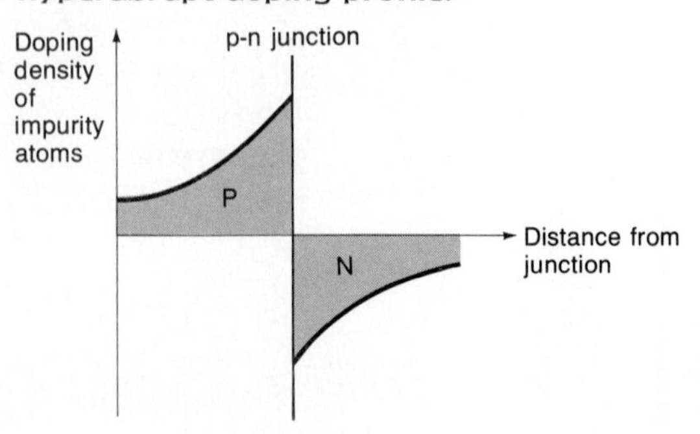

FIGURE 3-50 Varactor diode hyperabrupt doping profile.

The varactor diode is used in the voltage-controlled tuners found in many FM receivers and televisions. Its use has virtually eliminated the need for electromechanical tuning arrangements.

3-24 Diode Failure Modes

Obviously, we would like for solid-state components to be 100% reliable. However, this is unfortunately not the case. All solid-state devices have a certain probability of failure. To be good circuit designers and/or good circuit troubleshooters, we need to have some idea of the most likely failure modes for solid-state devices. This knowledge can help us anticipate failures in new circuit designs, and locate failures in a malfunctioning electronic circuit.

All the devices that we have studied so far—the silicon and germanium rectifier diodes, the silicon zener and avalanche diodes, the Schottky barrier diodes, and the varactor diodes will typically become short circuits or "leaky" when they fail. The probability of the devices becoming open circuits when they fail is not as great. As a rule of thumb, small-signal, rectifier, Schottky, and varactor diodes will become shorts, or exhibit excessive leakage current, 75% of the time when they fail. These diodes will fail as open circuits 25% of the time. Zener and avalanche diodes will become short circuits 60% of the time when they fail. These diodes will become open circuits 30% of the time, or go out of tolerance 10% of the time.

3-25 Diode Large-Signal Applications: Clippers

Diodes are often used to *clip* or *limit* voltage signals. One such application is shown in Fig. 3-51. The 1N4148 diodes are small-signal silicon units. The triangular symbol is used to represent an amplifier. (The amplifier will be introduced in Chapter 4.) The diode circuit in Fig. 3-51(a) is used to protect the amplifier's input from excessively large input signals.

In this example, we assume the amplifier's normal input signals are much less than 200 mV peak to peak. As can be seen in Fig. 3-51(b), the ac voltage signal across the diodes is too small to cause them to conduct. Therefore, they effectively act like open switches.

In Fig. 3-51(c) we see that a large (10-V p-p) signal has been applied to the amplifier's input. In this case, the diodes conduct to clip the signal across the amplifier's input to 700 mV peak. The amplifier's input is protected from input signals that attempt to exceed ± 700 mV.

The equivalent circuit when $v_s(t)$ is at its 5-V positive peak is depicted in Fig. 3-51(d). Observe that the knee-voltage model for the conducting diode (D_1) has been used. Since diode D_2 is reverse biased, it is modeled as an open switch. The circuit action during the negative half-cycles of input signal is similar. In this case D_1 is an open switch, and D_2 is replaced with its knee-voltage model.

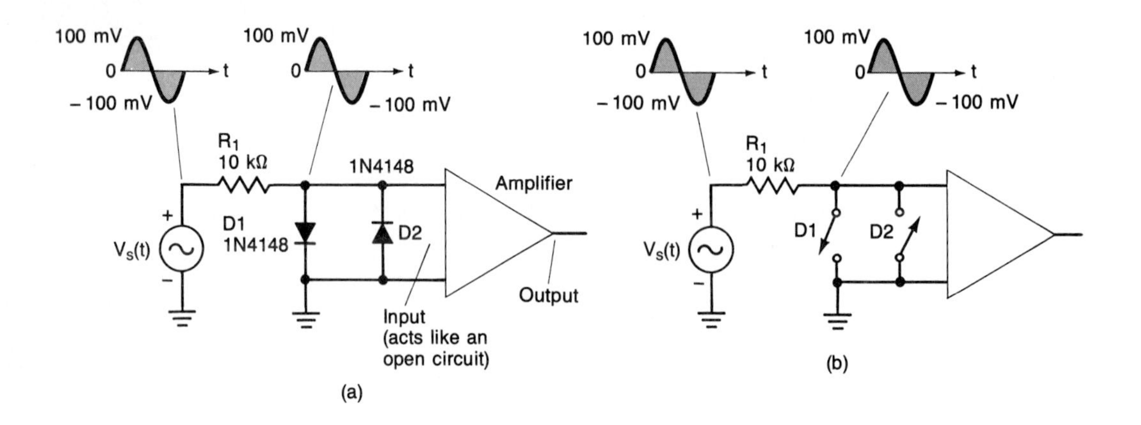

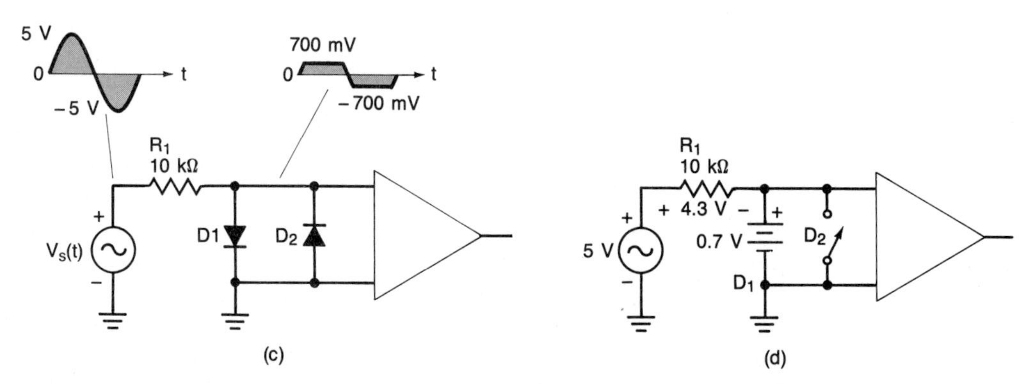

FIGURE 3-51 Diode clipper used to protect an amplifier's input: (a) circuit; (b) diodes act like open switches at low input levels; (c) limiting action; (d) equivalent circuit when $v_S(t)$ is at its 5-V peak.

Diode clippers can be biased to limit signals to much larger peak values [see Fig. 3-52(a)]. In this application, it is assumed that the normal input signals are on the order of 10 V peak. The circuit has been redrawn in Fig. 3-52(b). As can be seen in Fig. 3-52(c), when $v_s(t)$ is at its positive 10-V peak both diodes are nonconducting. Therefore, we replace them with open switches. Diode D_2 is reverse biased by the signal. (Its total reverse bias is 25 V.) Diode D_1 is also reverse biased. This is true because its cathode is *more positive* than its anode. (The reverse bias across D_1 is 5 V.)

If the input signal becomes too large (e.g., 16 V peak), the diodes will limit (or clip) the signal level to ± 15.7 V as shown in Fig. 3-52(d).

The clip points become obvious when we examine the equivalent circuits given in Fig. 3-52(e) and (f). Diode D_1 conducts on the positive half cycles when $v_s(t)$ attempts to exceed 15.7 V. Diode D_2 conducts when $v_s(t)$ attempts to become more negative than -15.7 V.

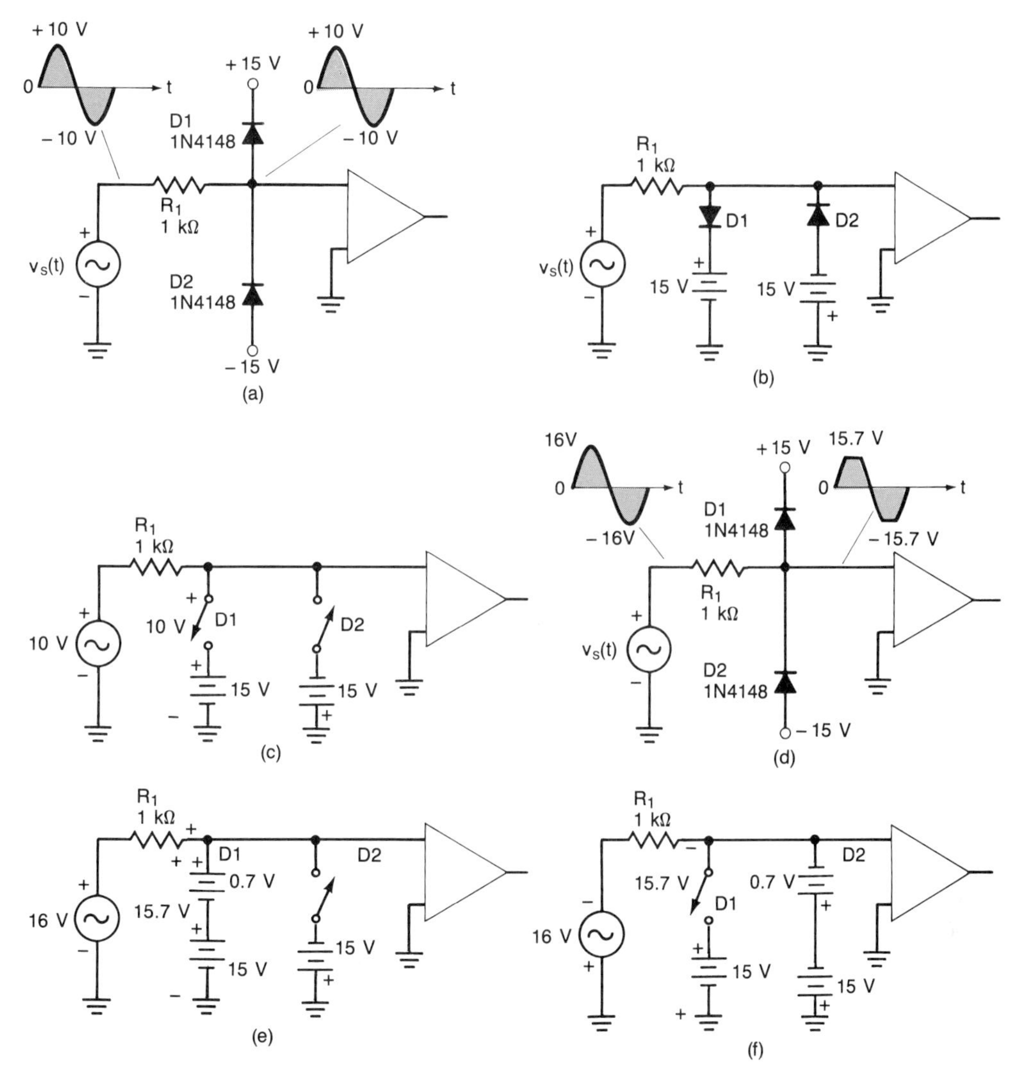

FIGURE 3-52 Biased clipper: (a) circuit without clipping; (b) equivalent circuit; (c) both diodes are off when $v_S(t)$ is at its 10-V peak; (d) circuit with clipping; (e) equivalent circuit when $v_S(t)$ is at its 16-V peak; (f) equivalent circuit when $v_S(t)$ is at its -16-V peak.

3-26 Clampers

A diode *clamper* circuit is shown in Fig. 3-53(a). The clamper is also referred to as a *dc restorer* and a (signal) *level shifter*. The latter term is probably the most descriptive.

As can be seen in Fig. 3-53(a), the input signal [$v_s(t)$] is pure ac. Specifically,

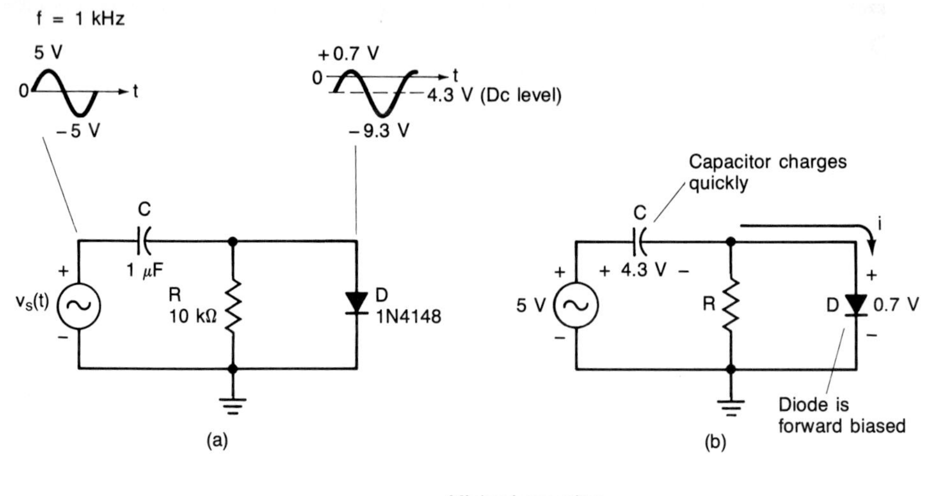

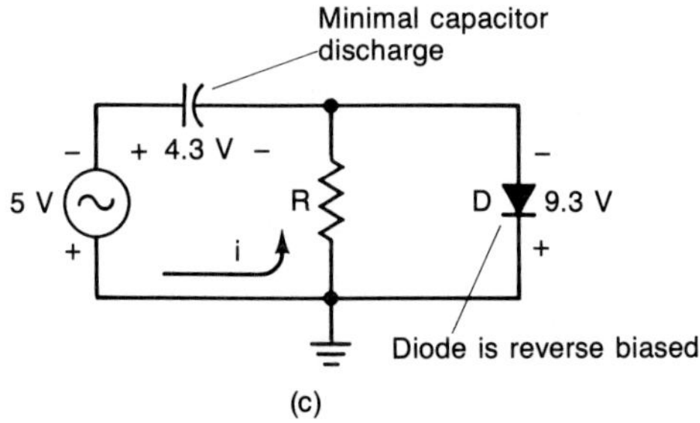

FIGURE 3-53 Diode clamper (dc level shifter): (a) circuit and waveforms; (b) capacitor charging; (c) diode is reverse biased and capacitor slightly discharges through R.

it has a dc level of zero. However, at the output of the circuit (across the diode), the ac signal is riding on a dc level of -4.3 V. The operation of the circuit is detailed in Fig. 3-53(b) and (c).

When $v_s(t)$ reaches its positive peak (5 V) the diode is forward biased, and the capacitor charges quickly to 4.3 V. The charging time constant is very small since the diode's forward resistance is very low. When $v_s(t)$ reaches its negative peak (-5 V), the diode is reverse biased. Consequently, the capacitor must discharge through the resistor. The discharge time constant (RC) is very large. In practice, we ensure that

$$\tau = RC \geq 10T$$

where T is the period of the highest signal frequency to be level shifted. (For Fig. 3-53(a) we have a T of 1 ms and a τ of 10 ms.) In Fig. 3-53(c) we see that the capacitor acts like a voltage source. Its voltage adds to v_s(t) and the total instantaneous voltage across the diode is -9.3 V.

PROBLEMS

Drill, Derivations, and Definitions

Section 3-2

3-1. A germanium diode has a reverse saturation current (I_S) of 1 μA at 25°C. Using the Shockley diode equation, compute the forward currents for $V_D = 0, 0.1, 0.2, 0.3, 0.31$, and 0.32 V. Also compute the reverse current for $V_D = -0.1, -1$, and -10 V. Construct a table similar to Table 3-1. Graph its *V–I* curve.

3-2. A silicon diode has a reverse saturation current (I_S) of 10 nA at 25°C. Construct a table of *V–I* values like Table 3-1. Use the same V_D values. Graph its *V–I* characteristic.

3-3. Repeat Prob. 3-1 for the same germanium diode at 100°C. Discuss the effect of temperature on its forward voltage drop. (*Hint*: Both V_T and I_S must be determined at 100°C.) Do not graph the curve.

3-4. A silicon diode has a reverse saturation current (I_S) of 10 μA at 125°C. Compute its I_S at 25°C.

3-5. Discuss the characteristics of a real diode that the Shockley diode equation does *not* take into account.

3-6. A silicon diode has a reverse current (I_R) of 0.025 mA at 25°C and an I_R of 0.120 mA at 100°C. Given that $I_R(T) = I_S(T) + I_{SL}$ (and I_{SL} does *not* vary with temperature), determine I_S (the reverse saturation current due to minority carriers) and I_{SL} (the surface leakage current) at 25°C. (Hint: Draw on Eq. 2-1.)

Section 3-3

3-7. Use the dc load-line approach to solve for the voltage across resistor R_D and the current through it [see Fig. 3-3(a)]. Assume that $R_D = 2\ k\Omega$, $R_L = 1\ k\Omega$, and V_S is 12 V.

3-8. Repeat Prob. 3-7 for $R_D = 200\ \Omega$, $R_L = 510\ \Omega$, and $V_S = 6$ V.

3-9. The diode shown in Fig. 3-4(a) has the *V–I* values indicated in Table 3-3. Using

TABLE 3-3
Diode's *V–I* Values

V_D (V)	I_D
0.75	73 mA
0.7	28 mA
0.6	4 mA
0.5	0.6 mA
0.4	≃0.1 mA
0.3	≃0 mA
0.2	≃0 mA
0.1	≃0 mA
0.0	0 mA
−0.1	−34 nA
−1.0	−40 nA
−10	−40 nA

TABLE 3-4
LED *V–I* Values

V_D (V)	I_D (mA)
0.0	0
1.0	≃0
1.5	0.6
1.7	1.3
1.9	3.1
2.1	7.3
2.2	11.2
2.3	17.1
2.4	26.1
2.5	40.0

these V–I values, *carefully* plot the diode's characteristic curve on graph paper. If $R_L = 150\ \Omega$ and $V_S = 3$ V, use a dc load line to determine the operating point (I_D and V_D).

3-10. Repeat Prob. 3-9 for $V_S = 6$ V.

3-11. Repeat Prob. 3-9 for $V_S = 6$ V and $R_L = 510\ \Omega$.

3-12. For the circuit of Fig. 3-54, use a dc load line to find the dc operating point. The diode's V–I characteristic data points are given in Table 3-3. Assume that $R_L = 10$ MΩ and $V_S = 3$ V.

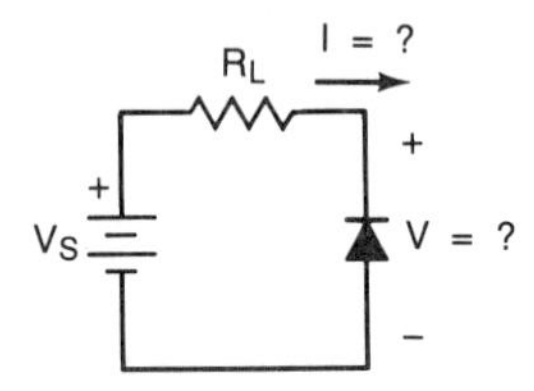

FIGURE 3-54 Diode circuit problem.

3-13. Repeat Prob. 3-12 for the same R_L and $V_S = 5$ V.

3-14. Given the zener diode and its idealized reverse V–I characteristic in Fig. 3-55, find its dc operating point using a dc load line.

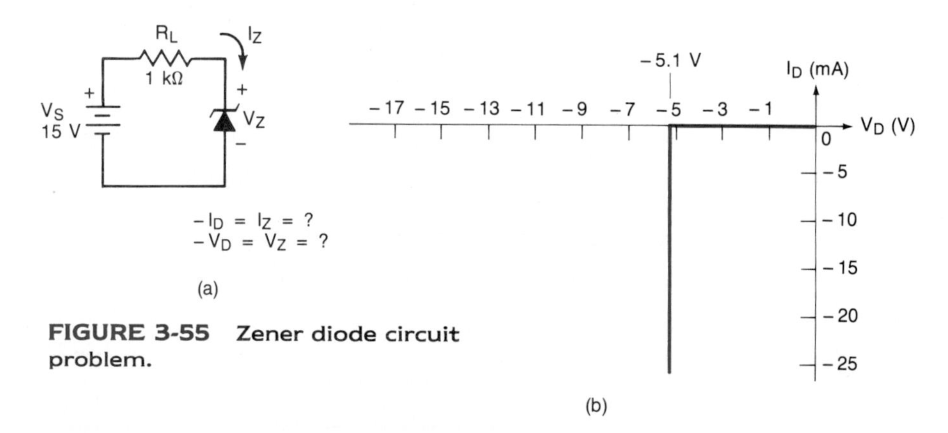

FIGURE 3-55 Zener diode circuit problem.

3-15. Repeat Prob. 3-14 if V_S is reduced to 3 V.

3-16. The forward V–I characteristic curve data for a green (GaP) LED are given in Table 3-4. Carefully plot the V–I characteristic on a sheet of graph paper. Solve the circuit shown in Fig. 3-56 by using the dc load-line approach.

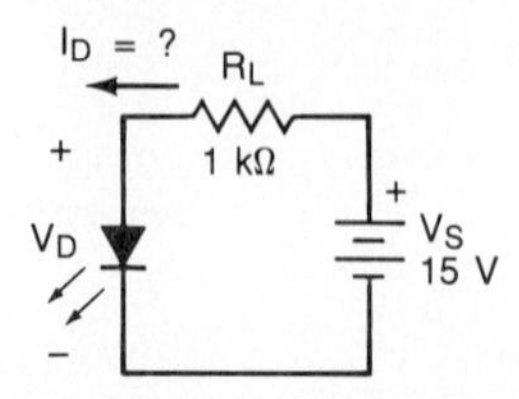

FIGURE 3-56 LED circuit problem.

Section 3-4

3-17. Repeat Prob. 3-9 using the *ideal diode* model. Compare the results. Which method yields the more accurate results?

3-18. Determine I_D and V_D for Fig. 3-4(a) with $V_S = 6$ V and $R_L = 150\ \Omega$ using the ideal diode model. Compare your results with those obtained in Prob. 3-10 using the dc load line. Which method yields the most accurate results?

Section 3-7

3-19. A silicon diode has a forward voltage drop of 0.75 V when a forward current of 40 mA flows through it. Sketch the dc models for this diode, which include (a) its knee voltage and (b) its dynamic resistance if its offset voltage V_O is 0.72 V.

Section 3-8

3-20. A silicon diode has a reverse current (I_R) of 15 nA when it has a reverse bias of 50 V. What is the ideal diode approximation of this diode when it is reverse biased? Draw its reverse current source model.

Section 3-9

3-21. Repeat Prob. 3-14 using the voltage source model for the zener diode. Compare the results.

Section 3-10

3-22. Arrive at the dynamic impedance model for the 1N5245 zener diode using the diode data given in Fig. 3-18. What is the voltage across the 1N5245 if the reverse breakdown current (I_Z) through it is 25 mA?

Section 3-11

3-23. If the diode in Fig. 3-20 is germanium, V_S is 4 V, and R_L is 1 kΩ, calculate the current through the diode.

Section 3-12

3-24. Repeat Prob. 3-23 if the diode is silicon.

Section 3-13

3-25. Given the circuit in Fig. 3-57, compute the voltage across R_L and the current through the diodes. Assume that the diodes are silicon. (*Hint*: Apply Thévenin's theorem.)

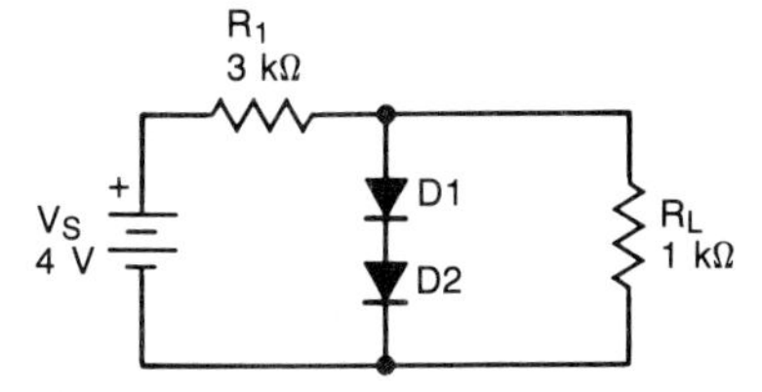

FIGURE 3-57 Diode circuit problem with loading.

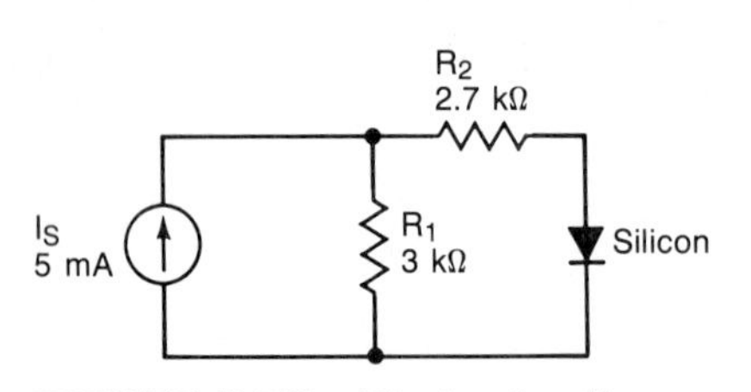

FIGURE 3-58 Diode circuit problem.

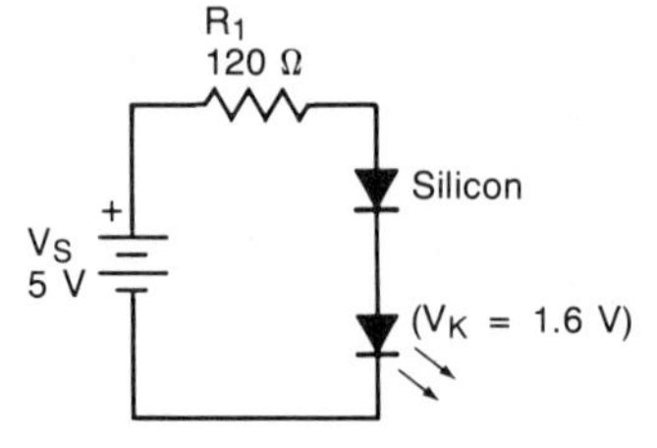

FIGURE 3-59 Diode/LED circuit problem.

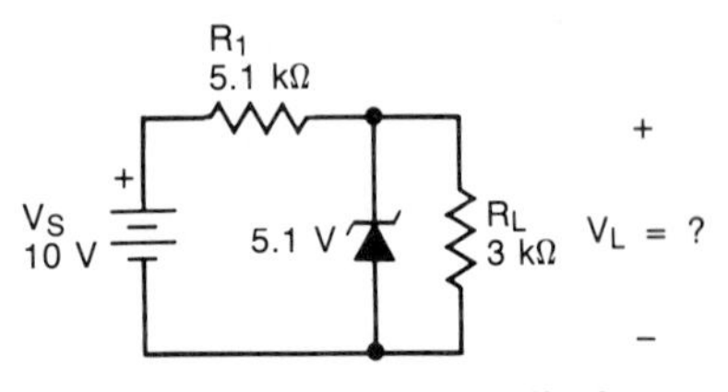

FIGURE 3-60 Zener diode circuit problem with loading.

3-26. Repeat Prob. 3-25 if the diodes are germanium and V_S is 5 V.

3-27. Given the circuit shown in Fig. 3-26(a), assume that V_S is 15 V, R_1 is 2 kΩ, R_2 is 3 kΩ, and R_3 is 5.1 kΩ. Determine the current through the silicon diode by using Thévenin's theorem.

3-28. Given the circuit shown in Fig. 3-58, use Thévenin's theorem to determine the current through the diode. (*Hint*: Remember that constant-current sources may be set to zero by replacing them with open circuits.)

3-29. Determine the current through the LED in Fig. 3-59.

3-30. Determine the voltage across R_L in Fig. 3-60. (*Hint*: Apply Thévenin's theorem.)

3-31. Repeat Prob. 3-30 for Fig. 3-60. Assume that R_L is increased to 10 kΩ.

Section 3-14

3-32. Graph a sine wave whose equation is

$$v(t) = 10 \sin 1250\pi t$$

3-33. Make an accurate sketch of a sine wave that has a peak-to-peak value of 100 mV and a period of 2 ms.

Section 3-15

3-34. Given the circuit shown in Fig. 3-29, write an equation for $v_S(t)$ if $V_S = 15$ V, and $v_s(t)$ is a 2-V-peak 20-kHz sine wave.

3-35. An audio signal produced by a crystal microphone has a peak value of 50 mV and a frequency of 10 kHz. The signal is sinusoidal and riding on a −7.5-V dc level. Write an equation that describes the total waveform, and make an accurate sketch.

Section 3-16

3-36. In your own words, define the superposition theorem. Explain why it may be used to perform a separate dc and ac analysis of an electronic circuit. Specifically, what are the constraints? Why is a dc analysis performed before an ac analysis?

3-37. Given the circuit shown in Fig. 3-29, with $V_S = 12$ V and

$$v_s(t) = 2 \sin 6000\pi t$$

draw the current waveform through R_L. R_L is 2 kΩ. Use the superposition theorem.

Section 3-17

3-38. Define r_{ac} and relate it to the slope of a line drawn tangent at the Q-point on a diode's V–I curve. Specifically, what happens to r_{ac} as I_D is increased?

3-39. A silicon diode is biased such that $V_D = 0.4$ V and $I_D = 0.1$ mA. If the diode's reverse saturation current is 2 nA, use Eq. 3-12 to determine its dynamic junction resistance r_j. Assume 25°C operation.

3-40. Repeat Prob. 3-39 if $V_D = 0.6$ V and $I_D = 10$ mA. (Note that $\eta = 1$ in this case.)

3-41. Repeat Prob. 3-40 using Eq. 3-15. How do the results compare?

3-42. A silicon diode is biased at 8 mA. Assuming that it is being operated at 25°C and has a bulk resistance of 1 Ω, compute its ac resistance.

3-43. Repeat Prob. 3-42 if the diode is germanium and is biased at 5 mA.

Section 3-19

3-44. Given the circuit shown in Fig. 3-37(a) and $R_1 = 60\ \Omega$, $R_2 = 12\ \text{k}\Omega$, $R_L = 10\ \text{k}\Omega$, $V_S = 15$ V, and that $v_s(t)$ is a 25-mV peak, 500-Hz sine wave, make a sketch of the ac waveform across R_L.

3-45. Repeat Prob. 3-44 if V_S is decreased to 10 V.

Section 3-20

3-46. Draw a schematic diagram of a half-wave rectifier, and describe how it works. Use the knee-voltage model in your explanation and include (large-signal) equivalent circuits. Assume a $v_s(t)$ of 5 V peak and a silicon diode.

Section 3-21

3-47. In your own words, briefly describe the term "diffusion capacitance."

3-48. A diode has a t_{rr} of 1 μs; what is the maximum frequency that it can rectify?

3-49. A diode has a t_{rr} of 10 ns; what is the maximum frequency that it can rectify?

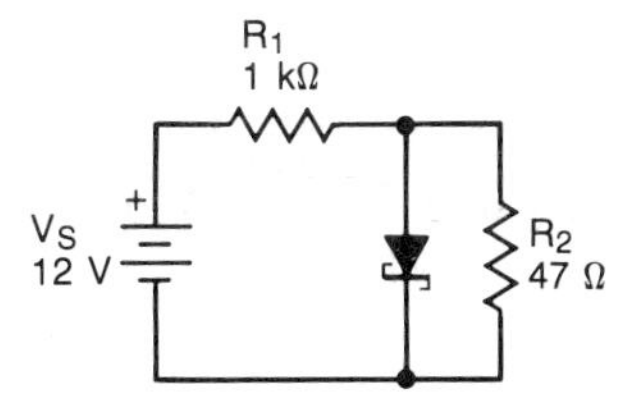

FIGURE 3-61 Schottky diode circuit problem.

3-50. Compute the current through the diode shown in Fig. 3-61. Take the diode's knee voltage into account.

3-51. Why is a Schottky diode referred to as a "hot carrier" diode? A "unipolar device"?

Section 3-23

3-52. A silicon diode has a linearly graded junction and a C_{j0} of 50 pF. Compute its junction capacitances at reverse biases of 0.1, 1, 10, and 20 V.

3-53. Repeat Prob. 3-52 if the diode has an abrupt junction.

3-54. Make a sketch of a linearly graded, an abrupt, and a hyperabrupt doping profile.

Section 3-25

3-55. The biased clipper in Fig. 3-52(a) has its ± 15 V supply voltages reduced to ± 6 V. What is the maximum peak-to-peak voltage that can appear across the amplifier's input?

3-56. Continue Prob. 3-55 by finding the reverse bias voltages across D_1 and D_2 if $v_s(t)$ is 3 V peak.

Section 3-26

3-57. The diode clamper in Fig. 3-53(a) has the diode reversed. Specifically, the anode now goes to ground. If $v_s(t)$ is a sine wave with a peak voltage of 10 V at a frequency of 5 kHz, find the dc level across the diode. Sketch the waveform that will appear across the diode.

Troubleshooting and Failure Modes

3-58. The LED in Fig. 3-56 is extinguished and the 1-kΩ resistor (rated at $\frac{1}{4}$ W) is warm. Determine the most likely circuit failure and explain your reasoning. The alternatives are: (a) the load resistor is shorted, (b) the load resistor is open, (c) the LED is open, and (d) the LED is shorted.

3-59. The voltage drop across the LED in Fig. 3-56 is 15 V, and the LED is extinguished. Using the alternatives cited in Prob. 3-58, indicate the most likely problem(s) and explain your rationale.

3-60. What is the probability that a reverse-biased rectifier will fail shorted or become "leaky"? (Refer to Section 3-24.)

3-61. The voltage (V_A) at point A in the circuit given in Fig. 3-62 is 3 V. The voltage at the anode of the diode is 0 V. The circuit is faulty. Determine the most likely failure and explain your reasoning. The alternatives are: (a) the diode is shorted, (b) the diode is open, (c) R_2 is open, (d) R_3 is open, and (e) R_3 is shorted. Determine the voltage at point A and at the anode of the diode when the circuit is operating properly.

FIGURE 3-62 Faulty diode circuit.

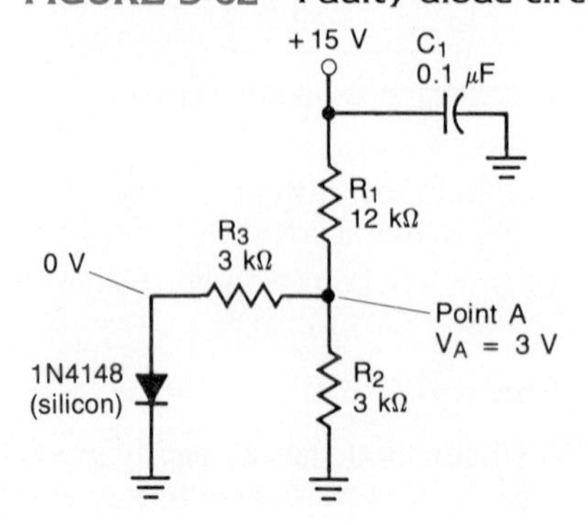

Design

3-62. A (new) zener diode is used in Fig. 3-55(a). It breaks down at 6.8 V. Determine the minimum (standard) value of R_L required to limit the current through the zener to slightly less than 40 mA. The voltage source V_S is still 14 V.

3-63. The LED in Fig. 3-56 is green (GaP) and drops approximately 1.8 V. Determine the minimum (standard) value of R_L required to limit the current through the LED to slightly less than 20 mA.

3-64. Repeat Prob. 3-63 if a red (GaAsP) LED is used. Its nominal voltage drop is 1.6 V.

3-65. A diode's reverse saturation current is almost always an undesirable phenomenon. However, in relatively low-temperature (e.g., less than 100°C) applications, it can be used as a temperature transducer. In neat, logical, step-by-step detail, solve Eq. 2-1 for temperature (T). Assume that the temperature range is from 0 to 100°C and that a 15-V supply is available. Sketch a simple circuit.

3-66. The voltage source in Fig. 3-60 is to be increased to 15 V. Determine the nearest $\pm 5\%$ tolerance standard resistor value for R_1 required to establish a current of slightly less than 25 mA through the 5.1-V zener diode. Assume that R_L is disconnected. Also determine the resistor's required power rating. What is the minimum value of R_L to ensure a V_L of 5.1 V if your R_1 is used?

Computer Problems

3-67. Write a BASIC computer program that computes a diode's current (I_D) by using the Shockley diode equation. It should accommodate both silicon and germanium diodes. It should prompt the user for the diode type (silicon or germanium), the temperature in degrees Celsius, I_S(25°C), and the diode voltage (V_D). It should output $I_S(T)$, V_T, and I_D. The program should then ask ''DO YOU WISH TO CONTINUE? (Y/N)''. If the user enters ''Y''es, the program should loop back to prompt the user for the next V_D. If the user selects ''N''o, the program should end.

3-68. Write a BASIC program that requests I_R (the diode total reverse current) at 25°C and its value at another temperature T. The program should then output I_S(25°C) and I_{SL}. Use your solution to Prob. 3-6 as a basis for the program's algorithm.

3-69. Combine your solutions to Probs. 3-67 and 3-68. Specifically, the program should satisfy the requirements of Prob. 3-67, but allow the user to enter I_R(25°C) and I_R at another temperature [$I_R(T)$].

4

Bipolar Junction Transistors

After Studying Chapter 4, You Should Be Able to:

- Describe the basic structure and operation of the npn and pnp BJTs.
- Define and use the α_{DC} and β_{DC} BJT parameters.
- Name and define the CE, CB, and CC BJT configurations.
- Generate and interpret the BJT V-I characteristic curves.
- Define the three BJT regions of operation: saturation, active, and cutoff.
- Describe BJT temperature effects and the leakage currents I_{CBO}, I_{CEO}, and I_{CES}.
- Find a BJT's Q-point by constructing and using its dc model and dc load line.
- Explain the BJT signal process.
- Draw the hybrid-pi BJT small-signal model and determine g_m, r_o, and r_π.
- Outline the basic BJT fabrication steps.

4-1 Role of the BJT in Electronics

In 1947, W. H. Brattain and J. Bardeen, both of Bell Labs, brought two closely spaced metallic needles into contact with the same germanium "base" wafer. The two needles served as electrodes, which were termed the *emitter* and *collector* terminals. A third, *base* terminal was also attached to the germanium crystal. Their experiments demonstrated that varying either the emitter or base dc terminal currents could cause a proportional variation in the collector voltage. Since an input (base or emitter) current could be used to control the output (collector) voltage, the gain (input-to-output transfer) of the device can be described as

$$\text{gain} = \frac{\text{output voltage}}{\text{input current}} \tag{4-1}$$

The units of the gain defined by Eq. 4-1 would be ohms. Hence J. R. Pierce (also of Bell Labs) described the device as a transfer resistor and coined the term *transistor*.

These first transistors were described as "point contact" devices. The point contact transistors were extremely difficult to produce commercially. However, in 1949 Willian Shockley theoretically described the *bipolar junction transistor*, or BJT.

The BJT truly revolutionized the electronics industry. Not only did it serve to render the vacuum tube obsolete, but the search for improved transistor manufacturing techniques laid the groundwork for modern integrated-circuit fabrication.

In a similar fashion, the linear integrated circuit is also pushing the discrete BJT into obsolescence in many applications. However, at present, we find that many new electronic circuit designs are a combination of both linear integrated circuits and discrete transistors. Consequently, we need to be reasonably comfortable with both of these electronic "tools." Further, an understanding of the discrete transistor can provide us with an intuitive understanding of the operation and the limitations of linear (and digital) integrated circuits.

4-2 Basic BJT Structure

Figure 4-1(a) illustrates an *npn* transistor. The *emitter* region is very heavily doped (n^+). The *base* region is very thin and lightly doped (p, or p^-) relative to the emitter region. The *collector* region is the largest and is moderately (n) to lightly (n^-) doped.

The *pnp* BJT is shown in Fig. 4-1(c). It is the complement of the *npn* BJT. Because the *npn* BJT offers a better high-frequency response, it is generally preferred

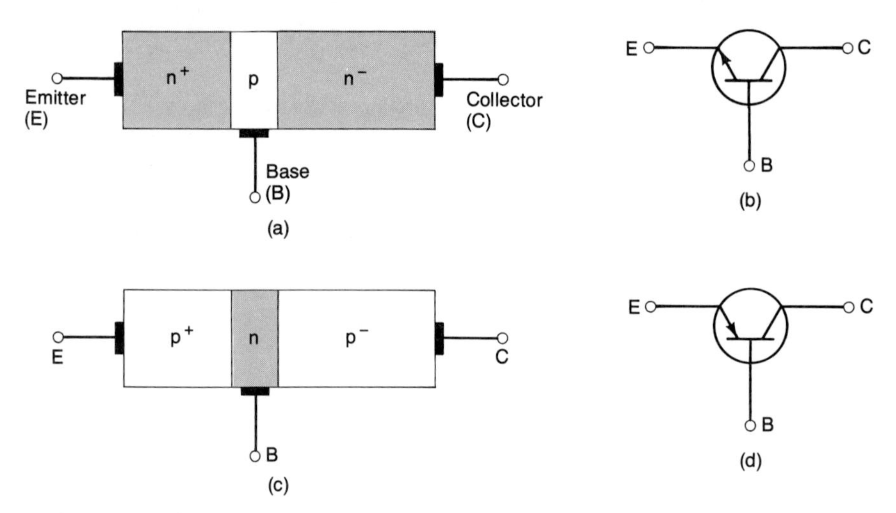

FIGURE 4-1 Npn and pnp BJT structures and symbols: (a) npn structure; (b) npn symbol; (c) pnp structure; (d) pnp symbol.

and more widely used than the *pnp* BJT. Consequently, we shall concentrate our discussions on the *npn* BJT. However, they may be readily extended to the *pnp* structure.

The schematic symbols for the *npn* and *pnp* BJTs have been depicted in Fig. 4-1(b) and (d), respectively. The only difference between the two symbols is that the arrow is *N*ot *P*ointing i*N* for the *NPN* transistor. In general, it should be remembered that the arrows on the schematic symbols of solid-state devices typically point to the *n*-type material.

4-3 Unbiased Transistors

Recall that when a semiconductor has been doped with pentavalent (donor) impurities, it becomes an *n*-type semiconductor. The additional free electrons tend to raise the Fermi (average electron energy) level within the crystal. Similarly, when a semiconductor has been doped with trivalent (acceptor) impurities, it becomes a *p*-type semiconductor. The additional holes tend to lower the Fermi level.

We also recall that when an *n*-type region is formed next to a *p*-type region, a barrier potential is produced at the *p*-*n* junction. The free electrons in the *n*-region will diffuse into the adjacent *p*-region to annihilate holes. Consequently, a layer of positive ions is formed on the *n*-side, and a layer of negative ions is formed on the *p*-side. This serves to create the barrier potential.

This same phenomenon occurs at both of the *p*-*n* junctions found in the *npn* transistor structure [see Fig. 4-2(a)]. Observe that the depletion regions (ionization layers) which are formed extend into the semiconductor crystal as a function of the doping levels. Specifically, it extends deeply into the lightly doped regions, and slightly into the heavily doped regions.

The formation of the barrier potential corresponds to the alignment of the Fermi

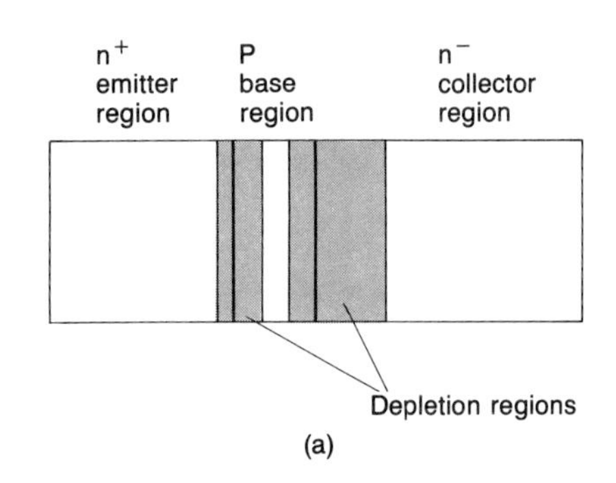

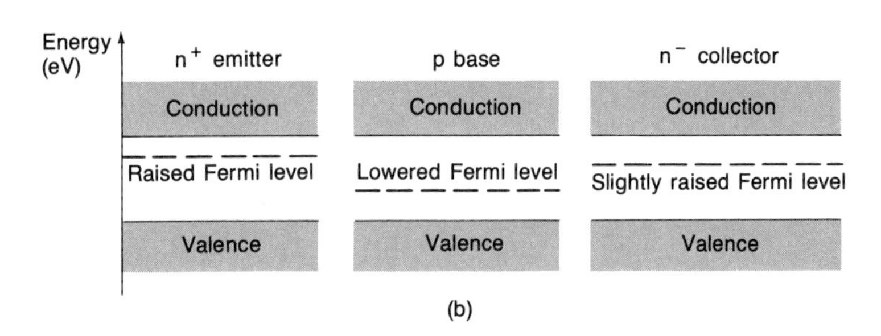

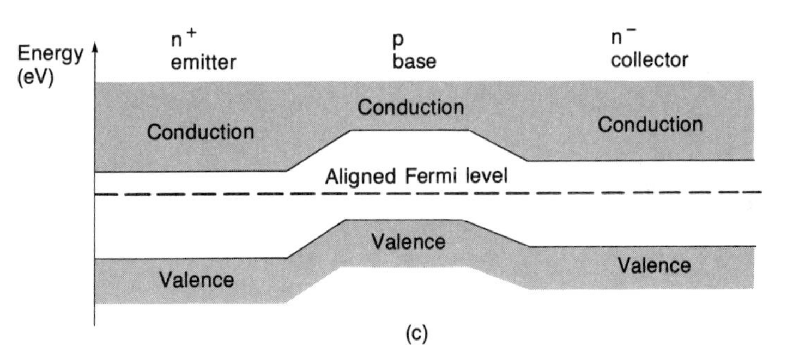

FIGURE 4-2 Unbiased transistor: (a) depletion regions; (b) isolated semiconductor crystals; (c) energy diagram of an npn transistor.

levels throughout the crystal. Recall that the Fermi-level alignment produces a distortion in the valence and conduction energy bands. Compare Fig. 4-2(b) and (c).

4-4 Operation of the BJT

To be used as an active (amplifying) device, the emitter–base *p-n* junction must be forward biased, while the collector–base *p-n* junction must be reverse biased. This is depicted in Fig. 4-3.

Since the emitter–base junction is forward biased, its depletion region is *narrow*. However, the relatively lightly doped base and collector regions produce a *wide*

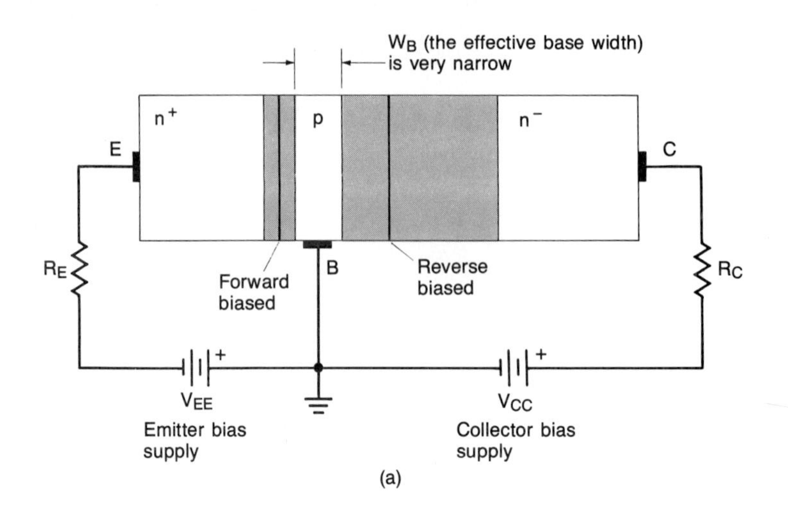

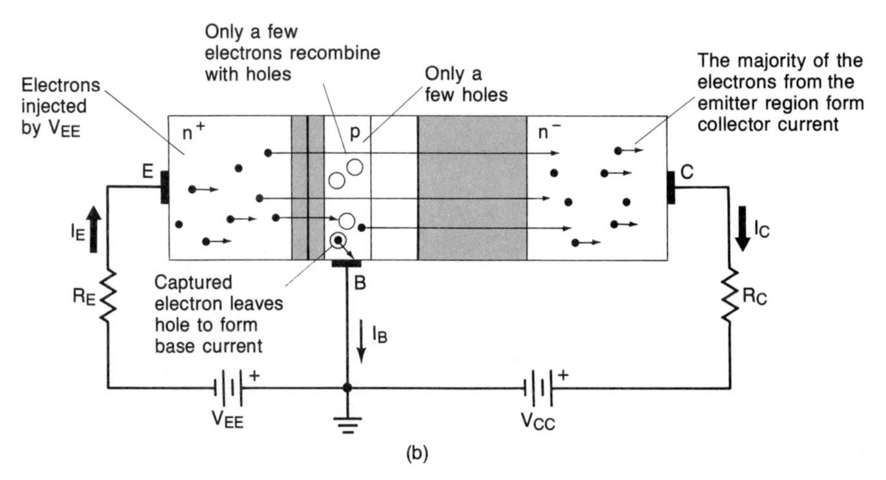

FIGURE 4-3 Basic operation of the BJT: (a) depletion regions and W_B; (b) charge carrier movement.

depletion region under reverse bias [see Fig. 4-3(a)]. Also observe that the *effective base width* W_B is the region between the two depletion regions.

The basic operation of the BJT can be understood by tracing the flow of electrons through it [refer to Fig. 4-3(b)]. Electrons are injected into the emitter region by the emitter bias supply V_{EE}. These conduction band electrons have enough energy to overcome the emitter–base barrier potential.

The emitter electrons enter the very thin, lightly doped base region. (Electrons are *minority* carriers in the *p*-type base region.) Because the base is so very lightly doped relative to the emitter region, only a few of the emitter electrons recombine with the holes doped into the base. Consequently, the base current I_B tends to be very *small*.

Because the effective base width W_B is so very thin, the high-energy minority carriers in the base region are close enough to the edge of the collector–base depletion region to be swept up or "collected" by it. Specifically, the negative electrons have enough energy to overcome the repulsion of the negative ions on the p side and the attraction of the positive ions on the n side.

Once the electrons have entered the collector region, they tend to drift toward the positive collector terminal under the influence of the collector bias supply V_{CC}. Ultimately, they flow through the external circuit and enter the positive terminal of V_{CC}.

The emitter current I_E, the collector current I_C, and the base current I_B (electron flows) have been indicated in Fig. 4-3(b). By Kirchhoff's current law, we arrive at

$$I_E = I_C + I_B \tag{4-2}$$

This is an extremely important relationship and should be committed to memory.

The base current (typically microamperes) is so small that we can state

$$I_E \simeq I_C \tag{4-3}$$

Equation 4-3 can be extremely useful when making a first-order approximation of the operation of a BJT in an electronic circuit.

The operation of the BJT may also be examined by means of its energy diagram. Recall from Chapter 2 that forward bias tends to align the valence and conduction energy bands, while reverse bias tends to increase their misalignment. Therefore, we arrive at the diagram presented in Fig. 4-4. The reader should compare Fig. 4-2(c) for the unbiased transistor with Fig. 4-4.

As can be seen in Fig. 4-4, electrons must go down a potential hill (lose energy) to travel from the emitter region to the base region. Further, the electrons must also go down a potential hill to move from the base region into the collector region. Observe that only a few of the electrons in the base region lose enough energy to

FIGURE 4-4 Energy diagram of an npn transistor in its active region.

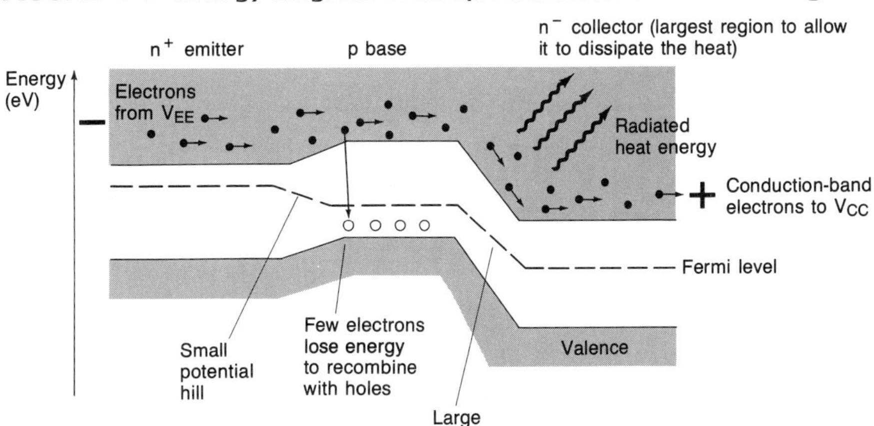

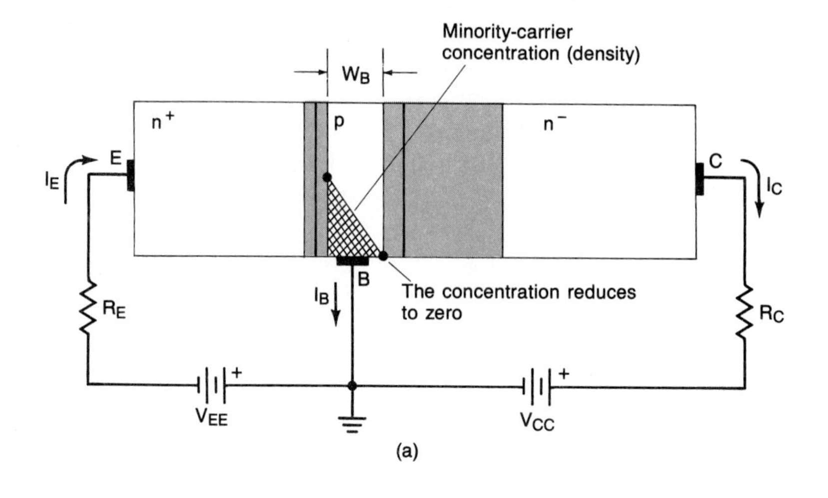

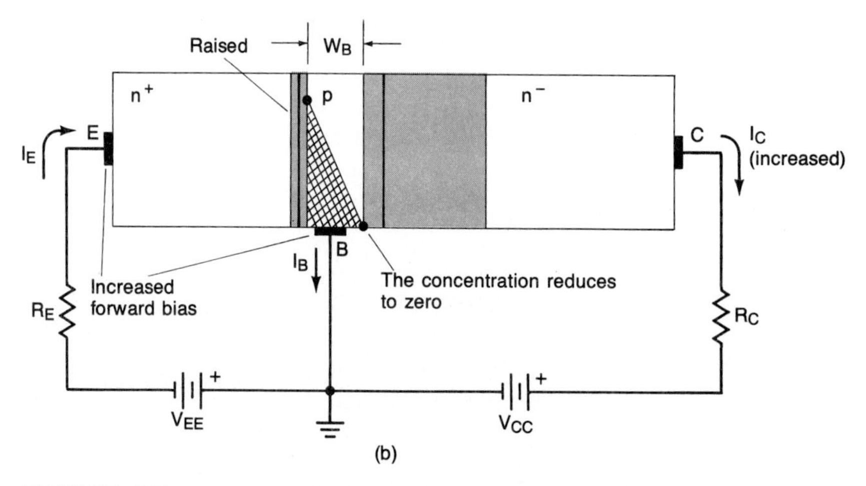

FIGURE 4-5 Minority-carrier concentration gradient in the base region: (a) concentration for a given emitter-base forward bias; (b) increased emitter-base forward bias increase the slope of the concentration gradient.

recombine with the (valence band) holes. Typically, at least 95% of the electrons have a long enough lifetime to diffuse into the collector region.

Also observe in Fig. 4-4 that the potential hill between the base and collector regions is very steep. Specifically, the electrons must lose a considerable amount of energy as they travel from the base region to the collector region. Typically, the electrons give up their energy in the form of heat. Consequently, the collector region must be able to dissipate this heat energy. It is for this reason that it is typically the largest of the three regions.

To conclude our discussion of the basic operation of the BJT, we present Fig. 4-5. Figure 4-5(a) illustrates the minority-carrier concentration gradient in the base region. The *slope* of the minority-carrier concentration is directly proportional to the emitter, base, and collector currents. If the emitter–base forward bias is increased, the slope of the gradient is increased. This implies that the emitter, base, and collector currents will be increased [see Fig. 4-5(b)]. The converse is also true. If the emitter–base forward bias is decreased, the slope of the gradient will be decreased, and all three of the currents will decrease.

Typically, either the base or emitter current is used to control the collector current.

4-5 BJT Configurations and Current Gain

As our studies progress, we shall see that the BJT can be used in three basic configurations: *common base*, *common emitter*, and *common collector* (see Fig. 4–6). Obviously, the term ''common'' implies that a particular transistor lead is connected to ground. Occasionally, this terminology may be misleading to the student. In many applications we may find that the ''common'' lead goes to the ac signal ground but *not* to the dc ground. Further, in some instances, *none* of the transistor leads may go directly to ground. To minimize the potential for confusion, we present Table 4-1. Specifically, to determine the BJT configuration, all we must do is determine the signal input, and output terminals of the BJT *with respect to ground.*

Manufacturers typically specify the *dc current gain* of a transistor on its data sheet. Current gain is defined by

$$\text{current gain} = \frac{\text{output current}}{\text{input current}} \tag{4-4}$$

The current gain for a BJT is usually specified for either its common-emitter configuration or its common-base configuration. The current gain for a BJT in its common-emitter configuration is given by

$$\boxed{h_{FE} \simeq \beta_{\text{DC}} = \frac{I_C}{I_B}} \tag{4-5}$$

The current gain for the BJT in its common-base configuration is given by

$$\boxed{h_{FB} \simeq \alpha_{\text{DC}} = \frac{I_C}{I_E}} \tag{4-6}$$

Equations 4-5 and 4-6 relate the *magnitudes* of I_E, I_C, and I_B.

TABLE 4-1
BJT Configurations

Configuration	Input Terminal	Output Terminal
Common base	Emitter	Collector
Common emitter	Base	Collector
Common collector	Base	Emitter

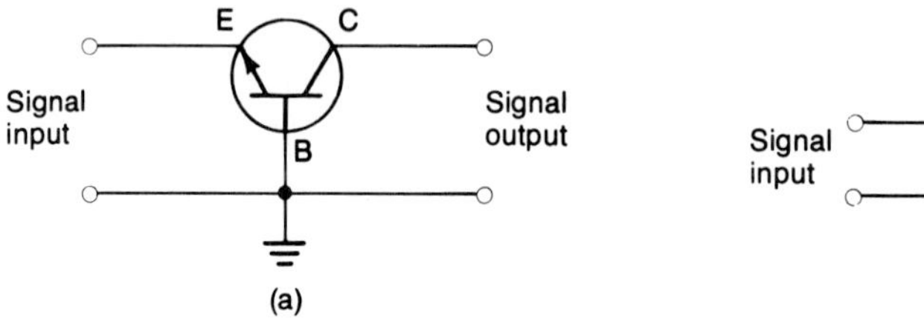

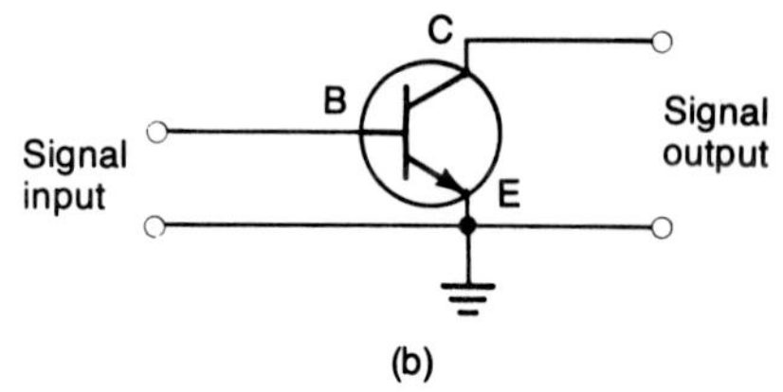

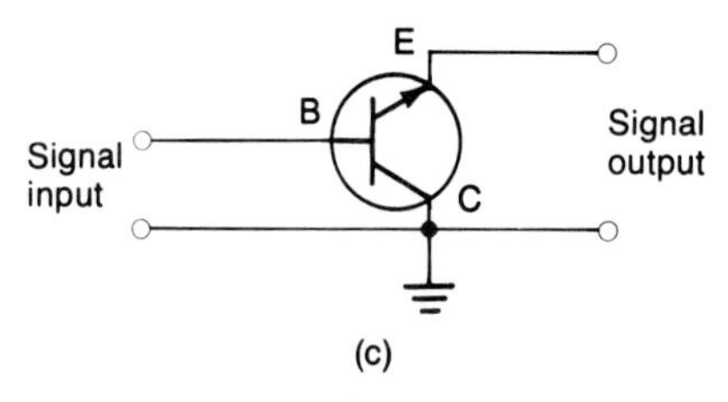

FIGURE 4-6 **BJT configurations: (a) common base; (b) common emitter; (c) common collector.**

β_{DC} (the dc beta) is also described as h_{FE}. The latter notation is based on an equivalent circuit for a BJT which is called the hybrid- or h-parameter model. The symbol h_{FE} stands for the *h*ybrid parameter *F*orward current gain of a BJT in its common *E*mitter configuration. Similarly, for α_{DC} (the dc alpha) h_{FB} is the h-parameter notation for the *F*orward current gain of a BJT in its common-*B*ase configuration. The h-parameter model will be discussed in more detail later.

By employing Eq. 4-2 and the definitions provided by Eqs. 4-5 and 4-6, it becomes possible to develop equations that allow us to find β_{DC} given α_{DC}, and vice versa. Hence, from Eq. 4-2,

$$I_C = I_E - I_B$$

and dividing both sides by I_C, we have

$$1 = \frac{I_E}{I_C} - \frac{I_B}{I_C} = \frac{1}{I_C/I_E} - \frac{1}{I_C/I_B}$$

and substituting in our definitions for α_{DC} and β_{DC}, we arrive at

$$1 = \frac{1}{\alpha_{DC}} - \frac{1}{\beta_{DC}} \tag{4-7}$$

Solving Eq. 4-7 for β_{DC} yields

$$\boxed{\beta_{DC} = \frac{\alpha_{DC}}{1 - \alpha_{DC}}} \tag{4-8}$$

Similarly, solving Eq. 4-7 for α_{DC} results in

$$\boxed{\alpha_{DC} = \frac{\beta_{DC}}{\beta_{DC} + 1}} \tag{4-9}$$

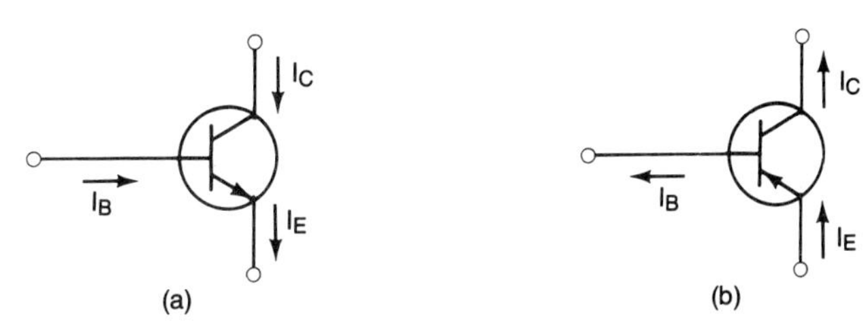

FIGURE 4-7 BJT conventional current directions: (a) npn; (b) pnp.

EXAMPLE 4-1

A silicon *npn* BJT is biased such that I_C is 1 mA and I_B is 10 μA. Calculate (a) its β_{DC}, (b) I_E, and (c) its α_{DC} [see Fig. 4-7(a)].

SOLUTION From Eq. 4-5:

$$\beta_{\mathrm{DC}} = \frac{I_C}{I_B} = \frac{1\ \mathrm{mA}}{10\mu\mathrm{A}} = 100$$

Application of Eq. 4-2 yields

$$I_E = I_C + I_B = 1\ \mathrm{mA} + 0.01\ \mathrm{mA} = 1.01\ \mathrm{mA}$$

Now Eq. 4-6 may be used.

$$\alpha_{DC} = \frac{I_C}{I_E} = \frac{1\ \mathrm{mA}}{1.01\ \mathrm{mA}} = 0.990$$

■

EXAMPLE 4-2

A silicon *pnp* BJT has a β_{DC} of 50 and an I_E of 1.5 mA. Determine I_C [see Fig. 4-7(b)].

SOLUTION From Eq. 4-9,

$$\alpha_{\mathrm{DC}} = \frac{\beta_{\mathrm{DC}}}{\beta_{\mathrm{DC}} + 1} = \frac{50}{50 + 1} = 0.980$$

By rearranging Eq. 4-6, we may solve for I_C.

$$I_C = \alpha_{\mathrm{DC}} I_E = (0.980)\,(1.5\ \mathrm{mA}) = 1.47\ \mathrm{mA}$$

■

Also note that in Fig. 4-7 the magnitudes and the directions of the conventional current flows for the *npn* and *pnp* BJTs have been defined. The reader should carefully observe the numerical results given in Examples 4-1 and 4-2. The base current is very small, the collector current is approximately equal to the emitter current, and α_{DC} is very close to unity. Typical values of β_{DC} range from about 10 to approximately 200, or more.

4-6 BJT Leakage Currents

Since the collector–base *p-n* junction is normally reverse biased, it is subject to the same reverse leakage currents that plague all reverse-biased diodes. The current components found in an *npn* BJT have been depicted in Fig. 4-8. Figure 4-8(a)

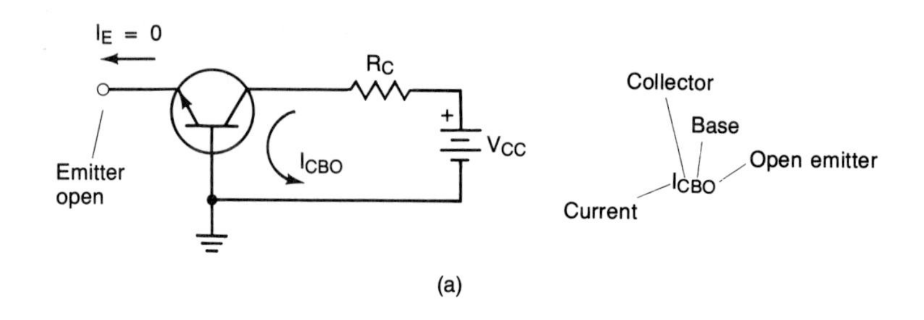

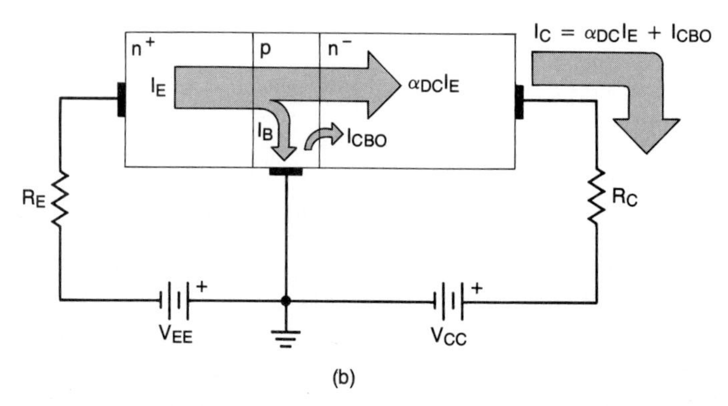

FIGURE 4-8 BJT current components: (a) conventional current definition of I_{CBO}; (b) electron flow current components.

emphasizes that a reverse leakage current flows from the collector through the collector-to-base junction when the emitter is left *open*. This current is denoted I_{CBO} and will be provided on some transistor data sheets.

In Fig. 4-8(b) all the current components, including I_{CBO}, have been shown. Their magnitudes are related by

$$\boxed{I_C = \alpha_{DC} I_E + I_{CBO}} \tag{4-10}$$

If I_{CBO} is approximately zero,

$$I_C \simeq \alpha_{DC} I_E$$

which is the *idealization* expressed previously by Eq. 4-6.

Equation 4-10 lends itself to the analysis of BJTs in the common-base configuration. By using Eq. 4-10 and several of the previously developed relationships, it becomes possible to derive a similar equation that promotes the analysis of a BJT in its common-emitter configuration:

$$I_C = \alpha_{DC} I_E + I_{CBO}$$

and since $I_E = I_C + I_B$,

$$\begin{aligned} I_C &= \alpha_{DC}(I_C + I_B) + I_{CBO} \\ &= \alpha_{DC} I_C + \alpha_{DC} I_B + I_{CBO} \end{aligned}$$

Solving for I_C, we have

$$I_C = \frac{\alpha_{DC}}{1 - \alpha_{DC}} I_B + \frac{I_{CBO}}{1 - \alpha_{DC}} \tag{4-11}$$

It can be shown that

$$\frac{1}{1 - \alpha_{DC}} = \beta_{DC} + 1$$

Substituting the relationship above and Eq. 4-8 into Eq. 4-11, we arrive at Eq. 4-12, which relates the magnitudes of the current components.

$$\boxed{I_C = \beta_{DC} I_B + (\beta_{DC} + 1)I_{CBO}} \tag{4-12}$$

The leakage current term is usually defined as I_{CEO}, and Eq. 4-12 may be written as

$$\boxed{I_C = \beta_{DC} I_B + I_{CEO}} \tag{4-13}$$

where $I_{CEO} = (\beta_{DC} + 1)I_{CBO}$.

The leakage current term I_{CEO} has been illustrated in Fig. 4-9(a). Quite often, manufacturers supply the value of I_{CEO} on their data sheets. In Fig. 4-9(b) the amplification of I_{CBO} by the factor $(\beta_{DC} + 1)$ has been more intuitively illustrated.

FIGURE 4-9 **Reverse leakage current I_{CEO}.**

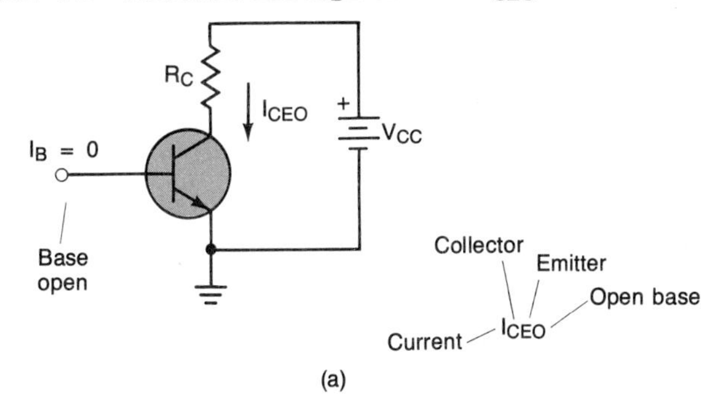

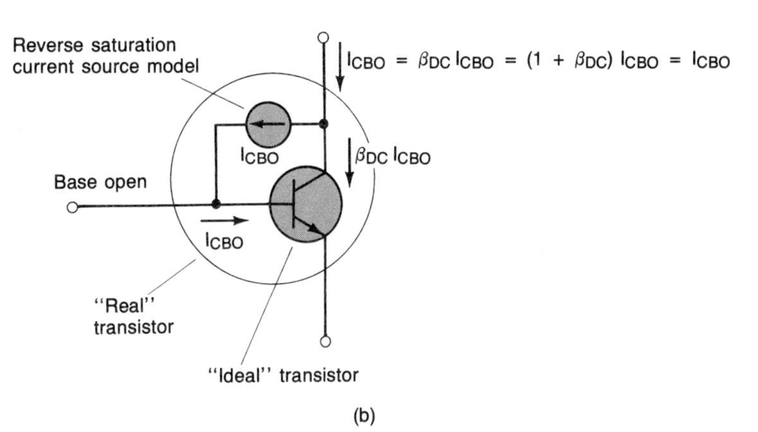

Leakage currents, such as I_{CBO} and I_{CEO}, are highly undesirable. Since leakage currents tend to increase as the ambient and/or junction temperature increases, they tend to shift the BJT Q-points as a function of temperature. In small-signal BJT voltage amplifiers, the Q-points may shift enough to cause a distortion in the output signal. In large-signal BJT power amplifiers, the leakage currents can contribute to a phenomenon known as *thermal runaway*, which can result in the destruction of power transistors. These considerations will be dealt with in our later work.

4-7 Transistor Convention

During the development of the V–I curves (and on manufacturers' data sheets), it is customary to apply the *transistor convention*. The transistor convention has been defined in Fig. 4-10(a).

Currents that flow *into* the transistor are called *positive*. Conversely, currents that flow *out* of the transistor are called *negative*. If the actual conventional current directions agree with the transistor convention, they are called positive. If not, they are called negative [see Fig. 4–10(b) and (c)].

To minimize the possibility of confusion during our later work, we shall only apply the transistor convention when we are dealing with V–I curves, graphical solutions, or interpreting a manufacturer's data sheet. Generally, we shall indicate conventional current directions, and assume that we are dealing with the magnitudes of these currents.

The voltages across the terminals of a transistor follow the usual double-subscript voltage convention. For example, V_{CE} is the voltage at the collector with respect to the emitter. If V_{CE} is 10 V, the collector is 10 V more positive

FIGURE 4-10 Transistor convention: (a) currents into a BJT are defined as positive; (b) an npn BJT has negative emitter current; (c) a pnp BJT has negative base and collector currents; (d) double-subscript voltage convention.

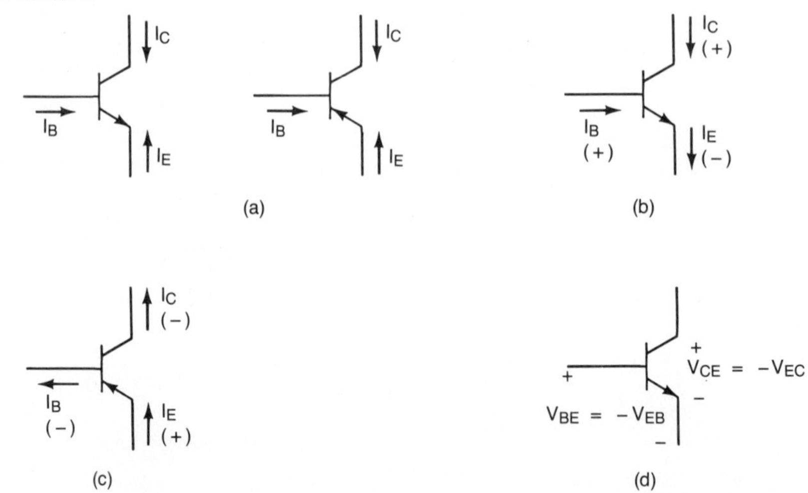

than the emitter. If V_{CE} is -10 V, the collector is 10 V more negative than the emitter. Therefore, we can also state that the emitter is 10 V more positive than the collector, or V_{EC} is $+10$ V.

4-8 BJT *V–I* Curves

We can generate three sets of *V–I* curves for the BJT: *input characteristics*, *output characteristics*, and *transfer characteristics*. Each of these three characteristic curves may be produced by using the circuit illustrated in Fig. 4-11.

The general procedure for generating the input curves for a BJT using the general circuit is as follows:

1. Set V_{22} to give a fixed value for V_2. V_2 is called the *parameter*.
2. Adjust V_{11} to give a convenient value of V_1 and measure the resulting I_1.
3. Repeat step 2 until the maximum desired value of I_1 is reached.
4. Readjust V_{22} to give a new value for V_2, and repeat steps 2 and 3.

Using the test circuit shown in Fig. 4-12(a) and following the procedure above, the input curves for an *npn* F JT may be generated. A typical set of input curves for a common-base BJT has been illustrated in Fig. 4-12(b). Observe that I_E is the input current, and it is labeled negative since it flows out of the emitter. The input voltage V_{EB} is also a negative quantity. This is true because the emitter is negative with respect to the base. The parameter is the collector-to-base voltage V_{CB}. This quantity is positive since the collector is positive with respect to the base.

Observe that the input curves for the BJT are very similar to those of a forward-biased rectifier diode. This is particularly true when the collector-to-base circuit is left open. The knee voltages of the input curves decrease as the collector-to-base voltage is increased. This is due to a phenomenon known as the *Early effect* or *base-width modulation*, which will be investigated in Section 4-16.

The output characteristic curves for the common-base *npn* BJT are developed in Fig. 4-13. The test circuit is given in Fig. 4-13(a). The emitter (input) current is the parameter. The collector-to-base voltage is varied and the corresponding collector currents are measured. Again, the transistor convention has been followed. The resulting *V–I* family of curves has been illustrated in Fig. 4-13(b).

Note that three regions have been indicated. Specifically, these are the *saturation*,

FIGURE 4-11 General test circuit to determine input, output, and transfer characteristic V-I curves.

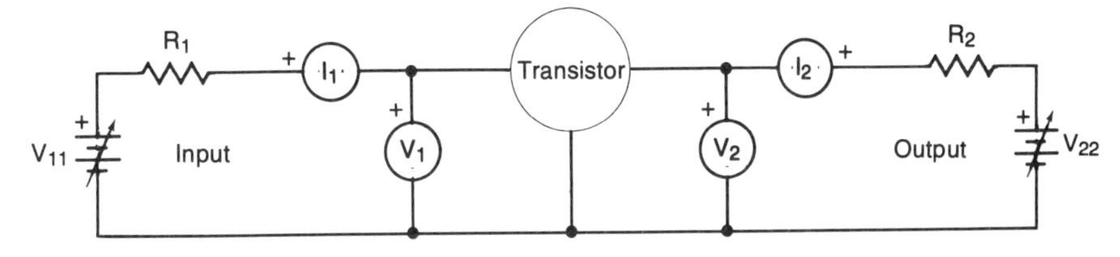

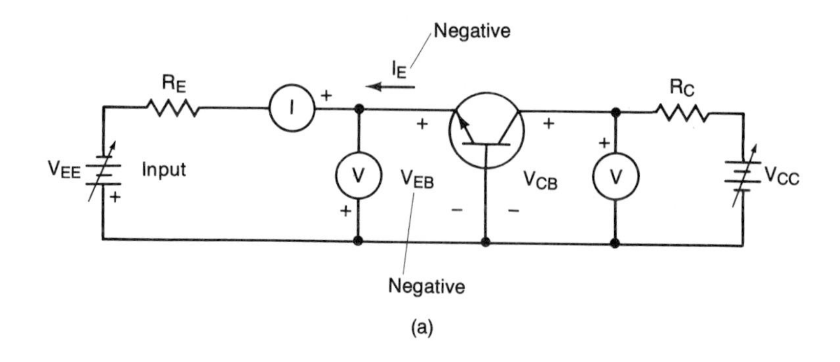

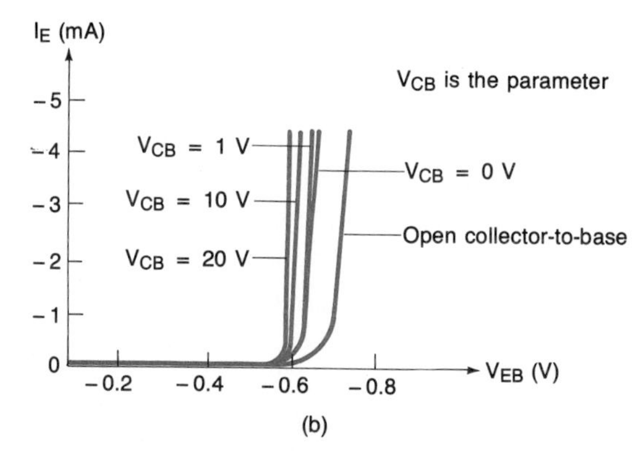

FIGURE 4-12 **Input curves for a silicon npn BJT in the common-base configuration: (a) test circuit; (b) typical curves.**

active, and *cutoff* regions. The region of operation for a BJT is determined by the biasing on its *p-n* junctions. These have been summarized in Table 4-2.

To be used as a linear amplifier, the BJT must be biased in its active region. The various regions of operation will be investigated more fully as our work progresses.

The test circuit and the input curves for an *npn* BJT in its common-emitter configuration are depicted in Fig. 4–14. Note that V_{CE} is the parameter.

The test circuit and the typical output curves for a common-emitter *npn* BJT have been given in Fig. 4-15 (page 133). In this case the parameter is the base (input) current. Once again, the saturation, active, and cutoff regions have been indicated.

In Fig. 4-16 (page 134) we have indicated the *transfer characteristic curve* for a silicon *npn* BJT in the common-emitter configuration. In this case the collector (output) current is plotted as a function of the controlling base-to-emitter (input) voltage. The collector-to-emitter voltage is the parameter. The reader should note that the three regions of operation have again been indicated. The *typical* saturation region has been shown in Fig. 4-16. However, the transistor will *not* saturate if V_{CE} is maintained at 10 V. This will become apparent in Chapter 6.

When V_{BE} is zero, the base is effectively *shorted* to the emitter terminal, and the collector current that flows is denoted as I_{CES}. Observe that I_{CES} is approximately equal to I_{CBO}.

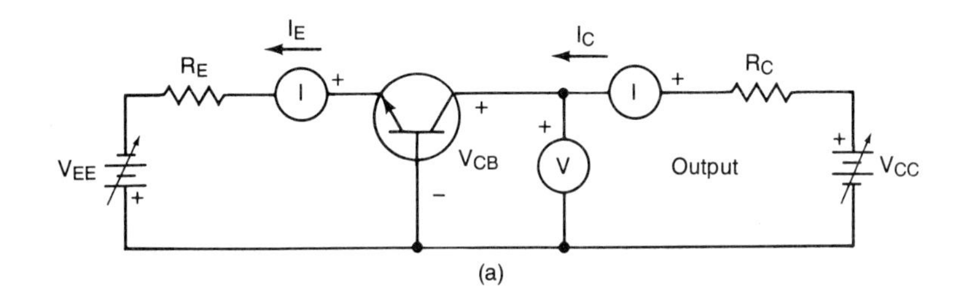

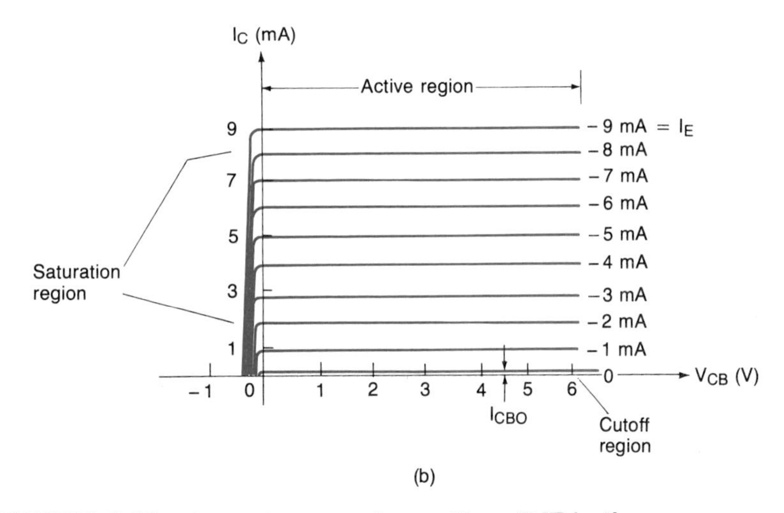

FIGURE 4-13 Output curves for a silicon BJT in the common-base configuration: (a) test circuit; (b) typical curves.

As the forward bias on the base–emitter *p-n* junction is increased, the collector current will rise exponentially. The typical *cut-in voltage* for a silicon transistor is 0.5 V (see Fig. 4-16). The cut-in voltage is usually defined to be the value of V_{BE} that produces a collector current that is 1% of its maximum value.

When V_{BE} is about 0.8 V, a common-emitter *npn* BJT will usually be in saturation. This value of V_{BE} is denoted as $V_{BE(\text{SAT})}$. The useful active region for a BJT is generally taken to be between the cut-in value for V_{BE} and $V_{BE(\text{SAT})}$. The BJT may

TABLE 4-2
BJT Regions of Operation and the Corresponding Biasing

Region of Operation	Emitter–Base Bias	Collector–Base Bias
Active	Forward	Reverse
Cutoff	Reverse	Reverse
Saturation	Forward	Forward

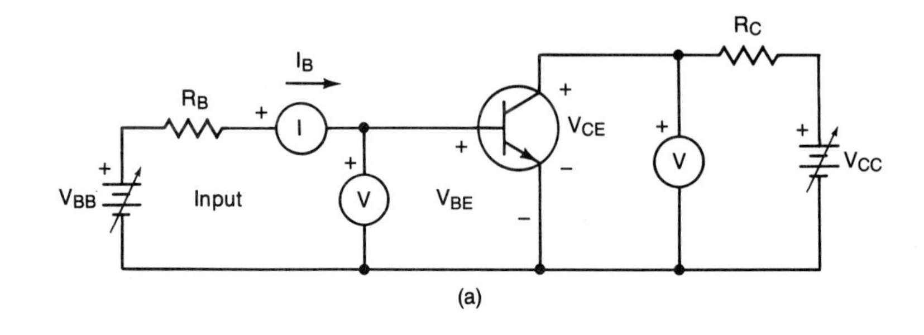

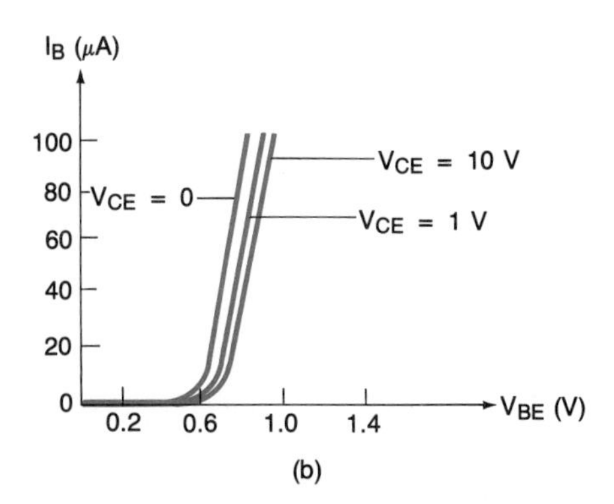

FIGURE 4-14 Common-emitter silicon npn BJT input curves: (a) test circuit; (b) typical input curves.

be used as a linear amplifier when it is biased in its active region and small signals are applied.

4-9 Temperature Effects

Recall that the electrical characteristics of semiconductors (including those used to make BJTs) are strongly affected by temperature. All of the BJT characteristics discussed thus far are strong functions of temperature.

> **As the temperature is *increased*, the collector leakage currents (I_{CBO}, I_{CEO}, and I_{CES}) tend to *increase*. The current gains (α_{DC} and β_{DC}) also tend to *increase* with temperature. The required base-to-emitter forward-bias voltage (V_{BE} or V_{EB}) to produce a given collector current tends to *decrease* with an *increase* in temperature. The base–emitter temperature coefficient is virtually identical to that of any forward-biased *p-n* junction.**

These temperature effects are reflected in the *V–I* curves of a BJT (see Fig. 4-17).

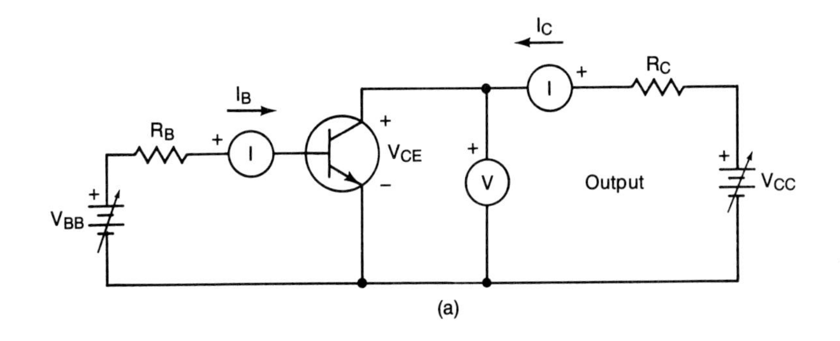

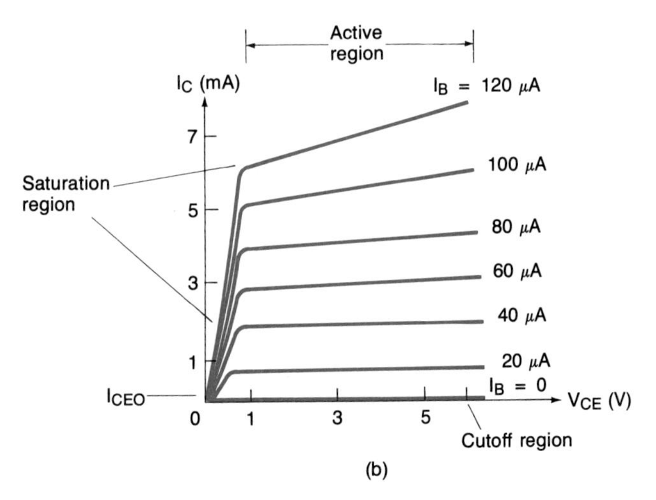

FIGURE 4-15 Common-emitter silicon npn BJT output curves: (a) test circuit; (b) typical curves.

EXAMPLE 4-3

The base-to-emitter voltage V_{BE} of a silicon BJT exhibits a temperature coefficient of approximately -2.2 mV/°C [see Fig. 4-17(a) and (c)]. If the transistor's V_{BE} is 0.7 V at 25°C, calculate its value at 50°C.

SOLUTION We can develop an equation for the V_{BE} of a silicon BJT at any temperature T in degrees Celsius.

$$V_{BE}(T) = V_{BE}(25°\text{C}) - (T - 25°\text{C})\,(2.2\text{ mV/°C})$$

Substituting in $T = 50°\text{C}$ gives us

$$\begin{aligned} V_{BE}(50°\text{C}) &= 0.7\text{ V} - (50°\text{C} - 25°\text{C})(2.2\text{ mV/°C}) \\ &= 0.7\text{ V} - 0.055\text{ V} = 0.645\text{ V} \end{aligned}$$

■

The collector leakage current I_{CBO} approximately doubles for each 8°C rise in temperature for silicon BJTs. If we slightly modify Eq. 2-1 which was introduced in Section 2-7, we arrive at

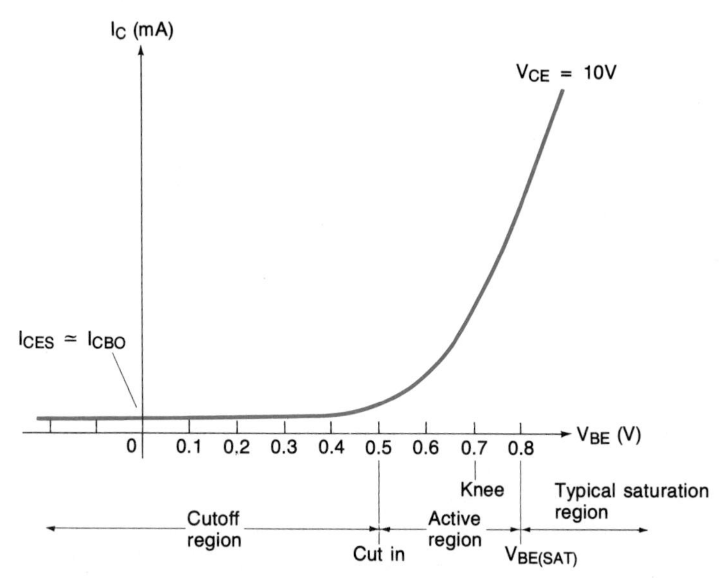

FIGURE 4-16 Transfer characteristic curve of a silicon npn BJT.

FIGURE 4-17 Temperature effects: (a) input curves; (b) output curves; (c) transfer curves.

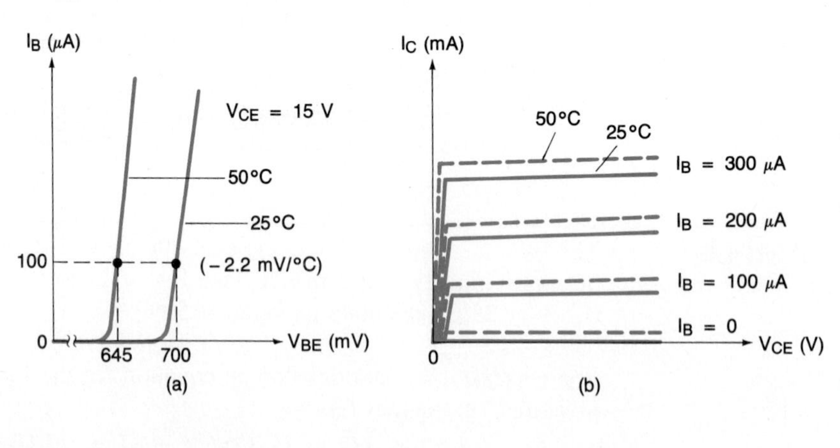

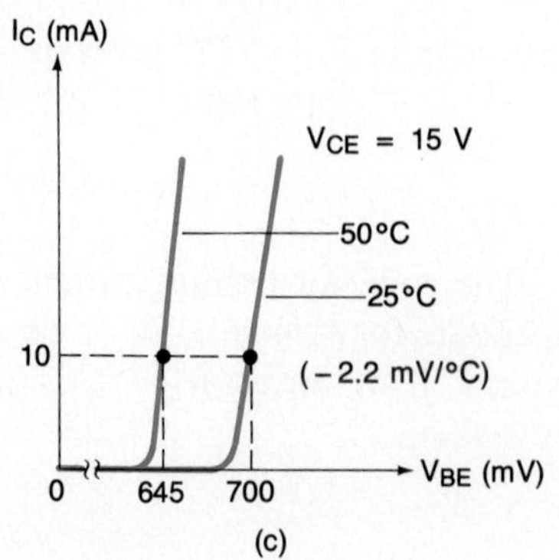

$$I_{CBO}(T) = [I_{CBO}(25°C)][2^{(T-25)/8}] \tag{4-14}$$

EXAMPLE 4-4

Given that a silicon *npn* BJT has an I_{CBO} of 10 nA at 25°C, determine its I_{CBO} at 50°C.

SOLUTION From Eq. 4-14,

$$I_{CBO}(50°C) = (10 \text{ nA})[2^{(50-25)/8}] = 87.2 \text{ nA}$$

■

The importance of the effects of temperature on the operation of the BJT cannot be overstated. Temperature variations can cause severe shifts in the Q-point of a BJT amplifier. This could possibly result in signal distortion. As we shall see, temperature effects cause the ac parameters of a BJT to change. Consequently, quantities such as a voltage amplifier's voltage gain and input impedance will also be affected by temperature. Fortunately, most of these effects can be compensated by using circuitry external to the BJT. These and other topics will be investigated more fully in our work to come.

4-10 BJT Analysis in Perspective

In Section 3-1 we discussed the approaches to the analysis of circuits that incorporate nonlinear circuit elements such as diodes. Specifically, we cited the use of nonlinear mathematics, graphical techniques, and the use of equivalent-circuit models. From the V–I curves developed in Section 4-8 it is clear that the BJT is also a nonlinear device. Consequently, the analysis alternatives offered in Section 3-1 also apply to the BJT.

For our purposes, we ignore the rigorous use of nonlinear mathematics. (However, the reader should recognize that very powerful nonlinear mathematical models, such as the Ebers–Moll model, have been developed and are widely accepted.) Graphical techniques, such as dc load lines and *bias lines*, are often used to determine the Q-points of BJTs. *Ac load lines* are also extremely useful for the analysis of *large-signal* effects in BJT amplifiers. However, most of our dc and (small-signal) ac analyses of BJT amplifiers will be developed around equivalent-circuit models.

Further, in Chapter 3 we saw that if we use piecewise *linear* dc models for the nonlinear device, and small-signal *linear* ac models, we may draw on the superposition theorem to separate the dc and ac analyses. This is exactly the same approach that we shall use with BJT amplifiers.

We shall demonstrate the importance and the techniques for the dc analysis of BJT amplifiers in the next section. However, most of our serious dc biasing is accomplished in Chapter 6.

4-11 Determining the *Q*-Point: Dc Load Lines

In Fig. 4-18(a) we have illustrated an *npn* BJT in the common-base configuration. To determine the transistor's dc operating point (*Q*-point), we may draw on the graphical (load-line) technique. (The reader may wish to review Section 3-3.)

However, it is necessary to apply the load-line technique *twice*—first to its input and then to its output. Since the BJT's input consists of the forward-biased emitter–base *p-n* junction, the load-line approach proves to be as inconvenient (if not more so) as it was for forward-biased diodes.

Consequently, we shall use the knee-voltage diode model by assuming that $|V_{EB}|$ is 0.3 V for germanium and 0.7 V for silicon transistors. Therefore, by applying Kirchhoff's voltage law around the BJT's input circuit [Fig. 4-18(a)] and solving for I_E, we have

$$-V_{EB} + I_E R_E - V_{EE} = 0$$

$$\boxed{I_E = \frac{V_{EE} + V_{EB}}{R_E}} \tag{4-15}$$

FIGURE 4-18 **Graphical analysis: (a) common-base npn silicon BJT; (b) input circuit analysis to find I_E; (c) output circuit analysis; (d) the load line is drawn to find the Q-point.**

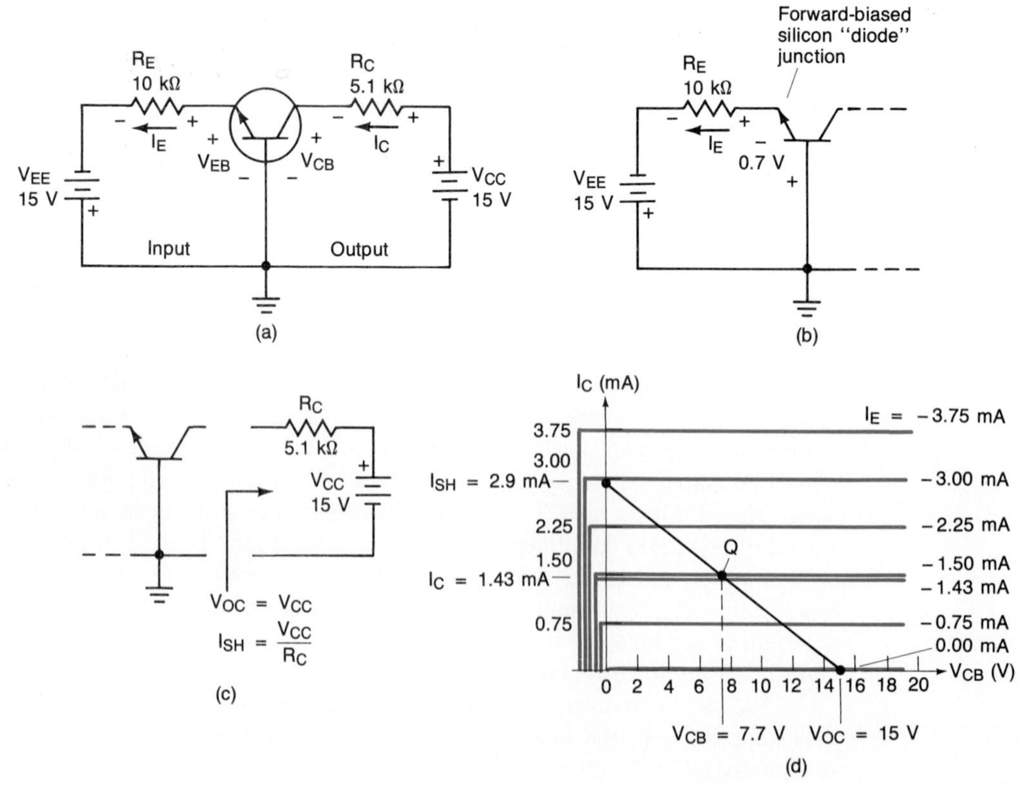

Substituting in the values given in Fig. 4-18(b),

$$I_E = \frac{V_{EE} + V_{EB}}{R_E} = \frac{15\text{V} + (-0.7\text{ V})}{10\text{ k}\Omega} = 1.43\text{ mA}$$

This is the value of I_E that determines the particular curve of interest in the family of output curves [see Fig. 4-18(d)]. Note that since I_E flows *out* of the emitter, it is a *negative* quantity according to the transistor convention.

To draw the load line on the output characteristic curves, we must determine the open-circuit voltage V_{OC} and the short-circuit current I_{SH} [see Fig. 4-18(c)]. Hence

$$V_{OC} = V_{CC} = 15\text{ V}$$

$$I_{SH} = \frac{V_{CC}}{R_C} = \frac{15\text{ V}}{5.1\text{ k}\Omega} = 2.94\text{ mA}$$

The intersection between the load line and the particular output characteristic curve determined above yields the Q-*point* [refer to Fig. 4-18(d)].

By projecting the *Q*-point over to the collector current axis, we can see that the collector current is approximately 1.43 mA. By projecting down to the collector–base voltage axis, we can also see that V_{CB} is approximately 7.7 V [see Fig. 4–18(d)].

As we mentioned in Chapter 3, graphical techniques are not always the most desirable alternative to the analysis of nonlinear circuit elements for two reasons. First, we must have access to the *V–I* curves for the device, or they must be generated. Second, our accuracy is limited by our neatness and the resolution of our graph. However, they do provide us with a "picture" of the operation of the device and the electronic circuit.

In Fig. 4-19(a) we have illustrated a dc load line with a *centered Q-point*. By "centered" we mean that I_C is one-half of I_{SH} and V_{CB} is one-half of V_{OC}. In Fig. 4-19(b) we can see that if I_E and V_{OC} are held constant, decreasing the value of R_C increases V_{CB}. The converse is also true, as shown in Fig. 4-19(c). Increasing R_C will decrease the value of V_{CB}.

Figure 4-20 illustrates the effects of varying V_{CC}. In Fig. 4-20(a) we can see that an increase in V_{CC} increases V_{CB}. Decreasing V_{CC} lowers the value of V_{CB} as shown in Fig. 4-20(b).

In Fig. 4-20(c) we can see that if I_E is lowered, then I_C will also be lowered, and the *Q*-point will again be centered. The significance of a centered *Q*-point is addressed in Chapter 6.

Figures 4-19 and 4-20 illustrate several important dc biasing relationships, and they should be studied carefully. In Fig. 4-21 we have illustrated the effects of changing I_E. A thorough examination will not only enhance our understanding of dc biasing, but also the effects of a signal on the BJT.

If I_E is increased, I_C will also increase. However, V_{CB} will decrease [refer to Fig. 4-21(a)]. A decrease in I_E will produce a decrease in I_C with an attendant increase in V_{CB}, as shown in Fig. 4-21(b).

If I_E is increased until I_C is approximately equal to I_{SH}, the transistor will be on the verge of saturation [see Fig. 4-21(c)]. Consequently, this maximum value of I_C is often termed $I_{C(SAT)}$. Hence

$$I_{C(SAT)} = I_{SH}$$

and

$$I_{C(SAT)} = \frac{V_{CC}}{R_C} \tag{4-16}$$

FIGURE 4-19 Effects of varying the collector resistor R_C.

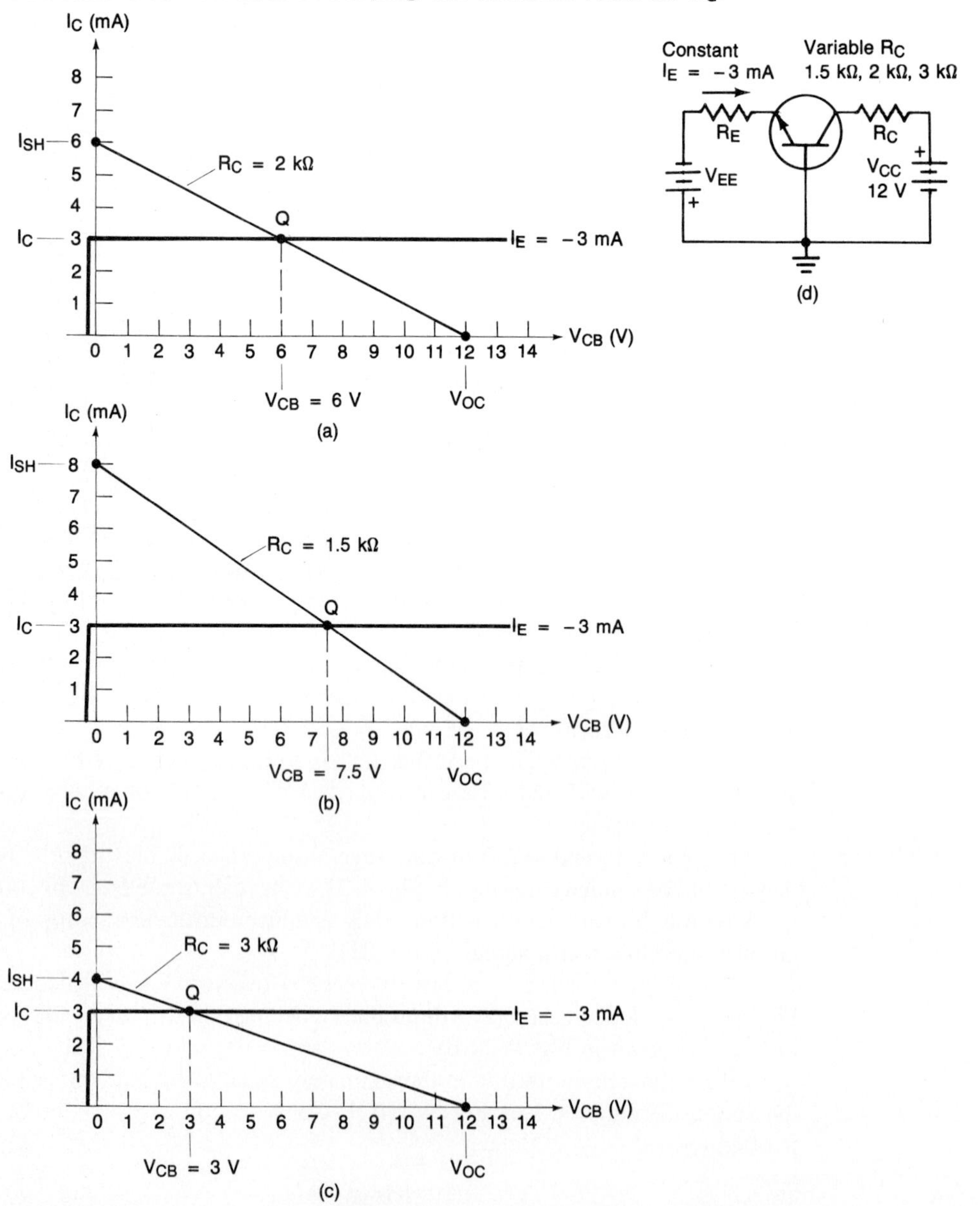

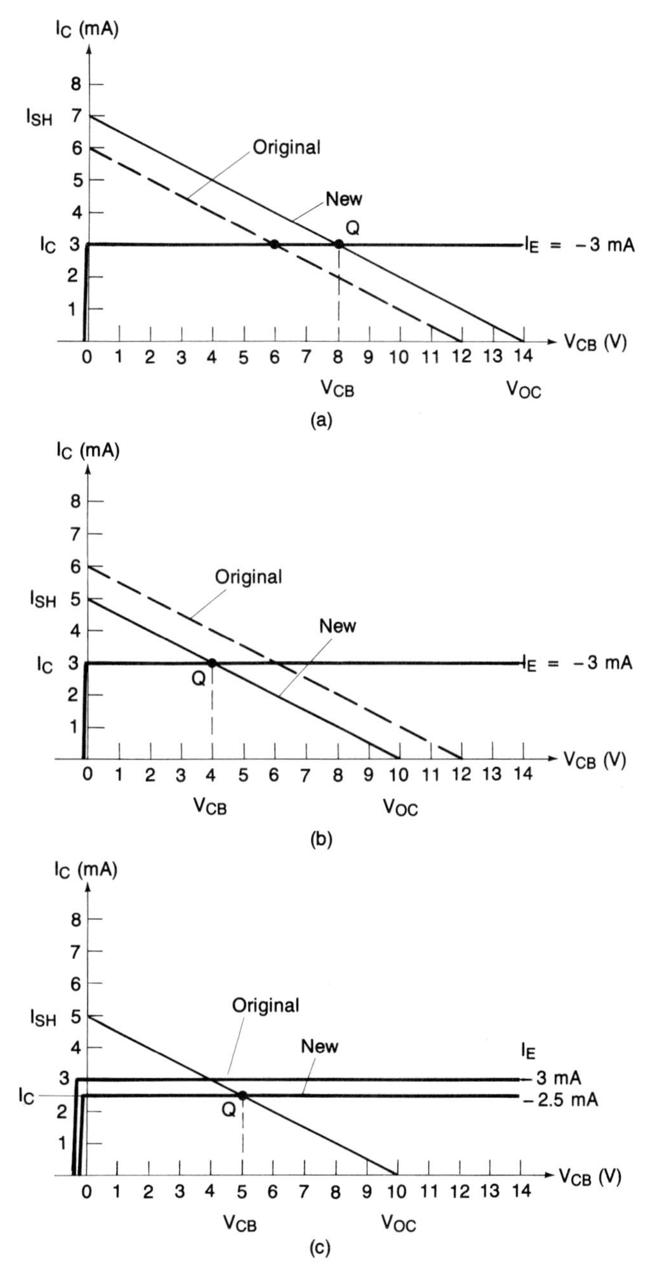

FIGURE 4-20 **Effects of the collector supply V_{CC} on the Q-point; (a) raising V_{CC} increases V_{CB}; (b) lowering V_{CC} reduces V_{CB}; (c) reducing the magnitude of I_E centers the Q-point.**

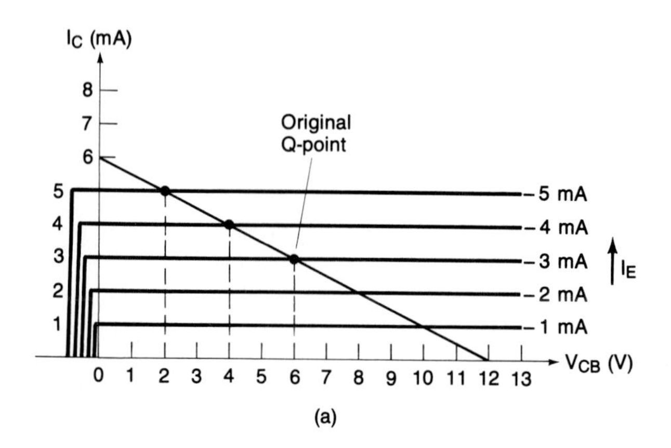

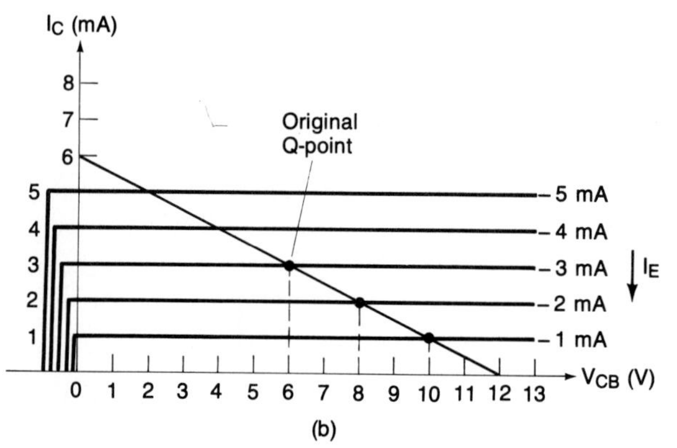

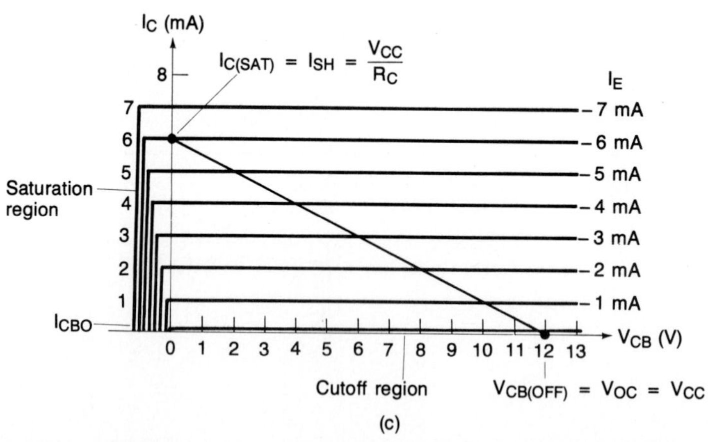

FIGURE 4-21 **Effects of I_E on the Q-point: (a) increasing I_E decreases V_{CB}; (b) decreasing I_E increases V_{CB}; (c) definition of $I_{C(SAT)}$ and $V_{CB(OFF)}$.**

Similarly, if I_E is reduced, V_{CB} will be increased to its maximum value V_{OC}. Since this region of operation is termed cutoff, this particular value of V_{CB} is denoted $V_{CB(OFF)}$. Also, from Fig. 4-21(c),

$$V_{CB(OFF)} = V_{OC}$$

and

$$V_{CB(\mathrm{OFF})} = V_{CC} \tag{4-17}$$

When a transistor is *saturated* its collector current will equal $I_{C(\mathrm{SAT})}$ and its V_{CB} will be (approximately) 0 V. When a transistor is in *cutoff*, its collector-to-base voltage will be $V_{CB(\mathrm{OFF})}$ and its collector current will be equal to I_{CBO}. Saturation and cutoff are *nonlinear regions of operation* for a BJT. Therefore, they form the boundary points for the operation of a BJT as a linear, active (amplifying) device.

4-12 BJT Dc Models

To eliminate the need for the graphical approach to determine the *Q*-point of a BJT, we may employ its dc piecewise linear model. The model utilizes a *current-controlled current source*.

The dc models for the *npn* and *pnp* BJTs are provided in Fig. 4-22(a) and (b), respectively. Conventional current directions have been indicated. In both cases,

$$I_C = \alpha_{\mathrm{DC}} I_E$$

Since α_{DC} has been shown to be approximately unity, we may use the approximation

$$I_C \simeq I_E$$

This approximation is indicated in Fig. 4-22.

To utilize the models given in Fig. 4-22(a) and (b), we shall use the knee-voltage models for the forward-biased emitter–base diodes, and assume that the collector currents are equal to the emitter currents. To expedite our solutions, we shall *not* take the time to redraw the BJT biasing circuits to include the BJT piecewise linear dc models indicated in Fig. 4-22(a) and (b). Instead, we will merely label our BJT modeling considerations directly on the original schematic diagrams [refer to Fig. 4-22(c) and (d)].

To analyze the BJT circuits given in Fig. 4-22(c) and (d), we may proceed as follows:

1. Analyze the input circuit by finding the emitter (input) current. This may be accomplished by applying Kirchhoff's voltage law and solving for I_E. For *both* Fig. 4-22(c) and (d),

$$I_E = \frac{V_{EE} - 0.7\ \mathrm{V}}{R_E} \tag{4-18}$$

2. Transfer across the device.

$$I_C \simeq I_E$$

3. Apply Kirchhoff's voltage law to the output (collector–base) circuit, and solve for V_{CB}. For the *npn* BJT given in Fig. 4-22(c), we have

$$V_{CB} = V_{CC} - I_C R_C \tag{4-19}$$

and for the *pnp* BJT given in Fig. 4-22(d),

$$V_{CB} = -V_{CC} + I_C R_C \tag{4-20}$$

FIGURE 4-22 BJT dc models: (a) npn piecewise linear model; (b) pnp piecewise linear model; (c) using the (silicon) npn model; (d) using the (silicon) pnp model.

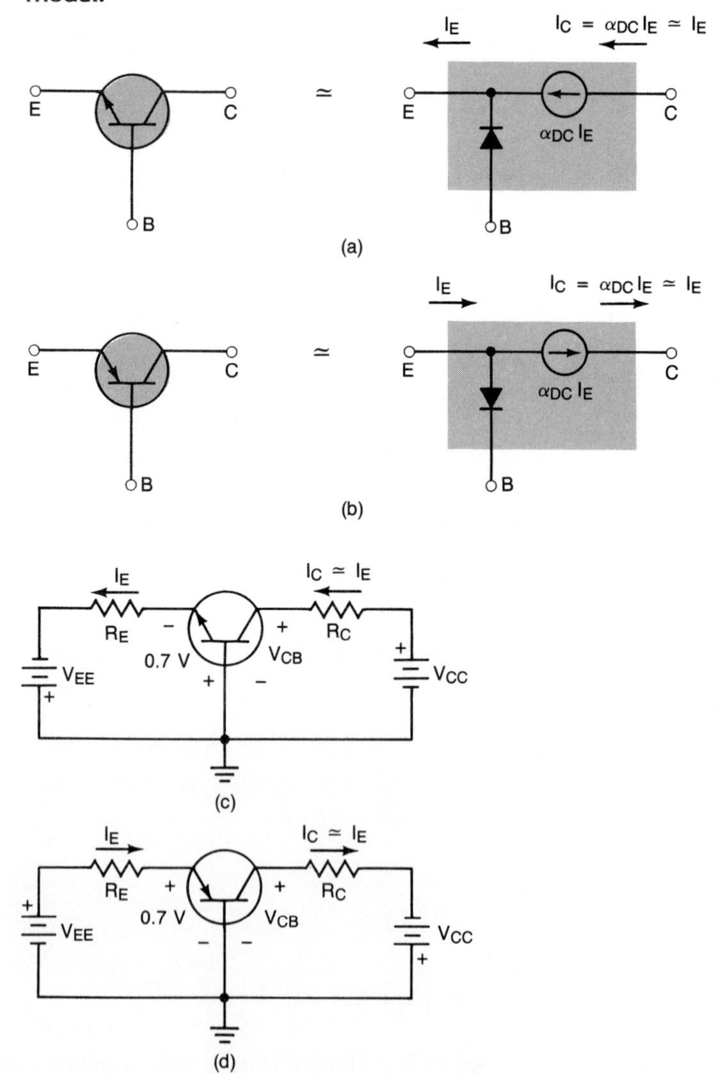

EXAMPLE 4-5

Repeat the analysis of the BJT common-base amplifier given in Fig. 4-18(a) using the dc piecewise linear model for the transistor. Also compute $I_{C(SAT)}$ and $V_{CB(OFF)}$.

SOLUTION First, we determine I_E by Eq. 4-18.

$$I_E = \frac{V_{EE} - 0.7\text{ V}}{R_E} = \frac{15\text{ V} - 0.7\text{ V}}{10\text{ k}\Omega} = 1.43\text{ mA}$$

(This was done previously.) To transfer across the device we merely assume that the collector current is approximately equal to the emitter current.

$$I_C \simeq I_E = 1.43\text{ mA}$$

and since the BJT is *npn*, we may use Eq. 4-19 to determine V_{CB}.

$$V_{CB} = V_{CC} - I_C R_C = 15\text{ V} - (1.43\text{ mA})(5.1\text{ k}\Omega) = 7.71\text{ V}$$

To determine $I_{C(SAT)}$ and $V_{CB(OFF)}$, we may draw on Eqs. 4-16 and 4-17, respectively.

$$I_{C(SAT)} = \frac{V_{CC}}{R_C} = \frac{15\text{ V}}{5.1\text{ k}\Omega} = 2.94\text{ mA}$$

$$V_{CB(OFF)} = V_{CC} = 15\text{ V}$$

By inspection of Fig. 4-18(d), we can see that our analytical solution agrees very closely with the load-line results. ■

EXAMPLE 4-6

The schematic diagram of a common-base BJT voltage amplifier has been provided in Fig. 4-23(a). Perform a dc analysis.

SOLUTION First, we recognize that capacitors C_1 and C_2 are dc blocking capacitors. From our work in Section 3-19, we may draw the dc equivalent circuit as indicated in Fig. 4-23(b). By Eq. 4-18 we have

$$I_E = \frac{V_{EE} - 0.7\text{ V}}{R_E} = \frac{15\text{ V} - 0.7\text{ V}}{15\text{ k}\Omega} = 0.953\text{ mA}$$

Now we may approximate I_C,

$$I_C \simeq 0.953\text{ mA}$$

and from Eq. 4-19,

$$V_{CB} = V_{CC} - I_C R_C = 15\text{ V} - (0.953\text{ mA})(6.8\text{ k}\Omega) = 8.52\text{ V}$$

To conclude our dc analysis, we shall determine $I_{C(SAT)}$ and $V_{CB(OFF)}$.

$$I_{C(SAT)} = \frac{V_{CC}}{R_C} = \frac{15\text{ V}}{6.8\text{ k}\Omega} = 2.21\text{ mA}$$

$$V_{CB(OFF)} = V_{CC} = 15\text{ V}$$ ■

The results of our dc analysis of Fig. 4-23(a) will be applied directly to our discussions in the next section. Therefore, the reader should study Example 4-6 carefully.

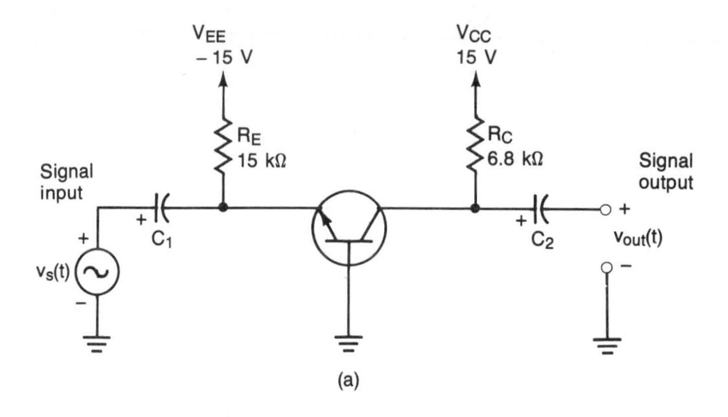

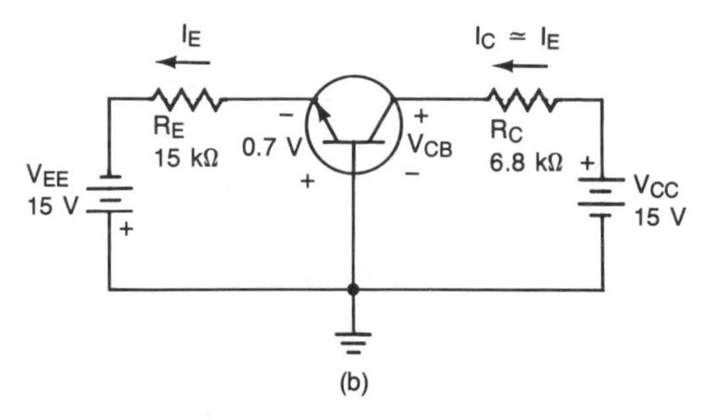

FIGURE 4-23 Dc analysis of a common-base voltage amplifier: (a) voltage amplifier; (b) dc equivalent circuit.

4-13 BJT Signal Process

Now that we have investigated the approaches to the dc analysis of a BJT, let us consider the effects of a signal on it. This will assist in our later work by allowing us to develop an intuitive ''feel'' for the operation of a BJT amplifier.

In Example 4-6 we performed a dc analysis of the common-base voltage amplifier given in Fig. 4-23(a). In Fig. 4-24 we see how an ac signal source $v_s(t)$ produces an ac signal in the BJT amplifier. Recall that the input and output coupling capacitors (C_1 and C_2) have a small capacitive reactance (X_C) to the ac signal. Specifically, we may treat them as short circuits to the ac signals and open circuits to the dc bias.

In Fig. 4-24(a) we can see that the ac voltage source develops an ac voltage across the forward-biased emitter–base diode. In Chapter 3 we saw that an ac signal applied to a forward-biased diode produces an ac voltage across it and an ac current through it. We also noted that the forward-biased diode must then possess an ac resistance. The forward-biased emitter–base ''diode'' behaves in much the same manner. Compare the *V–I* curve for the forward-biased diode given in Fig. 3-34 in Section 3-17 with the BJT input (emitter–base) *V–I* curves given in Fig. 4-12(b). The ac resistance of the emitter–base diode is very similar.

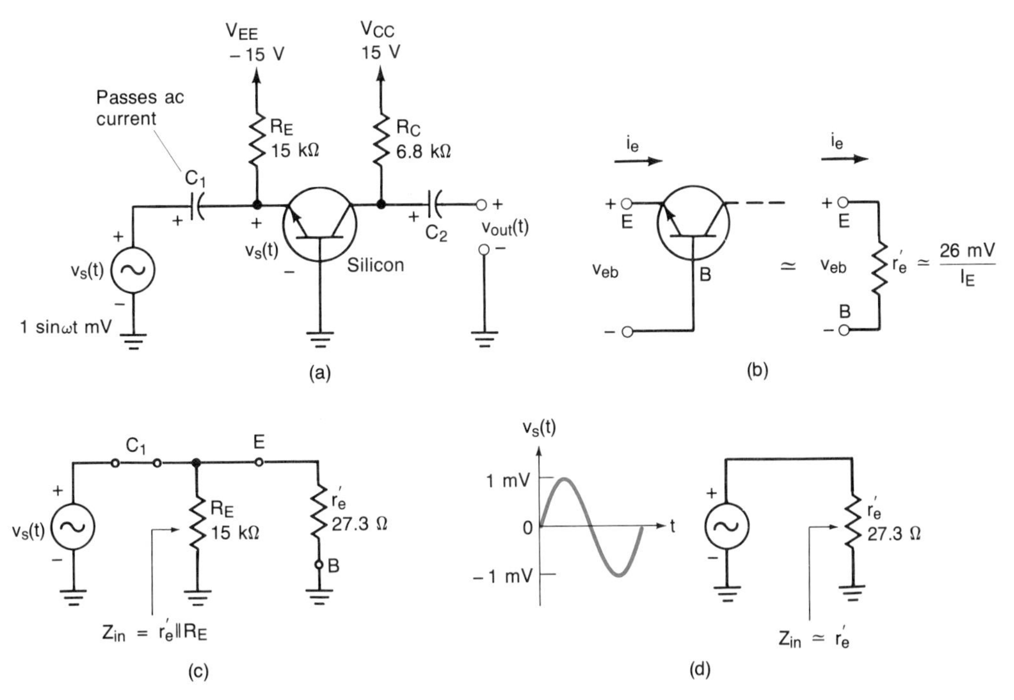

FIGURE 4-24 Ac analysis of the amplifier's input: (a) the ac signal source develops an ac voltage across the emitter-base p-n junction; (b) the small-signal ac equivalent of the emitter-base diode is r_e'; (c) ac equivalent of the amplifier's input; (d) simplified ac equivalent circuit.

In Section 3-17 we also developed Eq. 3-15 as an approximation of the diode's dynamic junction resistance r_j.

$$r_j \simeq \frac{26 \text{ mV}}{I_D} \tag{4-21}$$

The small-signal ac resistance of the emitter–base diode is called the *dynamic emitter resistance* r_e'. Its development is essentially the same as Eq. 3-15. Hence

$$r_e' \simeq \frac{26 \text{ mV}}{I_E} \tag{4-22}$$

Using the value of I_E determined in Example 4-6, we have

$$r_e' \simeq \frac{26 \text{ mV}}{I_E} = \frac{26 \text{ mV}}{0.953 \text{ mA}} \simeq 27.3\ \Omega$$

[see Fig. 4-24(b)].

In Fig. 4-24(c) we see the ac equivalent resistance of the amplifier's input circuit. The equivalent *input impedance* Z_{in} of the amplifier is

$$Z_{\text{in}} = r_e' \parallel R_E \tag{4-23}$$

and since $r_e' << R_E$,

$$Z_{\text{in}} \simeq r_e' \tag{4-24}$$

Therefore, the amplifier's input circuit may be simplified as depicted in Fig. 4-24(d).

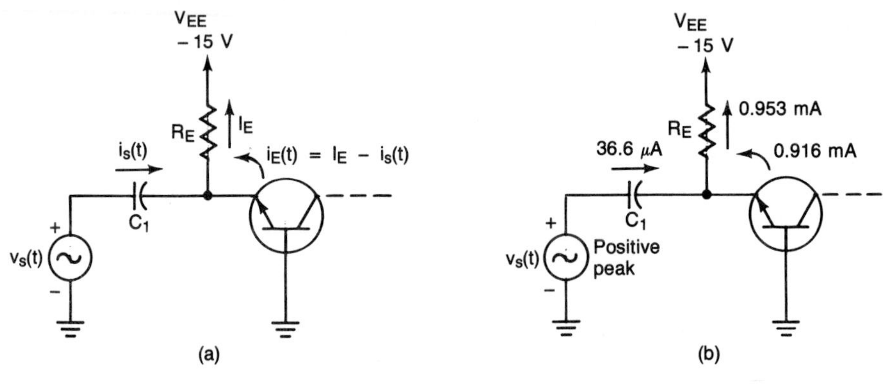

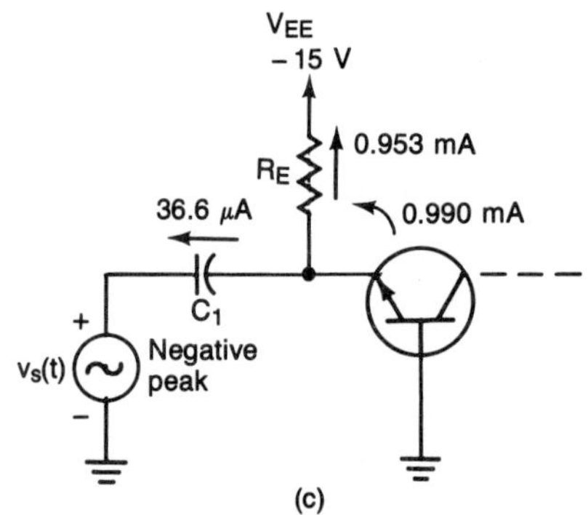

FIGURE 4-25 The signal source produces a variation in the emitter current: (a) the total instantaneous emitter current $i_E(t)$ is arrived at via Kirchhoff's current law; (b) $i_E(t)$ is reduced; (c) $i_E(t)$ is increased.

By inspection of the simplified ac equivalent input circuit, we can see that the instantaneous current drawn from $v_s(t)$ is

$$i_s(t) = \frac{v_s(t)}{r'_e} \tag{4-25}$$

and using the values given in Fig. 4-24(d), we may determine the peak current I_m drawn from the signal source

$$I_m = \frac{V_m}{r'_e} = \frac{1\text{ mV}}{27.3\ \Omega} = 36.6\ \mu\text{A}$$

The 1-mV-peak voltage source will supply a peak current of 36.6 μA to the input (emitter) of the BJT amplifier. This current will cause the emitter current to vary.

By applying Kirchhoff's current law at the emitter [Fig. 4-25(a)], we have

$$i_E(t) = I_E - i_s(t) \tag{4-26}$$

When the signal source causes a peak current of 36.6 μA to flow *into* the emitter as shown in Fig. 4-25(b), we have

$$i_E(t) = 0.953\text{ mA} - 36.6\ \mu\text{A} = 0.916\text{ mA}$$

Similarly, when the signal source causes an additional 36.6 μA to flow *out* of the emitter, as shown in Fig. 4-25(c),

$$i_E(t) = 0.953 \text{ mA} + 36.6\ \mu\text{A} = 0.990 \text{ mA}$$

As we have seen, the collector current is approximately equal to the emitter current. Hence

$$i_C(t) \simeq i_E(t) \tag{4-27}$$

Therefore, the collector current (which is controlled by the signal source) will vary from 0.916 to 0.990 mA [refer to Fig. 4-26(a)].

The instantaneous collector current determines the instantaneous magnitude of the collector-to-base (output) voltage $v_{CB}(t)$.

$$v_{CB}(t) = V_{CC} - i_C(t)R_C \tag{4-28}$$

Therefore, when $i_C(t)$ is 0.916 mA,

$$v_{CB}(t) = 15 \text{ V} - (0.916 \text{ mA})(6.8 \text{ k}\Omega) = 8.77 \text{ V}$$

FIGURE 4-26 BJT amplifier waveforms.

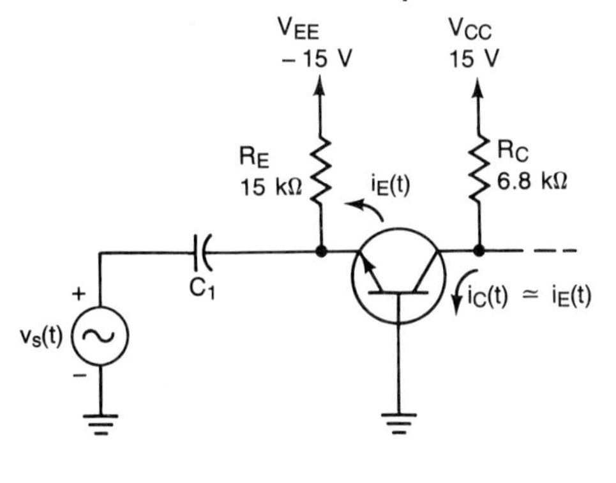

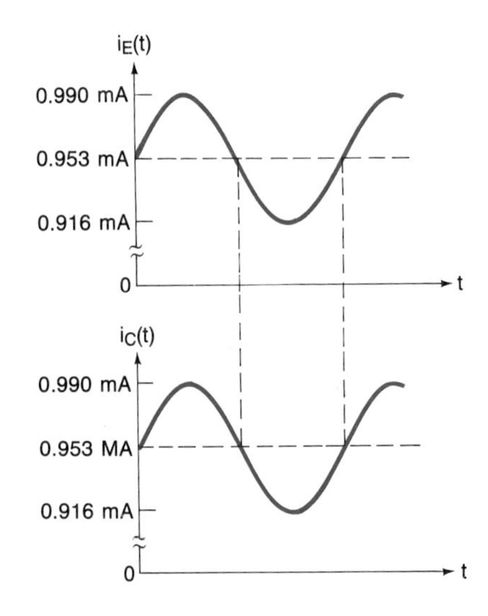

(a)

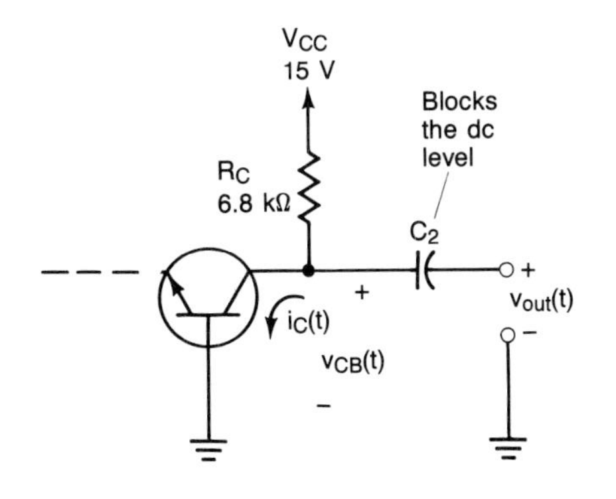

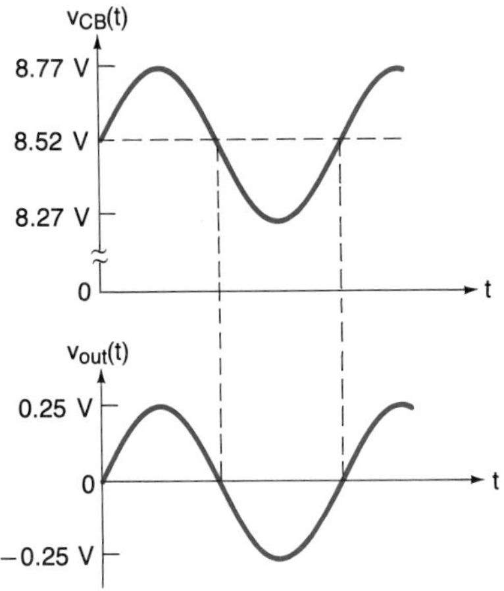

(b)

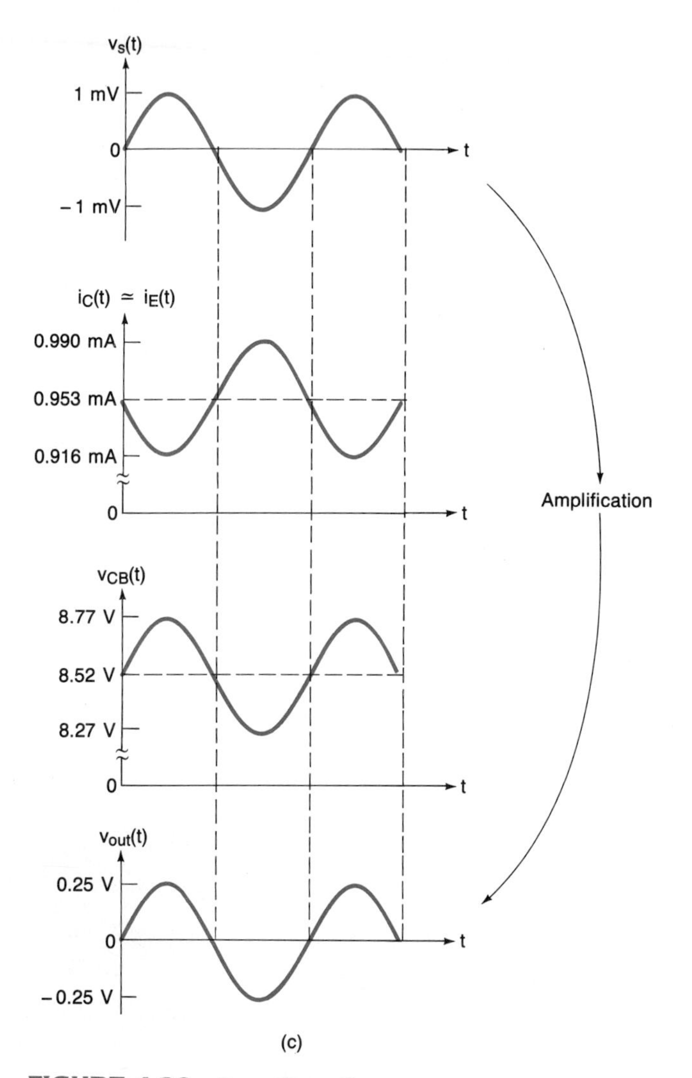

(c)

FIGURE 4-26 *(continued)*

and when $i_C(t)$ is 0.990 mA,

$$v_{CB}(t) = 15 \text{ V} - (0.990 \text{ mA})\,(6.8 \text{ k}\Omega) = 8.27 \text{ V}$$

The total collector-to-base voltage $v_{CB}(t)$ is illustrated in Fig. 4-26(b). Since capacitor C_2 blocks the dc level, an ac waveform appears across the output. The peak-to-peak ac output voltage is [from Fig. 4-26(b)]

$$v_{\text{out}}\,(\text{p-p}) = 8.77 \text{ V} - 8.27 \text{ V} = 0.50 \text{ V p-p}$$

The voltage gain of the amplifier is

$$Av = \frac{v_{\text{out}}(t)}{v_{\text{in}}(t)} = \frac{v_{\text{out}}(t)}{v_s(t)} = \frac{0.50 \text{ V p-p}}{2 \text{ mV p-p}} = 250$$

Carefully reflect on this result. The small-signal input voltage has produced a much larger output voltage. The transistor has made this possible. It is simply a controlling device. The signal has produced a change in the BJT's input which has produced a much larger change at its output.

A summary of the various waveforms has been provided in Fig. 4-26(c). The reader should study these waveforms and their phase relationships. By careful inspection of the collector current and the collector-to-base voltage waveforms, we see that as the collector current *increases*, the collector-to-base voltage *decreases*. Therefore, the BJT is described as having a *negative resistance* and it is classified as an *active device*.

Much attention has been directed toward the development of small-signal models for the BJT. In the next section we introduce some of these models and begin the development of the model that we shall use.

4-14 BJT Small-Signal Models

In Fig. 4-22 we indicated the dc models for the BJT. The BJT was considered to be a current-controlled current source. This is essentially the concept we used in the preceding section to describe the BJT signal process.

In Fig. 4-27 we have illustrated some of the more popular equivalent-circuit

FIGURE 4-27 BJT small-signal models: (a) r'_e model; (b) h-parameter model; (c) y-parameter model; (d) hybrid-pi model.

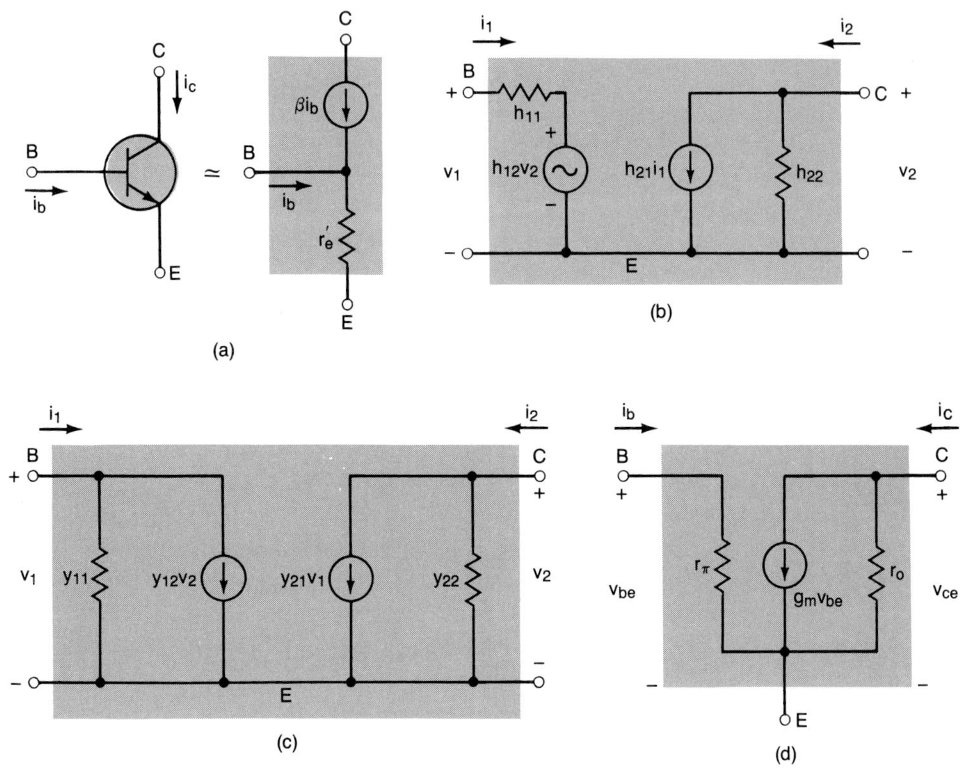

models for the BJT. The very simple r'_e-based model has been indicated in Fig. 4-27(a). The popularity of this model stems from the fact that it promotes an intuitive understanding of the ac operation of the BJT. This is the model we employed in the Section 4-12.

Figure 4-27(b) shows the *hybrid-* or *h-parameter model*. This model is obviously much more complex than the r'_e-based model. However, it is widely used to analyze the small-signal operation of the BJT. Manufacturers typically provide the values of the various parameters h_{11}, h_{12}, h_{21}, and h_{22} on their data sheets. The numerical subscripts we have indicated are generic. For instance, h_{21} is used to represent "the short-circuit forward current gain." If a BJT is in its common-base configuration h_{21} is approximately equal to its α_{ac} or α. In this case, h_{21} is called h_{fb}—hybrid parameter, *f*orward current gain, common *b*ase. If the BJT is in its common-emitter configuration, its current gain is approximately equal to its β_{ac} or β, and h_{21} becomes h_{fe}—*h*ybrid parameter, *f*orward current gain, common *e*mitter. The h-parameter model is generally used for signal analysis in the audio-frequency range (20 Hz to 20 kHz), although it may be used up to approximately 100 kHz. (Additional h-parameter considerations are provided in Appendix D.)

For analysis in the higher-frequency ranges (i.e., radio frequencies) the *short-circuit admittance* or *y-parameter model* is generally preferred [see Fig. 4-27(c)]. For example, if we were to conduct an analysis of a BJT found in an FM tuner, we might wish to employ the y-parameter model. Consequently, manufacturers quite often provide the y-parameters y_{11}, y_{12}, y_{21}, and y_{22} for high-frequency BJTs on their data sheets.

The h- and y-parameter models are very powerful. However, they prove to be difficult to use for most beginning students. These models are slightly abstract, and the values of each of the individual parameters depend on the BJT bias point, the operating temperature, and the signal frequency.

In Fig. 4-27(d) we see the simplified *hybrid-pi model*. This is the model on which we shall focus our attention. It is an attractive choice for several reasons. First, it promotes an intuitive understanding of the BJT. Second, it is easily modified to accommodate high-frequency analyses. Further, it is also extremely similar to the small-signal model commonly used for the analysis of the field-effect transistor (FET) that we pursue in Chapter 5. The hybrid-pi model may also be used to develop an ac equivalent circuit for the IC (integrated circuit) op amp.

4-15 Transconductance

As we have seen, the BJT may be thought of as a current-controlled current source. However, it may also be described as a *voltage-controlled current source* (refer to Fig. 4-16). The magnitude of the base-to-emitter (input) voltage controls the magnitude of the collector (output) current. Expressing this as a gain, we have

$$\text{gain} = \frac{\text{output current}}{\text{input voltage}} = \frac{i_C(t)}{v_{BE}(t)} \tag{4-29}$$

Obviously, Eq. 4-29 is a conductance. However, to emphasize that it denotes an input–output relationship, it is described as a *transfer conductance* or *transcon-*

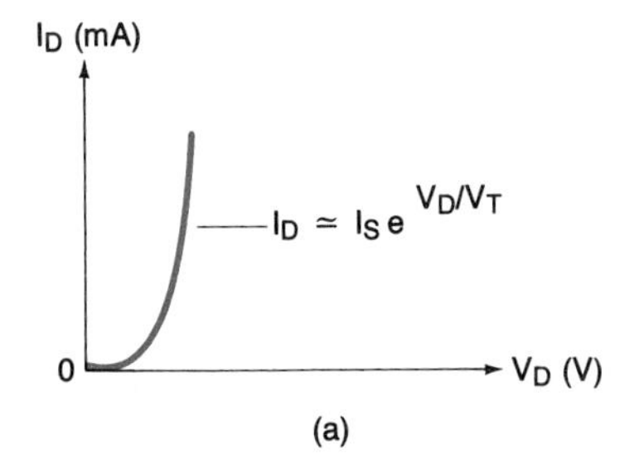

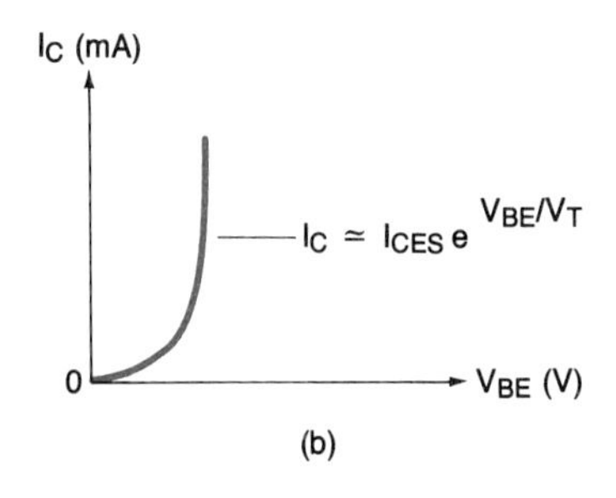

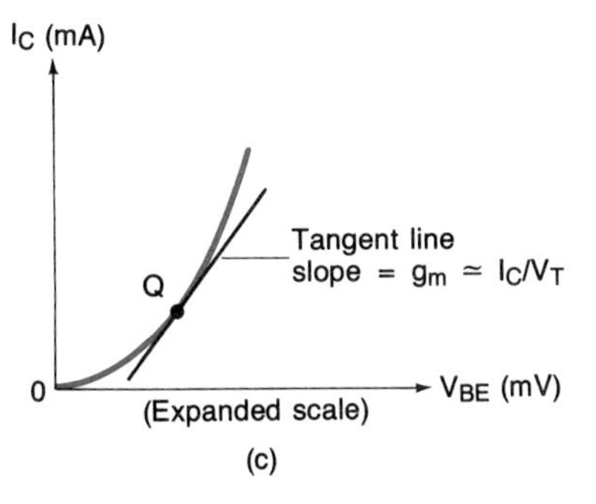

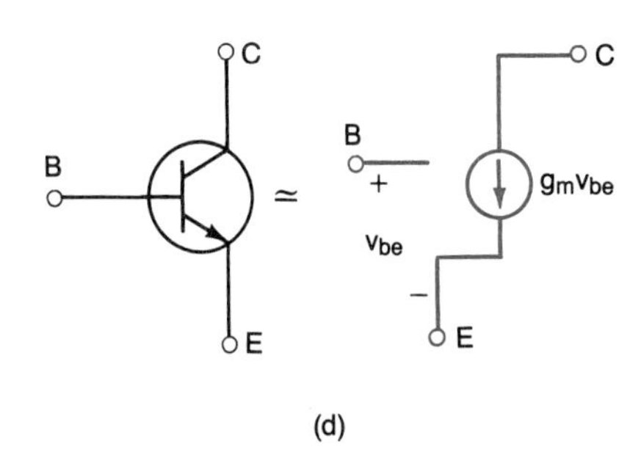

FIGURE 4-28 **BJT transconductance: (a) forward-biased diode V-I curve; (b) BJT transfer characteristic curve; (c) graphical definition of g_m; (d) "bare-bones" hybrid-pi model.**

ductance. The terms ''transconductance'' and *mutual conductance* are synonymous. Consequently, the symbol used for transconductance is g_m, and for the BJT,

$$g_m = \frac{i_C(t)}{v_{BE}(t)} \tag{4-30}$$

The transfer curve given in Fig. 4-16 has a shape that is very similar to the *V–I* curve of a forward-biased diode [see Fig. 4-28(a)]. Recall that we mathematically described the diode's *V–I* curve with the Shockley diode equation. Therefore, it should come as no great surprise that a similar exponential relationship may be used to describe the transfer curve [compare Fig. 4-28(a) and (b)]. Hence

$$\boxed{I_C \simeq I_{CES} e^{V_{BE}/V_T}} \tag{4-31}$$

where I_C = dc collector current
I_{CES} = collector leakage current with $V_{BE} = 0$ V
V_T = voltage equivalent of temperature

The voltage equivalent of temperature was originally defined by Eq. 3-2. It was also found to be approximately 26 mV at room temperature.

To develop an equation for g_m, we may employ calculus to take the derivative of Eq. 4-31,

$$g_m = \frac{dI_C}{dV_{BE}} = \frac{1}{V_T} I_{CES} e^{V_{BE}/V_T}$$

and if we substitute Eq. 4-31 into the result above, we have

$$\boxed{g_m \simeq \frac{I_C}{V_T}} \tag{4-32}$$

and at room temperature,

$$\boxed{\text{At 25°C:} \quad g_m \simeq \frac{I_C}{26 \text{ mV}}} \tag{4-33}$$

The small-signal transconductance of the BJT has been graphically defined in Fig. 4-28(c). Our ''bare-bones'' BJT model has been indicated in Fig. 4-28(d).

4-16 Early Effect and Base-Width Modulation

In Fig 4-15 we presented the common-emitter output characteristic curves. Turning back to Fig. 4-15, we can see that at higher values of I_C, an increase in V_{CE} produces an increase in I_C. However, at lower values of I_C, the effect of V_{CE} on I_C *appears* to be negligible. If a close analysis is made, it becomes evident that the *percentage* change in I_C due to V_{CE} is actually the *same* in both cases.

In 1952, J. Early of Bell Labs analyzed this effect, and this phenomenon was named after him. A survey of his efforts will help us understand why a BJT has an *output impedance* r_o and how we may determine its value. We shall then incorporate r_o into our hybrid-pi model of the BJT.

In Section 4-4 we explained the relationship between the minority-carrier concentration gradient in the base region and the transistor currents (refer to Fig. 4-5).

In Fig. 4-29(a) we see the minority-carrier concentration gradient at a relatively low value of collector-to-base bias V_{CB}. If V_{CB} is *increased*, the collector-to-base depletion region *widens* and the effective base width W_B is *decreased*. The decrease in W_B causes the *slope* of the concentration gradient to *increase*, which implies that I_C must *increase* [see Fig. 4-29(b)].

In Fig. 4-29(c) we see that the change in the slope of the concentration gradient due to changes in V_{CB} is minimal at low values of I_C. Figure 4-29(d) shows us that at larger values of I_C, a change in V_{CB} can produce a much larger change in the slope of the concentration gradient. Hence a resulting larger change in I_C occurs.

Early also observed that the slopes of the individual characterisitic curves are approximately constant. Further, if they are all projected backward [as shown in Fig. 4-30(a)], they all intersect a common point. This common point is termed the *Early voltage* V_A. Typical values of V_A range from about 150 to 250 V.

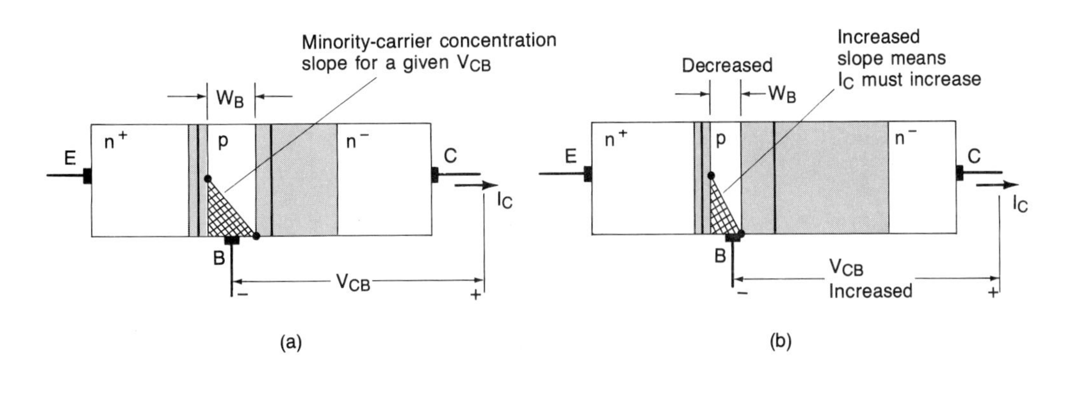

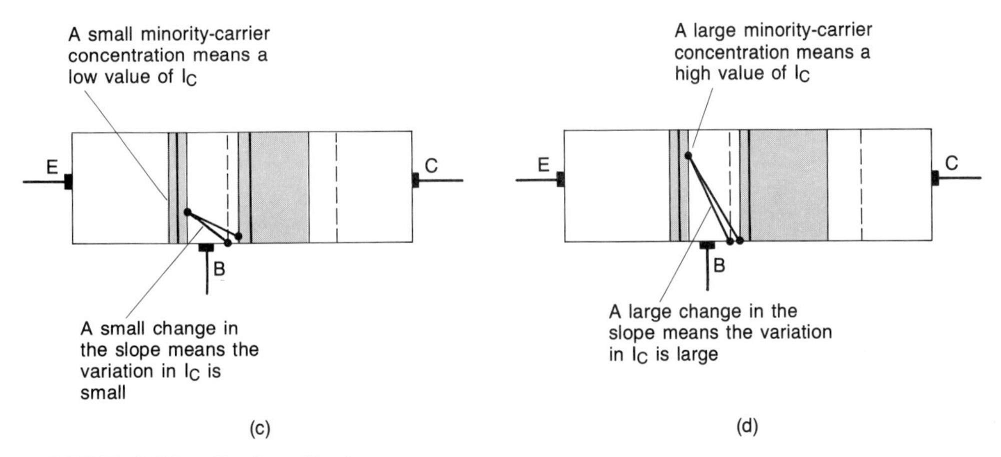

FIGURE 4-29 Early effect.

By inspection of Fig. 4-30(b), we can see that we may determine the dynamic resistance associated with each particular V–I curve of interest. This is the dynamic resistance or output impedance r_o of the BJT. Hence, from Fig. 4-30(b),

$$r_o = \frac{\Delta V_{CE}}{\Delta I_C} = \frac{V_{CE} - (-V_A)}{I_C - 0} = \frac{V_{CE} + V_A}{I_C}$$

and since typically $V_A >> V_{CE}$,

$$\boxed{r_o \simeq \frac{V_A}{I_C}} \tag{4-34}$$

where V_A is the Early voltage (typically 200 V).

Just as we found the dynamic resistance (r_O) of the diode in Section 3-7, now we have found the dynamic resistance (r_o) of the output of the BJT. A comparison between an ideal current source and one that includes a resistance has been provided

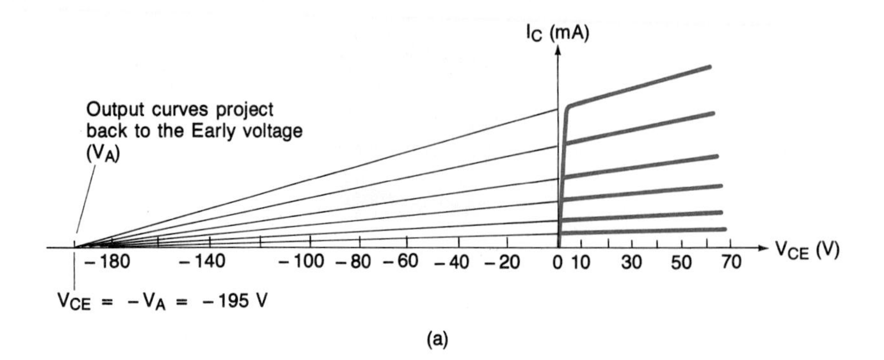

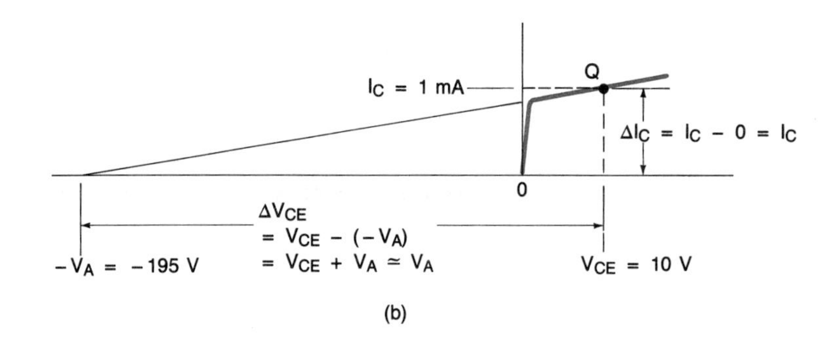

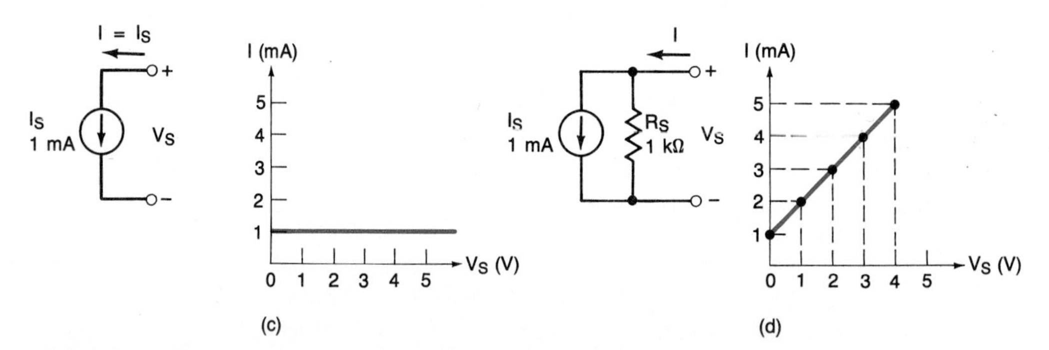

FIGURE 4-30 Determination and application of r_o in our BJT model: (a) finding the Early voltage; (b) graphical determination of r_o; (c) ideal constant-current source; (d) "real" constant-current source.

in Fig. 4-30(c) and (d), respectively. In Fig. 4-30(d) we may apply Kirchhoff's current law to arrive at

$$I = \frac{V_S}{R_S} + I_S = 1 \times 10^{-3} V_S + 1 \text{ mA}$$

From the result above and Fig. 4-30(d), it is clear that we may account for the upward slope in the BJT output characteristics by placing r_o in parallel with our

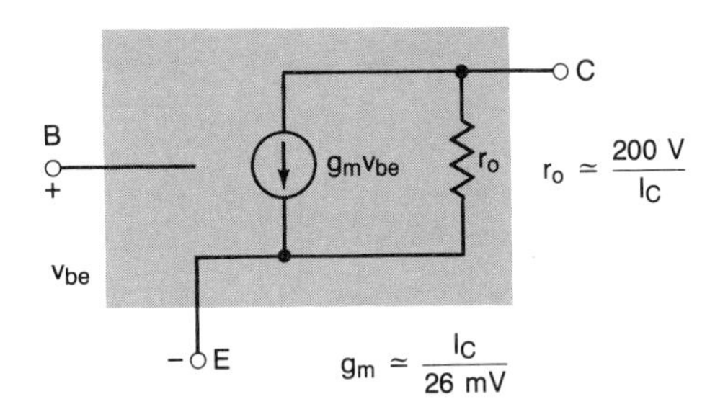

FIGURE 4-31 Addition of the BJT's output resistance r_o to our model.

constant-current source model of the BJT. Our ''improved'' small-signal BJT model has been shown in Fig. 4-31.

4-17 BJT Input Resistance

Now that we have seen that the BJT's output may be modeled as a voltage-controlled current source with a finite output resistance, it is time to consider the appropriate model for its *input*. In general, the input resistance of a device is defined as indicated in Fig. 4-32(a). As we saw in Section 4-13, the resistance ''looking into'' the emitter is r_e', and with reference to Fig. 4-32(a) and (b),

$$r_e' = \frac{v_{eb}}{i_e} \simeq \frac{26 \text{ mV}}{I_E} \tag{4-35}$$

where i_e is the magnitude of the instantaneous ac emitter current.

The resistance ''looking into'' the base terminal r_π,

$$r_\pi = \frac{v_{be}}{i_b} \tag{4-36}$$

Contrast Fig. 4-32(b) and (c). Figure 4-32(d) will serve as our reference in the work to follow.

To develop an equation for r_π, we shall first rewrite Eq. 4-2 using small-signal notation.

$$i_e = i_c + i_b \tag{4-37}$$

If we also employ small-signal notation to restate Eq. 4-5, and solve for i_c,

$$i_c = \beta i_b \tag{4-38}$$

where β is the *ac* current gain. The *h*-parameter equivalent uses lowercase subscripts and is denoted h_{fe}.

Substituting Eq. 4-38 into Eq. 4-37 yields

$$i_e = \beta i_b + i_b = i_b(\beta + 1) \tag{4-39}$$

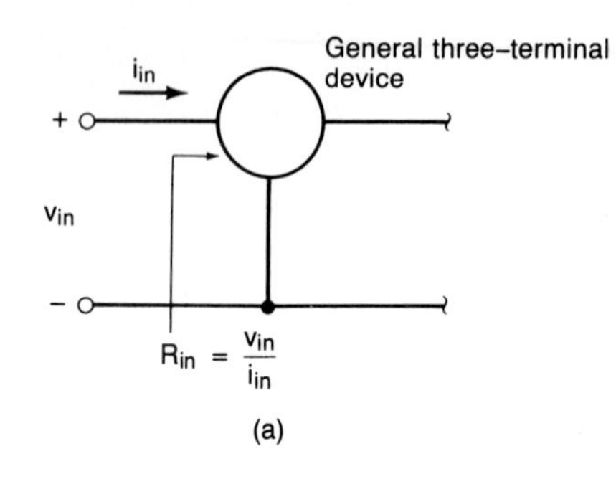

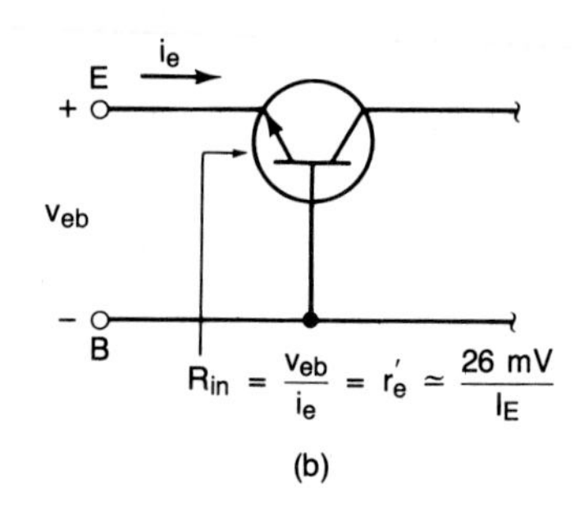

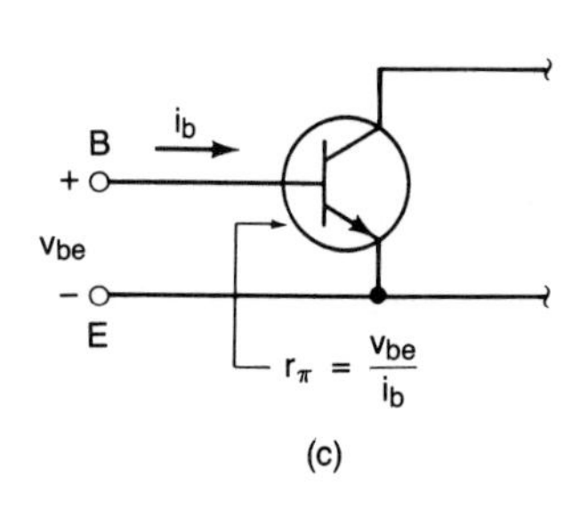

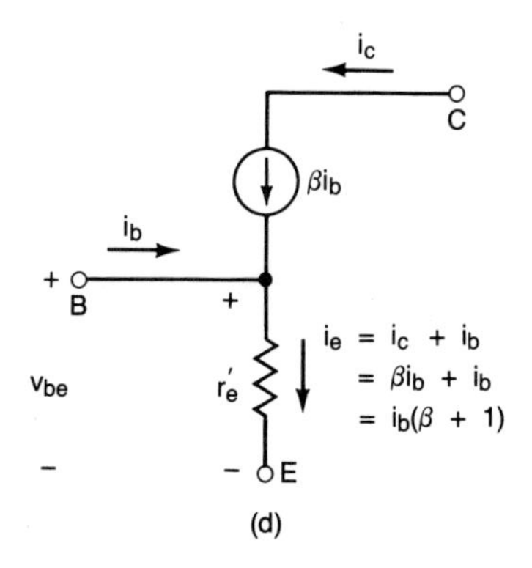

Npn or *pnp* BJTs:

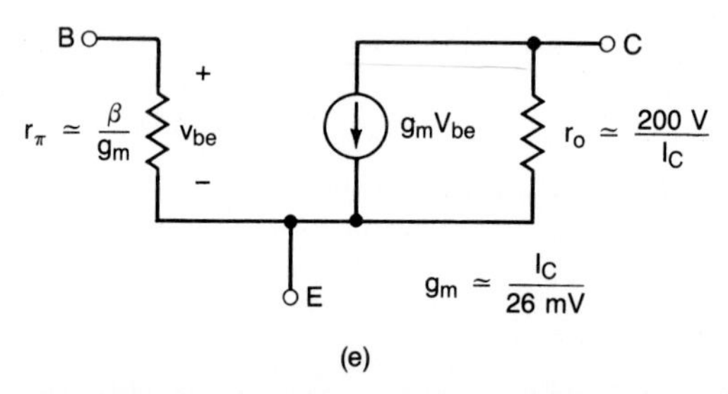

FIGURE 4-32 Development of r_π to complete the hybrid-pi BJT model: (a) definition of input resistance; (b) resistance "looking" into the emitter; (c) resistance "looking" into the base; (d) r_e' model used to develop r_π; (e) complete low-frequency, hybrid-pi model.

Using this result, and from Fig. 4-32(d), we see that

$$r_\pi = \frac{v_{be}}{i_b} = \frac{i_e r_e'}{i_b} = \frac{i_b(\beta + 1)r_e'}{i_b} = (\beta + 1)r_e'$$

and since $\beta >> 1$,

$$\boxed{r_\pi \simeq \beta r_e'} \qquad (4\text{-}40)$$

Let us reflect on this result. Since the magnitudes of the base-to-emitter and emitter-to-base voltages are equal, but the base current is much smaller (by a factor of about $1/\beta$), it makes sense that the resistance "looking into" the base is much larger (by a factor of about β) than the resistance "looking into" the emitter. Equation 4-40 expresses this observation.

To avoid calculating r'_e, which is not otherwise used in the hybrid-pi model, we may use the transconductance g_m to determine r_π. From Eqs. 4-33 and 4-35, if we assume that $I_C \simeq I_E$, we observe that

$$r'_e \simeq \frac{26 \text{ mV}}{I_C} = \frac{1}{I_C/26 \text{ mV}} = \frac{1}{g_m} \tag{4-41}$$

Substituting this into Eq. 4-40 yields

$$\boxed{r_\pi \simeq \frac{\beta}{g_m}} \tag{4-42}$$

Our completely developed hybrid-pi model has been given in Fig. 4-32(e). This shall be our BJT model for all of our small-signal, low-frequency analyses. As we shall see in Chapter 5, the small-signal model for the field-effect transistor will be virtually identical.

4-18 BJT Fabrication

To conclude our introduction to the BJT, we present a brief example of its fabrication. The reader may wish to review Sections 2-11 through 2-14. With that background, we see that Fig. 4-33 outlines the major steps involved in the fabrication of a discrete BJT.

In Fig. 4-33(a) we note that an n^- epitaxial layer has been grown upon an n^+ substrate. The richly doped substrate is used as a semiconductor extension of the metallic contact used to form the collector terminal of the transistor. The substrate is made just thick enough to minimize breakage during the handling that occurs during the fabrication process. The n^- epitaxial layer serves as the collector region.

The silicon dioxide (SiO_2) insulating layer is formed over the epitaxial layer as indicated in Fig. 4-33(b). By employing a precise photolithographic masking process, and a wet-chemistry etchant, a hole is produced in the SiO_2 layer.

Recall that the four possible acceptor dopants indicated in Table 1-1 were aluminum, boron, gallium, and indium. However, SiO_2 is only successful at masking boron. Consequently, boron is used to form the *p*-region by using gaseous diffusion [see Fig. 4-33(c)]. This *p*-type material serves as the base region for the BJT. Once again the crystal has a protective SiO_2 layer formed over it as shown in Fig. 4-33(d). A new hole is then etched in the SiO_2 layer [refer to Fig. 4-33(e)].

Since the *n*-type (donor) dopants antimony, arsenic, and phosphorus can all be blocked by SiO_2, any one of the three can be used to form the n^+ emitter region [see Fig. 4-33(f)]. By opening up an additional window to the base region, aluminum is then vacuum-deposited on the silicon wafer. Most of the aluminum is then etched

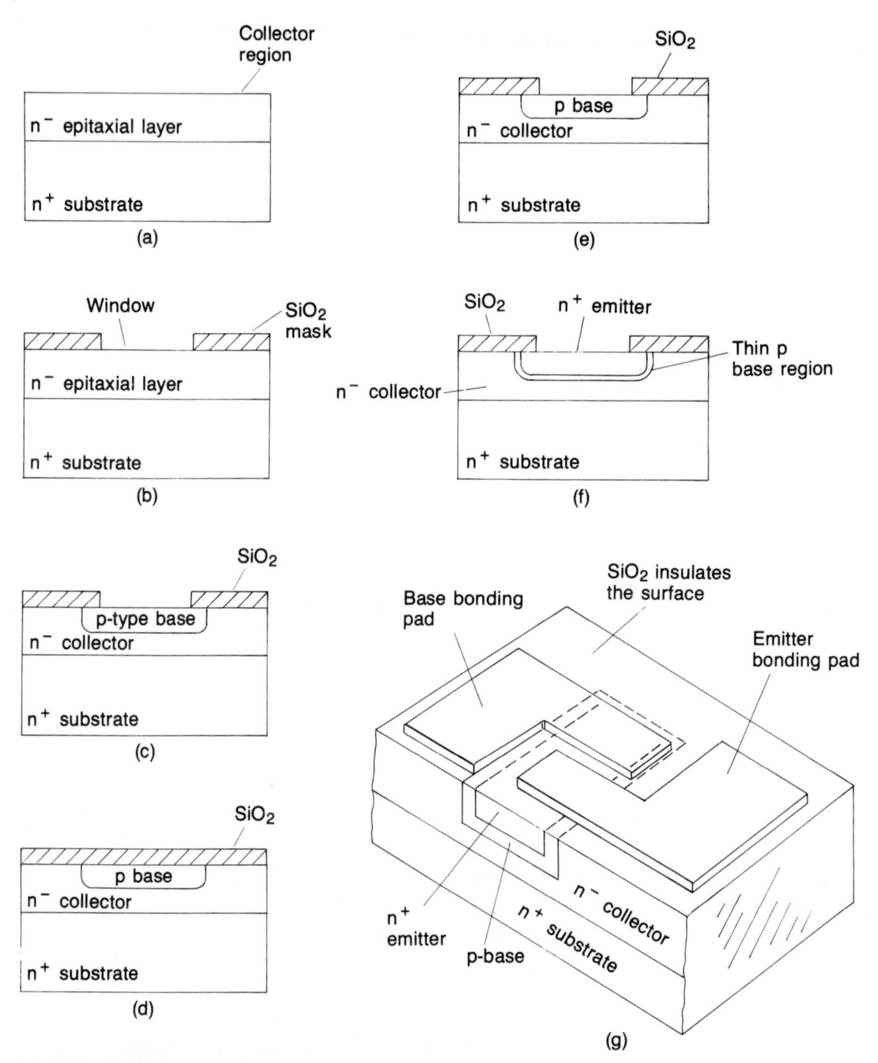

FIGURE 4-33 Fabrication of an npn (silicon) BJT.

away, leaving only the desired bonding pads for the base and emitter terminal connections. This is shown in Fig. 4-33(g). The two bonding pads are electrically insulated from one another by the SiO_2.

The completed transistor has been illustrated in Fig. 4-34(a). Observe that no bonding pad is formed for the collector. The collector region is electrically tied to the metallic header through the substrate. The actual collector terminal is butt welded to the outside of the header.

Note that *since the metallic case is welded to the header, the collector terminal is electrically in common with it.* The surface of the case serves as a heat radiator to assist in the removal of heat from the collector region. Similar metallic case styles and a plastic package have been shown in Fig. 4-34(b) and (c), respectively.

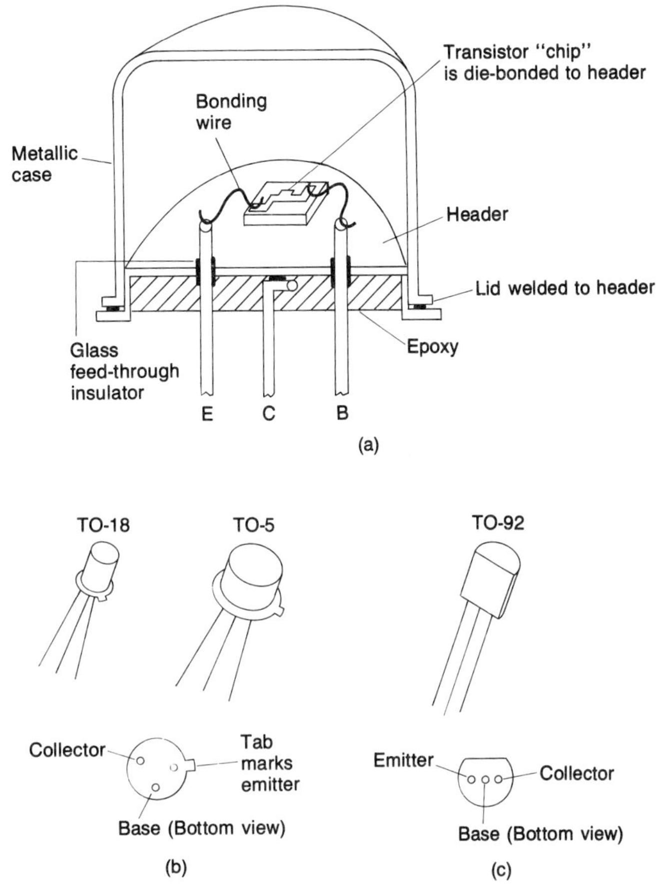

FIGURE 4-34 BJT case styles: (a) construction of a BJT in a TO-18 case style; (b) metallic case styles; (c) one-piece, injection-molded plastic package.

PROBLEMS

Drill, Derivations, and Definitions

Section 4-2

4-1. The ____ region of a BJT is the thinnest, the ____ region is the largest, and the ____ region is the most heavily doped.

4-2. Name the three BJT terminals. How many p-n junctions are used? Identify them.

Section 4-3

4-3. In an npn BJT ____ (positive, negative) ions exist on the emitter side of the p-n junction depletion region.

4-4. The depletion region extends more deeply into the ____ (base, collector) of the collector-base p-n junction.

Section 4-4

4-5. ____ (Holes, Electrons) are the minority carriers in the base region of a pnp BJT.

4-6. Define the term *effective base width* W_B.

4-7. In an npn BJT, electrons which cross the ____ (emitter-base, collector-base) p-n junction lose the most energy.

4-8. When a BJT is to be used as an amplifier, its emitter-base p-n junction is ____ (forward, reverse) biased and its collector-base p-n junction is ____ (forward, reverse) biased.

4-9. An *npn* BJT has an I_C of 2 mA and an I_B of 40 μA. Compute I_E. If I_B is doubled to 80 μA, determine the new values of I_C and I_E. (Hint: I_C will also double.)

4-10. A *pnp* BJT has an I_E of 1.5 mA and an I_C of 1.45 mA. Compute I_B. If I_E is doubled to 3 mA, find the new values of I_C and I_B. (Hint: I_B will also double.)

Section 4-5

4-11. Name the three basic BJT configurations. Name the input and output terminals for the three basic BJT configurations.

4-12. A BJT has an I_C of 1 mA and an I_B of 10 μA. Compute β_{DC} and α_{DC}.

4-13. A BJT has an I_C of 5 mA and an I_B of 20 μA. Compute its β_{DC} and α_{DC}.

4-14. If a BJT has an h_{FB} of 0.989, what is its h_{FE}?

4-15. If a BJT has an h_{FE} of 200, what is its h_{FB}?

Section 4-6

4-16. A BJT is in its common-base configuration. Its α_{DC} is 0.98 and its I_{CBO} is 25 nA. Compute its exact value of I_C when its I_E is 1 mA.

4-17. Repeat Prob. 4-16 for a BJT with an α_{DC} of 0.975 and an I_{CBO} of 50 nA when I_E is 3 mA.

4-18. A BJT has a β_{DC} of 100 and an I_{CBO} of 20 nA. Calculate its I_{CEO}.

4-19. A BJT has a β_{DC} of 250 and an I_{CBO} of 60 nA. Calculate its I_{CEO}.

4-20. A BJT is in its common-emitter configuration. Its β_{DC} is 125 and its I_{CEO} is 100 nA. Compute its exact value of I_C when its I_B is 8 μA.

4-21. Repeat Prob. 4-20 for a BJT with a β_{DC} of 50 and an I_{CEO} of 0.15 μA at an I_B of 100 μA.

4-22. Explain why I_{CEO} is larger than I_{CBO}. What is I_{CES}? How does I_{CES} compare to I_{CBO} and I_{CEO}?

4-23. In step-by-step detail, derive the relationship

$$\frac{1}{1 - \alpha_{DC}} = \beta_{DC} + 1$$

Use Eq. 4-8 to start your derivation.

4-24. Repeat Prob. 4-23, but use Eq. 4-9 to start your derivation.

Section 4-7

4-25. The transistor convention for defining current directions is also applied to linear and digital integrated circuits. The schematic symbol for an op amp integrated circuit is shown in Fig. 4-35. I_{OUT} in Fig. 4-35(a) is ____ (positive, negative) according to the transistor convention.

FIGURE 4-35 Op amp symbol and I_{OUT}.

4-26. Using the background provided in Prob. 4-25, I_{OUT} in Fig. 4-35(b) is ____ (positive, negative) according to the transistor convention.

Section 4-8

4-27. Make a neat sketch of the common-emitter output characteristic curves for a silicon *npn* BJT. Label the axes, indicate the parameter, and note the active, saturation, and cutoff regions.

4-28. Repeat Prob. 4-27 for a silicon *pnp* BJT.

Section 4-9

4-29. A silicon npn BJT has a measured V_{BE} of 0.65 V at 25°C. Determine its V_{BE} at 40°C and at 10°C.

4-30. A silicon pnp BJT has an I_{CBO} of 20 nA at 25°C. Find its I_{CBO} at 70°C.

4-31. A silicon npn BJT has an I_{CBO} of 45 nA at 60°C. Find its I_{CBO} at 25°C.

Section 4-11

4-32. Given the BJT shown in Fig. 4-36(a) and using the *V–I* curves shown in Fig. 4-36(b), use the load-line technique to determine I_C and V_{CB}.

4-33. Repeat Prob. 4-32 for the BJT circuit and the *V–I* curves given in Fig. 4-36(c) and (d), respectively.

Section 4-12

4-34. Using the BJT dc model, analyze the BJT circuit given in Fig. 4-36(a). Find I_C, V_{CB}, $V_{CB(OFF)}$, and $I_{C(SAT)}$.

4-35. Repeat Prob. 4-34 for the BJT circuit given in Fig. 4-36(c).

Section 4-13

4-36. The supply voltages in Fig. 4-24(a) have been reduced to ± 12V and R_E is lowered to 10 kΩ. Assume that $v_s(t)$ produces a 10-mV peak sine wave. Find I_E, I_C, and V_{CB}.

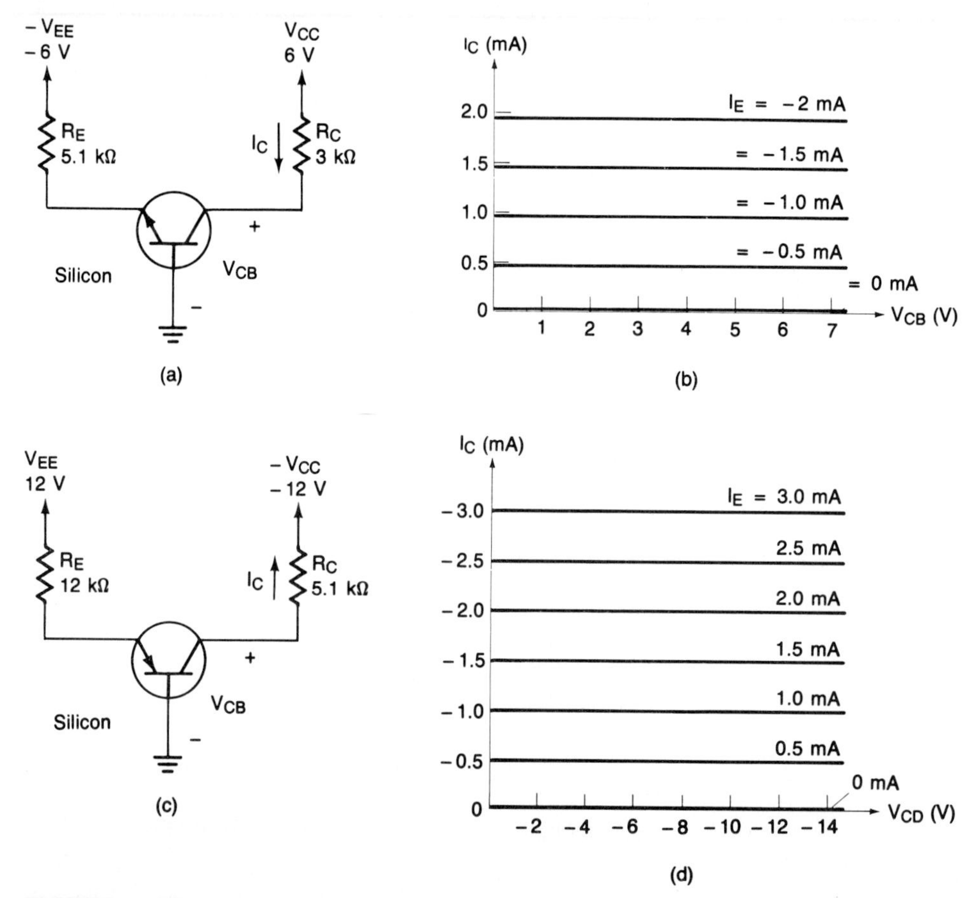

FIGURE 4-36 BJT circuit problems.

4-37. Continue Prob. 4-36 by finding r_e' and the peak current (I_m) drawn from $v_s(t)$.

4-38. Continue Prob. 4-37 by developing the waveforms associated with the amplifier. Label their positive and negative peak values. Use Fig. 4-26(c) as a guide.

Section 4-17

4-39. An *npn* BJT has a β of 100 at an I_C of 1 mA. Determine the values of g_m, r_o, and r_π. Make a sketch of its hybrid-pi model, and label its values.

4-40. Repeat Prob. 4-39 for a *pnp* BJT with a β of 60 at an I_C of 2 mA.

Section 4-18

4-41. Explain why boron is the preferred acceptor impurity used in the manufacture of silicon (diffused) BJTs.

Design

4-42. The *npn* BJT in Fig. 4-36(a) is to have an emitter current of 0.6 mA. Determine the nearest standard value for R_E. What is the nearest standard value of R_C required to give a V_{CB} of 3 V?

4-43. The *pnp* BJT given in Fig. 4-36(c) is to have an emitter current of 0.5 mA. Determine the nearest standard value required for R_E. Also find the nearest standard value of R_C required to yield a V_{CB} of approximately 6 V.

Troubleshooting and Failure Modes

Some of the common failure modes demonstrated by a small-signal BJT include a short between its collector and emitter, an open base, an open emitter, a degraded h_{FE}, and an excessive leakage current (i.e., I_{CBO}).

4-44. Determine the normal emitter-to-ground voltage of the silicon *npn* BJT given in Fig. 4-36(a). What would the voltage be if the emitter failed open?

4-45. Repeat Prob. 4-44 for the silicon *pnp* BJT given in Fig. 4-36(c).

4-46. The *pnp* BJT in Fig. 4-36(c) is excessively leaky. Specifically, its I_{CBO} has increased dramatically. Would this have an effect on V_{CB}? Will V_{CB} be too large, or too low? Explain.

4-47. A 510-Ω resistor was incorrectly installed for R_E in Fig. 4-36(a). Voltage measurements indicate a V_{EB} of -0.7 V, a V_{CE} of 0.2 V, and a V_{CB} of -0.5 V. Why is V_{CB} negative? What is the condition of the transistor - faulty, saturated, or cutoff?

Computer

4-48. Write a BASIC computer program that computes I_{CBO} as a function of temperature. It should prompt the user for the following: I_{CBO} at 25°C, the minimum temperature in °C, the maximum temperature in °C, and the step size (e.g., 1°C, 5°C, 10°C). The output should be tabular.

4-49. Write a program that will perform a dc analysis of a common-base amplifier [see Fig. 4-24(a)]. It should prompt the user for the following: V_{CC}, V_{EE}, R_E, R_C, α_{DC}, and the BJT type (silicon or germanium). The output should consist of V_{EB}, V_{CB}, I_E, I_C, $I_{C(SAT)}$, and $V_{CB(OFF)}$.

5

FIELD-EFFECT TRANSISTORS

After Studying Chapter 5, You Should Be Able to:

- Describe the basic structure and operation of the JFET.
- Name and define the CS, CD, and CG FET configurations, and explain their equivalence to the three BJT configurations.
- Generate and interpret FET V–I characteristic curves.
- Use the FET transfer equation.
- Explain temperature effects on an FET's V_p, I_{DSS}, and I_{GSS}.
- Describe the operation and characteristics of the DE- and E-MOSFET devices.
- Use the load line and transfer curve to find an FET's Q-point.
- Explain the FET signal process.
- Draw the hybrid-pi FET model, determine its g_m, r_o, and r_π, and relate them to the BJT.
- Describe the basic structure and operation of CMOS.

5-1 Role of the FET in Electronics

In Chapter 4 we introduced the BJT as a three-terminal active device. In this chapter we consider another class of three-terminal active devices called *field-effect transistors*.

The first commercially available field-effect transistors (FETs) were described as *junction* FETs (JFETs). The JFETs (first proposed by Shockley in 1952) were introduced after the BJTs. After the JFETs, another very similar family of field-effect transistors were developed—the *insulated-gate* FETs (IGFETs), which are more widely referred to as *metal-oxide-semiconductor* FETs (MOSFETs).

Discrete BJTs are presently being used in concert with linear integrated circuits in most new electronic designs. As we shall see, discrete FETs are also used in electronic designs when the required performance specifications so dictate. However, the primary thrust of the FET has been to enhance the design of linear, and particularly, digital integrated circuits. For example, the modern microprocessor simply could not have evolved without MOSFET technology.

Further, FET devices are continually arising which address new areas of application. Extremely high frequency (microwave) amplifiers are being designed around the *metal-semiconductor* FET (MESFET), and *gallium arsenide*–based FETs (GASFETs). High-power applications are being challenged by *vertical* channel MOSFETs (VMOS) and *insulated-gate transistors* (IGTs), which is a unique device that incorporates a MOSFET input and a BJT output.

The use of MOSFET technology is growing on almost a daily basis. In fact, it has been speculated that we are leaving the era of the bipolar transistor and entering the MOS technology age. At any rate, it is clear that we need to be well versed in the FET to master this evolving technology successfully. We begin our pursuit by considering the JFET.

5-2 Basic JFET Structure

In Fig. 5-1(a) we see the cross section of an *n-channel* JFET. Note that the JFET structure has two n^+ regions formed by gaseous diffusion: the *source* and the *drain*. Also observe that an n^- *channel* is doped in to connect the source and drain regions together. A *p*-type region is doped in above the channel, and the p^- substrate exists below the channel.

A two dimensional pictorial of the *n*-channel JFET has been provided in Fig. 5-1(b). The *p*-regions are connected together electrically. Its schematic symbol is provided in Fig. 5-1(c). The vertical bar represents the channel and *the arrow on*

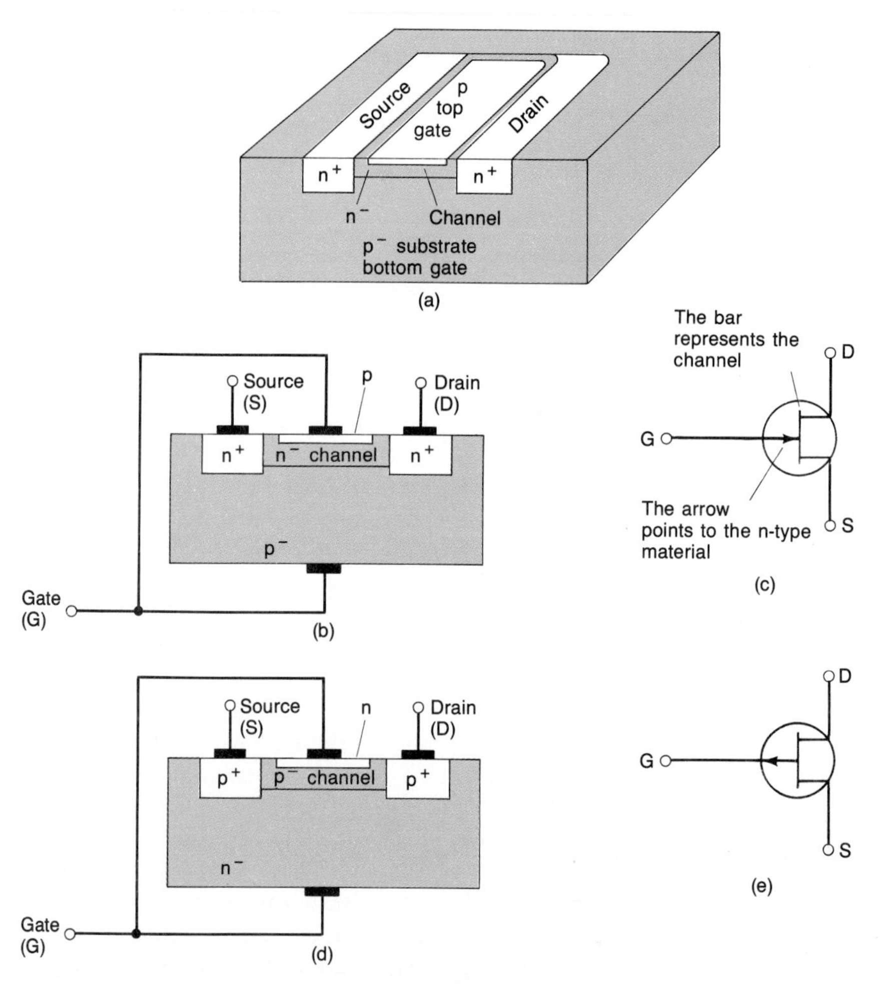

FIGURE 5-1 N- and p-channel JFETs: (a) diffused-channel, diffused-gate n-channel JFET; (b) two-dimensional pictorial of an n-channel JFET; (c) n-channel JFET symbol; (d) two-dimensional pictorial of an p-channel JFET; (e) p-channel JFET symbol.

schematic symbols points to the n-type material. Its complement is the *p-channel* JFET, whose cross section and schematic symbol are given in Fig. 5-1(d), and (e), respectively.

Note that the three JFET terminals are the *drain*, *source*, and *gate*.

The drain terminal is very similar to the collector terminal of the BJT. The gate is roughly analogous to the BJT's base terminal, and the source terminal is similar to the emitter terminal of the BJT.

These similarities will become apparent as our work progresses [see Fig. 5-2(a)].

We shall also see that the *n*-channel JFET is similar to the *npn* BJT and that the *p*-channel JFET has much in common with the *pnp* BJT. This is also emphasized in Fig. 5–2.

Just as the *npn* BJT is generally preferred because it has better high-frequency

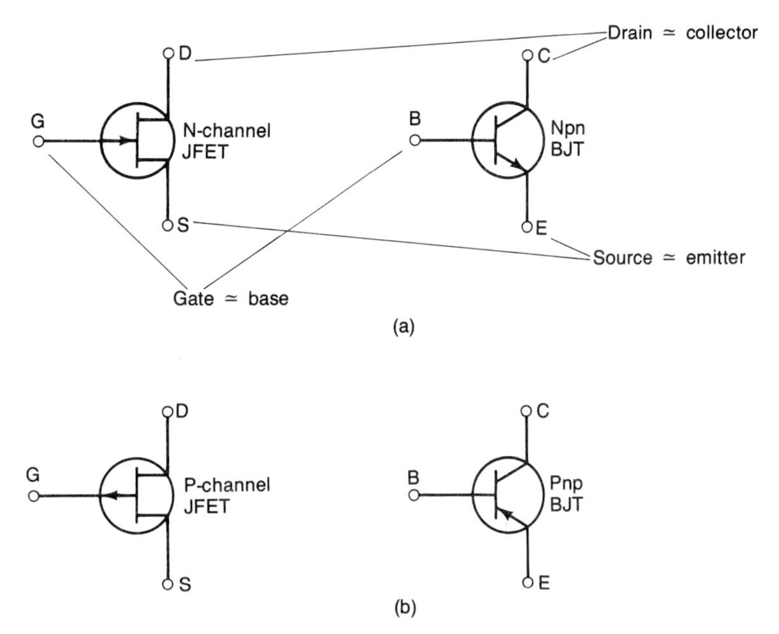

FIGURE 5-2 JFETs are similar to BJTs.

performance than the *pnp*, the *n*-channel JFET is favored over the *p*-channel JFET for the same reason. Consequently, we shall focus our discussions on the *n*-channel JFET. The thrust of our explanations may be extended to the *p*-channel device.

5-3 Operation of the JFET

As with any *p*-*n* junction, a depletion region forms at the interface. This is diagrammed in Fig. 5-3(a). The depletion regions extend deeply into the lightly doped n^- channel and the p^- substrate. Subsequently, the depletion region penetrates only slightly into the heavily doped n^+ source and drain regions.

In Fig. 5-3(b) we see that the source and the gate terminals have been tied to ground. An adjustable voltage source has been connected between the drain and ground. Its value will determine the drain-to-source voltage V_{DS}. Since the gate terminal has been tied directly to ground, V_{GS} will be 0 V.

Since the n^+ drain is positive and the (*p*-type) gate regions are tied to ground, the *p*-*n* junctions in the drain region are *reverse biased*. The encircled drain region in Fig. 5-3(b) has been enlarged in Fig. 5-4(a).

For small to moderate values of V_{DS} the reverse bias causes the depletion regions to widen as shown in Fig. 5-4(a). With the biasing as shown in Fig. 5-3(b), electrons will enter the channel through the source terminal (I_S), flow through the channel, and leave the JFET via the drain terminal (I_D). This has been noted in Fig. 5-4(a).

The channel can be dimensioned as shown in Fig. 5-4(b). It has a width W, a length L, and a thickness T. The channel possesses a certain resistance R which can be expressed by the familiar resistance formula $R = \rho L/A$, where ρ is the resistivity

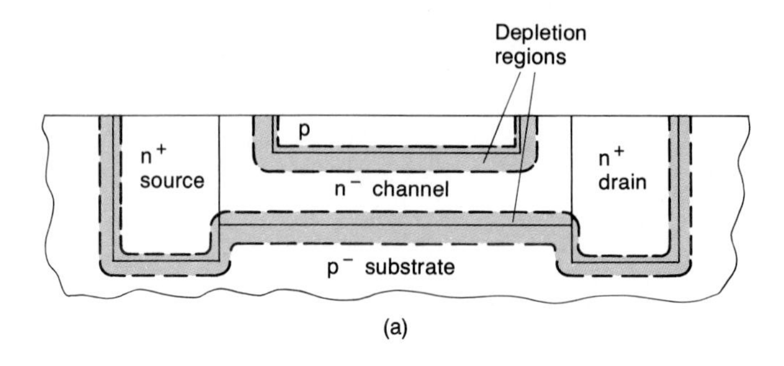

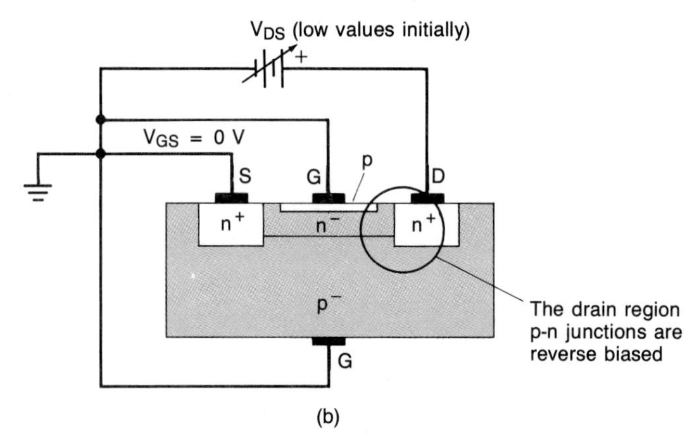

FIGURE 5-3 JFET depletion regions and the application of V_{DS}.

in ohm-meters, L is the length in meters, and A is the cross-sectional area in meters². Using our notation for the channel dimensions produces

$$R = \rho_c \frac{L}{WT} \tag{5-1}$$

The resistivity of the channel has been designated ρ_c. Its specific value is a function of the channel's doping level.

For low to moderate values of V_{DS} and the depletion regions indicated in Fig. 5-4(a), the channel's thickness T, and therefore its resistance, are essentially constant. Consequently, the JFET's V–I curve is linear, as indicated in Fig. 5-4(c). However, as V_{DS} is increased, the depletion regions widen and the V–I curve becomes nonlinear, as shown in Fig. 5-4(c). Specifically, the dynamic resistance of the channel begins to *increase*.

If V_{DS} is increased, the depletion regions widen to the point that they extend across the channel (see Fig. 5-5). Essentially, they have *pinched off* the channel. Appropriately, the value of V_{DS} that produces this condition is termed the *pinch-off voltage* V_P [refer again to Fig. 5-4(c)].

Recalling that the depletion region is devoid of majority charge carriers, it might appear that channel current should reduce to zero. However, this is *not* the case. High-energy conduction-band electrons that occur near the edge of the depletion region are swept across it and into the drain region to form the drain current. This is similar to the action in the collector–base region of the BJT.

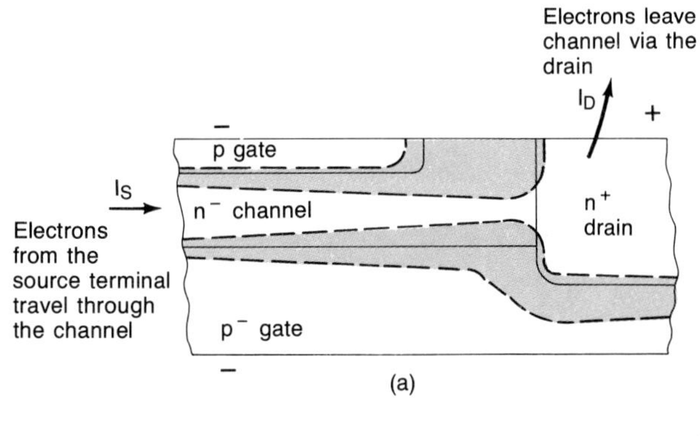

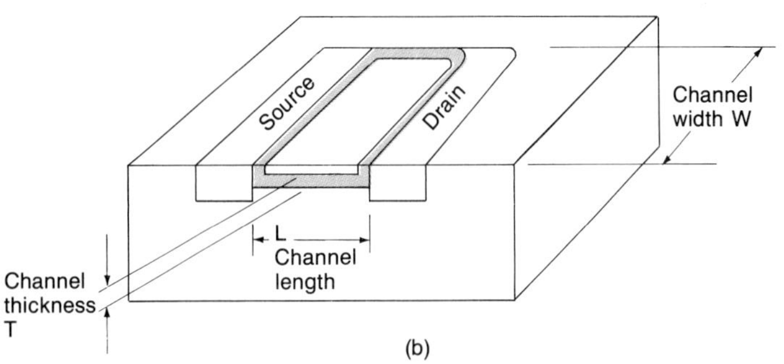

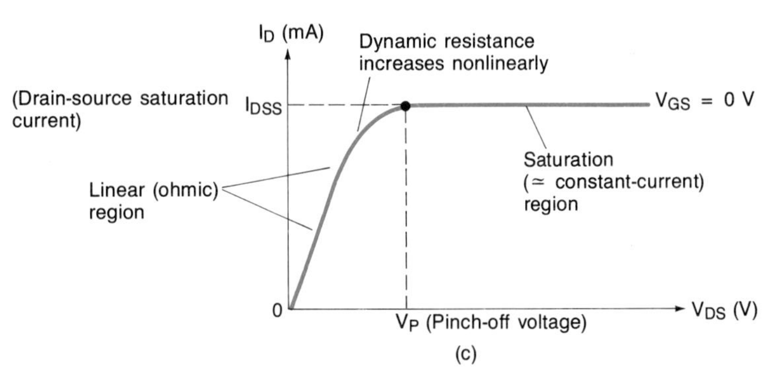

FIGURE 5-4 (a) JFET depletion regions; (b) channel dimensions; (c) V-I curve.

FIGURE 5-5 JFET in pinch-off.

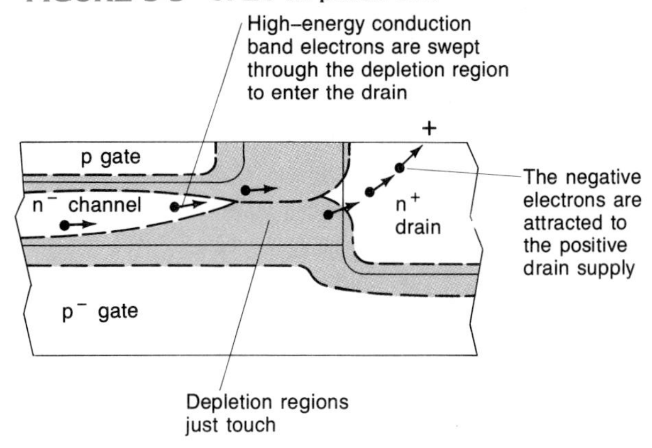

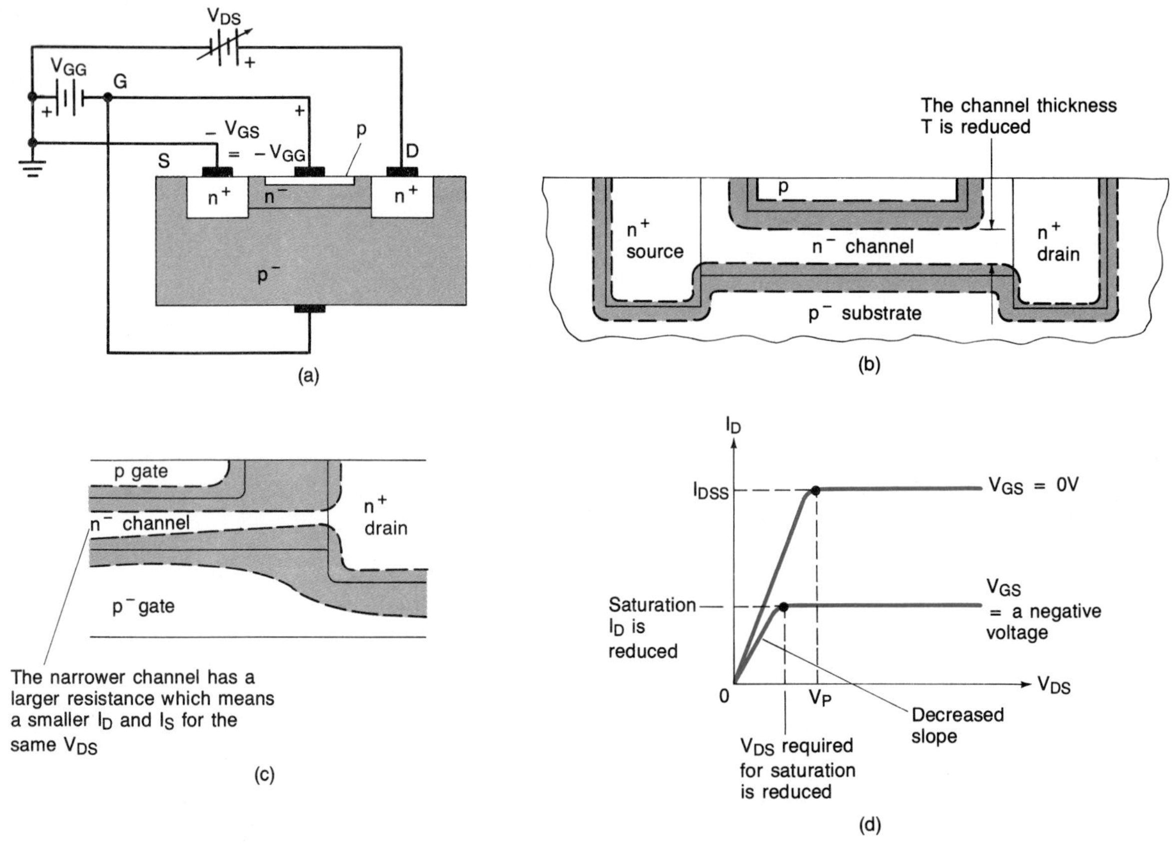

FIGURE 5-6 **Effects of a negative V_{GS}: (a) a negative V_{GS} is applied; (b) the negative gate voltage increases the depletion-region widths to reduce T (V_{DS} = 0 V); (c) the depletion regions with a negative V_{GS} and an applied V_{DS}; (d) V-I curves.**

In fact, when the JFET is in its pinch-off condition, the drain current I_D becomes essentially *constant*. When V_{GS} is zero (the gate is shorted to the source), the resulting drain current is called the *drain-to-source saturation current* I_{DSS}. I_{DSS} has been indicated in Fig. 5-4(c). Also note in Fig. 5-4(c) that two regions of operation have been indicated: the *linear* (or *ohmic*) *region* and the *saturation* (or *constant-current*) *region*. These will be more fully investigated as our work progresses.

In Fig. 5-6(a) we see that the gate-to-source voltage V_{GS} has been biased to a negative voltage. This also serves to reverse bias the gate-to-source *p-n* junction. The depletion regions along the channel will widen as illustrated in Fig. 5-6(b). Consequently, the thickness T of the channel will be reduced.

By inspection of Eq. 5-1 we can see that a *reduction* in T will *increase* the channel resistance R. Therefore, for low to moderate values of V_{DS}, the depletion regions will widen as indicated in Fig. 5-6(c). Compare Fig. 5-6(c) with 5-4(a). The increase in the channel resistance will produce the *V–I* curve with the slope shown in Fig. 5-6(d). The curve we developed when V_{GS} was zero volts has been included for reference.

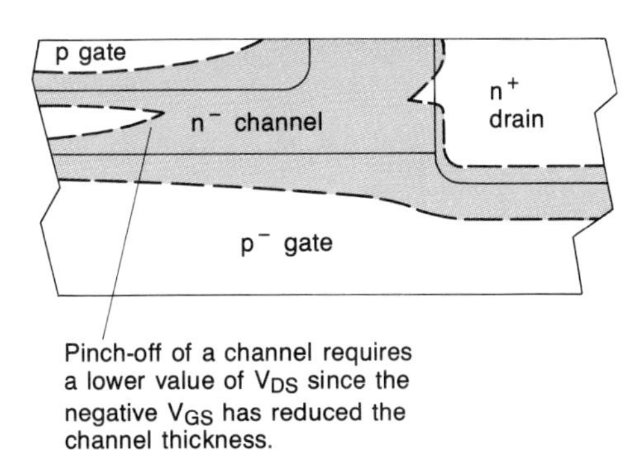

FIGURE 5-7 **JFET in pinchoff that has a negative gate-to-source bias.**

As V_{DS} is increased further, the channel will again become pinched off, as shown in Fig. 5-7, and the drain current will become constant. Contrast Figs. 5-5 and 5-7. The actual value of V_{DS} required to pinch off the channel is reduced from the value that was necessary when V_{GS} was zero volts. This occurs because of the reduced channel thickness produced by V_{GS}. This has been indicated in Figs. 5-6(d) and 5-7.

If we were to iterate this procedure, we could generate a family of *V–I* curves

FIGURE 5-8 **Typical common-source output V-I curves.**

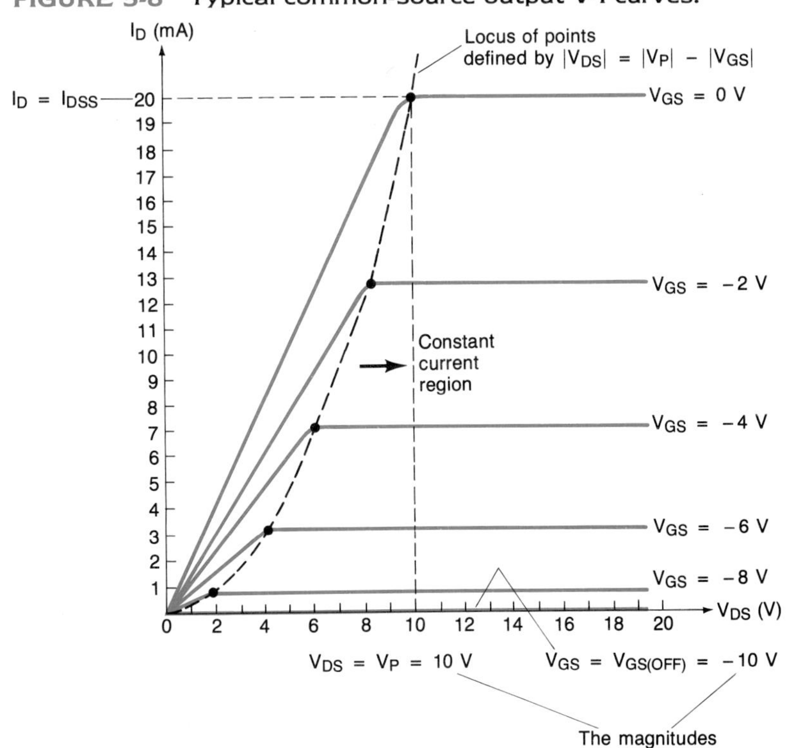

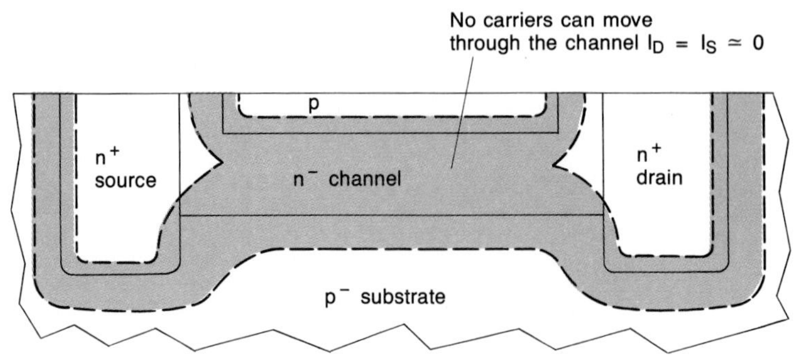

FIGURE 5-9 When $V_{GS} = V_{GS(OFF)}$ the JFET is in cutoff and no current can flow through the channel.

such as those given in Fig. 5-8. As we shall see in Section 5-4, these will be the common-source output characteristic *V–I* curves.

If V_{GS} were to be increased to a value termed $V_{GS(\text{OFF})}$, the depletion region widths will extend across the channel, as shown in Fig. 5-9. Carrier movement through the channel will be blocked and the drain and source currents will be reduced to approximately zero. The JFET is said to be *cut off*.

The magnitude of $V_{GS(\text{OFF})}$ will be equal to the magnitude of the pinch-off voltage (see Fig. 5-8). Hence

$$|V_{GS(\text{OFF})}| = |V_P| \tag{5-2}$$

Extending this concept, and by inspection of Fig. 5-8, we have

FIGURE 5-10 Simple comparison between the npn BJT and the n-channel JFET. (Conventional current directions have been indicated.)

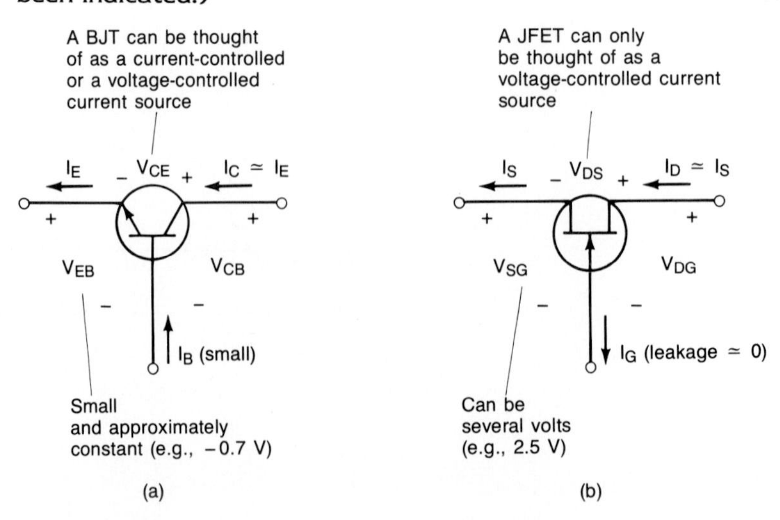

$$|V_{DS}| + |V_{GS}| = |V_P| \tag{5-3}$$

Let us reflect on the results of the discussion above. First, the JFET can be regarded as a *voltage-controlled current source*. Specifically, the magnitude of V_{GS} determines the magnitude of the drain current I_D (see Fig. 5-8).

Since V_{GS} reverse biases the gate-to-source *p-n* junction, the gate current I_G will be an extremely small reverse leakage current. Therefore, virtually all of the current that enters the source terminal will leave via the drain terminal. This is stated by

$$I_D \simeq I_S \tag{5-4}$$

This relationship should be memorized. To emphasize the similarities between the JFET and the BJT, we provide Fig. 5-10.

EXAMPLE 5-1

An *n*-channel JFET has an I_{DSS} of 20 mA and a $V_{GS(\text{OFF})}$ of -10 V. Assume that it is in pinch-off. What is its I_S when V_{GS} is zero volts? What value of V_{DS} is required to saturate the JFET if V_{GS} is -2 V?

SOLUTION When V_{GS} is equal to zero,

$$I_D = I_{DSS} = 20 \text{ mA}$$

and from Eq. 5-4,

$$I_S \simeq I_D = 20 \text{ mA}$$

The second question can be answered by employing Eqs. 5-2 and 5-3.

$$|V_P| = |V_{GS(\text{OFF})}| = 10 \text{ V}$$

and

$$|V_{DS}| = |V_P| - |V_{GS}| = 10 \text{ V} - 2 \text{ V} = 8 \text{ V}$$

■

5-4 FET Configurations and *V–I* Curves

Just as we saw that the BJT has three basic configurations, the same is true for the FETs (JFETs and MOSFETs). The *common-gate*, *common-source*, and *common-drain* configurations have been given in Fig. 5-11.

As with the BJT, the surest method for ascertaining the particular configuration is to determine the FET's signal input and output terminals with respect to ground. To assist us in our efforts, we have developed Table 5-1.

The three FET configurations have characteristics that are similar to the three corresponding BJT configurations. Therefore, we have indicated the BJT–FET correspondence in Table 5-2. (The reader should carefully study Tables 5-1 and 5-2 and compare them with Table 4-1. This will assist us in our later work.)

In Section 4-7 we introduced the transistor convention. This convention is applied to the FETs in exactly the same manner as it was to the BJTs. Further, we can also develop input, transfer, and output *V–I* characteristic curves for FETs. However,

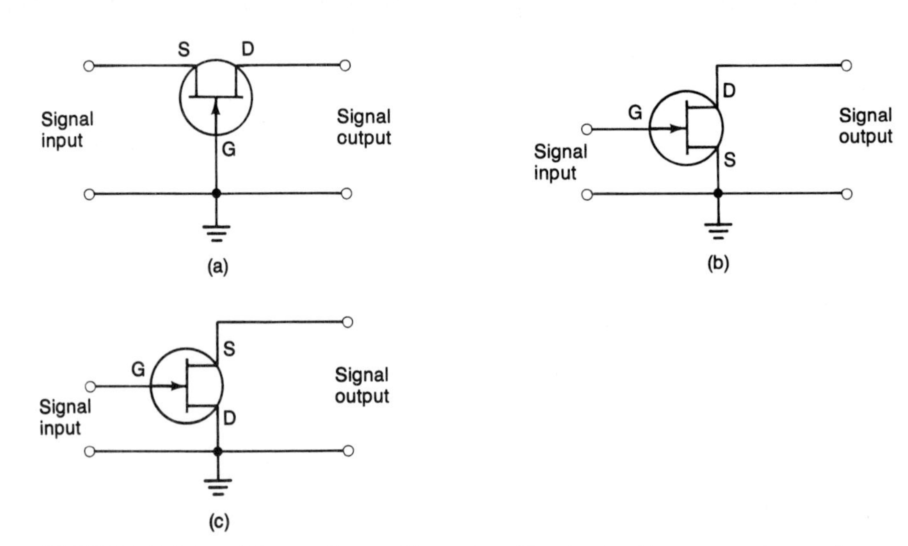

FIGURE 5-11 FET configurations (as illustrated by an n-channel JFET); (a) common gate; (b) common source; (c) common drain.

TABLE 5-1
FET Configurations

Configuration	Input Terminal	Output Terminal
Common gate	Source	Drain
Common source	Gate	Drain
Common drain	Gate	Source

TABLE 5-2
FET–BJT Correspondence

FET Configuration	BJT Configuration
Common gate	Common base
Common source	Common emitter
Common drain	Common collector

because the gate to source consists of a reverse-biased *p-n* junction, the gate current is approximately zero and its *V–I* characteristic is relatively uninteresting.

Subsequently, the most commonly developed *V–I* curves are the common-source output curves and the common-source transfer characteristic curves. The test circuit, typical output curves, and the corresponding transfer curve have been illustrated in Fig. 5-12.

5-5 FET Transfer Equation

In Section 4-14 we introduced Eq. 4-31, which provided an approximate mathematical description of the BJT transfer curve. Equation 5-5 is the equation for the transfer curve of the JFET.

$$I_D = I_{DSS}\left[1 - \frac{V_{GS}}{V_{GS(\mathrm{OFF})}}\right]^2 \tag{5-5}$$

The terms of Eq. 5-5 have all been defined previously. The only constraint on the equation is that the FET must be in its saturation region of operation. This is suggested

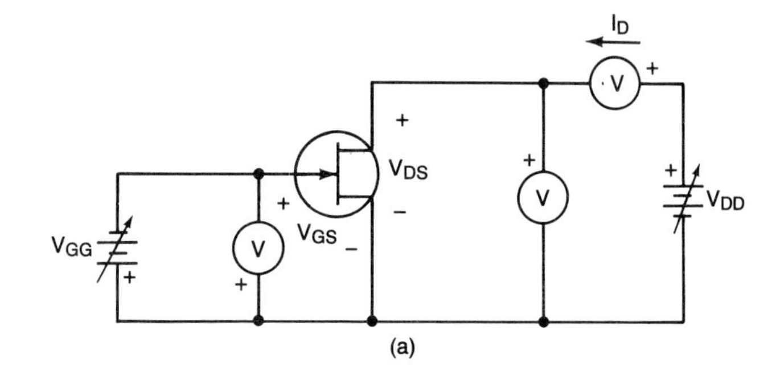

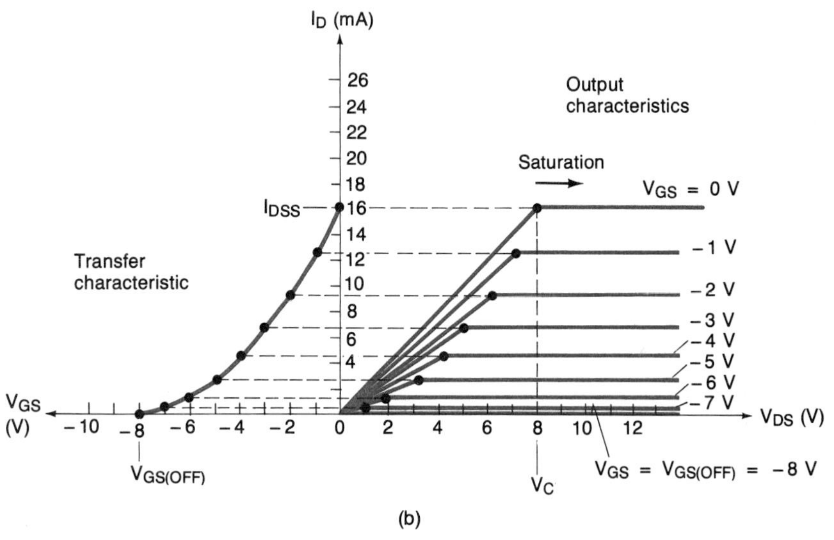

FIGURE 5-12 Common-source V-I curves: (a) test circuit (no current-limiting resistors are required); (b) V-I curves.

in Fig. 5-12(b) by the dashed lines that project from the output characteristic curves to the transfer characteristic curve. This poses no serious limitations since this is the normal region of operation for small-signal voltage amplifiers.

EXAMPLE 5-2

An n-channel JFET has a $V_{GS(\mathrm{OFF})}$ of -6 V and an I_{DSS} of 12 mA. Make a sketch of its transfer characteristic curve.

SOLUTION Using the given parameters and Eq. 5-5, we have

$$I_D = I_{DSS}\left[1 - \frac{V_{GS}}{V_{GS(\mathrm{OFF})}}\right]^2$$

$$= 12\ \mathrm{mA}\left(1 + \frac{V_{GS}}{6\ \mathrm{V}}\right)^2$$

We may obtain V–I coordinates by selecting values for V_{GS} between zero and $V_{GS(OFF)}$ volts, and substituting them into Eq. 5-5. For a V_{GS} of 0 V it is clear that I_D will be equal to I_{DSS} (12 mA). However, for a V_{GS} of -1 V,

$$I_D = 12 \text{ mA} \left(1 + \frac{-1 \text{ V}}{6 \text{ V}}\right)^2$$

$$= 8.33 \text{ mA}$$

Additional values may be computed in a similar fashion. Several of these have been summarized in Table 5-3, and they have been plotted in Fig. 5-13. ■

TABLE 5-3
Transfer Curve Values for Example 5-2

V_{GS} (V)	I_D (mA)
0	12
−1	8.33
−2	5.33
−3	3.00
−4	1.33
−5	0.33
−6	0

FIGURE 5-13 Transfer curve developed in Example 5-2.

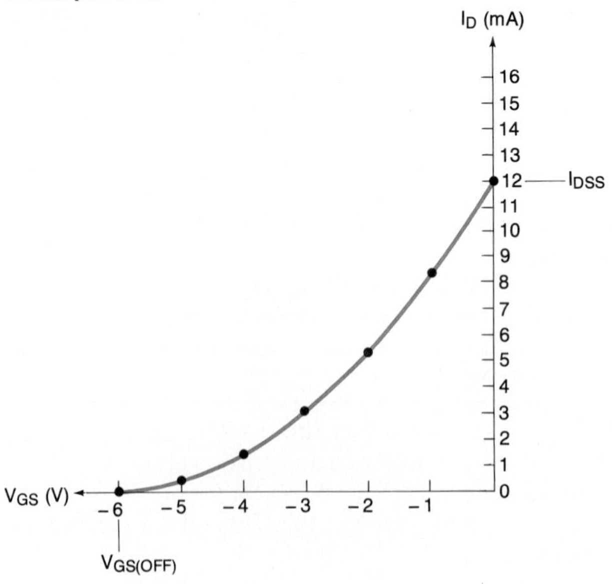

The reader should verify the values given in Table 5-3 and Fig. 5-13. We will be drawing upon Eq. 5-5 and transfer characteristic curves extensively in our work to come.

5-6 JFET Temperature Effects

The JFET, like all solid-state devices, is subject to the effects of temperature. The JFET (and BJT) parametric variations with temperature can cause us great concern when the Q-point must not drift excessively with temperature. (This is the usual case.) Fortunately, we can minimize the JFETs (and the BJTs) Q-point variations through the careful design of the bias circuit. To understand how the temperature sensitivity of JFETs can be minimized, we need to take a slightly closer look at its physical nature.

The JFET employs the electric field associated with a reverse-biased *p-n* junction to control the resistance (or conductivity) of a channel of semiconductor material. (Hence the term *field-effect* transistor is used.) Consequently, *no p-n* junctions are crossed by the carriers. Further, only majority carriers–either holes *or* electrons are involved in its operation. Therefore, the FET, like the Schottky diode, may be described as a *unipolar* device.

There are two primary mechanisms underlying the temperature sensitivity of the channel conductivity of a JFET:

1. Temperature *increases* cause a *decrease* in the depletion region width at the channel-gate *p-n* junction. This causes the channel thickness (T) to *increase*.
2. Carrier mobility decreases as temperature increases.

The fact that the channel thickness increases as the temperature increases causes I_D to increase. Another way of viewing the situation is to note that $V_{GS(\text{OFF})}$ increases in magnitude with temperature. Therefore, for a fixed value of V_{GS}, I_D will increase [see Fig. 5-14(a)]. As a rule of thumb, $V_{GS(\text{OFF})}$ *increases in magnitude with temperature. Specifically,* $V_{GS(\text{OFF})}$ $[V_p]$ *has a positive tempco of about* 2.2 *mV/°C.*

The second factor (carrier mobility decreases with increases in temperature) causes the channel conductivity to decrease with temperature increases. Recall that atoms in a solid tend to vibrate and that the amplitude of their vibrations is directly proportional to temperature. Therefore, the probability of collisions between the charge carriers and the atoms in the crystalline structure is increased. The net result is that the resistivity of the silicon has a *positive temperature coefficient*. Consequently, the drain current is *reduced* (assuming a constant V_{DS}) as the temperature is *increased*. The net result is that I_{DSS} *tends to decrease as the temperature increases* [see Fig. 5-14(b)].

Let us reflect on this. We have two distinct mechanisms affecting I_D as a function of temperature. Carrier mobility gives I_D a negative temperature coefficient, and the effect of $V_{GS(\text{OFF})}$ is to give I_D a positive temperature coefficient. Since both of these mechanisms occur simultaneously, it is possible to bias the JFET such that it will exhibit a *zero temperature coefficient*. In other words, I_D will *not* vary with temperature.

The value of V_{GS} that will give the I_D of both n- and p-channel JFETs a zero temperature coefficient is given by

$$|V_{GS}| \simeq |V_p| - 0.63 \text{ V} \tag{5-6}$$

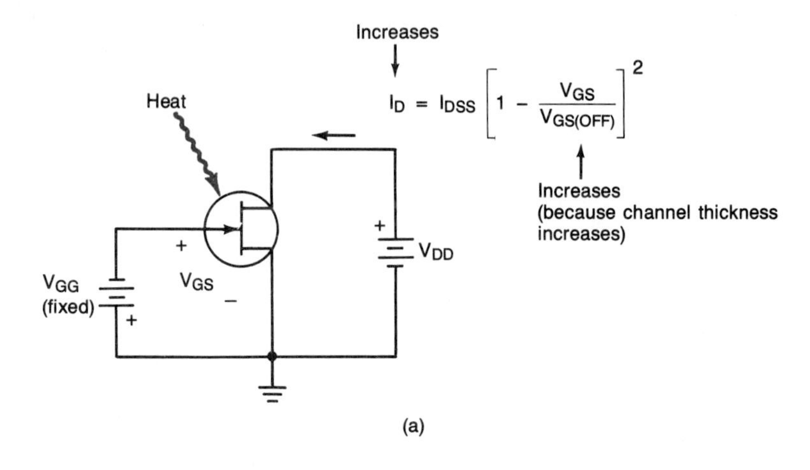

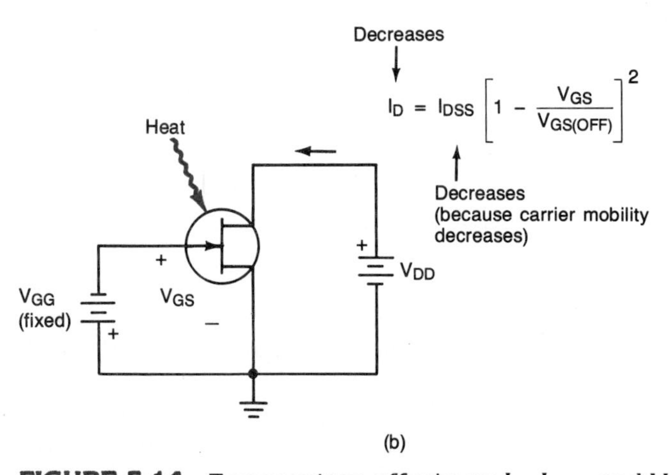

FIGURE 5-14 Temperature effects on I_D, I_{DSS}, and $V_{GS(OFF)}$.

where V_{GS} is the gate-to-source bias voltage to give I_D a zero temperature coefficient and V_p is the pinch-off voltage at 25°C.

In Fig. 5-15(a) we see the effect of temperature on the drain family of *V–I* curves. Note that the particular curve that complies with Eq. 5-6 does *not* change with temperature. Further, the *V–I* curves above it exhibit a negative temperature coefficient, while those below it exhibit a positive temperature coefficient.

On some data sheets manufacturers will provide additional curves such as those indicated in Fig. 5-15(b) and (c). The reader should verify the temperature coefficient of $V_{GS(\text{OFF})}$, V_p, and Eq. 5-6.

A third factor that affects the temperature sensitivity of JFETs is the minority-carrier generation within the gate–channel depletion region. This produces *gate leakage current*. This parameter is designated as I_{GSS} on FET data sheets. I_{GSS} stands for the gate-to-source leakage current with the drain shorted to the source (see Fig. 5-16).

I_{GSS} is simply the current through the reverse-biased *p-n* junction between the gate and source terminals. Just as we saw in Section 2-7, *the leakage current* (e.g., I_{GSS})

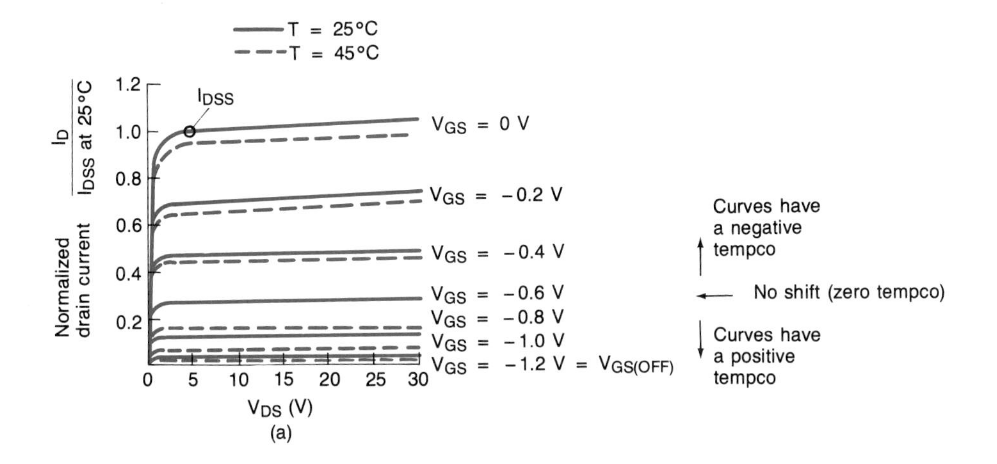

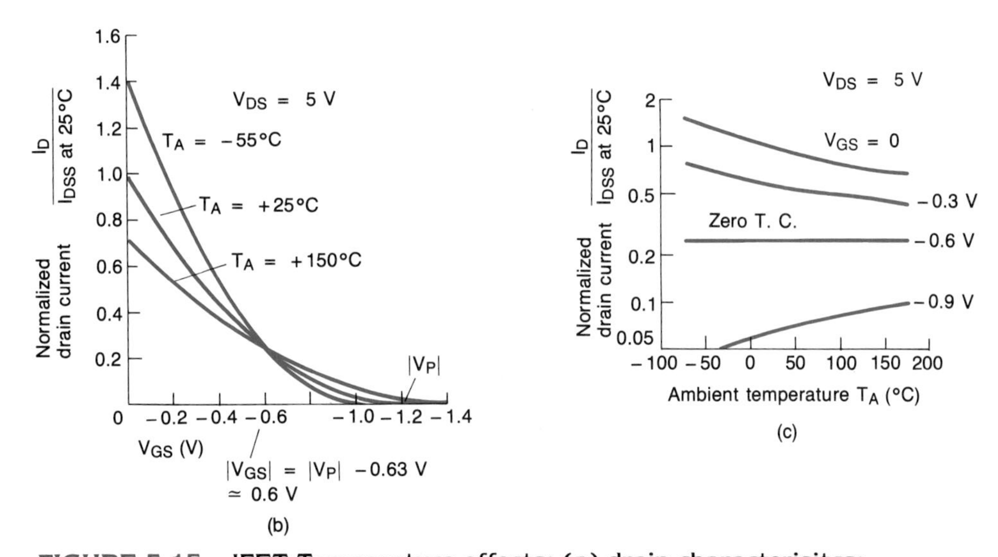

FIGURE 5-15 JFET Temperature effects: (a) drain characterisitcs; (b) transfer characteristics; (c) temperature coefficients with V_{GS} as a parameter.

FIGURE 5-16 Test circuit to determine I_{GSS}.

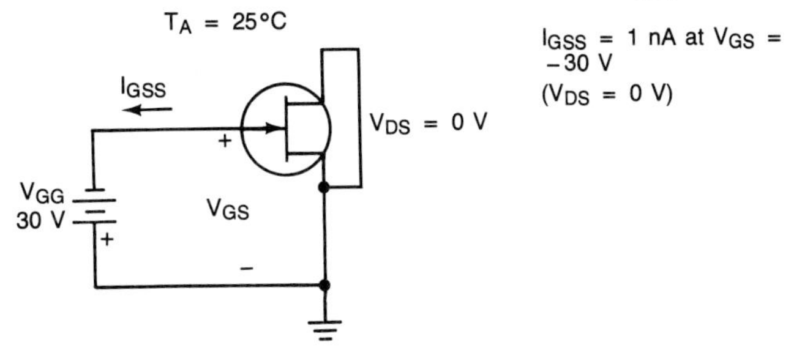

through a reverse-biased p-n junction approximately doubles for each 10°C *rise in temperature*. Therefore, we may use Eq. 2-1 to predict I_{GSS} at any given temperature.

EXAMPLE 5-3

An n-channel JFET has a pinch-off voltage of 3 V, an I_{DSS} of 20 mA, and an I_{GSS} of 5 nA at 25°C. What value of V_{GS} will give the JFET a zero tempco? What will the resulting drain current be at 50°C? What will its I_{GSS} be at that temperature?

SOLUTION We may use Eq. 5-6 to find the required value of V_{GS}.

$$|V_{GS}| = |V_p| - 0.63\text{ V} = 3\text{ V} - 0.63\text{ V} = 2.37\text{ V}$$

Hence V_{GS} should be -2.37 V. We may use Eq. 5-5 to find I_D at 50°C since the JFET will have a zero temperature coefficient.

$$I_D = I_{DSS}\left[1 - \frac{V_{GS}}{V_{GS(\text{OFF})}}\right]^2$$

$$= 20\text{ mA}\left(1 - \frac{2.37\text{ V}}{3\text{ V}}\right)^2 = 0.882\text{ mA}$$

To find I_{GSS}, we may use Eq. 2-1.

$$I_{GSS}(T) = [I_{GSS}\,(25°\text{C})][2^{(T-25)/10}]$$

$$= [5\text{ nA}][2^{(50-25)/10}] = 28.3\text{ nA}$$

■

5-7 Depletion–Enhancement Mode MOSFETs

Thus far we have seen that applying a negative V_{GS} to an n-channel JFET increases the depletion-region widths. The increase in the depletion regions reduces the thickness of the channel, which raises its resistance. The end result was a reduction in I_D.

If we were to reverse the polarity of V_{GG} to apply a *positive* V_{GS}, the p-n junctions between the gate and the channel would then be forward biased. Recalling that a forward bias decreases the width of a depletion region, we can see that the channel thickness would *increase* and the corresponding channel resistance would then *decrease*. Consequently, I_D would *increase* beyond the JFET's I_{DSS} value.

The normal operation of a JFET is in its depletion mode of operation. Some manufacturers classify this as *type A* operation. However, as we have just seen, it is also possible to *enhance* the conductivity of the JFET's channel.

For reasons that will become apparent later, we do *not* want the gate current to rise to an appreciable value. Therefore, the forward bias of the silicon p-n junction is usually restricted to a maximum of 0.5 V. A more conservative 0.2-V limit is indicated in Fig. 5-17.

As we shall see in Section 5-13, the *greater* I_D is compared to I_{DSS}, the *greater* the transconductance g_m will be. In Chapter 8 we shall see that the voltage gain is directly proportional to g_m. So, in general, the higher the g_m, the better, and we see one of the advantages of being able to enhance the channel.

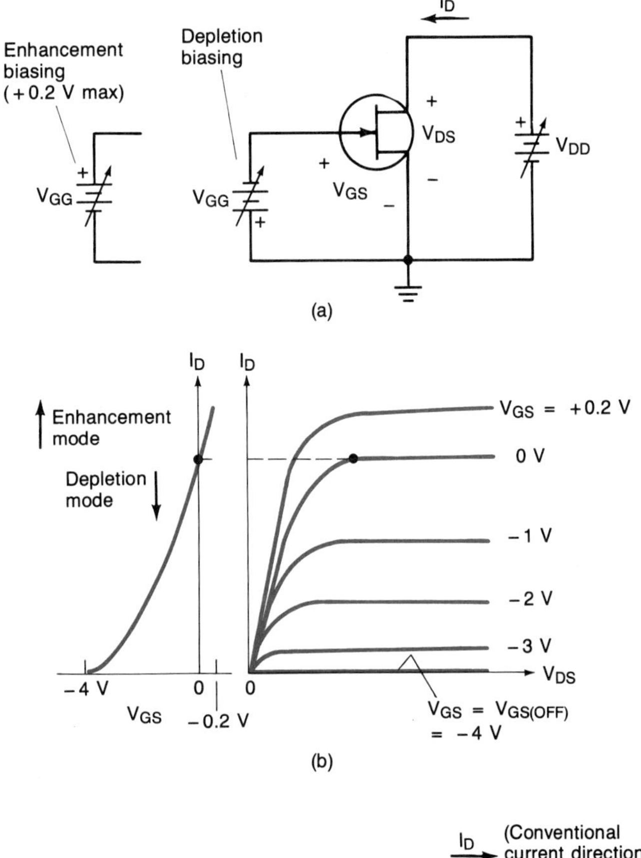

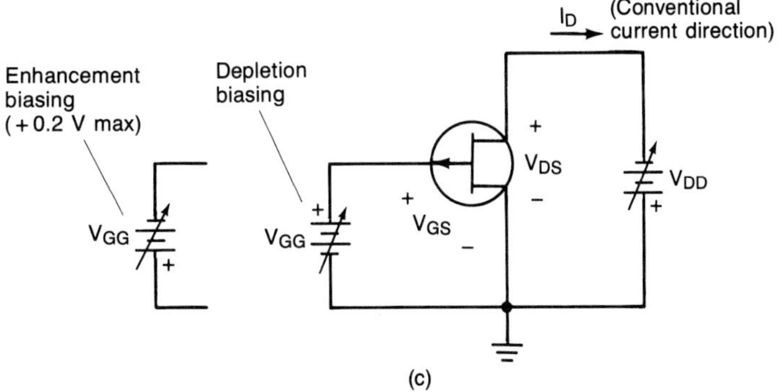

FIGURE 5-17 JFET depletion-enhancement operation: (a) n-channel JFET biasing; (b) n-channel transfer and output V-I characteristics; (c) p-channel JFET biasing.

In Fig. 5-17(a) we see the depletion–enhancement biasing of an *n*-channel JFET. The reader should contrast the required biasing of the *p*-channel JFET shown in Fig. 5-17(c). Also study the transfer and output *V–I* curves given in Fig. 5-17(b).

As its name suggests, the *depletion-enhancement MOSFET* (DE-MOSFET) was developed to be used in either or both the depletion and enhancement modes. Some

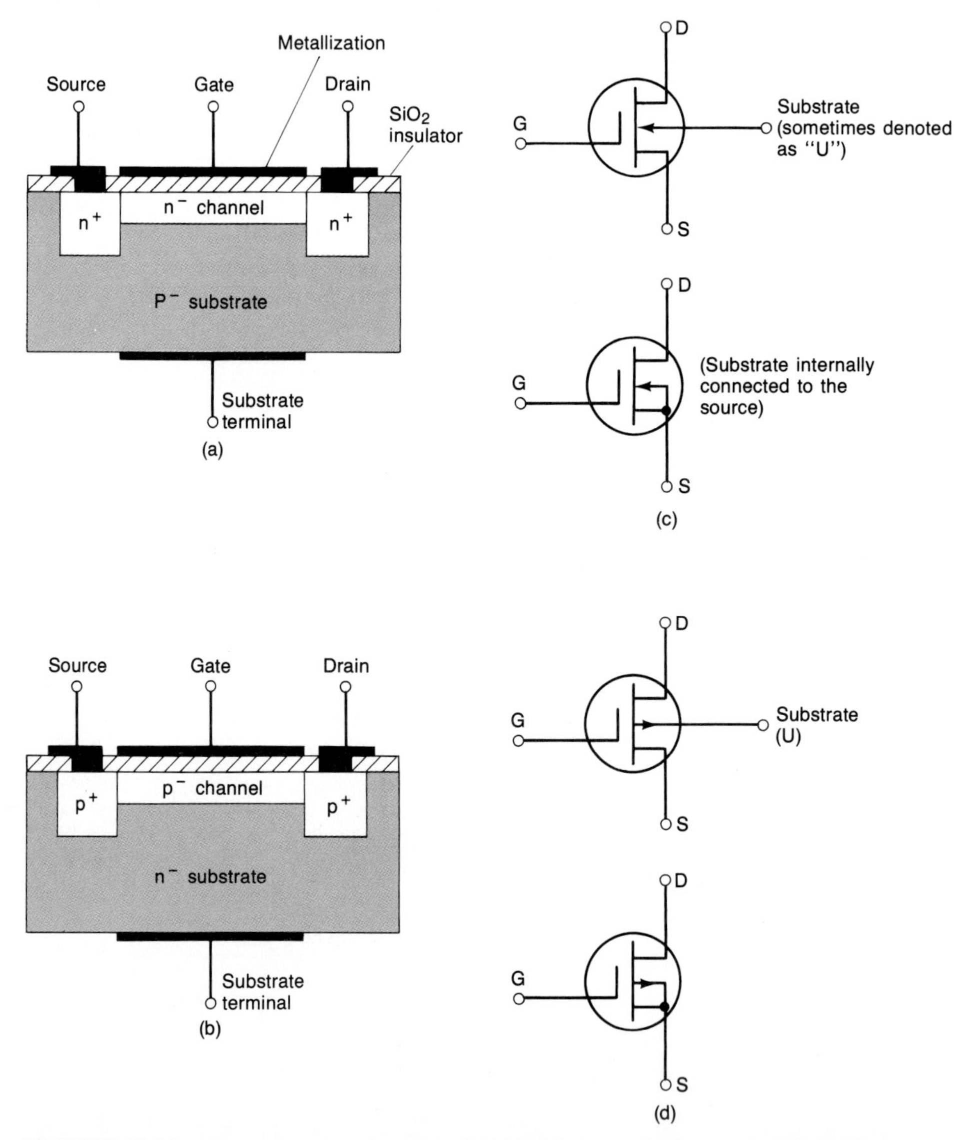

FIGURE 5-18 N- and p-channel DE-MOSFETs: (a) n-channel DE-MOSFET; (b) n-channel DE-MOSFET schematic symbols; (c) p-channel DE-MOSFETs; (d) p-channel DE-MOSFET schematic symbols.

manufacturers classify this as *type B* FET operation. The *n*-channel DE-MOSFET has been illustrated in Fig. 5-18(a), and its *p*-channel complement in Fig. 5-18(c). Their schematic symbols have been indicated in Fig. 5-18(b) and (d), respectively.

Note that some DE-MOSFETs have *four* terminals: the drain, gate, source, and *substrate*. In most applications, the substrate is merely tied to the source terminal. However, it may be used as another gate input terminal, or in the case of an *n*-channel DE-MOSFET, it may be tied to a negative supply voltage.

The internal action of an n-channel DE-MOSFET is described in Fig. 5-19. In Fig. 5-19(a) we see that V_{GS} is negative. Note that *no p-n* junction exists between the gate and the channel. The gate is electrically insulated by a very thin SiO_2 layer. The basic structure is very similar to a parallel-plate capacitor. The negative gate bias repels the negative electrons in the n^- channel. As a result, a depletion region forms to narrow the channel, reducing its conductivity, and lowering I_D. The process is very similar to the physical action in the JFET.

If we make V_{GS} positive, additional electrons will be drawn into the channel from the heavily doped n^+ source and drain regions. The additional electrons will increase the conductivity of the channel. The decrease in the channel's resistance will cause I_D to increase [see Fig. 5-19(b)].

FIGURE 5-19 **Operation of a DE-MOSFET: (a) depletion mode; (b) enhancement mode; (c) n-channel DE-MOSFET V-I curves; (d) p-channel DE-MOSFET V-I curves.**

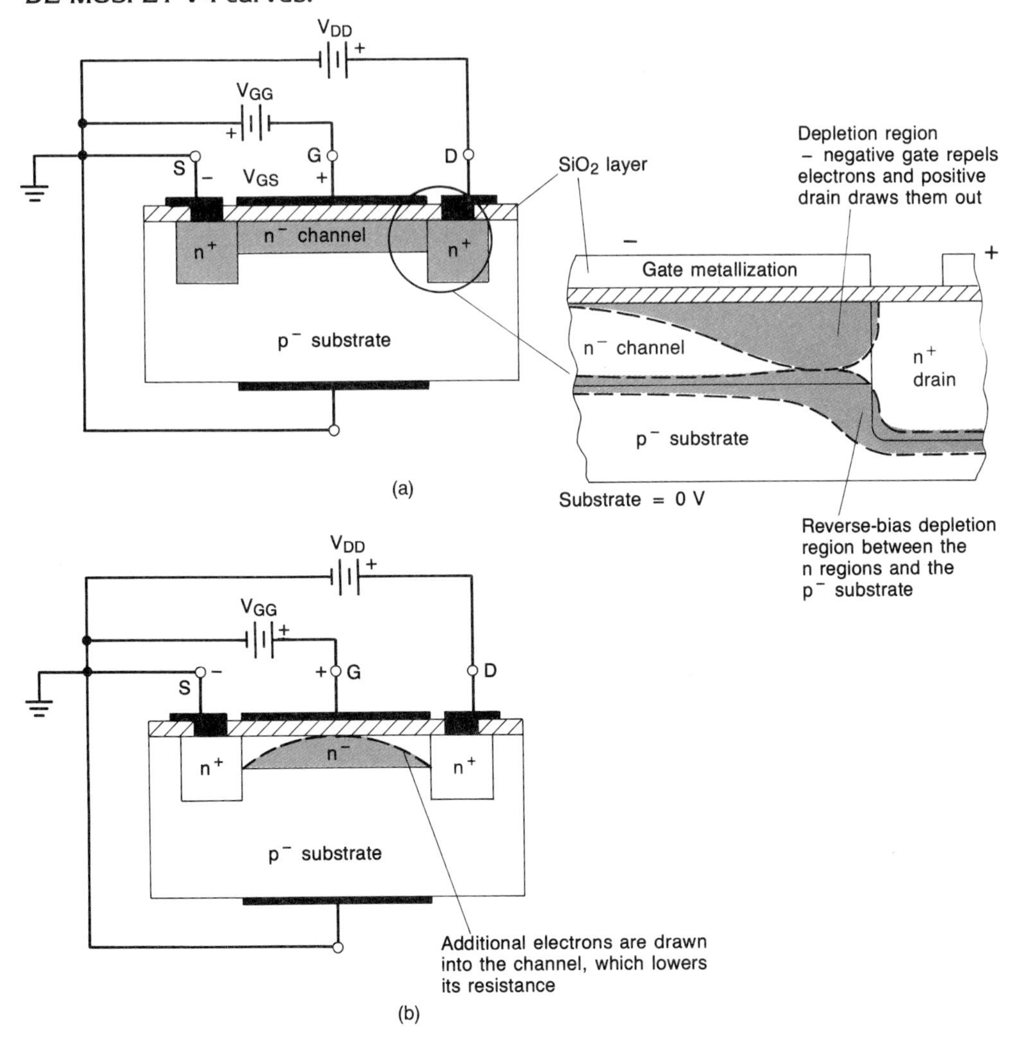

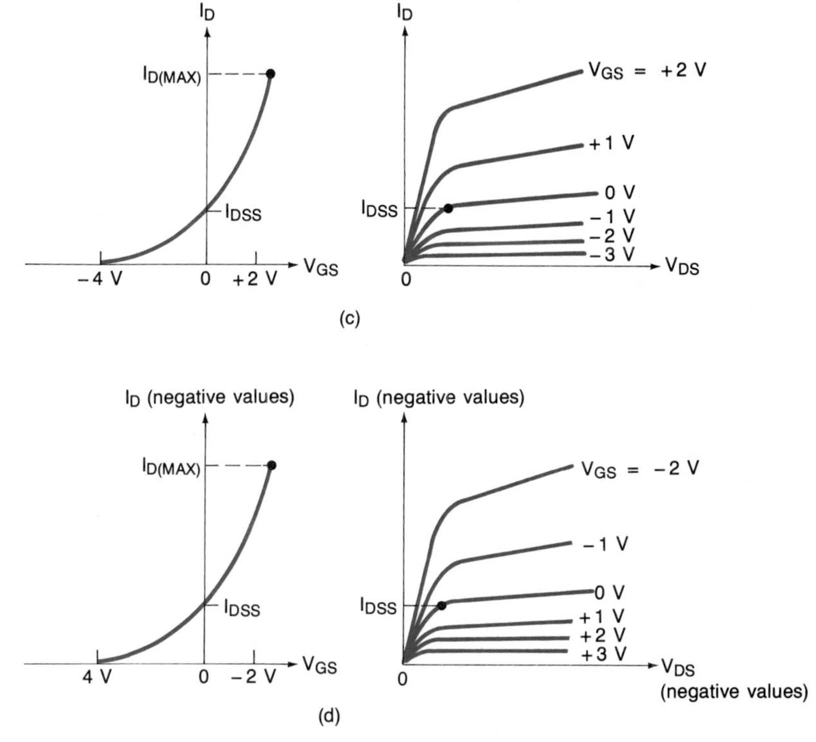

FIGURE 5-19 *(continued)*

Since we do not have to worry about turning on a *p-n* junction, V_{GS} can be much larger than the +0.5-V limitation we had for the *n*-channel JFET. However, we do have a maximum limit on $I_D - I_{D(\mathrm{MAX})}$. Refer to the *V–I* curves given in Fig. 5-19(c). The typical curves for a *p*-channel, DE-MOSFET are given in Fig. 5-19(d). The DE-MOSFET transfer curves may also be described by Eq. 5-5.

As a final comment we should note that the gate leakage current I_{GSS} of MOSFETs is much smaller than that of JFETs. The leakage current for JFETs is usually on the order of nanoamperes, while the leakage current for MOSFETs is typically on the order of picoamperes. Because the gate-to-source impedance of MOSFETs is so very large, it is possible that a large static charge can build on their gate terminals. If it exceeds about 100 V, the SiO_2 layer may break down, destroying the MOSFET. Consequently, MOSFETs and MOSFET integrated circuits must be handled with caution. It is for this reason that most MOSFETs are shipped in either conductive foam carriers or with shorting rings around their leads to prevent inadvertent destruction due to static electricity.

5-8 Enhancement MOSFETs

Figure 5-20(a) illustrates an *n*-channel *enhancement-only* MOSFET (E-MOSFET). Some manufacturers classify it as a *type C* FET. Its *p*-channel complement is shown in Fig. 5-20(c). Their schematic symbols are in Fig. 5-20(b) and (d), respectively.

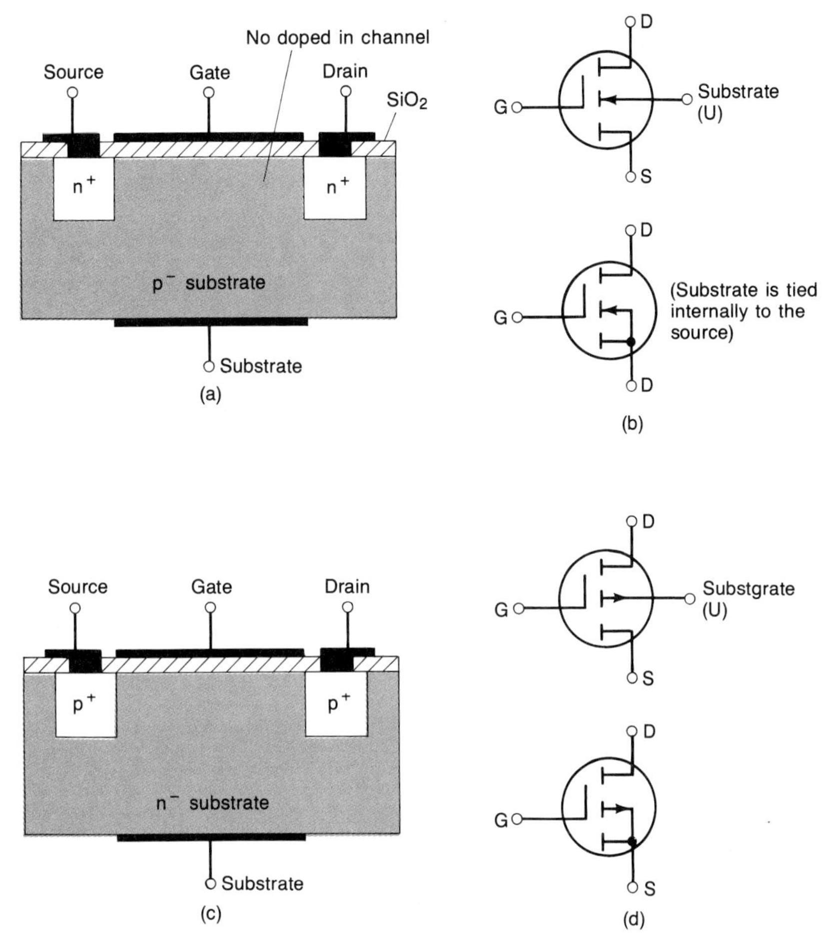

FIGURE 5-20 **Enhancement MOSFETs: (a) n-channel E-MOSFET; (b) n-channel schematic symbols; (c) p-channel E-MOSFET; (d) p-channel schematic symbols.**

Note that *no* channels are doped into these MOSFET structures. As we shall see, the channels are *electrically induced* into these MOSFETs. To help us differentiate between DE-MOSFETs and E-MOSFETs, the E-MOSFETs schematic symbols have dashed lines to represent the induced channels [refer to Fig. 5-20(b) and (d)].

When V_{GS} is zero volts, the E-MOSFET has an I_D of approximately zero. The reason why this is so can be seen in Fig. 5-21(a). Effectively, we have two back-to-back diode junctions in series with some equivalent resistance. With a short circuit between the gate and the source terminals, the drain *p-n* (diode) junctions are reverse biased. Consequently, only a very small reverse leakage current will flow between the drain and source terminals.

If the V_{GS} is made positive, an electric field is formed in the gate region. The positive gate serves to attract minority carriers (electrons) from the p^- substrate. When the gate becomes sufficiently positive, enough minority carriers exist along the surface of the substrate to form an *n*-type channel between the source and drain

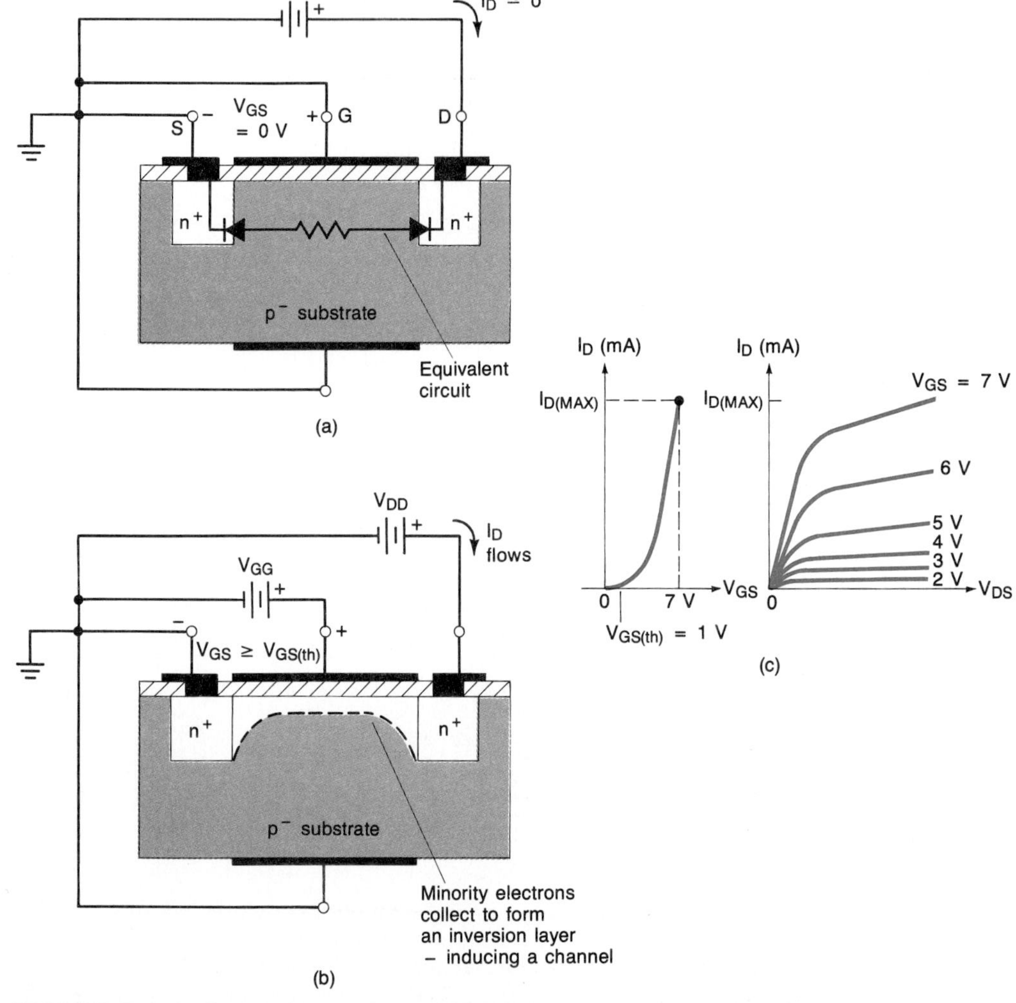

FIGURE 5-21 Operation of the E-MOSFET: (a) n-channel E-MOSFET with $V_{GS} = 0$ V; (b) n-channel E-MOSFET with $V_{GS} \geq V_{GS(th)}$; (c) V-I curves for an n-channel E-MOSFET.

regions. When this occurs, we have induced a channel by forming what is termed an *inversion layer* [see Fig. 5-21(b)].

The magnitude of V_{GS} required to form this inversion layer is termed the *gate-to-source threshold voltage* $V_{GS(\mathrm{th})}$. Manufacturers typically indicate $V_{GS(\mathrm{th})}$ on their E-MOSFET data sheets. As V_{GS} is increased beyond $V_{GS(\mathrm{th})}$, I_D will continue to increase. The *V–I* curves for an *n*-channel E-MOSFET are depicted in Fig. 5-21(c).

The equation for the transfer characteristic does *not* obey Eq. 5-5. However, it does follow a similar "square-law type" of relationship. The equation for the transfer characteristic of E-MOSFETs is given by

$$I_D = K\,[V_{GS} - V_{GS(\text{th})}]^2 \tag{5-7}$$

where I_D = drain current of an E-MOSFET
K = constant with units of amperes/volts^2
V_{GS} = gate-to-source bias voltage
$V_{GS(\text{th})}$ = gate-to-source threshold voltage

The constant K may be usually obtained *indirectly* from the manufacturer's data sheet. Consider Example 5-4.

EXAMPLE 5-4

An n-channel E-MOSFET has a $V_{GS(\text{th})}$ of 3.8 V and an I_D of 10 mA when V_{GS} is 8 V. Find its I_D when V_{GS} is 6 V.

SOLUTION First, we must use the given information and Eq. 5-7 to find K.

$$K = \frac{I_D}{[V_{GS} - V_{GS(\text{th})}]^2} = \frac{10\text{ mA}}{(8\text{ V} - 3.8\text{ V})^2}$$
$$= 5.67 \times 10^{-4}\text{ A/V}^2$$

Now that we have determined K, we may find I_D at the given V_{GS}.

$$I_D = K[V_{GS} - V_{GS(\text{th})}]^2$$
$$= (5.67 \times 10^{-4}\text{ A/V}^2)(6\text{ V} - 3.8\text{ V})^2$$
$$= 2.74\text{ mA}$$

■

5-9 FET Analysis in Perspective

In Section 4-10 we considered the alternatives to BJT circuit analysis. Most of our comments made in that section also apply, in general, to the FETs. However, there is one exception; most of our *dc analyses* will draw upon *graphical techniques*.

Admittedly, we have shunned, and will continue to shun, graphical techniques whenever possible. However, since the FETs are voltage-controlled current sources with a square-law relationship between their input voltages and their output currents, their biasing equations become unnecessarily complex with pure analytical methods. In the next section we introduce our graphical technique to determine the Q-point of a JFET voltage amplifier.

5-10 Determining the *Q*-Point

As we have seen, before an ac small-signal analysis can be performed, it is necessary to ascertain the Q-point. This was true for diodes and BJTs, and it is also true for FETs. The general procedure for FET amplifiers is very similar to the procedure we developed in Section 4-12 for the BJT. Let us review and expand on the procedure by contrasting the BJT with the FET.

1. Apply Kirchhoff's voltage law around the input circuit to determine the controlling input quantity. For BJTs, this generally means solving the resulting equation for either I_B or I_E. However, for FETs, we usually solve for V_{GS}.
2. Transfer across the device. For BJTs this is usually accomplished by using either β_{DC} or α_{DC}. However, with FETs, we shall use a graphical approach. Specifically, we shall plot the equation we wrote for V_{GS} in step 1 on the transfer characteristic curve for the particular FET.
3. Write a Kirchhoff's voltage law equation around the output circuit and solve for the output voltage. For BJTs this may be V_{CE} or V_{CB}. For FETs the output voltage may be V_{DS}.

   The general procedure for *any* three-terminal device is to analyze the input, transfer from the input to the output, and then analyze the output.

Now we focus our attention on the FET. Consider the JFET amplifier given in Fig. 5-22(a). Since the input and the output coupling capacitors are open circuits to dc, we arrive at the equivalent circuit shown in Fig. 5-22(b). Note that we have indicated the reverse leakage current I_{GSS}. Applying Kirchhoff's voltage law around the input circuit gives us

$$V_{GG} - I_{GSS} R_G + V_{GS} = 0$$

and solving for V_{GS}, we have

$$V_{GS} = -V_{GG} + I_{GSS} R_G$$

If we assume that I_{GSS} is small enough such that the voltage drop across R_G is negligible, then

$$V_{GS} \simeq -V_{GG} = -1 \text{ V}$$

This equation describes the *bias line* for the circuit. To transfer from the input circuit to the output, we merely plot the bias line on the FET's transfer characteristic curve. This is demonstrated in Fig. 5-22(c). The intersection between the bias line and the transfer *V–I* characteristic yields the drain current I_D. Hence

$$I_D = 12.25 \text{ mA}$$

We may complete our analysis by determining V_{DS}, the output voltage. By Kirchhoff's voltage law, and solving for V_{DS},

$$V_{DS} = V_{DD} - I_D R_D$$

Substituting in the values given in Fig. 5-22(a) yields

$$V_{DS} = 20 \text{ V} - (12.25 \text{ mA})(820 \ \Omega) = 9.96 \text{ V}$$

These bias values could have been obtained by using the dc load-line approach and the output characteristic curves. The load-line solution has been illustrated in Fig. 5-22(d). The reader should verify it. Notice the similarity between $I_{D(\text{SAT})}$ and $V_{DS(\text{OFF})}$ for the FET and $I_{C(\text{SAT})}$ and $V_{CE(\text{OFF})}$ for the BJT.

As we shall see in Chapter 6, the approach employing the transfer *V–I* curves and bias lines is the most expedient for a wide variety of problems. In the next section we shall see exactly how an input signal can cause the FET amplifier's operating point to vary about the *Q*-point we determined above.

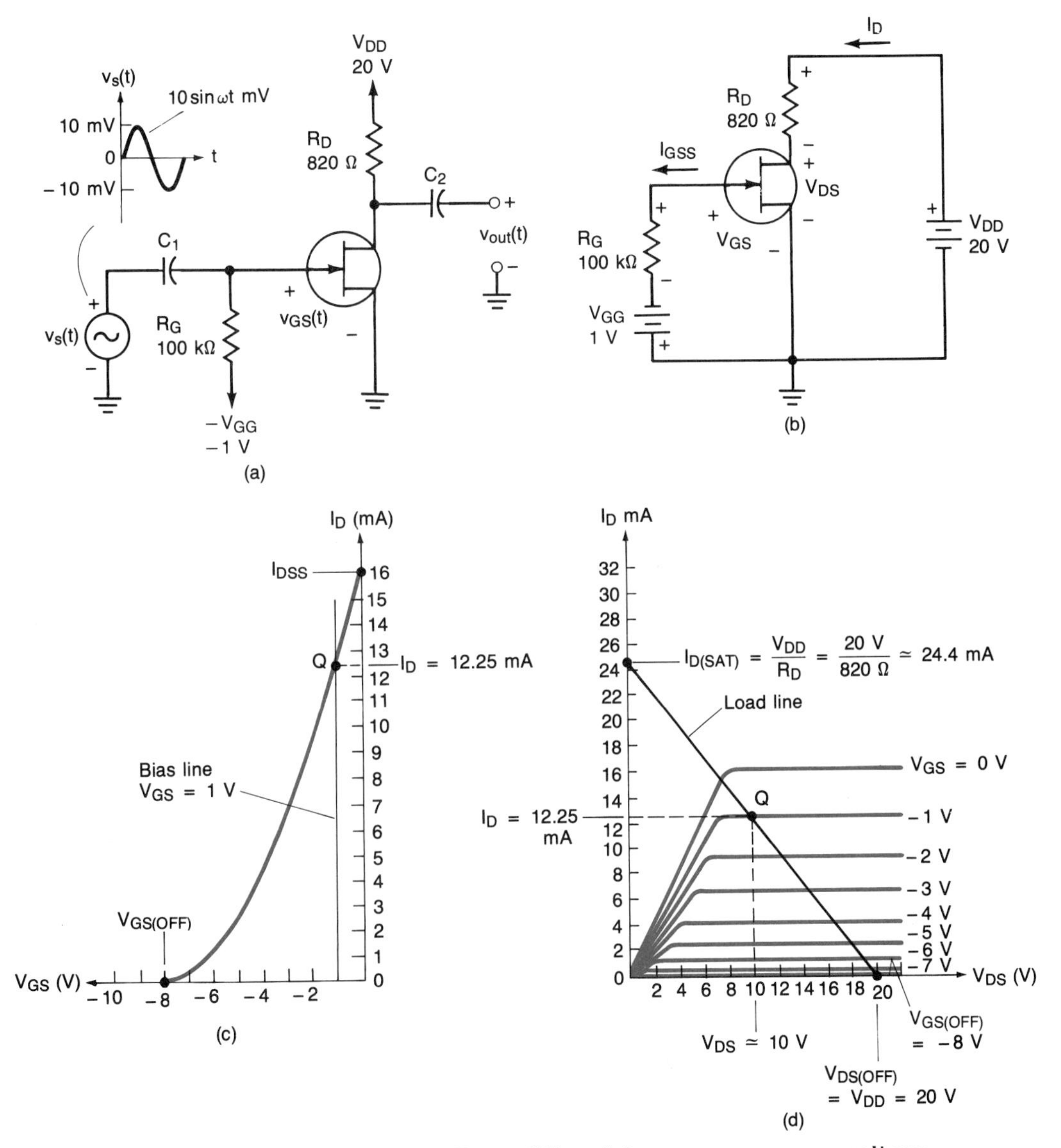

FIGURE 5-22 Dc analysis of a JFET amplifier: (a) common-source voltage amplifier; (b) dc equivalent circuit; (c) plotting the bias line; (d) load-line solution.

5-11 FET Signal Process

In Fig. 5-22(a) we see that the JFET voltage amplifier is being driven by a signal source, which is providing a 10-mV peak signal. The signal source adds to and subtracts from V_{GG}. This is made evident by the application of the superposition

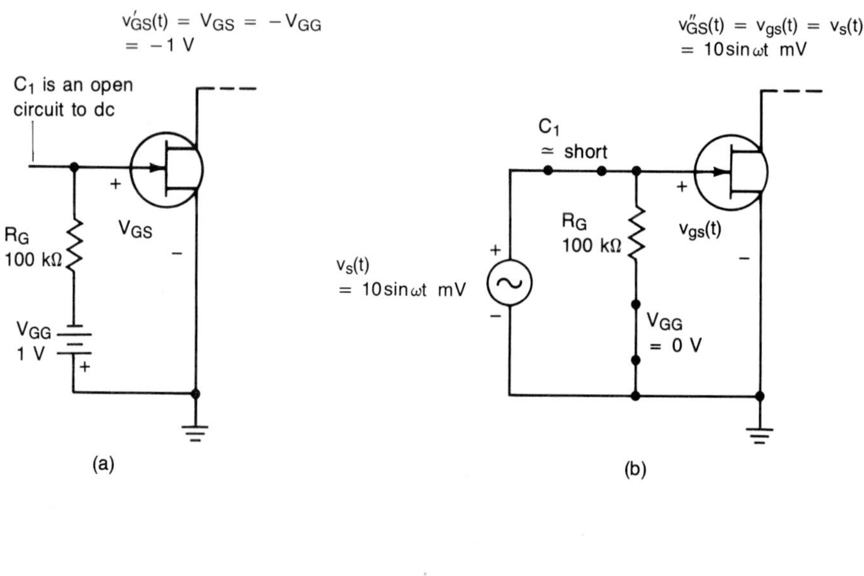

FIGURE 5-23 Analysis of the amplifier's input by superposition: (a) dc equivalent circuit; (b) ac equivalent circuit; (c) total gate-to-source voltage.

theorem and recalling that the coupling capacitors are open circuits to dc and short circuits to the ac signal [consider Fig. 5-23(a) and (b)].

The total instantaneous gate-to-source voltage $v_{GS}(t)$ is

$$
\begin{aligned}
v_{GS}(t) &= V_{GS} + v_{gs}(t) \\
&= -1 \text{ V} + 0.01 \sin \omega t \text{ V}
\end{aligned}
$$

The total instantaneous gate-to-source voltage has been illustrated in Fig. 5-23(c). Note that the signal is very small compared to the dc bias and that the gate-to-source voltage is always negative.

Now that we have analyzed the input, we must transfer across the FET to the output circuit. This can be accomplished via the transfer *V–I* characteristic curve [see Fig. 5-24(a)]. (Five significant figures have been used to illustrate the very small signal riding on the dc bias–not because we have an extraordinary level of accuracy.)

When $v_{GS}(t)$ becomes less negative, the total instantaneous drain current $i_D(t)$ increases, and as $v_{GS}(t)$ becomes more negative, i_D decreases. Since the *Q*-point is in the "linear" portion of the transfer curve, and because the signal is small, the FET's response is approximately linear. Specifically, the positive peak value of the

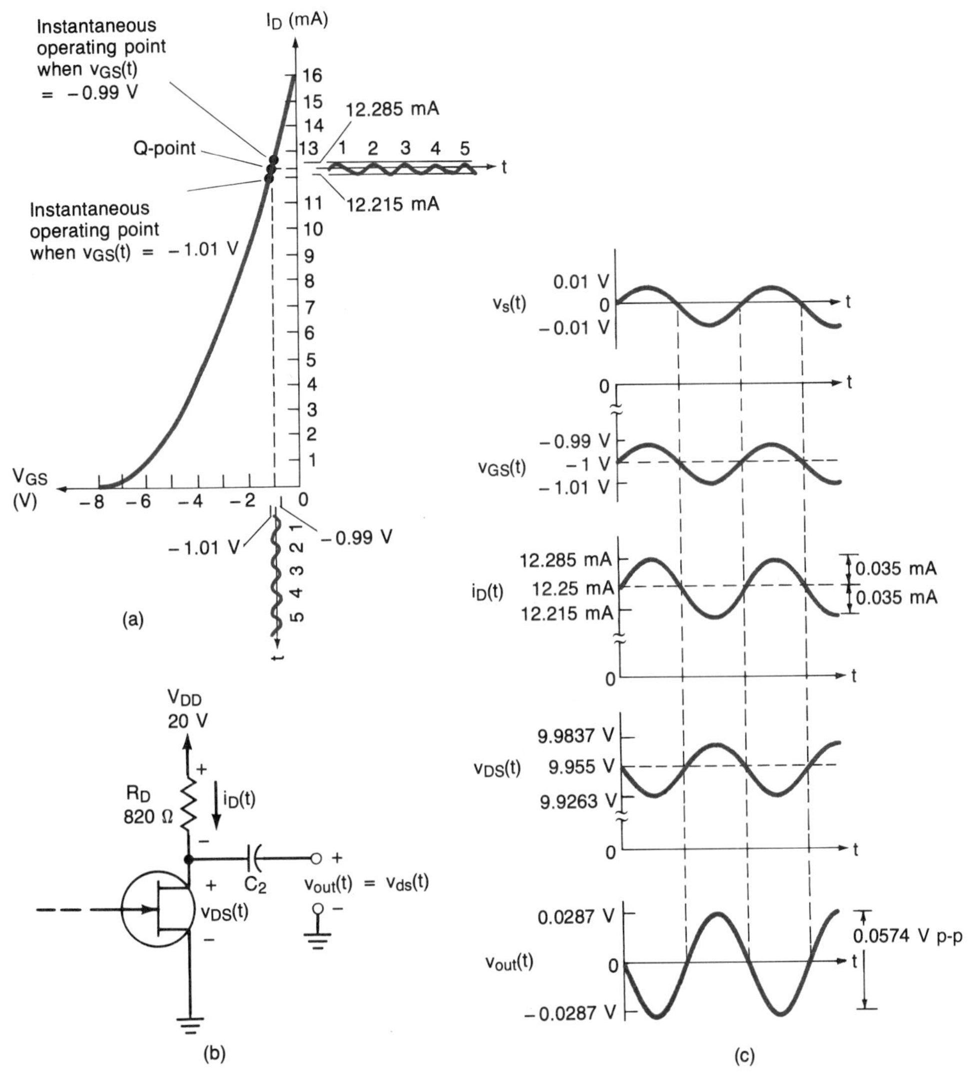

FIGURE 5-24 FET signal process: (a) transferring across the FET; (b) analyzing the output; (c) amplifer waveforms.

drain current and the negative peak are approximately equal in magnitude (about 0.035 mA).

The variations in $i_D(t)$ will produce a variation in the total instantaneous drain-to-source (output) voltage. By Kirchhoff's voltage law, and using the notation for total instantaneous quantities,

$$v_{DS}(t) = V_{DD} - i_D(t)R_D$$

[see Fig. 5-24(b)].

When the $i_D(t)$ is at its *maximum* value, $v_{DS}(t)$ will be at its *minimum* value. The converse is also true. The waveforms associated with the amplifier have been depicted in Fig. 5-24(c). These should be studied carefully.

Notice that the output voltage is 180° out of phase with the input voltage. This is not necessarily a problem, but it is a very important observation. This phase inversion is a characteristic of both the common-source and the common-emitter amplifiers. This will be demonstrated in Chapter 8.

The voltage gain of the amplifier can be found by taking the ratio of the peak-to-peak output voltage to the peak-to-peak input voltage, Hence

$$A_v = \frac{v_{\text{out}}(t)}{v_{\text{in}}(t)} = \frac{0.0574 \text{ V p-p} \angle -180^\circ}{0.02 \text{ V p-p}} = 2.87 \angle -180^\circ$$

The phasor notation is used to emphasize the input–output phase relationship. It is often abbreviated by using a negative sign. Thus

$$A_v = -2.87$$

Notice that the voltage gain of the common-source amplifier is considerably less than the voltage gain we obtained for the BJT amplifier in Section 4-12. This is *not* an unusual case. Generally, *BJTs provide much more voltage gain than FETs*. This will be elaborated in Chapter 8.

5-12 FET Small-Signal Models

To expedite our small-signal analyses of both FET and BJT amplifiers, we employ linear ac equivalent circuits. As we mentioned in Section 4-14, we shall use the hybrid-pi model for the BJT. One of our reasons for its selection was the fact that it is very similar to the model we shall use for the FETs.

Our selection of the hybrid-pi model in lieu of the h- or y- parameter models does present one disadvantage. Specifically, manufacturers tend to favor listing the h- and y-parameter values on their BJT data sheets. Similarly, we also find that manufacturers tend to publish y-parameter values for the FETs.

In Chapter 4 we saw how to estimate the parameters (g_m, r_o, and r_π) for our hybrid-pi BJT model. However, the FET parameters are not as easily estimated. Therefore, we shall be required to draw on the y-parameters given in the data sheets for our ac FET model. Consequently, we need to have some familiarity with their meaning. We shall continue to develop them as our work requires. However, for now, we merely present Table 5-4.

The table provides the notation used for the common-source FET y-parameters. For example, y_{fs} stands for the "short-circuit, forward transadmittance of the common-

TABLE 5-4
FET Short-Circuit y-Parameters

General Notation	Common Source	Admittance Type
y_{11}	y_{is}	Input
y_{12}	y_{rs}	Reverse transfer
y_{21}	y_{fs}	Forward transfer
y_{22}	y_{os}	Output

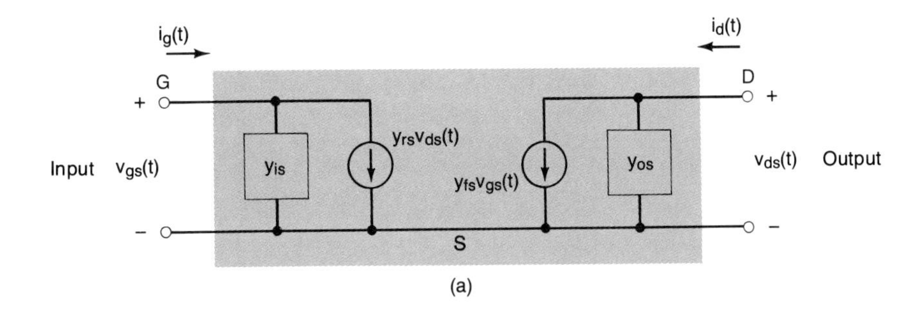

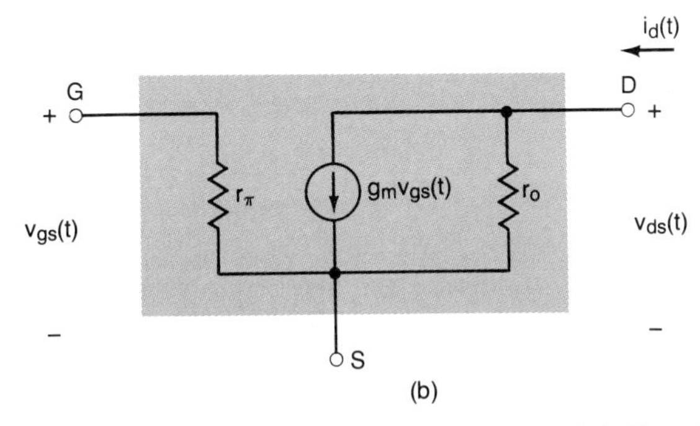

FIGURE 5-25 FET small-signal models: (a) the short-circuit admittance (y) parameter model favored by manufacturers; (b) the low-frequency small-signal hybrid-pi FET model.

source FET'' [see Fig. 5-25(a)]. Although this intimidating ``mouthful'' may seem to be quite impressive, we shall see that at low frequencies (i.e., 1 kHz) this simply stands for the transconductance g_m. This will be explained further in the next section.

5-13 Transconductance: JFETs and DE-MOSFETs

Although the BJT may be modeled quite easily as either a current-controlled current source or a voltage-controlled current source, there is little doubt that the FET is most appropriately modeled as the latter. Consequently, it becomes necessary to specify its transconductance g_m.

In Section 4-15 we derived the equations to describe the transconductance of the BJT. We shall develop the transconductance equation for the FET in a similar manner.

In general, the transconductance of the FET is (assuming that V_{DS} is held constant)

$$g_m = \frac{i_d(t)}{v_{gs}(t)} \tag{5-8}$$

The transconductance has been defined graphically on the transfer characteristic curve given in Fig. 5-26. Now since Eq. 5-5 describes the transfer curve mathematically, we may use it to develop an equation for g_m by taking its derivative. Hence

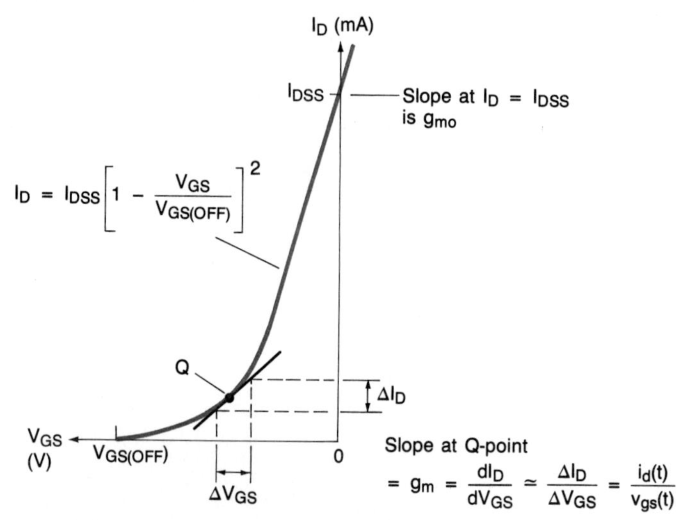

FIGURE 5-26 Graphical definition of the g_m for (n-channel) JFETs and DE-MOSFETs.

$$
\begin{aligned}
g_m &= \frac{dI_D}{dV_{GS}} = I_{DSS}\frac{d}{dV_{GS}}\left[1 - \frac{V_{GS}}{V_{GS(\text{OFF})}}\right]^2 \\
&= -\frac{2I_{DSS}}{V_{GS(\text{OFF})}}\left[1 - \frac{V_{GS}}{V_{GS(\text{OFF})}}\right]
\end{aligned}
\tag{5-9}
$$

Manufacturers typically specify the g_m for FETs at a V_{GS} of zero volts (see Fig. 5-26). We shall denote this tranconductance g_{m0}:

$$
g_{m0} = -\frac{2I_{DSS}}{V_{GS(\text{OFF})}} \tag{5-10}
$$

Substituting Eq. 5-10 into Eq. 5-9 yields

$$
\boxed{g_m = g_{m0}\left[1 - \frac{V_{GS}}{V_{GS(\text{OFF})}}\right]} \tag{5-11}
$$

where g_m = small-signal transconductance
g_{m0} = g_m when V_{GS} is zero volts
V_{GS} = gate-to-source bias voltage
$V_{GS(\text{OFF})}$ = gate-to-source cutoff voltage

By once again drawing on Eq. 5-5, we obtain

$$
I_D = I_{DSS}\left[1 - \frac{V_{GS}}{V_{GS(\text{OFF})}}\right]^2
$$

we may divide both sides by I_{DSS} and then take the square root of both sides.

$$\sqrt{\frac{I_D}{I_{DSS}}} = \left[1 - \frac{V_{GS}}{V_{GS(\text{OFF})}}\right]$$

and substituting the result above into Eq. 5-11, we arrive at the form offered by

$$g_m = g_{m0}\sqrt{\frac{I_D}{I_{DSS}}} \tag{5-12}$$

Equation 5-12 relates the magnitude of the g_m of a given FET to its Q-point (I_D). It will be quite useful in our later work. The only problem that remains is the determination of g_{m0}.

The notation "g_m" is generic in that it may be applied to either the FET or the BJT. However, as we mentioned in the preceding section, manufacturers tend to favor y-parameter notation. The forward transfer admittance may be designated as y_{fs}.

The term "admittance" is used in classical ac circuit theory to describe the ability of a network or a device to pass ac current flow. In general, it may be composed of a real part (conductance, or g), and an imaginary part (susceptance, or b). As we shall see, the FETs not only possess resistive effects but also *capacitive* effects. An admittance with a real part (g) and a capacitive susceptance (jb) will have the form

$$y = g + jb$$

Therefore, we should not be too surprised to find that the forward transfer admittance of an FET is

$$y_{fs} = g_{fs} + jb_{fs}$$

Fortunately, at low frequencies (i.e., 1 kHz), the capacitive effects are negligible and the magnitude of y_{fs} is approximately equal to the tranconductance g_m.

$$|y_{fs}| \simeq g_{fs} = g_m$$

EXAMPLE 5-5

A 2N5549 is an n-channel JFET that has a $|y_{fs0}|$ at 1 kHz of 15 mmho and an I_{DSS} of 60 mA. What is its g_m when its I_D is 40 mA?

SOLUTION Observe that the transconductance has been expressed in obsolete units (mhos). This is not unusual. Most manufacturers have not undergone the expense of updating their data sheets to use the SI units of siemens. Next, we note that $|y_{fs0}| = g_{fs0} = g_{m0} =$ 15 mS and apply Eq. 5-12.

$$g_m = g_{m0}\sqrt{\frac{I_D}{I_{DSS}}} = (1.5 \times 10^{-2})\sqrt{\frac{40\text{ mA}}{60\text{ mA}}}$$
$$= 12.25\text{ mS}$$

■

5-14 E-MOSFET Transconductance

Our work above applies primarily to the n- and p-channel JFETs and the DE-MOSFETs. The E-MOSFETs have a different equation for their drain current (see Fig. 5-27). Consequently, the equations for their transconductances have a slightly different form.

We may find the transconductance of the E-MOSFET by differentiating Eq. 5-7.

$$g_m = \frac{dI_D}{dV_{GS}} = \frac{d}{dV_{GS}} K[V_{GS} - V_{GS(\text{th})}]^2 \\ = 2K[V_{GS} - V_{GS(\text{th})}] \tag{5-13}$$

If we solve Eq. 5-7 for V_{GS}, we have

$$V_{GS} = \sqrt{\frac{I_D}{K}} + V_{GS(\text{th})}$$

and if we substitute this into Eq. 5-13 and exercise our algebra, we arrive at

$$\boxed{g_m = 2\sqrt{KI_D}} \tag{5-14}$$

where g_m = small-signal transconductance of an E-MOSFET
K = constant (amperes/volt^2)
I_D = drain bias current

Generally, manufacturers will specify the g_m of an E-MOSFET at a given I_D. Typically, we will need to know the g_m at a different I_D. To address this problem, we need to develop another relationship.

Therefore, we shall define g_{m1} as the transconductance when the drain current is I_{D1}, and g_{m2} occurs when the drain current is I_{D2}. By using Eq. 5-14, we may develop a ratio and solve for g_{m2}.

$$\boxed{g_{m2} = g_{m1}\sqrt{\frac{I_{D2}}{I_{D1}}}} \tag{5-15}$$

FIGURE 5-27 **Graphical definition of the g_m of an (n-channel) E-MOSFET.**

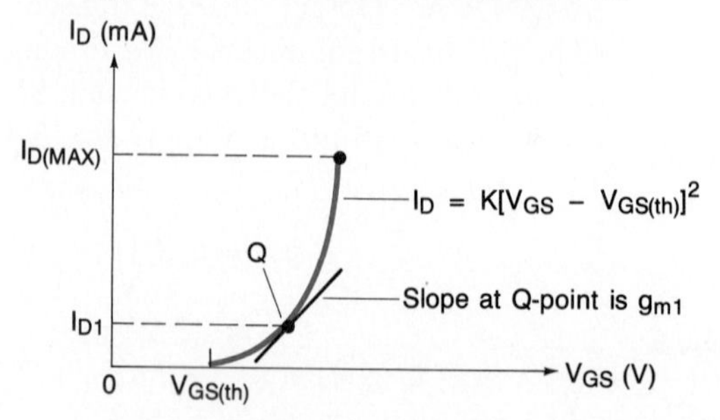

EXAMPLE 5-6

An n-channel E-MOSFET has a g_{fs} of 6 mS at an I_D of 10 mA and an $I_{D(\max)}$ of 50 mA. What is its g_m when I_D is 40 mA?

SOLUTION By Eq. 5-15,

$$g_{m2} = g_{m1}\sqrt{\frac{I_{D2}}{I_{D1}}} = (6 \times 10^{-3})\sqrt{\frac{40\text{ mA}}{10\text{ mA}}}$$
$$= 12\text{ mS} = g_m$$

■

5-15 FET Output and Input Impedance

Like the BJT, the FETs also exhibit an input and output impedance. Let us consider the output impedance r_o first. When a JFET or a DE-MOSFET just reach pinch-off, their depletion regions barely touch (refer to Fig. 5-5). However, as V_{DS} is increased, the depletion regions begin to move back along the channel as shown in Fig. 5-28(a). As can be seen, the channel length actually decreases slightly as V_{DS} increases. Consequently, the channel resistance decreases, and I_D will increase slightly as V_{DS} increases. This is very similar to the Early effect described in Section 4-16 for the BJTs. In fact, the FETs also have an Early voltage V_A, as shown in Fig. 5-28(b). Therefore, Eq. 4-34 may also be used for the FET if we merely substitute I_D for I_C.

We may use the short-circuit output admittance parameter y_{os} to provide an estimate of r_o (if it has been specified at a low frequency).

EXAMPLE 5-7

A 2N4119 is an n-channel JFET with a typical g_{os} specified as 0.8 μS at an I_D of 0.1 mA. Estimate its output resistance r_o at an I_D of 0.01 mA.

SOLUTION In general,

$$\frac{1}{|y_{os}|} \simeq \frac{1}{g_{os}} = r_o$$

Therefore, r_o is (at 0.1 mA) found by taking the reciprocal of g_{os}:

$$r_o = \frac{1}{g_{os}} = \frac{1}{0.8 \times 10^{-6}\text{ S}} = 1.25\text{ M}\Omega$$

Now since the Early voltage relationship also applies to the FET,

$$r_o = \frac{V_A}{I_D}$$

and solving for V_A and using the given set of conditions,

$$V_A = r_o I_D = (1.25\text{ M}\Omega)(0.1\text{ mA}) = 125\text{ V}$$

Now we may estimate r_o for an I_D 0.01 mA.

$$r_o = \frac{V_A}{I_D} = \frac{125\text{ V}}{0.01\text{ mA}} = 12.5\text{ M}\Omega$$

■

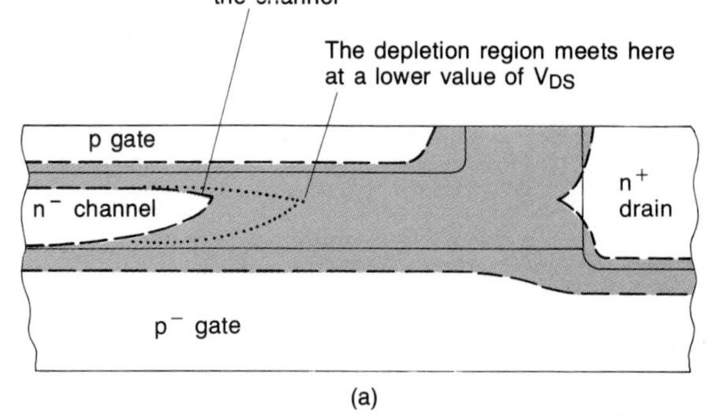

(a)

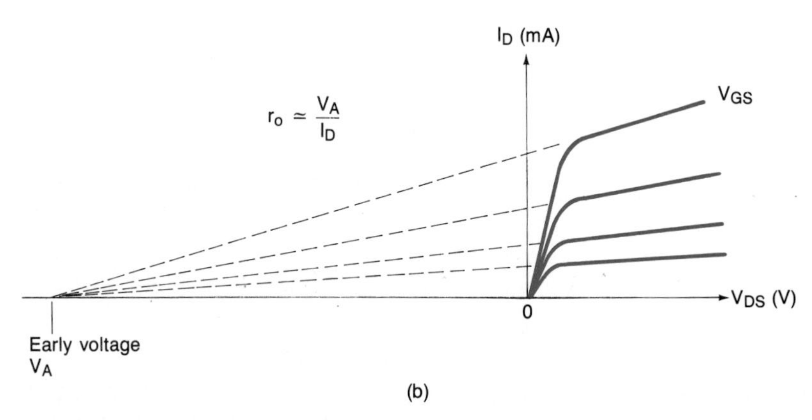

(b)

FIGURE 5-28 The r_o of FETs: (a) an increase in V_{DS} shortens the channel length, decreasing the channel resistance; (b) graphical determination of r_o by using the Early voltage.

FIGURE 5-29 Small-signal model for n- and p-channel JFETs, DE-MOSFETs, and E-MOSFETs.

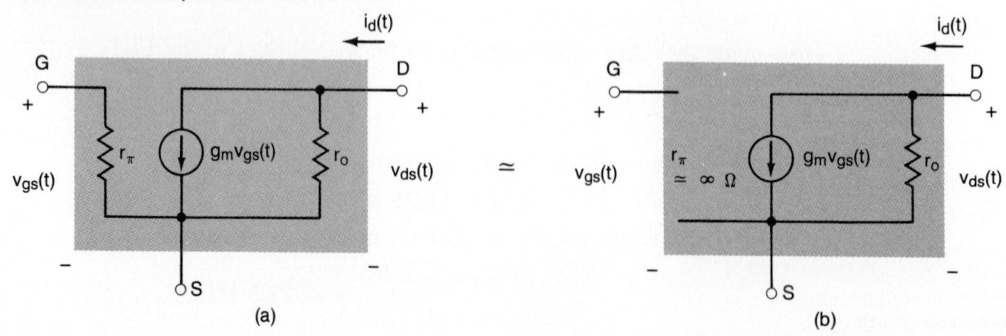

Example 5-7 illustrates exactly how we may estimate the r_o of an FET. However, we should emphasize that it is merely an estimate. If the manufacturer provides us with a graph of g_{os} values versus I_D, it should be used in lieu of our procedure. Generally, it will provide us with a closer value. For example, one of the manufacturers states that the typical value of g_{os} for a 2N4119 at an I_D of 0.01 mA is 28 μS. This equates to an r_o of 3.6 MΩ, and is probably closer to the truth than our 12.5-MΩ prediction.

The input impedance of the JFETs, as measured between the gate and the source is essentially the impedance of a reverse-biased diode–perhaps hundreds of megohms. The input impedance of MOSFETs is much higher–perhaps thousands of megohms! Consequently, we shall approximate the input impedance of both JFETs and MOSFETs as infinite ohms. This is stated as follows:

$$\text{For FETs:} \qquad r_\pi \simeq \infty\ \Omega \tag{5-16}$$

The small-signal ac model we shall use for FETs is shown in Fig. 5-29. We should emphasize that this model may be used for *n*- and *p*-channel JFETs, DE-MOSFETs, and E-MOSFETs.

5-16 Linear CMOS

When we have a *p*-channel E-MOSFET and an *n*-channel E-MOSFET configured as shown in Fig. 5-30(a), we have formed a new active-device combination which is described as a *complementary* MOSFET, or CMOS. Its fabrication has been depicted in Fig. 5-30(b). CMOS initially started out as a technology restricted primarily to digital integrated-circuit functions. However, its use has been extended to encompass a wide range of new products. Typical digital integrated-circuit applications include logic gates, counters, memories, and even microprocessors. Linear integrated-circuit applications include audio filters, phase-locked loops, and operational amplifiers. We also find CMOS used in analog–digital functions, including analog-to-digital converters (ADCs) and digital-to-analog converters (DACs).

Let us take a brief look at how the CMOS "building block" depicted in Fig. 5-30(a) operates. (This circuit is found in digital logic inverters and as the typical output stage of a CMOS op amp.) However, before we begin, let us recall two basic facts. First, a negative gate-to-source voltage is required to enhance a *p*-channel FET, and second, a positive gate-to-source voltage is required to enhance an *n*-channel FET.

In Fig. 5-31(a) we see that if $v_{\text{in}}(t)$ is 15 V, the *p*-channel E-MOSFET is nonconducting, or OFF, since its $v_{GS}(t)$ is zero, and the *n*-channel E-MOSFET is conducting, or ON. Therefore, $v_{\text{out}}(t)$ is approximately zero volts since the output point is effectively tied to ground by the *n*-channel E-MOSFET. Also note that $i_d(t)$ is zero.

If we set $v_{\text{in}}(t)$ to zero, as shown in Fig. 5-31(b), the *p*-channel E-MOSFET

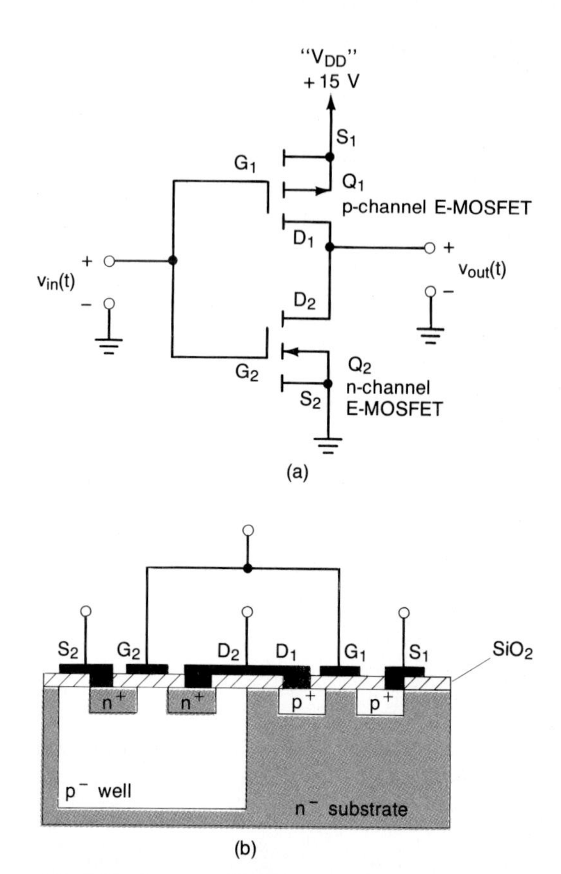

FIGURE 5-30 **CMOS: (a) schematic; (b) structure.**

will be ON. (Since its source is tied to $+15$ V, and its gate is at zero volts, the gate is effectively -15 V relative to the source terminal.) The *n*-channel E-MOSFET obviously has a $v_{GS}(t)$ of zero, and is OFF. Therefore, the output terminal is effectively tied to V_{DD} through the *p*-channel E-MOSFET. Again we note that $i_d(t)$ is zero.

In essence we have just described the operation of the CMOS inverter logic gate found in digital logic applications. We have also seen its primary advantage. Specifically, its current draw is zero when its output is either at 0 or 15 V. However, the CMOS pair will draw current as the output makes its transition from one extreme to the other.

The discussion above is applicable to us from a linear standpoint in that we have just described the ideal boundaries for the linear operation of a CMOS pair. In Fig. 5-31(c) we see a simple linear CMOS amplifier. Note that the gate terminals have been biased at $V_{DD}/2$.

In Fig. 5-31(d) we see the effect of applying an input signal voltage. As $v_{\text{in}}(t)$ goes positive, the *n*-channel E-MOSFET becomes more conductive, and the *p*-channel device becomes less conductive. The voltage divider action between the two FETs

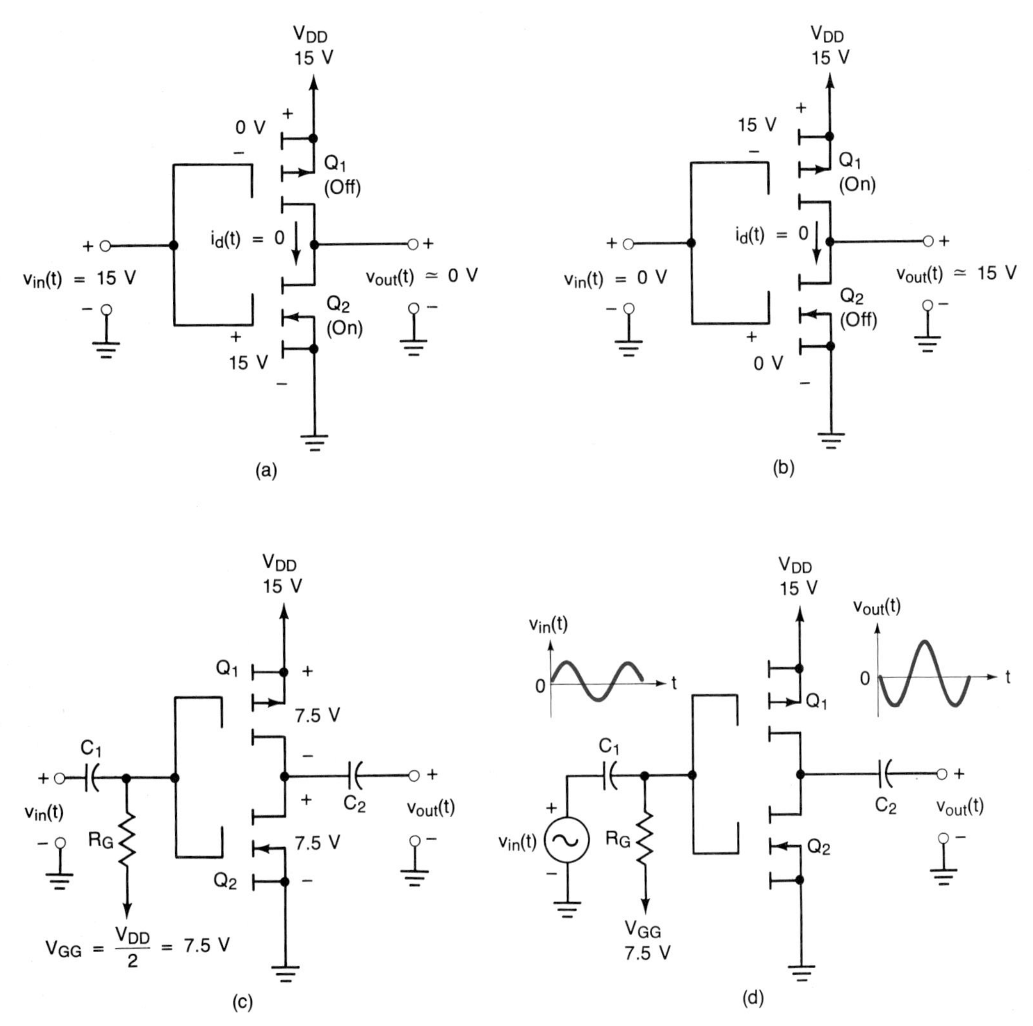

FIGURE 5-31 CMOS operation: (a) when v_{in} is positive Q_1 is off and Q_2 is on; (b) when v_{in} is 0 V Q_1 is on and Q_2 is off; (c) biasing CMOS for linear amplification; (d) applying a signal.

results in a decrease in $v_{out}(t)$. Similarly, as $v_{in}(t)$ goes negative, the gate-to-ground voltage decreases, the *p*-channel E-MOSFET becomes more conductive, the *n*-channel device becomes less conductive, and $v_{out}(t)$ increases.

Note that the input-output phase relationship is $-180°$ as indicated in Fig. 5-31(d). Since the input signal is applied to the gate terminals, and the output signal is taken at the drain terminals, *both* of the FETs are being used in their common-source configuration (refer to Table 5-1).

A summary of the various types of FETs has been provided in Fig. 5-32. Note that CMOS has been indicated as a combination of *n*- and *p*-channel E-MOSFETs.

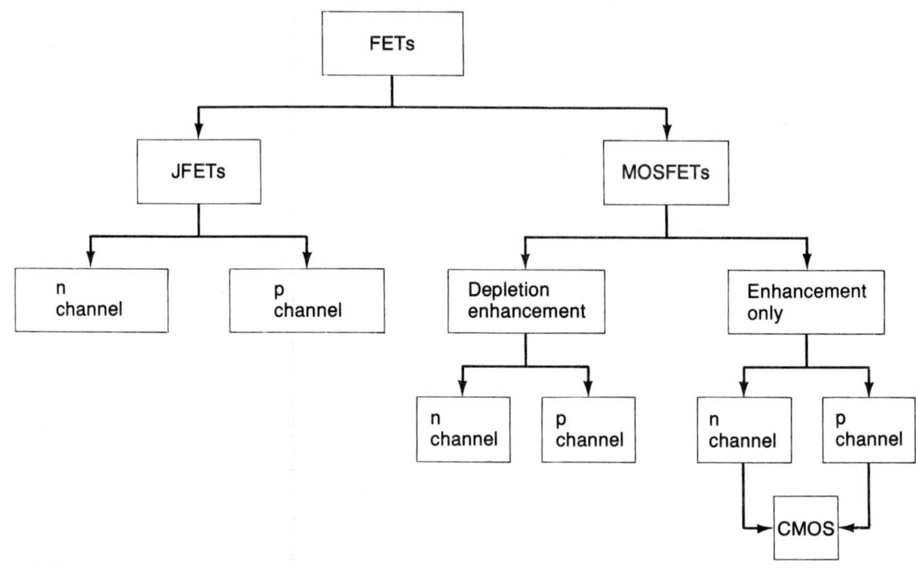

FIGURE 5-32 FET family.

PROBLEMS

Drill, Derivations, and Definitions

Section 5-2

5-1. The charge carriers in an *n*-channel JFET are ____ (holes, electrons).

5-2. As the charge carriers in a JFET move from the source region to the drain region they cross ____ (no, 1, 2) *p-n* junctions.

5-3. How do the three terminals of a JFET correspond to the BJT terminals?

Section 5-3

5-4. In our own words, explain why a depletion region forms at a *p-n* junction. How does its depth of penetration relate to doping levels?

5-5. What is the pinch-off voltage of an FET? If we apply Kirchhoff's voltage law around an FET, we arrive at $V_{DG} = V_{DS} - V_{GS}$. Relate the magnitude of the pinch-off voltage to V_{DG}, V_{GS}, and V_{DS}.

5-6. Name the two fundamental regions of operation for a JFET. Make a neat sketch of a typical *n*-channel FET's drain *V–I* characteristics, and indicate the two regions on it.

5-7. An *n*-channel JFET has a pinch-off voltage V_p of 5 V. What is the magnitude of its $V_{GS(\mathrm{OFF})}$? Explain the relationships between the channel depletion regions, V_p, and $V_{GS(\mathrm{OFF})}$.

5-8. An *n*-channel JFET has a $V_{GS(\mathrm{OFF})}$ of -8 V. If its V_{GS} is -3 V, what value of V_{DS} is required to place the JFET in pinch-off?

5-9. A *p*-channel JFET has a $V_{GS(\text{OFF})}$ of $+1.5$ V. If its V_{GS} is 1 V, what value of V_{DS} is required to place the JFET in pinch-off?

5-10. An FET has an I_D of 4 mA; what is its I_S? What is the ideal value of the gate current?

Section 5-4

5-11. Name the three FET configurations. What are their respective input and output terminals? How do the three FET configurations correspond to the three BJT configurations?

Section 5-5

5-12. A JFET has an I_{DSS} of 12 mA and a $V_{GS(\text{OFF})}$ of -3 V. Graph its transfer characteristic curve. Include a table of your data points.

5-13. A JFET has an I_{DSS} of 20 mA and a $V_{GS(\text{OFF})}$ of -8 V. Graph its transfer characteristic curve. Include a table of your data points.

Section 5-6

5-14. Why are FETs described as unipolar devices? Name another unipolar device.

5-15. What are the two basic mechanisms underlying the temperature sensitivity of I_D in JFETs? Describe them. What is I_{GSS}?

5-16. At low values of I_D, a *p*-channel JFET has a $V_{GS(\text{OFF})}$ of 1.2 V at 25°C. What is its $V_{GS(\text{OFF})}$ at 100°C?

5-17. Repeat Prob. 5-16 for an *n*-channel FET with a $V_{GS(\text{OFF})}$ of -8 V.

5-18. An *n*-channel JFET has a $V_{GS(\text{OFF})}$ of -5 V. What value of V_{GS} bias voltage is required to give the FET's drain current a zero temperature coefficient?

5-19. Repeat Prob. 5-18 if $V_{GS(\text{OFF})}$ is -1.5 V.

5-20. A JFET has an I_{GSS} of 15 nA at 25°C. What is its I_{GSS} at 75°C?

5-21. A JFET has an I_{GSS} of 10 nA at 25°C. What is its I_{GSS} at 125°C?

Section 5-7

5-22. *Negative* values of V_{GS} are required to _____ (enhance, deplete) the channels of both *n*-channel JFETs and DE-MOSFETs while *positive* values of V_{GS} are required to _____ (enhance, deplete) their channels.

5-23. When an FET is *enhanced*, the number of available charge carriers in its channel is _____ (increased, decreased).

5-24. Name the four terminals of a DE-MOSFET.

5-25. Is it possible to enhance a JFET? If so, what are the constraints on V_{GS}?

5-26. Can the substrate connection of a DE-MOSFET be biased to enhance the device? Is there a limit on the maximum enhancement bias between the substrate and source terminals? What do you think it is?

5-27. Why are MOSFETs typically shipped with their leads shorted together? Explain.

5-28. A 2N3796 is described by one manufacturer as being a low-power audio *n*-channel DE-MOSFET. Its typical I_{DSS} is 1.5 mA and its $V_{GS(\text{OFF})}$ is -3 V. Its maximum I_D

is 20 mA. Graph its transfer characteristic curve. Let V_{GS} be -3, -2.5, -2, -1.5, -1, -0.5, 0, 1, 2, 3, and 4 V. Include a table of your data points.

5-29. A 2N3797 DE-MOSFET is very similar to the 2N3796 given in Prob. 5-28. However, its typical $V_{GS(OFF)}$ is -5 V, and its typical I_{DSS} is 2.9 mA. Graph its transfer characteristic curve. Let V_{GS} be -5, -4, -3, -2, -1, 0, 1, 2, 3, and 4 V. Include a table of your data points.

Section 5-8

5-30. A Motorola 3N170 is an *n*-channel E-MOSFET. Its $V_{GS(th)}$ is typically 1 V, and produces an I_D of 10 μA. It has an $I_{D(ON)}$ of 10 mA for a V_{GS} of 10 V. Its $I_{D(max)}$ is 30 mA. Find K. Make a sketch of its transfer characteristic curve.

5-31. A Motorola 3N171 is an *n*-channel E-MOSFET. Its $V_{GS(th)}$ is typically 1.5 V, and produces an I_D of 10 μA. It has an $I_{D(ON)}$ of 10 mA for a V_{GS} of 10 V. Its $I_{D(max)}$ is 30 mA. Find K. Make a sketch of its transfer characteristic curve.

Section 5-10

5-32. The circuit in Fig. 5-22(a) has been modified. Specifically, $R_G = 200$ kΩ, $R_D = 1.2$ kΩ, $-V_{GG} = -2$ V, and V_{DD} remains at 20 V. The same JFET is used. This means it has the transfer characteristic shown in Fig. 5-22(c) and the drain V-I characteristics shown in Fig. 5-22(d). Find V_{GS}, I_D, V_{DS}, $I_{D(SAT)}$, and $V_{DS(OFF)}$. Assume an I_{GSS} of zero.

5-33. Repeat Prob. 5-32 if $-V_{GG}$ is changed to -2.5 V.

5-34. Repeat Prob. 5-32, but include the effects of I_{GSS} in determining V_{GS}. All component values are unchanged except that R_G has been increased to 10 MΩ. Assume an I_{GSS} of 10 nA.

Section 5-11

5-35. The amplifier in Fig. 5-22(a) has its $-V_{GG}$ changed to -0.5 V and the signal source amplitude is increased to -0.05 V peak. Make a sketch of the waveforms associated with the circuit. Use Fig. 5-24(c) as a guide. What is the voltage gain (A_V)?

5-36. In general, FET voltage amplifiers offer voltage gains that are ____ (greater than, less than) BJT voltage amplifiers.

Section 5-13

5-37. A 2N4341 is an *n*-channel JFET with the following specifications: $V_{GS(OFF)}$ is -6 V, I_{DSS} is 9 mA, and g_{fso} is 4000 μS. If the 2N4341 has an I_D of 8 mA, find its g_m. Repeat the analysis if I_D is lowered to 1 mA.

5-38. A 2N3578 is a *p*-channel JFET with the following specifications: $V_{GS(OFF)}$ is 4 V, I_{DSS} is -4.5 mA (the manufacturer is using the transistor convention–do not let this upset you), and g_{fso} is 3500 μS. Find the JFET's g_m if I_D is -4 mA. Repeat the analysis if I_D is -0.5 mA.

Section 5-14

5-39. An n-channel E-MOSFET has a g_{fs} of 1000 μS when V_{GS} is 10 V, a $V_{GS(\text{th})}$ of 3 V, an $I_{D(\text{ON})}$ of 3 mA when V_{GS} is 10 V, and an $I_{D(\text{MAX})}$ of 50 mA. Find its g_m when I_D is 10 mA, and its g_m when I_D is 40 mA.

5-40. Repeat Prob. 5-39 if the FET has a g_{fs} 1500 μS when V_{GS} is 10 V. All other factors are the same.

Section 5-15

5-41. An n-channel JFET has a $|y_{os}|$ of 50 μS when V_{GS} is 0 V, and an I_{DSS} of 20 mA. Find (a) its Early voltage, and (b) its r_o when I_D is 5 mA.

5-42. A 2N5647 is an n-channel JFET with a g_{os} of 5 μS at an I_D of 200 μA. Find its r_o at an I_D of 0.3 mA.

5-43. Draw the small-signal FET model for an n-channel JFET that utilizes g_m, r_o, and r_π. Can the same model be used for a p-channel E-MOSFET? Explain.

Section 5-16

5-44. Draw a neat representation of the structure of a CMOS pair. Label its components. Also draw the equivalent electrical schematic diagram.

5-45. The CMOS pair in Fig. 5-31(a) has a V_{DD} of 5 V. If $v_{\text{in}}(t)$ has an instantaneous value of 5 V, find $v_{gs1}(t)$, $v_{gs2}(t)$, and the approximate value of $v_{\text{out}}(t)$. Repeat the problem if $v_{\text{in}}(t)$ is 0 V.

5-46. The CMOS pair in Fig. 5-31(a) has a V_{DD} of 7.5 V. The source terminal of Q_2 goes to −7.5 V rather than ground. If $v_{\text{i}}(t)$ has an instantaneous value of +7.5 V, find $v_{gsI}(t)$, v_{gs2} (t), and the approximate value of $v_{\text{out}}(t)$. Repeat the problem if $v_{\text{in}}(t)$ is −7.5 V.

Design

5-47. Given the JFET voltage amplifier in Fig. 5-22(a), assume that the JFET has the transfer curve shown in Fig. 5-22(c). Determine the V_{GG} required to produce an I_D of 10 mA. Size R_D to provide a V_{DS} of approximately 10 V. (*Hint*: We can find V_{GG} by solving Eq. 5-5 for V_{GS} and assuming that V_{GS} will be equal to V_{GG}.)

5-48. Repeat Prob. 5-47 if I_D is to be 8 mA.

Troubleshooting and Failure Modes

Some of the more common failure modes for a JFET include a gate-to-drain or a gate-to-source short (about 40% of the time), an open drain or source terminal connection (about 10% of the time), or increased gate leakage current I_{GSS} (about 50% of the time). With these failure modes in mind, solve the problems below.

5-49. The (silicon) JFET in Fig. 5-22(a) has a short between its drain and gate terminals. What will the gate-to-ground voltage be? Explain.

5-50. The JFET in Fig. 5-22(a) has a "leaky" gate. I_{GSS} has increased to 5 μA. What will the gate-to-ground voltage be?

Computer

5-51. Write a BASIC program that will produce a table of I_D values for corresponding values of V_{GS}. The program should produce a selection menu to allow the user to select: (1) JFET or DE-MOSFET, or (2) E-MOSFET. If the first option is selected, the program should prompt the user for I_{DSS} and $V_{GS(\mathrm{OFF})}$. If the second option is selected, the program should prompt the user for $V_{GS(\mathrm{th})}$ and a value of I_D at a particular value of V_{GS}. (This will allow the program to calculate the constant K.) The program should compute 11 values of I_D. In the case of JFETs and DE-MOSFETs, the values of V_{GS} should range from zero to $V_{GS(\mathrm{OFF})}$. For E-MOSFETs the V_{GS} values should range from $V_{GS(\mathrm{th})}$ to the user-entered value of V_{GS}.

5-52. Modify the program you developed to satisfy the requirements of Prob. 5-51. Increase the menu options to three: (1) JFET, (2) DE-MOSFET, and (3) E-MOSFET. If option (2) is selected, the V_{GS} values shall go from $V_{GS(\mathrm{OFF})}$ through zero to $-V_{GS(\mathrm{OFF})}$ in 21 steps. Specifically, the DE-MOSFET is to be taken from cut-off through zero to a V_{GS} voltage with a magnitude equal to $V_{GS(\mathrm{OFF})}$, but with a polarity that will enhance the FET. The program must work for both *n*- and *p*-channel devices.

6

BJT, FET, and Integrated–Circuit Biasing

After Studying Chapter 6, You Should Be Able to:

- Explain the fundamental biasing problem.
- Analyze CE and CS fixed bias circuits.
- Analyze CE emitter-stabilized and CS self-bias circuits.
- Analyze collector voltage feedback bias and MOSFET (including CMOS) drain voltage feedback bias circuits.
- Analyze CE and CS voltage divider bias circuits.
- Analyze emitter bias BJT circuits and single- and dual-supply source bias circuits.
- Analyze a BJT differential pair bias circuit.
- Explain the operation of the current mirror.
- Describe op amp biasing using dual-polarity power supplies.

6-1 General Q-Point Considerations

In Chapters 4 and 5 we introduced the basic operation of the BJT and the FET, respectively. We also introduced the basic approaches to the dc bias analyses of these *potentially* active devices. We use the adjective "potentially" because if the dc bias is not correct, these devices will not serve as amplifiers. The required BJT and FET biasing constraints are simple.

A BJT must be biased in its active region of operation in order to behave as an active device. This is accomplished by ensuring that the base–emitter *p-n* junction is always forward biased, and the collector–base *p-n* junction is always reverse biased.

To serve as linear active devices, FETs must be in pinch-off. This is usually achieved by ensuring that the magnitude of V_{DS}, plus the magnitude of V_{GS}, always exceeds the magnitude of V_p.

The importance of a stable dc operating point, or Q-point, cannot be overstated. Even if the Q-point of a given BJT or FET remains in the linear active region of operation, we still want it to be stable or "locked" into position. The reason is simple. As we have seen many times before, the ac (signal) parameters of all solid-state devices are a strong function of the dc bias. Therefore, if we want the ac parameters of a solid-state device to be constant, a stable Q-point is mandatory.

In this chapter we shall see some of the bias circuits that have been developed to maintain a reasonably stable Q-point. A stable Q-point requires considerable prudence in the design of the biasing circuitry. This is true because we have *two* factors working against us. First, the tolerances associated with the dc parameters of discrete solid-state devices are rather large, and second, all of the dc parameters tend to vary with temperature. We discuss the tolerance of the dc parameters and provide some additional thermal considerations in the next section.

6-2 Device Parameter Variations Due to Tolerance and Temperature

Over the years, manufacturers have developed their manufacturing processes to provide us with a wide variety of very reliable solid-state components. Even so, we find that most solid-state components exhibit a wide variation in their various parameters. For example, the 2N4124 is a very widely used *npn* BJT. At an I_C of 2 mA, the minimum β_{DC} or h_{FE} is 120, and a maximum of 360. This means that an "off-the-

2N4123
2N4124

CASE 29-02, STYLE 1
TO-92 (TO-226AA)

GENERAL PURPOSE TRANSISTOR

NPN SILICON

MAXIMUM RATINGS

Rating	Symbol	2N4123	2N4124	Unit
Collector-Emitter Voltage	V_{CEO}	30	25	Vdc
Collector-Base Voltage	V_{CBO}	40	30	Vdc
Emitter-Base Voltage	V_{EBO}	5.0		Vdc
Collector Current — Continuous	I_C	200		mAdc
Total Device Dissipation @ T_A = 25°C Derate above 25°C	P_D	625 5.0		mW mW/°C
Total Device Dissipation @ T_C = 25°C Derate above 25°C	P_D	1.5 12		Watt mW/°C
Operating and Storage Junction Temperature Range	T_J, T_{stg}	−55 to +150		°C

THERMAL CHARACTERISTICS

Characteristic	Symbol	Max	Unit
Thermal Resistance, Junction to Case	$R_{\theta JC}$	83.3	°C/W
Thermal Resistance, Junction to Ambient	$R_{\theta JA}$	200	°C/W

ELECTRICAL CHARACTERISTICS (T_A = 25°C unless otherwise noted.)

Characteristic		Symbol	Min	Max	Unit
OFF CHARACTERISTICS					
Collector-Emitter Breakdown Voltage(1) (I_C = 1.0 mAdc, I_E = 0)	 2N4123 2N4124	$V_{(BR)CEO}$	 30 25	 — —	Vdc
Collector-Base Breakdown Voltage (I_C = 10 μAdc, I_E = 0)	 2N4123 2N4124	$V_{(BR)CBO}$	 40 30	 — —	Vdc
Emitter-Base Breakdown Voltage (I_E = 10 μAdc, I_C = 0)		$V_{(BR)EBO}$	5.0	—	Vdc
Collector Cutoff Current (V_{CB} = 20 Vdc, I_E = 0)		I_{CBO}	—	50	nAdc
Emitter Cutoff Current (V_{BE} = 3.0 Vdc, I_C = 0)		I_{EBO}	—	50	nAdc
ON CHARACTERISTICS					
DC Current Gain(1) (I_C = 2.0 mAdc, V_{CE} = 1.0 Vdc) (I_C = 50 mAdc, V_{CE} = 1.0 Vdc)	 2N4123 2N4124 2N4123 2N4124	h_{FE}	 50 120 25 60	 150 360 — —	—
Collector-Emitter Saturation Voltage(1) (I_C = 50 mAdc, I_B = 5.0 mAdc)		$V_{CE(sat)}$	—	0.3	Vdc
Base-Emitter Saturation Voltage(1) (I_C = 50 mAdc, I_B = 5.0 mAdc)		$V_{BE(sat)}$	—	0.95	Vdc
SMALL-SIGNAL CHARACTERISTICS					
Current-Gain — Bandwidth Product (I_C = 10 mAdc, V_{CE} = 20 Vdc, f = 100 MHz)	 2N4123 2N4124	f_T	 250 300	 — —	MHz
Output Capacitance (V_{CB} = 5.0 Vdc, I_E = 0, f = 100 MHz)		C_{obo}	—	4.0	pF
Input Capacitance (V_{BE} = 0.5 Vdc, I_C = 0, f = 100 kHz)		C_{ibo}	—	8.0	pF
Collector-Base Capacitance (I_E = 0, V_{CB} = 5.0 V, f = 100 kHz)		C_{cb}	—	4.0	pF
Small-Signal Current Gain (I_C = 2.0 mAdc, V_{CE} = 10 Vdc, f = 1.0 kHz)	 2N4123 2N4124	h_{fe}	 50 120	 200 480	—

MOTOROLA SEMICONDUCTORS SMALL-SIGNAL DEVICES

FIGURE 6-1 Partial data sheet for the 2N4124. (Reprinted with permission of Motorola, Inc.)

STATIC CHARACTERISTICS

FIGURE 9 – DC CURRENT GAIN

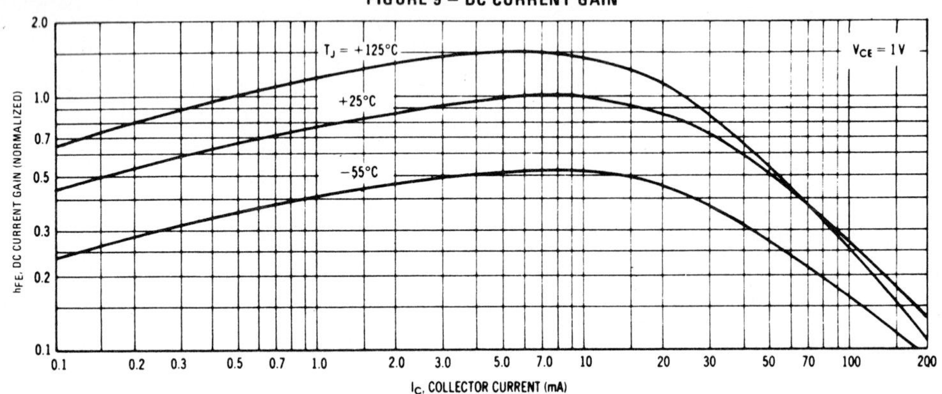

FIGURE 10 – COLLECTOR SATURATION REGION

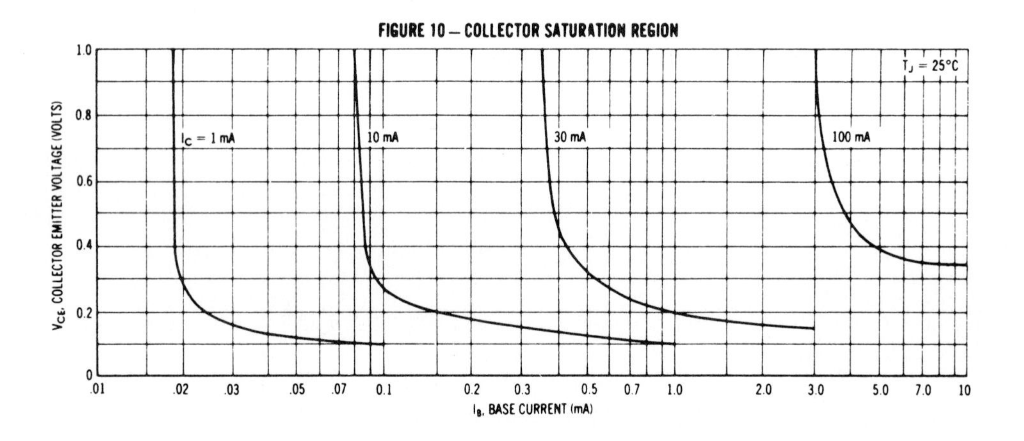

FIGURE 11 – "ON" VOLTAGES

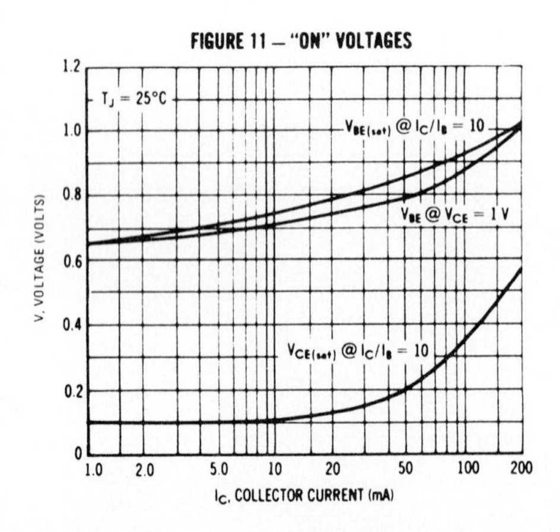

FIGURE 12 – TEMPERATURE COEFFICIENTS

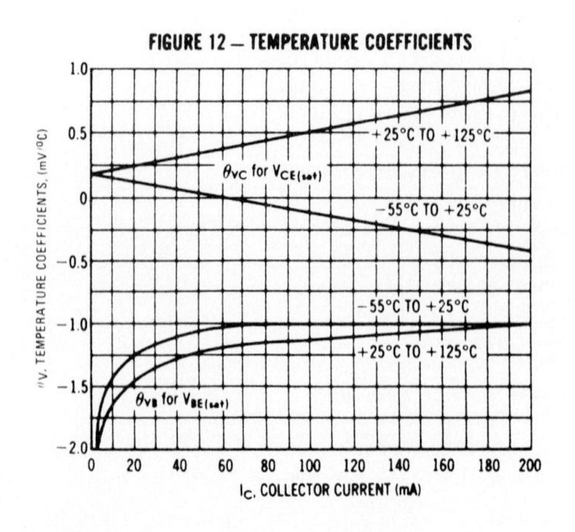

FIGURE 6-1 (continued)

shelf'' 2N4124 that we may use in our amplifier circuit (with an I_C of 2 mA) could have a dc current gain between 120 and 360. This three-to-one variation in β_{DC} can severely upset our Q-point if our bias circuitry does not incorporate some method to compensate for it. While a 3:1 variation in h_{FE} may seem to be quite large, some BJTs may have a 5:1, or even greater variation. A partial data sheet for the 2N4124 has been provided in Fig. 6-1.

To compound our problems further, the value of h_{FE} also tends to vary with the Q-point. A normalized graph of the variation of h_{FE} with I_C has been given in Figure 9 on the data sheet given in Fig. 6-1. Observe that h_{FE} also varies with temperature.

At this point it would seem that a bias circuit which relies on a *particular* value of h_{FE} has a hopelessly unpredictable Q-point. This is indeed true. Consequently, our primary objective in bias circuits is to make the Q-point as independent of the variations in β_{DC} or h_{FE} as possible. This is accomplished by designing the bias circuit to yield acceptable Q-point variations under the influence of a range in β_{DC} values. We shall denote the minimum and the maximum values of β_{DC} as $\beta_{DC(min)}$ and $\beta_{DC(max)}$, respectively. For example, if data are available, the minimum beta should be determined at the desired Q-point (I_C) and at the worst-case temperature. Consider Example 6-1.

EXAMPLE 6-1

The h_{FE} graph for the 2N4124 has been given in Figure 9 of Fig. 6-1. Determine the minimum beta [$\beta_{DC(min)}$] for a 2N4124 if we desire I_C to be 1 mA, V_{CE} to be on the order of 10 V, and the worst-case temperature to be $-40°C$.

SOLUTION Figure 6-1 lists the minimum h_{FE} as 120 when V_{CE} is 1 V, I_C is 2 mA, and the temperature is 25°C. Figure 9 on the data sheet also lists V_{CE} as being held at 1 V. Since we desire a V_{CE} on the order of 10 V, it appears that the data do not fit our application. However, since this is all the information we have, we are forced to use it. We also note that the manufacturer is using a logarithmic scale on both the vertical and horizontal axes. This allows the manufacturer to compress the data. The vertical (h_{FE}) axis has been normalized (to the h_{FE} that occurs when I_C is 7 mA at a temperature of 25°C). This allows the graph to be more of a universal representation. Let us locate the minimum h_{FE} value on Figure 9 of the data sheet. The graph indicates that at 25°C when I_C is 2 mA the normalized value of h_{FE} is approximately 0.88. If I_C is increased to about 7 mA, the normalized value of h_{FE} is unity. Let us determine h_{FE} at 25°C and 7 mA using the information given.

$$h_{FE}\,(I_C = 2\text{ mA}, T = 25°\text{C}) = 120 = 0.88 h_{FE}(I_C = 7\text{ mA}, T = 25°\text{C})$$

Therefore,

$$h_{FE}(I_C = 7\text{ mA}, T = 25°\text{C}) = \frac{120}{0.88} = 136$$

At a temperature of 25°C and an I_C of 1 mA, h_{FE} appears to be about 0.78 of 136. Hence

$$h_{FE}(I_C = 1\text{ mA}, T = 25°\text{C}) = (0.78)(136) = 106$$

Therefore, at an I_C of 1 mA and a temperature of 25°C, we can anticipate a minimum β_{DC} of 106. However, at lower temperatures we can expect an even lower minimum value of h_{FE}. We do not have a curve in Figure 9 for $-40°C$, but we do have one for $-55°C$. Therefore, we shall use it instead. When I_C is 1 mA at a temperature of $-55°C$, h_{FE} is approximately 0.4 of its normalized value (136). Therefore,

$$h_{FE}(I_C = 1 \text{ mA}, T = -55°\text{C}) = (0.4)(136) = 54.4$$

which completes the problem. ■

Like the BJT, the FET has a respectable tolerance in its parameters. The two most important JFET and DE-MOSFET (dc biasing) parameters are $V_{GS(\text{OFF})}$ and I_{DSS}. We must also take their maximum and minimum values into account. The partial data sheet for a 2N5458 *n*-channel JFET has been given in Fig. 6-2. One very important fact should be remembered when interpreting the data sheet:

FETS that have a *maximum* $V_{GS(\text{OFF})}$ will have a *maximum* value of I_{DSS}. The converse is also true.

As we mentioned in Chapter 5, we employ transfer characteristics and bias lines to determine the *Q*-points for FETs. This will be true in both the analysis and the design of FET bias circuits. The first step in both cases will be to plot the minimum and maximum transfer characteristic curves. This can be accomplished by drawing upon Eq. 5-5 to arrive at Eqs. 6-1 and 6-2.

For Minimum Transfer Curves:
$$I_D = I_{DSS(\text{min})}\left[1 - \frac{V_{GS}}{V_{GS(\text{OFF-min})}}\right]^2 \qquad (6\text{-}1)$$

For Maximum Transfer Curves:
$$I_D = I_{DSS(\text{max})}\left[1 - \frac{V_{GS}}{V_{GS(\text{OFF-max})}}\right]^2 \qquad (6\text{-}2)$$

EXAMPLE 6-2

A 2N5458 is an *n*-channel JFET. According to one manufacturer, $V_{GS(\text{OFF})}$ can range from -1 to -7 V and I_{DSS} can range from 2 to 9 mA (see Fig. 6-2). Plot its transfer characteristic curves.

SOLUTION By drawing on Eq. 6-1, we may plot the minimum transconductance curve. We shall use values for V_{GS} that range from 0 V to $V_{GS(\text{OFF})}$ in 0.2 V increments. As an example, let V_{GS} be -0.2 V,

$$I_D = I_{DSS(\text{min})}\left[1 - \frac{V_{GS}}{V_{GS(\text{OFF-min})}}\right]^2$$

$$= (2 \text{ mA})\left(1 - \frac{-0.2 \text{ V}}{-1 \text{ V}}\right)^2 = 1.28 \text{ mA}$$

In a similar fashion, we may determine the values for the maximum transfer curve. If V_{GS} is set to -1 V,

$$I_D = I_{DSS(\text{max})}\left[1 - \frac{V_{GS}}{V_{GS(\text{OFF-max})}}\right]^2$$

$$= (9 \text{ mA})\left(1 - \frac{-1 \text{ V}}{-7 \text{ V}}\right)^2 = 6.61 \text{ mA}$$

2N5457
2N5458
2N5459

CASE 29-02, STYLE 5
TO-92 (TO-226AA)

JFET
GENERAL PURPOSE

N-CHANNEL — DEPLETION

Refer to 2N4220 for graphs.

MAXIMUM RATINGS

Rating	Symbol	Value	Unit
Drain-Source Voltage	V_{DS}	25	Vdc
Drain-Gate Voltage	V_{DG}	25	Vdc
Reverse Gate-Source Voltage	V_{GSR}	−25	Vdc
Gate Current	I_G	10	mAdc
Total Device Dissipation @ T_A = 25°C Derate above 25°C	P_D	310 2.82	mW mW/°C
Junction Temperature Range	T_J	125	°C
Storage Channel Temperature Range	T_{stg}	−65 to +150	°C

ELECTRICAL CHARACTERISTICS (T_A = 25°C unless otherwise noted.)

Characteristic		Symbol	Min	Typ	Max	Unit
OFF CHARACTERISTICS						
Gate-Source Breakdown Voltage (I_G = −10 μAdc, V_{DS} = 0)		$V_{(BR)GSS}$	−25	—	—	Vdc
Gate Reverse Current (V_{GS} = −15 Vdc, V_{DS} = 0) (V_{GS} = −15 Vdc, V_{DS} = 0, T_A = 100°C)		I_{GSS}	 — —	 — —	 −1.0 −200	nAdc
Gate Source Cutoff Voltage (V_{DS} = 15 Vdc, I_D = 10 nAdc)	 2N5457 2N5458 2N5459	$V_{GS(off)}$	 −0.5 −1.0 −2.0	 — — —	 −6.0 −7.0 −8.0	Vdc
Gate Source Voltage (V_{DS} = 15 Vdc, I_D = 100 μAdc) (V_{DS} = 15 Vdc, I_D = 200 μAdc) (V_{DS} = 15 Vdc, I_D = 400 μAdc)	 2N5457 2N5458 2N5459	V_{GS}	 — — —	 −2.5 −3.5 −4.5	 — — —	Vdc
ON CHARACTERISTICS						
Zero-Gate-Voltage Drain Current* (V_{DS} = 15 Vdc, V_{GS} = 0)	 2N5457 2N5458 2N5459	I_{DSS}	 1.0 2.0 4.0	 3.0 6.0 9.0	 5.0 9.0 16	mAdc
SMALL-SIGNAL CHARACTERISTICS						
Forward Transfer Admittance Common Source* (V_{DS} = 15 Vdc, V_{GS} = 0, f = 1.0 kHz)	 2N5457 2N5458 2N5459	$\lvert y_{fs} \rvert$	 1000 1500 2000	 — — —	 5000 5500 6000	μmhos
Output Admittance Common Source* (V_{DS} = 15 Vdc, V_{GS} = 0, f = 1.0 kHz)		$\lvert y_{os} \rvert$	—	10	50	μmhos
Input Capacitance (V_{DS} = 15 Vdc, V_{GS} = 0, f = 1.0 MHz)		C_{iss}	—	4.5	7.0	pF
Reverse Transfer Capacitance (V_{DS} = 15 Vdc, V_{GS} = 0, f = 1.0 MHz)		C_{rss}	—	1.5	3.0	pF

*Pulse Test: Pulse Width ≤ 630 ms; Duty Cycle ≤ 10%.

FIGURE 6-2 Data sheet for the 2N5458. (Reprinted with permission of Motorola, Inc.)

This result and several other values have also been summarized in Table 6-1. The transfer curves have been illustrated in Fig. 6-3. ■

TABLE 6-1
Transfer Characteristic Curve Data

Minimum Values		Maximum Values	
V_{GS}(V)	I_D(mA)	V_{GS}(V)	I_D(mA)
0.0	2.00	0.0	9.00
−0.2	1.28	−1.0	6.61
−0.4	0.72	−2.0	4.59
−0.6	0.32	−3.0	2.94
−0.8	0.08	−4.0	1.65
−1.0	0.00	−5.0	0.73
		−6.0	0.18
		−7.0	0.00

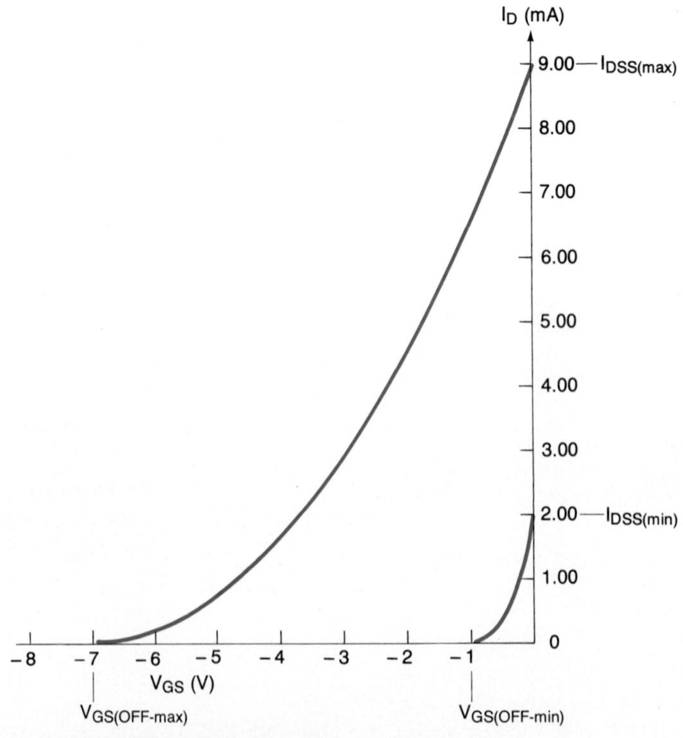

FIGURE 6-3 Minimum and maximum transfer curves for the 2N5458 JFET.

We shall focus our attention on the voltage breakdown mechanisms associated with BJTs and FETs in the next section. This is important since the breakdown voltages dictate the maximum dc supply voltages, such as V_{CC} and V_{DD}, which can be used with a given BJT or FET. Conversely, if the supply voltages are given, we must be able to ascertain if a given BJT or FET is suitable.

6-3 BJT and FET Breakdown Voltage Considerations

Both BJTs and FETs have maximum voltage ratings. These ratings are typically due to voltage breakdown mechanisms inherent in these devices. Since these breakdown voltages should never be exceeded or even approached, their magnitudes often dictate the suitability of a given device for a particular application. Consequently, these potential limitations merit our investigation. Let us consider the BJT first.

There are essentially *two* voltage breakdown mechanisms associated with the collector–base *p-n* junction: *avalanche breakdown* and *punch-through*. Avalanche voltage breakdown can best be understood by drawing an analogy between the avalanche diode (Section 2-8) and the collector–base junction. Their doping levels are comparable. Therefore, at large values of V_{CB}, it is quite possible that the reverse-biased collector–base junction may experience avalanche breakdown [see Fig. 6-4(a)]. When this occurs, the collector current I_C will tend to increase dramatically for relatively small further increases in V_{CB}. If I_C is not limited to a safe value, it is quite possible that the BJT will become severely damaged, or even destroyed. At the very least, normal transistor action will cease.

Punch-through (or reach-through) can also result in dangerously large values of I_C. BJT manufacturers are faced with a dilemma. To provide a BJT with a large current gain (i.e., β_{DC}), the base region must be very thin. However, as we saw in Section 4-16, V_{CB} modulates the effective width of the base region W_B. Specifically, as V_{CB} increases, W_B decreases. (This was called the Early effect.) If V_{CB} becomes sufficiently large, it is possible that the collector–base depletion region will extend across the base region until it touches the emitter–base depletion region. In other words, W_B is reduced to zero [see Fig. 6-4(b)].

When this occurs, I_C will flow through a heavily doped (low-resistance) FET-like "channel." Consequently, I_C will also dramatically increase as it did in the case of avalanche breakdown.

Either or both of these effects will limit the safe maximum voltage that may be impressed across the collector–base junction. Data sheets often express this rating as BV_{CBO}, or simply V_{CBO}. Again, the O subscript denotes that the third terminal (the emitter in this case) is left open during the measurement.

By inspection of Fig. 6-4(c) and the application of Kirchhoff's voltage law, we see that

$$V_{CE} = V_{CB} + V_{BE}$$

Therefore, it should come as no surprise that as V_{CE} increases, V_{CB} increases. Hence it is quite possible that a sufficiently large value of V_{CE} may also produce the conditions described above. Manufacturers denote this value of V_{CE} as BV_{CEO}, or V_{CEO}. A typical

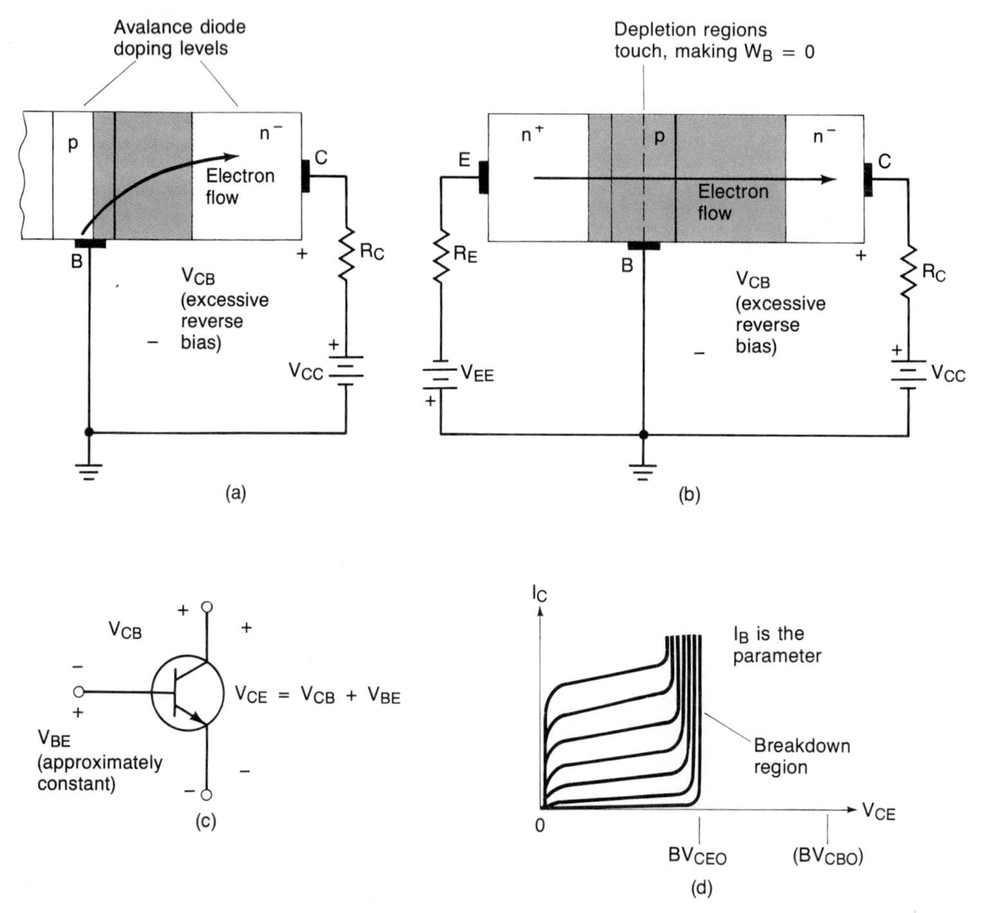

FIGURE 6-4 BJT breakdown mechanisms: (a) collector-base avalanche breakdown; (b) punch-through or reach-through; (c) voltage definitions; (d) CE V-I characteristics.

set of output V–I curves for a common-emitter BJT illustrating voltage breakdown has been shown in Fig. 6-4(d).

A third breakdown voltage parameter is also associated with the BJT – BV_{EBO} (or V_{EBO}). This refers to the maximum reverse voltage that may be impressed across the base–emitter p-n junction. The doping levels associated with the emitter and base regions are very similar to those encountered with zener diodes.

As a result, we should not be too surprised to learn that typical values of BV_{EBO} are often in the vicinity of 5 to 6 V. The primary breakdown mechanism associated with the base–emitter junction is the zener effect, in concert with the avalanche mechanism (see Fig. 6-5).

In general, the base–emitter junction will always be forward biased when we are dealing with small-signal voltage amplifiers. Therefore, the BV_{EBO} is of little consequence. However, we do need to pay particular attention to the collector voltage ratings. Specifically, they tend to limit the selection of BJTs for a given collector

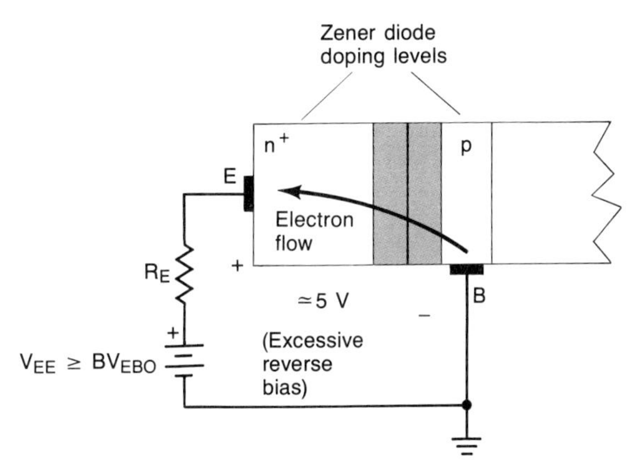

FIGURE 6-5 **Zener (and avalanche) reverse-voltage breakdown of the emitter-base p-n junction.**

supply voltage V_{CC}. Since BV_{CEO} is often less than, or equal to the value of BV_{CBO}, we offer the condition

$$V_{CC} \leq 0.8BV_{CEO} \tag{6-3}$$

which will provide at least a 20% safety margin.

The JFET maximum voltage ratings are dictated by the reverse voltage that will produce avalanche breakdown of the *p-n* junction which exists between the gate and the channel. The manufacturers typically provide two voltage ratings: the maximum drain-to-gate voltage V_{DG} and the maximum gate-to-source voltage V_{GS}. The voltage definitions for an *n*-channel JFET have been provided in Fig. 6-6(a). Hence, by applying Kirchhoff's voltage law around the transistor,

$$V_{DS} = V_{DG} + V_{GS}$$

As we saw in Chapter 5, V_{GS} assumes negative values when the *n*-channel JFET is being used in its depletion mode. Typical voltage values and their actual polarities for a small-signal JFET have been assigned in Fig. 6-6(b). As a result,

$$4\text{ V} = 7\text{ V} + (-3\text{ V})$$

and we see that Kirchhoff's voltage law is obeyed.

Both V_{GS} and V_{DG} serve to reverse bias the *p-n* junction between the gate and the channel. However, as we can see in Fig. 6-6(b), V_{DG} is the larger of the two. Consequently, *drain-to-gate* avalanche breakdown is the primary limiting factor in small-signal JFET operation. The effects of drain-to-gate breakdown can be seen in Fig. 6-6(c).

Recall that the JFET structure is symmetrical. Specifically, the roles of the drain and source terminals may be interchanged. Therefore, we typically find that for small-signal JFETs,

$$|BV_{GSS}| = |BV_{DGS}|$$

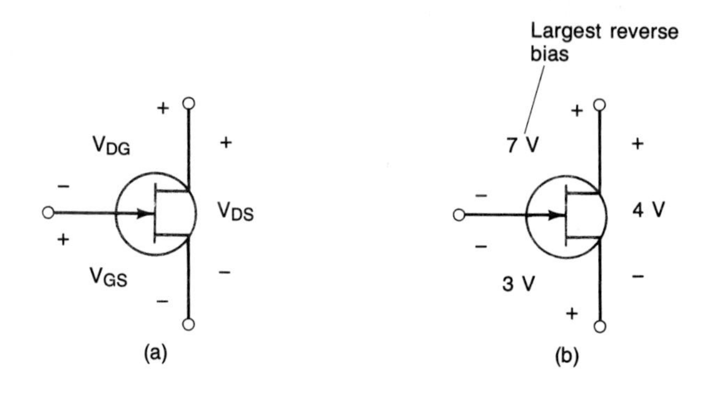

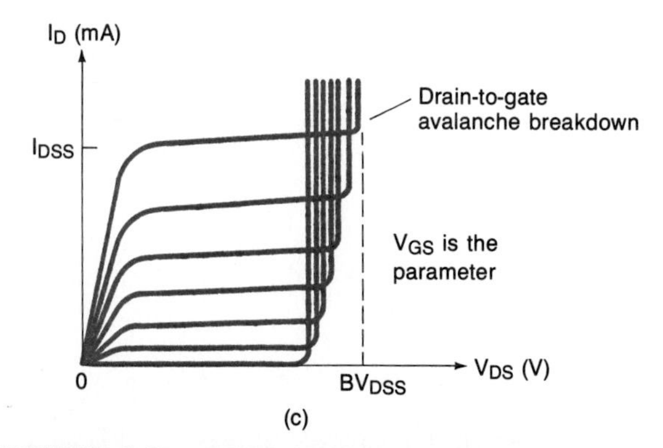

FIGURE 6-6 JFET voltage breakdown: (a) voltage definitions; (b) typical bias voltages; (c) drain-to-gate breakdown.

Observe that the third subscript is S. This means that the third terminal is shorted to ground during the measurement of these parameters. The JFET breakdown voltages have been defined in Fig. 6-7.

There are *two* constraints on the selection of the drain supply voltage V_{DD}. First, it must be large enough to permit operation of the JFET in its pinch-

FIGURE 6-7 JFET breakdown voltage measurements (typically, $|BV_{DGS}| = |BV_{GSS}| = |BV_{DSS}|$).

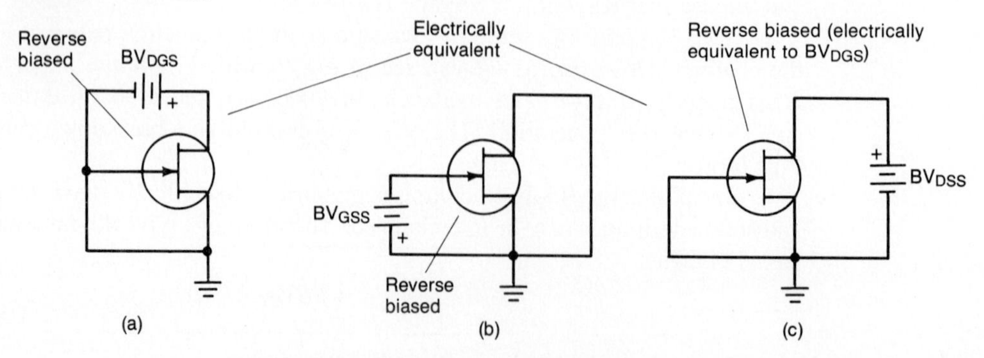

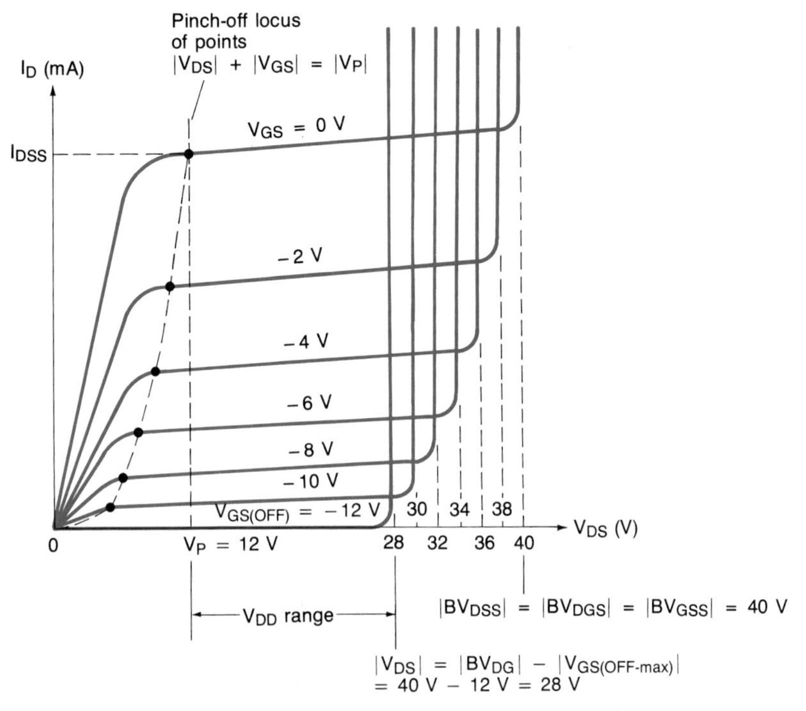

FIGURE 6-8 Range of V_{DD} values to ensure pinch-off and the avoidance of breakdown.

off (constant-current) region of operation. Second, it must not be large enough to allow the possibility of drain-to-gate breakdown.

These two constraints have been indicated on the drain characteristics given in Fig. 6-8. By inspection of it we can see that

$$|BV_{DG}| = |V_{DS}| + |V_{GS}|$$

and solving for $|V_{DS}|$,

$$|V_{DS}| = |BV_{DG}| - |V_{GS}|$$

Further analysis of the figure shows that drain-to-gate breakdown will not occur if

$$|V_{DS}| < |BV_{DG}| - |V_{GS(\text{OFF-max})}|$$

Obviously, if V_{DD} is selected to be a value less than or equal to the value of V_{DS} defined above, drain-to-gate breakdown cannot occur.

To ensure operation of a given JFET type in its constant-current region of operation, and to avoid the possibility of drain-to-gate breakdown, we provide

$$\boxed{1.2|V_{GS(\text{OFF-max})}| \leq V_{DD} \leq 0.8[|BV_{DG}| - |V_{GS(\text{OFF-max})}|]} \tag{6-4}$$

EXAMPLE 6-3

According to the manufacturer's data (Fig. 6-2), a 2N5458 *n*-channel JFET has a $V_{GS(\text{OFF-max})}$ of -7 V and a minimum of 25 V specified for its BV_{GSS}. What range of V_{DD} is suitable for this device?

SOLUTION Since

$$|BV_{DG}| = |V_{GSS}| = 25\text{ V},$$

we may employ Eq. 6-4.

$$1.2|V_{GS(\text{OFF-max})}| \leq V_{DD} \leq 0.8\,[|BV_{DG}| - |V_{GS(\text{OFF-max})}|]$$
$$1.2(7\text{ V}) \leq V_{DD} \leq 0.8(25\text{ V} - 7\text{ V})$$
$$8.4\text{ V} \leq V_{DD} \leq 14.4\text{ V}$$

■

EXAMPLE 6-4

Repeat Example 6-3 if the 2N5458 is to be used as a small-signal voltage amplifier with V_{GS} held at -0.5 V. Will a V_{DD} of 20 V be acceptable?

SOLUTION Equation 6-4 enforces unnecessarily harsh constraints on the upper limit for V_{DD} if V_{GS} is to be held at -0.5 V. We can state

$$V_{DD} \leq 0.8[|BV_{DG}| - |V_{GS}|]$$
$$= 0.8(25\text{ V} - 0.5\text{ V}) = 19.6\text{ V}$$

If we do not absolutely require a 20% safety margin, and we concede that most 2N5458s will probably have a capability greater than their 25-V rating, a V_{DD} of 20 V will be acceptable. ■

As a final comment, we should note that much of our discussion above also applies directly to the DE- and E-MOSFETs. However, the primary cause of breakdown in these devices is the electrical breakdown of the very thin SiO_2 insulating layer between the gate and the channel.

6-4 Midpoint Bias

To maintain small-signal linear operation, the operating point of an active device must not enter either cutoff or saturation. In fact, the characteristics of an active device may become nonlinear as the operating point *nears* either of these two boundaries. Consequently, it is good practice to establish the Q-point at the midpoint between saturation ($i_{C(\text{SAT})}$) and cutoff ($v_{CE(\text{OFF})}$). This is called *midpoint bias. Midpoint bias serves to provide the maximum possible symmetrical ac output voltage swing*. (This will be explained later when we investigate large-signal amplifiers.)

In Fig. 6-9(a) we see a common-emitter *npn* BJT that has been midpoint biased. Note that *no* load is connected across its output. Its Q-point has been indicated in Fig. 6-9(b). [In this case $i_{C(\text{SAT})}$ is equal to $I_{C(\text{SAT})}$ and $v_{CE(\text{OFF})}$ is equal to $V_{CE(\text{OFF})}$.] If a sinusoidal input signal is applied, the variation in base current will produce a sinusoidal variation in collector current. This, in turn, will result in a sinusoidal variation in the collector–emitter (output) voltage [see Fig. 6-9(c)]. If the BJT is *overdriven*, the operating point will enter saturation and cutoff. The output voltage

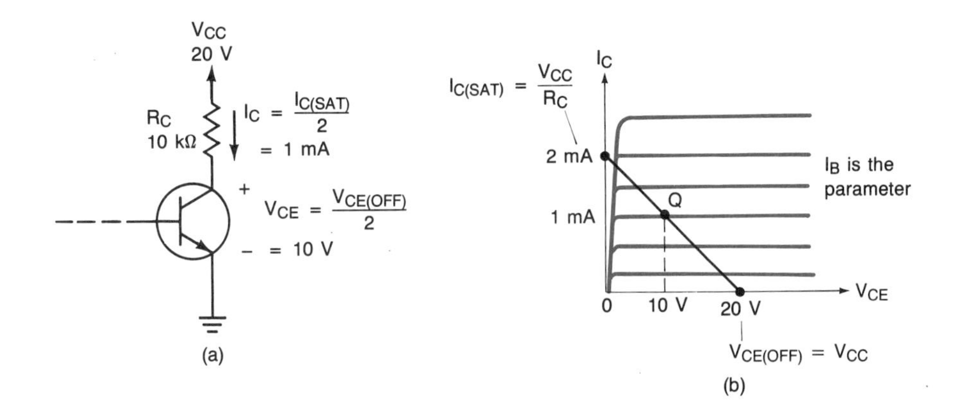

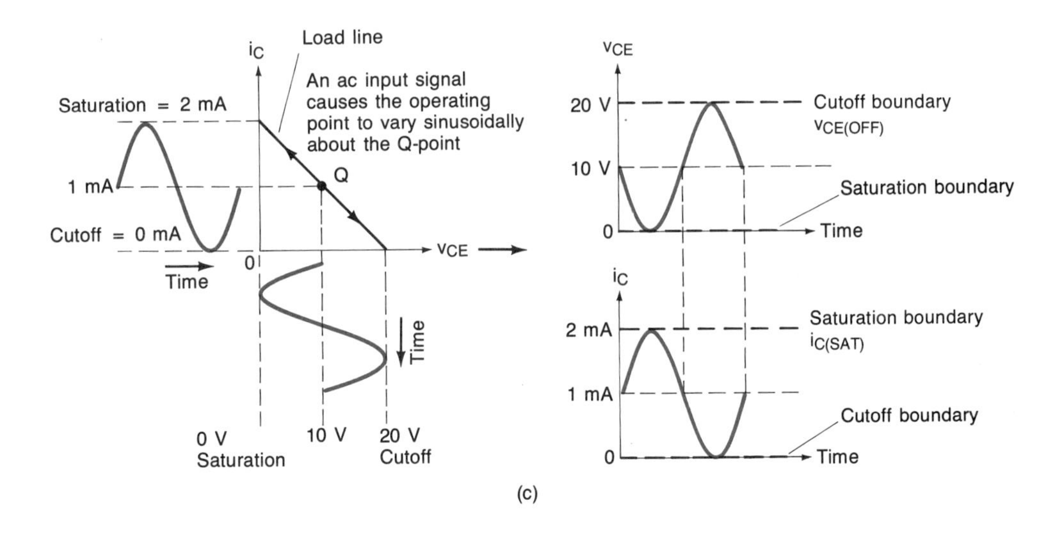

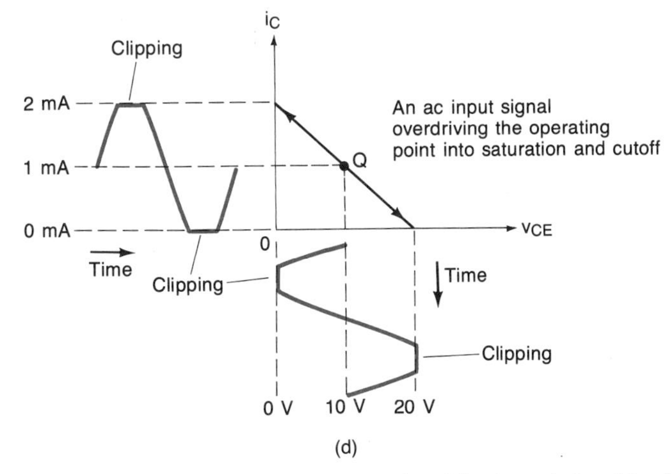

FIGURE 6-9 BJT signal process and midpoint biasing: (a) midpoint biased BJT; (b) load line; (c) signal process; (d) overdriving produces distortion.

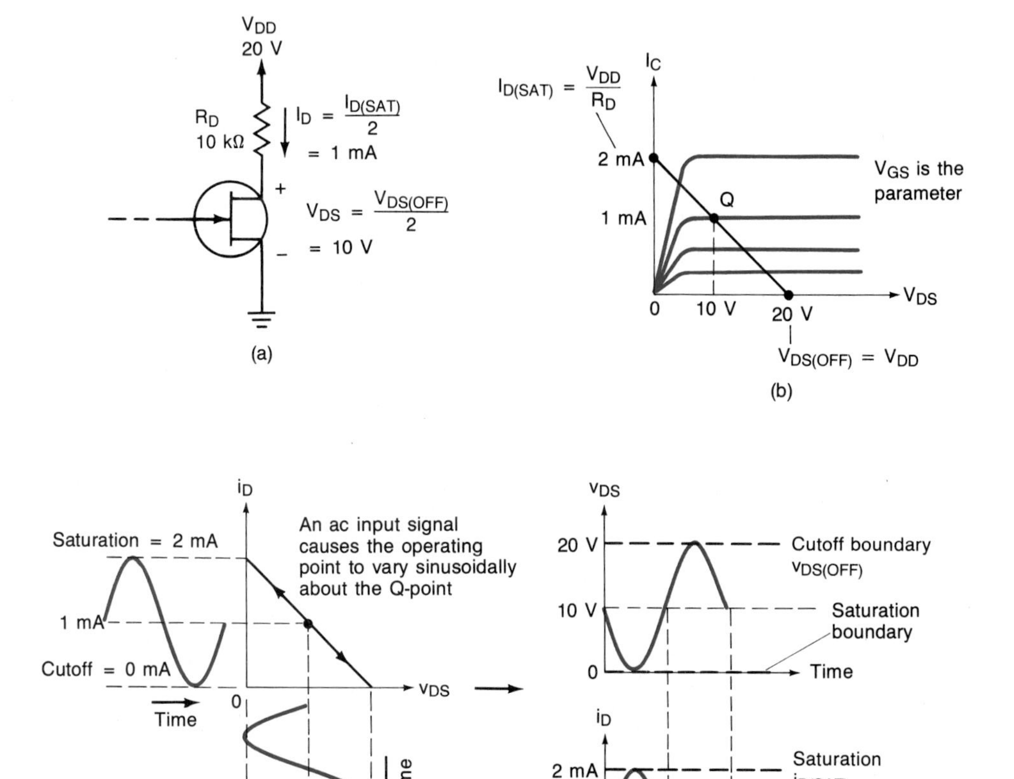

FIGURE 6-10 JFET signal process and midpoint biasing: (a) midpoint biased JFET; (b) dc load line; (c) signal process.

signal will then be "clipped" as illustrated in Fig. 6-9(d). Obviously, this distortion is highly undesirable. Exactly the same situation exists for the FETs (refer to Fig. 6-10).

In our work to follow, we shall introduce several of the classical BJT biasing circuits and their FET counterparts. To facilitate their comparison, we shall assume that *we are to use a* 2N4124 *npn BJT. The desired* I_C *is* 2 *mA, the collector supply voltage* V_{CC} *is* 20 *V, and we want midpoint bias.*

We shall also establish similar conditions for the various FET circuits. *We shall use a* 2N5458 *n-channel JFET,* I_D *is to be* 1 *mA,* V_{DD} *is* 20 *V, and we want midpoint bias.* In each case standard resistor values have been used. Consequently, the bias point may tend to become skewed more in one direction than the other. This is the normal situation in practical bias circuits. Our primary method of comparison will be to observe the percentage change in either the collector or drain quiescent currents. The ideal percentage change is *zero*. This will become clear as our work progresses.

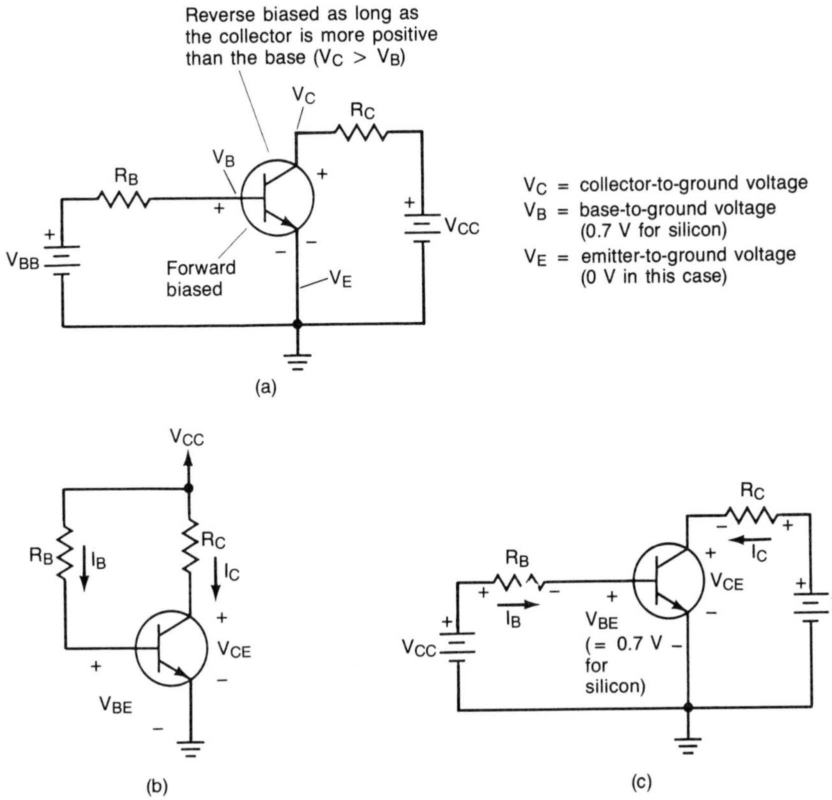

FIGURE 6-11 **CE fixed base bias: (a) biasing for linear operation; (b) using a single supply; (c) analyzing the circuit.**

A summary of the dc parameters for our devices is presented below. These specifications were extracted from the data sheets given in Figs. 6-1 and 6-2. For the 2N4124 *npn* BJT,

$BV_{CEO} = 25$ V	$I_{C(\max)} = 150$ mA
$BV_{CBO} = 30$ V	h_{FE} ($I_C = 2$ mA, $T = 25°$C) ranges
$BV_{EBO} = 5$ V	from 120 to 360 (3:1 variation)

For the 2N5458 *n*-channel JFET,

$BV_{DSS} = 25$ V	I_{DSS} ranges from 2 to 9 mA
$BV_{DGS} = 25$ V	(4.5:1 variation)
$BV_{GSS} = 25$ V	$V_{GS(OFF)}$ ranges from -1 to -7 V (7:1 variation)

6-5 *CE* Fixed Base Bias

In Fig. 6-11(a) we have an example of *common-emitter* (CE) *fixed base bias*. Notice that the base–emitter junction is forward biased. The collector–base junction will be reverse biased as long as the collector-to-ground voltage V_C is *more positive* than the base-to-ground voltage V_B. Recall that these are the necessary bias conditions for the BJT to be in its (linear) active region of operation.

If R_B is sized correctly, it is possible to use a V_{BB} that is equal to V_{CC}. If this is the case, and it usually is, the base and the collector circuits may share the collector supply voltage [see Fig. 6-11(b)].

To analyze this circuit, we proceed in exactly the same manner as we have done previously. First, we write an equation for the input. However, this time the controlling input quantity is I_B. Therefore, we shall solve our input equation for I_B. To transfer from the input circuit to the output, we shall employ β_{DC} to solve for I_C. Once we have ascertained I_C, we shall then apply Kirchhoff's voltage law and solve for the output voltage V_{CE}.

With reference to Fig. 6-11(c), we may use Kirchhoff's voltage law and solve for I_B.

$$I_B = \frac{V_{CC} - V_{BE}}{R_B} = \frac{V_{CC} - 0.7\text{ V}}{R_B} \tag{6-5}$$

Next, we may find I_C.

$$I_C = \beta_{DC} I_B \tag{6-6}$$

FIGURE 6-12 Analysis of fixed base bias: (a) saturation; (b) cutoff; (c) dc load line; (d) load line without V-I curves.

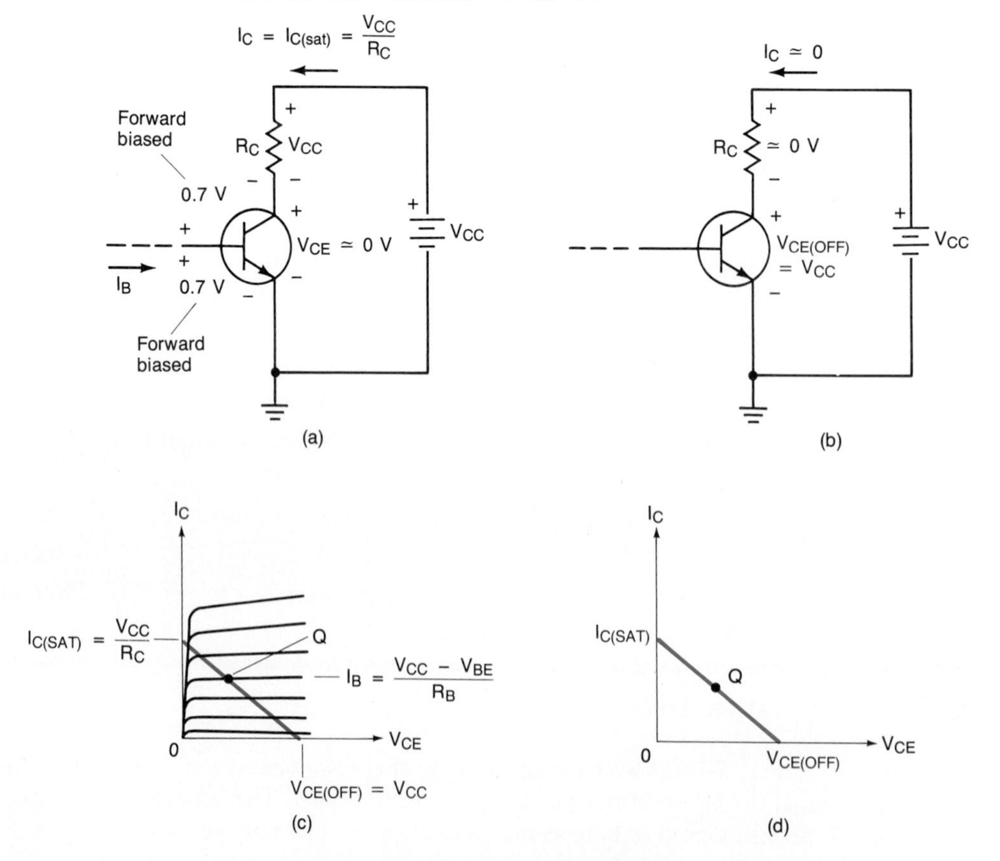

Now we may use Kirchhoff's voltage law to determine V_{CE}.

$$V_{CE} = V_{CC} - I_C R_C \tag{6-7}$$

Remember that the two ideal boundaries for linear operation are saturation and cutoff. For Fig. 6-12(a), we can see that $I_{C(\text{SAT})}$ is

$$I_{C(\text{SAT})} = \frac{V_{CC}}{R_C} \tag{6-8}$$

and in Fig. 6-12(b) we see that

$$V_{CE(\text{OFF})} = V_{CC} \tag{6-9}$$

The dc load line has been indicated in Fig. 6-12(c) to illustrate graphically the results of analysis. For clarity, we omit the characteristic curves in our discussions to emphasize the dc load line [see Fig. 6-12(d)].

EXAMPLE 6-5

A 2N4124 has an h_{FE} of 120. If it is used in the CE fixed-base-bias circuit shown in Fig. 6-13, determine I_B, I_C, V_{CE}, $I_{C(\text{SAT})}$, and $V_{CE(\text{OFF})}$.

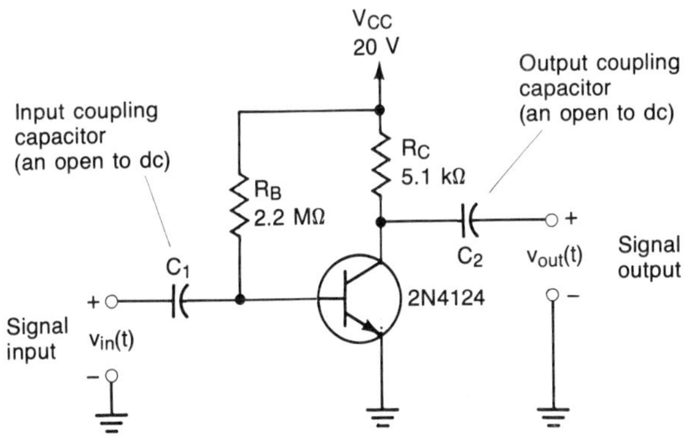

FIGURE 6-13 Fixed-base-bias circuit.

SOLUTION Following the procedure described above, we have

$$I_B = \frac{V_{CC} - 0.7\text{ V}}{R_B} = \frac{20\text{ V} - 0.7\text{ V}}{2.2\text{ M}\Omega} = 8.77\ \mu\text{A}$$

$$I_C = \beta_{DC} I_B = (120)(8.77\ \mu\text{A}) = 1.05\text{ mA}$$

$$V_{CE} = V_{CC} - I_C R_C = 20\text{ V} - (1.05\text{ mA})(5.1\text{ k}\Omega) = 14.6\text{ V}$$

$$V_{CE(\text{OFF})} = V_{CC} = 20\text{ V}$$

■

The fixed-bias circuit offers simplicity as its primary advantage. However, it is an extremely poor biasing arrangement as it does *not* provide any compensation for the parameter variations in a BJT due to temperature or tolerance.

Consider the example below.

EXAMPLE 6-6

Assume that the h_{FE} of the 2N4124 used in the circuit given in Fig. 6-13 increases to 360. (This could result from the replacement of the original 2N4124 with another 2N4124 with a higher h_{FE} or an increase in temperature.) Repeat the analysis requested in Example 6-5.

SOLUTION The base current will remain essentially constant.

$$I_B = 8.77\ \mu\text{A}$$

However, the collector current will tend to increase.

$$I_C = \beta_{DC} I_B = (360)(8.77\ \mu\text{A}) = 3.16\ \text{mA}$$

and

$$V_{CE} = V_{CC} - I_C R_C = 20\ \text{V} - (3.16\ \text{mA})(5.1\ \text{k}\Omega) = 3.88\ \text{V}$$

Obviously, $I_{C(SAT)}$ and $V_{CE(OFF)}$ are unaffected by the value of h_{FE}. ■

TABLE 6-2
BJT Fixed Bias

Parameter	Bias Values		Average	$\pm\%\ \Delta$
h_{FE}	120	360	240	±50
I_B	8.77 μA	8.77 μA	8.77 μA	±0
I_C	1.05 mA	3.16 mA	2.10 mA	±50
V_{CE}	14.6 V	3.88 V	9.27 V	±58
$I_{C(SAT)}$	3.92 mA	3.92 mA	—	—
$V_{CE(OFF)}$	20 V	20 V	—	—

The results of our analysis have been summarized in Table 6-2. The average values are determined by (maximum + minimum)/2. The % Δ values are determined by

$$\%\ \Delta = \frac{\frac{1}{2}(\Delta)}{\text{average}} \times 100\% = \frac{(\frac{1}{2})(\max - \min)}{(\frac{1}{2})(\max + \min)} \times 100\%$$

$$= \frac{\max - \min}{\max + \min} \times 100\%$$

and represent the percent change in the quantity about the average value. The only time the Q-point will be at its midpoint is when β_{DC} is halfway between 120 and 360, which is its average value, 240 (see Fig. 6-14).

As a final comment, we should point out that if Eq. 6-6 results in an I_C larger than $I_{C(SAT)}$, Eq. 6-7 will produce a *negative* value for V_{CE}. Some beginning students rigorously follow the procedure illustrated in Example 6-5 and are quite satisfied when negative values for V_{CE} occur. Obviously, this is *not* possible and is an indication that the BJT is in saturation.

6-6 *CS* Fixed Gate Bias

A common-source (CS) JFET that employs fixed bias is given in Fig. 6-15(a). In Section 5-10 we analyzed this biasing arrangement in detail. Its analysis follows

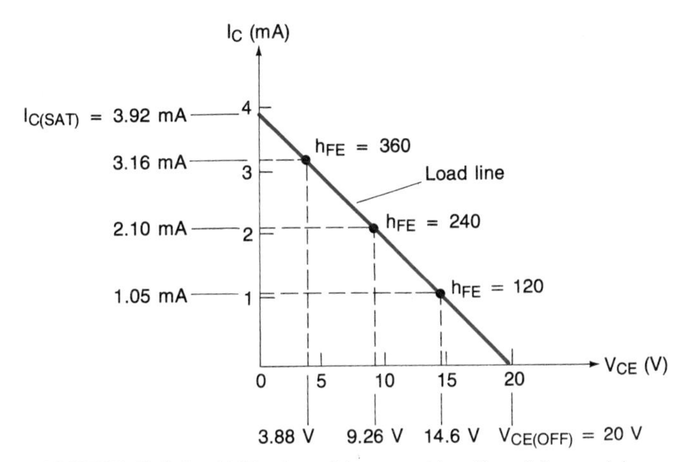

FIGURE 6-14 Effects of h_{FE} on the fixed base bias Q-point.

FIGURE 6-15 JFET fixed gate bias: (a) circuit; (b) bias line; (c) load line and Q-points.

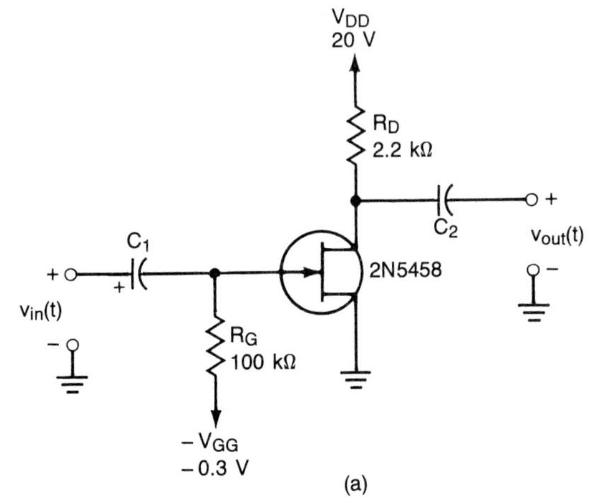

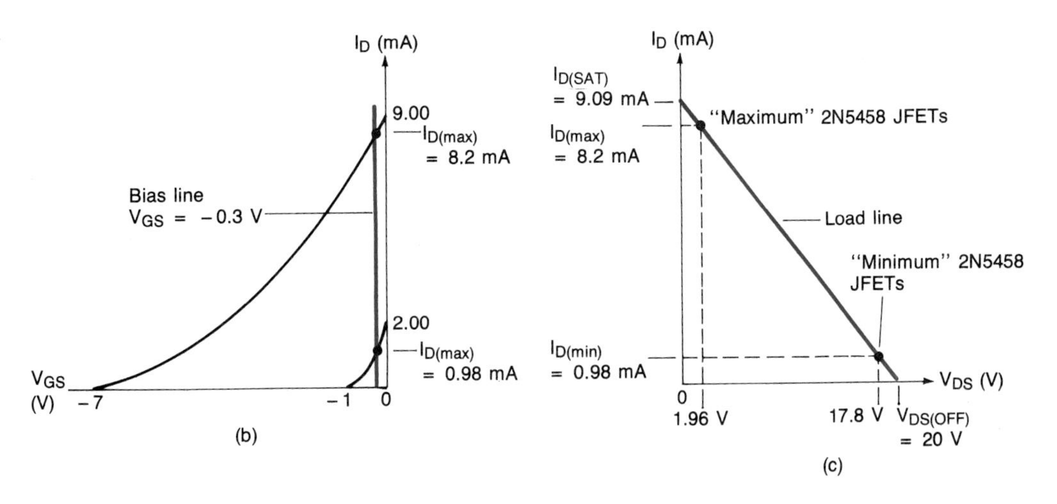

exactly the same strategy employed for the BJT with the exception that a graphical approach is used to transfer across the device.

To review, we first write the input equation and solve it for V_{GS}. We then employ the transfer characteristic curves and plot the bias line on it. This allows us to determine I_D. The analysis is completed by using Kirchhoff's voltage law to find V_{DS}.

To determine the effects of the device tolerance, we shall use the minimum and maximum transfer characteristic curves which were introduced in Section 6-2. This will allow us to determine the range of the possible values for I_D and V_{DS}.

EXAMPLE 6-7

The 2N5458 *n*-channel JFET analyzed in Example 6-2 is being used in the fixed-bias circuit given in Fig. 6-15(a). [The dc equivalent circuit is the same as the one shown in Fig. 5-22(b).] Determine V_{GS} and the minimum and maximum values for I_D and V_{DS}. Also find $I_{D(SAT)}$ and $V_{DS(OFF)}$.

SOLUTION The bias equation may be determined first.

$$V_{GS} = -V_{GG} = -0.3 \text{ V}$$

By plotting the bias line on the transfer characteristics as shown in Fig. 6-15(b), we may determine $I_{D(\min)}$ and $I_{D(\max)}$. By inspection,

$$I_{D(\min)} = 0.98 \text{ mA}$$
$$I_{D(\max)} = 8.2 \text{ mA}$$

With a little reflection, it should be clear that the minimum value of I_D will produce the maximum value for V_{DS}. Conversely, the maximum value of I_D will result in the minimum value of V_{DS}. Hence

$$V_{DS(\max)} = V_{DD} - I_{D(\min)}R_D$$
$$= 20 \text{ V} - (0.98 \text{ mA})(2.2 \text{ k}\Omega) = 17.8 \text{ V}$$

and

$$V_{DS(\min)} = V_{DD} - I_{D(\max)}R_D$$
$$= 20 \text{ V} - (8.2 \text{ mA})(2.2 \text{ k}\Omega) = 1.96 \text{ V}$$

To complete the problem, we find that

$$I_{D(SAT)} = \frac{V_{DD}}{R_D} = \frac{20 \text{ V}}{2.2 \text{ k}\Omega} = 9.09 \text{ mA}$$

and

$$V_{DS(OFF)} = V_{DD} = 20 \text{ V}$$

The load line is given in Fig. 6-15(c). ■

A summary of CS fixed bias is given in Table 6-3. Note that midpoint biasing with an I_D of 1 mA, is *not* possible for the 2N5458 in the fixed-bias configuration. Fixed bias does not compensate for variations in JFET device parameters.

Fixed bias is *not* an acceptable approach for either BJTs or FETs.

TABLE 6-3
JFET Fixed Bias

Parameter	Bias Values		Average	±% Δ
I_{DSS}	2 mA	9 mA	5.5 mA	±64
$V_{GS(\mathrm{OFF})}$	−1 V	−7 V	−4 V	±75
V_{GS}	−0.3 V	−0.3 V	−0.3 V	±0
I_D	0.98 mA	8.2 mA	4.59 mA	±79
V_{DS}	17.8 V	1.96 V	9.88 V	±80
$I_{D(\mathrm{SAT})}$	9.09 mA	9.09 mA	—	—
$V_{DS(\mathrm{OFF})}$	20 V	20 V	—	—

6-7 *CE* Emitter-Stabilized Bias

By adding a resistor R_E in series with the emitter, we arrive at the CE *emitter-stabilized* bias circuit shown in Fig. 6-16(a). If we apply Kirchhoff's voltage law around the input circuit as shown in Fig. 6-16(b),

$$-V_{CC} + I_B R_B + 0.7\text{ V} + (I_B + I_C)R_E = 0$$

Since we wish to solve for I_B, we must eliminate I_C. This can be accomplished by recalling that

$$I_C = \beta_{\mathrm{DC}} I_B$$

If we substitute this into our Kirchhoff's voltage law equation, we obtain

$$-V_{CC} + I_B R_B + 0.7\text{ V} + (I_B + \beta_{\mathrm{DC}} I_B)R_E = 0$$

If we solve the resulting equation for I_B, we have

$$I_B = \frac{V_{CC} - 0.7\text{ V}}{R_B + (1 + \beta_{\mathrm{DC}})R_E} \simeq \frac{V_{CC}}{R_B + \beta_{\mathrm{DC}} R_E} \tag{6-10}$$

The approximate relationship is valid as long as V_{CC} is much greater than 0.7 V and β_{DC} is much greater than unity. Although this is the usual case, we shall favor the more exact relationship.

To determine I_C, we again draw on β_{DC}.

$$I_C = \beta_{\mathrm{DC}} I_B$$

The output analysis yields

$$V_{CE} = V_{CC} - I_C R_C - (I_C + I_B)R_E \tag{6-11}$$
$$\simeq V_{CC} - (R_C + R_E)I_C$$

The approximation holds true as long as I_C is much larger than I_B. Again, this is the usual case.

In Fig. 6-16(a) we see the definitions of the terminal voltages with respect to ground. We may find the collector-to-ground voltage from

$$V_C = V_{CC} - I_C R_C \tag{6-12}$$

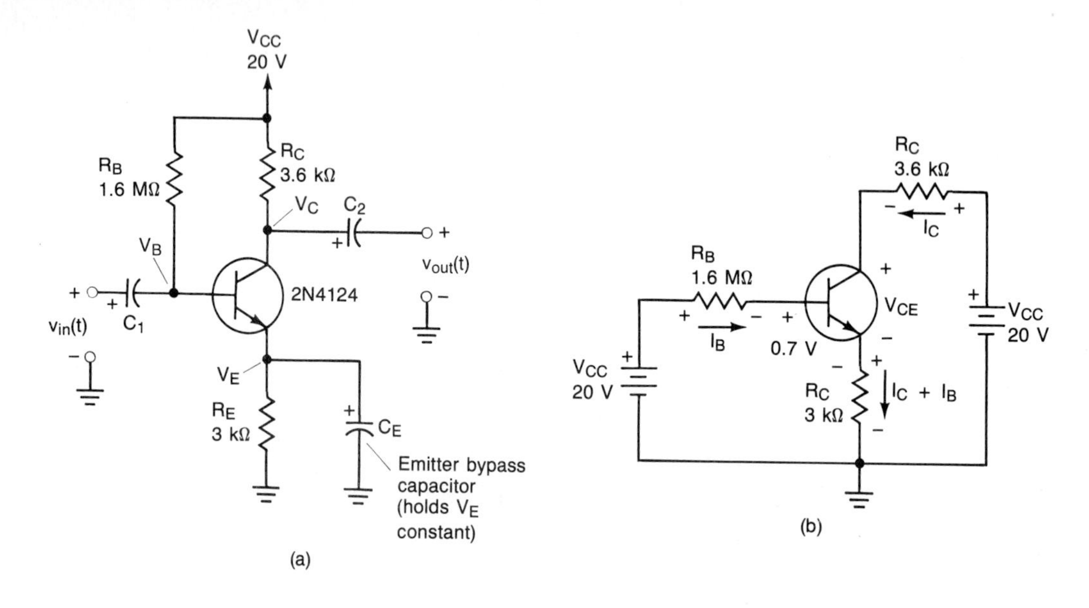

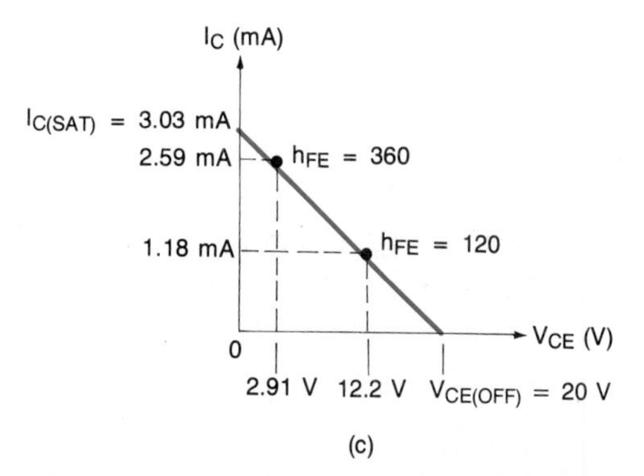

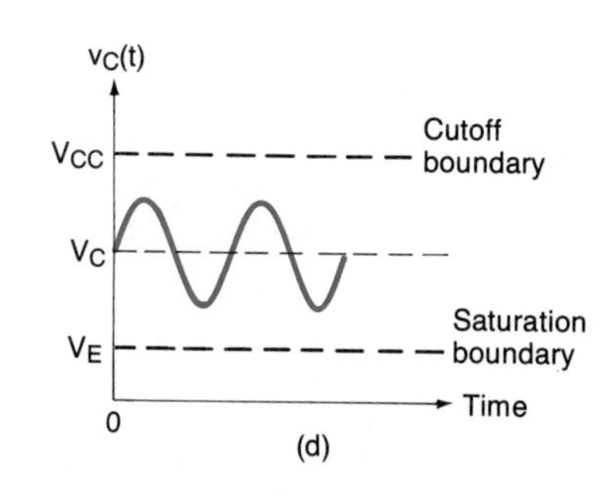

FIGURE 6-16 CE emitter-stabilized bias: (a) BJT circuit; (b) dc equivalent circuit; (c) dc load line; (d) cutoff and saturation limits.

and the emitter-to-ground voltage is given by

$$V_E = I_E R_E \simeq I_C R_E \tag{6-13}$$

The base-to-ground voltage is one base–emitter voltage drop higher than V_E.

$$V_B = V_E + 0.7\text{ V} \tag{6-14}$$

The saturation current $I_{C(\text{SAT})}$ is limited by R_C and R_E.

$$I_{C(\text{SAT})} = \frac{V_{CC}}{R_C + R_E} \tag{6-15}$$

and the cutoff voltages are

$$V_{CE(\text{OFF})} = V_{C(\text{OFF})} = V_{CC} \tag{6-16}$$

EXAMPLE 6-8

The BJT amplifier given in Fig. 6-16(a) uses emitter-stabilized bias. Assume that the 2N4124 has an h_{FE} of 120. Determine I_B, I_C, V_{CE}, V_C, V_E, V_B, $I_{C(\text{SAT})}$, $V_{CE(\text{OFF})}$, and $V_{C(\text{OFF})}$.

SOLUTION Observing the analysis outlined above, we proceed as follows:

$$I_B = \frac{V_{CC} - 0.7\text{ V}}{R_B + (1 + \beta_{\text{DC}})R_E} = \frac{20\text{ V} - 0.7\text{ V}}{1.6\text{ M}\Omega + (1 + 120)(3\text{ k}\Omega)}$$

$$= 9.83\ \mu\text{A}$$

$$I_C = \beta_{\text{DC}} I_B = (120)(9.83\ \mu\text{A}) = 1.18\text{ mA}$$

$$V_{CE} \simeq V_{CC} - I_C(R_C + R_E)$$

$$= 20\text{ V} - (1.18\text{ mA})(3.6\text{ k}\Omega + 3\text{ k}\Omega) = 12.2\text{ V}$$

$$V_C = V_{CC} - I_C R_C$$

$$= 20\text{ V} - (1.18\text{ mA})(3.6\text{ k}\Omega) = 15.8\text{ V}$$

$$V_E \simeq I_C R_E = (1.18\text{ mA})(3\text{ k}\Omega) = 3.54\text{ V}$$

$$V_B = V_E + 0.7\text{ V} = 3.54\text{ V} + 0.7\text{ V} = 4.24\text{ V}$$

$$I_{C(\text{SAT})} = \frac{V_{CC}}{R_C + R_E} = \frac{20\text{ V}}{3.6\text{ k}\Omega + 3\text{ k}\Omega} = 3.03\text{ mA}$$

$$V_{CE(\text{OFF})} = V_{C(\text{OFF})} = V_{CC} = 20\text{ V}$$

■

The addition of the emitter resistor serves to assist in the stabilization of the Q-point. This can be seen by close inspection of Eq. 6-10. If β_{DC} *increases* due to temperature, or because the transistor is replaced by another with a higher β_{DC}, the base current will be *reduced*. Consequently, the collector current will not increase as much as it would have in the case of fixed base bias. A more intuitive explanation also supports this observation. An increase in I_C increases the voltage drop across R_E. This reduces the voltage across R_B. Subsequently, I_B will be reduced to counter the increase in I_C. This process is referred to as *dc negative feedback*.

$$\overset{\uparrow}{I_C} = \overset{\uparrow}{\beta_{\text{DC}}}\overset{\downarrow}{I_B}$$

$\uparrow$ — the net increase in I_C is reduced

The circuit given in Fig. 6-16(a) has been reanalyzed using an h_{FE} of 360. The results are summarized in Table 6-4, and the load line is given in Fig. 6-16(c). The

TABLE 6-4
BJT Emitter-Stabilized Bias

Parameter	Bias Values		Average	±% Δ
h_{FE}	120	360	240	±50
I_B	9.83 μA	7.19 μA	8.51 μA	±16
I_C	1.18 mA	2.59 mA	1.88 mA	±37
V_{CE}	12.2 V	2.91 V	7.56 V	±61
$I_{C(\text{SAT})}$	3.03 mA	3.03 mA	—	—
$V_{CE(\text{OFF})}$	20 V	20 V	—	—

variation in I_C is $\pm37\%$ compared to $\pm50\%$ for fixed bias. However, the variation in V_{CE} has increased to $\pm61\%$. The circuit has improved the stability of I_C, but no improvement has been made on the variations in V_{CE}.

Obviously, the use of an emitter resistor to stabilize the Q-point is a step in the right direction. However, the emitter-stabilized bias circuit is usually not an acceptable solution in many problems. Observe [in Fig. 6-16(a)] that a capacitor has been placed in parallel with R_E. It is referred to as the *emitter bypass capacitor*. Its use allows dc negative feedback, but eliminates ac negative feedback by acting as a short circuit to the ac signal. Much more will be said about this in our later work. However, for now, we shall merely acknowledge that C_E serves to keep V_E constant and that V_E restricts our minimum output voltage swing [see Fig. 6-16(d)].

The effectiveness of the emitter-stabilized biasing arrangement is a direct function of the size of R_E. In practice we find that the size of R_E becomes prohibitively large in order to provide adequate biasing stabilization.

6-8 *CS* Source-Stabilized or Self-Bias

The FET equivalent to the BJT's emitter-stabilized bias circuit is the *source-stabilized* or *self-bias* circuit depicted in Fig. 6-17(a). The reason for the use of the term "self-bias" will be explained shortly.

By analyzing the input circuit as given in Fig. 6-17(b), Kirchhoff's voltage law produces

$$-I_{GSS}R_G + V_{GS} + I_D R_S = 0$$

We shall assume that the leakage current I_{GSS} produces a negligible voltage drop across R_G and solve for V_{GS} to obtain the bias line equation.

$$V_{GS} = -I_D R_S \tag{6-17}$$

By close inspection of this result the reason this FET circuit is described as "self-bias" becomes clear. The gate-to-ground voltage is held at zero volts. However, as I_D flows through R_S, the source is made *positive* with respect to ground, and a negative V_{GS} results.

If this bias equation is plotted on the minimum and maximum transfer characteristic curves, we may find the range of the possible values for I_D[see Fig. 6-17(c)]. Once this is accomplished, we may analyze the output circuit [see Fig. 6-17(b)].

$$V_{DS} = V_{DD} - (R_D + R_S)I_D \tag{6-18}$$

Observe the similarity between Eq. 6-18 and the output equation for the BJT's emitter-stabilized bias circuit (Eq. 6-11). With the exception of V_G, the equations for the terminal-to-ground voltages also strongly resemble those of the BJT [see Fig. 6-17(a)].

$$V_G = I_{GSS}R_G \simeq 0 \text{ V} \tag{6-19}$$

$$V_D = V_{DD} - R_D I_D \tag{6-20}$$

$$V_S = I_D R_S \tag{6-21}$$

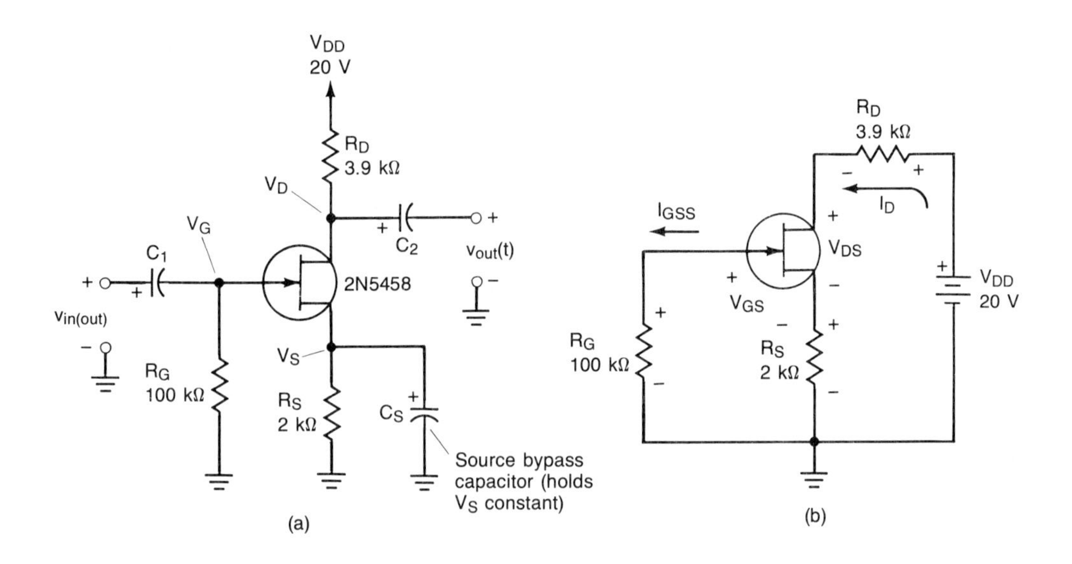

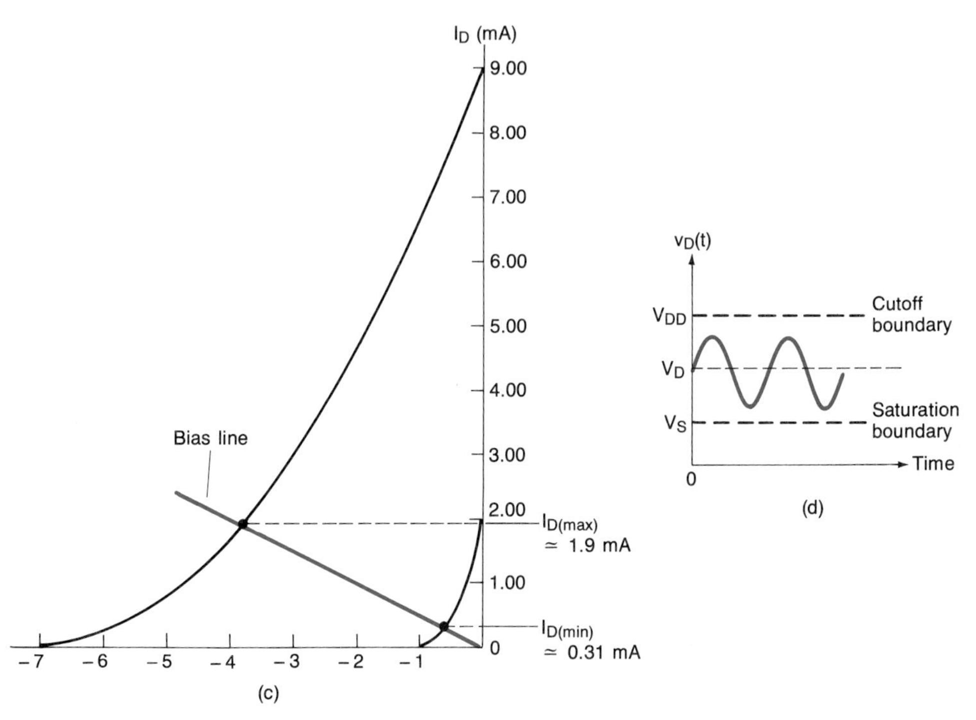

FIGURE 6-17 Source-stabilized (self-bias) JFET: (a) JFET circuit; (b) dc equivalent; (c) dc bias line; (d) cutoff and saturation limits.

EXAMPLE 6-9

Given the JFET amplifier in Fig. 6-17(a), determine the range of values for I_D, V_S, V_G, V_{GS}, V_{DS}, and V_D. Also ascertain $I_{D(\text{SAT})}$ and $V_{DS(\text{OFF})}$.

SOLUTION First, we notice that a *source* bypass capacitor C_S has been placed in parallel with R_S. It serves the same function as the emitter bypass capacitor. Therefore, the minimum value of instantaneous drain voltage $v_D(t)$ is limited to V_S. To determine I_D, we note that the bias equation is

$$V_{GS} = -I_D R_S = -(2000\ \Omega)I_D$$

By substituting in convenient values for I_D, we may plot the bias line on the transfer characteristic curves. For example, if we let I_D be 1 mA,

$$V_{GS} = -(2000\ \Omega)(1\text{ mA}) = -2\text{ V}$$

The resulting bias line has been plotted in Fig. 6-17(c). By inspection, we see that the intersections yield

$$I_{D(\text{min})} \simeq 0.31\text{ mA} \quad \text{and} \quad I_{D(\text{max})} \simeq 1.9\text{ mA}$$

Using these values, we may determine the range of possible voltage values for the circuit. When I_D is equal to $I_{D(\text{min})}$ [0.31 mA],

$$\begin{aligned}
V_{S(\text{min})} &= I_{D(\text{min})}R_S \\
&= (0.31\text{ mA})(2\text{ k}\Omega) = 0.62\text{ V} \\
V_G &= 0\text{ V} \\
V_{GS(\text{min})} &= -V_{S(\text{min})} = -0.62\text{ V} \\
V_{DS(\text{max})} &= V_{DD} - I_{D(\text{min})}(R_D + R_S) \\
&= 20\text{ V} - (0.31\text{ mA})(3.9\text{ k}\Omega + 2\text{ k}\Omega) = 18.2\text{ V} \\
V_{D(\text{max})} &= V_{DD} - I_{D(\text{min})}R_D \\
&= 20\text{ V} - (0.31\text{ mA})(3.9\text{ k}\Omega) = 18.8\text{ V}
\end{aligned}$$

and for an I_D of 1.9 mA,

$$\begin{aligned}
V_{S(\text{max})} &= I_{D(\text{max})}R_S \\
&= (1.9\text{ mA})(2\text{ k}\Omega) = 3.8\text{ V} \\
V_G &= 0\text{ V} \\
V_{GS(\text{max})} &= -V_{S(\text{max})} = -3.8\text{ V}
\end{aligned}$$

TABLE 6-5
JFET Source-Stabilized Bias

Parameter	Bias Values		Average	±% Δ
I_{DSS}	2 mA	9 mA	5.5 mA	±64
$V_{GS(\text{OFF})}$	−1 V	−7 V	−4 V	±75
V_{GS}	−0.62 V	−3.8 V	−2.21 V	±72
I_D	0.31 mA	1.9 mA	1.11 mA	±72
V_{DS}	18.2 V	8.79 V	13.5 V	±35
$I_{D(\text{SAT})}$	3.39 mA	3.39 mA	—	—
$V_{DS(\text{OFF})}$	20 V	20 V	—	—

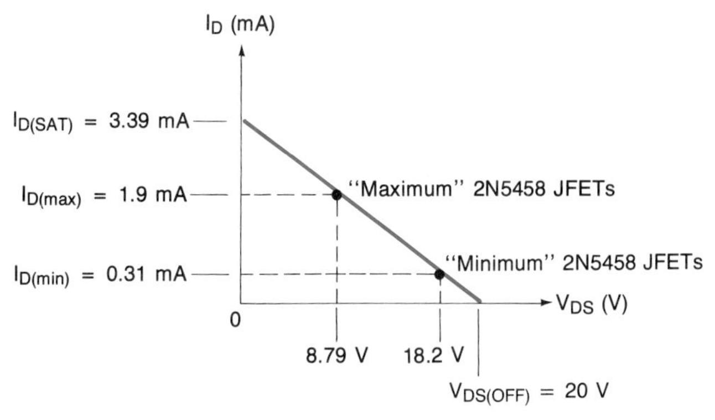

FIGURE 6-18 The dc load line for the JFET source-stabilized bias circuit.

$$V_{DS(\min)} = V_{DD} - I_{D(\max)}(R_D + R_S)$$
$$= 20\text{ V} - (1.9\text{ mA})(3.9\text{ k}\Omega + 2\text{ k}\Omega)$$
$$= 8.79\text{ V}$$
$$V_{D(\min)} = V_{DD} - I_{D(\max)}R_D$$
$$= 20\text{ V} - (1.9\text{ mA})(3.9\text{ k}\Omega) = 12.6\text{ V}$$

These two bias conditions have been summarized in Table 6-5. To complete the analysis, we find $I_{D(\text{SAT})}$ and $V_{DS(\text{OFF})}$.

$$I_{D(\text{SAT})} = \frac{V_{DD}}{R_D + R_S} = \frac{20\text{ V}}{3.9\text{ k}\Omega + 2\text{ k}\Omega} = 3.39\text{ mA}$$
$$V_{DS(\text{OFF})} = V_{DD} = 20\text{ V}$$

The load line is given in Fig. 6-18. ■

Again, as with the BJT, we see that the dc negative feedback has served to stabilize the bias. The variation in I_D has been reduced to $\pm 72\%$ as compared to $\pm 79\%$ for fixed gate bias. The variation in V_{DS} has been reduced to only $\pm 35\%$ compared to $\pm 80\%$ for fixed gate bias. However, the voltage-drop limitation imposed by R_S is just as severe as R_E.

The use of the source-stabilized FET biasing scheme is also generally undesirable if reasonably large signal swings are anticipated or we want to be absolutely sure of midpoint biasing.

The signal cutoff and saturation limits are shown in Fig. 6-17(d).

6-9 Collector Voltage Feedback Bias

The *collector voltage feedback* bias circuit has been depicted in Fig. 6-19(a). One of the primary advantages of this circuit is the lack of the emitter voltage limitation on the output voltage swing. Its simplicity is also an obvious plus. Let us proceed with its analysis.

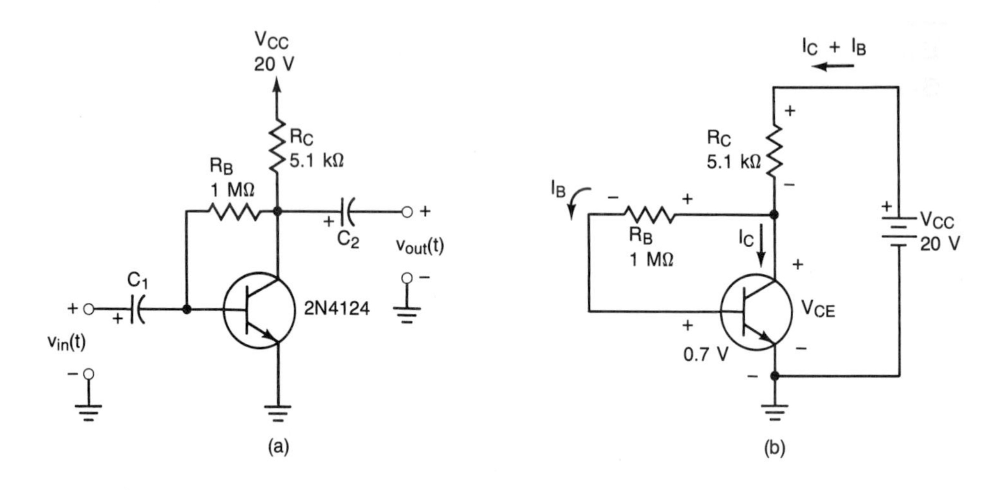

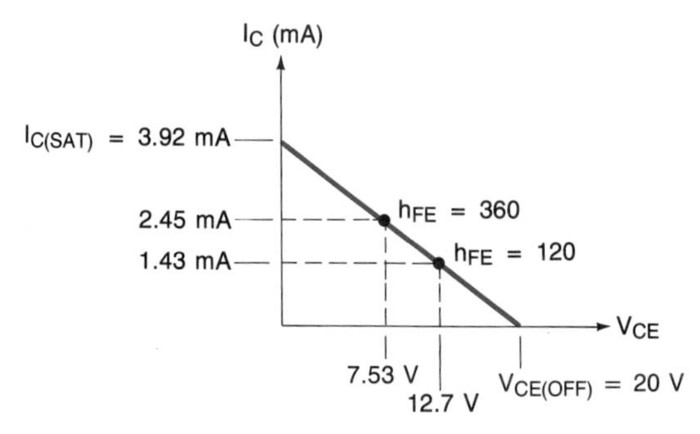

FIGURE 6-19 Collector voltage feedback bias: (a) BJT circuit; (b) dc equivalent circuit; (c) dc load line.

The input equation is obtained by applying Kirchhoff's voltage law around the input circuit as shown in Fig. 6-19(b). Its development is very similar to that for the emitter-stabilized bias circuit.

$$-V_{CC} + (I_C + I_B)R_C + I_B R_B + 0.7\text{ V} = 0$$

Using $I_C = \beta_{DC} I_B$ and solving for I_B yields

$$I_B = \frac{V_{CC} - 0.7\text{ V}}{R_B + (1 + \beta_{DC})R_C} \tag{6-22}$$

Compare this result with Eq. 6-10. Equation 6-22 also reflects the fact that the circuit will tend to counter variations in β_{DC}.

The output equation is based on the assumption that I_C is much greater than I_B [see Fig. 6-19(b)].

$$V_{CE} = V_C \simeq V_{CC} - I_C R_C \tag{6-23}$$

The rest of the equations describing this circuit are extremely similar to ones that we have developed previously. Consider Example 6-10.

EXAMPLE 6-10

Perform a dc analysis of the collector voltage feedback bias circuit given in Fig. 6-19(a). Assume that h_{FE} ranges from 120 to 360.

SOLUTION Following the normal procedure and assuming that h_{FE} is 120, we have

$$I_B = \frac{V_{CC} - 0.7\text{ V}}{R_B + (1 + \beta_{DC})R_C} = \frac{20\text{ V} - 0.7\text{ V}}{1\text{ M}\Omega + (1 + 120)(5.1\text{ k}\Omega)} = 11.9\ \mu\text{A}$$

$$I_C = \beta_{DC} I_B = (120)(11.9\ \mu\text{A}) = 1.43\text{ mA}$$

$$V_{CE} = V_C \simeq V_{CC} - I_C R_C$$

$$= 20\text{ V} - (1.43\text{ mA})(5.1\text{ k}\Omega) = 12.7\text{ V}$$

The collector saturation current is

$$I_{C(SAT)} = \frac{V_{CC}}{R_C} = \frac{20\text{ V}}{5.1\text{ k}\Omega} = 3.92\text{ mA}$$

By inspection of Fig. 6-19(b), we can see that $V_{CE(OFF)}$ is 20 V, V_B is 0.7 V, and of course, V_E must be 0 V. These results and the results of the analysis conducted when h_{FE} is 360 are given in Table 6-6. The load line is illustrated in Fig. 6-19(c). ■

TABLE 6-6
BJT Collector Voltage Feedback Bias

Parameter	Bias Values		Average	±% Δ
h_{FE}	120	360	240	±50
I_B	11.9 μA	6.79 μA	9.35 μA	±27
I_C	1.43 mA	2.45 mA	1.94 mA	±26
V_{CE}	12.7 V	7.53 V	10.1 V	±26
$I_{C(SAT)}$	3.92 mA	3.92 mA	—	—
$V_{CE(OFF)}$	20 V	20 V	—	—

From the results of our analysis, it is clear that this circuit configuration is superior to the emitter-stabilized bias circuit from both a bias stability and a signal swing standpoint. Once again we are using dc negative feedback to stabilize the Q-point. (Some ac negative feedback is also present, but it is not of immediate concern.)

The strategy behind the circuit is straightforward. The collector–emitter voltage V_{CE} is impressed across the base resistor R_B minus one base–emitter drop. If β_{DC} increases, I_C will increase. An increase in I_C will decrease V_{CE}. Therefore, the voltage across R_B will be reduced, I_B will also be reduced, and the increase in I_C due to β_{DC} will be countered.

6-10 Drain Voltage Feedback Bias

The voltage feedback bias technique *cannot* be applied to either the n- or the p-channel JFETs. This is true because we cannot enhance them any more than a few tenths of a volt. However, the technique can be applied to the DE-MOSFETs, the E-MOSFETs, or even a CMOS pair (see Fig. 6-20).

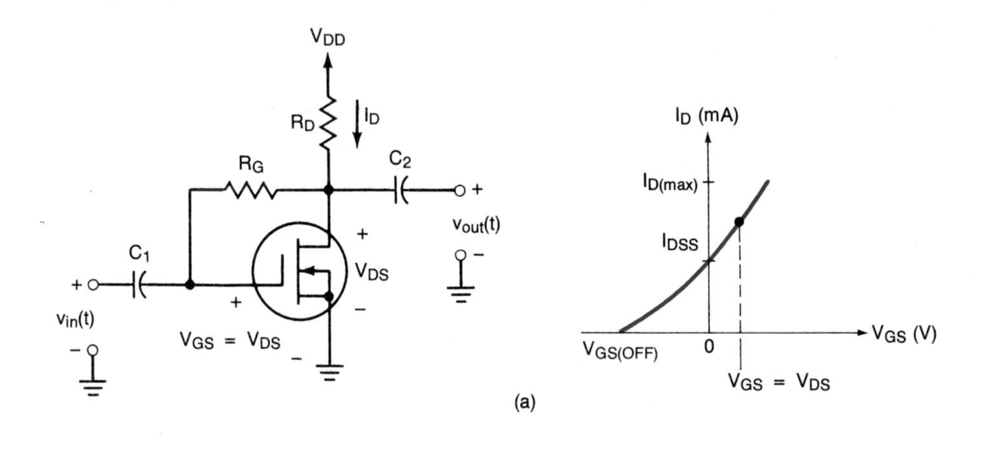

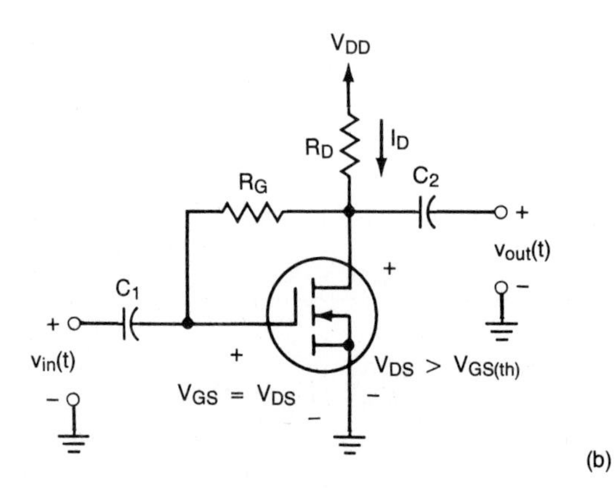

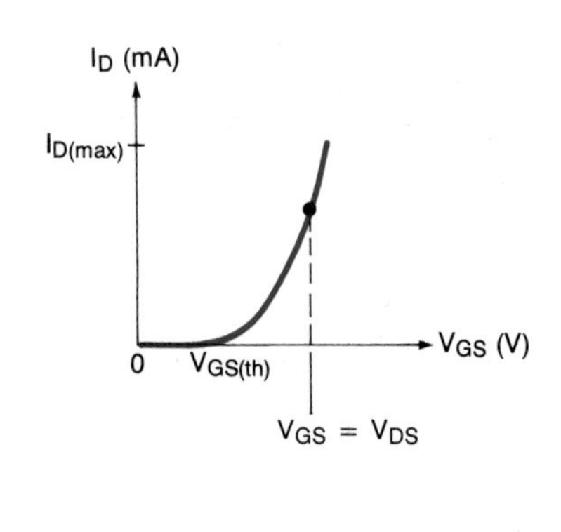

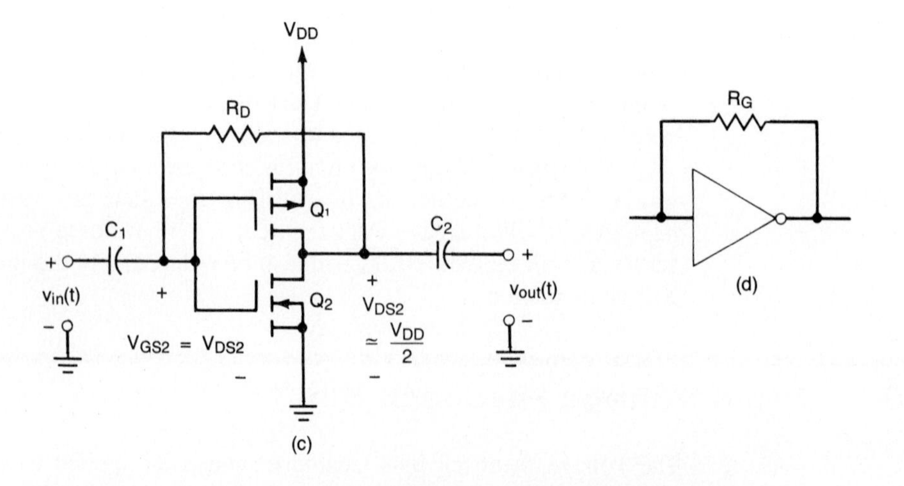

FIGURE 6-20 Drain voltage feedback bias: (a) DE-MOSFET; (b) E-MOSFET; (c) CMOS pair; (d) CMOS digital logic inverter with biasing for linear operation.

In Fig. 6-20(a) we see an *n*-channel DE-MOSFET being used in its enhancement mode of operation. Since the current drawn by the gate terminal is very nearly zero, no voltage is dropped across the gate resistor R_G and

$$V_{GS} = V_{DS}$$

Consequently, the bias line is given by

$$V_{GS} = V_{DD} - I_D R_D$$

A typical transconductance curve and a representative *Q*-point are also illustrated in Fig. 6-20(a).

An *n*-channel E-MOSFET with drain voltage feedback is provided in Fig. 6-20(b). Again, V_{GS} is equal to V_{DS}, and the bias equation is exactly the same as that for the DE-MOSFET. For proper operation of the E-MOSFET to occur, V_{DS} (and V_{GS}) must be greater than $V_{GS(\text{th})}$. This is emphasized in Fig. 6-20(b).

In Fig. 6-20(c) we have illustrated a CMOS pair. In Section 5-15 we introduced its principles of operation. Its operation as a linear amplifier was developed around the use of fixed bias. Specifically, we biased the gates at $V_{DD}/2$. This value of V_{GS} is typically well above the magnitude of the $V_{GS(\text{th})}$ for the two E-MOS devices. Therefore, both transistors will be into conduction. If the two MOSFETs have reasonably similar characteristics (the usual case), they will divide the drain supply voltage equally, as indicated in Fig. 6-20(c). Therefore, the use of drain voltage feedback will provide a V_{GS} of $V_{DD}/2$. We shall not delve into an explanation of the bias line at this point. This is because we are dealing with *nonlinear load resistors*. (Breathe a sigh of relief.)

Quite often, digital CMOS inverters are used in linear applications, such as amplifiers or oscillators, by employing drain voltage feedback bias [refer to Fig. 6-20(d)].

6-11 *CE* Voltage-Divider Bias

A circuit offering superior bias stability for a BJT when a single-polarity voltage supply is used is *voltage-divider bias* [see Fig. 6-21(a)]. The primary strategy behind this circuit rests in establishing a *bleeder current* through the voltage divider which effectively ''swamps out'' the *maximum* base current drawn by the BJT. A BJT will draw the maximum base current when that particular BJT possesses a *minimum* β_{DC} [see Fig. 6-21(b)].

This scheme effectively ''locks in'' the base-to-ground voltage V_B to a value that is *independent* of the particular β_{DC} possessed by the BJT. If I_C increases due to an increase in β_{DC}, I_E will also increase. An increase in I_E will raise the emitter-to-ground voltage V_E [see Fig. 6-21(c)].

From Fig. 6-21(c) we can see that

$$V_{BE} = V_B - V_E$$

Hence an *increase* in V_E will serve to *decrease* V_{BE}. In Section 4-14 we saw that the magnitude of V_{BE} controls the magnitude of I_C. Therefore, the decrease in V_{BE} will serve to reduce I_C—virtually canceling its increase due to β_{DC}.

The analysis of this circuit is easily accomplished after the application of

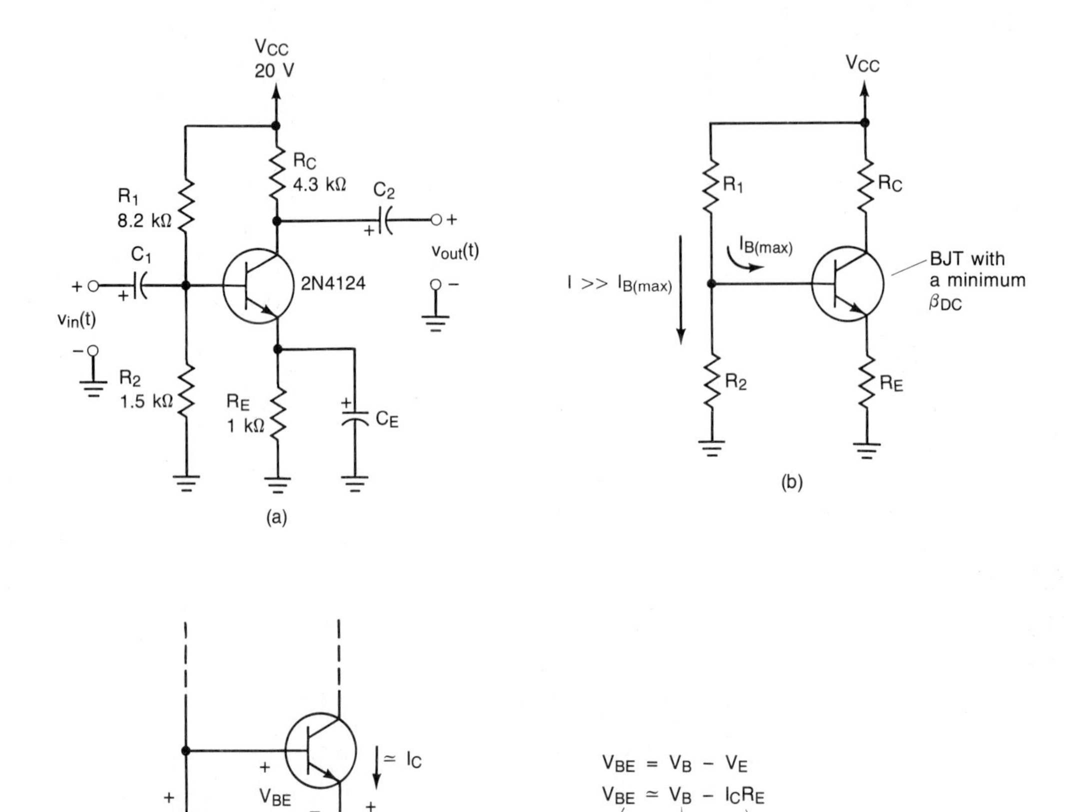

FIGURE 6-21 CE voltage-divider bias: (a) circuit; (b) the bleeder current I "swamps out" $I_{B(max)}$ to "lock in" V_B; (c) an increase in I_C will increase V_E to reduce V_{BE}.

Thévenin's theorem to simplify the base circuit as "seen" by the BJT's base terminal (refer to Fig. 6-22).

$$R_{TH} = R_1 \parallel R_2 \tag{6-24}$$

$$V_{TH} = \frac{R_2}{R_1 + R_2} V_{CC} \tag{6-25}$$

By applying Kirchhoff's voltage law around the equivalent input circuit, we arrive at the equation for I_B.

$$I_B = \frac{V_{TH} - 0.7 \text{ V}}{R_{TH} + (1 + \beta_{DC})R_E} \tag{6-26}$$

Observe its similarity to the equation we developed for the emitter-stabilized bias circuit (refer to Eq. 6-10). The balance of the analysis is merely a repetition of our previous efforts. Therefore, let us proceed with an example.

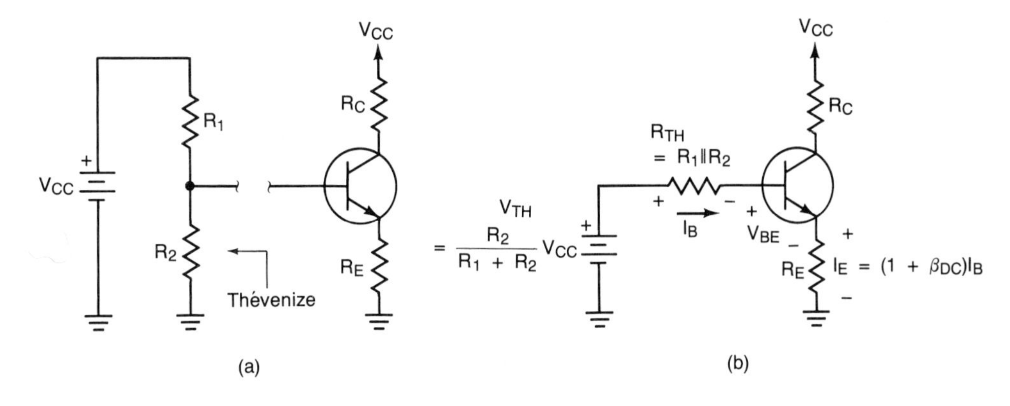

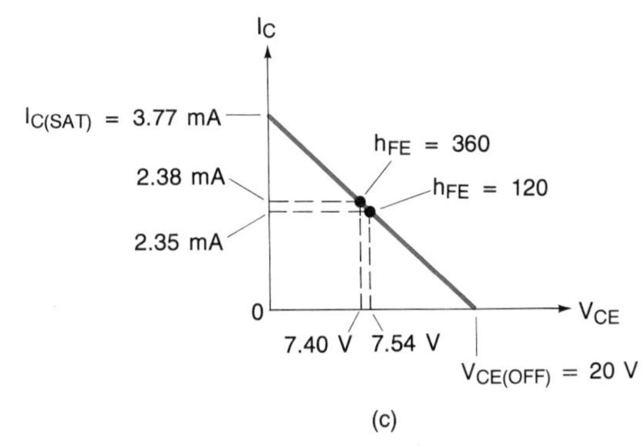

FIGURE 6-22 Determining I_B: (a) finding the Thévenin equivalent circuit; (b) analyzing the equivalent base circuit; (c) dc load line.

EXAMPLE 6-11

Perform a dc analysis of the voltage-divider bias circuit given in Fig. 6-21(a). Assume that h_{FE} varies from 120 to 360.

SOLUTION First, we must determine R_{TH} and V_{TH}.

$$R_{\text{TH}} = R_1 \parallel R_2 = 8.2\ \text{k}\Omega \parallel 1.5\ \text{k}\Omega = 1.268\ \text{k}\Omega$$

$$V_{\text{TH}} = \frac{R_2}{R_1 + R_2} V_{CC} = \frac{1.5\ \text{k}\Omega}{8.2\ \text{k}\Omega + 1.5\ \text{k}\Omega} 20\ \text{V} = 3.093\ \text{V}$$

and assuming an h_{FE} of 120, we may solve for I_B.

$$I_B = \frac{V_{\text{TH}} - 0.7\ \text{V}}{R_{\text{TH}} + (1 + \beta_{\text{DC}})R_E} = \frac{3.093\ \text{V} - 0.7\ \text{V}}{1.268\ \text{k}\Omega + (1 + 120)(1\ \text{k}\Omega)}$$
$$= 19.6\ \mu\text{A}$$

Further analysis yields

$$I_C = \beta_{\text{DC}} I_B = (120)(19.6\ \mu\text{A}) = 2.35\ \text{mA}$$
$$V_{CE} \simeq V_{CC} - I_C(R_C + R_E) = 20\ \text{V} - (2.35\ \text{mA})(4.3\ \text{k}\Omega + 1\ \text{k}\Omega)$$
$$= 7.54\ \text{V}$$
$$V_C = V_{CC} - I_C R_C = 20\ \text{V} - (2.35\ \text{mA})(4.3\ \text{k}\Omega) = 9.89\ \text{V}$$

$$V_E \simeq I_C R_E = (2.35 \text{ mA})(1 \text{ k}\Omega) = 2.35 \text{ V}$$

$$V_B = V_E + 0.7 \text{ V} = 2.35 \text{ V} + 0.7 \text{ V} = 3.05 \text{ V}$$

and the saturation and cutoff limits are

$$I_{C(\text{SAT})} = \frac{V_{CC}}{R_C + R_E} = \frac{20 \text{ V}}{4.3 \text{ k}\Omega + 1 \text{ k}\Omega} = 3.77 \text{ mA}$$

$$V_{CE(\text{OFF})} = V_{C(\text{OFF})} = V_{CC} = 20 \text{ V}$$

Repeating the calculations for an h_{FE} of 360 produces the results indicated in Table 6-7. The dc load line has been illustrated in Fig. 6-22(c). ■

TABLE 6-7
***CE* Voltage-Divider Bias**

Parameter	Bias Values		Average	±% Δ
h_{FE}	120	360	240	±50
I_B	19.60 μA	6.61 μA	13.1 μA	±49.6
I_C	2.35 mA	2.38 mA	2.37 mA	±0.63
V_{CE}	7.54 V	7.40 V	7.47 V	±0.94
$I_{C(\text{SAT})}$	3.77 mA	3.77 mA	—	—
$V_{CE(\text{OFF})}$	20 V	20 V	—	—

The voltage-divider bias circuit has provided an I_C that is "rock solid." Consequently, the ac signal parameters (which are affected by I_C) will also tend to be reasonably "locked in." I_B precisely counters the variation in h_{FE}, which results in less than a 1% variation in I_C and V_{CE}.

The superior performance of the BJT voltage-divider bias circuit makes it the most popular bias configuration when only a single-polarity supply voltage is available.

6-12 *CS* Voltage-Divider Bias

The voltage-divider technique may also be applied to the JFET, as shown in Fig. 6-23(a). The analysis proceeds in much the same manner as that employed for the BJT. Specifically, we apply Thévenin's theorem to the gate circuit to arrive at the equivalent circuit given in Fig.6-23(b).

$$R_{\text{TH}} = R_1 \parallel R_2 \tag{6-27}$$

$$V_{\text{TH}} = \frac{R_2}{R_1 + R_2} V_{DD} \tag{6-28}$$

An analysis of the equivalent circuit via Kirchhoff's voltage law and its solution for V_{GS} produces the bias equation.

$$V_{GS} = V_{\text{TH}} - I_D R_S \tag{6-29}$$

To plot the bias line on the transfer characteristic curves, we may select two convenient points. For example, if we let I_D be equal to zero, Eq. 6-29 reduces to

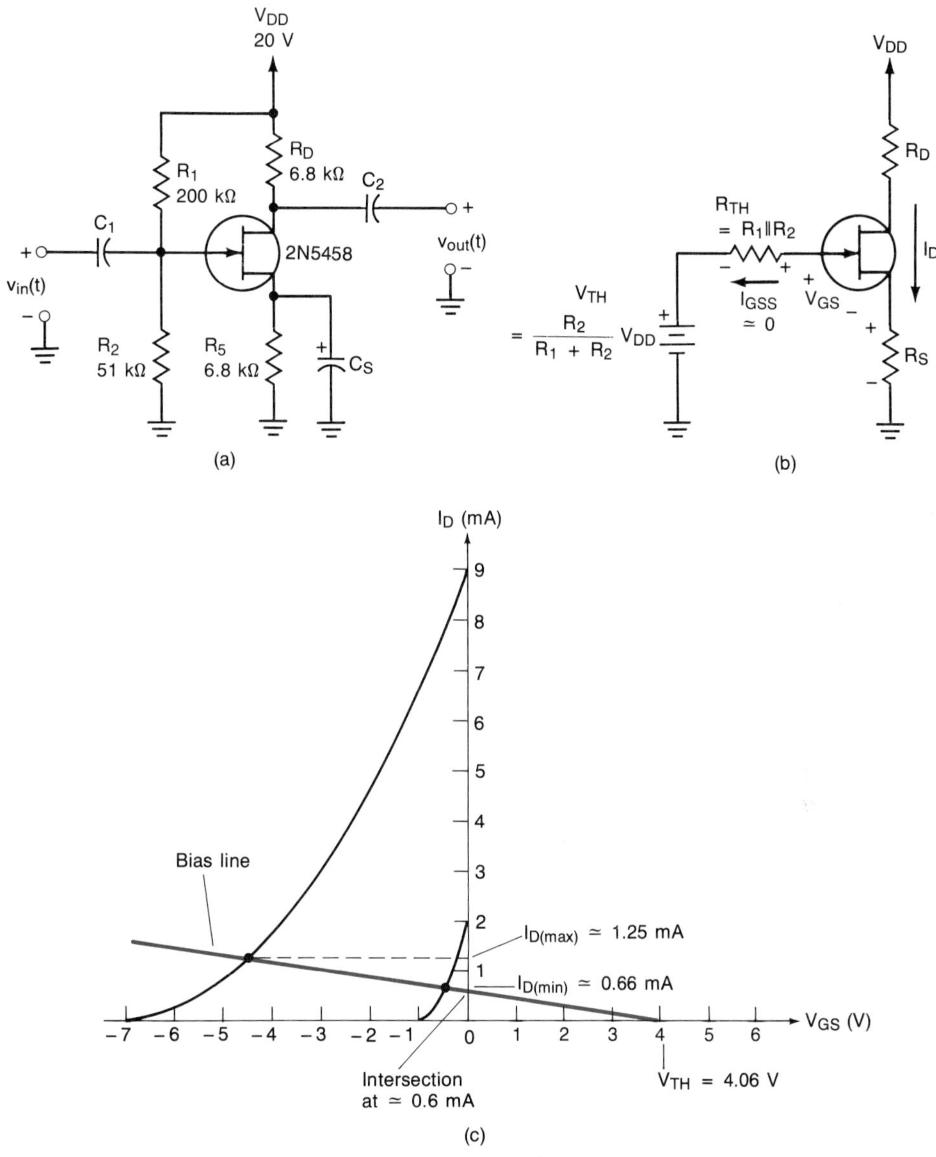

FIGURE 6-23 CS voltage-divider bias circuit: (a) circuit; (b) analyzing the equivalent gate circuit; (c) bias line.

$$V_{GS} = V_{\text{TH}}$$

Similarly, we may solve for the intercept on the I_D axis by setting V_{GS} equal to zero and solving Eq. 6-29 for I_D.

$$V_{\text{TH}} - I_D R_S = V_{GS} = 0$$

$$I_D = \frac{V_{\text{TH}}}{R_S}$$

Since the rest of the analysis steps are reasonably familiar to us at this point, let us proceed directly to an example.

EXAMPLE 6-12

Perform a complete dc analysis on the bias circuit given in Fig. 6-23(a).

SOLUTION The minimum and the maximum transfer curves for the 2N5458 have been plotted in Fig. 6-23(c). From Fig. 6-23(a) we can see that

$$R_{TH} = R_1 \| R_2 = 200\text{ k}\Omega \| 51\text{ k}\Omega = 40.6\text{ k}\Omega$$

$$V_{TH} = \frac{R_2}{R_1 + R_2} V_{DD} = \frac{51\text{ k}\Omega}{200\text{ k}\Omega + 51\text{ k}\Omega} 20\text{ V} = 4.06\text{ V}$$

Therefore, the bias equation is

$$V_{GS} = V_{TH} - I_D R_S = 4.06\text{ V} - (6.8\text{ k}\Omega)I_D$$

Next we solve for the intersection points. For $I_D = 0$ we obtain

$$V_{GS} = V_{TH} = 4.06\text{ V}$$

and for $V_{GS} = 0$ V,

$$I_D = \frac{V_{TH}}{R_S} = \frac{4.06\text{ V}}{6.8\text{ k}\Omega} = 0.597\text{ mA} \simeq 0.6\text{ mA}$$

These two points have been used to plot the bias line as illustrated in Fig. 6-23(c). The intersections between the bias line and the transfer curves indicate that

$$I_{D(\text{max})} = 1.25\text{ mA} \quad \text{and} \quad I_{D(\text{min})} = 0.66\text{ mA}$$

For 2N5458s with maximum transfer curves, an $I_{D(\text{max})}$ of 1.25 mA produces

$$V_{GS(\text{max})} = V_{TH} - I_{D(\text{max})}R_S = 4.06\text{ V} - (1.25\text{ mA})\,(6.8\text{ k}\Omega)$$
$$= -4.44\text{ V}$$

$$V_{S(\text{max})} = I_{D(\text{max})}R_S = (1.25\text{ mA})\,(6.8\text{ k}\Omega) = 8.5\text{ V}$$

$$V_{DS(\text{min})} = V_{DD} - I_{D(\text{max})}(R_D + R_S)$$
$$= 20\text{ V} - (1.25\text{ mA})\,(6.8\text{ k}\Omega + 6.8\text{ k}\Omega) = 3\text{ V}$$

$$V_{D(\text{min})} = V_{DD} - I_{D(\text{max})}R_D = 20\text{ V} - (1.25\text{ mA})\,(6.8\text{ k}\Omega)$$
$$= 11.5\text{ V}$$

The saturation and cutoff boundaries are

$$I_{D(\text{SAT})} = \frac{V_{DD}}{R_D + R_S} = \frac{20\text{ V}}{6.8\text{ k}\Omega + 6.8\text{ k}\Omega} = 1.47\text{ mA}$$

$$V_{DS(\text{OFF})} = V_{D(\text{OFF})} = V_{DD} = 20\text{ V}$$

The analysis for 2N5458s with a minimum transfer curve is included in Table 6-8. ■

TABLE 6-8
JFET Voltage-Divider Bias

Parameter	Bias Values		Average	±% Δ
I_{DSS}	2 mA	9 mA	5.5 mA	±64
$V_{GS(\text{OFF})}$	−1 V	−7 V	−4 V	±75
V_{GS}	−0.428 V	−4.44 V	−2.43 V	±82
I_D	0.66 mA	1.25 mA	0.955 mA	±31
V_{DS}	11.0 V	3.00 V	7 V	±57
$I_{D(\text{SAT})}$	1.47 mA	1.47 mA	—	—
$V_{DS(\text{OFF})}$	20 V	20 V	—	—

If we compare Tables 6-5 and 6-8, we see that the bias stability exceeds that of the source-stabilized circuit. However, it is not nearly as successful as BJT voltage-divider bias. The variation in I_D is reduced from ±72% to ±31%, but the variation in V_{DS} has *increased* from ±35% to ±57%.

6-13 *CE* Emitter Bias

If dual-polarity voltages are available, such as ±12 V, ±15 V, or ±20 V, it is possible to use a negative voltage source to bias the emitter of an *npn* BJT [see Fig.

FIGURE 6-24 CE emitter bias: (a) circuit; (b) analyzing the input; (c) analyzing the output; (d) the load line.

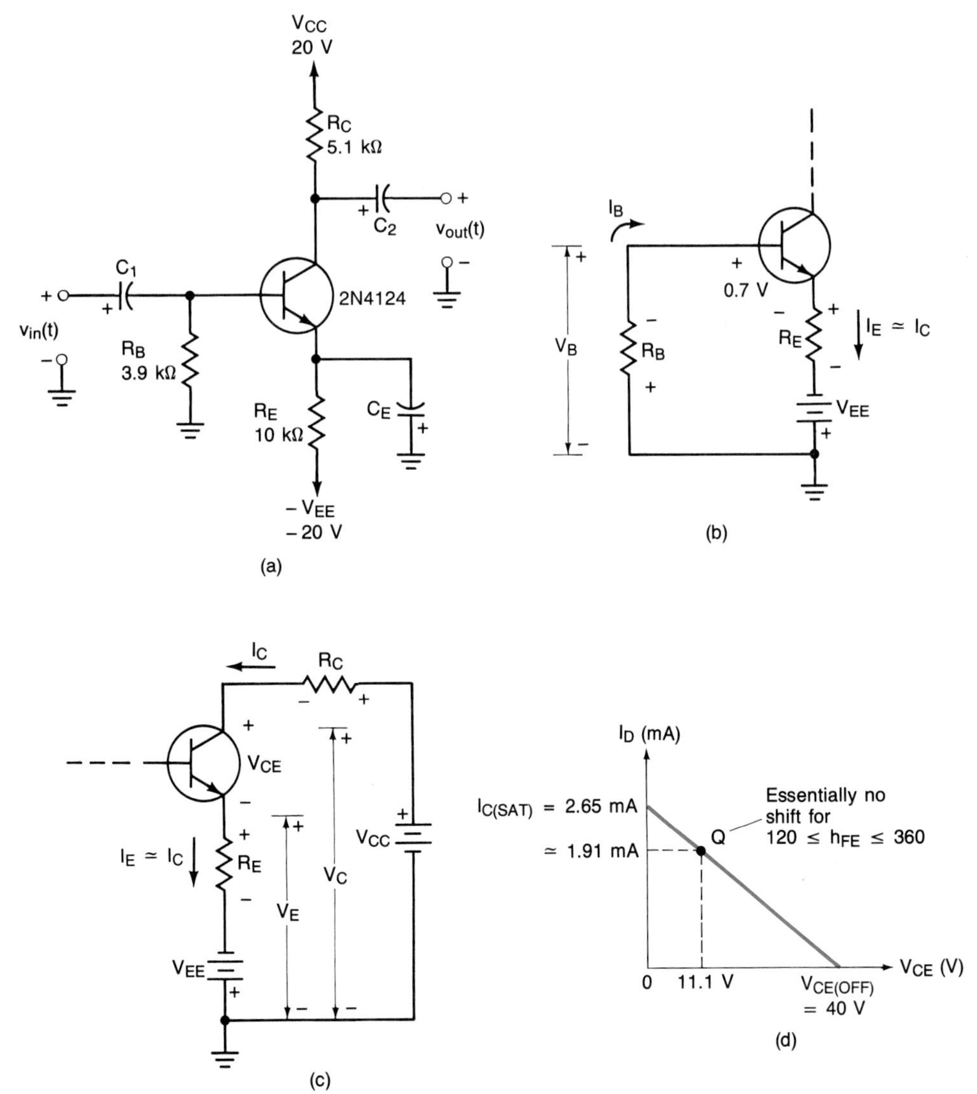

6-24(a)]. As we shall see, this particular scheme allows us to achieve an extremely stable Q-point. In most cases the bias stability exceeds that offered by the voltage-divider bias circuit.

From Fig. 6-24(b) and the application of Kirchhoff's voltage law, we may derive an equation for I_B.

$$I_B R_B + 0.7\text{ V} + I_E R_E - V_{EE} = 0$$

$$I_B R_B + 0.7\text{ V} + (1 + \beta_{DC}) I_B R_E - V_{EE} = 0$$

$$I_B = \frac{V_{EE} - 0.7\text{ V}}{R_B + (1 + \beta_{DC}) R_E} \tag{6-30}$$

The base-to-ground voltage V_B is

$$V_B = -I_B R_B \tag{6-31}$$

and assuming that I_B is very small, the voltage drop across R_B is negligible.

$$V_B = -I_B R_B \simeq 0\text{ V} \tag{6-32}$$

The emitter-to-ground [Fig. 6-24(c)] voltage is

$$V_E \simeq I_C R_E - V_{EE} \tag{6-33}$$

With these results and a little reflection, we see that the basic strategy behind this biasing approach is very similar to that of the voltage-divider biasing scheme. The base-to-ground voltage is again locked in, but this time at ground potential. The emitter resistor can assume much larger values (which enhances the bias stability).

We shall use Fig. 6-24(c) to assist us in the development of the equations for the collector–emitter and collector-to-ground voltages. Hence

$$-V_{CC} + I_C R_C + V_{CE} + I_E R_E - V_{EE} = 0$$

$$V_{CE} \simeq V_{CC} + V_{EE} - I_C(R_C + R_E) \tag{6-34}$$

$$V_C = V_{CC} - I_C R_C \tag{6-35}$$

The cutoff and saturation limits are

$$V_{CE(\text{OFF})} = V_{CC} + V_{EE} \tag{6-36}$$

$$V_{C(\text{OFF})} = V_{CC} \tag{6-37}$$

and

$$I_{C(\text{SAT})} = \frac{V_{CC} + V_{EE}}{R_C + R_E} \tag{6-38}$$

The reader is urged to verify the results above.

EXAMPLE 6-13

Given the BJT with emitter bias in Fig. 6-24(a), perform a complete dc analysis.

SOLUTION First, we shall use the minimum β_{DC} of 120 to determine the bias conditions. Hence

$$I_B = \frac{V_{EE} - 0.7\text{ V}}{R_B + (1 + \beta_{DC}) R_E} = \frac{20\text{ V} - 0.7\text{ V}}{3.9\text{ k}\Omega + (1 + 120)(10\text{ k}\Omega)}$$

$$= 15.9\ \mu\text{A}$$

$$I_C = \beta_{DC} I_B = (120)(15.9\ \mu\text{A}) = 1.91\text{ mA}$$

To determine V_{CE}, we must apply Eq. 6-34.

$$V_{CE} \simeq V_{CC} + V_{EE} - I_C(R_C + R_E)$$
$$= 20\text{ V} + 20\text{ V} - (1.91\text{ mA})(5.1\text{ k}\Omega + 10\text{ k}\Omega)$$
$$= 11.2\text{ V}$$

The terminal-to-ground voltages are

$$V_C = V_{CC} - I_C R_C = 20\text{ V} - (1.91\text{ mA})(5.1\text{ k}\Omega) = 10.3\text{ V}$$
$$V_B = -I_B R_B = -(15.9\ \mu\text{A})(3.9\text{ k}\Omega) = -0.062\text{ V} \simeq 0\text{ V}$$
$$V_E = -V_B - 0.7\text{ V} \simeq 0 - 0.7\text{ V} = -0.7\text{ V}$$

The nonlinear limits are

$$I_{C(\text{SAT})} = \frac{V_{CC} + V_{EE}}{R_C + R_E} = \frac{20\text{ V} + 20\text{ V}}{5.1\text{ k}\Omega + 10\text{ k}\Omega} = 2.65\text{ mA}$$
$$V_{CE(\text{OFF})} = V_{CC} + V_{EE} = 20\text{ V} + 20\text{ V} = 40\text{ V}$$
$$V_{C(\text{OFF})} = V_{CC} = 28\text{ V}$$

The analysis may be repeated in a similar fashion for BJTs with an β_{DC} of 360. The results are given in Table 6-9. The dc load line has been indicated in Fig. 6-24(d). ■

TABLE 6-9
***CE* Emitter Bias**

Parameter	Bias Values		Average	±% Δ
h_{FE}	120	360	240	±50
I_B	15.9 μA	5.34 μA	10.6 μA	±49.7
I_C	1.91 mA	1.92 mA	1.915 mA	±0.26
V_{CE}	11.2 V	11.0 V	11.1 V	±0.90
$I_{C(\text{SAT})}$	2.65 mA	2.65 mA	—	—
$V_{CE(\text{OFF})}$	40 V	40 V	—	—

Table 6-9 shows the superb stability of the emitter bias circuit. Its performance exceeds that offered by the voltage-divider bias circuit (Table 6-7). *The emitter bias circuit is almost totally unaffected by variations in* β_{DC}.

6-14 *CS* Source Bias

Extremely stable *Q*-points for FETs can be achieved by employing *CS source bias*. Specifically, the drain current is fixed by placing a constant-current source in series with the source terminal. Based on our work thus far, it is clear that a BJT with voltage-divider bias (Fig. 6-25) or emitter bias (Fig. 6-26) are logical choices.

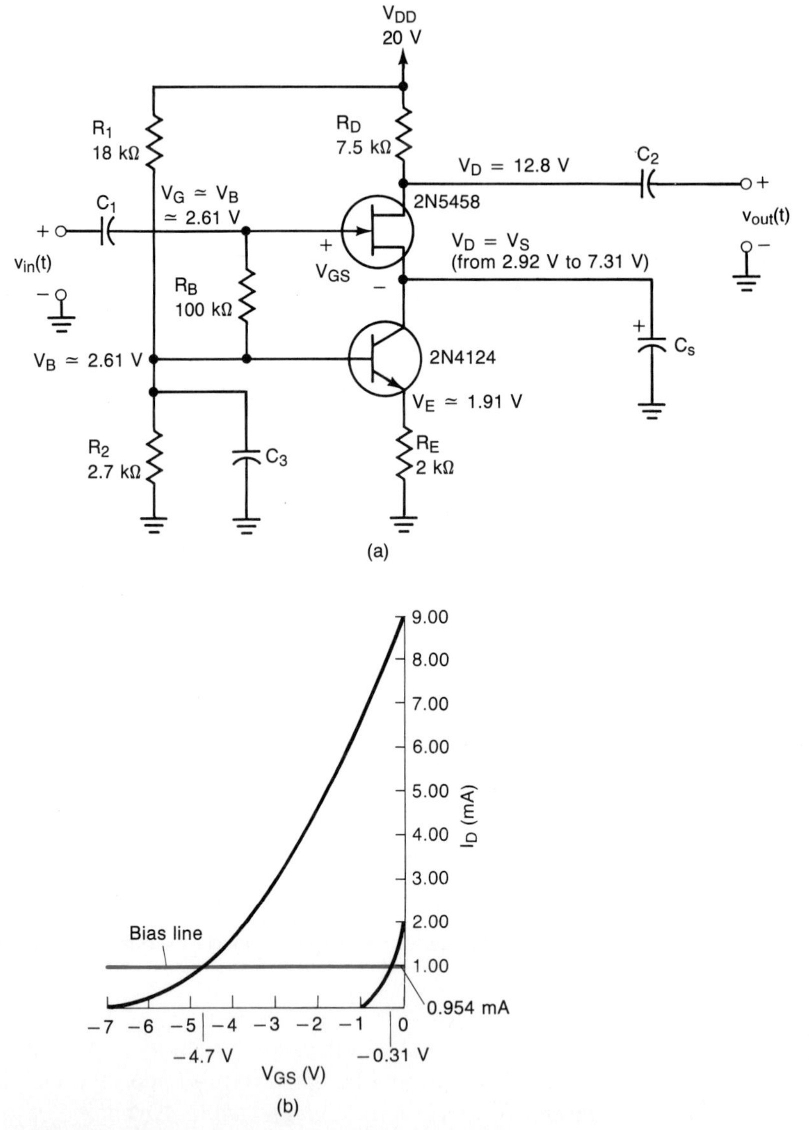

FIGURE 6-25 CS (single-supply) source bias: (a) circuit; (b) bias line.

The primary constraint on this biasing scheme is that the collector current I_C must be less than the $I_{DSS(\min)}$ for the FET.

In the case of the 2N5458, $I_{DSS(\min)}$ is 2 mA; therefore, I_C must be less than 2 mA for the bias circuit to work for all 2N5458 JFETs. An I_C of 1 mA would be an acceptable choice.

This biasing approach requires that we first analyze the BJT and then proceed to the FET. Consider Examples 6-14 and 6-15.

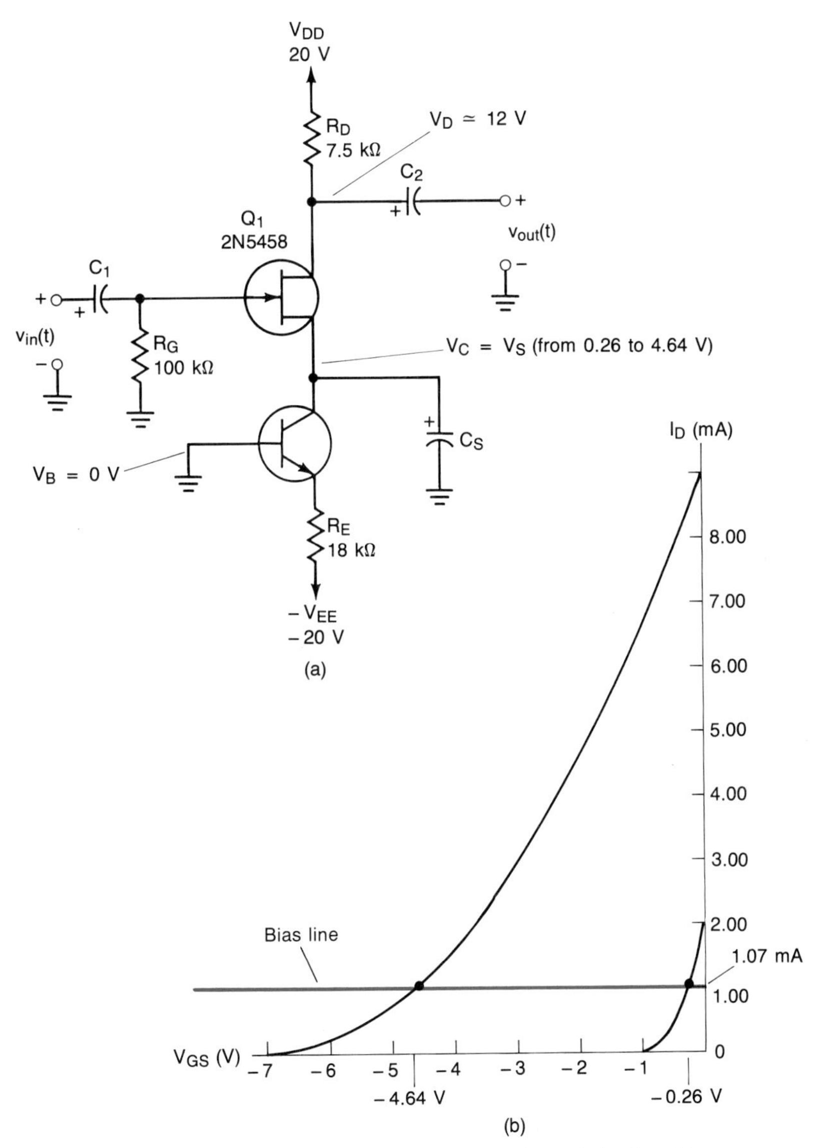

FIGURE 6-26 CS (dual-supply) source bias: (a) circuit; (b) bias line.

EXAMPLE 6-14

Perform a dc analysis of the circuit given in Fig. 6-25(a).

SOLUTION First, we note that the capacitors C_1 and C_2 are input and output coupling capacitors. Similarly, we recognize that C_S is a signal bypass capacitor. Capacitor C_3 is used for decoupling. These capacitors may be ignored during a dc analysis. Next, we shall analyze the BJT. We shall use approximations to expedite our analysis since we have two active devices to contend with.

The base-to-ground voltage may be approximated by simple voltage division between R_1 and R_2. (This is a good approximation if the bleeder current is much larger than the maximum anticipated base current.) Further, since the gate draws negligible current, the base-to-ground voltage and the gate-to-ground voltage are approximately equal.

$$V_B \simeq V_G \simeq \frac{R_2}{R_1 + R_2} V_{DD} = \frac{2.7\text{ k}\Omega}{18\text{ k}\Omega + 2.7\text{ k}\Omega} 20\text{ V} = 2.61\text{ V}$$

The emitter-to-ground voltage is

$$V_E = V_B - 0.7\text{ V} = 2.61\text{ V} - 0.7\text{ V} = 1.91\text{ V}$$

The emitter current (which determines the collector and drain currents) is

$$I_C = I_D \simeq I_E = \frac{V_E}{R_E} = \frac{1.91\text{ V}}{2\text{ k}\Omega} = 0.954\text{ mA}$$

The bias line has been plotted in Fig. 6-25(b). The intersections indicate that V_{GS} must range from approximately -0.31 to -4.7 V. To satisfy the FET, the source-to-ground voltage must be between 0.31 and 4.7 V *more positive* than V_G. Hence

$$\begin{aligned} V_{S(\text{min})} &= V_{C(\text{min})} = V_G + |V_{GS(\text{min})}| \\ &= 2.61\text{ V} + 0.31\text{ V} = 2.92\text{ V} \end{aligned}$$

and

$$\begin{aligned} V_{S(\text{max})} &= V_{C(\text{max})} = V_G + |V_{GS(\text{max})}| \\ &= 2.61\text{ V} + 4.7\text{ V} = 7.31\text{ V} \end{aligned}$$

The drain-to-ground voltage is

$$\begin{aligned} V_D &= V_{DD} - I_D R_D = 20\text{ V} - (0.954\text{ mA})\,(7.5\text{ k}\Omega) \\ &= 12.8\text{ V} \end{aligned}$$

As a final comment, we note that

$$V_{CE(\text{min})} = V_{C(\text{min})} - V_E = 2.92\text{ V} - 1.91\text{ V} = 1.01\text{ V}$$

Since $V_{CE(\text{min})} = 1.01\text{ V} \geq 0.7\text{ V}$, the BJT avoids saturation. If calculations reveal that $V_{CE(\text{min})}$ is less than 0.7 V, the BJT is in saturation and the circuit will not function properly. ■

EXAMPLE 6-15

Repeat Example 6-14 for the circuit given in Fig. 6-26(a).

SOLUTION First, we shall analyze the BJT. Observe that *no* base resistor is required. Consequently,

$$V_B = 0\text{ V}$$

By Kirchhoff's voltage law we may determine I_E to arrive at the approximate value of I_C.

$$I_C \simeq I_E = \frac{V_{EE} - 0.7\text{ V}}{R_E} = \frac{20\text{ V} - 0.7\text{ V}}{18\text{ k}\Omega} = 1.07\text{ mA}$$

The resulting bias line has been depicted in Fig. 6-26(b). The intersections indicate that V_{GS} can range from -0.26 to -4.64 V. This requires that V_S range from 0.26 to 4.64 V. Alternatively, we can state that V_C must range between these two extremes.

To ensure that the BJT remains in its linear range of operation, we should determine V_{CE}.

$$V_E = I_E R_E - V_{EE} = (1.07 \text{ mA})(18 \text{ k}\Omega) - 20 \text{ V}$$
$$= -0.74 \text{ V}$$

and V_{CE} must range from

$$V_{CE(\text{min})} = V_{C(\text{min})} - V_E = 0.26 \text{ V} - (-0.74 \text{ V}) = 1 \text{ V}$$
$$V_{CE(\text{max})} = V_{C(\text{max})} - V_E = 4.64 \text{ V} - (-0.74 \text{ V}) = 5.38 \text{ V}$$

These values appear to be reasonable. We shall finish our calculations by determining V_D.

$$V_D = V_{DD} - I_D R_D = 20 \text{ V} - (1.07 \text{ mA})(7.5 \text{ k}\Omega) = 12.0 \text{ V}$$ ■

6-15 Biasing a Differential Pair

BJTs and FETs may be configured as *differential amplifiers*. The advantages and characteristics of differential amplifiers will be investigated more fully in Chapter 8. However, for now, let us focus our attention on its dc analysis.

A BJT differential pair has been illustrated in Fig. 6-27(a). Observe that emitter bias has been used. The current that is established by the emitter resistor R_E and the emitter supply voltage V_{EE} is *shared* by the two BJTs, Q_1 and Q_2. The current that flows through R_E is referred to as the *tail current* I_T.

The electrical characteristics of the two BJTs are virtually identical in a good design. Consequently, the BJTs are described as being a *matched pair*. In fact, it is not unusual to see the BJTs constructed on the same piece of silicon in a single package. For example, the 2N2722 is described as a ''dual *npn* silicon transistor for small-signal, low-power differential amplifier applications.'' The dashed line around the BJTs in Fig. 6-27(a) is used to indicate that both BJTs share the same package. Because of their close matching we can state that

$$I_T = 2I_E$$

[see Fig. 6-27(a)].

The analysis of Fig. 6-27(a) is very similar to our previous efforts. If we assume that both halves of the differential circuit are matched, we need only analyze one-half of it. Therefore, let us analyze the base circuit of Q_1. By Kirchhoff's voltage law,

$$I_B R_B + 0.7 \text{ V} + 2I_E R_E - V_{EE} = 0$$
$$I_B R_B + 0.7 \text{ V} + 2(1 + \beta_{\text{DC}})I_B R_E - V_{EE} = 0$$

and solving for I_B gives us

$$I_B = \frac{V_{EE} - 0.7 \text{ V}}{R_B + 2(1 + \beta_{\text{DC}})R_E} \tag{6-39}$$

We could proceed in the normal fashion by next computing I_C and the corresponding terminal voltages. However, let us opt for a more approximate approach to expedite our analysis. First, we shall solve for I_T by applying Kirchhoff's voltage law, and assuming that I_B is negligible.

$$I_T \simeq \frac{V_{EE} - 0.7 \text{ V}}{R_E} \tag{6-40}$$

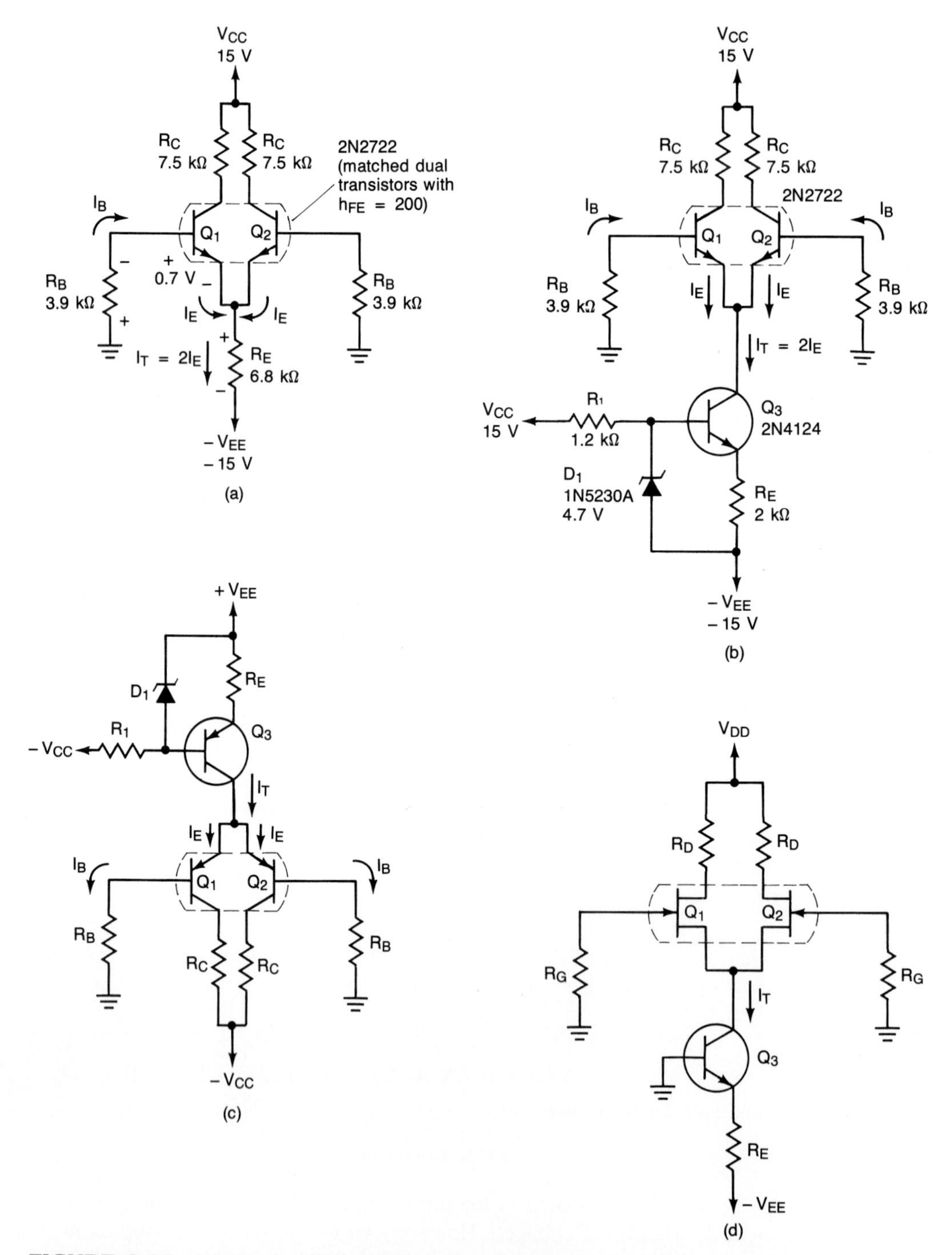

FIGURE 6-27 Biasing a differential pair: (a) differential BJT pair with emitter bias; (b) differential pair biased by a constant-current source; (c) pnp differential pair; (d) JFET differential pair.

If we use the values indicated in Fig. 6-27(a), we may conduct our analysis as follows:

$$I_T \simeq \frac{V_{EE} - 0.7 \text{ V}}{R_E} = \frac{15 \text{ V} - 0.7 \text{ V}}{6.8 \text{ k}\Omega} = 2.10 \text{ mA}$$

$$I_C \simeq I_E = \frac{I_T}{2} = \frac{2.10 \text{ mA}}{2} = 1.05 \text{ mA}$$

$$V_E = I_T R_E - V_{EE} = (2.10 \text{ mA})(6.8 \text{ k}\Omega) - 15 \text{ V} = -0.7 \text{ V}$$

$$V_C = V_{CC} - I_C R_C = 15 \text{ V} - (1.05 \text{ mA})(7.5 \text{ k}\Omega) = 7.12 \text{ V}$$

The reader should note that the bias values for Q_2 will be identical.

Employing a constant-current source [Fig. 6-27(b)] will improve the performance of the differential pair from both a biasing and a signal-processing standpoint. The 1N5230A is a 4.7-V zener diode (refer to the data sheet given in Fig. 3-18). Its function is to set the voltage across R_E equal to 4 V. This sets the tail current equal to 2 mA.

A differential pair may also be constructed from *pnp* BJTs [see Fig. 6-27(c)]. The operation is virtually identical to the *npn* BJT version. The reader should verify that the indicated current directions will satisfy the biasing requirements of the *pnp* BJTs.

Figure 6-27(d) illustrates that a differential pair may also be constructed by using FETs. Again, constant-current-source biasing has been used.

6-16 The Current Mirror

While discrete differential amplifiers are still being employed in new designs, its primary use is in the design of integrated-circuit operational amplifiers (op amps). In the design of op amps, circuit strategies are used that are not practical in discrete amplifiers. The new strategies are prompted by the fact that resistors utilize a great deal of "real estate," and precise matching of active devices is very practical since they share the same piece of silicon.

One such approach is the use of the *current mirror* to bias differential pairs [refer to Fig. 6-28(a)]. Observe that Q_1 has a short between its collector and base. Since only the base–emitter junction remains, Q_1 is called a "diode-connected" transistor. The current flowing through the resistor will be

$$I = \frac{V_S - V_{BE1}}{R} = \frac{15 \text{ V} - 0.650 \text{ V}}{10 \text{ k}\Omega} = 1.435 \text{ mA}$$

If we assume that Q_2's base current is negligibly small, then 1.435 mA is essentially the current flow through the (Q_1) diode. The operation of the current mirror may be described as follows:

1. Q_1 acts like a diode and a current is established through it.
2. Q_1 will develop a voltage drop (V_{BE1}) in response to the current.
3. The base–emitter of Q_2 is in parallel with Q_1 ($V_{BE1} = V_{BE2}$).
4. Q_2's collector current will be established in response to its V_{BE}.

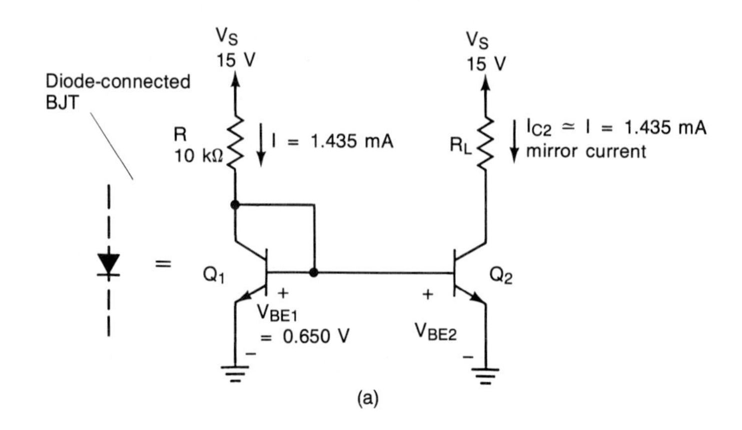

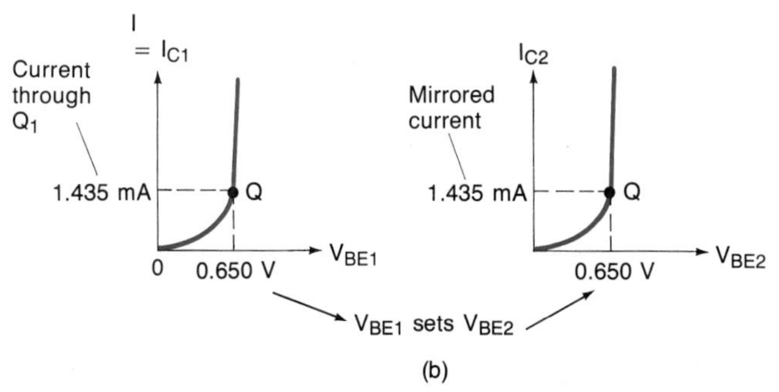

FIGURE 6-28 **Operation of the current mirror: (a) circuit; (b) V-I curves.**

5. Since the two transistors are matched, Q_2's collector current will be approximately equal to Q_1's current. Q_2's collector current is said to "mirror" the current through Q_1.

The circuit's operation is described further in Fig. 6-28(b).

Precise matching of the BJT transfer curves is absolutely essential. Recall that

$$I_C = I_{CES}e^{V_{BE}/V_T}$$

which was first defined in Eq. 4-31. By adding the appropriate subscripts, we may state for Q_1 and Q_2 in Fig. 6-28(a):

$$I_{C1} = I_{CES}e^{V_{BE1}/V_T}$$

$$I_{C2} = I_{CES}e^{V_{BE2}/V_T}$$

If we take a ratio, we obtain

$$\frac{I_{C2}}{I_{C1}} = \frac{I_{CES}e^{V_{BE2}/V_T}}{I_{CES}e^{V_{BE1}/V_T}} = e^{(V_{BE2} - V_{BE1})/V_T}$$

Taking the natural logarithm of both sides and solving for ΔV_{BE}, we have

$$\Delta V_{BE} = V_{BE2} - V_{BE1} = V_T \ln \frac{I_{C2}}{I_{C1}}$$

If we assume that I_{C2} is twice as large as I_{C1}, the corresponding mismatch in ΔV_{BE} is

$$\Delta V_{BE} = V_T \ln \frac{I_{C2}}{I_{C1}} = (26 \text{ mV}) \ln 2 = (26 \text{ mV})(0.693) = 18 \text{ mV}$$

Therefore, a difference of only 18 mV in the V_{BE} matching will produce a 100% error in our current mirror. This demonstrates the importance of matching the BJTs, and reinforces the futility of attempting to construct a current mirror from discrete BJTs.

We shall see exactly how the current mirror is employed to establish bias currents when we delve into the inner workings of the IC op amp. This is done in Chapter 9.

6-17 Biasing the Op Amp

The IC op amp is incredibly easy to use. This is one of the primary reasons it is selected for many applications. Initially, we shall regard it as merely another "active device"–just as the BJT or one of the many varieties of FETs.

FIGURE 6-29 **Biasing considerations for op amps: (a) schematic symbol; (b) maximum symmetrical output voltage swing; (c) BJT-input op amp; (d) FET-input op amp.**

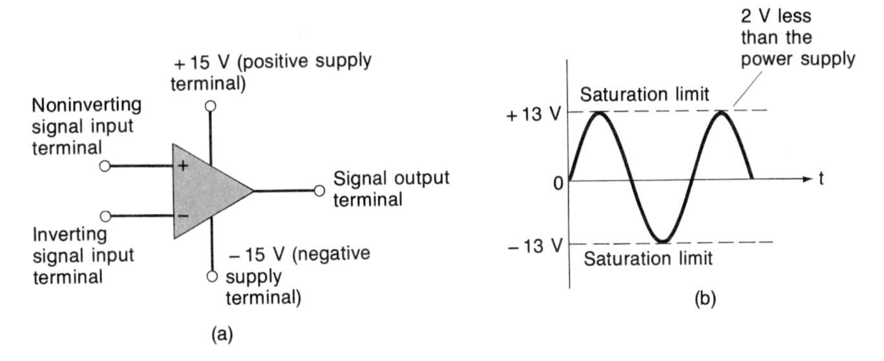

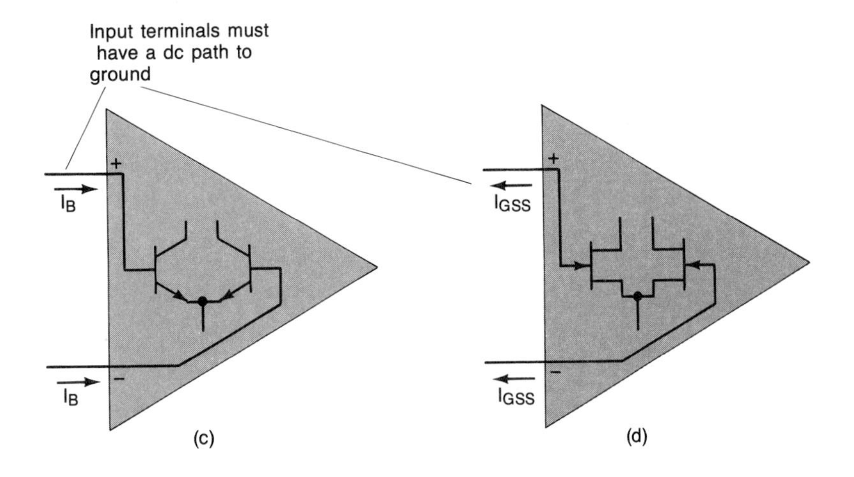

To bias an op amp we merely connect its positive and negative supply terminals to the power supplies [see Fig. 6-29(a)]. Note that the "+" terminal is the *noninverting signal input* terminal—*not* the positive supply terminal. Similarly, the "−" terminal is described as the *inverting signal input* terminal.

The op amp is described as being *direct coupled* since no internal dc blocking capacitors are used. This means that it may be used as a dc voltage amplifier. Its output voltage may be either positive or negative. As a rule of thumb, the magnitude of its peak output voltage swing is about *2 volts less* than the supply voltage. The maximum symmetrical output voltage swing for an op amp that has a ±15-V power supply has been depicted in Fig. 6-29(b).

There is one other extremely important biasing consideration for a successful application of op amps. The noninverting and inverting input terminals *must* have a dc return to ground. This should come as no great surprise. As we mentioned previously, the BJT and the FET differential pair is the fundamental building block of op amps [see Fig. 6-29(c) and (d)]. Much more will be said about this in our later work.

PROBLEMS

Drill, Derivations, and Definitions

Section 6-1

6-1. What bias conditions are required on the base–emitter and collector–base *p-n* junctions to ensure that a BJT remains in its active (linear) region of operation?

6-2. What bias conditions are required to ensure that a JFET or DE-MOSFET remains in its active (linear) region of operation?

Section 6-2

6-3. A 2N4124 BJT has a β_{DC} of 300 when I_C is 3 mA, V_{CE} is 10 V, and the ambient temperature is 25°C. Using Fig. 6-1, determine its β_{DC} at an I_C of 0.5 mA.

6-4. Repeat Prob. 6-3 if the β_{DC} is 160 when I_C is 3 mA, V_{CE} is 10 V, and the ambient temperature is 25°C. Also determine the β_{DC} at −55°C.

6-5. A 2N4341 *n*-channel JFET has a $V_{GS(OFF)}$ that ranges from −2 to −6 V and an I_{DSS} that ranges from 3 to 9 mA. Graph its minimum and maximum transfer characteristic curves. Present your data points in tabular form.

6-6. Repeat Prob. 6-5 for a 2N5459 *n*-channel JFET. Its data are given in Fig. 6-2. (Make a neat graph that can be used for later biasing problems.)

Section 6-3

6-7. Name the two voltage breakdown mechanisms associated with the collector–base *p-n* junction of a BJT. In your own words, describe them.

6-8. What is BV_{EBO}? In your own words, describe its cause and cite its typical range of values.

6-9. A BJT has a BV_{CEO} value of 30 V. What is its maximum collector supply voltage?

6-10. A BJT is to be used in an amplifier system with a V_{CC} of 15 V. What should its minimum BV_{CEO} rating be?

6-11. In your own words, explain the fundamental cause of voltage breakdown in a JFET. What is subject to voltage breakdown in a MOSFET?

6-12. An n-channel JFET has a $V_{GS(\text{OFF-max})}$ value of -3 V and a BV_{GSS} specified to be at least -30 V. What is the range of permissible V_{DD} values for this device?

6-13. A 2N5459 n-channel JFET has a $V_{GS(\text{OFF})}$ that ranges from -2 to -8 V and a BV_{GSS} of 25 V (minimum). What range of V_{DD} values is suitable for this device?

6-14. Repeat Prob. 6-13 if V_{GS} is to be fixed at -1 V.

Section 6-4

6-15. In your own words, explain what is meant by "midpoint bias." Why is it desirable? What is "clipping"?

Section 6-5

6-16. Given the BJT fixed-bias circuit shown in Fig. 6-13, determine I_B, I_C, V_{CE}, $I_{C(\text{SAT})}$, and $V_{CE(\text{OFF})}$. A 2N4123 is used. Use the range of β_{DC} values given in Fig. 6-1. V_{CC} is 15 V, R_C is 7.5 kΩ, and R_B is 1.5 MΩ.

6-17. Repeat Prob. 6-16 if R_B is reduced to 1 MΩ. Is there a problem? Explain.

Section 6-6

6-18. Analyze the JFET fixed-bias circuit given in Fig. 6-15(a). A 2N5459 is used. Determine the minimum and maximum values for I_D and V_{DS}. (You may wish to use a copy of the transfer curves you developed in Prob. 6-6.) Also find $I_{D(\text{SAT})}$ and $V_{DS(\text{OFF})}$. V_{DD} is 15 V, R_D is 1 kΩ, R_G is 150 kΩ, and V_{GG} is -1 V.

6-19. Repeat Prob. 6-18 if V_{DD} is 12 V. All other component values are the same.

Section 6-7

6-20. The BJT amplifier in Fig. 6-16(a) uses emitter-stabilized bias. Determine I_B, I_C, V_{CE}, V_C, V_B, V_E, $I_{C(\text{SAT})}$, $V_{CE(\text{OFF})}$, and $V_{C(\text{OFF})}$. The transistor is a 2N4123, V_{CC} is 15 V, R_C is 5.6 kΩ, R_B is 1 MΩ, and R_E is 1.5 kΩ. Use the β_{DC} data given in Fig. 6-1 for the 2N4123.

6-21. Repeat Prob. 6-20 if V_{CC} is 12 V. All other component values are the same.

6-22. In your own words, explain what is meant by "dc negative feedback." How is it employed in the emitter-stabilized bias circuit?

Section 6-8

6-23. What is the function of the emitter bypass capacitor? How does its function compare with that of the source bypass capacitor?

6-24. Given the source-stabilized bias circuit in Fig. 6-17(a), determine the range of values for I_D, V_S, V_G, V_{GS}, V_{DS}, and V_D. A 2N5459 is used. V_{DD} is 15 V, R_D is 3.9 kΩ, R_S is 2 kΩ, and R_G is 150 kΩ. (You may wish to use the transfer curves you developed in Prob. 6-6.)

6-25. Repeat Prob. 6-24 if V_{DD} is 12 V. All other values are unchanged.

Section 6-9

6-26. Perform a dc analysis of the collector voltage feedback bias circuit given in Fig. 6-19(a). Specifically, find I_B, I_C, V_{CE}, V_C, $I_{C(SAT)}$, and $V_{CE(OFF)}$. The transistor is a 2N4123, V_{CC} is 15 V, R_C is 7.5 kΩ, and R_B is 680 kΩ.

6-27. Repeat Prob. 6-26 if V_{CC} is 12 V. All other values are unchanged.

Section 6-10

6-28. Given the drain voltage feedback bias circuit in Fig. 6-20(a), determine V_{DS}, V_{GS}, and write the equation for its bias line. V_{DD} is 15 V, R_D is 7.5 kΩ, R_G is 1 MΩ, and I_D is 1 mA.

6-29. Repeat Prob. 6-28 if V_{DD} is 12 V. All other values are unchanged.

6-30. Explain how drain voltage feedback bias may be applied to an E-MOSFET; to a CMOS pair. Include the appropriate schematic diagrams.

Section 6-11

6-31. A BJT with voltage-divider bias has been given in Fig. 6-21(a). Find the range of values for I_B, I_C, V_E, V_B, V_{CE}, and V_C. Also find $I_{C(SAT)}$, $V_{C(OFF)}$, and $V_{CE(OFF)}$. A 2N4123 is used, V_{CC} is 15 V, R_C is 6.8 kΩ, R_E is 1.5 kΩ, R_1 is 13 kΩ, and R_2 is 2.2 kΩ.

6-32. Repeat Prob. 6-31 if V_{CC} is 12 V. All other values are unchanged.

Section 6-12

6-33. A JFET with voltage-divider bias has been given in Fig. 6-23(a). Determine the range of values for I_D, V_G, V_S, V_{DS}, V_D, $I_{D(SAT)}$, and $V_{D(OFF)}$. A 2N5459 is used. V_{DD} is 15 V, R_D is 3.6 kΩ, R_S is 6.8 kΩ, R_1 is 110 kΩ, and R_2 is 39 kΩ. (You may wish to use the transfer curves you developed in Prob. 6-6.)

6-34. Repeat Prob. 6-33 if V_{DD} is 12 V. All other values are unchanged.

Section 6-13

6-35. A BJT emitter bias circuit has been indicated in Fig. 6-24(a). Find the range of values for I_B, I_C, V_{CE}, V_C, V_B, and V_E. Also determine $I_{C(SAT)}$, $V_{CE(OFF)}$, and $V_{C(OFF)}$. A 2N4123 is used. V_{CC} is 15 V, $-V_{EE}$ is -15 V, R_C is 7.5 kΩ, R_E is 15 kΩ, and R_B is 3.3 kΩ.

6-36. Repeat Prob. 6-35 if V_{CC} is 12 V and $-V_{EE}$ is -12 V. All other values are unchanged.

Section 6-14

6-37. If a constant-current source is used to establish the I_D of an FET, what is the primary constraint on the current set by the constant-current source?

6-38. Given the constant-current source bias circuit in Fig. 6-25(a), determine V_B, V_G, V_E, I_C, I_D, $V_{S(min)}$, $V_{S(max)}$, and V_D. The JFET is a 2N5459 and the BJT is a 2N4123, V_{DD} is 15 V, R_D is 4.7 kΩ, R_E is 1.5 kΩ, R_G is 110 kΩ, R_1 is 13 kΩ, and R_2 is 2.2 kΩ. (You may wish to use the transfer curves you developed in Prob. 6-6.)

6-39. Repeat Prob. 6-38 for the constant-current source bias circuit given in Fig. 6-26(a). The JFET is a 2N5459 and the BJT is a 2N4123. V_{DD} is 15 V, $-V_{EE}$ is -15 V, R_D is 4.7 kΩ, R_E is 15 kΩ, and R_G is 110 kΩ. (You may wish to use the transfer curves you developed in Prob. 6-6.)

Section 6-15

6-40. Given a differential pair such as that shown in Fig. 6-27(a), with an R_E of 5.1 kΩ and an R_C of 5.1 kΩ, find I_T, I_E, I_C, V_E, and V_C. The 2N2722 dual transistors and supply voltages are unchanged.

6-41. Analyze the FET differential pair given in Fig. 6-27(d). R_E is 6.8 kΩ, R_D is 7.5 kΩ, and the power supply voltages are ± 15 V. Find I_T, I_{S1}, I_{D1}, and V_{D1}. Assume that V_{BE} is 0.7 V.

Section 6-16

6-42. In your own words, explain the operation of a current mirror using a "diode-connected" BJT. Why is a precise matching of the transfer curves required?

6-43. Refer to Fig. 6-28(a). R is increased to 20 kΩ, the emitters of both transistors are taken to a negative supply, and the supply voltages are ± 12 V, respectively, find the mirrored currents in Q_1 and Q_2. Assume that V_{BE1} is 0.62 V.

Section 6-17

6-44. Given the op amp shown in Fig. 6-29(a), assume that its power supply is ± 12 V. What is its maximum symmetrical ac output voltage swing? Make a sketch of its maximum sinusoidal output voltage.

6-45. Repeat Prob. 6-44 if the supplies are reduced to ± 6 V.

6-46. What is the fundamental biasing constraint on the input terminals of an op amp?

Troubleshooting

6-47. The emitter bypass capacitor in Fig. 6-21(a) has shorted. Determine V_B, V_E, and V_C.

6-48. The source bypass capacitor in Fig. 6-23(a) has shorted. Determine V_G, V_S, and V_D.

6-49. The base of the transistor in Fig. 6-24(a) has shorted to ground. Find V_B, V_E, and V_C. Will a dc bias check indicate that the circuit is faulty? Explain.

Design

A mastery of the strategies employed in electronic circuits can be achieved by careful analysis and drill. Once we understand the analysis and the constraints imposed on electronic circuits, we can intelligently modify the circuits to solve other problems. This leads us to the world of design. Problem 6-50 will direct the reader through the bias design of a BJT common-emitter voltage amplifier.

6-50. Design a BJT circuit such as the one shown in Fig. 6-21(a). V_{CC} is 12 V, I_C is to be approximately 1 mA, and the nearest upper standard 5%-tolerance resistor values are to be used.

(a) The V_E restricts the minimum output signal swing. Therefore, we should make it small. Nominally,

$$V_E = \frac{V_{CC}}{10}$$

Find V_E.

(b) I_E and I_C are approximately equal. This allows us to determine R_E.

$$R_E \simeq \frac{V_E}{I_C}$$

Find R_E.

(c) The boundaries on the maximum output signal swing are V_{CC} and V_E. Therefore, V_C should be centered between these two limits.

$$V_C = \frac{V_{CC}}{2} + \frac{V_E}{2}$$

Find V_C.

(d) Once we have determined V_C, we can find the voltage drop across the collector resistor.

$$V_{RC} = V_{CC} - V_C$$

Find V_{RC}.

(e) Now we can find R_C.

$$R_C = \frac{V_{RC}}{I_C}$$

Determine R_C, and pick the nearest upper standard value.

(f) The base-to-ground voltage will be one base–emitter voltage drop above V_E. Find V_B if the transistor is silicon.

(g) The BJT's $h_{FE(\text{min})}$ is 100. We can now find the maximum base current.

$$I_{B(\text{max})} = \frac{I_C}{h_{FE(\text{min})}}$$

Determine $I_{B(\text{max})}$.

(h) The bleeder current (I) through R_1 and R_2 should be much larger than $I_{B(\text{max})}$. Typically,

$$I = 10 I_{B(\text{max})}$$

Find I.

(i) Now that we have found V_B and I, we can find R_2.

$$R_2 = \frac{V_B}{I}$$

Determine the value of R_2 and pick the nearest upper standard value.

(j) The last step is to determine the value for R_1.

$$R_1 = \frac{V_{CC} - V_B}{I}$$

Find R_1.

Computer

6-51. Write a BASIC computer program that implements the design procedure described in Prob. 6-50. It should prompt the user for the target value for I_C, the available value of V_{CC}, and the minimum h_{FE}. It should automatically select standard ($\pm 5\%$) resistor values. Once it reports the selected resistor values, it should analyze the circuit using its selected resistor values.

7

Amplifier Models for Discrete and Integrated Circuits

After Studying Chapter 7, You Should Be Able to:

- Describe the basic voltage amplifier model.
- Explain input and output loading, and the differences between $A_{v(oc)}$, A_v, and A_{vs}.
- Describe the basic current amplifier model, and the meaning of $A_{i(sc)}$, A_i, and A_{is}.
- Find the current gain of a voltage amplifier.
- Find the power gain (A_p) of a voltage amplifier.
- Define and use decibels as they relate to voltage, current, and power gains.
- Define and use the relative decibel scales (dBm, dBV, and VU).
- Analyze cascaded voltage amplifier systems using the voltage amplifier models.
- Describe the op amp voltage-controlled voltage source model.

7-1 Basic Voltage Amplifier Model

In our work to come, we shall investigate small-signal amplifiers that employ BJTs, FETs, and integrated circuits. It will be our objective to find quantities such as the *input* and *output resistances* and the *voltage*, *current*, and *power gains*. Before our analyses become ''cluttered'' with circuitry particulars, models, and device nomenclature, it will be helpful to define these basic parameters and the concepts associated with *all* amplifiers.

All single-stage and multiple-stage electronic amplifiers may be represented by a single equivalent-circuit model. It does not matter whether we are dealing with integrated circuits, discrete BJTs, or FETs. The various equivalent-circuit models require that we invoke the *controlled-source concepts* taught in classical circuit analysis. The four fundamental controlled sources have been illustrated in Fig. 7-1. Observe that we have defined ac signal quantities by using lowercase letters. We shall drop the functional notation [e.g., v means $v(t)$ and i means $i(t)$].

We have already dealt with some of the controlled sources shown in Fig. 7-1. For example, we have treated the dc biasing of a BJT as a current-controlled current-source problem, and the dc biasing of an FET as a (nonlinear) voltage-controlled current-source problem. We have also developed small-signal models for *both* the BJT and the FET, which are voltage-controlled current sources. Now we shall apply the controlled source concept to the problem of modeling a *complete amplifier system*.

The *ideal* circuit model of a voltage amplifier is simply a voltage-controlled voltage source [see Fig. 7-2(a)]. A much more realistic, albeit slightly more complex model has been indicated in Fig. 7-2(b). This refined circuit model consists of an input resistance R_{in} (sometimes denoted Z_{in}), and the voltage-controlled voltage source with an output resistance R_{out} (sometimes denoted Z_{out}) in series. Note that the input (controlling) voltage is defined to be across R_{in}. The constant of proportionality between the input and output voltage is the open-circuit voltage gain $A_{v(oc)}$.

This model is a very good approximation at low signal frequencies. (It can easily be modified to accommodate most analyses at higher frequencies.) In Chapter 8 we shall see exactly how this equivalent circuit may be arrived at. Its use will simplify a host of problems.

The *ideal* voltage amplifier may be described as having the following attributes:

1. Infinite input resistance
2. Zero output resistance
3. Infinite voltage gain
4. Infinite bandwidth (a voltage gain that is not affected by the signal frequency)

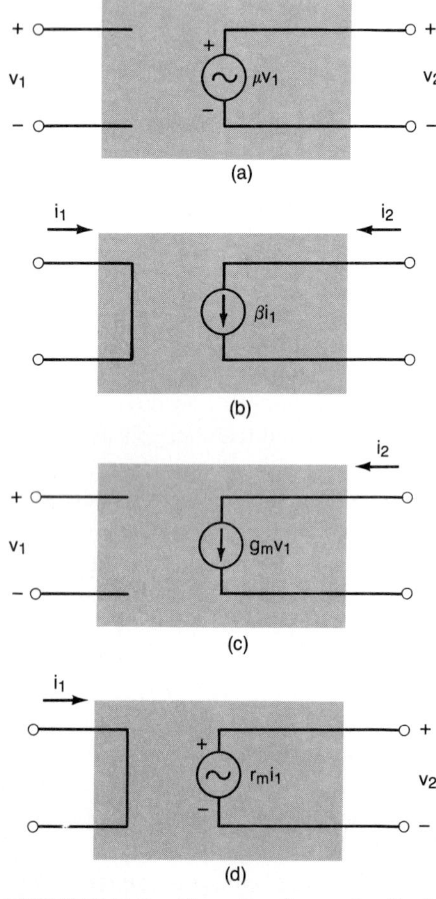

FIGURE 7-1 Four basic controlled sources: (a) voltage-controlled voltage source; (b) current-controlled current source; (c) voltage-controlled current source; (d) current-controlled voltage source.

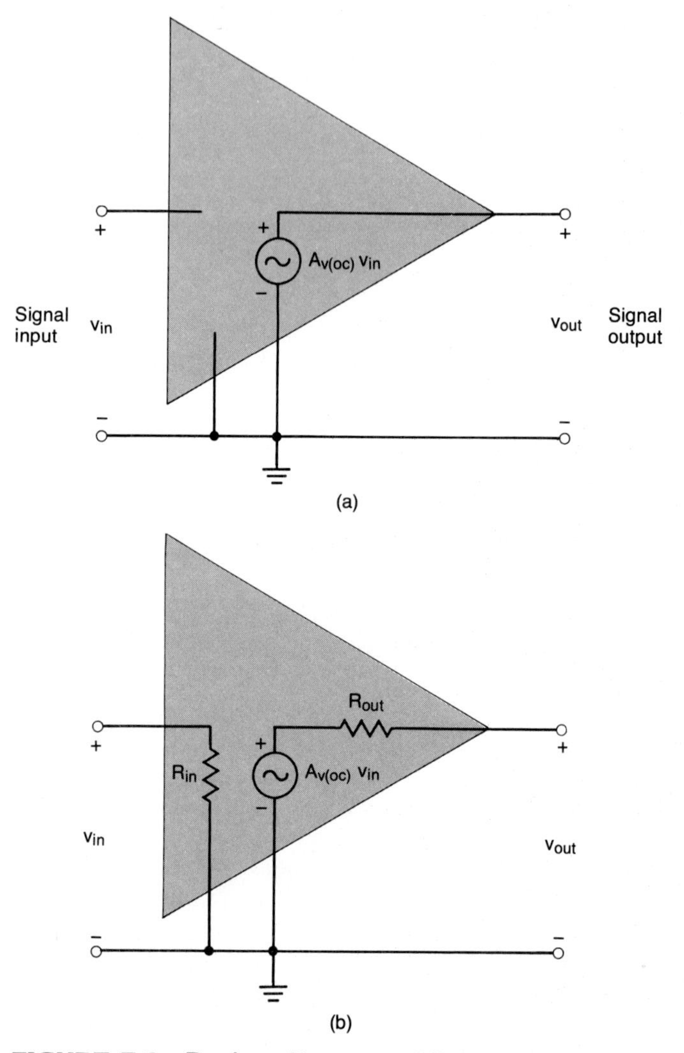

FIGURE 7-2 Basic voltage amplifier models: (a) ideal model; (b) more realistic model.

5. A zero output offset voltage (zero volts out with zero volts applied at the input)
6. No drift in these parameters when a change in temperature occurs

7-2 Unloaded Voltage Gain

In Fig. 7-3 we have an ideal signal source (zero output impedance) driving the input of a voltage amplifier that has been represented by our model. If *no load* is connected across the output of the amplifier, no current will be drawn from its output terminals,

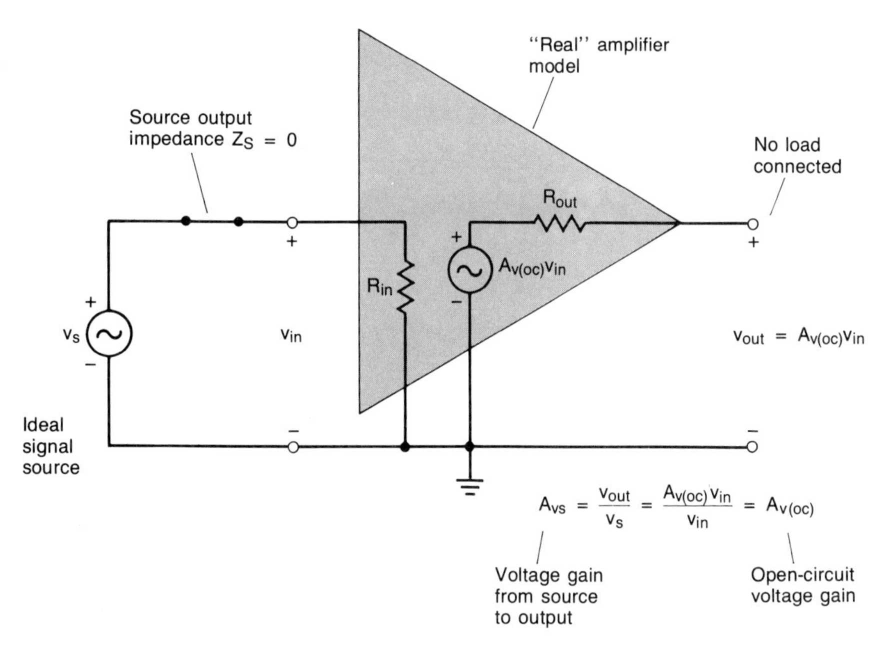

FIGURE 7-3 Voltage amplifier with no input or output loading.

and no voltage will be dropped across R_{out}. Therefore, the voltage at the output terminals will be

$$v_{out} = A_{v(oc)}v_{in}$$

If the signal source is ideal, its internal (resistive) impedance Z_S is zero. Consequently, any input current drawn by the amplifier will not cause a reduction in the terminal voltage of the signal source. Hence

$$v_{in} = v_s$$

and we find that the voltage gain is equal to $A_{v(oc)}$.

Observe that we have elected to use $A_{v(oc)}$ as the notation for the "open-circuit voltage gain." A_{vs} is used to represent the voltage gain from the signal source to the amplifier's output terminals.

The situation shown in Fig. 7-3 provides us with the maximum possible voltage gain ($A_{v(oc)}$). However, it is not realistic. Signal sources will have a nonzero output impedance, and R_{in} will tend to load down the signal source. Further, an amplifier is usually constructed to drive a given load, and connecting a load across its output will cause a reduction in its output voltage. Both input and output loading will tend to reduce the overall voltage gain. Input loading is explored in the next section.

7-3 Input Loading Effects

In Fig. 7-4 we see a voltage amplifier that has been represented by our model. Its output is not connected to a load, but its input is being driven by a signal source with

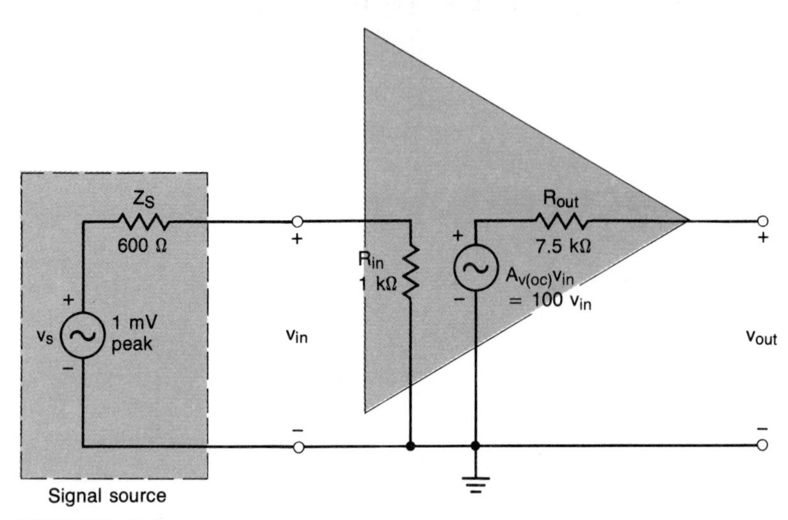

FIGURE 7-4 Input loading effects by R_{in} on the signal source.

an output impedance Z_S. The amplifier's input voltage v_{in} will be *less* than v_s because of the voltage division between Z_S and R_{in}. Hence

$$v_{in} = \frac{R_{in}}{R_{in} + Z_S} v_s \tag{7-1}$$

v_{out} is given by

$$v_{out} = A_{v(oc)} v_{in} = A_{v(oc)} \frac{R_{in}}{R_{in} + Z_S} v_s \tag{7-2}$$

and the (unloaded) voltage gain from the *signal source* to the output A_{vs} is

$$\boxed{A_{vs} = \frac{v_{out}}{v_s} = A_{v(oc)} \frac{R_{in}}{R_{in} + Z_S}} \tag{7-3}$$

EXAMPLE 7-1

Given the amplifier model and the parameters indicated in Fig. 7-4, find v_{in}, v_{out}, and the voltage gain from the signal source to the amplifier's output.

SOLUTION First, we shall find v_{in} by drawing on Eq. 7-1.

$$v_{in} = \frac{R_{in}}{R_{in} + Z_S} v_s = \frac{1\ \text{k}\Omega}{1\ \text{k}\Omega + 600\ \Omega}(1\ \text{mV}) = 0.625\ \text{mV}$$

Once we have found v_{in}, we may then find the open-circuit output voltage v_{out}.

$$v_{out} = A_{v(oc)} v_{in} = (100)\,(0.625\ \text{mV}) = 62.5\ \text{mV}$$

The ratio of the peak output voltage to the peak source voltage yields the voltage gain.

$$A_{vs} = \frac{v_{out}}{v_s} = \frac{62.5\ \text{mV}}{1\ \text{mV}} = 62.5$$

Alternatively, we could employ Eq. 7-3 directly.

$$A_{vs} = \frac{v_{out}}{v_s} = A_{v(oc)}\frac{R_{in}}{R_{in} + Z_S} = (100)\frac{1\ \text{k}\Omega}{1\ \text{k}\Omega + 600\ \Omega}$$
$$= 62.5$$

■

With a little reflection we can see why an infinite input resistance is desirable– all of v_s will then reach the input of the amplifier. Consequently, the overall voltage gain would not be affected by the value of Z_S. In practice, voltage amplifiers may have input resistances that range from several ohms to hundreds, or even thousands of megohms. The actual input resistance depends on the particular active device employed and the external circuit configuration. This will become obvious as our work progresses.

7-4 Output Loading Effects

In Fig. 7-5 we see an amplifier that has been represented by our model and has a load connected across its output terminals. In this case the output voltage produced by our controlled source will be divided between the output resistance and the load. Therefore, by voltage division,

$$v_{out} = \frac{R_L}{R_L + R_{out}} A_{v(oc)} v_{in} \tag{7-4}$$

and the voltage gain from the amplifier's input to its output is given by

$$A_v = \frac{v_{out}}{v_{in}} = \frac{R_L}{R_L + R_{out}} A_{v(oc)} \tag{7-5}$$

FIGURE 7-5 Output loading.

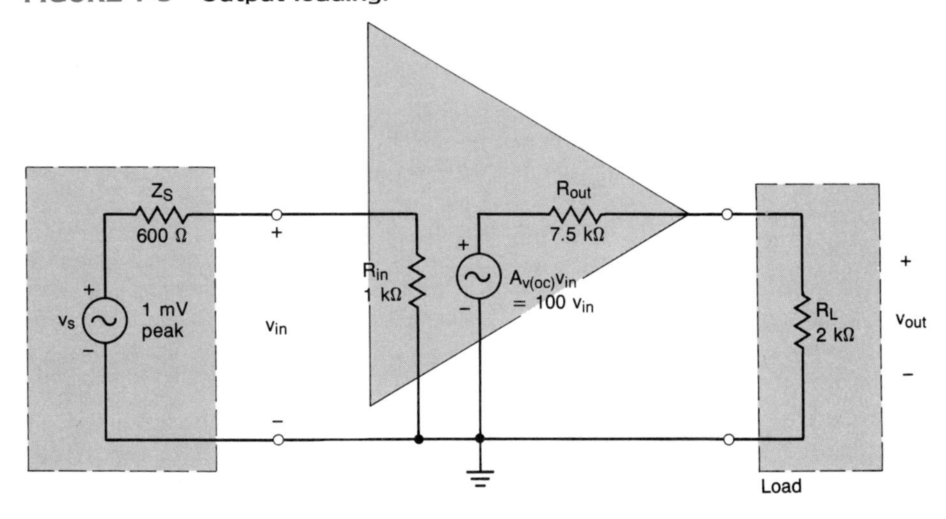

In this context we can see the need to distinguish between the unloaded (or open-circuit) voltage gain $A_{v(oc)}$, the voltage gain from the source to the output A_{vs}, and the loaded output voltage gain A_v. *A_v is the usual nomenclature applied to describe the voltage gain from an amplifier's input terminals to its loaded output terminals.*

If we substitute Eq. 7-1 for v_{in} into Eq. 7-4 and then solve for the overall voltage gain from the signal source to the loaded output, we obtain

$$A_{vs} = \frac{v_{out}}{v_s} = \frac{R_{in}}{R_{in} + Z_S} \frac{R_L}{R_L + R_{out}} A_{v(oc)} \tag{7-6}$$

EXAMPLE 7-2

Given the amplifier and the parameters indicated in Fig. 7-5, find v_{out}, the voltage gain from its input to its output and the overall voltage gain from the signal source to the output.

SOLUTION From our work in Example 7-1 we know that v_{in} is 0.625 mV, and application of Eq. 7-4 produces

$$v_{out} = \frac{R_L}{R_L + R_{out}} A_{v(oc)} v_{in}$$

$$= \frac{2\text{ k}\Omega}{2\text{ k}\Omega + 7.5\text{ k}\Omega} (100)\,(0.625\text{ mV}) = 13.2\text{ mV}$$

To find the voltage gain from the amplifier's input to its output, we shall merely find the ratio

$$A_v = \frac{v_{out}}{v_{in}} = \frac{13.2\text{ mV}}{0.625\text{ mV}} = 21.1$$

Now we shall demonstrate the application of Eq. 7-6 to find the overall voltage gain.

$$A_{vs} = \frac{v_{out}}{v_s} = \frac{R_{in}}{R_{in} + Z_S} \frac{R_L}{R_L + R_{out}} A_{v(oc)}$$

$$= \frac{1\text{ k}\Omega}{1\text{ k}\Omega + 600\ \Omega} \frac{2\text{ k}\Omega}{2\text{ k}\Omega + 7.5\text{ k}\Omega} (100) = 13.2$$

(The reader should note that we could have taken the ratio between v_{out} and v_s.) ■

With a little thought we can see why an output resistance of zero is desirable—no voltage division between R_{out} and R_L will occur. Consequently, the voltage gain will always be its maximum value and totally independent of R_L. As we shall see in our later work, an R_{out} of zero is impossible to achieve. However, it is possible to obtain output resistances that are only a fraction of an ohm.

7-5 Basic Current Amplifier Model

Most of our small-signal work will deal with voltage amplifiers. However, we occasionally encounter *current amplifiers*. The current amplifier is the *dual* of the voltage amplifier. ("Duality" means that all the voltage amplifier definitions and idealizations apply to the current amplifier. However, the terms ''voltage'' and ''current'' must be exchanged. For example, voltage gain becomes current gain, infinite input resistance becomes zero input resistance, zero output resistance becomes infinite output resistance, and so on.)

The *ideal* current amplifier exhibits the following attributes:

1. Zero input resistance
2. Infinite output resistance
3. Infinite current gain
4. Infinite bandwidth (a current gain that is not affected by the signal frequency)
5. A zero output offset current (zero current out with zero current directed into its input)
6. No drift in these parameters when a change in temperature occurs

The reader is invited to contrast these current amplifier attributes with those presented for the voltage amplifier in Section 7-1. Many of the same problems that plague voltage amplifiers are also found in current amplifiers. As an example, let us consider input and output loading effects.

As we shall see, input and output loading are due to current division between resistances. Therefore, we shall briefly review current division. Consider Fig. 7-6(a).

Resistances in parallel are more easily handled by considering their conductances. (Recall that conductances in parallel add.) Hence, by inspection of Fig. 7-6(a),

$$G_T = G_1 + G_2$$

The voltage across the parallel elements is given by

$$v = \frac{i_s}{G_T} = \frac{i_s}{G_1 + G_2} \tag{7-7}$$

and the current through G_2 is

$$i_2 = G_2 v = G_2 \frac{i_s}{G_1 + G_2} = \frac{G_2}{G_1 + G_2} i_s \tag{7-8}$$

Observe that Eq. 7-8 has a form that is the dual of the voltage-divider rule. Appropriately, it is called the current-divider rule.

The ideal current amplifier is a current-controlled current source such as that depicted in Fig. 7-1(b). A more realistic version has been indicated in Fig. 7-6(b). To facilitate our understanding of the current amplifier, we shall develop its relationships in analogous fashion to those of the voltage amplifier.

A current amplifier that is driving a short circuit has been given in Fig. 7-7(a). First we shall find its input current i_{in}.

$$i_{\text{in}} = \frac{G_{\text{in}}}{G_{\text{in}} + G_S} i_s \tag{7-9}$$

The output current is

$$i_{\text{out}} = A_{i(\text{sc})} i_{\text{in}} = A_{i(\text{sc})} \frac{G_{\text{in}}}{G_{\text{in}} + G_S} i_s \tag{7-10}$$

and the current gain from the signal source to the output is given by

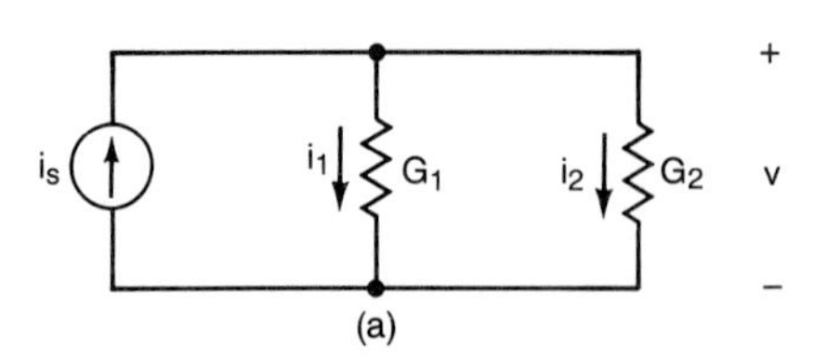

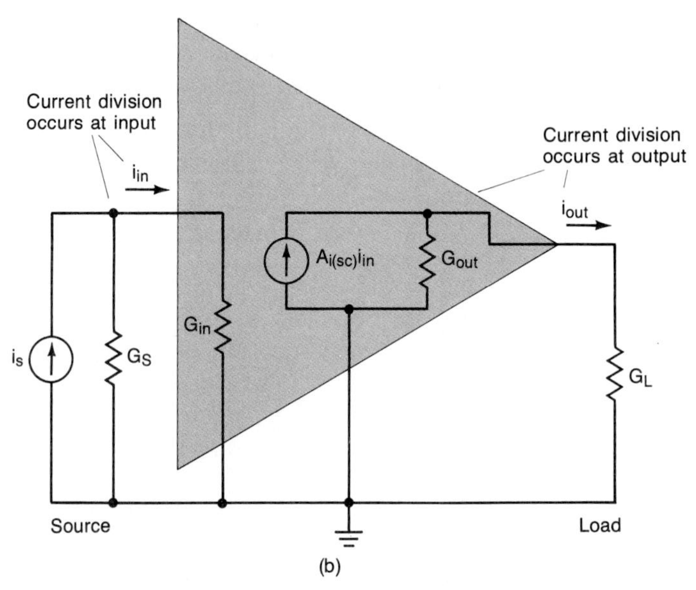

FIGURE 7-6 Current amplifier: (a) current division; (b) current amplifier with input and output loading.

$$A_{is} = \frac{i_{out}}{i_s} = A_{i(sc)} \frac{G_{in}}{G_{in} + G_S} \tag{7-11}$$

The effects of output loading occur when the load resistance is nonzero [see Fig. 7-7(b)]. The current division between G_{out} and G_L will cause a further reduction in the overall current gain. The output current division results in

$$i_{out} = \frac{G_L}{G_L + G_{out}} A_{i(sc)} i_{in} \tag{7-12}$$

If we substitute Eq. 7-9 into the result above and solve for the gain, we arrive at

$$A_{is} = \frac{i_{out}}{i_s} = \frac{G_{in}}{G_{in} + G_S} \frac{G_L}{G_L + G_{out}} A_{i(sc)} \tag{7-13}$$

Notice the similarity between Eqs. 7-13 and 7-6. The current amplifier relationships can be clarified by example.

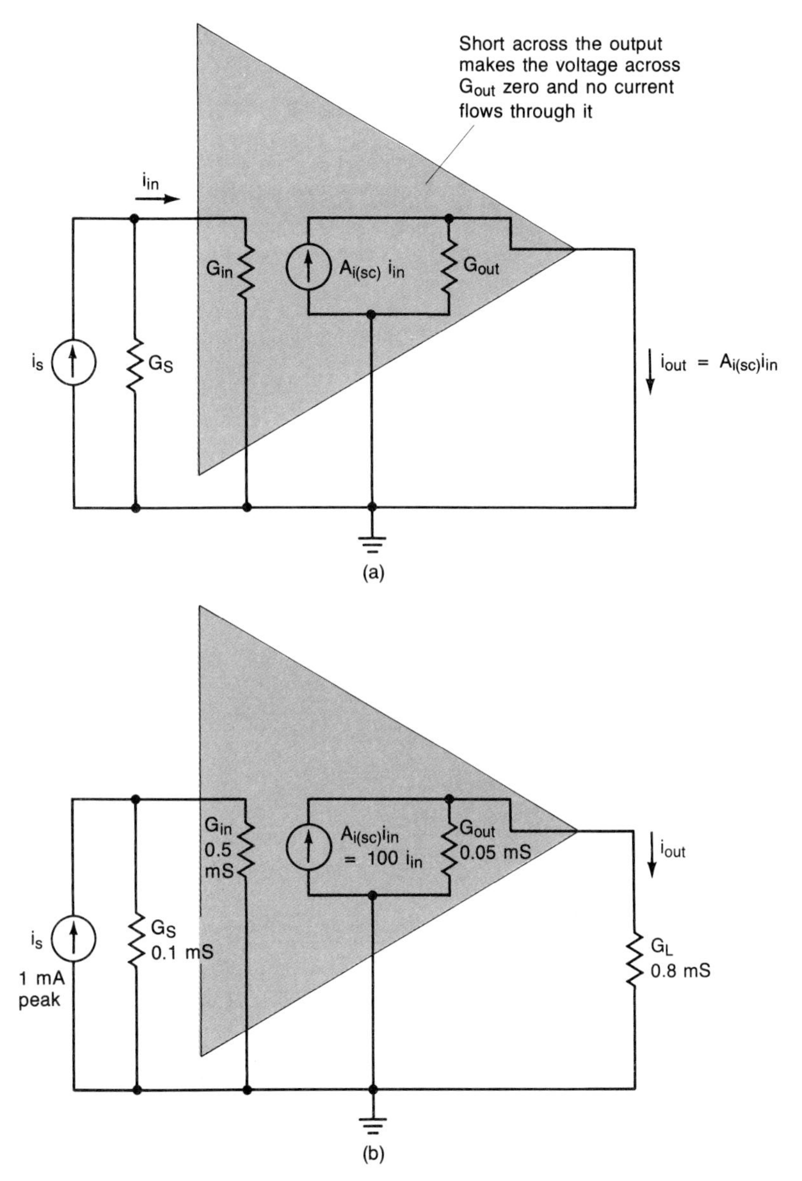

FIGURE 7-7 Current amplifier model: (a) driving a short; (b) driving a given load.

EXAMPLE 7-3

Given the current amplifier shown in Fig. 7-7(b), find i_{in}, i_{out}, the current gain from the amplifier's input to the load, and the overall current gain.

SOLUTION First, we shall find i_{in} by aplying Eq. 7-9.

$$i_{in} = \frac{G_{in}}{G_{in} + G_S} i_s = \frac{0.5 \text{ mS}}{0.5 \text{ mS} + 0.1 \text{ mS}} (10\ \mu\text{A}) = 8.33\ \mu\text{A}$$

The value of i_{out} may be found from Eq. 7-12 and then used to find A_i and A_{is}.

$$i_{out} = \frac{G_L}{G_L + G_{out}} A_{i(sc)} i_{in}$$

$$= \frac{0.8 \text{ mS}}{0.8 \text{ mS} + 0.05 \text{ mS}} (100)\,(8.33\ \mu\text{A}) = 0.784 \text{ mA}$$

$$A_i = \frac{i_{out}}{i_{in}} = \frac{0.784 \text{ mA}}{8.33\ \mu\text{A}} = 94.1$$

The ratio of the peak output current to the peak source current yields the overall current gain.

$$A_{is} = \frac{i_{out}}{i_s} = \frac{0.784 \text{ mA}}{10\ \mu\text{A}} = 78.4$$

An alternative approach requires the application of Eq. 7-13.

$$A_{is} = \frac{i_{out}}{i_s} = \frac{G_{in}}{G_{in} + G_S} \frac{G_L}{G_L + G_{out}} A_{i(sc)}$$

$$= \frac{0.5 \text{ mS}}{0.5 \text{ mS} + 0.1 \text{ mS}} \frac{0.8 \text{ mS}}{0.8 \text{ mS} + 0.05 \text{ mS}} (100) = 78.4$$

■

In addition to describing a voltage amplifier in terms of its voltage gain A_v, we occasionally must also specify its current gain A_i. The particular requirements are dictated by the amplifier's intended application.

7-6 Current Gain of a Voltage Amplifier

As we mentioned in Section 7-5, we may be interested in the current gain of a voltage amplifier—if the particular application so dictates. At the onset, we should mention that the ideal voltage amplifier has an infinite input resistance. Consequently, its input current is zero and its current gain is regarded as being indeterminant.

The simplest approach for finding the current gain of a voltage amplifier is to place its input in the basic form of our current amplifier model. This requires that we make a *source conversion*. Specifically, the signal source should be converted to its equivalent current source [see Fig. 7-8(a)]. Recall that a voltage source may be converted to an equivalent current source by dividing the source voltage by its source impedance. The source impedance is then placed in parallel with the resulting current source. This has been illustrated in Fig. 7-8(b).

By inspection of the output circuit, we can employ Ohm's law to arrive at an equation for i_{out}.

$$i_{out} = \frac{A_{v(oc)} v_{in}}{R_L + R_{out}} \tag{7-14}$$

The input voltage v_{in} is given by

$$v_{in} = i_{in} R_{in} \tag{7-15}$$

If we substitute this result into Eq. 7-14, we arrive at

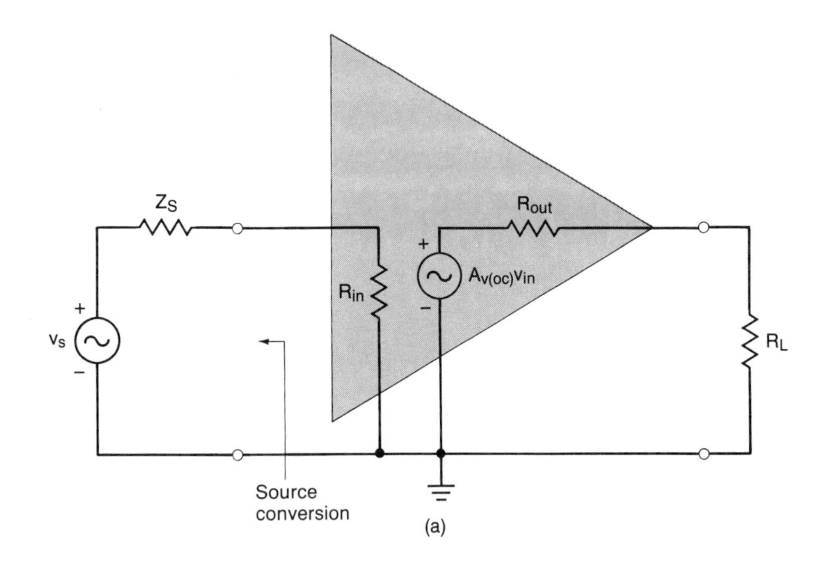

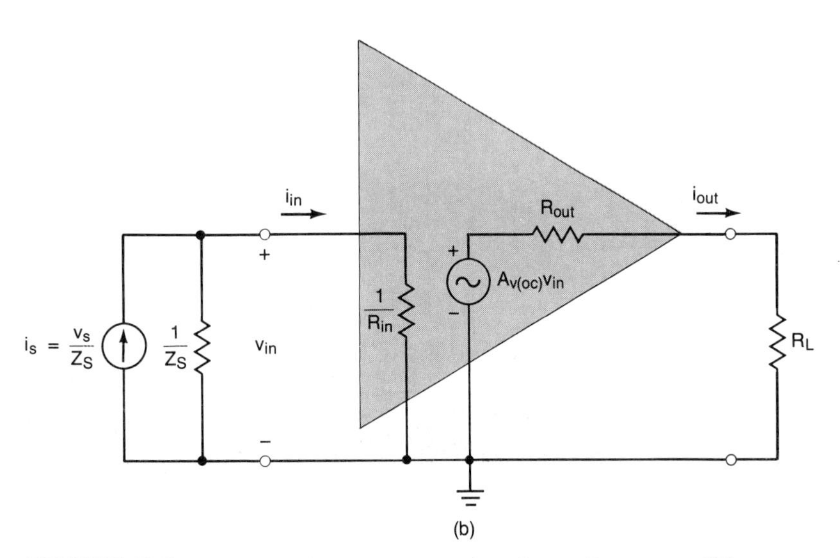

FIGURE 7-8 **Finding the current gain of a voltage amplifier: (a) source conversion; (b) the input circuit is converted into the basic form of a current amplifier.**

$$i_{out} = \frac{A_{v(oc)}}{R_L + R_{out}} i_{in} R_{in} = \frac{R_{in}}{R_L + R_{out}} A_{v(oc)} i_{in} \tag{7-16}$$

To find the current gain from the signal source to the output, we must apply current division at the input.

$$\begin{aligned} i_{in} &= \frac{1/R_{in}}{1/R_{in} + 1/Z_S} i_s = \frac{1/R_{in}}{(Z_S + R_{in})/R_{in}Z_S} i_s \\ &= \frac{1}{R_{in}} \frac{R_{in}Z_S}{Z_S + R_{in}} i_s = \frac{Z_S}{Z_S + R_{in}} i_s \end{aligned} \tag{7-17}$$

Substituting Eq. 7-17 into Eq. 7-16 yields

$$i_{out} = \frac{R_{in}}{R_L + R_{out}} A_{v(oc)} i_{in}$$

$$= \frac{R_{in}}{R_L + R_{out}} A_{v(oc)} \frac{Z_S}{Z_S + R_{in}} i_s$$

and solving for the current gain leads us to Eq. 7-18. (Note that i_s in Eq. 7-18 is *not* the current draw out of the source, but its maximum short-circuit output current.)

$$A_{is} = \frac{i_{out}}{i_s} = \frac{R_{in}}{R_L + R_{out}} A_{v(oc)} \frac{Z_S}{Z_S + R_{in}} \tag{7-18}$$

EXAMPLE 7-4

In Example 7-2, we found the voltage gain of the equivalent amplifier circuit shown in Fig. 7-5. Find its current gain.

SOLUTION We shall employ Eq. 7-18 to find the current gain.

$$A_{is} = \frac{i_{out}}{i_s} = \frac{R_{in}}{R_L + R_{out}} A_{v(oc)} \frac{Z_S}{Z_S + R_{in}}$$

$$= \frac{1\ \text{k}\Omega}{2\ \text{k}\Omega + 7.5\ \text{k}\Omega}(100)\frac{600\ \Omega}{600\ \Omega + 1\ \text{k}\Omega} = 3.95$$

■

7-7 Power Gain

In addition to describing an amplifier from a voltage or current gain perspective, power gain may also be used. The ratio of the (resistive) power delivered to the load to the (resistive) power provided by the signal source to the *amplifier's input* is defined to be the power gain.

With reference to Fig. 7-9 we can state that the power gain A_p is

$$A_p = \frac{P_{out}}{P_{in}} \tag{7-19}$$

When power levels are being analyzed we generally employ rms or effective values. The notation utilized for effective values requires that *capital letters* be used for voltages, currents, and power. Their subscripts, however, are *lowercase* letters. For example, the input voltage and current are denoted V_{in} and I_{in}, respectively.

By inspection of Fig. 7-9, we can see that

$$A_p = \frac{P_{out}}{P_{in}} = \frac{(V_{out})(I_{out})}{(V_{in})(I_{in})} = \frac{V_{out}}{V_{in}} \frac{I_{out}}{I_{in}}$$

and

$$A_p = A_v A_i \tag{7-20}$$

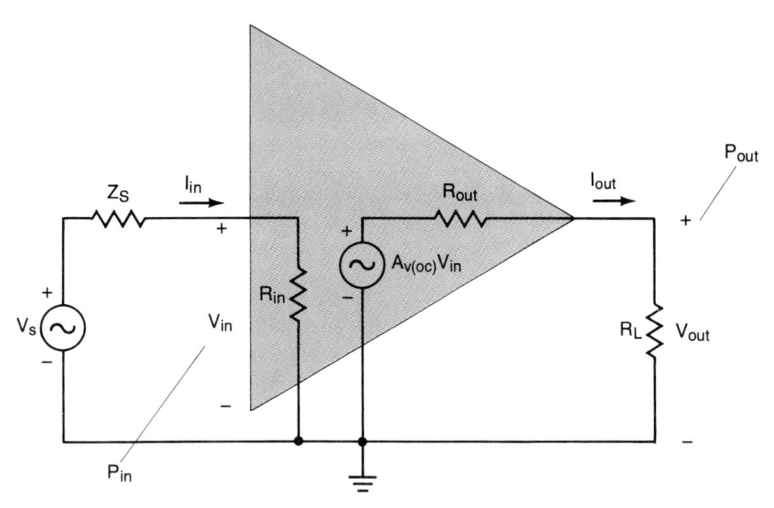

FIGURE 7-9 Power gain.

where A_v is the voltage gain from the amplifier's input to the load. This was given by Eq. 7-5.

$$A_v = \frac{R_L}{R_L + R_{out}} A_{v(oc)}$$

and the current gain from the amplifier's input to the load follows directly from Eq. 7-16.

$$A_i = \frac{i_{out}}{i_{in}} = \frac{R_{in}}{R_L + R_{out}} A_{v(oc)} \tag{7-21}$$

EXAMPLE 7-5

Given the voltage amplifier in Fig. 7-5, determine its power gain. What is the actual power delivered to the load?

SOLUTION First we shall find the voltage gain.

$$A_v = \frac{v_{out}}{v_{in}} = \frac{R_L}{R_L + R_{out}} A_{v(oc)}$$

$$= \frac{2\ \text{k}\Omega}{2\ \text{k}\Omega + 7.5\ \text{k}\Omega}(100) = 21.1$$

The current gain is

$$A_i = \frac{i_{out}}{i_{in}} = \frac{R_{in}}{R_L + R_{out}} A_{v(oc)}$$

$$= \frac{1\ \text{k}\Omega}{2\ \text{k}\Omega + 7.5\ \text{k}\Omega}(100) = 10.5$$

The power gain may be found by the product of the two gains.

$$A_p = A_v A_i = (21.1)\ (10.5) \simeq 222$$

In Example 7-2 we saw that the (sinusoidal) peak output voltage is 13.2 mV. Its rms value is

$$V_{out} = \frac{v_{out(peak)}}{\sqrt{2}} = 9.33 \text{ mV rms}$$

and the resulting output power is

$$P_{out} = \frac{V_{out}^2}{R_L} = \frac{(9.33 \text{ mV rms})^2}{2 \text{ k}\Omega} = 43.5 \text{ nW}$$

■

As we can see, even though the power gain is rather large (222), the output power level is extremely small (43.5 nW). However, the power level delivered by the signal source is only 0.195 nW. (The reader should verify this.) Typically, small-signal voltage amplifiers produce only a few milliwatts or less of output power. When several watts of output power are required, *large-signal* power amplifiers are used, and they are driven by voltage amplifiers. Power amplifiers are considered in Chapter 12.

7-8 Decibels

As we have seen, it is possible to describe amplifiers in terms of their voltage, current, and power gains. To simplify the analysis of electronic amplifiers and systems, a logarithmic measure called the *decibel* is used.

The origin of the decibel can be traced back to the telephone industry. Our ears respond to sound intensity in a nonlinear logarithmic fashion. Since many of the amplifiers were used to supply an audio output, the use of a logarithmic power output scale was deemed reasonable and convenient. The basic unit of power gain was termed the *bel* in honor of Alexander Graham Bell. Its definition is provided by

$$|A_p|_{bels} = \log \frac{P_{out}}{P_{in}} \text{ bels} \qquad (7\text{-}22)$$

Note that $|A_p|$ is read "the magnitude of the power gain."

For example, a power gain of 100 would be equivalent to

$$|A_p|_{bels} = \log \frac{P_{out}}{P_{in}} \text{ bels} = \log 100 \text{ bels} = 2 \text{ bels}$$

The bel proved to be too large for most work, so one-tenth of a bel–the decibel (dB)–was adopted. In this case, our power gain of 100 is equivalent to

$$|A_p|_{dB} = \frac{10 \text{ dB}}{\text{bel}} (|A_p|_{bels}) = \frac{10 \text{ dB}}{\text{bel}} (2 \text{ bels}) = 20 \text{ dB}$$

In short, to convert a power gain expressed in bels into decibels, we merely multiply the number of bels by 10. Symbolically,

$$|A_p|_{dB} = \frac{10 \text{ dB}}{\text{bel}} (|A_p|_{bels}) = \frac{10 \text{ dB}}{\text{bel}} \log \frac{P_{out}}{P_{in}} \text{ bel}$$

$$= 10 \log \frac{P_{out}}{P_{in}} \text{ dB}$$

and we arrive at Eq. 7-23. The units (dB) on its right-hand side are implied.

$$|A_p|_{dB} = 10 \log \frac{P_{out}}{P_{in}} \tag{7-23}$$

In practice, it is much easier to measure voltages than to measure power levels. Therefore, a relationship between the voltage gain and the power gain in decibels was developed. Its derivation proceeds as follows:

$$|A_p|_{dB} = 10 \log \frac{V_{out}^2/R_L}{V_{in}^2/R_{in}} = 10 \log \frac{V_{out}^2}{V_{in}^2}\frac{R_{in}}{R_L}$$

$$= 10 \log \left(\frac{V_{out}}{V_{in}}\right)^2 \frac{R_{in}}{R_L}$$

The result above can be simplified by recalling two of the fundamental properties of logarithms. First, the log of a quantity raised to a power is equal to that power times the log of the quantity. Second, the log of a product is equal to the sum of the logs of the individual terms of that product. Application of these two properties results in the following statement:

$$|A_p|_{dB} = 20 \log \frac{V_{out}}{V_{in}} + 10 \log \frac{R_{in}}{R_L} \tag{7-24}$$

Generally, when we are concerned about power gain, we are also interested in achieving maximum power transfer. If this is the case, impedances are matched. Specifically, the (resistive) source impedance Z_S, the input resistance R_{in}, the output resistance R_{out}, and the load resistance R_L will all be equal. (In telecommunications 600 Ω is typically used, and in radio frequency work 50 Ω is the standard.) Hence

$$R_{in} = R_L$$

and Eq. 7-24 reduces to

$$|A_p|_{dB} = 20 \log \frac{V_{out}}{V_{in}} + 10 \log 1$$

$$= 20 \log \frac{V_{out}}{V_{in}} + 0 = 20 \log \frac{V_{out}}{V_{in}} \tag{7-25}$$

In general, we can state:

For matched impedances: $$|A_p|_{dB} = 20 \log A_v \tag{7-26}$$

The magnitude of the voltage gain in decibels will be equal to the magnitude of the power gain in decibels *if the impedances are matched.*

In practice, we often express the magnitude of the voltage gain in decibels with a total *disregard* for the impedance levels. The only consequence of this is that the magnitude of the voltage gain in decibels will no longer be equal to the magnitude of the power gain in decibels. However, these two gains can be related if Eq. 7-24 is used.

However, if a simple statement of the voltage gain is required, we may use

$$|A_v|_{dB} = 20 \log A_v \tag{7-27}$$

In a similar fashion, the current gain in decibels can also be related to the power gain in decibels. Hence

$$\begin{aligned} |A_p|_{dB} &= 10 \log \frac{P_{out}}{P_{in}} = 10 \log \frac{I_{out}^2 R_L}{I_{in}^2 R_{in}} \\ &= 20 \log \frac{I_{out}}{I_{in}} + 10 \log \frac{R_L}{R_{in}} \end{aligned} \tag{7-28}$$

and if the impedances are matched, we arrive at

$$\text{For matched impedances:} \quad |A_p|_{dB} = 20 \log A_i \tag{7-29}$$

Merely to state the current gain in decibels, without concerning ourselves with its relationship to the power gain, we may use

$$|A_i|_{dB} = 20 \log A_i \tag{7-30}$$

EXAMPLE 7-6

Find the voltage gain, current gain, and the power gain in decibels for the amplifier given in Fig. 7-5.

SOLUTION In Example 7-5 we found that

$$A_v = \frac{v_{out}}{v_{in}} = 21.1 \qquad A_i = \frac{i_{out}}{i_{in}} = 10.5$$

and

$$A_p = A_v A_i \simeq 222$$

The voltage gain in decibels may be found from Eq. 7-27.

$$|A_v|_{dB} = 20 \log A_v = 20 \log(21.1) = 26.5 \text{ dB}$$

The current gain in decibels is given by Eq. 7-30.

$$|A_i|_{dB} = 20 \log A_i = 20 \log(10.5) = 20.4 \text{ dB}$$

The power gain in decibels is provided by Eq. 7-23.

$$|A_p|_{dB} = 10 \log A_p = 10 \log(222) = 23.5 \text{ dB}$$

Alternatively, we could have found the power gain in decibels by using either the voltage gain or the current gain, and their associated impedance levels (refer to Eqs. 7-24 and 7-25, respectively). Hence

$$\begin{aligned} |A_p|_{dB} &= 20 \log A_v + 10 \log \frac{R_{in}}{R_L} \\ &= 20 \log(21.1) + 10 \log \frac{1\ \text{k}\Omega}{2\ \text{k}\Omega} = 23.5 \text{ dB} \end{aligned}$$

or

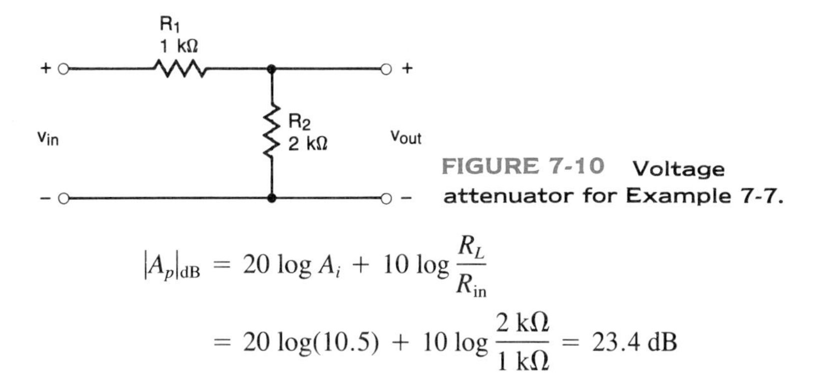

FIGURE 7-10 Voltage attenuator for Example 7-7.

$$|A_p|_{\text{dB}} = 20 \log A_i + 10 \log \frac{R_L}{R_{\text{in}}}$$

$$= 20 \log(10.5) + 10 \log \frac{2\ \text{k}\Omega}{1\ \text{k}\Omega} = 23.4\ \text{dB}$$

■

EXAMPLE 7-7

Find the voltage gain in decibels of the circuit given in Fig. 7-10.

SOLUTION In this case, we merely apply voltage division and solve for A_v.

$$A_v = \frac{v_{\text{out}}}{v_{\text{in}}} = \frac{R_2}{R_1 + R_2} = \frac{2\ \text{k}\Omega}{1\ \text{k}\Omega + 2\ \text{k}\Omega} = 0.667$$

(Since the voltage gain is less than unity, the circuit is classified as a *voltage attenuator*.) Now, from Eq. 7-27,

$$|A_v|_{\text{dB}} = 20 \log A_v = 20 \log(0.667) = -3.52\ \text{dB}$$

■

The examples above serve to illustrate a very important point. *Gains with a magnitude greater than unity will have positive decibel gains, and gains less than unity will be negative.*

In our work above we have seen how to convert a "straight" gain into its equivalent decibel gain. However, quite often, we have to convert a gain expressed in decibels into its equivalent "straight" gain. This merely involves a little algebra and taking the antilog of both sides of the resulting equation.

$$|A_v|_{\text{dB}} = 20 \log A_v$$

$$\log A_v = \frac{|A_v|_{\text{dB}}}{20}$$

$$\boxed{|A_v| = 10^{(|A_v|_{\text{dB}})/20}} \qquad (7\text{-}31)$$

where $|A_v|$ is the magnitude of the "straight" voltage gain and $|A_v|_{\text{dB}}$ is the magnitude of the voltage gain in decibels.

EXAMPLE 7-8

A voltage amplifier has a voltage gain of 40 dB and a current gain of 15 dB. Find its corresponding "straight" gains (ratios).

SOLUTION From Eq. 7-31 we can find the voltage gain.

$$|A_v| = 10^{(|A_v|_{\text{dB}})/20} = 10^{(40)/20} = 10^2 = 100$$

Similarly, we can find the current gain.

$$|A_i| = 10^{(|A_i|_{\text{dB}})/20} = 10^{(15)/20} = 10^{0.75} = 5.62$$

■

7-9 Relative Decibel Scales

As we have seen, the decibel scale is used to describe the ratio between an output quantity and a similar input quantity. In essence, we must always compare one quantity with reference to another quantity. In practice, this is *not* constrained to the ratio between output and input. This fact gives rise to two very popular *relative decibel scales*: dBm and dBV.

The dBm scale references power measurements to 1 mW dissipated in a 600-Ω resistive load. Its definition is provided by

$$|A_p|_{\text{dBm}} = 10 \log \frac{P}{1 \text{ mW}} \Bigg|_{R=600\ \Omega} \tag{7-32}$$

The statement above may be read: ''The magnitude of the power gain in dBm is equal to 10 times the log of the measured power P divided by 1 mW given that the impedance level is 600 Ω.'' Quite often, voltmeters will have a dBm scale, which will permit *voltage measurements* to be related to the dBm scale. Recalling that

$$P = \frac{V_r^2}{R}$$

$$V_r^2 = PR$$

$$V_r = \sqrt{PR}$$

and the voltage (V_r) corresponding to 1 mW in 600 Ω is

$$V_r = \sqrt{(1 \text{ mW})\,(600\ \Omega)} = 0.7746 \text{ V rms}$$

From our previous work it should be clear that if the impedance across which the voltage measurement is made is 600 Ω,

$$|A_p|_{\text{dBm}} = 20 \log \frac{V_r}{0.7746 \text{ V}} \Bigg|_{R=600\ \Omega} \tag{7-33}$$

If we are uninterested in relating our measurement to a power level, the impedance levels may be disregarded. Consequently, we arrive at

$$|A_v|_{\text{dBm}} = 20 \log \frac{V_r}{0.7746 \text{ V}} \tag{7-34}$$

The dBV measurement scale (also offered by many voltmeters) is much more straightforward - if we are totally uninterested in power levels. In this case, voltage measurements in decibels are referenced to 1 V (rms). No impedance level is included in its definition.

$$|A_v|_{\text{dBV}} = 20 \log \frac{V_r}{1 \text{ V}} = 20 \log V_r \tag{7-35}$$

EXAMPLE 7-9

If a voltage of 5 V rms appears across the output of an amplifier, what would a digital voltmeter read if it is configured to measure in dBm? In dBV?

SOLUTION This problem requires the use of Eqs. 7-34 and 7-35. The number of dBm would be

$$|A_v|_{dBm} = 20 \log \frac{V_r}{0.7746\ V} = 20 \log \frac{5\ V}{0.7746\ V} = 16.2\ dBm$$

and the number of dBV would be

$$|A_v|_{dBV} = 20 \log \frac{5\ V}{1\ V} = 20 \log 5 = 14.0\ dBV$$ ■

The relative decibel scales on a digital voltmeter can be a very powerful tool in the analysis of an amplifier circuit. For example, by measuring the output voltage in dBV and the input voltage in dBV, the voltage gain in decibels may be found by *subtracting* the two measurements. The argument for this proceeds as follows:

$$|A_v|_{dB} = 20 \log A_v = 20 \log \frac{V_{out}}{V_{in}} = 20 \log \frac{V_{out}/V_{ref}}{V_{in}/V_{ref}}$$

$$= 20 \log \frac{V_{out}}{V_{ref}} - 20 \log \frac{V_{in}}{V_{ref}}$$

$$= 20 \log \frac{V_{out}}{1\ V} - 20 \log \frac{V_{in}}{1\ V}$$

$$\boxed{|A_v|_{dB} = |A_v|_{dBV(out)} - |A_v|_{dBV(in)}} \qquad (7\text{-}36)$$

If we let V_{ref} be 0.7746 V, we can find $|A_v|_{dB}$ by working in dBms.

EXAMPLE 7-10

The voltage amplifier shown in Fig. 7-11 has an input of −40 dBV and an output of 6.02 dBV. What is its voltage gain in decibels? Its "straight" voltage gain? Its input voltage?

SOLUTION The gain in decibels is given by Eq. 7-36.

$$|A_v|_{dB} = |A_v|_{dBV(out)} - |A_v|_{dBV(in)}$$
$$= 6.02\ dBV - (-40\ dBV) = 46.02\ dB$$

The straight voltage gain may be found by using Eq. 7-31.

$$A_v = 10^{(|A_v|_{dB})/20} = 10^{46.02/20} = 200$$

In a similar fashion, we may find the rms value of the input voltage.

$$V_{in} = 10^{(|A_v|_{dBV(in)})/20} = 10^{-40/20} = 10^{-2} = 10\ mV\ rms$$ ■

FIGURE 7-11 Amplifier for Example 7-10.

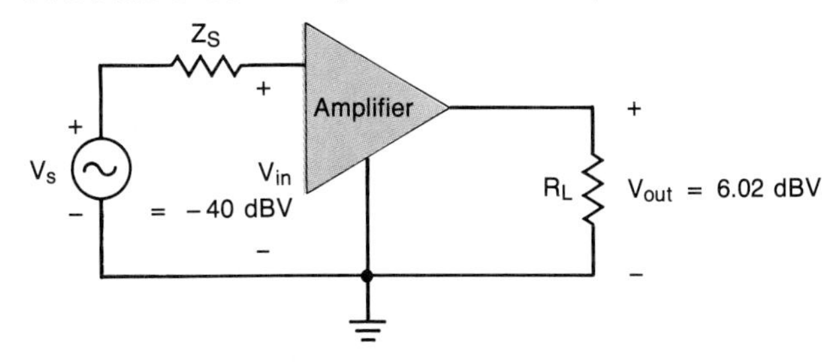

7-10 Cascaded Amplifier Systems

In Chapter 8 we develop equations for the analysis of the three fundamental BJT, FET, and *op amp configurations*. The three basic op amp configurations: the noninverting amplifier, the inverting amplifier, and the voltage follower also have a one-to-one correspondence with the BJT and FET configurations. These are listed in Table 7-1.

TABLE 7-1
BJT–FET–Op Amp Corresponding Amplifier Configurations

BJT	FET	Op amp
Common base	Common gate	Noninverting amp
Common emitter	Common source	Inverting amp
Common collector	Common drain	Voltage follower

The rationale behind the correspondence will become clear in Chapter 8. For example, we shall see that the common-base and common-gate amplifiers both exhibit relatively large voltage gains, and a 0° phase shift between the input and output voltages. However, they both tend to have very low input resistances and relatively large output resistances [see Fig. 7-12(a)]. The common-collector and common-drain amplifier configurations both have extremely large input resistances and very small output resistances. However, they offer voltage gains less than unity. Their input and output voltages are also in phase [see Fig. 7-12(b)]. To realize a voltage amplifier *system* whose attributes begin to approach those of the ideal amplifier (large input resistance, small output resistance, and a large voltage gain), we may use a *cascaded amplifier* arrangement [refer to Fig. 7-12(c)].

As we can see, when amplifiers are cascaded, the output of one stage is used to drive the input of the next stage. This results in an amplifier system with a performance which is superior to that which can be achieved via a single amplifier stage [see Fig. 7-12(d)].

Another benefit of a cascaded voltage amplifier arrangement is often a dramatic increase in the overall voltage gain. To see exactly how this occurs, inspect Fig. 7-13.

The overall (total) voltage gain A_{vT} is

$$A_{vT} = \frac{V_{\text{out}(3)}}{V_{\text{in}(1)}}$$

If we take the product of the individual *loaded* stage gains, we can arrive at the overall voltage gain. First, we see that

$$A_{v1}A_{v2}A_{v3} = \frac{V_{\text{out}(1)}}{V_{\text{in}(1)}}\frac{V_{\text{out}(2)}}{V_{\text{in}(2)}}\frac{V_{\text{out}(3)}}{V_{\text{in}(3)}}$$

and since the output of the first stage drives the input of the second stage, and the output of the second stage drives the input of the third stage, we may make the substitutions indicated below.

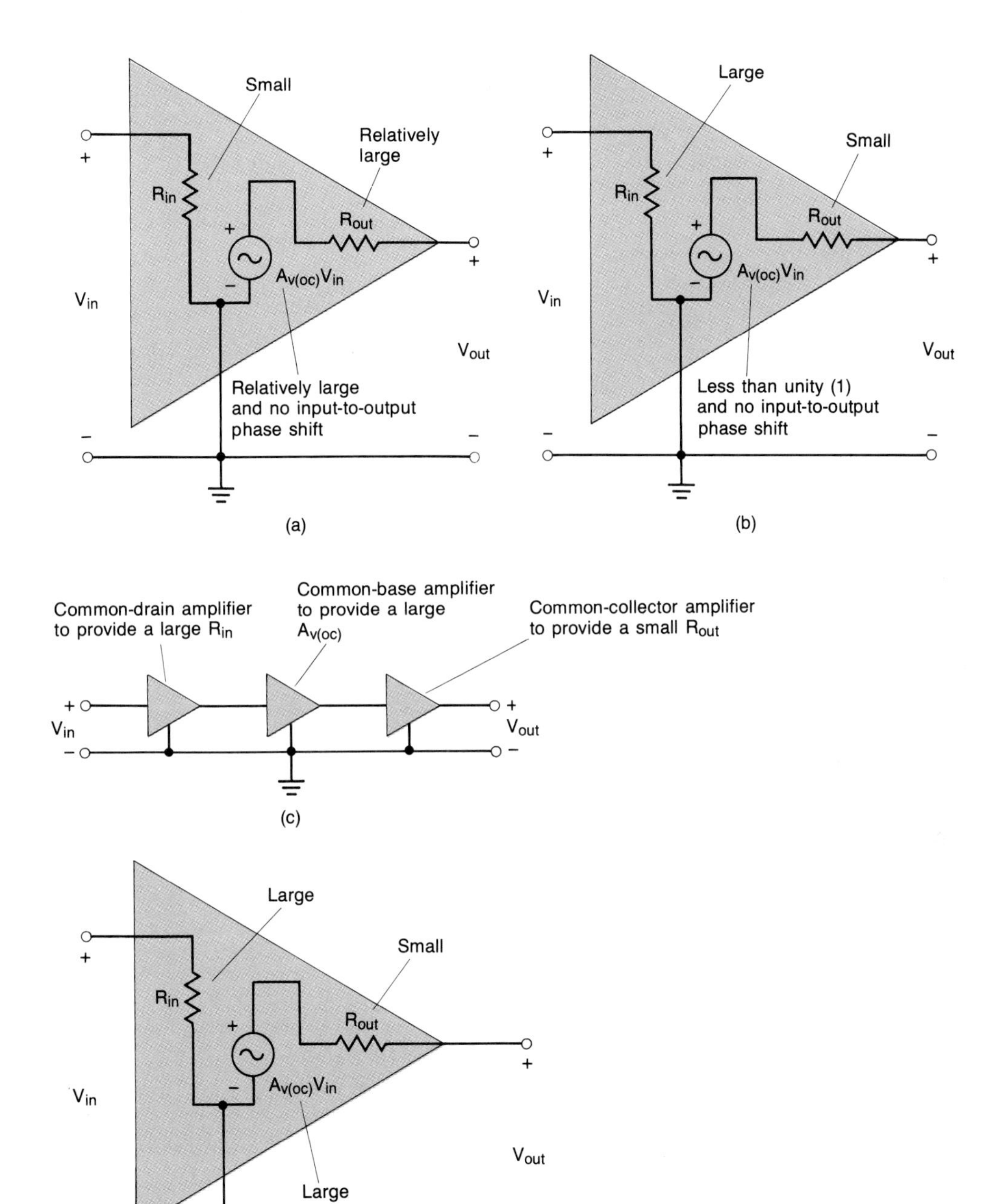

FIGURE 7-12 Cascading amplifier stages: (a) common-base or common-gate amplifier; (b) common-collector or common-drain amplifier; (c) cascaded amplifier system; (d) model of the cascaded amplifier system.

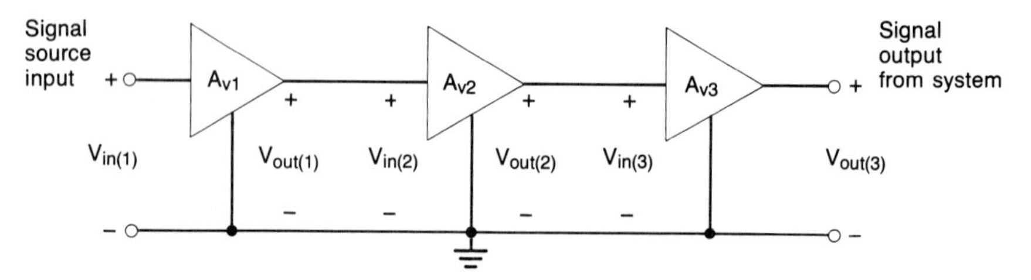

FIGURE 7-13 Cascaded voltage amplifier system.

$$A_{v1}A_{v2}A_{v3} = \frac{\cancel{V_{out(1)}}}{V_{in(1)}}\frac{\cancel{V_{out(2)}}}{\cancel{V_{out(1)}}}\frac{V_{out(3)}}{\cancel{V_{out(2)}}} = \frac{V_{out(3)}}{V_{in(1)}} = A_{vT}$$

Therefore, we can state

$$A_{vT} = A_{v1}A_{v2}A_{v3} \tag{7-37}$$

In general, for an amplifier with n stages, Eq. 7-38 defines the overall voltage gain.

$$A_{vT} = A_{v1}A_{v2}A_{v3} \cdots A_{v(n-1)}A_{vn} \tag{7-38}$$

If the loaded voltage gains are expressed as decibels, the overall voltage gain in decibels may be found by simple addition.

$$|A_{vT}|_{dB} = 20 \log A_{vT} = 20 \log (A_{v1}A_{v2} \cdots A_{vn})$$
$$= 20 \log A_{v1} + 20 \log A_{v2} + \cdots + 20 \log A_{vn}$$

$$|A_{vT}|_{dB} = |A_{v1}|_{dB} + |A_{v2}|_{dB} + \cdots + |A_{vn}|_{dB} \tag{7-39}$$

EXAMPLE 7-11

A two-stage voltage amplifier model has been illustrated in Fig. 7-14. The amplifier stages are identical and their output loads are identical. Find the overall voltage gain and express it in decibels.

SOLUTION The voltage gain of the first stage is

$$A_{v1} = \frac{V_{out(1)}}{V_{in(1)}} = \frac{R_{in(2)}}{R_{in(2)} + R_{out(1)}} A_{v(oc)1}$$
$$= \frac{1\text{ k}\Omega}{1\text{ k}\Omega + 5\text{ k}\Omega}(200) = 33.33$$

Similarly, the voltage gain of the second stage is

$$A_{v2} = \frac{V_{out(2)}}{V_{in(2)}} = \frac{R_L}{R_L + R_{out(2)}} A_{v(oc)2}$$
$$= \frac{1\text{ k}\Omega}{1\text{ k}\Omega + 5\text{ k}\Omega}(200) = 33.33$$

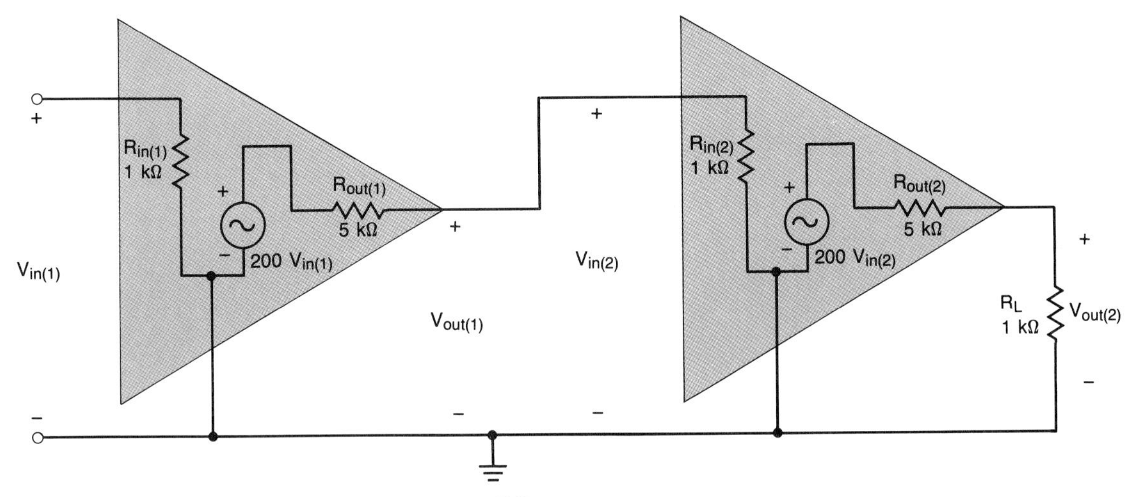

FIGURE 7-14 **Two-stage voltage amplifier.**

The overall voltage gain of the amplifier system is given by the product of the loaded voltage gains (Eq. 7-38).

$$A_{vT} = A_{v1}A_{v2} = (33.33)\,(33.33) = 1111$$

The voltage gain of the two stages in decibels is

$$|A_{v1}|_{\text{dB}} = 20 \log A_{v1} = 20 \log 33.33 = 30.46 \text{ dB}$$

and

$$|A_{v2}|_{\text{dB}} = |A_{v1}|_{\text{dB}} = 30.46 \text{ dB}$$

The overall voltage gain in decibels may be found through addition (Eq. 7-39).

$$|A_{vT}|_{\text{dB}} = |A_{v1}|_{\text{dB}} + |A_{v2}|_{\text{dB}} = 30.46 \text{ dB} + 30.46 \text{ dB}$$
$$= 60.92 \text{ dB}$$

Alternatively,

$$|A_{vT}|_{\text{dB}} = 20 \log A_{vT} = 20 \log 1111 = 60.91 \text{ dB}$$ ■

Example 7-11 illustrates one of the primary advantages of cascading amplifier stages - a dramatic increase in the overall voltage gain. The voltage gain of 33 possessed by each of the two stages has been increased to 1111 by cascading them. Also note that the input impedance of the second stage serves to load down the output of the first stage.

7-11 Op Amp Circuit Model

As we saw in Section 7-10, when amplifier stages are cascaded, very large voltage gains can be realized and the amplifier system input and output resistances can be made to approach those of the ideal voltage amplifier. Such is the case of the op amp. Several differential amplifier stages are cascaded to form a very high performance integrated-circuit amplifier.

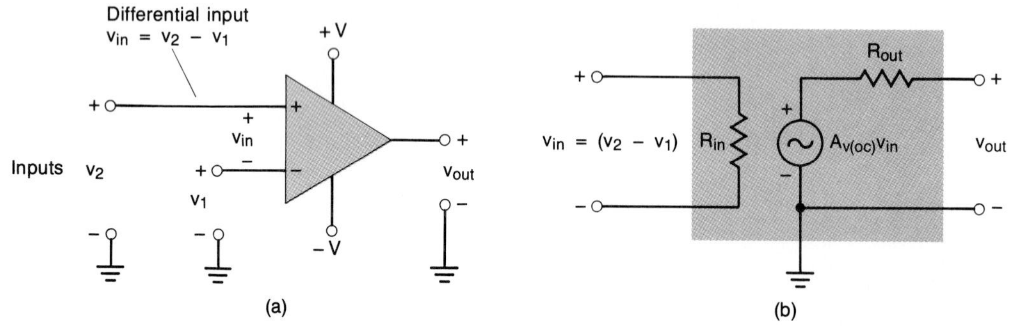

FIGURE 7-15 Op amp: (a) schematic symbol; (b) small-signal model.

We shall not delve into the theory of the signal processing of the differential pair until Chapter 8. For now, we shall merely introduce the characteristics of the signal input and output terminals of the op amp. In essence, we shall treat the op amp as another active device, like the BJT or the FET. However, we should acknowledge at the onset that op amps contain several transistors. (For example, the very popular 741 op amp contains 24 BJTs.)

The schematic symbol for an op amp has been repeated in Fig. 7-15(a). Recall that the "+" terminal is called noninverting input and the "−" terminal is the inverting input. As we shall see, the op amp behaves as a voltage-controlled voltage source. Notice that *two* input voltages have been defined in Fig. 7-15(a). This is due to the fact that the controlling voltage v_{in} is defined to be the *differential* (or difference) voltage across the op amp's two input terminals. This is emphasized in Fig. 7-15(b).

$$v_{in} = v_2 - v_1$$

The output voltage is directly proportional to the differential input voltage times the open-circuit voltage gain $A_{v(oc)}$. Hence

$$\boxed{v_{out} = A_{v(oc)}v_{in} = A_{v(oc)}(v_2 - v_1)} \tag{7-40}$$

The open-circuit voltage gain is, of course, ideally infinite. Real op amps typically have extremely large voltage gains. For example, the 741 op amp has a voltage gain that is specified to be a *minimum* of 50,000 and has a *typical* value of 200,000!

To define the op amp's characteristics graphically, we present Fig. 7-16. This figure depicts the op amp's input–output transfer characteristic curve. The particular values that have been indicated are for a typical 741 op amp with an $A_{v(oc)}$ of 200,000.

In Section 6-17 we pointed out that the maximum symmetrical output voltage swing of an op amp is limited to about 2 V less than its supply voltages. This fact is reflected in the op amp's transfer characteristic curve. Observe that the linear portion of the op amp's transfer characteristic curve occurs for differential input voltages of up to $\pm 65\ \mu V$. If v_{in} exceeds these boundaries, the op amp will go into saturation, and the output signal will be clipped.

Because of the large voltage gain associated with op amps, the resulting extremely small differential input voltage is often approximated as zero. This has been emphasized in Fig. 7-17(a).

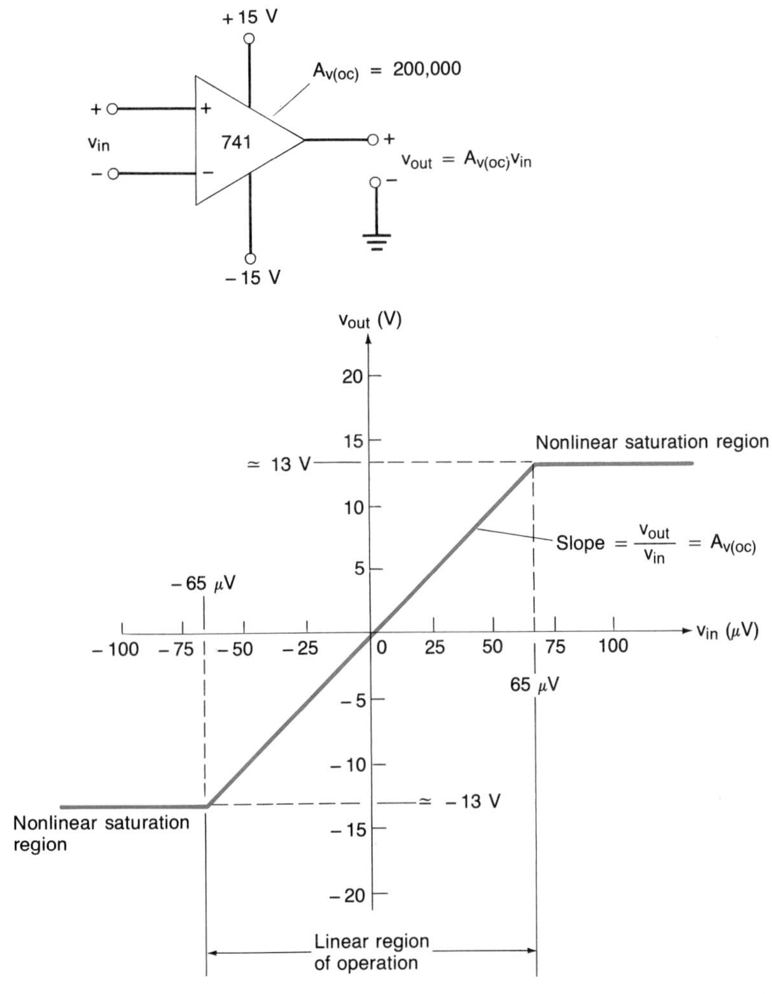

FIGURE 7-16 **Typical input-output transfer characteristic of a 741 op amp.**

To specify our op amp circuit model completely, we need to consider its R_{in} and R_{out} [refer to Fig. 7-17(b)]. The *differential input resistance* (R_{in}) is often referred to as simply the input resistance. It is the equivalent resistance as measured at either the inverting or the noninverting input with the other terminal connected to ground. For the 741 it is typically a relatively large 2 MΩ. However, for FET (differential) input op amps, values on the order of several gigaohms (10^9 Ω) are not uncommon. The typical R_{in} of a 741 op amp has been indicated in Fig. 7-17(b).

The output resistance (R_{out}) is the equivalent resistance that can be measured between the op amp's output and ground. For the 741 op amp, R_{out} is typically 75 Ω. This has also been indicated in Fig. 7-17(b).

Because of the large open-circuit voltage gain, the large input resistance, and the relatively small output resistance, we may idealize most op amps as shown in Fig. 7-17(c). This is the op amp model that we use for our work in Chapter 8.

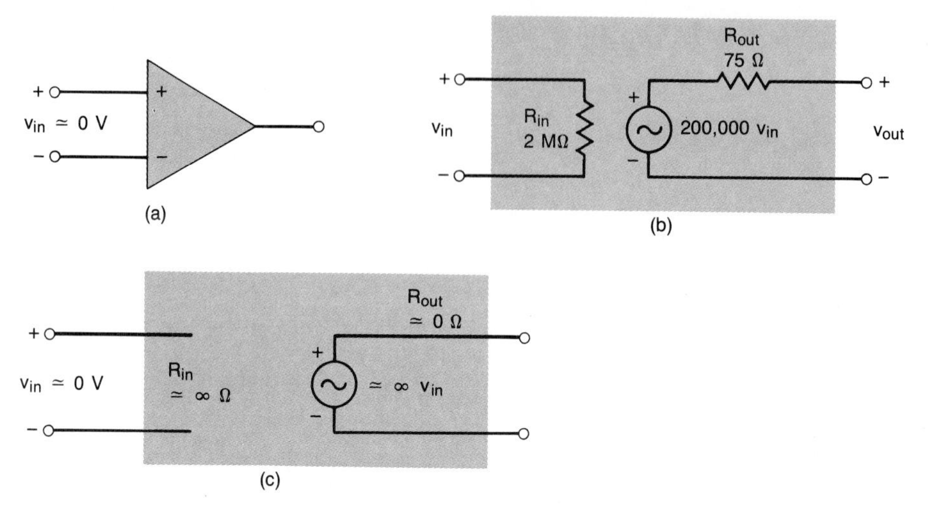

FIGURE 7-17 Op amp idealizations: (a) the differential input voltage of an op amp in its linear mode of operation is approximately zero; (b) typical equivalent-circuit values for a 741 op amp; (c) the ideal op amp model.

However, we shall continually refine our viewpoint of the op amp as our work progresses.

PROBLEMS

Drill, Derivations, and Definitions

Section 7-1

7-1. The controlled source in Fig. 7-1(a) has a μ of 30. If v_1 is 0.1 V peak, find v_2.

7-2. The controlled source in Fig. 7-1(b) has a β of 150. If i_1 is 100 μA peak, find i_2. Assume that a suitable load is connected across the source.

7-3. The controlled source in Fig. 7-1(c) has a g_m of 10 mS. If v_1 is 10 mV peak, find i_2. Assume that a suitable load is connected across the source.

7-4. The controlled source in Fig. 7-1(d) has an r_m of 1 kΩ. If i_1 is 1 mA peak, find v_2.

7-5. List the six characteristics associated with an ideal voltage amplifier.

Section 7-2

7-6. The amplifier in Fig. 7-2(b) has an $A_{v(oc)}$ of 100, an R_{in} of 10 kΩ, and an R_{out} of 3 kΩ. If v_{in} is 10 mV peak, find v_{out}.

7-7. The amplifier in Fig. 7-2(b) has a v_{in} of 5 mV peak, an R_{in} of 10 kΩ, an R_{out} of 5 kΩ, and a v_{out} of 2.5 V peak. What is its $A_{v(oc)}$?

Section 7-3

7-8. Given the amplifier model illustrated in Fig. 7-4, what is its input voltage v_{in}, its output voltage v_{out}, and its voltage gain A_{vs} from the signal source to its output? Assume that v_s is 5 mV peak, Z_s is 50 Ω, R_{in} is 2 kΩ, $A_{v(oc)}$ is 250, and R_{out} is 10 kΩ.

7-9. Repeat Prob. 7-8 if R_{in} is increased to 10 kΩ.

7-10. In your own words, explain why a voltage amplifier should have a large R_{in}.

Section 7-4

7-11. Repeat Prob. 7-8 for the amplifier circuit given in Fig. 7-5. Use the parameters given in Prob. 7-8. Assume that R_L is 5 kΩ.

7-12. Repeat Prob. 7-8 for the amplifier circuit given in Fig. 7-5. Use the parameters given in Prob. 7-8. Assume that R_L is 1 kΩ.

7-13. In your own words, explain why a voltage amplifier should have a small R_{out}.

Section 7-5

7-14. Explain what is meant by the term ''duality.'' Why is the current amplifier the dual of the voltage amplifier?

7-15. List the six characteristics of the ideal current amplifier.

7-16. Given the current amplifier indicated in Fig. 7-6(b), find the current gain A_{is} from the signal source to the load. Assume that G_S is 0.2 mS, G_{in} is 2 mS, $A_{i(sc)}$ is 200, G_{out} is 0.02 mS, and G_L is 0.15 mS. Find the peak load current if the signal source provides a peak current of 0.1 mA.

7-17. Repeat Prob. 7-16 if G_L is changed to 0.02 mS.

7-18. In your own words, explain why the G_{in} of a current amplifier should be large but its G_{out} should be small.

Section 7-6

7-19. Given the voltage amplifier indicated in Fig. 7-8(a), find its input-to-output voltage gain A_v, its source-to-output voltage gain A_{vs}, its input-to-output current gain A_i, and its source-to-output current gain A_{is}. Assume the following values: Z_S is 50 Ω, R_{in} is 3 kΩ, $A_{v(oc)}$ is 225, R_{out} is 6.8 kΩ, and R_L is 2.5 kΩ.

7-20. Repeat Prob. 7-19 if R_L is increased to 6.8 kΩ.

Section 7-7

7-21. Find the power gain of the amplifier described in Prob. 7-19.

7-22. Find the power gain of the amplifier described in Prob. 7-19. Assume that R_L is increased to 6.8 kΩ.

Section 7-8

7-23. An amplifier has a power gain A_p of 500; express its power gain in decibels. What is its gain in decibels if A_p is doubled? Halved? What are the respective changes in the dB values?

7-24. Repeat Prob. 7-23 for an amplifier with an A_p of 1000.

7-25. An amplifier has a voltage gain of 200 and a current gain of 150. Express these two gains in their decibel equivalents. Also express its power gain in decibels.

7-26. An amplifier has a voltage gain of 175 and a current gain of 50. Express these two gains in their decibel equivalents. Also express its power gain in decibels.

7-27. Given the attenuator circuit shown in Fig. 7-10. If R_1 is 5.1 kΩ and R_2 is 2.2 kΩ, find its voltage gain in decibels.

7-28. Repeat Prob. 7-27 if R_2 is increased to 100 kΩ.

7-29. An amplifier has a voltage gain of 43 dB and a current gain of 23 dB. Express these two gains as "straight" ratios.

7-30. Repeat Prob. 7-29 if the amplifier has a voltage gain of 41 dB and a current gain of 25 dB.

7-31. Show that the power gain in decibels may be found by the relationship

$$|A_p|_{\text{dB}} = \frac{|A_v|_{\text{dB}} + |A_i|_{\text{dB}}}{2}$$

Include all of the necessary algebraic steps.

Section 7-9

7-32. Define the relative decibel units dBm and dBV.

7-33. An amplifier has an ac output voltage of 2 V rms. What would a digital voltmeter indicate if it is configured to read in dBm? In dBV?

7-34. Repeat Prob. 7-33 if the amplifier's output voltage is 4 V rms.

7-35. An amplifier's input is −30 dBm and its output is 10 dBm. Find its voltage gain in decibels.

7-36. An amplifier's input is −43 dBV and its output is −2 dBV. Find its voltage gain in decibels.

Section 7-10

7-37. A four-stage cascaded voltage amplifier system has individual stage gains of 0.995, 150, 225, and 0.925. Find its total voltage gain. Also express it in decibels.

7-38. Repeat Prob. 7-37 if the individual gains are 0.821, 100, 95, and 120.

7-39. Express the individual stage gains given in Prob. 7-37 in their decibel equivalents and add them together to find the total gain in decibels. Your answer should be identical to that found in Prob. 7-37.

7-40. Express the individual stage gains given in Prob. 7-38 in their decibel equivalents and add them together to find the total gain in decibels. Your answer should be identical to that found in Prob. 7-38.

7-41. The cascaded amplifier in Fig. 7-14 has the following values: $R_{\text{in}(1)} = 10$ kΩ, $A_{v(\text{oc})1} = 150$, $R_{\text{out}(1)} = 5.1$ kΩ, $R_{\text{in}(2)} = 2$ kΩ, $A_{v(\text{oc})2} = 220$, $R_{\text{out}(2)} = 12$ kΩ, and $R_L =$ 10 kΩ. Find the overall voltage gain $[v_{\text{out}(2)}/v_{\text{in}(1)}]$ and express it in decibels.

7-42. The cascaded amplifier in Fig. 7-14 has the following values: $R_{\text{in}(1)} = 100$ kΩ, $A_{v(\text{oc})1} = 30$, $R_{\text{out}(1)} = 6.8$ kΩ, $R_{\text{in}(2)} = 3$ kΩ, $A_{v(\text{oc})2} = 180$, $R_{\text{out}(2)} = 5.6$ kΩ, and $R_L =$ 10 kΩ. Find the overall voltage gain $[v_{\text{out}(2)}/v_{\text{in}(1)}]$ and express it in decibels.

Section 7-11

7-43. An op amp has an $A_{v(oc)}$ of 100,000 and has a bias supply of ± 12 V. Make a neat sketch of its transfer characteristic curve.

7-44. Repeat Prob. 7-43 if $A_{v(oc)}$ is 150 dB and the bias supply voltage is ± 6 V.

7-45. Explain why the differential input voltage across the input terminals of an op amp may be approximated as zero.

Troubleshooting

When any analog electronic system is to be repaired, an oscilloscope should be used. This will quickly indicate severe dc bias problems, the presence or absence of a signal, and whether the signal is distorted. However, if we merely want to check gains in a reasonably functional system, a digital multimeter (DMM) with either a dBV or a dBm mode may be used.

7-46. An amplifier system is to have a gain of at least 100. A DMM indicates a reading of -23 dBV at its input and a reading of 25 dBV at its output. Is the amplifier's gain acceptable? What is its "straight ratio" voltage gain?

7-47. Repeat Prob. 7-46 if the amplifier's input reading is -45 dBm and its output reading is -8 dBm. (*Hint*: Working with dBm is identical to working in dBV.)

Design

7-48. An amplifier is driven by a signal source with a source resistance of 50 Ω. Determine the amplifier's required input resistance if no more than 1% of the (voltage) signal can be lost due to input loading effects.

7-49. An amplifier is to drive a load of 150 Ω. Determine the amplifier's required output resistance if at least 90% of its output voltage signal is to reach the load.

Computer

In our later work it will be necessary to have the computer determine gains in decibels. Decibel calculations require the use of common (base 10) logarithms. However, many personal computers will only compute natural (base *e*) logarithms. This problem can be overcome by drawing on the change-of-base formula taught in basic algebra courses.

$$\log_{10} A_v = \frac{\log_e A_v}{\log_e 10} = \frac{\log_e A_v}{2.302585}$$
$$= 0.4342945 \log_e A_v$$

7-50. Write a BASIC computer program that converts straight-ratio voltage gains into their decibel equivalents.

7-51. Write a BASIC computer program that prompts the user for Z_s, R_{in}, R_{out}, $A_{v(oc)}$, and R_L. It should then compute A_v, A_{vs}, A_i, A_{is}, and A_p. The gains should be expressed as straight ratios and in their decibel equivalents.

8

DISCRETE AND INTEGRATED AMPLIFIER CONFIGURATIONS

After Studying Chapter 8, You Should Be Able to:

- Determine the BJT and FET small-signal hybrid-pi model parameters (g_m, r_o, and r_π) at a given Q-point.
- Analyze the CE, CS, and op amp inverting amplifiers to find A_v, A_{vs}, A_i, A_{is}, and A_p.
- Explain the effect of removing the emitter bypass capacitor.
- Analyze the CB, CG, and op amp noninverting amplifiers to find A_v, A_{vs}, A_i, A_{is}, and A_p.
- Analyze the CC, CD, and op amp voltage follower circuits to find A_v, A_{vs}, A_i, A_{is}, and A_p.
- Analyze the discrete and op amp differential amplifiers.

8-1 Fundamental Voltage Amplifiers

In Section 7-10 we introduced three basic amplifier configurations: the noninverting amplifier, the inverting amplifier, and the voltage follower (see Table 7-1). A fourth amplifier category also exists - the differential amplifier. The differential amplifier was first introduced in Section 6-15. Each of these amplifier configurations is analyzed in detail in the sections to come. The procedure for the analysis of *any* amplifier is as follows:

1. Perform a dc analysis to find the Q-point.
2. Determine the small-signal (device) parameters at that Q-point.
3. Draw the ac equivalent circuit and determine R_{in}, R_{out}, and $A_{v(\text{oc})}$.
4. Use the equations developed for the general amplifier model to find A_v, A_{vs}, A_i, A_{is}, and A_p.

We have two primary goals. First, we want to emphasize the key characteristics of each of the amplifier configurations. Our second goal will be to demonstrate exactly how the configurations may be implemented by employing BJTs, FETs, or op amps. For convenience, a summary of the small-signal (device) models has been provided in Fig. 8-1.

Recall that each of the small-signal parameters has a value that is controlled by the dc operating point. To keep these parameters reasonably constant, we focus our attention on the more stable bias circuit arrangements. Specifically, we restrict our examples to those that employ voltage dividers. However, our analysis techniques may be directly extended to any of the other biasing circuit arrangements.

8-2 Determining the Device AC Parameters

In Section 4-13 we examined the BJT signal process by tracing the signal flow through a common-base BJT amplifier. Similarly, the signal process through a common-source FET amplifier was illustrated in Section 5-11.

The signal analysis of BJT, FET, and op amp voltage amplifiers may be simplified by drawing on the principle of superposition. The fundamental constraint is that the signals must be relatively small to ensure that the amplifiers remain linear. If this is the case, the dc and ac analyses may be conducted separately. The dc analysis allows us to find the dc bias voltages and currents. Finding the dc operating point allows us to ascertain the

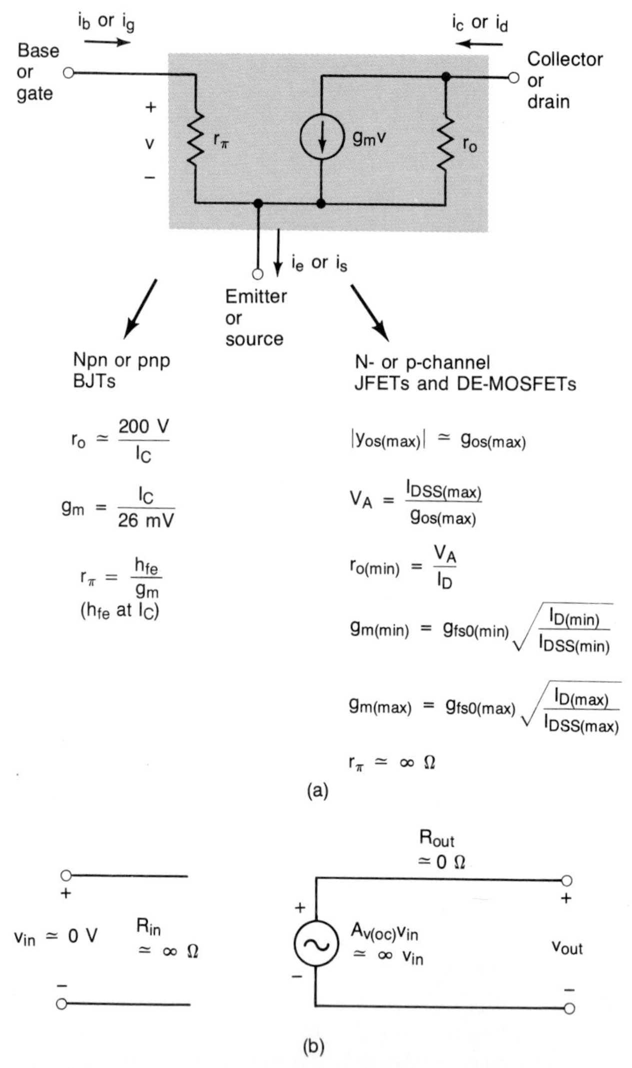

FIGURE 8-1 A summary of the small-signal models: (a) BJT/FET; (b) op amp (ideal) linear model.

magnitudes of the ac parameters possessed by a given active device at that *Q*-point.

In Fig. 8-2 we have illustrated the voltage-divider bias circuits for the 2N4124 BJT and the 2N5458 JFET. The dc analysis of these two circuits was accomplished in Sections 6-11 and 6-12, respectively. (The reader may wish to review those two sections.)

The common-emitter, common-source, common-base, common-gate, common-collector, and common-drain circuits have all been illustrated in Fig. 8-3. The only differences between them are the placement of the input and output coupling capacitors and the bypass capacitors. Since these capacitors are open circuits to dc, the dc

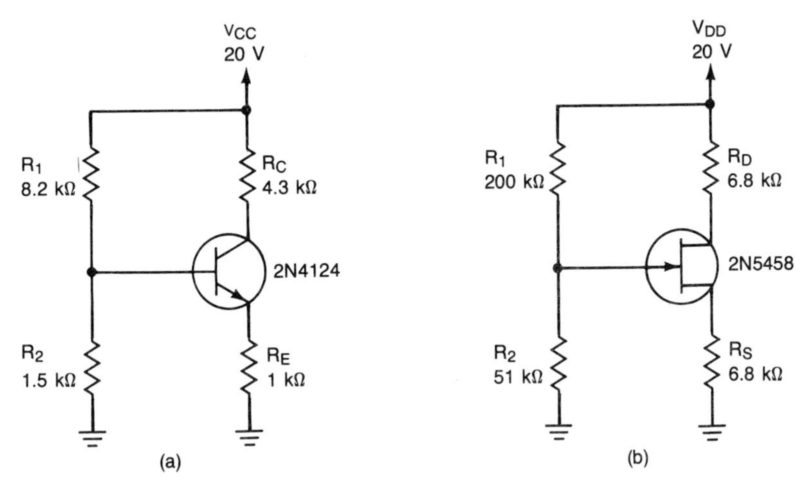

FIGURE 8-2 Dc bias circuits: (a) BJT; (b) JFET.

equivalent circuits and the resulting Q-points will be the same. *The dc bias values are the same - independent of the particular BJT, or FET signal configuration.* The obvious distinction between the various circuits rests in exactly where the input signal is injected and where the output signal is extracted. (The reader is invited to review Sections 4-5 and 5-4.)

Since the dc operating points of the three BJT circuits are the same and the three JFET circuits are the same, it is clear that their respective ac parameters will also remain unchanged. In fact, we shall see that the same small-signal device models (Fig. 8-1) may also be applied in each case.

Consequently, our next task is to determine the particular values of the parameters used in our transconductance equivalent-circuit models. The examples below will illustrate the procedures for finding g_m, r_o, and r_π for our 2N4124 BJT and our 2N5458 JFET.

EXAMPLE 8-1

Determine the values for the small-signal transconductance model of the 2N4124 BJT if its equivalent bias circuit is that depicted in Fig. 8-2(a).

SOLUTION Since the collector current is reasonably ''solid,'' we shall use its average value of 2.37 mA given in Table 6-7. The output resistance r_o may be found from Eq. 4-34.

$$r_o = \frac{V_A}{I_C} \simeq \frac{200 \text{ V}}{2.37 \text{ mA}} = 84.4 \text{ k}\Omega$$

The transconductance g_m is given by Eq. 4-33.

$$g_m = \frac{I_C}{26 \text{ mV}} = \frac{2.37 \text{ mA}}{26 \text{ mV}} = 91.2 \text{ mS}$$

The input impedance ''looking'' into the BJT's base–emitter terminals (r_π) is given by Eq. 4-42. However, before we can determine r_π, we must find the 2N4124's ac current gain $h_{fe}(\beta)$. This requires that we search its data sheet [Fig. 8-4(a)]. The ac

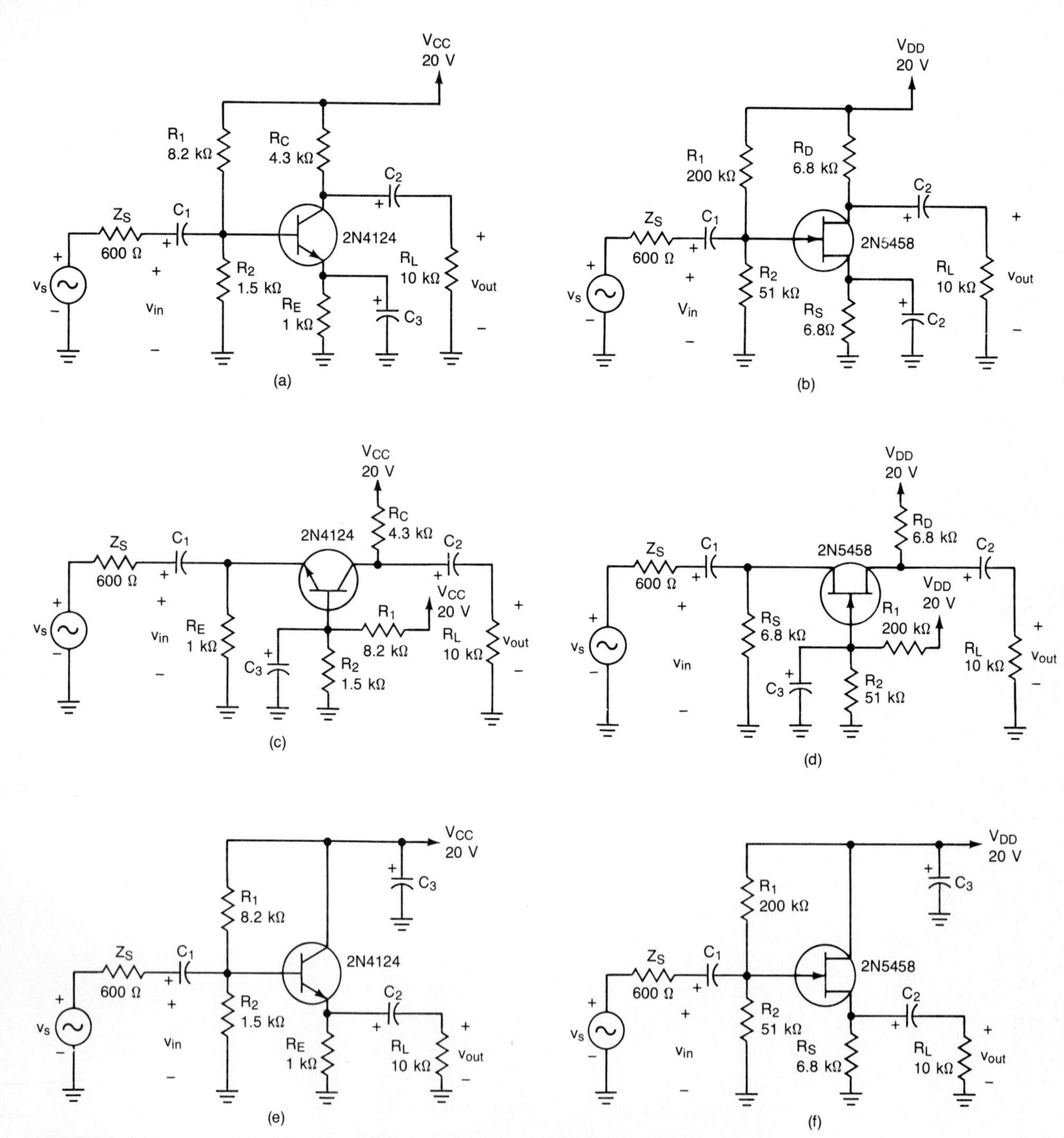

FIGURE 8-3 Amplifier configurations: (a) common-emitter; (b) common-source; (c) common-base; (d) common-gate; (e) common-collector; (f) common-drain.

SMALL-SIGNAL CHARACTERISTICS

Characteristic		Fig. No.	Symbol	Min	Max	Unit
High-Frequency Current Gain (I_C = 10 mAdc, V_{CE} = 20 Vdc, f = 100 MHz)	2N4123 2N4124		$\lvert h_{fe} \rvert$	2.5 3.0	– –	–
Current-Gain – Bandwidth Product (I_C = 10 mAdc, V_{CE} = 20 Vdc, f = 100 MHz)	2N4123 2N4124		f_T	250 300	– –	MHz
Output Capacitance (V_{CB} = 5 Vdc, I_E = 0, f = 100 kHz)		1	C_{ob}	–	4.0	pF
Input Capacitance (V_{BE} = 0.5 Vdc, I_C = 0, f = 100 kHz)		1	C_{ib}	–	8.0	pF
Small-Signal Current Gain (I_C = 2 mAdc, V_{CE} = 1 Vdc, f = 1 kHz)	2N4123 2N4124	5	h_{fe}	50 120	200 480	–
Noise Figure (I_C = 100 μAdc, V_{CE} = 5 Vdc, R_S = 1 k ohm, Noise Bandwidth = 10 Hz to 15.7 kHz)	2N4123 2N4124	3, 4	NF	– –	6.0 5.0	dB

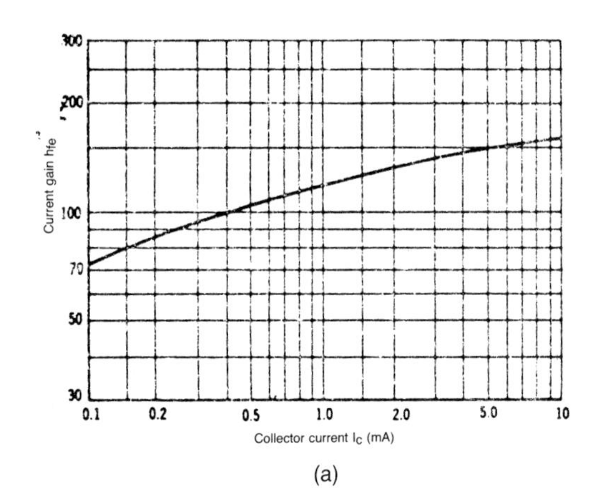

(a)

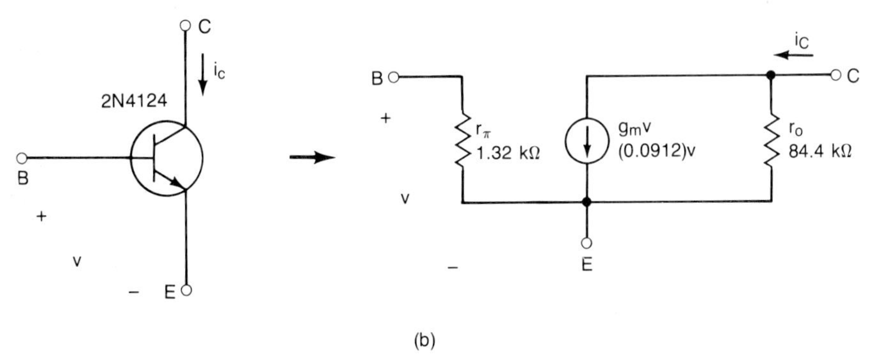

(b)

FIGURE 8-4 BJT and FET small-signal models: (a) 2N4124 data (Reprinted with permission of Motorola, Inc.); (b) 2N4124 small-signal model at its Q-point; (c) 2N5458 data (Reprinted with permission of Motorola, Inc.); (d) 2N5458 small-signal model at its Q-point.

ELECTRICAL CHARACTERISTICS (T_A = 25°C unless otherwise noted)

Characteristic		Symbol	Min	Typ	Max	Unit
OFF CHARACTERISTICS						
Gate-Source Breakdown Voltage (I_G = -10 μAdc, V_{DS} = 0)		BV_{GSS}	25	—	—	Vdc
Gate Reverse Current (V_{GS} = -15 Vdc, V_{DS} = 0)		I_{GSS}	—	—	1.0	nAdc
(V_{GS} = -15 Vdc, V_{DS} = 0, T_A = 100°C)			—	—	200	
Gate-Source Cutoff Voltage (V_{DS} = 15 Vdc, I_D = 10 nAdc)	2N5457	$V_{GS(off)}$	0.5	—	6.0	Vdc
	2N5458		1.0	—	7.0	
	2N5459		2.0	—	8.0	
Gate-Source Voltage (V_{DS} = 15 Vdc, I_D = 100 μAdc)	2N5457	V_{GS}	—	2.5	—	Vdc
(V_{DS} = 15 Vdc, I_D = 200 μAdc)	2N5458		—	3.5	—	
(V_{DS} = 15 Vdc, I_D = 400 μAdc)	2N5459		—	4.5	—	
ON CHARACTERISTICS						
Zero-Gate-Voltage Drain Current (1) (V_{DS} = 15 Vdc, V_{GS} = 0)	2N5457	I_{DSS}	1.0	3.0	5.0	mAdc
	2N5458		2.0	6.0	9.0	
	2N5459		4.0	9.0	16	
DYNAMIC CHARACTERISTICS						
Forward Transfer Admittance (1) (V_{DS} = 15 Vdc, V_{GS} = 0, f = 1 kHz)	2N5457	$\|y_{fs}\|$	1000	3000	5000	μmhos
	2N5458		1500	4000	5500	
	2N5459		2000	4500	6000	
Output Admittance (1) (V_{DS} = 15 Vdc, V_{GS} = 0, f = 1 kHz)		$\|y_{os}\|$	—	10	50	μmhos
Input Capacitance (V_{DS} = 15 Vdc, V_{GS} = 0, f = 1 MHz)		C_{iss}	—	4.5	7.0	pF
Reverse Transfer Capacitance (V_{DS} = 15 Vdc, V_{GS} = 0, f = 1 MHz)		C_{rss}	—	1.5	3.0	pF

(1) Pulse Test: Pulse Width ≤ 630 ms; Duty Cycle ≤ 10%

(2) Continuous package improvements have enhanced these guaranteed Maximum Ratings as follows: P_D = 1.0 W @ T_C = 25°C, Derate above 25°C – 8.0 mW/°C, T_J = -65 to +150°C, θ_{JC} = 125° C/W

(c)

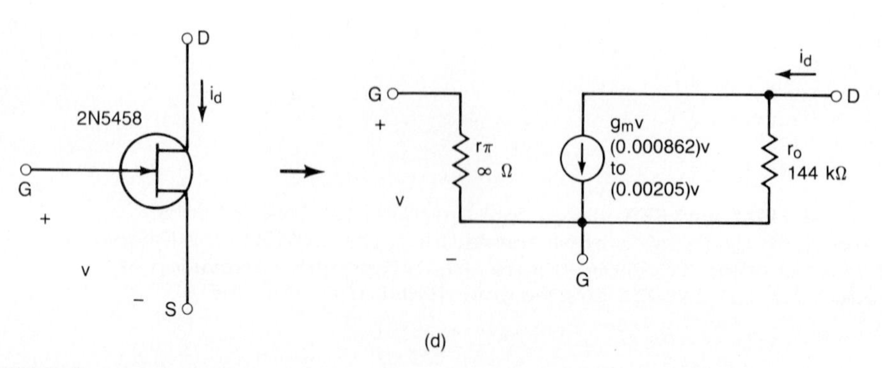

(d)

FIGURE 8-4 *(continued)*

current gain also varies with the Q-point in much the same fashion as the dc current gain h_{FE}. The data sheet indicates that h_{fe} can range from 120 to 480 under the conditions that I_C is 2 mA, V_{CE} is 10 V, the signal frequency is 1 kHz, and the ambient temperature is 25°C. To be conservative, we shall use the minimum value of 120. Hence

$$r_\pi = \frac{\beta}{g_m} = \frac{120}{91.2 \text{ mS}} = 1.32 \text{ k}\Omega$$

The 2N4124's model has been shown in Fig. 8-4(b). ■

EXAMPLE 8-2

Determine the values for the small-signal transconductance model of the 2N5458 JFET if its equivalent bias circuit is that depicted in Fig. 8-2(b). (Its bias values are given in Table 6-8.)

SOLUTION Since the drain current varies over a reasonably wide range of values, an average value is not extremely useful. Therefore, we must find the worst-case extremes for the values of its parameters. First, we shall find r_o. The manufacturer states that the maximum value of g_{os} is 50 μS when V_{DS} is 15 V, V_{GS} is 0 V, and f is 1 kHz [Fig. 8-4(c)]. No minimum value of g_{os} is specified.

In Section 6-2 we mentioned that FETs which have a *maximum* $V_{GS(OFF)}$ will also have a maximum I_{DSS}. These FETs will also possess a *maximum* g_{fs} and a *maximum* g_{os}. The converse is also true for FETs. Since the *maximum* g_{os} produces a *minimum* r_o, it poses the worst-case limitation. Therefore, we shall use its value in our FET models.

The manufacturer has specified that g_{os} has a *maximum* value of 50 μS when V_{GS} is 0 V. Therefore,

$$I_D = I_{DSS(\max)} = 9 \text{ mA}$$

and at this current

$$r_o = \frac{1}{g_{os}} = \frac{1}{50 \text{ μS}} = 20 \text{ k}\Omega$$

Following the procedure introduced in Example 5-7, we may now find the Early voltage V_A.

$$V_A = r_o I_D = (20 \text{ k}\Omega)(9 \text{ mA}) = 180 \text{ V}$$

Now we may find the *minimum* r_o at our $I_{D(\max)}$.

$$r_o = \frac{V_A}{I_{D(\max)}} = \frac{180 \text{ V}}{1.25 \text{ mA}} = 144 \text{ k}\Omega$$

This worst-case r_o has been used in our 2N5458 JFET model [see Fig. 8-4(d)].

The manufacturer also states that g_{fs0} (the forward transconductance when V_{GS} is 0 V) ranges from 1500 to 5550 μS. Following the procedure shown in Example 5-5, including the appropriate notation for $I_{D(\max)}$ and noting that $I_{DSS(\max)}$ is 9 mA,

$$g_{m(\max)} = g_{fs0(\max)} \sqrt{\frac{I_{D(\max)}}{I_{DSS(\max)}}} = (5500 \text{ μS}) \sqrt{\frac{1.25 \text{ mA}}{9 \text{ mA}}}$$
$$\simeq 2050 \text{ μS}$$

Similarly, since $I_{DSS(\min)}$ is 2 mA,

$$g_{m(\min)} = g_{fs0(\min)}\sqrt{\frac{I_{D(\min)}}{I_{DSS(\min)}}} = (1500\ \mu\text{S})\sqrt{\frac{0.66\text{ mA}}{2\text{ mA}}}$$
$$\simeq 862\ \mu\text{S}$$

These transconductance values have also been indicated in Fig. 8-4(d). The last step is the determination of r_π. From our work in Chapter 5 we shall assume that

$$r_\pi \simeq \infty$$

[see Fig. 8-4(d)]. ■

The reader should study Examples 8-1 and 8-2 carefully and verify the results. We shall draw on them for the balance of the chapter.

8-3 Inverting Voltage Amplifier

The ideal inverting voltage amplifier possesses all six of the attributes cited in Section 7-1. Its primary distinction is that its voltage gain will be *negative* - indicating that its output signal is 180° out of phase with reference to its input signal (see Fig. 8-5). This is not necessarily a problem, but it is an important observation.

As we shall see in the next few sections, the BJT in its common-emitter configuration and the FET in its common-source configuration will both be inverting voltage amplifiers. Further, the op amp can also be used as an inverting voltage amplifier if the input signal is tied to its inverting input terminal.

FIGURE 8-5 Inverting voltage amplifier.

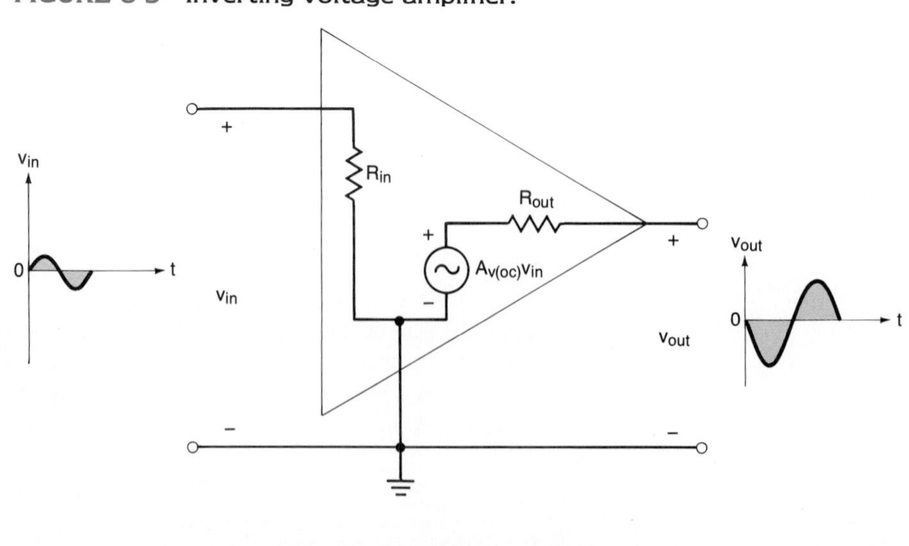

8-4 Common-Emitter and Common-Source Amplifiers

In Fig. 8-3(a) and (b) we depicted the common-emitter and common-source voltage amplifiers, respectively. Since we have already performed a dc analysis and determined the small-signal parameters, our next job is to determine the general amplifier parameters. Specifically, we shall find the input resistance R_{in}, the output resistance R_{out}, and the open-circuit voltage gain $A_{v(oc)}$. This will permit us to determine the voltage gains A_v and A_{vs}, the current gains A_i and A_{is}, and the power gain A_p (see Fig. 8-6). Although this may seem to be a formidable task, our work shall be essentially *halved* since our BJT and FET models are virtually identical.

The first step in the signal analysis of any amplifier is to draw its ac equivalent circuit and simplify it. Recalling that the coupling and bypass capacitors and the dc power supplies are essentially short circuits to the ac signal, we arrive at the ac equivalent circuits given in Fig. 8-7(a) and (b).

The next step involves replacing the BJT and the FET device symbols with their respective ac equivalent-circuit models. Since their models are virtually identical, *both* of the amplifiers may be represented as shown in Fig. 8-7(c).

FIGURE 8-6 Summary of the voltage amplifier relationships.

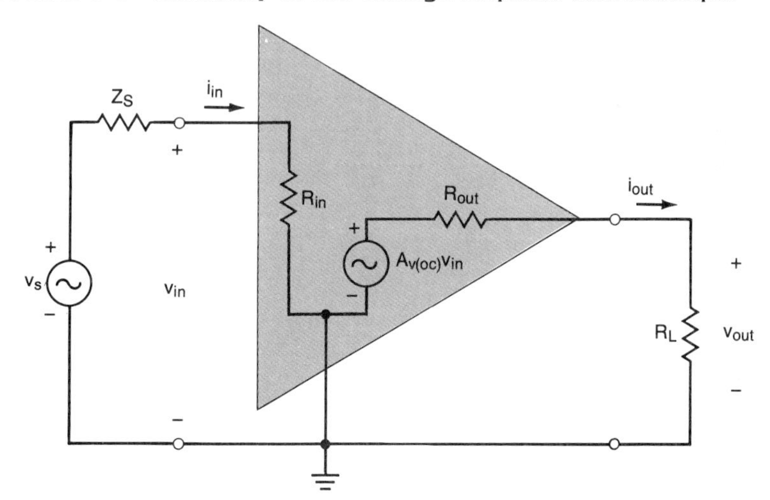

Once $A_{v(oc)}$, R_{in}, and R_{out} are known we may then find:

$$A_v = \frac{v_{out}}{v_{in}} = \frac{R_L}{R_L + R_{out}} A_{v(oc)}$$

$$A_{vs} = \frac{v_{out}}{v_s} = \frac{R_{in}}{R_{in} + Z_S} \frac{R_L}{R_L + R_{out}} A_{v(oc)} = \frac{R_{in}}{R_{in} + Z_S} A_v$$

$$A_i = \frac{i_{out}}{i_{in}} = \frac{R_{in}}{R_L + R_{out}} A_{v(oc)}$$

$$A_{is} = \frac{i_{out}}{i_s} = \frac{Z_S}{Z_S + R_{in}} \frac{R_{in}}{R_L + R_{out}} A_{v(oc)} = \frac{Z_S}{Z_S + R_{in}} A_i \text{ (Assumes a signal source conversion)}$$

$$A_p = \frac{P_{out}}{P_{in}} = A_v A_i$$

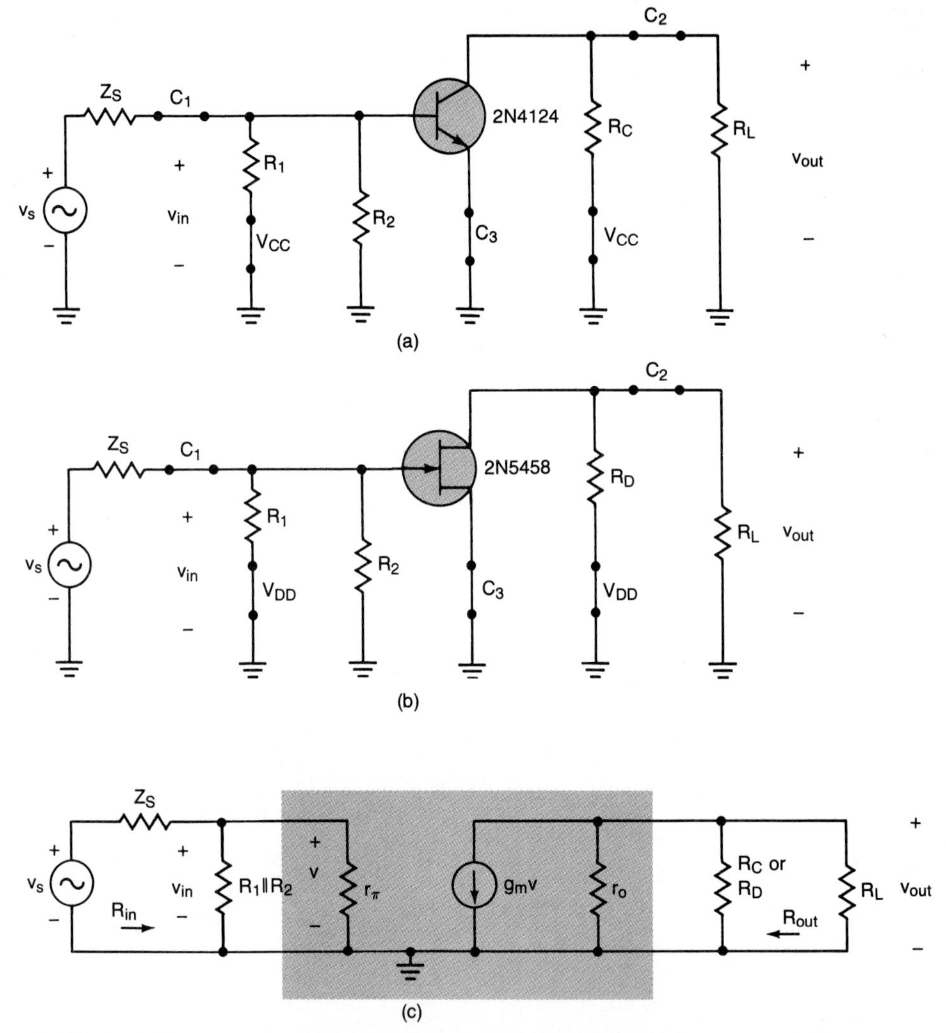

FIGURE 8-7 Ac equivalent circuits: (a) common-emitter ac equivalent circuit of Fig. 8-3(a); (b) common-source ac equivalent circuit of Fig. 8-3(b); (c) simplified ac equivalent circuit that includes the device model.

As summarized in Fig. 8-6, our first task is to determine R_{in}, R_{out}, and $A_{v(oc)}$. Then we can put our specific amplifiers into the general amplifier form. In Fig. 8-7(c) we can see that the input resistance R_{in} is formed by the parallel combination of R_1, R_2, and r_π. Hence we arrive at

$$\boxed{R_{in} = R_1 \parallel R_2 \parallel r_\pi} \tag{8-1}$$

EXAMPLE 8-3

Find the input resistance of the common-emitter amplifier given in Fig. 8-3(a).

SOLUTION From the ac equivalent circuit of Fig. 8-7(a) (Eq. 8-1), and recalling that r_π is 1.32 kΩ,

$$R_{in} = R_1 \,\|\, R_2 \,\|\, r_\pi = 8.2\ \text{k}\Omega \,\|\, 1.5\ \text{k}\Omega \,\|\, 1.32\ \text{k}\Omega = 647\ \Omega \quad ■$$

EXAMPLE 8-4

Repeat Example 8-3 for the FET amplifier given in Fig. 8-3(b).

SOLUTION Using exactly the same approach, and recalling that r_π is essentially infinite ohms,

$$R_{in} = R_1 \,\|\, R_2 \,\|\, r_\pi = R_1 \,\|\, R_2 \,\|\, \infty = R_1 \,\|\, R_2$$
$$= 200\ \text{k}\Omega \,\|\, 51\ \text{k}\Omega = 40.6\ \text{k}\Omega \quad ■$$

Let us reflect on these results. First, the input resistance of the FET amplifier is much larger than that of the BJT common-emitter stage. *The primary advantage of FET amplifiers is that they offer a much larger input resistance than that normally available from BJT amplifiers.* Second, the bias resistors are in parallel across the inputs of the voltage amplifiers. This poses a serious compromise in their selection. *To maintain good dc bias stability, the parallel combination of the bias resistors should be as* small *as possible. However, if a larger input resistance is required, the parallel combination should be as* large *as possible.* Therefore, practical designs require a judicious selection of the bias resistors.

The output resistance R_{out} of our amplifier has been defined in Fig. 8-7(c). Generally, it may be further defined as

$$R_{out} = \left.\frac{v_{out}}{i_{out}}\right|_{v_s=0,\ R_L=\infty} \tag{8-2}$$

If we apply the definition offered by Eq. 8-2 to the ac equivalent circuit given in Fig. 8-7(c), we arrive at the situation depicted in Fig. 8-8(a). Setting the signal source v_s to zero also sets the ac current source to zero. Therefore, the current source may be replaced by an open circuit [see Fig. 8-8(b)].

Be sure to note that we have disconnected our equivalent load resistance R_L. It is *not* considered to be a part of the amplifier. By inspection of Fig. 8-8(b) it is clear that in the case of the BJT amplifier, and for the JFET amplifier, we have,

$$R_{out} = r_o \,\|\, R_C \tag{8-3}$$

$$R_{out} = r_o \,\|\, R_D \tag{8-4}$$

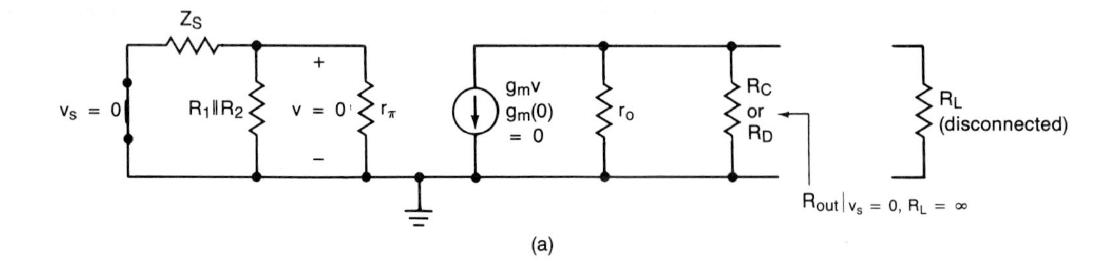

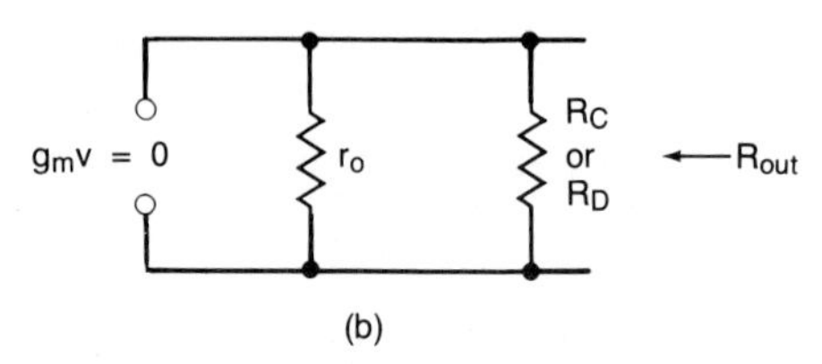

FIGURE 8-8 Determining R_{out}: (a) the signal source is set to zero and R_L is disconnected; (b) simplified output circuit.

EXAMPLE 8-5

Find the output resistance of the common-emitter amplifier given in Fig. 8-3(a).

SOLUTION Applying Eq. 8-3 gives us

$$R_{out} = r_o \parallel R_C = 84.4\ \text{k}\Omega \parallel 4.3\ \text{k}\Omega = 4.09\ \text{k}\Omega$$

and we note that

$$R_{out} \simeq R_C = 4.3\ \text{k}\Omega$$

■

EXAMPLE 8-6

Find the output resistance of the common-source amplifier given in Fig. 8-3(b).

SOLUTION Applying Eq. 8-4, we obtain

$$R_{out} = r_o \parallel R_D = 144\ \text{k}\Omega \parallel 6.8\ \text{k}\Omega = 6.49\ \text{k}\Omega$$

and we note again that

$$R_{out} \simeq R_D = 6.8\ \text{k}\Omega$$

■

In Examples 8-5 and 8-6, we can see that the r_o of the BJT and the JFET tend to be large compared to R_C and R_D, respectively. This is quite often the case. Therefore, we present the approximations below. For the BJT common-emitter amplifier,

$$\boxed{R_{out} \simeq R_C} \qquad (8\text{-}5)$$

and for the common-source amplifier,

$$\boxed{R_{out} \simeq R_D} \qquad (8\text{-}6)$$

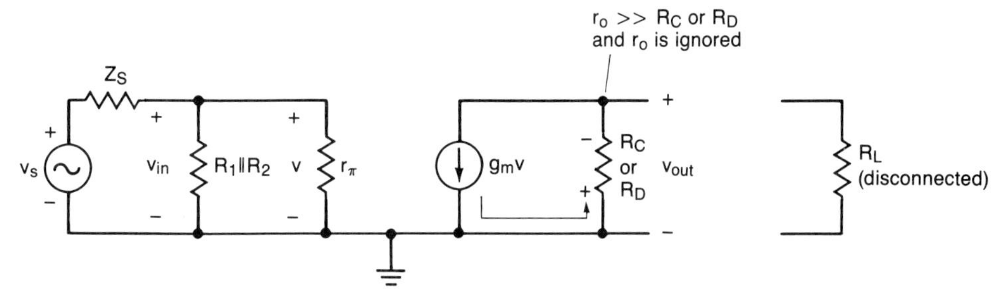

FIGURE 8-9 **Determination of $A_{v(oc)}$.**

These approximations have been reflected in the ac equivalent circuit given in Fig. 8-9.

Now let us find the open-circuit voltage gain $A_{v(oc)}$ (see Fig. 8-9). First, we note that

$$v = v_{in}$$

and by substitution, the voltage-controlled current source produces

$$g_m v = g_m v_{in}$$

The current source forces a current *up* through R_C or R_D. Therefore, the output voltage v_{out} is a *negative* quantity.

$$v_{out} = -g_m R_C v_{in} \quad \text{or} \quad v_{out} = -g_m R_D v_{in}$$

Solving for the voltage gain requires that we divide both sides by v_{in}. Hence, for the BJT,

$$\boxed{A_{v(oc)} = \frac{v_{out}}{v_{in}} = -g_m R_C} \tag{8-7}$$

and for the FET,

$$\boxed{A_{v(oc)} = \frac{v_{out}}{v_{in}} = -g_m R_D} \tag{8-8}$$

The negative signs on the voltage gains indicate the 180° phase shift that exists between v_{in} and v_{out}. Let us apply these results and demonstrate the use of Eqs. 8-7 and 8-8.

EXAMPLE 8-7

Calculate the open-circuit voltage gain of the BJT amplifier given in Fig. 8-3(a).

SOLUTION From Eq. 8-7,

$$A_{v(oc)} = \frac{v_{out}}{v_{in}} = -g_m R_C = -(91.2 \text{ mS})(4.3 \text{ k}\Omega) = -392 \quad ■$$

EXAMPLE 8-8

Find the open-circuit voltage gain of the JFET amplifier given in Fig. 8-3(b).

SOLUTION As we have seen, the g_m for the 2N5458 JFET ranges from 0.862 to 2.05 mS. Therefore, we must find the corresponding minimum and maximum voltage gains for our JFET amplifier. Modifying Eq. 8-8 slightly, we see that

$$A_{v(\text{oc-min})} = -g_{m(\text{min})}R_D = -(0.862 \text{ mS})(6.8 \text{ k}\Omega) = -5.86$$

and

$$A_{v(\text{oc-max})} = -g_{m(\text{max})}R_D = -(2.05 \text{ mS})(6.8 \text{ k}\Omega) = -13.9$$

■

From the results in Examples 8-7 and 8-8, we can see that because the transconductance of the BJT is much larger than that possessed by the JFET, the BJT amplifier provides much more voltage gain. This is *not* a unique situation.

BJT voltage amplifiers are capable of providing much more voltage gain than that exhibited by FET voltage amplifiers.

A summary of our work to this point has been provided in Fig. 8-10. Both the common-emitter and common-source amplifiers have been placed in the form of our general amplifier model.

Now we have simplified the circuits to the point where it becomes relatively easy to characterize our amplifiers more fully. Specifically, we can now determine A_v, A_{vs}, A_i, A_{is}, and A_p (refer to Fig. 8-6).

EXAMPLE 8-9

Find the loaded voltage gain A_v, the voltage gain from the signal source to the load A_{vs}, the loaded current gain A_i, the current gain from the signal source to the load A_{is}, and the power gain A_p for the common-emitter BJT amplifier given in Fig. 8-10(a).

SOLUTION Drawing from the summary provided in Fig. 8-6, we may proceed as follows:

$$A_v = \frac{v_{\text{out}}}{v_{\text{in}}} = \frac{R_L}{R_L + R_{\text{out}}} A_{v(\text{oc})} = \frac{10 \text{ k}\Omega}{10 \text{ k}\Omega + 4.3 \text{ k}\Omega}(-392) = -274$$

$$A_{vs} = \frac{v_{\text{out}}}{v_s} = \frac{R_{\text{in}}}{R_{\text{in}} + Z_S} A_v = \frac{647 \ \Omega}{647 \ \Omega + 600 \ \Omega}(-274) = -142$$

$$A_i = \frac{i_{\text{out}}}{i_{\text{in}}} = \frac{R_{\text{in}}}{R_L + R_{\text{out}}} A_{v(\text{oc})} = \frac{647 \ \Omega}{10 \text{ k}\Omega + 4.3 \text{ k}\Omega}(-392) = -17.7$$

$$A_{is} = \frac{i_{\text{out}}}{i_s} = \frac{Z_S}{Z_S + R_{\text{in}}} A_i = \frac{600 \ \Omega}{600 \ \Omega + 647 \ \Omega}(-17.7) = -8.52$$

$$A_p = A_v A_i = (-274)(-17.7) = 4850$$

■

EXAMPLE 8-10

Repeat Example 8-9 for the common-source JFET amplifier given in Fig. 8-10(b). Assume that the JFET amplifier has an $A_{v(\text{oc})}$ of -13.9.

SOLUTION We simply follow the same procedure as that outlined in Example 8-9.

$$A_v = \frac{v_{out}}{v_{in}} = \frac{R_L}{R_L + R_{out}} A_{v(oc)} = \frac{10\ \text{k}\Omega}{10\ \text{k}\Omega + 6.8\ \text{k}\Omega}(-13.9) = -8.27$$

$$A_{vs} = \frac{v_{out}}{v_s} = \frac{R_{in}}{R_{in} + Z_S} A_v = \frac{40.6\ \text{k}\Omega}{40.6\ \text{k}\Omega + 600\ \Omega}(-8.27) = -8.15$$

$$A_i = \frac{i_{out}}{i_{in}} = \frac{R_{in}}{R_L + R_{out}} A_{v(oc)} = \frac{40.6\ \text{k}\Omega}{10\ \text{k}\Omega + 6.8\ \text{k}\Omega}(-13.9) = -33.6$$

$$A_{is} = \frac{i_{out}}{i_s} = \frac{Z_S}{Z_S + R_{in}} A_i = \frac{600\ \Omega}{600\ \Omega + 40.6\ \text{k}\Omega}(-33.6) = -0.489$$

$$A_p = A_v A_i = (-8.27)(-33.6) = 278$$

■

The reader is encouraged to repeat Example 8-10 for a 2N5458 JFET with a minimum g_m [an $A_{v(oc)}$ of -5.86]. The results, plus a summary of Examples 8-9 and 8-10, have been provided in Table 8-1.

TABLE 8-1
Summary of the Common-Emitter and Common-Source Amplifiers

	2N4124	2N5458 (max)	2N5458 (min)
A_v	−274	−8.27	−3.49
A_{vs}	−142	−8.15	−3.44
A_i	−17.7	−33.6	−14.2
A_{is}	−8.52	−0.489	−0.207
A_p	4850	278	49.6

Once again, we see that the BJT offers much more voltage gain A_v than the FET. However, the voltage gain A_{vs} from the source to the load also tells quite a story. It is less than A_v because of the voltage division that occurs between the source resistance Z_S and the amplifier input resistance R_{in}. However, this gain reduction is far less severe in FET amplifiers because of their superior R_{in}. The loading effects are virtually negligible.

The (average) current gain A_i is, in general, larger for FETs. However, the current gain from the signal source to the load A_{is} is greatly reduced for the FET amplifier. This occurs because of the current-division problem between the source resistance R_S and the large R_{in} of the JFET amplifier.

We can also see in Table 8-1 that the BJT offers much more power gain A_p than the FET. Obviously, this occurs because both the A_v and A_i are moderately large for the BJT amplifier.

8-5 Removing the Emitter Bypass Capacitor

In Fig. 8-10(a) we see the emitter bypass capacitor C_3. Recall that it is included to keep the emitter-to-ground voltage constant. In other words, it serves to short, or bypass, the ac signal around the emitter resistor R_E. In so doing, the emitter bypass

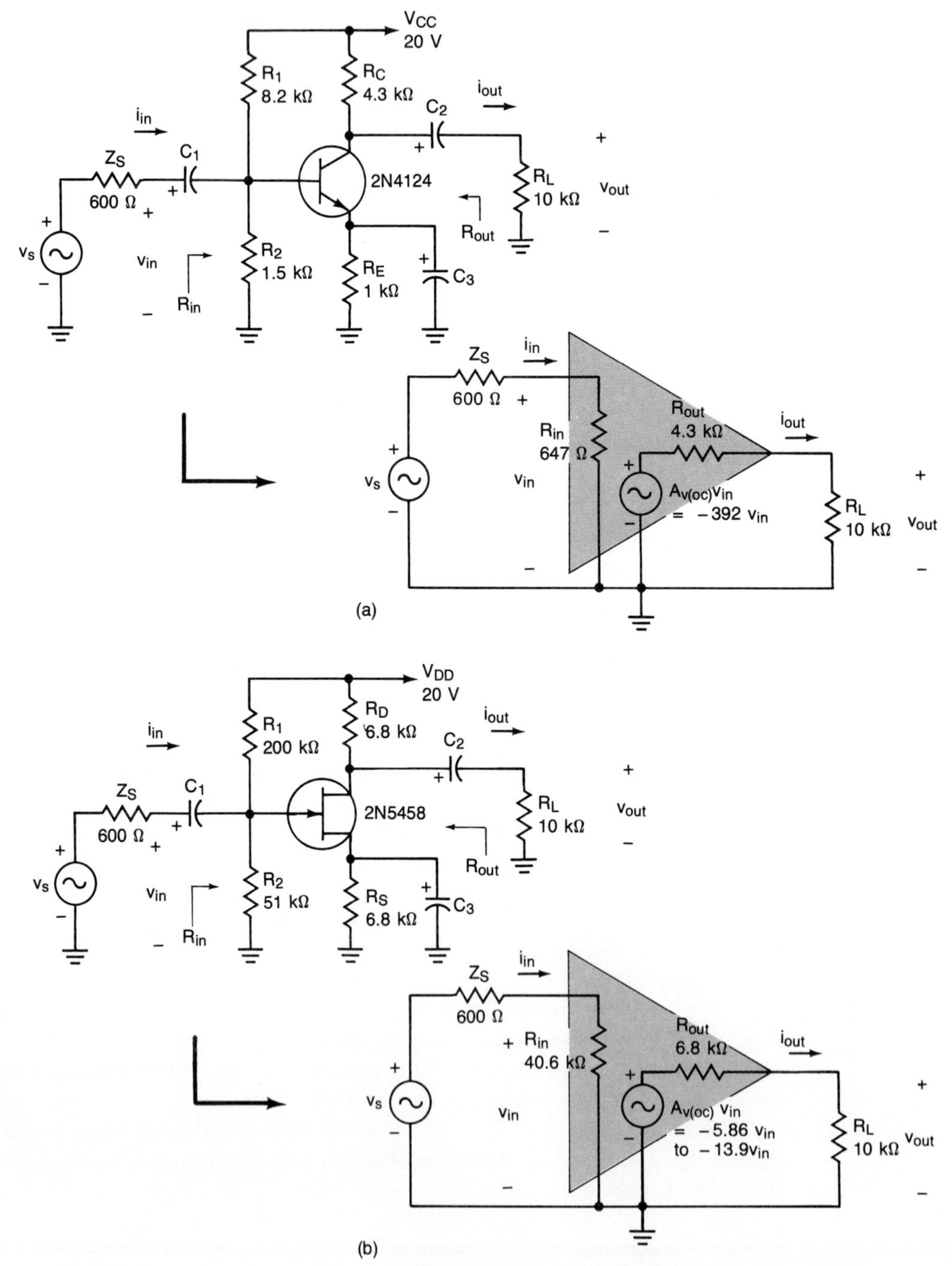

FIGURE 8-10 **BJT and FET amplifier models: (a) common-emitter; (b) common-source.**

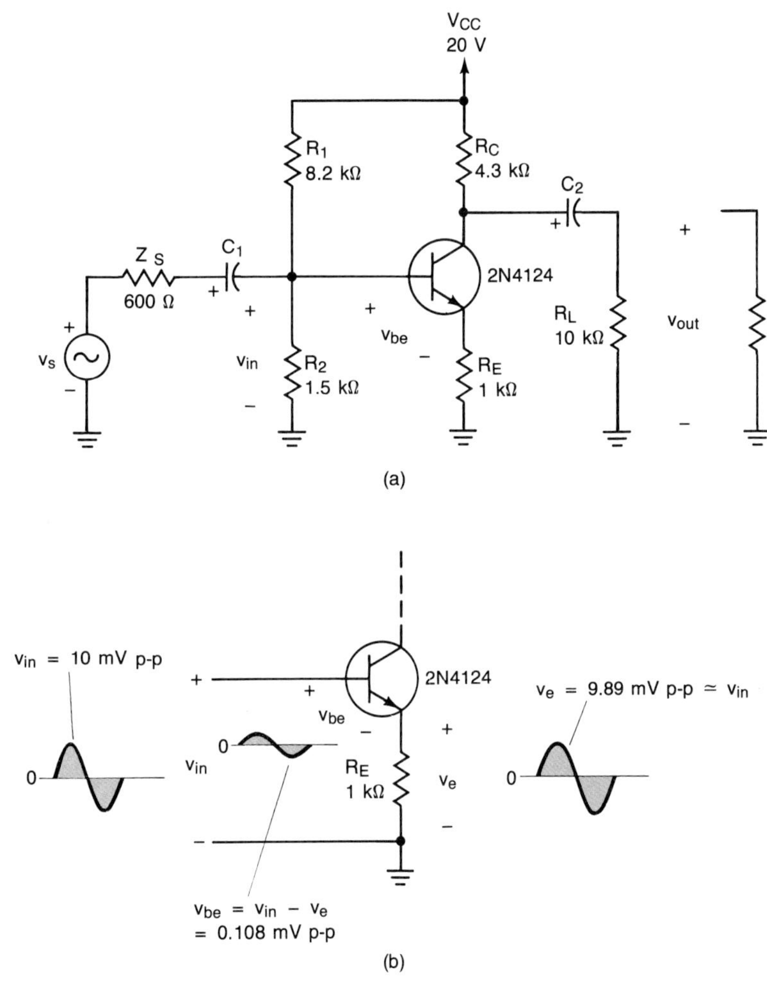

FIGURE 8-11 **Common-emitter amplifier with no emitter bypass capacitor: (a) amplifier; (b) signal relationships.**

capacitor is used to prevent *negative feedback* in the emitter circuit. [This is also the reason why the *source* bypass capacitor is used; see C_3 in Fig. 8-10(b).]

Very simply, negative feedback involves feeding back a portion of an amplifier's *output* signal to its *input* circuit. Further, the signal is fed back in such a manner that it *opposes* the input signal. We postpone an in-depth discussion of negative feedback until Chapter 10. For now, we merely focus our attention on the circuit given in Fig. 8-11(a).

As we shall see, the removal of the bypass capacitor results in an *increase* in the amplifier's *input resistance*, a *reduction* in its *voltage gain*, and an *increase* in its *output resistance* "looking into" the collector. Obviously, an increase in input resistance is a step in the right direction for a voltage amplifier. However, a reduction in the voltage gain and an increase in the output resistance are generally undesirable. Yet there are some advantages offered by these sacrifices. First, even though the voltage gain is reduced, it tends to be *more stable*. Specifically, it tends to be *less* dependent on the transistor's parameters (e.g., its g_m). Second, the increase in the collector-to-ground output resistance makes our approximation of R_{out} equal to R_C even *more realistic*. Before we delve into a rigorous analysis of Fig. 8-11(a), let us run through a more intuitive explanation of its operation.

When the input signal v_{in} goes positive, the transistor's base current i_b will increase. An increase in i_b will cause a proportional increase in the collector and emitter currents i_c and i_e, respectively. Since the emitter resistor is unbypassed, the voltage across it is free to increase (become more positive) with an increase in i_e. Similarly, when the input signal goes negative, the ac voltage across the emitter resistor will also go negative [see Fig. 8-11(b)].

Since very small base current i_b flows through the equivalent base-to-emitter resistance r_π, the voltage drop across it is also very small. The voltage drop across R_E will tend to be much larger since i_e flows through it. Consequently, the voltage developed across R_E will be approximately equal to v_{in} and in phase with it. Typical values have been indicated in Fig. 8-11(b). The voltage at the emitter essentially "follows" the input signal v_{in}.

The emitter circuit is often described as "pulling itself up by its own bootstraps." (Consequently, this technique is often referred to as *bootstrapping*.) This description reinforces the fact that the emitter-to-ground voltage v_e can *never* be equal to v_{in}, but always less than it.

In Fig. 8-12(a) we see a portion of the ac equivalent circuit when the emitter is bypassed. Obviously, the resistance "looking into" the base circuit $R_{in(base)}$ is equal to r_π.

$$R_{in(base)} = r_\pi = 1.32\ \text{k}\Omega$$

In Fig. 8-12(b) we see the situation that occurs when R_E is unbypassed. A quick look often tempts beginning students to state that $R_{in(base)}$ is merely equal to r_π plus R_E. This is *not* the case. It would be true if r_π and R_E were in *series* (experience the *same* current), but this does not occur. The base current flows through r_π, and the much larger emitter current flows through R_E.

Let us now see exactly what $R_{in(base)}$ is for Fig. 8-12(b). We first note that

$$i_b = \frac{v_{be}}{r_\pi} = \frac{0.108\ \text{mV p-p}}{1.32\ \text{k}\Omega} = 0.0818\ \mu\text{A p-p}$$

Therefore,

$$R_{in(base)} = \frac{v_{in}}{i_b} = \frac{10\ \text{mV p-p}}{0.0818\ \mu\text{A p-p}} = 122\ \text{k}\Omega$$

Obviously, this is a vast improvement over the $R_{in(base)}$ of 1.32 kΩ, which occurs when R_E is bypassed.

With this background in mind, we now develop a generalized equation for $R_{in(base)}$ (refer to Fig. 8-13). First we note that

$$v_{be} = i_b r_\pi \tag{8-9}$$

and $g_m v_{be} >> i_b$. Now by Kirchhoff's voltage law, and drawing on the relationships above, we arrive at

$$v_{in} \simeq i_b r_\pi + g_m v_{be} R_E = i_b r_\pi + g_m i_b r_\pi R_E = (1 + g_m R_E) r_\pi i_b \tag{8-10}$$

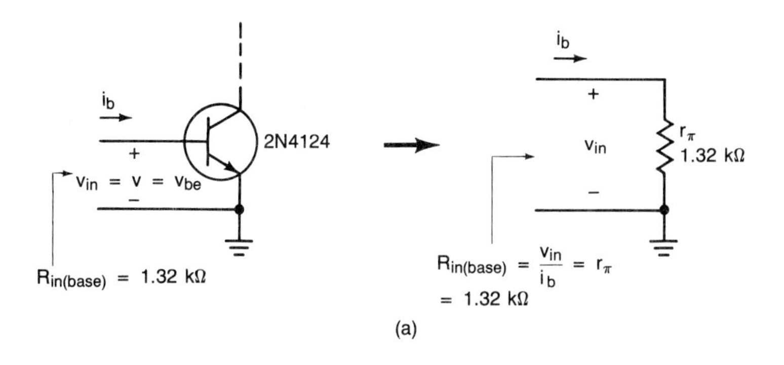

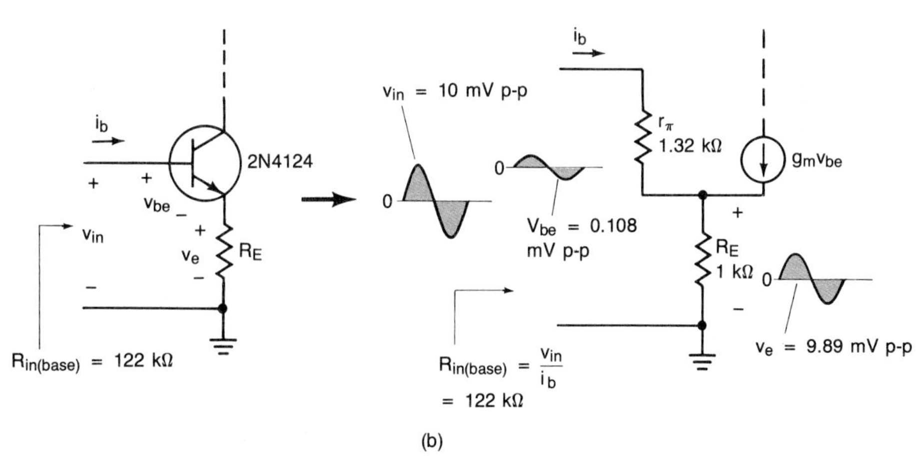

FIGURE 8-12 Finding $R_{in(base)}$: (a) emitter bypassed; (b) emitter unbypassed.

FIGURE 8-13 Partial ac equivalent circuit to determine $R_{in(base)}$.

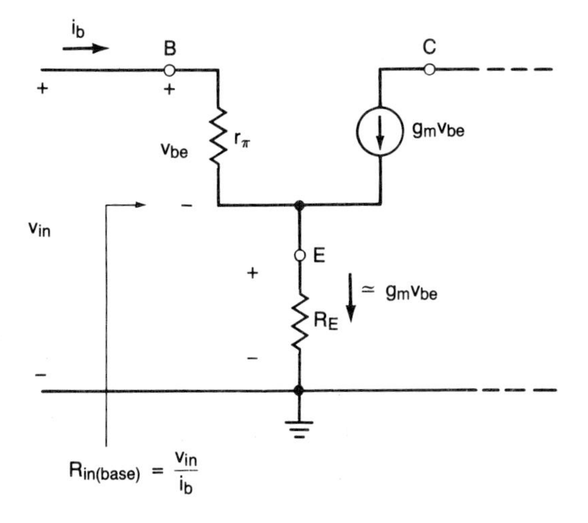

Dividing both sides by i_b yields the equation for $R_{in(base)}$.

$$R_{in(base)} = \frac{v_{in}}{i_b} = (1 + g_m R_E) r_\pi \tag{8-11}$$

Typically, $g_m R_E >> 1$, and therefore

$$\boxed{R_{in(base)} \simeq g_m R_E r_\pi} \tag{8-12}$$

A simpler form of Eq. 8-12 can be arrived at as follows:

$$R_{in(base)} \simeq g_m R_E r_\pi = g_m R_E \frac{h_{fe}}{g_m} = h_{fe} R_E$$

Most students find this form easier to remember.

By inspection of Fig. 8-11(a) and our previous work, it should be clear that R_{in} is given by

$$\boxed{R_{in} = R_1 \| R_2 \| R_{in(base)}} \tag{8-13}$$

EXAMPLE 8-11

Find the input resistance R_{in} of the common-emitter amplifier given in Fig. 8-11(a).

SOLUTION First we must find $R_{in(base)}$.

$$R_{in(base)} \simeq g_m R_E r_\pi = (91.2 \text{ mS})(1 \text{ k}\Omega)(1.32 \text{ k}\Omega) \simeq 120 \text{ k}\Omega$$

which agrees with our previous value of 122 kΩ. The input resistance is given by Eq. 8-13.

$$R_{in} = R_1 \| R_2 \| R_{in(base)} = 8.2 \text{ k}\Omega \| 1.5 \text{ k}\Omega \| 120 \text{ k}\Omega$$
$$= 1255 \ \Omega$$

■

This is a considerable improvement (+94%) over the R_{in} of 647 Ω that occurred when R_E was bypassed. *The main point to remember is that R_{in} can be greatly increased by removing the bypass capacitor.*

Now let us consider the unloaded voltage gain $A_{v(oc)}$ when R_E is unbypassed. Again, we shall initiate our analysis via an intuitive approach. In Fig. 8-14 we see that the input voltage essentially appears across R_E. Hence

$$v_e \simeq v_{in}$$

Therefore, the current through R_E is

$$i_e = \frac{v_e}{R_E} \simeq \frac{v_{in}}{R_E}$$

Since the emitter and collector currents are virtually the same, we can state that

$$i_c = i_e \simeq \frac{v_{in}}{R_E}$$

The output voltage is

$$v_{out} = -i_c R_C \simeq -\frac{v_{in}}{R_E} R_C$$

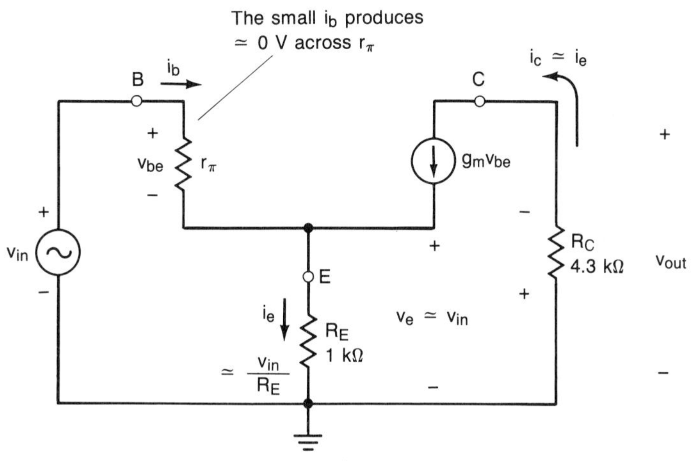

FIGURE 8-14 **Partial ac equivalent circuit to approximate $A_{v(oc)}$.**

and

$$A_{v(oc)} = \frac{v_{\text{out}}}{v_{\text{in}}} \simeq -\frac{R_C}{R_E} \tag{8-14}$$

Equation 8-14 *shows us that the open-circuit voltage gain is essentially determined by the ratio of the collector resistance to the emitter resistance. Obviously, the resulting voltage gain will be very stable since it does* not *depend on the BJT's* g_m. Further, our best guess as to a given amplifier's voltage gain will be more accurate since we will not have to draw on our approximation of g_m. Negative feedback has made this possible.

A more rigorous development of $A_{v(oc)}$ can be produced by referring back to Fig. 8-14. Since $g_m v_{be} >> i_b$,

$$\begin{aligned} v_{\text{in}} &= v_{be} + g_m v_{be} R_E \\ &= (1 + g_m R_E) v_{be} \end{aligned}$$

Solving for v_{be}, we have

$$v_{be} = \frac{v_{\text{in}}}{1 + g_m R_E} \tag{8-15}$$

The output voltage is

$$v_{out} = -g_m v_{be} R_C \tag{8-16}$$

Substituting Eq. 8-15 into Eq. 8-16 gives us

$$v_{\text{out}} = -\frac{g_m R_C}{1 + g_m R_E} v_{\text{in}} \tag{8-17}$$

Dividing both sides by v_{in} results in the expression for the voltage gain.

$$A_{v(oc)} = \frac{v_{out}}{v_{in}} = -\frac{g_m R_C}{1 + g_m R_E} \tag{8-18}$$

By inspection of Eq. 8-18, we can see that if

$$g_m R_E >> 1$$

then

$$A_{v(oc)} \simeq -\frac{g_m R_C}{g_m R_E} = -\frac{R_C}{R_E}$$

and we arrive at Eq. 8-14.

EXAMPLE 8-12

Find the $A_{v(oc)}$ of the common-emitter amplifier given in Fig. 8-11(a).

SOLUTION Recalling that g_m is 91.2 mS and drawing on Eq. 8-18, we obtain

$$A_{v(oc)} = -\frac{g_m R_C}{1 + g_m R_E} = -\frac{(91.2\text{ mS})(4.3\text{ k}\Omega)}{1 + (91.2\text{ mS})(1\text{ k}\Omega)} = -4.25$$

Equation 8-14 provides a very good approximation.

$$A_{v(oc)} \simeq -\frac{R_C}{R_E} = -\frac{4.3\text{ k}\Omega}{1\text{ k}\Omega} = -4.3$$

■

Generally, we may draw on the approximation offered by Eq. 8-14 for voltage gains of 10 or less without concerning ourselves about significant errors. At the very least, a quick inspection of the ratio of R_C to R_E can give us some indication of the approximate value of the voltage gain.

Now let us pursue the effects of the unbypassed R_E on the resistance ''looking into'' the collector. We shall refer to this as $R_{out(col)}$. Recall that the output resistance of an amplifier is defined under the condition that the signal source v_s is set equal to zero (refer to Eq. 8-2). This condition has been reflected in the unsimplified ac equivalent circuit given in Fig. 8-15(a). To ease our analysis, we shall find the Thévenin equivalent resistance of the input circuit [see Fig. 8-15(a) and (b)]. Hence

$$R_{TH} = Z_S \parallel R_1 \parallel R_2$$

Since our immediate interest is $R_{out(col)}$, we shall temporarily disconnect *both* R_L and R_C [see Fig. 8-15(c)]. Our ac equivalent circuit now reduces to that given in Fig. 8-15(d).

To excite our network we have added a fictitious voltage source for v_{out}. The equivalent resistance of the collector is defined by

$$R_{out(col)} = \frac{v_{out}}{i_c} \tag{8-19}$$

Application of Kirchhoff's voltage law around the output circuit of Fig. 8-15(d) produces

$$\begin{aligned} v_{out} &\simeq (i_c - g_m v_{be})r_o + i_c R_E \\ &\simeq i_c r_o - g_m r_o v_{be} + i_c R_E \end{aligned} \tag{8-20}$$

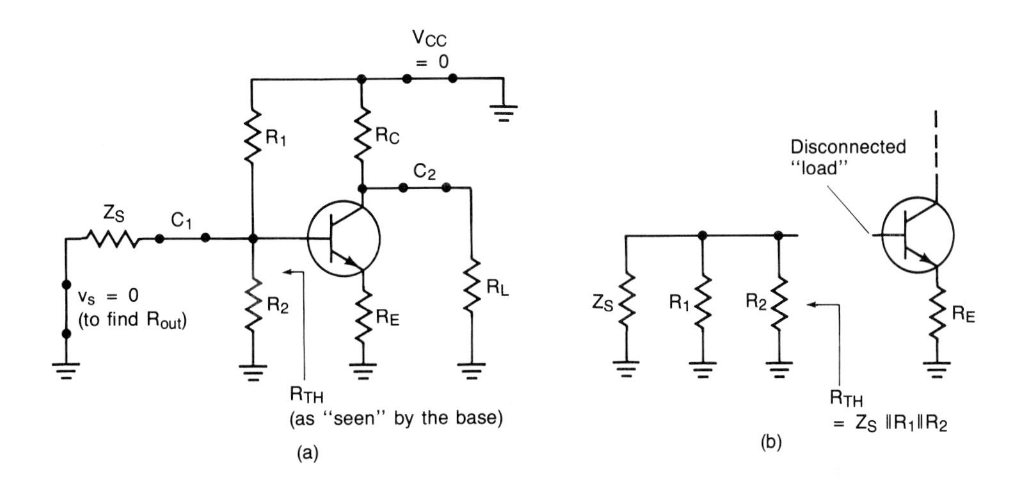

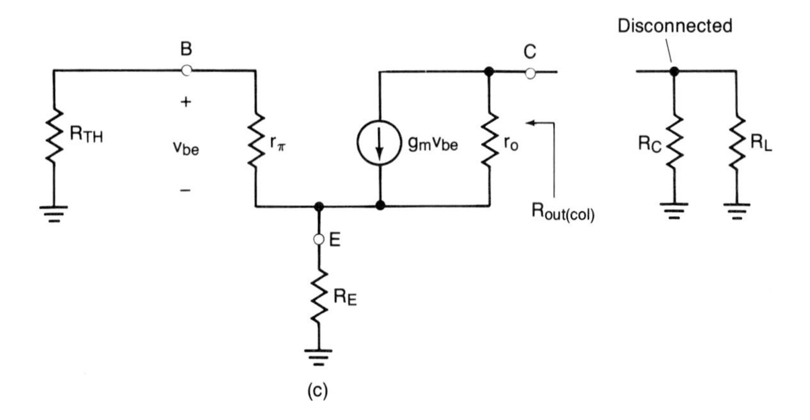

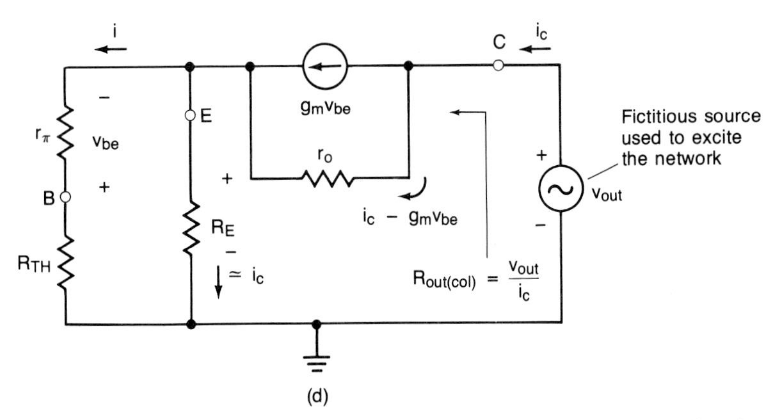

FIGURE 8-15 Finding $R_{out(col)}$: (a) unsimplified ac equivalent circuit; (b) simplifying the base circuit; (c) simplified ac equivalent circuit; (d) ac equivalent circuit for finding $R_{out(col)}$.

By current division the current i that flows through r_π is

$$i = \frac{R_E}{R_E + (r_\pi + R_{TH})} i_c \tag{8-21}$$

and v_{be} [as defined in Fig. 8-15(d)] will be a negative quantity.

$$v_{be} = -i r_\pi \tag{8-22}$$

Substitution of Eq. 8-21 into Eq. 8-22 produces

$$v_{be} = -\frac{R_E r_\pi}{R_E + (r_\pi + R_{TH})} i_c \tag{8-23}$$

Placing the result above into Eq. 8-20 provides us with our final equation for v_{out}.

$$\begin{aligned} v_{out} &= i_c r_o - g_m r_o \left[-\frac{R_E r_\pi}{R_E + (r_\pi + R_{TH})} i_c \right] + i_c R_E \\ &= \left[r_o + \frac{g_m r_o R_E r_\pi}{R_E + (r_\pi + R_{TH})} + R_E \right] i_c \end{aligned} \tag{8-24}$$

Dividing both sides of this result by i_c leads us to $R_{out(col)}$.

$$R_{out(col)} = \frac{v_{out}}{i_c} = r_o + \frac{g_m r_o r_\pi R_E}{R_E + r_\pi + R_{TH}} + R_E \tag{8-25}$$

In general, R_E tends to be small compared to the other two terms. Therefore, we arrive at the approximation offered by

$$\boxed{R_{out(col)} \simeq r_o + \frac{g_m r_o r_\pi R_E}{R_E + r_\pi + R_{TH}}} \tag{8-26}$$

EXAMPLE 8-13

Find the output resistance of the common-emitter amplifier given in Fig. 8-11(a).

SOLUTION To find $R_{out(col)}$, we must first find R_{TH}.

$$\begin{aligned} R_{TH} &= R_1 \,\|\, R_2 \,\|\, Z_S \\ &= 8.2\ \text{k}\Omega \,\|\, 1.5\ \text{k}\Omega \,\|\, 600\ \Omega \\ &= 407\ \Omega \end{aligned}$$

Now we may use Eq. 8-26 to find $R_{out(col)}$.

$$\begin{aligned} R_{out(col)} &\simeq r_o + \frac{g_m r_o r_\pi R_E}{R_E + r_\pi + R_{TH}} \\ &= 84.4\ \text{k}\Omega + \frac{(91.2\ \text{mS})(84.4\ \text{k}\Omega)(1.32\ \text{k}\Omega)(1\ \text{k}\Omega)}{1\ \text{k}\Omega + 1.32\ \text{k}\Omega + 407\ \Omega} \\ &= 3.81\ \text{M}\Omega \end{aligned}$$

Now we must merely recognize that $R_{out(col)}$ is in parallel with R_C. Thus

$$R_{out} = R_{out(col)} \,\|\, R_C = 3.81\ \text{M}\Omega \,\|\, 4.3\ \text{k}\Omega \simeq 4.3\ \text{k}\Omega$$

and we see that our approximation of R_{out} as R_C is even closer to the truth when R_E is unbypassed. ■

All of our work (notably Eqs. 8-12, 8-18, and 8-26) may be directly extended to the FET. However, unbypassed source resistance is typically *not* employed with FETs. This is true because the low voltage gains inherent in FET voltage amplifiers are often reduced to *less* than unity. Consequently, the FET voltage amplifiers often become voltage attenuators - an extremely undesirable situation. Therefore, we have not emphasized FETs in this section. *Partial* source resistor bypassing is possible [see Fig. 8-40]. Negative feedback (via an unbypassed source resistance) is often incorporated in FET *power amplifiers*, but more will be said about that in Chapter 12.

8-6 Op Amp as an Inverting Amplifier

The relative ease with which the op amp can be used as an inverting amplifier [Fig. 8-16(a)] is made possible by the liberal use of negative feedback. However, we can circumvent much of the feedback theory by invoking some of the *fundamental op amp idealizations*:

FIGURE 8-16 **Op amp inverting amplifier: (a) complete circuit; (b) amplifier circuit which includes the op amp model.**

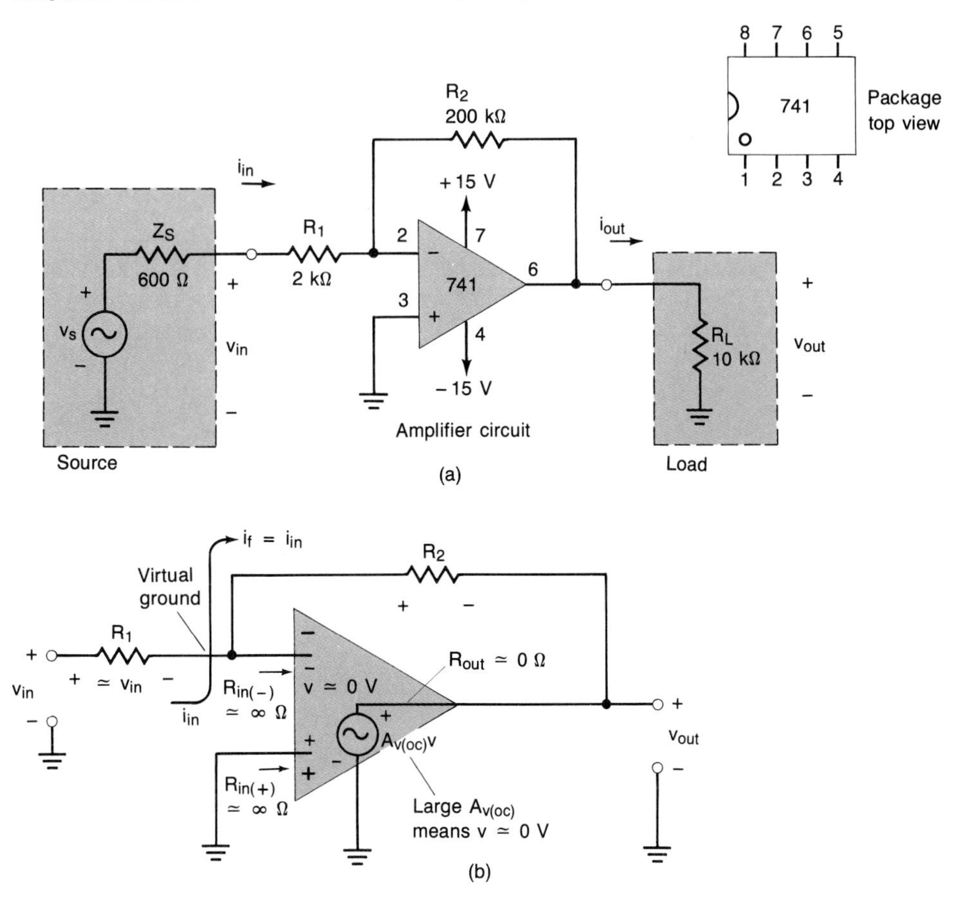

1. The input resistance of an op amp "looking" into either its inverting or noninverting input is infinite.
2. The open-circuit voltage gain of an op amp is infinite, and therefore its differential input voltage is zero.
3. The output resistance of an op amp is zero.

These idealizations are emphasized in Fig. 8-16(b). Note that since the differential input voltage is approximately zero and the noninverting input is tied to ground, the inverting input terminal of the op amp is described as being a *virtual ground* [see Fig. 8-16(b)].

The voltage gain equation may be derived by referencing Fig. 8-16(b). Since the inverting input terminal is at virtual ground, essentially all of v_{in} is impressed across R_1. The current that flows through R_1 is given by

$$i_{\text{in}} = \frac{v_{\text{in}}}{R_1} \tag{8-27}$$

Since the input resistance of the inverting input terminal ($R_{\text{in}(-)}$) is approximately infinite, the current through R_1 flows through R_2 (as a feedback current i_f).

$$i_f = i_{\text{in}} \tag{8-28}$$

By applying Kirchhoff's voltage law around the output loop, we obtain

$$v_{\text{out}} = -i_f R_2 + 0 = -i_f R_2 = -i_{\text{in}} R_2 \tag{8-29}$$

Substitution of Eq. 8-27 into Eq. 8-29 and solving for the voltage gain produce

$$\boxed{A_{v(\text{oc})} = \frac{v_{\text{out}}}{v_{\text{in}}} = -\frac{R_2}{R_1}} \tag{8-30}$$

As we can see, the voltage gain of an op amp inverting amplifier is given by the simple ratio of R_2 to R_1. This result is very similar to Eq. 8-14, which we developed for the common-emitter amplifier with no emitter bypass capacitor. Similarly, Eq. 8-30 is also an approximation—albeit an extremely accurate one. Once again negative feedback has provided us with a very stable voltage gain. As a final comment, we should point out that the open-circuit voltage gain given by Eq. 8-30 applies to the amplifier *circuit* given in Fig. 8-16(a). This should not be confused with the much larger (e.g., 25,000 or more) open-circuit voltage gain possessed by the particular integrated op amp. Both of these open-circuit voltage gains are *simultaneously* in force.

The input resistance of the op amp inverting voltage amplifier given in Fig. 8-16(a) is also incredibly easy to ascertain. By simply rearranging Eq. 8-27 we arrive at

$$\boxed{R_{\text{in}} = \frac{v_{\text{in}}}{i_{\text{in}}} = R_1} \tag{8-31}$$

Until feedback theory is seriously investigated in Chapters 10 and 11, we shall approximate the output resistance of Fig. 8-16(a) as zero.

$$\boxed{R_{\text{out}} \simeq 0\ \Omega} \qquad (8\text{-}32)$$

EXAMPLE 8-14

Find the loaded voltage gain A_v, the voltage gain from the signal source to the load A_{vs}, the loaded current gain A_i, the current gain from the signal source to the load A_{is}, and the power gain A_p for the op amp inverting amplifier circuit given in Fig. 8-16(a).

SOLUTION First, we shall place the circuit given in Fig. 8-16(a) in our general amplifier form (Fig. 8-17), where

$$A_{v(\text{oc})} = -\frac{R_2}{R_1} = -\frac{200\ \text{k}\Omega}{2\ \text{k}\Omega} = -100$$

$$R_{\text{in}} = R_1 = 2\ \text{k}\Omega$$

$$R_{\text{out}} \simeq 0\ \Omega$$

From the results above and Fig. 8-17, we may draw on the relationships summarized in Fig. 8-6. Hence

$$A_v = \frac{v_{\text{out}}}{v_{\text{in}}} = \frac{R_L}{R_L + R_{\text{out}}} A_{v(\text{oc})} = \frac{10\ \text{k}\Omega}{10\ \text{k}\Omega + 0.0\ \text{k}\Omega} A_{v(\text{oc})}$$

$$= A_{v(\text{oc})} = -100$$

$$A_{vs} = \frac{v_{\text{out}}}{v_s} = \frac{R_{\text{in}}}{R_{\text{in}} + Z_S} A_v = \frac{2\ \text{k}\Omega}{2\ \text{k}\Omega + 600\ \Omega}(-100)$$

$$= -76.9$$

$$A_i = \frac{i_{\text{out}}}{i_{\text{in}}} = \frac{R_{\text{in}}}{R_L + R_{\text{out}}} A_{v(\text{oc})} = \frac{2\ \text{k}\Omega}{10\ \text{k}\Omega + 0.0\ \text{k}\Omega}(-100)$$

$$= -20$$

FIGURE 8-17 General amplifier model for the op amp inverting amplifier circuit given in Fig. 8-16(a).

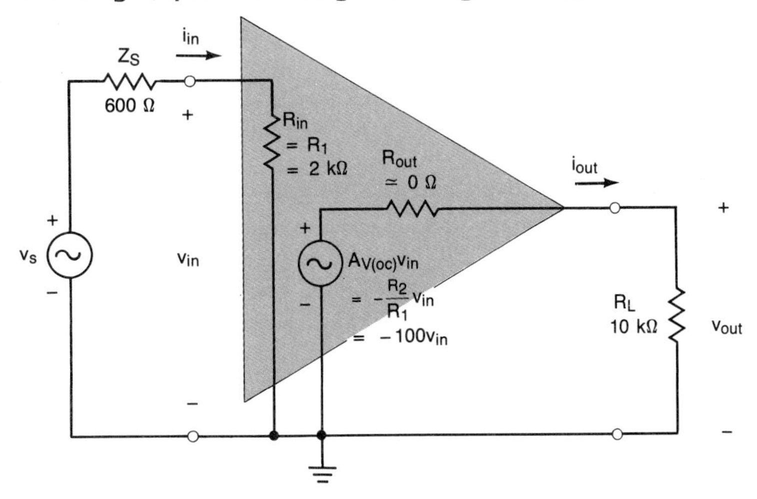

$$A_{is} = \frac{i_{out}}{i_s} = \frac{Z_S}{Z_S + R_{in}} A_i = \frac{600\ \Omega}{600\ \Omega + 2\ \text{k}\Omega}(-20)$$

$$= -4.62$$

$$A_p = A_v A_i = (-100)(-20) = 2000$$

■

As we can see, the procedure for characterizing an op amp inverting amplifier is identical to that required for the BJT and the FET inverting amplifiers. However, the initial dc analysis, and the determination of $A_{v(oc)}$, R_{in}, and R_{out} for the op amp inverting amplifier, are much easier. Now we can begin to appreciate why linear circuit designers prefer op amps to discrete circuit designs whenever possible.

The op amp is direct coupled internally. This means there are *no* coupling capacitors between the cascaded amplifiers inside the op amp. Consequently, the op amp will amplify dc as well as ac signals. For example, if the ac signal source in Fig. 8-16(a) is replaced with a 0.01 –V dc source, the output voltage (V_{OUT}) will be $A_v V_{IN} = (-100)(0.01\ \text{V}) = -1\ \text{V}$. Therefore, if the signal source produces an ac signal that rides on a dc level, an input coupling capacitor in series with the input resistor (R_1) may be required.

8-7 Noninverting Amplifier

As the name implies, the noninverting amplifiers demonstrate a voltage gain in which the input and output voltages are in phase. As we shall see, the BJT in its common-base configuration, the FET in its common-gate configuration, and of course, the op amp may all be used as noninverting voltage amplifiers. (The "follower" amplifiers are also noninverting, but we postpone our coverage of them until Sections 8-10 and 8-11.)

8-8 Common-Base and Common-Gate Amplifiers

The common-base and common-gate amplifiers typically exhibit a very low input resistance. Therefore, we find their use somewhat restricted in audio-frequency voltage amplifiers. However, an understanding of these amplifiers will be drawn upon in our analysis of the very important differential amplifier. We do find the common-base and common-gate amplifiers used in radio-frequency systems and in a special high-frequency amplifier referred to as the *cascode amplifier*. This configuration is employed in both discrete and integrated wide-bandwidth video amplifiers.

In Fig. 8-3(c) and (d), we saw the common-base and common-gate amplifiers, respectively. (The basic signal process of the common-base amplifier was explained in Section 4-13. The reader may wish to review that discussion at this point.)

The general ac equivalent circuits of Fig. 8-3(c) and (d) have been produced in Fig. 8-18(a) and (b). The base and gate bypass capacitors serve to tie those terminals to ac ground. Consequently, the R_1 and R_2 bias resistors are also shorted out of the circuit.

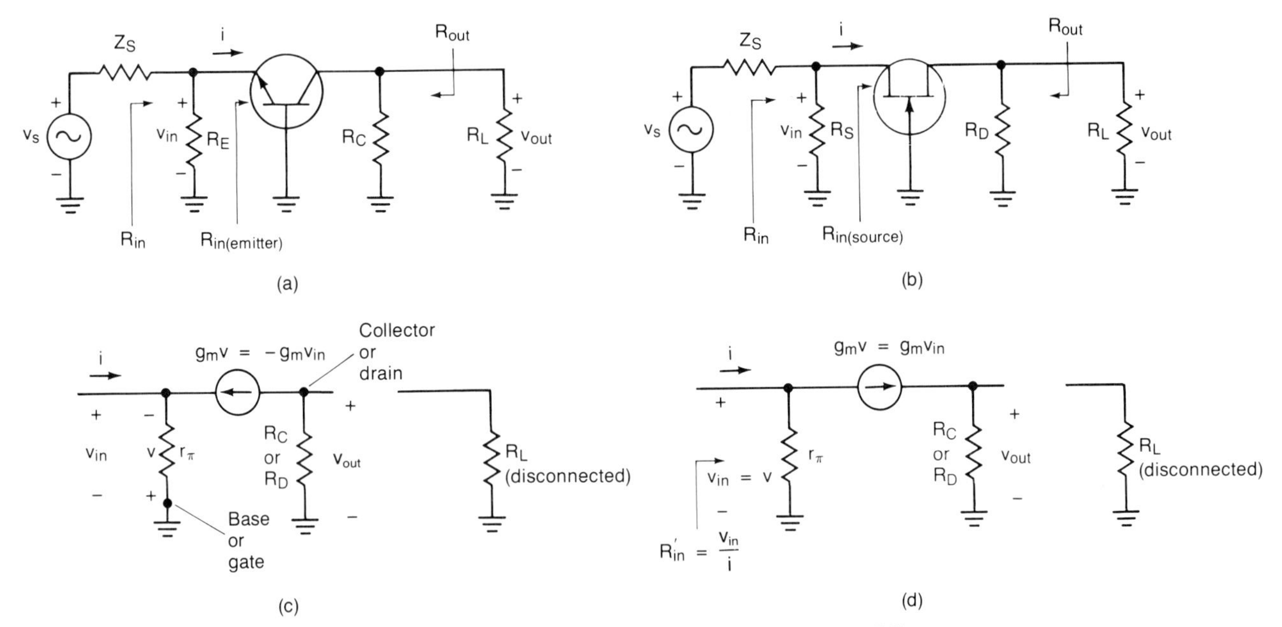

FIGURE 8-18 Analysis of the common-base and common-gate amplifiers: (a) common-base ac equivalent circuit; (b) common-gate ac equivalent circuit; (c) partial ac equivalent circuit; (d) partial ac equivalent circuit with v and the current source reversed.

First, we shall find the resistance "looking into" the BJT's emitter $R_{\text{in(emitter)}}$ and the FET's source $R_{\text{in(source)}}$. A partial ac equivalent circuit has been illustrated in Fig. 8-18(c). Observe that our controlling input voltage v is

$$v = -v_{\text{in}}$$

To simplify our discussion, we shall temporarily reverse the sense of our controlling input voltage v and, correspondingly, the direction of our controlled current source [see Fig. 8-18(d)].

By Kirchhoff's current law,

$$i = \frac{v_{\text{in}}}{r_\pi} + g_m v_{\text{in}} = \left(\frac{1}{r_\pi} + g_m\right) v_{\text{in}}$$

$$= \frac{1 + g_m r_\pi}{r_\pi} v_{\text{in}} \tag{8-33}$$

and solving for v_{in}/i, we arrive at our preliminary expression for $R_{\text{in(emitter)}}$ and $R_{\text{in(source)}}$ called R'_{in}.

$$R'_{\text{in}} = \frac{v_{\text{in}}}{i} = \frac{r_\pi}{1 + g_m r_\pi} \tag{8-34}$$

Typically, we find that $g_m r_\pi >> 1$. Hence for the BJT we have

$$\boxed{R_{\text{in(emitter)}} = \frac{r_\pi}{1 + g_m r_\pi} \simeq \frac{1}{g_m}} \tag{8-35}$$

and for the FET we have

$$R_{\text{in(source)}} = \frac{r_\pi}{1 + g_m r_\pi} \simeq \frac{1}{g_m} \tag{8-36}$$

It should be obvious at this point that the input resistance of the common-base amplifier is the parallel combination of $R_{\text{in(emitter)}}$ and R_E. Similarly, the input resistance of the common-gate amplifier may be found from the parallel combination of $R_{\text{in(source)}}$ and the source bias resistor R_S [see Fig. 8-18(a) and (b)].

The open-circuit voltage gain $A_{v(\text{oc})}$ for the BJT may be found by inspection of Fig. 8-18(d).

$$v_{\text{out}} = g_m R_C v_{\text{in}}$$

Dividing both sides by v_{in} produces

$$A_{v(\text{oc})} = \frac{v_{\text{out}}}{v_{\text{in}}} = g_m R_C \tag{8-37}$$

In a similar fashion, we arrive at Eq. 8-38 for the FET.

$$A_{v(\text{oc})} = \frac{v_{\text{out}}}{v_{\text{in}}} = g_m R_D \tag{8-38}$$

Note that these voltage gains are positive quantities, which denotes that the input and output voltages are in phase. The output resistance is again determined under the condition that v_s is set equal to zero. Simple inspection of Fig. 8-19 shows

FIGURE 8-19 Finding R_{out}: (a) ac equivalent circuit to find R_{out}; (b) simplified partial ac equivalent circuit.

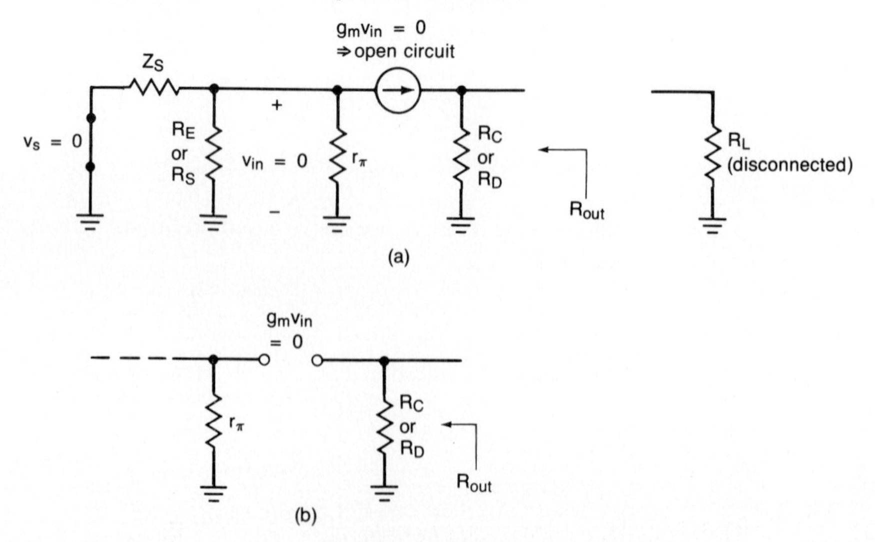

that the output resistance of the common-base amplifier and the common-gate amplifiers are given by Eqs. 8-39 and 40, respectively.

$$R_{out} = R_C \tag{8-39}$$

$$R_{out} = R_D \tag{8-40}$$

Note that the magnitudes of $A_{v(oc)}$ and R_{out} are equal to those of the common-emitter and common-source amplifiers.

EXAMPLE 8-15

Repeat the analysis requested in Example 8-14 for the common-base amplifier given in Fig. 8-3(c).

SOLUTION First, we recall that for the 2N4124 (at its bias point) we have a g_m of 91.2 mS and r_π of 1.32 kΩ. Hence, by Eq. 8-35,

$$R_{in(emitter)} = \frac{r_\pi}{1 + g_m r_\pi} = \frac{1.32\ \text{k}\Omega}{1 + (91.2\ \text{mS})(1.32\ \text{k}\Omega)}$$
$$= 10.9\ \Omega$$

which may be closely approximated as

$$R_{in(emitter)} \simeq \frac{1}{g_m} = \frac{1}{91.2\ \text{mS}} = 11.0\ \Omega$$

and R_{in} is

$$R_{in} = R_E \parallel R_{in(emitter)} = 1\ \text{k}\Omega \parallel 10.9\ \Omega \simeq 10.9\ \Omega$$

The open-circuit voltage gain may be found from Eq. 8-37.

$$A_{v(oc)} = g_m R_C = (91.2\ \text{mS})(4.3\ \text{k}\Omega) = 392$$

and from Eq. 8-39,

$$R_{out} = R_C = 4.3\ \text{k}\Omega$$

The complete analysis may now be conducted. Using the summary presented in Fig. 8-6, we obtain

$$A_v = \frac{v_{out}}{v_{in}} = \frac{R_L}{R_L + R_{out}} A_{v(oc)} = \frac{10\ \text{k}\Omega}{10\ \text{k}\Omega + 4.3\ \text{k}\Omega}(392)$$
$$= 274$$

$$A_{vs} = \frac{v_{out}}{v_s} = \frac{R_{in}}{R_{in} + Z_S} A_v = \frac{10.9\ \Omega}{10.9\ \Omega + 600\ \Omega}(274)$$
$$= 4.89$$

$$A_i = \frac{i_{out}}{i_{in}} = \frac{R_{in}}{R_L + R_{out}} A_{v(oc)} = \frac{10.9\ \Omega}{10\ \text{k}\Omega + 4.3\ \text{k}\Omega}(392)$$
$$= 0.299$$

$$A_{is} = \frac{i_{out}}{i_s} = \frac{Z_S}{Z_S + R_{in}} A_i = \frac{600\ \Omega}{600\ \Omega + 10.9\ \Omega}(0.299)$$
$$= 0.294$$

$$A_p = A_v A_i = (274)(0.299) = 81.9$$

■

Several important observations can be made from Example 8-15. First, the extremely low R_{in} tends to severely load down the signal source. Second, the low input resistance also tends to reduce the current gain greatly. Consequently, the power gain of the common-base amplifier is respectibly less than that available from the common-emitter amplifier (refer to Example 8-9).

EXAMPLE 8-16

Repeat Example 8-15 for the common-gate amplifier given in Fig. 8-3(d).

SOLUTION We follow the procedure demonstrated in Example 8-15. Using the *maximum* parameters for the 2N5458 FET (a g_m of 2.05 mS and an r_π that is approximately infinite) gives us

$$R_{\text{in(source)}} = \frac{r_\pi}{1 + g_m r_\pi} \simeq \frac{1}{g_m} = \frac{1}{2.05\text{ mS}} = 488\ \Omega$$

$$R_{\text{in}} = R_S \,\|\, R_{\text{in(source)}} = 6.8\text{ k}\Omega \,\|\, 488\ \Omega = 455\ \Omega$$

$$A_{v(\text{oc})} = g_m R_D = (2.05\text{ mS})(6.8\text{ k}\Omega) = 13.9$$

$$R_{\text{out}} = R_D = 6.8\text{ k}\Omega$$

These values allow us to complete the analysis.

$$A_v = \frac{v_{\text{out}}}{v_{\text{in}}} = \frac{R_L}{R_L + R_{\text{out}}} A_{v(\text{oc})} = \frac{10\text{ k}\Omega}{10\text{ k}\Omega + 6.8\text{ k}\Omega}(13.9) = 8.27$$

$$A_{vs} = \frac{v_{\text{out}}}{v_s} = \frac{R_{\text{in}}}{R_{\text{in}} + Z_S} A_v = \frac{455\ \Omega}{455\ \Omega + 600\ \Omega}(8.27) = 3.57$$

$$A_i = \frac{i_{\text{out}}}{i_{\text{in}}} = \frac{R_{\text{in}}}{R_L + R_{\text{out}}} A_{v(\text{oc})} = \frac{455\ \Omega}{10\text{ k}\Omega + 6.8\text{ k}\Omega}(13.9) = 0.376$$

$$A_{is} = \frac{i_{\text{out}}}{i_s} = \frac{Z_S}{Z_S + R_{\text{in}}} A_i = \frac{600\ \Omega}{600\ \Omega + 455\ \Omega}(0.376) = 0.214$$

$$A_p = A_v A_i = (8.27)(0.376) = 3.11$$

■

TABLE 8-2
Summary of the Common-Base and Common-Gate Amplifiers

	2N4124	2N5458 (max)	2N5458 (min)
A_v	274	8.27	3.49
A_{vs}	4.89	3.57	2.17
A_i	0.299	0.376	0.346
A_{is}	0.294	0.214	0.130
A_p	81.9	3.11	1.21

The reader is encouraged to repeat the analysis for a 2N5458 FET with a minimum set of parameters [refer to Fig. 8-4(d)]. A summary of the results is presented in Table 8-2. Compare Table 8-2 with Table 8-1 for the common-emitter and common-source amplifiers.

8-9 Op Amp Noninverting Amplifier

As we saw in Section 8-8, the low input resistance of the common-base and common-gate configurations severely limits the application of these noninverting amplifiers. However, the op amp may be used as a noninverting amplifier, which demonstrates an extremely *high* input resistance. Therefore, we find it extensively used in lieu of its discrete counterparts.

The noninverting op amp configuration may be formed by slightly modifying the inverting op amp circuit. Specifically, the input signal is connected to the op amp's noninverting input terminal, and its inverting input terminal is taken to ground through R_1. Compare Fig. 8-20 with Fig. 8-16(a).

The analysis of the noninverting amplifier can be simplified by once again drawing on the three op amp idealizations cited in Section 8-6. These idealizations have been illustrated in Fig. 8-20. As we can see, since the op amp's differential input voltage is approximately zero, v_{in} appears across R_1. Therefore, the current through R_1 is

$$i = \frac{v_{in}}{R_1} \tag{8-41}$$

Since the op amp's inverting input terminal is assumed to have infinite input resistance, it draws no current. Consequently, the current through R_2 will be equal to the current through R_1. If we apply Kirchhoff's voltage law around the output circuit, we can state

$$v_{out} = iR_2 + iR_1 \tag{8-42}$$

FIGURE 8-20 **Op amp noninverting amplifier.**

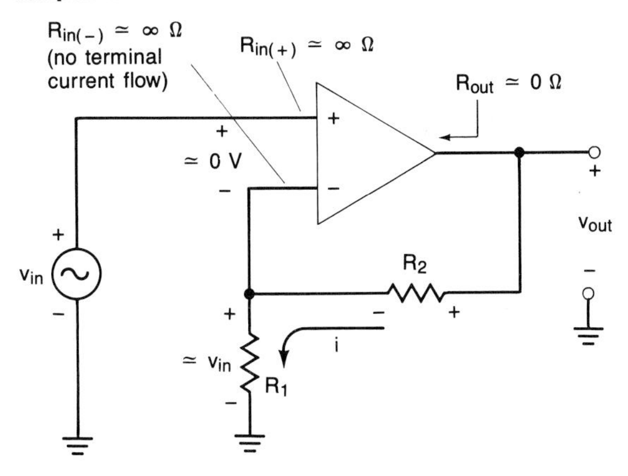

Substitution of Eq. 8-41 into Eq. 8-42 produces

$$v_{out} = \frac{R_2}{R_1} v_{in} + \frac{R_1}{R_1} v_{in} = \left(\frac{R_2}{R_1} + 1\right) v_{in} \tag{8-43}$$

Dividing both sides by v_{in} yields the equation for the open-circuit voltage gain.

$$A_{v(oc)} = \frac{v_{out}}{v_{in}} = 1 + \frac{R_2}{R_1} \tag{8-44}$$

The noninverting amplifier circuit has a voltage gain equation which is extremely similar to that of the inverting amplifier. Compare Eq. 8-44 with Eq. 8-30. Once again, negative feedback has produced an extremely stable voltage gain that depends on the ratio of R_2 to R_1.

Since the input resistance of the amplifier circuit is equal to the input resistance of the op amp's noninverting input terminal, we can state that

$$R_{in} \simeq \infty\ \Omega \tag{8-45}$$

The R_{in} given by Eq. 8-45 is, of course, an idealization, but it does provide a useful approximation in many applications.

There is an extremely important limitation on the circuit illustrated in Fig. 8-20. *The noninverting input terminal must be provided with a dc bias current path to ground.* This simple fact was first pointed out in Section 6-17. Therefore, the signal source must be *direct coupled* (no dc blocking capacitors) in order to achieve this requirement. Further, this direct-coupled signal source may *not* be disconnected while power is applied to the op amp. *If the dc bias path is interrupted while power is applied to the op amp, its noninverting terminal will be unconnected, or "floating." When this occurs the op amp's output will unpredictably wander to either the positive or negative power supply "rail."* The output will return to its proper level when the input connection is restored. However, some op amps may experience a more serious condition known as "latch-up."

When an op amp experiences latch-up its output will be in saturation at either its positive or its negative power supply "rail," and remain there. The output will be effectively unresponsive to any input signals. This condition is extremely undesirable, but generally nondestructive to op amps, with a feature known as output short-circuit protection. An op amp can resume normal operation after it has experienced latch-up. Power should be removed, a dc bias current path for its input terminals must be provided, and its power supply voltages must then be restored.

One way to avoid latch-up problems (or an unpredictable op amp output) is to ensure that the op amp always has a dc bias path to ground. This may be achieved by the addition of a bias resistor R_3 (see Fig. 8-21). When R_3 is used, the signal source is not required to provide the dc bias path to ground.

The primary trade-off by the use of R_3 is that the input resistance of the noninverting amplifier circuit is lowered to approximately the value of R_3. Hence, when R_3 is used, R_{in} is

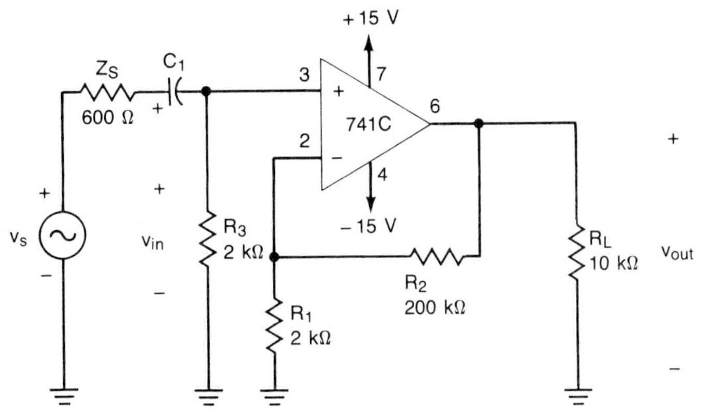

FIGURE 8-21 Op amp noninverting amplifier for Example 8-17.

$$\boxed{R_{in} \simeq R_3} \tag{8-46}$$

Without further explanation at this point, we should mention that the optimum choice for R_3 is to make it equal to the parallel combination of R_1 and R_2. This stems from a consideration known as *bias current compensation*. (This is discussed further in the problems at the end of the chapter.) Hence, for Fig. 8-21,

$$\boxed{R_3 = R_1 \parallel R_2} \tag{8-47}$$

The output resistance of the noninverting amplifier may also be approximated as zero [just as it was for the op amp inverting amplifier (see Fig. 8-20)].

EXAMPLE 8-17

Repeat Example 8-15 for the op amp noninverting amplifier given in Fig. 8-21.

SOLUTION First we shall find $A_{v(\text{oc})}$, R_{in}, and R_{out}.

$$A_{v(\text{oc})} = 1 + \frac{R_2}{R_1} = 1 + \frac{200\ \text{k}\Omega}{2\ \text{k}\Omega} = 101$$

$$R_{in} = R_3 = 2\ \text{k}\Omega$$

$$R_{out} = 0\ \Omega$$

Completing the analysis, we obtain

$$A_v = \frac{v_{out}}{v_{in}} = \frac{R_L}{R_L + R_{out}} A_{v(\text{oc})} = A_{v(\text{oc})}$$

$$= 101$$

$$A_{vs} = \frac{v_{out}}{v_s} = \frac{R_{in}}{R_{in} + Z_S} A_v = \frac{2\ \text{k}\Omega}{2\ \text{k}\Omega + 600\ \Omega}(101)$$

$$= 77.7$$

$$A_i = \frac{i_{out}}{i_{in}} = \frac{R_{in}}{R_L + R_{out}} A_{v(oc)} = \frac{2\ \text{k}\Omega}{10\ \text{k}\Omega + 0\ \Omega}(101)$$
$$= 20.2$$
$$A_{is} = \frac{i_{out}}{i_s} = \frac{Z_S}{Z_S + R_{in}} A_i = \frac{600\ \Omega}{600\ \Omega + 2\ \text{k}\Omega}(20.2)$$
$$= 4.66$$
$$A_p = A_v A_i = (101)(20.2) = 2040$$

■

A quick comparison of these results with the summary given in Table 8-2 clearly indicates the superiority of the op amp noninverting amplifier. However, we remind the reader that the discrete noninverting amplifiers do hold a position of importance in some high-frequency applications.

8-10 Emitter and Source Followers

Both common-collector BJTs and common-drain FETs may be used in amplifier circuits referred to as *followers*. These follower circuits exhibit a large input resistance, a small output resistance, and a voltage gain of approximately unity. Further, the output voltage tends to be in phase with the input voltage—hence the term "follower."

At first glance, one might be tempted to dismiss these circuits as useless because of their unity voltage gain. However, their input and output resistances promote their use as input and output *buffer* stages in *amplifier systems*.

The common-collector (emitter follower) and the common-drain (source follower) [amplifier circuits] were first illustrated in Fig. 8-3(e) and (f), respectively. Their individual ac equivalent circuits have been depicted in Fig. 8-22(a) and (b) and their generalized ac equivalent in Fig. 8-22(c).

In Section 8-5 we saw the effects of removing the emitter (or source) bypass capacitor. The main thrust of that discussion may be extended directly to the follower circuits. For example, the input resistance $R_{in(base)}$ is a strong function of the unbypassed emitter resistance. Since the load resistance in parallel with R_E forms the equivalent unbypassed emitter resistance, we must modify Eq. 8-12 slightly. Hence, for the BJT,

$$\boxed{R_{in(base)} \simeq g_m(R_E \parallel R_L)r_\pi} \qquad (8\text{-}48)$$

Similarly, for the FET we have

$$\boxed{R_{in(gate)} \simeq g_m(R_S \parallel R_L)r_\pi} \qquad (8\text{-}49)$$

Since the r_π for the FET is so large, we may approximate Eq. 8-49 further as

$$\boxed{R_{in(gate)} \simeq \infty\ \Omega} \qquad (8\text{-}50)$$

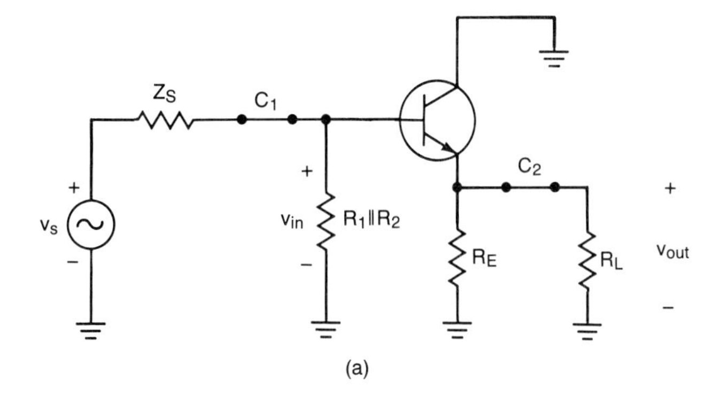

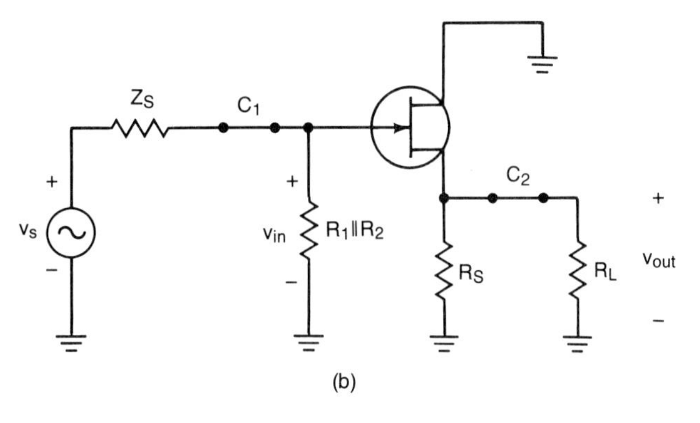

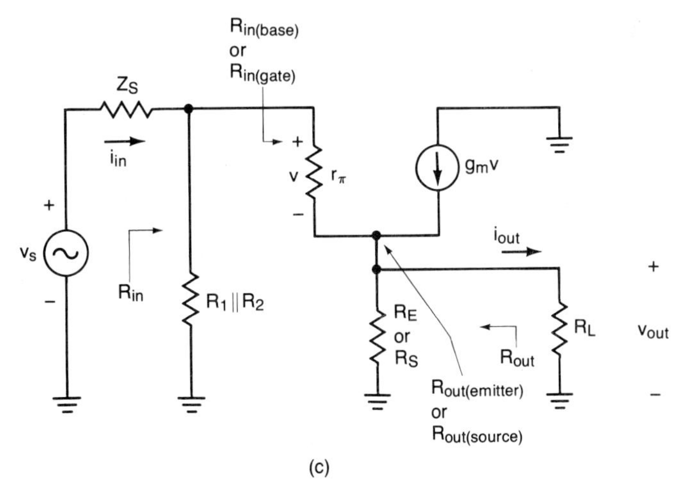

FIGURE 8-22 Follower ac equivalent circuits: (a) common-collector (emitter follower); (b) common-drain (source follower); (c) generalized ac equivalent circuit.

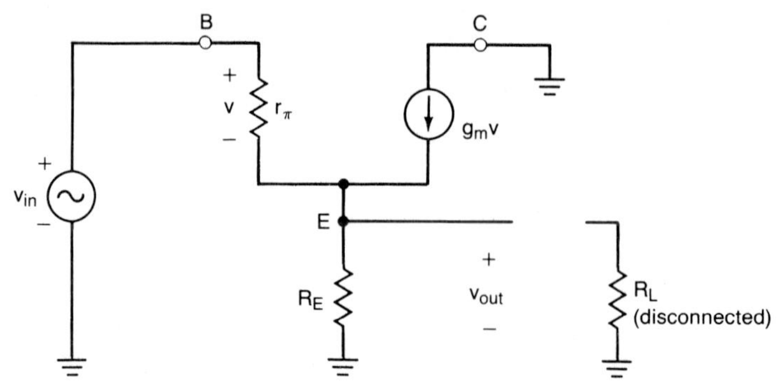

FIGURE 8-23 Partial ac equivalent circuit used to find $A_{v(oc)}$.

The equation for $A_{v(oc)}$ may be found by analyzing the partial BJT ac equivalent circuit given in Fig. 8-23. First we shall apply Kirchhoff's voltage law around the input loop.

$$v_{in} = v + v_{out}$$

Solving for v yields

$$v = v_{in} - v_{out} \tag{8-51}$$

The output voltage may be written as

$$v_{out} \simeq g_m v R_E \tag{8-52}$$

Substituting Eq. 8-51 into Eq. 8-52 gives us

$$\begin{aligned} v_{out} &\simeq g_m v R_E = g_m(v_{in} - v_{out})R_E \\ &= g_m R_E v_{in} - g_m R_E v_{out} \end{aligned}$$

Solving for v_{out} yields

$$\begin{aligned} v_{out} + g_m R_E v_{out} &= g_m R_E v_{in} \\ v_{out}(1 + g_m R_E) &= g_m R_E v_{in} \\ v_{out} &= \frac{g_m R_E}{1 + g_m R_E} v_{in} \end{aligned}$$

and dividing both sides by v_{in} produces $A_{v(oc)}$. Hence, for the BJT

$$A_{v(oc)} = \frac{v_{out}}{v_{in}} = \frac{g_m R_E}{1 + g_m R_E} \tag{8-53}$$

Similarly, for the FET,

$$A_{v(oc)} = \frac{v_{out}}{v_{in}} = \frac{g_m R_S}{1 + g_m R_S} \tag{8-54}$$

By inspection of Eq. 8-53 we can see that if $g_m R_E >> 1$,

$$A_{v(oc)} \simeq \frac{g_m R_E}{g_m R_E} = 1$$

[In a similar fashion, it is also possible for $A_{v(oc)}$ of the source follower to approach unity.]

The equivalent circuit given in Fig. 8-24(a) may be used to determine R_{out}. To simplify our work R_E will temporarily be disconnected [see Fig. 8-24(b)]. Note that the application of Eq. 8-2 has again required that we excite out network with v_{out}. Also note that we have simplified the base circuit by finding the equivalent resistance R_{TH} "seen" by the base terminal.

Since a negative control voltage $(-v)$ is produced, we have taken the liberty of temporarily reversing its definition and the direction of our current source [see Fig. 8-24(c)].

By applying Kirchhoff's current law at the emitter node, we write

$$i_{out} = \frac{v_{out}}{r_\pi + R_{TH}} + g_m v \tag{8-55}$$

Since r_π and R_{TH} are in series, we may apply voltage division to arrive at an equation for v.

FIGURE 8-24 Finding $R_{out(emitter)}$.

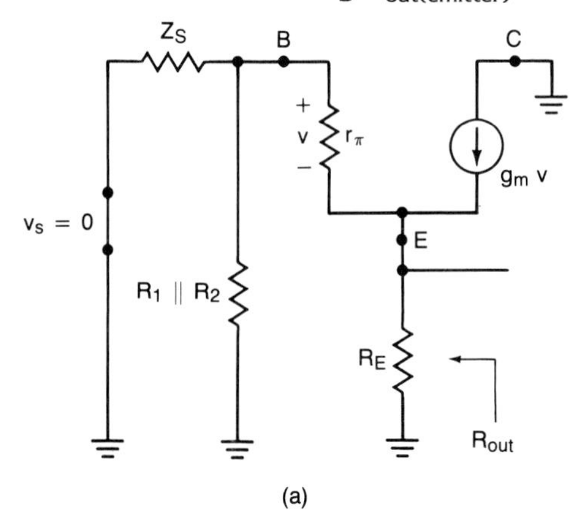

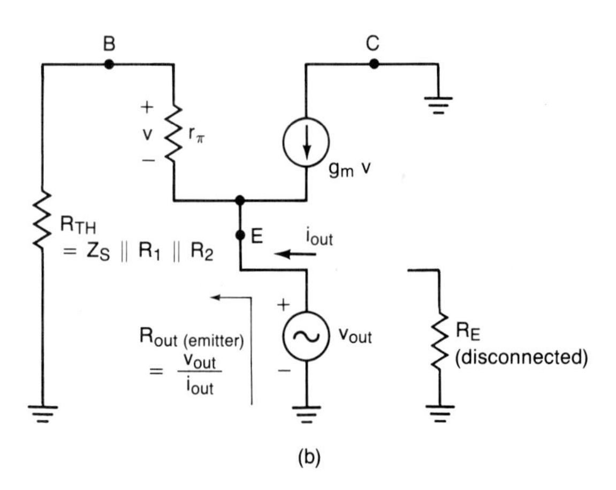

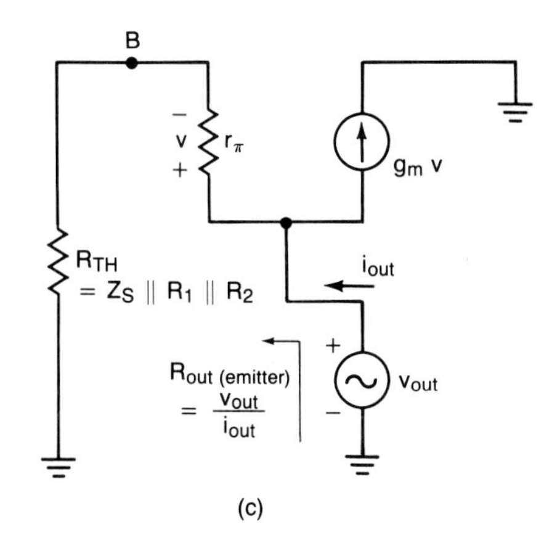

$$v = \frac{r_\pi}{r_\pi + R_{TH}} v_{out} \tag{8-56}$$

Substituting Eq. 8-56 into Eq. 8-55 and solving for the ratio of v_{out} to i_{out} results in Eq. 8-57.

$$\begin{aligned} i_{out} &= \frac{v_{out}}{r_\pi + R_{TH}} + g_m \frac{r_\pi}{r_\pi + R_{TH}} v_{out} \\ &= \left(\frac{1}{r_\pi + R_{TH}} + \frac{g_m r_\pi}{r_\pi + R_{TH}} \right) v_{out} \\ &= \frac{1 + g_m r_\pi}{r_\pi + R_{TH}} v_{out} \end{aligned}$$

$$\boxed{R_{out(emitter)} = \frac{v_{out}}{i_{out}} = \frac{r_\pi + R_{TH}}{1 + g_m r_\pi}} \tag{8-57}$$

The analogous FET relationship (and taking into account that r_π approaches infinity) is given by

$$\boxed{R_{out(source)} = \frac{r_\pi + R_{TH}}{1 + g_m r_\pi} \simeq \frac{r_\pi}{g_m r_\pi} = \frac{1}{g_m}} \tag{8-58}$$

EXAMPLE 8-18

Find the loaded voltage gain A_v, the voltage gain from the signal source to the load A_{vs}, the current gain A_i, the current gain from the signal source to the load A_{is}, and the power gain A_p for the emitter follower given in Fig. 8-3(e).

SOLUTION Once again recalling that g_m is 91.2 mS and r_π is 1.32 kΩ, we may use Eqs. 8-48, 8-53, and 8-57 to assist us in placing our emitter follower in the general amplifier form.

$$\begin{aligned} R_{in(base)} &\simeq g_m(R_E \parallel R_L) r_\pi \\ &= (91.2\ \text{mS})(1\ \text{k}\Omega \parallel 10\ \text{k}\Omega)(1.32\ \text{k}\Omega) = 109\ \text{k}\Omega \\ R_{in} &= R_1 \parallel R_2 \parallel R_{in(base)} = 8.2\ \text{k}\Omega \parallel 1.5\ \text{k}\Omega \parallel 109\ \text{k}\Omega \\ &= 1253\ \Omega \\ A_{v(oc)} &= \frac{g_m R_E}{1 + g_m R_E} = \frac{(91.2\ \text{mS})(1\ \text{k}\Omega)}{1 + (91.2\ \text{mS})(1\ \text{k}\Omega)} = 0.989 \end{aligned}$$

and we see that $A_{v(oc)}$ is very close to unity. Next we find the output resistance.

$$\begin{aligned} R_{TH} &= R_1 \parallel R_2 \parallel Z_S = 8.2\ \text{k}\Omega \parallel 1.5\ \text{k}\Omega \parallel 600\ \Omega = 407\ \Omega \\ R_{out(emitter)} &= \frac{r_\pi + R_{TH}}{1 + g_m r_\pi} = \frac{1.32\ \text{k}\Omega + 407\ \Omega}{1 + (91.2\ \text{mS})(1.32\ \text{k}\Omega)} \\ &= 14.2\ \Omega \\ R_{out} &= R_{out(emitter)} \parallel R_E = 14.2\ \Omega \parallel 1\ \text{k}\Omega \simeq 14.2\ \Omega \end{aligned}$$

Continuing the analysis, we have

$$A_v = \frac{v_{out}}{v_{in}} = \frac{R_L}{R_L + R_{out}} A_{v(oc)} = \frac{10\text{ k}\Omega}{10\text{ k}\Omega + 14.2\ \Omega}(0.989)$$
$$= 0.988$$
$$A_{vs} = \frac{v_{out}}{v_s} = \frac{R_{in}}{R_{in} + Z_S} A_v = \frac{1253\ \Omega}{1253\ \Omega + 600\ \Omega}(0.988)$$
$$= 0.668$$
$$A_i = \frac{i_{out}}{i_{in}} = \frac{R_{in}}{R_L + R_{out}} A_{v(oc)} = \frac{1253\ \Omega}{10\text{ k}\Omega + 14.2\ \Omega}(0.989)$$
$$= 0.124$$
$$A_{is} = \frac{i_{out}}{i_s} = \frac{Z_S}{Z_S + R_{in}} A_i = \frac{600\ \Omega}{600\ \Omega + 1253\ \Omega}(0.124)$$
$$= 0.0402$$
$$A_p = A_v A_i = (0.988)(0.124) = 0.123$$

■

The low voltage and current gains have produced an extremely low power gain. Clearly, the primary advantage offered by the emitter follower are the capabilities of a moderately large R_{in} and a small R_{out}. (This is demonstrated further in Example 8-20.) Therefore, it serves as an excellent choice as an input and/or output stage of an amplifier system (refer back to Fig. 7-12).

EXAMPLE 8-19

Repeat Example 8-18 for the drain follower given in Fig. 8-3(f).

SOLUTION For a 2N5458 with a *maximum* set of parameters (a g_m of 2.05 mS and an r_π of infinity),

$$R_{in(gate)} \simeq g_m(R_S \| R_L)r_\pi \simeq \infty$$
$$R_{in} = R_1 \| R_2 \| R_{in(gate)} = R_1 \| R_2 = 200\text{ k}\Omega \| 51\text{ k}\Omega$$
$$= 40.6\text{ k}\Omega$$
$$A_{v(oc)} = \frac{g_m R_S}{1 + g_m R_S} = \frac{(2.05\text{ mS})(6.8\text{ k}\Omega)}{1 + (2.05\text{ mS})(6.8\text{ k}\Omega)} = 0.933$$

The source terminal output resistance $R_{out(source)}$ and the resulting R_{out} will be approximated.

$$R_{out(source)} \simeq \frac{1}{g_m} = \frac{1}{2.05\text{ mS}} = 488\ \Omega$$
$$R_{out} = R_{out(source)} \| R_S = 488\ \Omega \| 6.8\text{ k}\Omega = 455\ \Omega$$

The analysis may be finished in the usual manner.

$$A_v = \frac{v_{out}}{v_{in}} = \frac{R_L}{R_L + R_{out}} A_{v(oc)} = \frac{10\text{ k}\Omega}{10\text{ k}\Omega + 455\ \Omega}(0.933)$$
$$= 0.892$$
$$A_{vs} = \frac{v_{out}}{v_s} = \frac{R_{in}}{R_{in} + Z_S} A_v = \frac{40.6\text{ k}\Omega}{40.6\text{ k}\Omega + 600\ \Omega}(0.892)$$
$$= 0.879$$

$$A_i = \frac{i_{\text{out}}}{i_{\text{in}}} = \frac{R_{\text{in}}}{R_L + R_{\text{out}}} A_{v(\text{oc})} = \frac{40.6\text{ k}\Omega}{10\text{ k}\Omega + 455\ \Omega}(0.933)$$
$$= 3.62$$
$$A_{is} = \frac{i_{\text{out}}}{i_s} = \frac{Z_S}{Z_S + R_{\text{in}}} A_i = \frac{600\ \Omega}{600\ \Omega + 40.6\text{ k}\Omega}(3.62)$$
$$= 0.0527$$
$$A_p = A_v A_i = (0.892)(3.62) = 3.23$$

The summary of Examples 8-18 and 8-19, and the results of the analysis of a 2N5458 with a minimum set of parameters is given in Table 8-3. As we can see, the FETs offer a superior R_{in} but a larger R_{out}. They do exhibit a much larger current gain and, consequently, power gain than does the BJT.

TABLE 8-3
Summary of the Common-Collector and Common-Drain Amplifiers

	2N4124	2N5458 (max)	2N5458 (min)
A_v	0.988	0.892	0.777
A_{vs}	0.668	0.879	0.766
A_i	0.124	3.62	3.15
A_{is}	0.0402	0.0527	0.0459
A_p	0.123	3.23	2.45

One of our primary objectives has been to illustrate that the same basic BJT and FET bias circuits can be used in any one of the three basic amplifier configurations. This has been demonstrated throughout the chapter. However, we should point out that the dc biasing can be designed to maximize the performance of a given amplifier. Consider Example 8-20.

EXAMPLE 8-20

Find R_{in}, R_{out}, and A_v for the emitter follower given in Fig. 8-25. Express the voltage gain in decibels.

FIGURE 8-25 Emitter follower for Example 8-20.

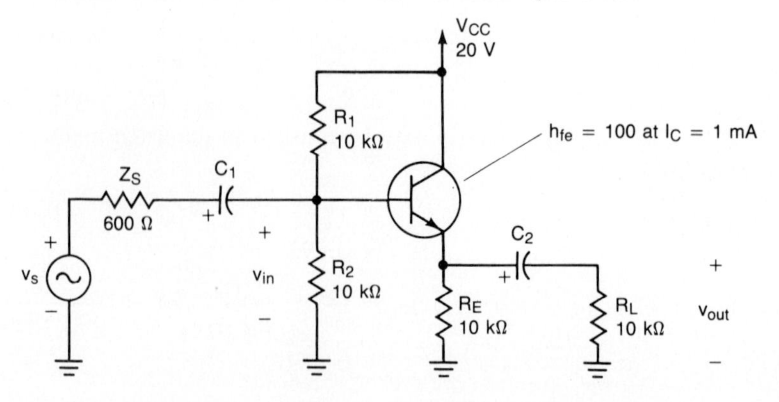

SOLUTION To expedite the analysis we shall draw upon approximations. First, we must locate the Q-point.

$$V_B = \frac{R_2}{R_1 + R_2} V_{CC} = \frac{10\ \text{k}\Omega}{20\ \text{k}\Omega}(20\ \text{V}) = 10\ \text{V}$$

$$V_E = V_B - 0.7\ \text{V} = 10\ \text{V} - 0.7\ \text{V} = 9.3\ \text{V}$$

$$I_C \simeq I_E = \frac{V_E}{R_E} = \frac{9.3\ \text{V}}{10\ \text{k}\Omega} = 0.93\ \text{mA}$$

Now we shall approximate the BJT's small-signal parameters. (Observe that the BJT's h_{fe} is 100 as noted in Fig. 8-25.) We shall not trouble ourselves with r_o.

$$g_m = \frac{I_C}{26\ \text{mV}} = \frac{0.93\ \text{mA}}{26\ \text{mV}} = 35.8\ \text{mS}$$

$$r_\pi = \frac{h_{fe}}{g_m} = \frac{100}{35.8\ \text{mS}} = 2.80\ \text{k}\Omega$$

The general parameters R_{in} and $A_{v(\text{oc})}$ are

$$\begin{aligned} R_{\text{in(base)}} &= g_m(R_E \parallel R_L)r_\pi \\ &= (35.8\ \text{mS})(10\ \text{k}\Omega \parallel 10\ \text{k}\Omega)(2.80\ \text{k}\Omega) = 501\ \text{k}\Omega \\ R_{\text{in}} &= R_1 \parallel R_2 \parallel R_{\text{in(base)}} \simeq R_1 \parallel R_2 = 10\ \text{k}\Omega \parallel 10\ \text{k}\Omega \\ &= 5\ \text{k}\Omega \end{aligned}$$

$$A_{v(\text{oc})} = \frac{g_m R_E}{1 + g_m R_E} = \frac{(35.8\ \text{mS})(10\ \text{k}\Omega)}{1 + (35.8\ \text{mS})(10\ \text{k}\Omega)} = 0.997$$

The output resistance R_{out} is

$$R_{\text{TH}} = R_1 \parallel R_2 \parallel Z_S = 5\ \text{k}\Omega \parallel 600\ \Omega = 536\ \Omega$$

$$\begin{aligned} R_{\text{out(emitter)}} &= \frac{r_\pi + R_{\text{TH}}}{1 + g_m r_\pi} = \frac{2.80\ \text{k}\Omega + 536\ \Omega}{1 + (35.8\ \text{mS})(2.80\ \text{k}\Omega)} \\ &= 33\ \Omega \end{aligned}$$

$$R_{\text{out}} = R_{\text{out(emitter)}} \parallel R_E \simeq R_{\text{out(emitter)}} = 33\ \Omega$$

and A_v, and its value in decibels can now be determined.

$$\begin{aligned} A_v &= \frac{R_L}{R_L + R_{\text{out}}} A_{v(\text{oc})} = \frac{10\ \text{k}\Omega}{10\ \text{k}\Omega + 33\ \Omega}(0.997) \\ &= 0.994 \end{aligned}$$

$$|A_v|_{\text{dB}} = 20 \log A_v = 20 \log 0.994 = -0.0523\ \text{dB}$$

■

By inspection of Fig. 8-25 and Example 8-20, we can see that R_1 and R_2 are equal in value. Therefore, the base-to-ground voltage V_B, and approximately the emitter-to-ground voltage V_E, are about one-half of the collector supply voltage V_{CC}. This Q-point allows the emitter voltage swing to assume much larger values than the circuit given in Fig. 8-3(e). Further, R_{in} has been increased because R_1 and R_2 are larger.

8-11 Op Amp Voltage Follower

The op amp voltage follower has been shown in Fig. 8-26(a). Since the op amp's differential input voltage is approximately zero, we can see that the v_{out} must be approximately equal to v_{in}, and the circuit will have an $A_{v(oc)}$ of approximately unity.

If the signal source does not provide the noninverting input terminal with a bias current path to ground, and/or a dc blocking capacitor is used, we must again install a bias resistor R_3. Observe in Fig. 8-26(b) that R_3 has been included to ensure a dc bias path to ground for the noninverting input terminal. For bias current compensation (which will be explored later) the resistance in the feedback circuit R_2 should be equal to R_3. This has also been noted in Fig. 8-26(b).

The addition of R_2 does *not* affect the unity voltage gain of the circuit. This is true because the high input resistance of the inverting input terminal does not promote a significant current flow through R_2. Alternatively, we may turn to Eq. 8-44. If R_1 is infinite (an open), then $A_{v(oc)}$ reduces to unity.

FIGURE 8-26 Op amp voltage follower.

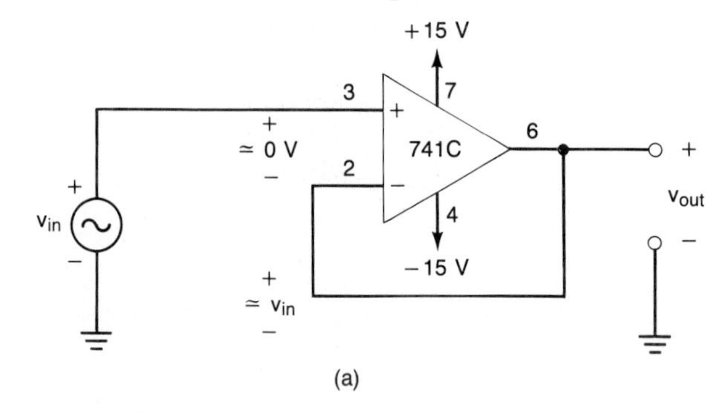

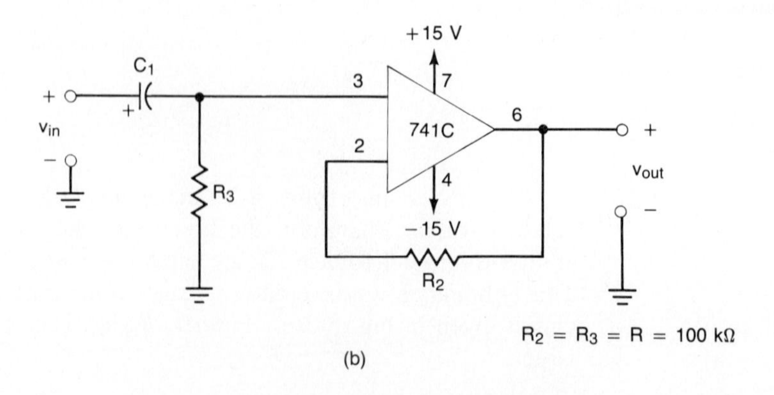

8-12 Discrete Differential Amplifier

Discrete differential amplifier circuits [Fig. 8-27(a)] are occasionally found in the ''front-end'' vertical amplifiers in wide-bandwidth oscilloscopes, as part of the error amplifier circuitry in power supply regulators, and as the driver stages in audio power amplifiers. Observe that we have defined an inverting input and a noninverting input.

FIGURE 8-27 **Discrete differential amplifier: (a) basic circuit; (b) the noninverting input signal is set to zero; (c) the equivalent resistance "seen" by Q_1's emitter; (d) since the output resistance of Q_1's emitter is $1/g_m$, $v_e = v_{in1}/2$.**

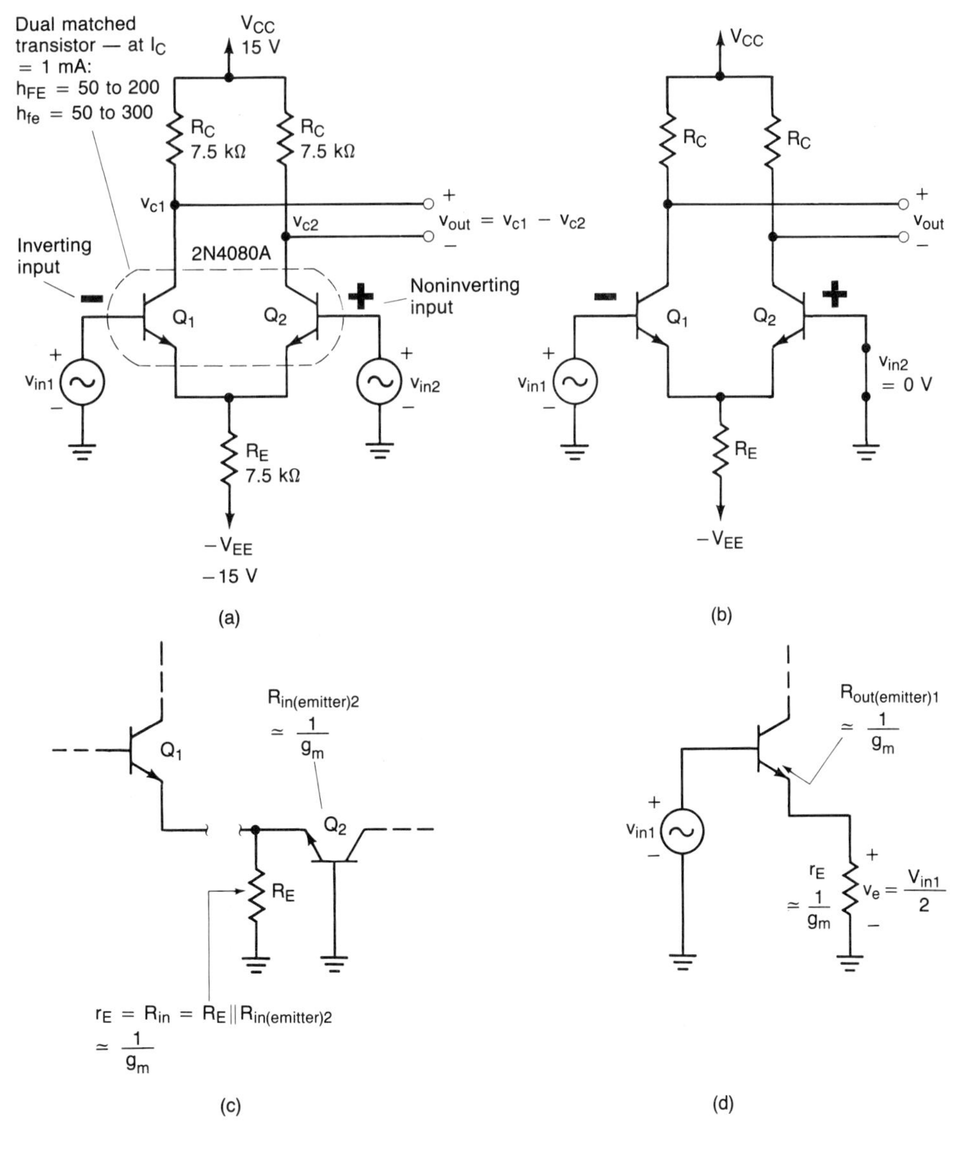

As we shall see, an understanding of the differential amplifier will require us to draw on the fundamental concepts associated with common-emitter, common-base, and common-collector amplifiers. First we shall derive an equation for $A_{v(\text{oc})}$ by taking a rather intuitive approach.

If we call on the principle of superposition, we may temporarily set v_{in2} to zero [see Fig. 8-27(b)]. The equivalent ac resistance "seen" by the emitter of transistor Q_1 will be called r_E. It is composed of the parallel combination of the resistance "looking" into the emitter (input terminal) of Q_2 and R_E [see Fig. 8-27(c)]. Calling upon Eq. 8-35, we see that

$$R_{\text{in(emitter)2}} = \frac{r_\pi}{1 + g_m r_\pi} \simeq \frac{1}{g_m}$$

and

$$r_E = R_E \parallel R_{\text{in(emitter)2}} \simeq R_{\text{in(emitter)2}}$$

As we have seen (Eq. 8-57), the output resistance of a BJT's emitter is

$$R_{\text{out(emitter)1}} = \frac{r_\pi + R_{\text{TH}}}{1 + g_m r_\pi}$$

and in this example $R_{\text{TH}} = 0$. Hence

$$R_{\text{out(emitter)1}} = \frac{r_\pi}{1 + g_m r_\pi} \simeq \frac{1}{g_m}$$

Since the output resistance of Q_1's emitter is equal to the equivalent emitter-to-ground resistance "looking" into Q_2's input, v_e will be equal to one-half of v_{in1} [see Fig. 8-27(d)].

By drawing on Eq. 8-18 for the voltage gain of a common-emitter amplifier with an unbypassed emitter resistance, and using r_E to represent the equivalent ac emitter resistance, we see that

$$v_{c1} = -\frac{g_m R_C}{1 + g_m r_E} v_{\text{in1}} = -\frac{g_m R_C}{1 + g_m(1/g_m)} v_{\text{in1}}$$
$$= -\frac{g_m R_C}{2} v_{\text{in1}}$$

[refer to Fig. 8-28(a)].

Transistor Q_2 acts like a common-base amplifier. Q_2's input voltage is v_e. Therefore, the voltage at Q_2's collector may be found by employing Eq. 8-37.

$$v_{c2} = g_m R_C v_e = g_m R_C \frac{v_{\text{in1}}}{2} = \frac{g_m R_C}{2} v_{\text{in1}}$$

The output voltage is merely the difference between these two collector-to-ground voltages. Hence

$$v_{\text{out}} = v_{c1} - v_{c2} = -\frac{g_m R_C}{2} v_{\text{in1}} - \frac{g_m R_C}{2} v_{\text{in1}}$$
$$= -g_m R_C v_{\text{in1}}$$

These results, and typical signals have been depicted in Fig. 8-28(b).

If, by superposition, we now set v_{in1} equal to zero and reactivate v_{in2}, we may now analyze its contribution to form v_{out}. By symmetry it should be clear that

$$v_{\text{out}} = v_{c1} - v_{c2} = g_m R_C v_{\text{in2}}$$

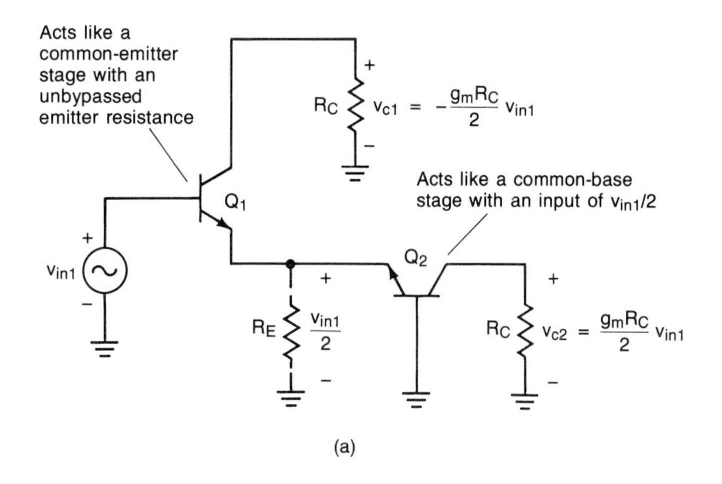

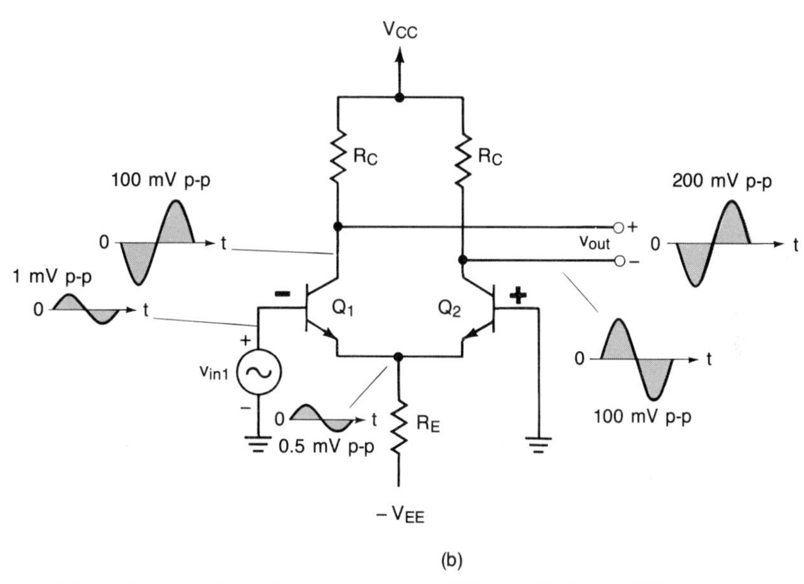

FIGURE 8-28 **Signal process in a differential amplifier: (a) partial ac equivalent circuit; (b) signal process.**

If both sources act simultaneously, we may sum the contributions of $v_{\text{in}1}$ and $v_{\text{in}2}$. Hence

$$\begin{aligned} v_{\text{out}} &= v_{c1} - v_{c2} = g_m R_C v_{\text{in}2} - g_m R_C v_{\text{in}1} \\ &= g_m R_C (v_{\text{in}2} - v_{\text{in}1}) \end{aligned}$$

At this point, we may now define the open-circuit differential voltage gain $A_{vd(\text{oc})}$ offered by the amplifier given in Fig. 8-27(a).

$$A_{vd(\text{oc})} = \frac{v_{c1} - v_{c2}}{v_{\text{in}2} - v_{\text{in}1}} = \frac{v_{\text{out}}}{v_{\text{in}2} - v_{\text{in}1}} = g_m R_C \tag{8-59}$$

The input resistance at *either the inverting or the noninverting input terminals* may be found by using Eq. 8-11, which we developed for the common-emitter amplifier when its emitter resistor is unbypassed. Letting r_E represent the ac emitter resistance, we have

$$R_{in} = R_{in(base)} = (1 + g_m r_E) r_\pi$$

and since $r_E \simeq 1/g_m$,

$$R_{in} = \left(1 + g_m \frac{1}{g_m}\right) r_\pi = 2r_\pi$$

$$\boxed{R_{in} = 2r_\pi} \qquad (8\text{-}60)$$

Using our BJT model, the ac equivalent circuit of the differential amplifier may be drawn [see Fig. 8-29(a)]. If we set the input signals to zero, we may find the output resistance [see Fig. 8-29(b)]. By inspection, we can see that if the input signals are set equal to zero, the controlled current sources also go to zero (become open circuits). Therefore, the equivalent resistance from one output terminal to the other is R_{out}.

$$\boxed{R_{out} = 2R_C} \qquad (8\text{-}61)$$

EXAMPLE 8-21

Find R_{in}, R_{out}, and $A_{vd(oc)}$ for the differential amplifier given in Fig. 8-27(a).

SOLUTION The BJTs are reasonably well matched since a 2N4080A *dual* silicon transistor is used. We shall minimize the dc analysis. Finding the tail current I_T will lead us to the collector currents.

$$I_T = \frac{V_{EE} - 0.7\text{ V}}{R_E} = \frac{15\text{ V} - 0.7\text{ V}}{7.5\text{ k}\Omega} = 1.91\text{ mA}$$

$$I_{E1} = I_{E2} = \frac{I_T}{2} = \frac{1.91\text{ mA}}{2} = 0.955\text{ mA}$$

$$I_{C1} = I_{C2} = I_{E1} = I_{E2} = 0.955\text{ mA}$$

Now we shall estimate the small-signal parameters possessed by the BJTs.

$$g_m = \frac{I_C}{26\text{ mV}} = \frac{0.955\text{ mA}}{26\text{ mV}} = 36.7\text{ mS}$$

$$r_\pi = \frac{h_{fe}}{g_m} = \frac{50}{36.7\text{ mS}} = 1.36\text{ k}\Omega$$

Now we may characterize the amplifier.

$$R_{in} = 2r_\pi = (2)(1.36\text{ k}\Omega) = 2.72\text{ k}\Omega$$

$$R_{out} = 2R_C = (2)(7.5\text{ k}\Omega) = 15\text{ k}\Omega$$

$$A_{vd(oc)} = \frac{v_{out}}{v_{in2} - v_{in1}} = g_m R_C = (36.7\text{ mS})(7.5\text{ k}\Omega)$$
$$= 275$$

■

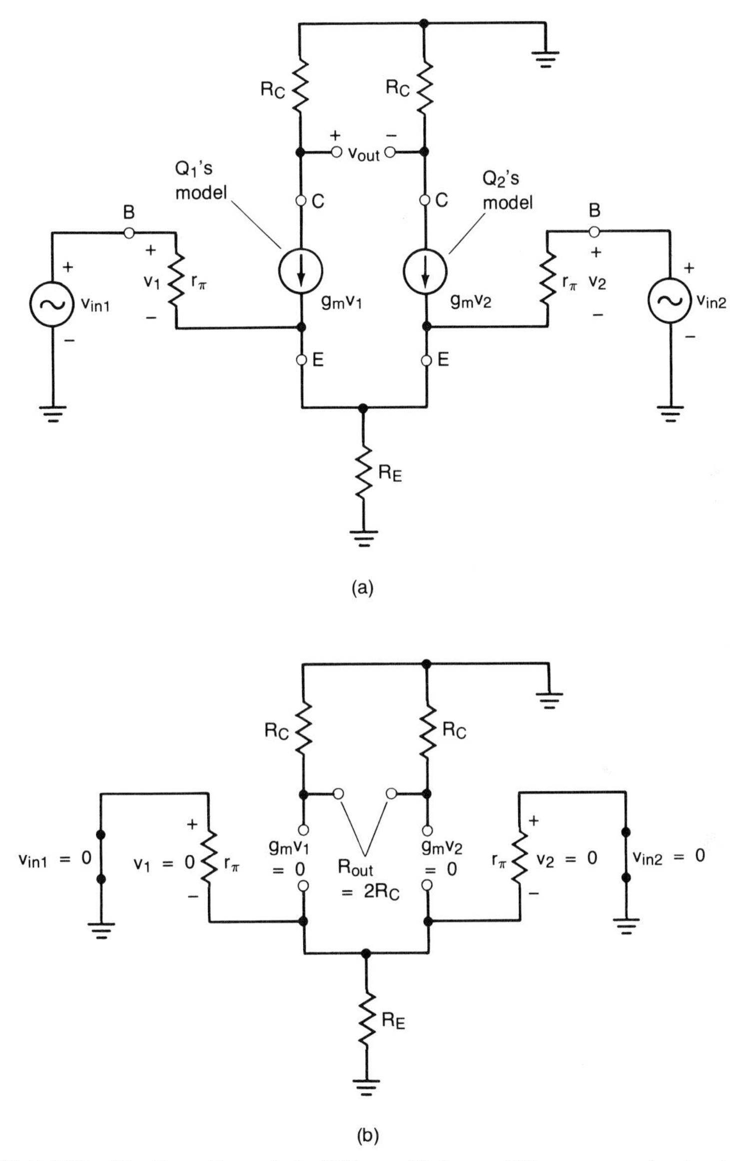

FIGURE 8-29 **Finding R_{out}: (a) differential amplifier ac equivalent circuit; (b) setting $v_{in1} = v_{in2} = 0$ to find R_{out}.**

In Fig. 8-30(a) we see the differential amplifier in action when v_{in2} is greater than v_{in1}. Similarly, in Fig. 8-30(b) we see that if v_{in1} is greater than v_{in2}, the output amplitude is proportional to the difference between the two signals but shifted 180° in phase. Figure 8-30(c) illustrates that if the two input voltages are *equal* in magnitude, and *in phase* the output will be *zero*.

Although the example illustrated in Fig. 8-30(c) may seem trivial at this point, do *not* treat it lightly. The fact that a differential amplifier will tend to cancel out any equal (common) signals at its input terminals illustrates its *common-mode rejec-*

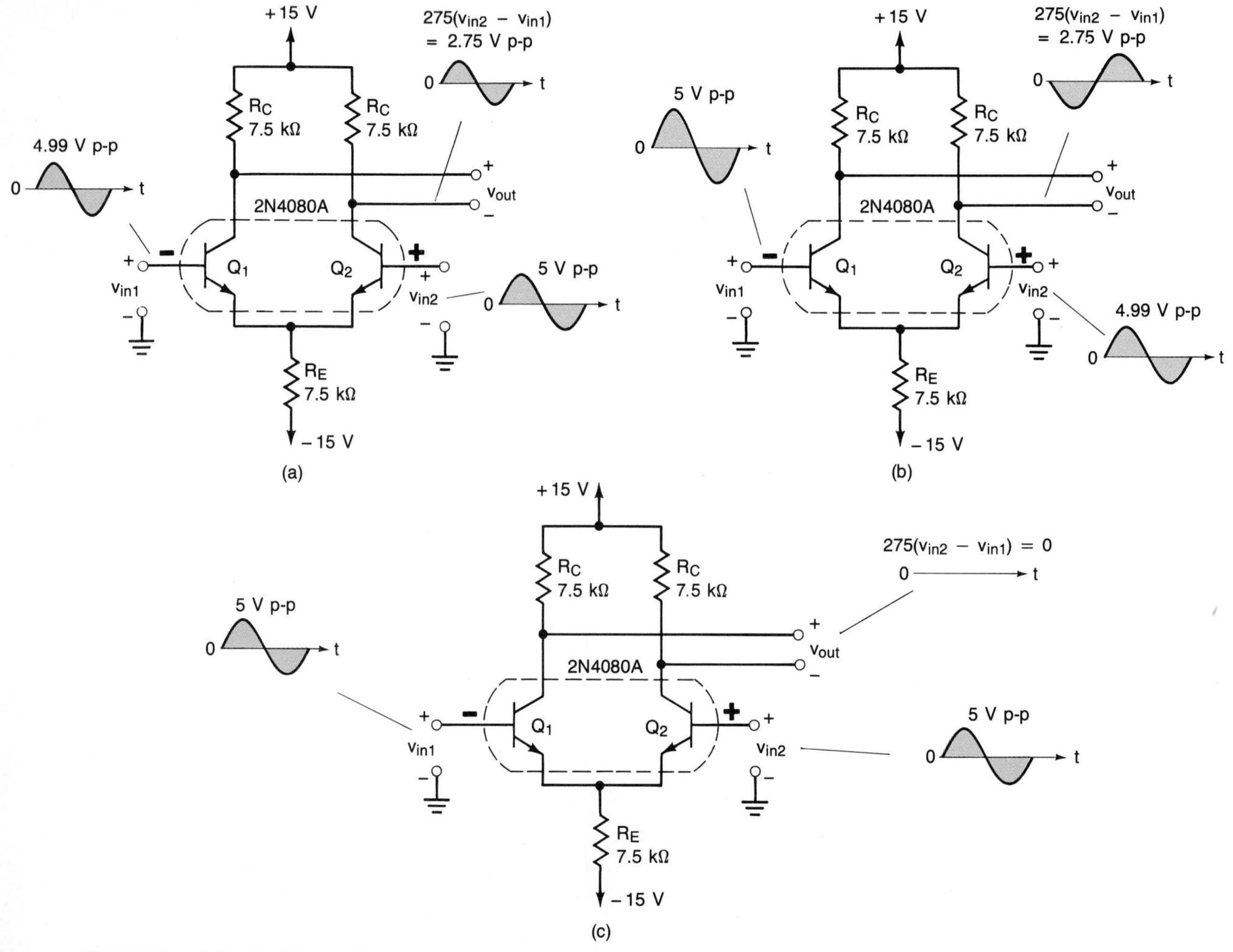

FIGURE 8-30 Differential output signals: (a) $v_{in2} > v_{in1}$; (b) $v_{in1} > v_{in2}$; (c) $v_{in1} = v_{in2}$.

tion. This is a very powerful feature and has served to launch the differential amplifier into many instrumentation applications. The reason is simple. In electrically ''noisy'' environments, the ground connection at the signal source will *not* be at the same potential as the ground connection at the amplifier's input [see Fig. 8-31(a)]. Consequently, both the signal and the 60-Hz noise are amplified. If the signal source is allowed to ''float'' (no ground connection), 60-Hz noise may be capacitively and/or magnetically coupled into it [see Fig. 8-31(b)]. However, the 60-Hz *common-mode noise voltage* is rejected by the differential amplifier.

The ratio of the output voltage to the common mode input voltage $v_{in(cm)}$ of a differential amplifier is defined as the *common-mode voltage gain* $A_{v(cm)}$.

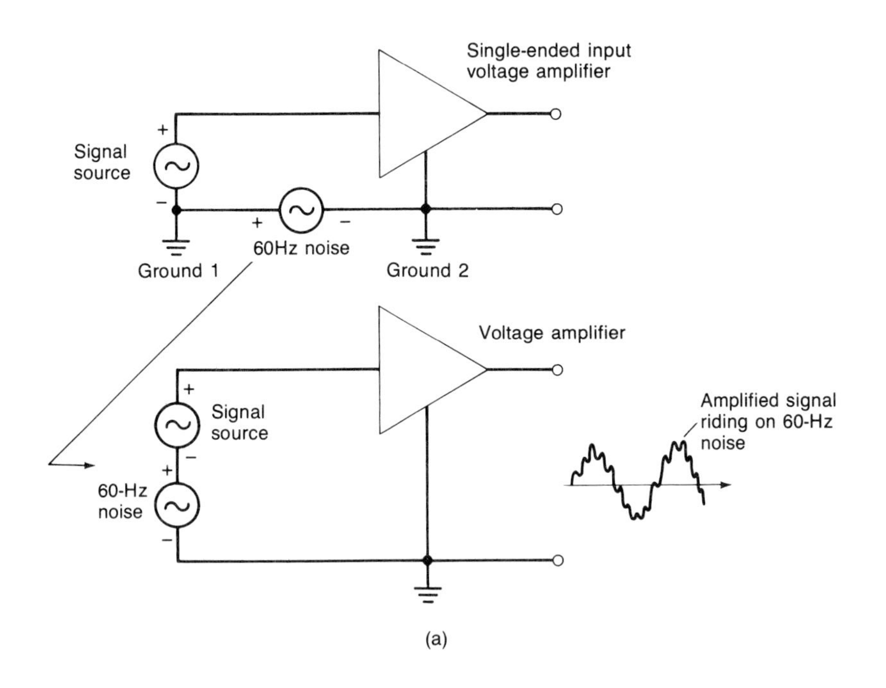

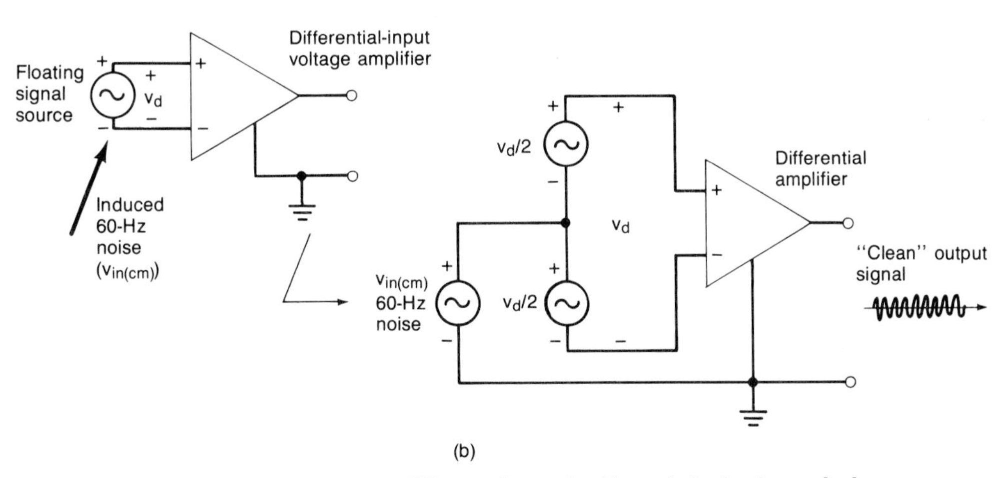

FIGURE 8-31 Differential amplifier noise rejection: (a) single-ended amplifier offers no noise rejection; (b) differential amplifier eliminates 60-Hz noise.

$$A_{v(cm)} = \frac{v_{out}}{v_{in(cm)}} \tag{8-62}$$

For the idealized differential amplifiers shown in Fig. 8-30, the common-mode voltage gain $A_{v(cm)}$ is *zero*.

Any imbalance between the left- and right-hand sides of the differential amplifier given in Fig. 8-30 will result in a nonzero $A_{v(\mathrm{cm})}$. The degree to which $A_{v(\mathrm{cm})}$ approaches zero depends on how well the resistors and transistors are matched, and how well they track one another with temperature changes. Practically, we work to ensure that

$$A_{v(\mathrm{cm})} << 1$$

From a common-mode rejection standpoint, the differential amplifier is indeed a very powerful tool. The differential input *and* the differential output make extremely small values of $A_{v(\mathrm{cm})}$ possible. However, most applications require that the differential amplifier provide an output signal that is single-ended (ground-referenced) [see Fig. 8-32(a)].

FIGURE 8-32 Single-ended output differential: (a) circuit; (b) input signals; (c) representing the differential and common-mode input signals.

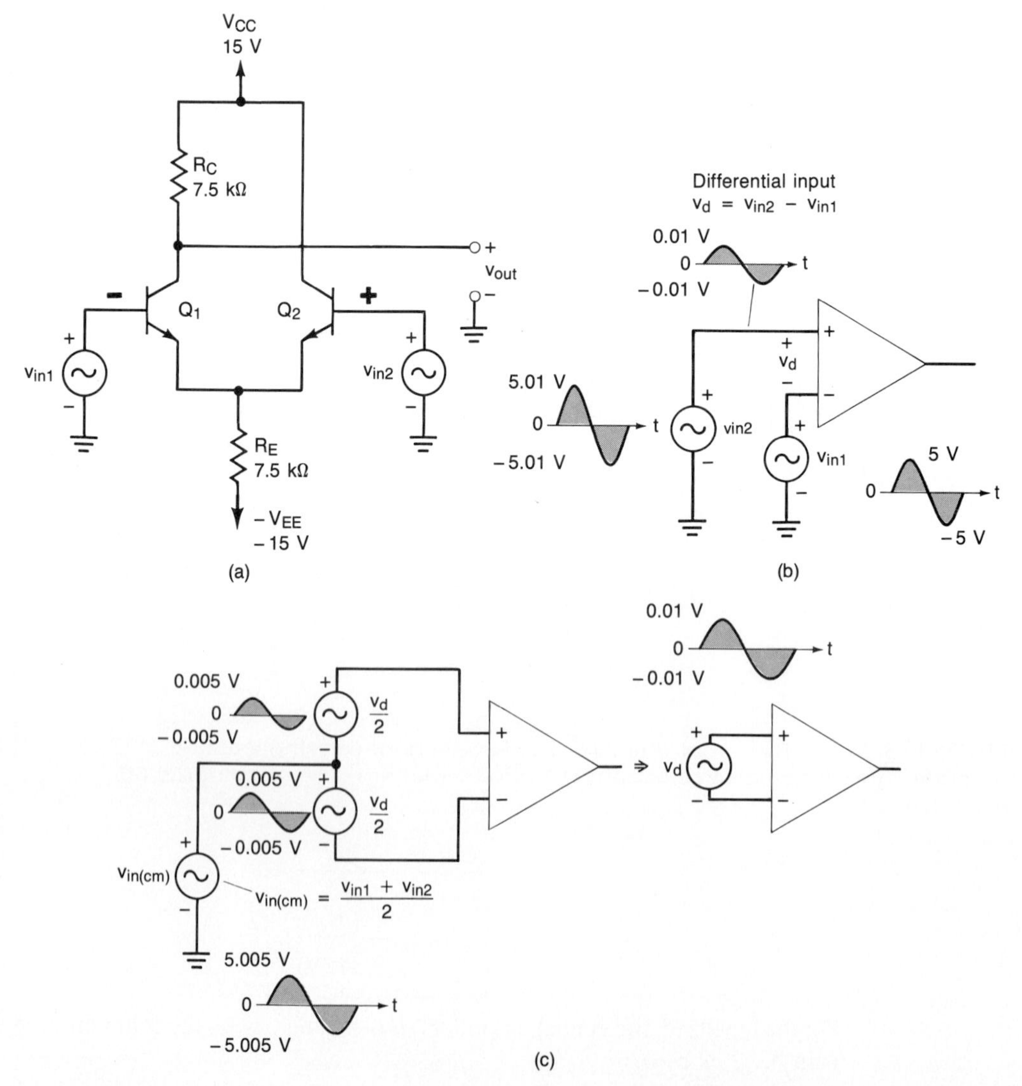

A single-ended output is required to ease the interface requirements between the differential amplifier and the rest of the electronic system. Unfortunately, a single-ended output reduces $A_{vd(oc)}$ by one-half. This should be evident from our previous work. Hence

$$A_{vd(oc)} = \frac{v_{out}}{v_{in2} - v_{in1}} = \frac{g_m R_C}{2} \tag{8-63}$$

Further, a single-ended output also serves to degrade the common-mode rejection. However, by understanding the differential and common-mode signal process, we shall see how to remedy this reduction in common-mode rejection.

In Fig. 8-32(b) we have represented a differential amplifier with signals at its inverting and noninverting inputs, and the resulting difference, or differential, input signal between its two input terminals. If we temporarily ignore common-mode effects, we may represent the differential input signal as indicated in Fig. 8-32(c).

In Fig. 8-33(a) we have represented a portion of the differential amplifier and the ac differential signal path. It is easy to see that R_E is not involved in the signal flow. This serves to reinforce the reason R_E does not appear in our equation for $A_{vd(oc)}$. Be sure to note that the ac differential current flow has been indicated, *not* the total instantaneous current flow.

Figure 8-33(b) illustrates the effects of the common-mode signal. In this case we can see that the ac common-mode signal does flow through R_E, and

$$v_e = 2i_e R_E$$

Since our output signal is taken at the collector of Q_1, we are not really interested in Q_2. Therefore, we may simplify our work by *splitting the circuit into two parts* [refer to Fig. 8-33(c)]. Note that splitting the circuit requires that we double the value of the R_E's in order to keep v_e the same. Hence

$$v_e = i_e(2R_E) = 2i_e R_E$$

To find the common-mode voltage gain $A_{v(cm)}$, we note that the analysis is virtually identical to our work for the common-emitter amplifier with an unbypassed emitter resistance [refer to Eq. 8-18 and see Fig. 8-33(d)].

$$A_{v(cm)} = \frac{v_{out}}{v_{in(cm)}} = -\frac{g_m R_C}{1 + 2g_m R_E} \tag{8-64}$$

Since $2g_m R_E >> 1$,

$$A_{v(cm)} = \frac{v_{out}}{v_{in(cm)}} \simeq -\frac{R_C}{2R_E} \tag{8-65}$$

One final parameter is quite often used to describe how well a differential amplifier (such as our discrete amplifier or an integrated op amp) works is the *common-mode rejection ratio* (CMRR).

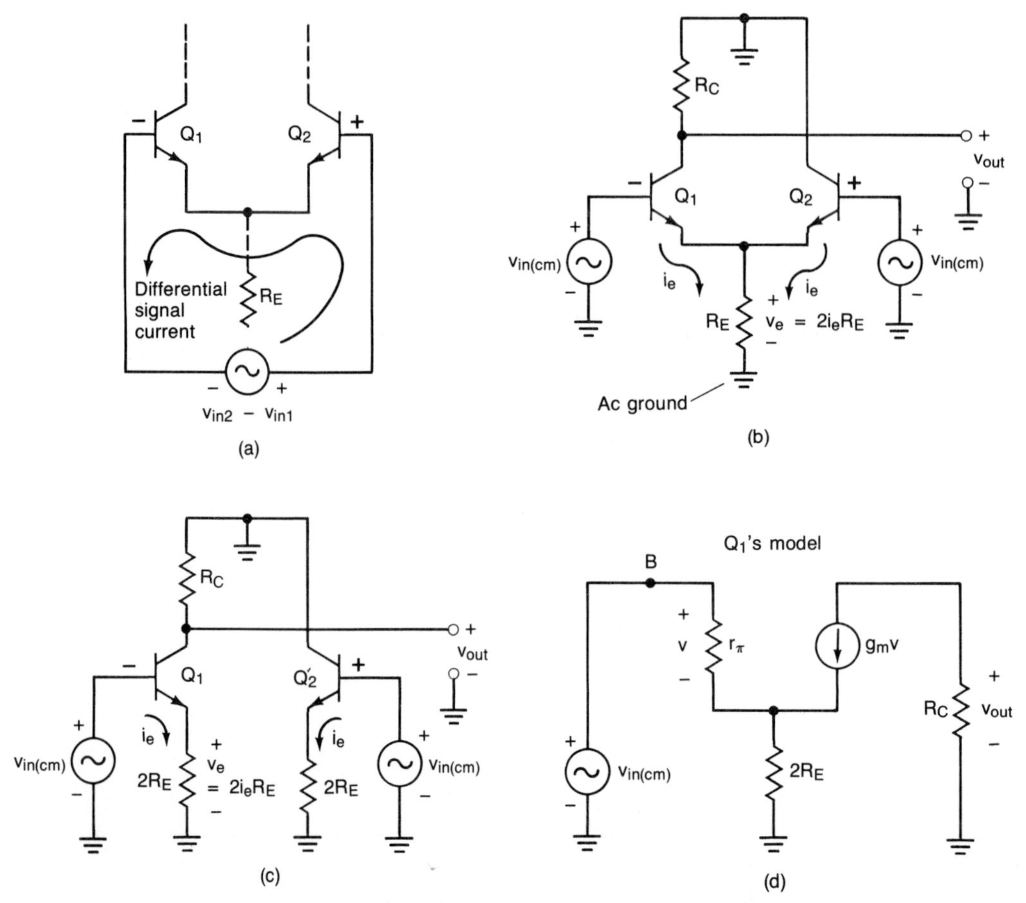

FIGURE 8-33 Finding $A_{v(cm)}$: (a) differential signal; (b) common-mode signal flow; (c) splitting the circuit is made possible by doubling R_E; (d) simplified ac equivalent circuit to find $A_{v(cm)}$.

$$\text{CMRR} = \left| \frac{A_{vd}}{A_{v(\text{cm})}} \right| \tag{8-66}$$

Quite often it is expressed in decibels.

$$|\text{CMRR}|_{\text{dB}} = 20 \log \text{CMRR} \tag{8-67}$$

Generally, we want the CMRR to be very large.

EXAMPLE 8-22

Find R_{in}, R_{out}, $A_{vd(\text{oc})}$, $A_{v(\text{cm})}$, and the CMRR both as a straight ratio and in decibels for Fig. 8-32(a).

SOLUTION The collector currents are unaffected by the removal of Q_2's collector resistor and are the same as they were in Example 8-21 ($I_{C1} = I_{C2} = 0.955$ mA). Therefore, the BJT parameters are also the same ($g_m = 36.7$ mS and $r_\pi = 1.36$ kΩ). Now we may characterize the amplifier.

$$R_{in} = 2r_\pi = 2(1.36\text{ k}\Omega) = 2.72\text{ k}\Omega$$

$$R_{out} = R_C = 7.5\text{ k}\Omega$$

$$A_{vd(oc)} = \frac{v_{out}}{v_{in2} - v_{in1}} = \frac{g_m R_C}{2} = \frac{(36.7\text{ mS})(7.5\text{ k}\Omega)}{2} = 138$$

$$A_{v(cm)} = \frac{v_{out}}{v_{in(cm)}} = -\frac{R_C}{2R_E} = -\frac{7.5\text{ k}\Omega}{(2)(7.5\text{ k}\Omega)} = -0.5$$

and we can see that the differential amplifier will attenuate common-mode signals by one-half.

$$\text{CMRR} = \left|\frac{A_{vd(oc)}}{A_{v(cm)}}\right| = \left|\frac{138}{-0.5}\right| = 275$$

$$|\text{CMRR}|_{dB} = 20 \log \text{CMRR} = 20 \log 275 = 48.8\text{ dB}$$ ■

In Chapter 6 we saw how a constant-current source could be used to bias a differential amplifier. Recall that this technique results in a very stable dc operating point. In addition, Example 8-23 will serve to illustrate that the CMRR is tremendously improved by this biasing scheme. Consider Fig. 8-34(a).

The three transistors are available as part of an RCA CA3046 transistor array. The BJTs are all constructed on the same piece of silicon, which is housed in a 14-pin dual-in-line (DIP) package [see Fig. 8-34(b)]. The corresponding pin numbers have been indicated in Fig. 8-34(a).

EXAMPLE 8-23

Find the CMRR and the magnitude of the CMRR in decibels of the differential amplifier given in Fig. 8-34(a).

SOLUTION The tail current is formed by the collector current of Q_3. The 5.6-V zener diode is used to establish Q_3's emitter current and place its emitter at −7.1 V with respect to ground. Using approximations yields

$$I_T = I_{C3} \simeq \frac{V_Z - 0.7\text{ V}}{R_E} = \frac{5.6\text{ V} - 0.7\text{ V}}{4.7\text{ k}\Omega} = 1.04\text{ mA}$$

The collector currents of Q_1 and Q_2 are both equal to one-half of I_T.

$$I_{C1} = I_{C2} = \frac{I_T}{2} = \frac{1.04\text{ mA}}{2} = 0.520\text{ mA}$$

Now we may determine Q_1's collector-to-ground voltage.

$$V_{C1} = V_{CC} - I_{C1}R_C = 12\text{ V} - (0.52\text{ mA})(12\text{ k}\Omega) = 5.76\text{ V}$$

The reader is encouraged to verify the balance of the dc values indicated in Fig. 8-34. For Q_3,

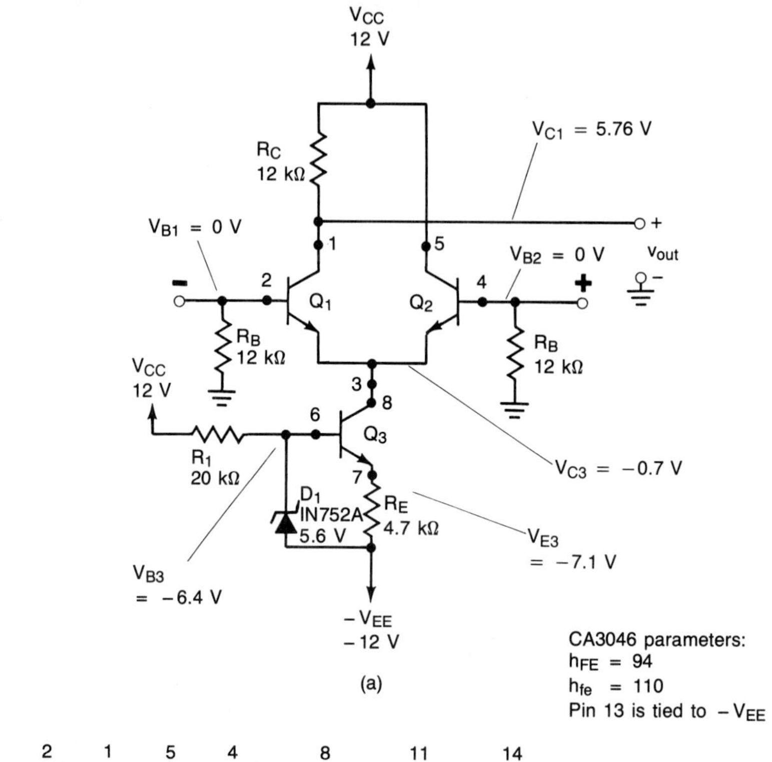

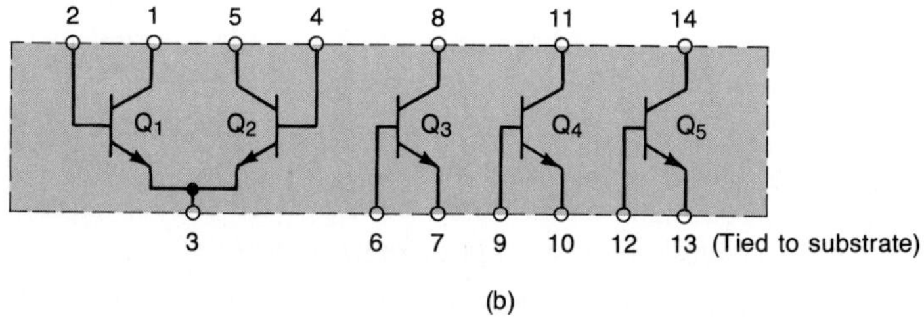

FIGURE 8-34 **Circuit and data for Example 8-23: (a) differential amplifier with constant-current source bias; (b) RCA CA3046 transistor array.**

$$g_{m3} = \frac{I_{C3}}{26 \text{ mV}} = \frac{1.04 \text{ mA}}{26 \text{ mV}} = 40 \text{ mS}$$

$$r_{\pi 3} = \frac{h_{fe}}{g_{m3}} = \frac{110}{40 \text{ mS}} = 2.75 \text{ k}\Omega$$

$$r_{o3} = \frac{200 \text{ V}}{I_{C3}} = \frac{200 \text{ V}}{1.04 \text{ mA}} = 192 \text{ k}\Omega$$

and for Q_1 and Q_2,

$$g_{m1} = g_{m2} = \frac{I_{C1}}{26 \text{ mV}} = \frac{0.520 \text{ mA}}{26 \text{ mV}} = 20 \text{ mS}$$

$$r_{\pi 1} = r_{\pi 2} = \frac{h_{fe}}{g_{m1}} = \frac{110}{20 \text{ mS}} = 5.5 \text{ k}\Omega$$

Now we may determine $A_{vd(\text{oc})}$.

$$A_{vd(oc)} = \frac{g_{m1}R_C}{2} = \frac{(20 \text{ mS})(12 \text{ k}\Omega)}{2}$$
$$= 120$$

Before we can determine $A_{v(cm)}$ we must first find the equivalent ac emitter resistance r_E. This happens to be equal to the resistance "looking into" the collector of Q_3. Refer back to our discussion of $R_{out(col)}$ in Section 8-5. By drawing on Eq. 8-26 and noting that the R_{TH} of Q_3's base circuit is zero,

$$r_E = R_{out(col)3} \simeq r_{o3} + \frac{g_{m3}r_{o3}r_{\pi 3}R_E}{R_E + r_{\pi 3}}$$
$$= 192 \text{ k}\Omega + \frac{(40 \text{ mS})(192 \text{ k}\Omega)(2.75 \text{ k}\Omega)(4.7 \text{ k}\Omega)}{4.7 \text{ k}\Omega + 2.75 \text{ k}\Omega}$$
$$= 13.5 \text{ M}\Omega$$

Now by applying Eq. 8-65 we may find $A_{v(cm)}$.

$$A_{v(cm)} = \frac{v_{out}}{v_{in(cm)}} = -\frac{R_C}{2r_E} = -\frac{12 \text{ k}\Omega}{(2)(13.5 \text{ M}\Omega)}$$
$$= -4.44 \times 10^{-4}$$

From Eqs. 8-66 and 8-67,

$$\text{CMRR} = \left|\frac{A_{vd(oc)}}{A_{v(cm)}}\right| = \left|\frac{120}{-4.44 \times 10^{-4}}\right| = 2.70 \times 10^5$$
$$|\text{CMRR}|_{dB} = 20 \log \text{CMRR} = 20 \log 2.70 \times 10^5 = 109 \text{ dB}$$

Obviously, the use of a constant-current source has greatly improved the CMRR. This is one of the reasons this technique is favored in differential amplifiers. This is particularly true in the case of integrated-circuit op amp designs.

8-13 Op Amp Differential Amplifier

In Fig. 8-35(a) we see an op amp differential amplifier circuit. Recall that the gain on the inverting input terminal is $-R_2/R_1$ (refer to Eq. 8-30). However, the gain on the noninverting input terminal is (from Eq. 8-44) $1 + R_2/R_1$. Therefore, *to provide equal gains on both of the input terminals, it is necessary to attenuate the noninverting input signal via a resistive voltage divider*.

To analyze the op amp differential amplifier we may use the principle of superposition. First, we shall set v_{in2} to zero [see Fig. 8-35(b)]. The parallel equivalent resistance between the noninverting input and ground does *not* affect the voltage gain. Subsequently, from Eq. 8-30,

$$v'_{out} = -\frac{R_2}{R_1} v_{in1} \tag{8-68}$$

If we now set v_{in1} to zero, we arrive at Fig. 8-35(c). Since the noninverting input terminal of the op amp does not draw appreciable current, the resistors associated with that input are effectively in series. Therefore, by voltage division,

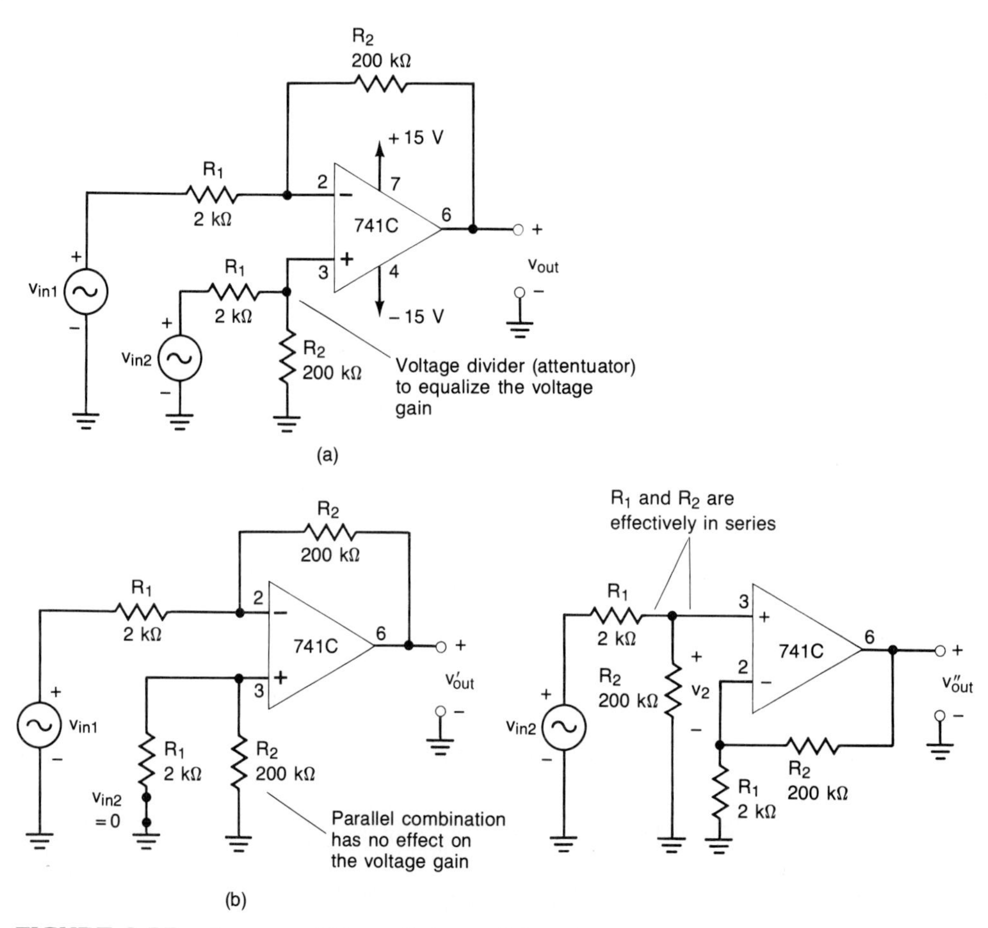

FIGURE 8-35 Op amp differential amplifier: (a) circuit; (b) setting v_{in2} to 0; (c) setting v_{in1} to 0.

$$v_2 = \frac{R_2}{R_1 + R_2} v_{in2} \tag{8-69}$$

Since v_2 is the voltage that appears at the noninverting input terminal,

$$v''_{out} = \left(1 + \frac{R_2}{R_1}\right) v_2 \tag{8-70}$$

and by substituting Eq. 8-69 into Eq. 8-70, we obtain

$$\begin{aligned} v''_{out} &= \left(1 + \frac{R_2}{R_1}\right) \frac{R_2}{R_1 + R_2} v_{in2} \\ &= \frac{R_1 + R_2}{R_1} \frac{R_2}{R_1 + R_2} v_{in2} \\ &= \frac{R_2}{R_1} v_{in2} \end{aligned} \tag{8-71}$$

Summation of Eqs. 8-68 and 8-71 produces the equation for v_{out}:

$$v_{out} = v'_{out} + v''_{out} = -\frac{R_2}{R_1} v_{in1} + \frac{R_2}{R_1} v_{in2}$$

$$= \frac{R_2}{R_1} (v_{in2} - v_{in1})$$

The differential voltage gain is

$$A_{vd(oc)} = \frac{v_{out}}{v_{in2} - v_{in1}} = \frac{R_2}{R_1} \tag{8-72}$$

The primary disadvantage of the op amp differential amplifier circuit is that the input resistance of the inverting input $R_{in(-)}$ and the input resistance of the noninverting

FIGURE 8-36 **Finding the input resistances: (a) "looking" into the inverting input; (b) "looking" into the noninverting input.**

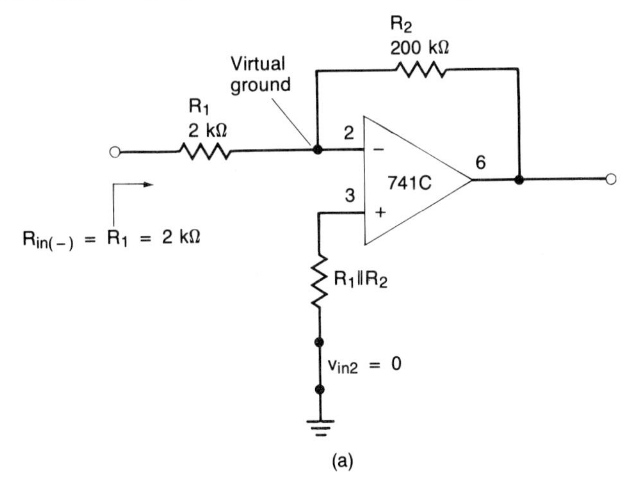

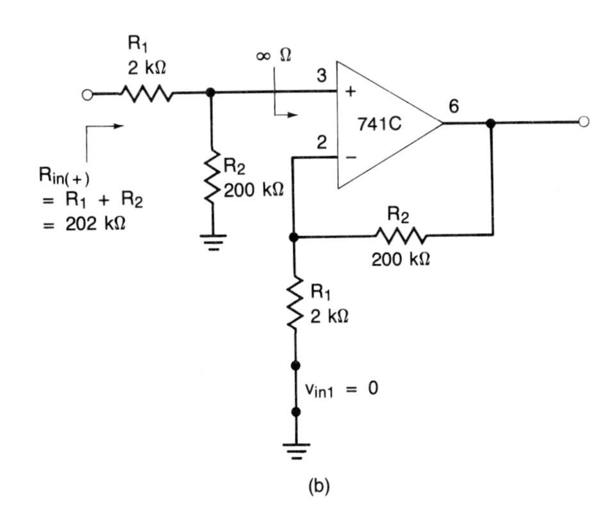

input $R_{in(+)}$ are *not* equal. Consequently, their loading effects upon their respective signal sources are *not* equal. This can degrade the CMRR.

From Fig. 8-36(a) we can see that

$$R_{in(-)} = R_1 \tag{8-73}$$

and from Fig. 8-36(b),

$$R_{in(+)} = R_1 + R_2 \tag{8-74}$$

The output resistance will again be assumed to be approximately equal to zero.

EXAMPLE 8-24

Find $A_{vd(oc)}$, $R_{in(-)}$, $R_{in(+)}$, and R_{out} for the op amp differential amplifier circuit given in Fig. 8-35(a).

SOLUTION From Eqs. 8-72, 8-73, and 8-74,

$$A_{vd(oc)} = \frac{v_{out}}{v_{in2} - v_{in1}} = \frac{R_2}{R_1} = \frac{200\ \text{k}\Omega}{2\ \text{k}\Omega} = 100$$

$$R_{in(-)} = R_1 = 2\ \text{k}\Omega$$

$$R_{in(+)} = R_1 + R_2 = 2\ \text{k}\Omega + 200\ \text{k}\Omega = 202\ \text{k}\Omega$$

$$R_{out} \simeq 0\ \Omega$$

■

FIGURE 8-37 Op amp voltage followers used to buffer the inputs of an op amp differential amplifier.

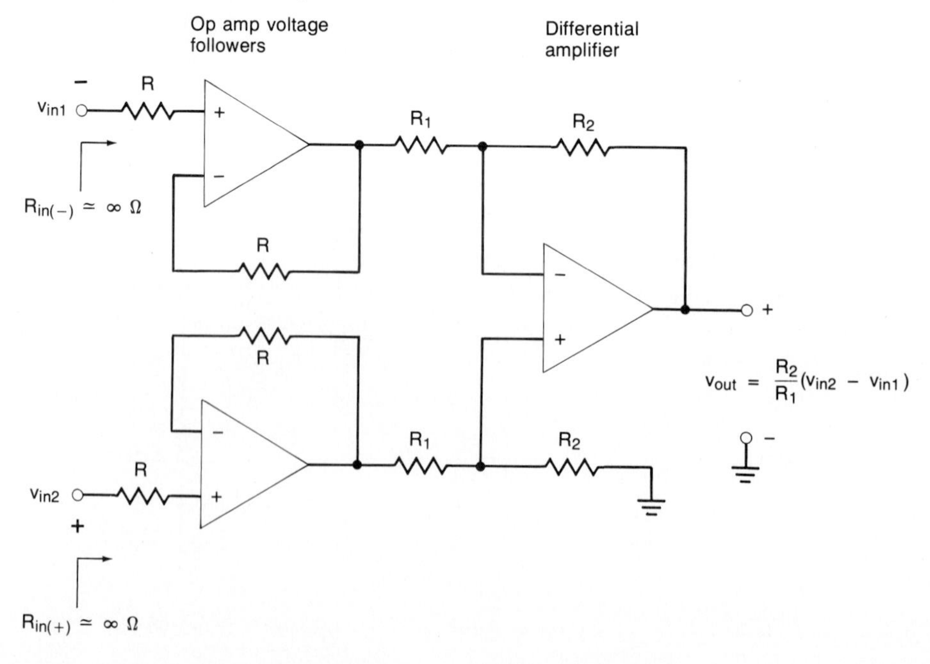

We can see that the mismatch between $R_{in(-)}$ and $R_{in(+)}$ can be quite significant. We can minimize the mismatch by employing op amp voltage followers to drive the inverting and the noninverting inputs (see Fig. 8-37).

As we have seen, the differential amplifier may be used to amplify the difference between its two inputs, v_{in1} and v_{in2}. However, there is an extremely important constraint on v_{in1} and v_{in2}. Specifically, their total instantaneous values must *always* be between the positive and the negative power supply rails. This is described as the op amp's *common-mode input voltage range. If the inputs are not within the common-mode voltage input range, the output of some op amps will tend to latch up. Further, if the inputs are too large, an op amp's input may be destroyed!* Consider Fig. 8-38.

FIGURE 8-38 The differential inputs must be within the common-mode input voltage range: (a) inputs within the ±15-V common-mode input range; (b) inputs outside of the common-mode input voltage range.

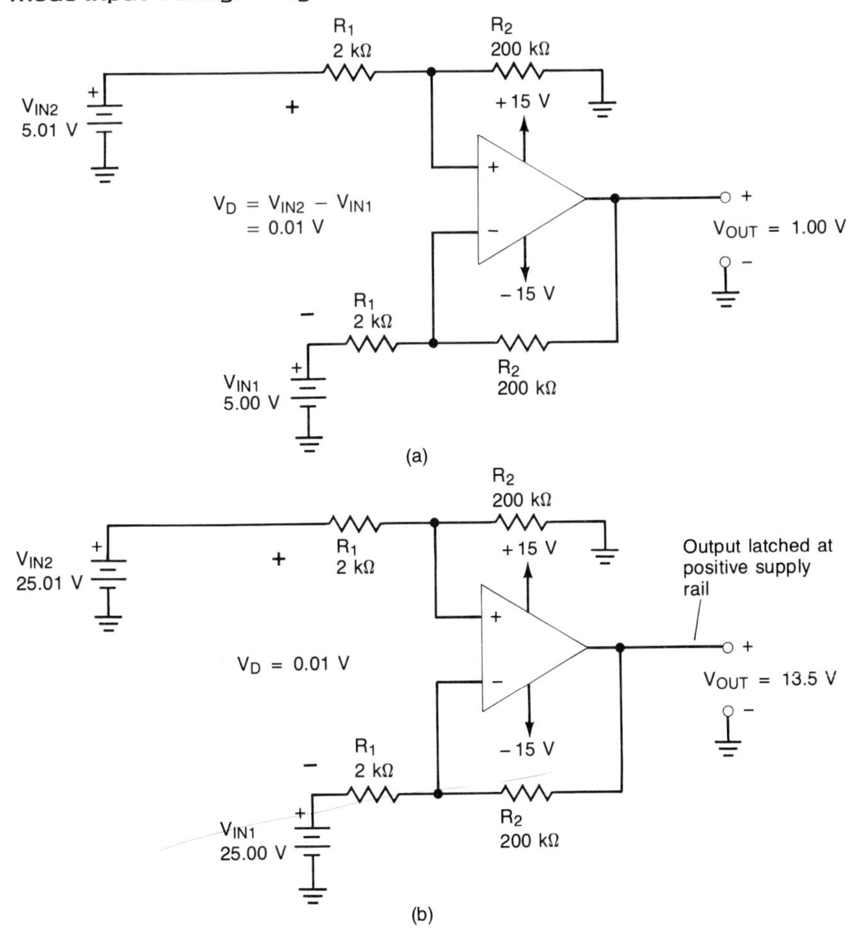

PROBLEMS

Drill, Derivations, and Definitions

Section 8-1

8-1. From memory, list the four basic amplifier configurations.

8-2. The small-signal ac parameters are *independent* of the Q-point. True False

The reader is advised to perform the analysis requested in Probs. 8-3 and 8-4. Many of the following problems require the results of these problems and those of Probs. 8-7 and 8-9. To assist you in checking your work, key answers will be provided for dependent problems.

Section 8-2

8-3. Given the BJT bias circuit in Fig. 8-39(a), perform a dc analysis. Specifically, find I_B, I_C, V_{CE}, V_C, V_E, V_B, $I_{C(SAT)}$, and $V_{CE(OFF)}$. Present your answers in tabular form. (ANS: $I_{C(min)} = 0.938$ mA; $I_{C(max)} = 0.966$ mA.)

FIGURE 8-39 Bias circuits: (a) npn BJT; (b) n-channel JFET.

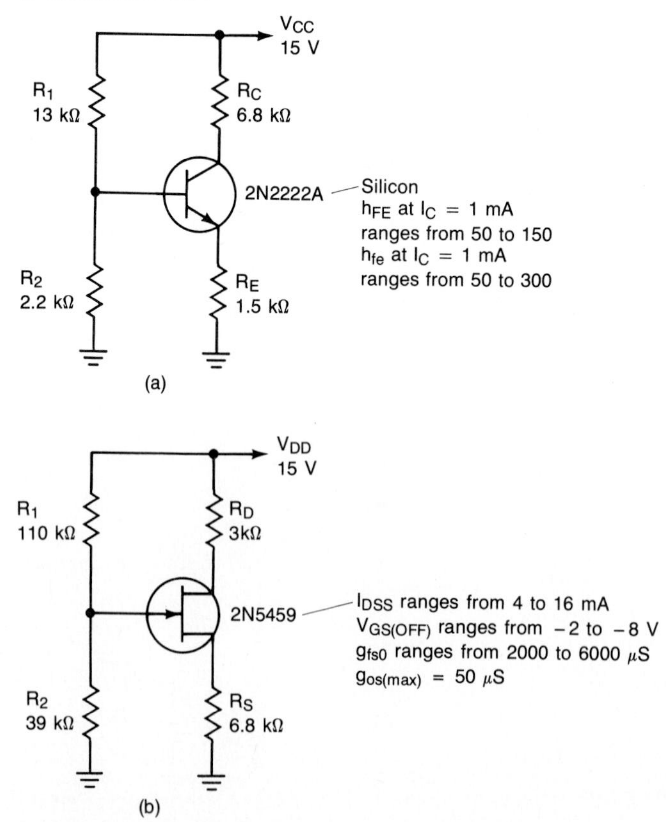

8-4. Determine I_G, I_D, V_{DS}, V_D, V_S, V_G, $I_{D(SAT)}$, and $V_{DS(OFF)}$ for the FET bias circuit given in Fig. 8-39(b). Present your answers in tabular form. (ANS: $I_{D(min)} = 0.745$ mA; $I_{D(max)} = 1.41$ mA.)

8-5. Explain why the dc bias point of an active device such as a BJT or an FET is unaffected by the particular signal configuration if *RC* coupling is used.

8-6. What factors determine the particular amplifier configuration of an active device? Give an example.

8-7. Find the average value of I_C for the 2N2222A BJT bias circuit analyzed in Prob. 8-3 [refer to Fig. 8-39(a)]. Use the average value to determine r_o, g_m, and r_π. Use $h_{fe(min)}$. (ANS: $r_\pi = 1.366$ kΩ.)

8-8. Repeat Prob. 8-7 using $h_{fe(max)}$.

8-9. Determine the worst-case (minimum) value of r_o for the 2N5459 JFET given in Fig. 8-39(b). Also find $g_{m(max)}$, $g_{m(min)}$, and r_π. Use the I_D values determined in Prob. 8-4. (ANS: $g_{m(min)} = 868$ μS.)

Section 8-4

8-10. The bias circuit shown in Fig. 8-39(a) is to be used as a common-emitter amplifier [see Fig. 8-3(a)] with a Z_S of 600 Ω and an R_L of 10 kΩ. Find its R_{in}, R_{out}, and $A_{v(oc)}$. Continue the analysis by determining A_v, A_{vs}, A_i, A_{is}, and A_p. Assume that r_o is large enough to be ignored.

8-11. Repeat Prob. 8-10 for the 2N5459 JFET bias circuit shown in Fig. 8-39(b) if it is to be used as a common-source amplifier [see Fig. 8-3(b)]. The Z_S is 600 Ω and the R_L is 10 kΩ. Use the maximum g_m.

8-12. Repeat Prob. 8-10 for the 2N5459 JFET common-source amplifier described in Prob. 8-11 if the JFET has the minimum g_m.

8-13. Compare the typical values of A_v, A_i, R_{in}, and R_{out} for the common-emitter and the common-source amplifiers.

Section 8-5

8-14. In your own words, explain the term "negative feedback."

8-15. If the emitter bypass capacitor of a common-emitter amplifier is removed, what happens to A_v, R_{in}, and R_{out}? Explain.

8-16. What is meant by the term "bootstrapping"? Explain.

8-17. If the emitter bypass capacitor of the amplifier described in Prob. 8-10 is removed, find $A_{v(oc)}$, $R_{in(base)}$, R_{in}, and R_{out}. Complete the analysis by finding A_v, A_{vs}, A_i, A_{is}, and A_p.

8-18. Why is it generally undesirable to have a common-source FET voltage amplifier without a source bypass capacitor?

8-19. To improve voltage gain stability without turning a JFET amplifier into a voltage attenuator, it is possible to *partially* bypass its source resistor (see Fig. 8-40). Find $A_{v(oc)}$ with R_{S1} and R_{S2} completely bypassed (no source resistance), partially bypassed (no R_{S2} in the ac equivalent circuit), and unbypassed (both R_{S1} and R_{S2} appear in the ac equivalent circuit). Assume a g_m of 0.868 mS. (Hint: Use Eq. 8-18 as a *basis* for your solutions.)

8-20. Repeat Prob. 8-19, but assume a g_m of 1.781 mS.

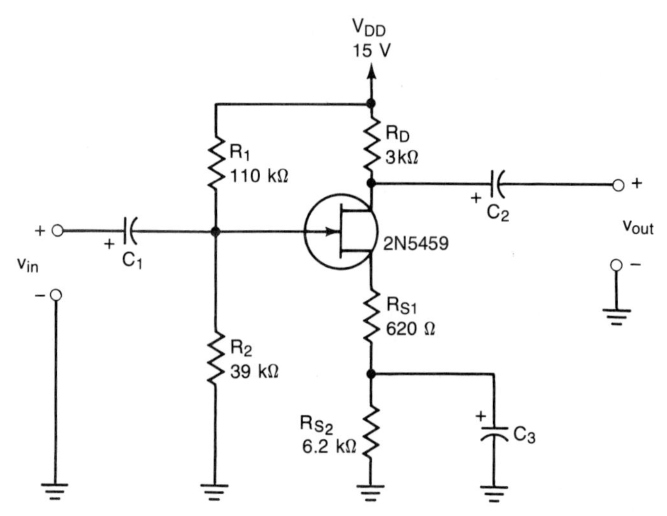

FIGURE 8-40 JFET common-source amplifier with partial source bypassing.

Section 8-6

8-21. List the three fundamental op amp idealizations.

8-22. Explain what is meant by the term ''virtual ground'' as it relates to op amps.

8-23. Given the op amp inverting amplifier in Fig. 8-16(a), assume that R_2 is 150 kΩ and R_1 is 20 kΩ. Determine $A_{v(oc)}$, R_{in}, and R_{out}. Complete the analysis by determining A_v, A_{vs}, A_i, A_{is}, and A_p.

8-24. The op amp inverting amplifier in Fig. 8-16(a) has an R_1 of 10 kΩ and an R_2 of 47 kΩ. The ac source is replaced with a dc source that makes V_{IN} equal to -1 V. Find V_{OUT}. Repeat the problem if V_{IN} is 2 V.

8-25. The signal source in Fig. 8-16(a) produces an ac signal that rides on a dc level. Specifically, $v_S = -2 + 0.01 \sin \omega t$. Find v_{OUT}. Is there a problem? What would be a solution?

Section 8-8

8-26. A common-base amplifier [Fig. 8-3(c)] uses the bias circuit given in Fig. 8-39(a). Assume that Z_S is 600 Ω and R_L is 10 kΩ. Find its R_{in}, R_{out}, and $A_{v(oc)}$. Complete the analysis by determining A_v, A_{vs}, A_i, A_{is}, and A_p.

8-27. Repeat Prob. 8-26 for a common-gate amplifier [Fig. 8-3(d)] which uses the bias circuit given in Fig. 8-39(b). Assume that Z_S is 600 Ω, R_L is 10 kΩ, and the JFET has its maximum g_m.

8-28. Repeat Prob. 8-27 if the JFET has its minimum g_m of 0.868 mS.

Section 8-9

8-29. Explain the term ''latch-up'' as it relates to op amps. What is the function of R_3 in Fig. 8-21?

8-30. Given the noninverting amplifier in Fig. 8-21, assume that R_2 is 150 kΩ and that both R_1 and R_3 are 2 kΩ. Determine R_{in}, R_{out}, and $A_{v(\text{oc})}$. Also find A_v, A_{vs}, A_i, A_{is}, and A_p.

8-31. The input coupling capacitor (C_1) in Fig. 8-21 is removed and replaced by a short circuit. The ac signal source is replaced by a dc source that sets the input voltage (V_{IN}) equal to $-0.01\ V$. Find V_{OUT}. Repeat the problem if V_{IN} is $0.1\ V$.

Section 8-10

8-32. What are the primary advantages offered by either the emitter or the source follower? What are their ideal voltage gains in decibels?

8-33. The bias circuit given in Fig. 8-39(a) is to be used as an emitter follower [Fig. 8-3(e)]. R_C is replaced by a short circuit. Assume that Z_S is 600 Ω and R_L is 10 kΩ. Determine R_{in}, $A_{v(\text{oc})}$, and R_{out}. Complete the analysis by finding A_v, A_{vs}, A_i, A_{is}, and A_p.

8-34. Repeat Prob. 8-33 for a source follower [Fig. 8-3(f)], which uses the bias circuit shown in Fig. 8-39(b). R_D is replaced by a short circuit. Assume that Z_S is 600 Ω and R_L is 10 kΩ. Use the minimum g_m.

8-35. Repeat Prob. 8-34 if the JFET has its maximum g_m of 1.781 mS.

Section 8-11

8-36. Given the op amp voltage follower in Fig. 8-26(b), assume that R_2 and R_3 are both 220 kΩ. Find R_{in}, R_{out}, and $A_{v(\text{oc})}$.

Section 8-12

8-37. The discrete differential amplifier illustrated in Fig. 8-27(a) has its resistor values changed to 12 kΩ and its supplies to ±12 V. Perform an approximate dc analysis to ascertain the tail current I_T and the collector currents. Determine g_m and r_π. Assume that the two BJTs are perfectly matched and have an h_{fe} of 100. Calculate A_{vd}, R_{in}, and R_{out}.

8-38. Repeat Prob. 8-37 if the supply voltages are increased to ±16 V. All other component values are unchanged.

8-39. Explain the term "common-mode rejection."

8-40. Define the term "common-mode noise."

8-41. If the BJTs and resistors in a discrete differential amplifier are *not* perfectly matched, will the common-mode voltage gain $A_{v(\text{cm})}$ increase or decrease? Explain. What is the ideal value for $A_{v(\text{cm})}$?

8-42. Given the discrete differential amplifier in Fig. 8-32(a), observe that it has a single-ended output. Find its R_{in}, R_{out}, $A_{vd(\text{oc})}$, $A_{v(\text{cm})}$, and its CMRR. Express the CMRR both as a straight ratio and in decibels. The resistors are both equal to 12 kΩ, the voltage supplies are ±12 V, and the BJTs each have an h_{fe} of 100.

8-43. A constant-current source has been used to bias the differential amplifier given in Fig. 8-34(a). The zener diode has been changed to a 6.8-V unit, R_1 is 1.5 kΩ, and R_E is 5.6 kΩ. Repeat the analysis requested in Prob. 8-42.

Section 8-13

8-44. Given the op amp differential amplifier circuit shown in Fig. 8-35(a), assume that R_1 is 1 kΩ and R_2 is 150 kΩ. Determine the circuit's $A_{vd(oc)}$, R_{out}, $R_{in(+)}$, and $R_{in(-)}$.

8-45. Repeat Prob. 8-44 if R_1 is 10 kΩ and R_2 is 47 kΩ.

8-46. Explain the primary disadvantage of the basic op amp differential amplifier circuit. Suggest a remedy and draw its schematic diagram.

8-47. In your own words, explain the term "common-mode input voltage range." What might happen to an op amp if one or both of its input signals are outside the common-mode input voltage range?

Troubleshooting and Failure Modes

An amplifier can exhibit a number of failure modes. For instance, the gain can be too low, the output signal may be absent, and/or the output signal can be distorted. When this occurs the next step is to check the dc biases.

8-48. The JFET amplifier given in Fig. 8-10(b) is malfunctioning. The voltage gain is only -0.9 instead of its normal range from -5.86 to -13.9. No distortion is evident and the dc biases are normal. Select the most likely cause(s) from the following alternatives: (a) the source bypass capacitor is shorted, (b) the source bypass capacitor is open, (c) the output coupling capacitor is open, and (d) the input coupling capacitor is open. Explain your reasoning.

8-49. The JFET amplifier in Fig. 8-10(b) has no output signal. A dc bias check indicates that the drain voltage is 10 V. A second measurement indicates that V_{DS} is 0 V. Select the most likely cause(s) from the following alternatives: (a) the JFET is shorted between its drain and source terminals, (b) R_2 is shorted, (c) R_2 is open, and (d) the source bypass capacitor is open. Explain your reasoning.

Design

All op amps require an input bias current that flows from the external circuitry connected to its inverting and noninverting input terminals. This can cause an offset voltage error at its output terminal. This problem can be minimized by obeying one simple rule:

To compensate for bias currents, the equivalent resistance to ground "seen" by the noninverting input terminal must be equal to that "seen" by the inverting input terminal.

This has been detailed in Fig. 8-41. Assume that all the designs below must include bias current compensation.

8-50. Design an op amp inverting amplifier that has an R_{in} of 10 kΩ and an A_v of -5.1.

8-51. Design an op amp noninverting amplifier to provide an $A_{v(oc)}$ of 100, have a R_{in} of 1 kΩ, and operate from a ± 15-V power supply. It is to amplify a small (ac) input signal that is riding on a 3-V dc bias level.

8-52. Design a common-emitter voltage amplifier using a 2N2222A [see Fig. 8-39(a)]. *No* emitter bypass capacitor is to be used. The amplifier should meet the following requirements: $A_{v(oc)} = -10$, $R_{out} = 10$ kΩ, $R_{in} \geq 1.5$ kΩ, $I_c \simeq 1$ mA, and operate from a 20-V dc supply. Draw the schematic diagram and include input and output coupling capacitors.

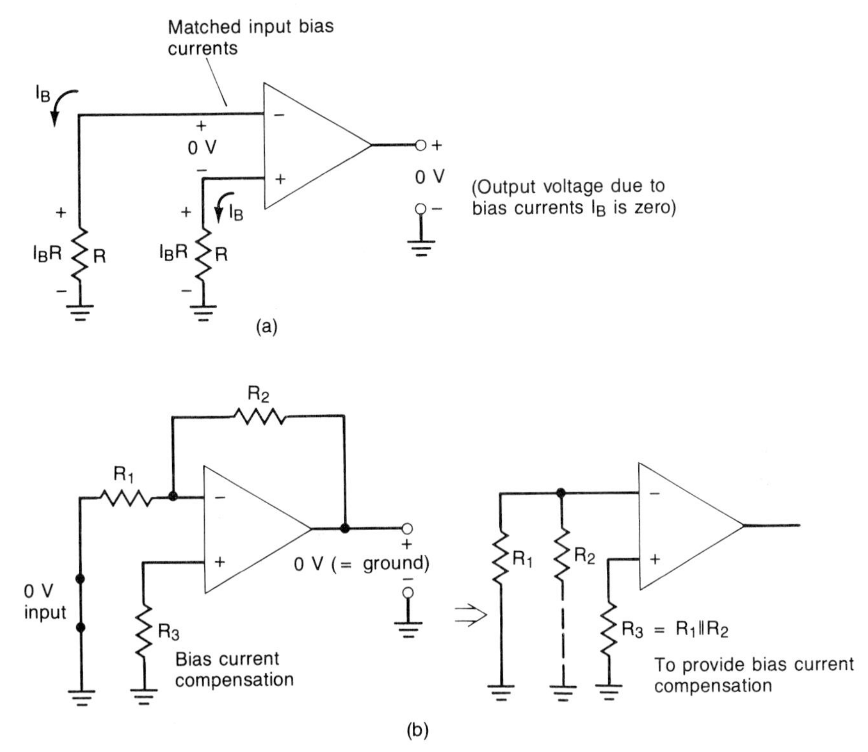

FIGURE 8-41 Bias current compensation: (a) the equivalent resistances to ground must be equal to produce an output voltage of 0 V; (b) bias current compensation for an inverting amplifier.

Computer

8-53. Write a BASIC program that analyzes a common-emitter voltage amplifier. It should request g_m, r_π, R_C, R_E, R_1, R_2, Z_S, and R_L. It should then compute R_{in}, R_{out}, $A_{v(oc)}$, A_v, A_{vs}, A_i, A_{is}, and A_p. The gains should be expressed as straight ratios and in their decibel equivalents.

8-54. Write a BASIC program that performs the dc bias design described in Prob. 6-51 and also incorporates the ac analysis features described in Prob. 8-53. If you have previously written the programs to satisfy Prob. 6-51 and 8-53, all you must do is merge the two solutions into one program.

9

Cascaded Amplifiers and Frequency Response

After Studying Chapter 9, You Should Be Able to:

- Analyze a discrete cascaded amplifier using approximations to find A_v, A_{vs}, A_i, A_{is}, and A_p.
- Analyze an integrated circuit cascaded amplifier to find A_v, A_{vs}, A_i, A_{is}, and A_p.
- Describe the op amp transconductance model.
- Determine the corner frequencies and develop a graph of the frequency response of the simple RC low-pass circuit.
- Determine the corner frequencies and develop a graph of the frequency response of the simple RC high-pass filter circuit.
- Use the Bode approximations to determine frequency response.
- Develop the low-frequency response of a discrete, RC-coupled, single-stage amplifier.
- Analyze a single-supply inverting and noninverting op amp amplifier circuit and determine its frequency response.

- Determine the BJT and FET device capacitances using data sheet information.
- Employ Miller's theorem.
- Determine the high-frequency roll-off in the discrete BJT and FET amplifiers.
- Analyze and explain the high-frequency roll-off associated with the frequency-compensated op amp.
- Apply the techniques studied in this chapter to the cascade amplifier.

9-1 Discrete Cascaded Amplifier System

In Fig. 9-1(a) we have a cascaded voltage amplifier system. A common-source FET is used as the input stage to provide a high input resistance, and therefore minimize the loading on the signal source. The second stage is a common-emitter amplifier. Its function is to provide most of the system gain. An emitter follower has been selected as the output stage to give the amplifier system a low output resistance. This serves to minimize output loading effects.

FIGURE 9-1 Cascaded amplifier system: (a) circuit; (b) ac equivalent circuit.

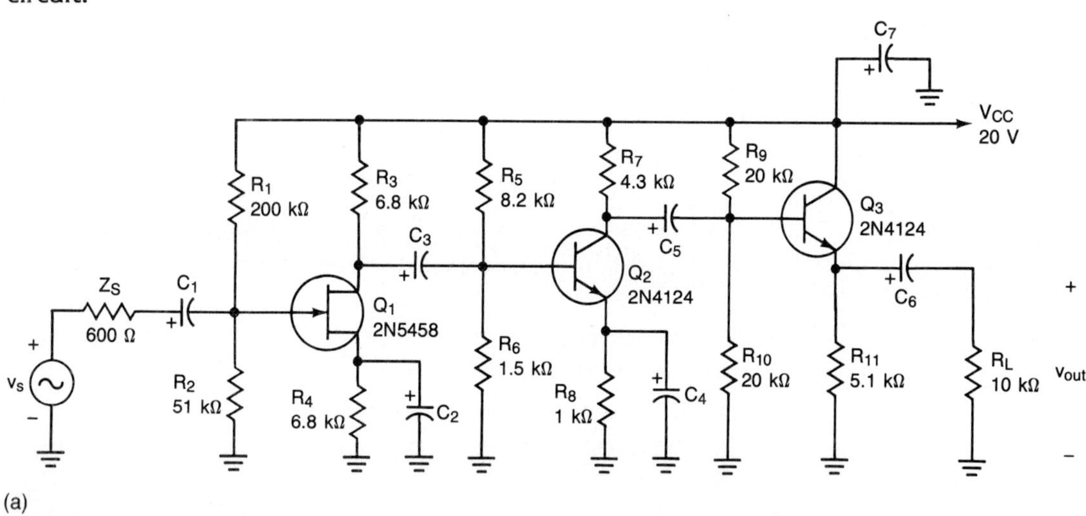

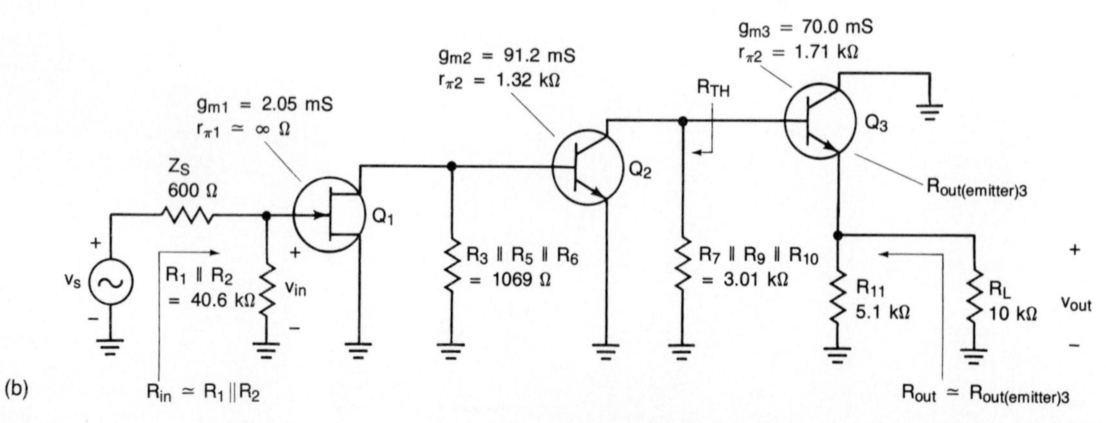

The first step in our investigation is to perform a dc analysis. Next, the ac parameters for each of the active devices must be found from their respective data sheets. The ac equivalent circuit must then be drawn [see Fig. 9-1(b)]. This allows us easily to ascertain the equivalent load and input resistances of each of the individual stages. If we then apply the relationships developed in Chapter 8, we may find the R_{in}, $A_{v(\mathrm{oc})}$, and R_{out} for the amplifier system. The final step in the analysis requires that we apply the equations derived in Chapter 7.

Before we begin our analysis let us make some general observations. First, the input and output coupling capacitors may be shared between stages. Specifically, in Fig. 9-1(a) we note that C_3 and C_5 are both input *and* output coupling capacitors. Therefore, they are often simply referred to as coupling capacitors or dc blocking capacitors. Capacitor C_2 is Q_1's source bypass capacitor and C_4 is Q_2's emitter bypass capacitor.

Capacitor C_7 is a power supply decoupling capacitor. Its function is to bypass the ac signal around the dc power supply. This is extremely important in multiple-stage amplifier systems. If one of the amplifier stages draws significant current, it is quite possible that power supply variations will result. If this occurs, the bias of all the other amplifier stages will vary. The bias variation will follow the signal being processed by the stage producing the supply fluctuations. This can result in distortion and/or *oscillations*. (When oscillations are produced, the amplifier system actually generates a signal.) This highly undesirable situation can be avoided by proper power supply bypassing.

We also see in Fig. 9-1(a) that the familiar R_E, R_S, R_C, and so on, notation has not been used. This is the normal case in "real world" schematics. The component reference designators will typically have the form illustrated in the figure.

Now that our fundamental observations have been made, we shall proceed to the dc analysis. A quick inspection of the component values reveals that we have already performed the dc analysis for the first two stages in Chapter 6. The results are summarized in Tables 6-7 and 6-8. The dc bias analysis of the third stage can quickly be completed by drawing on approximations. Hence

$$V_{B3} \simeq \frac{R_{10}}{R_9 + R_{10}} V_{CC} = \frac{20\ \mathrm{k}\Omega}{20\ \mathrm{k}\Omega + 20\ \mathrm{k}\Omega}(20\ \mathrm{V}) = 10\ \mathrm{V}$$

$$V_{E3} = V_{B3} - 0.7\ \mathrm{V} = 10\ \mathrm{V} - 0.7\ \mathrm{V} = 9.3\ \mathrm{V}$$

$$I_{C3} \simeq I_{E3} = \frac{V_{E3}}{R_{11}} = \frac{9.3\ \mathrm{V}}{5.1\ \mathrm{k}\Omega} = 1.82\ \mathrm{mA}$$

From our previous work, we have seen that the maximum g_m for the 2N5458 JFET (Q_1) is 2.05 mS, and its r_π is approximately infinite. These have been noted in Fig. 9-1(b). We have also determined previously that the g_m for the 2N4124 BJT (Q_2) is 91.2 mS, and its r_π is 1.32 kΩ. These values have also been indicated in Fig. 9-1(b).

At a collector current of 1.82 mA, the h_{fe} of the 2N4124 is approximately 120. The rest of the parameters for Q_3 may be determined in the usual manner.

$$g_{m3} = \frac{I_{C3}}{26\ \mathrm{mV}} = \frac{1.82\ \mathrm{mA}}{26\ \mathrm{mV}} = 70.0\ \mathrm{mS}$$

$$r_{\pi 3} = \frac{h_{fe}}{g_{m3}} = \frac{120}{70.0 \text{ mS}} = 1.71 \text{ k}\Omega$$

These results have also been included in Fig. 9-1(b).

The input resistance of our amplifier system is equal to the input resistance of the common-source amplifier.

$$R_{\text{in}} = R_1 \parallel R_2 \parallel r_{\pi 1} \simeq R_1 \parallel R_2 = 200 \text{ k}\Omega \parallel 51 \text{ k}\Omega = 40.6 \text{ k}\Omega$$

The output resistance is equal to the output resistance of the emitter follower [see Fig. 9-1(b)]. The first step in its determination is to find the Thévenin equivalent resistance ''seen'' by the base terminal of Q_3. Hence

$$R_{\text{TH}} \simeq R_7 \parallel R_9 \parallel R_{10} = 4.3 \text{ k}\Omega \parallel 20 \text{ k}\Omega \parallel 20 \text{ k}\Omega = 3.01 \text{ k}\Omega$$

$$R_{\text{out}} = R_{\text{out(emitter)3}} \parallel R_{11} \simeq R_{\text{out(emitter)3}}$$

$$= \frac{r_{\pi 3} + R_{\text{TH}}}{1 + g_{m3} r_{\pi 3}} = \frac{1.71 \text{ k}\Omega + 3.01 \text{ k}\Omega}{1 + (70.0 \text{ mS})(1.71 \text{ k}\Omega)} = 39.1 \ \Omega$$

The next step in our analysis is to find the open-circuit voltage gain $A_{v(\text{oc})}$ for the amplifier system. To accomplish this, it is necessary first to find the individual *loaded* voltage gains for the first two amplifier stages. Our efforts can be considerably reduced if we digress slightly in order to make some extremely useful observations.

Let us first consider the common-source voltage amplifier. We recall that

$$A_{v(\text{oc})} = -g_m R_D \tag{9-1}$$

and

$$R_{\text{out}} \simeq R_D \tag{9-2}$$

We also know that the loaded voltage gain is given by

$$A_v = \frac{R_L}{R_L + R_{\text{out}}} A_{v(\text{oc})} \tag{9-3}$$

If we substitute Eqs. 9-1 and 9-2 into Eq. 9-3, we obtain

$$A_v = \frac{R_L}{R_L + R_{\text{out}}} A_{v(\text{oc})} = \frac{R_L}{R_L + R_D}(-g_m R_D)$$

$$= -g_m \frac{R_L R_D}{R_L + R_D} = -g_m (R_L \parallel R_D)$$

If we let r_D represent the ac equivalent resistance connected between the drain and ground (e.g., the parallel combination of R_D and R_L) we arrive at

$$\boxed{A_v = -g_m r_D} \tag{9-4}$$

where r_D is the ac equivalent drain resistance of the common-source amplifier.

In a similar fashion, we can develop Eq. 9-5 for the common-emitter amplifier.

$$\boxed{A_v = -g_m r_C} \tag{9-5}$$

where r_C is the ac equivalent collector resistance of the common-emitter amplifier. Consider Fig. 9-2(a). The equivalent resistances r_D and r_C have been identified. Specifically,

$$
\begin{aligned}
r_D &= R_3 \parallel R_5 \parallel R_6 \parallel r_{\pi 2} \\
&= 6.8\ \text{k}\Omega \parallel 8.2\ \text{k}\Omega \parallel 1.5\ \text{k}\Omega \parallel 1.32\ \text{k}\Omega = 591\ \Omega \\
r_C &= R_7 \parallel R_9 \parallel R_{10} = 4.3\ \text{k}\Omega \parallel 20\ \text{k}\Omega \parallel 20\ \text{k}\Omega = 3.01\ \text{k}\Omega
\end{aligned}
$$

and from Eqs. 9-4 and 9-5,

$$
\begin{aligned}
A_{v1} &= -g_{m1}r_D = -(2.05\ \text{mS})(591\ \Omega) = -1.21 \\
A_{v2} &= -g_{m2}r_C = -(91.2\ \text{mS})(3.01\ \text{k}\Omega) = -275
\end{aligned}
$$

The open-circuit voltage gain of the third stage may be found by using Eq. 8-53.

$$
A_{v3(\text{oc})} = \frac{g_{m3}R_{11}}{1 + g_{m3}R_{11}} = \frac{(70.0\ \text{mS})(5.1\ \text{k}\Omega)}{1 + (70.0\ \text{mS})(5.1\ \text{k}\Omega)} = 0.997
$$

FIGURE 9-2 Continuing the analysis of the cascaded amplifier: (a) finding r_D and r_C; (b) amplifier model.

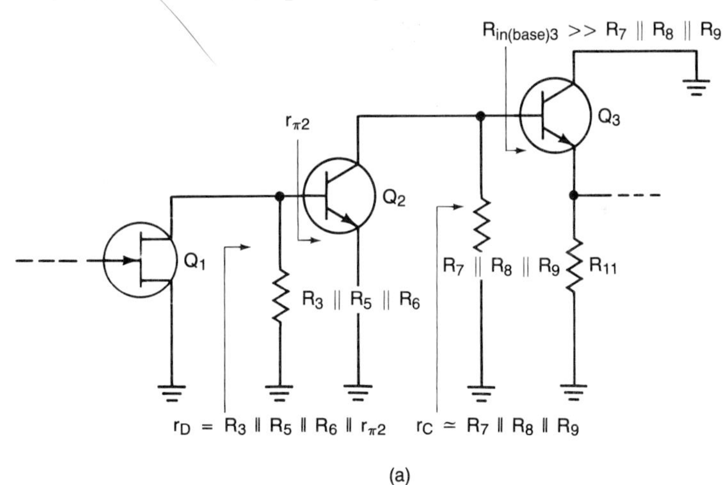

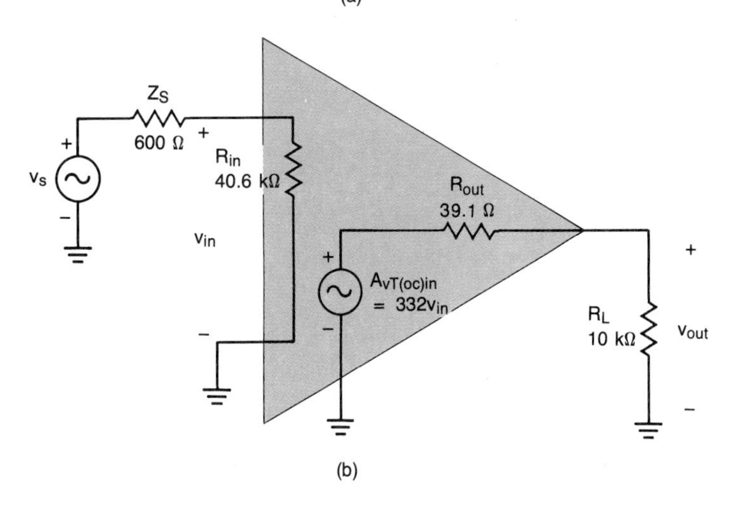

The open-circuit voltage gain of the cascaded amplifier may be found by applying Eq. 7-38.

$$A_{vT(oc)} = A_{v1}A_{v2}A_{v3(oc)} = (-1.21)(-275)(0.997) = 332$$

At this point we have reduced the cascaded amplifier system to the form of our general amplifier model [see Fig. 9-2(b)].

The balance of the analysis may be conducted in the usual manner.

$$A_{vT} = \frac{v_{out}}{v_{in}} = \frac{R_L}{R_L + R_{out}} A_{vT(oc)} = \frac{10\ \text{k}\Omega}{10\ \text{k}\Omega + 39.1\ \Omega}(332) = 331$$

$$A_{vsT} = \frac{v_{out}}{v_s} = \frac{R_{in}}{R_{in} + Z_S} A_{vT} = \frac{40.6\ \text{k}\Omega}{40.6\ \text{k}\Omega + 600\ \Omega}(331) = 326$$

$$A_{iT} = \frac{i_{out}}{i_{in}} = \frac{R_{in}}{R_L + R_{out}} A_{vT(oc)} = \frac{40.6\ \text{k}\Omega}{10\ \text{k}\Omega + 39\ \Omega}(332) = 1343$$

$$A_{isT} = \frac{i_{out}}{i_s} = \frac{Z_S}{Z_S + R_{in}} A_{iT} = \frac{600\ \Omega}{600\ \Omega + 40.6\ \text{k}\Omega}(1343) = 19.6$$

$$A_{pT} = A_{vT}A_{iT} = (331)(1343) = 4.45 \times 10^5$$

If we compare the results above with our work in Chapter 8, we immediately notice that the cascaded amplifier offers superior performance compared to any of the single-stage discrete amplifiers. In the next section we investigate a cascaded amplifier system that uses IC (integrated circuit) op amps.

9-2 Op Amp Cascaded Amplifier System

In Fig. 9-3(a) the pin connections for the 747 dual op amp integrated circuit have been illustrated. The 14-pin package contains two separate op amps. The op amps share common positive (V^+) and negative (V^-) power supply connections. *The offset null terminals are used for balancing the outputs of the two op amps. Specifically, by using an external potentiometer connected to the offset null terminals, it is possible to adjust the dc output of the op amps to precisely zero volts when zero volts appears at their respective input terminals. This is very important in most dc amplifier applications.* However, it does not immediately concern us. Therefore, these terminals are left unconnected in Fig. 9-3(b).

Observe that the IC pin connections have been indicated in Fig. 9-3(b). Also note that the op amps have been decoupled from the power supplies via capacitors C_2 and C_3. Let us proceed with the analysis.

As we have seen, the input resistance of the noninverting op amp amplifier is approximately equal to dc return resistor R_1 in Fig. 9-3(b). Hence

$$R_{in} = R_1 = 43\ \text{k}\Omega$$

The output resistance of the amplifier system is equal to the output resistance of the second stage, and

$$R_{out} \simeq 0\ \Omega$$

Now let us find the overall voltage gain. As we have seen, since the output resistance of the op amp amplifier circuits is approximately zero, the open-circuit

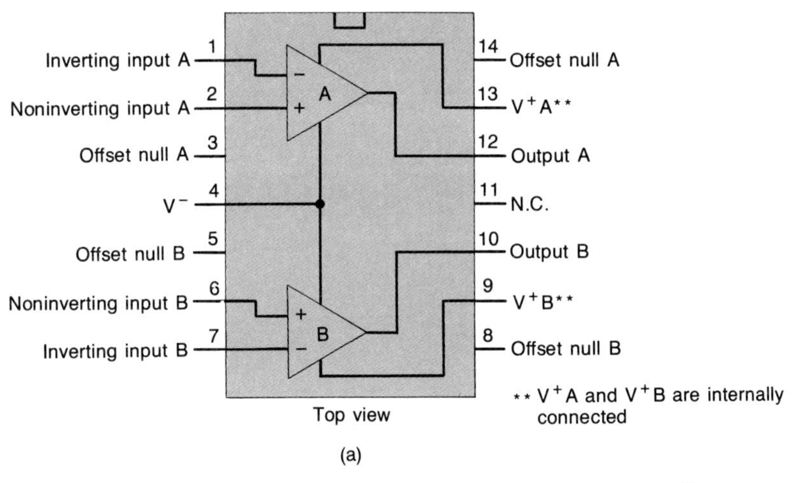

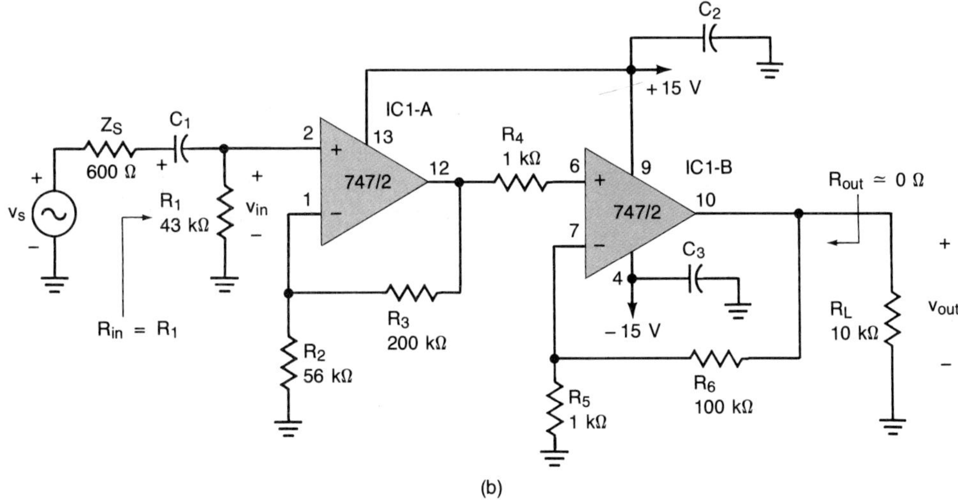

FIGURE 9-3 **Cascaded amplifier using the 747 dual op amp: (a) 14-pin dual inline package (DIP); (b) circuit.**

voltage gains $A_{v(\text{oc})}$ are essentially equal to the loaded voltage gains A_v. From Eq. 8-44 we recall that

$$A_{v1} = A_{v1(\text{oc})} = 1 + \frac{R_3}{R_2} = 1 + \frac{200\text{ k}\Omega}{56\text{ k}\Omega} = 4.57$$

$$A_{v2} = A_{v2(\text{oc})} = 1 + \frac{R_6}{R_5} = 1 + \frac{100\text{ k}\Omega}{1\text{ k}\Omega} = 101$$

and the total voltage gain A_{vT} is

$$A_{vT} = A_{vT(\text{oc})} = A_{v1}A_{v2(\text{oc})} = (4.57)(101) = 462$$

At this point, we may readily complete the analysis. Hence

$$A_{vsT} = \frac{v_{\text{out}}}{v_s} = \frac{R_{\text{in}}}{R_{\text{in}} + Z_S} A_{vT} = \frac{43\text{ k}\Omega}{43\text{ k}\Omega + 600\ \Omega}(462)$$
$$= 456$$

$$A_{iT} = \frac{i_{out}}{i_{in}} = \frac{R_{in}}{R_L + R_{out}} A_{vT(oc)} = \frac{43\ \text{k}\Omega}{10\ \text{k}\Omega + 0\ \Omega}(462)$$
$$= 1987$$
$$A_{isT} = \frac{i_{out}}{i_s} = \frac{Z_S}{Z_S + R_{in}} A_{iT} = \frac{600\ \Omega}{600\ \Omega + 43\ \text{k}\Omega}(1987)$$
$$= 27.3$$
$$A_{pT} = A_{vT}A_{iT} = (462)(1987) = 918 \times 10^3$$

Obviously, the analysis of the op amp cascaded amplifier is considerably less strenuous than the analysis of the discrete cascaded amplifier. Further, its performance is, in general, superior.

9-3 Cascaded Amplifiers Found in Integrated Circuit Op Amps

The IC op amp is a simply marvelous tool. It is a physically small ''block of voltage gain'' which (through the use of negative feedback) allows us to construct amplifiers with outstanding characteristics and with a minimum of external components. At this point it would seem quite appropriate to examine the basic structure of the IC op

FIGURE 9-4 Fundamental block diagram of the IC op amp.

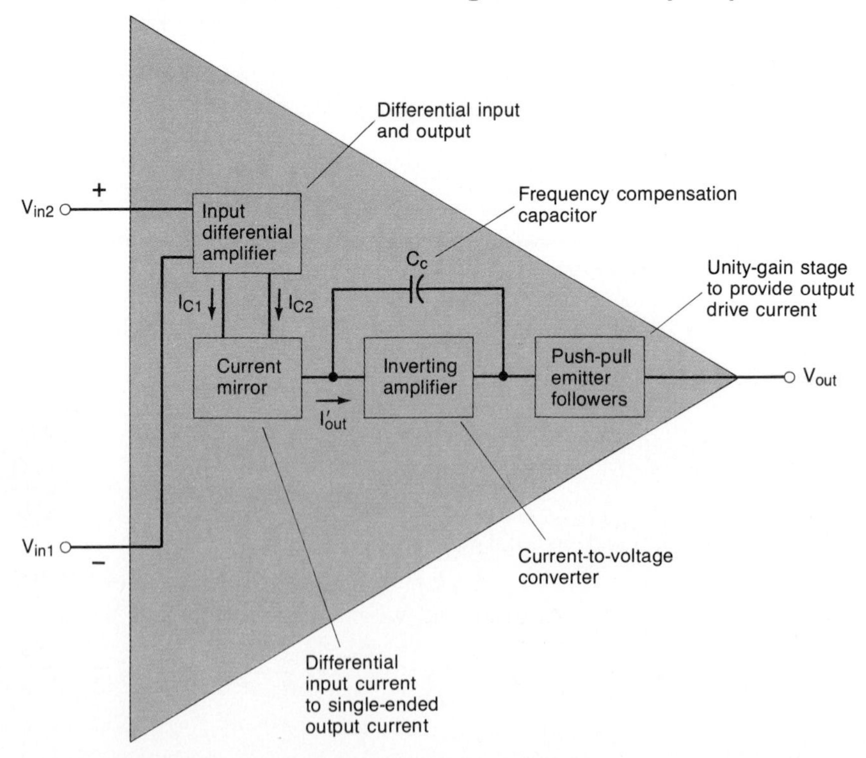

amp. This will serve to dispell some of its mystery and help us to understand more fully its applications and limitations.

In Fig. 9-4 we see the general block diagram of an IC op amp. *The op amp circuitry has been broken down into four basic cascaded stages: an input differential amplifier with a differential output, a current mirror, an inverting current-to-voltage amplifier, and an emitter-follower output stage.* In the text to follow, we investigate the function and operation of each of these stages. The equivalent circuits we shall investigate are extremely rudimentary and are merely representative.

A typical *pnp* differential input stage has been depicted in Fig. 9-5(a). Observe that it is biased by a constant-current source. The actual constant-current source found

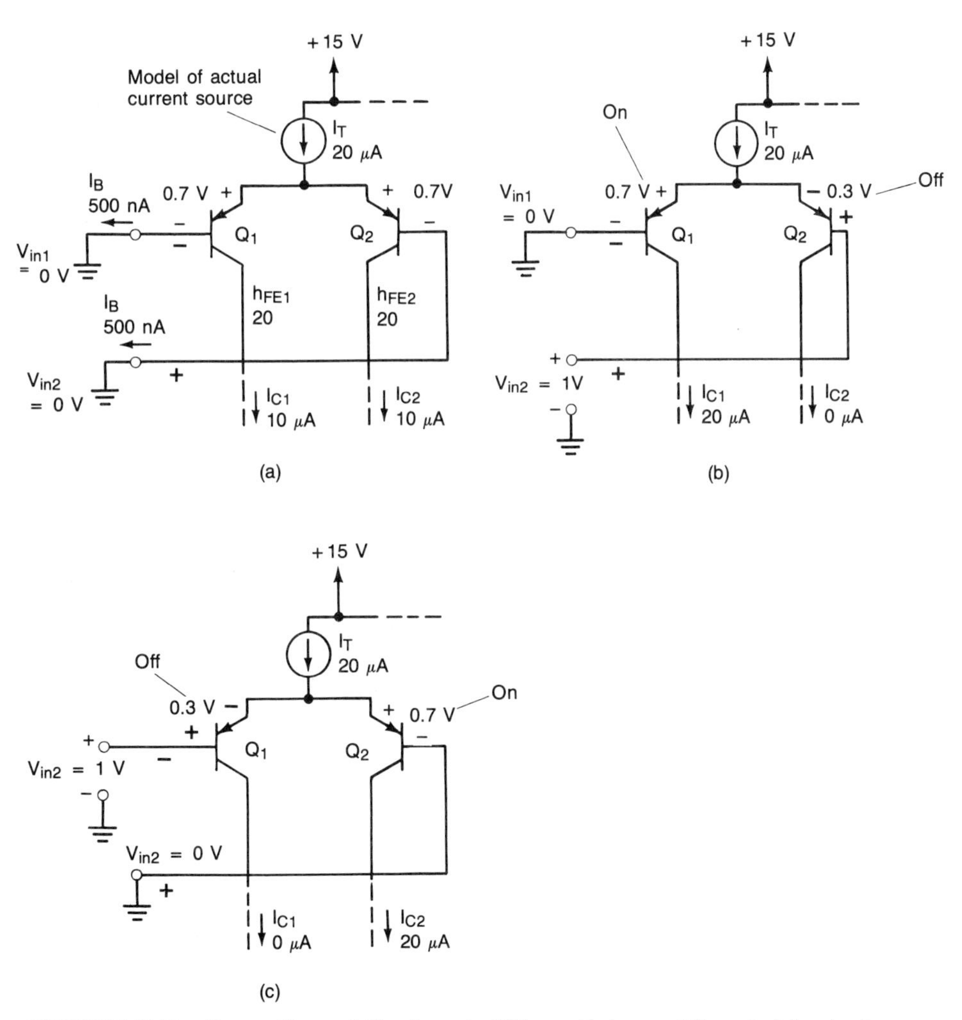

FIGURE 9-5 **Operation of the input differential amplifier: (a) typical bias values for the 741 op amp have been indicated in our pnp version; (b) noninverting input more positive than the inverting input makes $I_{C1} > I_{C2}$; (c) inverting input more positive than the noninverting input makes $I_{C2} > I_{C1}$.**

in most IC op amps is rather complex. Therefore, we shall simplify the discussion by representing it by the ideal current source symbol indicated in Fig. 9-5.

The tail current established by the constant-current source is usually a relatively small value. For example, in the case of the 741 op amp, 20 μA is used. This is done to keep the input bias current I_B small. The 741 op amp has a *maximum* I_B of 500 nA. This suggests that the h_{FE}'s of its input transistors have a minimum value of 20. We shall also use this h_{FE} for our *pnp* input transistors [see Fig. 9-5(a)].

Constant-current source biasing is used for three main reasons. First, it provides an extremely stable Q-point, as was discussed in Chapter 6. Second, transistors require less real estate than resistors on the silicon substrate. The third reason was explained in Section 8-12. Constant-current source biasing improves the common-mode rejection of a differential pair. This is true when the differential amplifier's output is either single-ended or differential.

To understand how the differential amplifier processes both ac and dc inputs, we offer the two worst-case conditions depicted in Fig. 9-5(b) and (c). In Fig. 9-5(b) we see that if $V_{\text{IN}2}$ is 1 V *more* positive than $V_{\text{IN}1}$, I_{C1} will be equal to I_T, and I_{C2} will be zero. Specifically, Q_2 will be in cutoff and Q_1 will be fully conducting. The converse is illustrated in Fig. 9-5(c).

The situations shown in Fig. 9-5(b) and (c) represent a nonlinear mode of operation - saturation. The main point to note is that the tail current is shared by the two transistors and exchanges between their collectors.

In Fig. 9-6 we have added the current mirror. (The current mirror was introduced in Section 6-16.) Its function is to convert the differential output current of the first stage into a single-ended output current. Let us see how this occurs.

The collector current that flows through Q_1 is directed into the collector of diode-connected Q_3. Transistor Q_3 will develop a base-to-emitter voltage V_{BE} in response to its forced collector current. Since the base-emitter terminals of Q_3 and Q_4 are parallel, Q_4 will have the same V_{BE}. Assuming that Q_4's transfer characteristic curve matches Q_3's, it will experience the same collector current. From Fig. 9-6(a) and Kirchhoff's current law,

$$I'_{\text{OUT}} = I_{C2} - I_{C1}$$

For the purposes of illustration, we have again illustrated a nonlinear (saturation) condition in Fig. 9-6(b). In this case we can see that 20 μA flows *into* Q_4's collector. Transistor Q_4 is described as *sinking* current. Similarly, if I_{C1} were zero and I_{C2} were 20 μA, 20 μA would be *leaving* Q_4's collector, and it would be described as *sourcing* current.

The third basic stage of the op amp serves as an inverting current-to-voltage converter. To understand this circuit, the reader is directed to Fig. 9-7(a). Assume that the amplifier is a high-gain inverting amplifier and it has a relatively large input resistance. Therefore, virtually all of the current that is directed into its input terminal will flow through the feedback impedance Z_f connected between its input and output. Because of its large voltage gain, its input terminal will be a virtual ground, and

$$v_{\text{out}} = -iZ_f$$

A single BJT can offer extremely large voltage gains (in excess of 1000) if a constant-current source is placed in series with its collector [see Fig. 9-7(b)].

Using these concepts, we now refer to Fig. 9-7(c). Transistor Q_5 serves as a

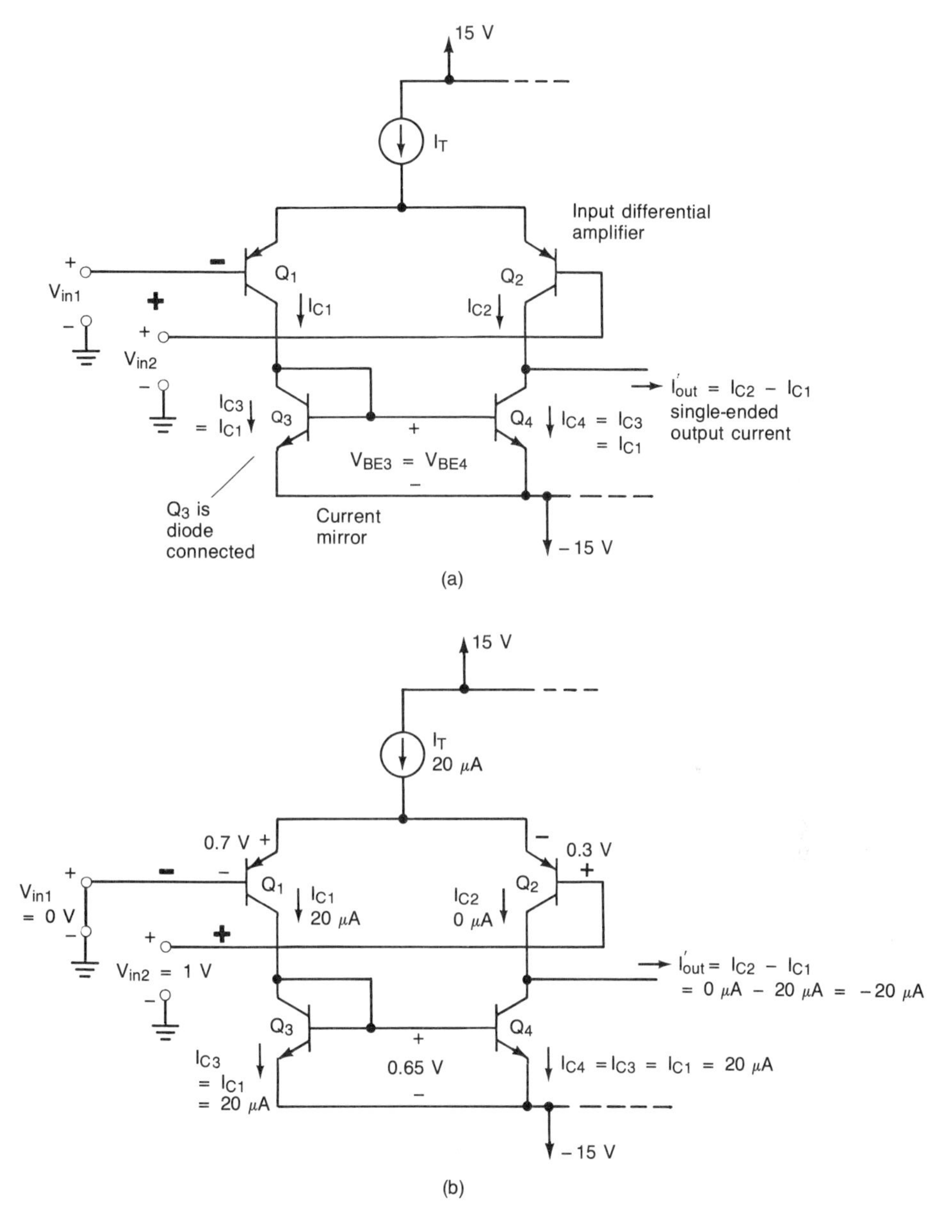

FIGURE 9-6 Basic operation of the current mirror: (a) the output of the differential amplifier is directed into the current mirror; (b) currents when the noninverting input is large enough to produce saturation.

high-gain inverting amplifier since a constant-current source I_{BIAS} is placed in series with its collector. It is employed as a current-to-voltage converter. Its feedback element is a capacitor C_c. The capacitor is used to provide *frequency compensation*. It is integrated on the silicon chip and has a magnitude on the order of picofarads. For example, the compensation capacitor C_c used in the 741 op amp is 30 pF. As we

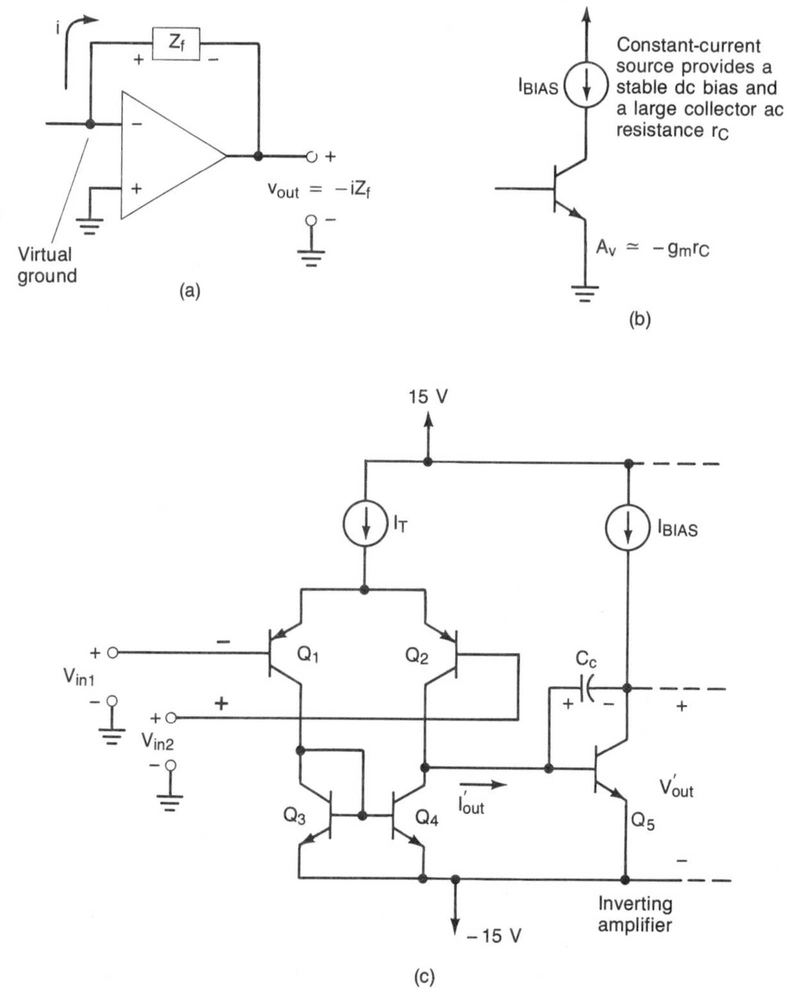

FIGURE 9-7 Operation of the inverting amplifier stage: (a) active (inverting) current-to-voltage converter; (b) using a constant-current source to obtain a large A_v; (c) inverting amplifier stage.

shall see in Section 10-13, the primary function of this capacitor is to prevent the op amp from going into oscillation.

Let us trace through the operation of the circuit given in Fig. 9-7(c). The differential input voltage at the base terminals of Q_1 and Q_2 produces a proportional differential output current. This differential output current is injected into the current mirror (Q_3 and Q_4). The current mirror converts its differential input current into a single-ended output current I'_{OUT}.

The output current I'_{OUT} is directed into the base circuit of Q_5. Since Q_5 behaves as a large-voltage-gain inverting amplifier, it serves to convert I'_{OUT} to a voltage V'_{OUT}. We will have more to say about this later.

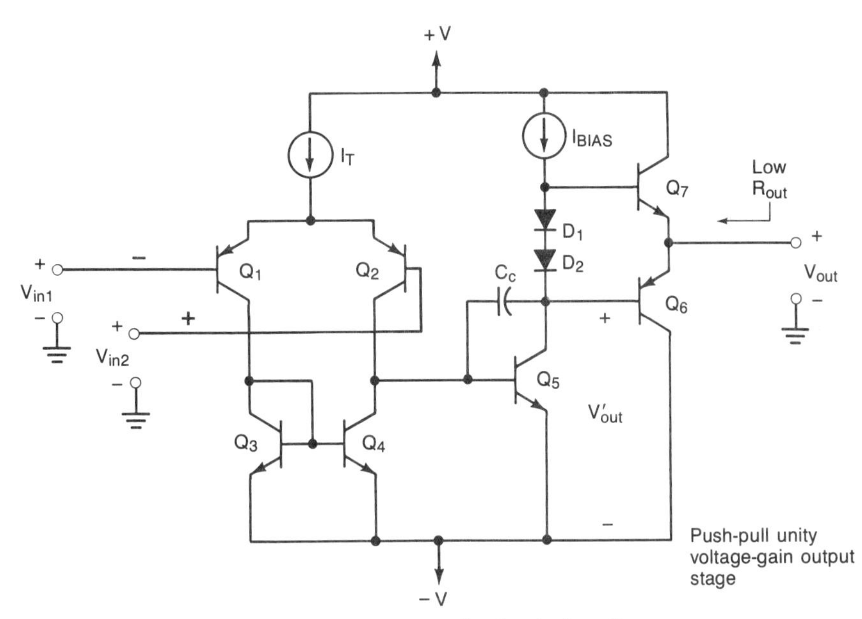

FIGURE 9-8 Complete op amp equivalent circuit.

The equivalent output stage has been included in Fig. 9-8. Its configuration is often referred to as being *push-pull*. This is due to the fact that the two transistors Q_6 and Q_7 alternate in conduction. Specifically, the *npn* BJT Q_7 conducts when V'_{OUT} is positive, and the *pnp* BJT Q_6 conducts when V'_{OUT} is negative. Diodes D_1 and D_2 are forward biased by the current source I_{BIAS}. These diodes are actually diode-connected BJTs. Their function is to provide bias stabilization for the output BJTs Q_6 and Q_7. The specifics associated with push-pull amplifiers are addressed more fully in Chapter 12.

For now we merely observe that the input signal to the output stage V'_{OUT} is tied to the base circuits of the BJTs and the output signals are taken from their emitters. From Table 4-1 we recall that this allows us to classify the output transistors as being common-collector amplifiers (emitter followers). Recall that emitter followers exhibit a voltage gain of approximately unity. Further, they also provide a low output resistance (see Fig. 9-8).

The primary function of the output stage is to give the op amp a low output resistance and additional drive capability. By ''drive'' we mean the current supplied to the load. For example, the 741 op amp can supply a peak load current of approximately ± 10 mA.

9-4 Op Amp Transconductance Model

With our greater understanding of the IC op amp's internal circuitry, it is now possible to develop a small-signal *transconductance model*. This will provide us with an even greater insight into the (ac) operation of the op amp in our later work. Consider Fig.

9-9(a). The two fundamental op amp functional blocks have been illustrated. To remind us that the op amp's output is buffered, we have illustrated its (unity-gain) output stage in phantom.

The input differential pair consists of *pnp* BJTs. Consequently, as their respective base terminals become more *positive*, their collector currents will *decrease*. The magnitude of their collector currents can be related to their input voltages via transconductance. Hence

$$i_{c1} = -g_m v_{\text{in}1} \quad \text{and} \quad i_{c2} = -g_m v_{\text{in}2} \tag{9-6}$$

The minus signs denote the phase inversions, and we have assumed that the transconductances of the two transistors (Q_1 and Q_2) are both equal to g_m.

As we saw in Fig. 9-8, the input differential amplifier drives a current mirror

FIGURE 9-9 Op amp transconductance model: (a) IC op amp equivalent circuit; (b) basic op amp ac model; (c) low-frequency BJT/FET small-signal model.

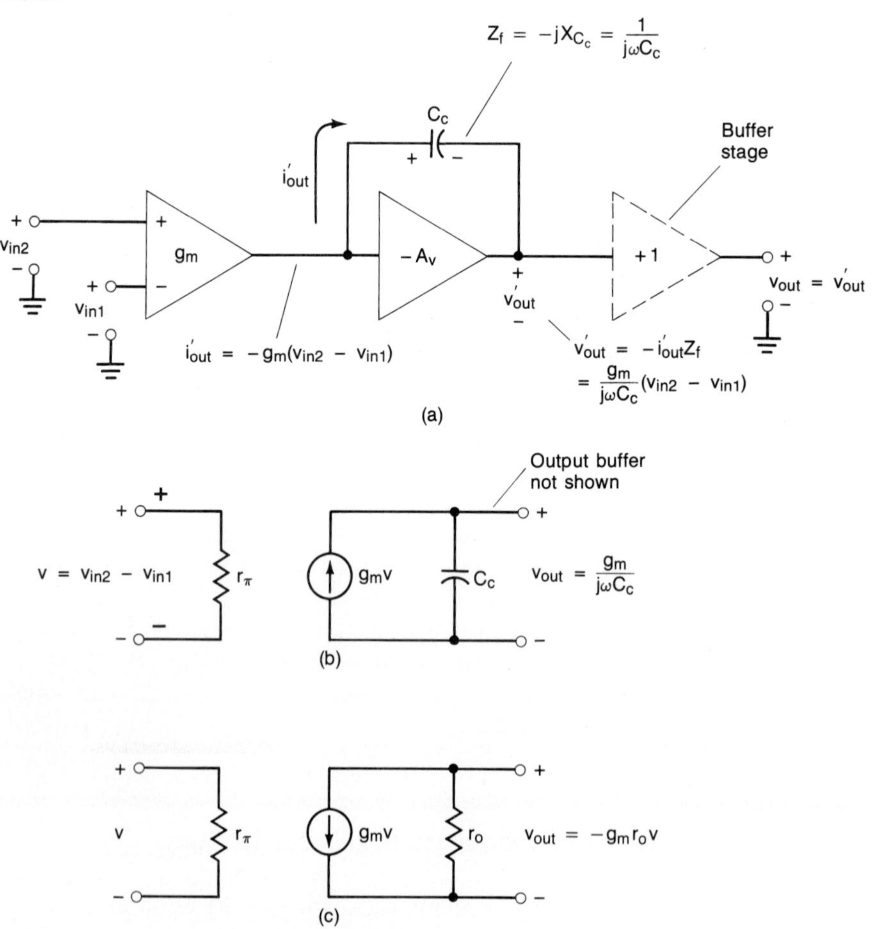

which, in turn, produces a proportional single-ended output current. Therefore, we may describe the output current i'_{out} in terms of transconductance. Thus

$$\begin{aligned} i'_{\text{out}} &= i_{c2} - i_{c1} = -g_m v_{\text{in2}} - (-g_m v_{\text{in1}}) \\ &= -g_m(v_{\text{in2}} - v_{\text{in1}}) \end{aligned} \tag{9-7}$$

Since the second stage consists of an inverting current-to-voltage converter, we may write

$$\begin{aligned} v'_{\text{out}} &= -i'_{\text{out}}Z_f = -\left[-g_m(v_{\text{in2}} - v_{\text{in1}})\right]\frac{1}{j\omega C_c} \\ &= \frac{g_m}{j\omega C_c}(v_{\text{in2}} - v_{\text{in1}}) \end{aligned} \tag{9-8}$$

Since the output of the inverting current-to-voltage converter is directed into the input of the push-pull, emitter-follower output stage, we may take its $(+1)$ voltage gain times v'_{out} to obtain v_{out}.

$$v_{\text{out}} = (+1)(v'_{\text{out}}) = \frac{g_m}{j\omega C_c}(v_{\text{in2}} - v_{\text{in1}}) \tag{9-9}$$

To appreciate this result more fully, the reader is directed to Fig. 9-9(b). The op amp's transconductance model is shown. The r_π for the op amp is its *differential input resistance*. This parameter is typically available on the op amp's data sheet. For example, the 741C op amp has a differential input resistance of at least 300 kΩ, with its typical value being 2 MΩ [see Fig. 9-38(b)].

The transconductance model of the IC op amp is very similar to our BJT and FET low-frequency models. Contrast Fig. 9-9(b) with the model shown in Fig. 9-9(c). Note that we are *not* suggesting that there are *no* reactive (capacitive) components in the high-frequency BJT and FET models. We are merely attempting to illustrate the similarities.

If we pause to reflect on Eq. 9-9 and Fig. 9-9(b), it immediately becomes obvious that the ac voltage gain of the op amp will *decrease* as the signal frequency *increases*. Recall that

$$\omega = 2\pi f \quad \text{and} \quad A_v = \frac{v_{\text{out}}}{v_{\text{in2}} - v_{\text{in1}}} = \frac{g_m}{j\omega C_c}$$

This variation in voltage gain as a function of frequency is a very significant observation and serves to lead us to the importance of an amplifier's *frequency response*. In the balance of this chapter we focus our attention on frequency effects and their analysis.

9-5 Frequency Domain

As we saw in Section 9-4, the ac voltage gain of the op amp tends to decrease as the signal frequency is increased. In other words, the voltage gain may be regarded as a function of frequency. This is true not only for the IC op amp but for *all*

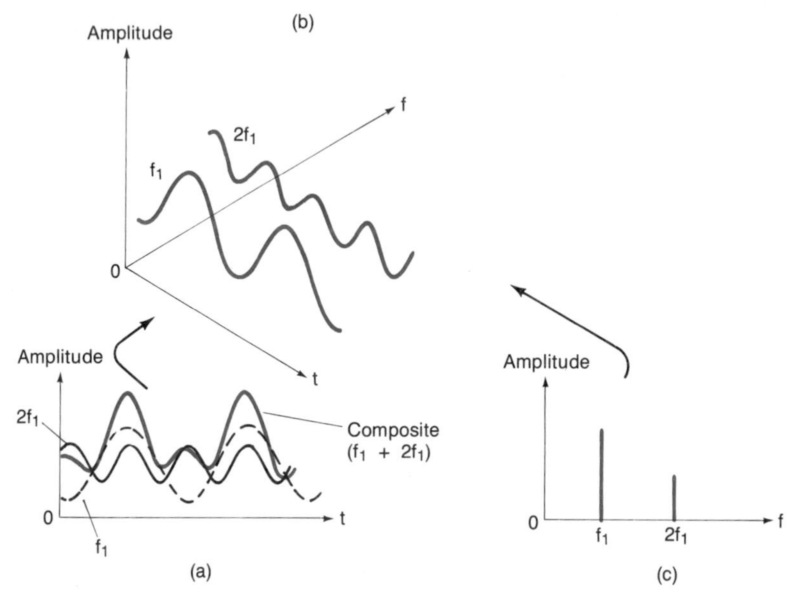

FIGURE 9-10 Frequency domain: (a) time-domain viewpoint; (b) two viewpoints; (c) frequency-domain viewpoint.

amplifiers. In addition to voltage gain, the current and power gains, and the input and output impedances are also functions of frequency.

This frequency dependence of the various amplifier parameters requires that we shift our perspective from the familiar time domain to the *frequency domain*. The time domain may be viewed as a plot of amplitude versus time. The frequency domain is a plot of amplitude versus frequency (see Fig. 9-10).

In Fig. 9-10(a) we see a time-domain representation of a complex waveform. In general, a "complex waveform" is a waveshape that is formed by more than one sinusoidal component. In this example we have a complex waveform with a sinusoidal component at its fundamental frequency f_1 and a sinusoidal component at twice the fundamental frequency $2f_1$ (a second harmonic). On an oscilloscope we would see the composite complex waveform. This would be a time-domain representation. In this case it would be up to us to determine the harmonic content of the complex waveform [see Fig. 9-10(b)].

In Fig. 9-10(c) we see a frequency-domain representation. This type of display is available from an electronic instrument called a *spectrum analyzer*. Obviously, it is much easier to ascertain the harmonic content of the complex waveform with a frequency-domain representation. Frequency-domain representations are useful not only in the determination of spectral content, but will also assist us in our analysis of the frequency response of amplifier systems.

Most of our analysis of frequency effects can be accomplished by understanding the simple *RC* low-pass and *RC* high-pass filter circuits. The fundamentals of these circuits are presented in the next three sections.

9-6 *RC* Low-pass Filters

In Fig. 9-11(a) we see a generalized ac network. Using the notation for rms values (capital V and lowercase subscripts) we may apply voltage division to arrive at V_{out}.

$$V_{out} = \frac{Z_2}{Z_1 + Z_2} V_{in} \tag{9-10}$$

and solving for the voltage gain, we have

$$A_v = \frac{V_{out}}{V_{in}} = \frac{Z_2}{Z_1 + Z_2} \tag{9-11}$$

With this review, let us investigate the low-pass *RC* filter shown in Fig. 9-11(b).

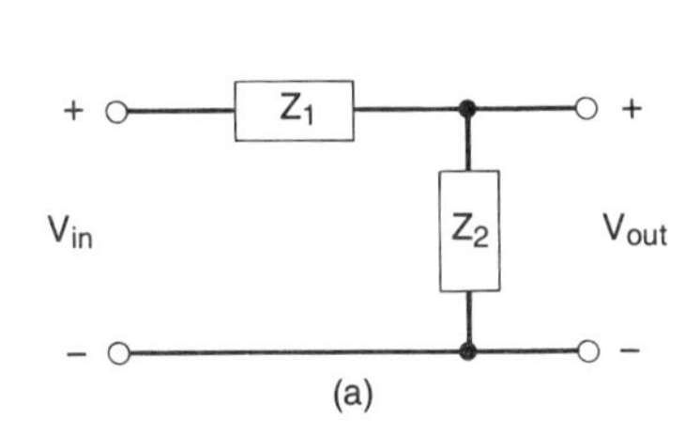

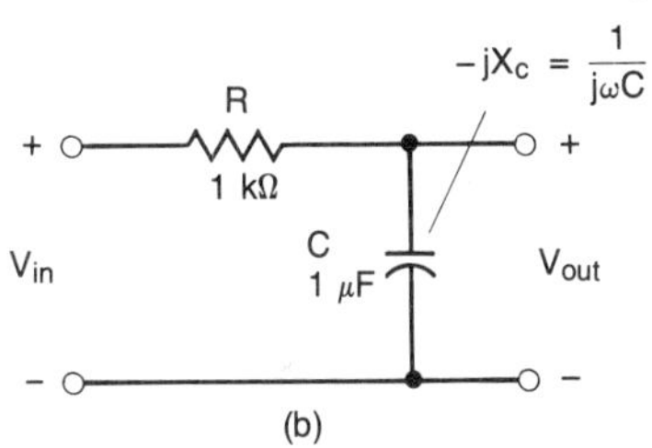

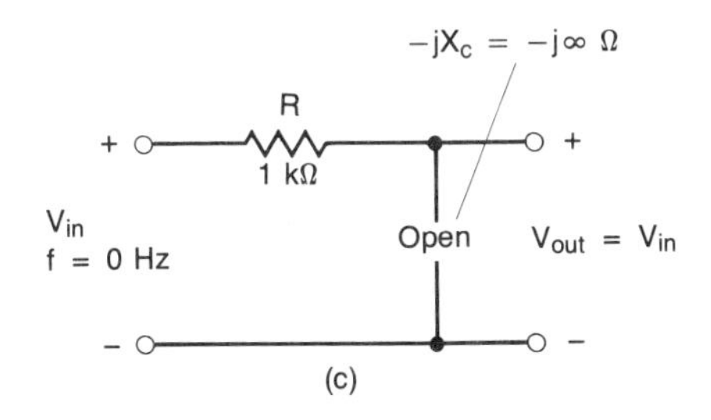

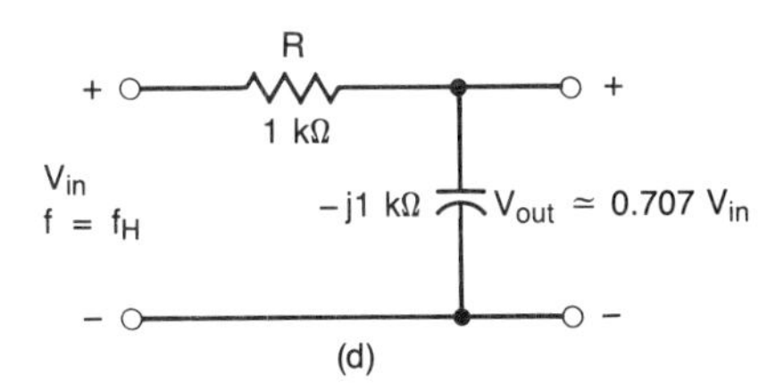

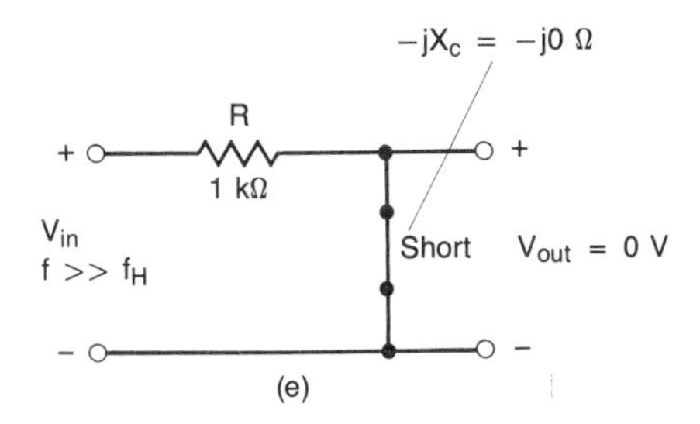

FIGURE 9-11 Operation of the RC low-pass filter: (a) generalized network; (b) filter circuit; (c) the low-pass filter has unity (0 dB) voltage gain at dc (0 Hz); (d) at the upper corner frequency $|X_c| = R$ and the output voltage is reduced to $0.707V_{in}$; (e) at very high frequencies $V_{out} \simeq 0$ V.

At low frequencies the capacitive reactance X_c is very large. In fact, at dc (0 Hz), we assume that X_c is infinite. This is equivalent to stating that the capacitor is an open circuit [refer to Fig. 9-11(c)]. In this case the voltage gain is unity.

$$A_v = \frac{V_{out}}{V_{in}} = \frac{V_{in}}{V_{in}} = 1$$

and in decibels,

$$|A_v|_{dB} = 20 \log 1 = 0 \text{ dB}$$

As the signal frequency increases, X_c decreases. At *one particular* frequency the magnitude of the capacitive reactance will be equal to the resistance [see Fig. 9-11(d)]. The voltage gain at this particular frequency is (by Eq. 9-11)

$$A_v = \frac{Z_2}{Z_1 + Z_2} = \frac{-jX_c}{R - jX_c} = \frac{-j1000\ \Omega}{1000\ \Omega - j1000\ \Omega}$$

$$= \frac{1000 \angle -90^\circ}{\sqrt{1000^2 + 1000^2} \angle -\tan^{-1}(1000/1000)}$$

$$= \frac{1000 \angle -90^\circ}{1414 \angle -45^\circ} = 0.707 \angle -45^\circ$$

The corresponding gain in decibels at this frequency is

$$|A_v|_{dB} = 20 \log 0.707 = -3.01 \text{ dB} \simeq -3 \text{ dB}$$

The voltage gain has been reduced from 0 dB to −3.01 dB at this particular frequency. Consequently, this is referred to as the "3-dB down point." Since the only time this occurs is when the magnitude of the capacitive reactance is equal to the resistance, we may easily determine this high-frequency point f_H. (As we shall see, f_H is often referred to as the *upper corner frequency*.) Hence

$$R = X_c = \frac{1}{2\pi f_H C}$$

and solving for f_H in terms of R and C, we have

$$\boxed{f_H = \frac{1}{2\pi RC}} \qquad (9\text{-}12)$$

For our example in Fig. 9-11(b),

$$f_H = \frac{1}{2\pi RC} = \frac{1}{2\pi(1\text{ k}\Omega)(1\ \mu\text{F})} = 159 \text{ Hz}$$

As the frequency is increased beyond f_H, the capacitive reactance will continue to decrease. Specifically, as f approaches infinity, X_c approaches zero.

$$A_v = \frac{-j0\ \Omega}{1000\ \Omega - j0\ \Omega} = 0$$

and the voltage gain in decibels approaches negative infinity.

$$|A_v|_{dB} = -\infty \text{ dB}$$

This represents infinite attenuation. Of course, −∞ dB is a *theoretical* limit, but it does serve to give us the general idea.

Because the low-pass filter circuit is so fundamental to much of our *high-*

frequency analysis, we shall develop some general relationships. This will ultimately serve to simplify our analyses. Referring again to Fig. 9-11(b), we may restate the voltage gain.

$$A_v = \frac{V_{out}}{V_{in}} = \frac{1/j\omega C}{R + 1/j\omega C} = \frac{1/j\omega C}{(j\omega RC + 1)/j\omega C} = \frac{1}{j\omega C}\,\frac{j\omega C}{1 + j\omega RC}$$
$$= \frac{1}{1 + j\omega RC} \tag{9-13}$$

The radian frequency ω_H is equal to 2π times the upper corner frequency f_H. Now since the f_H occurs when the magnitude of the capacitive reactance is equal to the resistance,

$$|X_c| = \frac{1}{\omega_H C} = R$$

and

$$\frac{1}{\omega_H} = RC$$

If we substitute in this result into Eq. 9-13, we obtain

$$A_v = \frac{1}{1 + j\omega RC} = \frac{1}{1 + j(\omega/\omega_H)}$$

Recalling that $\omega = 2\pi f$ and $\omega_H = 2\pi f_H$, we obtain

$$A_v = \frac{1}{1 + j(2\pi f/2\pi f_H)} = \frac{1}{1 + j(f/f_H)}$$

$$\boxed{A_v = \frac{1}{1 + j(f/f_H)}} \tag{9-14}$$

This relationship may seem to be a bit more abstract, but it tends to be a lot less cumbersome than working with reactances and impedances. Equation 9-14 is described as being *normalized* about the upper corner frequency f_H. As we shall see, all *RC* low-pass filters demonstrate the same voltage-gain response about their particular f_H. Equation 9-14 describes that response. Since the equation is so powerful, it merits our serious attention.

The rectangular (real and imaginary) form of Eq. 9-14 may be converted to polar (magnitude and phase angle) form.

$$A_v = \frac{1}{1 + j(f/f_H)} = \frac{1}{\sqrt{1 + (f/f_H)^2}\angle \tan^{-1}(f/f_H)}$$
$$= \frac{1}{[1 + (f/f_H)^2]^{1/2}}\angle -\tan^{-1}(f/f_H) \tag{9-15}$$

Generally, it is much more convenient to work with frequency response plots when their amplitude (the magnitude of A_v) is expressed in decibels.

$$|A_v| = \frac{1}{[1 + (f/f_H)^2]^{1/2}}$$

$$|A_v|_{dB} = 20 \log \frac{1}{[1 + (f/f_H)^2]^{1/2}} \tag{9-16}$$

If we employ some of the basic logarithm identities, we may simplify Eq. 9-16.

$$|A_v|_{dB} = 20 \log 1 - 20 \log[1 + (f/f_H)^2]^{1/2}$$
$$= 0 \text{ dB} - 10 \log[1 + (f/f_H)^2]$$

$$\boxed{|A_v|_{dB} = -10 \log[1 + (f/f_H)^2]} \tag{9-17}$$

The corresponding phase angle θ may be extracted from Eq. 9-15.

$$\boxed{\theta = -\tan^{-1}(f/f_H)} \tag{9-18}$$

Let us now see exactly how to use Eqs. 9-17 and 9-18. Consider the case in which the signal frequency is much smaller than f_H.

$$f = 0.01 f_H$$

From Eq. 9-17,

$$|A_v|_{dB} = -10 \log[1 + (f/f_H)^2]$$
$$= -10 \log[1 + (0.01 f_H/f_H)^2]$$
$$= -10 \log[1 + (0.01)^2] = -0.0004 \text{ dB} \simeq 0 \text{ dB}$$

and from Eq. 9-18,

$$\theta = -\tan^{-1}(f/f_H) = -\tan^{-1}(0.01 f_H/f_H)$$
$$= -\tan^{-1} 0.01 = -0.573° \simeq 0°$$

If we proceed in a similar fashion for other values of f, we arrive at the values summarized in Table 9-1. The reader is encouraged to verify the results.

TABLE 9-1
***RC* Low-Pass Filter Frequency Response**

f	$\|A_v\|$	θ
$0.01f_H$	-0.0004 dB $\simeq 0$ dB	$-0.573° \simeq 0°$
$0.1f_H$	-0.0864 dB $\simeq 0$ dB	$-5.71° \simeq 0°$
f_H	-3.01 dB $\simeq -3$ dB	$-45°$
$10f_H$	-20.04 dB $\simeq -20$ dB	$-84.3° \simeq -90°$
$100f_H$	-40 dB	$-89.4° \simeq -90°$
$1000f_H$	-60 dB	$-89.9° \simeq -90°$

Be sure to note the approximations indicated in Table 9-1. We shall draw on these in our frequency response plots, which are discussed in the next section.

9-7 Bode Approximations

The complete frequency response graph includes both magnitude and phase shift information plotted as a function of frequency. As an example, consider Fig. 9-12(a). The data points for the graph were taken from Table 9-1. Notice that the horizontal axis is scaled logarithmically. The factor-of-10 change in frequency between divisions (e.g., from f_H to $10f_H$) is referred to as a *decade*. Since the logarithm of zero is undefined, zero does *not* appear on the horizontal frequency axis.

The vertical axis is linearly scaled, but the magnitude of the voltage gain has been plotted in decibels. The right-hand vertical axis has been linearly scaled in degrees.

By using the approximations cited in Table 9-1, we may rather easily draw the

FIGURE 9-12 Complete frequency responses of the RC low-pass filter: (a) actual response; (b) Bode approximation.

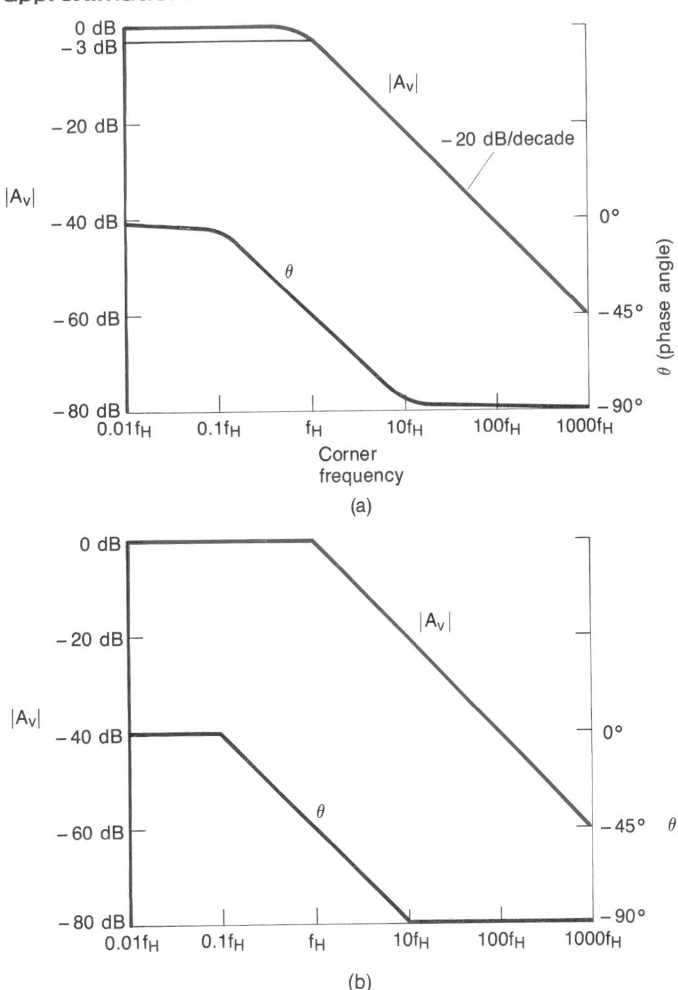

frequency response plot for the simple *RC* low-pass filter. These are referred to as the *Bode approximations*. The Bode plot (named after Hendrick W. Bode of Bell Labs) consists of straight-line approximations of the actual smoothly varying response curve [see Fig. 9-12(b)]. The rules for drawing a Bode plot are extremely simple.

1. Determine f_H by using Eq. 9-12.
2. For all frequencies below f_H, the gain is approximated as zero decibels.
3. The gain decreases at a rate of -20 dB per decade as the frequency is increased beyond f_H.
4. The phase angle is zero degrees for all frequencies at least one decade below f_H.
5. The phase angle becomes negative, and decreases linearly passing through $-45°$ at f_H as frequency increases.
6. At one decade above f_H the phase angle reaches $-90°$, and remains $-90°$ for all frequencies beyond that point.

EXAMPLE 9-1

Draw the frequency response curve of the *RC* low-pass filter given in Fig. 9-13(a) using Bode approximations. Is the phase angle leading or lagging? Explain.

SOLUTION First, we shall use Eq. 9-12 to determine f_H.

$$f_H = \frac{1}{2\pi RC} = \frac{1}{2\pi(1\ \text{k}\Omega)(100\ \text{pF})} = 1.59\ \text{MHz}$$

Observe that the relatively small capacitance (100 pF) has resulted in a relatively high corner frequency. Using the steps outlined above, the Bode plot has been illustrated in Fig. 9-13(b). In response to the question regarding the nature of the phase angle, we must state that the output *lags* the input [see the phasor diagram given in Fig. 9-13(c)]. Consequently, the *RC* low-pass filter is sometimes referred to as a *lag network*. ■

9-8 *RC* High-Pass Filters

Equally important in the analysis of the frequency response of amplifiers is the high-pass *RC* filter circuit [see Fig. 9-14(a)]. Before we begin a serious analysis, let us take a more qualitative look at the circuit.

At low frequencies the capacitive reactance of the capacitor will be very large. Consequently, it will drop most of V_{in}, and the voltage gain will be much, much less than unity. As the frequency is increased, X_c will continue to decrease. At one particular frequency, the magnitude of X_c will be equal to R. This will be defined as the *lower corner frequency* f_L. As the frequency increased beyond this point, X_c will approach 0 Ω (a short circuit), and the voltage gain will approach unity (0 dB).

With this survey in mind, our immediate objective is to develop a normalized gain equation in much the same fashion as we derived Eq. 9-14. By applying voltage division, we may readily solve for the voltage gain.

$$A_v = \frac{V_{\text{out}}}{V_{\text{in}}} = \frac{R}{R + 1/j\omega C}$$

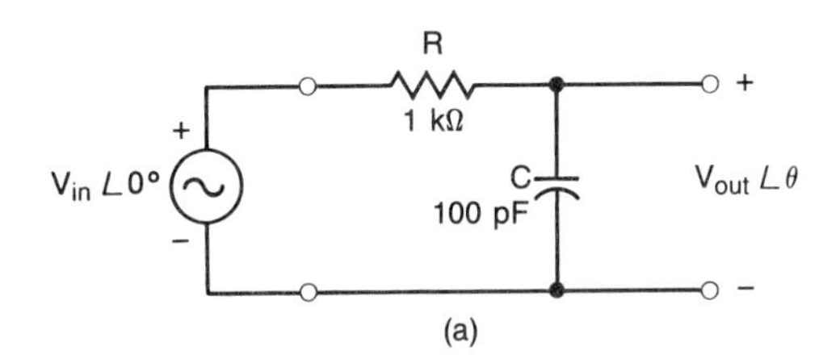

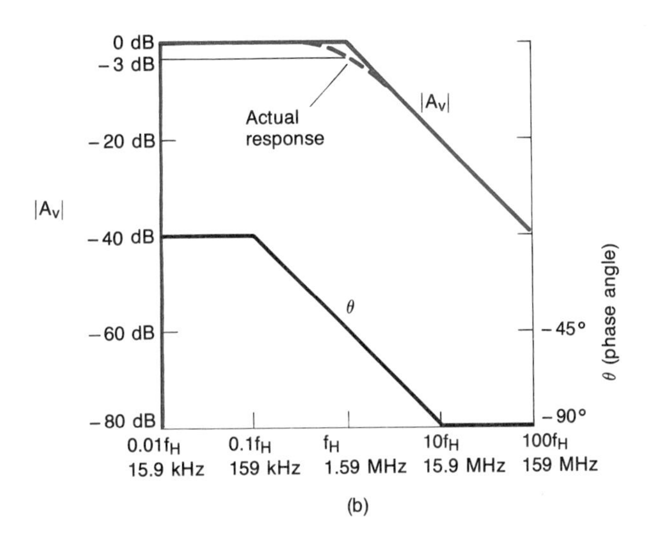

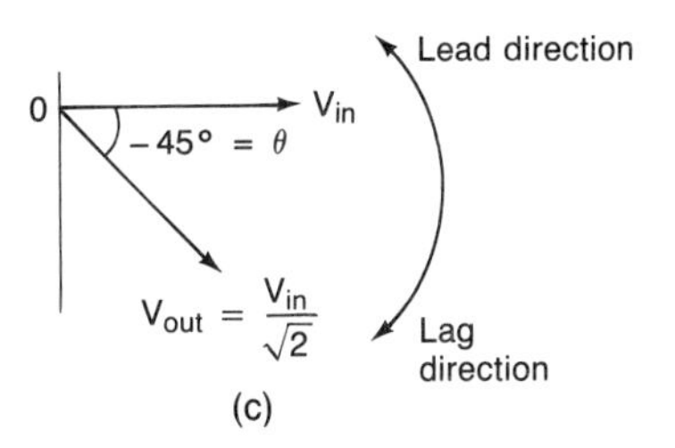

FIGURE 9-13 RC low-pass filter for Example 9-1: (a) circuit; (b) Bode approximation; (c) phasor diagram at the corner frequency (V_{out} lags V_{in}).

Next, in our quest for a normalized expression, we shall divide the numerator and each of the denominator terms by R.

$$A_v = \frac{R/R}{R/R + 1/j\omega RC} = \frac{1}{1 - j(1/\omega RC)} = \frac{1}{1 - j(1/\omega)(1/RC)}$$

As we noted previously, the corner frequency of the high-pass filter also occurs when the magnitude of the capacitive reactance is equal to the resistance. Hence

$$\omega_L = \frac{1}{RC}$$

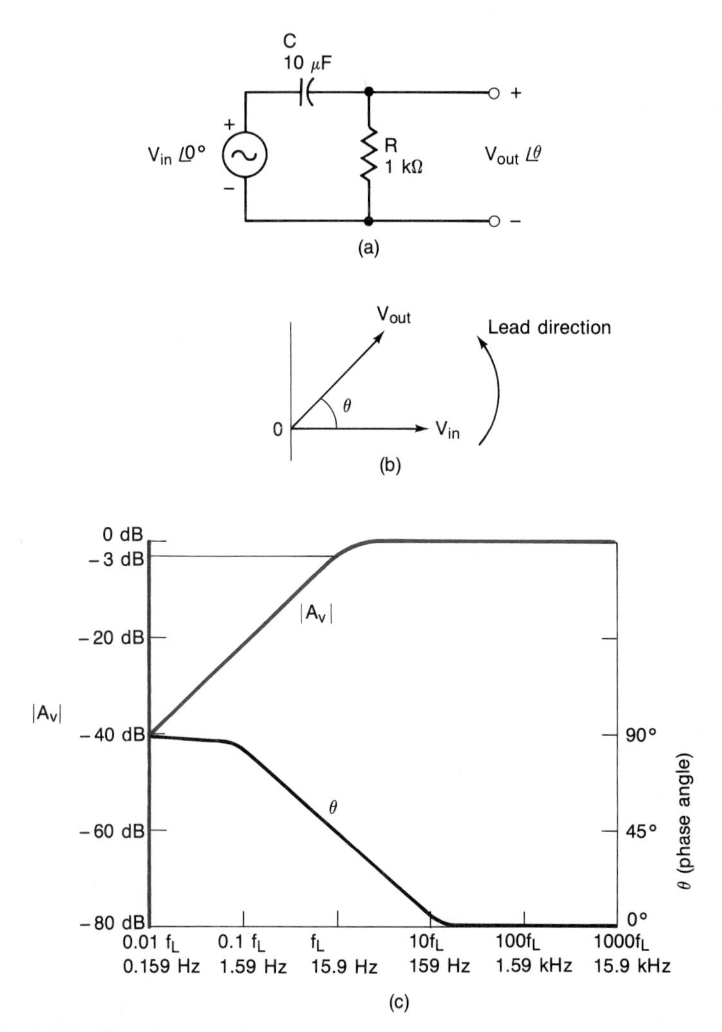

FIGURE 9-14 **RC high-pass filter: (a) circuit; (b) V_{out} leads V_{in}; (c) frequency response.**

Substituting this into our expression for the voltage gain, we arrive at the following result:

$$A_v = \frac{V_{\text{out}}}{V_{\text{in}}} = \frac{1}{1 - j(\omega_L/\omega)}$$

Since $\omega_L = 2\pi f_L$ and $\omega = 2\pi f$,

$$A_v = \frac{V_{\text{out}}}{V_{\text{in}}} = \frac{1}{1 - j(2\pi f_L/2\pi f)}$$

$$\boxed{A_v = \frac{1}{1 - j(f_L/f)}} \qquad (9\text{-}19)$$

The rectangular form of our frequency-normalized voltage gain equation may be converted to its polar form. This will allow us to ascertain its magnitude and the corresponding phase shift at any particular frequency of interest.

$$A_v = \frac{1}{1 - j(f_L/f)} = \frac{1}{\sqrt{1 + (f_L/f)^2} \angle -\tan^{-1}(f_L/f)}$$
$$= \frac{1}{[1 + (f_L/f)^2]^{1/2}} \angle \tan^{-1}(f_L/f) \qquad (9\text{-}20)$$

The magnitude of the voltage gain of the highpass filter is given by Eq. 9-21.

$$|A_v| = \frac{1}{[1 + (f_L/f)^2]^{1/2}}$$
$$|A_v|_{dB} = 20 \log \frac{1}{[1 + (f_L/f)^2]^{1/2}} \qquad (9\text{-}21)$$

Application of basic logarithmic identities produces Eq. 9-22.

$$|A_v|_{dB} = 20 \log 1 - 20 \log[1 + (f_L/f)^2]^{1/2}$$
$$= 0 \text{ dB} - 10 \log[1 + (f_L/f)^2]$$

$$\boxed{|A_v|_{dB} = -10 \log[1 + (f_L/f)^2]} \qquad (9\text{-}22)$$

The corresponding phase angle θ may be found by using

$$\boxed{\theta = \tan^{-1}(f_L/f)} \qquad (9\text{-}23)$$

In this case the phase angle is leading. Consequently, the high-pass RC filter is occasionally referred to as a *lead network*. Refer to the phasor diagram shown in Fig. 9-14(b).

EXAMPLE 9-2

Determine the complete frequency response of the high-pass filter given in Fig. 9-14(a).

SOLUTION First, we must determine f_L. Since this occurs when the magnitude of X_c is equal to the resistance R, we may draw on Eq. 9-12. All that is required is to change the subscript of the frequency f from H to L. Hence

$$f_L = \frac{1}{2\pi RC} = \frac{1}{2\pi(1 \text{ k}\Omega)(10 \text{ }\mu\text{F})} = 15.9 \text{ Hz}$$

To generate the frequency response, we shall again opt for decade multiples of the corner frequency. For example, if we let the frequency be low compared to f_L (e.g., let f equal $0.01f_L$), we may proceed as follows:

$$|A_v|_{dB} = -10 \log[1 + (f_L/f)^2]$$
$$= -10 \log[1 + (f_L/0.01f_L)^2]$$
$$= -10 \log 10{,}001 = -40 \text{ dB}$$

and from Eq. 9-23,

$$\theta = \tan^{-1}(f_L/f) = \tan^{-1}(f_L/0.01f_L)$$
$$= \tan^{-1} 100 = 89.4° \simeq 90°$$

If we continue in a similar fashion for other values of f, we arrive at the values summarized in Table 9-2. The frequency response has been graphed in Fig. 9-14(c). ■

TABLE 9-2
***RC* High-Pass Filter Frequency Response**

f	$\lvert A_v \rvert$	θ
$0.01f_L$	−40 dB	89.4° ≃ 90°
$0.1f_L$	−20 dB	84.3° ≃ 90°
f_L	−3.01 dB ≃ −3 dB	45°
$10f_L$	0 dB	5.71° ≃ 0°
$100f_L$	0 dB	0.573° ≃ 0°
$1000f_L$	0 dB	0.0573° ≃ 0°

To expedite the analysis of the high-pass *RC* filter, it is possible to use the Bode approximations. The procedure is very similar to that cited for the low-pass filter in Section 9-7.

1. Determine f_L by using Eq. 9-12. (Of course, we must change the subscript.)
2. For all frequencies above f_L, the gain is approximated as zero decibels.
3. The gain decreases at a rate of −20 db/decade as the frequency is decreased below f_L.
4. The phase angle is zero degrees for all frequencies at least one decade above f_L.
5. The phase angle becomes positive, and its magnitude increases linearly passing through 45° at f_L as frequency decreases.
6. At one decade below f_L the phase angle reaches 90°, and remains 90° for all frequencies below that point.

The Bode plot for the high-pass filter analyzed in Example 9-2 has been illustrated in Fig. 9-15. Contrast Fig. 9-15 with Fig. 9-14(c).

9-9 Low-Frequency Roll-Off in the BJT and FET Amplifiers

In our previous analysis of amplifiers, we have assumed that the capacitive reactances of the input and output coupling capacitors and the bypass capacitors have been small enough to be regarded as short circuits. Since $X_c = 1/2\pi fC$ this requires that the capacitance C be *large enough to provide a small X_c at the lowest frequency of interest*. In the audio-frequency range (20 Hz to 20 kHz) the lowest frequency is, of course, 20 Hz.

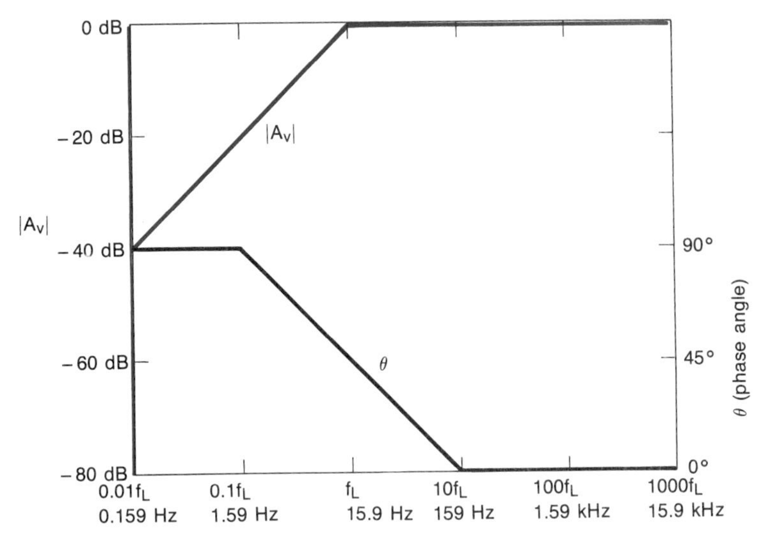

FIGURE 9-15 Bode approximation of the RC high-pass filter response given in Fig. 9-14(c).

As the signal frequency is decreased, at some point the capacitive reactances will become significant. Consider Fig. 9-16(a) and (b). Notice that we have assigned values to the capacitors.

Examine Fig. 9-17. As the capacitive reactance of the input capacitor C_1 increases, the voltage-divider action between it, and the input impedance of the amplifier will become increasingly significant. Specifically, C_1 will drop more and more of the input signal. This will cause a reduction in the actual signal reaching the input of the amplifier. Since the net input voltage V'_{in} is reduced, the output voltage of the amplifier will decrease (resulting in a lower voltage gain).

The same effect occurs at the output. In this case we have ac voltage division between C_2 and the load R_L. As the capacitive reactance of C_2 increases, less of the voltage produced at collector V'_{out} will actually reach the load. This will also promote a reduction in the voltage gain. This is also depicted in Fig. 9-17.

The emitter bypass capacitor will also produce a reduction in the voltage gain as the signal frequency decreases. Recall that the emitter bypass capacitor was included to prevent negative feedback at the emitter with its attendant reduction in voltage gain. As the signal frequency decreases, the bypassing becomes less effective, and the voltage gain will decrease. The *lower limit* on the unloaded voltage gain is

$$A_{v(\text{oc})} = -\frac{g_m R_C}{1 + g_m R_E}$$

which was developed when *no* emitter bypass capacitor was used.

These same effects also occur in the case of FET amplifiers. Even the op amp voltage amplifier will experience these same low-frequency limitations when capacitive input and output coupling and bypassing are utilized. This is explained in the next section.

Now that we have a basic "feel" for the low-frequency limitations on capacitively coupled amplifiers, let us proceed with the analysis (see Fig. 9-18). This will

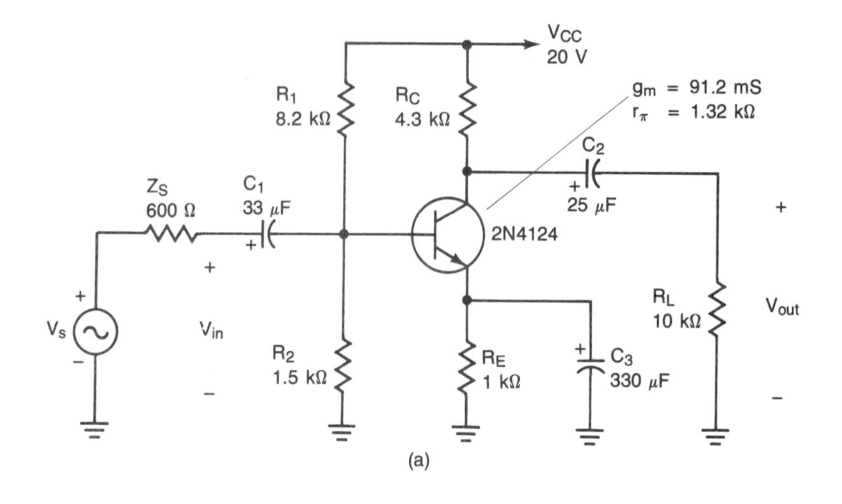

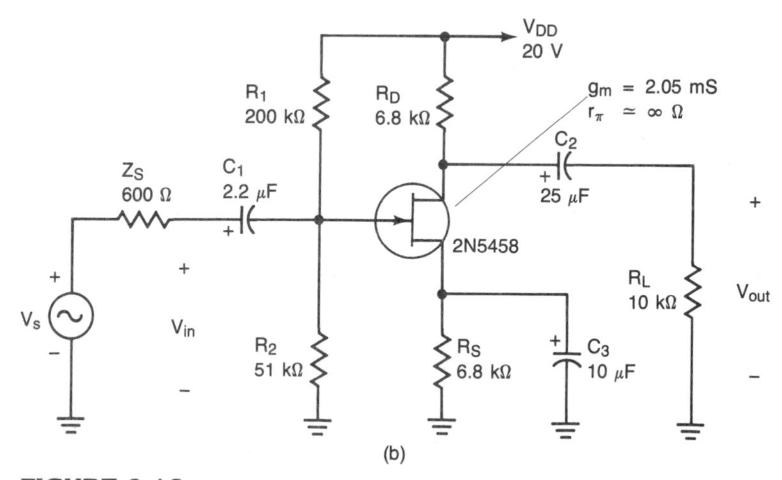

FIGURE 9-16 **Inverting amplifiers to be analyzed: (a) common-emitter; (b) common-source.**

require us to draw on our work with high-pass *RC* filters. Initially, we shall disregard the effects of the emitter and source bypass capacitors. (This allows a simplified approach, but results in an approximation.)

First, we shall analyze the input circuit. Since the capacitive reactance of C_1 is significant at low frequencies, we may write

$$V_{\text{in}} = \frac{R_{\text{in}}}{R_{\text{in}} + Z_S + 1/j\omega C_1} V_s$$

Dividing each of the numerator and denominator terms by $(R_{\text{in}} + Z_S)$ yields

$$\begin{aligned} V_{\text{in}} &= \frac{R_{\text{in}}/(R_{\text{in}} + Z_S)}{1 + 1/j\omega(R_{\text{in}} + Z_S)C_1} V_s \\ &= \frac{R_{\text{in}}}{R_{\text{in}} + Z_S} \frac{1}{1 + 1/j\omega(R_{\text{in}} + Z_S)C_1} V_s \end{aligned}$$

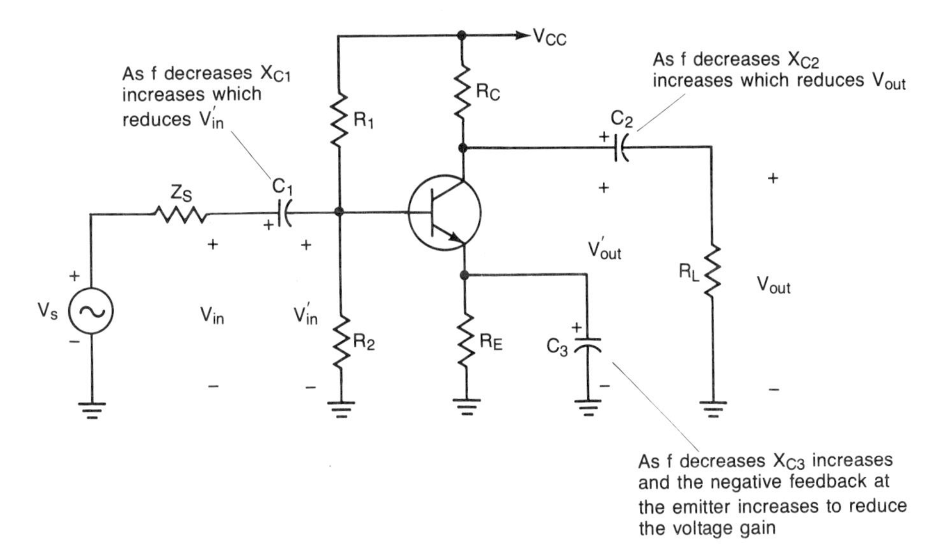

FIGURE 9-17 Low-frequency effects in the common-emitter amplifier.

Defining the low-end corner frequency due to C_1, as ω_{L1},

$$\omega_{L1} = \frac{1}{(R_{\text{in}} + Z_S)C_1}$$

We may substitute in this definition to obtain the normalized frequency form of the equation.

$$V_{\text{in}} = \frac{R_{\text{in}}}{R_{\text{in}} + Z_S} \frac{1}{1 + (1/j\omega)\,[1/(R_{\text{in}} + Z_S)C_1]} V_s$$

$$= \frac{R_{\text{in}}}{R_{\text{in}} + Z_S} \frac{1}{1 - j(\omega_{L1}/\omega)} V_s$$

FIGURE 9-18 Low-frequency equivalent circuit.

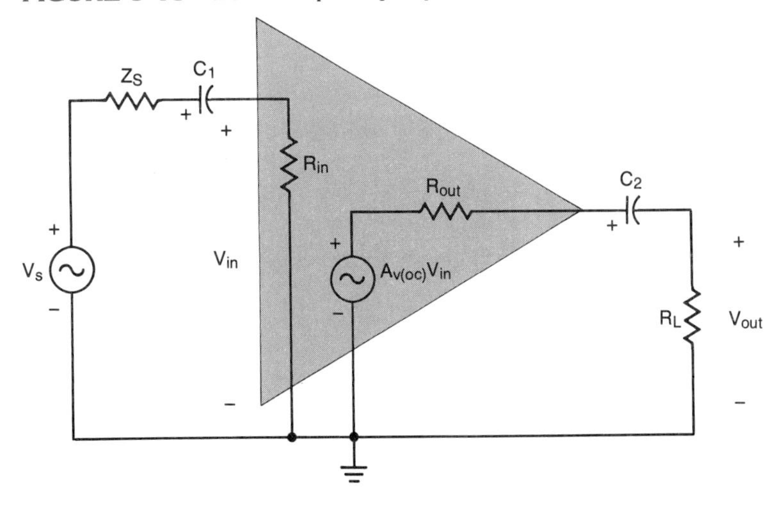

and canceling the 2π's contained in ω and ω_{L1}, we arrive at

$$V_{in} = \frac{R_{in}}{R_{in} + Z_S} \frac{1}{1 - j(f_{L1}/f)} V_s \tag{9-24}$$

where

$$f_{L1} = \frac{1}{2\pi(R_{in} + Z_S)C_1}$$

We may perform a similar analysis of the output circuit of the amplifier model given in Fig. 9-18. Hence

$$\begin{aligned} V_{out} &= \frac{R_L}{R_L + R_{out} + 1/j\omega C_2} A_{v(oc)} V_{in} \\ &= \frac{R_L}{R_L + R_{out}} \frac{1}{1 + 1/j\omega(R_L + R_{out})C_2} A_{v(oc)} V_{in} \end{aligned}$$

and if we define the corner frequency due to C_2 to be ω_{L2},

$$\omega_{L2} = \frac{1}{(R_L + R_{out})C_2}$$

If we substitute in this relationship, we may develop the normalized frequency form of the equation for V_{out}. The required steps are the same as those used to arrive at Eq. 9-24.

$$V_{out} = \frac{R_L}{R_L + R_{out}} \frac{1}{1 - j(f_{L2}/f)} A_{v(oc)} V_{in} \tag{9-25}$$

where

$$f_{L2} = \frac{1}{2\pi(R_L + R_{out})C_2}$$

By combining Eqs. 9-24 and 9-25, we obtain

$$\begin{aligned} V_{out} &= \frac{R_L}{R_L + R_{out}} \frac{1}{1 - j(f_{L2}/f)} A_{v(oc)} V_{in} \\ &= \frac{R_L}{R_L + R_{out}} \frac{1}{1 - j(f_{L2}/f)} A_{v(oc)} \frac{R_{in}}{R_{in} + Z_S} \frac{1}{1 - j(f_{L1}/f)} V_s \end{aligned} \tag{9-26}$$

Admittedly, Eq. 9-26 appears to be most formidable. Its use and meaning will become more apparent as our work progresses. For now, we shall merely note that it is similar to our development of A_{vs} at the higher frequencies in Chapter 8. However, it also reflects that the low-frequency voltage gain has two distinct corner frequencies f_{L1} and f_{L2}. These corner frequencies are produced by capacitors C_1 and C_2, respectively. Now let us include the effects of the emitter bypass capacitor. As we shall see, the emitter bypass capacitor will produce a third unique corner frequency in our amplifier's low-frequency voltage gain.

In Fig. 9-19 we have illustrated a common-emitter amplifier with an emitter bypass capacitor. Our intent is to solve for $A_{v(oc)}$ and substitute it into Eq. 9-26. This will provide us with the complete description of the low-frequency voltage gain.

Since the emitter-to-ground impedance Z_e is significant at low frequencies, we may write the equation for $A_{v(oc)}$ as

$$A_{v(\text{oc})} = -\frac{g_m R_C}{1 + g_m Z_e} \tag{9-27}$$

The emitter-to-ground impedance may be approximated by

$$Z_e \simeq \frac{1}{j\omega C_3} \| R_E \simeq \frac{1}{j\omega C_3} \tag{9-28}$$

We may obtain an equation for $A_{v(\text{oc})}$ by substituting Eq. 9-28 for Z_e into Eq. 9-27.

$$A_{v(\text{oc})}[f] = -\frac{g_m R_C}{1 + g_m Z_e} \simeq -\frac{g_m R_C}{1 + g_m(1/j\omega C_3)}$$

$$= -\frac{g_m R_C}{1 - j(g_m/\omega C_3)} \tag{9-29}$$

Observe that we have emphasized that the open-circuit voltage gain is a function of the signal frequency by using the notation $A_{v(\text{oc})}[f]$. If we define the corner frequency as

$$\omega_{L3} = \frac{g_m}{C_3}$$

The normalized frequency form of Eq. 9-29 is given by

$$A_{v(\text{oc})}[f] = -\frac{g_m R_C}{1 - j(1/\omega)(g_m/C_3)} = -\frac{g_m R_C}{1 - j(\omega_{L3}/\omega)}$$

$$= -\frac{g_m R_C}{1 - j(f_{L3}/f)} \tag{9-30}$$

where $f_{L3} \simeq g_m/2\pi C_3$.

The low-frequency open-circuit voltage gain $A_{v(\text{oc})}[f]$ may be written in terms of the *midband open-circuit* voltage gain $A_{v(\text{oc})}$ by recalling that

$$A_{v(\text{oc})} = -g_m R_C$$

FIGURE 9-19 Low-frequency common-emitter amplifier with an emitter bypass capacitor.

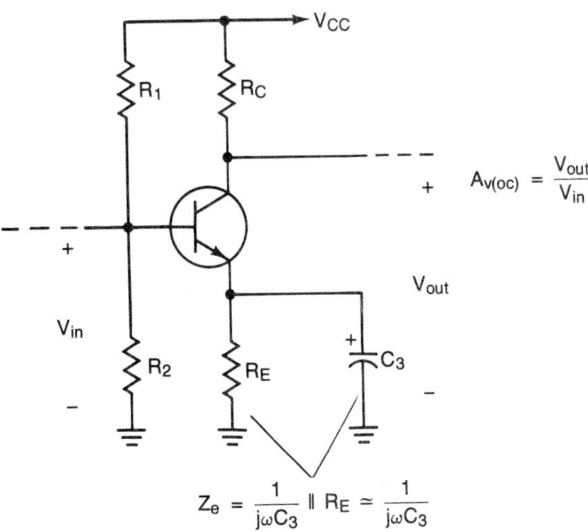

and substituting this into Eq. 9-30.

$$A_{v(\text{oc})}[f] = \frac{-g_m R_C}{1 - j(f_{L3}/f)} = \frac{A_{v(\text{oc})}}{1 - j(f_{L3}/f)} \tag{9-31}$$

By taking a similar approach, we may solve Eq. 9-26 for the voltage gain from the source to the load as a function of frequency $A_{vs}[f]$ in terms of A_{vs}. Recalling that

$$A_{vs} = \frac{V_{\text{out}}}{V_s} \tag{9-32}$$

we may solve Eq. 9-26. Hence

$$A_{vs}[f] = \frac{R_L}{R_L + R_{\text{out}}} A_{v(\text{oc})}[f] \frac{R_{\text{in}}}{R_{\text{in}} + Z_S} \frac{1}{1 - j(f_{L1}/f)} \frac{1}{1 - j(f_{L2}/f)} \tag{9-33}$$

Observe that we have emphasized that the open-circuit voltage gain is also a function of frequency. If we substitute Eq. 9-31 into Eq. 9-33, we obtain the approximation

$$A_{vs}[f] \simeq \frac{R_L}{R_L + R_{\text{out}}} \frac{A_{v(\text{oc})}}{1 - j(f_{L3}/f)} \frac{R_{\text{in}}}{R_{\text{in}} + Z_S} \frac{1}{1 - j(f_{L1}/f)} \frac{1}{1 - j(f_{L2}/f)} \tag{9-34}$$

As we have seen before, the loaded voltage gain may be defined as

$$A_{vs} = \frac{R_L}{R_L + R_{\text{out}}} A_{v(\text{oc})} \frac{R_{\text{in}}}{R_{\text{in}} + Z_S} \tag{9-35}$$

If we substitute the definition above into Eq. 9-34, we obtain our ultimate approximation.

$$\boxed{A_{vs}[f] \simeq \frac{A_{vs}}{[1 - j(f_{L1}/f)][1 - j(f_{L2}/f)][1 - j(f_{L3}/f)]}} \tag{9-36}$$

Let us pause to reflect on Eq. 9-36. It illustrates that the low-end roll-off in the voltage gain as a function of frequency has three distinct corner frequencies. The three corner frequencies f_{L1}, f_{L2}, and f_{L3} are produced by the input coupling capacitor C_1, the output coupling capacitor C_2, and the emitter bypass capacitor C_3, respectively. Equation 9-36 can be extended directly to the FET and the op amp voltage amplifiers. All that is required is that we find A_{vs} (the midband source-to-load voltage gain) and find the corner frequency associated with each coupling or bypass capacitor that is employed.

Since we have considerable experience in finding the A_{vs} for amplifier circuits, let us concentrate on the determination of the corner frequencies. In the case of the common-emitter amplifier, we have seen

$$f_{L1} = \frac{1}{2\pi(R_{\text{in}} + Z_S)C_1}$$

$$f_{L2} = \frac{1}{2\pi(R_L + R_{\text{out}})C_2}$$

$$f_{L3} = \frac{g_m}{2\pi C_3} = \frac{1}{2\pi(1/g_m)C_3}$$

In each case the corner frequency is found by knowing the capacitance, and *the Thévenin equivalent resistance* "seen" by each capacitance. (This technique is an approximation since we have ignored any possible interaction between the corner frequencies produced by C_1, C_2, and C_3.) This has been demonstrated in Fig. 9-20. Obviously, this observation can greatly simplify our frequency response analysis of amplifier circuits.

The actual plotting of the complete frequency response (both the magnitude in decibels and the phase angle) can also be rather easily accomplished by drawing on the Bode approximations. Let us see exactly how and why this may be done.

The magnitude of $A_{vs}[f]$ in decibels is

$$|A_{vs}[f]|_{\text{dB}} = 20 \log \left| \frac{A_{vs}}{[1 + (f_{L1}/f)^2]^{1/2} [1 + (f_{L2}/f)^2]^{1/2} [1 + (f_{L3}/f)^2]^{1/2}} \right|$$

FIGURE 9-20 Using the Thévenin equivalent resistances to determine the corner frequencies: (a) f_{L1}; (b) f_{L2}; (c) f_{L3}.

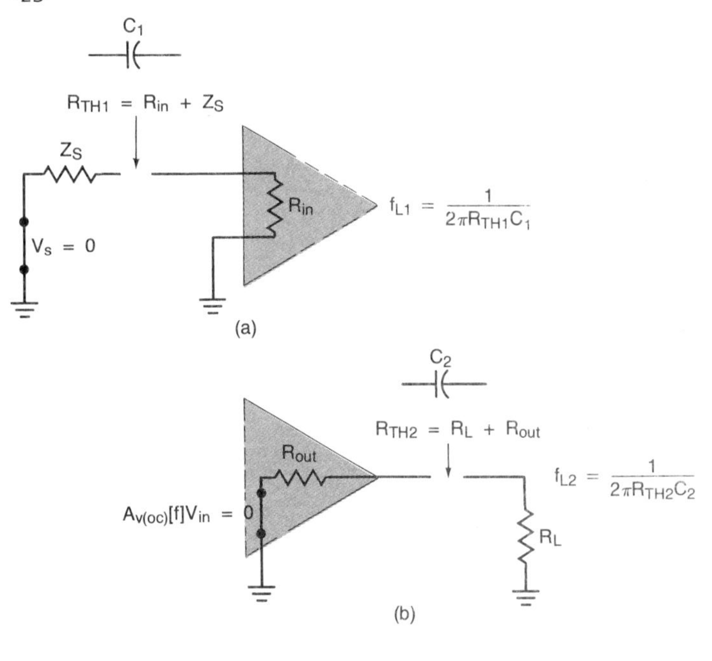

The relationship above may be simplified by once again calling on some of the basic logarithmic identities.

$$|A_{vs}|_{\text{dB}} = 20 \log A_{vs} - 10 \log[1 + (f_{L1}/f)^2] - 10 \log[1 + (f_{L2}/f)^2] - 10 \log[1 + (f_{L3}/f)^2]$$

As we can see, the *composite amplitude response* is given by the *sum of the individual terms*. This implies that a *graphical addition* is possible. This will be demonstrated in the next example.

Now let us direct our attention to the determination of the composite phase angle, which we shall denote as ϕ (lowercase phi). Since the denominator terms in Eq. 9-36 are all multiplications, their individual phase angles are additive. Bringing their sum up into the numerator reverses their signs from negative to positive. Letting θ represent the phase angle of the midband voltage gain (e.g., 0° for a noninverting amplifier and 180° for an inverting amplifier), we arrive at

$$\phi = \theta + \tan^{-1}\frac{f_{L1}}{f} + \tan^{-1}\frac{f_{L2}}{f} + \tan^{-1}\frac{f_{L3}}{f}$$

Therefore, the *composite phase angle* ϕ is also a *sum of the individual terms*. This again implies that a graphical addition is possible.

EXAMPLE 9-3

Determine the three corner frequencies for the low-end frequency roll-off of the common-emitter amplifier given in Fig. 9-16(a).

SOLUTION If we had never seen this circuit before, we would first perform a dc analysis. Using the results, we would then determine the small-signal parameters. The g_m and r_π for the 2N4124 have been indicated in Fig. 9-16(a). The next step entails the determination of R_{in}, R_{out}, and $A_{v(\text{oc})}$ for the midfrequency range.

$$R_{\text{in}} = R_1 \| R_2 \| r_\pi = 8.2\ \text{k}\Omega \| 1.5\ \text{k}\Omega \| 1.32\ \text{k}\Omega = 647\ \Omega$$

$$R_{\text{out}} \simeq R_C = 4.3\ \text{k}\Omega$$

$$A_{v(\text{oc})} = -g_m R_C = -(91.2\ \text{mS})(4.3\ \text{k}\Omega) = -392 = 392\ \angle -180^\circ$$

The midband source-to-load voltage gain may be found in the usual fashion.

$$A_{vs} = \frac{R_L}{R_L + R_{\text{out}}} A_{v(\text{oc})} \frac{R_{\text{in}}}{R_{\text{in}} + Z_S}$$

$$= \frac{10\ \text{k}\Omega}{10\ \text{k}\Omega + 4.3\ \text{k}\Omega}(392\ \angle -180^\circ)\frac{647\ \Omega}{647\ \Omega + 600\ \Omega}$$

$$= 142\ \angle -180^\circ$$

$$|A_{vs}|_{\text{dB}} = 20 \log A_{vs} = 20 \log 142 = 43.0\ \text{dB}$$

$$\theta = -180^\circ$$

Now we shall determine the three corner frequencies.

$$R_{\text{TH1}} = Z_S + R_{\text{in}} = 600\ \Omega + 647\ \Omega = 1247\ \Omega$$

$$f_{L1} = \frac{1}{2\pi R_{\text{TH1}} C_1} = \frac{1}{2\pi(1247\ \Omega)(33\ \mu\text{F})} = 3.87\ \text{Hz} \approx 4\ \text{Hz}$$

$$R_{\text{TH2}} = R_L + R_{\text{out}} = 10\ \text{k}\Omega + 4.3\ \text{k}\Omega = 14.3\ \text{k}\Omega$$

$$f_{L2} = \frac{1}{2\pi R_{\text{TH2}}C_2} = \frac{1}{2\pi(14.3\text{ k}\Omega)(25\ \mu\text{F})} = 0.445\text{ Hz} \approx 0.4\text{ Hz}$$

$$R_{\text{TH3}} \simeq \frac{1}{g_m} = \frac{1}{91.2\text{ mS}} \simeq 11.0\ \Omega$$

$$f_{L3} = \frac{1}{2\pi R_{\text{TH3}}C_3} = \frac{1}{2\pi(11.0\ \Omega)(330\ \mu\text{F})} = 43.8\text{ Hz} \approx 40\text{ Hz}$$

■

Now let us find the frequency response for the amplifier analyzed in Example 9-3. (To simplify, we have made rough approximations to place the corners one full decade apart.) The complete frequency response could be plotted by solving Eq. 9-36 for the magnitude of $A_{vs}[f]$ in decibels and the corresponding phase angle for each particular frequency of interest. However, this is rather cumbersome without the aid of a computer. Therefore, we shall opt for the use of Bode approximations for each of the terms and graphically add the results.

In Fig. 9-21(a) we have illustrated the amplitude plot of A_{vs} (43 dB) and the high-pass filter response associated with the f_{L3} ($\approx$ 40 Hz) corner frequency. The composite may be found by graphical addition and is indicated in Fig. 9-21(b). As we can see, one decade below the f_{L3} corner frequency, the composite response is reduced from 43 dB to 23 dB.

The high-pass filter response associated with the f_{L1} ($\approx$ 4 Hz) corner frequency has been added to the previous resultant as shown in Fig. 9-21(c). The second corner frequency causes the roll-off in the voltage gain to increase to -40 dB per decade. Similarly, the f_{L2} response has been added in to form the complete amplitude response in Fig. 9-21(d).

The phase shift ϕ of the amplifier may also be sketched via the Bode approximations. The procedure is very similar to that for the amplitude response (see Fig. 9-22).

The 180° phase shift associated with the BJT and the phase shift produced by the f_{L3} corner frequency have been shown in Fig. 9-22(a). In Fig. 9-22(b) we see that the phase shift due the f_{L1} corner frequency causes the amplifier's phase shift to change from $-180°$ in a positive direction toward $-90°$. As can be seen in Fig. 9-22(c), the f_{L1} corner frequency causes the phase shift to increase another $+90°$.

The effects of the f_{L2} corner frequency [Fig. 9-22(c)] on the total phase shift have been included in Fig. 9-22(d). Since it is often desirable to know both the total phase shift ϕ and the amplitude $|A_{vs}[f]|$ at each particular frequency, the complete response has been illustrated.

EXAMPLE 9-4

Determine the midband and the low-end frequency response of the common-source amplifier given in Fig. 9-16(b).

SOLUTION The approach is identical to that of the common-emitter amplifier. First we perform the midband analysis.

$$R_{\text{in}} = R_1 \parallel R_2 = 200\text{ k}\Omega \parallel 51\text{ k}\Omega = 40.6\text{ k}\Omega$$

$$R_{\text{out}} \simeq R_D = 6.8\text{ k}\Omega$$

$$A_{v(\text{oc})} = -g_m R_D = -(2.05\text{ mS})(6.8\text{ k}\Omega) = 13.9\ \underline{/-180°}$$

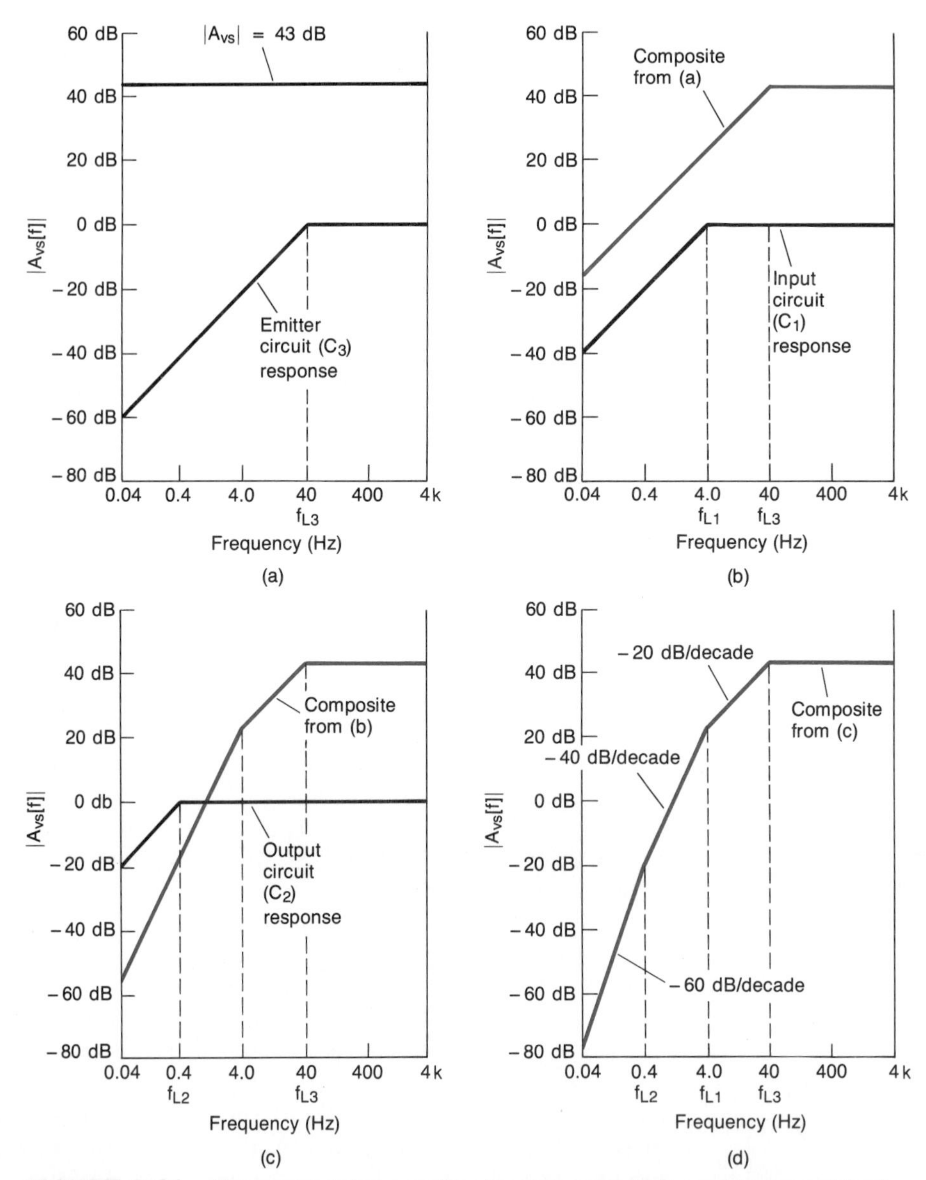

FIGURE 9-21 **The complete amplitude response is found by graphical addition: (a) A_{vs} and C_3; (b) A_{vs}, C_3, and C_1; (c) A_{vs}, C_3, C_1, and C_2; (d) complete response.**

$$
\begin{aligned}
A_{vs} &= \frac{R_L}{R_L + R_{\text{out}}} A_{v(\text{oc})} \frac{R_{\text{in}}}{R_{\text{in}} + Z_S} \\
&= \frac{10\ \text{k}\Omega}{10\ \text{k}\Omega + 6.8\ \text{k}\Omega} (13.9 \underline{\angle -180^\circ}) \frac{40.6\ \text{k}\Omega}{40.6\ \text{k}\Omega + 600\ \Omega} \\
&= 8.15 \underline{\angle -180^\circ}
\end{aligned}
$$

The magnitude of the midband voltage gain may be determined in decibels.

$$|A_{vs}|_{\text{dB}} = 20 \log A_{vs} = 20 \log 8.15 = 18.2 \text{ dB}$$

$$\theta = -180°$$

Now we shall determine the three corner frequencies.

$$R_{\text{TH1}} = Z_S + R_{\text{in}} = 600\ \Omega + 40.6\ \text{k}\Omega = 41.2\ \text{k}\Omega$$

$$f_{L1} = \frac{1}{2\pi R_{\text{TH1}} C_1} = \frac{1}{2\pi(41.2\ \text{k}\Omega)(2.2\ \mu\text{F})} = 1.76\ \text{Hz}$$

$$R_{\text{TH2}} = R_L + R_{\text{out}} = 10\ \text{k}\Omega + 6.8\ \text{k}\Omega = 16.8\ \text{k}\Omega$$

$$f_{L2} = \frac{1}{2\pi R_{\text{TH2}} C_2} = \frac{1}{2\pi(16.8\ \text{k}\Omega)(25\ \mu\text{F})} = 0.379\ \text{Hz}$$

$$R_{\text{TH3}} \simeq \frac{1}{g_m} = \frac{1}{2.05\ \text{mS}} = 488\ \Omega$$

$$f_{L3} = \frac{1}{2\pi R_{\text{TH3}} C_3} = \frac{1}{2\pi(488\ \Omega)(10\ \mu\text{F})} = 32.6\ \text{Hz}$$

■

The low-end frequency response is taken to be limited by the 3-dB down point produced by the *largest* corner frequency. In Example 9-4, 32.6 Hz (f_{L3}) would be the limit. In this respect f_{L3} is described as being the *dominant corner frequency*. *It is important to note that f_{L1} and f_{L2} are at least one full decade below f_{L3}. This ensures that f_{L3} is indeed dominant.*

9-10 Single-Supply Op Amps

In our previous work we have operated our op amps with dual-polarity supply voltages (e.g., ±12 V or ±15 V). However, in some applications, dual-polarity power supply voltages are unavailable and it may still be desirable to use an IC op amp. In Fig. 9-23(a) we see exactly how this may be accomplished.

Observe that we have illustrated a 741 op amp that is being operated with a single 20-V power supply. Also note that input and output coupling (dc blocking) capacitors C_1 and C_2 have been used. Capacitor C_3 may be described as a decoupling capacitor. Its function will become apparent shortly.

In Fig. 9-23(b) the dc equivalent circuit has been illustrated. Resistors R_1 and R_2 form a voltage divider that is used to bias the noninverting input terminal to one-half of the supply voltage (10 V). Since capacitor C_3 is an open circuit to dc, the dc equivalent circuit is essentially a (unity-gain) voltage follower. Therefore, the output of the op amp will also be biased to approximately one-half of the supply voltage (10 V).

When an op amp is biased by dual-polarity supply voltages, it exhibits the transfer characteristic curve depicted in Fig. 9-24(a). However, when it is biased only by a single positive-supply voltage, its negative supply terminal is taken to ground

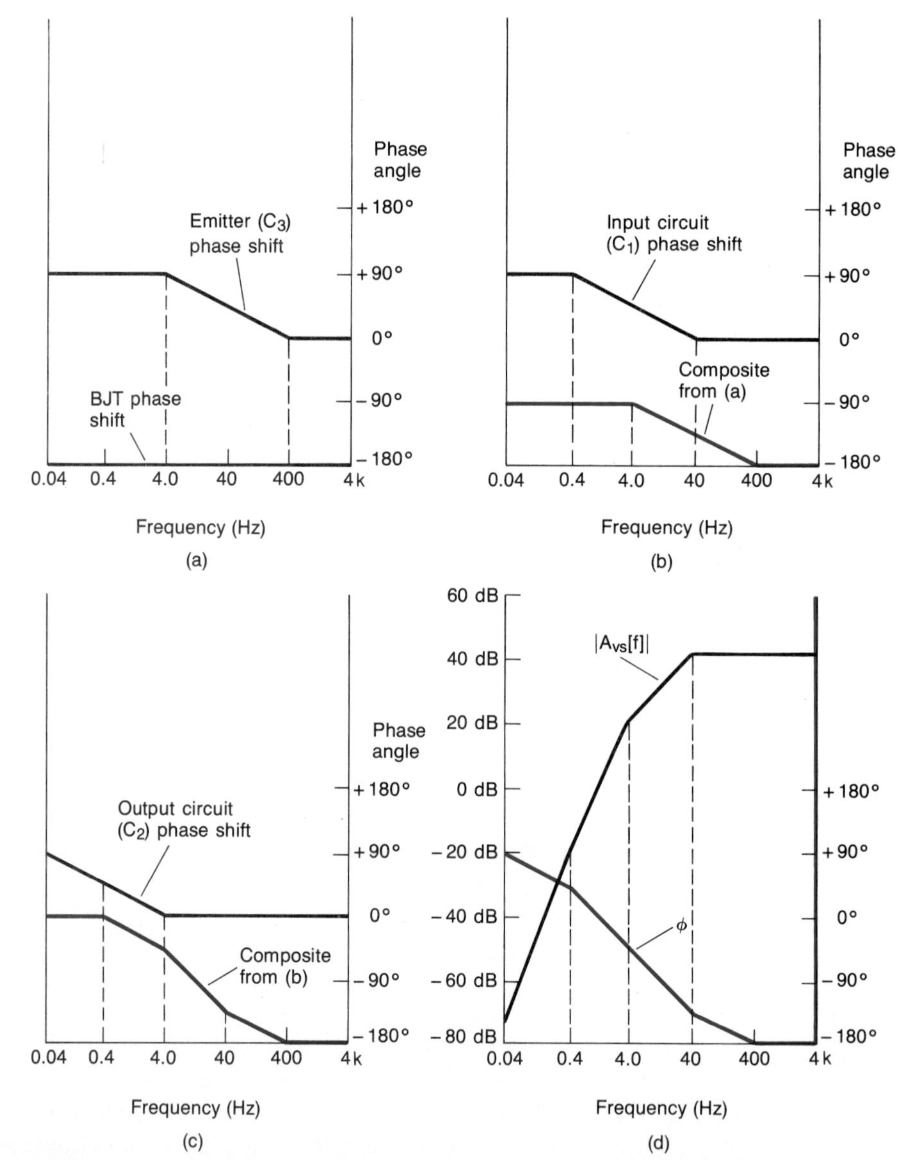

FIGURE 9-22 The total phase shift is found by graphical addition: (a) BJT and C_3; (b) BJT, C_3, and C_1; (c) BJT, C_3, C_1, and C_2; (d) complete amplitude and phase response.

[pin 4 in Fig. 9-23(b)] and its transfer characteristic curve is restricted to the first and second quadrants [refer to Fig. 9-24(b)].

In the midband frequency range, the capacitors become short circuits. In this case, we have the ac equivalent circuit shown in Fig. 9-24(c). Obviously, the circuit then forms a noninverting amplifier configuration. The difference between the dc voltage gain and the ac voltage gain is made possible by C_3.

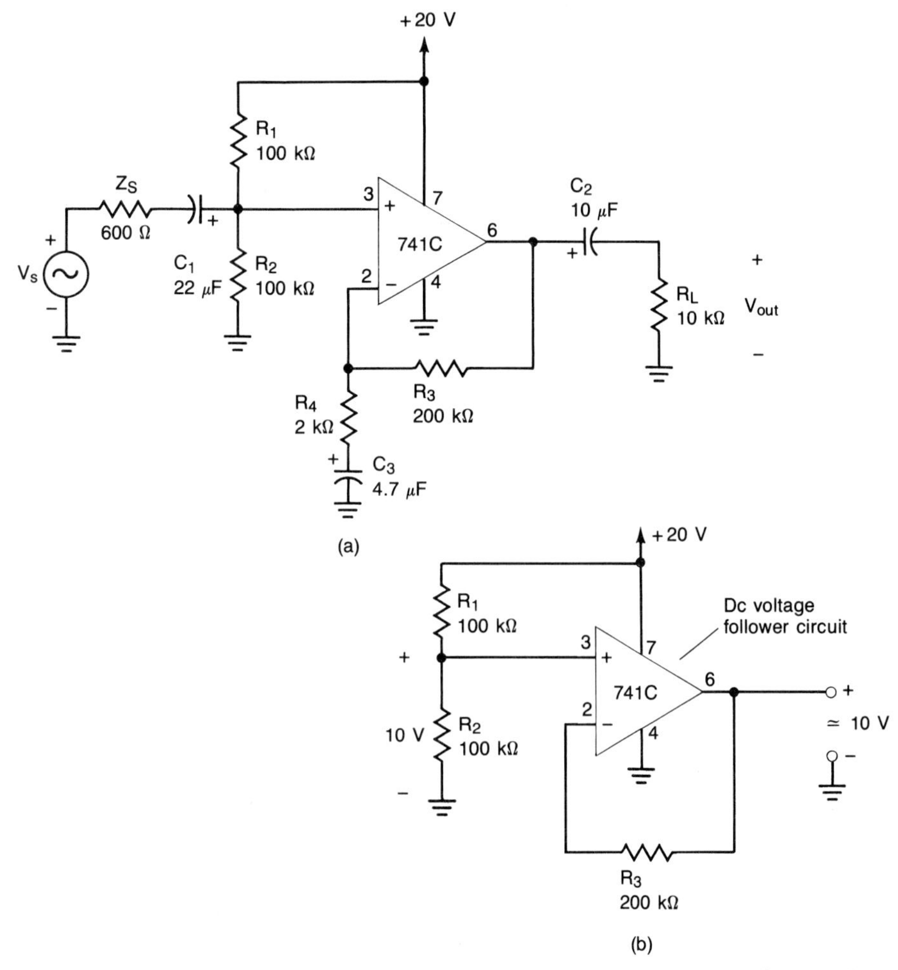

FIGURE 9-23 **Single-supply noninverting amplifier: (a) circuit; (b) dc equivalent.**

EXAMPLE 9-5

Find the midband voltage gain in decibels and the corner frequencies that determine the low-end roll-off of the single-supply amplifier given in Fig. 9-23(a). Which one is dominant?

SOLUTION The midband analysis may be accomplished by referring to Fig. 9-24(c).

$$R_{in} \simeq R_1 \parallel R_2 = 100 \text{ k}\Omega \parallel 100 \text{ k}\Omega = 50 \text{ k}\Omega$$

$$A_{v(oc)} = A_v = 1 + \frac{R_3}{R_4} = 1 + \frac{200 \text{ k}\Omega}{2 \text{ k}\Omega} = 101 \underline{/0^\circ}$$

$$A_{vs} = \frac{R_{in}}{R_{in} + Z_S} A_v = \frac{50 \text{ k}\Omega}{50 \text{ k}\Omega + 600 \ \Omega}(101 \underline{/0^\circ}) = 99.8 \underline{/0^\circ}$$

$$|A_{vs}|_{dB} = 20 \log A_{vs} = 20 \log (99.8) = 40 \text{ dB}$$

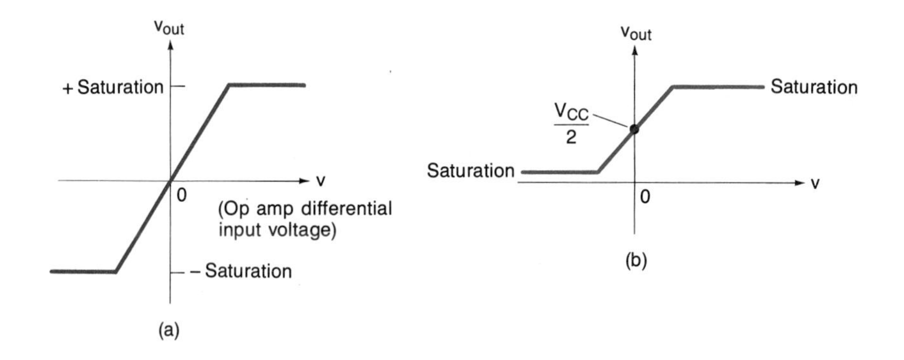

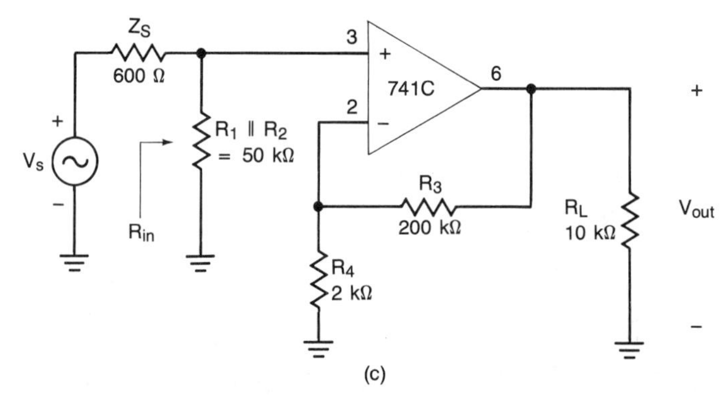

FIGURE 9-24 Dc and ac analysis: (a) transfer characteristic when biased by a dual-polarity supply voltage; (b) single-supply transfer characteristic; (c) ac (midband) equivalent circuit.

The determination of the corner frequencies requires that we first find the Thévenin equivalent resistance ''seen'' by each of the three capacitors (see Fig. 9-25).

$$R_{\text{TH1}} = Z_S + R_1 \| R_2 = 600\ \Omega + 50\ \text{k}\Omega = 50.6\ \text{k}\Omega$$

$$f_{L1} = \frac{1}{2\pi R_{\text{TH1}} C_1} = \frac{1}{2\pi(50.6\ \text{k}\Omega)(22\ \mu\text{F})} = 0.143\ \text{Hz}$$

$$R_{\text{TH2}} = R_L = 10\ \text{k}\Omega$$

$$f_{L2} = \frac{1}{2\pi R_{\text{TH2}} C_2} = \frac{1}{2\pi(10\ \text{k}\Omega)(10\ \mu\text{F})} = 1.59\ \text{Hz}$$

$$R_{\text{TH3}} = R_4 = 2\ \text{k}\Omega$$

$$f_{L3} = \frac{1}{2\pi R_{\text{TH3}} C_3} = \frac{1}{2\pi(2\ \text{k}\Omega)(4.7\ \mu\text{F})} = 16.9\ \text{Hz}$$

The amplifier's f_{L3} is dominant. Therefore, the amplifier's low-end roll-off will occur at 16.9 Hz. ■

As a final comment, we should point out that we do *not* have the necessary skills at this point to draw the frequency response graph for the amplifier. This is

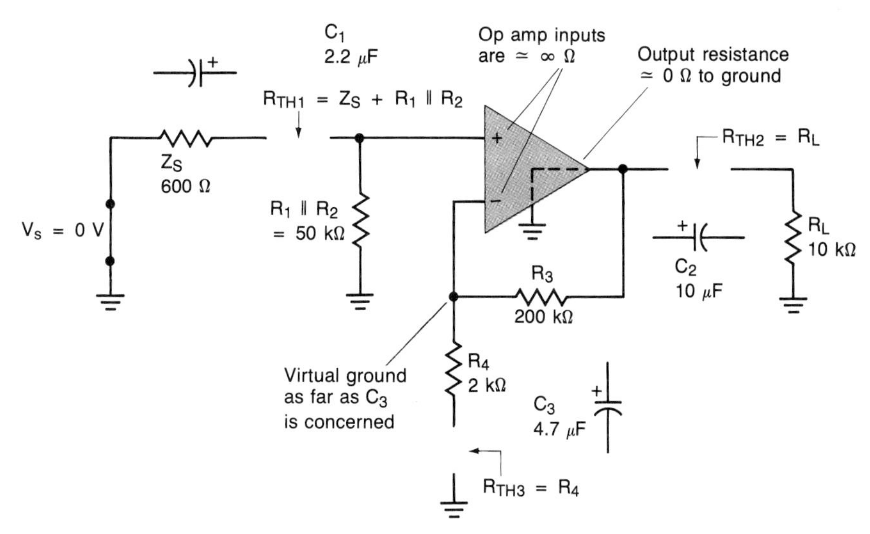

FIGURE 9-25 **Finding the frequency response of the single-supply op amp.**

true because the amplifier's gain equation contains a *pole* at 16.9 Hz and a *zero* at 0.168 Hz. Poles and zeros are covered in Chapter 10.

9-11 BJT and FET Device Capacitances

The basic BJT/FET *low-frequency* model has been repeated in Fig. 9-26(a). Both the BJT and FET exhibit small (picofarad) capacitive effects [see Fig. 9-26(b) and (c)]. At high frequencies the capacitive reactances of these small capacitances become low enough to be regarded as significant. In fact, these device capacitances essentially determine the high-frequency roll-off in the gain of amplifiers. Consequently, they merit our serious attention.

The existence of the capacitive effects should not be a startling revelation. All forward-biased *p-n* junctions demonstrate a capacitive effect known as storage or diffusion capacitance. The normally forward-biased base–emitter *p-n* junction demonstrates diffusion capacitance C_{be}. The capacitive effect associated with a reverse-biased *p-n* junction is called junction capacitance. Since the collector-to-base *p-n* junction is normally reverse biased it exhibits junction capacitance [see C_{cb} in Fig. 9-26(b)]. Similarly, the capacitances C_{dg} and C_{gs} are also junction capacitances since the *p-n* junctions associated with a JFET are also typically reverse biased.

The capacitances C_{ce} and C_{ds} are often termed "header capacitances." These capacitances are associated with the physical construction of the active devices on a header. Since these capacitances are generally less than a picofarad, we shall ignore them.

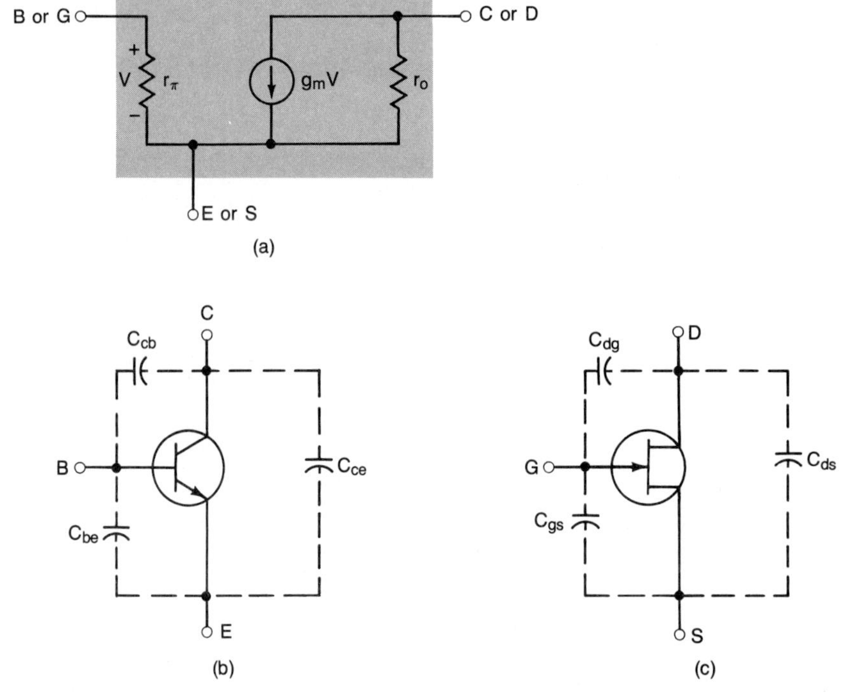

FIGURE 9-26 BJT and FET device capacitances: (a) BJT/FET low-frequency model; (b) BJT capacitances; (c) FET capacitances.

9-12 BJT High-Frequency Model

Let us first devote our attentions to the high-frequency model of the BJT. This equivalent circuit has been illustrated in Fig. 9-27(a). Three components have been added to the low-frequency model given in Fig. 9-26(a). The symbol r'_{bb} denotes the *base-spreading resistance*. The base-spreading resistance represents the resistive path between the ohmic contact used to form the base terminal connection and the active base region of the BJT. Typical values of r'_{bb} range between approximately 30 and 700 Ω. The two other components are C_{be} and C_{cb}, which were discussed previously.

To determine r'_{bb} we need to understand r_π more fully, and how it relates to h_{ie} [see Appendix D]. The symbol h_{ie} is one of the four hybrid parameters first introduced in Chapter 4. The *h* denotes that it is one of the four hybrid parameters. The *i* indicates *i*nput impedance and the *e* stands for common *e*mitter [see Fig. 9-27(b)].

As we can see, h_{ie} represents the sum of the dynamic junction resistance r_π and r'_{bb}. Therefore, we can state

$$r'_{bb} = h_{ie} - r_\pi \tag{9-37}$$

where h_{ie} and r_π are determined at the same *Q*-point.

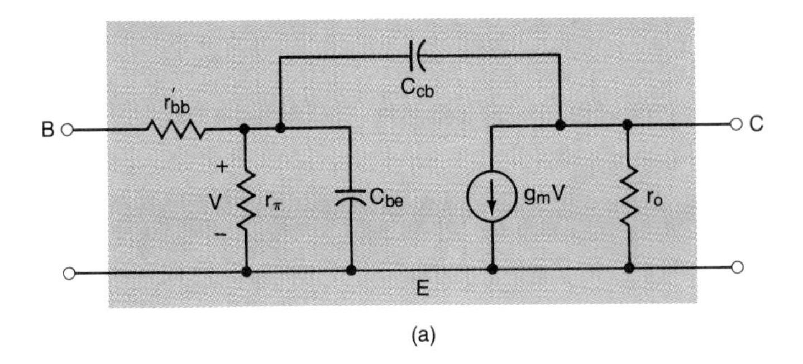

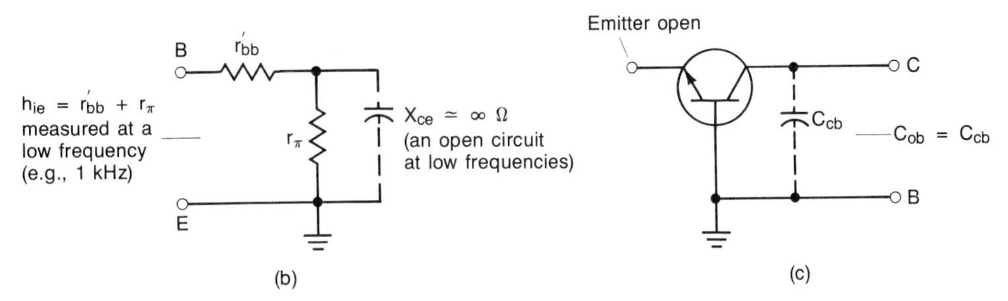

FIGURE 9-27 **BJT high-frequency parameters r'_{bb} and C_{cb}: (a) high-frequency hybrid-pi model; (b) h_{ie}; (c) collector-to-base output capacitance.**

The capacitance C_{cb} is relatively easy to determine if the manufacturer has been kind enough to supply us with C_{ob}. The capacitance C_{ob} is the collector-to-base (output) capacitance for a BJT in its common-base configuration as measured with the emitter (input) terminal left open [unconnected-see Fig. 9-27(c)].

$$C_{cb} = C_{ob} \tag{9-38}$$

where C_{ob} is measured at our Q-point.

The last BJT parameter to be determined is the diffusion capacitance C_{be}. It is very difficult to measure C_{be} directly since it is shunted by r_π. Consequently, manufacturers typically specify its value by indirect means. Manufacturers typically elect to specify either the BJT's *current gain–bandwidth product* f_T, or the *common-emitter current gain cutoff* (corner) *frequency* f_{hfe}. Our next objective will be to develop the relationships between f_{hfe}, f_T, and C_{be}.

Recall that the h-parameter h_{fe} is described as the common-emitter short-circuit current gain. Specifically, it is the ratio of the collector current to the base current as measured when an *ac* short circuit is placed between the collector and emitter terminals [see Fig. 9-28(a)].

In Fig. 9-28(b) we have depicted the BJT's high-frequency equivalent circuit with a short circuit placed across its output. Our next goal is to derive an equation for h_{fe}. The input admittance Y_{in} of the network given in Fig. 9-28(b) is

$$Y_{in} = \frac{1}{r_\pi} + j\omega(C_{be} + C_{cb}) = \frac{1 + j\omega r_\pi(C_{be} + C_{cb})}{r_\pi}$$

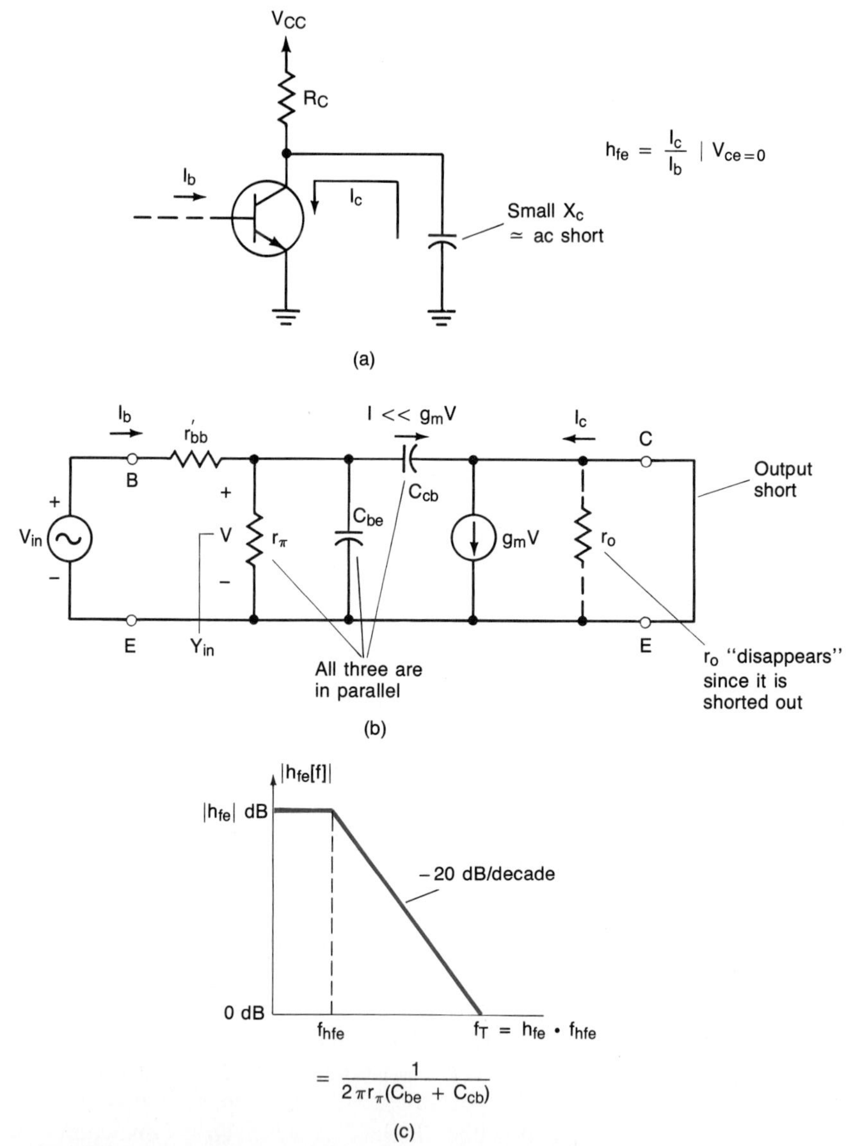

FIGURE 9-28 Finding $h_{fe}[f]$: (a) partial circuit to measure h_{fe}; (b) hybrid-pi high-frequency equivalent circuit to find $h_{fe}[f]$; (c) Bode approximation.

The input impedance Z_{in} is the reciprocal of the admittance.

$$Z_{\text{in}} = \frac{1}{Y_{\text{in}}} = \frac{r_\pi}{1 + j\omega r_\pi(C_{be} + C_{cb})}$$

The controlling voltage V may now be defined.

$$V = Z_{\text{in}} I_b = \frac{r_\pi}{1 + j\omega r_\pi(C_{be} + C_{cb})} I_b$$

The (ac) current that flows through C_{cb} is much less than the current produced by the controlled source.

$$I << g_m V$$

Therefore, we can write

$$I_c = g_m V = \frac{g_m r_\pi}{1 + j\omega r_\pi(C_{be} + C_{cb})} I_b$$

Recalling that

$$r_\pi = \frac{h_{fe}}{g_m}$$

we may substitute this into our equation for I_c. Hence

$$I_c = \frac{g_m(h_{fe}/g_m)}{1 + j\omega r_\pi(C_{be} + C_{cb})} I_b = \frac{h_{fe}}{1 + j\omega r_\pi(C_{be} + C_{cb})} I_b$$

If we divide both sides of the equation by I_b, we arrive at our equation for the short-circuit current gain as a function of frequency $h_{fe}[f]$.

$$h_{fe}[f] = \frac{I_c}{I_b} = \frac{h_{fe}}{1 + j\omega r_\pi(C_{be} + C_{cb})} \tag{9-39}$$

Next we define the corner frequency associated with $h_{fe}[f]$. to be

$$\omega_{hfe} = \frac{1}{r_\pi(C_{be} + C_{cb})}$$

$$\frac{1}{\omega_{hfe}} = r_\pi(C_{be} + C_{cb})$$

Substituting this into Eq. 9-39, we obtain the normalized frequency form of our equation for $h_{fe}[f]$.

$$h_{fe}[f] = \frac{h_{fe}}{1 + j(\omega/\omega_{hfe})} = \frac{h_{fe}}{1 + j(f/f_{hfe})} \tag{9-40}$$

The bode approximation of Eq. 9-40 has been presented in Fig. 9-28(c). At this point we develop a relationship between C_{be} and f_{hfe}. Since the corner frequency has been defined as

$$f_{hfe} = \frac{1}{2\pi r_\pi(C_{be} + C_{cb})} \tag{9-41}$$

we may solve for C_{be} to arrive at Eq. 9-42.

$$2\pi r_\pi(C_{be} + C_{cb}) = \frac{1}{f_{hfe}}$$

$$\boxed{C_{be} = \frac{1}{2\pi\, r_\pi f_{hfe}} - C_{cb}} \tag{9-42}$$

Observe in Fig. 9-28(c) that the frequency at which $h_{fe}[f]$ is equal to unity is f_T. If we let

$$f = f_T = h_{fe} f_{hfe}$$

and substitute this into Eq. 9-40, we obtain

$$h_{fe}[f] = \frac{h_{fe}}{1 + j(f_T/f_{hfe})} = \frac{h_{fe}}{1 + j(h_{fe}f_{hfe}/f_{hfe})} = \frac{h_{fe}}{1 + jh_{fe}}$$

$$\simeq \frac{h_{fe}}{jh_{fe}} = 1 \angle -90°$$

By drawing on Eq. 9-41, we may develop our relationship for C_{be} in terms of f_T.

$$f_T = h_{fe}f_{hfe} = \frac{h_{fe}}{2\pi r_\pi(C_{be} + C_{cb})} = \frac{h_{fe}/r_\pi}{2\pi(C_{be} + C_{cb})}$$

Now recalling that

$$r_\pi = \frac{h_{fe}}{g_m}$$

we may state

$$g_m = \frac{h_{fe}}{r_\pi}$$

Substituting this into our equation for f_T produces

$$f_T = \frac{g_m}{2\pi(C_{be} + C_{cb})} \qquad (9\text{-}43)$$

Solving Eq. 9-43 for C_{be} leads us to

$$\boxed{C_{be} = \frac{g_m}{2\pi f_T} - C_{cb}} \qquad (9\text{-}44)$$

EXAMPLE 9-6

Given the BJT amplifier shown in Fig. 9-29(b), find the BJT's high-frequency parameters C_{cb}, r'_{bb}, and C_{be}.

SOLUTION As we have seen, the collector current of the 2N4124 is 2.37 mA, which produces a g_m of 91.2 mS. Inspection of the BJT's data sheet [Fig. 9-29(a)] reveals that h_{fe} is a minimum of 120, C_{ob} is a maximum of 4 pF, h_{ie} is approximately 1.75 kΩ, and f_T is 300 MHz. Since r_π has been shown to be 1.32 kΩ, we may use Eq. 9-37 to find the base spreading resistance r'_{bb}.

$$r'_{bb} = h_{ie} - r_\pi = 1.75\ \text{k}\Omega - 1.32\ \text{k}\Omega = 430\ \Omega$$

Bias corrections for the collector-to-base capacitance C_{ob} have been given in Fig. 9-29(a).

$$C_{cb} = C_{ob} = 1.9\ \text{pF}$$

Equation 9-44 permits us to determine the base-to-emitter diffusion capacitance C_{be}.

$$C_{be} = \frac{g_m}{2\pi f_T} - C_{cb} = \frac{91.2\ \text{mS}}{2\pi(300\ \text{MHz})} - 1.9\ \text{pF} = 46.5\ \text{pF}$$

These results have been illustrated in Fig. 9-29(c). ■

SMALL SIGNAL CHARACTERISTICS

High-Frequency Current Gain (I_C = 10 mAdc, V_{CE} = 20 Vdc, f = 100 MHz)	2N4123 2N4124		$\|h_{fe}\|$	2.5 3.0	– –	–
Current-Gain – Bandwidth Product (I_C = 10 mAdc, V_{CE} = 20 Vdc, f = 100 MHz)	2N4123 2N4124		f_T	250 300	– –	MHz
Output Capacitance (V_{CB} = 5 Vdc, I_E = 0, f = 100 kHz)		1	C_{ob}	–	4.0	pF
Input Capacitance (V_{BE} = 0.5 Vdc, I_C = 0, f = 100 kHz)		1	C_{ib}	–	8.0	pF
Small-Signal Current Gain (I_C = 2 mAdc, V_{CE} = 1 Vdc, f = 1 kHz)	2N4123 2N4124	5	h_{fe}	50 120	200 480	–
Noise Figure (I_C = 100 μAdc, V_{CE} = 5 Vdc, R_S = 1 k ohm, Noise Bandwidth = 10 Hz to 15.7 kHz)	2N4123 2N4124	3 4	NF	– –	6.0 5.0	dB

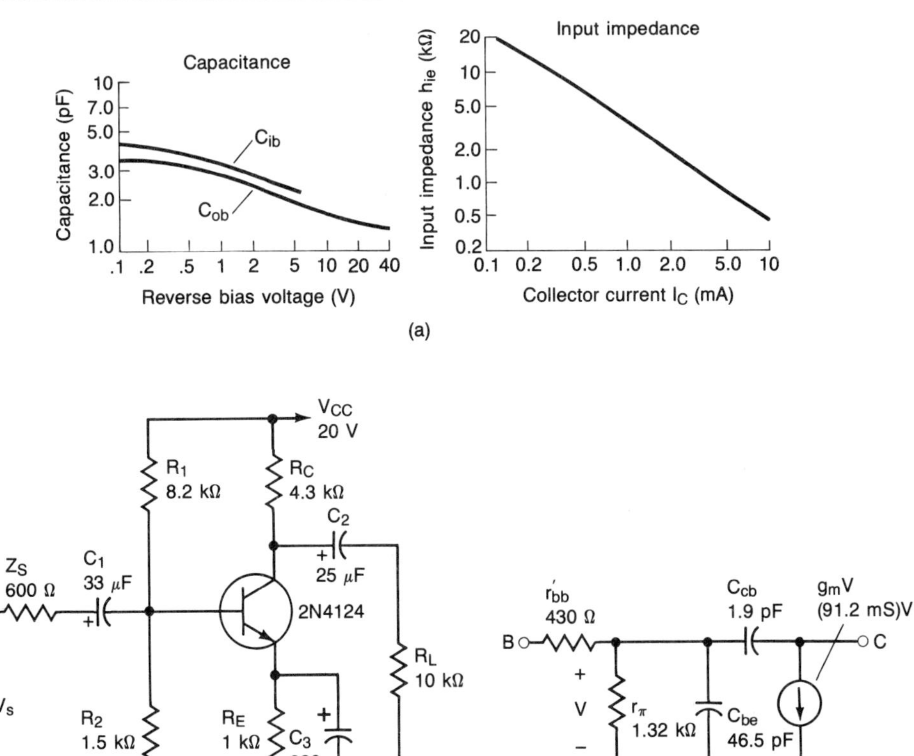

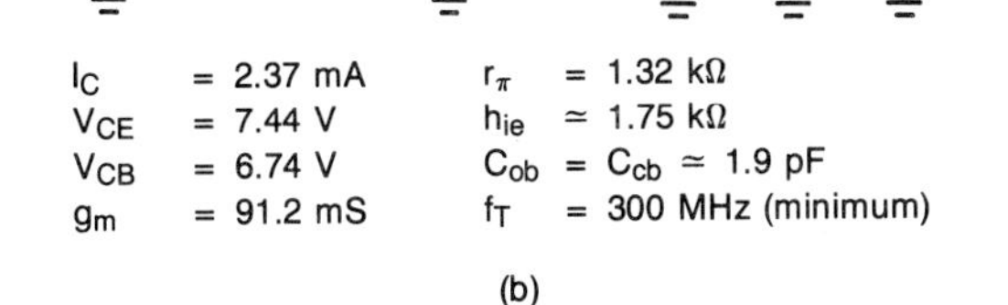

FIGURE 9-29 High-frequency model of the 2N4124 BJT: (a) partial data sheet (reprinted with permission of Motorola, Inc.); (b) amplifier; (c) model.

9-13 FET High-Frequency Model

The capacitances C_{gs} and C_{dg} associated with FETs are relatively easy for manufacturers to specify, as they are shunted by reverse-biased (high-resistance) *p-n* junctions. The FET capacitances have been repeated in Fig. 9-30(a).

The FET capacitances are typically measured under (ac) short-circuit conditions. For example, C_{iss} is the input capacitance with an ac short placed between the drain to source [see Fig. 9-30(b)]. Since C_{gs} is in parallel with C_{dg} and capacitances in parallel add:

$$C_{iss} = C_{gs} + C_{dg}$$

Occasionally, data sheets also list C_{oss} [see Fig. 9-30(c)]. In this case,

$$C_{oss} = C_{ds} + C_{dg}$$

One additional capacitance C_{rss} is also typically specified. C_{rss} is called the reverse transfer capacitance. It is equal to C_{dg}. Hence

$$C_{rss} = C_{dg}$$

FIGURE 9-30 **FET capacitances: (a) device capacitances; (b) input capacitance C_{iss}; (c) output capacitance C_{oss}.**

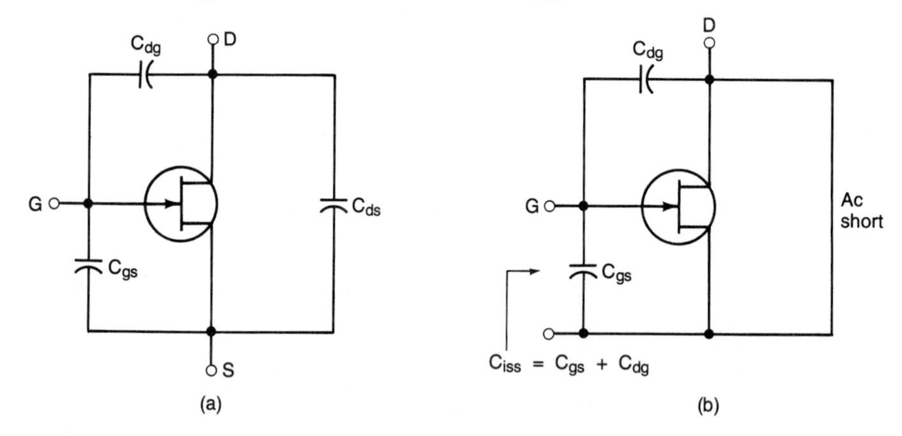

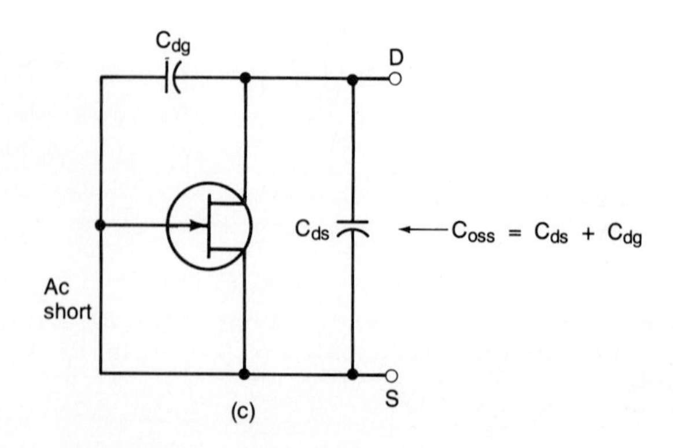

To relate these data sheet parameters to the device capacitances C_{dg}, C_{gs}, and C_{ds} required in our high-frequency FET model, we present the equations cited below.

$$C_{dg} = C_{rss} \tag{9-45}$$

$$C_{gs} = C_{iss} - C_{rss} \tag{9-46}$$

$$C_{ds} = C_{oss} - C_{rss} \tag{9-47}$$

EXAMPLE 9-7

Given the common-source FET amplifier shown in Fig. 9-31(b), find the FET's high-frequency parameters C_{gs} and C_{dg}.

SOLUTION An investigation of the manufacturer's data sheet [Fig. 9-31(a)] yields the typical and maximum values of C_{iss} (4.5 pF typical, 7.0 pF maximum) and C_{rss} (1.5 pF typical, 3.0 pF maximum) for a V_{DS} of 15 V and a V_{GS} of 0 V as the given dc bias conditions. The range of dc bias values for our circuit was summarized in Table 6-8. A quick inspection soon reveals that the bias values are not the same as the manufacturer's test conditions. However, since they are all that we have, we are forced to use them. (Corrections can be made, but we shall not delve into them.) To be conservative, we shall use the maximum values. Hence, by Eq. 9-45,

$$C_{dg} = C_{rss} = 3.0 \text{ pF}$$

and by Eq. 9-46,

$$C_{gs} = C_{iss} - C_{rss} = 7.0 \text{ pF} - 3.0 \text{ pF} = 4 \text{ pF}$$

The model has been depicted in Fig. 9-31(c). ■

9-14 Miller's Theorem

As we saw in Figs. 9-29 and 9-31, the high-frequency BJT and FET models both include a feedback capacitance connected between their input and output terminals. The analysis of amplifier circuits employing a feedback impedance between their input and output terminals can be greatly simplified by drawing on *Miller's theorem*.

In Fig. 9-32(a) we see our generalized amplifier model with a feedback impedance Z_f connected between its input and output. The loaded voltage gain

$$A_v = \frac{V_{\text{out}}}{V_{\text{in}}} = \frac{R_L}{R_L + R_{\text{out}}} A_{v(\text{oc})}$$

will be slightly modified because an additional current I_f is drawn from the amplifier's output. At this point we shall *not* concern ourselves with the determination of A_v.

From Fig. 9-32(a),

$$I_f = \frac{V_{\text{out}} - V_{\text{in}}}{Z_f}$$

DYNAMIC CHARACTERISTICS

Forward Transfer Admittance (1) (V_{DS} = 15 Vdc, V_{GS} = 0, f = 1 kHz)	2N5457	$\|y_{fs}\|$	1000	3000	5000	μmhos
	2N5458		1500	4000	5500	
	2N5459		2000	4500	6000	
Output Admittance (1) (V_{DS} = 15 Vdc, V_{GS} = 0, f = 1 kHz)		$\|y_{os}\|$	—	10	50	μmhos
Input Capacitance (V_{DS} = 15 Vdc, V_{GS} = 0, f = 1 MHz)		C_{iss}	—	4.5	7.0	pF
Reverse Transfer Capacitance (V_{DS} = 15 Vdc, V_{GS} = 0, f = 1 MHz)		C_{rss}	—	1.5	3.0	pF

(1) Pulse Test: Pulse Width ≤ 630 ms; Duty Cycle ≤ 10%

(a)

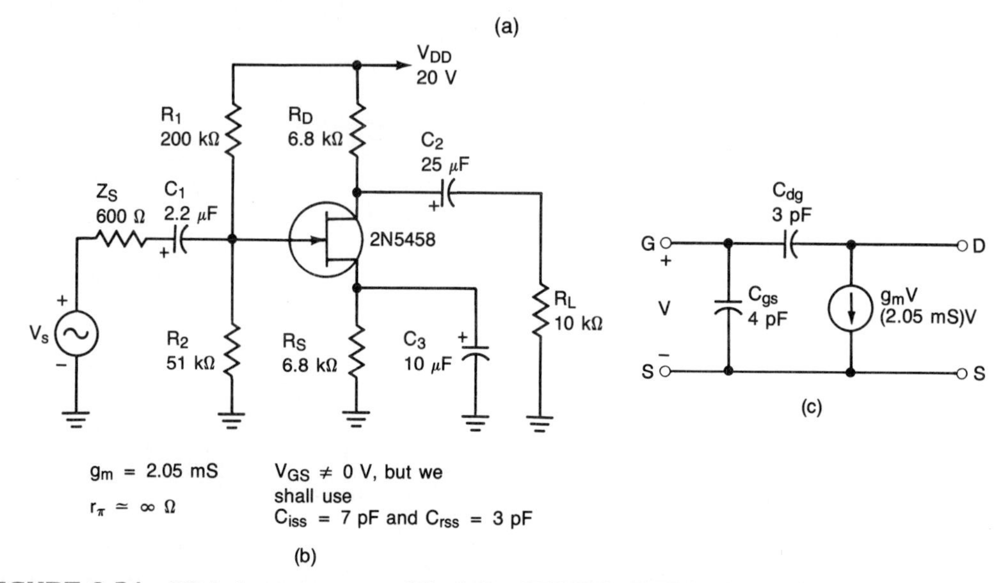

FIGURE 9-31 High-frequency model of the 2N5458 JFET: (a) partial data sheet (reprinted with permission of Motorola, Inc.); (b) amplifier; (c) model.

Since $V_{\text{in}} = V_{\text{out}}/A_v$, we may substitute this into the relationship above.

$$I_f = \frac{V_{\text{out}} - V_{\text{out}}/A_v}{Z_f} = \frac{(1 - 1/A_v)V_{\text{out}}}{Z_f}$$

$$= \frac{(A_v - 1)/A_v}{Z_f} V_{\text{out}} = \frac{V_{\text{out}}}{A_v Z_f/(A_v - 1)}$$

Let us reflect on this result. If a feedback impedance Z_f is connected between an amplifier's input and output terminals, an additional current I_f will be drawn from the amplifier's output. If we remove Z_f and connect an impedance Z_{fo} between the amplifier's output and ground, *exactly the same current results*. The value of this equivalent output impedance is given by

$$Z_{fo} = \frac{A_v Z_f}{A_v - 1} \tag{9-48}$$

This observation has been illustrated in Fig. 9-32(b). If $A_v >> 1$,

$$Z_{fo} \simeq \frac{A_v}{A_v} Z_f = Z_f$$

FIGURE 9-32 Miller's theorem: (a) output analysis; (b) output equivalent; (c) input analysis; (d) input equivalent; (e) basic intent of Miller's theorem.

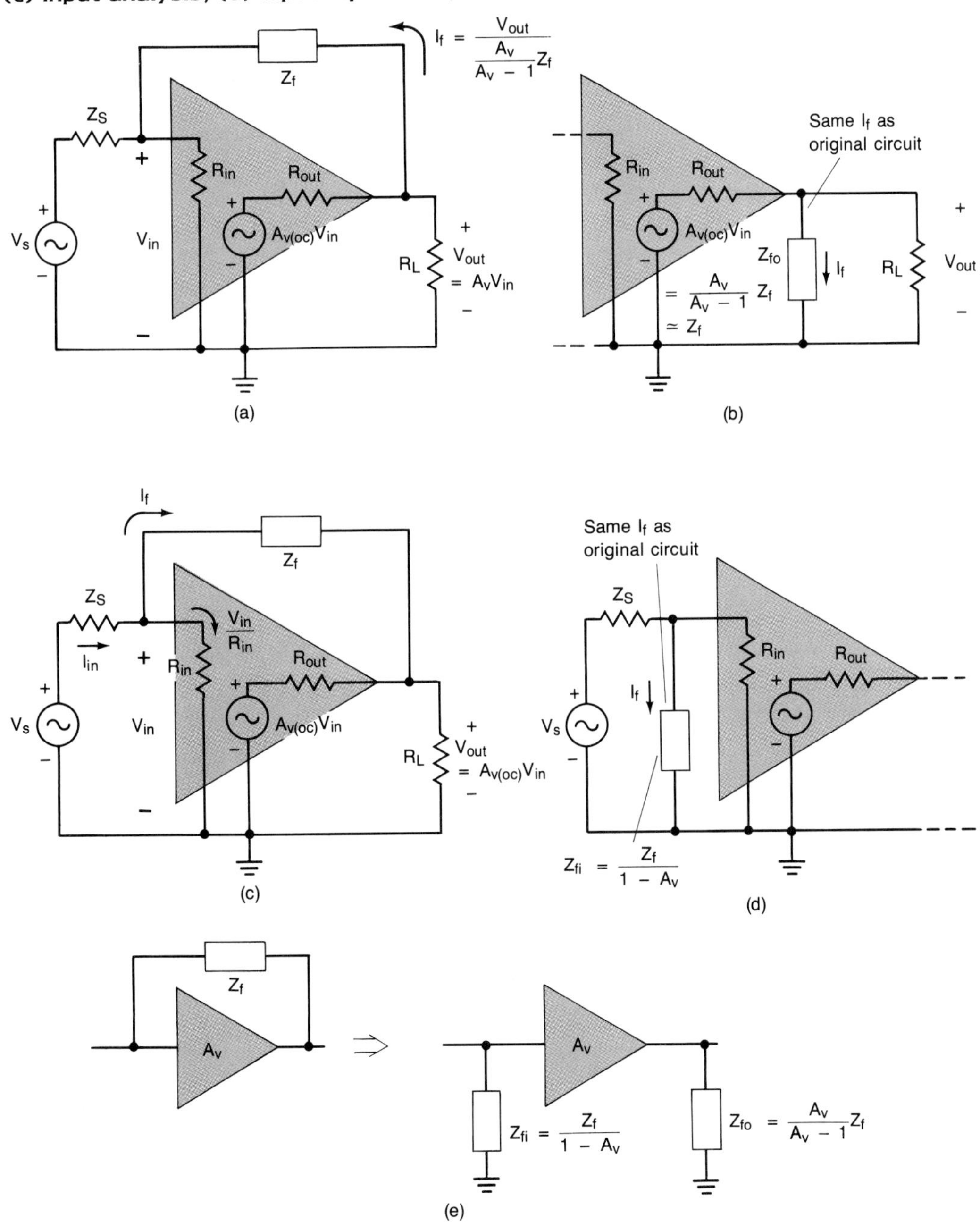

In brief, we can analyze an amplifier with a feedback impedance Z_f merely by removing it and connecting an equivalent impedance across the amplifier's output.

To include the effects of Z_f, we must slightly modify our equation for A_v. To simplify our notation, we define the *ac equivalent load impedance* Z_L as the parallel combination of Z_{fo} and R_L. Thus

$$A_v = \frac{V_{\text{out}}}{V_{\text{in}}} = \frac{Z_L}{Z_L + R_{\text{out}}} A_{v(\text{oc})} \tag{9-49}$$

where $Z_L = Z_{fo} \parallel R_L$. In practice, the feedback impedance is often so large that it is a reasonable approximation to ignore the Z_{fo} loading effects on the output. In this event we may find A_v in the usual manner and *disregard* Eq. 9-49.

The feedback impedance also affects an amplifier's input circuit - but to a *much larger degree*. Now that we know how to approximate A_v, let us analyze the input circuit. Consider Fig. 9-32(c).

An additional current component I_f is drawn from the signal source. This will affect the amplifier's equivalent input impedance. Let us see why this is so by first finding the amplifier's equivalent input admittance Y_{in}.

$$I_{\text{in}} = \frac{V_{\text{in}}}{R_{\text{in}}} + \frac{V_{\text{in}} - V_{\text{out}}}{Z_f} = \frac{V_{\text{in}}}{R_{\text{in}}} + \frac{V_{\text{in}} - A_v V_{\text{in}}}{Z_f}$$

$$= \left(\frac{1}{R_{\text{in}}} + \frac{1 - A_v}{Z_f}\right) V_{\text{in}}$$

$$Y_{\text{in}} = \frac{I_{\text{in}}}{V_{\text{in}}} = \frac{1}{R_{\text{in}}} + \frac{1 - A_v}{Z_f}$$

Recalling that admittances in parallel add, the result above suggests that we have two equivalent input admittances connected across the amplifier's input circuit. The reciprocal of the admittance term due to the feedback through Z_f produces the equivalent impedance across the amplifier's input Z_{fi}. Hence

$$Z_{fi} = \frac{1}{(1 - A_v)/Z_f} = \frac{Z_f}{1 - A_v}$$

$$\boxed{Z_{fi} = \frac{Z_f}{1 - A_v}} \tag{9-50}$$

where A_v includes the effects of output shunting by Z_f [see Fig. 9-32(d)]. The "bottom line" has been emphasized in Fig. 9-32(e).

9-15 High-Frequency Roll-Off in the BJT and FET Amplifiers

To perform a high-frequency analysis with relatively minimal effort, we need to eliminate the feedback capacitance (e.g., C_{cb} or C_{dg}). We shall do this by using Miller's theorem. Consider Fig. 9-33(a). The feedback impedance Z_f is a capacitive reactance X_{cf}.

$$Z_f = -jX_{cf} = \frac{1}{j\omega C_f}$$

By Miller's theorem the equivalent input impedance is, from Eq. 9-50,

$$Z_{fi} = \frac{Z_f}{1 - A_v} = \frac{jX_{cf}}{1 - A_v} = \frac{1/j\omega C_f}{1 - A_v} = \frac{1}{j\omega(1 - A_v)C_f}$$

From this result it is clear that the feedback capacitance C_f produces a much larger equivalent capacitance across the amplifier's input. Specifically,

$$\boxed{C_{fi} = (1 - A_v)C_f} \tag{9-51}$$

Similarly, we may use Miller's theorem to find the equivalent capacitance across the amplifier's output. From Eq. 9-48,

$$Z_{fo} = \frac{A_v Z_f}{A_v - 1} \simeq Z_f = \frac{1}{j\omega C_f}$$

and

$$\boxed{C_{fo} \simeq C_f} \tag{9-52}$$

These results have been summarized in Fig. 9-33(b).

FIGURE 9-33 **Applying Miller's theorem: (a) impedances; (b) equivalent capacitances.**

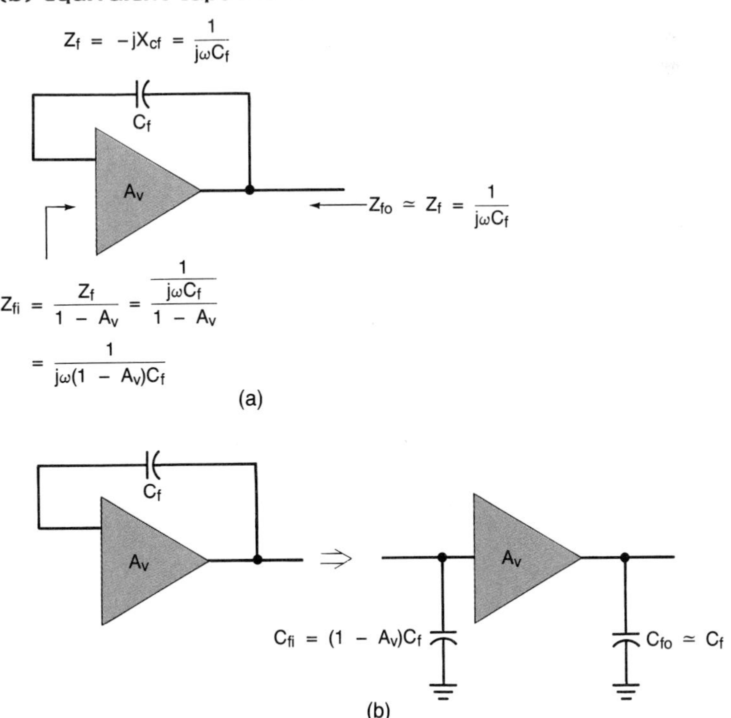

Now let us pursue the high-frequency analysis of our common-emitter amplifier. It has been repeated in Fig. 9-34(a). The high-frequency equivalent circuit has been given in Fig. 9-34(b). The capacitances C_{w1} and C_{w2} are used to represent stray wiring capacitances. All of the 2N4124's high-frequency parameters have been indicated.

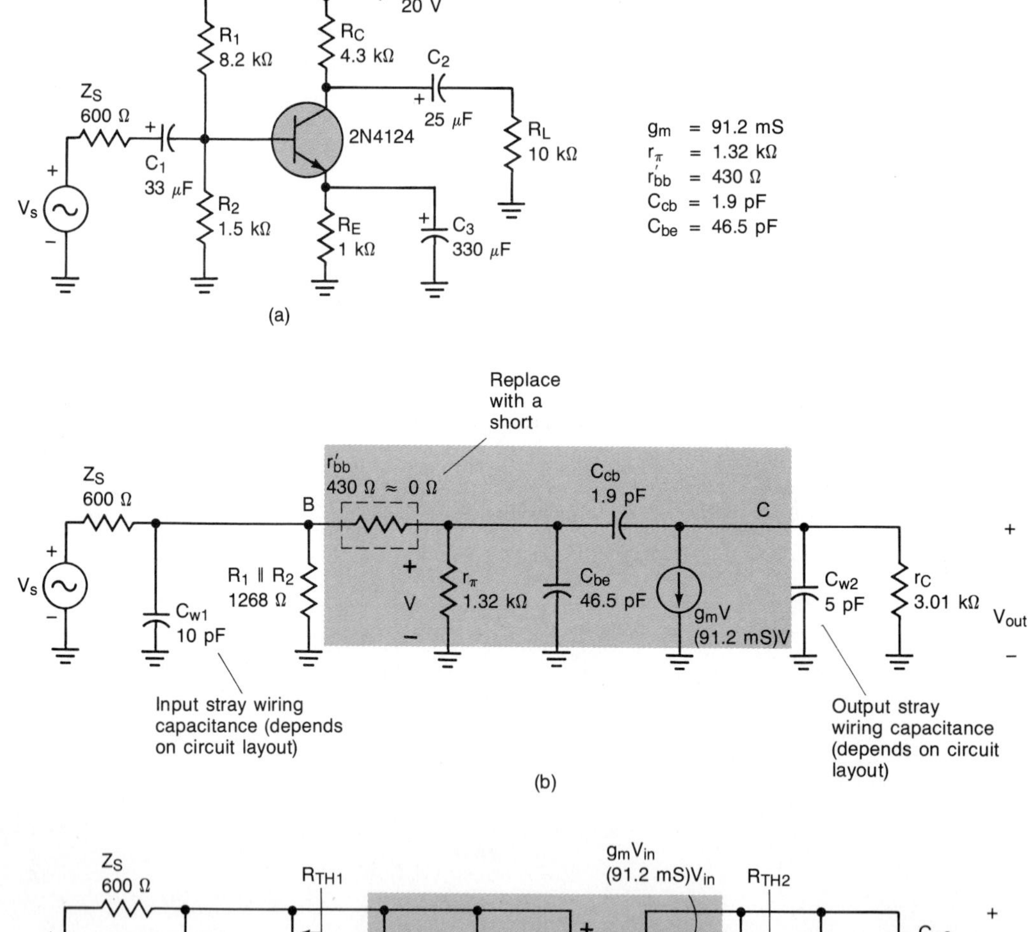

FIGURE 9-34 High-frequency analysis of the common-emitter amplifier: (a) schematic diagram; (b) high-frequency ac equivalent circuit; (c) simplification by Miller's theorem.

Observe that r'_{bb} has been shown in a dashed-line box. To simplify our work we shall ignore it. This will have the effect of raising the value of our prediction of the input high-end roll-off frequency and raise the values of A_v and A_{vs}. *Rigorous analyses should include* r'_{bb}.

To apply Miller's theorem, we need to know the voltage gain A_v "across" the feedback impedance. This is the voltage gain from the active base region (across r_π) to the load. Then we can find C_{fi} and C_{fo} by using Eqs. 9-51 and 9-52, respectively. Hence,

$$A_v = \frac{V_{\text{out}}}{V} = -g_m r_C = -(91.2\text{mS})(3.01\text{ k}\Omega) = -275$$

Note that we have used the ac collector resistance r_C to shorten our efforts. Continuing the analysis, we see that

$$\begin{aligned} C_{fi} &= (1 - A_v)C_f = (1 - A_v)C_{cb} \\ &= (1 + 275)(1.9\text{ pF}) = 524\text{ pF} \end{aligned}$$

and

$$C_{fo} \simeq C_{cb} = 1.9\text{ pF}$$

These results have been used to simplify the high-frequency equivalent circuit [refer to Fig. 9-34(c)]. The equivalent capacitances C_{in} and C_{out} have been defined in the figure. With a little reflection, it is clear that the capacitances C_{in} and C_{out} effectively form low-pass filters. Consequently, as the signal frequency increases, their capacitive reactances decrease, which serves to reduce the overall voltage gain of the amplifier.

To determine the input and output corner frequencies, we must first find the Thévenin equivalent resistances "seen" by C_{in} and C_{out}. The reader is again directed to Fig. 9-34(c).

$$\begin{aligned} R_{\text{TH1}} &= Z_S \,\|\, R_1 \,\|\, R_2 \,\|\, r_\pi = 600\ \Omega \,\|\, 1260\ \Omega \,\|\, 1.32\text{ k}\Omega \\ &= 311\ \Omega \end{aligned}$$

and

$$f_{H1} = \frac{1}{2\pi R_{\text{TH1}} C_{\text{in}}} = \frac{1}{2\pi(311\ \Omega)(580.5\text{ pF})} = 882\text{ kHz}$$

Similarly,

$$R_{\text{TH2}} = r_C = 3.01\text{ k}\Omega$$

$$f_{H2} = \frac{1}{2\pi\, R_{\text{TH2}}\, C_{\text{out}}} = \frac{1}{2\pi(3.01\text{ k}\Omega)(6.9\text{ pF})} = 7.66\text{ MHz}$$

Clearly, the input circuit corner frequency f_{H1} is dominant. This is true because of the large equivalent input capacitance given by Miller's theorem. Quite often, this is referred to as the "Miller effect."

From our extensive analysis of an amplifier's low-frequency response the reader should be willing to accept that the high-frequency voltage gain may be described by

$$A_{vs}[f] = \frac{A_{vs}}{[1 + j(f/f_{H1})][1 + j(f/f_{H2})]} \tag{9-53}$$

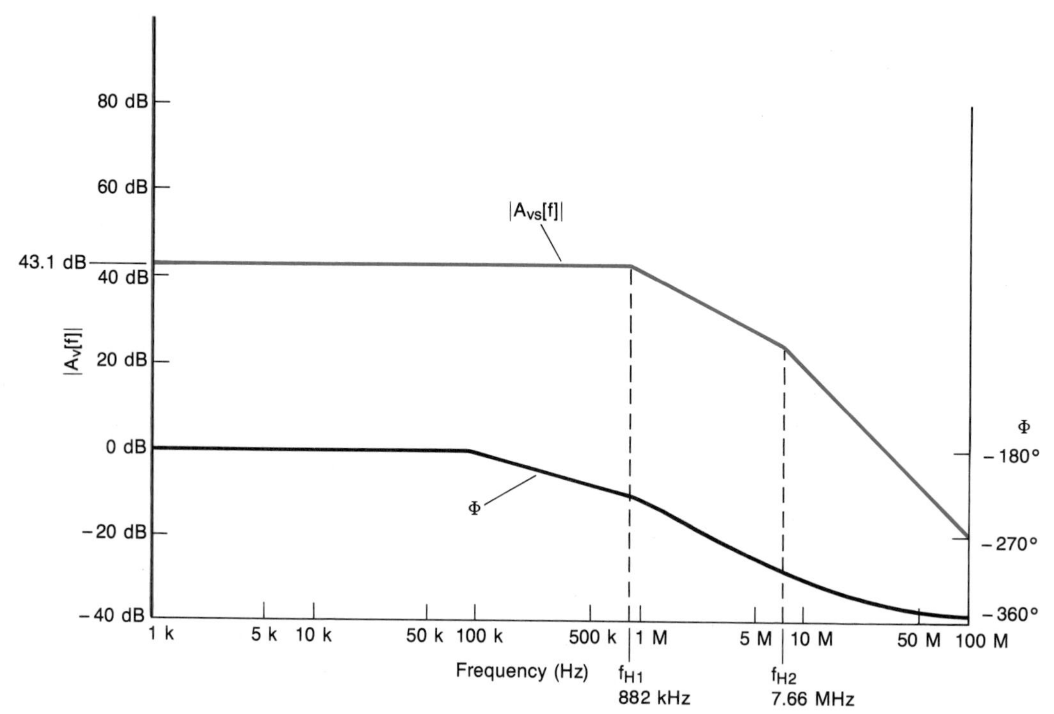

FIGURE 9-35 High-frequency response of the BJT amplifier given in Fig. 9-34(a).

The magnitude in decibels of Eq. 9-53 is given by

$$|A_{vs}[f]|_{\text{dB}} = 20 \log A_{vs} - 10 \log[1 + (f/f_{H1})^2] - 10 \log[1 + (f/f_{H2})^2] \tag{9-54}$$

The corresponding phase angles may be found from

$$\phi = \theta - \tan^{-1}\frac{f}{f_{H1}} - \tan^{-1}\frac{f}{f_{H2}} \tag{9-55}$$

Without delving into the mathematical proof, it should be reasonably clear that Bode approximations may be used to sketch an amplifier's high-frequency response. This has been accomplished in Fig. 9-35. From Fig. 9-34(c) and recalling that the capacitances are open circuits in the midband, we have

$$R_{\text{in}} = R_1 \parallel R_2 \parallel r_\pi = 1268\ \Omega \parallel 1.32\ \text{k}\Omega = 647\ \Omega$$

$$A_{vs} = \frac{R_{\text{in}}}{R_{\text{in}} + Z_S} A_v = \frac{647\ \Omega}{647\ \Omega + 600\ \Omega}(-275) = -143$$

$$|A_{vs}|_{\text{dB}} = 20 \log A_{vs} = 20 \log 143 = 43.1\ \text{dB}$$

EXAMPLE 9-8

Given the common-source JFET amplifier shown in Fig. 9-36(a), perform a high-frequency analysis to determine f_{H1} and f_{H2}.

SOLUTION The high-frequency parameters C_{dg} and C_{gs} determined in Example 9-7 have been repeated in Fig. 9-36(a). The midband ac equivalent circuit has been

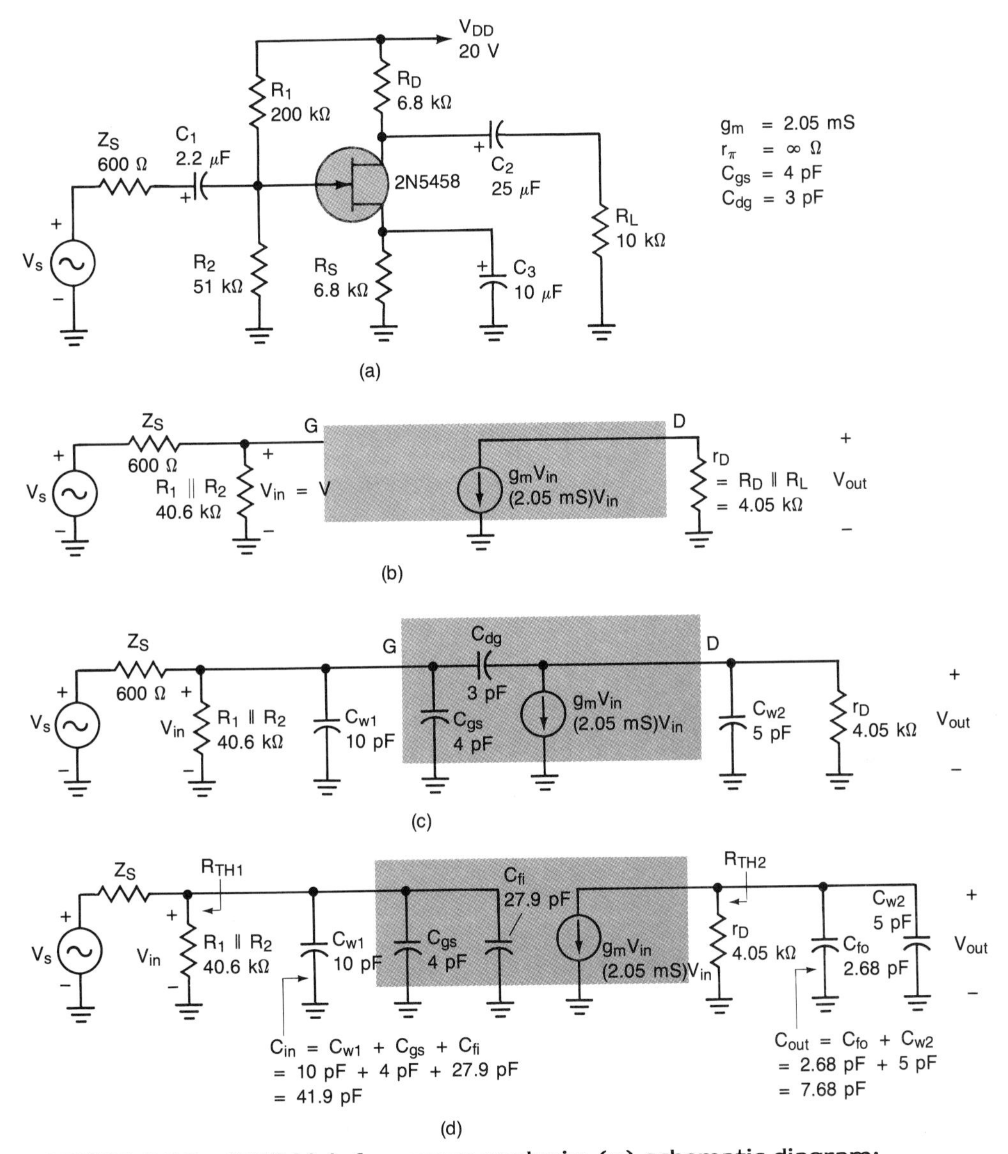

FIGURE 9-36 **JFET high-frequency analysis: (a) schematic diagram; (b) midband ac equivalent circuit; (c) high-frequency ac equivalent circuit; (d) simplification by using Miller's theorem.**

drawn in Fig. 9-36(b). Using the ac equivalent drain resistance r_D, we may easily find A_v.

$$A_v = \frac{V_{\text{out}}}{V_{\text{in}}} = -g_m r_D = -(2.05 \text{ mS})(4.05 \text{ k}\Omega) = -8.30$$

The high-frequency ac equivalent circuit [Fig. 9-36(c)] includes stray wiring capacitances (C_{w1} and C_{w2}), C_{dg} and C_{gs}. By applying Miller's theorem, we may find C_{fi} and C_{fo}.

$$C_{fi} = (1 - A_v)C_f = (1 - A_v)C_{dg}$$
$$= (1 + 8.30)(3 \text{ pF}) = 27.9 \text{ pF}$$

Since the gain A_v is relatively small, we may *not* approximate C_{fo}. Hence

$$C_{fo} = \frac{A_v C_f}{A_v - 1} = \frac{A_v C_{dg}}{A_v - 1} = \frac{(-8.30)(3 \text{ pF})}{-8.30 - 1} = 2.68 \text{ pF}$$

Using these results, we arrive at the simplified circuit given in Fig. 9-36(d). Note that C_{in} and C_{out} have been defined. The Thévenin equivalent resistance associated with C_{in} is used to determine f_{H1}.

$$R_{\text{TH1}} = Z_S \| R_1 \| R_2 = 600 \ \Omega \| 40.6 \text{ k}\Omega = 591 \ \Omega$$

$$f_{H1} = \frac{1}{2\pi R_{\text{TH1}} C_{\text{in}}} = \frac{1}{2\pi(591 \ \Omega)(41.9 \text{ pF})} = 6.43 \text{ MHz}$$

Similarly, for the output we have

$$R_{\text{TH2}} = r_D = 4.05 \text{ k}\Omega$$

$$f_{H2} = \frac{1}{2\pi R_{\text{TH2}} C_{\text{out}}} = \frac{1}{2\pi(4.05 \text{ k}\Omega)(7.68 \text{ pF})} = 5.12 \text{ MHz}$$ ■

The complete frequency response of an amplifier should encompass the frequencies for which both the low-end and high-end amplitude roll-off occurs. This has been illustrated in Fig. 9-37. The *bandwidth* (BW) of an amplifier is defined to be between the two 3-dB down points. In general,

$$\boxed{\text{BW} = f_H - f_L \simeq f_H} \tag{9-56}$$

FIGURE 9-37 Complete amplifier spectral response.

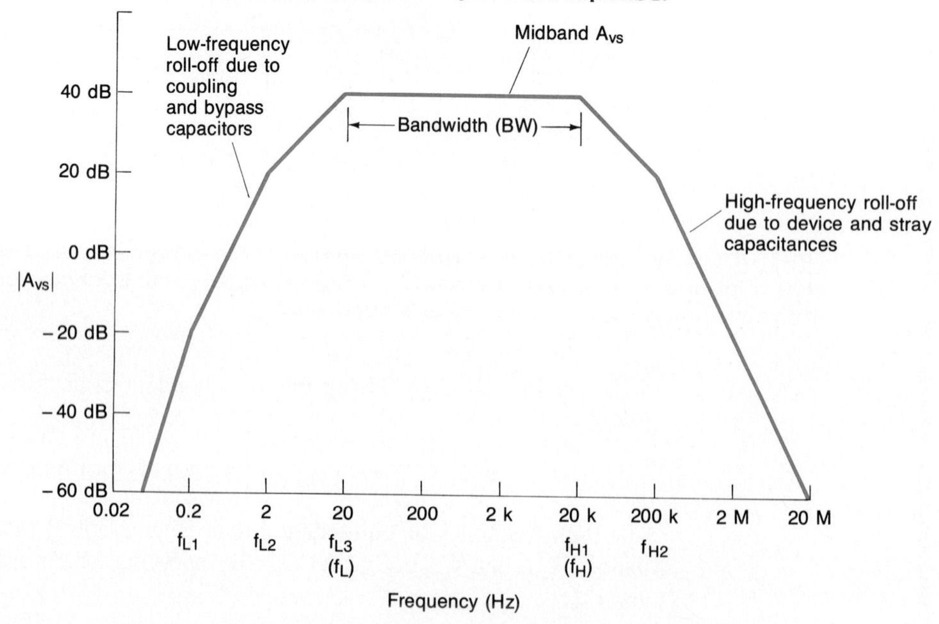

where f_H is high-end corner frequency and f_L is the low-end corner frequency.

The f_H for an amplifier is equal to its *lowest* high-end corner frequency - if its other high-end corner frequencies are at least one decade greater. Otherwise, an interaction will occur that could place f_H at a frequency *below* the lowest high-end corner. Similarly, the f_L is equal to the *highest* low-end corner frequency - if its other low-end corner frequencies are at least one decade below it. Otherwise, interaction can occur that would place f_L *above* its dominant low-end corner frequency.

Since f_H is typically much greater than f_L, the bandwidth is approximately equal to f_H. The bandwidth of our BJT amplifier is approximately 865 kHz and the bandwidth of our JFET amplifier is approximately 3.68 MHz. Typically, FET amplifiers offer much better high-end frequency responses than do BJT amplifiers.

9-16 High-Frequency Roll-Off in the Frequency-Compensated Op Amp

In Section 9-4 we developed the op amp transconductance model. Specifically, we saw that (Eq. 9-9)

$$v_{\text{out}} = \frac{g_m}{j\omega C_c}(v_{\text{in2}} - v_{\text{in1}})$$

and solving for the ac voltage gain as a function of frequency yields

$$A_v[f] = \frac{v_{\text{out}}}{v_{\text{in2}} - v_{\text{in1}}} = \frac{g_m}{j\omega C_c} = \frac{g_m}{j2\pi f C_c} \quad (9\text{-}57)$$

As we have seen previously, negative feedback has a stabilizing effect on an amplifier's voltage gain. This fact is reflected in Eq. 9-57. Specifically, the ac voltage gain is under relatively tight control because of the local negative feedback produced by the feedback capacitor C_c (refer to Fig. 9-4). The dc voltage gain is *not* as well defined. This is because the feedback capacitor is essentially an open circuit to dc. Therefore, the high-voltage-gain inverting amplifier is operated without feedback as far as dc is concerned. When no feedback is used, an amplifier is described as being "open loop."

The frequency response of an op amp without any external feedback has been depicted in Fig. 9-38. A_{VO} is the dc voltage gain. The use of the frequency compensation capacitor C_c produces the corner frequency f_H and causes the voltage gain to roll off at a rate of -20 dB/decade.

The frequency that lowers the op amp's open loop (no external feedback) voltage gain to unity (0 dB) is called f_T. This notation is referred to as the *transition frequency* or the *gain–bandwidth product*. The term "transition frequency" is appropriate since the op amp makes a transition from amplification to attenuation for signal frequencies beyond f_T. The term "gain-bandwidth product" is probably the most useful description (see Fig. 9-38).

The *current* gain–bandwidth product for the BJT and the gain–bandwidth product for an op amp are very similar. The reader is invited to compare Figs. 9-28(c) and 9-38(a).

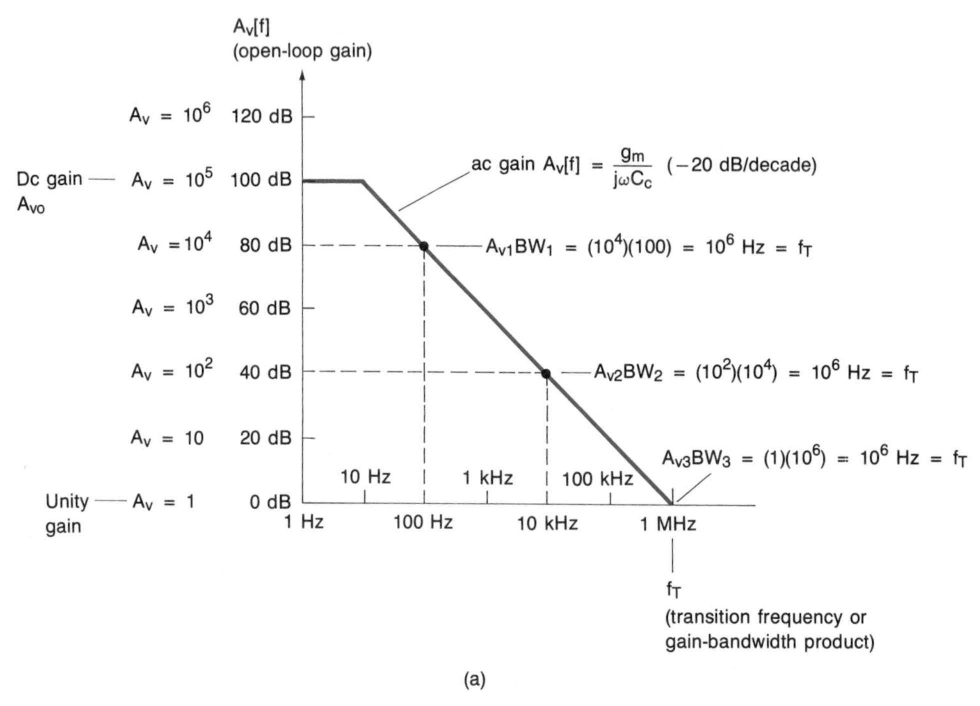

FIGURE 9-38 Open-loop frequency response of a frequency-compensated op amp: (a) typical values for the 741 op amp.

EXAMPLE 9-9

A 741 op amp has a C_c of 30 pF and an f_T of 1 MHz. Find its equivalent g_m. [The 741's partial data sheet is given in Fig. 9-38(b).]

SOLUTION since $A_v[f]$ is unity at f_T, we may draw on Eq. 9-57 to solve for g_m.

$$|A_v[f]| = \frac{g_m}{2\pi f_T C_c} = 1$$

$$g_m = 2\pi f_T C_c = 2\pi(1 \text{ MHz})(30 \text{ pF}) = 0.188 \text{ mS}$$

■

PROBLEMS

Drill, Derivations, and Definitions

Section 9-1

9-1. A common-emitter BJT amplifier using a 2N3904 silicon *npn* transistor has been illustrated in Fig. 9-39(a). The pertinent transistor parameters have been summarized

LM741/LM741A/LM741C/LM741E Operational Amplifier

General Description

The LM741 series are general purpose operational amplifiers which feature improved performance over industry standards like the LM709. They are direct, plug-in replacements for the 709C, LM201, MC1439 and 748 in most applications.

The amplifiers offer many features which make their application nearly foolproof: overload protection on the input and output, no latch-up when the common mode range is exceeded, as well as freedom from oscillations.

The LM741C/LM741E are identical to the LM741/LM741A except that the LM741C/LM741E have their performance guaranteed over a 0°C to +70°C temperature range, instead of −55°C to +125°C.

Schematic and Connection Diagrams (Top Views)

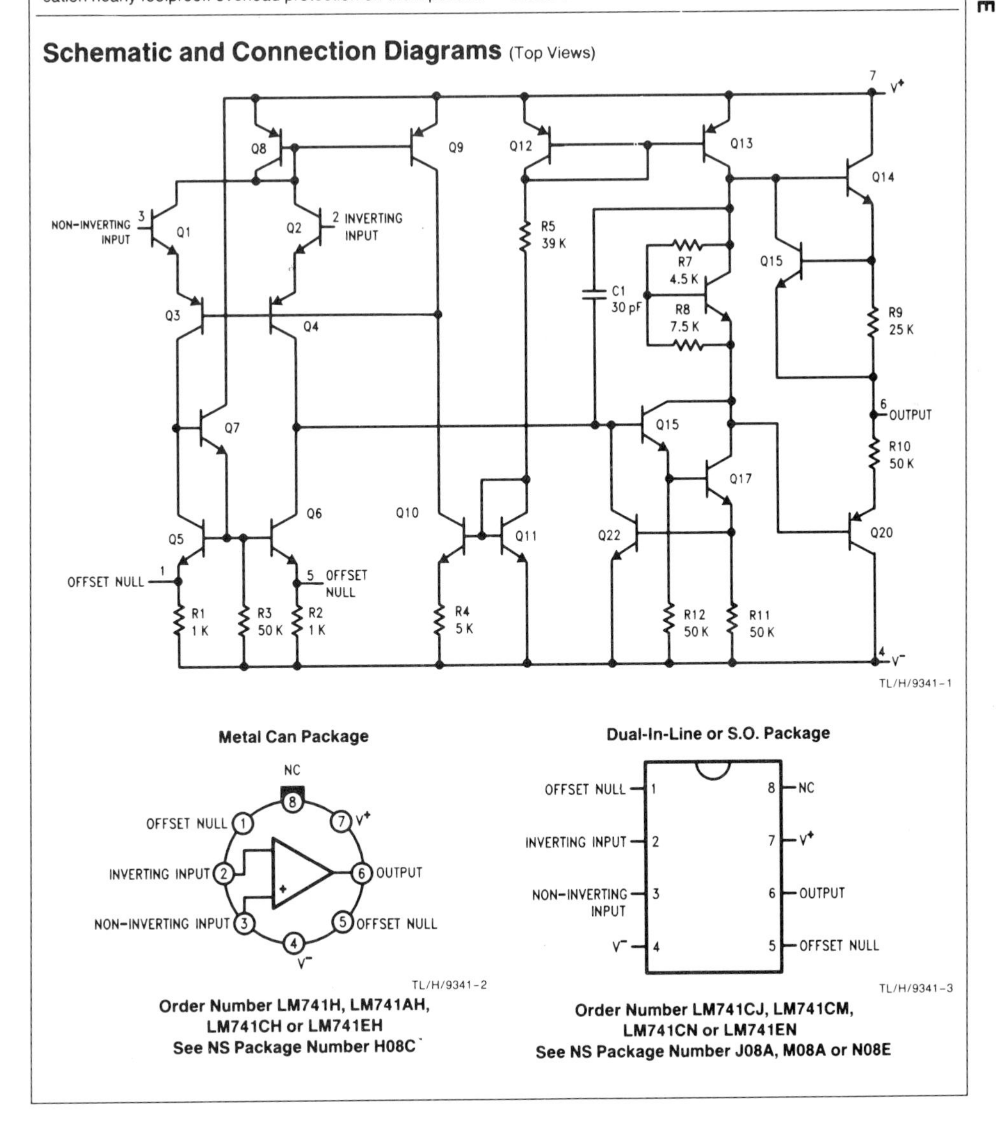

Order Number LM741H, LM741AH, LM741CH or LM741EH
See NS Package Number H08C

Order Number LM741CJ, LM741CM, LM741CN or LM741EN
See NS Package Number J08A, M08A or N08E

FIGURE 9-38 *(continued)* (b) 741C op amp data sheet (reprinted with permission of National Semiconductor).

LM741/LM741A/LM741C/LM741E

Absolute Maximum Ratings

If Military/Aerospace specified devices are required, contact the National Semiconductor Sales Office/Distributors for availability and specifications.
(Note 5)

	LM741A	LM741E	LM741	LM741C
Supply Voltage	±22V	±22V	±22V	±18V
Power Dissipation (Note 1)	500 mW	500 mW	500 mW	500 mW
Differential Input Voltage	±30V	±30V	±30V	±30V
Input Voltage (Note 2)	±15V	±15V	±15V	±15V
Output Short Circuit Duration	Indefinite	Indefinite	Indefinite	Indefinite
Operating Temperature Range	−55°C to +125°C	0°C to +70°C	−55°C to +125°C	0°C to +70°C
Storage Temperature Range	−65°C to +150°C	−65°C to +150°C	−65°C to +150°C	−65°C to +150°C
Junction Temperature	150°C	100°C	150°C	100°C
Soldering Information				
N-Package (10 seconds)	260°C	260°C	260°C	260°C
J- or H-Package (10 seconds)	300°C	300°C	300°C	300°C
M-Package				
Vapor Phase (60 seconds)	215°C	215°C	215°C	215°C
Infrared (15 seconds)	215°C	215°C	215°C	215°C

See AN-450 "Surface Mounting Methods and Their Effect on Product Reliability" (Appendix D) for other methods of soldering surface mount devices.

Electrical Characteristics (Note 3)

Parameter	Conditions	LM741A/LM741E			LM741			LM741C			Units
		Min	Typ	Max	Min	Typ	Max	Min	Typ	Max	
Input Offset Voltage	T_A = 25°C $R_S \le$ 10 kΩ $R_S \le$ 50Ω	 	 0.8	 3.0		 1.0 	 5.0 		 2.0 	 6.0 	 mV mV
	$T_{AMIN} \le T_A \le T_{AMAX}$ $R_S \le$ 50Ω $R_S \le$ 10 kΩ			 4.0 			 6.0			 7.5	 mV mV
Average Input Offset Voltage Drift				15							μV/°C
Input Offset Voltage Adjustment Range	T_A = 25°C, V_S = ±20V	±10				±15			±15		mV
Input Offset Current	T_A = 25°C		3.0	30		20	200		20	200	nA
	$T_{AMIN} \le T_A \le T_{AMAX}$			70		85	500			300	nA
Average Input Offset Current Drift				0.5							nA/°C
Input Bias Current	T_A = 25°C		30	80		80	500		80	500	nA
	$T_{AMIN} \le T_A \le T_{AMAX}$			0.210			1.5			0.8	μA
Input Resistance	T_A = 25°C, V_S = ±20V	1.0	6.0		0.3	2.0		0.3	2.0		MΩ
	$T_{AMIN} \le T_A \le T_{AMAX}$, V_S = ±20V	0.5									MΩ
Input Voltage Range	T_A = 25°C							±12	±13		V
	$T_{AMIN} \le T_A \le T_{AMAX}$				±12	±13					V
Large Signal Voltage Gain	T_A = 25°C, $R_L \ge$ 2 kΩ V_S = ±20V, V_O = ±15V V_S = ±15V, V_O = ±10V	 50 			 50	 200		 20	 200		 V/mV V/mV
	$T_{AMIN} \le T_A \le T_{AMAX}$, $R_L \ge$ 2 kΩ, V_S = ±20V, V_O = ±15V V_S = ±15V, V_O = ±10V V_S = ±5V, V_O = ±2V	 32 10			 25 			 15 			 V/mV V/mV V/mV

Electrical Characteristics (Note 3) (Continued)

Parameter	Conditions	LM741A/LM741E			LM741			LM741C			Units
		Min	Typ	Max	Min	Typ	Max	Min	Typ	Max	
Output Voltage Swing	V_S = ±20V $R_L \ge$ 10 kΩ $R_L \ge$ 2 kΩ	 ±16 ±15									 V V
	V_S = ±15V $R_L \ge$ 10 kΩ $R_L \ge$ 2 kΩ				 ±12 ±10	 ±14 ±13		 ±12 ±10	 ±14 ±13		 V V
Output Short Circuit Current	T_A = 25°C $T_{AMIN} \le T_A \le T_{AMAX}$	10 10	25 	35 40		25 			25 		mA mA
Common-Mode Rejection Ratio	$T_{AMIN} \le T_A \le T_{AMAX}$ $R_S \le$ 10 kΩ, V_{CM} = ±12V $R_S \le$ 50 kΩ, V_{CM} = ±12V	 80	 95		 70 	 90 		 70 	 90 		 dB dB
Supply Voltage Rejection Ratio	$T_{AMIN} \le T_A \le T_{AMAX}$, V_S = ±20V to V_S = ±5V $R_S \le$ 50Ω $R_S \le$ 10 kΩ	 86 	 96 		 77	 96		 77	 96		 dB dB
Transient Response Rise Time Overshoot	T_A = 25°C, Unity Gain		 0.25 6.0	 0.8 20		 0.3 5			 0.3 5		 μs %
Bandwidth (Note 4)	T_A = 25°C	0.437	1.5								MHz
Slew Rate	T_A = 25°C, Unity Gain	0.3	0.7			0.5			0.5		V/μs
Supply Current	T_A = 25°C					1.7	2.8		1.7	2.8	mA
Power Consumption	T_A = 25°C V_S = ±20V V_S = ±15V		 80 	 150 		 50	 85		 50	 85	 mW mW
LM741A	V_S = ±20V T_A = T_{AMIN} T_A = T_{AMAX}			 165 135							 mW mW
LM741E	V_S = ±20V T_A = T_{AMIN} T_A = T_{AMAX}			 150 150							 mW mW
LM741	V_S = ±15V T_A = T_{AMIN} T_A = T_{AMAX}					 60 45	 100 75				 mW mW

Note 1: For operation at elevated temperatures, these devices must be derated based on thermal resistance, and T_j max. (listed under "Absolute Maximum Ratings"). $T_j = T_A + (\theta_{jA} P_D)$.

Thermal Resistance	Cerdip (J)	DIP (N)	TO-5 (H)	SO-8 (M)
θ_{jA} (Junction to Ambient)	100°C/W	100°C/W	150°C/W	195°C/W
θ_{jC} (Junction to Case)	N/A	N/A	80°C/W	N/A

Note 2: For supply voltages less than ±15V, the absolute maximum input voltage is equal to the supply voltage.

Note 3: Unless otherwise specified, these specifications apply for V_S = ±15V, −55°C ≤ T_A ≤ +125°C (LM741/LM741A). For the LM741C/LM741E, these specifications are limited to 0°C ≤ T_A ≤ +70°C.

Note 4: Calculated value from: BW (MHz) = 0.35/Rise Time(μs).

Note 5: For military specifications see RETS741X for LM741 and RETS741AX for LM741A.

LM741/LM741A/LM741C/LM741E

(b)

FIGURE 9-38 *(continued)*

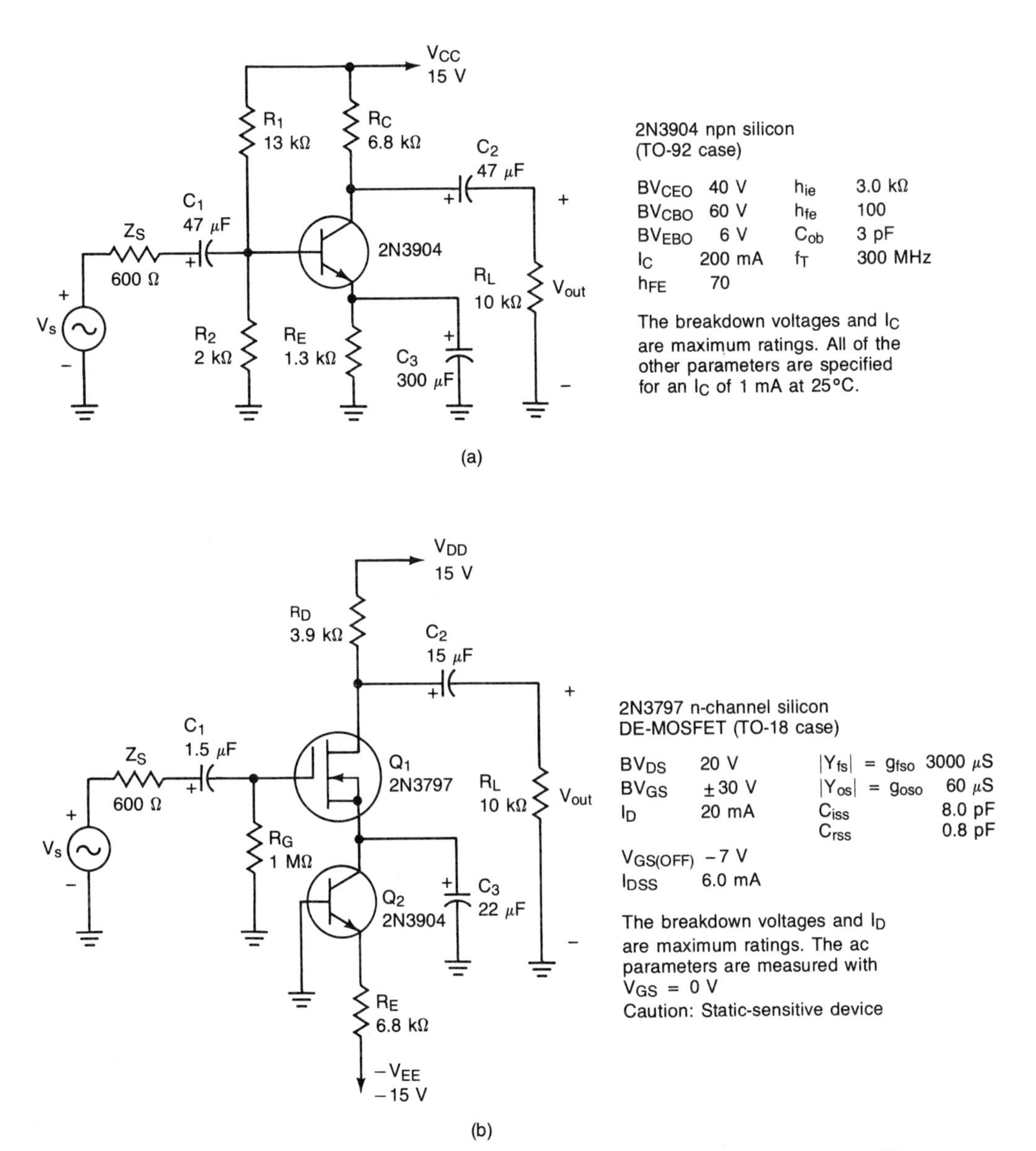

FIGURE 9-39 Single-stage amplifiers to be analyzed: (a) common-emitter; (b) common-source.

in the figure. Draw the dc equivalent circuit. Perform an approximate dc analysis to determine V_B, V_E, I_E, I_C, V_C, and V_{CE}.

9-2. A common-source FET amplifier using a 2N3797 *n*-channel DE-MOSFET has been illustrated in Fig. 9-39(b). The pertinent transistor parameters have been summarized in the figure. Draw the dc equivalent circuit. Determine I_D and use the FET's transconductance equation to find V_{GS}. Complete the analysis by finding V_G, I_S, V_D, V_{DS}, and V_{CE}.

9-3. Determine the small-signal parameters g_m, r_o, and r_π for the 2N3904 BJT at the Q-point found in Prob. 9-1. Verify that r_o is large enough to be ignored.

9-4. Determine the small-engine parameters g_m, r_o, and r_π for the 2N3797 MOSFET at the Q-point found in Prob. 9-2. Verify that r_o is large enough to be ignored.

9-5. Draw the ac equivalent circuit of the common-emitter amplifier given in Fig. 9-39(a). Find its R_{in}, $A_{v(\text{oc})}$, and R_{out}. Use these results to find A_v and A_{vs}.

9-6. Draw the ac equivalent circuit of the common-source amplifier given in Fig. 9-39(b). Find its R_{in}, $A_{v(\text{oc})}$, and R_{out}. Use these results to find A_v and A_{vs}.

9-7. What is meant by the term "ac equivalent collector resistance" (r_C)? Find r_C for the common-emitter amplifier given in Fig. 9-39(a) and use it to determine A_v. How does its value compare with the A_v found in Prob. 9-5?

9-8. What is meant by the term "ac equivalent drain resistance" (r_D)? Find r_D for the common-source amplifier given in Fig. 9-39(b) and use it to determine A_v. How does its value compare with the A_v found in Prob. 9-6?

9-9. A discrete cascaded amplifier system has been illustrated in Fig. 9-40(a). Draw and simplify the amplifier's ac equivalent circuit. Using r_D and r_C, find the individual loaded voltage gains for the first two stages. Find the open-circuit voltage gain for the third stage. Using these results, find the open-circuit voltage gain $A_{vT(\text{oc})}$ for the amplifier system. Complete the analysis by determining R_{in}, R_{out}, A_{vT}, A_{vsT}, A_{iT}, A_{isT}, and A_{pT} for the amplifier system.

Section 9-2

9-10. A cascaded amplifier system employing op amps has been given in Fig. 9-40(b). Find R_{in}, R_{out}, and $A_{vT(\text{oc})}$ for the amplifier system. Using these results, determine A_{vT}, A_{vsT}, A_{iT}, A_{isT}, and A_{pT} for the amplifier system.

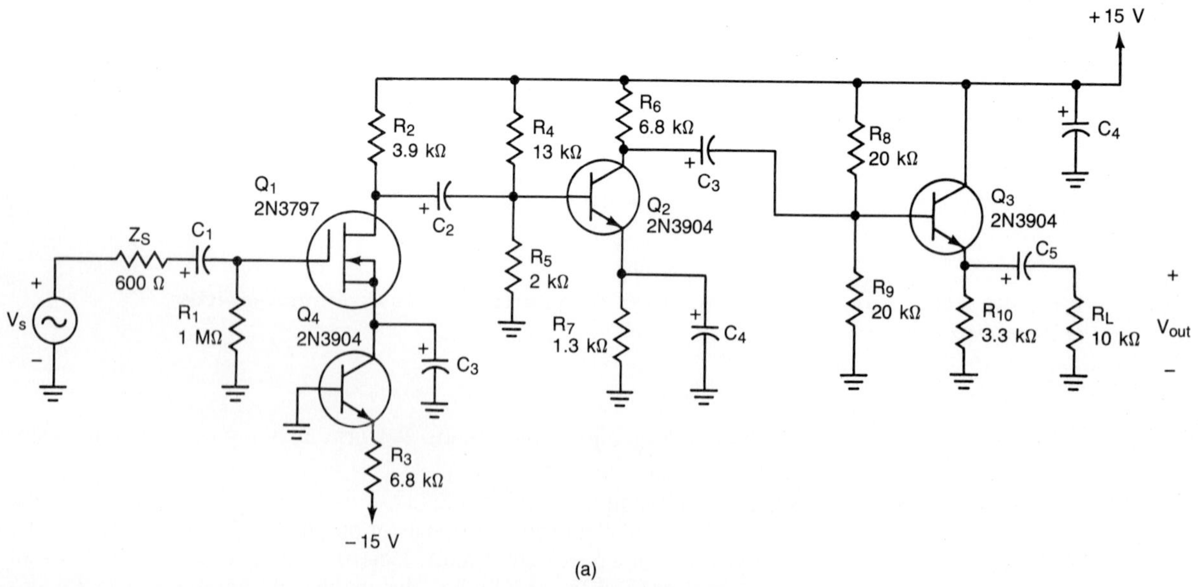

FIGURE 9-40 Cascaded amplifiers to be analyzed: (a) discrete; (b) integrated circuit.

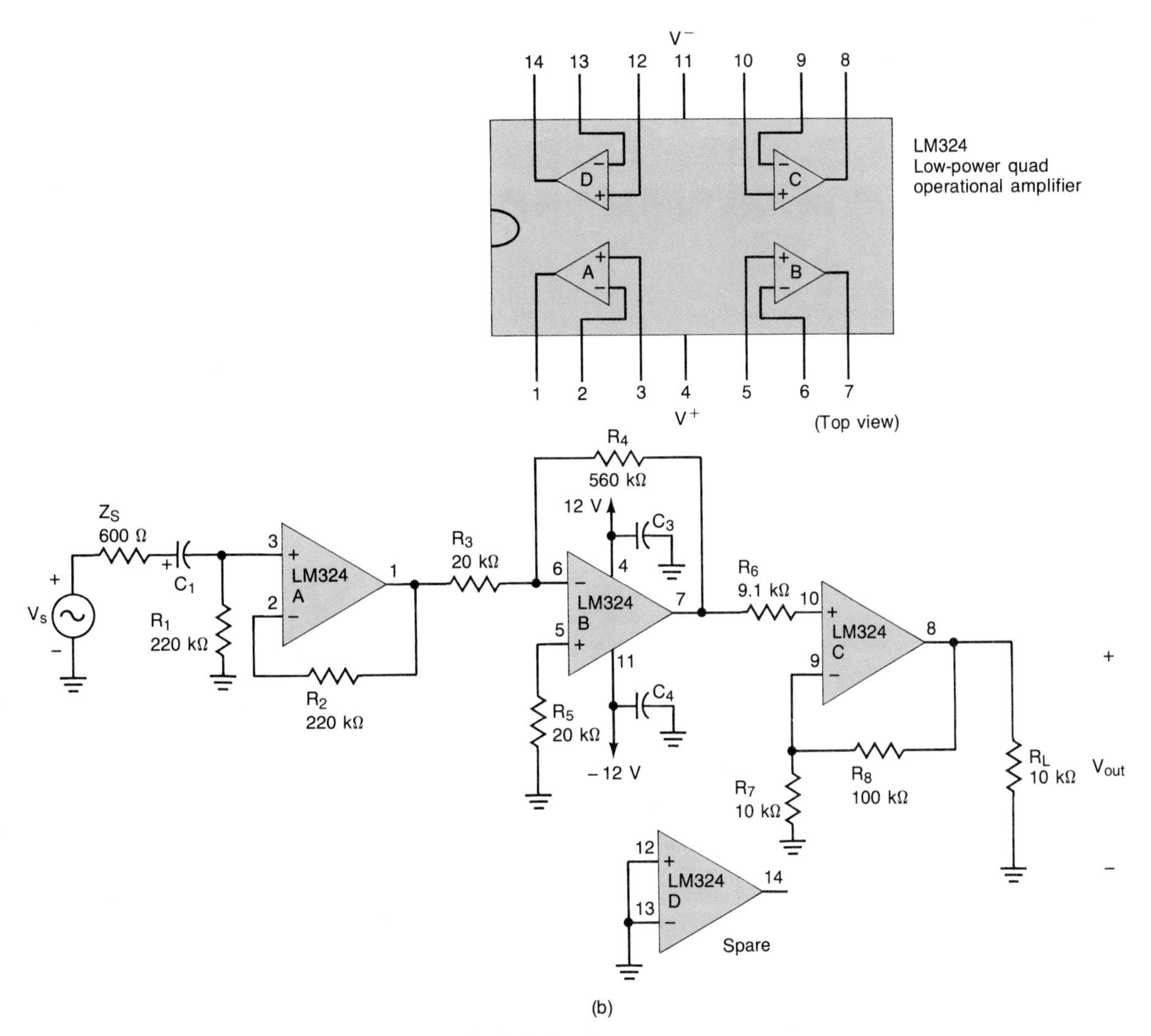

FIGURE 9-40 (continued)

9-11. Repeat Prob. 9-10 for the cascaded amplifier system given in Fig. 9-40(b). The LM324 quad op amp is still used. The rest of the components are changed as follows: R_1 and R_2 are 150 kΩ, R_3 is 10 kΩ, R_4 is 100 kΩ, R_5 is 9.1 kΩ, R_6 is 5.1 kΩ, R_7 and R_8 are 10 kΩ, and the supply voltages are ±15 V.

9-12. The input coupling capacitor (C1) is removed and replaced by a short circuit in Fig. 9-40(b). The ac signal source is replaced by a dc source. The component values are the same as those specified in Prob. 9-11. If the voltage at pin 3 is −0.5 V with respect to ground, what are the voltages at pins 1, 7, and 8?

9-13. The input coupling capacitor (C1) is removed and replaced by a short circuit in Fig. 9-40(b). The ac signal source is replaced by a dc source. The component values are the same as specified in Prob. 9-11. If the voltage at pin 3 is 0.60 V with respect to ground, what are the voltages at pins 1, 7, and 8?

9-14. Repeat Prob. 9-13 if the voltage at pin 3 is increased to 1 V.

Section 9-3

9-15. Draw the block diagram of the typical circuitry found in an op amp. Label each of the blocks.

9-16. In your own words, briefly explain the operation of each of the four fundamental stages found in the typical op amp.

Section 9-4

9-17. Which two stages in the op amp determine the op amp's transconductance?

9-18. As the signal frequency increases the ac voltage gain of an op amp will tend to ______ (increase, decrease).

Section 9-5

9-19. Explain what is meant by the term "frequency domain." What are its advantages over a time-domain representation?

9-20. What is the basic difference between an oscilloscope and a spectrum analyzer?

Section 9-7

9-21. Find the corner frequency for the low-pass filter given in Fig. 9-13(a), and use the Bode approximations to graph its amplitude and phase response as a function of frequency. Draw your Bode plot on five-cycle semilog paper. Assume that C is 100 pF and R is 2 kΩ.

9-22. Repeat Prob. 9-21 if C is 1 μF and R is 10 kΩ.

Section 9-8

9-23. Find the corner frequency for the high-pass filter given in Fig. 9-14(a), and use the Bode approximations to graph its amplitude and phase response as a function of frequency. Draw your Bode plot on five-cycle semilog paper. Assume that C is 0.68 μF and R is 10 kΩ.

9-24. Repeat Prob. 9-23 if C is 1000 pF and R is 2 kΩ.

Section 9-9

9-25. Draw the low-frequency equivalent circuit of the common-emitter amplifier shown in Fig. 9-39(a). Determine the Thévenin equivalent resistance "seen" by each of the capacitors C_1, C_2, and C_3. Find the corner frequency produced by each of these capacitors. Draw the amplifier's low-end frequency response by using the Bode approximations. Use five-cycle semilog paper to graph your results. (Note that this is a continuation of the analyses requested in Probs. 9-1, 9-3, 9-5, and 9-7.)

9-26. Repeat Prob. 9-25 for the common-source amplifier given in Fig. 9-39(b). (Note that this is a continuation of the analyses requested in Probs. 9-2, 9-4, 9-6, and 9-8.)

Section 9-10

9-27. An op amp that is being biased by a single-polarity power supply has been illustrated in Fig. 9-41(a). Draw its dc equivalent circuit and find its dc output voltage. Draw its midband ac equivalent circuit, and determine its R_{in}, R_{out}, and A_{vs}. Determine its three corner frequencies.

9-28. Repeat Prob. 9-27 for the op amp circuit given in Fig. 9-41(b). In this case there are only two corner frequencies.

9-29. As the base-emitter forward bias increases, the base-emitter diffusion capacitance _______ (increases, decreases).

9-30. Base-emitter (diffusion) capacitance is normally _______ (larger, smaller) than collector-base (junction) capacitance.

9-31. Header capacitances (e.g., C_{ce} or C_{ds}) are normally very _______ (large, small).

FIGURE 9-41 Single-supply op amp amplifiers to be analyzed: (a) noninverting; (b) inverting.

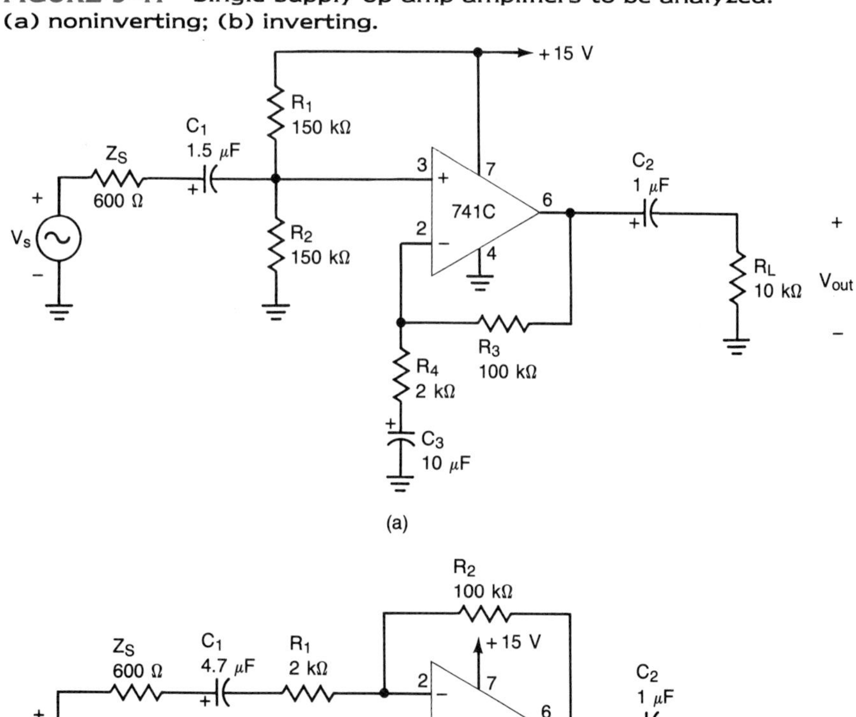

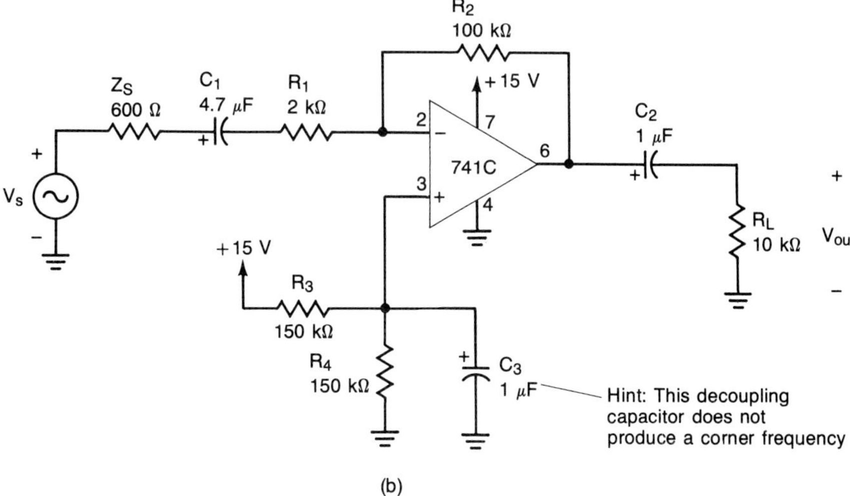

Section 9-12

9-32. Determine the base-spreading resistance r'_{bb}, the collector-to-base junction capacitance C_{cb}, and the base-to-emitter diffusion capacitance C_{be} for the 2N3904 BJT given in Fig. 9-39(a). (Note that this is a continuation of Probs. 9-1, 9-3, 9-5, 9-7, and 9-25.)

9-33. Draw the midband ac equivalent circuit of the common-emitter amplifier given in Fig. 9-39(a) which includes r'_{bb}. Taking the effects of r'_{bb} into account, recompute the amplifier's midband voltage gain A_v. (*Hint*: Voltage division occurs between r'_{bb} and r_π.) How does it compare with the A_v found in Prob. 9-7?

Section 9-13

9-34. Determine capacitances C_{dg} and C_{gs} for 2N3797 DE-MOSFET given in Fig. 9-39(b). In this case the device capacitances are *not* junction capacitances. The MOSFET's capacitances are the result of the capacitive effects between the gate metallization and the channel through the insulating SiO_2 layer. (Note that this is a continuation of Probs. 9-2, 9-4, 9-6, 9-8, and 9-26.)

Section 9-15

9-35. Draw the high-frequency ac equivalent circuit of the common-emitter amplifier given in Fig. 9-39(a). Indicate the device capacitances C_{cb} and C_{be}. Ignore the base-spreading resistance r'_{bb}. Employ Miller's theorem to simplify the circuit. Determine the corner frequencies produced by the equivalent input and output capacitances. Graph the amplifier's high-end frequency response using the Bode approximations on five-cycle semilog paper.

9-36. Draw the high-frequency ac equivalent circuit of the common-source amplifier given in Fig. 9-39(b). Indicate the device capacitances C_{dg} and C_{gs}. Use Miller's theorem to simplify the circuit. Determine the corner frequencies produced by the equivalent input and output capacitances. Graph the amplifier's high-end frequency response using the Bode approximations on five-cycle semilog paper.

9-37. Define the term ''bandwidth.'' If an amplifier has an upper corner frequency f_H of 30 kHz and a lower corner frequency f_L of 100 Hz, what is the amplifier's bandwidth BW? If its midband A_{vs} is 30 dB, find its A_{vs} at 30 kHz.

9-38. A *cascode* amplifier is shown in Fig. 9-42. The first stage (Q_1) is a common-emitter amplifier and the second stage is a common-base amplifier. Find V_{B1}, V_{E1}, I_{C1}, I_{C2}, V_{B2}, and V_{E2}. The two BJTs are identical. Find g_m and r_π. Determine the midband A_v for Q_1 and Q_2. Using Miller's theorem explain why the circuit has a superior high-end frequency response. Find the high-end corner frequencies associated with Q_1's input and Q_2's output. Compare your results with the common-emitter stage given in Fig. 9-34(a) [analyzed in Prob. 9-35].

Section 9-16

9-39. What is the function of the frequency-compensation capacitor found in many op amps?

9-40. The LM324 is described as a quad op amp. In order to place four frequency-compensated op amps in the same IC package, the size of C_c was reduced from the 30-pF value found in the 741 op amp down to 5 pF. The f_T of the LM324 op amps

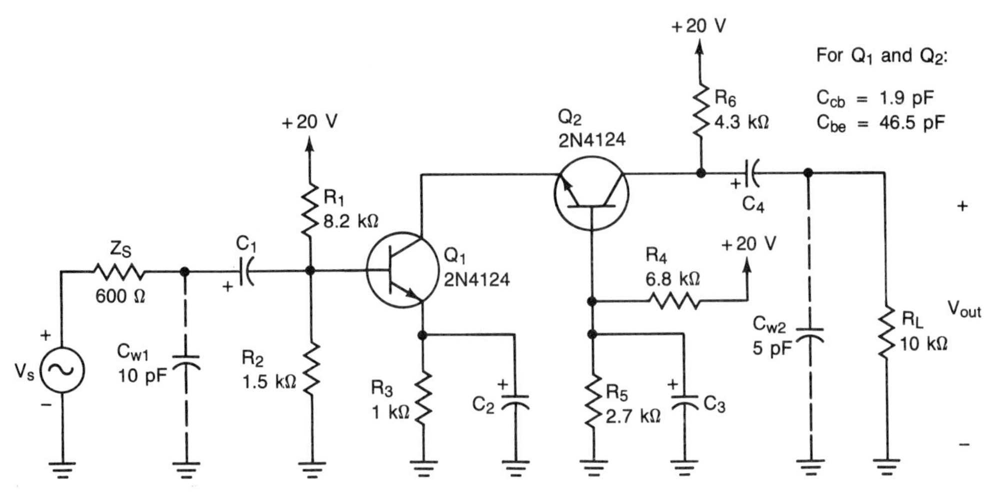

FIGURE 9-42 Cascode amplifier.

is 1 MHz. Find their equivalent g_m. Compare your answer with the g_m found for the 741 op amp in Example 9-9.

Troubleshooting and Failure Modes

9-41. The amplifier system shown in Fig. 9-1(a) is faulty. The ac signal appears at the base of Q_3 but does not appear across R_L. The dc biases associated with Q_3 appear to be normal. Select the possible failure(s) from the following alternatives and explain your reasoning: (a) capacitor C_5 is open, (b) capacitor C_6 is open, (c) transistor Q_3 has a collector-to-emitter short, and (d) the load R_L is shorted to ground.

9-42. Further investigation of the faulty amplifier described in Prob. 9-41 reveals that at *low* frequencies a signal begins to appear at the emitter of Q_3. What is the problem with the amplifier system? Explain.

9-43. Capacitor C_3 in Fig. 9-41(a) has shorted. Determine the voltages at pins 3, 2, and 6 relative to ground.

Design

The low-end frequency response of an amplifier is controlled by making one of the corner frequencies *dominant*. In the case of a BJT amplifier this is accomplished by sizing the emitter bypass capacitor to produce a corner frequency at the desired low-end roll-off point. The coupling capacitors are then selected to produce corner frequencies that are at least one decade below the corner frequency produced by the emitter bypass capacitor.

9-44. Design a common-emitter BJT voltage amplifier such as that given in Fig. 9-16(a). V_{CC} is 12 V and I_C is to be approximately 1.5 mA. The (silicon) BJT has an h_{fe} of 110 and an h_{FE} of 80. Determine the standard values required for R_1, R_2, R_C, and R_E. The source resistance is 50 Ω and the load resistance is 100 kΩ. Size the emitter bypass capacitor such that the low-end roll-off occurs at about 20 Hz. Also size the input and output coupling capacitors to produce corner frequencies which are at approximately 2 Hz. Use standard capacitor values.

9-45. Repeat Prob. 9-44 if V_{CC} is 18 V and I_C is to be approximately 2 mA.

9-46. Design a noninverting op amp amplifier with a single supply voltage of 20 V, an R_{in} of at least 100 kΩ, and an ac voltage gain of 101. If the amplifier is to have a bandwidth of 20 kHz, what is the op amp's minimum gain–bandwidth product? The amplifier's low-end roll-off frequency must be at least 20 Hz, with a source resistance of 100 Ω.

Computer

9-47. Write a BASIC program that will perform a low-frequency analysis of an amplifier that has three corner frequencies. The program should prompt the user for f_{L1}, f_{L2}, f_{L3}, and the amplifier's midband voltage gain in decibels. The program should automatically generate a table of 55 values for A_v in decibels and the corresponding phase angles in degrees. The frequencies of analysis should start one decade *above* the highest low-end corner frequency (f_{L1}) and decrement in multiples of f_{L1}. For example, if f_{L1} is 1000 Hz, f_{L2} is 100 Hz, and f_{L3} is 10 Hz, the analysis frequencies should be 0.01, 0.02, 0.03, . . . , 0.1, 0.2, 0.3, . . . , 1, 2, 3, . . . , 10, 20, 30, . . . , 100, 200, 300, . . . , 1 k, 2 k, 3 k, . . . , 8 k, 9 k, and 10 kHz.

9-48. Write a BASIC program that will perform a high-frequency analysis of an amplifier that has three corner frequencies. The program should prompt the user for f_{H1}, f_{H2}, f_{H3}, and the amplifier's midband voltage gain in decibels. The program should automatically generate a table of 55 values for A_v in decibels and the corresponding phase angle in degrees. The frequencies of analysis should start one decade *below* the lowest high-end corner frequency (f_{H1}) and increment in multiples of f_{H1}. For example, if f_{H1} is 20 Hz, f_{H2} is 100 Hz, and f_{H3} is 1 kHz, the analysis frequencies should be 2, 4, 6, 8, 10, 20, 40, . . . , 100, 120, 140, 160, 180, 200, . . . , 2 k, 4 k, 6 k, . . . , 20 k, 40 k, 60 k, . . . , 200 k, 400 k, . . . , 800 k, 1 M, 1.2 M, 1.6 M, 1.8 M, and 2 MHz.

10

NEGATIVE FEEDBACK

After Studying Chapter 10, You Should Be Able to:

- Describe the four basic amplifiers: voltage, transconductance, transresistance, and current.
- Explain the effect of voltage-series negative feedback on input and output resistance.
- Explain the effect of voltage-series negative feedback on bandwidth and nonlinear distortion.
- Describe the significance of loop gain frequency response on input and output resistances, and the % THD.
- Explain how a negative feedback amplifier can go into oscillation.
- Apply the Bode approximation to determine closed-loop stability by finding phase margins.
- Describe and analyze dominant-pole and lag-lead frequency compensation circuits.

10-1 General Feedback Considerations

The negative-feedback concept occurred to Harold S. Black (of Bell Labs) on his way to work one morning. He later patented its application to electronic amplifiers in 1927. Since that time its use has been extended beyond amplifier theory into the realm of automatic control systems, including those found in modern robots.

We have used *dc negative feedback* in our dc bias circuits to achieve superior bias stability. Our first exposure to *ac negative feedback* occurred when we first removed the emitter bypass capacitor found in the common-emitter amplifier. In this case the negative feedback at the emitter increased the input and output resistances. It also produced a reduction in the amplifier's voltage gain. This was deemed an equitable trade-off, as the resulting voltage gain was much more stable. Specifically, it was approximately determined by the ratio of the collector resistance to the emitter resistance. This obviously minimizes the effects of active device tolerance and changing temperature. We also made extensive use of negative feedback in our op amp circuits. Since the op amp is direct coupled, dc and ac negative feedback occur simultaneously.

As we shall see, negative feedback generally improves the input and output resistances of amplifier circuits, stabilizes the gain, improves bandwidth, and reduces nonlinear distortion. Simply, negative feedback allows us to realize accurate, predictable, linear systems.

The many benefits offered through the use of negative feedback might tempt one to conclude that we should use as much of it as possible. However, there is a limit. If too much negative feedback is used, an amplifier or control system might break into *oscillations*. When this occurs an amplifier actually generates an output signal and may become totally unresponsive to its input. A control system may shatter into pieces, and a robot may run amok.

The benefits and pitfalls associated with negative feedback merit our serious attention. By studying the negative-feedback theories and principles, we can greatly extend our knowledge of electronic amplifiers, amplifier systems, and control systems in general.

10-2 Four Basic Amplifiers

Although we have expended much of our previous efforts on the voltage amplifier, there are four fundamental (small-signal) amplifier categories: *voltage*, *current*, *transconductance*, and *transresistance*. Their respective models have been summarized in Fig. 10-1.

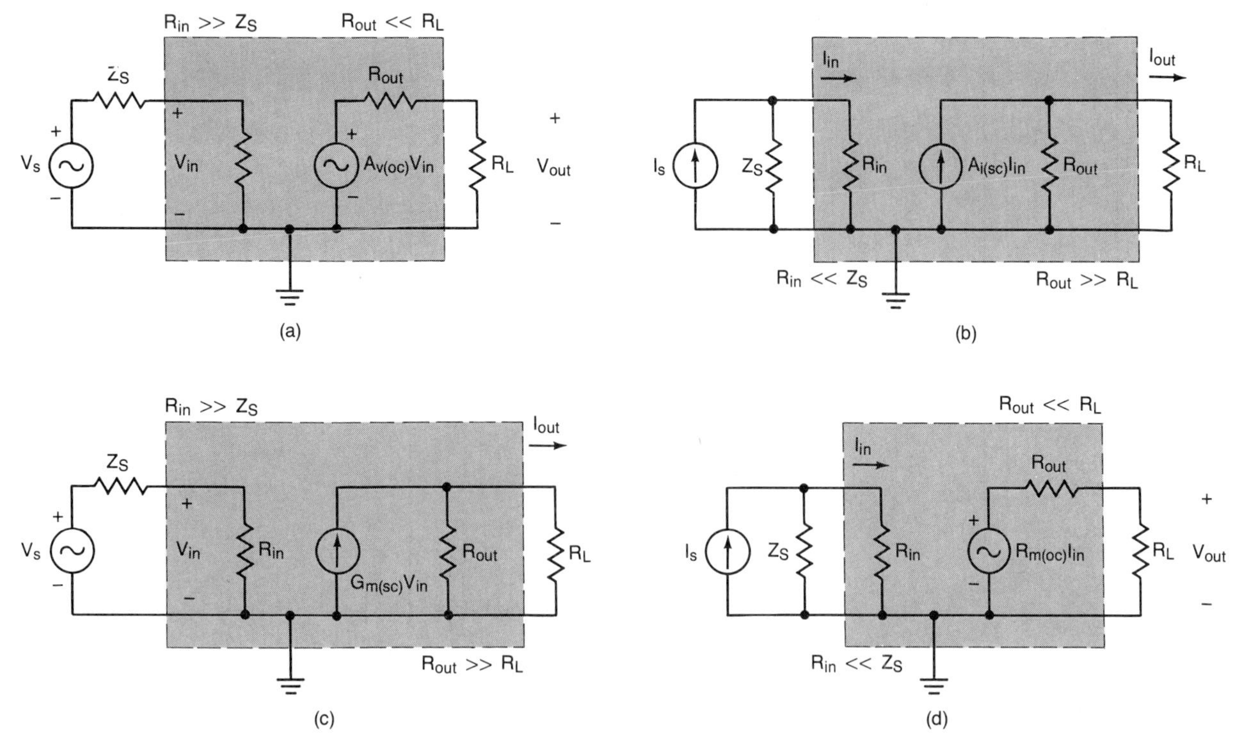

FIGURE 10-1 Four amplifier models: (a) voltage; (b) current; (c) transconductance (voltage-to-current); (d) transresistance (current-to-voltage).

Figure 10-1(a) illustrates our voltage amplifier. Recall that its ideal input resistance R_{in} is infinite and its ideal output resistance R_{out} is zero. In general, we can state that

$$R_{\text{in}} >> Z_S \qquad (\text{ideally, } R_{\text{in}} = \infty\ \Omega)$$
$$R_{\text{out}} << R_L \qquad (\text{ideally, } R_{\text{out}} = 0\ \Omega)$$

If these two conditions are met,

$$A_{vs} = \frac{V_{\text{out}}}{V_s} = \frac{R_{\text{in}}}{R_{\text{in}} + Z_S} A_{v(\text{oc})} \frac{R_L}{R_L + R_{\text{out}}}$$
$$\simeq \frac{R_{\text{in}}}{R_{\text{in}}} A_{v(\text{oc})} \frac{R_L}{R_L} = A_{v(\text{oc})}$$

The current amplifier was introduced in Section 7-5. Its model has been repeated in Fig. 10-1(b). Its ideal input resistance R_{in} is zero, and its ideal output resistance R_{out} is infinite. Its general constraints are

$$R_{\text{in}} << Z_S \qquad (\text{ideally, } R_{\text{in}} = 0\ \Omega)$$
$$R_{\text{out}} >> R_L \qquad (\text{ideally, } R_{\text{out}} = \infty\ \Omega)$$

These conditions promote the following approximation:

$$A_{is} = \frac{I_{\text{out}}}{I_s} = \frac{Z_S}{R_{\text{in}} + Z_S} A_{i(\text{sc})} \frac{R_{\text{out}}}{R_L + R_{\text{out}}}$$
$$\simeq \frac{Z_S}{Z_S} A_{i(\text{sc})} \frac{R_{\text{out}}}{R_{\text{out}}} = A_{i(\text{sc})}$$

and we achieve the maximum possible gain.

The transconductance amplifier is often referred to as a *voltage-to-current converter*. Very simply, an input voltage signal is used to produce a proportional output current [refer to Fig. 10-1(c)]. In this case, the ideal input and output resistances are infinite, and we can state that

$$R_{\text{in}} >> Z_S \qquad (\text{ideally, } R_{\text{in}} = \infty\ \Omega)$$
$$R_{\text{out}} >> R_L \qquad (\text{ideally, } R_{\text{out}} = \infty\ \Omega)$$

These conditions lead us to

$$G_{ms} = \frac{I_{\text{out}}}{V_s} = \frac{R_{\text{in}}}{R_{\text{in}} + Z_S} G_{m(\text{sc})} \frac{R_{\text{out}}}{R_L + R_{\text{out}}}$$
$$\simeq \frac{R_{\text{in}}}{R_{\text{in}}} G_{m(\text{sc})} \frac{R_{\text{out}}}{R_{\text{out}}} = G_{m(\text{sc})}$$

where G_{ms} is the mutual conductance, or transconductance from the signal source to the load, and $G_{m(\text{sc})}$ is the maximum possible transconductance as measured from the amplifier's input to its output with the load short-circuited.

The transresistance amplifier is also referred to as a *current-to-voltage converter*. Its model has been given in Fig. 10-1(d). Its function is to provide an output voltage that is proportional to its input current. Consequently, its ideal input and output resistances are zero.

$$R_{\text{in}} << Z_S \qquad (\text{ideally, } R_{\text{in}} = 0\ \Omega)$$
$$R_{\text{out}} << R_L \qquad (\text{ideally, } R_{\text{out}} = 0\ \Omega)$$

Hence

$$R_{ms} = \frac{V_{\text{out}}}{I_s} = \frac{Z_S}{R_{\text{in}} + Z_S} R_{m(\text{oc})} \frac{R_L}{R_L + R_{\text{out}}}$$
$$\simeq \frac{Z_S}{Z_S} R_{m(\text{oc})} \frac{R_L}{R_L} = R_{m(\text{oc})}$$

where R_{ms} is the mutual resistance, or transresistance from the signal source to the load, and $R_{m(\text{oc})}$ is the maximum transresistance as measured from the amplifier's input to its open-circuited output.

10-3 Feedback Amplifiers

To optimize the four amplifier configurations, negative feedback is used. The feedback arrangements best suited to stabilize the gains of our four basic amplifiers have been shown in Fig. 10-2. The feedback arrangements have been categorized as *voltage-series*, *current-shunt*, *current-series*, and *voltage-shunt*.

In both the voltage and transresistance amplifiers, we are interested in delivering

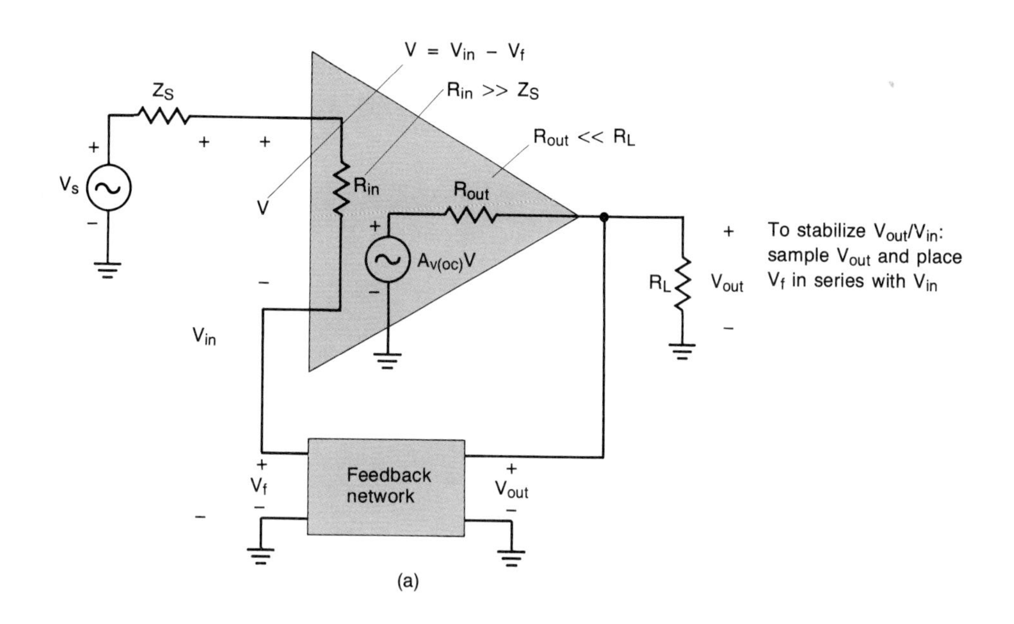

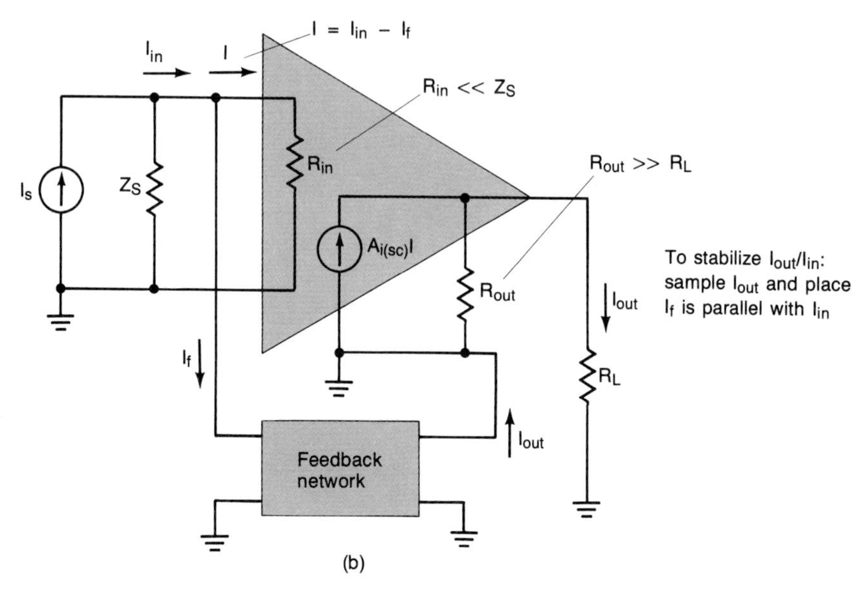

FIGURE 10-2 Four basic feedback arrangements: (a) voltage amplifier with voltage-series negative feedback; (b) current amplifier with current-shunt negative feedback; (c) transconductance amplifier with current-series negative feedback; (d) transresistance amplifier with voltage-shunt negative feedback.

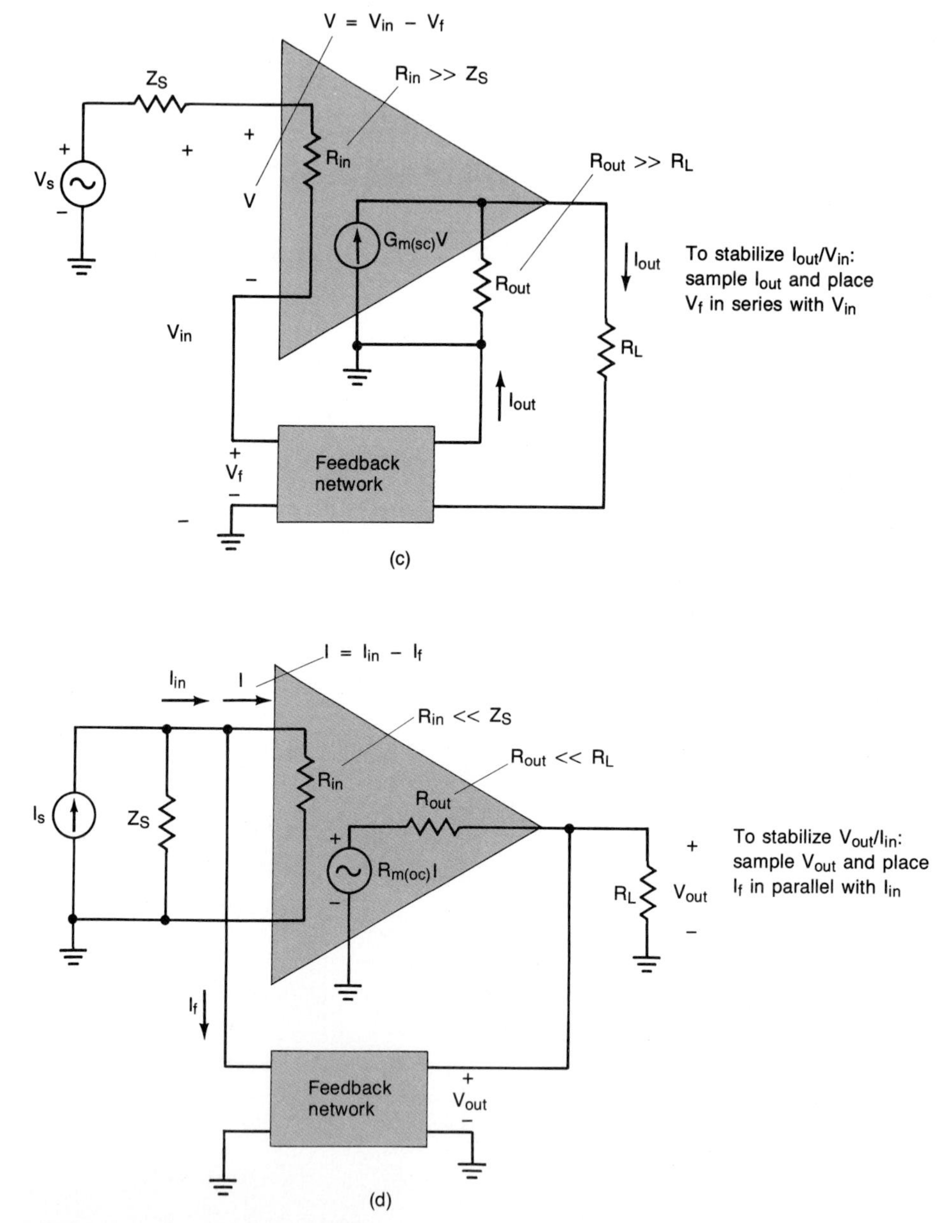

FIGURE 10-2 *(continued)*

an output voltage V_{out} across the load. Therefore, the input of the feedback network is connected across the load. Figure 10-2(a) and (d) show *output voltage sampling*. In both the current and transconductance amplifiers, the output quantity is a current I_{out}. Therefore, the input of the feedback network is arranged to monitor I_{out} by placing it in series with the load. Figure 10-2(b) and (c) show *output current sampling*.

Both voltage and transconductance amplifiers are driven by voltage signals. Therefore, the output of the feedback network is placed in series with the signal source [see Fig. 10-2(a) and (c)]. The current and transresistance amplifiers are driven by current signals. Therefore, the output of the feedback network is placed in shunt (parallel) with the amplifier's input [see Fig. 10-2(b) and (d)].

10-4 Voltage-Series Negative Feedback

The noninverting op amp amplifier configuration (Fig. 10-3) utilizes voltage-series negative feedback. The input of the feedback network is connected across the amplifier's output. The output of the feedback network is in series with the signal source. In many respects the noninverting op amp amplifier is the most suitable choice for demonstrating classical feedback theory. We shall use it to illustrate the feedback concepts in this chapter. Chapter 11 will move us into the other topologies.

In the analysis of this feedback amplifier, we shall make the usual assumptions:

1. The input resistance of the op amp is large enough to be regarded as infinite.
2. The output resistance of the op amp is small enough to be regarded as zero.

Using these assumptions, we may develop the voltage gain of the amplifier with negative feedback. We shall call this the *closed-loop voltage gain* $A_{v(\text{cl})}$. By Kirchhoff's voltage law,

$$V_{\text{in}} = V + V_f \tag{10-1}$$

where V is the op amp's differential (controlling) input voltage and V_f is the feedback voltage.

The gain of the feedback network is called the *feedback factor* β. It is defined by

$$\beta = \frac{V_f}{V_{\text{out}}} \tag{10-2}$$

Solving for V_f produces the following result:

$$V_f = \beta V_{\text{out}} \tag{10-3}$$

Substitution of Eq. 10-3 into Eq. 10-1 yields

$$V_{\text{in}} = V + V_f = V + \beta V_{\text{out}} \tag{10-4}$$

The output voltage V_{out} is determined by the op amp's loaded, *open-loop* (no feedback) *voltage gain* $A_{v(\text{ol})}$, and its differential input voltage V. Specifically,

$$V_{\text{out}} = A_{v(\text{ol})}V$$

Solving this relationship for V and placing it into Eq. 10-4 allows us to solve for $A_{v(\text{cl})}$.

$$\frac{V_{out}}{A_{v(ol)}} + \beta V_{out} = V_{in}$$

$$V_{out}\left[\frac{1}{A_{v(ol)}} + \beta\right] = V_{out}\frac{1 + \beta A_{v(ol)}}{A_{v(ol)}} = V_{in}$$

$$\boxed{A_{v(cl)} = \frac{V_{out}}{V_{in}} = \frac{A_{v(ol)}}{1 + \beta A_{v(ol)}}} \tag{10-5}$$

As we shall see, this is an extremely important result. *The term* $\beta A_{v(ol)}$ *is called the loop gain*. If the loop gain $\beta A_{v(ol)}$ is much greater than 1, we arrive at the approximation

$$A_{v(cl)} = \frac{A_{v(ol)}}{1 + \beta A_{v(ol)}} \simeq \frac{A_{v(ol)}}{\beta A_{v(ol)}}$$

$$\boxed{A_{v(cl)} \simeq \frac{1}{\beta}} \tag{10-6}$$

where $\beta A_{v(ol)} >> 1$.

In Fig. 10-4(a) we see the 741C op amp in its noninverting amplifier configuration. The feedback factor β may be found by finding the voltage gain of the feedback network. By voltage division we can see that

$$\beta = \frac{V_f}{V_{out}} = \frac{R_1}{R_1 + R_2} \tag{10-7}$$

FIGURE 10-3 Voltage-series negative feedback of the op amp noninverting amplifier.

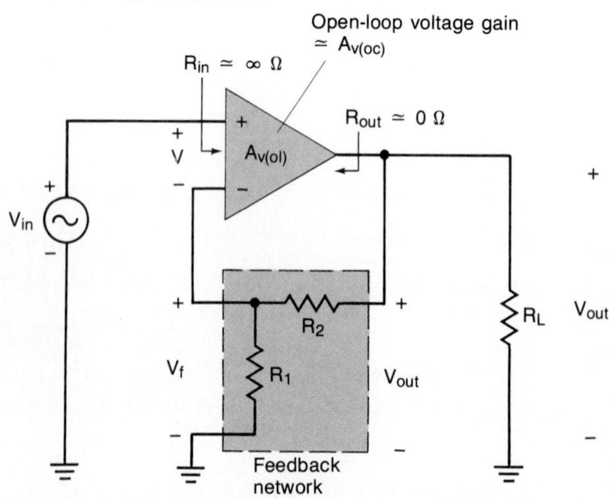

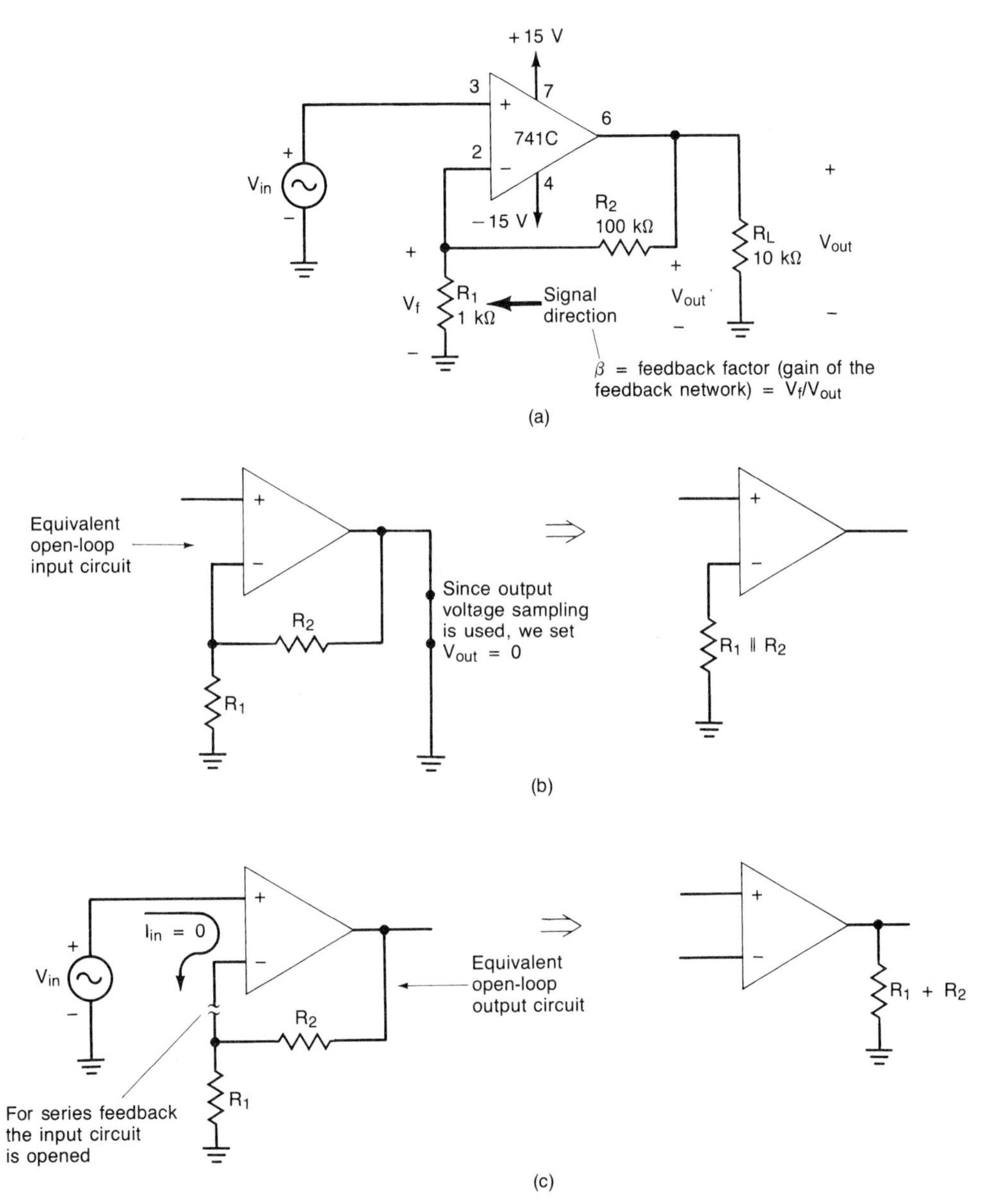

FIGURE 10-4 Practical example of voltage-series negative feedback: (a) amplifier; (b) open-loop input circuit; (c) open-loop output circuit; (d) open-loop amplifier.

This result is based on the assumption that the differential input resistance of the op amp is virtually infinite.

If we use the relationship above and the approximation cited in Eq. 10-6, we obtain

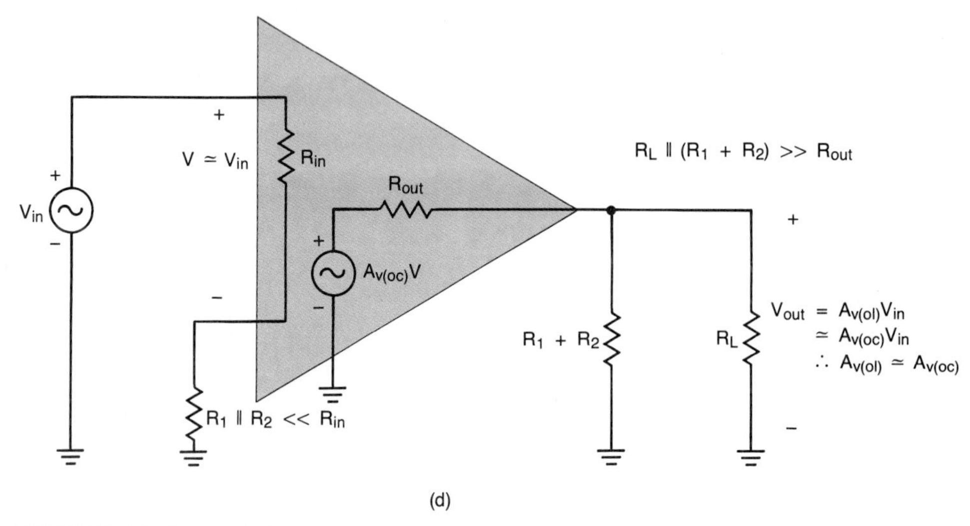

FIGURE 10-4 *(continued)*

$$A_{v(\text{cl})} \simeq \frac{1}{\beta} = \frac{1}{R_1/(R_1 + R_2)} = \frac{R_1 + R_2}{R_1}$$

$$= \frac{R_1}{R_1} + \frac{R_2}{R_1} = 1 + \frac{R_2}{R_1} \tag{10-8}$$

Classical feedback theory has led us to the same gain equation that we developed previously by more intuitive means (refer to Eq. 8-44).

Although the approximations offered by Eqs. 10-6 and 10-8 are suitable for most applications, we occasionally need additional accuracy. When this is the case, we turn to Eq. 10-5. To employ it, we need to find the open-loop voltage gain $A_{v(\text{ol})}$. *Even though $A_{v(\text{ol})}$ is the amplifier's voltage gain without feedback, it should include the loading effects produced by the feedback network.* In general, $A_{v(\text{ol})}$ may be determined by opening up the feedback network. The rules for doing this are as follows:

To determine the equivalent input circuit:

1. Set $V_{\text{out}} = 0$ for output voltage sampling. This may be accomplished by shorting the output to ground.
2. Set $I_{\text{out}} = 0$ for output current sampling. This may be accomplished by opening up the output circuit.

To determine the equivalent output circuit:

1. Set $V_{\text{in}} = 0$ for shunt feedback. This may be accomplished by shorting the input to ground.

2. Set $I_{in} = 0$ for series feedback. This may be accomplished by opening the input loop.

Applying these rules to our noninverting op amp amplifier [Fig. 10-4(a)] which employs voltage-series negative feedback leads us to Fig. 10-4(b) and (c). The final open-loop equivalent circuit has been given in Fig. 10-4(d).

If we retain our assumptions that the op amp's input resistance is approximately infinite, and that the op amp's output resistance is approximately zero, the loading effects will be negligible. Therefore, the open-loop voltage gain will be approximately equal to the op amp's differential voltage gain. Hence

$$A_{v(ol)} \simeq A_{v(oc)}$$

Our primary objective has been to introduce and illustrate the rules - even though the op amp is marvelous enough to approximate the ideal voltage amplifier.

EXAMPLE 10-1

The 741C op amp used in Fig. 10-4(a) has a differential voltage gain that can vary between 20,000 and 100,000. Assuming that its 300 kΩ (minimum) input resistance is large enough to be ignored, find its range of voltage gains and its ideal voltage gain.

SOLUTION From Eq. 10-7,

$$\beta = \frac{R_1}{R_1 + R_2} = \frac{1\text{ k}\Omega}{1\text{ k}\Omega + 100\text{ k}\Omega} = 0.\dot{0}0990$$

Assuming that $A_{v(oc)}$ is approximately equal to $A_{v(ol)}$, we may use Eq. 10-5 to find the range of voltage gains.

$$A_{v(cl)} = \frac{A_{v(ol)}}{1 + \beta A_{v(ol)}} = \frac{20{,}000}{1 + (0.00990)(20{,}000)} = 100.5$$

$$A_{v(cl)} = \frac{A_{v(ol)}}{1 + \beta A_{v(ol)}} = \frac{100{,}000}{1 + (0.00990)(100{,}000)} = 100.9$$

From Eq. 10-8 we may find the ideal closed-loop voltage gain.

$$A_{v(cl)} \simeq \frac{1}{\beta} = 1 + \frac{R_2}{R_1} = 1 + \frac{100\text{ k}\Omega}{1\text{ k}\Omega} = 101$$ ■

Obviously, Eq. 10-8 is an accurate approximation of the voltage gain, and we can see the stabilizing effect offered by negative feedback. The variation in the open-loop voltage gain (from 20,000 to 100,000) has been considerably reduced (100.5 to 100.9). We may conclude that negative feedback allows us to realize a much more stable, accurate voltage gain.

10-5 Effect of Series Negative Feedback on Input Resistance

As we saw in Example 10-1, voltage-series negative feedback stabilizes the voltage-gain variations normally found in an open-loop amplifier. When the feedback signal is in *series* with the input circuit, it also serves to *raise the amplifier's input resistance*. Consider Fig. 10-5(a).

The phasing of the feedback signal serves to reduce the net voltage signal across R_{in}. This causes an attendant reduction in the current i_{in} drawn from the signal source. Therefore, the resistance "seen" by the signal source must be higher.

The rationale for this conclusion has been illustrated in Fig. 10-5(a). If we assume that the op amp has an open-loop voltage gain of 20,000, its closed-loop voltage gain is 100.4925. (This gain was arrived at in Example 10-1, and approximated

FIGURE 10-5 Input resistance with voltage-series negative feedback: (a) signal relationships; (b) developing R_{inf}.

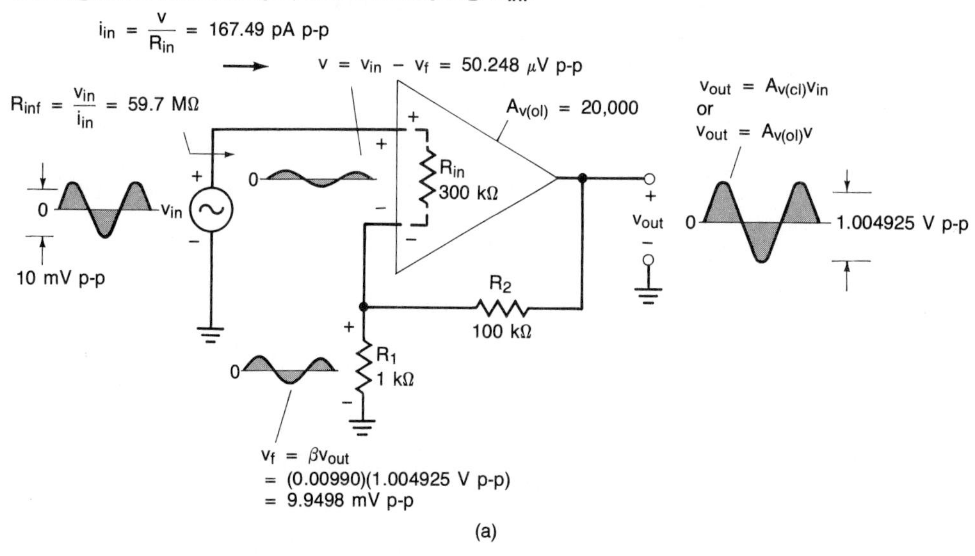

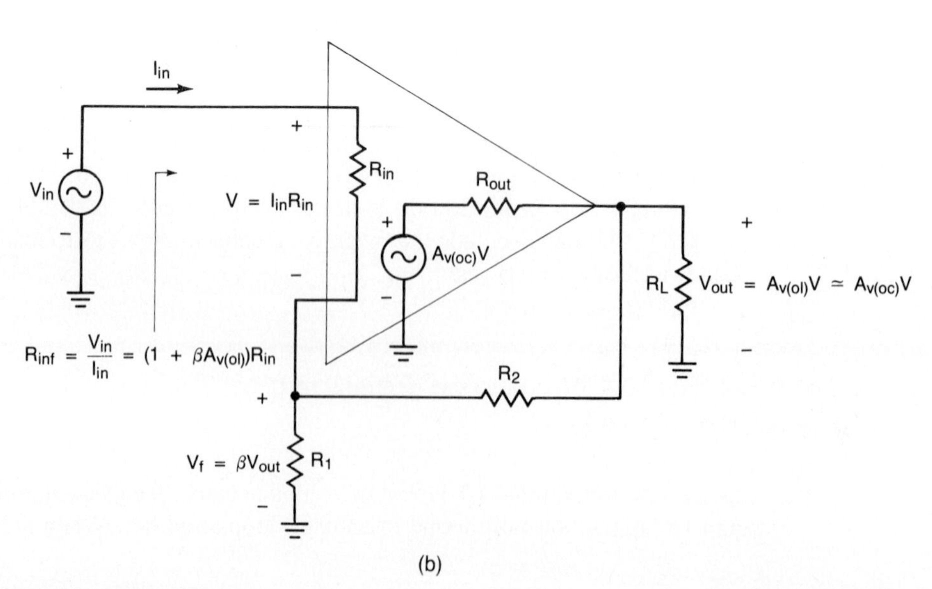

as 100.5.) Therefore, a 10-mV peak-to-peak input voltage produces 1.004925 V peak to peak at the output. The feedback voltage across R_1 is 9.9498 mV peak to peak since the feedback factor β is 0.00990 (which was also determined in Example 10-1). The voltage developed across the op amp's differential input resistance R_{in} is the difference between v_{in} and v_f. This voltage will be 50.248 μV peak to peak. This voltage develops a current flow through R_{in}, which is 167.49 pA peak to peak. The resulting input resistance with feedback R_{inf} is the ratio of v_{in} to i_{in}, or 59.7 MΩ! As a final comment, *notice that the output voltage may be found by taking the product of* v_{in}, *and the closed-loop voltage gain, or by finding the product of the op amp's differential input voltage and the open-loop voltage gain.*

Using feedback theory it is now possible to generalize how much the input resistance will be increased. Consider Fig. 10-5(b). The input resistance with feedback will be denoted R_{inf} and is given by the ratio of V_{in} to I_{in}.

We shall again assume that $A_{v(ol)}$ is approximately equal to $A_{v(oc)}$. From Kirchhoff's voltage law,

$$V_{in} = I_{in}R_{in} + \beta V_{out} \tag{10-9}$$

The output voltage is determined by the open-loop voltage gain and the signal V across R_{in}. Hence

$$V_{out} = A_{v(ol)}V = A_{v(ol)}I_{in}R_{in} \tag{10-10}$$

If we substitute Eq. 10-10 into Eq. 10-9, we obtain Eq. 10-11.

$$V_{in} = I_{in}R_{in} + \beta A_{v(ol)}I_{in}R_{in} = (1 + \beta A_{v(ol)})I_{in}R_{in}$$

$$\boxed{R_{inf} = \frac{V_{in}}{I_{in}} = [1 + \beta A_{v(ol)}]R_{in}} \tag{10-11}$$

As we can see, the input resistance R_{inf} with series negative feedback is equal to a factor of 1 plus the loop gain times the input resistance without feedback R_{in}. Consider the following examples.

EXAMPLE 10-2

Determine the *minimum* input resistance of the noninverting op amp amplifier given in Fig. 10-4(a).

SOLUTION From Example 10-1 we recall that the 741C op amp has an open-loop voltage gain that ranges from 20,000 to 100,000. The minimum R_{inf} will occur with 741C's that possess a minimum voltage gain. Since the 741C has a minimum differential input resistance of 300 kΩ, we shall also employ this minimum in Eq. 10-11 to determine R_{inf}. In Example 10-1 we saw that the feedback factor β is 0.00990. Hence

$$\begin{aligned} R_{inf} &= [1 + \beta A_{v(ol)}]\, R_{in} \\ &= [1 + (0.00990)(20{,}000)][300\ \text{k}\Omega] = 59.7\ \text{M}\Omega \end{aligned}$$

This result agrees with the more intuitive development presented in Fig. 10-5(a). ■

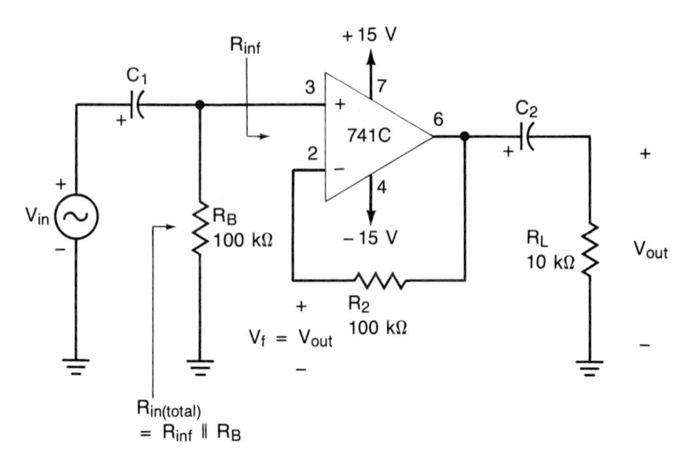

FIGURE 10-6 Voltage follower for Example 10-3.

EXAMPLE 10-3

The 741C is being used as a voltage follower in Fig. 10-6. Find the input resistance "looking" into the op amp's input terminals, and the total input resistance presented to the signal source.

SOLUTION The voltage follower uses 100% feedback. Specifically, we can see that

$$V_f = V_{\text{out}}$$

which means that the feedback factor β is unity. Using the parameters given in Example 10-2, and Eq. 10-11 allows us to find the minimum R_{inf} "looking into" the noninverting input terminal.

$$\begin{aligned} R_{\text{inf}} &= [1 + \beta A_{v(\text{ol})}]\, R_{\text{in}} \\ &= [1 + (1)(20{,}000)][300\text{ k}\Omega] = 6.00\text{ G}\Omega \end{aligned}$$

The total input resistance presented to the source is the R_{inf} in parallel with the dc return resistor R_B. Hence

$$R_{\text{in(total)}} = R_{\text{inf}} \parallel R_B = 6\text{ G}\Omega \parallel 100\text{ k}\Omega \simeq 100\text{ k}\Omega$$ ■

If the signal source is coupled directly to the op amp's input, the use of R_B is not required. (This assumes, of course, that the signal source is *never* disconnected from the op amp while power is applied. Recall that a floating input may promote latch-up or a wandering output.) In these cases we can readily see that the voltage follower provides an extremely large input resistance. Consequently, it is not uncommon to find a voltage follower as the input stage of a high-performance, cascaded amplifier system. Negative feedback has made this high input resistance possible.

10-6 Effect of Output Voltage Negative Feedback on Output Resistance

Quite often, the load resistance across the output of an amplifier is *not* constant. Specifically, an amplifier may be required to drive a dynamic load, or it may be desirable to have a given amplifier drive a range of possible fixed loads. In either case, the output voltage of an amplifier will be affected by a change in the load resistance. For example, if the load resistance were to decrease, more current will be drawn from the amplifier's output. This will cause the voltage drop across the amplifier's output resistance to increase. Therefore, the output voltage will decrease.

Output voltage negative feedback will tend to minimize this effect. Consider Fig. 10-7(a). If an increase in I_{out} occurs, V_{out} will decrease. This will decrease the feedback voltage V_f. Therefore, more of V_{in} will be impressed across R_{in} (V will increase). This, in turn, will cause V_{out} to increase. In other words, negative feedback tends to counter output voltage changes. This has the same effect as *lowering* the amplifier's output resistance.

By using the feedback network to monitor an amplifier's output voltage, the amplifier's output resistance with feedback R_{of} will be less than the amplifier's open-loop output resistance R_{out} [see Fig. 10-7(b)].

FIGURE 10-7 Output resistance with voltage-series negative feedback: (a) signal process; (b) R_{of}; (c) equivalent circuit to find R_{of}.

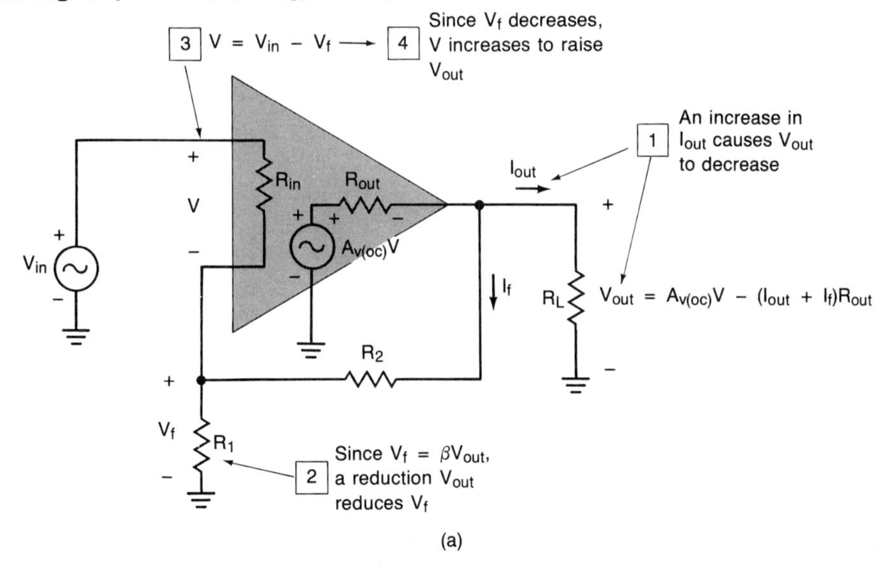

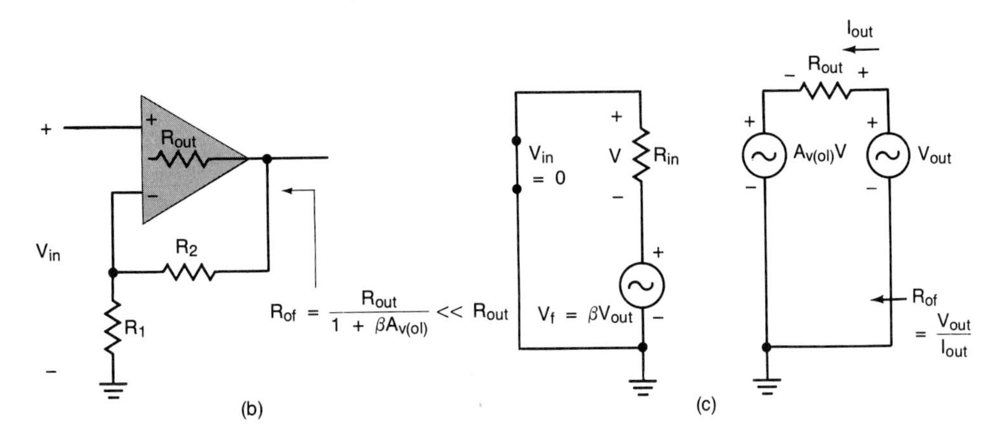

Observe that we shall again assume that the loaded open-loop voltage gain $A_{v(\text{ol})}$ is approximately equal to $A_{v(\text{oc})}$. To find the output resistance we set V_{in} equal to zero and place a fictitious voltage source V_{out} across the output of the feedback amplifier.

In Fig. 10-7(c) we can see that

$$V = -V_f = -\beta V_{\text{out}} \tag{10-12}$$

If we apply Kirchhoff's voltage law around the output loop and substitute Eq. 10-12 into the result, we arrive at

$$\begin{aligned} V_{\text{out}} &= I_{\text{out}} R_{\text{out}} + A_{v(\text{ol})} V \\ &= I_{\text{out}} R_{\text{out}} - \beta A_{v(\text{ol})} V_{\text{out}} \end{aligned} \tag{10-13}$$

The output resistance with feedback R_{of} may be found by solving for the ratio of V_{out} to I_{out}. Thus

$$V_{\text{out}} + \beta A_{v(\text{ol})} V_{\text{out}} = I_{\text{out}} R_{\text{out}}$$

$$V_{\text{out}}(1 + \beta A_{v(\text{ol})}) = I_{\text{out}} R_{\text{out}}$$

$$\boxed{R_{of} = \frac{V_{\text{out}}}{I_{\text{out}}} = \frac{R_{\text{out}}}{1 + \beta A_{v(\text{ol})}}} \tag{10-14}$$

Equation 10-14 indicates that the output resistance with feedback may be found by taking the amplifier's open-loop output resistance and dividing it by a quantity equal to 1 plus the loop gain. This can have a dramatic effect.

EXAMPLE 10-4

Find the maximum R_{of} of the noninverting op amp amplifier circuit given in Fig. 10-4(a). The 741C has a minimum $A_{v(\text{oc})}$ of 20,000 and an open-loop output resistance of 75 Ω. (This low output resistance is made possible by the use of an emitter-follower push-pull output stage.)

SOLUTION The solution merely requires direct application of Eq. 10-14. Recalling that the feedback factor β is 0.00990, we have

$$\begin{aligned} R_{of} &= \frac{R_{\text{out}}}{1 + \beta A_{v(\text{ol})}} = \frac{75\ \Omega}{1 + (0.00990)(20{,}000)} \\ &= 0.377\ \Omega \end{aligned}$$

■

As we can see, the use of negative feedback has provides an output resistance that is negligibly small. With an R_{out} this small, the output loading effects are virtually nonexistent.

10-7 Negative-Feedback Effects on Bandwidth

The use of negative feedback tends to extend the bandwidth of an amplifier. This is true for each of the four fundamental amplifiers (voltage, transresistance, transconductance, and current) with their associated feedback topologies (voltage-series, voltage-shunt, current-series, and current-shunt, respectively). Since we have initially been dealing with the voltage amplifier and voltage-series negative feedback, we shall use it as the vehicle to explain this effect.

In Fig. 10-8(a) we see the open-loop voltage gain of the 741C op amp plotted as a function of frequency. We shall denote the open-loop voltage gain as a function of frequency by $A_{v(\text{ol})}[f]$. Since the op amp is direct coupled, its frequency response extends down to dc (0 Hz), which cannot be indicated on our logarithmic frequency axis. The op amp's corner frequency is determined by its small (30-pF) frequency-

FIGURE 10-8 Negative feedback extends the bandwidth: (a) 741C open-loop response; (b) closed-loop response.

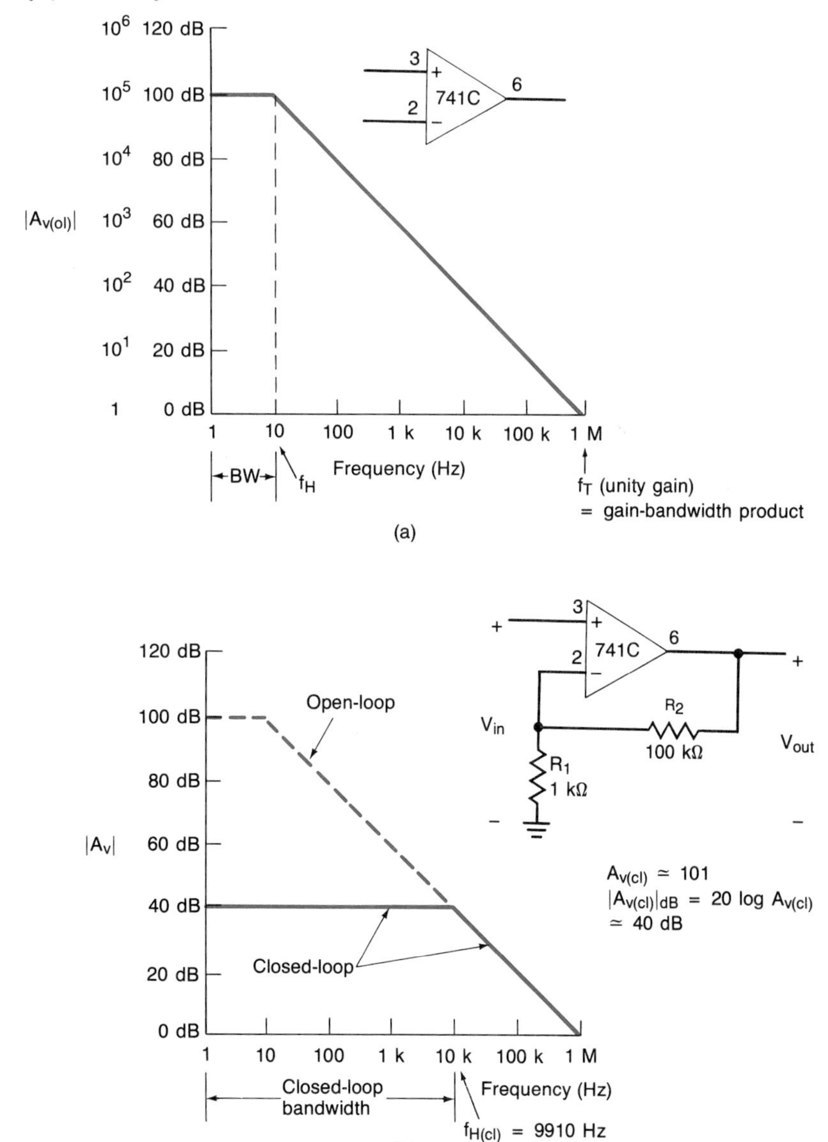

compensation capacitor. This dominant corner frequency limits the open-loop bandwidth to only 10 Hz!

Let us proceed to develop an analytical description of the effects of negative feedback on this very limited open-loop bandwidth. From Eq. 10-5 we recall that

$$A_{v(\text{cl})} = \frac{A_{v(\text{ol})}}{1 + \beta A_{v(\text{ol})}}$$

From our work in Chapter 9, it should be clear that the open-loop voltage gain as a function of frequency $A_{v(\text{ol})}[f]$ may be described by

$$A_{v(\text{ol})}[f] = \frac{A_{v(\text{ol})}}{1 + j(f/f_H)} \tag{10-15}$$

$A_{v(\text{ol})}$ is the dc voltage gain, and f_H is the dominant corner frequency (e.g., 10 Hz for the 741C op amp). We may arrive at an equation for the closed-loop voltage gain as a function of frequency $A_{v(\text{cl})}[f]$ merely by substituting Eq. 10-15 into Eq. 10-5. Hence

$$A_{v(\text{cl})}[f] = \frac{A_{v(\text{ol})}[f]}{1 + \beta A_{v(\text{ol})}[f]} = \frac{\dfrac{A_{v(\text{ol})}}{1 + jf/f_H}}{1 + \beta \dfrac{A_{v(\text{ol})}}{1 + jf/f_H}}$$

If we multiply the numerator and denominator terms by $(1 + jf/f_H)$, we arrive at the following result:

$$A_{v(\text{cl})}[f] = \frac{A_{v(\text{ol})}}{[1 + j(f/f_H)] + \beta A_{v(\text{ol})}} = \frac{A_{v(\text{ol})}}{1 + \beta A_{v(\text{ol})} + j(f/f_H)}$$

Dividing the numerator and denominator terms by $(1 + \beta A_{v(\text{ol})})$ leads us to

$$A_{v(\text{cl})}[f] = \frac{\dfrac{A_{v(\text{ol})}}{1 + \beta A_{v(\text{ol})}}}{1 + \dfrac{jf/f_H}{1 + \beta A_{v(\text{ol})}}} = \frac{\dfrac{A_{v(\text{ol})}}{1 + \beta A_{v(\text{ol})}}}{1 + j\dfrac{f}{(1 + \beta A_{v(\text{ol})})f_H}}$$

$$\boxed{A_{v(\text{cl})}[f] = \frac{A_{v(\text{cl})}}{1 + j[f/f_{H(\text{cl})}]}} \tag{10-16}$$

where $A_{v(\text{cl})}[f]$ = closed-loop voltage gain as a function of frequency

$A_{v(\text{cl})} = A_{v(\text{ol})}/[1 + \beta A_{v(\text{ol})}]$

$f_{H(\text{cl})}$ = closed-loop upper corner frequency, which is $[1 + \beta A_{v(\text{ol})}]f_H$

The closed-loop upper corner frequency $f_{H(\text{cl})}$ is increased by a factor of 1 plus the loop gain.

In Section 9-16 we learned that f_T is called the gain–bandwidth product. In Fig. 10-8(a) we see that the f_T of the 741C op amp is 1 MHz. The gain–bandwidth product has been defined by

$$f_T = A_{v(\text{ol})} f_H \tag{10-17}$$

If we find the product of $A_{v(\text{cl})}$ and $f_{H(\text{cl})}$, we see that it too will be equal to f_T.

$$A_{v(\text{cl})} f_{H(\text{cl})} = \frac{A_{v(\text{ol})}}{\cancel{1 + \beta A_{v(\text{ol})}}} [\cancel{1 + \beta A_{v(\text{ol})}}] f_H$$
$$= A_{v(\text{ol})} f_H$$

$$\boxed{f_T = A_{v(\text{ol})} f_H = A_{v(\text{cl})} f_{H(\text{cl})}} \qquad (10\text{-}18)$$

This is an extremely useful equation, as the examples below will demonstrate.

EXAMPLE 10-5

The 741C op amp has the open-loop frequency response curve given in Fig. 10-8(a). Observe that the open-loop voltage gain $A_{v(\text{ol})}$ is 100 dB (100,000) at low frequencies. Further, the open-loop corner frequency f_H is 10 Hz, and the gain–bandwidth product is 1 MHz. Find the closed-loop corner frequency $f_{H(\text{cl})}$ for the noninverting amplifier circuit given in Fig. 10-8(b).

SOLUTION Recalling that the feedback factor β is 0.00990, and drawing on Eq. 10-16, we obtain

$$f_{H(\text{cl})} = [1 + \beta A_{v(\text{ol})}] f_H$$
$$= [1 + (0.00990)(1 \times 10^5)][10 \text{ Hz}] = 9910 \text{ Hz}$$ ■

EXAMPLE 10-6

Repeat Example 10-5 using the ideal closed-loop voltage gain and Eq. 10-18.

SOLUTION As we saw from Eq. 10-8,

$$A_{v(\text{cl})} \simeq \frac{1}{\beta} = 1 + \frac{R_2}{R_1} = 101$$

and from Eq. 10-18,

$$f_{H(\text{cl})} = \frac{f_T}{A_{v(\text{cl})}} = \frac{1 \text{ MHz}}{101} = 9901 \text{ Hz}$$ ■

As we can see, the use of the idealized voltage gain and the gain-bandwidth product are in close agreement with the result indicated in Example 10-5. Consequently, the procedure illustrated in Example 10-6 is the most commonly used approach - particularly when op amps are used. The closed-loop frequency response is given in Fig. 10-8(b).

Recall that (from Eq. 9-56) an amplifier's bandwidth is given by

$$\text{BW} = f_H - f_L$$

Since an op amp is direct coupled, $f_L = 0$ Hz *and therefore, for a frequency-compensated op amp,* BW $= f_H$.

By using the gain-bandwidth product for a given op amp and the closed-loop voltage gain, it is relatively easy to determine the closed-loop bandwidth of a given frequency-compensated amplifier. For example, if a 741C op amp (with an f_T of 1 MHz) is used in a noninverting amplifier configuration with a closed-loop voltage gain $A_{v(\text{cl})}$ of 100, the bandwidth will be

$$\text{BW} = \frac{f_T}{A_{v(\text{cl})}} = \frac{1 \text{ MHz}}{100} = 10 \text{ kHz}$$

In a similar fashion, we can determine the bandwidth for any given closed-loop voltage gain (see Table 10-1).

TABLE 10-1
Closed-Loop Bandwidth of an Amplifier with a Gain–Bandwidth Product of 1 MHz

Closed-Loop Voltage Gain	Bandwidth
10,000	100 Hz
1,000	1 kHz
100	10 kHz
10	100 kHz
1	1 MHz

In each case, the product of the closed-loop voltage gain and the bandwidth is equal to the f_T of the amplifier. With a little thought, we can see that the more negative feedback we have, the wider the bandwidth will be. The respective bandwidths have been illustrated in Fig. 10-9.

FIGURE 10-9 Relationship between closed-loop gain and bandwidth.

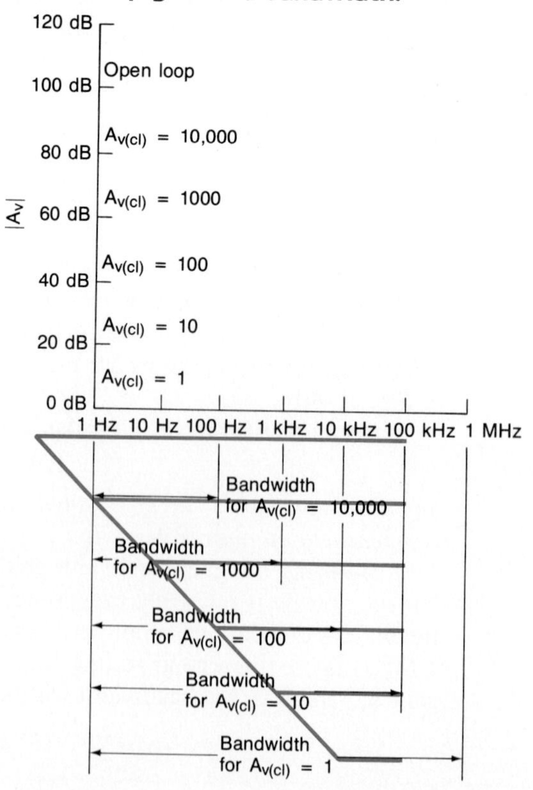

The gain–bandwidth product of a given amplifier can severely restrict its maximum closed-loop voltage gain. Consider Example 10-7.

EXAMPLE 10-7

Determine the maximum possible closed-loop voltage gain of a 741C op amp that is to be used as an audio amplifier.

SOLUTION A 741C has an f_T of 1 MHz. Since the audio-frequency range extends to 20 kHz, the maximum closed-loop voltage gain is

$$A_{v(\mathrm{cl})} = \frac{f_T}{\mathrm{BW}} = \frac{1\ \mathrm{MHz}}{20\ \mathrm{kHz}} = 50$$

■

10-8 Negative-Feedback Effects on Nonlinear Distortion

In our previous work we have constrained our amplifiers to operate on small signals. This was necessary to ensure that our BJT, FET, and op amp *linear* models were valid. Consider the BJT. Its transfer characteristic curve has been illustrated in Fig. 10-10(a).

For small operating point excursions, sinusoidal variations in v_{BE} produce approximately sinusoidal variations in i_C. Since

$$v_C = -\ i_C R_C$$

the collector (output) voltage will also be approximately sinusoidal, and the output response is described as being linear.

The large-signal condition has been depicted in Fig. 10-10(b). In this case the sinusoidal variation in v_{BE} produces a nonsinusoidal i_C. Therefore, the proportional output voltage v_C will also be distorted. When an amplifier changes the shape of the input waveform in this fashion we say that it has produced *amplitude distortion*. This is always associated with the nonlinear transfer characteristic of all real amplifiers.

Nonsinusoidal periodic waveforms may be represented as a sum of sinusoidal components. The amplitudes and frequencies associated with each of the individual sinusoidal components may be determined by a technique taught in calculus called the Fourier series expansion. The sinusoidal components will be at frequencies that are multiples (harmonics) of the fundamental frequency. This is the frequency-domain perspective of nonlinear amplitude distortion.

The time- and frequency-domain representations of a linear system have been illustrated in Fig. 10-11(a) and (b), respectively. Similarly, we see the representations of a nonlinear system in Fig. 10-11(c) and (d). Observe that a 1-kHz sinusoidal input has produced a nonsinusoidal output that contains a component at the fundamental (1 kHz), the second harmonic (2 kHz), the third harmonic (3 kHz), and so on. We shall not concern ourselves with the determination of the amplitudes of the individual harmonics (V_1, V_2, V_3, respectively) at this point. However, it is important to note that the amplitudes of the individual harmonics tend to diminish as their order increases.

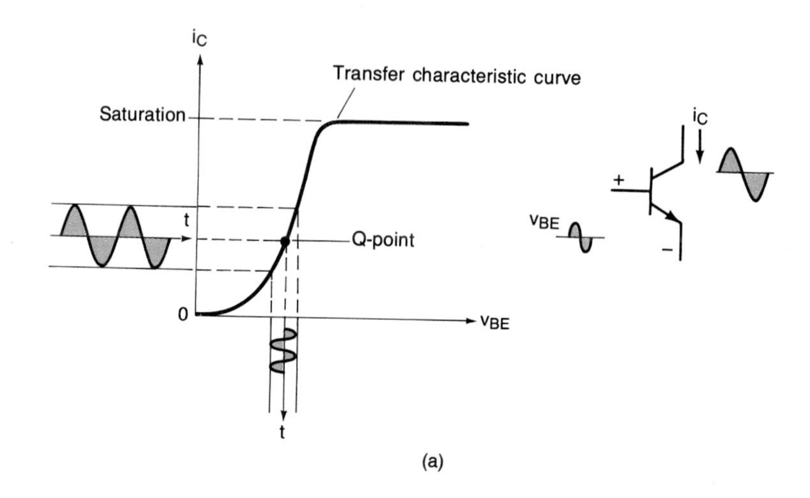

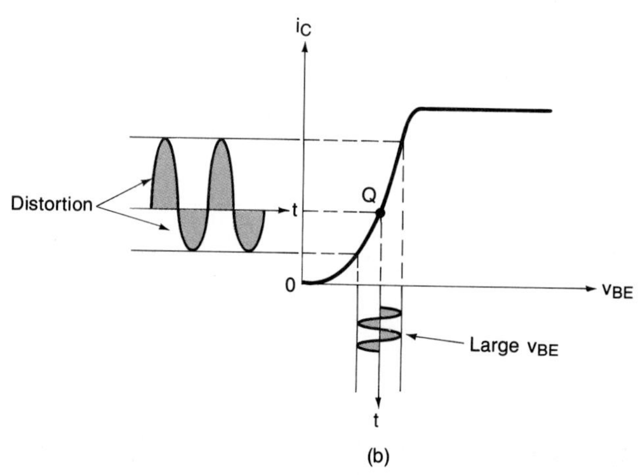

FIGURE 10-10 Nonlinear amplitude distortion: (a) approximately linear small-signal response; (b) nonlinear large-signal response.

Specifically, the second harmonic is generally larger than the third, the third is generally larger than the fourth, and so on.

Typically, we characterize the linearity (or nonlinearity) of a given amplifier by referring to its *total harmonic distortion* THD. *The THD is defined as the root sum of squares (RSS) of the individual distortion components.*

The percent of second harmonic distortion ($\% D_2$) is given by the ratio of amplitude of the second harmonic V_2 to the amplitude of the fundamental V_1. In a similar fashion, the percent of third harmonic distortion ($\% D_3$) may be found from the ratio of V_3 to V_1. THD may be found by using these distortion components in Eq. 10-19.

$$\% D_2 = \frac{V_2}{V_1} \times 100\% \qquad \text{percent of second harmonic distortion}$$

$$\% D_3 = \frac{V_3}{V_1} \times 100\% \qquad \text{percent of third harmonic distortion}$$

$$\% D_n = \frac{V_n}{V_1} \times 100\% \qquad \text{percent of } n\text{th harmonic distortion}$$

$$\boxed{\% \text{THD} = \sqrt{(\% D_2)^2 + (\% D_3)^2 + \cdots + (\% D_n)^2}} \qquad (10\text{-}19)$$

FIGURE 10-11 Time- and frequency-domain representations: (a) time-domain representation of a linear amplifier; (b) frequency-domain representation of a linear amplifier; (c) time-domain representation of a nonlinear amplifier; (d) frequency-domain representation of a nonlinear amplifier.

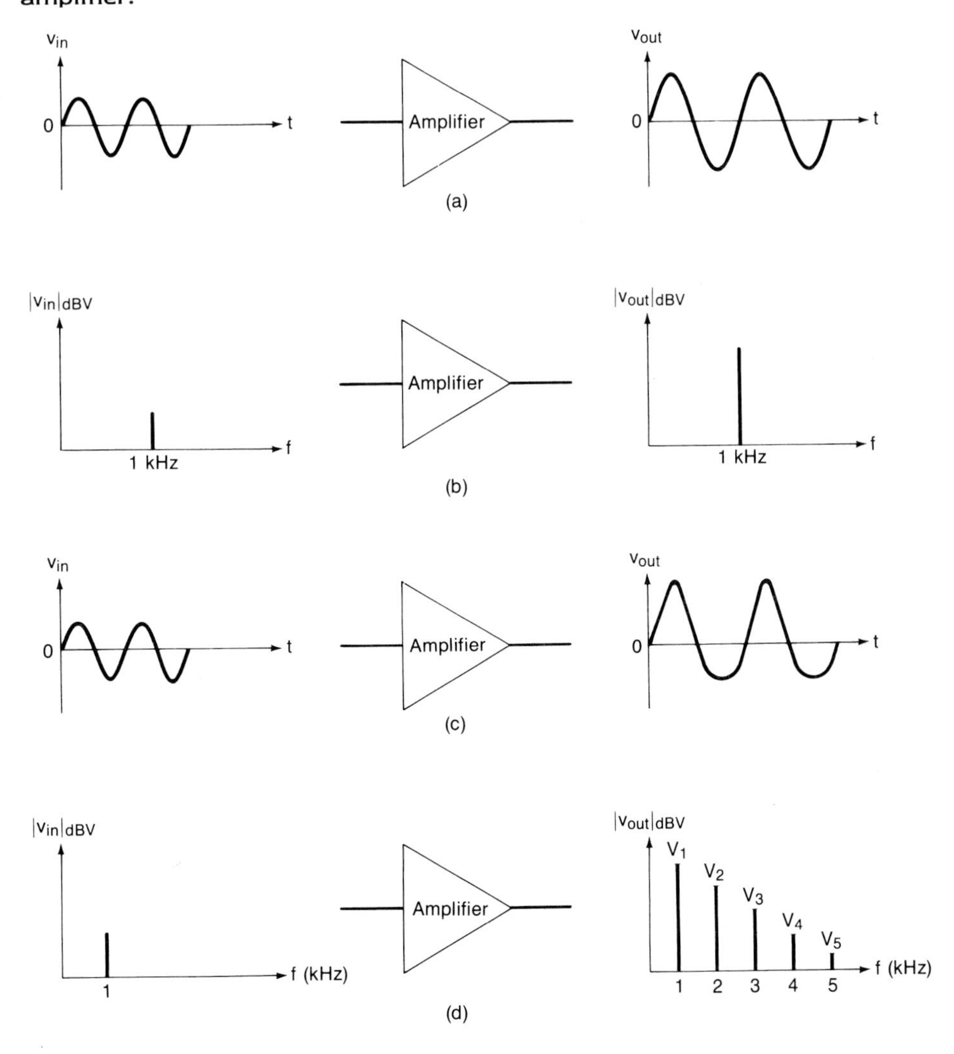

EXAMPLE 10-8

The amplifier in Fig. 10-11(d) has the following spectral content in its output: $V_1 = 4$ V rms, $V_2 = 0.4$ V rms, $V_3 = 0.2$ V rms, and $V_4 = 0.1$ V rms. Find the percent of total harmonic distortion % THD at 1 kHz.

SOLUTION First we must find the individual harmonic components, and then from Eq. 10-19 we may find the % THD.

$$\% D_2 = \frac{V_2}{V_1} \times 100\% = \frac{0.4 \text{ V rms}}{4 \text{ V rms}} \times 100\% = 10\%$$

$$\% D_3 = \frac{V_3}{V_1} \times 100\% = \frac{0.2 \text{ V rms}}{4 \text{ V rms}} \times 100\% = 5\%$$

$$\% D_4 = \frac{V_4}{V_1} \times 100\% = \frac{0.1 \text{ V rms}}{4 \text{ V rms}} \times 100\% = 2.5\%$$

$$\begin{aligned} \% \text{ THD} &= \sqrt{(\% D_2)^2 + (\% D_3)^2 + \cdots + (\% D_n)^2} \\ &= \sqrt{(10\%)^2 + (5\%)^2 + (2.5\%)^2} \\ &= \sqrt{131.25} = 11.5\% \end{aligned}$$

■

A % THD of 11.5% is quite excessive and could easily be detected by the average listener. As we shall see, *negative feedback will reduce nonlinear amplitude distortion*. Let us see how this occurs. From Eq. 10-19 we find that

$$\begin{aligned} \% \text{ THD} &= \sqrt{(\% D_2)^2 + (\% D_3)^2 + \cdots + (\% D_n)^2} \\ &= \sqrt{\left(\frac{V_2}{V_1} \times 100\%\right)^2 + \left(\frac{V_3}{V_1} \times 100\%\right)^2 + \cdots + \left(\frac{V_n}{V_1} \times 100\%\right)^2} \end{aligned}$$

If we square the individual terms and factor, we see that

$$\begin{aligned} \% \text{ THD} &= \sqrt{\frac{V_2^2}{V_1^2}(100\%)^2 + \frac{V_3^2}{V_1^2}(100\%)^2 + \cdots + \frac{V_n^2}{V_1^2}(100\%)^2} \\ &= \sqrt{\frac{(100\%)^2}{V_1^2}(V_2^2 + V_3^3 + \cdots + V_n^2)} \\ &= \sqrt{\frac{V_2^2 + V_3^2 + \cdots + V_n^2}{V_1^2}} \times 100\% \end{aligned}$$

We shall define V_{THD} as

$$V_{\text{THD}} = \sqrt{V_2^2 + V_3^2 + \cdots + V_n^2}$$

By substitution we obtain

$$\boxed{\% \text{ THD} = \frac{V_{\text{THD}}}{V_i} \times 100\%} \tag{10-20}$$

where % THD = percent of total harmonic distortion

V_{THD} = equivalent harmonic rms voltage = $(V_2^2 + V_3^2 + \cdots + V_n^2)^{1/2}$

V_1 = rms value of the fundamental

The equivalent rms value of any nonsinusoidal waveform is given by the square root of the sum of the squares of the rms values of the individual spectral (sinusoidal) components.

EXAMPLE 10-9

Repeat Example 10-8 by first finding the rms value of the harmonic content in the output voltage waveform, and then applying Eq. 10-20.

SOLUTION First we shall find V_{THD}.

$$V_{THD} = \sqrt{V_2^2 + V_3^2 + \cdots + V_n^2}$$

$$= \sqrt{(0.4 \text{ V rms})^2 + (0.2 \text{ V rms})^2 + (0.1 \text{ V rms})^2}$$

$$= 0.458 \text{ V rms}$$

and from Eq. 10-20,

$$\% \text{ THD} = \frac{V_{THD}}{V_1} \times 100\% = \frac{0.458 \text{ V rms}}{4 \text{ V rms}} \times 100\% = 11.5\%$$

■

Thus the individual harmonic components produced by the nonlinear transfer characteristic curve may be lumped together into one equivalent distortion voltage V_{THD}. Since this distortion is produced within the amplifier, we may model the situation as depicted in Fig. 10-12(a).

To gain a more intuitive feel for the effects of negative feedback on nonlinear amplitude distortion, consider Fig. 10-12(b). The fundamental and its second harmonic have been added together to form a single, distorted output waveform.

In Fig. 10-12(b) we see that a fraction of this distorted output voltage is developed across the feedback resistor R_1. Since the voltage v is the difference between v_{in} and v_f, we see that it is *predistorted* in a manner which opposes, or counters, the distortion produced within the amplifier. Consequently, we see in Fig. 10-12(b) that the output signal distortion is reduced.

In an effort to quantify this discussion, return to Fig. 10-12(a). Application of Kirchhoff's voltage law around the output produces the following statement:

$$V_{out} = V_{THD} + A_{v(ol)}V$$

and since

$$V = V_{in} - V_f = V_{in} - \beta V_{out}$$

we may substitute this into our equation for V_{out}.

$$V_{out} = V_{THD} + A_{v(ol)}(V_{in} - \beta V_{out})$$

$$= V_{THD} + A_{v(ol)}V_{in} - \beta A_{v(ol)}V_{out}$$

Solving for V_{out} gives us

$$V_{out} + \beta A_{v(ol)}V_{out} = A_{v(ol)}V_{in} + V_{THD}$$

$$V_{out}(1 + \beta A_{v(ol)}) = A_{v(ol)}V_{in} + V_{THD}$$

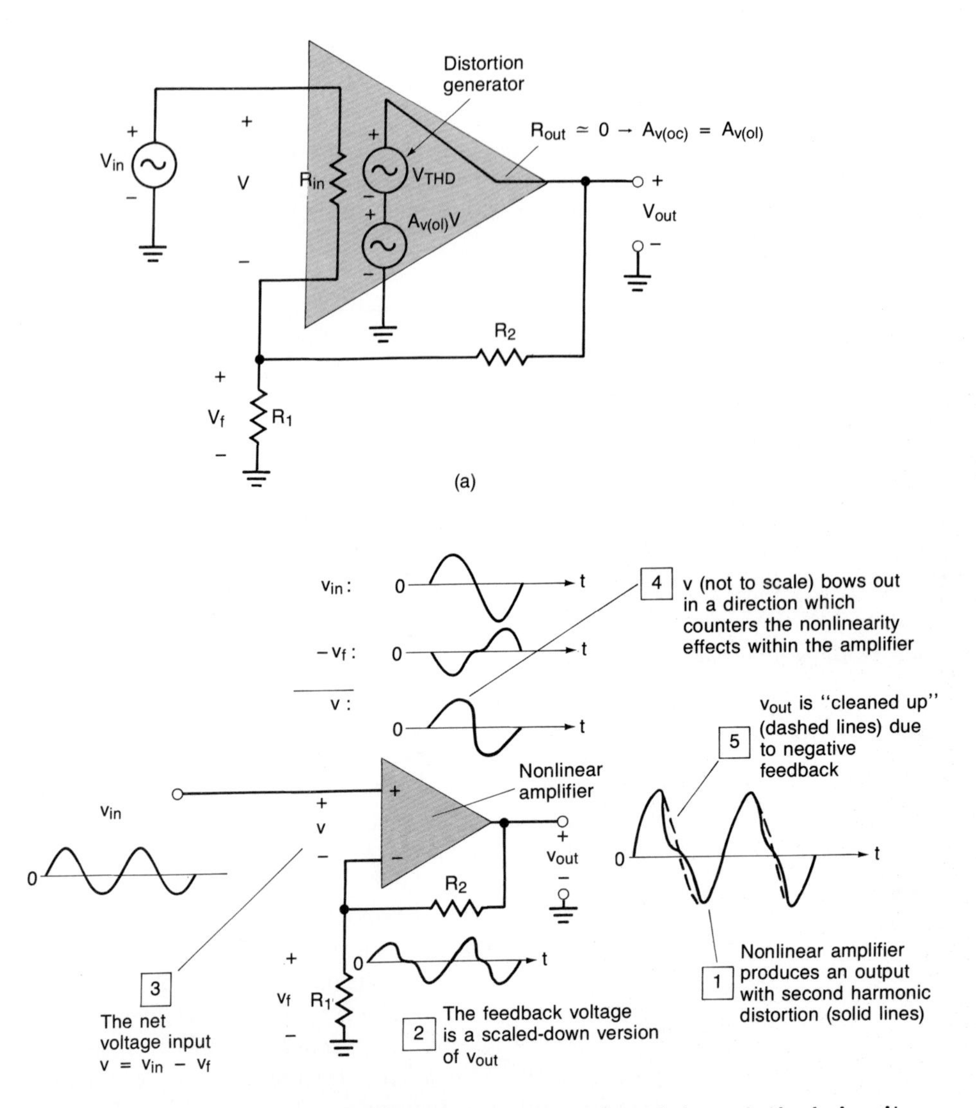

FIGURE 10-12 Negative feedback and distortion: (a) analytical circuit model; (b) negative feedback reduces distortion.

If we divide both sides by 1 plus the loop gain, we arrive at the following equation for V_{out}:

$$V_{out} = \underbrace{\frac{A_{v(ol)}}{1 + \beta A_{v(ol)}} V_{in}}_{\text{ideal component}} + \underbrace{\frac{V_{THD}}{1 + \beta A_{v(ol)}}}_{\text{distortion component}}$$

As we can see, the output contains an ideal and a distortion component. The distortion voltage with feedback $V_{THD(cl)}$ has been stated in

$$V_{THD(cl)} = \frac{V_{THD}}{1 + \beta A_{v(ol)}} \quad (10\text{–}21)$$

If we divide both sides of Eq. 10-21 by the rms value of the fundamental V_1 and find the percentage, we obtain Eq. 10-22 for the percent of THD with negative feedback $\%\ THD_{(cl)}$.

$$\%\ THD_{(cl)} = \frac{V_{THD(cl)}}{V_1} \times 100\% = \frac{V_{THD}/V_1}{1 + \beta A_{v(ol)}} \times 100\%$$

$$\boxed{\%\ THD_{(cl)} = \frac{\%\ THD}{1 + \beta A_{v(ol)}}} \quad (10\text{-}22)$$

EXAMPLE 10-10

Given that the noninverting op amp amplifier in Fig. 10-4(a) has an open-loop % THD of 11.5% at 1 kHz, an $A_{v(ol)}$ of 100,000, and a β of 0.00990, find the $\%\ THD_{(cl)}$.

SOLUTION By direct application of Eq. 10-22, we may find the closed-loop distortion.

$$\%\ THD_{(cl)} = \frac{\%\ THD}{1 + \beta A_{v(ol)}} = \frac{11.5\%}{1 + (0.00990)(100{,}000)}$$
$$= 0.0116\%$$

■

Obviously, negative feedback has served to reduce the nonlinear distortion greatly. Most of us have trouble in detecting % THD levels below 1% in audio amplifier systems.

10-9 Summary of Voltage-Series Negative Feedback

As we have seen, voltage-series negative feedback serves to stabilize the voltage gain, increase the input resistance, decrease the output resistance, increase the bandwidth, and reduce the nonlinear distortion. The primary sacrifice to achieve these benefits is a reduction in the maximum available (open-loop) voltage gain. A summary of our work thus far has been presented in Table 10-2.

TABLE 10-2
Summary of Voltage-Series Negative Feedback as Used with Voltage Amplifiers

Parameter	Equation	Comments
$A_{v(cl)}$	$A_{v(ol)}/[1 + \beta A_{v(ol)}]$	Decreased
R_{inf}	$[1 + \beta A_{v(ol)}]R_{in}$	Increased
R_{of}	$R_{out}/[1 + \beta A_{v(ol)}]$	Decreased
$BW_{(cl)}$	$[1 + \beta A_{v(ol)}]BW$	Increased
$\%\ THD_{(cl)}$	$\%\ THD/[1 + \beta A_{v(ol)}]$	Decreased

The bandwidth and distortion relationships may be generalized to *all* feedback amplifiers. However, the closed-loop input and output resistances depend on the feedback topology. This will become apparent as our work progresses.

10-10 Significance of Frequency Response, % THD, and Loop Gain

In Chapter 9 we developed Eq. 9-57 for the ac (open-loop) voltage gain of a frequency-compensated op amp [see Fig. 10-13(a)].

$$A_v[f] = \frac{g_m}{j2\pi f C_c}$$

The ac response [Fig. 10-13(b)] is determined by the op amp's g_m and C_c. The unity-gain frequency (f_T) is tightly controlled by the manufacturers. (The op amp is described as being frequency compensated for unity gain. The significance of this description will become apparent in the next two sections.)

From our extensive work in Chapter 9, it should be obvious that C_c will produce a corner frequency f_H [see Fig. 10-13(c)]. We can generalize the response illustrated in Fig. 10-13(c) by using the amplifier model shown in Fig. 10-14(a). An IC op amp, or a discrete amplifier, can be forced to exhibit the response shown in Fig. 10-14(b). Therefore, the amplifier model given in Fig. 10-14(a) is a much more universal representation.

In Chapter 9 we saw that an amplifier which has the frequency response illustrated in Fig. 10-14(b) may be described by

$$A_{v(\mathrm{ol})}[f] = \frac{A_{v(\mathrm{ol})}}{1 + j(f/f_H)} \tag{10-23}$$

We are reminded of the corresponding magnitude in decibels, and the phase angle by Eqs. 10-24 and 10-25, respectively.

$$|A_{v(\mathrm{ol})}[f]|_{\mathrm{dB}} = 20 \log A_{v(\mathrm{ol})} - 10 \log|1 + (f/f_H)^2| \tag{10-24}$$

$$\phi = -\tan^{-1}\frac{f}{f_H} \tag{10-25}$$

For a factor-of-10 increase in the signal frequency (a decade) beyond f_H, the amplitude given by Eq. 10-24 decreases, or rolls off, at a rate of -20 dB/decade. For instance, if we let f equal $10f_H$,

$$\begin{aligned}|A_{v(\mathrm{ol})}[f]|_{\mathrm{dB}} &= 20 \log A_{v(\mathrm{ol})} - 10 \log|1 + (10f_H/f_H)^2| \\ &= 20 \log A_{v(\mathrm{ol})} - 10 \log|1 + 100| \\ &= 20 \log A_{v(\mathrm{ol})} - 10 \log 101 \\ &\simeq 20 \log A_{v(\mathrm{ol})} - 20 \text{ dB}\end{aligned}$$

Alternatively, the roll-off may be described in terms of *octaves*. An octave is a factor-of-2 change in frequency. If we let f in Eq. 10-24 equal $2f_H$, $4f_H$, $8f_H$, and so on, we see that the gain rolls off at a rate of approximately -6 dB/octave. Both of these descriptions have been indicated in Fig. 10-14(b). These descriptions are used interchangeably (e.g., -20 dB/decade $=$ -6 dB/octave and -40 dB/decade $=$

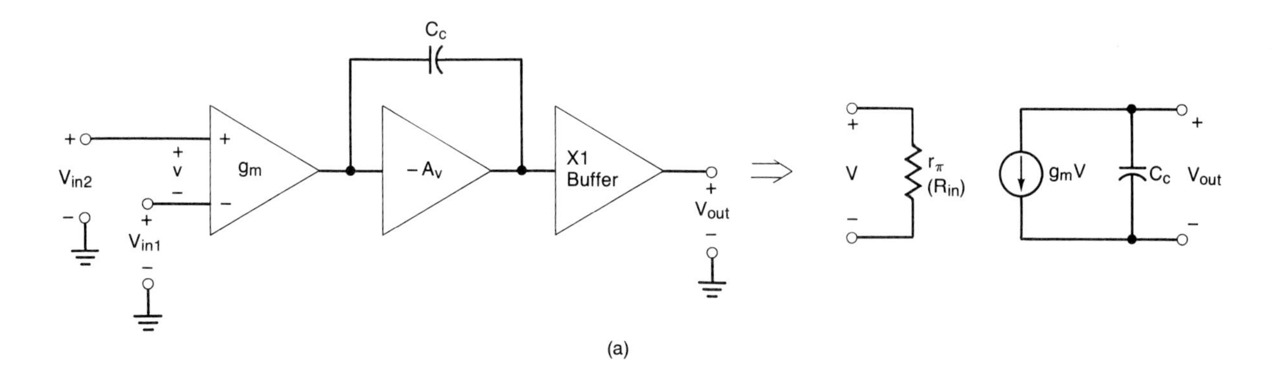

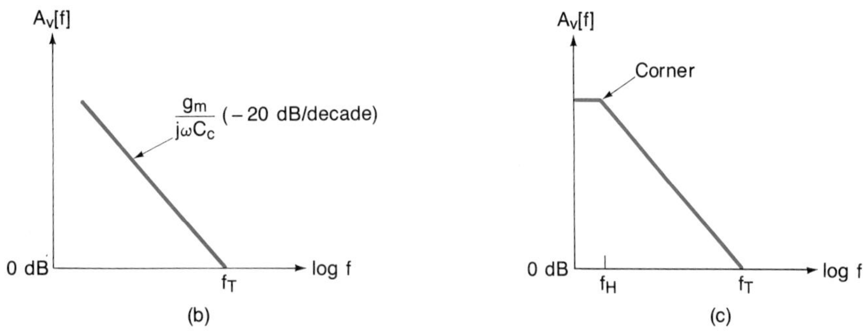

FIGURE 10-13 Frequency-compensated op amp: (a) block diagram and transconductance model; (b) frequency response defined by the gain equation; (c) C_c produces a corner frequency.

−12 dB/octave). However, we shall continue to favor expressing gain changes in dB per decade.

Now let us consider the effects of negative feedback [see Fig. 10-15(a)]. The feedback factor is (from Eq. 10-7)

$$\beta = \frac{R_1}{R_1 + R_2} = \frac{1\ \text{k}\Omega}{1\ \text{k}\Omega + 20\ \text{k}\Omega} = 0.04762$$

The closed-loop gain is (from Eq. 10-5)

$$A_{v(\text{cl})} = \frac{A_{v(\text{ol})}}{1 + \beta A_{v(\text{ol})}} = \frac{100{,}000}{1 + (0.04762)(100{,}000)} = 21$$

The corresponding closed-loop gain in decibels is

$$|A_{v(\text{cl})}|_{\text{db}} = 20 \log A_{v(\text{cl})} = 20 \log 21 = 26.4\ \text{dB}$$

The corner frequency with negative feedback was defined by Eq. 10-16. In this case,

$$\begin{aligned} f_{H(\text{cl})} &= (1 + \beta A_{v(\text{ol})}) f_H \\ &= [1 + (0.04762)(100{,}000)][10\ \text{Hz}] = 47.6\ \text{kHz} \end{aligned}$$

The closed-loop frequency response has been illustrated in Fig. 10-15(b). Note that the closed-loop response [including the closed-loop corner frequency $f_{H(\text{cl})}$] has been superimposed on the open-loop gain response. In Section 10-7 we saw that the closed-loop response may be described by Eq. 10-16 (which has been repeated below).

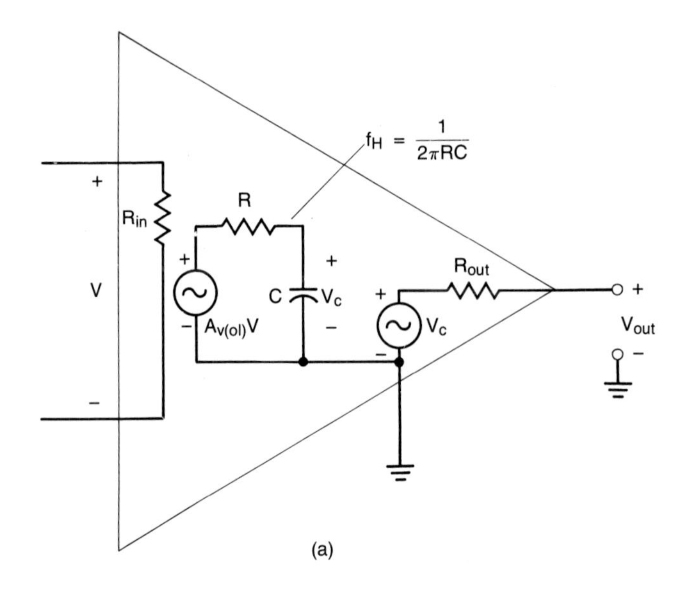

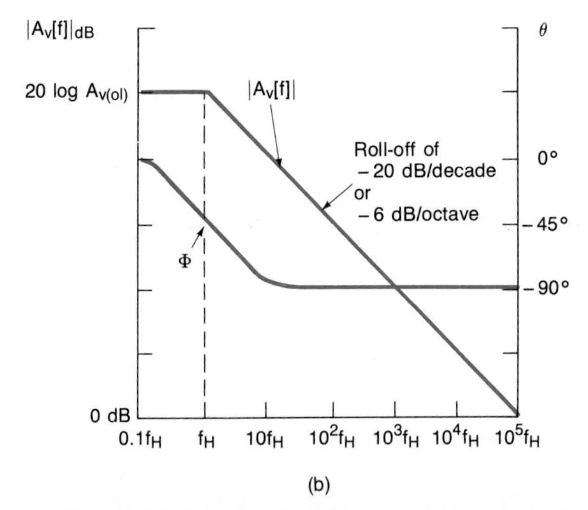

FIGURE 10-14 Frequency-compensated (for unity gain) amplifier: (a) universal model for discrete and integrated-circuit amplifiers; (b) frequency response.

$$A_{v(cl)}[f] = \frac{A_{v(cl)}}{1 + j[f/f_{H(cl)}]}$$

In other words, *the closed-loop response has the same shape as the open-loop response, and ultimately rolls off at* -20 *dB/decade* (*or* -6 *dB/octave*).

Let us review the response curves given in Fig. 10-15(b) from a loop-gain perspective. At low frequencies we have seen that

$$A_{v(cl)} = \frac{A_{v(ol)}}{1 + \beta A_{v(ol)}}$$

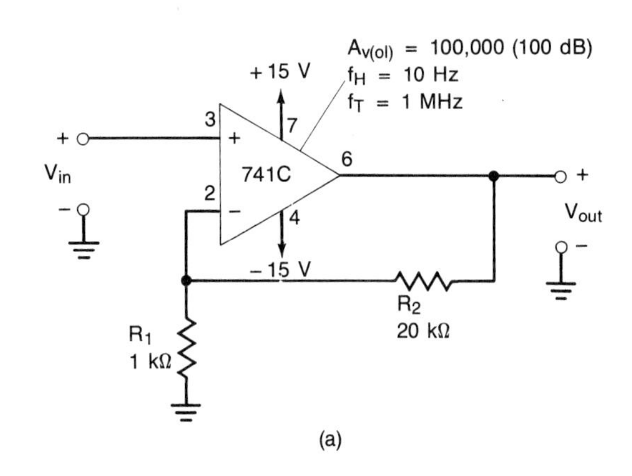

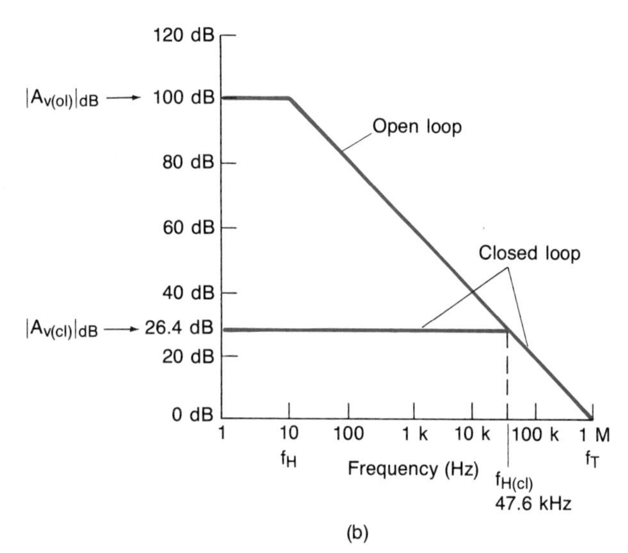

FIGURE 10-15 Closed-loop frequency response: (a) circuit; (b) open- and closed-loop frequency response.

and in decibels,

$$
\begin{aligned}
|A_{v(\text{cl})}|_{\text{dB}} &= 20 \log A_{v(\text{cl})} = 20 \log \frac{A_{v(\text{ol})}}{1 + \beta A_{v(\text{ol})}} \\
&= 20 \log A_{v(\text{ol})} - 20 \log |1 + \beta A_{v(\text{ol})}| \\
&\simeq 20 \log A_{v(\text{ol})} - 20 \log \beta A_{v(\text{ol})}
\end{aligned}
$$

The relationship above indicates that *the closed-loop voltage gain in decibels is approximately equal to the open-loop gain in decibels minus the loop gain in decibels*. In Fig. 10-15(a) the loop gain in decibels is

$$
\begin{aligned}
|\beta A_{v(\text{ol})}|_{\text{dB}} &= 20 \log \beta A_{v(\text{ol})} \\
&= 20 \log (0.04762)(100{,}000) \simeq 73.6 \text{ dB}
\end{aligned}
$$

and

$$
\begin{aligned}
|A_{v(\text{cl})}|_{\text{dB}} &= |A_{v(\text{ol})}|_{\text{dB}} - |\text{loop gain}|_{\text{dB}} \\
&= 20 \log A_{v(\text{ol})} - 20 \log \beta A_{v(\text{ol})} \\
&= 100 \text{ dB} - 73.6 \text{ dB} = 26.4 \text{ dB}
\end{aligned}
$$

The relationships between the open-loop gain, the closed-loop gain, and the loop gain have been noted in Fig. 10-16(a). To emphasize the relationships further, Fig. 10-16(b) has been provided.

The more loop gain used in the amplifier, the greater the bandwidth (BW) will be. Signal frequencies within the bandwidth will all be amplified uniformly. Signal

FIGURE 10-16 Relationships between the open-loop, loop, and closed-loop gains.

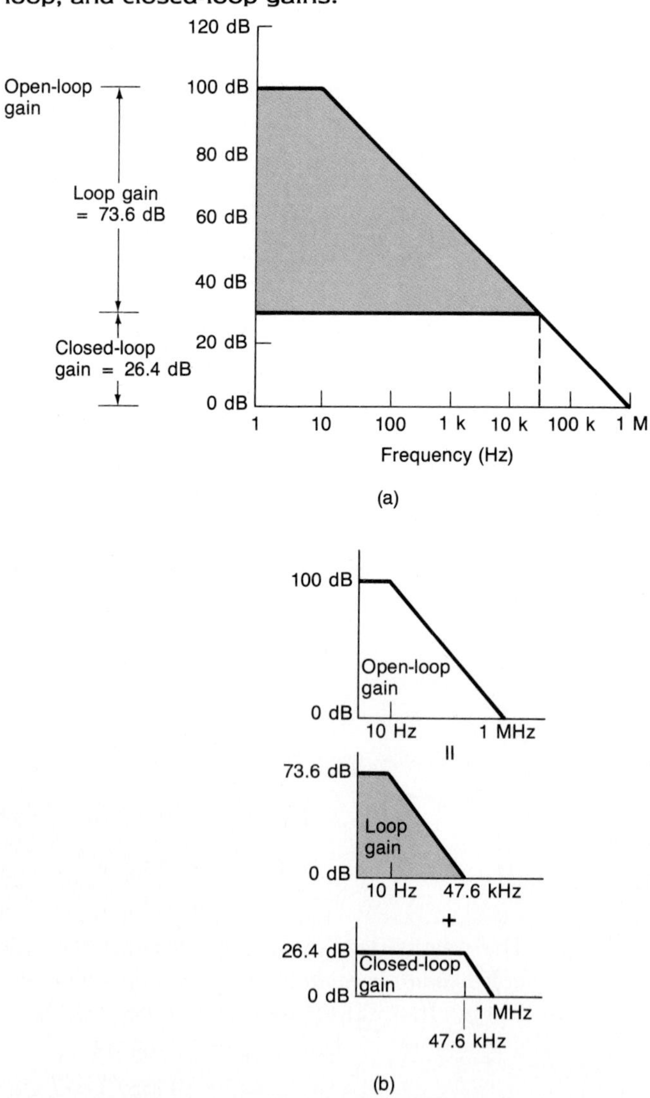

frequencies beyond the corner frequency will be amplified by a *lesser* amount. This can constitute *frequency distortion*. This type of distortion may not be as readily apparent to the reader as amplitude distortion. Therefore, we shall elaborate.

Recall that all periodic complex waveforms may be represented as a sum of sine waves. Consider Fig. 10-17(a). A square wave has been illustrated. If the Fourier series expansion were applied to it, the result below is obtained.

$$\begin{aligned} v(t) = {} & 31.83 \sin 2\pi(1000)t + 10.61 \sin 2\pi(3000)t \\ & + 6.366 \sin 2\pi(5000)t + 4.547 \sin 2\pi(7000)t \\ & + 3.537 \sin 2\pi(9000)t + \cdots + \frac{4V_m}{n\pi} \sin 2\pi nft \end{aligned} \tag{10-26}$$

FIGURE 10-17 **Bandwidth and frequency distortion: (a) square wave; (b) superposition representation.**

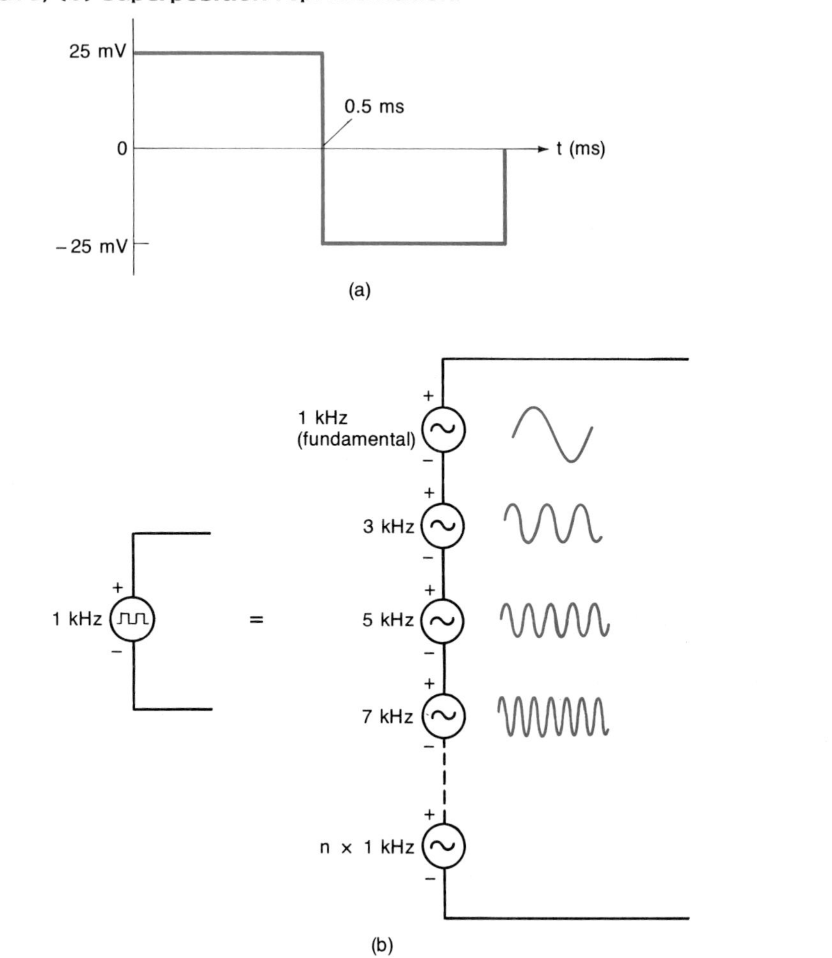

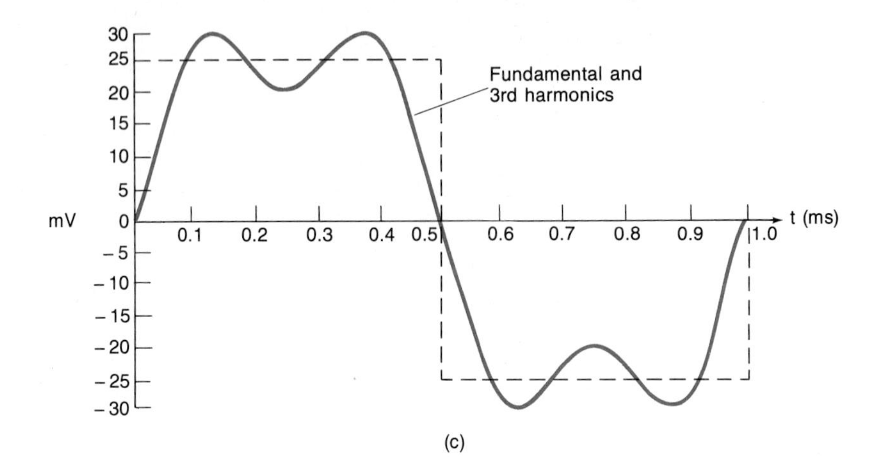

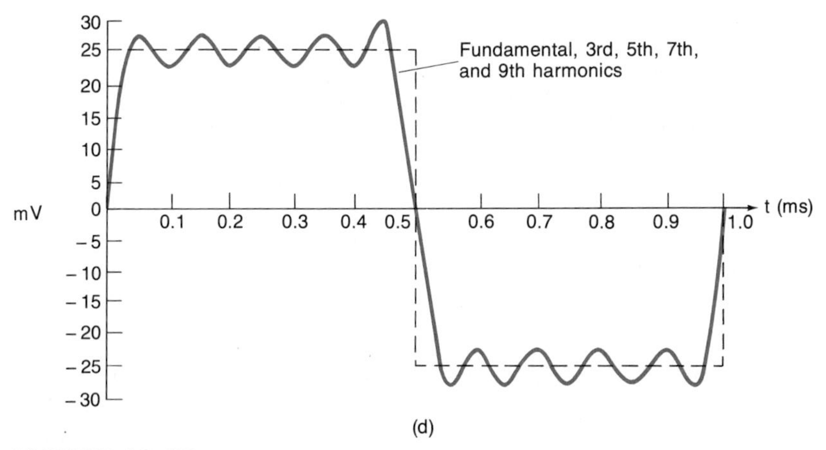

FIGURE 10-17 *(continued)* **(c) fundamental and third harmonic; (d) fundamental, third, fifth, seventh, and ninth harmonics.**

where V_m is the peak value of the square wave and n is the order of the harmonic (odd numbers only). Note that the peak values are in millivolts.

The expansion lists the fundamental, third, fifth, seventh, and ninth harmonics, but may be extended out to an infinite number of terms. The greater the number of harmonics, the closer the series will approximate the square wave given in Fig. 10-17(a).

In Fig. 10-17(b) this concept is demonstrated electrically. Specifically, we see that sinusoidal voltage sources placed in series may be used to synthesize a square wave - provided that each of the sinusoidal sources has the correct amplitude, frequency, and phase relationship. In Fig. 10-17(c) the sum of the fundamental and third harmonics is given. In Fig. 10-17(d) we see the sum that includes components up to the ninth harmonic. Figure 10-17(d) obviously offers a much closer approximation.

If the frequency response (bandwidth) of a given amplifier is severely limited, the higher harmonics of complex waveforms will be attenuated. Consequently, a square-wave input [Fig. 10-17(a)] might produce an output in which only the fundamental and third harmonics are significant. The result would be an output waveform such as that shown in Fig. 10-17(c). If the bandwidth of the amplifier were increased (e.g., by negative feedback) such that only the harmonics above the ninth were lost, the output waveform would be similar to that given in Fig. 10-17(d).

Now we can truly appreciate the importance of a wide bandwidth - frequency distortion is reduced. *Negative feedback (a large loop gain) can serve to minimize frequency distortion.*

Now let us consider the effects of the loop gain on the total harmonic distortion (% THD). Assume that the % THD of the 741C [Fig. 10-15(a)] is measured to be 10% at 10 kHz [refer to Fig. 10-18(a)].

FIGURE 10-18 Loop gain and % THD: (a) open-loop gain of the 741C; (b) loop gain; (c) loop-gain response.

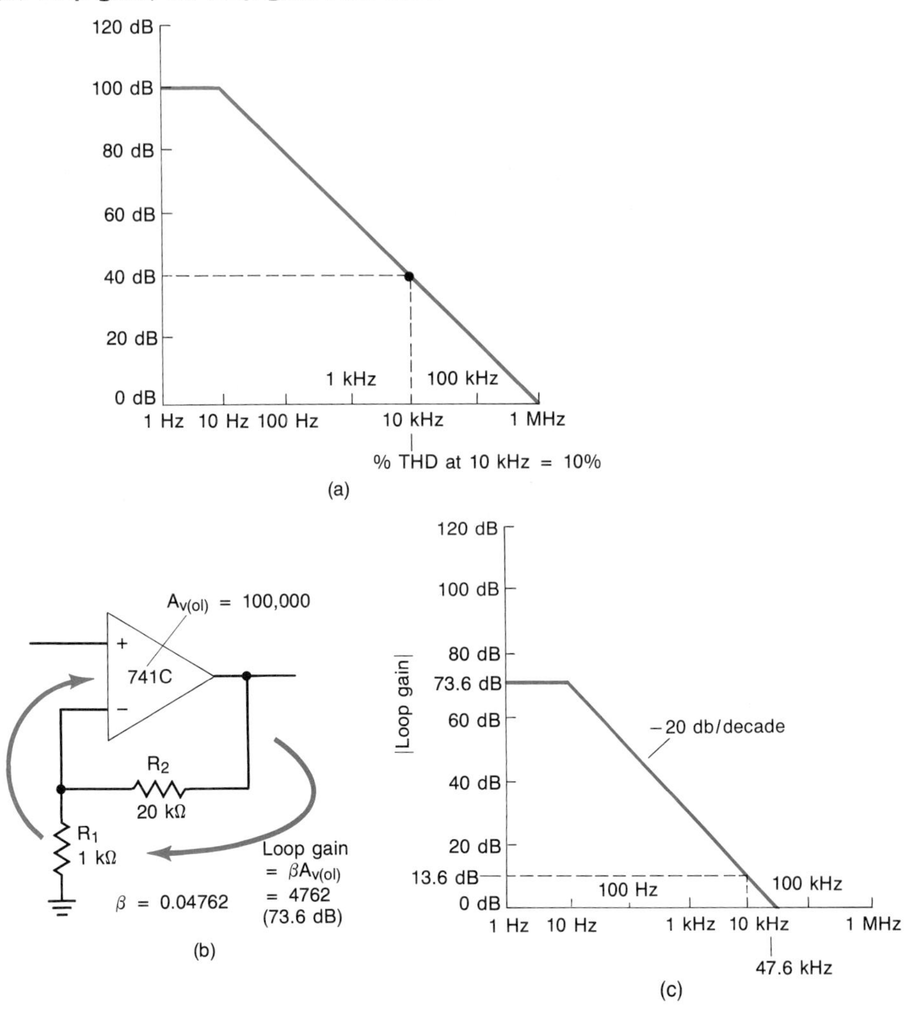

If we use negative feedback to provide a closed-loop gain of 21 (26.4 dB), the corresponding loop gain is

$$\beta A_{v(\mathrm{ol})} = (0.04762)(100{,}000) = 4762$$

or 73.6 dB [see Fig. 10-18(b)]. In this case, the total harmonic distortion will be (from Eq. 10-22)

$$\%\ \mathrm{THD}_{(\mathrm{cl})} = \frac{\%\ \mathrm{THD}}{1 + \beta A_{v(\mathrm{ol})}} \simeq \frac{\%\ \mathrm{THD}}{\beta A_{v(\mathrm{ol})}} = \frac{10\%}{4762} = 0.0021\%$$

Once again we see that negative feedback has significantly reduced the harmonic distortion. However, this is true only at low frequencies (below the open-loop corner f_H). The loop gain varies with frequency. Consider Fig. 10-18(c). At 10 kHz the loop gain is only 13.6 dB. This corresponds to

$$\beta A_{v(\mathrm{ol})}[f] = 4.762$$

Therefore, the % THD at 10 kHz is

$$\%\ \mathrm{THD}_{(\mathrm{cl})} \simeq \frac{\%\ \mathrm{THD}}{\beta A_{v(\mathrm{ol})}} = \frac{10\%}{4.762} = 2.1\%$$

The reduction in the loop gain allows more of the open-loop distortion to appear at higher frequencies.

Because the loop gain varies with frequency, the % THD of voltage amplifiers should be stated at a given closed-loop gain and the frequency at which it was measured. (*Power amplifiers* should also include a statement of the output power level.)

To minimize both frequency and nonlinear distortion, the loop gain should be large. This can be achieved by increasing the negative feedback, the open-loop gain, and/or the open-loop bandwidth.

Manufacturers of IC op amps typically opt to extend the open-loop bandwidth. Specifically, they provide circuit designers with op amps which are either *decompensated* or *uncompensated.*

Decompensated op amps employ lower values of internal frequency-compensation capacitors. Uncompensated op amps do not include any internal frequency compensation capacitors. These op amps, cascaded op amps, and discrete device amplifier systems require that we understand the need for frequency compensation and how to apply it. This is necessary if we are to avoid *oscillations.*

10-11 Closed-Loop Stability: The Oscillation Problem

Negative feedback enhances amplifier performance. However, as we cautioned in Section 10-1, too much negative feedback can produce instability. Specifically, the amplifier might break into oscillations. Let us see exactly how and why this occurs.

In Chapter 9 we saw that the gain of an amplifier tends to decrease as the signal frequency increases. The frequency roll-off is produced by the capacitive effects within the amplifier. Recall that the two major sources of these capacitive effects are the various BJT and FET device capacitances, and stray (wiring) capacitances.

Because of these various capacitances, it is not unusual to find two, three, or more high-frequency breakpoints (or corner frequencies). (The reader is invited to refer back to Section 9-13.) We found two high-frequency breakpoints in both the single-stage BJT and FET amplifiers. Subsequently, the possibility of several corner frequencies in a cascaded amplifier system (including the cascaded amplifiers found in the IC op amp) should not be too astounding. Examples have been provided in Fig. 10-19.

In Fig. 10-19(a) we see an open-loop response with two corner frequencies. Note that for frequencies beyond f_{H2} the roll-off rate increases to -40 dB/decade (-12 dB/octave). In Fig. 10-19(b) we see an open-loop response with three corner frequencies.

Figure 10-19(b) also indicates the situation if negative feedback is used to provide an ideal closed-loop voltage gain of 40 dB (100). Once again, all that is required is that we superimpose the closed-loop gain on the open-loop response. In this case, it is extremely important to note that the gain rolls off at a rate of -60 dB/decade beyond the intersection between the ideal closed-loop gain and the open-loop response.

When we are dealing with amplifiers with more than one open-loop corner frequency, we shall generally opt for graphical analyses. This approach tends to be much less cumbersome when compared to more mathematical approaches.

This becomes apparent when we describe the open-loop responses mathematically. For example, the response shown in Fig. 10-19(a) would be defined by

$$A_{v(\mathrm{ol})}[f] = \frac{A_{v(\mathrm{ol})}}{(1 + jf/f_{H1})(1 + jf/f_{H2})} \tag{10-27}$$

FIGURE 10-19 Frequency responses with multiple corner frequencies: (a) 2nd-order; (b) 3rd order.

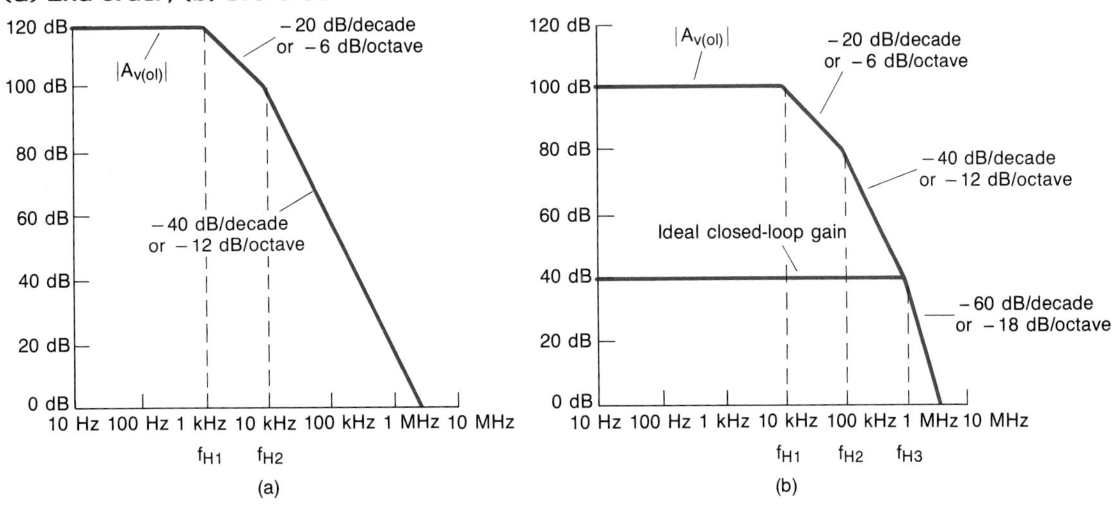

The corresponding magnitude in decibels and phase information for our noninverting amplifier are provided by

$$|A_{v(\mathrm{ol})}[f]|_{\mathrm{dB}} = 20 \log A_{v(\mathrm{ol})} - 10 \log[1 + (f/f_{H1})^2] - 10 \log[1 + (f/f_{H2})^2] \quad (10\text{-}28)$$

$$\phi = -\tan^{-1}\frac{f}{f_{H1}} - \tan^{-1}\frac{f}{f_{H2}} \quad (10\text{-}29)$$

Similarly, the response depicted in Fig. 10-19(b) is described by

$$A_{v(\mathrm{ol})}[f] = \frac{A_{v(\mathrm{ol})}}{(1 + jf/f_{H1})(1 + jf/f_{H2})(1 + jf/f_{H3})} \quad (10\text{-}30)$$

$$\begin{aligned} |A_{v(\mathrm{ol})}[f]|_{\mathrm{dB}} &= 20 \log A_{v(\mathrm{ol})} - 10 \log|1 + (f/f_{H1})^2| \\ &\quad - 10 \log|1 + (f/f_{H2})^2| \\ &\quad - 10 \log|1 + (f/f_{H3})^2| \end{aligned} \quad (10\text{-}31)$$

$$\phi = -\tan^{-1}\frac{f}{f_{H1}} - \tan^{-1}\frac{f}{f_{H2}} - \tan^{-1}\frac{f}{f_{H3}} \quad (10\text{-}32)$$

Equations 10-27 and 10-29 suggest that the maximum possible phase shift with two corner frequencies is $-180°$. Equations 10-30 and 10-32 indicate that the maximum possible phase shift with three corner frequencies is $-270°$.

The phase angle of the feedback voltage (or current) is *critical* [see Fig. 10-20(a)]. The input signal v_{in} and the output signal v_{out} are in phase. Specifically, there is 0° of phase shift across the amplifier. This is true only for signal frequencies at least one decade below the lowest closed-loop corner frequency.

If the amplifier has the open-loop response shown in Fig. 10-19(b) and the signal frequency were to be increased sufficiently, the phase shift will change. In fact, if the signal frequency were to be increased to 333 kHz, the phase shift between v_{in} and v_{out} would be $-180°$! [See Fig. 10-20(b).]

Consider what has happened. The feedback voltage v_f has changed sign. Instead of being *negative* (opposing v_{in}) it is now *positive* (additive with v_{in}). This is termed *positive feedback*. If it becomes sufficiently large, the amplifier will begin to generate a signal, or *oscillate*. Examine Fig. 10-21.

All active and passive devices will generate small levels (typically, nanovolts or less) of white electrical noise. White noise is a random generation of electrical signals that encompass the frequency spectrum from essentially dc (0 Hz) to extremely high (many gigahertz) frequencies. The frequency-domain representation of white noise has been included in Fig. 10-21(a).

We shall not delve into the various noise mechanisms in detail. However, it should be reasonably clear that thermal energy will produce a random motion in free electrons. This random electron motion serves to produce a random current. If this random current exists in a resistor (or any resistance), a random (noise) voltage will be developed across that resistor.

At some frequency f_o (e.g., 333 kHz) the phase shift across that amplifier will be $-180°$. This particular frequency will be contained in the white noise [refer to Fig. 10-21(a)].

Suppose that 1 μV peak to peak of noise voltage at f_o appears across the amplifier's input. This gives rise to the situation shown in Fig. 10-21(b). At a frequency of f_o

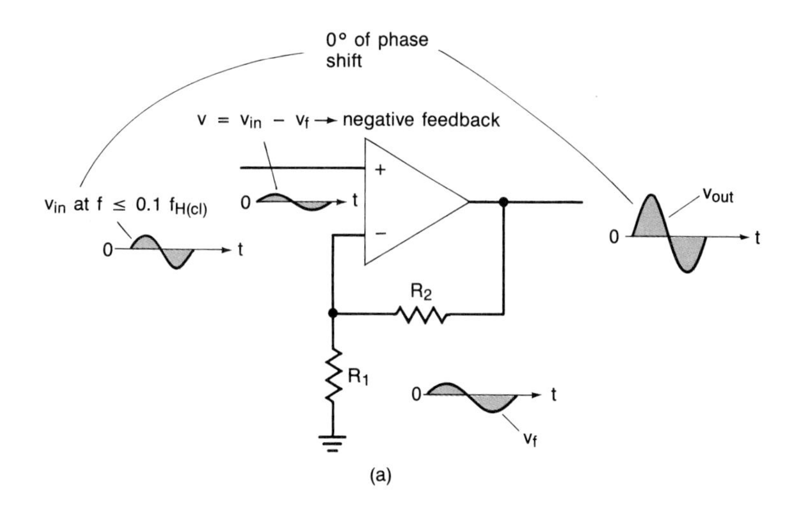

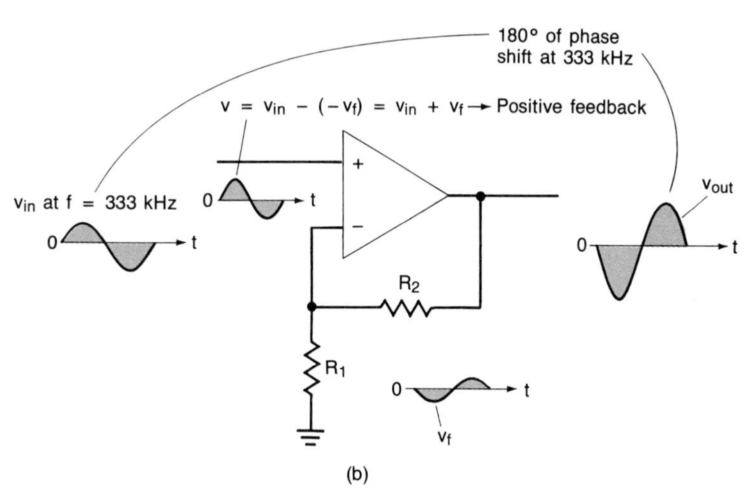

FIGURE 10-20 **The phase shift of the feedback signal is critical: (a) $f \leq 0.1 f_{H(cl)}$; (b) $f = 333$ kHz.**

the amplifier's open-loop voltage gain has been assumed to drop off to 234 [see Fig. 10-21(c)]. Therefore,

$$\begin{aligned} v_{out} &= A_{v(ol)}[f]\, v = (234 \angle -180^\circ)(1\ \mu V_{p\text{-}p}) \\ &= 234\ \mu V_{p\text{-}p} \angle -180^\circ \end{aligned}$$

as indicated in Fig. 10-21(b). The feedback factor β is

$$\beta = \frac{R_1}{R_1 + R_2} = \frac{1\ k\Omega}{1\ k\Omega + 99\ k\Omega} = 0.01$$

Since

$$\begin{aligned} v_f &= \beta v_{out} = (0.01)(234\ \mu V_{p\text{-}p} \angle -180^\circ) \\ &= 2.34\ \mu V_{p\text{-}p} \angle -180^\circ \end{aligned}$$

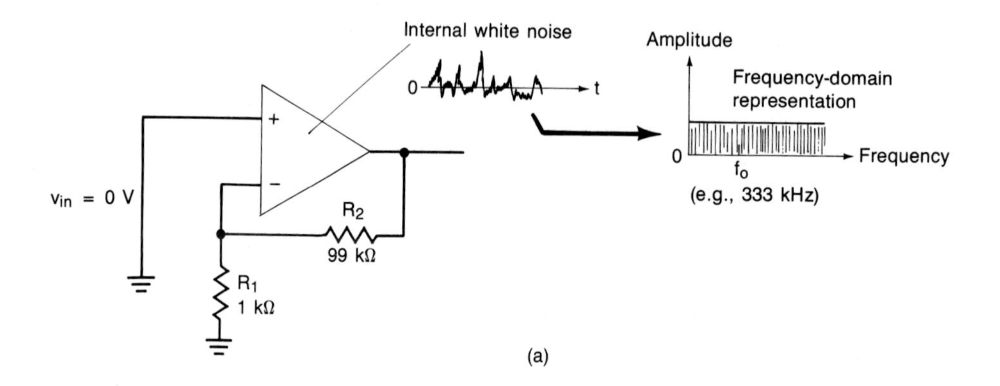

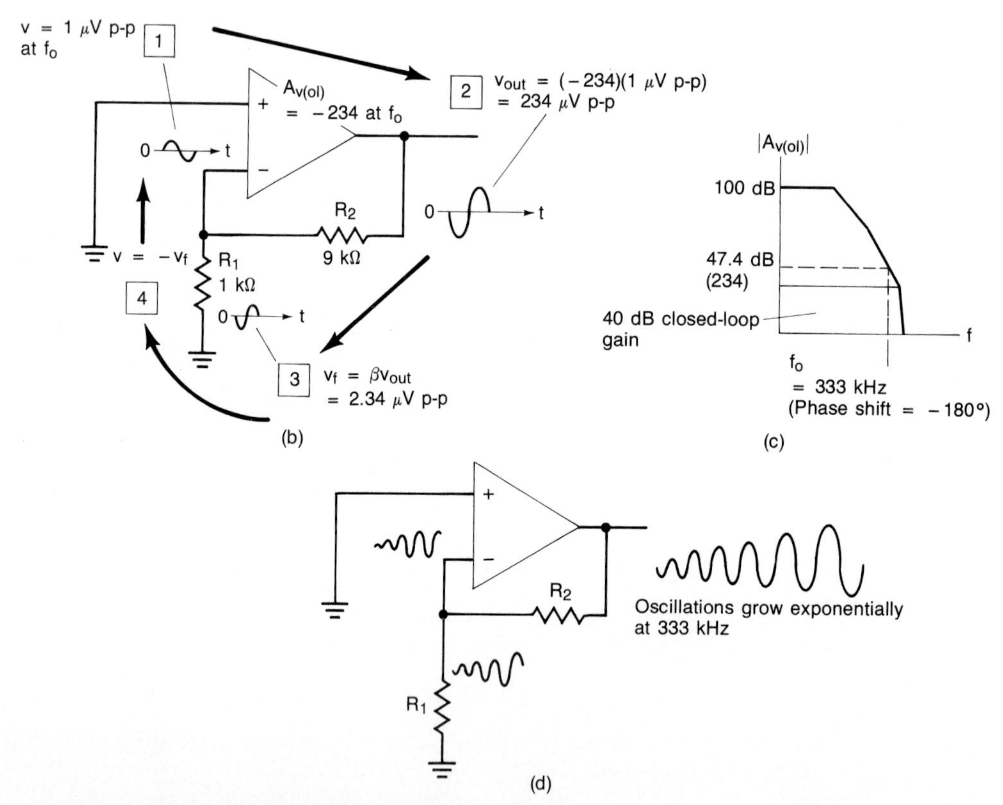

FIGURE 10-21 **Oscillation buildup: (a) white noise contains 333 kHz; (b) oscillation process; (c) Bode plot; (d) oscillations grow exponentially.**

and from Fig. 10-21(b) we see that

$$v = -v_f$$

Therefore, the voltage appearing across the input will *increase* to 2.34 μV peak to peak. This net input voltage will produce an output of

$$v_{out} = A_{v(ol)}[f]v = (234 \angle -180^\circ)(2.34\ \mu V_{p\text{-}p})$$
$$= 548\ \mu V_{p\text{-}p} \angle -180^\circ$$

The increase in v_{out} will increase v_f.

$$v_f = \beta v_{out} = (0.01)(548\ \mu V_{p\text{-}p} \angle -180°)$$
$$= 5.48\ \mu V_{p\text{-}p} \angle -180°$$

and the new input voltage will be 5.48 μV peak to peak. The process continues and v_{out}, v_f, and v will grow exponentially. This has been depicted in Fig. 10-21(d).

The key behind this instability is that the phase shift must be 180° *and* the loop gain must be greater than unity. The phase shift requirement should be obvious. The argument for the loop gain requirement has been developed below.

In the previous example we saw that in each case the feedback voltage v_f was greater than the amplifier's input voltage v. Hence

$$v_f > v$$
$$\beta v_{out} > v$$
$$\beta A_{v(ol)}[f] v > v$$

Dividing both sides by v yields

$$\beta A_{v(ol)}[f] > 1$$

To summarize, for oscillations to build up we must satisfy two conditions:

1. $\phi = \pm 180°$
2. $|\beta A_{v(ol)}[f]| > 1$

In real amplifiers, the output signal builds up until the amplifier approaches or enters saturation. When an amplifier nears saturation its voltage gain will begin to decrease. Consequently, when an amplifier oscillates, its peak-to-peak output voltage is still generally limited by its power supply rails.

To sustain *oscillations in an amplifier, it is only necessary that the magnitude of the loop gain be equal to unity. Therefore, we may slightly modify the two conditions cited above.*

1. $\phi = \pm 180°$
2. $|\beta A_{v(ol)}[f]| = 1$ (0 dB)

These two conditions are called the *Barkhausen criteria for sustained oscillations*.

To determine if a given closed-loop amplifier system will oscillate, we must look at the Bode plot of its loop gain. It is very important to note that the Barkhausen criteria are based on *loop gain*, *not* just on the *open-loop gain*. In the example above, the feedback elements were resistive. Therefore, β was a real number with a phase shift of 0°. However, in general, the feedback elements may be reactive. Hence

$$|\beta[f]A_{v(ol)}[f]| \angle \phi = |\beta[f]| \angle \theta_1\ |A_{v(ol)}[f]| \angle \theta_2$$
$$= |\beta[f]|\,|A_{v(ol)}[f]| \angle \theta_1 + \theta_2$$

From the above we can see that we must look at the loop gain, *not* just the open-loop gain. The magnitude of the loop gain is equal to the product of the feedback factor and the open-loop gain at each particular frequency, and the total phase angle is given by the sum of their respective phase angles.

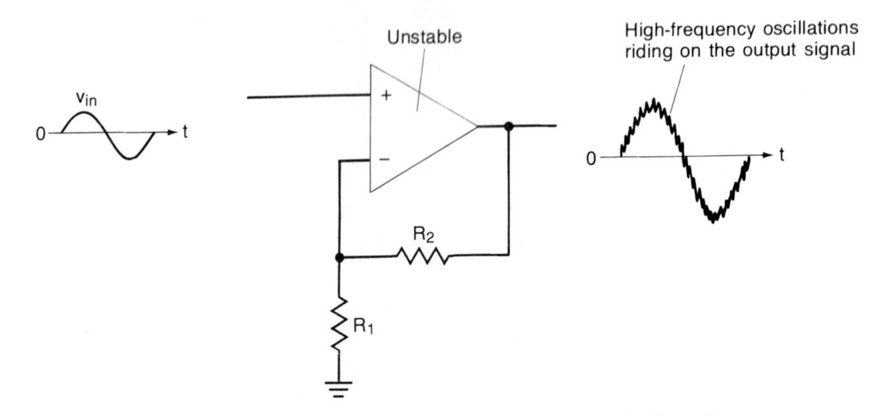

FIGURE 10-22 **Unstable amplifier with an input signal.**

In Fig. 10-21 no input signal source was shown. This was done to emphasize that an unstable amplifier is capable of generating its own output signal. If an input signal is applied, the output will be a combination of the amplified input signal and the generated signal. An example has been illustrated in Fig. 10-22. If the oscillation frequency is sufficiently high, an output waveform may appear to be "fuzzy" when displayed on an oscilloscope.

The oscillations are undesirable. They sap energy from our power supply, produce additional (unnecessary) heat dissipation, and may be coupled into other parts of the electronic system.

10-12 Stability Criteria

If the feedback network is resistive, the feedback factor does not change with frequency and has no phase angle associated with it. Consequently, the loop gain will have the same corner frequencies and phase angles as those demonstrated by the open-loop gain. Therefore, the shape of the loop-gain response will be the same as that of the open-loop gain (refer to Fig. 10-16).

To determine whether or not a given amplifier will oscillate, we may use either of two approaches. We can find the magnitude of the loop gain at which the phase angle is 180°, or we find the value of the phase angle at which the loop gain is unity (0 dB). We favor the latter approach.

Figure 10-23 represents a loop-gain response with three corner frequencies. They occur at 1, 10, and 100 kHz. The maximum phase shift with three corner frequencies is −270°. In Fig. 10-23(a) we see that when the phase angle reaches −180°, the loop gain is a negative dB value (−21.8 dB). This means that the loop gain is less than unity. This indicates that the amplifier will be stable. The magnitude of the difference between the gain at −180° and 0 dB indicates how stable the circuit is. This quantity is called the *gain margin*. The gain margin in this case is 21.8 dB.

In Fig. 10-23(b) we see the alternative approach. The loop gain crosses the horizontal axis (0 dB) at the *crossing frequency* f_c. The 0-dB crossing frequency f_c has a corresponding *crossing angle* θ_c of −125°. Since the crossing angle is *less* than

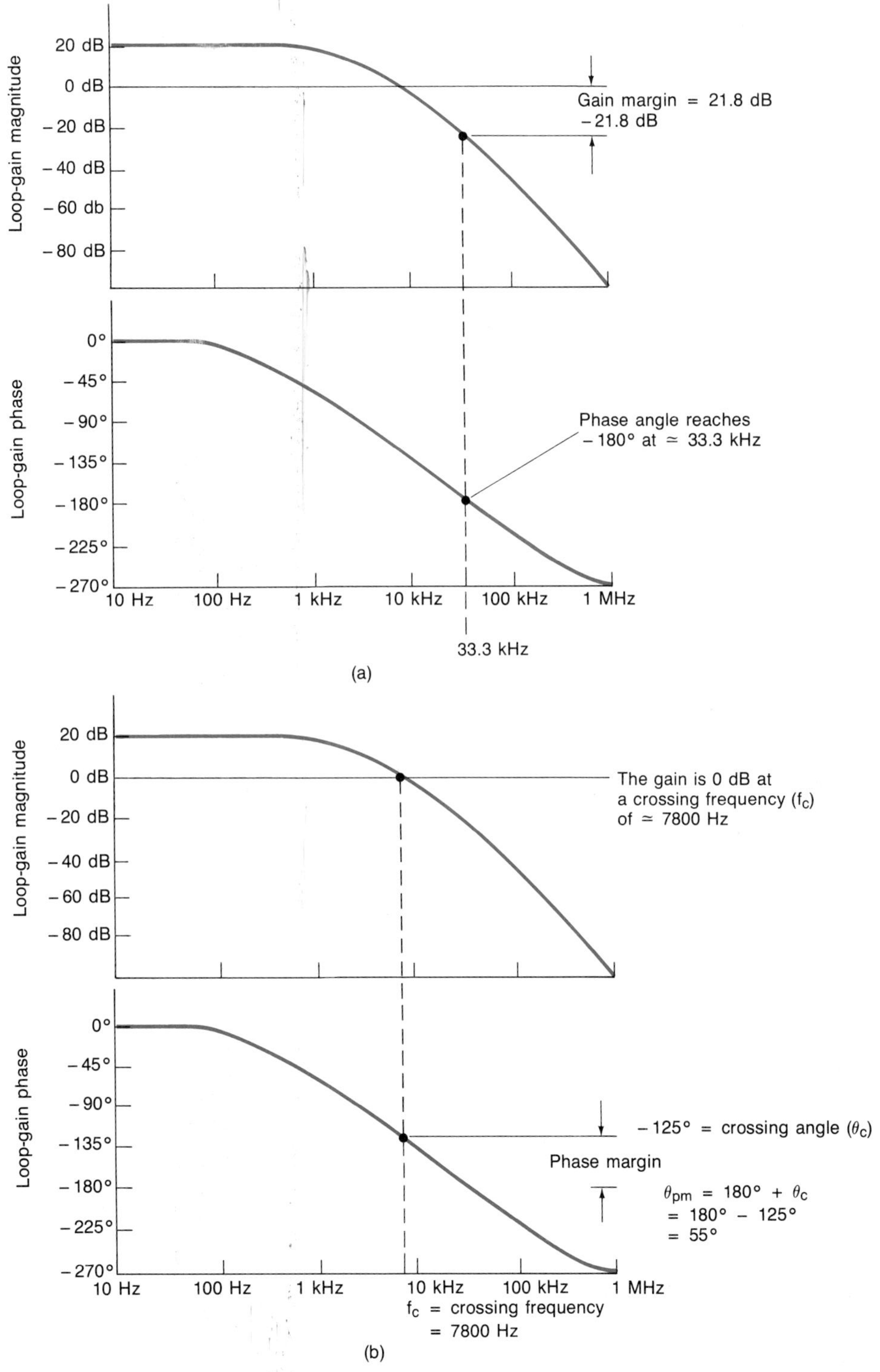

FIGURE 10-23 Stability criteria: (a) finding the gain margin; (b) finding the phase margin.

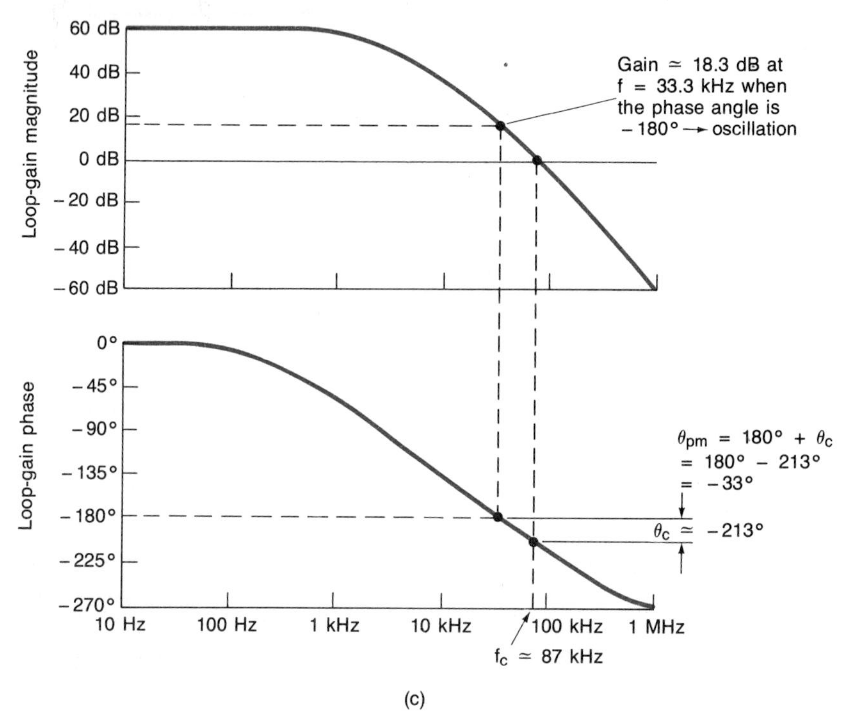

(c)

FIGURE 10-23 *(continued)* (c) a negative phase margin indicates that oscillations will occur.

the required $-180°$, the circuit is stable and will not oscillate. As a measure of the relative stability of a given amplifier circuit, we define the term *phase margin* θ_{pm}.

$$\theta_{pm} = 180° + \theta_c \tag{10-33}$$

where θ_{pm} is the phase margin and θ_c is the crossing angle.

In Fig. 10-23(b) the phase margin θ_{pm} is

$$\theta_{pm} = 180° + \theta_c = 180° - 125° = 55°$$

The circuit will be stable since the phase angle is not $-180°$, and its corresponding phase margin is a *positive value. A positive phase margin always indicates that a circuit or system will be stable.*

The closer the phase margin gets to 0°, the closer the circuit comes to oscillating. For example, if the crossing angle θ_c is 180°, the phase margin is

$$\theta_{pm} = 180° + \theta_c = 180° - 180° = 0°$$

and the amplifier will be unstable.

In Fig. 10-23(c) we see a loop-gain response with the same three corner frequencies. The low-frequency loop gain has been increased to 60 dB. In this case θ_c is $-213°$, and the phase margin is

$$\theta_{pm} = 180° + \theta_c = 180° - 213° = -33°$$

A negative *phase margin indicates an unstable situation*. Therefore, the amplifier will oscillate. If we inspect Fig. 10-23(c), we can see that the decibel value of the loop gain is positive (greater than unity) when the phase shift is $-180°$.

EXAMPLE 10-11

Given the op amp amplifier in Fig. 10-24(a) and its open-loop response in Fig. 10-24(b), find the closed-loop gain, sketch the loop gain response, find θ_c, and determine if the amplifier is stable.

SOLUTION The ideal closed-loop gain is

$$A_{v(\mathrm{cl})} = 1 + \frac{R_2}{R_1} = 1 + \frac{1\ \mathrm{M}\Omega}{100\ \Omega} = 10{,}001$$

and in decibels,

$$\begin{aligned} |A_{v(\mathrm{cl})}|_{\mathrm{dB}} &= 20 \log A_{v(\mathrm{cl})} \\ &= 20 \log 10{,}001 \\ &= 80\ \mathrm{dB} \end{aligned}$$

FIGURE 10-24 Circuit and responses for Example 10-11: (a) circuit; (b) op amp's open-loop gain; (c) ideal closed-loop gain; (d) loop gain.

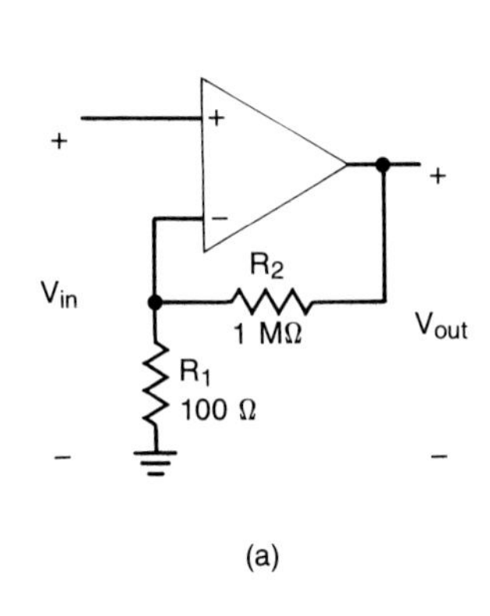

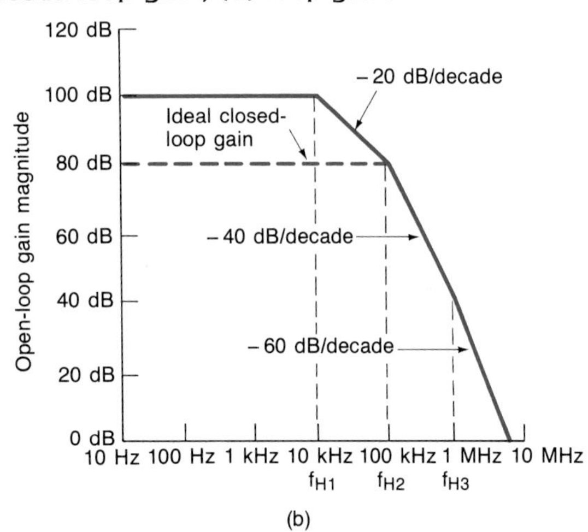

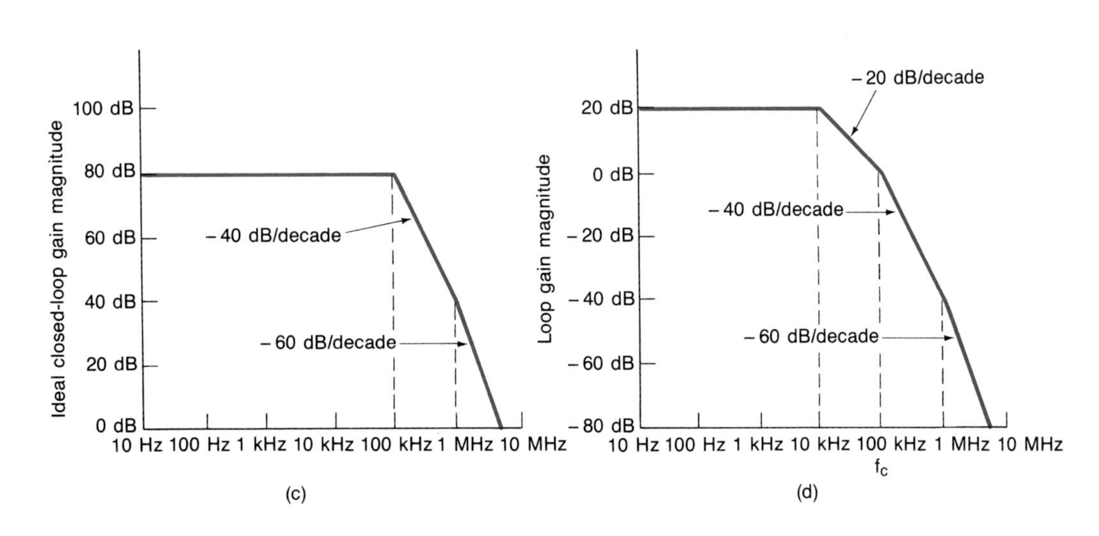

The *ideal* closed-loop gain response has been sketched in Fig. 10-24(c). For reference, it has also been superimposed as a dashed line on the amplifier's open-loop response in Fig. 10-24(b). The loop-gain plot has been sketched in Fig. 10-24(d). No phase response has been given. However, in Fig. 10-24(d) we see that f_c is 100 kHz, and that the corner frequencies occur at 10 kHz, 100 kHz, and 1 MHz. We may find the crossing angle θ_c by applying Eq. 10-23, and evaluating it at f_c.

$$\begin{aligned}\theta_c &= -\tan^{-1}\frac{f_c}{f_{H1}} - \tan^{-1}\frac{f_c}{f_{H2}} - \tan^{-1}\frac{f_c}{f_{H3}} \\ &= -\tan^{-1}\frac{100\text{ kHz}}{10\text{ kHz}} - \tan^{-1}\frac{100\text{ kHz}}{100\text{ kHz}} - \tan^{-1}\frac{100\text{ kHz}}{1\text{ MHz}} \\ &= -84.3^\circ - 45^\circ - 5.7^\circ = -135^\circ\end{aligned}$$

The phase margin may be found from Eq. 10-24.

$$\theta_{pm} = 180^\circ + \theta_c = 180^\circ - 135^\circ = 45^\circ$$

The positive phase margin indicates that the amplifier will not oscillate. ■

EXAMPLE 10-12

The same op amp is to be used in the amplifier circuit given in Fig. 10-25(a). The op amp's open-loop response has been repeated in Fig. 10-25(b). Repeat the analysis requested in Example 10-11.

SOLUTION The ideal closed-loop gain is

$$A_{v(cl)} = 1 + \frac{R_2}{R_1} = 1 + \frac{100\text{ k}\Omega}{100\ \Omega} = 1001$$

and in decibels,

$$\begin{aligned}|A_{v(cl)}|_{dB} &= 20\log A_{v(cl)} \\ &= 20\log 1001 \\ &= 60\text{ dB}\end{aligned}$$

The ideal closed-loop response has been illustrated in Fig. 10-25(c) and shown as a dashed line in Fig. 10-25(b). The loop-gain response has been illustrated in Fig. 10-25(d). The 0-dB crossing frequency f_c is approximately 300 kHz. The corresponding crossing angle θ_c is

$$\begin{aligned}\theta_c &= -\tan^{-1}\frac{f_c}{f_{H1}} - \tan^{-1}\frac{f_c}{f_{H2}} - \tan^{-1}\frac{f_c}{f_{H3}} \\ &= -\tan^{-1}\frac{300\text{ kHz}}{10\text{ kHz}} - \tan^{-1}\frac{300\text{ kHz}}{100\text{ kHz}} - \tan^{-1}\frac{300\text{ kHz}}{1\text{ MHz}} \\ &= -88.1^\circ - 71.6^\circ - 16.7^\circ = -176.4^\circ\end{aligned}$$

and the phase margin is

$$\theta_{pm} = 180^\circ + \theta_c = 180^\circ - 176.4^\circ = 3.6^\circ$$

The phase angle is not quite -180°, and the phase margin is not quite 0°. However, with the advent of stray capacitances and component tolerances, chances are that the circuit will oscillate. Consequently, this circuit would be deemed *marginally stable*. ■

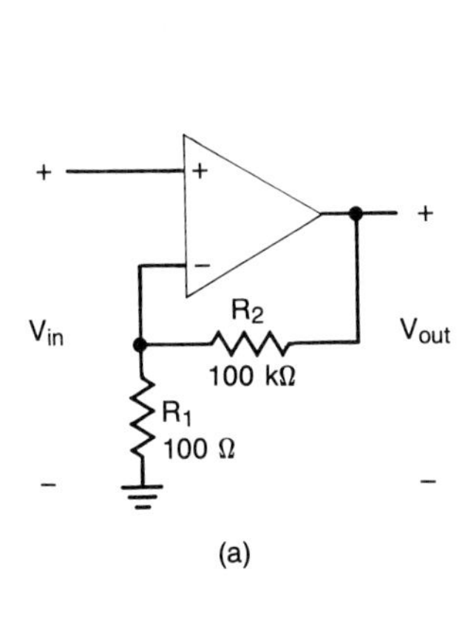

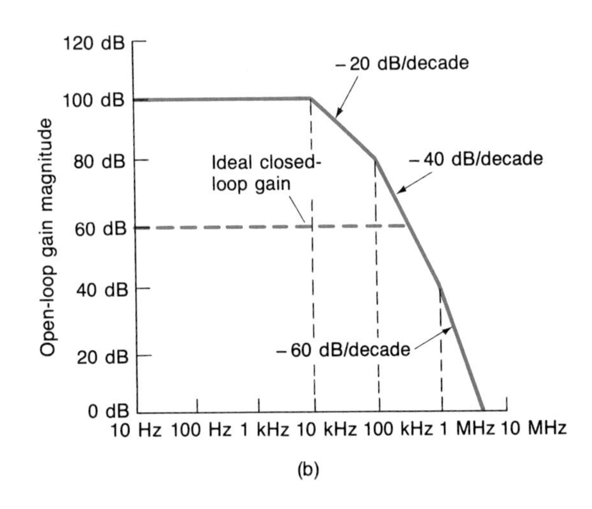

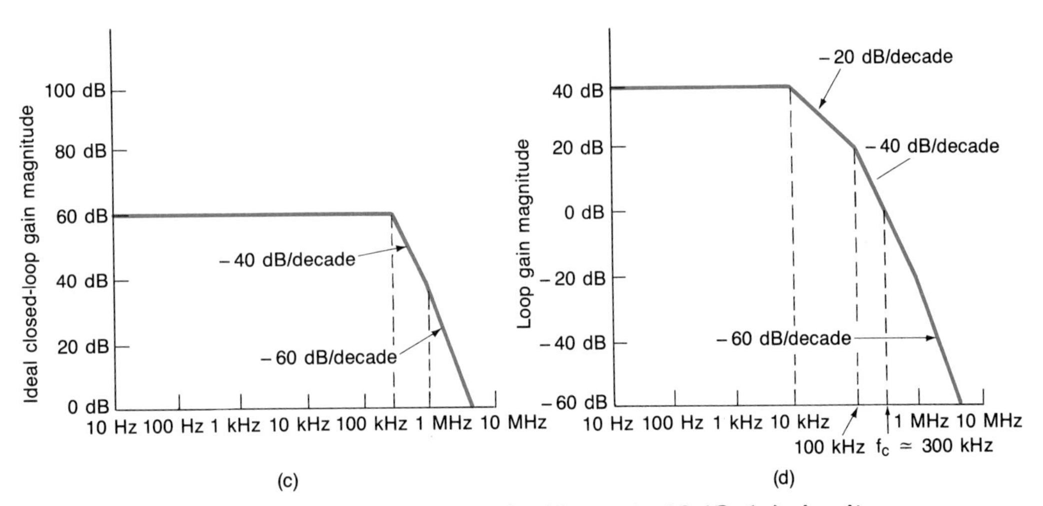

FIGURE 10-25 **Circuit and responses for Example 10-12: (a) circuit; (b) op amp's open-loop gain; (c) ideal closed-loop gain; (d) loop gain.**

EXAMPLE 10-13

Repeat the analysis if the same op amp is used in the amplifier circuit given in Fig. 10-26(a).

SOLUTION Following the same procedure, we have

$$A_{v(\text{cl})} = 1 + \frac{R_2}{R_1} = 1 + \frac{9\ \text{k}\Omega}{1\ \text{k}\Omega} = 10$$

$$|A_{v(\text{cl})}|_{\text{dB}} = 20 \log A_{v(\text{cl})}$$
$$= 20 \log 10 = 20\ \text{dB}$$

The open-loop and ideal closed-loop gains have been indicated in Fig. 10-26(b) and (c). The loop gain is given in Fig. 10-26(d). The crossing frequency is 2.08 MHz.

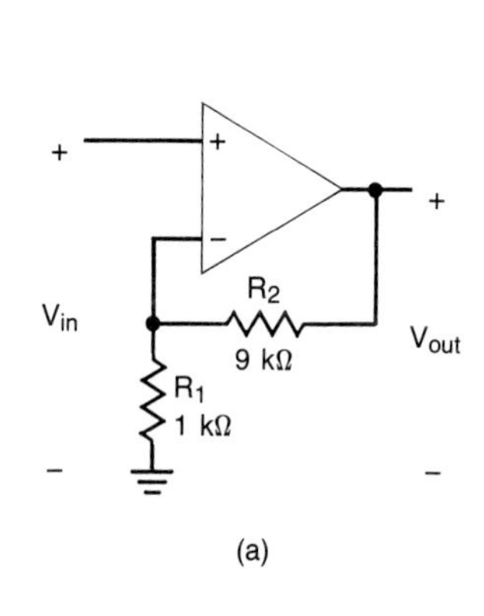

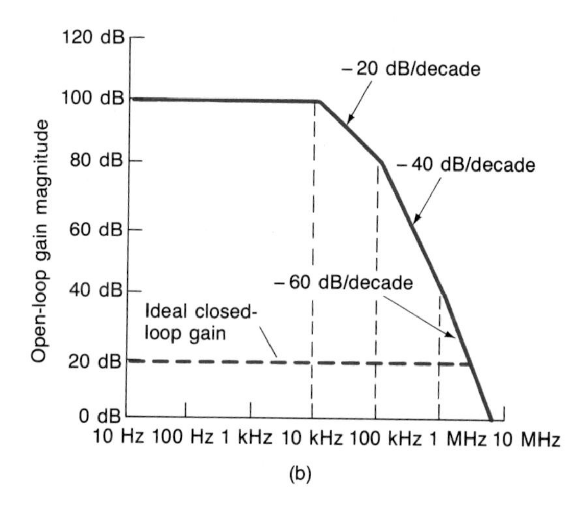

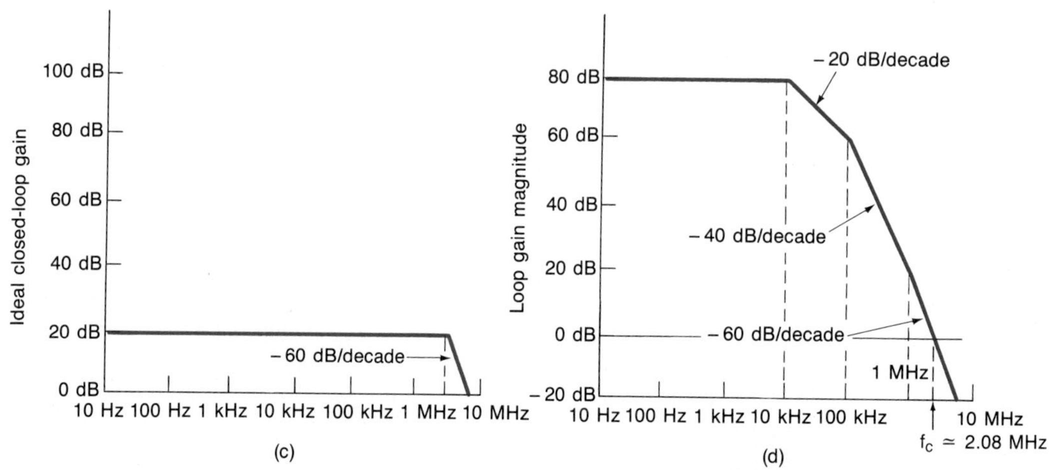

FIGURE 10-26 Circuit and responses for Example 10-13: (a) circuit; (b) op amp's open-loop gain; (c) ideal closed-loop gain; (d) loop gain.

$$
\begin{aligned}
\theta_c &= -\tan^{-1}\frac{f_c}{f_{H1}} - \tan^{-1}\frac{f_c}{f_{H2}} - \tan^{-1}\frac{f_c}{f_{H3}} \\
&= -\tan^{-1}\frac{2.08\ \text{MHz}}{10\ \text{kHz}} - \tan^{-1}\frac{2.08\ \text{MHz}}{100\ \text{kHz}} - \tan^{-1}\frac{2.08\ \text{MHz}}{1\ \text{MHz}} \\
&= -89.7^\circ - 87.2^\circ - 64.3^\circ = -241.2^\circ
\end{aligned}
$$

and

$$\theta_{pm} = 180^\circ + \theta_c = 180^\circ - 241.2^\circ = -61.2^\circ$$

The negative phase margin indicates that the circuit will oscillate. ■

We should point out that Examples 10-11, 10-12, and 10-13 are approximations since we have not taken the time to draw the exact response curves. This will yield some error in the results. However, let us reflect on these three examples. The same

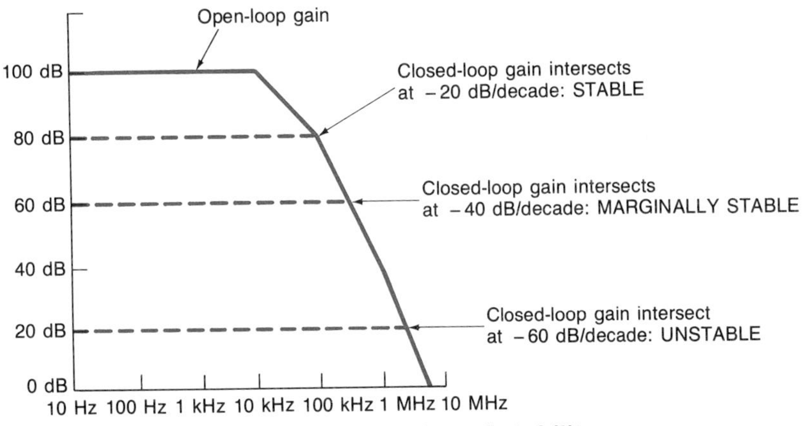

FIGURE 10-27 Quick determination of stability.

op amp has been used in three different amplifier circuits. In the first case the circuit was stable. In the second case the circuit was marginally stable. The third amplifier circuit was unstable. By referring to Figs. 10-24(d), 10-25(d), and 10-26(d), we can make some extremely useful observations. A quick determination of an amplifier's stability can be made by looking at the slope of the loop-gain curve as it passes through 0 dB. Specifically, we can see that:

1. −20 dB/decade crossing: absolutely stable
2. −40 dB/decade crossing: marginally stable
3. −60 dB/decade: unstable

If we understand the relationships between the open-loop response, the ideal closed-loop response, and loop gain, we can further ease our efforts. The loop-gain crossing frequency f_c is equal to the ideal closed-loop corner frequency $f_{H(\text{cl})}$. Consequently, *the slope of the open-loop response curve at the intersection between the closed-loop gain and the open-loop response gives us the slope of loop gain at its 0-dB crossing*. This has been illustrated in Fig. 10-27. It is important to note that this quick inspection applies only when purely resistive feedback elements are used.

Now we can truly begin to appreciate the primary pitfall associated with the use of negative feedback. Negative feedback results in a very predictable gain, improved input and output resistances, a wider bandwidth, and reduced nonlinear distortion. In fact, the more negative the feedback (the greater the loop gain), the greater the improvements. However, as Fig. 10-27 clearly indicates, the more negative feedback we use, the greater the possibility of oscillation.

The intersection between the ideal closed-loop gain and the open-loop response can indeed be used to determine whether or not a closed-loop system will oscillate. However, the actual closed-loop frequency response is not easy to predict when the open-loop response has two or more corner frequencies. Quite often a "peaking" effect will be observed (see Fig. 10-28). The closed-loop response can be predicted by using a graphical technique based on a *Nichol's chart*. This will be left for more advanced work.

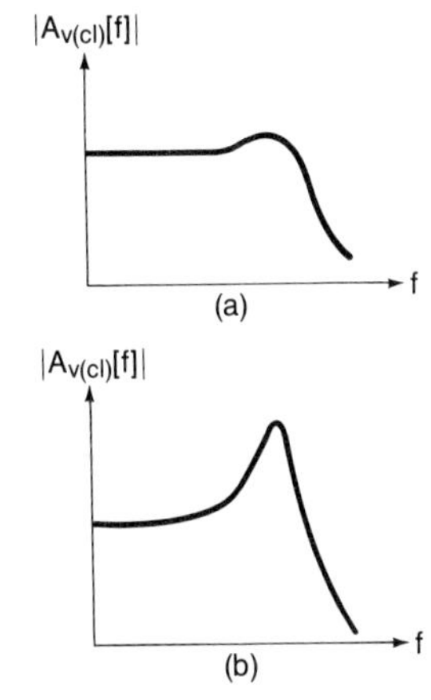

FIGURE 10-28 Typical closed-loop responses for amplifiers with two or more open-loop corner frequencies: (a) slight peaking when the loop gain θ_c is $-90°$; (b) significant peaking when the loop gain θ_c is $-170°$; (c) gain goes to infinity (producing oscillations) if $\theta_c \leq -180°$.

In "real" amplifier systems the frequency at which the loop-gain phase shift reaches $-180°$ will *not* be the actual frequency of oscillation. The actual oscillation frequency will be (unpredictably) lower. This is true because the active devices will enter saturation. This causes the amplifier to "slow down" or become "sluggish."

10-13 Frequency Compensation

To enjoy the benefits of negative feedback, we must understand how to "tame" an amplifier or amplifier system to prohibit uncontrolled oscillations. This may be accomplished by employing *frequency compensation.*

Frequency-compensation circuits are designed to reshape the magnitude and phase plots of the loop gain such that the magnitude is less than unity when the phase angle reaches 180°. There are three general approaches:

1. *Lag compensation* (also called "dominant-pole compensation")
2. *Lag-lead compensation* (also called "pole-zero compensation")
3. *Lead compensation*

We shall focus our immediate attention on the first two approaches.

10-14 Lag (Dominant-Pole) Compensation

In classical network theory, a $j\omega$ term in the denominator of a gain equation is called a *pole*, and a $j\omega$ term in the numerator of a gain equation is called a *zero*. The characteristics of poles and zeros will be illustrated as our work progresses. Their complete mathematical ramifications will be left for more advanced work.

Lag or dominant-pole compensation has been illustrated in Fig. 10-29. This is the technique used by op amps that have been frequency compensated for unity gain (e.g., the 741 op amp). The general approach is to use a capacitance (e.g., C_c) to produce a corner frequency f_H which is much lower than the open-loop corner frequencies. Consider Fig. 10-29(a).

The open-loop corner frequencies (without C_c) are 1, 3, and 10 MHz. If a dominant corner (pole) frequency is designed into the open-loop response (e.g., at 10 Hz for the 741 op amp by using a C_c of 30 pF), we obtain the open-loop response shown in Fig. 10-29(b). For any closed-loop response down to unity (0 dB), the intersection between the closed-loop gain and the open-loop response has a slope of -20 dB/decade. Therefore, the amplifier will be stable for closed-loop gains as low as unity [see Fig. 10-29(c)].

With a little reflection we can see that the basic approach is to insert a pole such that f_{H1} occurs at an open-loop gain of 0 dB. Another way of viewing the situation is to state that f_T (the unity-gain frequency) is tightly controlled.

10-15 Lag-Lead (Pole-Zero) Compensation

The second approach (lag–lead or pole–zero compensation) may be accomplished by employing the network depicted in Fig. 10-30(a). To understand this circuit, let us first make some intuitive observations. At low frequencies, the capacitive reactance will be very large and the voltage gain will be approximately unity [see Fig. 10-30(b)]. As the frequency is increased, the capacitive reactance will decrease. If X_c is much larger than R_3, we have the approximate situation shown in Fig. 10-30(c). In essence, we have a low-pass filter. Therefore, the voltage gain will roll off at a rate of -20 dB/decade. When the frequency is increased further, the capacitive reactance becomes negligibly small, and we have the situation shown in Fig. 10-30(d). The phase shift will become 0°, and the maximum attenuation (minimum voltage gain) is given by the voltage division between R and R_3.

With this background in mind, let us proceed with a more rigorous analysis. By applying voltage division to the lag–lead circuit [Fig. 10-30(a)] we may arrive at its voltage gain as a function of frequency.

$$A_v[f] = \frac{V_{\text{out}}}{V_{\text{in}}} = \frac{R_3 + 1/j\omega C}{R + R_3 + 1/j\omega C} = \frac{(1 + j\omega R_3 C)/j\omega C}{[1 + j\omega(R + R_3)C]/j\omega C}$$

$$= \frac{1 + j\omega R_3 C}{1 + j\omega(R + R_3)C} \qquad (10\text{-}34)$$

FIGURE 10-29 Op amp that has been frequency-compensated for unity gain: (a) op amp equivalents.

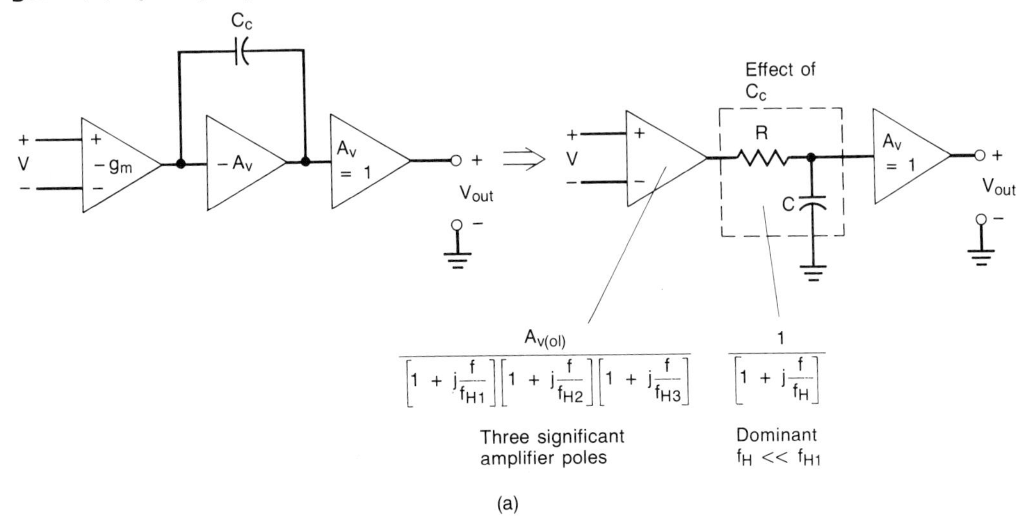

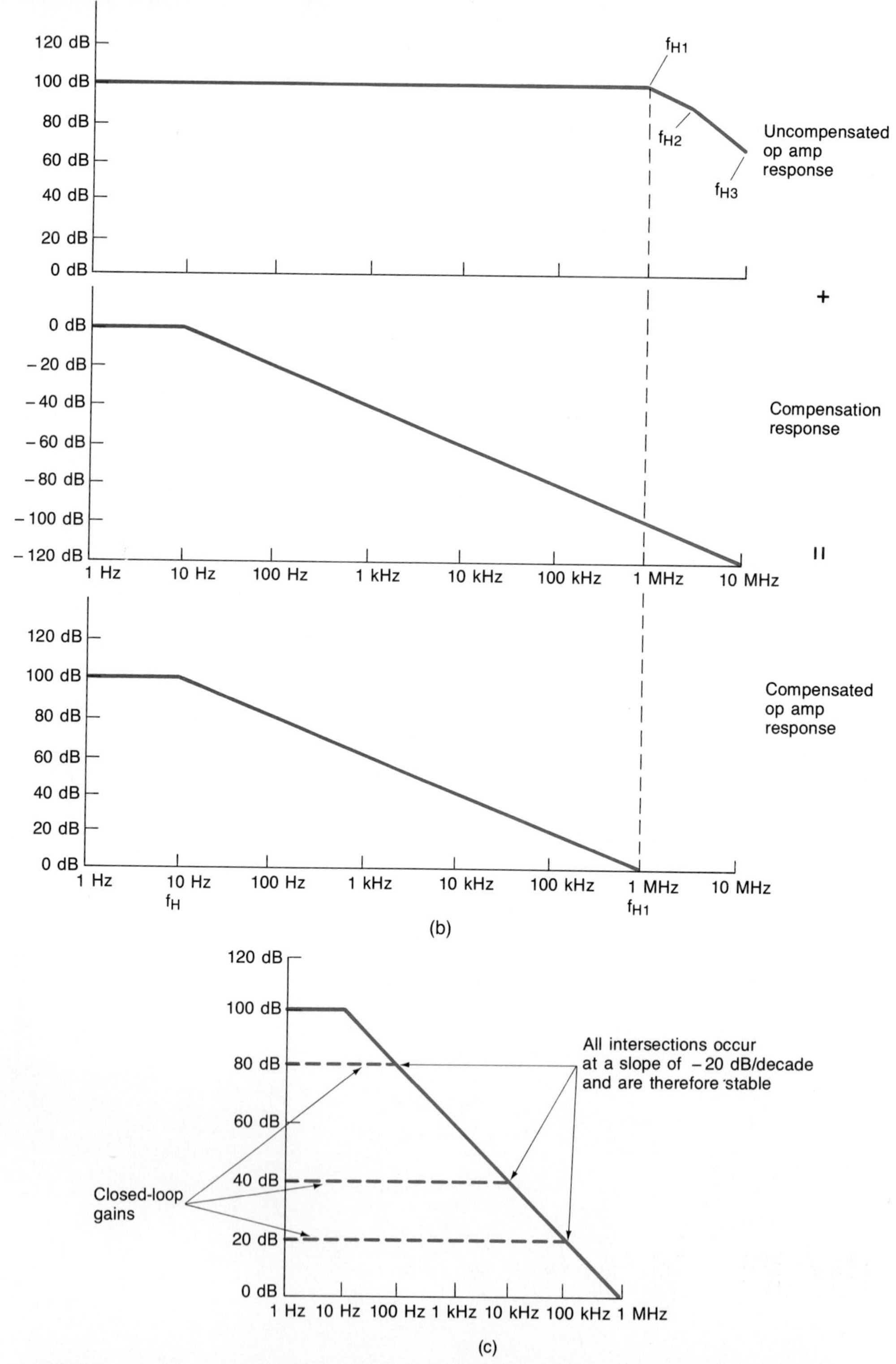

FIGURE 10-29 (*continued*) (b) the (lag or dominant-pole) compensation network provides an op amp open-loop response that rolls off at −20 dB/decade down to 0 dB; (c) closed-loop gains as low as 0 dB will result in stable operation.

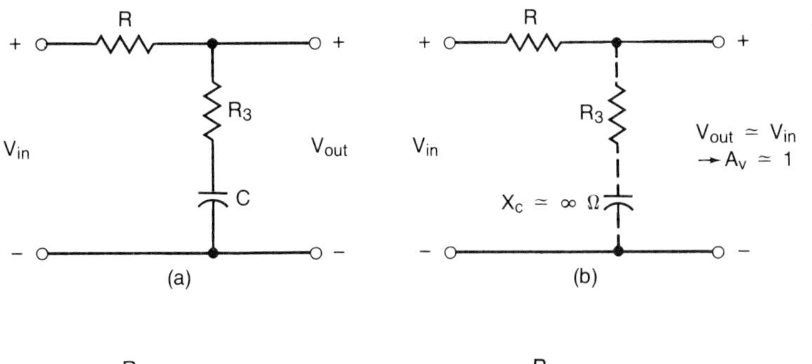

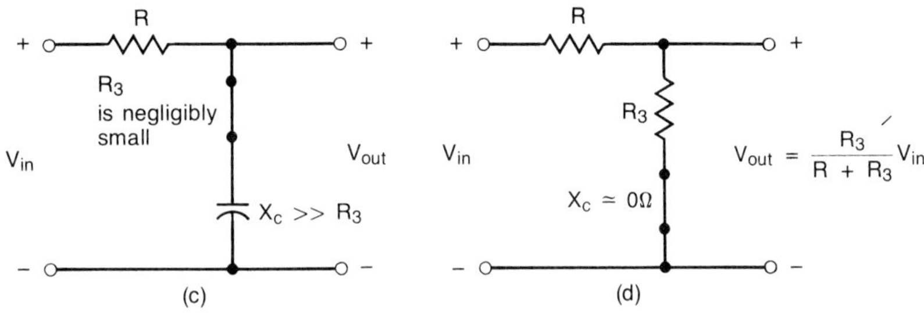

FIGURE 10-30 Lag-lead network: (a) circuit; (b) low-frequency equivalent; (c) at slightly higher frequencies the circuit behaves as a low-pass filter; (d) at high frequencies the circuit behaves as a resistive attenuator.

To further clarify the terms ''pole'' and ''zero,'' they have been emphasized in our result.

a ''zero'' term

$$A_v[f] = \frac{1 + j\omega R_3 C}{1 + j\omega(R + R_3)C}$$

a ''pole'' term

To be consistent with the terminology found in classical network analysis, we shall define ω_z and ω_p as the zero and pole frequencies, respectively. Hence

$$\omega_z = \frac{1}{R_3C} \rightarrow R_3C = \frac{1}{\omega_z}$$

and

$$\omega_p = \frac{1}{(R + R_3)C} \rightarrow (R + R_3)C = \frac{1}{\omega_p}$$

Substituting these definitions into Eq. 10-34 results in Eq. 10-35.

$$A_v[f] = \frac{V_{\text{out}}}{V_{\text{in}}} = \frac{1 + j(\omega/\omega_z)}{1 + j(\omega/\omega_p)}$$

$$\boxed{A_v[f] = \frac{V_{\text{out}}}{V_{\text{in}}} = \frac{1 + jf/f_z}{1 + jf/f_p}} \tag{10-35}$$

where

$$f_z = \frac{1}{2\pi R_3 C} \quad \text{and} \quad f_p = \frac{1}{2\pi(R + R_3)C}$$

The magnitude in decibels and the phase angle in degrees may be found in the usual fashion.

$$A_v[f] = \frac{\sqrt{1 + (f/f_z)^2}}{\sqrt{1 + (f/f_p)^2}} \angle \tan^{-1}(f/f_z) - \tan^{-1}(f/f_p)$$

Hence

$$|A_v[f]|_{\text{dB}} = 10 \log|1 + (f/f_z)^2| - 10 \log|1 + (f/f_p)^2| \tag{10-36}$$

$$\phi = \tan^{-1}\frac{f}{f_z} - \tan^{-1}\frac{f}{f_p} \tag{10-37}$$

EXAMPLE 10-14

Given the network shown in Fig. 10-30(a), assume that R is 5.6 kΩ, R_3 is 1.2 kΩ, and C is 0.06 μF, and plot the magnitude and gain as a function of frequency.

SOLUTION First we find the two corner frequencies.

$$f_z = \frac{1}{2\pi R_3 C} = \frac{1}{2\pi(1.2\ \text{k}\Omega)(0.06\ \mu\text{F})} = 2.21\ \text{kHz}$$

$$f_p = \frac{1}{2\pi(R + R_3)C} = \frac{1}{2\pi(5.6\ \text{k}\Omega + 1.2\ \text{k}\Omega)(0.06\ \mu\text{F})} = 390\ \text{Hz}$$

and the gain equation in decibels is

$$|A_v[f]|_{\text{dB}} = 10 \log|1 + (f/f_z)^2| - 10 \log|1 + (f/f_p)^2|$$
$$= 10 \log\left|1 + \left(\frac{f}{2.21\ \text{kHz}}\right)^2\right| - 10 \log\left|1 + \left(\frac{f}{390\ \text{Hz}}\right)^2\right|$$

The corresponding phase angles in degrees are given by

$$\phi = \tan^{-1}\frac{f}{f_z} - \tan^{-1}\frac{f}{f_p} = \tan^{-1}\frac{f}{2.21\ \text{kHz}} - \tan^{-1}\frac{f}{390\ \text{Hz}}$$

Evaluating the gain and phase equations at various frequencies produces the plots shown in Fig. 10-31(a). ■

Observe that the Bode approximation of the amplitude response has been given in Fig. 10-31(a). The Bode approximation can be arrived at through graphical addition.

In Fig. 10-31(b) we see that a pole produces a response that rolls off at a rate of −20 dB/decade for frequencies beyond f_p. Figure 10-31(c) represents the Bode approximation of a zero. For frequencies beyond f_z, the slope of the response shows an increase at a rate of +20 dB/decade. As we can see in Fig. 10-31(d), the graphical addition of these two terms produces the Bode approximation of the complete amplitude response. Compare the Bode approximation and the actual response shown in Fig. 10-31(a). Now let us see exactly how the lag–lead network can be used to prevent oscillations.

In Fig. 10-32(a) we see the open-loop response of an amplifier that is to be used with a closed-loop gain of 100 (40 dB). Since the closed-loop gain intersects

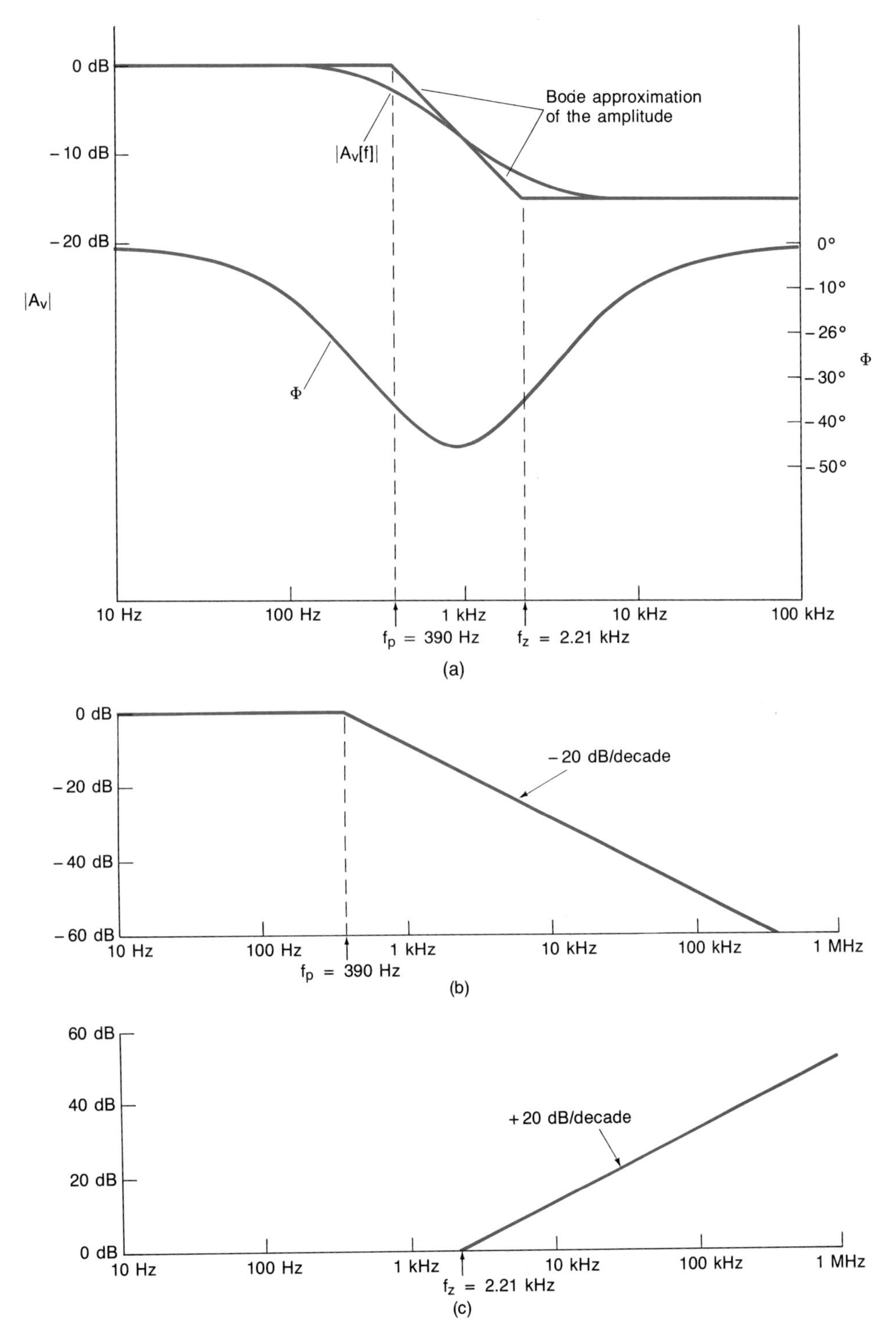

FIGURE 10-31 **Lag-lead response: (a) complete frequency response; (b) pole; (c) zero; (d) sum.**

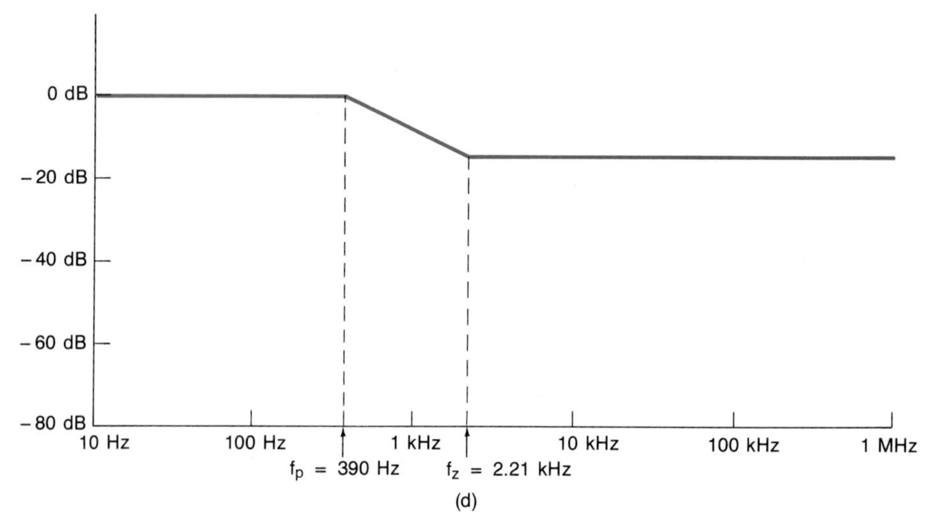

FIGURE 10-31 *(continued)*

the open-loop response at its −40 dB/decade slope (point A), we have a marginally stable amplifier.

EXAMPLE 10-15

Find the phase margin of the uncompensated amplifier given in Fig. 10-32(a).

SOLUTION The intersection between the open- and closed-loop responses at point A gives us the loop gain's 0-dB crossing frequency f_c. By inspection, we can see that it is approximately 90 kHz. We can also see that the open-loop corner frequencies occur at 2.2, 40, and 200 kHz. With these observations, we can now find the crossing angle θ_c.

$$\begin{aligned}\theta_c &= -\tan^{-1}\frac{f_c}{f_{H1}} - \tan^{-1}\frac{f_c}{f_{H2}} - \tan^{-1}\frac{f_c}{f_{H3}} \\ &= -\tan^{-1}\frac{90\text{ kHz}}{2.2\text{ kHz}} - \tan^{-1}\frac{90\text{kHz}}{40\text{ kHz}} - \tan^{-1}\frac{90\text{ kHz}}{200\text{ kHz}} \\ &= -88.6^\circ - 66.0^\circ - 24.2^\circ = -178.8^\circ\end{aligned}$$

The phase margin is

$$\theta_{\text{pm}} = 180^\circ + \theta_c = 180^\circ - 178.8^\circ = 1.2^\circ$$

The small positive phase margin indicates that the amplifier circuit is extremely close to oscillation. Frequency compensation is mandatory. ■

In Fig. 10-32(b) we see the addition of our lag–lead network to the amplifier circuit. Recalling that the overall voltage gain of a cascaded system is the product of the individual (loaded) voltage gains, we find that

$$A_{v(\text{ol})c}[f] = \frac{V_{\text{out}}}{V_{\text{in}}} = \frac{V_1}{V_{\text{in}}}\frac{V_{\text{out}}}{V_1} = A_{v(\text{ol})}[f]A_v[f] \qquad (10\text{-}38)$$

where $A_{v(ol)c}[f]$ is the *compensated* open-loop voltage gain as a function of frequency. Substituting in the expressions for our gains produces

$$A_{v(ol)c}[f] = \frac{A_{v(ol)}}{[1 + j(f/f_{H1})][1 + j(f/f_{H2})][1 + j(f/f_{H3})]} \frac{1 + j(f/f_z)}{1 + j(f/f_p)} \tag{10-39}$$

FIGURE 10-32 Lag-lead frequency compensation: (a) the closed-loop response is marginally stable ($\Theta_{pm} = 1.1°$); (b) addition of a lag-lead network to alter the open-loop response.

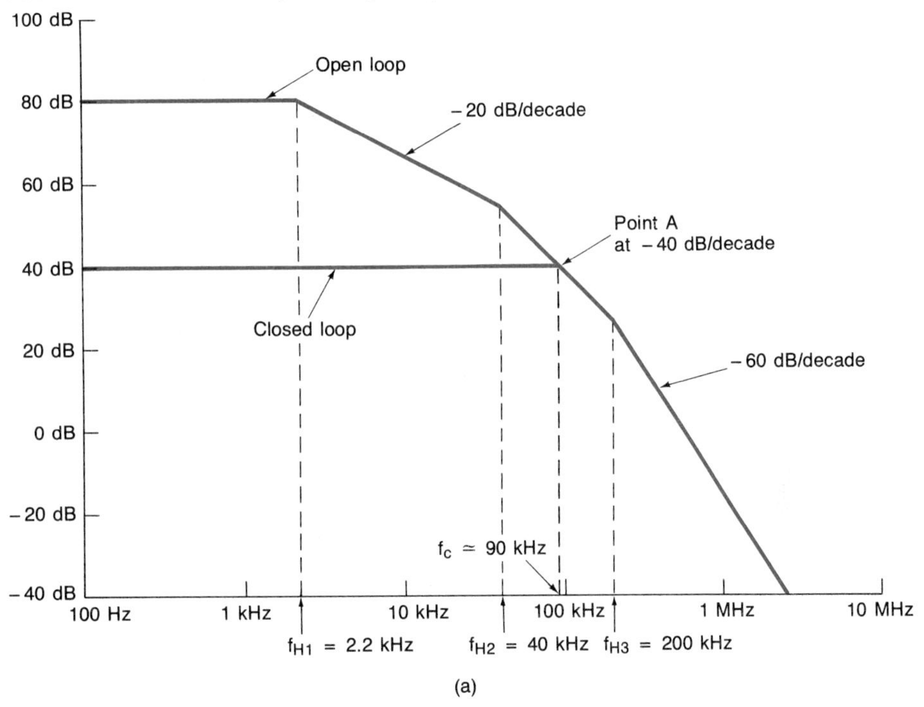

(a)

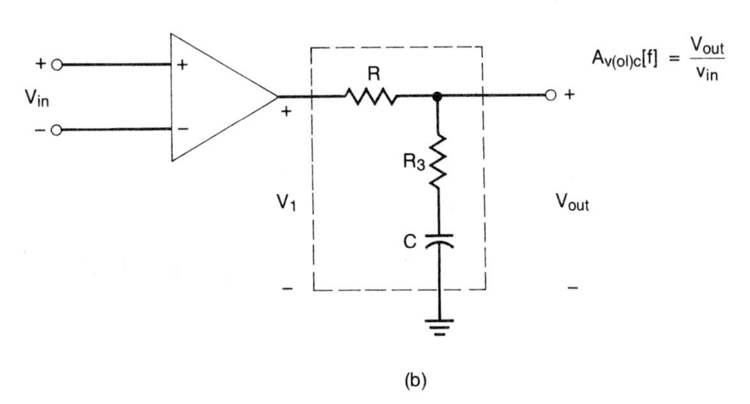

(b)

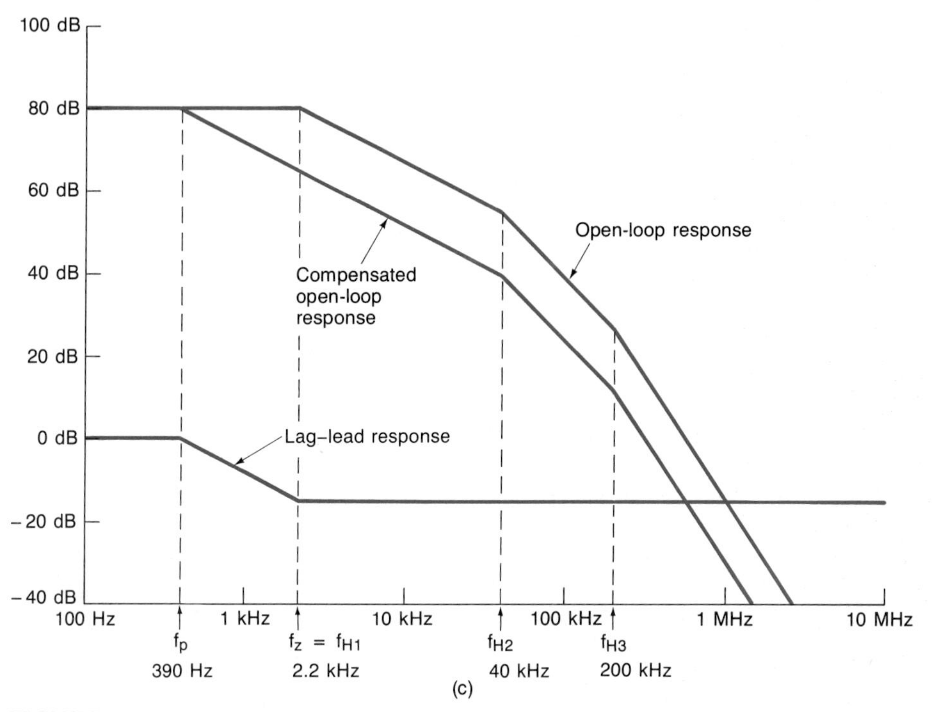

FIGURE 10-32 (*continued*) (c) the compensated open-loop response is the graphical sum of the lag-lead and open-loop responses; (d) compensated closed-loop gain.

The usual strategy is to set the zero of the lag-lead compensation network equal to the lowest corner (pole) frequency of the amplifier. Hence

$$f_z = f_{H1}$$

Therefore,

$$A_{v(\text{ol})c}[f] = \frac{A_{v(\text{ol})}}{[\cancel{1 + j(f/f_{H1})}][1 + j(f/f_{H2})][1 + j(f/f_{H3})]} \frac{\cancel{1 + j(f/f_z)}}{1 + j(f/f_p)}$$

$$= \frac{A_{v(\text{ol})}}{[1 + j(f/f_p)][1 + j(f/f_{H2})][1 + j(f/f_{H3})]} \tag{10-40}$$

Consider what has happened. We have used the compensation network to cancel f_{H1} and replace it with f_p. We have reshaped the open-loop response. This becomes even more apparent if we employ graphical addition [see Fig. 10-32(c)].

By controlling the high-frequency attenuation of the compensation network, the point of intersection between the closed-loop gain and the compensated open-loop response can be adjusted to occur at f_{H2}. The high-frequency attenuation of the compensation network is produced by voltage division between R and R_3. These resistors are selected to reduce the uncompensated open-loop gain at f_{H2} to the desired closed-loop gain. This is precisely the approach taken to arrive at the response indicated in Fig. 10-32(d). The crossing frequency of the loop gain has been set equal to f_{H2}. For comparison purposes, the uncompensated response has also been indicated in Fig. 10-32(d).

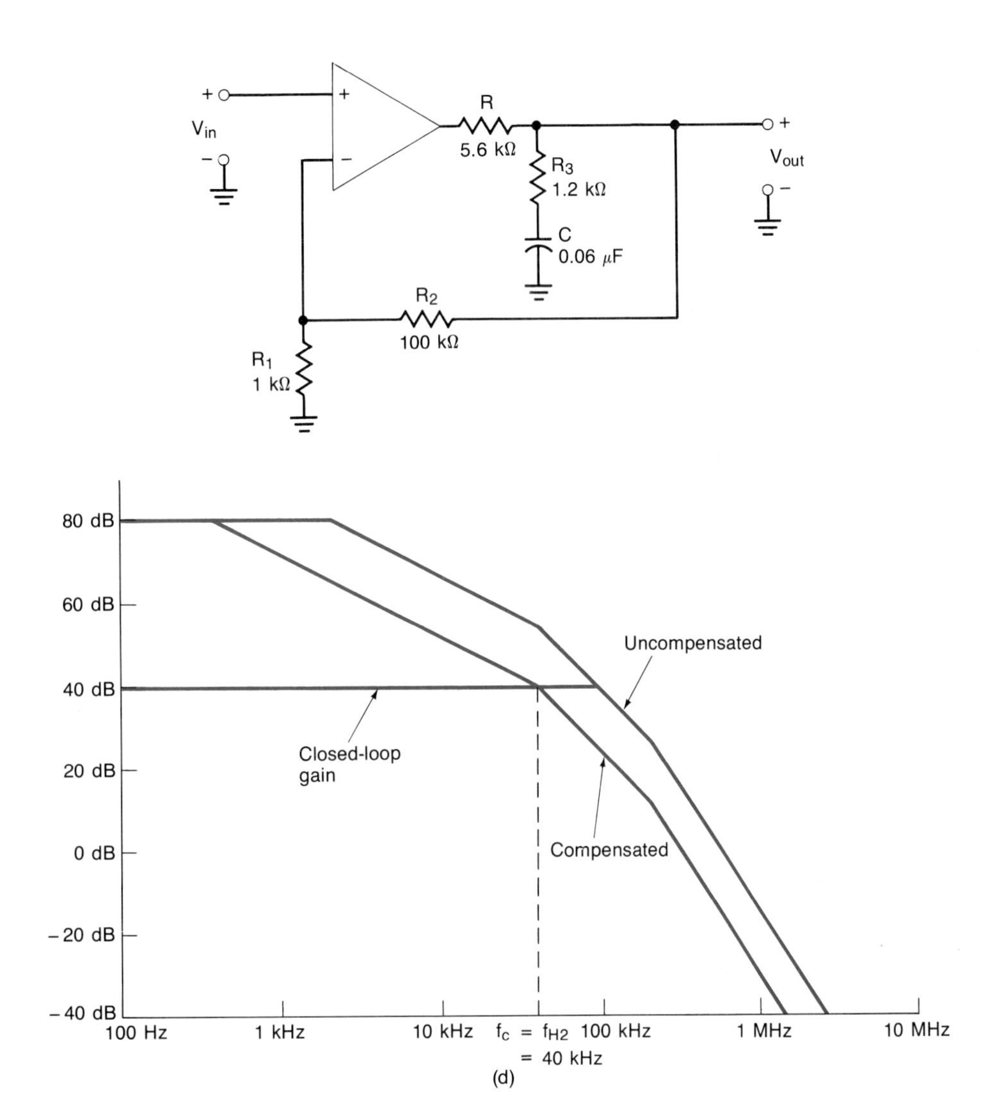

FIGURE 10-32 *(continued)*

EXAMPLE 10-16

Find the phase margin of the frequency-compensated amplifier given in Fig. 10-32(d).

SOLUTION The intersection between the closed-loop gain and the compensated open-loop gain yields the crossing frequency f_c. By inspection, we see that f_c is equal to f_{H2} (40 kHz), and f_{H1} is equal to our new pole at 390 Hz.

$$\begin{aligned}\theta_c &= -\tan^{-1}\frac{f_c}{f_{H1}} - \tan^{-1}\frac{f_c}{f_{H2}} - \tan^{-1}\frac{f_c}{f_{H3}}\\ &= -\tan^{-1}\frac{40\text{ kHz}}{390\text{ Hz}} - \tan^{-1}\frac{40\text{ kHz}}{40\text{ kHz}} - \tan^{-1}\frac{40\text{ kHz}}{200\text{ kHz}}\\ &= -89.4^\circ - 45.0^\circ - 11.3^\circ = -145.7^\circ\end{aligned}$$

Now we can find the phase margin.

$$\theta_{pm} = 180° + \theta_c = 180° - 145.7° = 34.3°$$

The positive phase margin indicates stable operation. ■

Now let us see how the values for the compensation network were arrived at. Consider Example 10-17.

EXAMPLE 10-17

Given that R is 5.6 kΩ, determine the required values of R_3 and C to produce the compensated response shown in Fig. 10-32(d).

SOLUTION First, we shall determine the required high-frequency attenuation to be provided by the compensation network. The open-loop gain at f_{H2} ($A_{v(ol)}[f_{H2}]$) is 55 dB, and the desired closed-loop gain $A_{v(cl)}$ is 40 dB. Therefore, the required high-frequency attenuation of the compensation network is

$$\text{attenuation} = A_{v(ol)}[f_{H2}] - A_{v(cl)} = 55 \text{ dB} - 40 \text{ dB}$$
$$= 15 \text{ dB}$$

Therefore, A_v must be -15 dB. The corresponding ''straight-ratio'' voltage gain is

$$A_v = 10^{|A_v|_{dB}/20} = 10^{-15/20} = 10^{-0.75} = 0.1778$$

Since the attenuation is determined by voltage division between R and R_3, we may solve for R_3. Hence

$$A_v = \frac{R_3}{R + R_3}$$

and

$$R_3 = \frac{A_v R}{1 - A_v} = \frac{(0.1778)(5.6 \text{ k}\Omega)}{1 - 0.1778} = 1.21 \text{ k}\Omega$$

The nearest standard value is 1.2 kΩ. The zero frequency f_z is set equal to f_{H1} (2.2 kHz), and in Eq. 10-35 we saw that

$$f_z = \frac{1}{2\pi R_3 C}$$

and solving for C gives us

$$C = \frac{1}{2\pi f_z R_3} = \frac{1}{2\pi(2.2 \text{ kHz})(1.2 \text{ k}\Omega)} = 0.0603 \text{ }\mu\text{F}$$

The nearest standard value is 0.06 μF. To complete any design, we should analyze our circuit by using the selected standard values. Hence

$$f_p = \frac{1}{2\pi(R + R_3)C} = \frac{1}{2\pi(5.6 \text{ k}\Omega + 1.2 \text{ k}\Omega)(0.06 \text{ }\mu\text{F})}$$
$$= 390 \text{ Hz}$$

which agrees with our result in Example 10-14, and

$$f_z = \frac{1}{2\pi R_3 C} = \frac{1}{2\pi(1.2 \text{ k}\Omega)(0.06 \text{ }\mu\text{F})} = 2.21 \text{ kHz}$$

This is very close to the required value of 2.2 kHz, considering that standard component values have been used. ■

The value of R was given in Example 10-17. If this were not done, we would select a nominal value (e.g., a few kilohms) and iterate through the procedure until reasonable values of R_3 and C are obtained. Generally, in the case of IC op amps, the value of R is *not* given. It will be the output resistance of a intermediate amplifier stage within its internal cascaded amplifier system. Consider Fig. 10-33.

FIGURE 10-33 Lag-lead frequency compensation of an op amp.

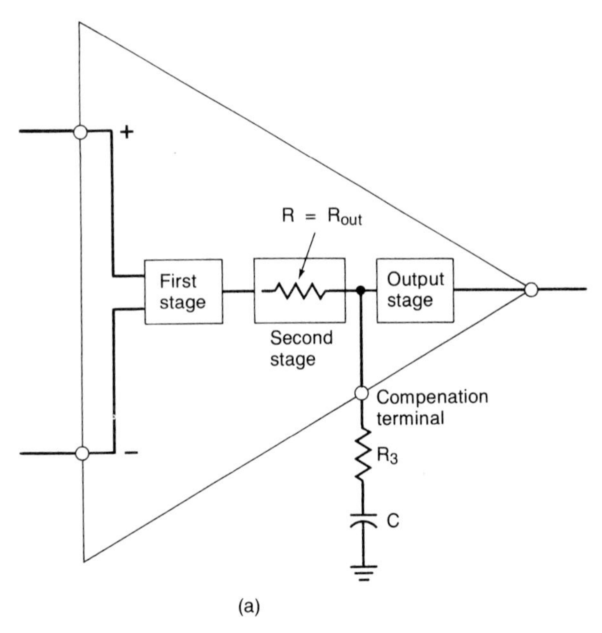

(a)

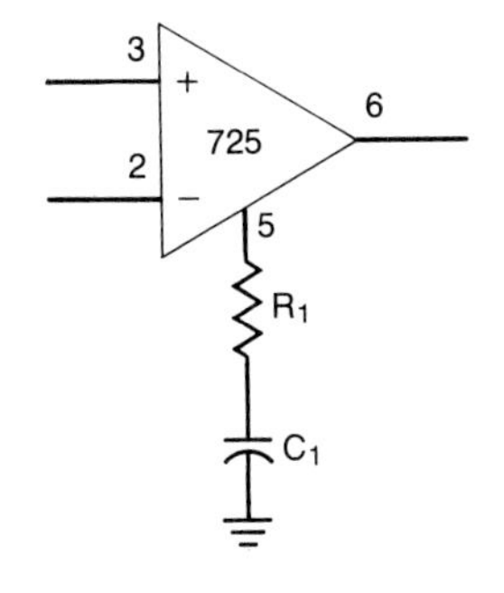

$A_{v(cl)}$	R_1	C_1
10,000	10 kΩ	50 pF
1,000	470 Ω	0.001 μF
100	47 Ω	0.01 μF

(b)

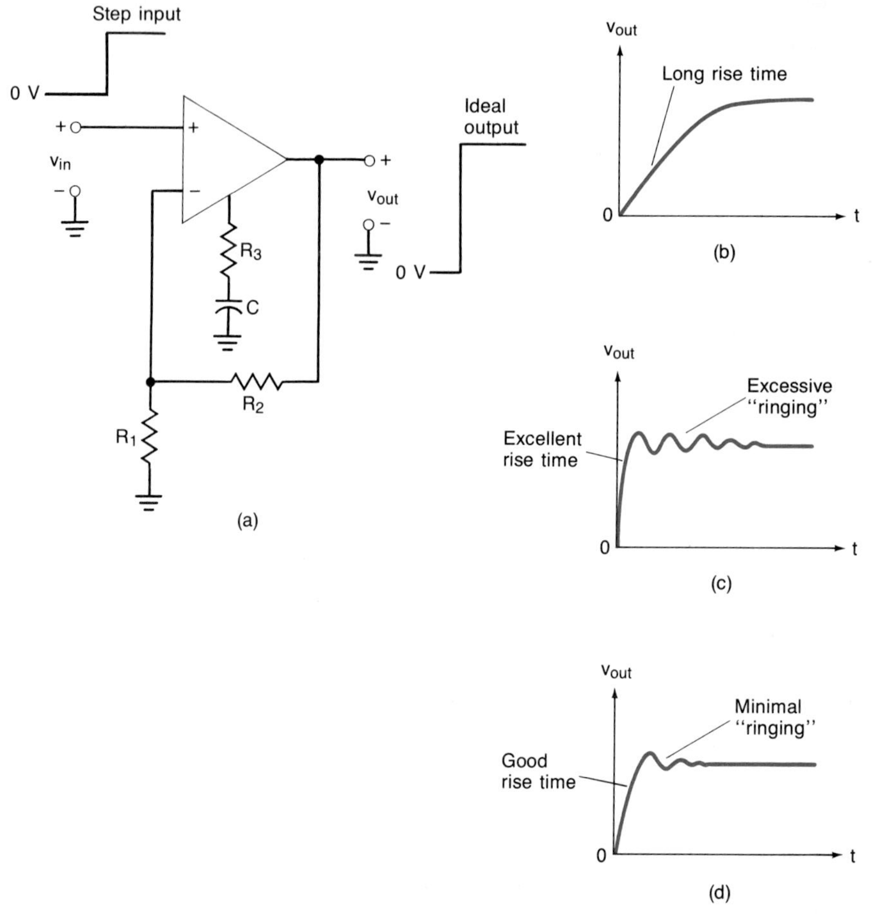

FIGURE 10-34 Transient response and Θ_{pm}: (a) compensated amplifier; (b) large Θ_{pm}; (c) small Θ_{pm}; (d) $\Theta_{pm} = 45°$.

In many cases the manufacturer will supply us with the required values of the external compensation components [see Fig. 10-33(b)]. In general, we should not deviate too far from the recommendations.

The phase margin and the transient (large-signal) response of an amplifier system are intimately related. From a stability standpoint, it would seem that a large phase margin is desirable. However, if the phase margin is too large, the amplifier's response to a step input [Fig. 10-34(a)] will be severely limited. Specifically, its rise time will be too long [see Fig. 10-34(b)]. If the phase margin is too small, the amplifier's response to a step input will exhibit "ringing" [see Fig. 10-34(c)]. *The optimum phase margin is* 45° [refer to Fig. 10-34(d)]. The transient response in this case demonstrates a good rise time with minimal ringing.

PROBLEMS

Drill, Derivations, and Definitions

Section 10-2

10-1. Name the two types of small-signal amplifiers that are controlled by an input voltage. In general, what are the constraints on their input resistances?

10-2. Name the two types of small-signal amplifiers that are controlled by an input current. In general, what are the constraints on their input resistances?

10-3. Name the two types of small-signal amplifiers that deliver an output voltage. In general, what are the constraints on their output resistances?

10-4. Name the two types of small-signal amplifiers that deliver an output current. In general, what are the constraints on their output resistances?

Section 10-3

10-5. When is output voltage sampling appropriate in a feedback amplifier? Output current sampling?

10-6. When is it appropriate to feed back a signal in series with an amplifier's input? In shunt?

10-7. Which feedback topology is best suited for a voltage amplifier? A current amplifier? Explain.

10-8. Which feedback topology is best suited for a transconductance amplifier? A transresistance amplifier? Explain.

Section 10-4

10-9. The LM318 is a high-speed (wide-bandwidth) operational amplifier. It is available in an eight-pin DIP and is pin-for-pin compatible with the 741C op amp. If it is used to replace the 741C in Fig. 10-4(a), find its minimum, maximum, and ideal closed-loop voltage gains. The LM318's open-loop voltage gain ranges from 25,000 to 200,000, and its minimum input resistance is 500 kΩ.

10-10. Repeat Prob. 10-9 if a higher-grade LM118 is used. The LM118's voltage gain ranges from 50,000 to 200,000, and its minimum input resistance is 1 MΩ.

Section 10-5

10-11. Feeding back a signal in series with the input circuit of an amplifier tends to raise its input resistance. Assume that an LM318 is used in Fig. 10-4(a). Find the minimum R_{inf} if R_{in} is 500 kΩ (minimum) and the minimum $A_{v(ol)}$ is 25,000.

10-12. Repeat Prob. 10-11 if an LM118 is used. Assume that R_{in} is 1 MΩ and $A_{v(ol)}$ is 50,000.

10-13. What is the magnitude of the feedback factor β when an op amp is being used as a voltage follower? Is the loop gain large, or small? Explain.

10-14. An LM318 is to be used as a voltage follower in a circuit such as that shown in Fig. 10-6. Assume that R_B and R_2 are 200 kΩ. R_L is 5.1 kΩ, R_{in} is 500 kΩ, and $A_{v(ol)}$ is 25,000. Find $A_{v(cl)}$, R_{inf}, and the total input resistance $R_{in(total)}$ presented to the signal source. How do the exact answers compare to our normal approximations?

10-15. Repeat Prob. 10-14 if an LM118 is used. In this case, R_B and R_2 are 470 kΩ, R_L is 5.1 kΩ, R_{in} is 500 kΩ, and $A_{v(ol)}$ is 50,000.

Section 10-6

10-16. When the input of the feedback network is connected across an amplifier's output, the output resistance is lowered. Assume that an LM318 is used in the noninverting amplifier circuit given in Fig. 10-4(a). Find R_{of} if R_{out} is 25 Ω and $A_{v(ol)}$ is 25,000.

10-17. Repeat Prob. 10-16 if an LM118 is used with an R_{out} of 25 Ω and an $A_{v(ol)}$ of 50,000 is used.

Section 10-7

10-18. An LM216 operational amplifier has internal frequency compensation for unity voltage gain. Its gain–bandwidth product is 100 kHz, and its open-loop voltage gain is 105 dB. Make a sketch of the Bode approximation of its amplitude response. What is its open-loop bandwidth?

10-19. An LM258 is a dual operational amplifier. Specifically, it contains two independent op amps in a single 14-pin DIP. The op amps are internally frequency compensated for unity voltage gain. If the gain–bandwidth product is 1 MHz and the open-loop voltage gain of these op amps is 110 dB, make a sketch of the Bode approximation of their ampltitude response. Find their open-loop bandwidth.

10-20. An op amp has an open-loop voltage gain of 110 dB and a gain–bandwidth product of 10 MHz. Find its closed-loop bandwidth when $A_{v(cl)}$ is 100. What is its open-loop bandwidth?

10-21. Repeat Prob. 10-20 if the closed-loop gain $A_{v(cl)}$ is reduced to 10.

Section 10-8

10-22. In your own words, explain how amplitude distortion (without clipping) is produced in an amplifier. Illustrate your answer with a sketch.

10-23. What is meant by the term ''total harmonic distortion'' (THD)? How is it defined?

10-24. An amplifier is driven by a 1-kHz sine wave. Its output contains the following spectral content: 3 V rms at 1 kHz, 0.5 V rms at 2 kHz, 0.2 V rms at 3 kHz, and 0.04 V rms at 4 kHz. Find the % THD at 1 kHz.

10-25. An amplifier is driven by a 10-kHz sine wave. Its output contains the following spectral content: 5 V rms at 10 kHz, 0.4 V rms at 20 kHz, 0.062 V rms at 30 kHz, and 0.004 V rms at 40 kHz. Find the % THD at 10 kHz.

10-26. Find the total rms value (V_{THD}) of the harmonic content contained in the output waveform of the amplifier described in Prob. 10-24.

10-27. Find the total rms value (V_{THD}) of the harmonic content contained in the output waveform of the amplifier described in Prob. 10-25.

10-28. Explain how negative feedback reduces amplitude distortion in an amplifier.

10-29. Given that a noninverting op amp amplifier has an open-loop % THD of 5% at 1 kHz, an $A_{v(\text{ol})}$ of 20,000, and an ideal closed-loop voltage gain of 20, find the closed-loop total harmonic distortion % $\text{THD}_{(\text{cl})}$.

10-30. Repeat Prob. 10-29 if the ideal closed-loop gain is reduced to 10.

Section 10-10

10-31. What is an octave? What frequency would be one octave above 15 kHz? What frequency would be two octaves below 15 kHz?

10-32. A noninverting amplifier is frequency compensated for unity gain, has an open-loop gain of 110 dB, a gain–bandwidth product of 10 MHz, and a closed-loop gain of 40 dB. Sketch its open-loop, closed-loop, and loop-gain frequency response plots.

10-33. A noninverting amplifier is frequency compensated for unity gain, has an open-loop gain of 80 dB, a gain–bandwidth product of 100 kHz, and a closed-loop voltage gain of 60 dB. Sketch its open-loop, closed-loop, and loop-gain frequency response plots.

10-34. What is frequency distortion in an amplifier? How does it differ from amplitude distortion?

10-35. A square wave has a frequency of 10 kHz, a positive peak of 10 V, and a negative peak of -10 V. Use Eq. 10-26 to write the Fourier series expansion out to the fifth harmonic. Graph this approximation.

10-36. Repeat Prob. 10-35 for a 2-kHz square wave with a positive peak of 5 V and a negative peak of -5 V.

10-37. Explain why the % $\text{THD}_{(\text{cl})}$ tends to increase as the signal frequency is raised.

10-38. The closed-loop total harmonic distortion % $\text{THD}_{(\text{cl})}$ of an amplifier increases with frequency as shown in Fig. 10-35(a). Find the % $\text{THD}_{(\text{cl})}$ at 1 kHz and at 10 kHz.

FIGURE 10-35 Frequency effects on closed-loop parameters: (a) closed-loop distortion; (b) closed-loop output resistance.

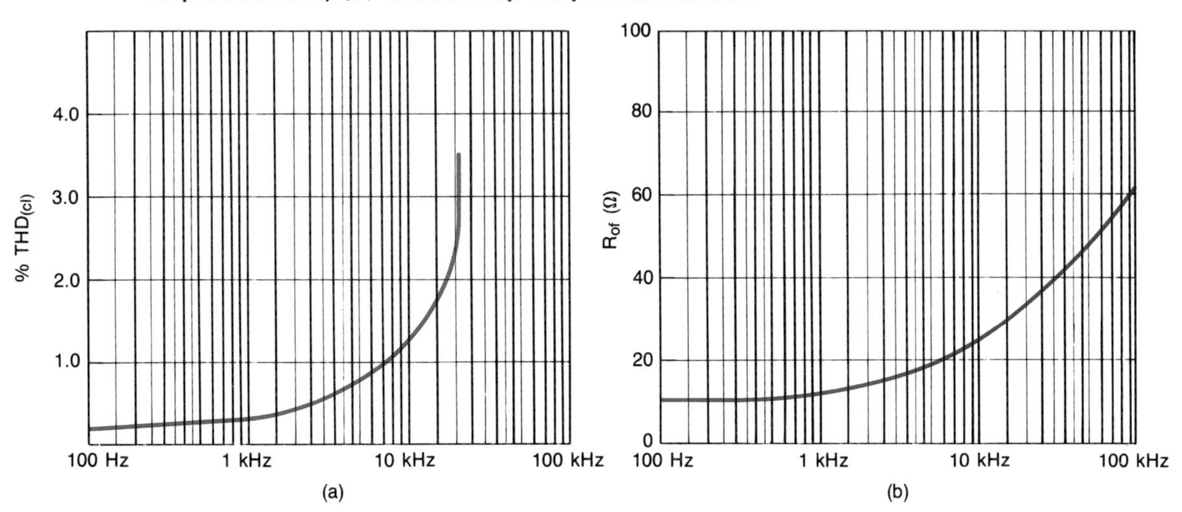

10-39. Explain why the closed-loop output resistance R_{of} of a noninverting op amp amplifier circuit will tend to increase as the signal frequency is raised.

10-40. The closed-loop output resistance of an amplifier increases as shown in Fig. 10-35(b). Determine R_{of} at 100 Hz, 10 kHz, and 100 kHz.

10-41. Describe the fundamental differences between a decompensated op amp and an uncompensated op amp.

Section 10-11

10-42. Explain what is meant by positive feedback. Describe how it may produce oscillation in an amplifier.

10-43. What is "white" noise?

10-44. What are the two Barkhausen criteria for sustained oscillations in a feedback amplifier? Why is oscillation in an amplifier undesirable?

FIGURE 10-36 Loop gain responses: (a) Prob. 10-46; (b) Prob. 10-47.

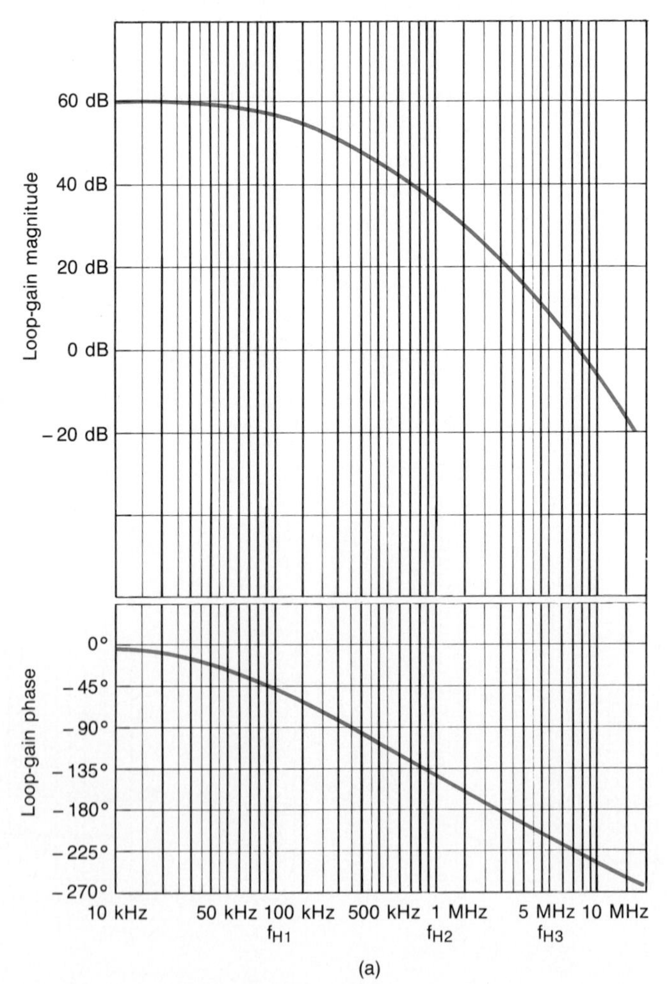

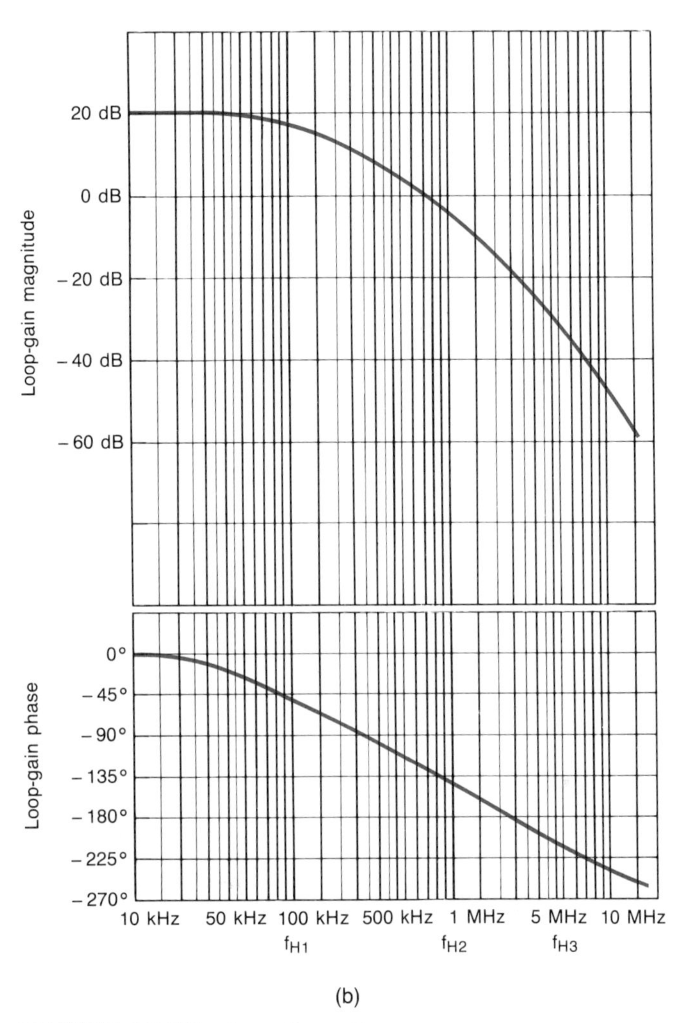

FIGURE 10-36 *(continued)*

Section 10-12

10-45. In your own words, explain the difference between the terms "gain margin" and "phase margin."

10-46. Given the loop-gain response curves in Fig. 10-36(a), determine the gain margin, the crossing frequency, the crossing angle θ_c, and the phase margin θ_{pm}. Is the circuit stable?

10-47. Repeat Prob. 10-46 for the loop-gain response curves given in Fig. 10-36(b).

10-48. An op amp's open-loop response is shown in Fig. 10-37(a). The amplifier circuit is given in Fig. 10-37(b). Determine the closed-loop gain and sketch the loop-gain response. Also find θ_c and θ_{pm}. Is the circuit stable?

10-49. Repeat Prob. 10-48 if R_2 is changed to 100 kΩ.

10-50. Repeat Prob. 10-48 if R_2 is changed to 10 kΩ.

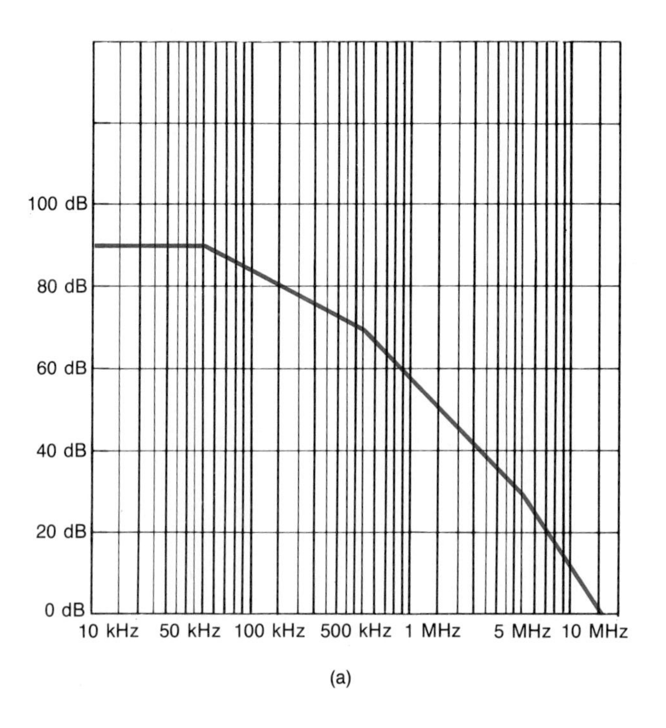

(a)

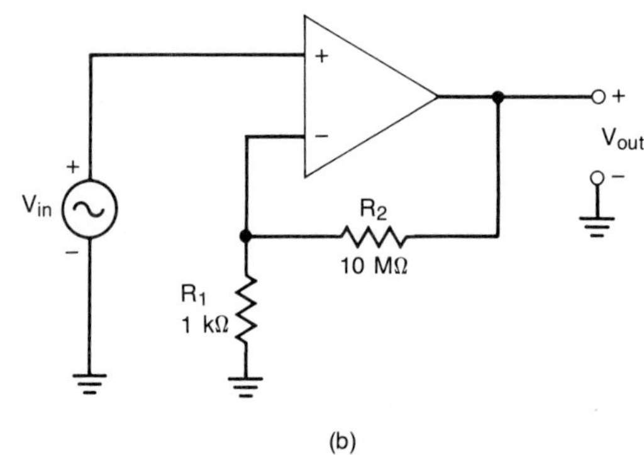

(b)

FIGURE 10-37 Response and circuit for Probs. 10-48 through 10-50: (a) open-loop response; (b) circuit.

10-51. A quick determination of an amplifier's stability can be made by looking at the slope of the loop-gain response as it passes through 0 dB. State the three classifications absolutely stable, marginally stable, and unstable in terms of the loop-gain slope through 0 dB. Express your answers in terms of dB per decade.

10-52. Repeat Prob. 10-51, but express your answers in terms of dB per octave.

Section 10-13

10-53. What is the purpose of frequency compensation? What is the basic approach?

10-54. Name the three basic types of frequency compensation. Which type is normally found in op amps that have been internally frequency compensated for unity voltage gain?

Section 10-14

10-55. Explain why an op amp that has been frequency compensated for unity voltage gain is stable for closed-loop gains as low as 0 dB.

10-56. An LM149 op amp is an example of a decompensated op amp. Its internal frequency-compensation capacitor is only 1 pF. Decompensation is used to provide an op amp with a relatively wide bandwidth. The only trade-off is that the op amp will not be stable for closed-loop voltage gains less than 5. It is described as having a gain–bandwidth product of 4 MHz for gains of 5 or greater. Determine its closed-loop bandwidth if $A_{v(\mathrm{cl})}$ is equal to 100.

10-57. For the LM149 described in Prob. 10-56, determine the frequency at which its open-loop gain will start to roll off at a rate of -40 dB/decade. Its $A_{v(\mathrm{ol})}$ is 50,000 and its slope rolls off at -40 dB/decade for an open-loop gain of 5. A graphical solution is acceptable.

Section 10-15

10-58. A lag–lead network has been depicted in Fig. 10-30(a). Assume that R is 6.8 kΩ, R_3 is 1.5 kΩ, and C is 0.05 μF, and plot the gain and phase as a function of frequency. Use five-cycle semilog paper.

10-59. Repeat Prob. 10-58 if R is 18 kΩ, R_3 is 1 kΩ, and C is 0.008 μF.

10-60. An op amp has a low-frequency open-loop gain of 100 dB and corner frequencies at 100 kHz, 1 MHz, and 5 MHz. It is to be used in an amplifier with a closed-loop gain of 60 dB. Sketch the frequency response of the open-loop gain, and superimpose the closed-loop gain. Use five-cycle semilog paper and start at 100 Hz. Determine the crossing frequency, the crossing angle, and the phase margin. Is the amplifier stable?

10-61. Repeat Prob. 10-60 for the same op amp if the closed-loop gain is reduced to 40 dB.

10-62. The op amp of Prob. 10-60 has been compensated by using a lag–lead network [refer to Fig. 10-32(d)]. In this case, R is 9.1 kΩ, R_3 is 1 kΩ, and C is 0.0015 μF. Sketch the compensated open-loop response and the closed-loop response on semilog paper. You may wish to employ the same graph paper used to solve Prob. 10-60. Determine the crossing angle and the phase margin. Is the amplifier stable?

10-63. The op amp of Prob. 10-61 has been compensated by using a lag–lead network [refer to Fig. 10-32(d)]. In this case, R is 9.1 kΩ, R_3 is 91 Ω, and C is 0.018 μF. Sketch the compensated open-loop response and the closed-loop response on semilog paper. You may wish to employ the same graph paper used to solve Prob. 10-60. Determine the crossing angle and the phase margin. Is the amplifier stable?

10-64. What is the optimum phase margin for a good transient response? What constitutes a ‘‘good’’ transient response?

Troubleshooting and Failure Modes

The most common failure mode for an op amp amplifier is for its output to latch at either its positive or negative supply rail. When investigating such a failure the first measurements should be the voltages present at the op amp's power supply pins. If the negative supply pin is open, the negative supply pin will be pulled *toward* the positive supply potential. In this case the op amp's output will be latched at the positive supply rail. If the supply pins have their proper voltages, a latched output can mean an open feedback loop, a relatively large dc input voltage, one of the input pins does not have a dc bias path to ground, or the op amp's input has exceeded its common-mode input voltage range. If none of these problems exist, the op amp itself has failed and should be replaced. With this discussion in mind, answer the questions below.

10-65. The op amp circuit shown in Fig. 10-4(a) is malfunctioning. The voltage readings at the various pins are as follows: pin 2 is at -12.9 mV, pin 3 is at 0 V, pin 4 is at -14.9 V, pin 6 is at -13 V, and pin 7 is at -4 V. Select the most likely failure from the following list of alternatives: (a) the positive supply connection is open, (b) the noninverting input is floating, (c) the feedback loop is open, and (d) R_1 is shorted. Explain why, or why not, the various alternatives are likely.

10-66. Repeat Prob. 10-65 if the voltage readings are the following: pin 2 is at 0 V, pin 3 is at 0 V, pin 4 is at -15.1 V, pin 6 is at 13.5 V, and pin 7 is at 15 V.

Design

10-67. It is desired to have a noninverting op amp amplifier with a bandwidth of at least 20 kHz and a closed-loop voltage gain of 100. Specify the op amp's required minimum f_T.

10-68. An op amp is uncompensated and has corner frequencies at 200 kHz, 2 MHz, and 10 MHz. Its open-loop voltage gain is 100 dB. It is to be used in a noninverting amplifier that has a closed-loop voltage gain of 40 dB. Design a lag–lead compensation network (if necessary) with an R of 10 kΩ. Find values for R_1, R_2, R_3, and C [refer to Fig. 10-32(d)].

Computer

10-69. Write a BASIC program that will find the voltage gain in decibels and the phase angle in degreees for a lag–lead network [see Fig. 10-30(a)]. The program should prompt the user for the zero frequency, the pole frequency, and the desired frequency of interest. The program should continue to repeat until the user elects to terminate its execution (e.g., ''ANOTHER FREQUENCY (Y/N)?'') is responded to with a ''N''o.

10-70. Modify the program written to satisfy the requirements of Prob. 10-69. The program is to automatically generate an analysis from the user-specified minimum and maximum frequencies in 20 equal steps.

11

Additional Negative Feedback Concepts and Applications

After Studying Chapter 11, You Should Be Able to:

- Explain the differences and relative merits between local negative feedback and overall negative feedback in a cascaded amplifier system.
- Analyze and explain the operation of active negative feedback as it relates to a bootstrapped voltage follower.
- Define the characteristics and use of current-series negative feedback in transconductance (voltage-to-current) amplifiers.
- Analyze op amp feedback loops that include the base-emitter junction of an output boost transistor.
- Define the characteristics and use of voltage-shunt negative feedback in transresistance (current-to-voltage) amplifiers.
- Define the characteristics and use of current-shunt negative feedback in current amplifiers.

11-1 Cascaded Amplifiers with Negative Feedback

In Chapter 9 we saw the advantages offered by the use of cascaded amplifiers. Now we consider cascaded amplifier systems from a negative-feedback standpoint. Suppose that we desire a precision amplifier with a closed-loop gain of 121. The op amp we intend to use has an open-loop gain of 10,000. Our first choice might be the single-stage noninverting amplifier depicted in Fig. 11-1(a). Without hesitation, we see that the ideal closed-loop gain is given by

$$A_{v(\text{cl})} = 1 + \frac{R_2}{R_1} = 1 + \frac{120\ \text{k}\Omega}{1\ \text{k}\Omega} = 121$$

However, let us scrutinize the situation more closely. If we first find the feedback factor β, we may determine a more realistic value for the closed-loop voltage gain.

$$\beta = \frac{R_1}{R_1 + R_2} = \frac{1\ \text{k}\Omega}{1\ \text{k}\Omega + 120\ \text{k}\Omega} = 0.008264$$

$$A_{v(\text{cl})} = \frac{A_{v(\text{ol})}}{1 + \beta A_{v(\text{ol})}} = \frac{10{,}000}{1 + (0.008264)(10{,}000)} = 119.6$$

The gain of 119.6 is approximately 1.16% below our desired gain of 121. This departure from the ideal may be directly attributed to a low value of loop gain. To achieve greater precision, the cascaded system shown in Fig. 11-1(b) may be used. Both amplifiers are assumed to possess an open-loop gain of 10,000. By inspection, we can see that the ideal voltage gains of each of the two stages is

$$A_{v(\text{cl})} = 1 + \frac{R_2}{R_1} = 1 + \frac{10\ \text{k}\Omega}{1\ \text{k}\Omega} = 11$$

and since the overall (total) gain $A_{v(\text{cl})T}$ is the product of the individual stage gains,

$$A_{v(\text{cl})T} = (A_{v(\text{cl})1})\,(A_{v(\text{cl})2}) = (11)(11) = 121$$

If we again strive for more accuracy in our assessment, we see that

$$\beta = \frac{R_1}{R_1 + R_2} = \frac{1\ \text{k}\Omega}{1\ \text{k}\Omega + 10\ \text{k}\Omega} = 0.09091$$

$$A_{v(\text{cl})1} = \frac{A_{v(\text{ol})}}{1 + \beta A_{v(\text{ol})}} = \frac{10{,}000}{1 + (0.09091)(10{,}000)} = 10.988 = A_{v(\text{cl})2}$$

$$A_{v(\text{cl})T} = A_{v(\text{cl})1}A_{v(\text{cl})2} = (10.988)(10.988) = 120.7$$

The total gain is much closer (only -0.248%) to our ideal gain of 121. The conclusion at this point is clear.

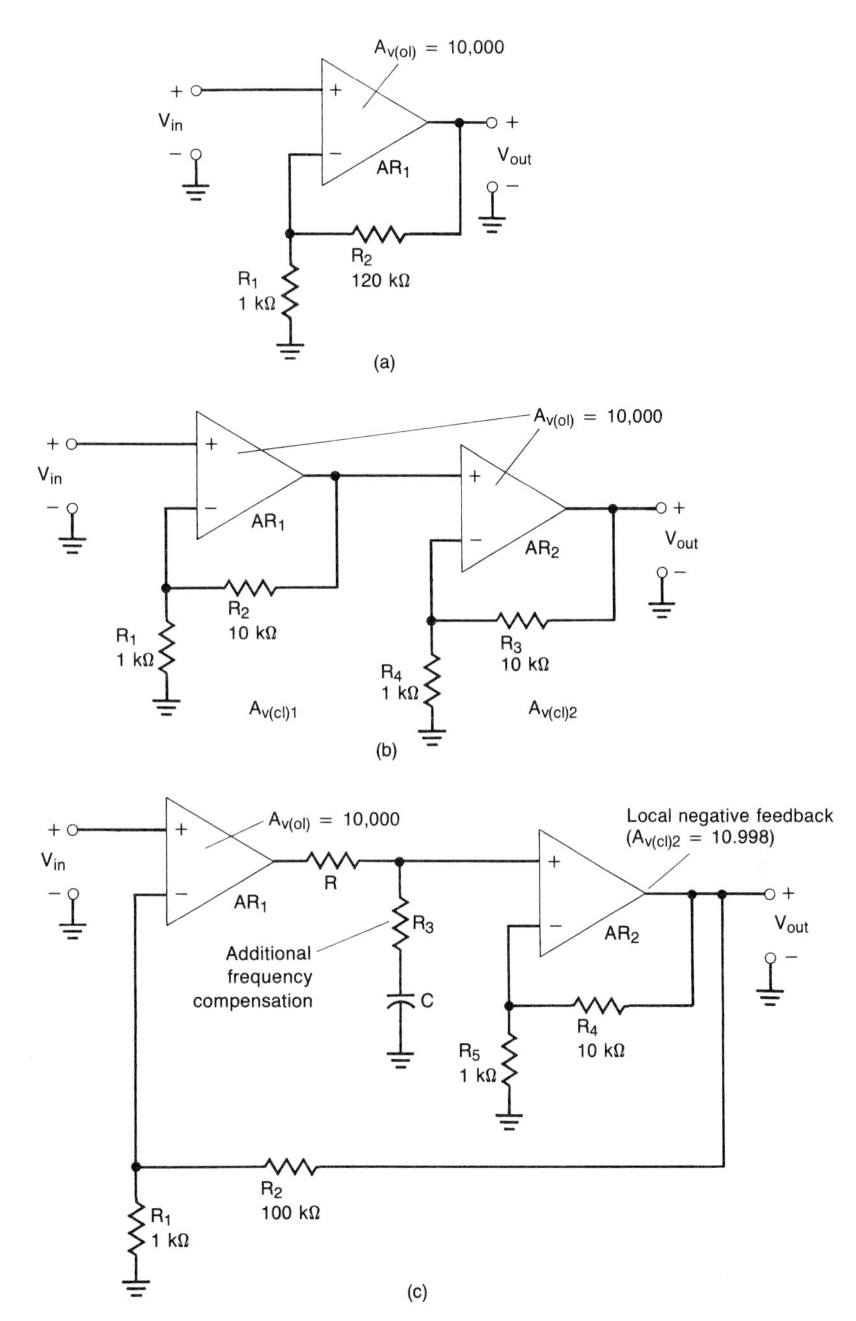

FIGURE 11-1 Negative feedback and cascaded systems: (a) noninverting amplifier has a gain of 119.6; (b) cascaded noninverting amplifiers provide a gain of 120.7; (c) local and overall negative feedback provide a gain of 120.9.

To achieve a given gain, it is better to use two stages with a lot of negative feedback than one single stage with less negative feedback.

Now consider Fig. 11-1(c).

Local negative feedback is used to stabilize the gain of the second stage. However, the main feedback loop has been wrapped around *both* amplifier stages. In this case the total (open-loop) voltage gain may be determined as follows:

$$A_{v(\mathrm{ol})T} = A_{v(\mathrm{ol})}A_{v(\mathrm{cl})2} = (10{,}000)(10.988) = 109{,}880$$

The feedback factor is the same as it was for Fig. 11-1(a). Specifically, β is 0.008264. The overall (total) closed-loop voltage gain is

$$A_{v(\mathrm{cl})T} = \frac{A_{v(\mathrm{ol})T}}{1 + \beta A_{v(\mathrm{ol})T}} = \frac{109{,}880}{1 + (0.008264)(109{,}880)}$$

$$= 120.9$$

The gain of 120.9 is a mere 0.826% below our desired gain of 121. Figure 11-1(c) offers the best gain precision. This is true in many cases. Superior performance results when the negative feedback loop is wrapped around the entire system in concert with local negative feedback. However, we should point out that loop stability is an extremely important consideration. We must ensure that the system does not oscillate.

Superior gain stability results when local negative feedback is used in conjunction with a feedback loop wrapped around the entire system.

However, before we decide which configuration to use, we must also consider amplifier internal noise and economics.

11-2 Active Negative Feedback

With the advent of low-cost, high-performance IC op amps, circuit techniques have been developed which were impractical when only discrete active devices were available. An example of this is the use of *active negative feedback*. Specifically, it is now economically feasible to use amplifiers as an integral part of the feedback network.

As we have seen (Example 10-3), the voltage follower exhibits an extremely large input resistance. Recall that the ideal voltage amplifier has an infinite input resistance. Therefore, a voltage follower would seem to be the ideal choice as the input stage in a cascaded amplifier system. However, the op amp's noninverting input terminal must always have a dc bias path to ground. This is ensured by the use of R_B in Fig. 11-2(a). Unfortunately, R_B tends to lower the input resistance of the amplifier. This can pose a serious limitation.

Figure 11-2(b) represents a unique solution. The circuit may be described as a "bootstrapped voltage follower." As we shall see, the circuitry ensures an input bias current to the noninverting input terminal of amplifier AR_1, and simultaneously preserves the extremely large input resistance associated with the use of voltage followers.

Amplifier AR_1 provides the main signal path, and amplifier AR_2 provides the active negative-feedback path. Both amplifiers are being used in their voltage-follower

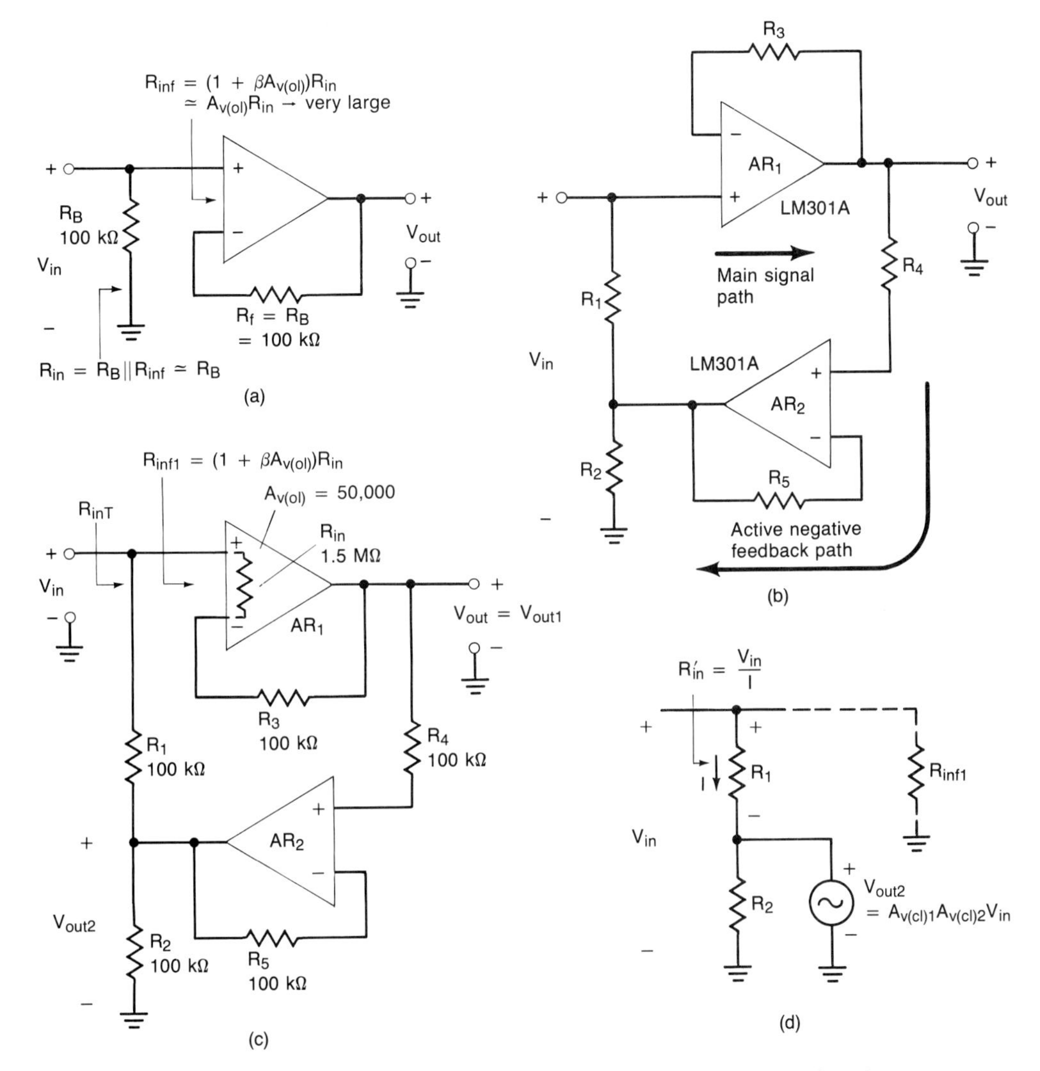

FIGURE 11-2 Analyzing the bootstrapped voltage follower: (a) voltage follower; (b) bootstrapped voltage follower; (c) finding R_{inf1}; (d) input model.

mode of operation. Resistors R_1 and R_2 are used to provide the bias current path for AR_1's noninverting input terminal. Resistors R_3, R_4, and R_5 are used to provide bias current compensation. (Bias current compensation was introduced in Chapter 8.)

The op amps are relatively inexpensive LM301A operational amplifiers. According to their manufacturer's data sheet, they have a minimum differential input resistance R_{in} of 1.5 MΩ and a minimum open-loop voltage gain $A_{v(ol)}$ of 50,000. Let us analyze the circuit to determine the total input resistance R_{inT}.

Since AR_1 is a voltage follower, it uses 100% of local negative feedback (β is unity). Its input resistance with negative feedback R_{inf1} is

$$
\begin{aligned}
R_{\text{in}f1} &= (1 + \beta A_{v(\text{ol})})R_{\text{in}} \\
&= [1 + (1)(50{,}000)][1.5\ \text{M}\Omega] \\
&= 75 \times 10^9\Omega = 75\ \text{G}\Omega
\end{aligned}
$$

[see Fig. 11-2(c)]. The closed-loop voltage gain of AR_1 is

$$A_{v(\text{cl})1} = \frac{A_{v(\text{ol})}}{1 + \beta A_{v(\text{ol})}} = \frac{50{,}000}{1 + (1)(50{,}000)} = 0.99998$$

and

$$V_{\text{out1}} = A_{v(\text{cl})1}V_{\text{in}} = 0.99998V_{\text{in}}$$

The output of AR_1 is directed into the input of AR_2. The closed-loop voltage gain of AR_2 will also be minimum of 0.99998. Therefore, V_{out2} is

$$V_{\text{out2}} = A_{v(\text{cl})2}V_{\text{out1}} = A_{v(\text{cl})2}A_{v(\text{cl})1}V_{\text{in}}$$

Substituting in our values yields

$$V_{\text{out2}} = (0.99998)(0.99998)V_{\text{in}} = 0.99996V_{\text{in}}$$

With these results, we can now find $R_{\text{in}T}$. A model of the input circuit has been given in Fig. 11-2(d). To simplify our efforts, we shall temporarily ignore the effects of the input resistance of the noninverting input terminal of AR_1. We shall call the input resistance due to R_1 loading effects R'_{in}. By applying Kirchhoff's voltage law, we may find the expression for R'_{in}. Hence

$$
\begin{aligned}
V_{\text{in}} &= IR_1 + A_{v(\text{cl})1}A_{v(\text{cl})2}V_{\text{in}} \\
V_{\text{in}} - A_{v(\text{cl})1}A_{v(\text{cl})2}V_{\text{in}} &= IR_1 \\
V_{\text{in}}(1 - A_{v(\text{cl})1}A_{v(\text{cl})2}) &= IR_1 \\
R'_{\text{in}} &= \frac{V_{\text{in}}}{I} = \frac{R_1}{1 - A_{v(\text{cl})1}A_{v(\text{cl})2}}
\end{aligned}
$$

Substituting in the values, we may determine R'_{in}.

$$R'_{\text{in}} = \frac{R_1}{1 - A_{(\text{cl})1}A_{v(\text{cl})2}} = \frac{100\ \text{k}\Omega}{1 - 0.99996} = 2.5\ \text{G}\Omega$$

The total input resistance $R_{\text{in}T}$ is given by the parallel combination of $R_{\text{in}f1}$ and R'_{in}.

$$
\begin{aligned}
R_{\text{in}T} &= R'_{\text{in}} \parallel R_{\text{in}f1} = 2.5\ \text{G}\Omega \parallel 75\ \text{G}\Omega \\
&= 2.42\ \text{G}\Omega
\end{aligned}
$$

Consider this result. We have taken relatively inexpensive op amps with an input resistance of 1.5 MΩ, ensured a dc bias path to ground, and achieved an input resistance of 2.42 GΩ! This input resistance level is virtually an open circuit. Active negative feedback has made this possible. If high-quality op amps with even larger input resistances are used, input resistances on the order of tera (10^{12}) ohms can be achieved.

The dc equivalent input resistance of our bootstrapped voltage follower is also incredibly large. The input dc bias current of the LM301A ranges from a typical value of 70 nA to a maximum value of 250 nA. Consider Fig. 11-3(a). With a V_{IN} of 10 V, the static resistance R_{IN} has a *minimum* value of

$$R_{\text{IN}} = \frac{V_{\text{IN}}}{I_{\text{IN}}} \simeq \frac{V_{\text{IN}}}{I_B} = \frac{10\ \text{V}}{250\ \text{nA}} = 40\ \text{M}\Omega$$

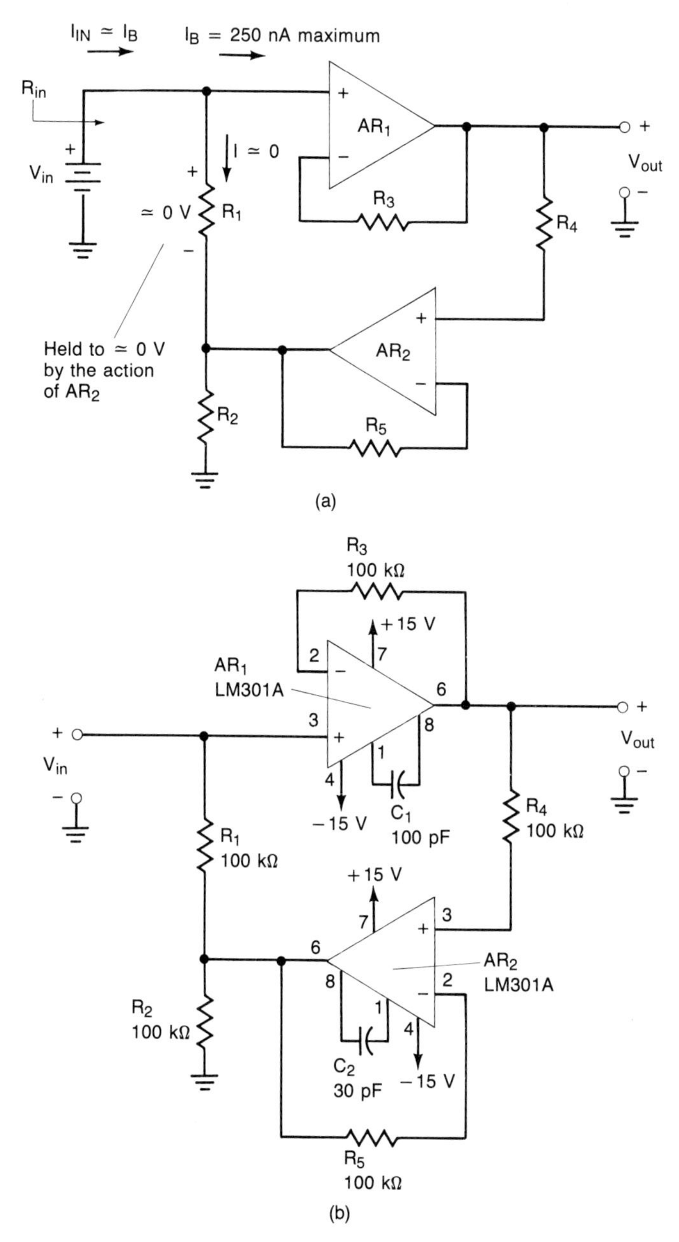

FIGURE 11-3 Bootstrapped voltage follower: (a) finding the dc input resistance R_{IN}; (b) complete schematic.

Experimental laboratory data indicated a dc input resistance R_{IN} of well over 100 MΩ.

The complete electrical schematic has been given in Fig. 11-3(b). The LM301A op amp utilizes dominant pole frequency compensation via an external capacitor. *For active negative feedback to work, the feedback amplifier AR_2 must be "faster" than* *the main signal path amplifier AR_1.* The speed of an amplifier is inversely proportional to the size of its frequency compensation capacitance. According to the manufacturer, C_c must be at least 30 pF to ensure stability for unity gain. Consequently, C_c for AR_2 is 30 pF, and AR_1 has been slowed down by using C_c of 100 pF.

We shall see another example of active negative feedback in Chapter 15. It is an extremely powerful technique.

11-3 Current-Series Negative Feedback

In Section 10-2 the transconductance amplifier (voltage-to-current converter) was introduced [refer to Fig. 10-2(c)]. Current-series negative feedback [Fig. 10-3(c)] is the optimum topology for a transconductance amplifier. Recall that the basic intent is to use an input voltage signal to provide an output current signal.

In Fig. 11-4 we see the idealized transconductance amplifier. The input and output resistances have been assumed to be infinite. Let us develop the feedback equation for the transconductance amplifier with negative feedback. The closed-loop transconductance shall be called $G_{m(\text{cl})}$.

By applying Kirchhoff's voltage law around the input circuit,

$$V_{\text{in}} = V + V_f \qquad (11\text{-}1)$$

In Fig. 11-4 we also see that

$$I_{\text{out}} = G_{m(\text{ol})}V \rightarrow V = \frac{I_{\text{out}}}{G_{m(\text{ol})}}$$

and

$$V_f = I_{\text{out}}R_f$$

FIGURE 11-4 Ideal transconductance amplifier with current-series negative feedback.

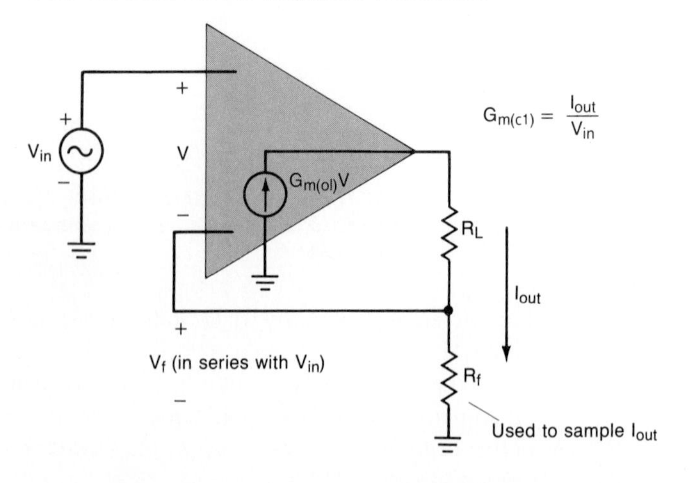

If we substitute these relationships into Eq. 11-1, we arrive at Eq. 11-2.

$$V_{in} = \frac{I_{out}}{G_{m(ol)}} + I_{out}R_f = \left[\frac{1}{G_{m(ol)}} + R_f\right]I_{out}$$

$$= \frac{1 + R_f G_{m(ol)}}{G_{m(ol)}} I_{out}$$

and solving for I_{out}/V_{in} yields $G_{m(cl)}$.

$$\boxed{G_{m(cl)} = \frac{I_{out}}{V_{in}} = \frac{G_{m(ol)}}{1 + R_f G_{m(ol)}}} \tag{11-2}$$

where $G_{m(cl)}$ = closed-loop transconductance
$G_{m(ol)}$ = open-loop transconductance
R_f = feedback factor (a resistance)

This result has a form that is extremely similar to our equation for $A_{v(cl)}$ [refer to Eq. 10-5].

Observe that the feedback factor is a resistance R_f and the loop gain is $R_f G_{m(ol)}$. Now let us look at a transconductance amplifier which has been designed around an op amp [see Fig. 11-5(a)].

The circuit appears to be a noninverting amplifier at first glance. However, notice that the load resistance R_L occupies the position normally held by "R_2" and our feedback element is R_f. Our objective is to control the current I_{out} through the load R_L. R_f is used to sense I_{out} and develop a feedback voltage which is in series with the input circuit.

Before we begin a rigorous analysis, let us first exercise our intuition. As we can see in Fig. 11-5(a), with a large open-loop voltage gain, the differential input voltage V is approximately zero. Therefore, V_{in} effectively appears across R_f. If we assume that the inverting input terminal draws negligible current, R_f is effectively in series with R_L. Hence

$$I_{out} \simeq \frac{V_{in}}{R_f}$$

The resulting transconductance with negative feedback is (approximately)

$$G_{m(cl)} = \frac{I_{out}}{V_{in}} \simeq \frac{1}{R_f} \tag{11-3}$$

This same approximation can be arrived at by setting the loop gain [$R_f G_{m(ol)}$] in Eq. 11-2 to be much larger than unity.

Now let us pursue the circuit from the more classical approach. In Section 10-4 we saw that to analyze a feedback amplifier, we must find the equivalent input circuit, the equivalent output circuit, and the feedback factor.

To find the equivalent open-loop input circuit when output current sampling is used, we must open the output circuit [see Fig. 11-5(b)]. To find the equivalent open-loop output circuit when the feedback signal is in series with the input circuit, we must also open the input circuit [see Fig. 11-5(c)]. The equivalent open-loop circuit has been given in Fig. 11-5(d). Let us proceed with the analysis.

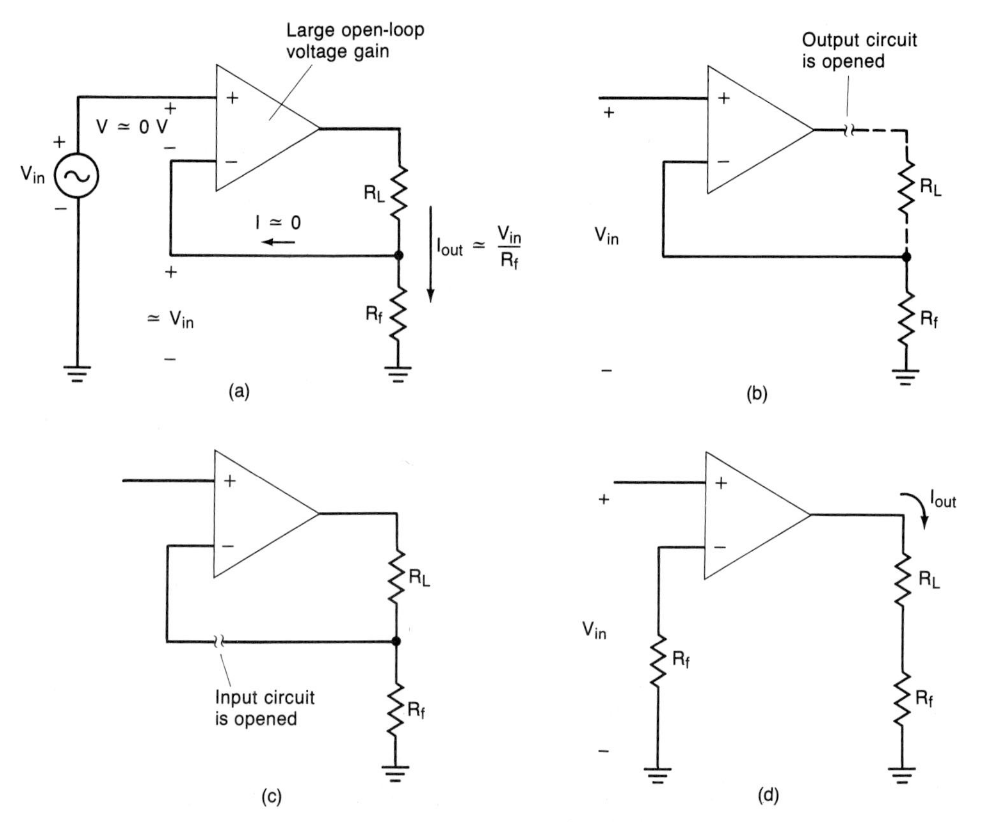

FIGURE 11-5 Op amp transconductance amplifier: (a) approximate analysis; (b) finding the open-loop input circuit; (c) finding the open-loop output circuit; (d) equivalent open-loop circuit.

Because op amps typically have a large R_{in}, we can state that

$$R_{\text{in}} >> R_f$$

and

$$V \simeq V_{\text{in}}$$

This has been noted in Fig. 11-6(a). The output current I_{out} is given by Ohm's law.

$$I_{\text{out}} = \frac{A_{v(\text{ol})}V}{R_L + R_f} \simeq \frac{A_{v(\text{ol})}V_{\text{in}}}{R_L + R_f}$$

The open-loop transconductance is given by

$$G_{m(\text{ol})} = \frac{I_{\text{out}}}{V_{\text{in}}} = \frac{A_{v(\text{ol})}}{R_L + R_f} \tag{11-4}$$

From Fig. 11-6 and assuming that the input resistance of the inverting input terminal is very large, we may state that

$$V_f = I_{\text{out}}R_f$$

and

$$R_f = \frac{V_f}{I_{\text{out}}} \tag{11-5}$$

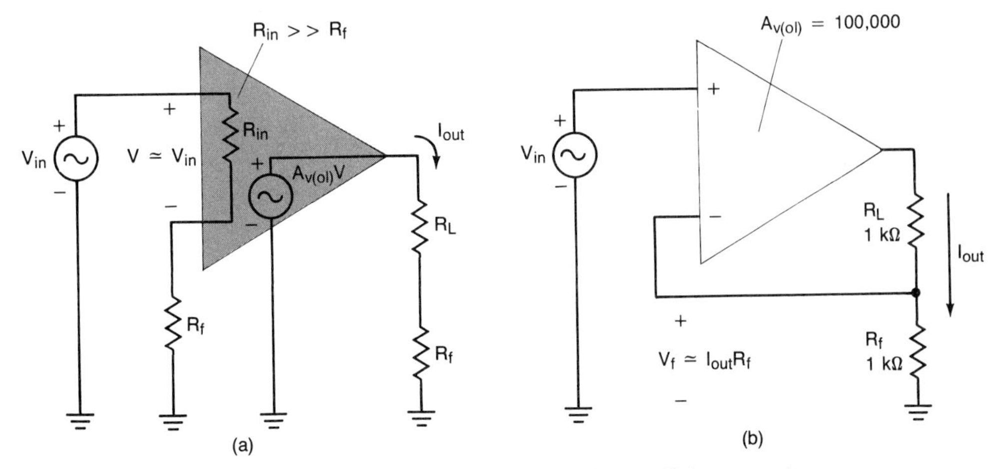

FIGURE 11-6 Analyzing the circuit: (a) open-loop; (b) example.

If we place Eqs. 11-4 and 11-5 into the general form given by Eq. 11-2, we obtain

$$G_{m(\text{cl})} = \frac{G_{m(\text{ol})}}{1 + R_f G_{m(\text{ol})}} = \frac{A_{v(\text{ol})}/(R_L + R_f)}{1 + R_f[A_{v(\text{ol})}/(R_L + R_f)]} \tag{11-6}$$

Once again, we can see that if the loop gain is much greater than unity, we arrive at the approximation offered by

$$G_{m(\text{cl})} = \frac{\dfrac{A_{v(\text{ol})}}{(R_L + R_f)}}{1 + R_f\left[\dfrac{A_{v(\text{ol})}}{(R_L + R_f)}\right]} \simeq \frac{\dfrac{A_{v(\text{ol})}}{(R_L + R_f)}}{R_f\left[\dfrac{A_{v(\text{ol})}}{(R_L + R_f)}\right]} = \frac{1}{R_f} \tag{11-7}$$

This is precisely the same result that we developed by our more intuitive approach (refer to Eq. 11-3).

EXAMPLE 11-1

Find the closed-loop transconductance $G_{m(\text{cl})}$ of the circuit given in Fig. 11-6(b).

SOLUTION From Eq. 11-4 we first find the open-loop transconductance $G_{m(\text{ol})}$.

$$G_{m(\text{ol})} = \frac{A_{v(\text{ol})}}{R_L + R_f} = \frac{100{,}000}{1\text{ k}\Omega + 1\text{ k}\Omega} = 50\text{ S}$$

By inspection, we see that R_f is 1 kΩ, and from Eq. 11-6,

$$G_{m(\text{cl})} = \frac{G_{m(\text{ol})}}{1 + R_f G_{m(\text{ol})}} = \frac{50\text{ S}}{1 + (1\text{ k}\Omega)(50\text{ S})} = 0.99998\text{ mS}$$

$$\simeq 1\text{ mS}$$

Another way of viewing this result is that the gain is 1 mA per volt of input. ■

Voltage-to-current conversion has many applications in electronic systems. For example, the horizontal and vertical deflection coils, and the focusing coils found in television sets, and video monitors are excited by current signals. The current signals

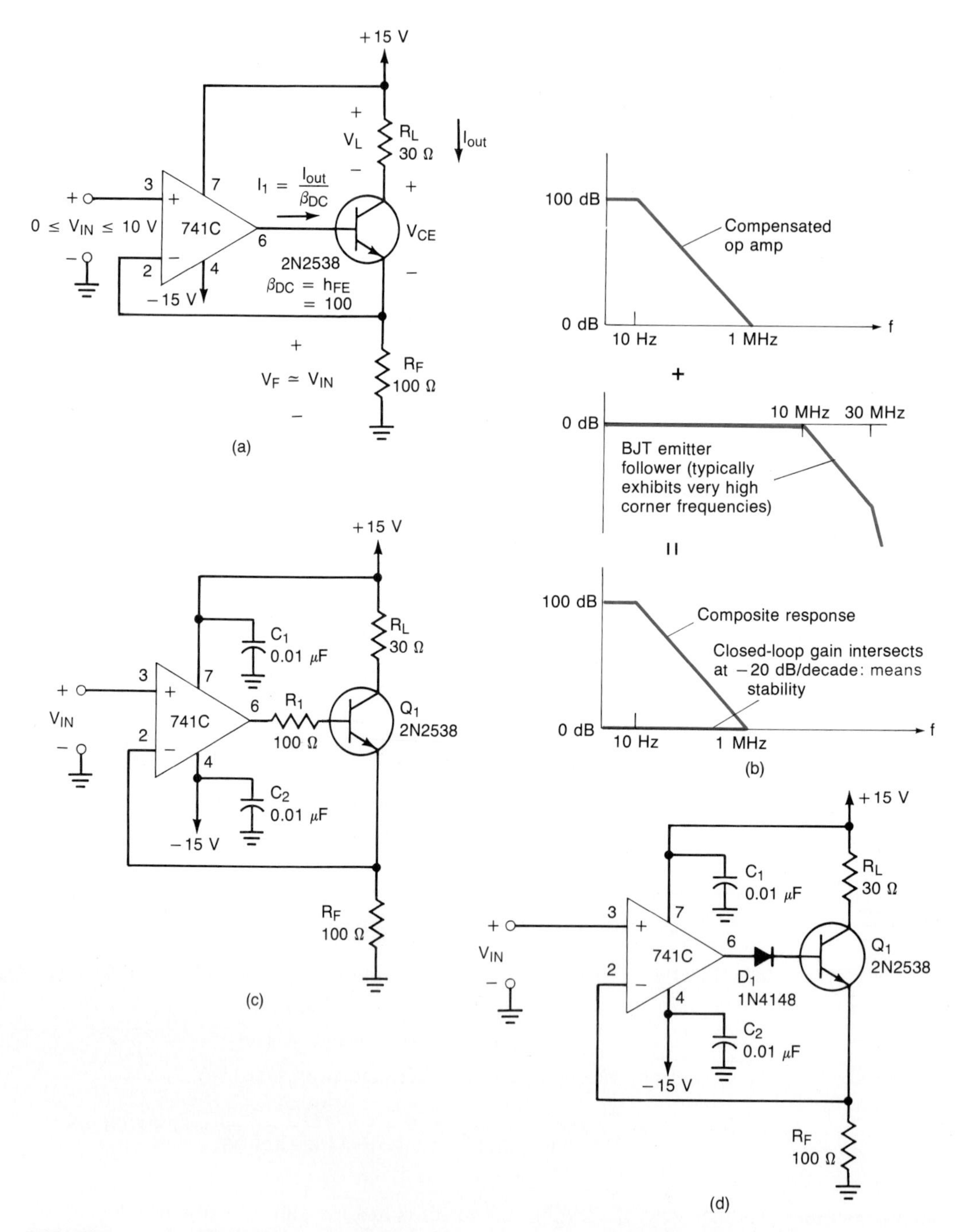

FIGURE 11-7 "Boosted" op amp transconductance amplifier: (a) basic circuit; (b) finding the stability; (c) adding R_1, C_1, and C_2 to avoid oscillations; (d) protecting the BJT and avoiding oscillations.

are used to produce magnetic fields around the coils, which, in turn, are used to control the electron beam directed toward the screen. The required current signals are often produced by voltage-to-current converters. In process control systems, signals are often transmitted through lines as currents (typically, 4 to 20 mA) because current signals are inherently more immune (than voltage signals) to electrical noise, and line resistance effects are rendered insignificant. Once again, the usual approach is to use a voltage-to-current converter.

Typically, *the peak current drawn from the output of most general-purpose op amps should not exceed* 10 *mA*. In the event that larger output currents are required, the cascaded arrangement illustrated in Fig. 11-7(a) may be used. This technique is a very popular method of boosting the maximum available output current from an op amp circuit. Therefore, it merits our serious attention.

Observe that the feedback loop has been wrapped around both stages. Consequently, the voltage across R_F (the emitter voltage of the BJT) will be approximately equal to V_{IN}. Therefore,

$$V_F \simeq V_{\mathrm{IN}}$$

and

$$I_E \simeq \frac{V_{\mathrm{IN}}}{R_F}$$

The output current I_{OUT} through the load is the collector current of the BJT. Hence

$$I_{\mathrm{OUT}} = I_C = \alpha_{\mathrm{DC}} I_E \simeq I_E$$

The output voltage of the op amp V_1 will be one base–emitter drop (approximately 0.7 V) higher than the emitter voltage. Therefore,

$$\begin{aligned} V_1 &= V_F + 0.7 \text{ V} \\ &\simeq V_{\mathrm{IN}} + 0.7 \text{ V} \end{aligned}$$

(The op amp works to make the voltage at its two inputs equal.)

The current drawn from the op amp's output I_1 will be reduced by the BJT's current gain.

$$I_1 = \frac{I_C}{\beta_{\mathrm{DC}}} = \frac{I_{\mathrm{OUT}}}{\beta_{\mathrm{DC}}}$$

The voltage across the load will be

$$V_L = I_{\mathrm{OUT}} R_L$$

and the BJT's collector-to-emitter voltage will be

$$V_{CE} = V_{CC} - V_L - V_{\mathrm{IN}}$$

EXAMPLE 11-2

Analyze the transconductance amplifier given in Fig. 11-7(a) if its input voltage is 10 V dc.

SOLUTION We observe that R_L is 30 Ω, h_{FE} is 100, V_{CC} is 15 V, and use the approximations cited above.

$$V_F \simeq V_{\mathrm{IN}} = 10 \text{ V}$$

$$I_E \simeq \frac{V_{\mathrm{IN}}}{R_F} = \frac{10 \text{ V}}{100\ \Omega} = 100 \text{ mA}$$

$$I_{OUT} \simeq I_E = 100 \text{ mA}$$
$$V_1 = V_{IN} + 0.7 \text{ V} = 10 \text{ V} + 0.7 \text{ V} = 10.7 \text{ V}$$
$$I_1 = \frac{I_{OUT}}{\beta_{DC}} = \frac{100 \text{ mA}}{100} = 1 \text{ mA}$$
$$V_L = I_{OUT}R_L = (100 \text{ mA})(30\Omega) = 3 \text{ V}$$
$$V_{CE} = V_{CC} - V_L - V_{IN}$$
$$= 15 \text{ V} - 3 \text{ V} - 10 \text{ V} = 2 \text{ V}$$

■

The circuit has a gain of 10 mA of output current for each volt of input. It will provide this gain as long as both the op amp and the BJT remain in their linear regions of operation. For example, the BJT cannot maintain a constant collector current for values of V_{CE} too close to 0 V, and the op amp's output voltage cannot exceed approximately V_{CC} minus 2 V.

If we assume that the maximum input voltage V_{IN} is 10 V, R_L can range from 0 to 50 Ω. For example, if R_L is 50 Ω and I_{OUT} is 100 mA, then

$$V_L = I_{OUT}R_L = (0.1 \text{ A})(50 \, \Omega) = 5 \text{ V}$$

and

$$V_{CE} = V_{CC} - V_L - V_{IN}$$
$$= 15 \text{ V} - 5 \text{ V} - 10 \text{ V} = 0 \text{ V}$$

The transistor will be saturated. An further increases in R_L will cause a reduction in the 100-mA output current. Since R_L can range from 0 to 50 Ω, V_L can range from 0 to 5 V. Our output will source a current of 100 mA for V_L's that range from 0 to 5 V. This is called the *compliance* of our current source. Specifically, the compliance is 5 V.

There are two other concerns that require our attention if we are to ensure reliable operation of our circuit. The first is the stability of our circuit, and the second deals with the protection of our BJT.

The feedback loop is used to enclose both the op amp and our BJT. As far as the feedback loop is concerned, the BJT is a common-collector (emitter-follower) amplifier. The op amp's output is directed into the BJT's (input) base terminal, and the feedback voltage is taken from the BJT's (output) emitter terminal.

If the frequency response of the loop gain is examined, no instability will be obvious. Since the closed-loop voltage gain from the op amp's noninverting input terminal to the BJT's emitter is approximately 0 dB, the circuit appears to be stable. Consider Fig. 11-7(b).

However, the amplifier's wide bandwith (1 MHz) makes the circuit susceptible to small stray capacitances and inductances. (The most troublesome stray inductance typically originates within the power supply lines.) These stray reactances can cause (unpredictable) additional phase shifts. Consequently, the possibility of high-frequency oscillations exists. This potential oscillation problem can be avoided by installing bypass capacitors at the supply terminals of the op amp, and placing a small resistance in series with the BJT's base terminal [refer to Fig. 11-7(c)].

Another precaution is often necessary. If V_{IN} becomes negative, the circuit will not function. More important, the base-emitter *p-n* junction of the BJT will become

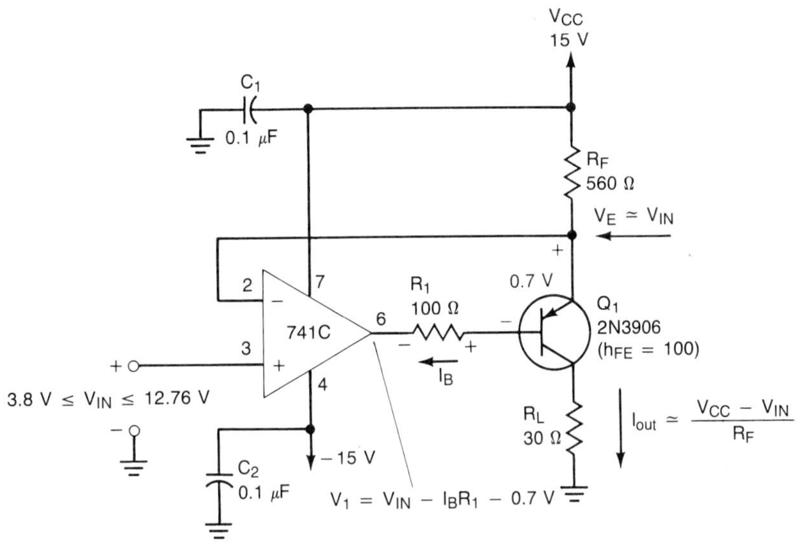

FIGURE 11-8 Transconductance amplifier with a ground-referenced load.

reverse biased. The maximum reverse-bias rating of a BJT base-emitter junction is typically limited to 5 V. If V_{IN} becomes negative, the BJT's base-emitter junction will become reverse biased and the BJT will enter cutoff. When this occurs the feedback loop will be opened up, and the large open-loop voltage gain of the op amp will cause its output to hit the negative rail. (The op amp's output will saturate at about −13 V.) Consequently, the large reverse bias will cause zener breakdown of the BJT's base–emitter junction. Without sufficient current limiting, the BJT will be destroyed.

The usual soltuion to this potential failure is to place a diode in series with the BJT's base terminal [see Fig. 11-7(d)]. The diode's small forward resistance is often sufficient to eliminate the need for the small resistance placed in series with the base terminal. It should be clear that the addition of this diode requires that the op amp's output voltage be 1.4 V higher than V_{IN}.

A "floating" (not referenced to ground) load is often undesirable. If the *npn* BJT is replaced by a *pnp* BJT, the circuit operation is essentially unaffected, but the load may now be referenced to ground (see Fig. 11-8). The pertinent equations have been indicated in the figure. The reader will be directed through the analysis of this circuit in the problems at the end of the chapter.

11-4 Voltage-Shunt Negative Feedback

Voltage-shunt negative feedback is the optimum topology for a transresistance amplifier. Recall that the function of the transresistance amplifier is to provide an output voltage that is proportional to the input current. Therefore, it is often referred to as a *current-to-voltage converter*. The closed-loop transresistance will be denoted by $R_{m(\text{cl})}$ [see Fig. 11-9(a)].

Both the input and output resistances have been idealized as zero ohms. By applying Kirchhoff's current law to the input, we may write

$$I_{\text{in}} = I + I_f \tag{11-8}$$

Further inspection of Fig. 11-9(a) yields that

$$V_{\text{out}} = -R_{m(\text{ol})}I \rightarrow I = -\frac{V_{\text{out}}}{R_{m(\text{ol})}} \tag{11-9}$$

and since R_{in} is zero ohms, V_{out} is impressed across the feedback conductance G_f. Hence

$$I_f = -G_f V_{\text{out}} \tag{11-10}$$

If we substitute the relationships offered by Eqs. 11-9 and 11-10 into Eq. 11-8, we obtain Eq. 11-11. Thus

$$I_{\text{in}} = I + I_f = -\frac{V_{\text{out}}}{R_{m(\text{ol})}} - G_f V_{\text{out}} = -\left[\frac{1}{R_{m(\text{ol})}} + G_f\right]V_{\text{out}}$$

$$= -\frac{1 + G_f R_{m(\text{ol})}}{R_{m(\text{ol})}} V_{\text{out}}$$

$$\boxed{R_{m(\text{cl})} = \frac{V_{\text{out}}}{I_{\text{in}}} = -\frac{R_{m(\text{ol})}}{1 + G_f R_{m(\text{ol})}}} \tag{11-11}$$

Despite the negative sign, the general form of Eq. 11-11 is similar to that for our voltage amplifier with voltage-series negative feedback and that of our transconductance amplifier with current-series negative feedback. The feedback factor is a conductance G_f and the loop gain is $G_f R_{m(\text{ol})}$. Once again, we see that if the loop gain is much greater than unity, the gain is approximately equal to a circuit constant. In this case,

$$R_{m(\text{cl})} \simeq -\frac{1}{G_f} = -R_f$$

Figure 11-9(b) represents the op amp version of a transresistance amplifier. Notice that the polarity of the controlling differential input voltage has been reversed [see Fig. 11-9(c)]. Consequently, the polarity of the op amp's controlled source has also been reversed.

Our first step is to find the equivalent open-loop circuit. Since output voltage sampling is used, the equivalent input circuit may be found by shorting the output to ground. Because the feedback signal is in parallel with the input circuit, the equivalent open-loop output circuit may be found by shorting the input to ground. The entire equivalent open-loop circuit has been depicted in Fig. 11-9(d).

In Fig. 11-9(d) we note that

$$R_f << R_{\text{in}}$$

Therefore, virtually all of I_{in} flows through R_f. Hence

$$V = I_{\text{in}} R_f$$

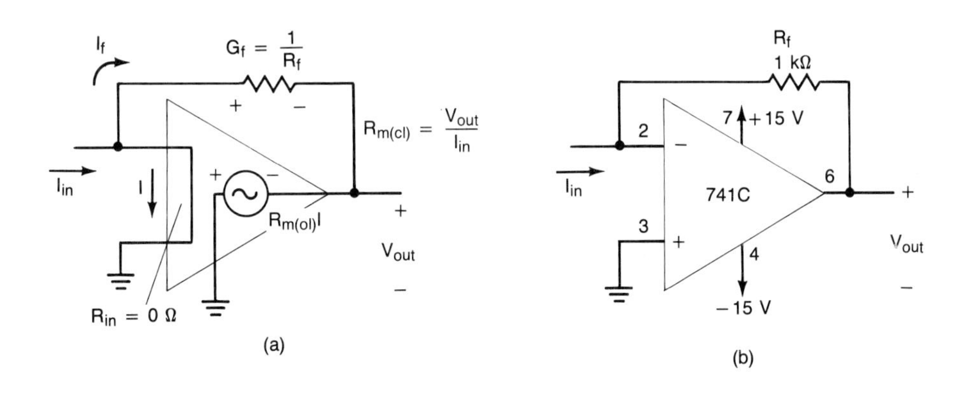

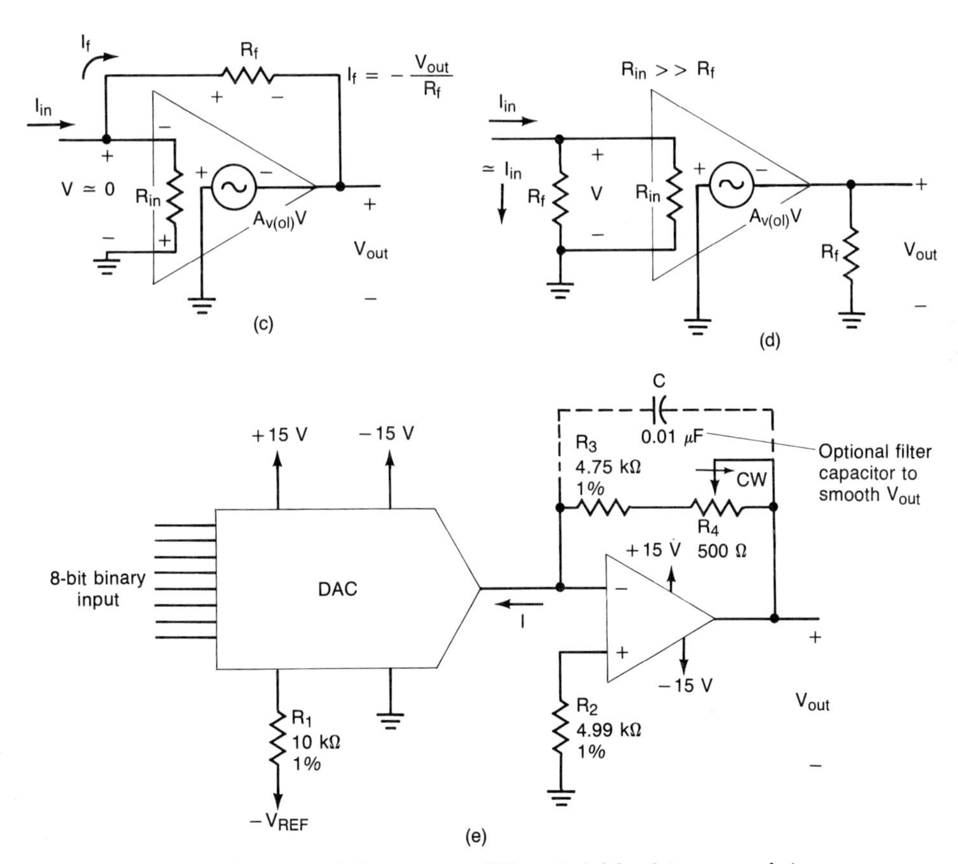

FIGURE 11-9 Transresistance amplifier: (a) ideal transresistance amplifier with voltage-shunt negative feedback; (b) op amp transresistance amplifier; (c) op amp equivalent circuit; (d) equivalent open-loop circuit; (e) converting a DAC's current output to a voltage.

The expression for the output voltage may be determined by substituting the result above into the op amp's output equation.

$$V_{\text{out}} = -A_{v(\text{ol})}V = -A_{v(\text{ol})}R_f I_{in}$$

Solving for the ratio of V_{out} to I_{in} produces the open-loop transresistance $R_{m(\text{ol})}$.

$$R_{m(\text{ol})} = \frac{V_{\text{out}}}{I_{\text{in}}} = -A_{v(\text{ol})}R_f \qquad (11\text{-}12)$$

Since the open-loop voltage gain of an op amp is so very large, we shall approximate the differential input voltage as zero. Therefore, virtually all of V_{out} will be impressed across R_f. From Fig. 11-9(c) we write

$$I_f = -\frac{V_{\text{out}}}{R_f}$$

and the feedback factor is

$$\frac{I_f}{V_{\text{out}}} = -\frac{1}{R_f} \qquad (11\text{-}13)$$

if we place $R_{m(\text{ol})}$ and the feedback factor $-1/R_f$ into the general feedback equation form, we obtain

$$R_{m(\text{cl})} = \frac{V_{\text{out}}}{I_{\text{in}}} = \frac{R_{m(\text{ol})}}{1 + G_f R_{m(\text{ol})}} = \frac{-A_{v(\text{ol})}R_f}{1 + [-1/R_f][-A_{v(\text{ol})}R_f]}$$

$$= -\frac{A_{v(\text{ol})}R_f}{1 + (1/R_f)A_{v(\text{ol})}R_f} \simeq -\frac{1}{1/R_f} = -R_f \qquad (11\text{-}14)$$

The approximation in Eq. 11-14 is valid as long as the loop gain [in this case simply $A_{v(\text{ol})}$] is much greater than unity. This is the usual case.

EXAMPLE 11-3

Find the gain of the op amp circuit given in Fig. 11-9(b). The op amp's open loop voltage gain is 100,000.

SOLUTION We shall apply Eq. 11-14 to determine the circuit's closed-loop transresistance.

$$R_{m(\text{cl})} = -\frac{A_{v(\text{ol})}R_f}{1 + (1/R_f)A_{v(\text{ol})}R_f} = -\frac{(100{,}000)(1\ \text{k}\Omega)}{1 + (1/1\ \text{k}\Omega)(100{,}000)(1\ \text{k}\Omega)}$$

$$= -999.99\ \Omega$$

and ideally,

$$R_{m(\text{cl})} \simeq -R_f = -1000\ \Omega$$

Obviously, the approximation provides us with a very accurate prediction. Notice that the units of transresistance are ohms. The gain could also be described as being 1000 V per ampere or 1 V per milliampere. The latter is generally a more accurate description of op amp transresistance amplifiers. ■

An example of a practical application of the current-to-voltage converter has been given in Fig. 11-9(e). Many of the currently available high-speed digital-to-analog converters (DACs) provide a *current output*. The op amp transresistance amplifier is used to convert the output current of the DAC into a voltage. This is necessary because of the limited compliance of DACs (particularly, CMOS DACs). We shall not detail the operation of the DAC, but merely point out this widespread application.

The most common application of voltage-shunt negative feedback is the op amp inverting voltage amplifier [refer to Fig. 11-10(a)]. In this case we are using an op amp *transresistance* amplifier as a *voltage* amplifier. To relate this circuit's voltage gain to that of the noninverting voltage amplifier, we shall develop an equation that has the same form as Eq. 10-5. For clarity, we shall express our results in terms of the op amp's open-circuit voltage gain $A_{v(oc)}$. To ease our efforts, we shall draw on Miller's theorem [see Fig. 11-10(b)]. By voltage division,

$$V = \frac{\dfrac{R_2}{[1 + A_{v(oc)}]}}{R_1 + \dfrac{R_2}{[1 + A_{v(oc)}]}} V_{in} = \frac{\dfrac{R_2}{[1 + A_{v(oc)}]}}{\dfrac{[R_1(1 + A_{v(oc)}) + R_2]}{[1 + A_{v(oc)}]}} V_{in}$$

$$= \frac{R_2}{R_1 + R_2 + R_1 A_{v(oc)}} V_{in} \tag{11-15}$$

FIGURE 11-10 The op amp inverting voltage amplifier uses voltage-shunt negative feedback: (a) circuit; (b) simplification by Miller's theorem.

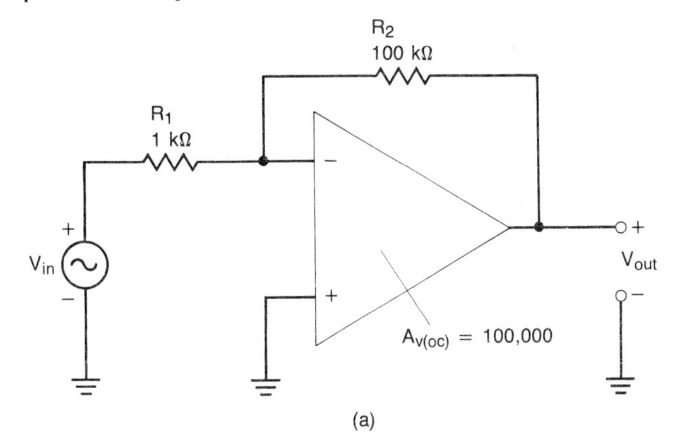

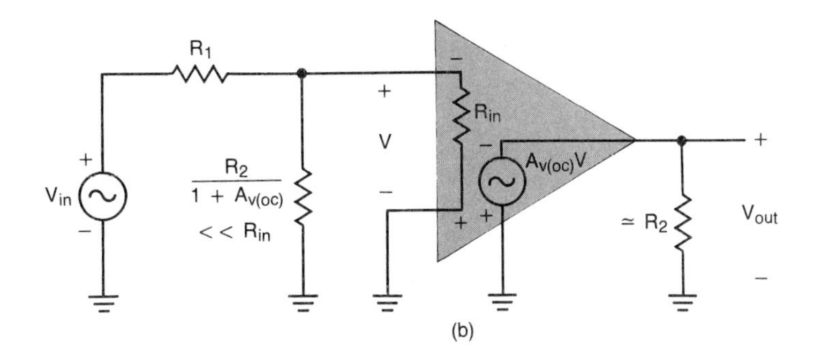

By inspection of Fig. 11-10(b) and substitution of Eq. 11-15,

$$V_{out} = -A_{v(oc)}V = -\frac{R_2 A_{v(oc)}}{R_1 + R_2 + R_1 A_{v(oc)}} V_{in}$$

Dividing each of the numerator and denominator terms by $(R_1 + R_2)$ yields

$$V_{out} = -\frac{\dfrac{R_2 A_{v(oc)}}{(R_1 + R_2)}}{1 + \dfrac{R_1 A_{v(oc)}}{(R_1 + R_2)}} V_{in} = -\frac{\dfrac{R_2 A_{v(oc)}}{(R_1 + R_2)}}{1 + \left(\dfrac{R_1}{R_2}\right)\left[\dfrac{R_2 A_{v(oc)}}{(R_1 + R_2)}\right]} V_{in}$$

$$A_{v(cl)} = \frac{V_{out}}{V_{in}} = -\frac{\dfrac{R_2 A_{v(oc)}}{(R_1 + R_2)}}{1 + \left(\dfrac{R_1}{R_2}\right)\left[\dfrac{R_2 A_{v(oc)}}{(R_1 + R_2)}\right]} \tag{11-16}$$

If we do not let the minus sign trouble us, we can see that Eq. 11-16 has been placed in the general form of Eq. 10-5. This has been emphasized below.

$$A_{v(cl)} = \frac{V_{out}}{V_{in}} = -\frac{R_2 A_{v(oc)}/(R_1 + R_2)}{1 + [R_1/R_2][R_2 A_{v(oc)}/(R_1 + R_2)]}$$

open-loop gain terms $(A_{v(ol)})$

feedback factor β

If the loop gain is much greater than unity, we arrive at our familiar approximation of the inverting amplifier's closed-loop voltage gain.

$$A_{v(cl)} \simeq -\frac{1}{\beta} = -\frac{1}{R_1/R_2} = -\frac{R_2}{R_1} \tag{11-17}$$

EXAMPLE 11-4

Determine the closed-loop gain of the inverting amplifier given in Fig. 11-10(a).

SOLUTION From Eq. 11-16,

$$A_{v(cl)} = \frac{V_{out}}{V_{in}} = -\frac{R_2 A_{v(oc)}/[R_1 + R_2]}{1 + [R_1/R_2][R_2 A_{v(oc)}/(R_1 + R_2)]}$$

$$= -\frac{(100\text{ k}\Omega)(100{,}000)/(1\text{ k}\Omega + 100\text{ k}\Omega)}{1 + [1\text{ k}\Omega/100\text{ k}\Omega][(100\text{ k}\Omega)(100{,}000)/(1\text{ k}\Omega + 100\text{ k}\Omega)]} = -99.9$$

The ideal closed-loop gain is given by Eq. 11-17.

$$A_{v(cl)} \simeq -\frac{R_2}{R_1} = -\frac{100\text{ k}\Omega}{1\text{ k}\Omega} = -100$$

Obviously, there is a close agreement between this approximation and our more accurate prediction of -99.9. ■

11-5 Op amp Inverting Summer

Another application of voltage-shunt negative feedback has been illustrated in Fig. 11-11. This circuit is called an *inverting summer*. Essentially, it may be used to add two or more voltages together. This is necessary when we wish to offset a dc signal, add a dc offset to an ac signal, or add ac signals together. The latter application is found in audio equipment called *mixers*.

Given that the op amp's inverting input terminal is a virtual ground, we can state that

$$I_1 = \frac{V_1}{R_1} \qquad I_2 = \frac{V_2}{R_2} \qquad I_3 = \frac{V_3}{R_3}$$

This is depicted in Fig. 11-11(a). Assuming that no current flows into the op amp's inverting input terminal, I_1, I_2, and I_3 are all directed into the op amp's feedback resistor R_f. Hence

$$I_f = I_1 + I_2 + I_3 = \frac{V_1}{R_1} + \frac{V_2}{R_2} + \frac{V_3}{R_3}$$

FIGURE 11-11 **Op amp inverting summer: (a) basic circuit; (b) example.**

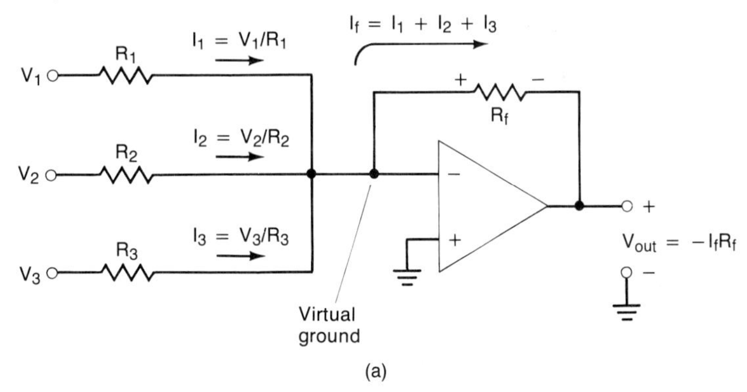

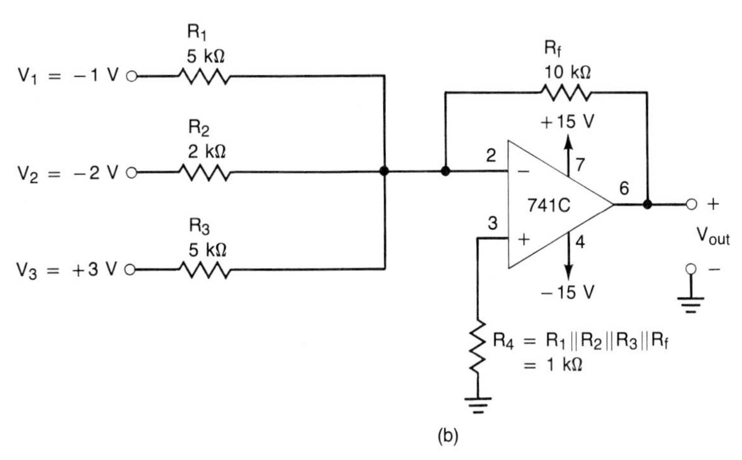

The output voltage is equal to the voltage developed across R_f. This leads us to

$$\begin{aligned} V_{\text{out}} &= -I_f R_f = -\left(\frac{V_1}{R_1} + \frac{V_2}{R_2} + \frac{V_3}{R_3}\right) R_f \\ &= -\left(\frac{R_f}{R_1} V_1 + \frac{R_f}{R_2} V_2 + \frac{R_f}{R_3} V_3\right) \end{aligned} \tag{11-18}$$

Equation 11-18 tells us that we cannot only add V_1, V_2, and V_3 together, but we can independently weight their contribution to the output voltage. However, if $R_1 = R_2 = R_3 = R$, then

$$\begin{aligned} V_{\text{out}} &= -\left(\frac{R_f}{R} V_1 + \frac{R_f}{R} V_2 + \frac{R}{R} V_3\right) \\ &= -\frac{R_f}{R}(V_1 + V_2 + V_3) \end{aligned} \tag{11-19}$$

and each input voltage has the *same* weight. Let us predict the (dc) output voltage of the summer shown in Fig. 11-11(b). From Eq. 11-18,

$$\begin{aligned} V_{\text{OUT}} &= -\left(\frac{R_f}{R_1} V_1 + \frac{R_f}{R_2} V_2 + \frac{R_f}{R_3} V_3\right) \\ &= -\left[\frac{10\text{ k}\Omega}{5\text{ k}\Omega}(-1\text{ V}) + \frac{10\text{ k}\Omega}{2\text{ k}\Omega}(-2\text{ V}) + \frac{10\text{ k}\Omega}{5\text{ k}\Omega}(3\text{ V})\right] \\ &= 6\text{ V} \end{aligned}$$

11-6 Current-Shunt Negative Feedback

When the objective is to achieve a stabilized current gain, current-shunt negative feedback is used. The output current is sampled by a series element and fed back in parallel with the input circuit [see Fig. 11-12(a)].

Assuming that the inverting input terminal has a very large input resistance, virtually all of I_{in} flows through R_2. The current that flows through R_1 will be equal to the sum of I_{out} and I_{in}. Observe that resistors R_1 and R_2 are connected across the op amp's differential input. Therefore, by Kirchhoff's voltage law,

$$\begin{aligned} V &\simeq I_{\text{in}} R_2 + (I_{\text{in}} + I_{\text{out}}) R_1 = I_{\text{in}} R_2 + I_{\text{in}} R_1 + I_{\text{out}} R_1 \\ &= I_{\text{in}}(R_1 + R_2) + I_{\text{out}} R_1 \end{aligned}$$

Since the open-loop voltage gain of the op amp is so very large, V will be approximately zero. Therefore, the inverting input terminal is a virtual ground. Hence

$$I_{\text{in}}(R_1 + R_2) + I_{\text{out}} R_1 \simeq 0$$

and solving for the closed-loop current gain $A_{i(\text{cl})}$ yields

$$\begin{aligned} I_{\text{out}} R_1 &= -I_{\text{in}}(R_1 + R_2) \\ I_{\text{out}} &= -\frac{R_1 + R_2}{R_1} I_{\text{in}} = -\left(1 + \frac{R_2}{R_1}\right) I_{\text{in}} \end{aligned}$$

$$\boxed{A_{i(\text{cl})} = \frac{I_{\text{out}}}{I_{\text{in}}} = -\left(1 + \frac{R_2}{R_1}\right)} \tag{11-20}$$

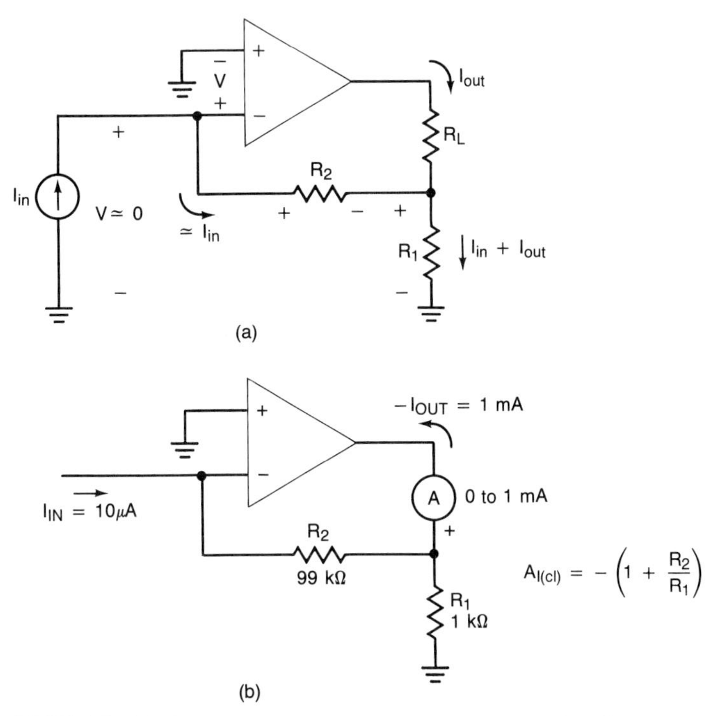

FIGURE 11-12 Current-shunt negative feedback and the op amp current amplifier: (a) basic circuit; (b) example.

Because the load is floating in Fig. 11-12(a), the circuit is not very popular. However, it does serve floating loads quite nicely. Consider the next example.

EXAMPLE 11-5

A 1-mA (full-scale) D'Arsonval ammeter is to be driven by a current that ranges from 0 to 10 μA. Will the circuit given in Fig. 11-12(b) suffice?

SOLUTION The ammeter is a floating load therefore the scheme is sound. The current gain is

$$A_{I(\text{cl})} = \frac{I_{\text{OUT}}}{I_{\text{IN}}} = -\left(1 + \frac{R_2}{R_1}\right) = -\left(1 + \frac{99\ \text{k}\Omega}{1\ \text{k}\Omega}\right) = -100$$

and if I_{IN} is 10 μA,

$$I_{\text{OUT}} = A_{I(\text{cl})} I_{\text{IN}} = -(100)(10\ \mu\text{A}) = -1\ \text{mA}$$

The gain is adequate. Since the current gain is negative, I_{OUT} flows *into* the op amp's output terminal. Consequently, the negative terminal of the ammeter must connect to the op amp's output [refer to Fig. 11-12(b)]. ■

11-7 Summary

In each of the negative-feedback arrangements, the feedback amplifier assumes characteristics that approach the ideal situation. The gain is stabilized, the input and output resistances are improved, the bandwidth is increased, and the distortion is reduced.

We have not mathematically proven these statements in all cases, but the reader should be readily willing to accept these facts based on our extensive work in Chapter 10. A summary has been provided in Table 11-1.

TABLE 11-1
Summary of the Closed-Loop Amplifiers

	Type of Feedback			
Parameter	Voltage-Series	Current-Shunt	Voltage-Shunt	Current-Series
R_{inf}	Increased	Decreased	Decreased	Increased
R_{of}	Decreased	Increased	Decreased	Increased
$BW_{(cl)}$	Increased	Increased	Increased	Increased
$\%\ THD_{(cl)}$	Decreased	Decreased	Decreased	Decreased
Gain	$A_{v(cl)}$	$A_{i(cl)}$	$R_{m(cl)}$	$G_{m(cl)}$

PROBLEMS

Drill, Derivations, and Definitions

Section 11-1

11-1. An amplifier system is to provide a closed-loop voltage gain of 1000. If two cascaded amplifiers are to be used, and their individual closed-loop voltage gains are to be equal, compute their value. In general, will two cascaded stages [Fig. 11-1(b)] offer more precision than a single stage [Fig. 11-1(a)] with less negative feedback? Base your answer on the loop-gain considerations. Assume that $A_{v(ol)}$ is 20,000.

11-2. Repeat Prob. 11-1 if the overall system gain is to be 2000.

11-3. Explain why the amplifier system given in Fig. 11-1(c) is superior to the one given in Fig. 11-1(b). What is the purpose of R, R_3, and C?

11-4. What is meant by the term ''local negative feedback''?

Section 11-2

11-5. What is meant by the term ''active negative feedback''? What is the primary constraint on the frequency compensation of the active feedback element?

11-6. A bootstrapped voltage follower, such as that depicted in Fig. 11-3(b), has R_1 through R_5 equal to 470 kΩ. The minimum open-loop voltage gain of the op amps is 75,000, and their minimum differential input resistances are 2 MΩ. Find the circuit's total input resistance R_{inT}.

11-7. Repeat Prob. 11-6 if the op amps have minimum input resistances of 5 MΩ and a minimum open-loop voltage gains of 100,000.

Section 11-3

11-8. A transconductance amplifier (Fig. 11-4) has a $G_{m(ol)}$ which ranges from 15 to 30 S. If R_f is 100 Ω and R_L is 1 kΩ, find the minimum and maximum vlaues of the closed-loop transconductance. What is the ideal closed-loop transconductance?

11-9. Repeat Prob. 11-8 if $G_{m(ol)}$ ranges from 20 to 40 S, R_f is 200 Ω, and R_L remains 1 kΩ.

11-10. Find the closed-loop transconductance of the op amp circuit given in Fig. 11-6(b). Assume that $A_{v(ol)}$ is 20,000, R_L is 1 kΩ, and R_f is 2 kΩ. What is the ideal closed-loop transconductance?

11-11. Repeat Prob. 11-10 if $A_{v(ol)}$ remains at 20,000, R_L is 2 kΩ, and R_f is 10 kΩ.

11-12. Repeat the analysis conducted in Example 11-2 if V_{IN} is reduced to 2 V dc.

11-13. Repeat the analysis conducted in Example 11-2 if V_{IN} is reduced to 1 V dc.

11-14. What is meant by the term "compliance"?

11-15. What is the function of R_1, C_1, and C_2 in Fig. 11-7(c)? Will R_1 raise, or lower, the circuit's compliance? Explain.

11-16. Why is diode D_1 required in Fig. 11-7(d)? Specifically, what function(s) does it serve? Will its presence raise, or lower, the circuit's output compliance? By how much?

11-17. A voltage-to-current converter with a ground-referenced load has been illustrated in Fig. 11-8. Assume that V_{IN} ranges from 3.8 to 12.76 V. Find V_E, V_1, V_L, V_{CE}, and I_{OUT} at the two extremes.

11-18. A diode has *not* been placed in series with the base of transistor Q_1 (Fig. 11-8). Explain why a diode is *not* necessary.

Section 11-4

11-19. An ideal transresistance amplifier has been shown in Fig. 11-9(a). Given that the open-loop transresistance can range from 10 to 20 kΩ, and that R_f is 1 kΩ, find the minimum and maximum closed-loop transresistance. What is the ideal closed-loop transresistance?

11-20. Repeat Prob. 11-19 if $R_{m(ol)}$ ranges from 50 to 100 kΩ and R_f is 10 kΩ.

11-21. Find the closed-loop transresistance of the op amp circuit given in Fig. 11-9(b). The op amp's open-loop voltage gain is 10,000 and R_f is 2 kΩ. What is the ideal transresistance?

11-22. Repeat Prob. 11-21 if $A_{v(ol)}$ is 25,000 and R_f is 10 kΩ.

11-23. What is a DAC? What is the function of the op amp circuit given in Fig. 11-9(e)?

11-24. Ideally, what is the gain (in V/mA) of the op amp circuit given in Fig. 11-9(e) if R_4 is adjusted fully clockwise (CW)?

11-25. Ideally, what is the gain (in V/mA) of the op amp circuit given in Fig. 11-9(e) if R_4 is adjusted fully counterclockwise (CCW)?

11-26. Determine the closed-loop voltage gain of the op amp inverting amplifier given in Fig. 11-10(a) if $A_{v(oc)}$ is 25,000, R_1 is 2 kΩ, and R_2 is 100 kΩ. What is the ideal closed-loop voltage gain?

11-27. Repeat Prob. 11-26 if $A_{v(oc)}$ is increased to 120,000.

11-28. An inverting (op amp) voltage amplifier [Fig. 11-10(a)] has *less bandwidth than a noninverting amplifier using the same op amp* [which means the same $A_{v(oc)}$ and f_T]. Given the (inverting) op amp amplifier relationships

$$f_{H(\text{cl})} = (1 + \beta A_{v(\text{ol})})f_H = \left[1 + \frac{R_1}{R_1 + R_2} A_{v(\text{oc})}\right] f_H$$

$$A_{v(\text{cl})} \simeq \frac{1}{\beta} \quad \text{and} \quad \beta = \frac{R_1}{R_2}$$

show that

$$f_{H(\text{cl})} \simeq \frac{\beta}{\beta + 1} f_T$$

in step-by-step detail.

11-29. An op amp has an $A_{v(\text{oc})}$ of 100,000 and an f_T of 1 MHz. Find its approximate closed-loop bandwidth if it is to be used as a voltage follower. Find its approximate closed-loop bandwidth if it is to be used as a unity-gain inverting amplifier.

Section 11-5

11-30. An inverting summer [Fig. 11-11(b)] has an R_f of 20 kΩ, an R_1 and R_2 of 5 kΩ, and an R_3 of 2 kΩ. Determine its output voltage if V_1, V_2, and V_3 are all 0.2 V.

Section 11-6

11-31. Given the current amplifier shown in Fig. 11-12(a), assume that R_1 is 1 kΩ, R_2 is 10 kΩ, and R_L is 100 Ω. Find the ideal current gain. What is the gain in decibels?

11-32. Repeat Prob. 11-31 if R_2 is increased to 200 kΩ.

11-33. Distinguish between the terms "sinking" and "sourcing" as they relate to constant-current sources.

Design

11-34. A voltage-to-current converter is to drive a ground-referenced load. The circuit configuration shown in Fig. 11-8 is acceptable. A V_{IN} of -10 V is to produce 25 mA down through the load, and a V_{IN} of -1 V is to produce 16 mA down through the load. Determine the required value of R_F. If the op amp's maximum output current is to be no greater than 10 mA, what is the minimum h_{FE} of transistor Q_1? Select the minimum acceptable value of V_{CC} from the following "standard" voltages: 6, 10, 12, 15, 20, and 25 V.

11-35. A current-to-voltage converter such as that shown in Fig. 11-9(e) is to be designed. When I is 0.3 mA the ouput voltage to be $+10$ V. When I is 0 mA the output voltage is to be 0 V. Standard 5%-tolerance resistors are to be specified. Determine the values of R_2 and R_3. Include a potentiometer (R_4) to trim the circuit gain. Also provide a bias current compensation resistor [such as R_2 in Fig. 11-9(e)] in series with the op amp's noninverting input terminal.

11-36. It is desired to convert a voltage signal that ranges from 0 to 5 V into a current signal that ranges from 4 to 20 mA. Complete the design shown in Fig. 11-13.

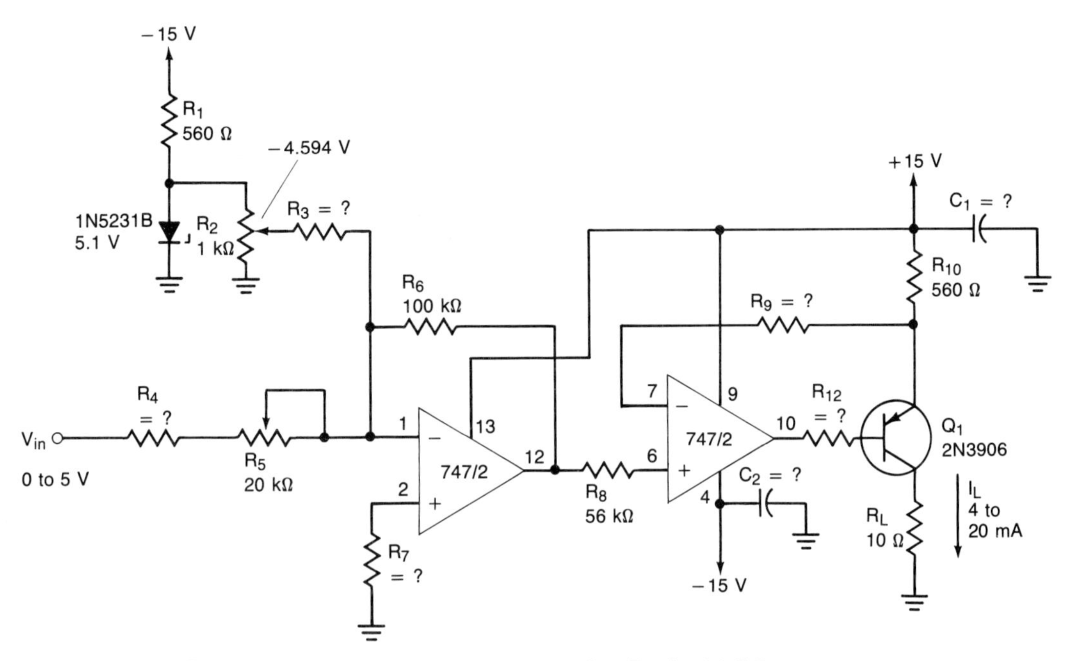

FIGURE 11-13 Voltage-to-current converter for Prob. 11-36.

Troubleshooting

As we saw in Section 11-1, superior performance can be achieved when local and overall negative feedback are used [refer to Fig. 11-1(c)]. Troubleshooting such a system can be difficult. A *system* failure can result from a failure in virtually any *one* component in the system. The overall negative feedback will upset all the stages found in the system. Carefully analyzing our measurements can often lead us to the problem. However, in some cases it may be necessary to break (open) the overall feedback loop.

11-37. Refer to Fig. 11-1(c). The following dc measurements have been taken: V_{IN} is 0 V, the output of AR_1 is +13 V, the voltage across R_1 is −0.129 V, the voltage at the noninverting input of AR_2 is +13 V, the voltage at the output of AR_2 is −13 V, and the voltage across R_5 is −1.18 V. Select the most probable failure(s) from the following list of alternatives: (a) resistor R_1 has opened, (b) op amp AR_1 has failed, (c) op amp AR_2 has failed, and (d) resistor R_5 has opened. Explain your reasons for accepting or rejecting each of the alternatives.

11-38. Refer to Fig. 11-1(c). The following dc measurements have been taken: V_{IN} is 0 V the output of AR_1 is −13 V, the voltage across R_1 is +0.129 V, the voltage at the noninverting input of AR_2 is −13 V, the voltage at the output of AR_2 is +13 V, and the voltage across R_5 is +1.18 V. Select the most probable failure(s) from the following list of alternatives: (a) resistor R_1 has opened, (b) op amp AR_1 has failed, (c) op amp AR_2 has failed, and (d) resistor R_5 has opened. Explain your reasons for accepting or rejecting each of the alternatives.

Computer

11-39. Write a BASIC program that will analyze an op amp inverting amplifier. It should prompt the user for the op amp parameters and the values of the gain-setting resistors. It should provide the ideal values of R_{in}, R_{out}, and $A_{v(cl)}$, and the more exact values. (*Hint*: Use Miller's theorem for R_{inf} and R_{of}.)

11-40. Write a BASIC program that will design an inverting op amp amplifier circuit. Specifically, it should prompt the user for the required $A_{v(cl)}$, R_{inf}, and R_{of}. It should then compute the nearest 5%-tolerance standard resistor values. It should also determine the required value of $A_{v(ol)}$ for the op amp for a less-than-0.1% gain error when the resistors are at their nominal values.

DISCRETE AND INTEGRATED POWER AMPLIFIERS

After Studying Chapter 12, You Should Be Able to:

- Use ac load lines to predict the maximum (undistorted) peak-to-peak ac output voltage of CE and CC BJT amplifiers.
- Define the terms ''slew rate,'' ''TIM,'' and ''power bandwidth.''
- Analyze a class A power amplifier with an op amp driver to predict its $P_{out(ac)}$, $P_{in(dc)}$, and % η.
- Explain the classes (A, AB, B, and C) of operation.
- Analyze a class AB push-pull power amplifier to determine its various waveforms, bias values, $P_{out(ac)}$, $P_{in(dc)}$, and % η.
- Explain the SOA limits of a bipolar power transistor.
- Explain the operation of the Darlington pair and quasi-complementary symmetry.
- Describe the differences between VMOS and DMOS.
- Analyze the V_{BE} multiplier circuit.

- Analyze power amplifiers with inductive loads.
- Analyze overcurrent and open-circuit output protection circuits.
- Analyze thermal circuits and size heat sinks.
- Explain the operation of single-supply IC push-pull power amplifiers.
- Analyze a power amplifier with a common-source push-pull output stage.
- Explain the frequency compensation considerations associated with the common-source push-pull output stage.
- Discuss the relationship between power bandwidth and voltage gain.
- Explain the operation of the constant-current diode.

12-1 Role of the Power Amplifier

We begin our study of a new class of amplifiers in this chapter - the *power amplifier*. The power amplifier is used to boost the signals produced by our electronics to the point where they may drive respectable loads such as loudspeakers, motors, solenoids, and relays. Many audio and industrial control amplifier systems may be represented by the general block diagram in Fig. 12-1.

A *transducer* is used to convert one form of energy into another type. For example, a microphone is used to convert acoustical energy into electrical energy. Conversely, a loudspeaker is used to convert electrical energy into acoustical energy. A motor is a transducer that is used to convert electrical energy into mechanical energy. As we can see in Fig. 12-1, transducers appear at the input and output of the general amplifier system.

The input transducer produces small electrical (typically, voltage) signals that are used to drive the input of the first stage. As we can see, the first stage is a voltage amplifier. Its function is to present a high resistance to the input transducer to minimize loading effects and to increase the amplitude of the voltage signal. Successive voltage amplifier stages are used to provide further increases in the amplitude of the voltage signal. As the signal progresses through the system, its amplitude increases to the point where it can no longer be classified as a "small signal." Specifically, the operating points of the active devices make much larger excursions. Under these conditions, our small-signal models *no longer apply*. This requires that we develop some additional techniques to assist us in our analysis of these *large-signal* amplifiers.

The power amplifier that drives the output transducer demands even more consideration than the small-signal voltage amplifiers that we have focused on thus far. For example, we now have to consider if the output signals will be large enough to produce clipping if the operating point enters saturation and/or cutoff. This will require a graphical analysis called the *ac load line*. Clipping produces severe distortion and must be avoided.

FIGURE 12-1 General amplifier system.

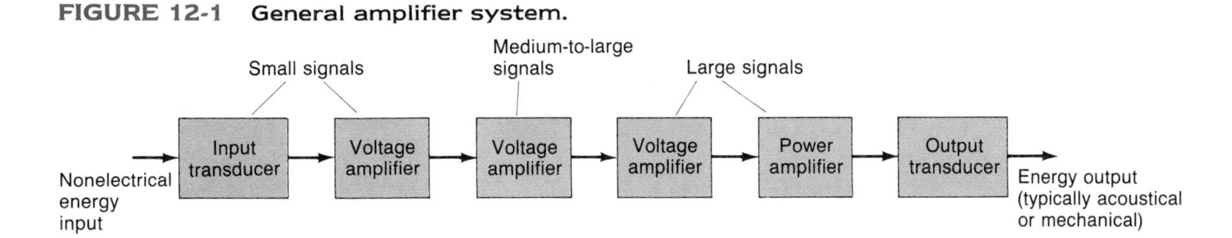

12-2 Common-Emitter AC Load Line

A common-emitter BJT voltage amplifier has been shown in Fig. 12-2. If we were to perform a *small-signal analysis*, a dc analysis would be the required first step. The dc bias values have been given in Fig. 12-2. (The reader should verify the indicated dc bias values.) Knowing that I_C is 0.98 mA and that V_{CE} is 5.49 V, allows us to ascertain the BJT's small-signal ac parameters. We shall assume that the BJT's data sheet specifies an h_{fe} of approximately 100.

$$g_m = \frac{I_C}{26\text{ mV}} = \frac{0.98\text{ mA}}{26\text{ mV}} = 37.7\text{ mS}$$

$$r_\pi = \frac{h_{fe}}{g_m} = \frac{100}{37.7\text{ mS}} = 2.65\text{ k}\Omega$$

and

$$r_o = \frac{V_A}{I_C} = \frac{200\text{ V}}{0.98\text{ mA}} = 204\text{ k}\Omega$$

Assuming that r_o is large enough to be ignored, the small-signal analysis may be continued in the fashion detailed below.

$$A_{v(oc)} = -g_m R_C = -(37.7\text{ mS})(8.2\text{ k}\Omega) = -309$$

$$A_v = \frac{V_{out}}{V_{in}} = -g_m r_C = -g_m(R_C \parallel R_L)$$

$$= -(37.7\text{ mS})(8.2\text{ k}\Omega \parallel 10\text{ k}\Omega) = -170$$

$$R_{in} = r_\pi \parallel R_1 \parallel R_2 = 2.65\text{ k}\Omega \parallel 13\text{ k}\Omega \parallel 2.2\text{ k}\Omega = 1.10\text{ k}\Omega$$

and

$$R_{out} = R_C = 8.2\text{ k}\Omega$$

This concludes the "bare-bones" small-signal analysis of the amplifier. However, suppose that the input signal could become sufficiently large to cause us concern about the possibility of output clipping. In particular, it is possible that our loaded

FIGURE 12-2 **Common-emitter voltage amplifier.**

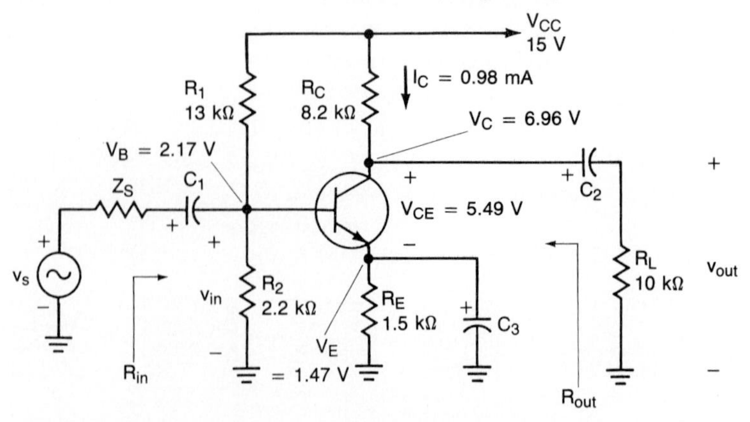

voltage gain of 170 might be large enough to allow the input signal to produce output distortion. This poses a *large-signal* question. Specifically, what is the maximum, undistorted, peak-to-peak output voltage swing? In Chapter 6 we addressed this problem, but *no load was connected across the output of the amplifier* (see Fig. 12-3).

The maximum positive output excursion is limited by V_{CC} (15 V), and the maximum negative-going excursion is limited by the emitter-to-ground voltage V_E (1.47 V). Recall that V_E is held constant by the emitter bypass capacitor C_3. *The minimum peak excursion limits the maximum peak-to-peak output voltage swing.* Therefore, in this example,

$$v_{\text{out(p-p max)}} = 2(5.49 \text{ V}) = 10.98 \text{ V}$$

Any further increases in v_{out} will result in clipping of the negative half-cycle of the output voltage waveform. In Fig. 12-3 the maximum input signal is *approximately*

$$v_{\text{in(p-p max)}} \simeq \frac{v_{\text{out(p-p max)}}}{|A_{v(\text{oc})}|} = \frac{10.98 \text{ V p-p}}{309} = 35.5 \text{ mV p-p}$$

FIGURE 12-3 Maximum peak-to-peak output voltage of the unloaded common-emitter amplifier.

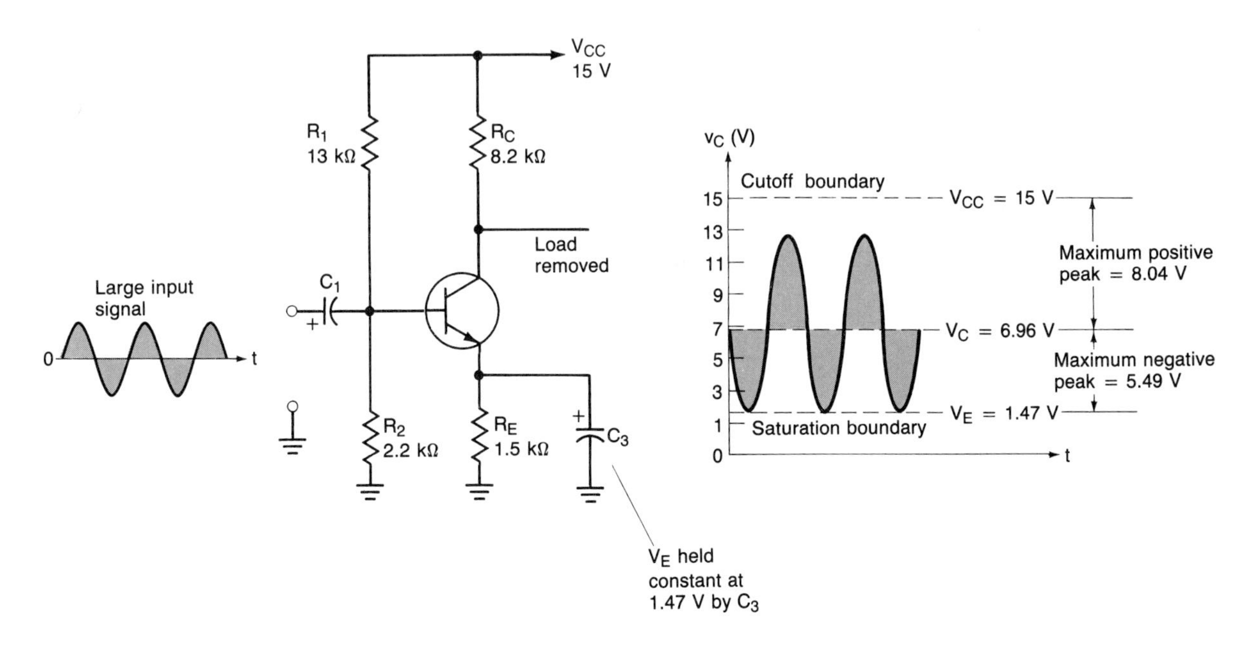

The calculation above applies when R_L is *not* connected. When the 10-kΩ load resistor *is* connected, the voltage gain will be reduced (from −309 to −170) and the maximum peak-to-peak output voltage swing will *also be reduced*. The easiest way to predict what the reduced voltage swing will be, and to visualize why this is so, is to invoke a graphical analysis technique that is called the *ac load line*. Before we delve into this method, let us first review the dc load-line procedure.

Recall that the dc load line is a graphical technique for performing a dc analysis. Let us plot the dc load line for the BJT amplifier given in Fig. 12-3. First we must find $V_{CE(\text{OFF})}$ and $I_{C(\text{SAT})}$.

$$V_{CE(\text{OFF})} = V_{CC} = 15 \text{ V}$$

$$I_{C(\text{SAT})} = \frac{V_{CC}}{R_C + R_E} = \frac{15 \text{ V}}{8.2 \text{ k}\Omega + 1.5 \text{ k}\Omega} = 1.55 \text{ mA}$$

These two points are then plotted upon the BJT's output characteristic curves. The base current must then be determined to locate the Q-point (see Fig. 12-4). The actual BJT curve has been omitted for clarity.

When the BJT's operating point is caused to vary by an input signal, ac voltages and currents are produced. By inspection of Fig. 12-3, we recall that the emitter bypass capacitor serves to effectively short the emitter to ground. Therefore, the emitter resistor does *not* appear in the (midband) ac equivalent circuit. Consequently, it will *not* serve to limit the peak ac current $i_{c(\text{sat})}$ (refer to Fig. 12-5).

As we can see in Fig. 12-4, if the dc operating point moves up, I_C increases and V_{CE} decreases. *The collector-to-emitter voltage must decrease to zero in order for the collector current to increase to* $I_{C(\text{SAT})}$. Exactly the *same* situation exists in

FIGURE 12-4 Dc load line.

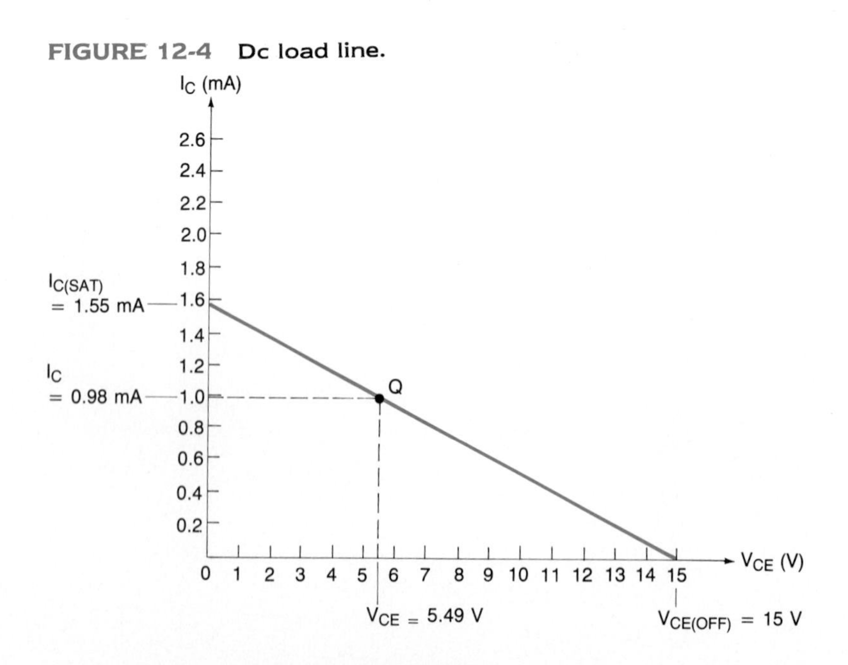

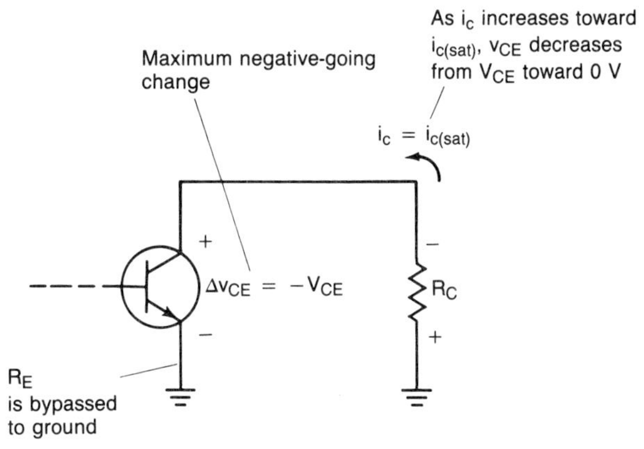

FIGURE 12-5 **Output ac equivalent circuit of Fig. 12-3.**

the case of the ac load line. Therefore, the maximum negative-going change in the collector-to-emitter voltage is given by

$$\Delta v_{CE} = 0 - V_{CE} = -V_{CE}$$

or the magnitude of the negative peak voltage $v_{ce(\text{neg. pk})}$ is

$$|v_{ce(\text{neg. pk})}| = V_{CE}$$

(see Fig. 12-5).

Consequently, the peak ac current $i_{c(\text{sat})}$ for the amplifier shown in Fig. 12-3 is given by

$$i_{c(\text{sat})} = \frac{|v_{ce(\text{neg. peak})}|}{R_C} = \frac{V_{CE}}{R_C}$$

$$i_{c(\text{sat})} = \frac{V_{CE}}{R_C} \tag{12-1}$$

For our example (Fig. 12-3),

$$i_{c(\text{sat})} = \frac{V_{CE}}{R_C} = \frac{5.49\text{ V}}{8.2\text{ k}\Omega} = 0.670\text{ mA}$$

This is the *peak ac current*. However, the *total instantaneous peak collector current* $i_{C(\text{SAT})}$ is the sum of the peak ac collector current, and the dc collector current I_C. Hence, for Fig. 12-3,

$$i_{C(\text{SAT})} = I_C + \frac{V_{CE}}{R_C} \tag{12-2}$$

and in our example $i_{C(\text{SAT})}$ is

$$i_{C(\text{SAT})} = I_C + \frac{V_{CE}}{R_C} = 0.98\text{ mA} + \frac{5.49\text{ V}}{8.2\text{ k}\Omega} = 1.65\text{ mA}$$

This result forms one of the endpoints for our ac load line (see Fig. 12-6).

In a similar fashion, the *increase* in the ac collector-to-emitter (output) voltage must be accompanied by a *decrease* in the collector current. Specifically, the change in the collector current from I_C to zero will develop a positive peak voltage across

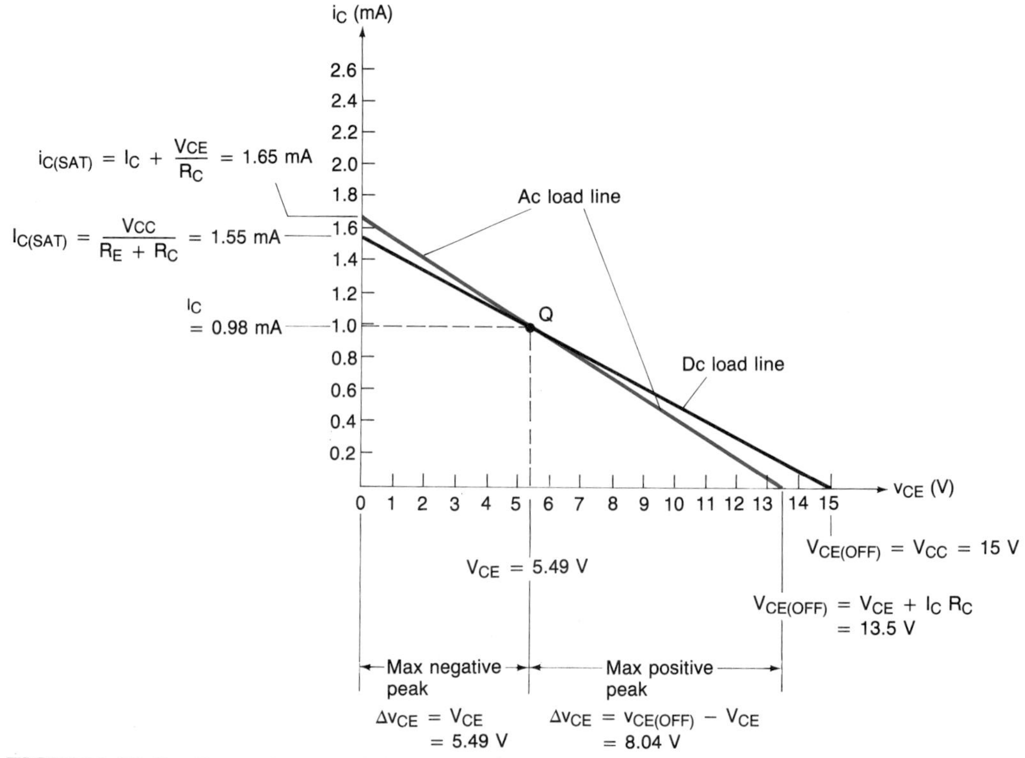

FIGURE 12-6 Dc and ac load lines for the amplifier given in Fig. 12-3.

R_C. This positive peak voltage corresponds to the ac collector-to-emitter cutoff voltage $v_{ce(\text{off})}$. Hence

$$v_{ce(\text{pos. pk})} = v_{ce(\text{off})} = \Delta i_c R_C = (I_C - 0)R_C$$
$$= I_C R_C$$

Again, for our example (Fig. 12-3),

$$v_{ce(\text{off})} = I_C R_C = (0.98 \text{ mA})(8.2 \text{ k}\Omega) = 8.04 \text{ V}$$

The total instantaneous collector-to-emitter voltage v_{CE} must increase beyond V_{CE}. Therefore, $v_{CE(\text{OFF})}$ is given by

$$v_{CE(\text{OFF})} = V_{CE} + I_C R_C \tag{12-3}$$

For our example,

$$v_{CE(\text{OFF})} = V_{CE} + I_C R_C = 5.49 \text{ V} + (0.98 \text{ mA})(8.2 \text{ k}\Omega)$$
$$= 13.5 \text{ V}$$

This forms the second endpoint for our ac load line. It has been indicated in Fig. 12-6.

Inspection of Fig. 12-6 demonstrates that the ac load line allows us to reach the same conclusion that was arrived at more intuitively in Fig. 12-3. In both cases, the maximum negative peak voltage is 5.49 V, and the maximum positive peak voltage is 8.04 V. The power behind the ac load line rests in the fact that it can be readily extended to the less obvious situations which arise when we connect R_L across the amplifier's output. If we allow r_L to represent the equivalent ac load across the amplifier, we can extend Eq. 12-2 to represent the general situation (see Eq. 12-4). By inspection of Fig. 12-7, we can see that the change in the collector-to-emitter voltage forms a peak collector current $i_{c(\text{sat})}$ which is limited by r_L. In this case *r_L is formed by the parallel combination of R_C and R_L.*

$$i_{C(\text{SAT})} = I_C + \frac{V_{CE}}{r_L} \tag{12-4}$$

Similarly, the use of r_L allows us to rewrite Eq. 12-3 as indicated in

$$v_{CE(\text{OFF})} = V_{CE} + I_C r_L \tag{12-5}$$

Now let us consider the effects of R_L when it is connected across the output of our amplifier (refer to Fig. 12-2). As we can see in Fig. 12-7, r_L is

$$r_L = R_C \parallel R_L = 8.2\ \text{k}\Omega \parallel 10\ \text{k}\Omega = 4.51\ \text{k}\Omega$$

From Eqs. 12-4 and 12-5, we can find $i_{C(\text{SAT})}$ and $v_{CE(\text{OFF})}$ for our example.

$$i_{C(\text{SAT})} = I_C + \frac{V_{CE}}{r_L} = 0.98\ \text{mA} + \frac{5.49\ \text{V}}{4.51\ \text{k}\Omega} = 2.20\ \text{mA}$$

and

$$v_{CE(\text{OFF})} = V_{CE} + I_C r_L = 5.49\ \text{V} + (0.98\ \text{mA})(4.51\ \text{k}\Omega) = 9.91\ \text{V}$$

These two points have been used to draw the ac load line as illustrated in Fig. 12-8.

By inspection of Fig. 12-8, we can see that the maximum positive peak output signal is 4.42 V. Therefore, the maximum undistorted output voltage is 8.84 V peak to peak. Let us contrast this result. When R_L is not connected, the maximum peak-to-peak output voltage is 10.98 V, but when R_L is connected, the maximum peak-to-peak output swing is reduced to 8.84 V.

In general, the maximum positive peak output voltage is given

$$v_{\text{out(pos. peak max)}} = I_C r_L \tag{12-6}$$

and the maximum negative peak output signal is

$$v_{\text{out(neg. peak max)}} = V_{CE} \tag{12-7}$$

FIGURE 12-7 **Output ac equivalent circuit for the amplifer given in Fig. 12-2.**

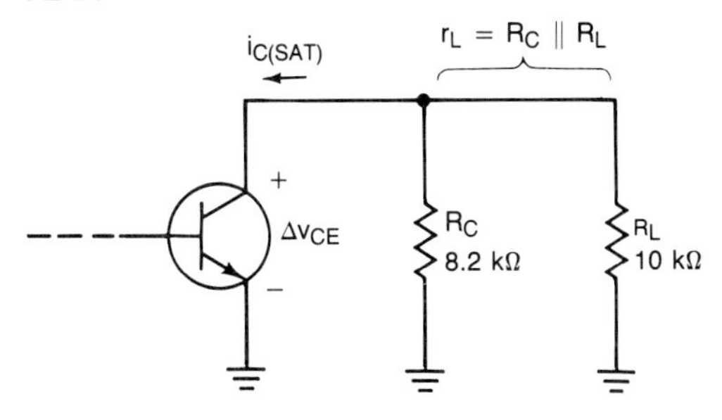

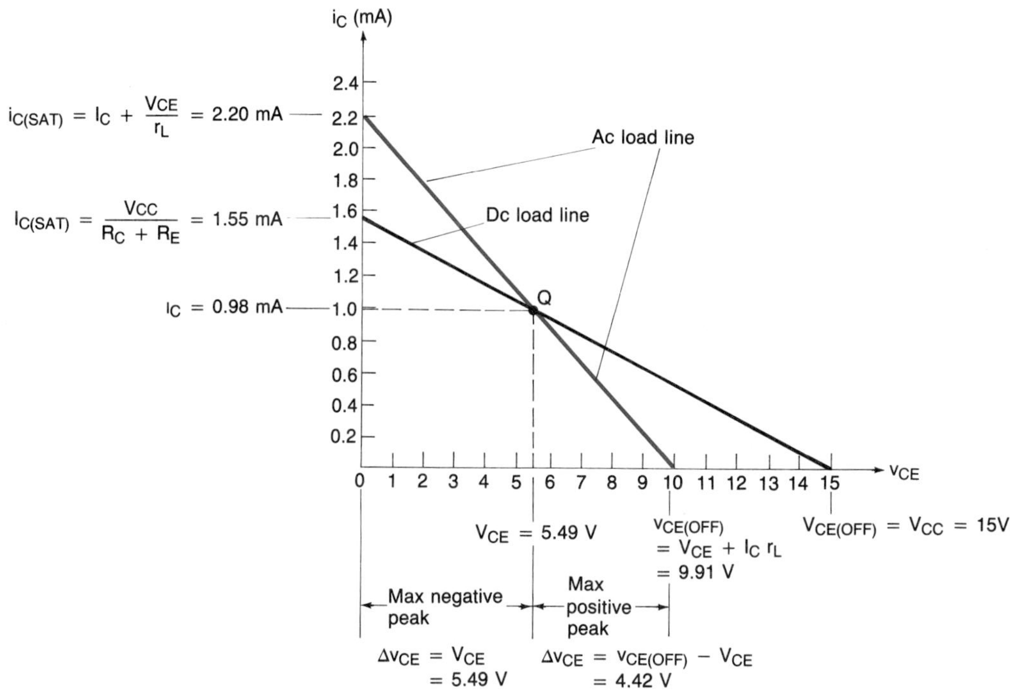

FIGURE 12-8 Dc and ac load lines for the loaded amplifier given in Fig. 12-2.

The maximum peak-to-peak output signal is given by two times the smaller of the two. A summary has been given by

$$v_{\text{out(p-p max)}} = 2\,I_C r_L \Big|_{I_C r_L \leq V_{CE}} \tag{12-8}$$

or

$$v_{\text{out(p-p max)}} = 2V_{CE} \Big|_{V_{CE} \leq I_C r_L} \tag{12-9}$$

12-3 Common-Collector AC Load Line

Our study of the ac load line in Section 12-2 may readily be extended to the common-collector or emitter-follower BJT stage. However, before we begin our study of the ac load line for the emitter follower, let us digress slightly to understand the role of the emitter follower in power amplifiers.

In Fig. 12-1 we saw the block diagram of the general amplifier system. The general strategy in most amplifier system designs is to use voltage amplifiers to increase the voltage swing of the input transducer. Quite often, the role of the power amplifier is to increase the amplitude of the attendant *current swing*. In other words, the voltage gain of the power amplifier stage is typically small. However, its *current gain* is often quite large. These are the attributes possessed by the common-collector (or the common-drain) amplifier. Therefore, emitter followers (or source followers) are quite often used in power amplifiers.

Another benefit offered by the emitter follower (and the source follower) is a low output resistance. Recall that maximum power transfer from a source to a load occurs when the internal resistance of the source matches the resistance of the load. In the case of the 4-, 8-, or 16-Ω loudspeaker, the emitter or source follower would seem to be the obvious choice. We should point out that a transformer could be used to match the high output impedance of a common-emitter stage to a low load resistance. However, transformers are bulky, expensive, and have a relatively poor frequency response. Consequently, most modern audio power amplifier designs have phased out the use of transformers.

A common-collector amplifier has been illustrated in Fig. 12-9. We shall perform a dc analysis, a small-signal ac analysis, draw the dc load line, and then draw the ac load line.

The dc analysis of this circuit is very similar to that for the common-emitter amplifier. Thus

FIGURE 12-9 Analyzing the emitter follower: (a) circuit; (b) circuit used to determine $A_{v(oc)}$ in Chapter 8; (c) shortcut to determine the loaded voltage gain A_v.

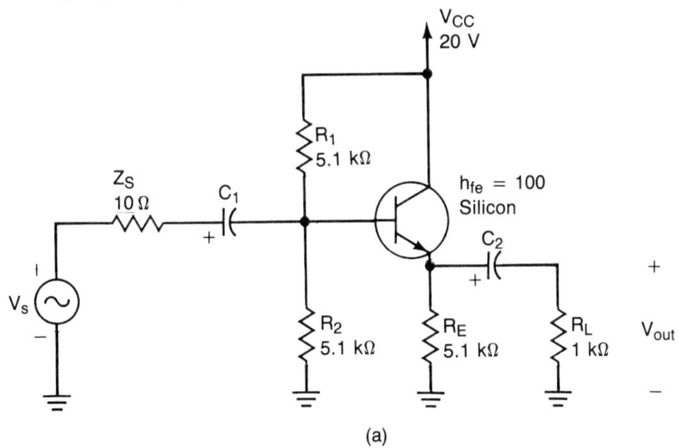

$$A_{v(oc)} = \frac{V_{out}}{V_{in}} = \frac{g_m R_E}{1 + g_m R_E}$$

V_{in} R_E V_{out} R_L (disconnected)

(b)

$$A_v = \frac{V_{out}}{V_{in}} = \frac{g_m r_L}{1 + g_m r_L}$$

V_{in} $r_L = R_E \parallel R_L$ V_{out}

(c)

$$V_B = \frac{R_2}{R_1 + R_2} V_{CC} = \frac{5.1\ \text{k}\Omega}{5.1\ \text{k}\Omega + 5.1\ \text{k}\Omega}(20\ \text{V}) = 10\ \text{V}$$

$$V_E = V_B - 0.7\ \text{V} = 10\ \text{V} - 0.7\ \text{V} = 9.3\ \text{V}$$

$$I_C \simeq I_E = \frac{V_E}{R_E} = \frac{9.3\ \text{V}}{5.1\ \text{k}\Omega} = 1.82\ \text{mA}$$

$$V_{CE} = V_{CC} - V_E = 20\ \text{V} - 9.3\ \text{V} = 10.7\ \text{V}$$

By inspection, we note that the collector-to-ground voltage V_C is equal to V_{CC} (20 V).

The parameters for the BJT have been computed below. Note in Fig. 12-9(a) that the BJT's h_{fe} is 100.

$$g_m = \frac{I_C}{26\ \text{mV}} = \frac{1.82\ \text{mA}}{26\ \text{mV}} = 70.0\ \text{mS}$$

$$r_\pi = \frac{h_{fe}}{g_m} = \frac{100}{70.0\ \text{mS}} = 1.43\ \text{k}\Omega$$

and

$$r_o = \frac{V_A}{I_C} = \frac{200\ \text{V}}{1.82\ \text{mA}} = 110\ \text{k}\Omega$$

Once again, we shall assume that an r_o of 110 kΩ is large enough to be ignored. (To appreciate fully our small-signal ac analysis of the emitter follower below, the reader might find it beneficial to review Section 8-10.)

The input resistance for the emitter follower may be computed by employing Eq. 8-48.

$$\begin{aligned} R_{\text{in(base)}} &\simeq g_m(R_E \parallel R_L)r_\pi \\ &= (70\ \text{mS})(5.1\ \text{k}\Omega \parallel 1\ \text{k}\Omega)(1.43\ \text{k}\Omega) = 83.7\ \text{k}\Omega \\ R_{\text{in}} &= R_1 \parallel R_2 \parallel R_{\text{in(base)}} = 5.1\ \text{k}\Omega \parallel 5.1\ \text{k}\Omega \parallel 83.7\ \text{k}\Omega \\ &= 2.47\ \text{k}\Omega \end{aligned}$$

To determine the voltage gain, we must employ Eq. 8-53. We shall calculate the *loaded* voltage gain merely by substituting the ac equivalent load resistance r_L (the parallel combination of R_E and R_L) in for R_E. The rational for this substitution has been developed in Fig. 12-9(b) and (c). Hence

$$\begin{aligned} A_v &= \frac{V_{\text{out}}}{V_{\text{in}}} = \frac{g_m r_L}{1 + g_m r_L} = \frac{g_m(R_E \parallel R_L)}{1 + g_m(R_E \parallel R_L)} \\ &= \frac{(70\ \text{mS})(5.1\ \text{k}\Omega \parallel 1\ \text{k}\Omega)}{1 + (70\ \text{mS})\ (5.1\ \text{k}\Omega \parallel 1\ \text{k}\Omega)} = 0.983 \end{aligned}$$

To determine the output resistance, we draw on Eq. 8-57. First, we must find the Thévenin equivalent resistance "seen" by the base terminal of the BJT. Thus

$$\begin{aligned} R_{\text{TH}} &= Z_S \parallel R_1 \parallel R_2 = 10\ \Omega \parallel 5.1\ \text{k}\Omega \parallel 5.1\ \text{k}\Omega \\ &\simeq 10\ \Omega \end{aligned}$$

Now we may compute R_{out}.

$$\begin{aligned} R_{\text{out(emitter)}} &= \frac{r_\pi + R_{\text{TH}}}{1 + g_m r_\pi} \\ &= \frac{1.43\ \text{k}\Omega + 10\ \Omega}{1 + (70\ \text{mS})(1.43\ \text{k}\Omega)} = 14.2\ \Omega \simeq R_{\text{out}} \end{aligned}$$

The dc load line may be plotted once we find $I_{C(SAT)}$ and $V_{CE(OFF)}$.

$$I_{C(SAT)} = \frac{V_{CC}}{R_E} = \frac{20\text{ V}}{5.1\text{ k}\Omega} = 3.92\text{ mA}$$

$$V_{CE(OFF)} = V_{CC} = 20\text{ V}$$

The dc load line has been depicted in Fig. 12-10(a). Note that the Q-point found above has been indicated on it.

FIGURE 12-10 Emitter-follower large-signal analysis: (a) load lines; (b) ac equivalent circuit; (c) interpreting where output clipping will occur.

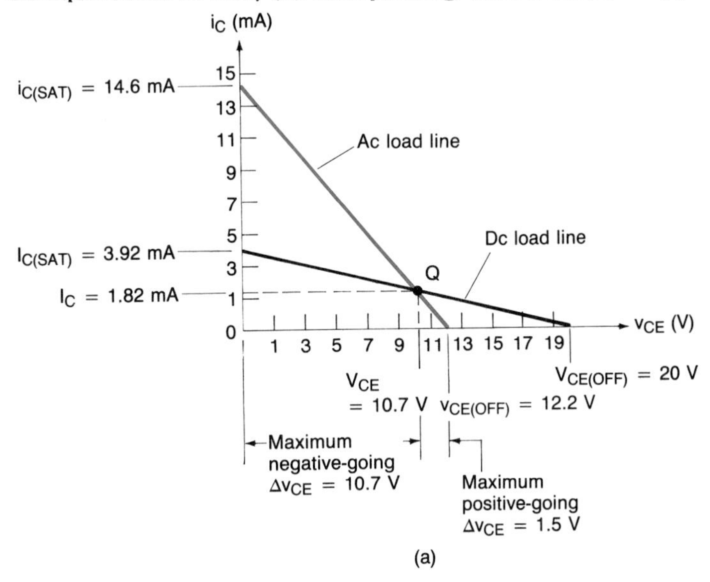

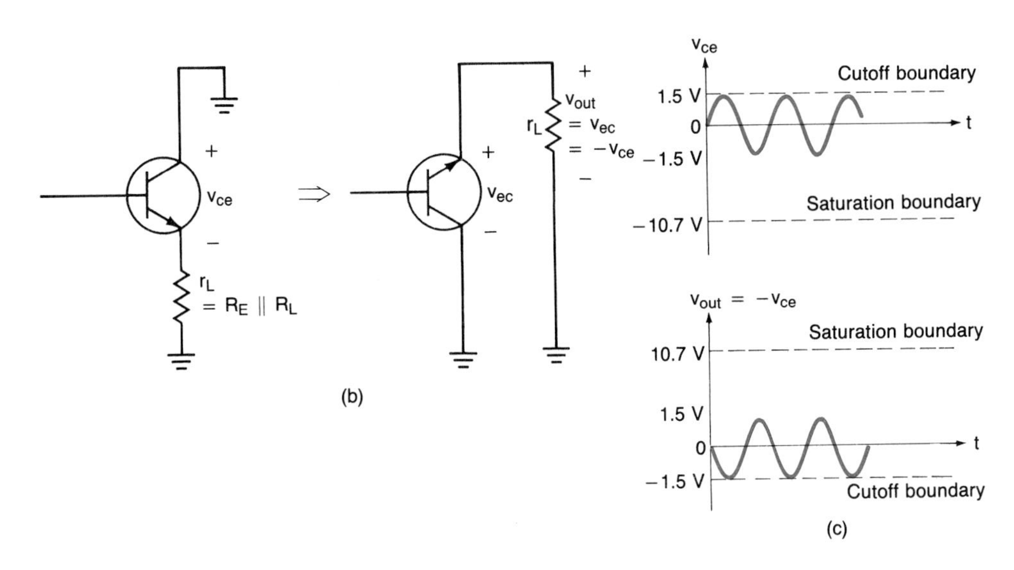

To find $i_{C(SAT)}$ and $v_{CE(OFF)}$, we may use Eqs. 12-4 and 12-5, respectively. We again note that r_L is the parallel combination of R_E and R_L [refer to Fig. 12-10(b)].

$$r_L = R_E \parallel R_L = 5.1 \text{ k}\Omega \parallel 1 \text{ k}\Omega = 836 \ \Omega$$

$$i_{C(SAT)} = I_C + \frac{V_{CE}}{r_L} = 1.82 \text{ mA} + \frac{10.7 \text{ V}}{836 \ \Omega} = 14.6 \text{ mA}$$

$$v_{CE(OFF)} = V_{CE} + I_C r_L = 10.7 \text{ V} + (1.82 \text{ mA})(836 \ \Omega) = 12.2 \text{ V}$$

The corresponding ac load line has been illustrated in Fig. 12-10(a).

As can be seen in Fig. 12-10(a), the collector-to-emitter voltage has a maximum positive-going peak of only 1.5 V while the negative-going peak can reach 10.7 V before saturation occurs. These are statements of the changes in the collector-to-emitter voltage, *not* the output voltage. The changes in the collector-to-emitter voltage will produce a corresponding change in the output (load) voltage. However, the *sense* of the change is reversed. As v_{CE} *decreases*, the voltage across the load *increases*. Conversely, as v_{CE} *increases*, the voltage across the load *decreases*. With a little thought, we can see that the ac load line in Fig. 12-10(a) supports this. It has also been reflected in the ac equivalent circuit given in Fig. 12-10(b). Therefore,

$$v_{out} = v_{ec} = -v_{ce}$$

Clipping of the *positive* peak of the output voltage will occur if it attempts to exceed 10.7 V. Clipping of the negative peak of the output voltage will occur at 1.5 V, and this is the limitation [see Fig. 12-10(c)].

EXAMPLE 12-1

Find the maximum peak-to-peak output voltage of the emitter follower given in Fig. 12-11(a).

SOLUTION We observe that emitter bias has been used. First, we must perform a dc analysis (refer to Section 6-13). The small base current produces a negligibly small voltage drop across R_B. The base-to-emitter voltage drop is also small compared to V_{EE}. We can simplify our efforts by using approximations.

$$V_B \simeq 0 \text{ V}$$

$$V_E \simeq 0 \text{ V}$$

$$I_C \simeq I_E \simeq \frac{V_{EE}}{R_E} = \frac{15 \text{ V}}{15 \text{ k}\Omega} = 1 \text{ mA}$$

$$V_{CE} \simeq V_{CC} = 15 \text{ V}$$

The dc load line may be drawn once we have determined $I_{C(SAT)}$ and $V_{CE(OFF)}$. Hence

$$I_{C(SAT)} = \frac{V_{CC} + V_{EE}}{R_E} = \frac{15 \text{ V} + 15 \text{ V}}{15 \text{ k}\Omega} = 2 \text{ mA}$$

$$V_{CE(OFF)} = V_{CC} + V_{EE} = 15 \text{ V} + 15 \text{ V} = 30 \text{ V}$$

Now we may develop the ac load line.

$$r_L = R_E \parallel R_L = 15 \text{ k}\Omega \parallel 10 \text{ k}\Omega = 6 \text{ k}\Omega$$

$$i_{C(SAT)} = I_C + \frac{V_{CE}}{r_L} = 1 \text{ mA} + \frac{15 \text{ V}}{6 \text{ k}\Omega} = 3.50 \text{ mA}$$

$$v_{CE(OFF)} = V_{CE} + I_C r_L = 15 \text{ V} + (1 \text{ mA})(6 \text{ k}\Omega) = 21 \text{ V}$$

The load lines have been given in Fig. 12-11(b). As we can see, the maximum positive-going output voltage is limited to less than 15 V, and the maximum negative-going output voltage is restricted to less than 6 V. Therefore, we can state that output clipping will not occur as long as the output voltage is less than 12 V peak to peak. ■

12-4 Slew Rate, TIM, and Power Bandwidth

In Section 12-3 we saw how to predict the clipping that will result in the distortion of a large output signal. The ac load line helps us anticipate, and hopefully avoid, the occurrence of clipping.

The ac load line for the large-signal stages found in a power amplifier must be adjusted to avoid clipping when the load is connected.

FIGURE 12-11 Emitter follower for Example 12-1: (a) circuit; (b) load lines.

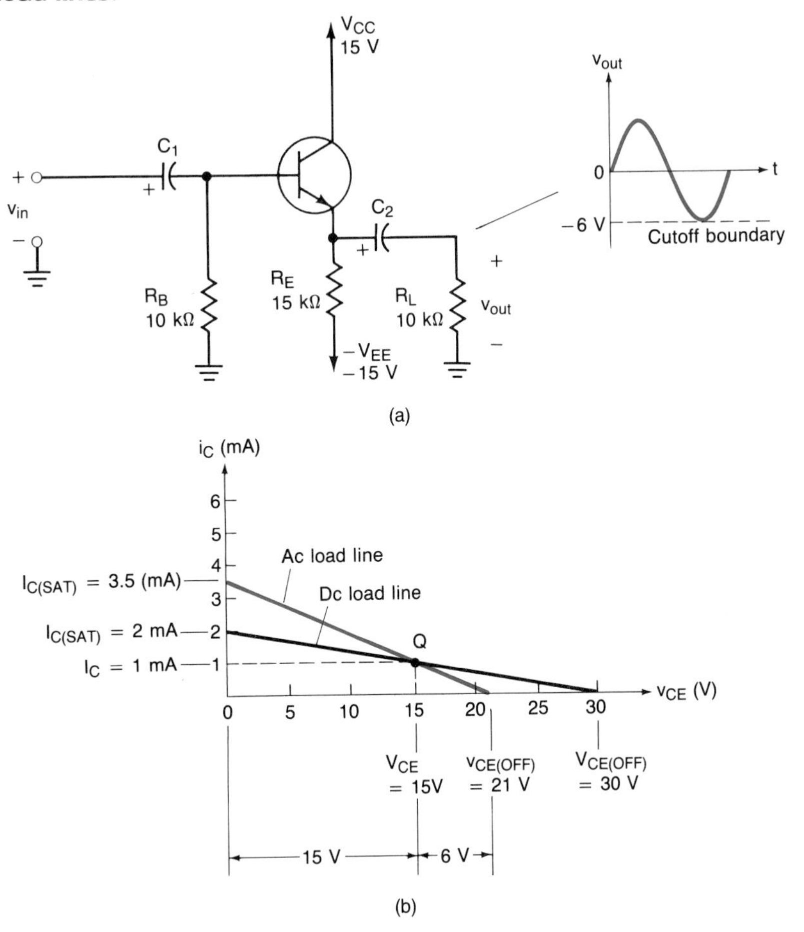

The larger the signal swing, the more severe the nonlinearity in the gain of a power amplifier system will be. This gives rise to total harmonic distortion (THD). Very simply, for a pure sinusoidal input signal, harmonics of it will appear at the output. The problem becomes even more acute when two or more sinusoidal signals are simultaneously presented at the input. The output will not only contain harmonics, but sum and difference frequencies as well.

For example, if 3 kHz and 10 kHz appear at the input, the output will contain the fundamental frequencies (3 kHz and 10 kHz), harmonics (6 kHz, 9 kHz, 12 kHz, . . . , and 20 kHz, 30 kHz, 40 kHz, . . .), and the sum and difference components of the fundamentals and the harmonics. In this example we would have 7 kHz and 13 kHz produced by the fundamentals, 14 kHz and 26 kHz produced by the second harmonics, 21 kHz and 39 kHz produced by the third harmonics, and so on. The output will literally be "splattered" with unwanted frequencies. This is termed *intermodulation distortion* (IM). (Most listeners can tolerate some THD, but find IM very annoying.) Both THD and IM can be minimized by using negative feedback. Therefore, negative feedback is an integral part of all "real" power amplifiers.

Negative feedback (and a large loop gain) is used to minimize THD and IM in a power amplifier.

Less obvious, but equally important, is the *slew-rate* limitation of an amplifier. The slew rate becomes very important when an amplifier is expected to operate on large signals at relatively high frequencies. As we shall see, the slew-rate limitation of a large-signal amplifier is another source of potentially severe distortion.

The maximum rate of change in the output voltage of an amplifier system is called its *slew rate* (SR) [see Fig. 12-12(a)]. Capacitances (which by definition oppose a change in voltage) in an amplifier serve to limit its slew rate. In the case of operational amplifiers and power amplifiers, the frequency-compensation capacitors serve as the primary source of slew-rate limiting.

To understand the reasons why the change in the output voltage of an amplifier may be relatively "sluggish," let us first consider the IC op amp. Our simplified op amp equivalent circuit has been given in Fig. 12-12(b).

In Section 9-3 we saw that the maximum current out of the current mirror (Q_3 and Q_4) is equal to the tail current. Recall that the 741 op amp has a tail current I_T of 20 μA. Therefore, the current mirror has a maximum total instantaneous output of $\pm I_T$ or ± 20 μA in this instance [see Fig. 12-12(b)].

The output of the current mirror is directed into the input of the inverting current-to-voltage converter stage. The frequency-compensation capacitor C_c limits how rapidly the output voltage can change. Recall that for any capacitor,

$$i_c = C \frac{dv_C}{dt}$$

which means that the effective current "through" a capacitor is proportional to the rate of change in the voltage across it. In Fig. 12-12(b),

$$i'_{\text{out(max)}} = \pm I_T = \pm C_c \frac{dv_c}{dt} = \mp C_c \frac{dv_{\text{out}}}{dt}$$

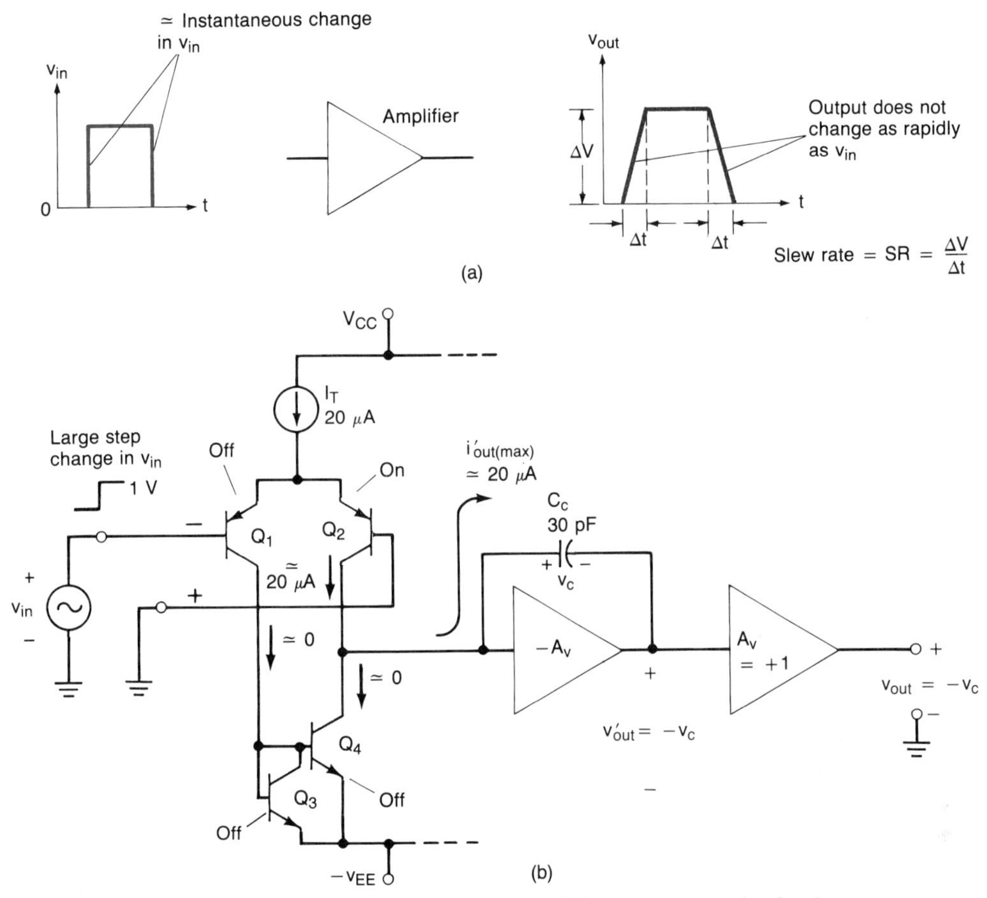

FIGURE 12-12 Slew rate: (a) its meaning; (b) op amp equivalent circuit.

If we solve for the rate of change in the voltage with respect to time, the slew rate is given by

$$\begin{aligned} \text{SR} &= \frac{dv_{\text{out}}}{dt} = \frac{\pm I_T}{C_c} = \frac{\pm 20\ \mu\text{A}}{30\ \text{pF}} = \pm 667\ \text{kV/s} \\ &= \pm 0.67\ \text{V}/\mu\text{s} \end{aligned}$$

Typically, the slew rate is expressed in volts per microsecond as demonstrated above.

To reach the slew-rate limitation, the input transistors (Q_1 and Q_2) must be fully switched. Generally, the differential input voltage must be greater than ±120 mV for a bipolar op amp (which contains only bipolar transistors) and greater than ±1 to ±3 V for an FET-input op amp.

Obviously, we want our op amps (and our power amps) to have as large a slew rate as possible. One way to achieve this is to *increase* the input stage biasing current I_T and to *decrease* the size of C_c.

$$SR = \frac{i_{T(\max)}}{C_c}$$

(increase $i_{T(\max)}$ to raise SR; decrease C_c to raise SR)

However, both of these actions will tend to *raise* the unity-gain frequency f_T of the op amp. Recall that

$$A_v = \frac{g_m}{j\omega C_c}$$

Since the magnitude of an op amp's voltage gain reduces to unity at f_T,

$$|A_v| = \frac{g_m}{2\pi f_T C_c} = 1 \rightarrow f_T = \frac{g_m}{2\pi C_c}$$

and for any bipolar transistor,

$$g_m = \frac{I_C}{26 \text{ mV}}$$

Therefore, increasing the tail current will raise I_C, which, in turn, will increase g_m. An increase in g_m and/or a decrease in C_c will increase f_T.

$$f_T = \frac{g_m}{2\pi C_c}$$

(g_m: increasing I_T to improve the slew rate raises g_m, which also increases f_T)

(C_c: decreasing C_c to improve the slew rate also increases f_T)

The disadvantage of increasing f_T is simply an increase in the possibility of oscillation at low closed-loop voltage gains. Ideally, we would like to increase the first stage biasing current to improve the slew rate without increasing g_m. The obvious solution is to employ active devices which inherently offer a low value of g_m. From our earlier work we recall that JFETs demonstrate this characteristic. Consequently, *op amps that employ JFET inputs* (Fig. 12-13) *have been developed to provide much larger slew rates than those available from standard bipolar op amps.*

The slew-rate limitation poses some serious closed-loop ramifications. If the input signal changes rapidly and the output does not change quickly enough, the amplifier's negative feedback signal will also not change quickly enough. Consequently, during rapid input changes, the amplifier effectively operates in the *open-loop mode*.

In the case of power amplifiers, the slew rate is an extremely important parameter. If the slew rate is too low, slew-induced distortion can occur. This type of distortion is often called *transient intermodulation* (TIM) *distortion* (see Fig. 12-14).

The TIM occurs when the input changes faster than the output can follow. As a result, the negative feedback signal cannot change quickly enough. This condition indicates that the amplifier is essentially operating in its open-loop mode, and that the feedback signal is *trying* to correct the output waveform. Very large overshoots and undershoots can occur under these conditions.

A "good" high-fidelity 100 watts per channel stereo power amplifier driving an 8-Ω load might have a slew rate on the order of 50 V/μs to eliminate TIM. In general, the higher the output power levels, the greater the slew

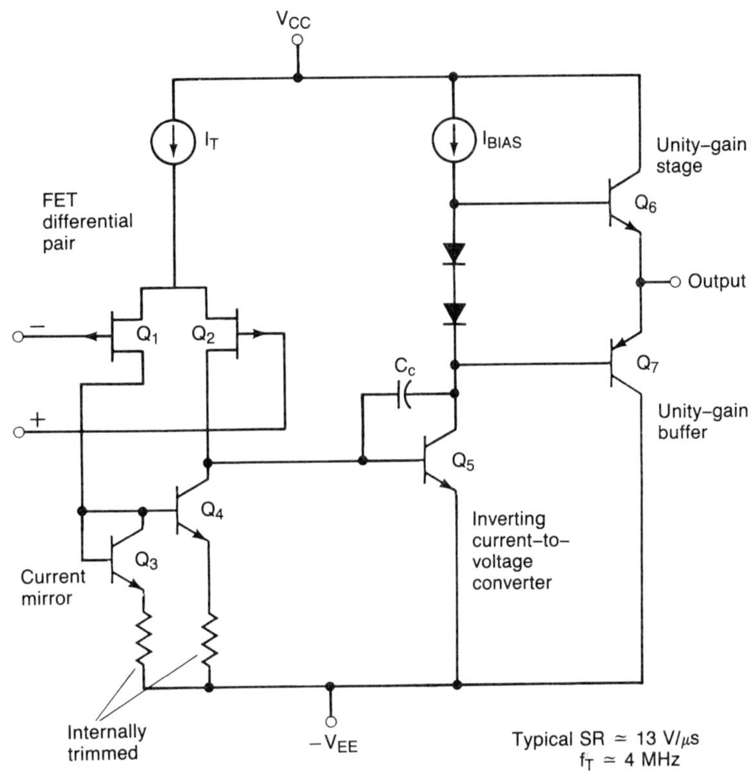

FIGURE 12-13 A bipolar-FET (BiFET) op amp typically offers very good slew rates.

rate should be. The optimum specification for power amplifiers should be normalized to the output levels. This requires that the slew rate for power amplifiers be specified in units of volts per microsecond per volt of output. In this case we could state that the slew rate should be from 0.5 to 1 V/μs per peak output volt for a high-performance power amplifier.

The slew rate is often referenced to a strictly sinusoidal output by means of another parameter called the *power bandwidth*. For a sinusoidal output voltage, we can state that

$$v_{out} = V_m \sin \omega t$$

When large sinusoidal *steady-state* signals are involved, the slew-rate limitation of an amplifier may produce distortion in the waveform. Note that we are concerning ourselves with possible distortion in steady-state signals in this case, *not* transient (TIM) problems. To determine when this will occur, we need to find the maximum rate of change in a sinusoidal output voltage. This is accomplished mathematically by finding the derivative of the sinusoidal output voltage.

$$\frac{dv_{out}}{dt} = \frac{d}{dt}(V_m \sin \omega t) = \omega V_m \cos \omega t$$

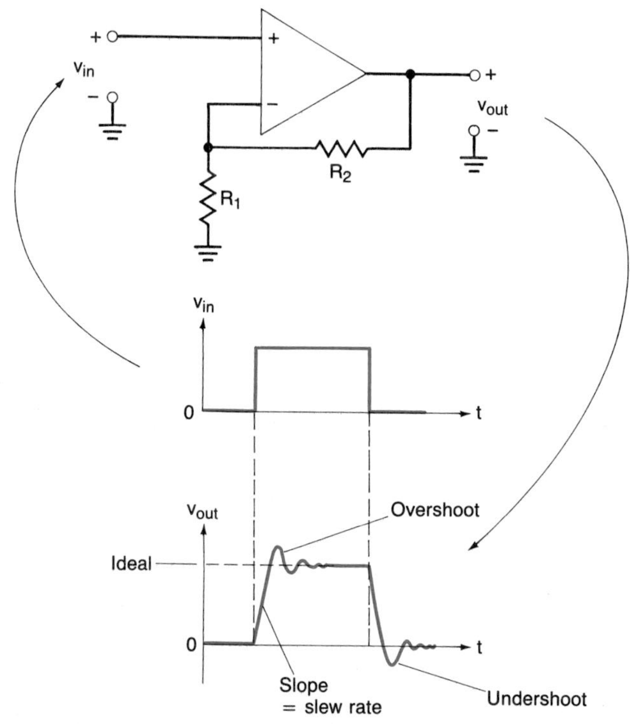

FIGURE 12-14 **Transient intermodulation (TIM) distortion.**

The maximum rate of change in a sinusoidal waveform occurs at its zero crossings [see Fig. 12-15(a)]. If we let $t = 0$,

$$\frac{dv_{out}}{dt} = \omega V_m \cos \omega(0) = \omega V_m(1) = \omega V_m$$

To avoid slew-rate distortion in the output voltage waveform, the maximum rate of change in the output must be *less* than the slew rate of the amplifier. Since the maximum rate of change in the sinusoidal output voltage waveform is directly proportional to its peak value V_m and frequency f_{max}, we may develop Eq. 12-10.

$$\left.\frac{dv_{out}}{dt}\right|_{max} = \omega_{max} V_m \leq \text{SR}$$

$$\omega_{max} \leq \frac{\text{SR}}{V_m} \rightarrow f_{max} \leq \frac{\text{SR}}{2\pi V_m}$$

$$\boxed{f_{max} = \frac{\text{SR}}{2\pi V_m}} \tag{12-10}$$

where f_{max} = power bandwidth
SR = slew rate
V_m = peak output voltage

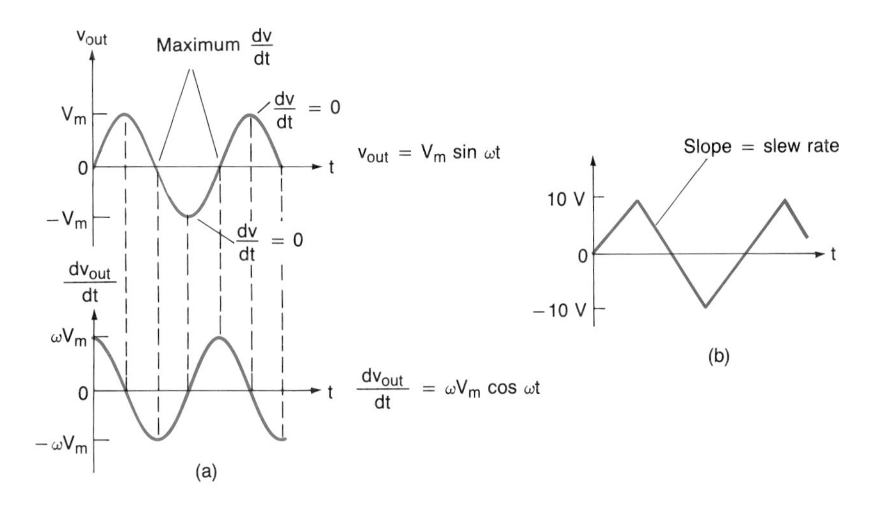

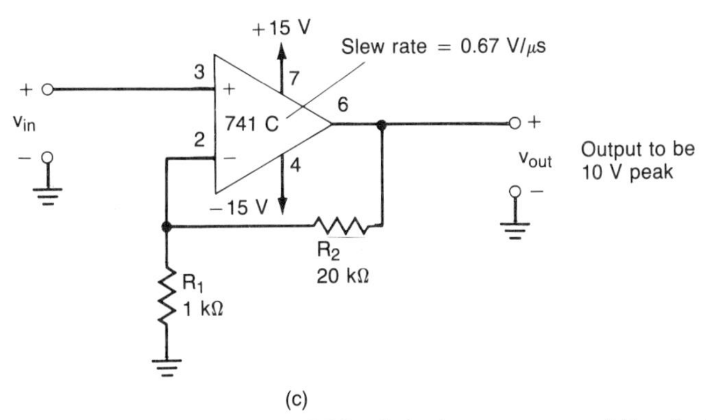

FIGURE 12-15 **Power bandwidth: (a) sine wave and its derivative; (b) amplifer output with a sinusoidal input when its power bandwidth is exceeded; (c) circuit for Example 12-2.**

If the frequency of the output signal for a given peak output voltage V_m exceeds the power bandwidth f_{max}, distortion will result. The output waveform will begin to assume a triangular appearance [see Fig. 12-15(b)].

We should point out that the severe distortion depicted in Fig. 12-15(b) occurs only when the signal frequency is far beyond the power bandwidth. Specifically, f_{max} tells us the point at which the distortion just begins to occur.

EXAMPLE 12-2

Given the 741 op amp circuit shown in Fig. 12-15(c), find its power bandwidth if the peak output voltage is to be 10 V and the 741's slew rate is 0.67 V/μs.

SOLUTION The solution follows directly from Eq. 12-10.

$$f_{max} = \frac{SR}{2\pi V_m} = \frac{0.67\ \text{V}/\mu\text{s}}{2\pi(10\ \text{V})} = 10.7\ \text{kHz}$$

Obviously, the 741's poor slew rate severely limits its application in high-fidelity, large-signal audio-amplifier systems. ■

12-5 Class A Power Amplifiers

A class A power amplifier is defined as a power amplifier in which the collector (or drain) current flows for 360° *of the input signal.* Except for the term "power," all the small- and large-signal amplifiers we have studied thus far could be categorized as "class A" by this definition. An extremely simple class A power amplifier has been given in Fig. 12-16(a). The circuit is described as a series-fed class A power amplifier. The term "series-fed" is derived from the fact that the load R_L is connected in series with the output transistor.

At first glance, the circuit could be mistaken for a voltage amplifier. However, note that I_C is *one ampere*. Voltage amplifiers typically have quiescent collector currents on the order of a few milliamperes. (The nominal I_C for a discrete voltage

FIGURE 12-16 Series-fed class A power amplifier: (a) power amplifier and its dc equivalent; (b) load lines; (c) waveforms.

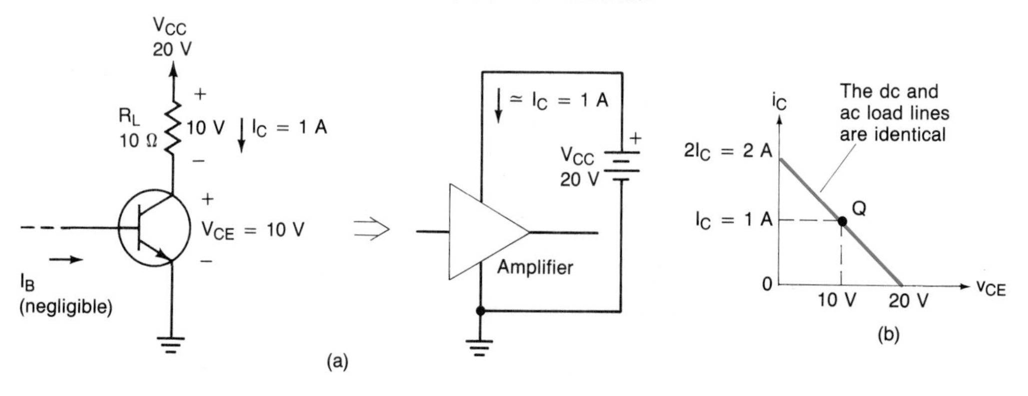

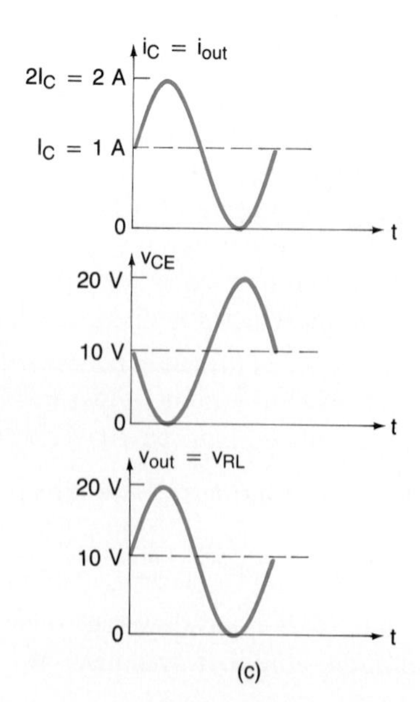

amplifier is 1 mA.) This is done to keep r_π (and therefore R_{in}) reasonably large. However, in power amplifiers, we not only need large output voltage swings, but large current swings as well. This requires a large I_C.

Since large power levels are involved, the *efficiency* of the amplifier becomes a major concern. Recall that the efficiency of any system is a measure of how well the system converts its applied input power into useful output power. The power losses in a system lower its efficiency. The power losses in an electronic power amplifier are due to the power dissipations within the active and passive devices.

Power dissipations in our active devices produce temperature rises. Semiconductors hate heat. It degrades their performance, reduces their reliability, and can result in their destruction if their maximum junction temperature T_j (typically, 150 to 175°C) is exceeded.

In an electronic power amplifier, its primary source of input power is its dc power supply. The dc input power to the amplifier shown in Fig. 12-16(a) is given by

$$P_{in(dc)} = V_{CC} I_C \tag{12-11}$$

This assumes that the base current is negligibly small compared to the collector current.

In Fig. 12-16(b) we see the load lines. Note that since R_L is in series with the BJT, the ac and dc load lines are the same. By inspection, we can see that midpoint bias has been used. The peak total instantaneous collector current is $2I_C$. Therefore, the peak ac (output) current through the load $I_{out(max)}$ is

$$I_{out(max)} = I_C$$

[see Fig. 12-16(c)]. Its corresponding rms value is

$$I_{out} = \frac{I_{out(max)}}{\sqrt{2}} = \frac{I_C}{\sqrt{2}}$$

Similarly, the peak ac (output) voltage across the load is

$$V_{out(max)} = V_{CE}$$

and its corresponding rms value is

$$V_{out} = \frac{V_{out(max)}}{\sqrt{2}} = \frac{V_{CE}}{\sqrt{2}}$$

The ac (output) power $P_{out(ac)}$ developed in the load is given by the product of the rms value of the output voltage developed across it, and the rms value of the current through it. Hence

$$P_{out(ac)} = \frac{V_{CE}}{\sqrt{2}} \frac{I_C}{\sqrt{2}} = \frac{V_{CE} I_C}{2} \tag{12-12}$$

Since midpoint biasing has been used, we can restate Eq. 12-11.

$$P_{in(dc)} = V_{CC} I_C = 2V_{CE} I_C \tag{12-13}$$

The efficiency η is given by the ratio of the ac output power to the dc input power. Expressing the efficiency as a percentage, and from Eqs. 12-12 and 12-13, we have

$$\begin{aligned} \%\,\eta &= \frac{P_{out(ac)}}{P_{in(dc)}} \times 100\% = \frac{(V_{CE} I_C)/2}{2V_{CE} I_C} \times 100\% \\ &= \frac{V_{CE} I_C}{2} \frac{1}{2V_{CE} I_C} \times 100\% = \frac{1}{4} \times 100\% = 25\% \end{aligned}$$

$$\text{The ideal class A efficiency:} \qquad \% \eta = 25\% \tag{12-14}$$

The ideal (maximum) efficiency of a class A power amplifier is 25%. Let us consider exactly what this means. If we want a 100-W class A power amplifier, we must provide it with 400 W of dc input power! This means that 300 W of power is lost within the amplifier as heat.

EXAMPLE 12-3

Find the dc input power, the ac output power, and the efficiency of the series-fed, class A power amplifier given in Fig. 12-16(a).

SOLUTION From Eq. 12-11 we can find the dc input power.

$$P_{\text{in(dc)}} = I_C V_{CC} = (1\text{ A})(20\text{ V}) = 20\text{ W}$$

The ac output power may be determined by applying Eq. 12-12.

$$P_{\text{out(ac)}} = \frac{V_{CE} I_C}{2} = \frac{(10\text{ V})(1\text{ A})}{2} = 5\text{ W}$$

The efficiency $\% \eta$ is

$$\% \eta = \frac{P_{\text{out(ac)}}}{P_{\text{in(dc)}}} \times 100\% = \frac{5\text{ W}}{20\text{ W}} \times 100\% = 25\%$$

■

While an efficiency of 25% is extremely low, real (nonideal) class A power amplifiers have even lower efficiencies. Typical values of efficiency may be less than 5% and as large as approximately 20%.

Besides inefficiency, the series-fed class A power amplifier offers a second big disadvantage - it allows a dc bias current to flow through the load. This can produce additional (unnecessary) load dissipation and contribute to poor dynamic response in some loads (e.g., a loudspeaker).

12-6 A Practical Class A Power Amplifier

The poor efficiency of the class A power amplifier limits it to low-power (typically, milliwatt) applications. A simple circuit using an op amp driver has been given in Fig. 12-17(a). First, we note that the amplifier is being operated from a single-polarity, 12-V power supply. The op amp is an LM324. It is described as a low-power (low power consumption) quad (four op amps in a single package) op amp. It is designed to operate from a single-supply voltage, and includes internal, unity-gain frequency compensation.

A "single-supply" op amp provides output voltage swings near ground (0 V) *and very close to* V_{CC}. In the case of the LM324, the maximum positive output voltage is V_{CC} minus 1.5 V and a minimum output voltage of 0 to 20 mV. In our particular circuit, the op amp's maximum positive output voltage is

$$V_{CC} - 1.5\text{ V} = 12\text{ V} - 1.5\text{ V} = 10.5\text{ V}$$

As we shall see, this will not pose a serious limitation.

A single-supply op amp does not preclude the necessity for biasing up its input. This is still required if linear operation is to be expected. In Fig. 12-17(b), the noninverting input is biased to 7.91 V by R_1 and R_2. (The basic operation of the single-supply op amp amplifier was discussed in Section 9-10.)

The feedback loop has been wrapped around both the op amp and the BJT. (This technique was explained in Section 11-3.) The output transistor is a low-cost TIP41C *npn* power transistor which has been mounted on a heat sink. (The characteristics of power transistors and heat sinking are discussed later in this chapter.)

The dc equivalent circuit of this class A amplifier has been given in Fig. 12-17(b). Again, we note that the feedback loop includes both the op amp and the

FIGURE 12-17 Practical 250-mW class A power amplifier: (a) power amplifier (with driver); (b) dc analysis; (c) ac equivalent circuits; (d) load lines; (e) maximum output waveform.

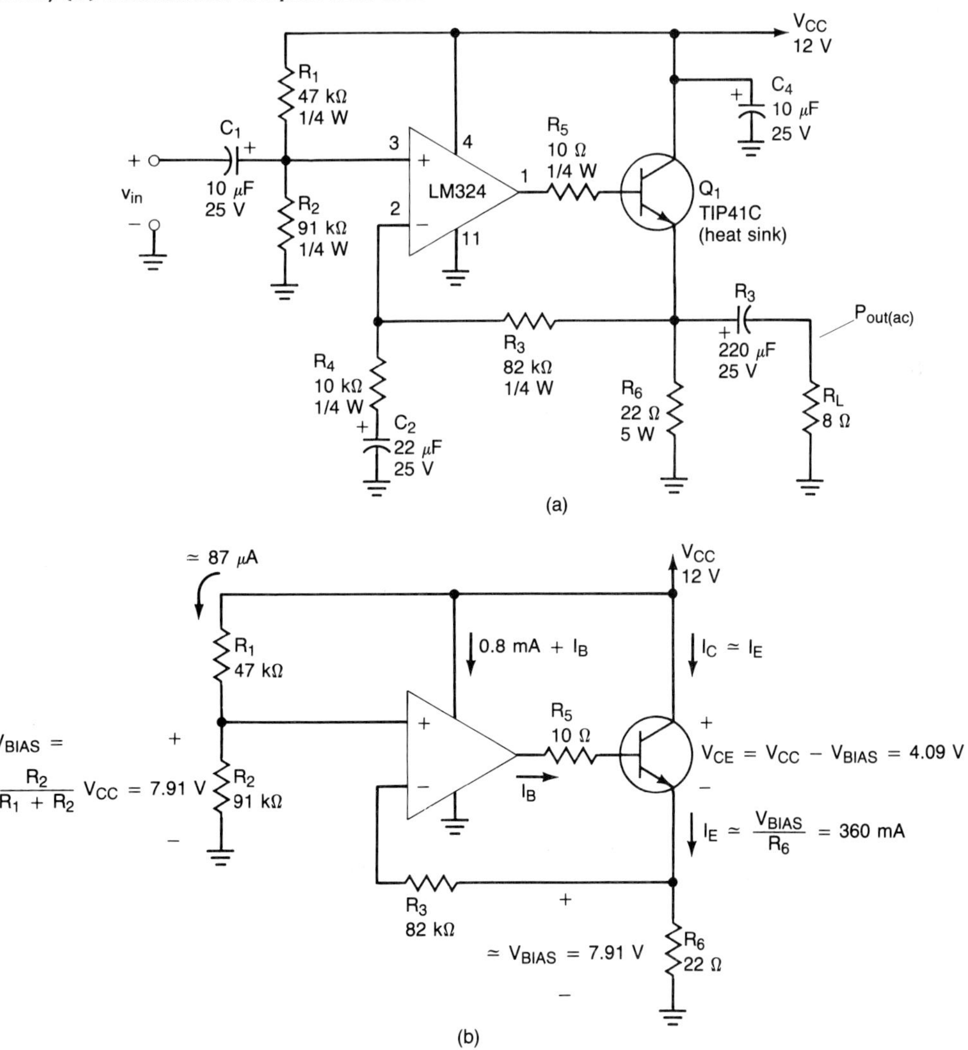

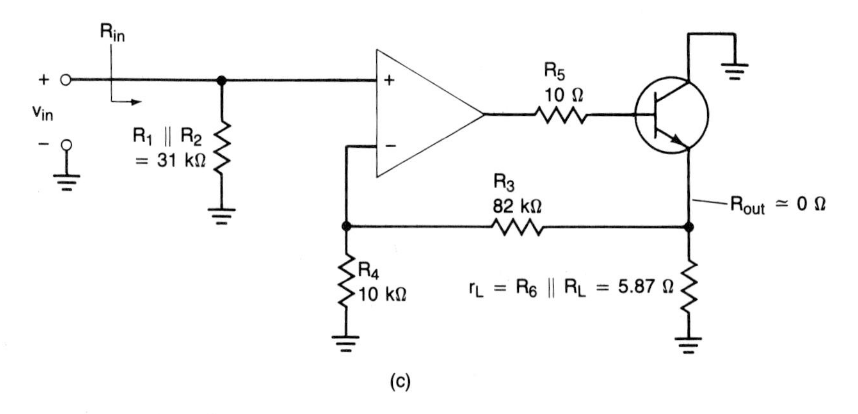

(c)

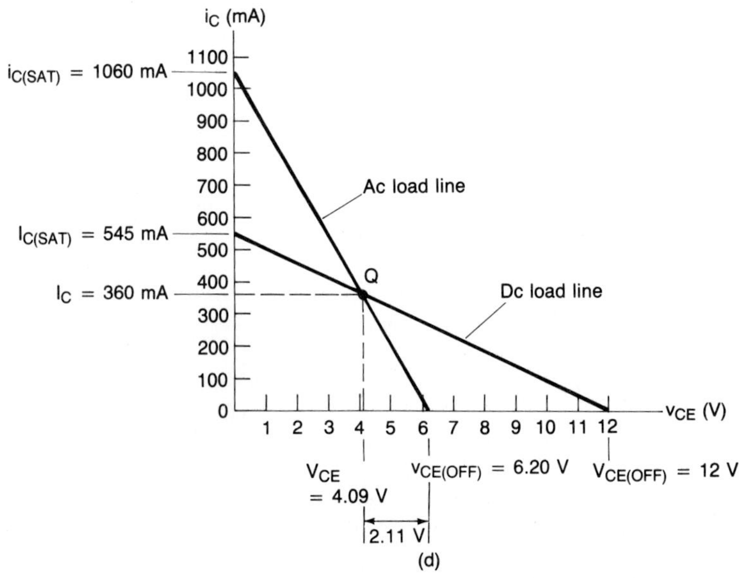

(d)

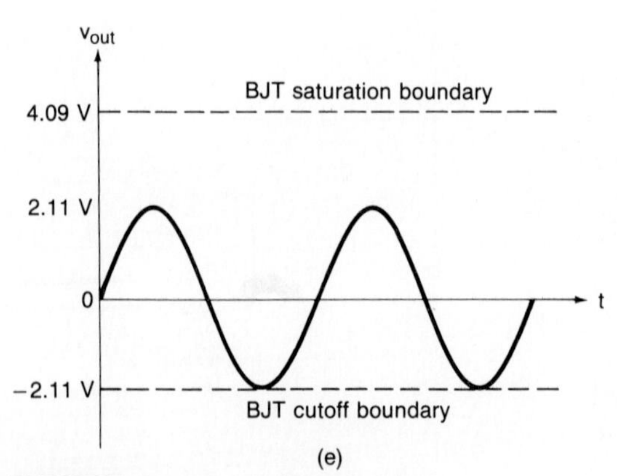

(e)

FIGURE 12-17 ***(continued)***

BJT. The dc blocking action of capacitor C_2 produces a voltage-follower circuit. Since the op amp "works" to keep its noninverting and inverting inputs at the same potential, V_{BIAS} appears across the BJT's emitter resistor R_6. The dc emitter current is

$$I_E = \frac{V_{BIAS}}{R_6} = \frac{7.91\ \text{V}}{22\ \Omega} \simeq 360\ \text{mA}$$

and

$$I_C \simeq I_E = 360\ \text{mA}$$

The BJT's dc collector current is rather large, but this is typical for class A power amplifiers. The dc load line and Q-point have been indicated in Fig. 12-17(d). The reader should verify it.

The LM324 op amp draws a nominal bias current of only 0.8 mA. The bias resistors will only pull another 87 μA. The BJT's base current is supplied by the op amp's output. This current is ultimately drawn from the op amp's power supply terminal [see Fig. 12-17(b)].

Since these currents are so very small compared to the collector current of 360 mA, we shall ignore them. Consequently, the amplifier's input power is

$$P_{\text{in(dc)}} \simeq V_{CC} I_C = (12\ \text{V})(360\ \text{mA}) = 4.32\ \text{W}$$

Because of the op amp driver (with its large open-loop voltage gain and large input resistance), we can easily approximate the amplifier's ac parameters. The ac equivalent circuit has been provided in Fig. 12-17(c). Hence

$$R_{\text{in}} \simeq R_1 \parallel R_2 = 47\ \text{k}\Omega \parallel 91\ \text{k}\Omega \simeq 31\ \text{k}\Omega$$

$$A_{v(\text{cl})} \simeq 1 + \frac{R_3}{R_4} = 1 + \frac{82\ \text{k}\Omega}{10\ \text{k}\Omega} = 9.2$$

$$R_{\text{out}} \simeq 0\ \Omega$$

To find the efficiency, we must first develop the ac load line for the emitter-follower output transistor. As can be seen in Fig. 12-17(c), the ac load resistance r_L is given by the parallel combination of the emitter resistor R_6 and R_L.

$$r_L = R_6 \parallel R_L = 22\ \Omega \parallel 8\ \Omega = 5.87\ \Omega$$

Equations 12-4 and 12-5 give us the endpoints for the ac load line.

$$i_{C(\text{SAT})} = I_C + \frac{V_{CE}}{r_L} = 360\ \text{mA} + \frac{4.09\ \text{V}}{5.87\ \Omega} \simeq 1.06\ \text{A}$$

$$v_{CE(\text{OFF})} = V_{CE} + I_C r_L = 4.09\ \text{V} + (360\ \text{mA})(5.87\ \Omega) = 6.20\ \text{V}$$

The ac load line has been drawn in Fig. 12-17(d). Observe that the peak positive-going collector-to-emitter voltage is 2.11 V, and the peak negative-going collector–emitter voltage is 4.09 V. From Section 12-3 we recall that this means that the maximum *positive* peak load voltage is 4.09 V, and the maximum *negative* peak load voltage is 2.11 V [see Fig. 12-17(e)].

Since the maximum undistorted peak output voltage across the load is 2.11 V, the equivalent rms output voltage is

$$V_{\text{out}} = \frac{V_{\text{out(peak)}}}{\sqrt{2}} = \frac{2.11\ \text{V}}{\sqrt{2}} = 1.49\ \text{V rms}$$

Therefore, the peak ac power developed in the load is

$$P_{out(ac)} = \frac{V_{out}^2}{R_L} = \frac{(1.49\ V)^2}{8\ \Omega} \simeq 278\ mW$$

The efficiency of the amplifier is

$$\%\ \eta = \frac{P_{out(ac)}}{P_{in(dc)}} \times 100\% = \frac{0.278\ W}{4.32\ W} \times 100\% = 6.44\%$$

Obviously, this ''practical'' class A power amplifier demonstrates an efficiency that is far below the ideal maximum of 25% for series fed. [The maximum efficiency for Fig. 12-17(a) with R_6 and R_L equal in value is 8.33% (see Prob. 12-33).] The primary power losses occur in the emitter resistor and the BJT. The *maximum* power loss in the BJT occurs when *no* signal is applied. Hence

$$P_C = V_{CE}I_C = (4.09\ V)(360\ mA) = 1.47\ W$$

12-7 Classes of Amplifiers: The Quest for Improved Efficiency

Even when we are dealing with relatively small (e.g., mW) power levels, power losses demand our attention. For example, in battery-operated systems, we simply cannot afford to waste power. Very few consumers are pleased at the prospect of frequently replacing or recharging batteries. If a system is to use a line-operated dc power supply, power losses are still highly undesirable. Power losses produce temperature rises in both the active and passive devices.

Temperature rises reduce the reliability of electronic systems. In fact, as a rough rule of thumb, the operating life of bipolar transistors approximately doubles for each 10°C reduction in their (collector-base) junction temperature.

Therefore, improving the efficiency of a power amplifier is an extremely important goal. As a means to this end, various *classes* of amplifier circuits have been developed. These have been illustrated in Fig. 12-18.

In Fig. 12-18(a), a general amplifier circuit has been provided. The class of operation can be changed merely by adjusting the base bias supply V_B to position the Q-point. As we can see, the class A power amplifier [Fig. 12-18(b)] permits the collector current to flow for 360° of the input signal. The class AB amplifier [Fig. 12-18(c)] has its Q-point lowered such that collector current flows for less than 360°, but more than 180°. The class B amplifier [Fig. 12-18(d)] allows the collector current to flow for only 180° of the input signal. The class C amplifier [Fig. 12-18(e)] has collector currents that flow for *less* than 180° of the input signal. Other categories exist, but we shall not consider them at this time.

The class AB amplifier is the scheme most often used for audio-frequency power amplifiers. The distortion in the collector current is eliminated by using transistor *complementary pairs*. A complementary transistor pair consists of an *npn* and a *pnp* bipolar transistor which have closely matched electrical specifications. (Alternatively,

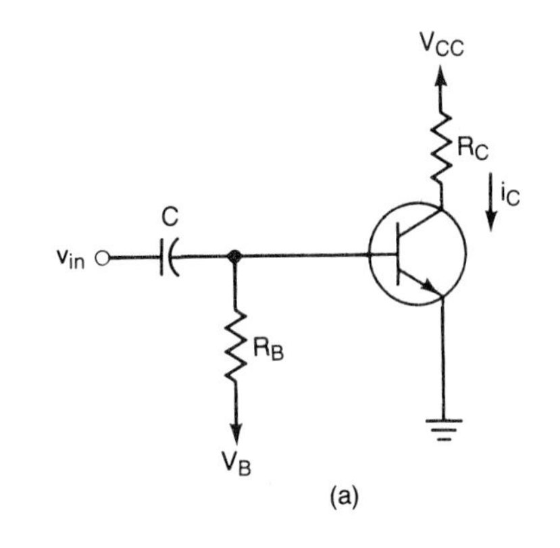

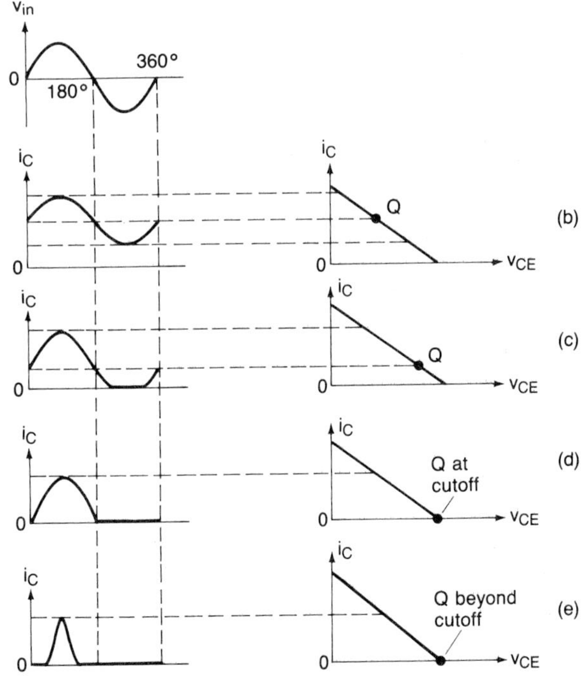

FIGURE 12-18 Classes of amplifier operation: (a) general circuit; (b) class A; (c) class AB; (d) class B; (e) class C.

the pair may be an *n*-channel and a *p*-channel MOSFET.) The use of complementary pairs is demonstrated in the next section.

The severe distortion in the collector current of the class C power amplifier can be minimized by using resonant *LC* tuned circuits. Consequently, even though the class C amplifier tends to be much more efficient than the class A, AB, and B amplifiers, it use is restricted to radio-frequency (RF) power amplifiers. We shall not address the class C amplifier further.

12-8 Class B and AB Power Amplifiers

The class B power amplifier serves as the basis for the class AB configuration.

The class AB amplifier is the most widely used audio power amplifier configuration. Class AB push-pull complementary symmetry power amplifiers offer lower transistor power dissipation, improved efficiency, and lower distortion than is achievable via the class A scheme. The relative importance of the class AB push-pull complementary symmetry amplifier requires that we devote much of our studies toward it.

We shall present our preliminary discussion of it, and then digress to the types of output devices and their characteristics. Once we understand the output device considerations, we can then direct our studies to the circuitry typically ''nested'' within power amplifiers to optimize performance. To understand the class AB power amplifier, we shall first direct our attention to the class B power amplifier.

The essence of class B operation can be seen in Fig. 12-19(a). No dc forward bias is applied to the base of the BJT. Its base terminal is simply tied to ground through R_B. Therefore, the BJT is biased at cutoff. Consequently, when *no* signal is applied, the BJT's collector current is zero. Since the collector current is zero, the voltage drop across R_L will also be zero. The voltage drop across the BJT (V_{CE}) will be equal to V_{CC}. Even though the BJT's collector-to-emitter voltage is equal to V_{CC}, the BJT will *not* dissipate any power.

$$P_C = V_{CE}I_C = V_{CC}(0\text{A}) = 0\text{ W}$$

A no-signal power dissipation of zero is an obvious big advantage of the class B power amplifier over the class A version.

When the input signal goes negative, the BJT be will biased further into cutoff. However, when the input signal goes positive, the BJT's base-emitter *p-n* junction will become forward biased and collector current will flow. As a result, the positive half-cycle of the input signal will produce a positive half-cycle of output voltage across R_L [refer again to Fig. 12-19(a)].

Obviously, a single-stage class B power amplifier stage produces too much distortion to be of much use in audio applications. The typical remedy is to employ a complementary transistor pair in a class B push-pull arrangement [see Fig. 12-19(b)]. Q_1 is an *npn* BJT and Q_2 is a *pnp* BJT whose electrical dc and ac specifications closely match those possessed by Q_1. As we mentioned previously, this constitutes a complementary pair.

On the positive half-cycle of the input signal, the base-emitter of Q_1 is forward biased, while the base-emitter of Q_2 is reverse biased. Therefore, Q_1 will conduct and Q_2 will remain in cutoff. Conversely, on the negative half-cycle of the input, Q_1 will be in cutoff while Q_2 is forward biased into conduction. The operation (and circuit) of the upper (*npn*) stage is symmetrical with the lower (*pnp*) stage. Now we can see where the term ''complementary symmetry'' comes from.

In Fig. 12-19(c), we note that when Q_1 conducts, it *pushes* current through R_L. When Q_2 conducts, it *pulls* current from ground through R_L [see Fig. 12-19(d)]. Now the description ''class B push-pull'' operation should be clear.

Let us examine the operation of the circuit more closely. Both transistors have

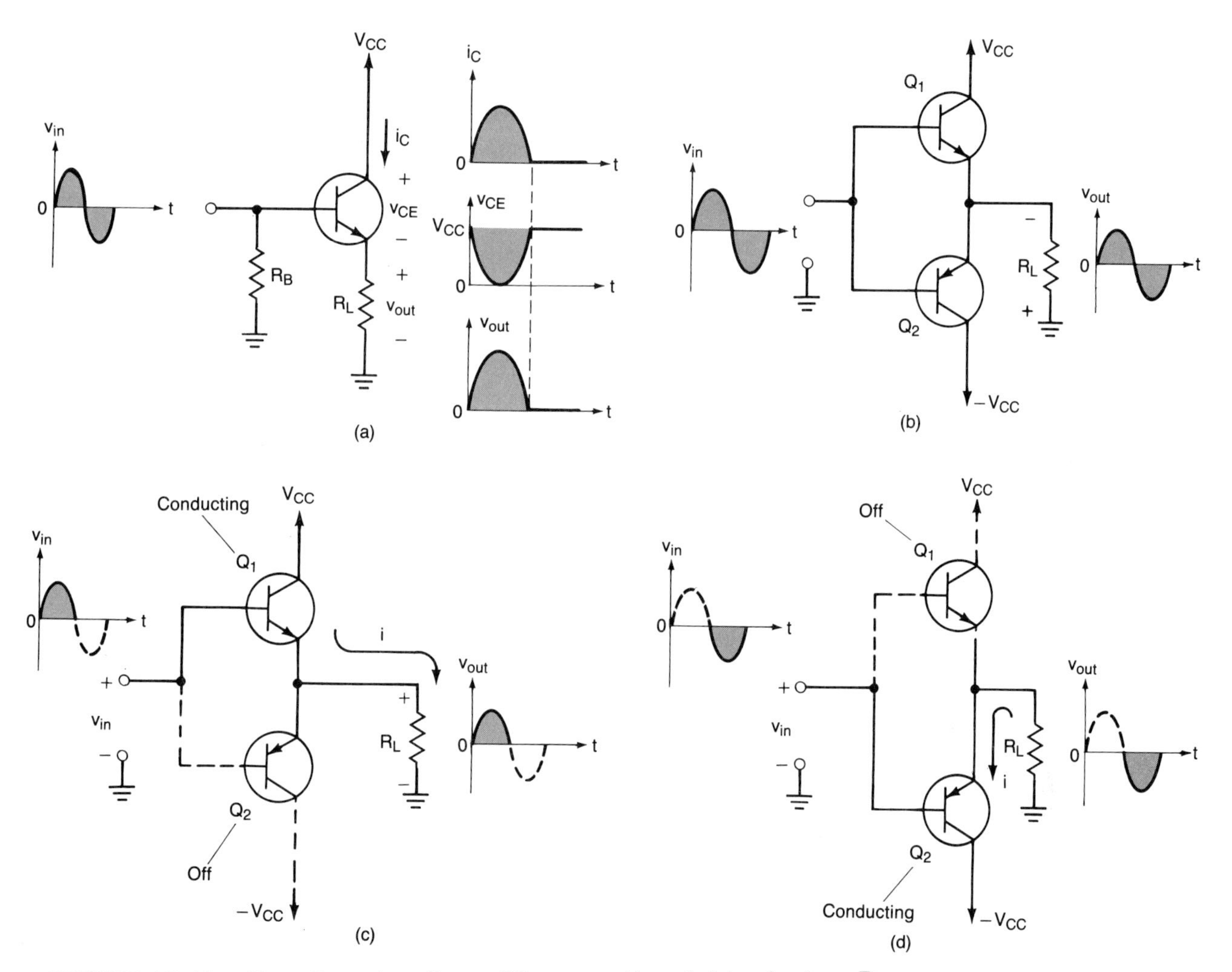

FIGURE 12-19 Class B push-pull amplifier operation: (a) basic class B amplifier; (b) push-pull class B amplifier using complementary symmetry; (c) when v_{in} is positive Q_1 conducts; (d) when v_{in} is negative Q_2 conducts.

no forward bias when no signal is applied. Therefore, their respective collector currents and power dissipations are zero when the circuit is idling. This is an ideal situation. However, in real class B push-pull power amplifiers this scheme will produce *crossover distortion* [see Fig. 12-20(a)].

Recall that silicon transistors must have at least 0.5 to 0.6 V of forward base–emitter bias before they will go into conduction. Since the forward bias for the bipolar transistors is produced by the input signal, *both* of the transistors will be nonconducting, or OFF, when the input signal is approximately ± 0.5 V. This forms a "dead band" in the input and produces crossover distortion in the output.

To avoid crossover distortion, it is necessary to add a small amount of forward

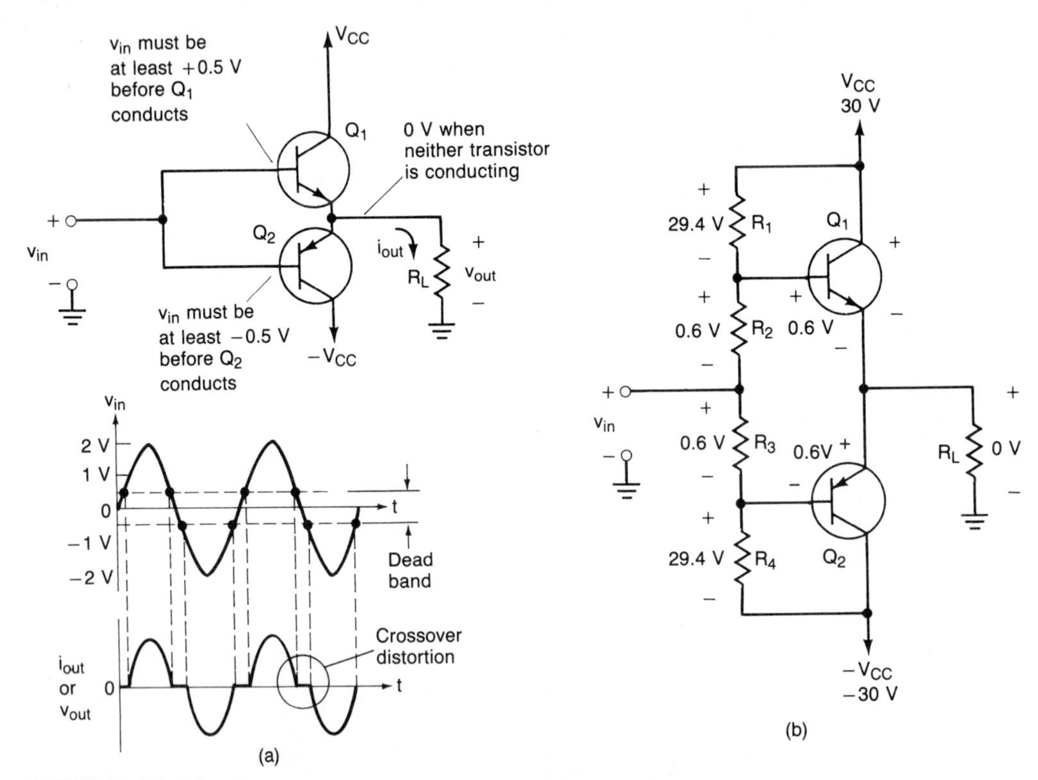

FIGURE 12-20 Crossover distortion and its elimination: (a) crossover distortion; (b) dc bias for class AB operation to eliminate crossover distortion.

bias to take the BJTs to the verge of conduction, or slightly beyond. This does slightly lower the efficiency of the circuit, but it alleviates the crossover distortion problem. Technically, the operation of the BJTs lies between class B and class A. Therefore, the circuit operation is often referred to as being class AB [see Fig. 12-20(b)].

Let us now investigate the operation of the complementary symmetry, push-pull, class AB power amplifier. The circuit has been repeated in Fig. 12-21(a).

During the positive half-cycle of the input signal, transistor Q_1 goes further into conduction. The positive input signal drives transistor Q_2 into cutoff. As the collector current of Q_1 increases, its collector-to-emitter voltage decreases. The decrease in the collector-to-emitter voltage of Q_1 serves to make the voltage across the load increase in a positive-going direction. With a little reflection, we can see that as the load voltage becomes more positive, the emitter of transistor Q_2 also becomes more positive. Therefore, the *collector-to-emitter voltage* of Q_2 becomes more *negative*. Most beginners are amazed to find a sinusoidal collector-to-emitter voltage across the BJTs since they only pass current for approximately one-half cycle. The rest of the circuit action has been summarized by the waveforms included in Fig. 12-21. The reader should study the waveforms carefully.

The average dc voltage across the amplifier is $2V_{CC}$. If we neglect the small

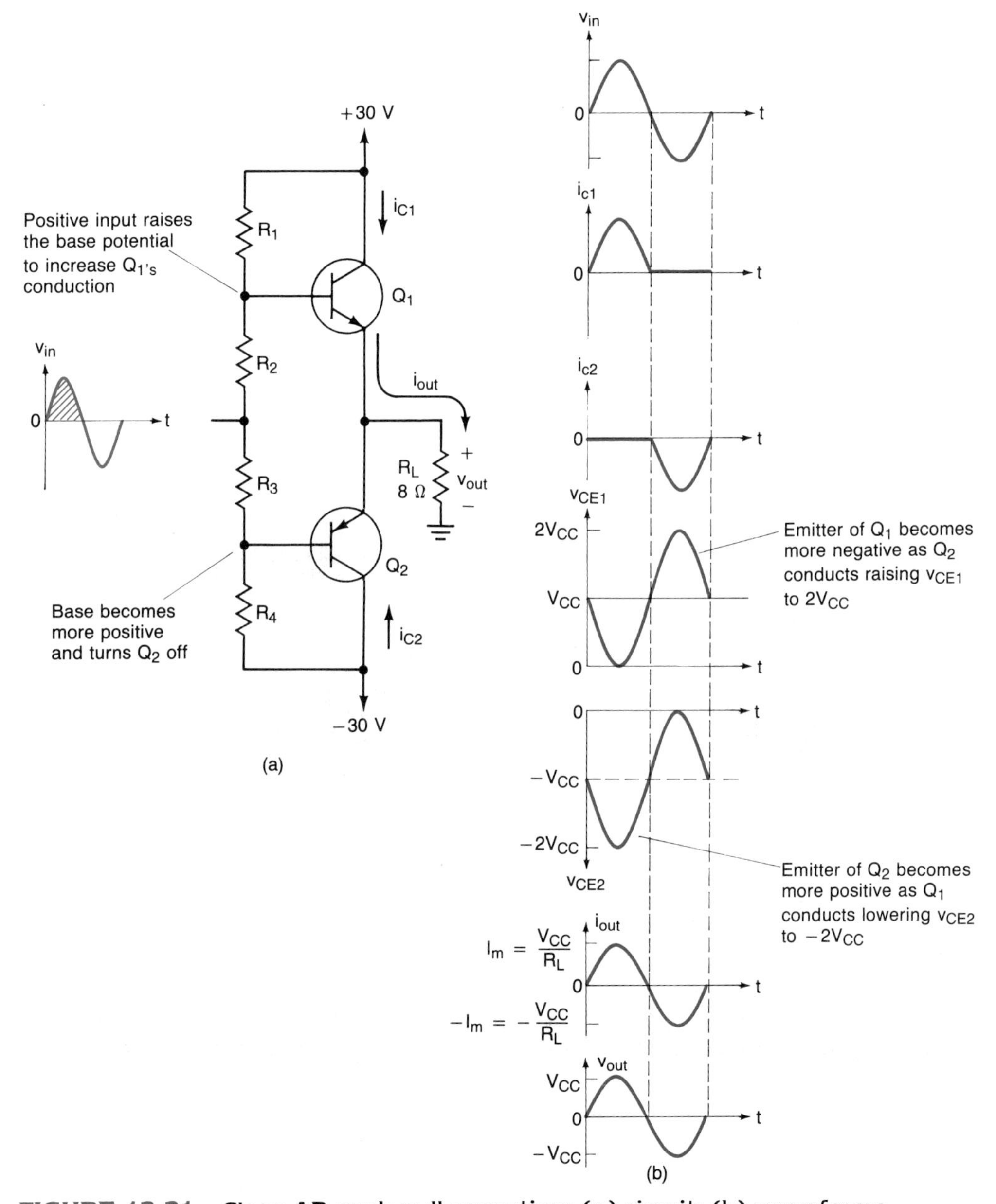

FIGURE 12-21 **Class AB push-pull operation: (a) circuit; (b) waveforms.**

bias current through the BJTs, we can easily determine the average current drawn from the dc power supply. The BJTs alternate in conduction. Their corresponding collector current waveforms have been repeated in Fig. 12-22(a). *The average (or dc) current drawn from the dc power supply is directly proportional to the output signal swing.*

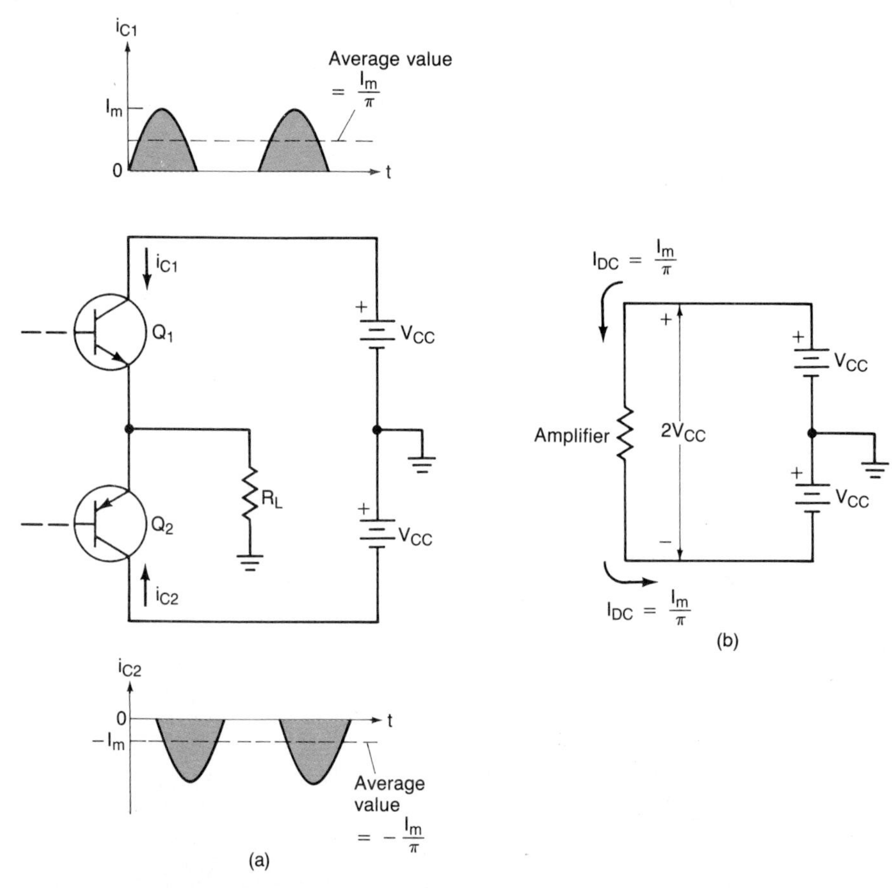

FIGURE 12-22 Finding the dc input power to the class AB push-pull power amplifier: (a) average current; (b) equivalent circuit.

The average dc current I_{DC} can be shown (Chapter 13) to be

$$I_{\mathrm{DC}} = \frac{I_m}{\pi} \simeq 0.318 I_m \tag{12-15}$$

The average dc current drawn by the *npn* BJT from the positive supply is equal to the magnitude of the dc current drawn by the *pnp* BJT from the negative supply. Therefore, we may model the situation as depicted in Fig. 12-22(b).

The dc input power $P_{\mathrm{in(dc)}}$ to the class AB push-pull power amplifier is

$$P_{\mathrm{in(dc)}} = 2V_{CC}\frac{I_m}{\pi} = \frac{2V_{CC}I_m}{\pi}$$

Inspection of Fig. 12-21 reveals that the ac output power delivered to the load is

$$P_{\text{out(ac)}} = \frac{V_{CC}}{\sqrt{2}} \frac{I_m}{\sqrt{2}} = \frac{V_{CC} I_m}{2}$$

The ideal efficiency of the class B (class AB) amplifier is

$$\% \eta = \frac{P_{\text{out(ac)}}}{P_{\text{in(dc)}}} \times 100\% = \frac{(V_{CC} I_m)/2}{(2V_{CC} I_m)/\pi} \times 100\%$$

$$= \frac{V_{CC} I_m}{2} \frac{\pi}{2V_{CC} I_m} \times 100\% = \frac{\pi}{4} \times 100\% = 78.5\%$$

The ideal class B efficiency: $\% \eta = 78.5\%$ (12-16)

Obviously, this represents a vast improvement over the series-fed class A power amplifier with its ideal efficiency of only 25%. Equation 12-16 does not specifically cover the class AB amplifier. However, if the bias current through the output transistors is kept relatively small, the ideal class AB efficiency will also approach 78.5%.

EXAMPLE 12-4

Find the efficiency of the class AB push-pull power amplifier given in Fig. 12-21(a). Assume that the load is an 8-Ω (resistive) loudspeaker and that the no-signal biasing current is negligible.

SOLUTION First, we note that the collector supply voltage is ±30 V. The peak ac voltage across the 8-Ω load is therefore 30 V. The peak current through the load is

$$I_m = \frac{V_{CC}}{R_L} = \frac{30 \text{ V}}{8\ \Omega} = 3.75 \text{ A}$$

The dc input power is

$$P_{\text{in(dc)}} = \frac{2V_{CC} I_m}{\pi} = \frac{(60 \text{ V})(3.75 \text{ A})}{\pi} = 71.62 \text{ W}$$

Now we find the rms value of the ac output power.

$$P_{\text{out(ac)}} = \frac{V_{CC} I_m}{2} = \frac{(30 \text{ V})(3.75 \text{ A})}{2} = 56.25 \text{ W}$$

The efficiency is

$$\% \eta = \frac{P_{\text{out(ac)}}}{P_{\text{in(dc)}}} \times 100\% = \frac{56.25 \text{ W}}{71.62 \text{ W}} \times 100\% \simeq 78.5\% \ ■$$

Before we develop the additional circuitry normally found in power amplifiers, we need to understand the characteristics of the power semiconductors. Many possibilities exist in the type of output devices employed. The major classifications include bipolar power transistors, Darlington power transistors, and field-effect power transistors. The particular type of device employed dictates the specific types of circuitry needed to support (and protect) its proper operation.

12-9 Bipolar Power Transistors and SOA

The dc beta of bipolar power transistors tends to fall off as the collector current increases (see Fig. 12-23). This tendency points to a very disheartening fact: At high current levels, we can expect the B_{DC} to fall off. This is precisely the time when we need as high a B_{DC} as possible. Low B_{DC} values increase the demands on the circuitry driving the base. More will be said about this later.

To appreciate more fully the demands placed on a BJT power output device, let us consider a brief example. If a class AB push-pull power amplifier is to deliver 100 W to an 8-Ω (resistive) speaker, we find that it must provide a peak output voltage of 40 V, and a peak current of 5 A. In this instance,

$$I_m = \frac{V_{CC}}{R_L} = \frac{40\text{ V}}{8\ \Omega} = 5\text{ A}$$

$$P_{\text{out(ac)}} = \frac{V_{CC}I_m}{2} = \frac{(40\text{ V})(5\text{ A})}{2} = 100\text{ W}$$

The output transistors must be able to block collector-to-emitter voltages of 80 V ($2V_{CC}$) and pass the 5 A of peak collector current. These values are in sharp contrast with the low-level signals encountered in our study of voltage amplifiers. BJTs designed to handle these large voltage and current signals are necessarily different from those designed to handle small signals.

Since bipolar power transistors must be able to pass large collector currents, their collector-base *p-n* junctions must be able to dissipate large power levels. Therefore, the collector-base *p-n* junctions of power BJTs must have much larger areas than those found in small-signal BJTs.

The large collector-base area also increases the collector-base junction capacitance C_{ob}. The values of C_{ob} typically range from 100 to 2000 pF. As we saw in Section 9-12, the gain-bandwidth product f_T of a BJT is inversely proportional to C_{ob} (and C_{be}). Consequently, the f_T for small-signal BJTs can be several hundred megahertz. However, the f_T for power BJTs may be as low as 0.5 MHz, with 1 to 50 MHz being typical values.

FIGURE 12-23 The h_{FE} of a power transistor falls off as I_C increases.

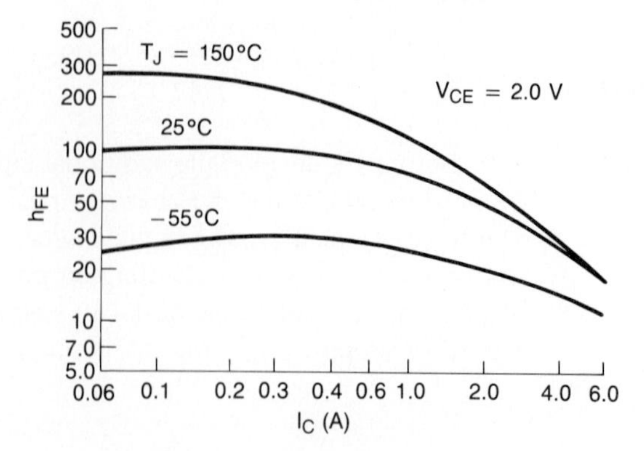

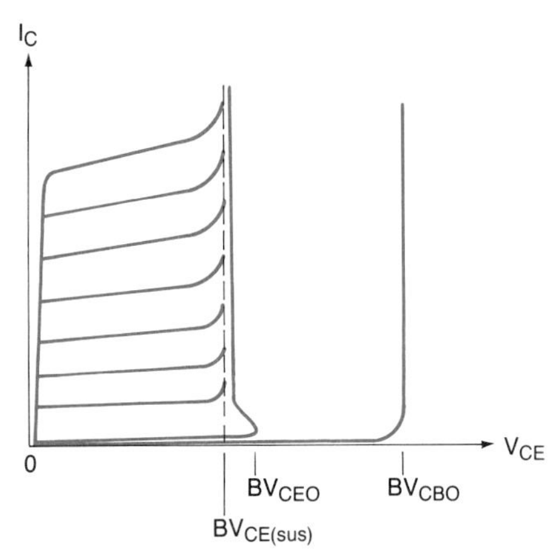

FIGURE 12-24 BJT breakdown voltage ratings.

Very large (perhaps several hundred volts) breakdown voltage ratings are often required for power devices. (The voltage ratings for a BJT have been summarized in Fig. 12-24.) Generally, the largest breakdown voltage rating is BV_{CBO}. Recall that the value of BV_{CBO} corresponds to the avalanche breakdown voltage of the collector-base *p-n* junction as measured with the emitter lead open. The value of BV_{CEO} is generally less than BV_{CBO} and is measured with the base lead open. The third voltage rating, $BV_{CE(\mathrm{sus})}$, is the *collector-to-emitter sustaining voltage*. It represents the maximum collector-to-emitter voltage with the base current held constant. Obviously, the BJT should not be operated in circuits in which its voltage ratings will be exceeded or even approached. Numerous other voltage ratings are often provided on power BJT data sheets. These are explained by the test conditions specified on the data sheet and/or in application notes.

The large geometries of the BJT power devices also result in wide effective base widths. Therefore, the current gains are also much lower than those found in small-signal BJTs. Typical values of β_{DC} range from 20 to 70. At large values of I_C it is not unusual to find that β_{DC} has dropped to 10 or less.

The large collector–base junction area found in power BJTs also promotes large values of collector leakage current I_{CBO}. Values of I_{CBO} in a power transistor can be as large as several milliamperes.

Recall that the leakage current I_{CBO} (and I_{CEO}) is extremely temperature sensitive. The collector-base *p-n* junction temperature is not only a function of the ambient temperature, but the self-generated temperature rise as well. As collector current in a BJT increases, its collector-base power dissipation increases. An increase in power dissipation produces an increase in the junction temperature. An increase in the junction temperature will cause I_{CBO} to increase. The increase in I_{CBO} increases the collector current, and the cycle repeats. This phenomenon is called *thermal runaway*. If the junction temperature is not kept low, and/or the collector current is not limited, the BJT may destroy itself. This is a serious problem that is particularly acute in power amplifiers. We address this concern later.

Current ratings in power BJTs are also very important. Typically, the manufacturer will specify the maximum collector current $I_{C(\max)}$ that a BJT can pass. The value of $I_{C(\max)}$ may be limited by the minimum value of β_{DC}, or by the current capability of the BJT's collector-base junction, and/or that of the internal bonding wires. It should not be exceeded.

Another very important limitation in bipolar power transistors is a mechanism called the *second breakdown effect*. It is often denoted as $I_{S/b}$ on data sheets. This phenomenon is energy dependent and thus a function of time as well as voltage and current. Basically, second breakdown results from "hot spots" formed within the BJT's collector-base region due to nonuniform current densities. If a localized area in a BJT is hotter than surrounding areas, its resistivity will decrease. Therefore, it will tend to "hog" more of the current. This, in turn, will increase its temperature even more, prompting a further increase in the current density. If this process is allowed to continue, that area of the BJT will be destroyed. This reduction in the junction area will reduce the current-handling capability of the device. Typically, this will cause other hotspots to form, leading to the ultimate failure of the device. As we shall see, most manufacturers will provide curves that show the limitations imposed by second breakdown.

Additional power BJT ratings include the maximum collector–base junction temperature T_j, and maximum power dissipation P_C. The maximum T_j typically ranges from 150 to 200°C. Both T_j and P_C are a function of the transistor package or case style. Examples have been provided in Fig. 12-25.

In an effort to remove heat from the collector-base *p-n* junction, the collector is thermally (and electrically) bonded to the metallic mounting tab (flange) of the

FIGURE 12-25 Examples of Joint Electron Device Engineering Council (JEDEC) power transistor case styles: (a) JEDEC TO-3 (T_j maximum typically 200°C); (b) JEDEC TO-220 (T_j maximum typically 150°C).

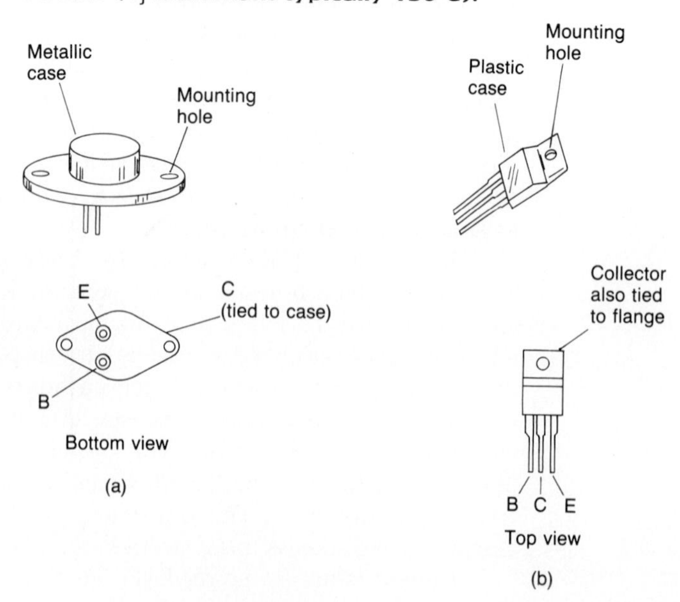

TO-220 package and the case of the TO-3 package. We shall investigate this further in Section 12-17 when we consider heat sinking.

The power dissipated in the collector is given by

$$P_C = V_{CE} I_{C(\text{AVG})} \tag{12-17}$$

Note that $I_{C(\text{AVG})}$, the average dc collector current, has been used in lieu of I_C. This allows us to extend Eq. 12-17 to all (e.g., class A and class AB) power amplifiers. The maximum power dissipation of a BJT has been plotted on the common-emitter output characteristics of a BJT in Fig. 12-26(a). *In general, the operating points of*

FIGURE 12-26 Maximum power and the SOA: (a) the maximum power curve has been indicated on the common-emitter V-I characteristics; (b) the SOA is indicated on a log-log plot.

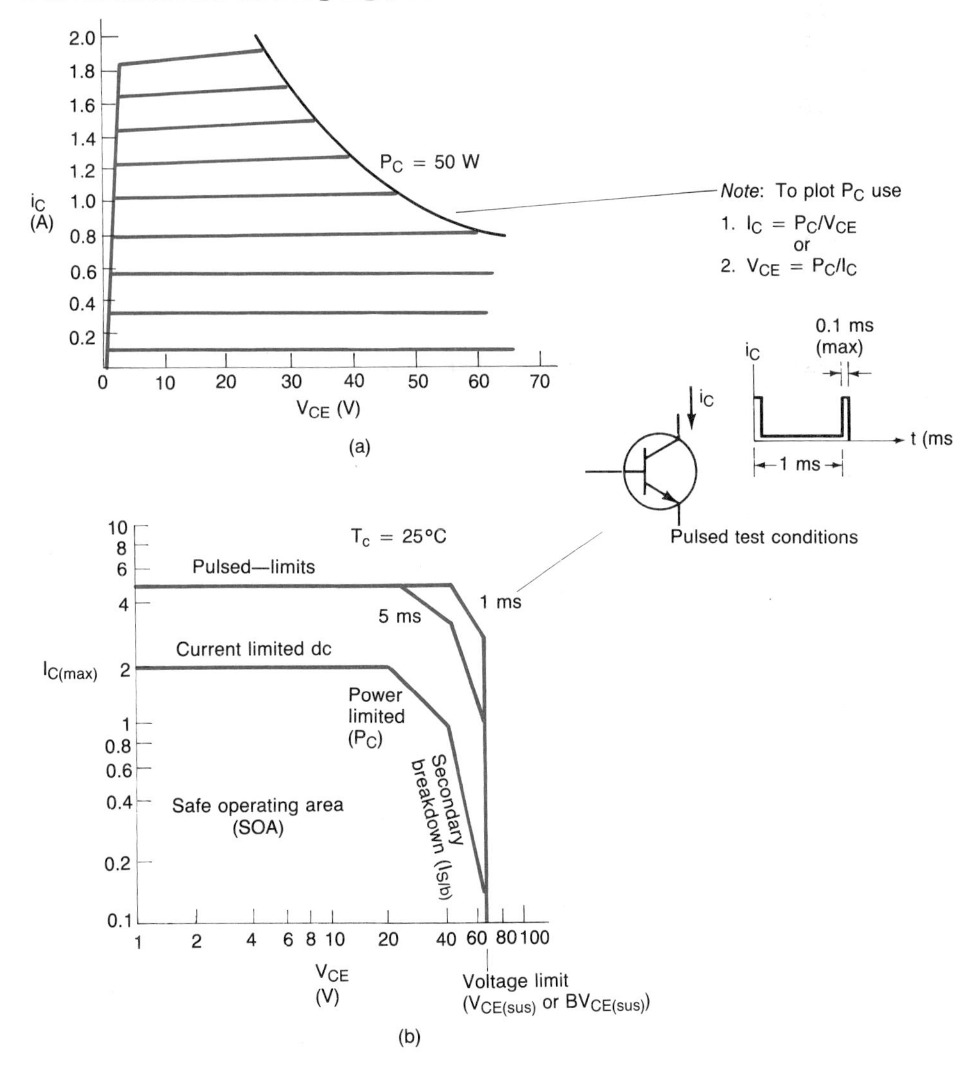

the BJT should be restricted to the left *of the maximum collector power curve*. It is extremely important to note that the maximum collector power is often specified for a junction temperature of 25°C. If the junction temperature is increased due to an increase in the ambient temperature and/or self-heating effects, the transistor will have to be derated. More will be said about this later.

The boundaries imposed by the maximum collector power dissipation, the maximum collector current, $BV_{CE(\text{sus})}$, and $I_{S/b}$ have been depicted graphically in Fig. 12-26(b). The area within the curve is called the *safe operating area* (SOA). We must work to ensure that the BJT operation is always confined within the SOA.

Observe that a log-log plot has been used to show the SOA. Also note that the SOA has been specified at a *case temperature* T_c of 25°C. In general, the junction temperature will be higher than T_c. T_c is specified since it is easily monitored.

12-10 Biasing and Drive Requirements in a Class AB Power Amplifier

A practical class AB push-pull power amplifier using an op amp driver has been given in Fig. 12-27(a). To provide bias stability (e.g., to avoid thermal runaway) several precautions have been taken. First, we see that small emitter resistors have been used. From our work in Chapter 6, we recall that emitter resistors tend to improve bias stability. The larger their value, the greater the bias stability will be. However, the emitter resistors are in series with the load. Therefore, they must be small enough to pass the peak current demanded by the load. They must also be kept small to minimize their power dissipation. *While emitter resistors enhance the bias stability, their use also lowers the efficiency of the amplifier.* (An equation for sizing R_E is given in Prob. 12-83.)

Diodes D_1 and D_2 are used to provide additional *temperature compensation*. Recall that the forward voltage drop across a silicon diode decreases at the rate of approximately 2 mV for each 1°C rise in temperature. Similarly, the forward base–emitter voltage drop across a silicon transistor also decreases at the same rate. Therefore, the forward bias required to turn on the output transistors decreases as their temperature increases. If a fixed base-emitter bias is used, the collector currents will increase with temperature. The diodes are used to adjust the base-emitter forward bias automatically as a function of temperature. For optimal results, the temperature coefficients of the diodes and the BJTs should be matched, and the diodes should be thermally bonded to the output transistors.

Since matching temperature coefficients between the (discrete) devices is, at best, difficult to achieve, a negative-feedback loop has been wrapped around the power amplifier. In this case, the op amp works to keep the output at zero volts when no signal is applied. The circuit is extremely stable.

Typically, the output of an IC op amp will be within millivolts of zero when no input signal is applied. Quite often, this is sufficient. However, note in Fig. 12-27(a) that a potentiometer has been connected to the op amp. Many op amps have a pair of *offset null* terminals. In the case of the 741 op amp, the manufacturer suggests that a 10-kΩ potentiometer be used with its rotor connected to the negative supply voltage. The potentiometer is adjusted while monitoring the output of the amplifier. By carefully "trimming" in its value, the output may be precisely adjusted to within microvolts of zero when no input signal is applied.

Also note that decoupling capacitors C_1 and C_2 have been included. These are mandatory if spurious oscillations due to interstage coupling is to be avoided. The power transistors quite often draw respectable current spikes from the power supply.

If potentiometer R_2 is carefully adjusted to balance the output to zero volts, and we assume that the forward-biased diodes D_1 and D_2 each drop 0.7 V, we obtain the dc values given in Fig. 12-27(b). The reader should study the results carefully.

The base-emitter voltage drops across the transistors have been shown to be 0.6 V. This is a guess. Typically, they will range from about 0.5 to nearly 0.7 V. Their exact values are difficult to predict. In practice, it is not unusual to find power amplifiers with adjustable (via a trimpot) bias networks.

By inspection of Fig. 12-27(b), and assuming that the base currents are neg-

FIGURE 12-27 "Practical" class AB power amplifier: (a) complete schematic; (b) dc equivalent circuit.

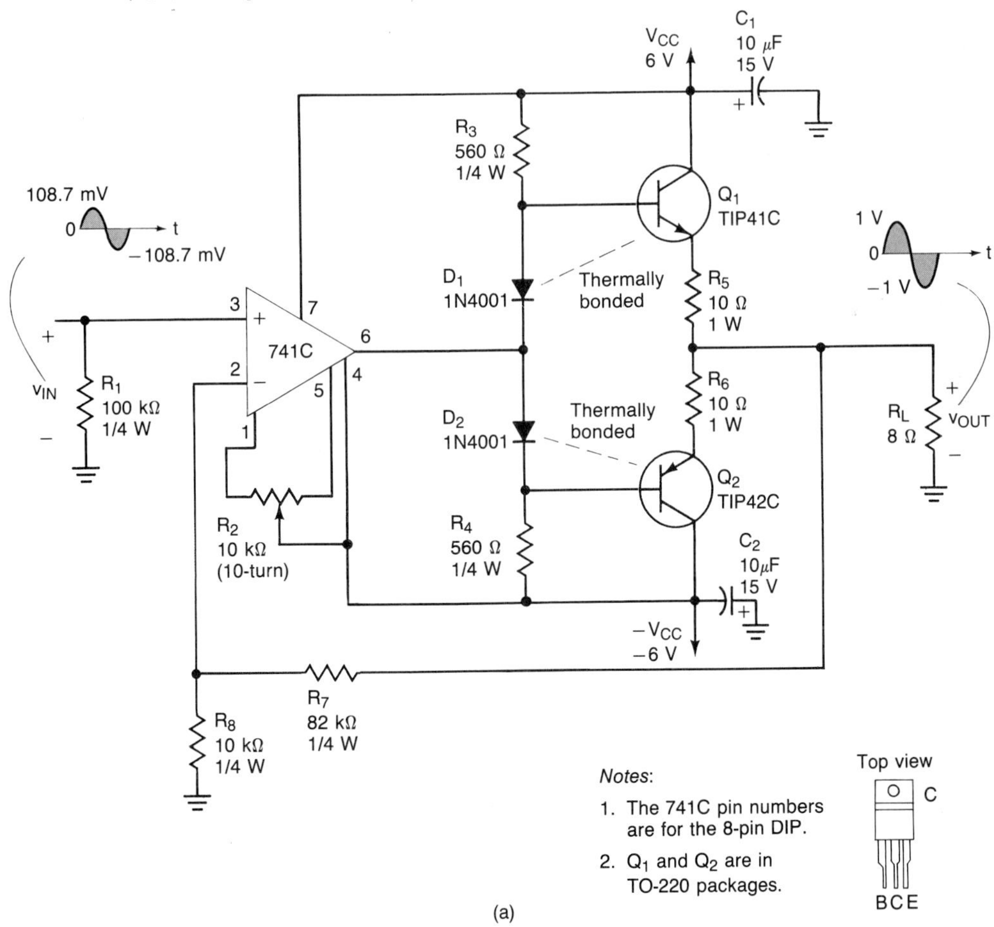

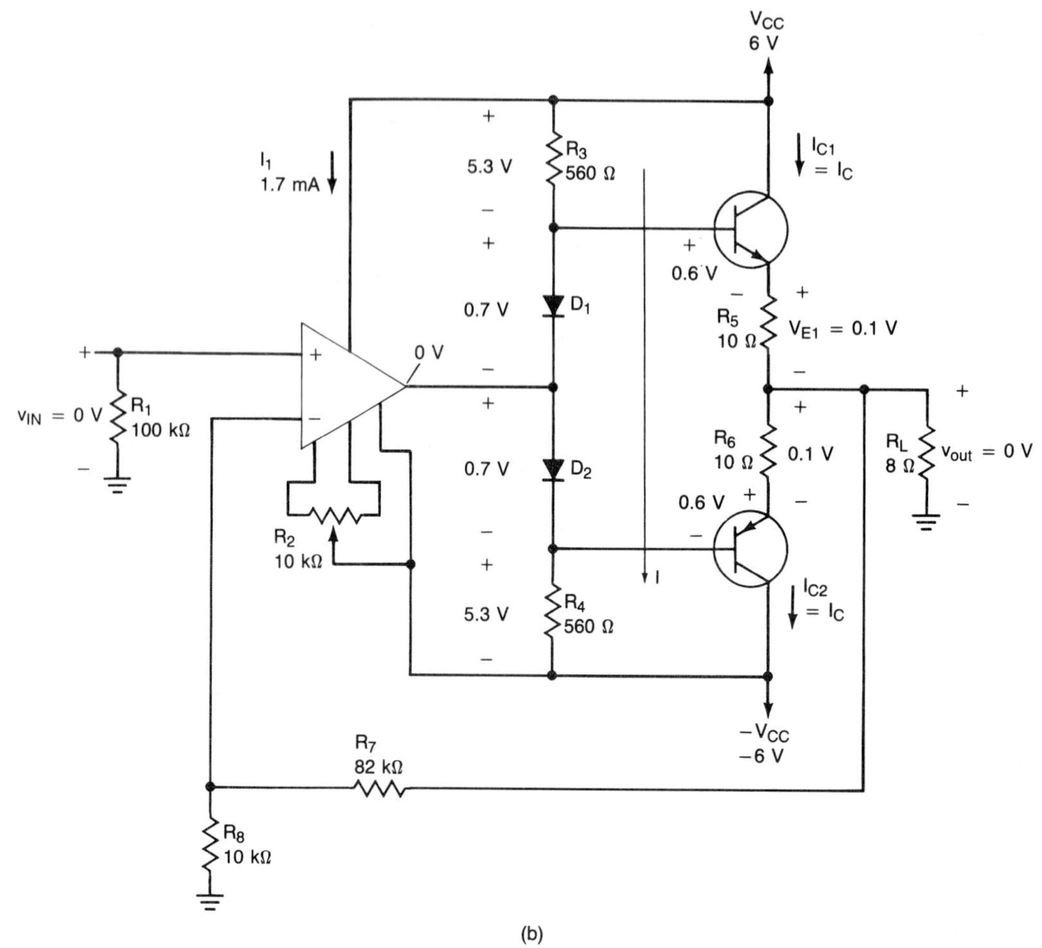

FIGURE 12-27 *(continued)*

ligible, we may find the bias current I. This current flows down through R_3, D_1, D_2, and R_4.

$$I \simeq \frac{V_{R3}}{R_3} = \frac{5.3\ \text{V}}{560\ \Omega} = 9.46\ \text{mA}$$

The collector current I_{C1} through transistor Q_1 is

$$I_{C1} = I_{C2} \simeq \frac{V_{E1}}{R_5} = \frac{0.1\ \text{V}}{10\ \Omega} = 10\ \text{mA}$$

By symmetry, and since there is 0 V across R_L, the collector currents of the two output transistors are equal.

We now know the dc bias current I, and the approximate dc collector current. To find the total current drawn from the dc power supply, we must also determine the op amp's required bias current. According to the manufacturer's data sheet, the no-signal (standby) current for the 741C op amp is typically 1.7 mA. It has been indicated in Fig. 12-27(b).

The total idling current for the power amplifier is given by Kirchhoff's current law.

$$I_{IN(standby)} = I_1 + I + I_C$$
$$= 1.7 \text{ mA} + 9.46 \text{ mA} + 10 \text{ mA} = 21.2 \text{ mA}$$

Since the total dc voltage impressed across the amplifier is 12 V, the amplifier's standby power dissipation is

$$P_{IN(standby)} = 2V_{CC}I_{IN(standby)}$$
$$= (12 \text{ V})(21.2 \text{ mA}) = 254 \text{ mW}$$

Admittedly, this is not much standby power. However, ideally we do not want to throw away any power. The ideal standby power is zero. The only way that we can reduce our standby power further is to lower I_1, I, and/or I_C.

The only way we can reduce the op amp's standby current is to replace the 741C with a low-power op amp (such as the LM324). The diode bias current I can be reduced by increasing the size of R_3 and R_4. However, we must ensure that enough bias current flows down through the diodes such that the op amp's output signal *never* turns them off. If this occurs, distortion will result. This potential problem will become more apparent as our discussion continues.

Reducing the quiescent current through the output transistors also poses drawbacks. The lower its value, the more crossover distortion effects (and other nonlinearities) become apparent.

We are *not* suggesting that the push-pull power amplifier given in Fig. 12-27(a) is an optimal design. We are merely pointing out the many considerations involved when striving for efficiency. Let us proceed with the analysis.

The negative feedback loop has been wrapped around the entire amplifier system. Therefore, its closed-loop voltage gain is determined rather easily.

$$A_{v(cl)} = \frac{V_{out}}{V_{in}} + 1 + \frac{R_7}{R_8} = 1 + \frac{82 \text{ k}\Omega}{10 \text{ k}\Omega} = 9.2$$

If the input voltage is 108.7 mV peak, the peak output voltage will be

$$v_{OUT} = A_{v(cl)}v_{IN} = (9.2)(108.7 \text{ mV peak}) = 1 \text{ V peak}$$

[see Fig. 12-27(a)].

The following equation allows us to determine the ac power developed in the load R_L.

$$P_{out(ac)} = \frac{(V_{out(peak)}/\sqrt{2})^2}{R_L} = \frac{V^2_{out(peak)}}{2R_L} \qquad (12\text{-}18)$$

For Fig. 12-27(a) we find that

$$P_{out(ac)} = \frac{V^2_{out(peak)}}{2R_L} = \frac{(1 \text{ V})^2}{2(8 \ \Omega)} = 0.0625 \text{ W}$$

The peak load current will be

$$I_{out(peak)} = \frac{V_{out(peak)}}{R_L} = \frac{1 \text{ V}}{8 \ \Omega} = 125 \text{ mA peak}$$

The signal current flowing through the output transistors will raise their average collector currents beyond I_C. This is indicated in

$$I_{C(\text{AVG})} = \frac{I_{\text{out(peak)}}}{\pi} + I_C \quad (12\text{-}19)$$

In our example, Eq. 12-19 yields

$$I_{C(\text{AVG})} = \frac{I_{\text{out(peak)}}}{\pi} + I_C = \frac{125 \text{ mA}}{\pi} + 10 \text{ mA} = 49.8 \text{ mA}$$

Under these conditions, the total dc input current is

$$I_{\text{in}} \simeq I_1 + \text{I} + \text{I}_{C(\text{AVG})} = 1.7 \text{ mA} + 9.46 \text{ mA} + 49.8 \text{ mA} \simeq 61 \text{ mA}$$

and the dc input power is

$$P_{\text{in(dc)}} = I_{\text{IN}}(2V_{CC}) = (61 \text{ mA})(12 \text{ V}) = 732 \text{ mW}$$

The efficiency may now be determined.

$$\% \, \eta = \frac{P_{\text{out(ac)}}}{P_{\text{in(dc)}}} \times 100\% = \frac{62.5 \text{ mW}}{732 \text{ mW}} \times 100\% = 8.54\%$$

This efficiency is considerably less than the theoretical 78.5%, but it is still somewhat better than the class A power amplifier analyzed in Section 12-6. Rest assured that we shall greatly improve on this situation.

To understand completely the power amplifier given in Fig. 12-27(a), we need to develop its various waveforms. This will give us respectable insight into the operation of this relatively popular amplifier configuration.

Recall that when Q_1 is conducting, transistor Q_2 is off. Consequently, during the positive half-cycle, we have the approximate situation shown in Fig. 12-28(a). Since the peak output current flows through Q_1's emitter resistor, it should not prove too surprising that the positive peak emitter voltage v_{E1} is defined by

$$v_{E1(\text{peak})} = i_{\text{OUT(peak)}}(R_E + R_L) \quad (12\text{-}20)$$

In our example, v_{E1} is

$$\begin{aligned} v_{E1(\text{peak})} &= i_{\text{OUT(peak)}}(R_E + R_L) \\ &= (0.125 \text{ A})(10 \,\Omega + 8\Omega) = 2.25 \text{ V} \end{aligned}$$

During the negative half-cycle, transistor Q_1 is nonconducting. Therefore, its emitter current and the voltage drop across its emitter resistor are both zero. Consequently, the voltage at the emitter of transistor Q_1 will be equal to the peak negative output voltage [see Fig. 12-28(b)].

By symmetry, the waveform at the emitter of transistor Q_2 is easily determined. Its emitter waveform will have a *negative* peak of 2.25 V and a *positive* peak of 1 V [refer to Fig. 12-28(c)].

Now let us find the base-to-ground voltage waveforms. During the positive half-cycle when Q_1's emitter voltage reaches 2.25 V, its base voltage will also be at its positive peak. Since its base-emitter *p-n* junction is forward biased, its base voltage will be one base-emitter drop more positive than its emitter.

$$v_{B1} = v_{E1} + 0.7 \text{ V} \quad (12\text{-}21)$$

In our example, Q_1's base voltage will be

$$v_{B1} = V_{E1} + 0.7 \text{ V} = 2.25 \text{ V} + 0.7 \text{ V} = 2.95 \text{ V}$$

The op amp's output voltage v_1 will be one diode (D_1) drop less than the base-to-ground voltage v_{B1}. Hence

$$v_1 = v_{B1} - v_{D1} \quad (12\text{-}22)$$

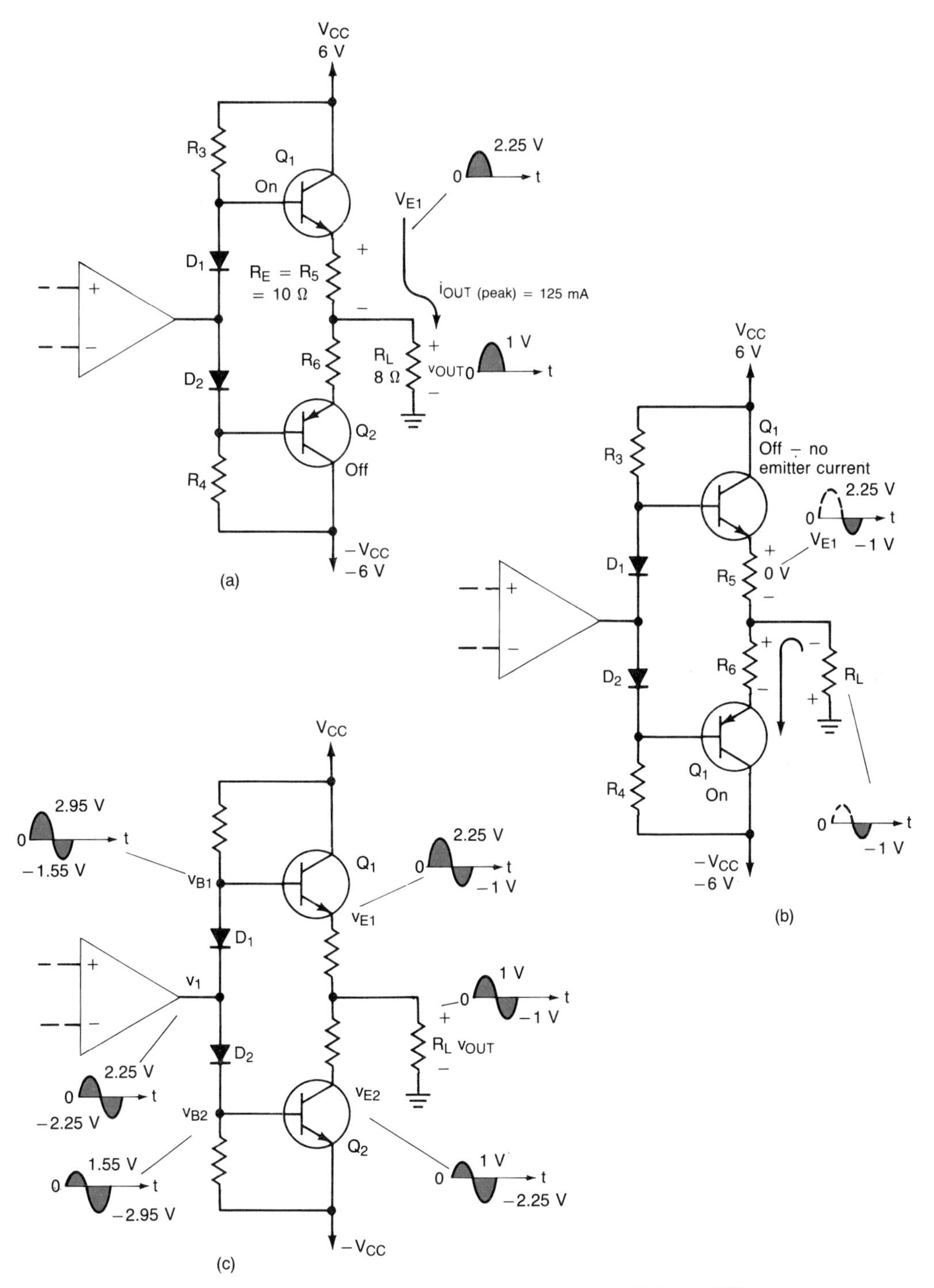

FIGURE 12-28 Amplifier's waveforms: (a) developing Q_1's emitter waveform during the positive half-cycle; (b) Q_1's emitter during the negative half-cycle; (c) waveform summary.

and for our amplifier we have

$$v_1 = v_{B1} - v_{D1} = 2.95\text{ V} - 0.7\text{ V} = 2.25\text{ V}$$

[refer to Fig. 12-28(c)].

Once again, by symmetry, we can see in Fig. 12-28(c) that during the negative half-cycle the base of transistor Q_2 will be one base-emitter drop *less* than the emitter voltage.

$$v_{B2} = v_{E2} - 0.7\text{ V} \tag{12-23}$$

Thus we see that

$$v_{B2} = v_{E2} - 0.7\text{ V} = -2.25\text{ V} - 0.7\text{ V} = -2.95\text{ V}$$

During the negative half-cycle, the output of the op amp will obviously be negative. However, its peak negative output voltage will be one diode (D_2) drop more positive than Q_2's base-to-ground voltage.

$$v_1 = v_{B2} + v_{D2} \tag{12-24}$$

As can be seen in Fig. 12-28(c),

$$v_1 = v_{B2} + v_{D2} = -2.95\text{ V} + 0.7\text{ V} = -2.25\text{ V}$$

The peak negative base-to-ground voltage of Q_1 will be one diode (D_1) drop more positive than the op amp's peak negative output voltage. This is stated by

$$v_{B1} = v_1 + v_{D1} \tag{12-25}$$

In our example, Eq. 12-25 produces

$$v_{B1} = -2.25\text{ V} + 0.7\text{ V} = -1.55\text{ V}$$

Now that we have found all the peak values, we can see how each of the waveforms in Fig. 12-28(c) was derived. Laboratory results agree very closely with the indicated values. Now that we understand the proper operation of the class AB power amplifier, let us investigate the causes of signal clipping.

As we mentioned previously, clipping may be produced by any (or all) of three possible sources. The output transistors, the bias diodes (D_1 and D_2), and the output of the op amp all pose potential clipping problems. Let us see why. First, we consider the output transistors.

If Q_1 is very close to saturation (fully ON) and Q_2 is in cutoff (OFF), we have the *approximate* situation shown in Fig. 12-29(a). (The approximation improves as larger values of V_{CC} are used.) By inspection,

$$i_{C(\text{SAT})} = \frac{V_{CC}}{R_E + R_L} = \frac{V_{CC}}{R_5 + R_L} = \frac{6\text{ V}}{10\ \Omega + 8\ \Omega} = 333.3\text{ mA}$$

The corresponding peak output voltage will be

$$\begin{aligned} v_{\text{OUT(peak)}} &= i_{C(\text{SAT})} R_L \\ &= (333.3\text{ mA})(8\ \Omega) = 2.67\text{ V} \end{aligned}$$

As a quick approximation, we may use simple voltage division between R_E and R_L. Specifically,

$$v_{\text{OUT(peak)}} = \frac{R_L}{R_L + R_E} V_{CC} = \frac{8\ \Omega}{8\ \Omega + 10\ \Omega}(6\text{ V}) = 2.67\text{ V}$$

This is perfectly legitimate-provided that we understand what we are doing.

By symmetry we can see that the peak negative output voltage is -2.67 V. These limits have been indicated in Fig. 12-29(b). If the amplifier is overdriven, the output will be clipped as shown in Fig. 12-29(c).

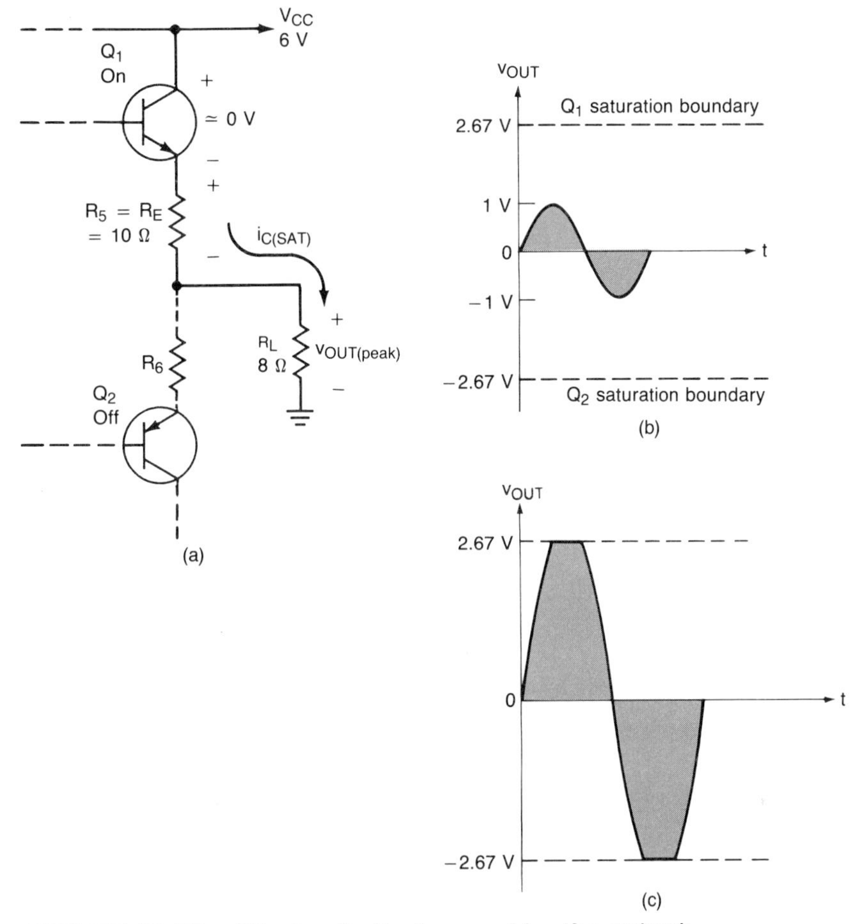

FIGURE 12-29 **Clipping limits imposed by the output transistors: (a) finding $i_{C(SAT)}$; (b) saturation boundaries; (c) output distortion produced by Q_1 and Q_2.**

Now let us investigate the possibility of diode clipping. The bias current I down through the diodes was previously determined to be 9.46 mA. As long as the bias diodes (D_1 and D_2) remain sufficiently forward biased, no appreciable distortion will occur. Consider the situation shown in Fig. 12-30(a).

During the positive half-cycle, the output of the op amp goes positive. The positive-going output voltage of the op amp serves to forward bias transistor Q_1. The ac base current may be regarded to flow as indicated in Fig. 12-30(a). However, the total instantaneous bias current will continue to flow *down* through the diode bias network. However, as the output voltage of the op amp swings positive, the net voltage across R_3 is reduced, and we must ensure that enough current flows down through R_3 to meet the base current requirements of Q_1. The total instantaneous base current i_B may be determined as follows:

$$i_{\text{OUT}} = i_E \simeq i_C$$

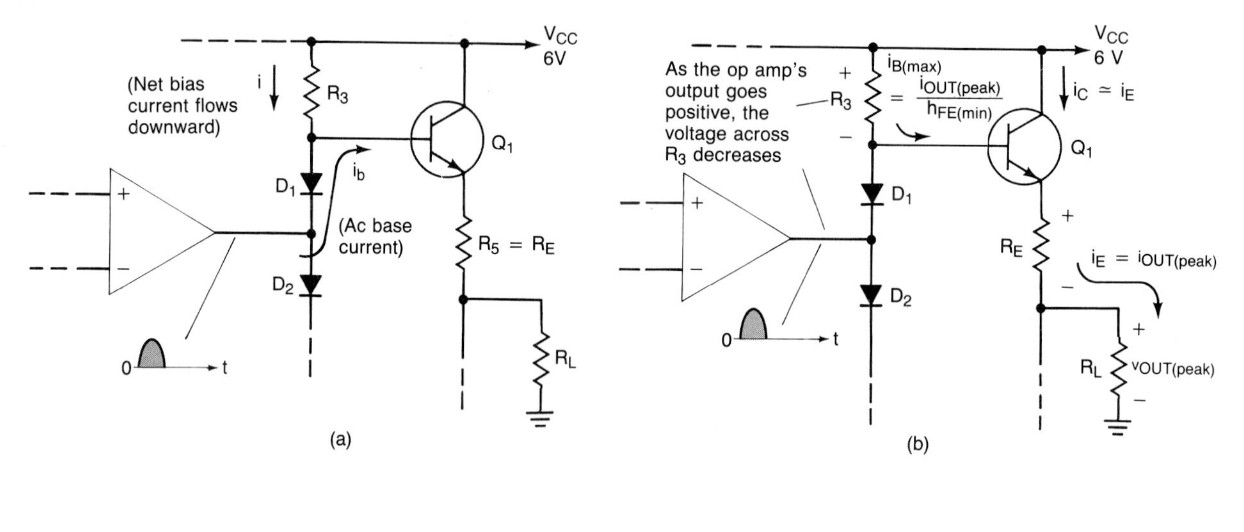

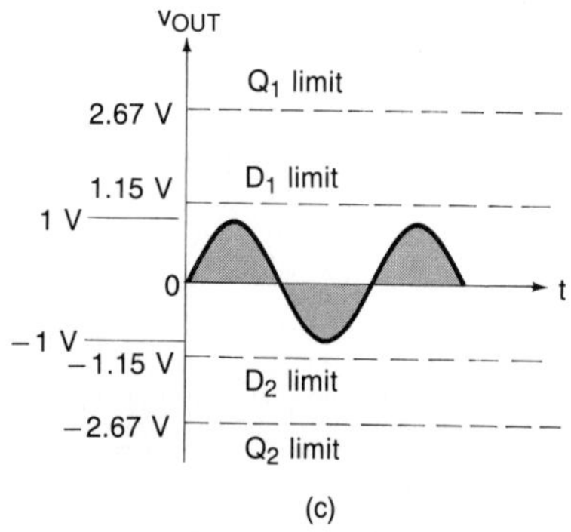

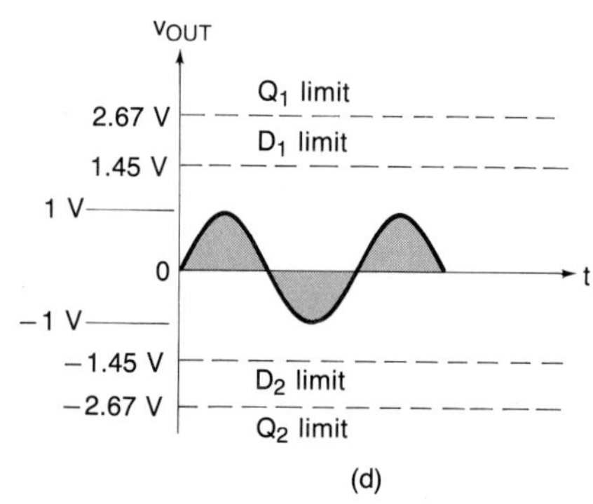

FIGURE 12-30 Clipping limits imposed by the bias diodes D_1 and D_2; (a) ac base current flows into the BJT, but the net bias current flows downward ($i_b < i$); (b) the total instantaneous base current flows down through R_3; (c) $h_{FE} = 30$ clipping limits; (d) $h_{FE} = 50$ clipping limits.

and since $h_{FE} = i_C/i_B$,

$$i_B = \frac{i_C}{h_{FE}} \simeq \frac{i_{\text{OUT}}}{h_{FE}}$$

The *minimum* h_{FE} for the TIP41C (and the TIP42C) bipolar transistors is 30. The *maximum* base current occurs for BJTs with the *minimum* h_{FE}.

$$i_{B(\text{max})} \simeq \frac{i_{\text{OUT}}}{h_{FE(\text{min})}}$$

However, since the total instantaneous base current must be provided through R_3, let us consider the problem carefully.

$$v_{\text{OUT(peak)}} = V_{CC} - i_{B(\text{max})}R_3 - V_{BE1} - i_{\text{OUT(peak)}}R_E$$

[see Fig. 12-30(b)]. Since

$$i_{\text{OUT(peak)}} = \frac{v_{\text{OUT(peak)}}}{R_L}$$

and

$$i_{B(\max)} = \frac{i_{\text{OUT(peak)}}}{h_{FE(\min)}} = \frac{v_{\text{OUT(peak)}}/R_L}{h_{FE(\min)}} = \frac{v_{\text{OUT(peak)}}}{h_{FE(\min)}R_L}$$

Then by substitution we can see that

$$v_{\text{OUT(peak)}} = V_{CC} - i_{B(\max)}R_3 - V_{BE1} - i_{(\text{OUT(peak)}}R_E$$

$$= V_{CC} - \frac{v_{\text{OUT(peak)}}}{h_{FE(\min)}R_L} R_3 - V_{BE1} - \frac{v_{\text{OUT(peak)}}}{R_L} R_E$$

Solving for the peak output voltage yields

$$v_{\text{OUT(peak)}} = \frac{V_{CC} - V_{BE1}}{1 + R_3/h_{FE(\min)}R_L + R_E/R_L} \qquad (12\text{-}26)$$

In our example,

$$v_{\text{OUT(peak)}} = \frac{V_{CC} - V_{BE1}}{1 + R_3/h_{FE(\min)}R_L + R_E/R_L}$$

$$= \frac{6\text{ V} - 0.7\text{ V}}{1 + 560\ \Omega/(30)(8\ \Omega) + 10\ \Omega/8\ \Omega} = 1.16\text{ V}$$

Do not allow Eq. 12-26 to become baffling. The larger $h_{FE(\min)}$ and R_L become, the greater the peak output voltage will become. Similarly, the peak output voltage can be increased by reducing the value of R_E and/or R_3. Consider the consequences.

We want a very large value of $h_{FE(\min)}$, but standard bipolar transistors simply will not provide us with large values of $h_{FE(\min)}$. We do not have much to say about R_L. Its value is dictated by the nature of the problem at hand. If we lower the value of R_E, we sacrifice bias stability and risk thermal runaway. Lowering R_3 raises the bias current I and increases our standby power dissipation.

As we can see in Fig. 12-30(c), it is quite possible that output clipping could result from the diodes rather than the output transistors. It all depends on the h_{FE} of the BJTs. For example, if we are fortunate enough to have BJTs with an h_{FE} of 50, Eq. 12-26 indicates that

$$v_{\text{OUT(peak)}} = \frac{V_{CC} - V_{BE1}}{1 + R_3/h_{FE(\min)}R_L + R_E/R_L}$$

$$= \frac{6\text{ V} - 0.7\text{ V}}{1 + 560\ \Omega/(50)(8\ \Omega) + 10\ \Omega/8\ \Omega} = 1.45\text{ V}$$

[see Fig. 12-30(d)].

The key point to remember is that the greater the h_{FE} is, the less the base current (or *drive requirement*) will be. However, we must always assume that our transistors will have the *minimum* h_{FE}.

The third potential source of clipping depends on the op amp's output. Most general-purpose op amps can deliver peak output currents which range from 10 mA to a maximum of 30 mA. Generally, distortion will begin to appear as the output current *nears* the op amp's current limit.

To determine the peak output current demanded from the op amp, it is necessary to draw the ac equivalent circuit [see Fig. 12-31(a)]. Recall that the ac resistance (the

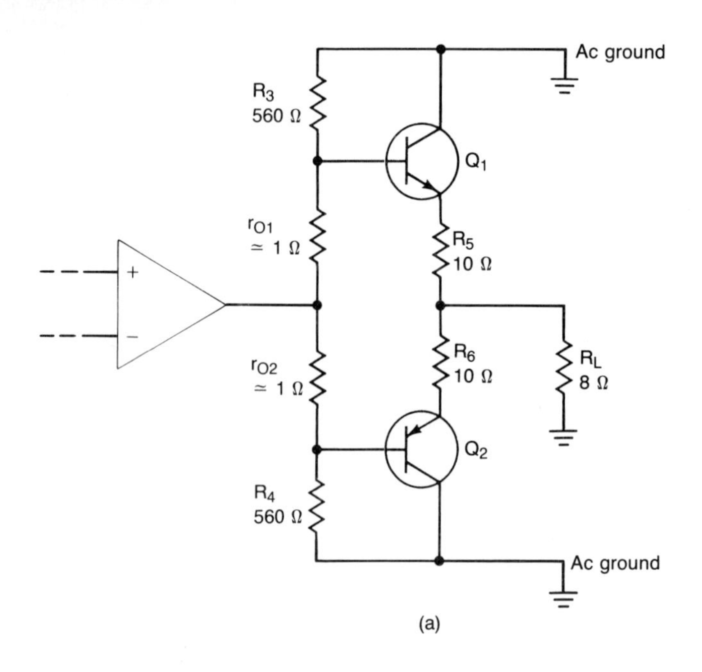

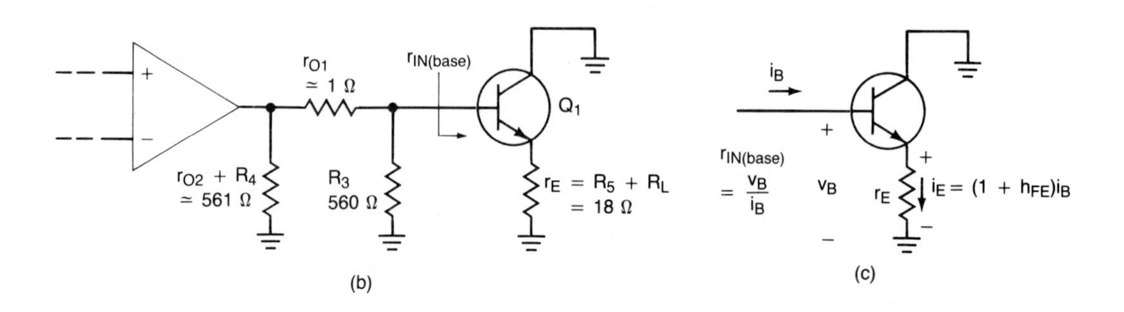

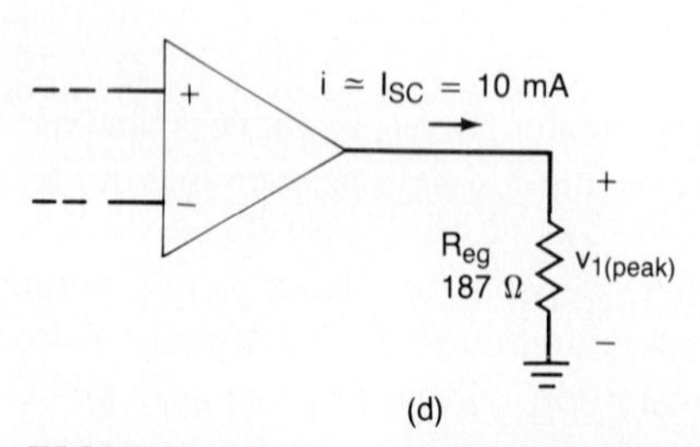

FIGURE 12-31 Finding the equivalent resistance across the op amp's output: (a) ac equivalent circuit; (b) simplified ac equivalent circuit; (c) finding $r_{IN(base)}$; (d) finding the maximum op amp output voltage.

dynamic resistance) of a diode is r_O (Chapter 3). At high current levels, the value of r_O approaches the bulk resistance of the diode. Assuming that the bulk resistance is 1 Ω, we arrive at the ac equivalent shown in Fig. 12-31(a).

Since only one output transistor is in conduction at a given time (e.g., Q_2 is in cutoff during the positive half-cycle), the ac equivalent circuit may be simplified as shown in Fig. 12-31(b). To determine the ac equivalent resistance "looking" into the base of transistor Q_1, we must digress slightly.

Referring to Fig. 12-31(c), the argument proceeds as follows:

$$\begin{aligned} i_E &= i_C + i_B = h_{FE}i_B + i_B \\ &= (h_{FE} + 1)i_B \end{aligned}$$

Further,

$$v_B \simeq v_E = i_E r_E = (h_{FE} + 1)i_B r_E$$

and

$$\boxed{r_{\text{IN(base)}} = \frac{v_B}{i_B} \simeq (h_{FE} + 1)r_E} \tag{12-27}$$

where $r_{\text{IN(base)}}$ = large-signal BJT input resistance
h_{FE} = dc current gain
r_E = ac equivalent emitter resistance

If we again use the minimum h_{FE} of 30, we may find the minimum $r_{\text{IN(base)}}$ for Q_1. Thus, from Fig. 12-31(b),

$$r_E = R_5 + R_L = 10\ \Omega + 8\ \Omega = 18\ \Omega$$

and from Eq. 12-27,

$$\begin{aligned} r_{\text{IN(base)}} &\simeq (h_{FE} + 1)r_E = (30 + 1)(18\ \Omega) \\ &= 558\ \Omega \end{aligned}$$

The equivalent resistance across the output of the op amp may now be found [see Fig. 12-31(b) and (d)].

$$\begin{aligned} R_{\text{eq}} &= (R_4 + r_{O2}) \parallel (r_{O1} + R_3 \parallel r_{\text{IN(base)}}) \\ &= (561\ \Omega) \parallel (1\ \Omega + 560\ \Omega \parallel 558\ \Omega) = 187\ \Omega \end{aligned}$$

The 741C op amp can supply between 10 and 30 mA of output current. This is typically denoted as the short-circuit output current I_{SC} on its data sheet. If we take the worst case, the output of a given 741C op amp may clip if its output current exceeds 10 mA peak. Given that the equivalent resistance across its output is 187 Ω, this translates to a peak output voltage of

$$v_{1\text{(peak)}} = I_{SC}R_{\text{eq}} = (10\text{ mA})(187\ \Omega) = 1.87\text{ V peak}$$

In this case, the output of the op amp will clip at only 1.87 V peak [see Fig. 12-32(a)]. The base, emitter, and output voltage waveforms were developed in exactly the same manner as presented above. The reader should verify their peak values.

Obviously, as shown in Fig. 12-32(b), the op amp poses the severest limitation. Keep in mind, however, that Fig. 12-32(b) represents the worst-case situations. Specifically, the h_{FE} of the transistors has been assumed to be the minimum (30), and the op amp's maximum output current has also been assumed to be at its minimum

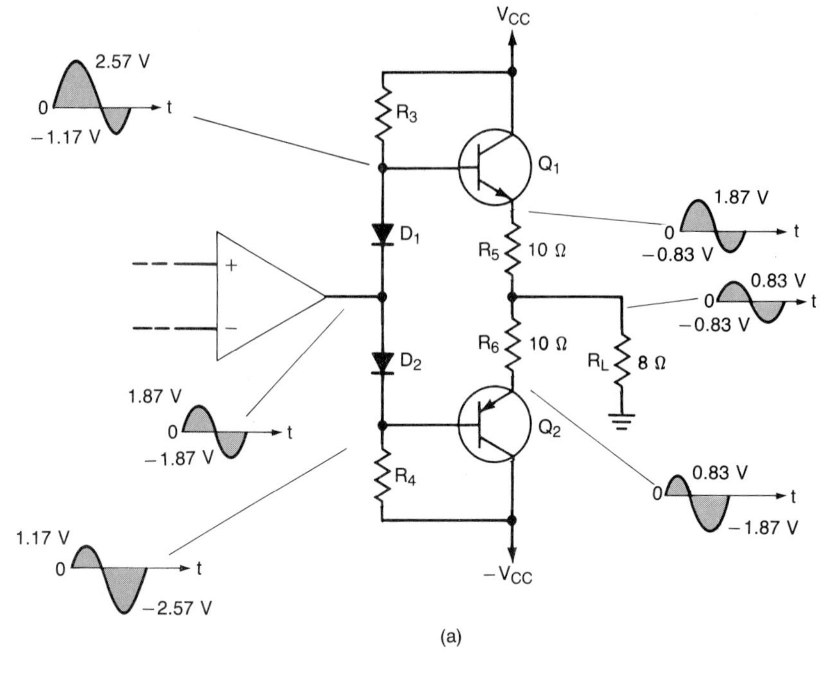

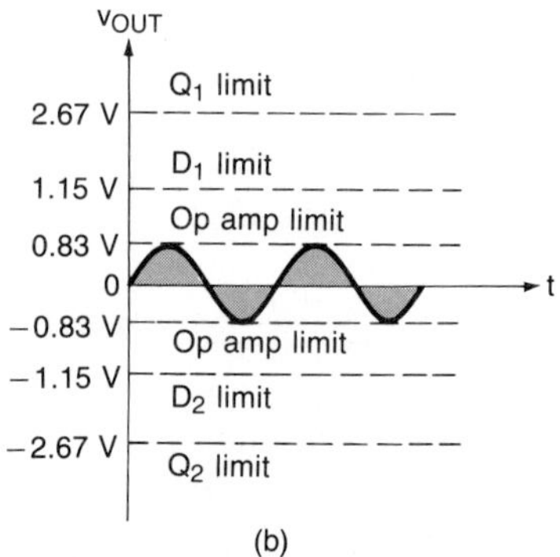

FIGURE 12-32 **Worst-case output limits if I_{SC} = 10 mA and h_{FE} = 30; (a) maximum peak-to-peak waveforms; (b) output limits.**

(10 mA). In the real world, we must face up to these facts if we are to develop and/or troubleshoot amplifier designs successfully.

The two most serious limitations - the diode bias and the op amp's peak output current - are directly related to the h_{FE} of the output transistors. The lower the h_{FE}, the more base current our output transistors require and the more acute the clipping problem becomes. The input (base) current is often termed the output transistors' input *drive requirement*.

Obviously, we want to minimize the drive requirement. For this reason, several alternative output devices (and new technologies) have been developed. In the next three sections we learn about these devices and see how they may be used to improve the performance of our power amplifiers.

12-11 Darlington Pairs

There are two fundamental approaches to improving the performance of the power amplifier given in Fig. 12-27(a). First, we can include an intermediate driver stage to interface between the output of the op amp and the input of the power transistors. Second, we can use output devices that require less input drive.

The *Darlington configuration* using discrete transistors represents the first alternative [see Fig. 12-33(a)]. The two-transistor combination may be thought of as a single, high-performance transistor with the three basic terminals. These have been emphasized in Fig. 12-33(a) by B, C, and E.

By inspection of Fig. 12-33(a), we can see that

$$i_{C1} = h_{FE1} i_{B1}$$

and

$$i_{E1} \simeq i_{C1} = h_{FE1} i_{B1}$$

Since the emitter of Q_1 is connected to the base of Q_2,

$$i_{B2} = i_{E1} \simeq h_{FE1} i_{B1}$$

and

$$\begin{aligned} i_{C2} &= h_{FE2} i_{B2} = h_{FE2}(h_{FE1} i_{B1}) \\ &= h_{FE1} h_{FE2} i_{B1} \end{aligned}$$

Since the base of our composite transistor is the base of Q_1 and the collector is essentially that of Q_2, the total current gain $h_{FE(\text{total})}$ is given by

$$\boxed{h_{FE(\text{total})} = \frac{i_{C2}}{i_{B1}} = h_{FE1} h_{FE2}} \tag{12-28}$$

where $h_{FE(\text{total})}$ = current gain of the Darlington pair
h_{FE1} = current gain of the input BJT
h_{FE2} = current gain of the output BJT

Consider this result. The overall current gain is equal to the *product* of the individual BJT gains. For example, if Q_1 is a 2N3904 transistor with a minimum h_{FE} of 100 and Q_2 is our TIP41C with a minimum h_{FE} of 30, the total gain is (from Eq. 12-28)

$$h_{FE(\text{total})} = h_{FE1} h_{FE2} = (100)(30) = 3000$$

This very high current gain indicates that much less base (drive) current is required. The *pnp* complement Darlington using the 2N3906 and TIP42C transistors has been indicated in Fig. 12-33(b).

Since the Darlington pair may be thought of as a high-beta transistor with three terminals, we may quite easily modify the power amplifier given in Fig. 12-27(a) [see Fig. 12-33(c)]. Note that four bias diodes (D_1 through D_4) are now required to forward bias the base-emitter *p-n* junctions of each of the four output transistors.

Several enhancements are now possible. Since the base drive has been reduced, the bias resistors R_3 and R_4 have been increased to reduce the bias current I. This offers two immediate advantages. First, lowering I reduces the idling current and the

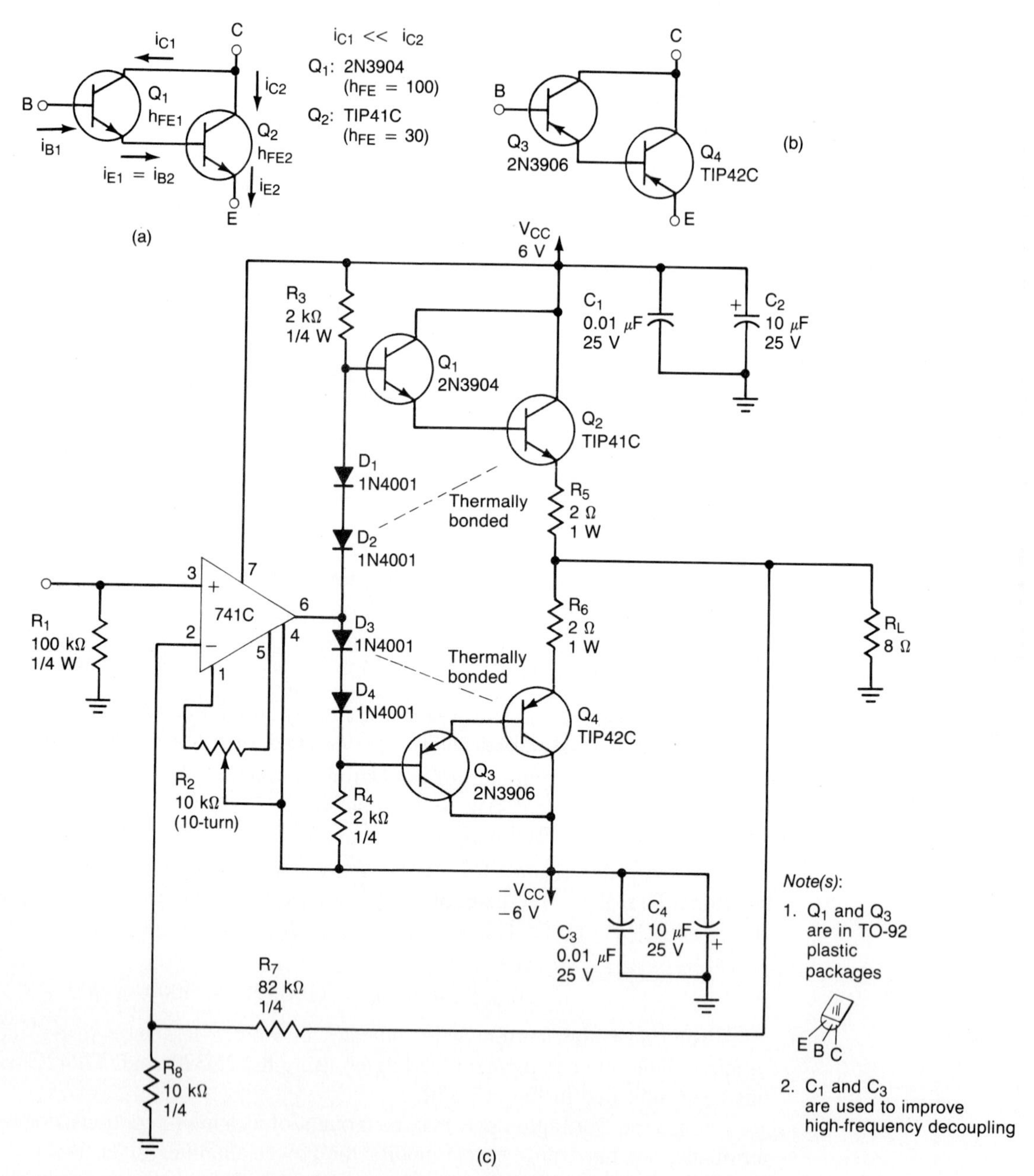

FIGURE 12-33 Darlington pair and its application: (a) npn; (b) pnp; (c) 640-mW class AB push-pull power amplifier.

standby power dissipation. Second, increasing the bias resistors also reduces the loading on the output of the op amp.

EXAMPLE 12-5

Given the power amplifier in Fig. 12-33(c), assume that v_{OUT} is 3.2 V peak. Determine the ac output power, the peak ac output current, the approximate dc input power, and the efficiency.

SOLUTION The ac output power is

$$P_{out(ac)} = \frac{v^2_{OUT(peak)}}{2R_L} = \frac{(3.2\text{ V})^2}{(2)(8\ \Omega)} = 0.64\text{ W}$$

The peak output current is

$$i_{OUT(peak)} = \frac{v_{OUT(peak)}}{R_L} = \frac{3.2\text{ V}}{8\ \Omega} = 400\text{ mA}$$

To find the dc input power, we must first find the approximate average collector current.

$$I_C \simeq \frac{i_{OUT(peak)}}{\pi} = \frac{400\text{ mA}}{\pi} = 127\text{ mA}$$

The dc input power is

$$P_{in(dc)} \simeq 2V_{CC}I_C = (2)(6\text{ V})(127\text{ mA}) = 1.53\text{ W}$$

and the efficiency is

$$\%\ \eta = \frac{P_{out(ac)}}{P_{in(dc)}} \times 100\% = \frac{0.64\text{ W}}{1.53\text{ W}} \times 100\% = 42\%$$

■

Before we proceed further, let us ponder the results of this example. By adding two transistors (and two diodes), we have lowered the drive requirement, which has allowed us to squeeze out much more output power. As an added bonus, the efficiency has been vastly improved. (Notice that we did not include the BJTs' Q-point current in our efficiency calculations. Laboratory work indicated the BJT no-signal current was only 14 mA, which is relatively small compared to our average collector currents of 127 mA. The measured efficiency was 37.4% and included the diode bias current as well as the BJT Q-point current.)

EXAMPLE 12-6

Sketch the emitter waveforms for Q_2 and Q_4, and the base waveforms for Q_1 and Q_3 [see Fig. 12-33(c)]. Also sketch the op amp's output waveform.

SOLUTION Following the same basic procedure outlined in Section 12-10, we first find Q_2's positive peak.

$$v_{E2} = i_{OUT(peak)}(R_5 + R_L)$$
$$= (400\text{ mA})(2\ \Omega + 8\ \Omega) = 4\text{ V}$$

The negative peak will be -3.2 V. By symmetry, we now know that Q_4's emitter waveform will have a positive peak of 3.2 V and a negative peak of 4 V. Transistor

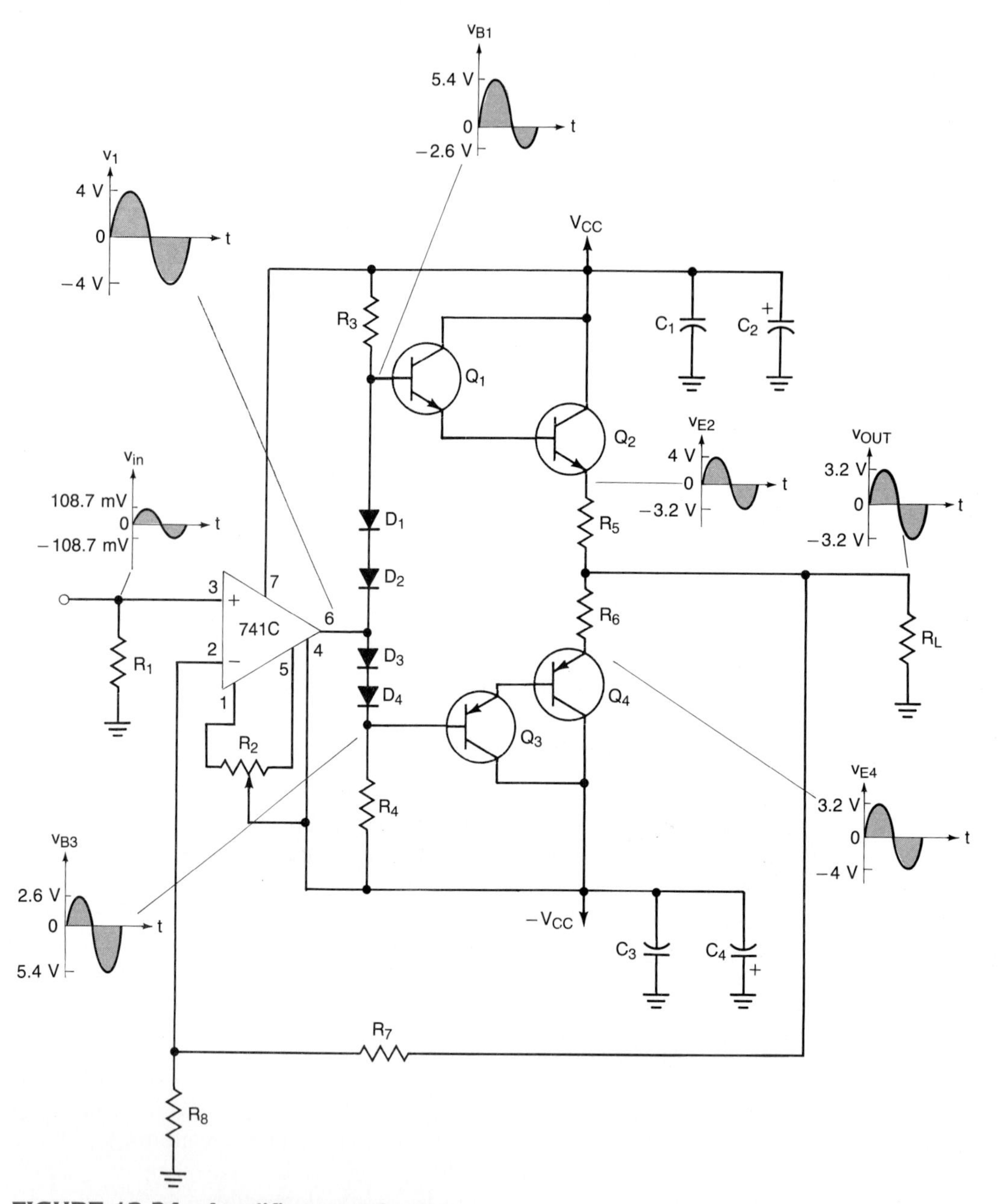

FIGURE 12-34 Amplifier waveforms.

Q_1 will have a base waveform with a positive peak voltage 1.4 V (two base–emitter drops) more positive than Q_2's emitter waveform.

$$v_{B1} = v_{E2} + 1.4\text{ V} = 4\text{ V} + 1.4\text{ V} = 5.4\text{ V}$$

Simultaneously, the op amp will have a positive peak output voltage that is two diode drops less than v_{B1}.

$$v_1 = v_{B1} - 1.4\text{ V} = 5.4\text{ V} - 1.4\text{ V} = 4\text{ V}$$

By symmetry, its peak negative output will be -4 V. This negative peak allows us to find quite easily the negative peak base voltage v_{B1}. It will be two diode drops more positive. Hence

$$v_{B1} = v_1 + 1.4 \text{ V} = -4 \text{ V} + 1.4 \text{ V} = -2.6 \text{ V}$$

Now we can see that Q_3's base voltage will have a positive peak of 2.6 V and a negative peak of -5.4 V. These various waveforms have been indicated in Fig. 12-34. ■

EXAMPLE 12-7

Determine the maximum peak-to-peak output voltage that can be provided by the power amplifier given in Fig. 12-33(c). Assume that the total h_{FE} of the Darlington pairs has a minimum value ($h_{FE(\text{total-min})}$) of 3000.

SOLUTION First, we shall consider the output transistors. (We temporarily ignore base drive requirements and apply voltage division.)

$$v_{\text{OUT(peak)}} = \frac{R_L}{R_L + R_5} V_{CC} = \frac{8\ \Omega}{8\ \Omega + 2\ \Omega}(6 \text{ V}) = 4.8 \text{ V}$$

Therefore, the output transistors can (ideally) produce a maximum output voltage of 9.6 V peak to peak. Now let us consider the bias diodes. A slight modification of Eq. 12-26 greatly simplifies our task.

$$v_{\text{OUT(peak)}} = \frac{V_{CC} - (V_{BE1} + V_{BE2})}{1 + R_3/h_{FE(\text{total-min})}R_L + R_5/R_L}$$

$$= \frac{6 \text{ V} - 1.4 \text{ V}}{1 + 2\text{ k}\Omega/(3000)(8\ \Omega) + 2\ \Omega/8\ \Omega} = 3.45 \text{ V}$$

Under the worst-case considerations, the bias diodes will permit a maximum output voltage of 6.9 V peak to peak. The op amp's I_{SC} will be assumed to be 10 mA. The ac equivalent circuit across its output has been illustrated in Fig. 12-35. Hence

$$r_{\text{IN(base)}} \simeq (1 + h_{FE(\text{total-min})})(R_5 + R_L)$$

$$= (1 + 3000)(2\ \Omega + 8\ \Omega) \simeq 30 \text{ k}\Omega$$

and by inspection, and neglecting the diode resistances,

$$R_{\text{eq}} \simeq R_4 \parallel R_3 \parallel r_{\text{IN(base)}}$$

$$= 2\text{ k}\Omega \parallel 2\text{ k}\Omega \parallel 30\text{ k}\Omega = 968\ \Omega$$

The peak output voltage is

$$v_{1\text{(peak)}} = I_{\text{SC}}R_{\text{eq}} = (10 \text{ mA})(968\ \Omega) = 9.68 \text{ V}$$

Consider this result carefully. The equivalent resistance across the output of the op amp is sufficiently large such that the peak current limitation will *not* be reached. However, with a V_{CC} of 6 V, the op amp's output will saturate at about 4 V. Therefore, the op amp limits the maximum peak-to-peak output voltage to 6.4 V peak to peak as shown in Fig. 12-34. ■

The usefulness of the Darlington configuration has prompted many manufacturers to provide *monolithic power Darlington transistors*. The transistors are three-terminal devices with the equivalent schematic diagram shown in Fig. 12-36(a). The resistances (R_1 and R_2) serve to improve the transistor's switching performance (from

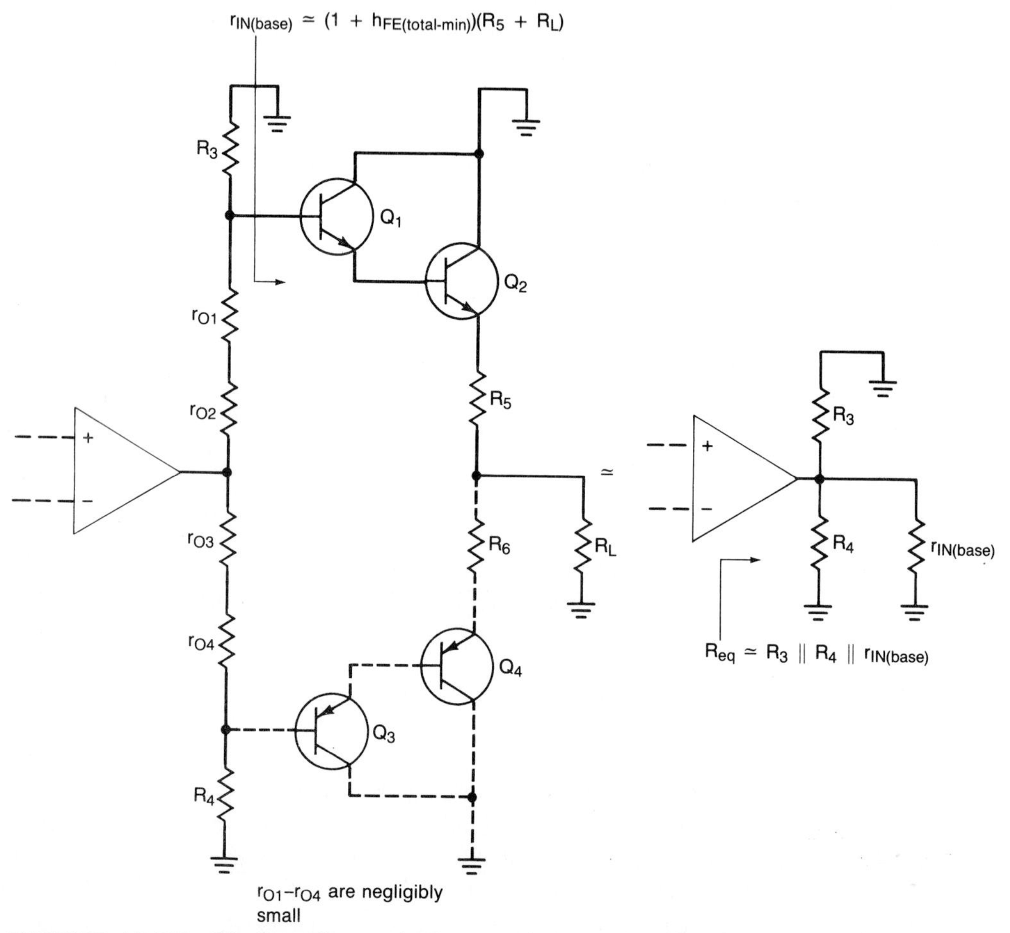

FIGURE 12-35 Finding the equivalent resistance loading the op amp's output.

on to off), and also minimize the effects of leakage current. The diode D is a "parasitic" device that does not impair the performance of the transistor since it is normally reverse biased. In fact, it helps protect the transistor from reverse voltage transients that typically occur when the transistor is used to switch inductive loads. However, it is not an extremely fast diode and therefore cannot always be relied on for good transient suppression. The actual fabrication of the monolithic power Darlington transistor has been detailed in Fig. 12-36(b). The transistor is described as being a "double-epitaxial, single-diffused device."

The electrical schematic of a *pnp* Darlington power transistor has been shown in Fig. 12-36(c). Typically, the details (e.g., R_1, R_2, and diode D) are not included in their schematic symbol [see Fig. 12-36(d)].

In this case the TIP120 and the TIP125 are used as the complementary output transistors. The minimum $h_{FE(\text{total})}$ is specified by the manufacturer as 1000. Although

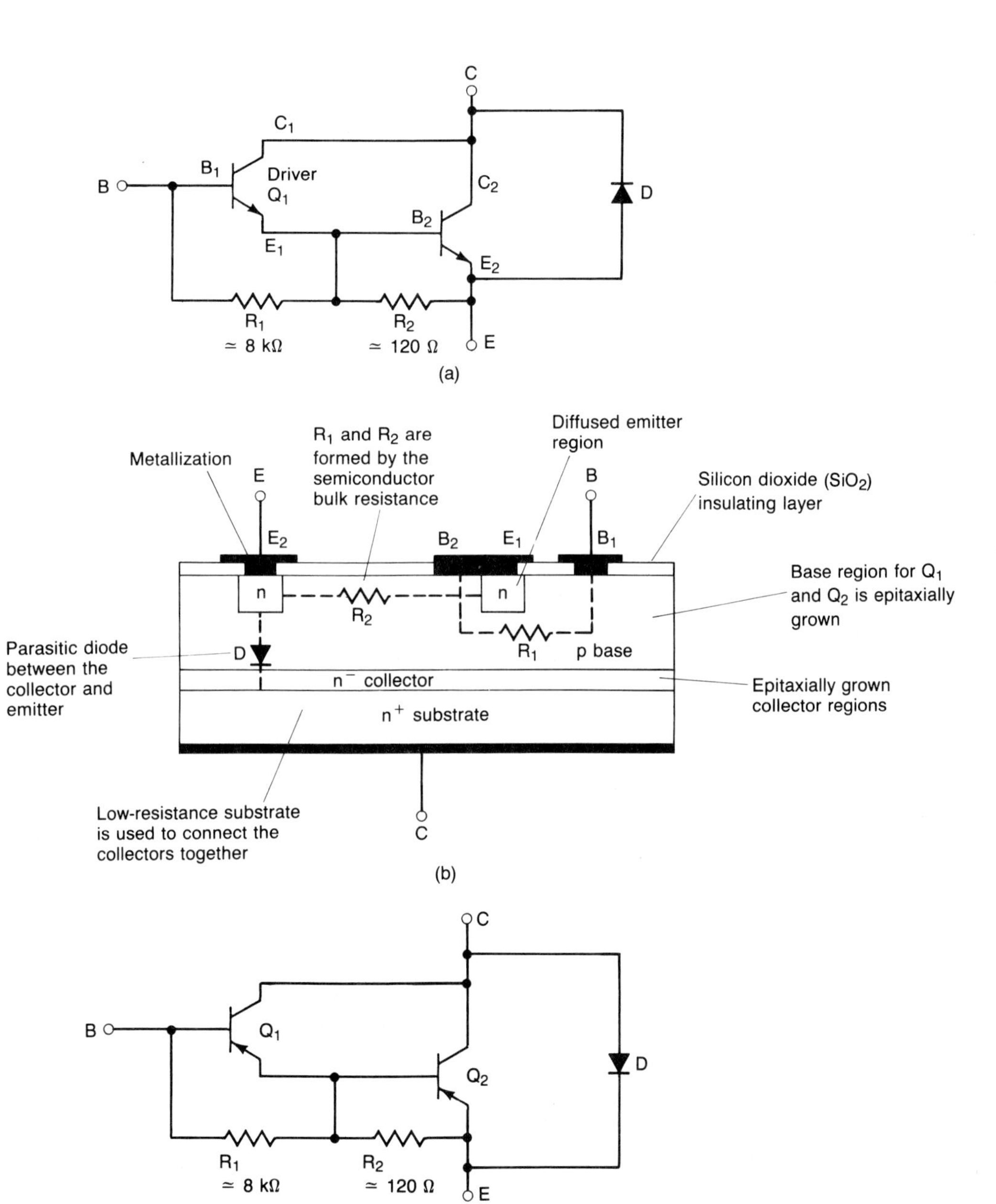

FIGURE 12-36 Monolithic Darlington transistor: (a) TIP120 (npn) equivalent circuit; (b) its structure; (c) TIP125 (pnp) equivalent circuit; (d) push-pull power amplifier with Darlington output transistors.

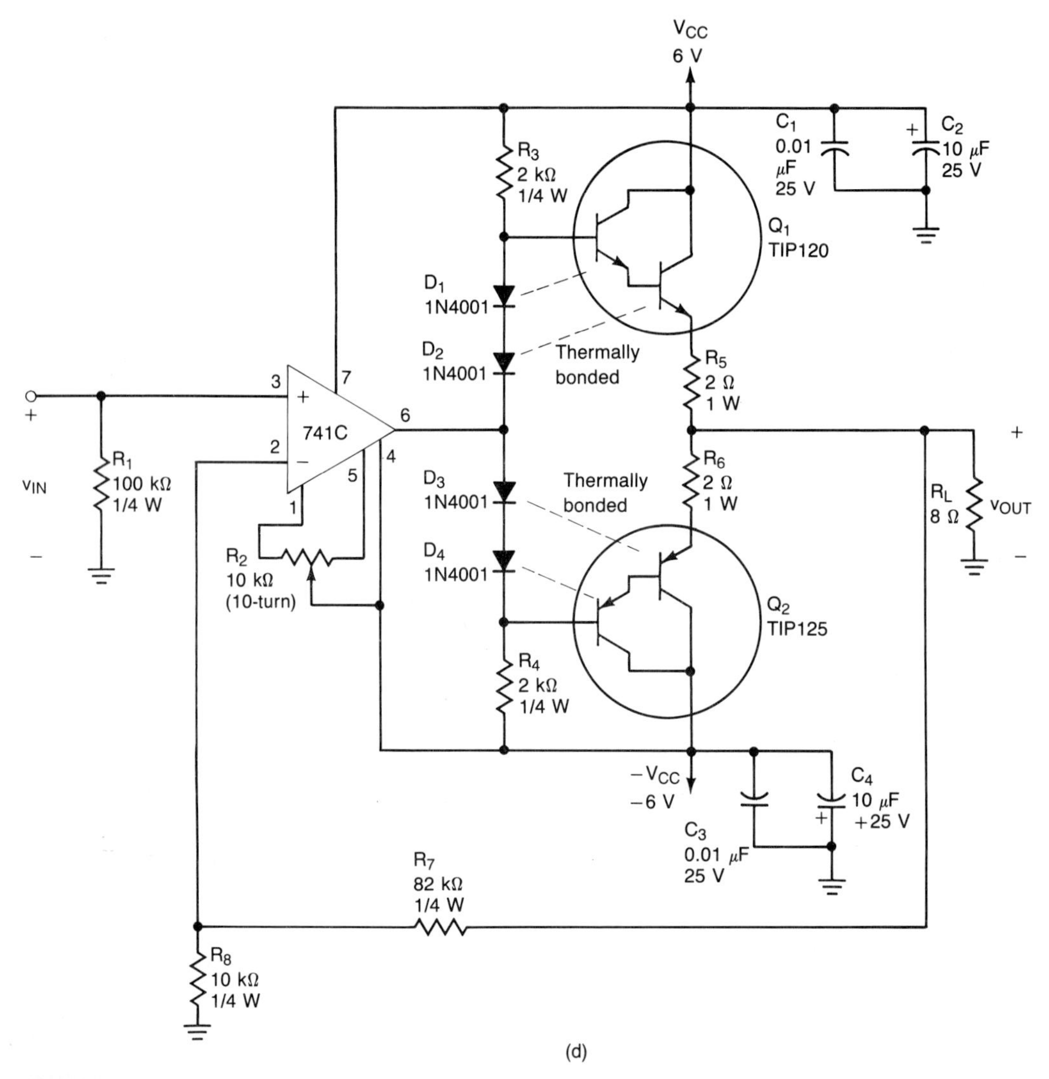

FIGURE 12-36 *(continued)*

this is a lower minimum than the 3000 we achieved via our discrete approach, we have reduced the circuit complexity by using only two transistors as opposed to four. [In the problems at the end of the chapter, the reader will be asked to analyze the power amplifier given in Fig. 12-36(d).]

12-12 Quasi-complementary Symmetry

Quite often - particularly when high power output levels are required, it is difficult to find closely matched complementary pairs. Further, the reliability of silicon *npn* BJTs exceeds that of the silicon *pnp* BJTs at elevated junction temperatures. Specifically,

$$\text{failure ratio} = \frac{\text{failure rate at } T_j}{\text{failure rate at } 25°\text{C}}$$

For a junction temperature T_j of 175°C, silicon BJTs demonstrate a failure ratio of about 10:1. For silicon *pnp* BJTs at the same temperature, the failure ratio is

FIGURE 12-37 Quasi-complementary symmetry: (a) quasi-pnp Darlington pair; (b) the pnp Darlington pair has been replaced with Q_3 and Q_4.

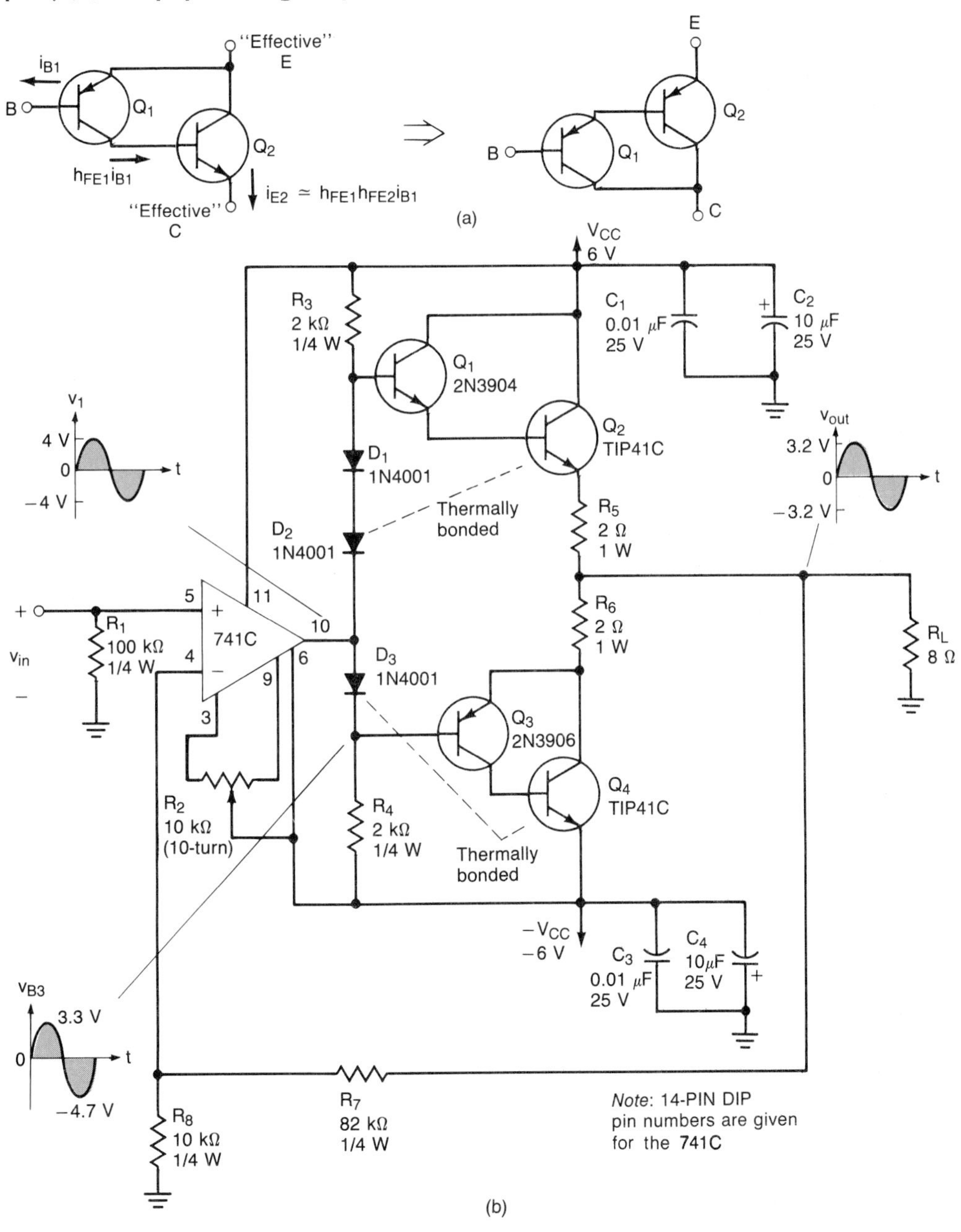

approximately 19:1. Consequently, circuit designers are often prompted to search for alternatives to the standard complementary output transistors. One approach is to utilize quasi-complementary pairs. Very simply, by using a *pnp–npn* BJT combination, we may derive an equivalent *pnp* power Darlington transistor [see Fig. 12-37(a)].

The total current gain of the quasi-*pnp* Darlington is also equal to the product of the individual transistor current gains. The configuration may be thought of as a single *pnp* Darlington power transistor with the equivalent collector, emitter, and base terminals indicated in Fig. 12-37(a). It may be used in our power amplifier as shown in Fig. 12-37(b). The resulting configuration is called a *quasi-complementary symmetry power amplifier*. Observe that only three bias diodes are required.

The operation of the amplifier is virtually identical to that of the amplifier given in Fig. 12-34. The only waveform that will be altered is the one that appears at the base of the transistor Q_3. This is true because we have eliminated the need for one of the bias diodes (D_4). Consequently, the peak values for v_{B3} are only *one* diode drop less than the peak output voltage appearing at the op amp's output. The positive peak is

$$v_{B3} = v_1 - 0.7 \text{ V} = 4 \text{ V} - 0.7 \text{ V} = 3.3 \text{ V}$$

and the negative peak voltage is

$$v_{B3} = v_1 - 0.7 \text{ V} = -4 \text{ V} - 0.7 \text{ V} = -4.7 \text{ V}$$

[see Fig. 12-37(b)].

We do encounter an additional constraint in the quasi-complementary symmetry power amplifier. Consider the voltage drop across R_3. It must always be large enough to provide sufficient reverse collector-to-base bias for Q_1. We handled this problem indirectly by ensuring that enough current flows through R_3 to meet Q_1's peak input current demands.

By inspection of Fig. 12-37(b), we see that the voltage drop across resistor R_4 must supply Q_3's collector-to-base reverse bias *and* the forward base-emitter drop required by Q_4. Consequently, the minimum voltage drop across R_4 is required to be at 0.7 V larger than the drop across R_3. Hence, if output clipping occurs, the Q_3-Q_4 bias network is probably the culprit.

12-13 Power MOSFETs: VMOS and DMOS

The bipolar and Darlington power transistors enjoy widespread use in power applications. However, the bipolar power transistor requires relatively "hefty" input drive circuitry, and the Darlington typically exhibits relatively large collector-to-emitter saturation voltages [$V_{CE(\text{SAT})}$].

In Fig. 12-38 we see a TIP41C power transistor and a TIP120 Darlington power transistor that have just reached saturation. The h_{FE} of the TIP41C is assumed to be 10 and the h_{FE} of the TIP120 is assumed to be 1000. Notice their respective base currents. We have seen the need for a low input drive requirement, but we have not yet investigated the need for a low saturation (ON) voltage.

As we can see in Fig. 12-38, the TIP41C with its $V_{CE(\text{SAT})}$ of 1.5 V at an $I_{C(\text{SAT})}$ of 6 A dissipates 9 W. However, the TIP120 with a $V_{CE(\text{SAT})}$ of 4 V at an $I_{C(\text{SAT})}$ of

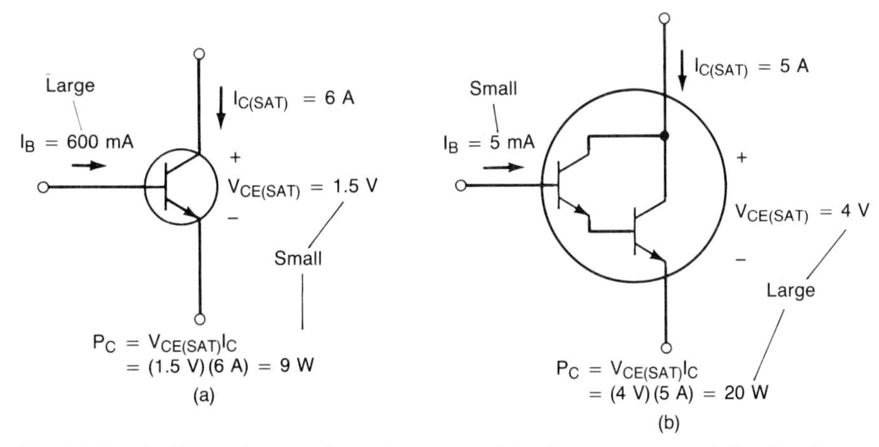

FIGURE 12-38 Comparison between bipolar power and Darlington (bipolar) power transistors: (a) TIP41C (h_{FE} = 10); (b) TIP120 (h_{FE} = 1000).

only 5 A dissipates a respectable *20 W* of collector power! Recall that the ideal $V_{CE(\mathrm{SAT})}$ is zero. If this were achievable, the collector dissipation would be zero when the BJT is in saturation. In contrast, the TIP120 throws away 20 W of power.

The ideal power transistor should require zero input drive current and have an output saturation voltage of zero. Its frequency response should also be extremely good (dc to light would be nice). In their quest for the ideal power transistor, many manufacturers have developed *power MOSFETs*.

The first commercially available power MOSFETs appeared in 1976. Power MOSFETs are characterized by very small input drive current, freedom from secondary breakdown effects (which greatly improves its SOA), and a better high-frequency response than BJTs - particularly in switching applications. However, they are not clearly superior to the bipolar transistors in all applications. Power MOSFETs have their own set of limitations. For example, their input capacitances must be charged and discharged by their drive circuits. Further, power MOSFETs contain a parasitic bipolar transistor, which may cause problems in some switching applications. The MOSFET's saturation voltage tends to increase more than a similar bipolar transistor as temperature is increased. However, their inherent advantages often merit their serious consideration in many power applications. This relatively new and continually emerging technology deserves our attention.

In Chapter 5 we introduced the JFET, the DE-MOSFET, and the E-MOSFET, but these were *small-signal devices* that are *not* designed to accommodate large power levels. Power MOSFETs are typically enhancement-only MOSFETs (E-MOSFETs), which are usually designed for *vertical* drain current flow as opposed to *horizontal* current flow. Consider Fig. 12-39(a).

Fig. 12-39(a) illustrates the familiar small-signal, *n*-channel E-MOSFET. Its key limitation in power applications is its channel length. The longer the channel, the greater the transconductance. However, as the length of the channel increases, so does its resistance.

When the small-signal MOSFET is turned fully ON, or saturated, its channel

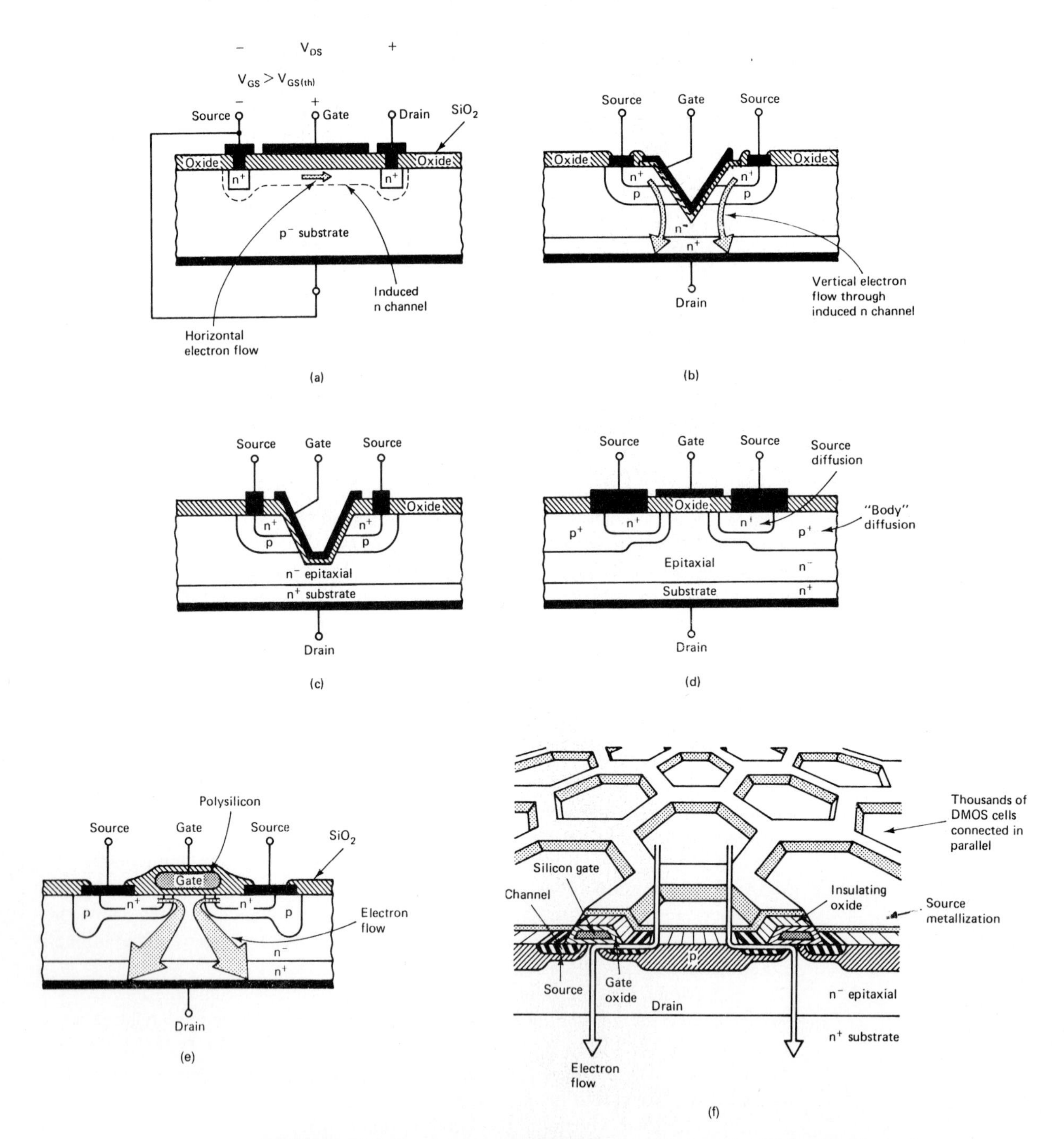

FIGURE 12-39 Power MOSFET structures: (a) small-signal, n-channel E-MOSFET; (b) V-grooved power MOSFET (VMOS); (c) U-grooved power MOSFET; (d) double-diffused power MOSFET (DMOS) with metallized gate; (e) DMOS power MOSFET with a polysilicon gate; (f) surface geometry of a DMOS HEXFET produced by International Rectifier.

resistance ($r_{DS(\text{ON})}$) may be on the order of 30 to 50 Ω. If such a transistor could pass 1 A, it would drop 30 V and dissipate 30 W of power! Therefore, the large values of $r_{DS(\text{ON})}$ prohibit the small-signal structure from being employed in power applications.

To develop E-MOSFETs that can handle large power levels [e.g., have a small $r_{DS(\text{ON})}$], manufacturers have developed new MOSFET geometries. An example is given in Fig. 12-39(b). The drain has been located at the *bottom* of the device while the source is located on the top. This arrangement is designed to promote a *vertical* flow of drain current. The V-grooved gate (which supports the acronym ''VMOS'') in concert with this geometry produces transconductances which are very large (typically, *hundreds* to *thousands* of millisiemens) and a short channel length. Because of the reduced channel length, the $r_{DS(\text{ON})}$ for V-groove MOSFETs is generally small, often *less than* 1 Ω.

The operation of the *n*-channel VMOS device is essentially the same as the small-signal, *n*-channel E-MOSFET. With a V_{GS} of zero, no appreciable drain current flows. As V_{GS} becomes positive, and increases beyond the threshold voltage $V_{GS(\text{th})}$, drain current will begin to flow.

The V-groove gate configuration produces a significant limitation. The point of the sharp V-shaped groove produces intense electric fields, which limits the MOSFET's breakdown voltage rating. The breakdown voltage ratings of VMOS power transistors tend to be much lower than those offered by bipolar transistors. Bipolar transistors are available with breakdown voltage ratings of several hundred volts. (Some BJTs have ratings as high as 1200 V.) However, VMOS transistors typically have voltage ratings less than 100 V. To elevate voltage ratings the U-grooved structure was devised [see Fig. 12-39(c)].

Further improvements in power MOSFET voltage and power ratings have been made possible by the development of the *double-diffused* power MOSFET (DMOS) [refer to Fig. 12-39(d)]. The terminology ''double-diffused'' comes from the sequential manner in which the *p*-doped ''body'' diffusion is followed by the second, highly doped n^+ source diffusion.

The DMOS power FET offers very large voltage ratings (hundreds of volts), a low $r_{DS(\text{ON})}$, and very large transconductances. The DMOS power FETs are respectable competitors to the power BJTs in many high-power applications.

The small-signal MOSFETs and early DMOS had gate structures that were produced by depositing metal on the insulating (SiO_2) layer. The latest technology utilizes a conductive *polysilicon gate* which is diffused on the SiO_2 layer. The polysilicon is highly doped to lower its resistivity [see Fig. 12-39(e)]. This technique minimizes gate overlapping (which reduces its distance from the source metallization) and results in lower interelectrode capacitance.

Most commercially available power MOSFETs are either VMOS or DMOS. However, there is an abundance of trade names, which can, at times, be overwhelming. For example, International Rectifier refers to its DMOS as HEXFETs [Fig. 12-39(f)], Siemens calls its devices SIPMOS, and RCA has its line of TMOS. Typically, these various trade names reflect differences in the device surface geometries. Functionally, they are DMOS or VMOS.

Another interesting point to consider is that power MOSFET structures are composed of *cells*. Typically, thousands of these cells are all connected in parallel

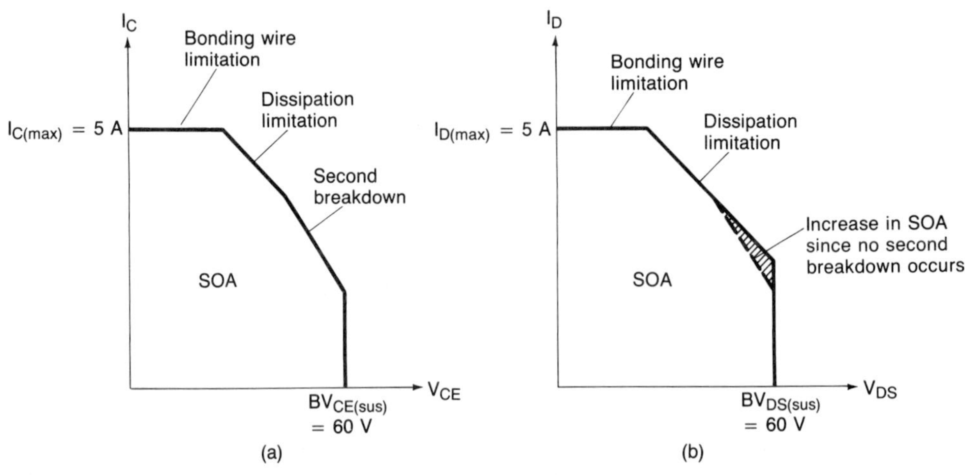

FIGURE 12-40 SOA comparison between bipolar and field-effect power transistors: (a) BJT; (b) MOSFET.

to form the power MOSFET device. This is illustrated by the basic HEXFET structure given in Fig. 12-39(f).

The HEXFET surface is composed of hexagonal cells all connected in parallel by the continuous sheet of source and drain metallization. The polysilicon gates are also tied together. The usual density of the cells in a power MOSFET structure exceeds 500,000 per square inch. This parallel arrangement is used to increase the current-handling capability of the power MOSFET device. Further, this parallel structure reduces $r_{DS(ON)}$ and promotes very large transconductances. The price to be paid for these enhancements is rather large gate-to-source capacitances.

For the purposes of comparison, the safe operating area of a bipolar power transistor and a similar power MOSFET have been depicted in Fig. 12-40. The bipolar power transistor's SOA has been shown in Fig. 12-40(a). The power MOSFET is not limited by secondary breakdown effects. Consequently, its SOA is improved [compare Fig. 12-40(a) and (b)].

In Section 5-6 we investigated JFET temperature effects. Specifically, we saw that I_D has a positive temperature coefficient (tempco) at low values of I_D and a negative tempco at high values of I_D. All FETs - junction, MOS, VMOS, and DMOS - exhibit this same effect. It is this fundamental mechanism that allows MOSFET cells and power MOSFET devices to be connected in parallel. MOSFETs tend to ''current share'' as opposed to ''current hogging.'' They do *not* exhibit the second breakdown mechanism.

As we saw in Section 4-9, the BJT's collector current tends to increase with temperature. Consequently, I_C demonstrates a positive tempco. Virtually all bipolar transistor failures can be traced directly to the BJT's positive temperature coefficient. It contributes to the thermal stresses produced by current hogging, secondary breakdown, and thermal runaway. By virture of their I_D negative tempcos, VMOS and DMOS devices are not as prone to these effects. This will be qualified in the next section.

The characteristics of power MOSFETs will become more apparent as our work progresses. In audio power amplifiers power MOSFETs offer lower distortion, simplified drive circuitry, and less stringent protection requirements.

12-14 Bias Stability and the V_{BE} Multiplier

Both the bipolar and MOS power transistors are temperature sensitive. The collector current of a bipolar transistor exhibits a positive tempco as illustrated in Fig. 12-41(a). Careful study of the transfer curve reveals that the required base-to-emitter voltage decreases at a rate of approximately -2.2 mV/°C in order to maintain a given collector current.

The transfer curve of a power MOSFET has been shown in Fig. 12-41(b). At *low* values of drain current, we see that the drain current also has a positive tempco. However, at higher values of I_D, the drain current actually *decreases* as the temperature increases. Also note that the drain current does have a zero temperature coefficient (0TC) point (refer to Section 5-6). If the MOSFET were biased at the 0TC point, its drain current would *not* change with temperature.

Typically, the class AB push-pull output transistors in an audio power amplifier have a quiescent collector or drain current of less than 150 mA. The precise value is dictated by the standby power and distortion requirements. Since the drain current of most power MOSFETs demonstrates a positive tempco at low values of I_D, they require the *same* considerations as bipolar output transistors.

Figure 12-41(b) indicates that the forward threshold voltage [$V_{GS(\text{th})}$] decreases as the temperature increases in order to maintain a given drain current. The temperature coefficient of the $V_{GS(\text{th})}$ for power MOSFETs can range from approximately -2.5 to -6 mV/°C.

FIGURE 12-41 BJT and MOSFET transfer curves: (a) to maintain a given I_C, V_{BE} decreases at a rate of approximately -2.2 mV/°C; (b) at low values of I_D, the required value of V_{GS} to maintain a given I_D decreases at a rate which ranges from approximately -2.5 to -6 mV/°C.

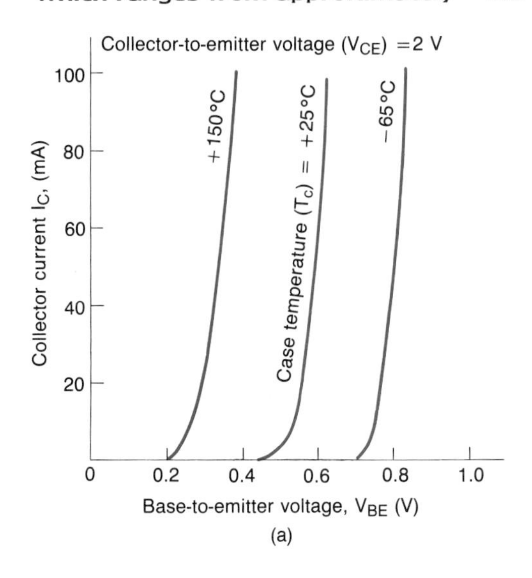

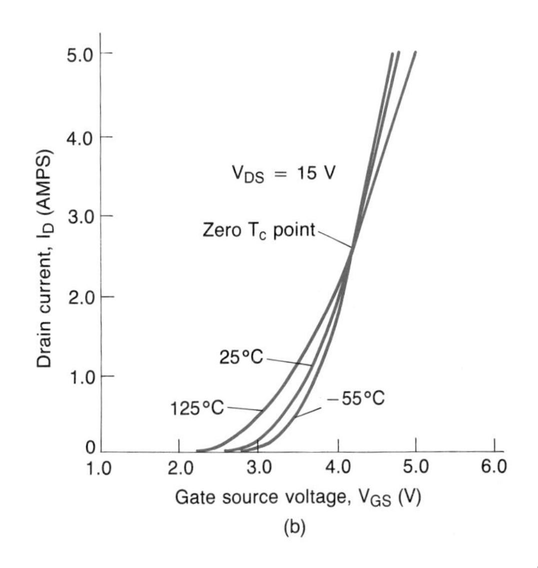

The conclusion to be drawn is that if a stable quiescent current (I_C or I_D) is to be maintained, we must provide a bias circuit that *decreases the bias voltage as the temperature increases*. Previously, we employed bias diodes to accomplish temperature compensation. However, a more elegant (and adjustable) approach is to use a circuit described as a V_{BE} *multiplier* [see Fig. 12-42(a)].

The V_{BE} multiplier consists of transistor Q_1 and resistors R_1, R_2, and R_3. To work correctly, a constant-current source is required. One option is to use the scheme depicted in Fig. 12-42(a). Transistor Q_2, resistors R_4 and R_5, and diodes D_1 and D_2 are used to establish a constant current I_{C2}.

Darlington output transistors (Q_3 and Q_4) have also been indicated. Emitter resistors R_6 and R_7 are used to assist their bias stability. The drive signal injection point has also been indicated for reference. We shall not concern ourselves with the signal-processing aspects of Fig. 12-42(a) at this point. Let us see how the V_{BE} multiplier works. Typical component values have been indicated in Fig. 12-42(a).

First, we review the operation of the constant-current source. It has been redrawn in Fig. 12-42(b). Ground has been indicated in phantom since it really does not enter into the analysis. The negative supply voltage provides the forward bias for diodes D_1 and D_2, and the base-emitter junction of transistor Q_2. The current I is

$$I = \frac{V_{CC} - V_{D1} - V_{D2}}{R_4} \simeq \frac{25\text{ V} - 1.4\text{ V}}{4.7\text{ k}\Omega} = 5.02\text{ mA}$$

This current produces a bias voltage across diodes D_1 and D_2 of approximately 1.4 V. Assuming that Q_2's base-emitter voltage drop is 0.7 V, we can now find the emitter current.

$$I_{E2} = \frac{(V_{D1} + V_{D2}) - V_{BE2}}{R_5} \simeq \frac{1.4\text{ V} - 0.7\text{ V}}{56\ \Omega} = 12.5\text{ mA}$$

If we assume further that the base current I_{B2} is negligibly small,

$$I_{C2} \simeq I_{E2} = 12.5\text{ mA}$$

The constant-current source will continue to provide a constant current as long as the voltage drop across it is at least 1.4 V [refer to Fig. 12-42(b)]. For voltage drops less than 1.4 V, Q_2 will begin to enter saturation, and its collector current will decrease.

Diodes D_1 and D_2 provide dc bias with a measure of temperature compensation. (A slight overcompensation can exist since *two* diodes are used.) The circuit provides a reasonably constant current over a relatively wide range of ambient temperatures. Now let us look at the operation of the V_{BE} multiplier. It has been redrawn in Fig. 12-42 (c).

For proper operation to occur, the base current must be small relative to the bleeder current I_1, and I_1 must be small compared to the collector current I_{C1}. Mathematically,

$$I_{B1} < I_1 < I_{C1}$$

Therefore, the total current flow through the circuit is approximately equal to I_{C1}. From Fig. 12-42(c) we note that

$$I_1 \simeq \frac{V_{BE1}}{R_2 + R_3} \tag{12-29}$$

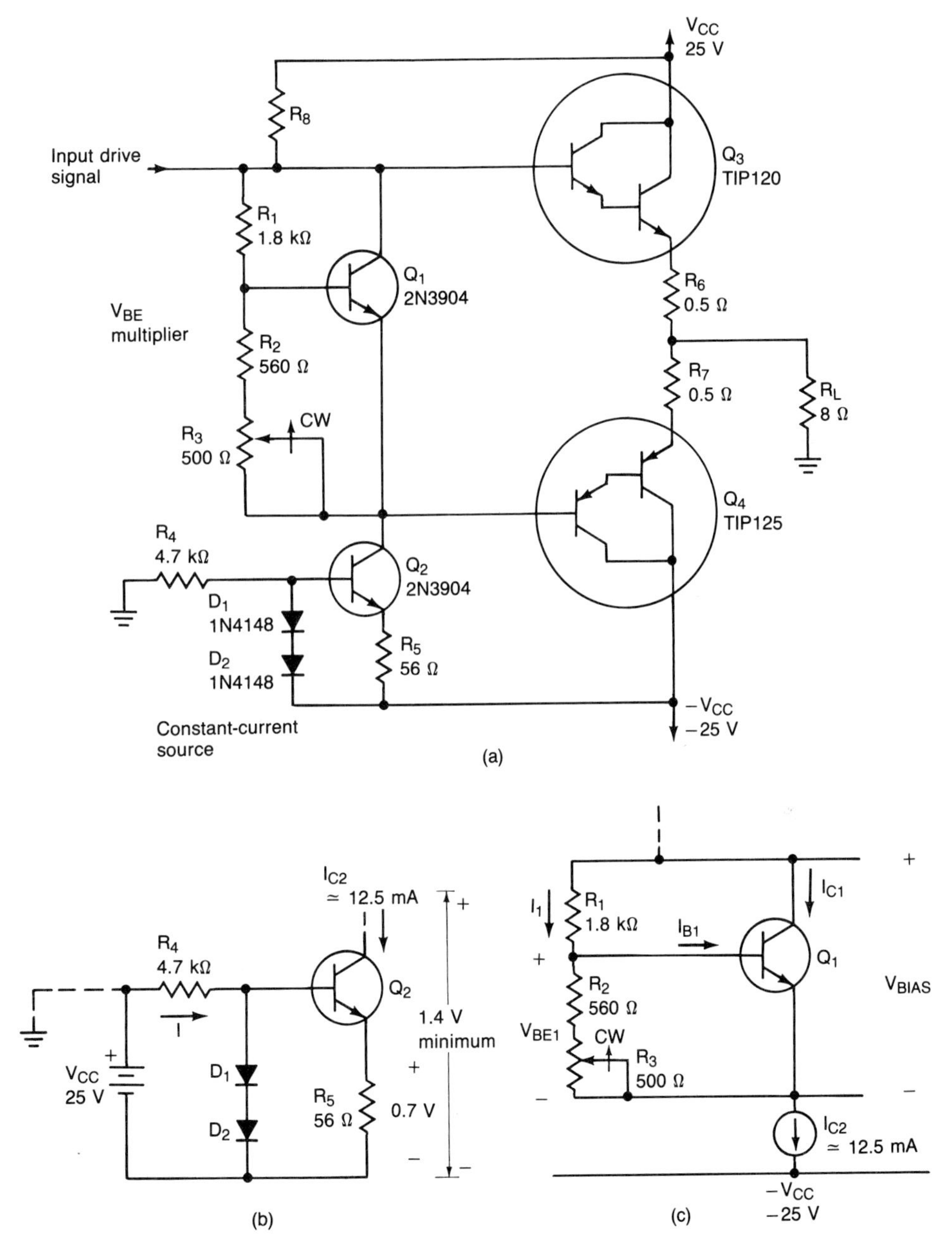

FIGURE 12-42 V_{BE} multiplier: (a) using a V_{BE} multiplier to bias complementary Darlington output transistors; (b) current source; (c) analyzing the V_{BE} multiplier.

Since I_{B1} is small compared to I_1, we can find the voltage drops across R_1, R_2, and R_3. The sum of the voltage drops is equal to V_{BIAS}.

$$V_{BIAS} \simeq I_1(R_1 + R_2 + R_3) \tag{12-30}$$

Substitution of Eq. 12-29 into Eq. 12-30 produces Eq. 12-31.

$$V_{BIAS} \simeq (R_1 + R_2 + R_3)I_1 = \frac{R_1 + R_2 + R_3}{R_2 + R_3} V_{BE1}$$

$$= \left(\frac{R_1}{R_2 + R_3} + 1\right) V_{BE1}$$

$$\boxed{V_{BIAS} \simeq \left(1 + \frac{R_1}{R_2 + R_3}\right) V_{BE1}} \tag{12-31}$$

From the result above, we can readily see why the circuit is called a "V_{BE} multiplier." The collector-to-emitter voltage of Q_1 (V_{BIAS}) will be greater than V_{BE1}.

EXAMPLE 12-8

Find the minimum and maximum values of V_{BIAS} for the V_{BE} multiplier given in Fig. 12-42(c). Assume that V_{BE1} is 0.7 V.

SOLUTION With R_3 adjusted fully clockwise (CW), R_3 is approximately zero, and from Eq. 12-31,

$$V_{BIAS} \simeq \left(1 + \frac{R_1}{R_2 + R_3}\right) V_{BE1} = \left(1 + \frac{1.8\ \text{k}\Omega}{560\ \Omega}\right)(0.7\ \text{V})$$

$$= 2.95\ \text{V}$$

If R_3 is adjusted fully counterclockwise, its resistance is 500 Ω. Hence

$$V_{BIAS} \simeq \left(1 + \frac{R_1}{R_2 + R_3}\right) V_{BE1} = \left(1 + \frac{1.8\ \text{k}\Omega}{1060\ \Omega}\right)(0.7\ \text{V})$$

$$= 1.89\ \text{V}$$

■

This example demonstrates that V_{BIAS} may be adjusted from 1.89 to 2.95 V. These values seem reasonable if we remind ourselves that we have *four* base-emitter *p-n* junctions (Q_3 and Q_4) to forward bias. Now let us see how V_{BIAS} changes with temperature. This requires that we apply some calculus. Disinterested (and/or intimidated) readers may skip over the mathematics.

Taking the derivative of Eq. 12-31 with respect to temperature provides us with

$$\frac{dV_{BIAS}}{dT} = \frac{d}{dT}\left[\left(1 + \frac{R_1}{R_2 + R_3}\right) V_{BE1}\right]$$

$$= \left(1 + \frac{R_1}{R_2 + R_3}\right)\frac{dV_{BE1}}{dT} \tag{12-32}$$

The derivative suggests that the approximation given by Eq. 12-33 may be used.

$$\frac{\Delta V_{BIAS}}{\Delta T} \simeq \left(1 + \frac{R_1}{R_2 + R_3}\right)\frac{\Delta V_{BE1}}{\Delta T} \tag{12-33}$$

The term $\Delta V_{BE1}/\Delta T$ is approximately -2.2 mV/°C for a single silicon bipolar transistor. In Fig. 12-42(a), $\Delta V_{\mathrm{BIAS}}/\Delta T$ will depend on the setting of potentiometer R_3. If R_3 is fully clockwise, V_{BIAS} is 2.95 V and

$$\frac{\Delta V_{\mathrm{BIAS}}}{\Delta T} \simeq \left(1 + \frac{R_1}{R_2 + R_3}\right)\frac{\Delta V_{BE1}}{\Delta T}$$

$$= \left(1 + \frac{1.8\ \mathrm{k}\Omega}{560\ \Omega}\right)\frac{-2.2\ \mathrm{mV}}{1^\circ\mathrm{C}} = -9.27\ \mathrm{mV/^\circ C}$$

and if R_3 is fully counterclockwise, V_{BIAS} is 1.89 V and

$$\frac{\Delta V_{\mathrm{BIAS}}}{\Delta T} \simeq \left(1 + \frac{R_1}{R_2 + R_3}\right)\frac{\Delta V_{BE1}}{\Delta T}$$

$$= \left(1 + \frac{1.8\ \mathrm{k}\Omega}{1060\ \Omega}\right)\frac{-2.2\ \mathrm{mV}}{1^\circ\mathrm{C}} = -5.94\ \mathrm{mV/^\circ C}$$

Since the larger bias voltage also results in larger negative temperature coefficients, the V_{BE} multiplier finds widespread application in power amplifiers with bipolar, Darlington, and MOSFET power transistors. For optimal operation, transistor Q_1 of the V_{BE} multiplier should be mounted on the *same* heat sink as the output transistors. In this way they will experience nearly the same junction temperatures.

Referring to Fig. 12-42(a), the operation of the V_{BE} multiplier can easily be understood. As the temperature increases, the collector current of Q_1 would normally increase. However, the constant-current source (Q_2) keeps Q_1's collector current constant. Therefore, V_{BE1} is forced to decrease with temperature. The corresponding decrease in V_{BIAS} serves to keep the quiescent collector currents of the output devices relatively constant.

The output transistor idling current should be adjusted with R_3 initially set fully counterclockwise. Resistor R_3 should then be slowly adjusted clockwise until the minimum quiescent collector current that yields acceptable distortion is reached. The idling current can be monitored by measuring the voltage drops across the emitter resistors. If caution is not used, the transistors may experience thermal runaway.

Now let us investigate the operation of a power amplifier that incorporates the V_{BE} multiplier. One such example is the circuit given in Fig. 12-43(a). At first glance, the circuit may appear to be quite frightening. However, its fundamental operation is quite similar to some of our previous investigations. Consider the amplifier studied earlier in Fig. 12-33(c).

The 741 op amp is used as the driver stage in both cases. Bias diodes D_1 and D_2 in Fig. 12-33(c) have been replaced by a V_{BE} multipler (Q_1) in Fig. 12-43(a). Similarly, another V_{BE} multiplier (Q_2) has been used to replace bias diodes D_3 and D_4 in Fig. 12-33(c).

The bias resistor R_4 in Fig. 12-33(c) has been replaced by a constant-current source in Fig. 12-43(a). Bias resistor R_3 in Fig. 12-33(c) has been repeated in Fig. 12-43(a), but *two* resistors (R_2 and R_3) have been used and a *bootstrapping capacitor* C_1 has been added. The function of C_1 will be explained shortly.

Also notice that the negative feedback loop has been wrapped around the entire amplifier system [see resistors R_{14} and R_{15} in Fig. 12-43(a)]. Capacitor C_4 blocks the dc bias such that *100% dc negative feedback* is used. This greatly assists the dc bias

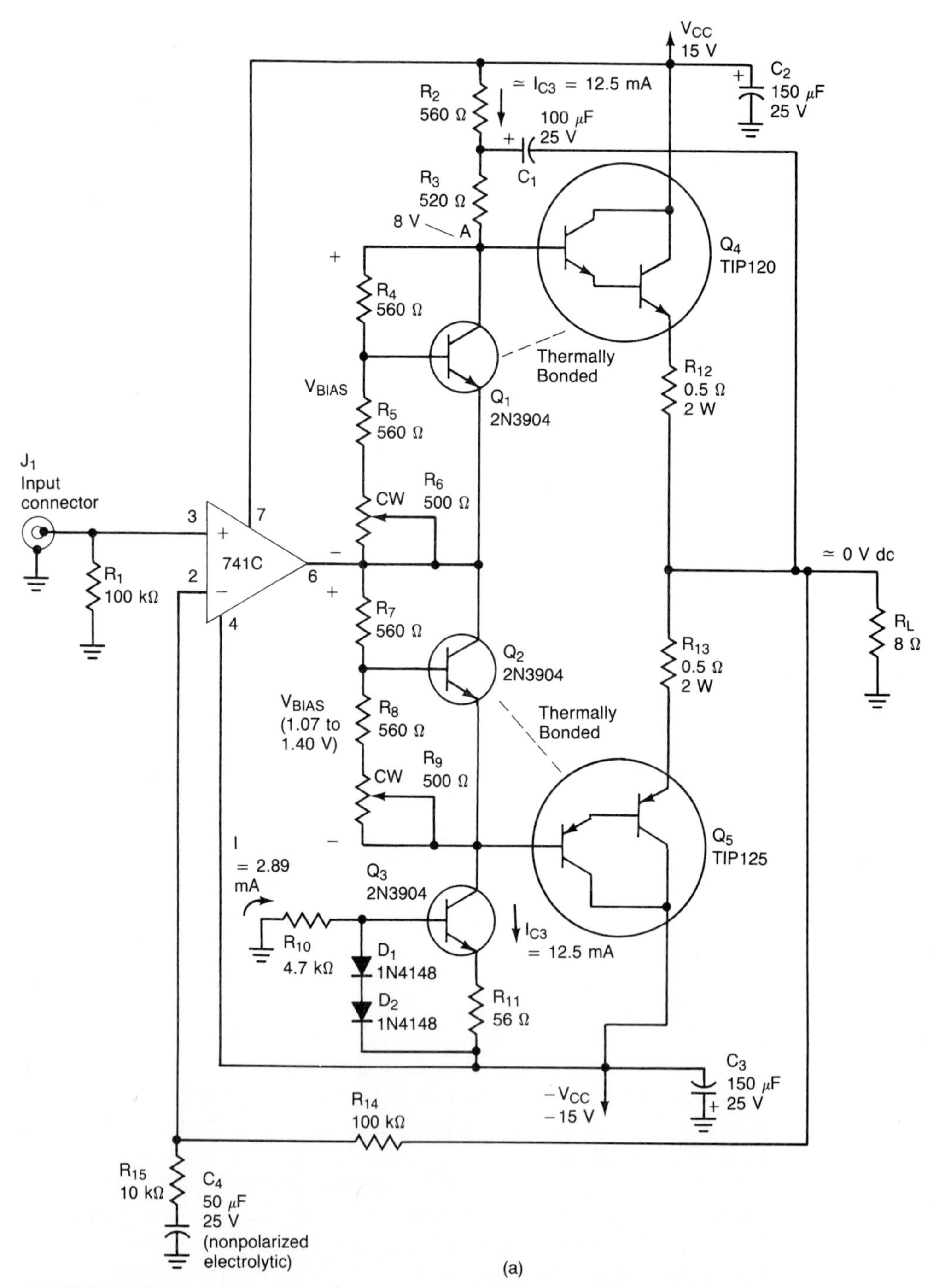

FIGURE 12-43 Practical $6\frac{1}{4}$-W power amplifier: (a) schematic; (b) waveforms.

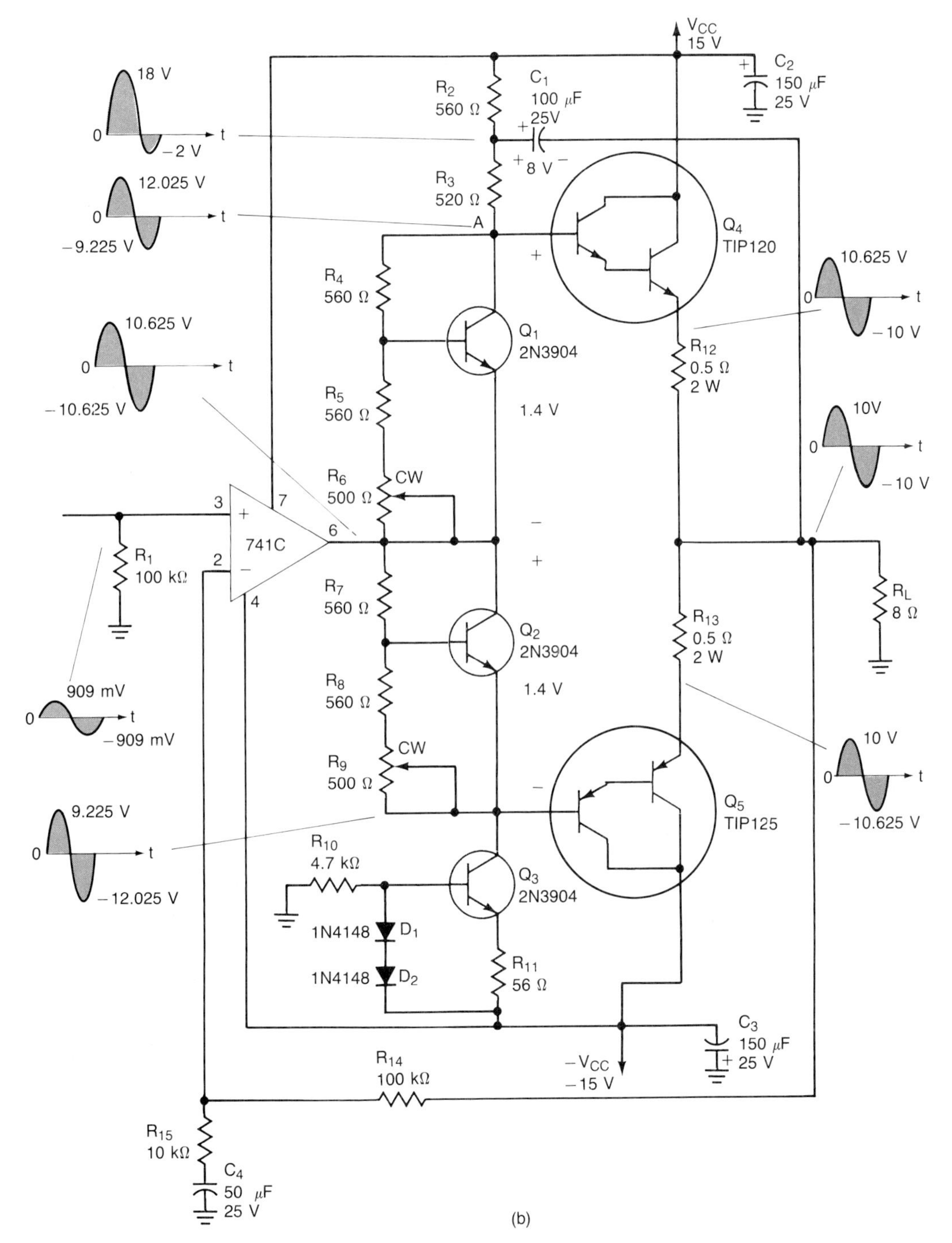

FIGURE 12-43 *(continued)*

stability. However, the amplifier's low-frequency response will no longer extend to dc (0 Hz), but this is not a great price to pay in an *audio* amplifier system.

EXAMPLE 12-9

Perform a dc analysis of the V_{BE} multipliers shown in Fig. 12-43(a). Specifically, determine the approximate current through the diodes D_1 and D_2, the collector current of Q_3, and the adjustment range of the V_{BE} multipliers.

SOLUTION The current I [Fig. 12-43(a)] through the diodes is

$$I \simeq \frac{V_{CC} - 1.4\text{ V}}{R_{10}} = \frac{15\text{ V} - 1.4\text{ V}}{4.7\text{ k}\Omega} = 2.89\text{ mA}$$

The emitter current of Q_3 is

$$I_{E3} = \frac{(V_{D1} + V_{D2}) - V_{BE3}}{R_{11}} \simeq \frac{1.4\text{ V} - 0.7\text{ V}}{56\ \Omega} = 12.5\text{ mA}$$

and $I_{C3} \simeq I_{E3} \simeq 12.5$ mA. Since the two V_{BE} multipliers are identical, we only need to make one analysis. Using Eq. 12-31 and assuming that R_6 is fully counterclockwise,

$$V_{\text{BIAS}} \simeq \left(1 + \frac{R_4}{R_5 + R_6}\right) V_{BE1} = \left(1 + \frac{560\ \Omega}{1060\ \Omega}\right)(0.7\text{ V})$$
$$\simeq 1.07\text{ V}$$

Similarly, if potentiometer R_6 is adjusted fully clockwise,

$$V_{\text{BIAS}} \simeq \left(1 + \frac{R_4}{R_5 + R_6}\right) V_{BE1} = \left(1 + \frac{560\ \Omega}{560\ \Omega}\right)(0.7\text{ V})$$
$$\simeq 1.40\text{ V}$$

■

To provide 6.25 W into 8 Ω, the peak output voltage must be 10 V [see Fig. 12-43(b)]. The balance of the waveforms have also been indicated in the figure. The emitter waveforms are determined in exactly the same fashion as described previously. The positive peaks are very nearly equal to the magnitude of the negative peaks. This is true because the emitter resistors are relatively small compared to the 8-Ω load.

The base waveforms are determined in the same manner as done previously. Consider the base of Q_4. The positive peak is two base-emitter voltage drops larger than the emitter voltage.

The V_{BE} multipliers act as constant-voltage sources. In Fig. 12-43(b) they have each been adjusted to 1.4 V. Therefore, the base of Q_4 must always be 1.4 V more positive than the op amp's output voltage.

At this point it should be clear that the determination of the output, emitter, base, and op amp output voltage waveforms proceeds in exactly the same manner as our previous efforts. The only waveform that we have not encountered before is the one that appears at the junction between R_2 and R_3. Observe that its positive peak is 18 V!

This large positive peak is made possible by the bootstrap capacitor C_1. In the no-signal, or dc steady-state condition, the voltage at point A in Fig. 12-43(b) is approximately

$$V_A \simeq V_{CC} - I_{C3}R_2 = 15\text{ V} - (12.5\text{ mA})(560\ \Omega)$$
$$= 8\text{ V}$$

and since the output is approximately 0 V, the capacitor will charge to 8 V. If the signal frequency is relatively high (in this example a few hertz), the capacitor will not appreciably charge or discharge. Therefore, it will maintain a constant 8 V across its terminals. Consequently, the capacitor will behave as a voltage source, and the voltage at point A will always be 8 V *more positive* than the output voltage.

Recall that the bias circuit produces clipping if the voltage across the bias resistors drops too low. This prohibits the bias network from supplying the peak base current demanded by the output transistors. The bootstrap capacitor minimizes this problem by keeping the voltage drop (and therefore the available current) across R_3 large.

This arrangement constitutes *positive feedback*. However, because emitter followers have voltage gains less than unity, potential instability (oscillations) are avoided.

However, spurious parasitic oscillations are always a possibility. These may be avoided, or cured by careful circuit layout, decoupling the op amp, inserting 100-Ω resistors in series with the Q_4 and Q_5 base terminals, and/or installing small (e.g., 47-pF) capacitors between the collector–base terminals of the output transistors (Q_4 and Q_5). To further enhance the circuit's operation, the 741C may be replaced with an LF411 FET-input op amp. This will dramatically improve the circuit's slew rate to approximately 10 V/μs (see Fig. 12-49).

EXAMPLE 12-10

Find the maximum undistorted output voltage as determined by the output transistors [Q_4 and Q_5 in Fig. 12-43(a)]. Assume that their $V_{CE(\text{SAT})}$ is 1 V.

SOLUTION Since $V_{CE(\text{SAT})}$ is 1 V, the net voltage applied to the emitter-load circuit is

$$V_{CC} - V_{CE(\text{SAT})} = 15\text{ V} - 1\text{ V} = 14\text{ V}$$

and by voltage division,

$$v_{\text{OUT(peak)}} = \frac{R_L}{R_E + R_L}[V_{CC} - V_{CE(\text{SAT})})] = \frac{8\ \Omega}{0.5\ \Omega + 8\ \Omega}(14\text{ V})$$
$$= 13.2\text{ V}$$

[see Fig. 12-44(a)]. ■

EXAMPLE 12-11

In Fig. 12-43(a) the constant-current source must have at least 1.4 V across it to operate correctly. Determine its restriction on the maximum peak output voltage.

SOLUTION The collector of Q_3 must be at a potential of at least -13.6 V to be 1.4 V more positive than $-V_{CC}$. Therefore, the peak emitter voltage of Q_5 will be

$$v_{E5(\text{peak})} = -V_{CC} + 1.4\text{ V} + v_{BE5}$$
$$= -15\text{ V} + 1.4\text{ V} + 1.4\text{ V} = -12.2\text{ V}$$

The peak output voltage is

$$v_{OUT(peak)} = \frac{R_L}{R_E + R_L} v_{E5(peak)} = \frac{8\ \Omega}{0.5\ \Omega + 8\ \Omega}(-12.2\ \text{V})$$
$$= -11.5\ \text{V}$$

[see Fig. 12-44(b)]. ■

FIGURE 12-44 Finding the maximum output voltage: (a) the clipping point due to $V_{CE(SAT)}$ and the emitter resistor; (b) the clipping point due to the constant-current source; (c) peak output limitation due to the op amp; (d) peak currents during the positive peak output voltage; (e) peak currents during the negative peak output voltage.

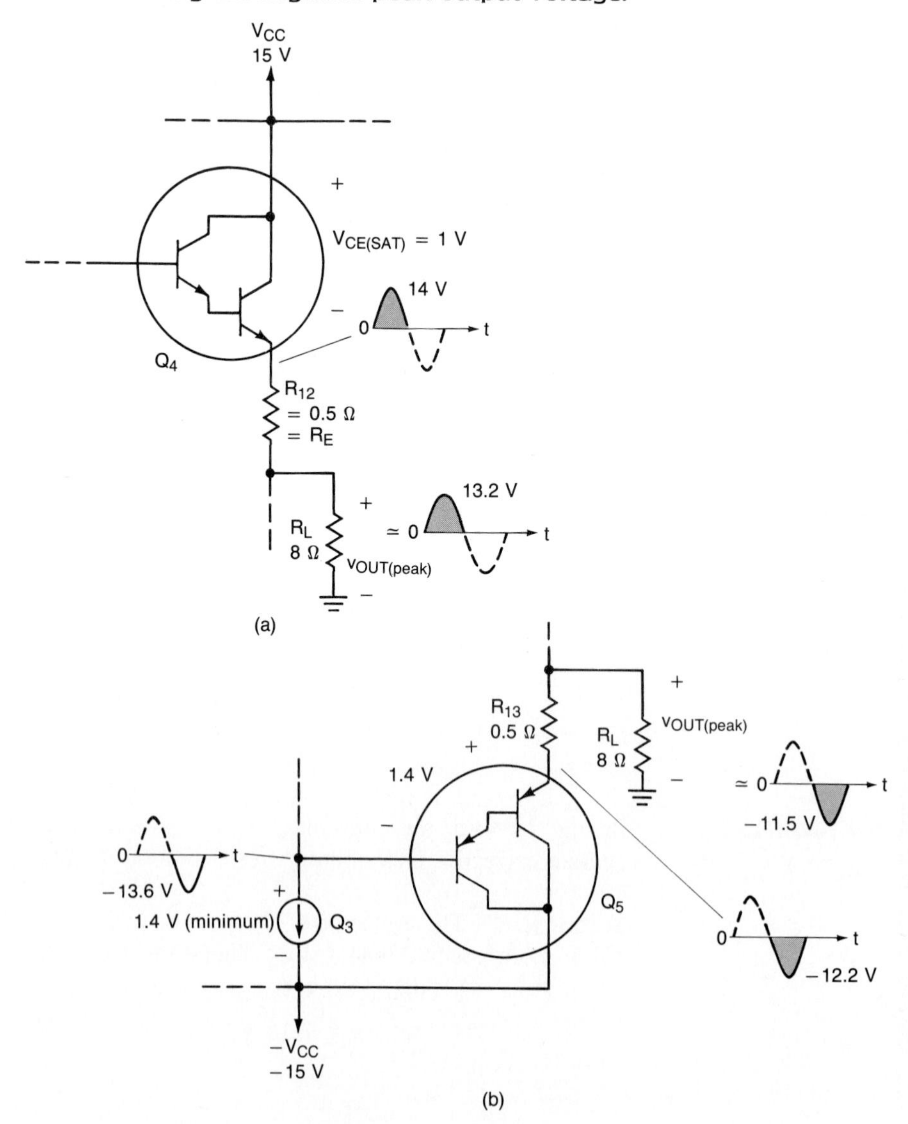

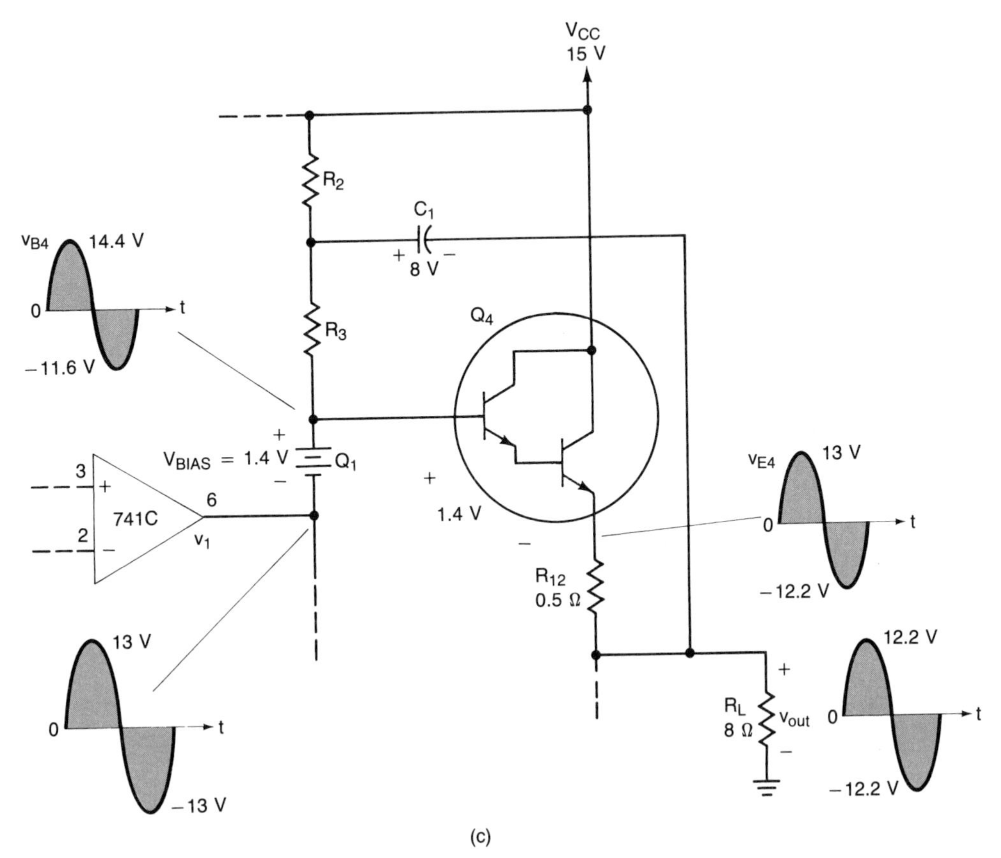

FIGURE 12-44 *(continued)*

EXAMPLE 12-12

Determine the output limitation imposed by the bias resistors R_2 and R_3 in Fig. 12-43(a). Conduct the analysis with and without bootstrap capacitor C_1.

SOLUTION If the bootstrap capacitor is *not* used, we may apply a slight modification of Eq. 12-26 as we did in Example 12-7.

$$v_{\mathrm{OUT(peak)}} = \frac{V_{CC} - V_{BE4}}{1 + (R_2 + R_3)/h_{FE(\mathrm{min})}R_L + R_{12}/R_L}$$

$$= \frac{15\ \mathrm{V} - 1.4\ \mathrm{V}}{1 + 1080\ \Omega/(1000)(8\ \Omega) + 0.5\ \Omega/8\ \Omega} = 11.4\ \mathrm{V}$$

If bootstrap capacitor C_1 is used, point A in Fig. 12-43(b) will be about 8 V higher than the output. Therefore, the emitter can reach V_{CC} minus $V_{CE(\mathrm{SAT})}$, and the peak output voltage found in Example 12-10 applies.

$$V_{\mathrm{OUT(peak)}} = 13.2\ \mathrm{V}$$

■

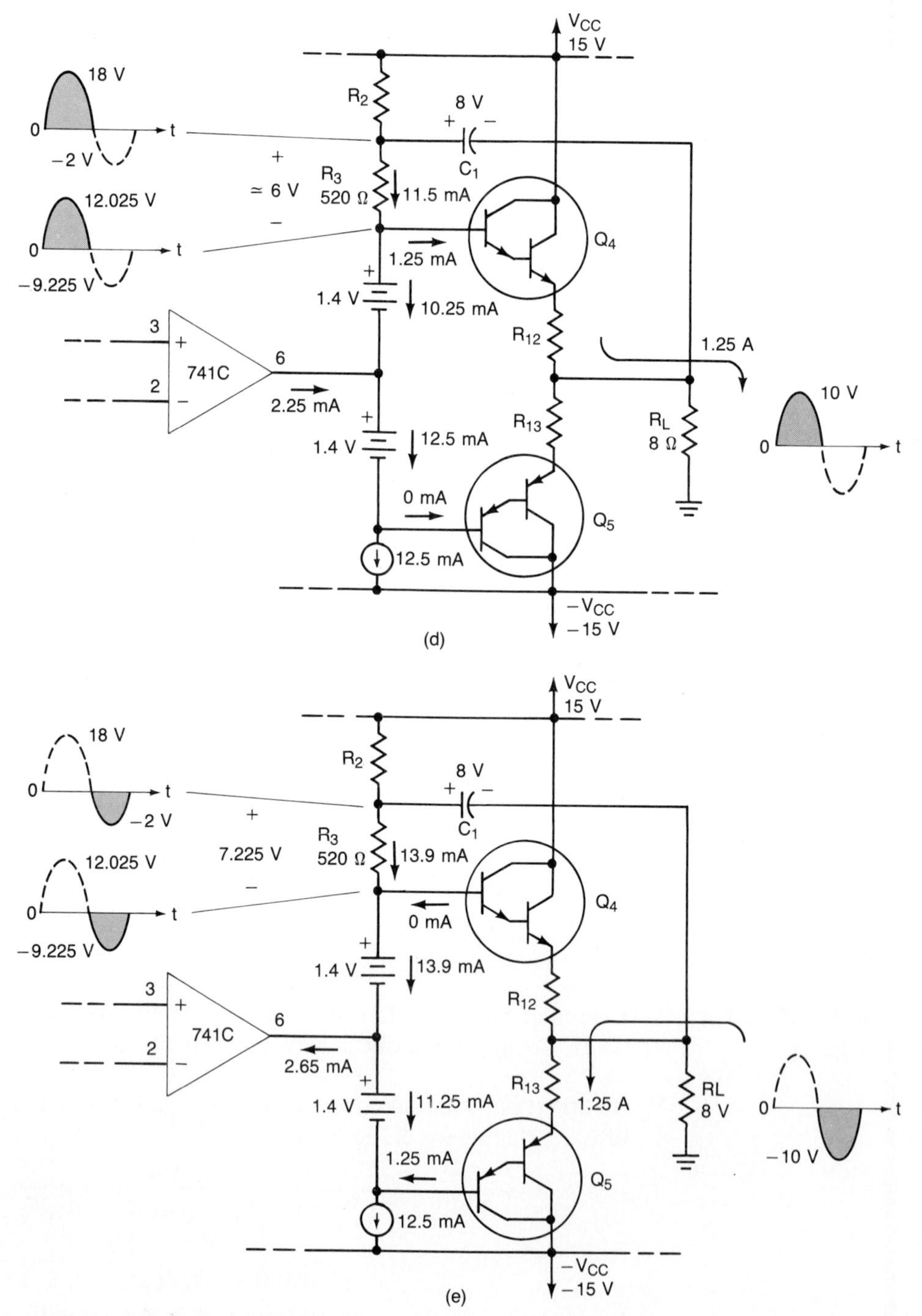

FIGURE 12-44 *(continued)*

EXAMPLE 12-13

Determine the maximum peak output voltage swing as limited by the op amp in Fig. 12-43(a).

SOLUTION Assuming that the op amp's peak output voltage v_1 is 13 V, the base of Q_4 will be

$$v_{B4} = v_1 + V_{\text{BIAS}} = 13\text{ V} + 1.4\text{ V} = 14.4\text{ V}$$

and the peak emitter voltage is

$$v_{E4(\text{peak})} = v_{B4} - v_{BE4} = 14.4\text{ V} - 1.4\text{ V} = 13\text{ V}$$

The corresponding peak output voltage is

$$v_{\text{OUT(peak)}} = \frac{R_L}{R_E + R_L} v_{E4(\text{peak})} = \frac{8\ \Omega}{0.5\ \Omega + 8\ \Omega}(13\text{ V})$$
$$= 12.2\text{ V}$$

[see Fig. 12-44(c)] ■

The op amp's output current limitation (typically 10 mA peak) is *not* a factor. This is true because of the low input drive current requirement of the Darlington output transistors and the use of bootstrap capacitor C_1 [refer to Fig. 12-44(d)].

During the positive half-cycle, the voltage drop across R_3 is approximately 6 V. By Ohm's law, 11.5 mA will flow down through R_3. As we can see, the peak base current into Q_4 is 1.25 mA. By Kirchhoff's current law, 10.25 mA must flow down through the upper V_{BE} multiplier (Q_1). Since transistor Q_5 is nonconducting during the positive half-cycle, 12.5 mA flows down through the lower V_{BE} multiplier (Q_2). Kirchhoff's current law indicates that 2.25 mA is required to flow out of the op amp's output terminal.

The negative half-cycle has been depicted in Fig. 12-44(e). By following the same approach, we can see that during the peak negative half-cycle, 2.65 mA is required to flow into the op amp's output.

Obviously, the demands on the op amp's output are well below its 10 mA peak rating. Also note that bootstrapping has kept the voltage drop across R_3 relatively constant. Therefore, the current flow through it is also relatively constant.

EXAMPLE 12-14

Determine the approximate efficiency of the power amplifier given in Fig. 12-43(a).

SOLUTION If we assume that the output voltage is 10 V peak,

$$i_{\text{OUT(peak)}} = \frac{v_{\text{OUT(peak)}}}{R_L} = \frac{10\text{ V}}{8\ \Omega} = 1.25\text{ A}$$

and

$$I_C = \frac{i_{\text{OUT(peak)}}}{\pi} = \frac{1.25\text{ A}}{\pi} = 398\text{ mA}$$

The average dc input power is

$$P_{\text{in(dc)}} \simeq 2V_{CC} I_C = (2)(15\text{ V})(398\text{ mA}) = 11.94\text{ W}$$

The efficiency (for 6.25 W of ac output power) is

$$\%\ \eta = \frac{P_{\text{out(ac)}}}{P_{\text{in(dc)}}} \times 100\% = \frac{6.25\text{ W}}{11.64\text{ W}} \times 100\% = 52.4\%$$ ■

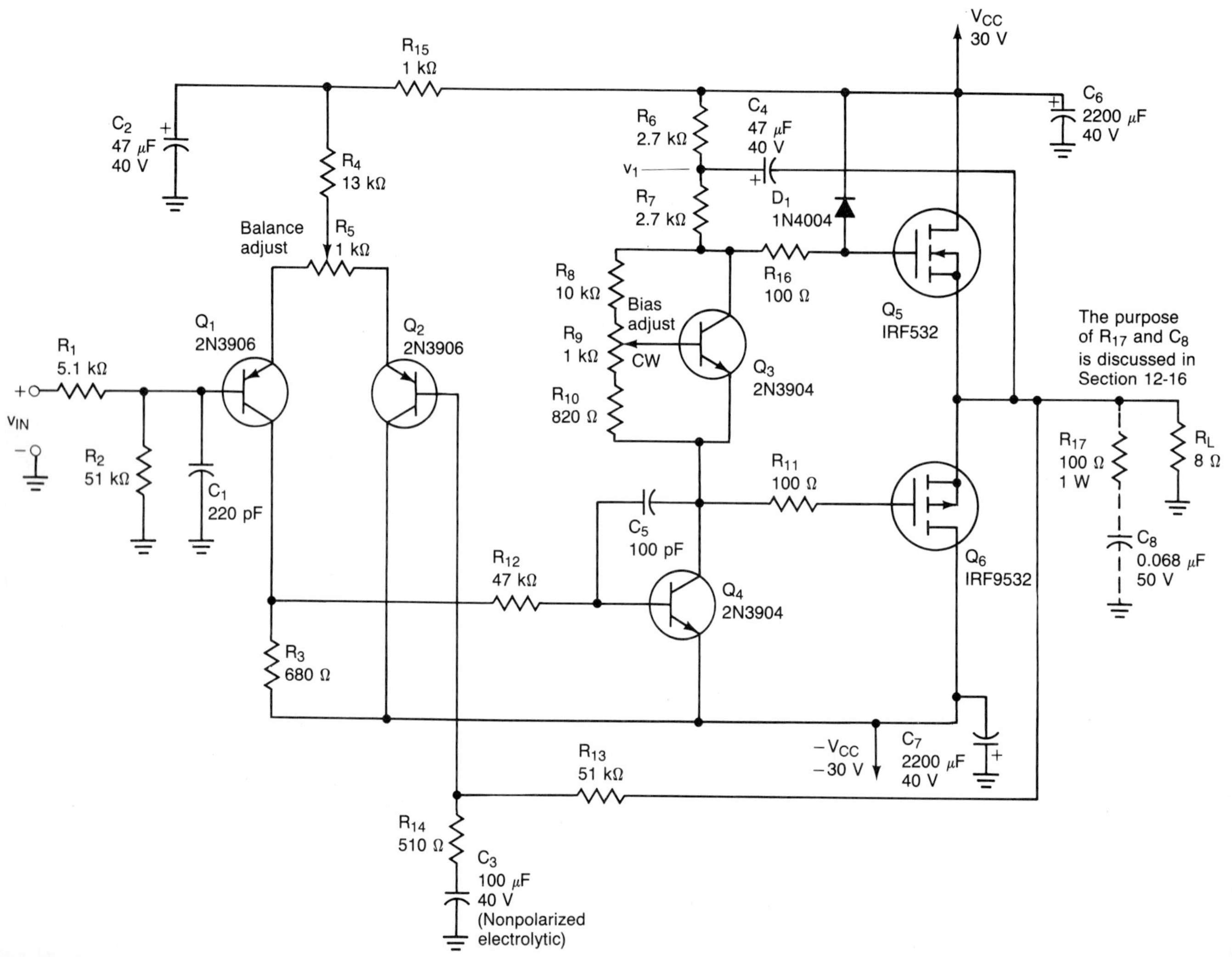

FIGURE 12-45 **Discrete 30-W power amplifier using HEXFETs and a V_{BE} multiplier.**

Consider this result. The power amplifier given in Fig. 12-33(c) gave us 0.64 W with an efficiency of 42%. By raising the supply voltages, lowering the emitter resistors, using bootstrapping, and employing V_{BE} multipliers, we have achieved 6.25 W at an efficiency of 52%! We are starting to get really serious.

In Fig. 12-45 a power amplifier that uses power MOSFETs has been illustrated. Notice that a V_{BE} multiplier (Q_3) has been used to bias the output transistors (Q_5 and Q_6). The output devices are HEXFETs manufactured by International Rectifier. Let us temporarily avoid the rigors of analysis and "walk through" the circuit qualitatively.

The amplifier is capable of delivering 30.25 W into an 8-Ω load with an efficiency of 57.6%. The input stage consists of a discrete differential amplifier (Q_1 and Q_2) which employs *pnp* transistors. Resistor R_2 provides a dc return to ground for the base of Q_1 and works in concert with R_1 and C_1 to suppress high-frequency input noise. (With a little reflection the reader should recognize that we have a low-pass filter.)

The input signal is directed into the noninverting input of the differential amplifier. The use of a differential input stage (whether it be integrated or discrete) easily allows us to wrap the feedback loop (R_{13}, R_{14}, and C_3) around the entire amplifier system. Emitter biasing of the differential amplifier is accomplished via emitter resistor R_4. The R_{15}-C_2 combination serves as another low-pass filter to keep supply noise (e.g., the ripple discussed in Chapter 13) from disturbing the sensitive input stage. The potentiometer R_5 is used to balance the differential pair. This is useful in setting up the bias for the maximum undistorted output voltage.

The output of the differential amplifier is directed into the base of transistor Q_4. Transistor Q_4 serves as the driver stage. It is a class A voltage amplifier. Since it is class A, it keeps the average current through the V_{BE} multiplier (Q_3) relatively constant. Capacitor C_5 provides the amplifier with dominant pole frequency compensation. This minimizes the possibility of uncontrolled oscillations.

Capacitor C_4 is used to provide bootstrapping. This not only provides additional drive for output transistor Q_5, but also serves to keep the average collector current through Q_4 relatively constant.

Diode D_1 is used to limit the gate voltage of Q_5 to one diode drop greater than $+V_{DD}$ (refer to Section 3-25). This protects the input of Q_5 in case it is overdriven. Resistor R_{16} serves two purposes. First, it limits the current through diode D_1, and second, it prevents parasitic oscillations. Similarly, the function of R_{11} is to prevent parasitic oscillations in Q_6.

Also notice the absence of source resistors. These are not always required for thermal stability in MOSFET power devices. The reader will be led through a detailed analysis of this circuit in the problems at the end of the chapter.

EXAMPLE 12-15

The n-channel IRF532 has a $V_{GS(\text{th})}$ that can range from 2 to 4 V. The p-channel IRF9532 has a $V_{GS(\text{th})}$ that can range from -2 to -4 V. Is the V_{BE} multiplier given in Fig. 12-45 adequate?

SOLUTION Study the V_{BE} multiplier circuit carefully. Notice that the potentiometer configuration has been altered in this design. Nevertheless, from Eq. 12-31 we can see that if R_9 is adjusted fully counterclockwise,

$$V_{\text{BIAS(min)}} \simeq \left(1 + \frac{R_8}{R_9 + R_{10}}\right) V_{BE3}$$

$$= \left(1 + \frac{10\ \text{k}\Omega}{1820\ \Omega}\right)(0.7\ \text{V}) = 4.5\ \text{V}$$

and if R_9 is fully clockwise,

$$V_{\text{BIAS(max)}} \simeq \left(1 + \frac{R_8 + R_9}{R_{10}}\right) V_{BE3}$$

$$= \left(1 + \frac{11\ \text{k}\Omega}{820\ \Omega}\right)(0.7\ \text{V}) = 10.1\ \text{V}$$

Since the V_{BE} multiplier is used to provide V_{GS} bias for both Q_5 and Q_6, it must be able to supply from 4 to 8 V. The V_{BE} multiplier appears to be adequate. (Recall that MOSFET gate current is essentially zero. Therefore, there will be no dc voltage drops across R_{11} and R_{16}.) ■

12-15 Reactive Loads: The Real World

One of the most important aspects of power amplifier design is to ensure that the output devices are always operated within their safe operating area (SOA) boundaries. This can be verified by plotting the ac load line on the derated SOA of the output device under consideration.

In Fig. 12-46(a) we see the output transistors for a 20-W power amplifier. The output transistors are being used to drive an 8-Ω load. As we can see in Fig. 12-46(a), 20 W of power into an 8-Ω load requires a peak load voltage of 17.9 V and a peak load (collector) current of 2.24 A. The corresponding collector-to-emitter voltage and collector current waveforms have been generated in Fig. 12-46(b). For simplicity, the effect of the very small emitter resistances and the idle current has been ignored.

Study these waveforms carefully. If no signal is applied, or when it crosses zero, the collector current is also (approximately) zero. Under these conditions, the transistor is in cutoff, and its collector-to-emitter voltage will be equal to V_{CC} (25 V). As the signal goes positive, the collector current will increase to 2.24 A peak, and the collector-to-emitter voltage will decrease (to a minimum of 7.1 V in this example).

During the negative half-cycle, Q_1 will be in cutoff, and its collector current will be zero. However, because transistor Q_2 is conducting, the load will go to -17.9 V peak. Since Q_1's emitter is at -17.9 V and its collector is tied to V_{CC} (25 V), its collector-to-emitter voltage will reach 42.9 V.

By inspection of Fig. 12-46(b), we can see that v_{CE} is a sine wave that has been shifted by 180° and rides on a dc level. Mathematically,

$$v_{CE} = V_{CC} - V_m \sin \omega t \tag{12-34}$$

where v_{CE} = total instantaneous collector-emitter voltage
V_m = peak ac output voltage
ωt = angle (radians)

The collector current is discontinuous. Therefore,

$$i_C = I_m \sin \omega t \;\Big|\; 0 \le \omega t \le \pi \qquad = 0 \;\Big|\; \pi < \omega t < 2\pi \tag{12-35}$$

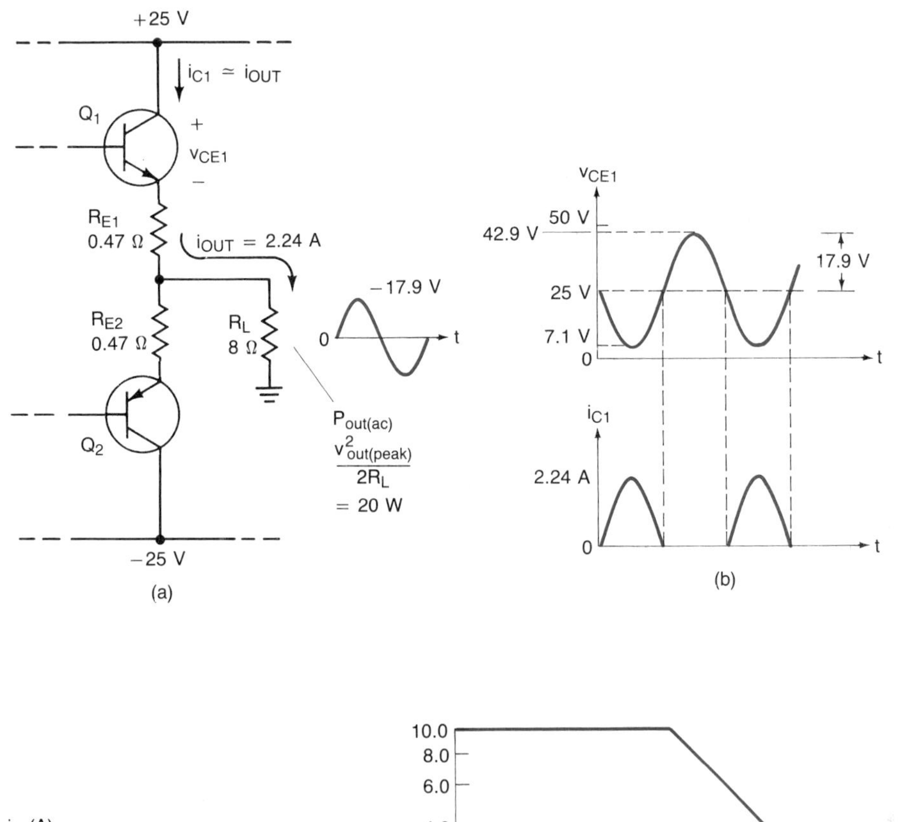

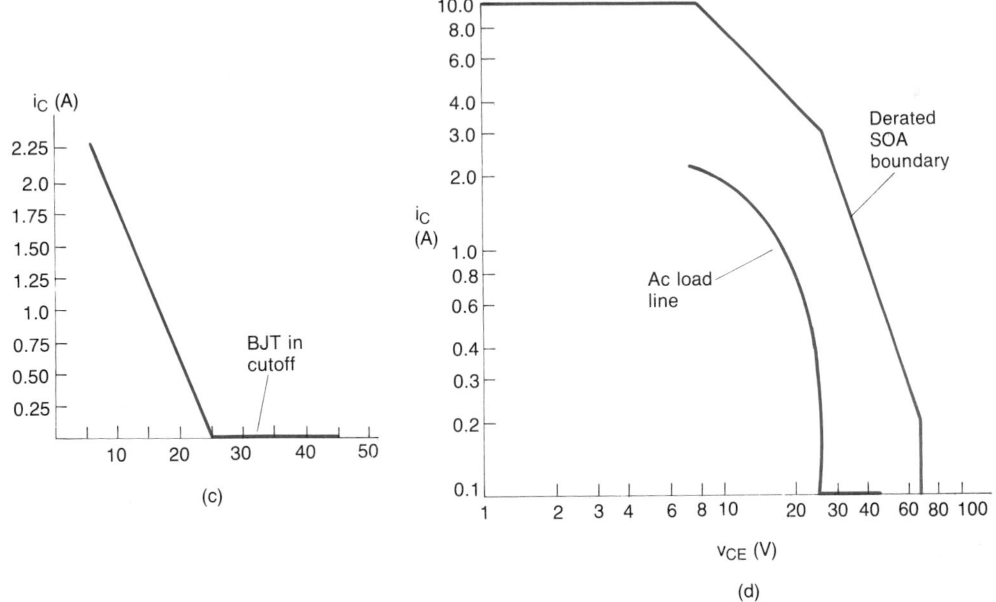

FIGURE 12-46 Ac load-line analysis when R_L is 8 Ω (resistive): (a) output of a 20-W power amplifer; (b) waveforms; (c) ac load line plotted using the data points given in Table 12-1; (d) ac load line plotted on the BJT's SOA graph.

where i_C = total instantaneous collector current
I_m = peak ac output current
ωt = angle (radians)

Since most of us prefer to work in degrees rather than radians, we can rewrite Eqs. 12-34 and 12-35 as follows:

$$\begin{aligned} v_{CE} &= V_{CC} - V_m \sin\theta \\ i_C &= I_m \sin\theta \quad \Big| \; 0° \le \theta \le 180° \\ &= 0 \quad \Big| \; 180° < \theta < 360° \end{aligned} \tag{12-36}$$

where θ is the angle in degrees.

In Fig. 12-46(b) we see that

$$v_{CE} = 25 - 17.9 \sin\theta \text{ volts}$$

and

$$\begin{aligned} i_C &= 2.24 \sin\theta \text{ amperes} \,|\, 0° \le \theta \le 180° \\ &= 0 \,|\, 180° < \theta < 360° \end{aligned}$$

These two expressions not only give us a way of precisely graphing the waveforms, they also describe the ac load line. Specifically, we can find v_{CE} and the corresponding i_C, and plot the results.

EXAMPLE 12-16

Draw the ac load line for transistor Q_1 in Fig. 12-46(a) using Eq. 12-36.

SOLUTION As an example, we shall determine v_{CE} and i_C when θ is 45°.

$$v_{CE} = 25 - 17.9 \sin\theta = 25 - 17.9 \sin 45° = 12.3 \text{ V}$$

$$i_C = 2.24 \sin\theta = 2.24 \sin 45° = 1.58 \text{ A}$$

The rest of the calculations are very similar and have been summarized in Table 12-1. The points have been plotted to form the ac load line as depicted in Fig. 12-46(c). ■

TABLE 12-1
AC Load-Line Data Points

θ (deg)	v_{CE} (V)	i_C (A)
0	25.0	0
45	12.3	1.58
90	7.1	2.24
135	12.3	1.58
180	25.0	0
225	37.7	0
270	42.9	0
315	37.7	0
360	25.0	0

If the ac load line is plotted on the SOA of the transistor, we arrive at Fig. 12-46(d). Recall from Section 12-9 that the SOA is typically plotted on a log-log graph. This changes the shape of our load line as indicated. This should not bother us; we are not looking for signal relationships but are verifying that our ac load line lies within the SOA boundaries. All things considered, this is a relatively painless method for verifying reliable transistor operation.

The loads we have worked with thus far have been *resistive*. However, in the "real world" our power amplifiers must drive motors, solenoids, and loudspeakers. These loads are typically combinations of resistance and *inductance*. Inductive loads have been known to give technicians and engineers severe headaches, and have caused the "death" of countless numbers of output transistors. Although not as common, but just as troublesome, capacitive loads have also caused a great deal of grief.

Reactive loads cause phase shifts between the voltage and current. The magnitude of the load and the phase shift are both functions of frequency.

In Fig. 12-47(a) we see the power amplifier output stage given in Fig. 12-46(a), but the resistive 8-Ω load has been replaced by an 8-Ω loudspeaker. Loudspeakers are anything but resistive. A simple diagram of a permanent-magnet loudspeaker has been illustrated in Fig. 12-47(b).

The signal drives the voice coil (an inductive electromagnet). The signal produces magnetic flux lines around the voice coil which alternately attract and repulse the permanent magnet located behind the speaker cone. As the speaker cone moves, the air around it is moved and sound is produced.

The rated 8 Ω of speaker impedance typically occurs at only one frequency (usually 400 Hz). The actual speaker impedance tends to vary with frequency, and tends to be a combination of resistance and inductive reactance over most of the audio frequency range. Recall that in a *pure inductance*, the current lags the voltage across it by 90°. *In the case of loudspeakers, the worst-case phase angle is normally* 60°.

If the same peak output voltage, and current are maintained, the waveforms for v_{CE}, and i_C are as shown in Fig. 12-47(c). The phase shift causes increased power dissipation in Q_1. With a pure resistive load, v_{CE} is at its minimum when i_C is at its maximum. Conversely, when v_{CE} is at its maximum, i_C is at its minimum. This situation promotes minimum power dissipation in the output transistor.

However, an inductive load can severely alter this relationship [see Fig. 12-47(c)]. When v_{CE} is equal to V_{CC} (25 V), the collector current is 1.94 A (0.866 of its peak value). Since the BJT's dissipation can be greatly increased, the ac load line may make significant excursions outside the SOA. Obviously, we need to plot it.

The analytical expressions for v_{CE} and i_C [which is lagging by π/3 radians or 60°, as shown in Fig. 12-47(c)] have been given by

$$
\begin{aligned}
v_{CE} &= V_{CC} - V_m \sin \omega t \\
i_C &= I_m \sin(\omega t - \pi/3) \quad \Big| \quad \pi/3 \le \omega t \le 4\pi/3 \\
&= 0 \quad \Big| \quad 0 < \omega t < \pi/3 \quad \text{or} \quad 4\pi/3 < \omega t < 2\pi
\end{aligned}
\tag{12-37}
$$

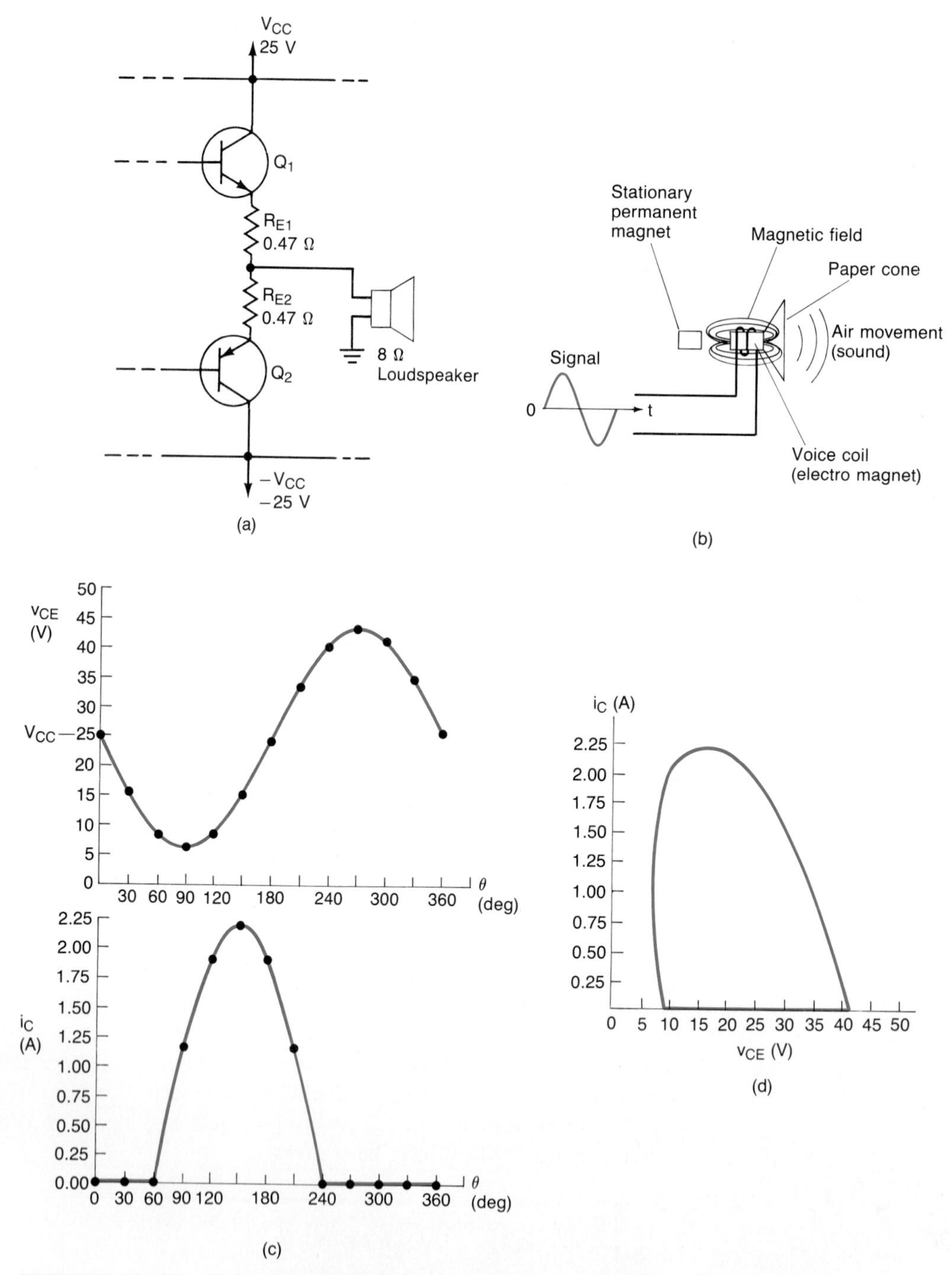

FIGURE 12-47 Inductive ac load line: (a) power amplifier output stage driving a loudspeaker; (b) permanent-magnet loudspeaker; (c) waveforms with an inductive (60° lagging) load; (d) load line plotted using the Table 12-2 data points; (e) ac (inductive) load line plotted on the SOA graph.

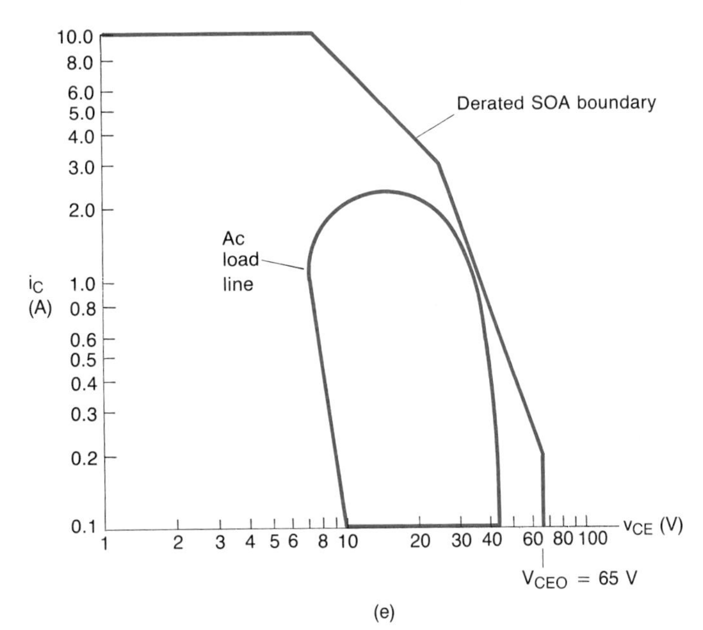

FIGURE 12-47 *(continued)*

and if we express the sine waves in terms of degrees, we obtain

$$\boxed{\begin{aligned} v_{CE} &= V_{CC} - V_m \sin\theta \\ i_C &= I_m \sin(\theta - 60^\circ) \;\Big|\; 60^\circ \le \theta \le 240^\circ \\ &= 0 \;\Big|\; 0^\circ < \theta < 60^\circ \quad \text{or} \quad 240^\circ < \theta < 360^\circ \end{aligned}} \tag{12-38}$$

In our example, Eq. 12-38 yields

$$\begin{aligned} v_{CE} &= 25 - 17.9 \sin\theta \\ i_C &= 2.24 \sin(\theta - 60^\circ) \,|\, 60^\circ \le \theta \le 240^\circ \\ &= 0 \,|\, 0 < \theta < 60^\circ \text{ or } 240^\circ < \theta < 360^\circ \end{aligned}$$

EXAMPLE 12-17

Draw the ac load line for the power amplifier given in Fig. 12-47(a). Assume that V_m is 17.9 V, I_m is 2.24 A, and the load is resistive–inductive such that it produces a 60° phase shift.

SOLUTION Sample calculations will be conducted for a θ of 0° and 90°. For a θ of 0°,

$$v_{CE} = 25 - 17.9 \sin\theta = 25 - 17.9 \sin 0^\circ = 25 \text{ V}$$

$$i_C = 0$$

and

$$v_{CE} = 25 - 17.9 \sin 90° = 7.1 \text{ V}$$

$$i_C = 2.24 \sin(\theta - 60°) = 2.24 \sin(90° - 60°) = 1.12 \text{ A}$$

The rest of the calculations are similar and have been summarized in Table 12-2.

TABLE 12-2
Inductive AC Load-Line Data Points

θ (deg)	v_{CE} (V)	i_C (A)
0	25.00	0
30	16.00	0
60	9.50	0
90	7.10	1.12
120	9.50	1.94
150	16.00	2.24
180	25.00	1.94
210	33.95	1.12
240	40.50	0
270	42.90	0
300	40.50	0
330	33.95	0
360	25.00	0

The ac load data points given in Table 12-2 have been plotted in Fig. 12-47(d). Observe that the load line is elliptical in nature. Its shape is altered further when it is plotted on the log-log graph of the SOA [Fig. 12-47(e)]. Observe that the inductive ac load line comes dangerously close to the BJT's SOA boundary in its second breakdown region. *The main point to remember is that inductive loads can greatly increase the power dissipation of the output transistors. The inductive effects simply cannot be ignored.* ■

12-16 Output Protection

When the output of a power amplifier, or any electronic system, is connected to the "outside world," it may be subjected to harsh treatment. Consider the audio power amplifier. We can go to great lengths to ensure that the amplifier will provide a given output power into a specific load, maintain a negligible distortion level, exhibit an impressive bandwidth, and have nearly an ideal efficiency. However, since its output is taken off the confines of the printed circuit board, we cannot guarantee that it will be handled kindly. For example, it may be connected to an incorrect load (e.g., a 4-Ω speaker was used instead of an 8-Ω), it may be left open (someone forgot to connect the speaker), or it may be short circuited (the speaker, or the wires connecting it, failed). For maximum reliability the power amplifier should be "forgiving." We must protect the output.

Typically, the output devices used in a power amplifier are one of the most expensive components in the amplifier system. Let us see how to protect them from an overload (too much output current) or a shorted load. The most common approach to short-circuit or overload protection is to sense the output current via the emitter or source resistor, and use the resulting voltage to turn on a bipolar transistor. The transistor is used to divert the input drive signal from the output device. Consider Fig. 12-48(a).

Transistors Q_3 and Q_4 provide the output protection. Observe that their base–emitter *p-n* junctions are connected across the emitter resistors. During normal operation, both of these transistors are OFF. In Fig. 12-48(a) we can see that if the load current is 1 A peak, Q_3's base-emitter voltage is 0.47 V. Since silicon BJTs do not start conducting until their V_{BE} is approximately 0.5 V, Q_3 is just on the verge of conducting. Therefore, its collector current will be approximately zero. If the load current is increased such that the voltage drop across R_{E1} is 0.6 V, transistor Q_3 will conduct [see Fig. 12-48(b)].

At this point any further increases in the base drive to output transistor Q_1 will be diverted from its base terminal by Q_3 [refer to Fig. 12-48(c)]. Even though the base drive to Q_1 has been increased to 50 mA, the output current will essentially be restricted to 1.28 A. Therefore, we have achieved *current limiting*. The maximum (or short-circuit) current is given by

$$I_{SC} = \frac{0.6\ \text{V}}{R_E} \tag{12-39}$$

where I_{SC} is the short-circuit current.

A current limit of 1.28 A is too low if we desire 20 W of output power. From Fig. 12-46(a), we recall that this requires a peak load current of 2.24 A. To use the basic current limiting scheme shown above, we must reduce the base-emitter bias of our overload protection transistors. One option is to lower the emitter resistors. However, this will reduce the bias stability of our output transistors. A second option has been depicted in Fig. 12-48(d).

A voltage divider (R_1 and R_2) has been used to reduce the base-emitter voltage applied to Q_3. In practice, the voltage-divider resistance ($R_1 + R_2$) should be much larger than the emitter resistance, and the base current drawn by Q_3 should be negligibly small compared to the current flow through R_1 and R_2. The values indicated in Fig. 12-48(d) satisfy these requirements.

By applying voltage division, Ohm's law, and assuming that V_{BE} must be 0.6 V to turn on Q_3, we obtain Eq. 12-40.

$$I_{SC}R_E = \frac{R_2}{R_1 + R_2} = 0.6\ \text{V}$$

$$I_{SC} = \frac{0.6\ \text{V}}{R_E}\,\frac{R_1 + R_2}{R_2}$$

$$I_{SC} = \frac{0.6\ \text{V}}{R_E}\left(1 + \frac{R_1}{R_2}\right) \tag{12-40}$$

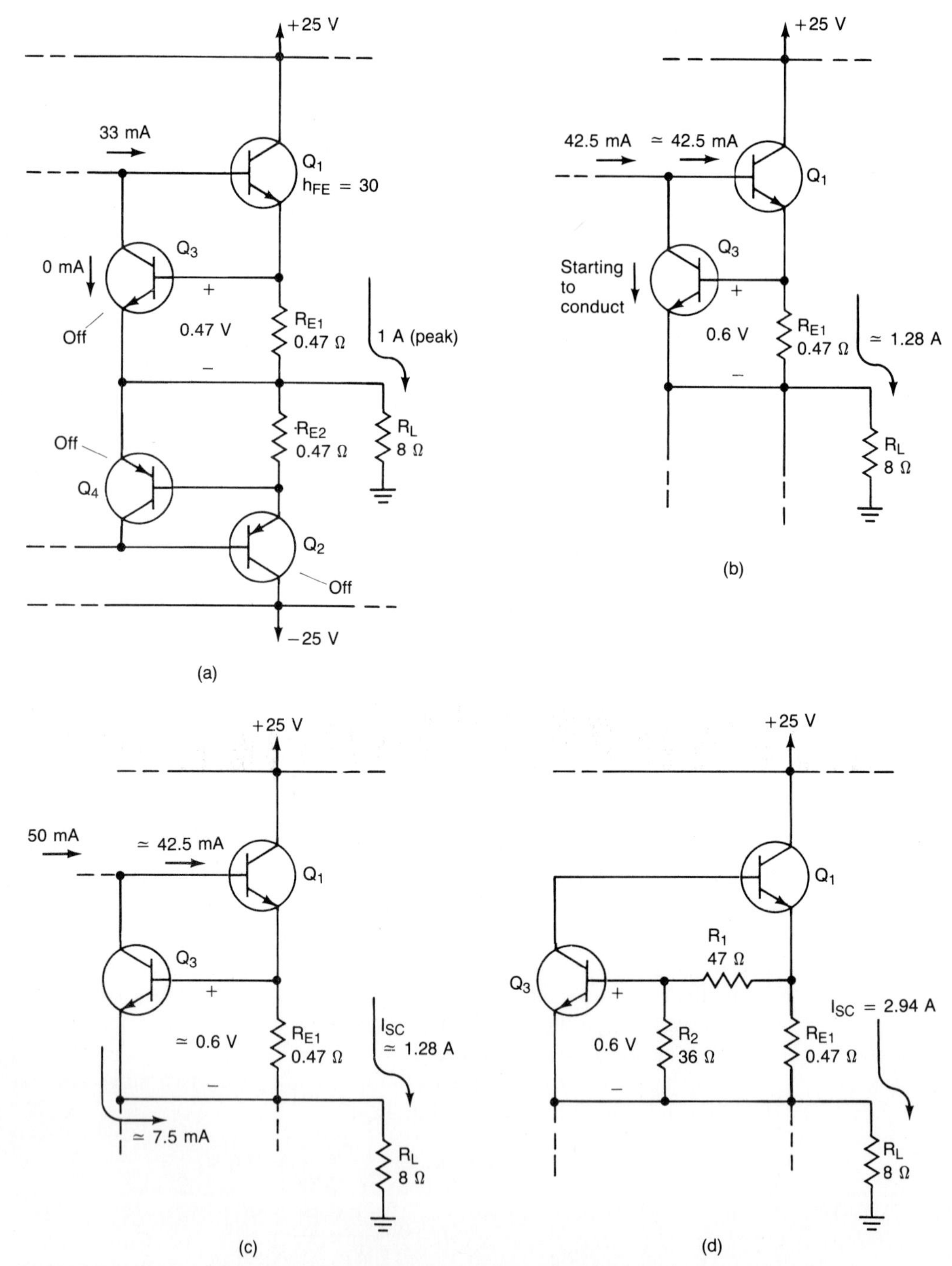

FIGURE 12-48 Short-circuit (and overload) protection: (a) when 1 A flows through R_{E1}, V_{BE3} is only 0.47 V and Q_3 remains off; (b) when V_{BE} reaches 0.6 V, Q_3 will start to divert current from Q_1's base; (c) the output current is limited; (d) a voltage divider has been used to raise the current limit to 2.94 A.

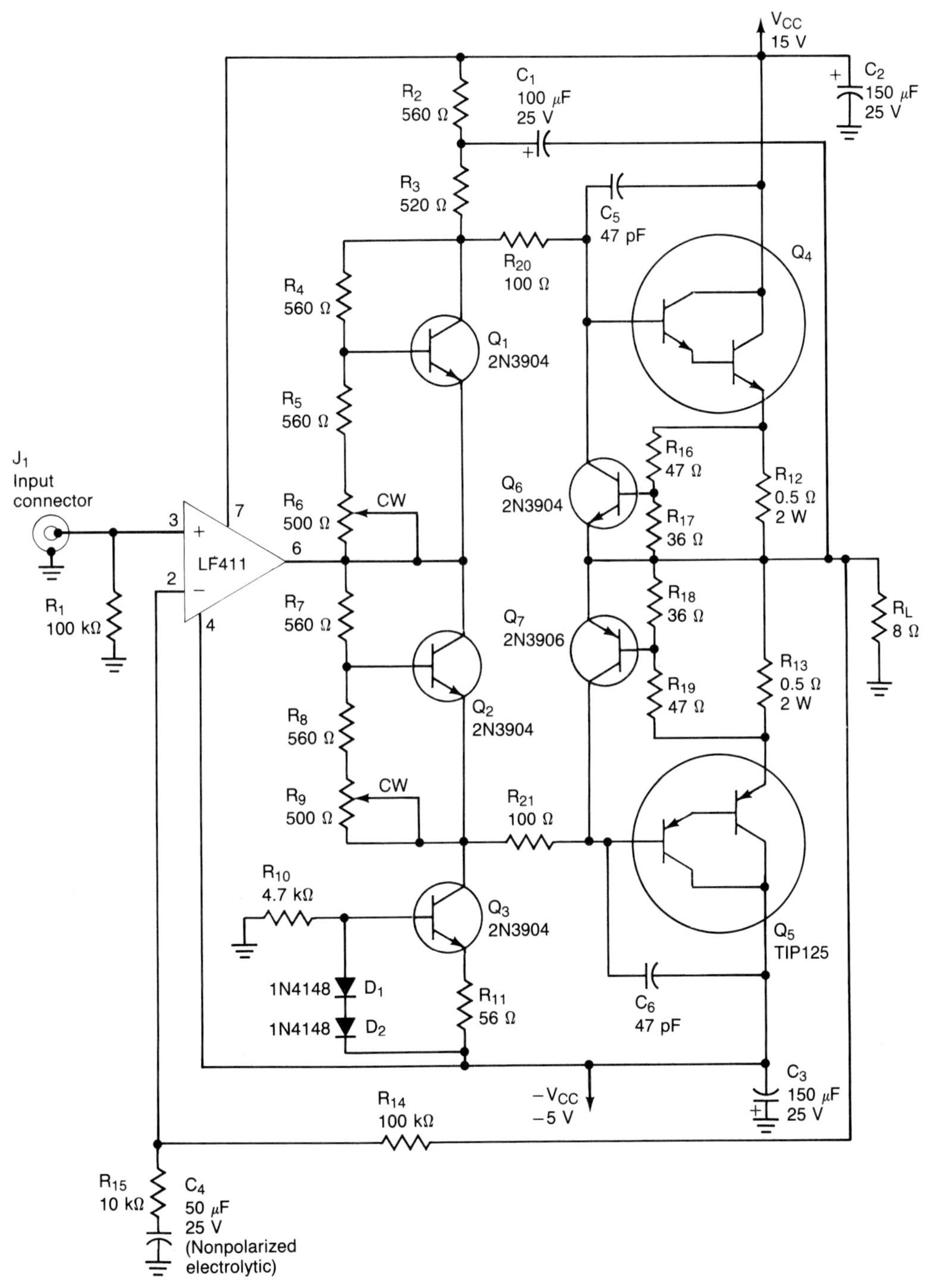

FIGURE 12-49 Transistors Q_6 and Q_7 have been added to provide output short-circuit protection. (R_{20}, R_{21}, C_5, and C_6 are used to prevent parasitic oscillations.)

In Fig. 12-48(d) we obtain

$$I_{SC} = \frac{0.6\text{ V}}{R_E}\left(1 + \frac{R_1}{R_2}\right) = \frac{0.6\text{ V}}{0.47\ \Omega}\left(1 + \frac{47\ \Omega}{36\ \Omega}\right) = 2.94\text{ A}$$

and we see that the circuit will limit the current to 2.94 A, which is above the normal maximum peak current of 2.24 A. The various voltages and currents have been indicated in Fig. 12-48(d).

To put the technique in perspective, current limiting can be added to the power amplifier shown in Fig. 12-43 (see Fig. 12-49). Again, we emphasize that the overload protection transistors are normally OFF. Therefore, the normal signal analysis is unaffected by their presence. The value of I_{SC} for Fig. 12-49 will be determined by the reader in the problems at the end of the chapter. (Also note that we have included R_{20}, R_{21}, C_5, and C_6 to eliminate parasitic oscillations as discussed in Section 12-14. The LF411 FET-input op amp is used to provide an excellent slew rate.)

Now that we have discussed overload and short-circuit protection, we need to consider the effects of a power amplifier with *no load connected*. Our first thought might be that if no load is connected, no damage will be done. In the case of the power amplifier given in Fig. 12-49, this is indeed true. However, this is not always the case. If the frequency compensation of the amplifier is designed with the assumption that the rated load is connected, this may *not* be the case (refer back to Fig. 12-45).

Recall that C_5 provides dominant pole frequency compensation. However, if the load is disconnected, the open-loop voltage gain will increase dramatically. Under these circumstances, the negative feedback may be large enough to cause high-frequency oscillations. The uncontrolled oscillations can destroy the output transistors if their power dissipation becomes excessive. Under these conditions, a series network such as R_{17} and C_8 shown in dashed lines in Fig. 12-45 should be added across the output.

Under normal conditions, the series *RC* (R_{17} and C_8) network will not affect amplifier performance. However, if R_L is disconnected, it will act as the amplifier's load, which serves to limit the open-loop voltage gain. Therefore, oscillations will be avoided. As a final comment, we should also mention that as an added bonus, the network reduces high-frequency noise at the output.

12-17 Thermal Analysis and Heat Sinking

In power amplifiers, the output devices are called upon to control large current and voltage swings. Consequently, their power dissipations tend to be vary large. The large power dissipations produce a rise in the junction temperature (T_j). In the case of bipolar transistors, T_j refers to collector–base *p-n* junction temperature. However, T_j is treated by manufacturers as a generic term for the chip temperature of power MOSFETs and ICs. Typically, the maximum allowable junction temperatures range from 125 to 200°C.

Elevated temperatures are undesirable for many reasons: performance is degraded, reliability is impaired, and device destruction becomes a real possibility. *Our fundamental problem is to remove heat from the semiconductor chip in order to keep T_j as low as possible.* This problem is compounded by economics and space restrictions.

To keep our semiconductors operating at sufficiently low temperatures, we use *heat sinks*. Heat sinks are available in a wide variety of shapes and sizes. Two examples have been provided in Fig. 12-50(a). Fins are used to increase the surface area (and therefore the emissivity) of many heat sinks.

A typical heat-sink configuration has been shown in Fig. 12-50(b). In this example, a power transistor in a TO-3 case style has been indicated. Since the collector (or drain) is electrically tied to the case, and the heat sink is normally mounted to the chassis (ground), an electrical insulator must be placed between the transistor and the heat sink. The most common insulating material is mica. To promote good heat transfer from the transistor's case to the heat sink, it is necessary to place silicon grease on both sides of the insulator.

To size the required heat sink properly, we need to understand the heat transfer problem quantitatively. The most common approach is to use a *thermal circuit*, which is analogous to an electrical circuit [see Fig. 12-50(c)]. In the analogy, the power dissipation (P_D) may be regarded as a current source, temperature drops (ΔT) are equivalent to voltage drops, and *thermal resistance* (θ) is the opposition the thermal energy (heat flux) flowing between two points. With these definitions in mind, the thermal circuit given in Fig. 12-50(c) begins to make sense.

Absolute zero ($-273°$C) is our reference. The entire thermal circuit is "lifted" up from "ground" by ambient temperature T_A which has been shown as a variable-voltage source. We have three thermal resistances to contend with: the thermal resistance between the junction and the case (θ_{jc}), the thermal resistance between the case and the heat sink (θ_{cs}), and the thermal resistance between the heat sink and ambient air (θ_{sa}). The sum of these three thermal resistance is the total thermal resistance from the junction to ambient (θ_{ja}).

$$\boxed{\theta_{ja} = \theta_{jc} + \theta_{cs} + \theta_{sa}} \qquad (12\text{-}41)$$

where θ_{ja} = junction-to-ambient thermal resistance
θ_{jc} = junction-to-case thermal resistance
θ_{cs} = case-to-sink thermal resistance
θ_{sa} = heat sink-to-ambient thermal resistance

If we remember that temperature is analogous to voltage, the junction temperature (T_j), the case temperature (T_c), and the heat-sink temperature (T_{hs}) defined in Fig. 12-49(c) should be clear. Since the actual junction temperature T_j is our most important consideration, we must apply "thermal Ohm's law" and "thermal Kirchhoff's voltage law" to obtain

$$\boxed{T_j = P_D\theta_{ja} + T_A} \qquad (12\text{-}42)$$

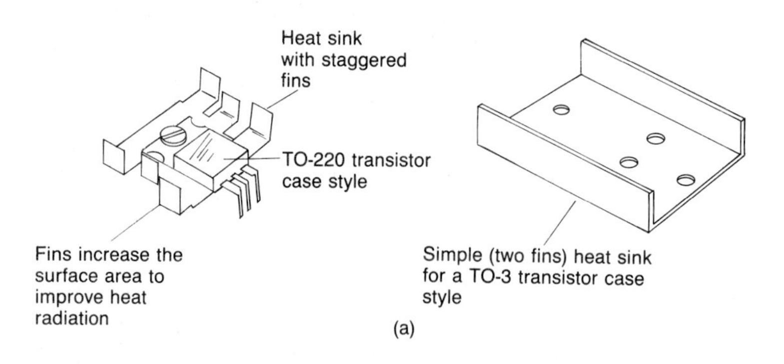

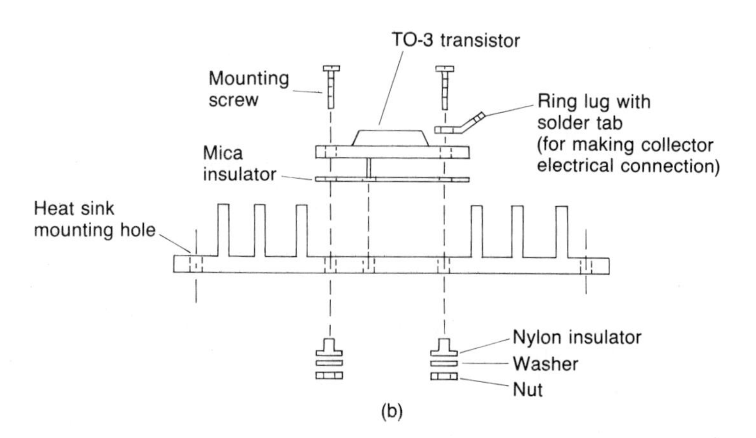

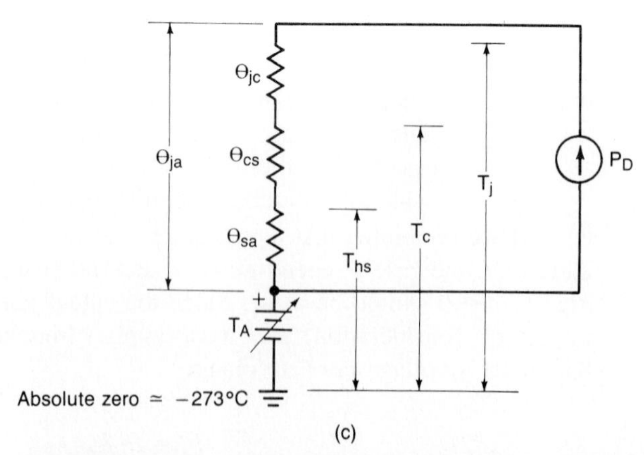

FIGURE 12-50 Heat sinks and thermal analysis: (a) heat sink examples; (b) mounting a TO-3 case style; (c) thermal equivalent circuit.

In the specification of heat sinks, we often require the thermal resistance from junction to ambient. This is easily determined if we merely rearrange Eq. 12-42 as shown in

$$\theta_{ja} = \frac{T_j - T_A}{P_D} \tag{12-43}$$

From Eq. 12-43 it should be obvious that thermal resistance has units of °C/W.

The heat-sinking problem requires that we ensure that the thermal resistance from junction to ambient (θ_{ja}) is low enough to keep the transistor under consideration from overheating. The ideal value of θ_{ja} is zero. Since θ_{ja} is actually the sum of three separate thermal resistances (Eq. 12-41), let us briefly consider each of them in turn.

The construction of a power transistor determines its θ_{jc}. Occasionally, the θ_{jc} for a power device may *not* be specified on its data sheet. In these instances, the data sheet usually includes a power derating curve [see Fig. 12-51].

As we can see, the power derating curve for a 2N3719 has been provided. The curve relates the maximum power dissipation rating to the transistor's *case temperature*. Consider this carefully. A 2N3719 transistor is billed as a 150-W power transistor. However, this rating applies only if the case temperature T_c is held at 25°C. If the transistor heats up, its maximum power rating is reduced. For example, if the case temperatrure is allowed to reach 75°C, the transistor can dissipate only 107 W!

The slope of the derating curve yields the derating factor D. The reciprocal of the derating factor D produces θ_{jc}. This has been shown in Fig. 12-51. Some of the typical θ_{jc} values have been given in Fig. 12-52.

FIGURE 12-51 **Finding θ_{jc} for the 2N3716 power transistor.**

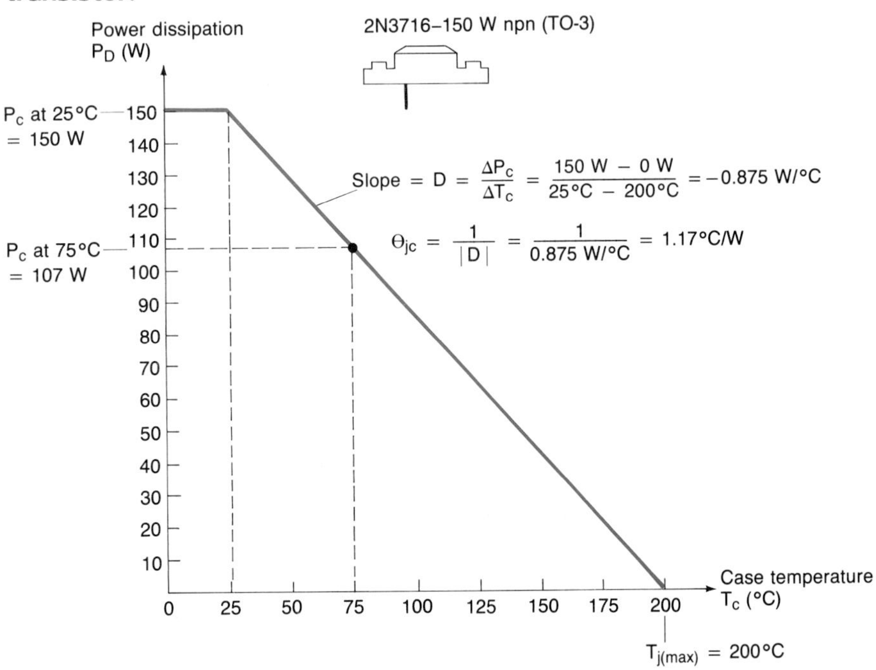

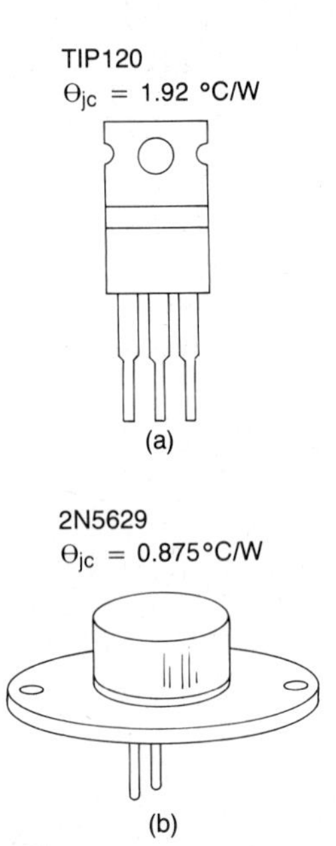

FIGURE 12-52 Examples of junction-to-case thermal resistance (θ_{jc}): (a) the TO-220 case style has θ_{jc} values from 1.7 to 5.0°C/W; (b) the TO-3 case style has θ_{jc} values from 0.5 to 6.0°C/W.

The first thermal resistance encountered external to the transistor is through the semiconductor-to-heat sink interface. *The thermal control measures taken at this point can have a major impact on the operating junction temperatures for a given power dissipation* P_D. For example, a TO-202 BJT case-to-sink interface with *no mica insulator and coated with silicon grease* can have a θ_{cs} as low as 0.9°C/W. However, if the same transistor is mounted to its heat sink *without* using silicon grease, θ_{cs} can be as high as 20°C/W.

EXAMPLE 12-18

A TO-202 case-style transistor is supposed to be mounted directly on its heat sink with silicon grease. However, the technician inadvertently installed the transistor without applying silicon grease. If the transistor dissipates 30 W, how much hotter will it operate?

SOLUTION The problem can be resolved fairly easily if we use the thermal circuit given in Fig. 12-50(c). If we assume that θ_{cs} is 0.9°C/W when no insulator is used and silicon grease is employed,

$$T_c = P_D\theta_{cs} + T_{hs}$$
$$= (30\text{ W})(0.9°\text{C/W}) + T_{hs} = 27°\text{C} + T_{hs}$$

Therefore, if the instructions were followed correctly, the case temperature will only be 27°C hotter than the heat sink. Now if we assume that the incorrectly mounted transistor has a θ_{cs} of 20°C/W,

$$T_c = P_D\theta_{cs} + T_{hs}$$
$$= (30\text{ W})(2°\text{C/W}) + T_{hs} = 60°\text{C} + T_{hs}$$

The transistor's case will be 60°C hotter than the heat sink. If the case is 60°C hotter, the junction temperature will also be 60°C hotter. The technician's error has produced a significant (33°C) difference. ■

The importance of the interface between the device and the heat sink was emphasized in Example 12-18. Typical θ_{cs} values are provided in Table 12-3.

TABLE 12-3
Typical Values of θ_{cs} (°C/W)

Case Style	Direct Contact	Direct Contact with Grease	Indirect Contact with Grease and Mica Insulator
TO-3	0.5–0.7	0.3–0.5	0.4–0.6
TO-202	1.5–2.0	0.9–1.2	1.2–1.7
TO-220	1.0–1.3	0.6–0.8	0.8–1.1

The last parameter to consider is the thermal resistance of the heat sink (θ_{sa}). There can *never* be a single θ_{sa} value for a given heat sink, only a value derived from a given *set* of conditions. The thermal resistance of a heat sink tends to vary as the temperature changes. Specifically, the heat-sink thermal resistance tends to *decrease* as the temperature of the heat sink is *raised*. Furthermore, airflow can have a dramatic effect (even at low velocities) by increasing the heat transfer efficiency from the heat sink to ambient air. For these reasons, heat-sink θ_{sa} numbers should be treated with care. Find out the test conditions and/or obtain a graph of θ_{sa} versus temperature.

EXAMPLE 12-19

A silicon power transistor has a heat sink with a θ_{sa} of 1.5°C/W. The transistor is rated at 120 W (25°C) and has a θ_{jc} of 1.25°C/W. A mounting insulator is used which has a θ_{cs} of 0.4°C/W. What is the maximum power the transistor can dissipate if the maximum ambient temperature is 40°C and T_j is rated at 175°C?

SOLUTION First, we must determine θ_{ja} by using Eq. 12-41.

$$\theta_{ja} = \theta_{jc} + \theta_{cs} + \theta_{sa}$$
$$= 1.25°\text{C/W} + 0.4°\text{C/W} + 1.5°\text{C/W} = 3.15°\text{C/W}$$

From Eq. 12-42 we recall that

$$T_j = P_D\theta_{ja} + T_A$$

and solving for P_D allows us to complete the problem.

$$P_{D(\max)} = \frac{T_{j(\max)} - T_{A(\max)}}{\theta_{ja}} = \frac{175°\text{C} - 40°\text{C}}{3.15°\text{C/W}} = 42.8\text{W}$$ ■

EXAMPLE 12-20

A TIP120 *npn* power transistor is rated at 65 W (at 25°C). The transistor has a maximum T_j of 150°C and a θ_{jc} of 1.92°C/W. It is to dissipate 10 W in a power amplifier and the maximum ambient temperature is 60°C. Determine the θ_{sa} of its heat sink. Assume that the transistor is to be mounted to the heat sink with a mica insulator using silicon grease.

SOLUTION First, we use Eq. 12-43 to determine the maximum allowable junction-to-ambient thermal resistance.

$$\theta_{ja(\max)} = \frac{T_{j(\max)} - T_{A(\max)}}{P_{D(\max)}} = \frac{150°\text{C} - 60°\text{C}}{10\ \text{W}} = 9°\text{C/W}$$

By rearranging Eq. 12-41, we may solve for the maximum heat-sink thermal resistance. (θ_{cs} is obtained from Table 12-3.)

$$\begin{aligned}\theta_{sa(\max)} &= \theta_{ja(\max)} - \theta_{jc(\max)} - \theta_{cs(\max)} \\ &= 9°\text{C/W} - 1.92°\text{C/W} - 1.1°\text{C/W} = 5.98°\text{C/W}\end{aligned}$$

Therefore, the heat sink must have a $\theta_{sa} < 5.98°\text{C/W}$.

If we are to specify the heat sink completely, space, economics, and manufacturer's lead time must be considered. Manufacturers of heat sinks include Aavid Engineering, Inc., IERC, Staver, Thermalloy, and Wakefield. ■

12-18 IC Power Amplifiers

In most of the power amplifier circuits presented thus far, we have opted to utilize IC op amps in the "front end." This was done because low-cost general-purpose op amps often represent the most cost-effective approach. Similarly, for many low-power (e.g., typically less than 10 W) applications, the *IC power amplifier* can represent a feasible alternative to discrete (or IC-discrete) power amplifier designs. We need to be familiar with this technology.

The internal "workings" of integrated audio power amplifiers do not differ significantly from the traditional integrated op amps studied thus far. They typically employ the standard linear op amp "tricks" such as the use of current mirrors, active loads, and internal frequency compensation. The major differences occur in the output stages, where class AB power amplifiers are used. The geometry of the IC layout is usually designed to promote thermal stability across the chip.

In Fig. 12-53(a) the typical input stage has been shown. As we can see, a differential input stage with current-source biasing, a current mirror, and an inverting current-to-voltage converter has been used. These stages are used to establish the transconductance of the audio power amplifier. A simplified output stage has been given in Fig. 12-53(b).

It is difficult to fabricate high-quality *pnp* bipolar power transistors in monolithic ICs. Therefore, many IC designers elect to use quasi-complementary symmetry. (Q_9 and Q_{10} behave as a single *pnp* output transistor.) Transistor Q_5 (the current-to-voltage converter) also serves as a class A driver. Also note that diode biasing has been used to eliminate crossover distortion. The normal idle current typically ranges from 1 to 15 mA.

Since the output of the audio power amplifier is used to drive external loads, most integrated circuits include output protection. Current-limiting circuitry (via Q_7 and Q_{11}) has been added to the output stage as shown in Fig. 12-53(c). In addition to output protection, most integrated audio power amplifiers also provide *thermal shutdown*. The typical circuitry used has also been illustrated in Fig. 12-53(c).

Diode D_3 is an avalanche diode. Recall from Chapter 2 that avalanche diodes exhibit a positive temperature coefficient. As their junction temperature increases, their breakdown voltage also increases. The base-emitter bias voltage of a BJT

FIGURE 12-53 IC power amplifer: (a) equivalent input stage determines the transconductance; (b) adding the output stage; (c) the complete equivalent circuit includes IC protection circuitry.

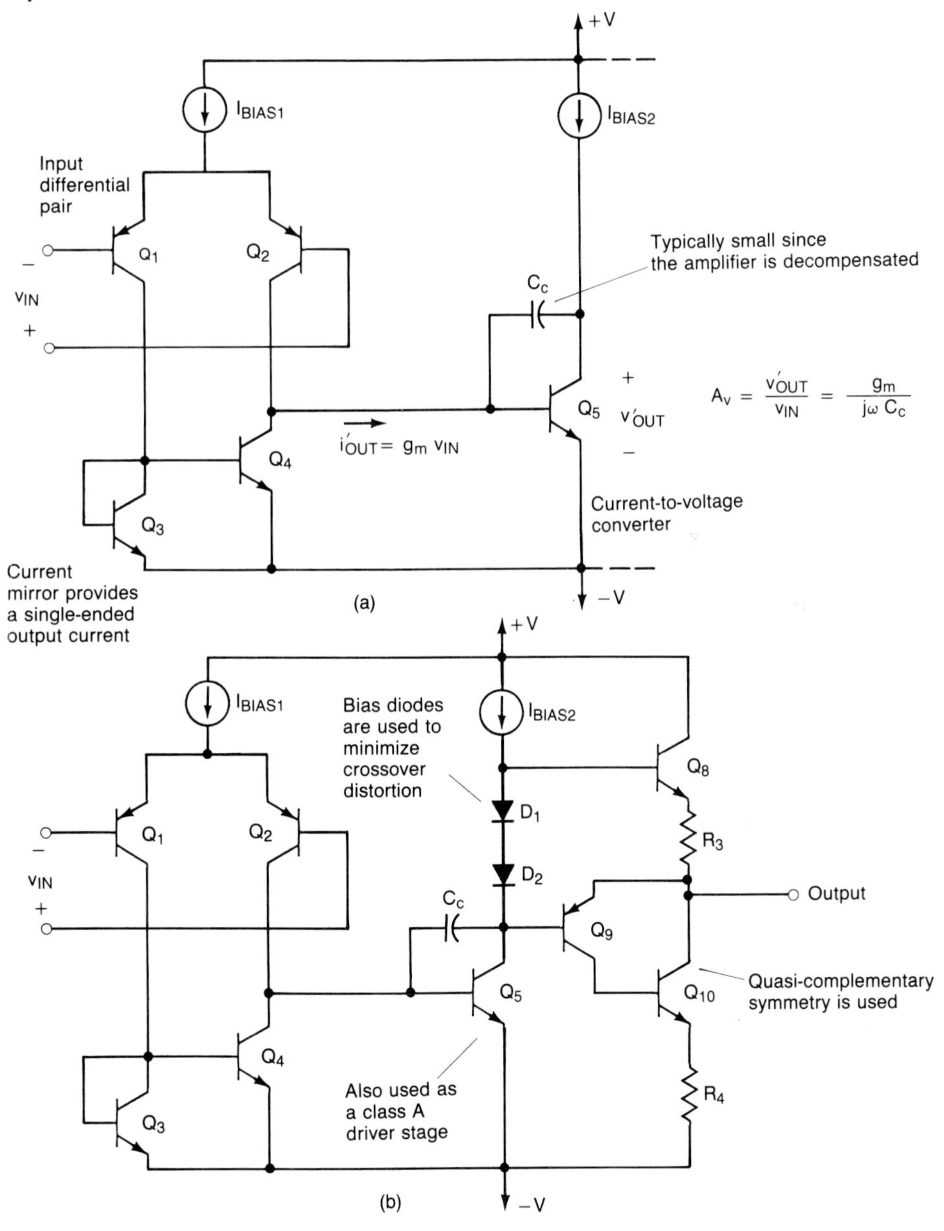

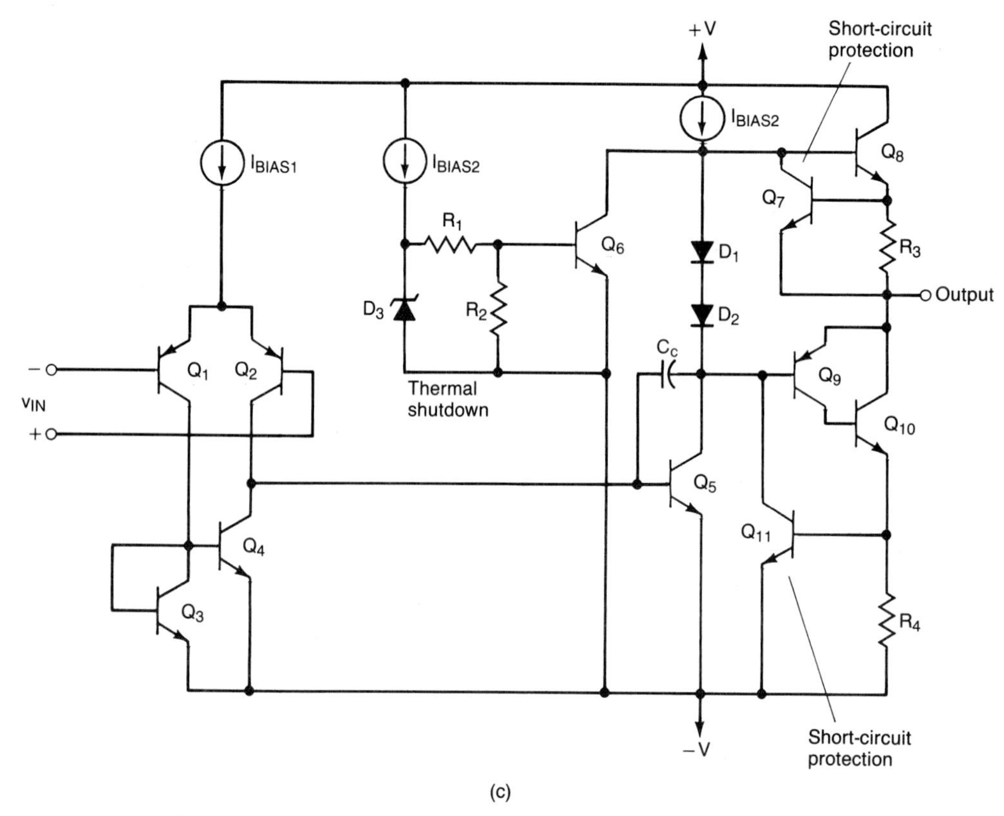

FIGURE 12-53 *(continued)*

demonstrates a negative temperature coefficient. Therefore, as the temperature of the chip rises, the avalanche voltage of D_3 increases, and the base-emitter threshold voltage drop of Q_6 decreases. The resistor ratio of R_1 and R_2 assists in determining the temperature at which transistor Q_6 begins to conduct. Typically, this occurs at about 150°C. The conduction of Q_6 is used to divert the dc bias and the input signal from the output transistors - effectively shutting down the IC.

The addition of thermal shutdown to integrated audio power amplifiers (and other high-power ICs) greatly enhances their reliability. If heat sinking is inadequate in a discrete design, the output transistors will typically fail. However, in a thermally protected IC, inadequate heat sinking results in reduced drive to the load, to restrict the chip temperature to a safe value.

In many of the power amplifier circuits discussed thus far, we have indicated the use of split power supplies (e.g., ± 15 V). This was necessary to ensure that the output circuit was balanced to keep a dc bias voltage from appearing across the load. However, the use of a single supply voltage is also possible. One such approach has been indicated in Fig. 12-54(a). In this case the output of the amplifier is typically set to a dc voltage that is equal to one-half of the supply voltage. This is accomplished by using a voltage divider at the noninverting input of the op amp driver, and wrapping a unity-gain (to dc) feedback loop around the amplifier system [see R_7, R_8, and C_3 in Fig. 12-54(a)].

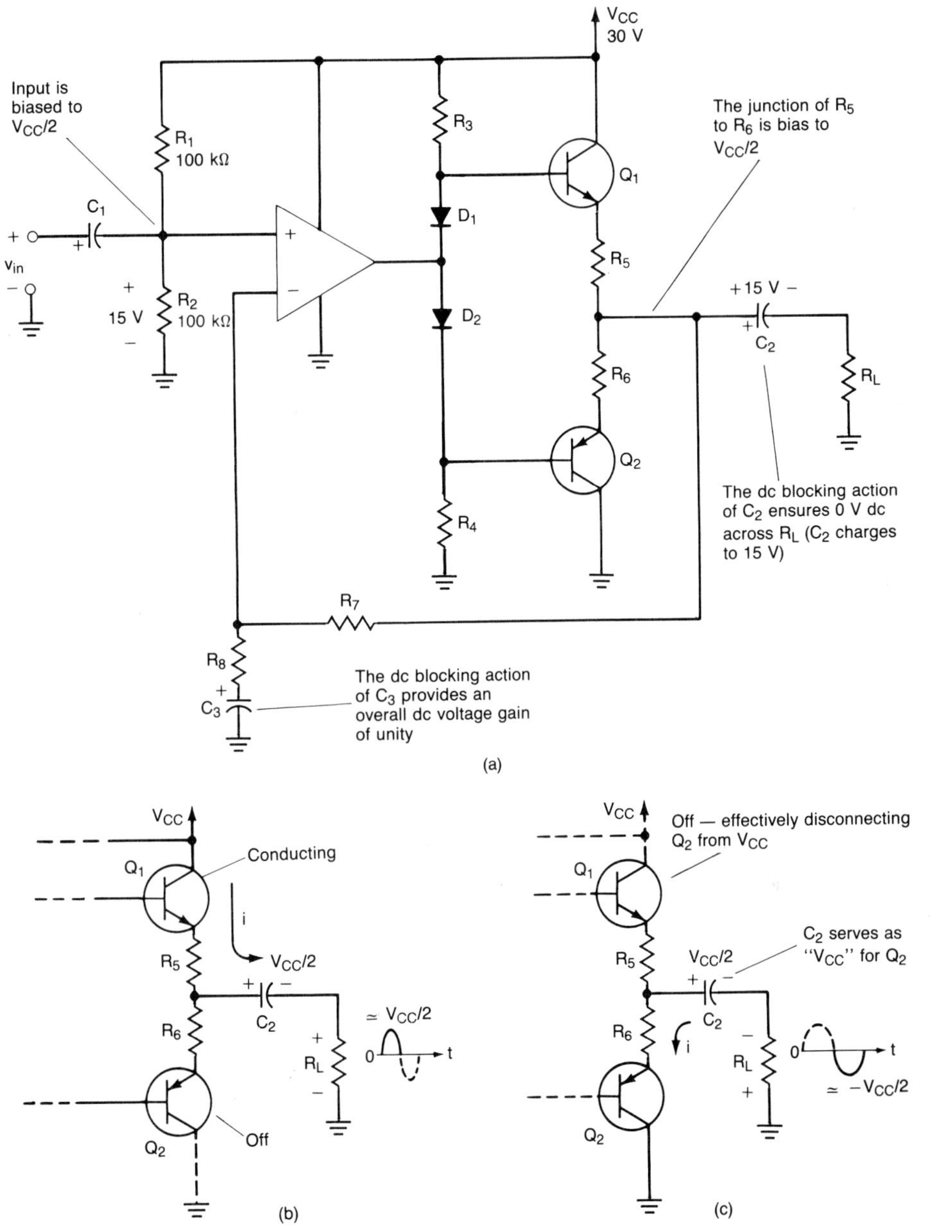

FIGURE 12-54 Single-supply power amplifier: (a) dc bias; (b) positive half-cycle; (c) negative half-cycle.

To keep the dc voltage across the load at zero, an output coupling (dc blocking) capacitor (C_2) is used. Obviously, this approach limits the low-end frequency response, but it does eliminate the need for dual power supplies. The push-pull action of the output stage is essentially unchanged. Consider Fig. 12-54(b).

During the positive half-cycle, transistor Q_1 conducts and transistor Q_2 is OFF.

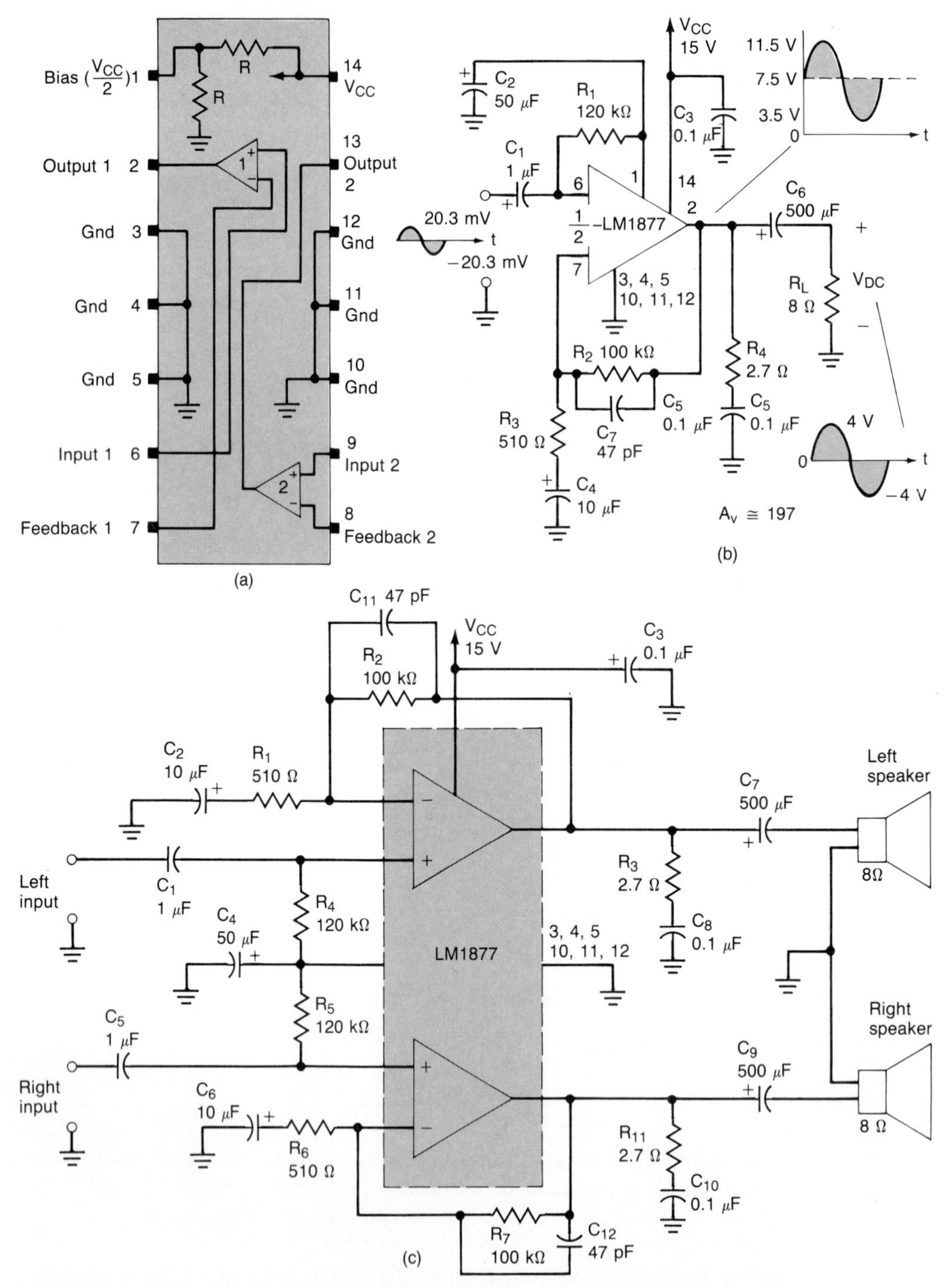

FIGURE 12-55 Using the LM1877 integrated-circuit power amplifer: (a) layout of the LM1877; (b) single-channel, 1-W audio amplifier; (c) 2-W (1 W per channel) stereo power amplifier using the LM1877.

During the negative half-cycle, transistor Q_1 is OFF and transistor Q_2 conducts. Observe that the output coupling capacitor charges to $V_{CC}/2$ [as indicated in Fig. 12-54(a)]. Since transistor Q_2 is effectively disconnected from V_{CC} (Q_1 is OFF), its source of "V_{CC}" is provided by the output coupling capacitor [see Fig. 12-54(c)].

To establish a bias voltage reference at $V_{CC}/2$, a simple voltage divider is used. The LM1877 dual audio power amplifier also employs this scheme [see pin 1 in Fig. 12-55(a)]. The LM1877 contains two separate power amplifiers (ideal for stereo applications) and is capable of delivering 2 W per channel into 8-Ω loads. However, the % THD is approximately 10% when 2 W is being delivered. Therefore, a more conservative design of 1 W is recommended since the % THD at this level is only around 0.055%.

The LM1877 is decompensated. Consequently, it is stable only for closed-loop voltage gains greater than 10. Its supply voltages can range from 6 V to an absolute maximum of 26 V. The LM1877 has a typical power bandwidth of 65 kHz when delivering 1.5 W to an 8-Ω load. Its typical slew rate is 2 V/μs.

A 1-W power amplifier has been shown in Fig. 12-55(b). The noninverting input has been biased to 7.5 V by connecting it to the pin 1 reference terminal through R_1. Capacitor C_2 is used for decoupling. The input and output coupling capacitors are C_1 and C_6, respectively. The output is biased to 7.5 V because of the action of feedback resistors R_2 and R_3, and dc blocking capacitor C_4. Capacitor C_7 has been added to reduce the amplifier's gain at high frequencies. This also tends to prohibit high-frequency oscillation. The R_4-C_5 network is used to promote high-frequency stability in the event that the 8-Ω load is disconnected.

A simple stereo power amplifier that uses both halves of the LM1877 has been given in Fig. 12-55(c). Although we are not dealing with an extremely high-performance power amplifier, it should be clear that we have a low-cost design which sports a minimal component count.

A number of monolithic and hybrid integrated power amplifiers are available. The LM1877 is a simple example of this growing technology. The price and performance of integrated power amplifiers are becoming extremely competitive with discrete designs.

12-19 Common-Source Output Stage

In Fig. 12-56(a) we see the complete electrical schematic of a high-performance, 15-watt audio power amplifier. A quick inspection soon reveals that the amplifier design incorporates several unique features. Do not let the circuit become overwhelming. Before we are through, the subtle details and the rationale behind the design will become clear.

A partial circuit diagram has been indicated in Fig. 12-56(b). As we can see, the gates of the MOSFETs are being driven from the op amp's supply terminals. This unconventional approach becomes clear when we recall that the output circuit of most monolithic op amps consists of a push-pull stage. Consequently, the resistors in series with the op amp's supply terminals serve as collector resistors. The resulting phase relationships have also been illustrated in Fig. 12-56(b).

The op amp's output stage has been converted to a pair of common-emitter push-pull drivers. The voltage at the supply terminals of the op amp is used not only to drive the output stage, but to provide dc bias as well.

By inspection of Fig. 12-56(b), we can also see that the gate terminals of the output transistors are driven, and the output signal is extracted from their drain terminals. Therefore, the output transistors are being used in their common-source configuration.

Let us reflect briefly at this point. Instead of using the op amp's output terminal to drive a pair of push-pull source followers, we are using the op amp's supply terminals to drive the output stage. The primary advantage in this approach is that we are obtaining some additional voltage gain from the op amp's output stage. Further,

FIGURE 12-56 **15-W class AB push-pull audio power amplifier using a common-source complementary pair: (a) schematic; (b) partial schematic.**

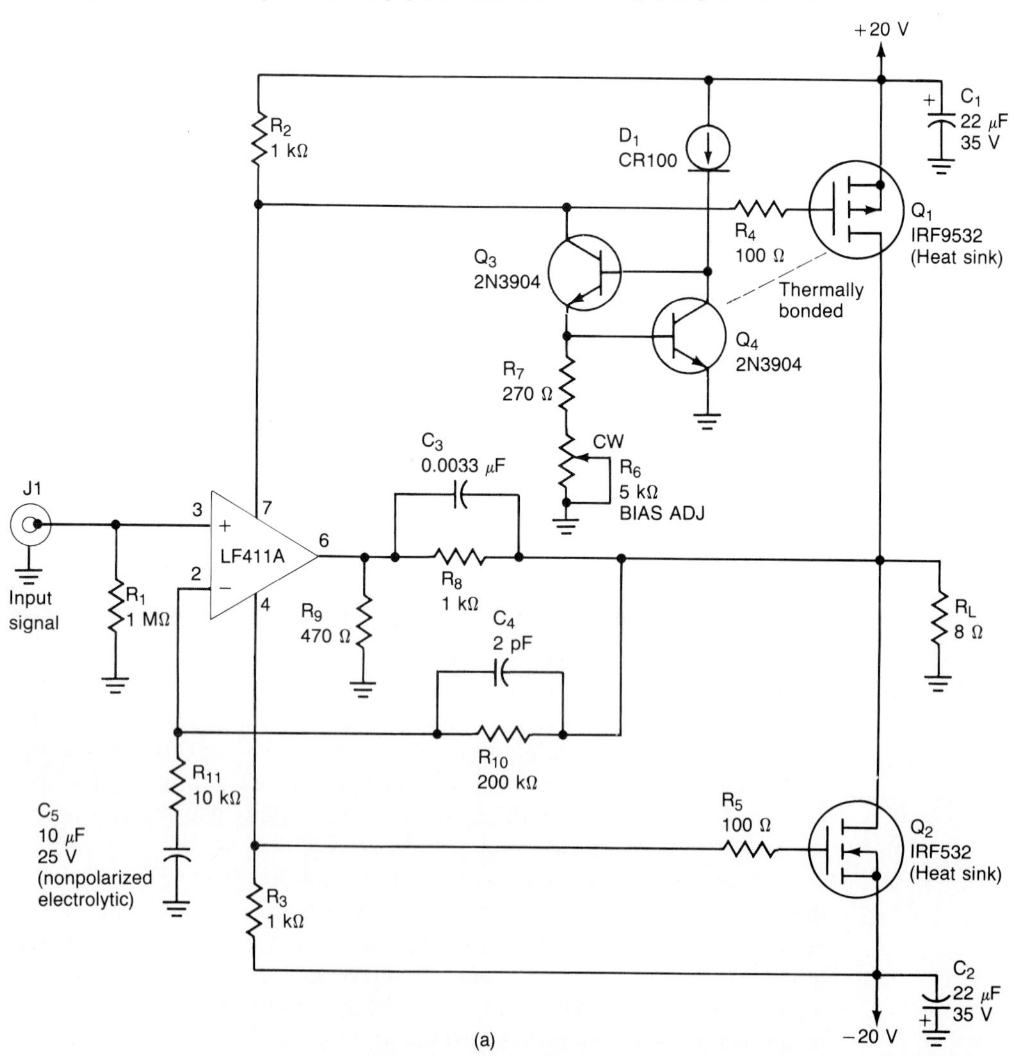

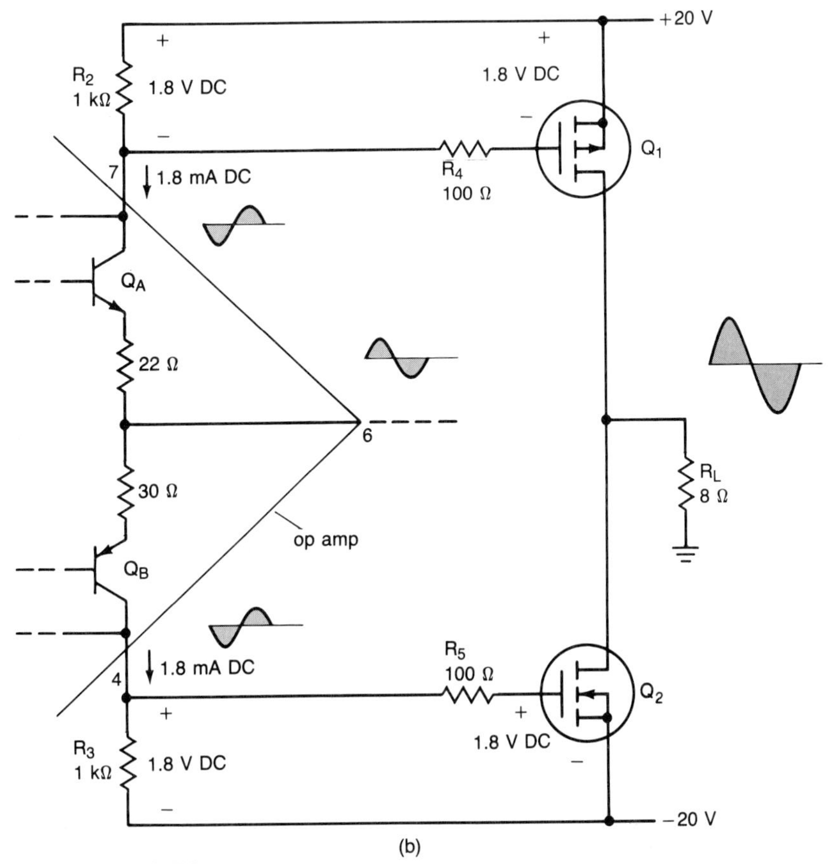

FIGURE 12-56 *(continued)*

the output of most monolithic op amps can only swing to within about 2 V of the supply rails. (The LF411A can typically swing to within 1.5 V of the supply rails. However, according to the manufacturer, the worst-case limitation is 3 V.) The use of the source-follower configuration in the output stage also suffers from inherent disadvantages. First, its voltage gain is less than unity and second, it has a substantial dc threshold offset voltage. (Recall that the only way we could boost up the output swing was through the use of a bootstrapping capacitor.)

The amplifier depicted in Fig. 12-56(a) does not suffer from these disadvantages. Therefore, its output voltage can swing much closer to the supply rails. As we shall see, this will allow us to obtain a larger ac output power, and obtain a greater efficiency. Now that we have highlighted the basic strategy behind Fig. 12-56(a), let us proceed with the analysis.

12-20 Common-Source Dc Analysis

The output transistors are n- and p-channel complementary DMOS power transistors. Their primary dc electrical specifications have been given in Table 12-4.

The LF411A op amp has a JFET input stage. Therefore, the dc bias current

TABLE 12-4
Dc Electrical Ratings

Parameter	IRF532 (*n*-channel)	IRF9532 (*p*-channel)
$V_{DS(\max)}$	100 V	− 100 V
$I_{D(\max)}$ @ 100°C	8.0A	− 6.5 A
$P_{D(\max)}$ @ 25°C	75 W	75 W
D (Derating)	− 0.60 W/°C	− 0.60 W/°C
$V_{GS(\text{th})}$	2.0 to 4.0 V	− 2.0 to − 4.0 V

drawn by its input terminals is low (50 pA), and its slew rate is high (10 V/μs). Its gain-bandwidth product ranges from 3 to 4 MHz. Its maximum supply votlage is ± 22 V. Therefore, it is quite comfortable with the ± 20 V supply indicated in Fig. 12-56(a). Its nominal supply current is 1.8 mA.

The supply current drawn by the op amp develops a dc voltage across the resistors (R_2 and R_3) in series with the supply terminals [see Fig. 12-56(b)]. By Ohm's law we can see that 1.8 V will be developed across the resistors. This establishes the gate-to-source dc bias voltages for the MOSFETs.

In Table 12-4 we can see that the threshold voltage $V_{GS(\text{th})}$ for the MOSFETs ranges from 2 to 4 V. Therefore, the MOSFETs are biased toward, but not beyond, their thresholds.

Temperature compensation and some means of adjusting the dc bias is necessary. The bias should be adjusted to minimize crossover distortion without requiring excessive dc idling currents. Although a V_{BE} multiplier could be used, an alternative scheme employing Q_3, Q_4, R_6, R_7, and D_1 has been indicated [see Figs. 12-56(a) and 12-57(a)].

The circuit can best be described as a temperature-compensating current source. Notice that transistor Q_4 is thermally bonded to output transistor Q_1. Transistor Q_4 is biased by a constant current. In this case, the constant current is produced by a *constant-current diode* D_1.

The constant-current diode is a two-terminal device. Its operation has been detailed in Fig. 12-57(b). As we can see, it is a self-biased JFET that is biased beyond pinchoff. The CR100 constant-current diode has a nominal current of 1.0 mA but it can range from 0.9 to 1.1 mA.

The base-emitter voltage of transistor Q_4 (V_{BE4}) is used to establish the emitter current of transistor Q_3.

$$I_{E3} = \frac{V_{BE4}}{R_6 + R_7} \tag{12-44}$$

The collector current of Q_3 is approximately equal to its emitter current.

$$I_{C3} \simeq I_{E3}$$

As we can see in Fig. 12-57(a), Q_3's collector current flows through R_2. Therefore, the current through R_2 consists of the op amp's bias current (I_1) and I_{C3}. This current sets the voltage drop across R_2 which establishes the gate-to-source bias voltage of transistor Q_1. (Recall that no appreciable voltage drop will be developed across the gate resistors R_4 and R_5.)

$$V_{GS1} = -(I_1 + I_{C3})R_2 \tag{12-45}$$

FIGURE 12-57 Dc analysis: (a) partial schematic to study the dc biasing; (b) constant-current diode (0.900 to 1.10 mA).

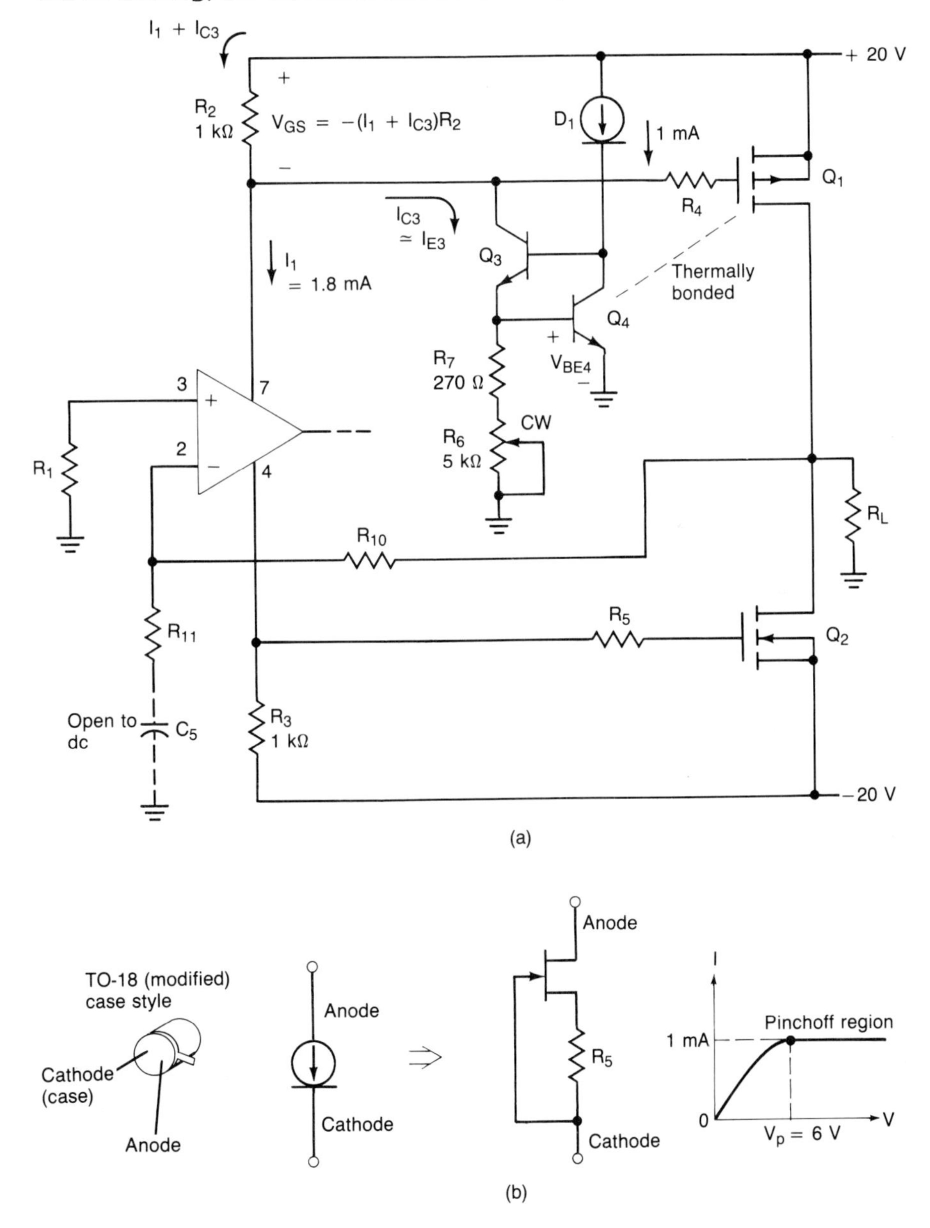

A quick inspection of Fig. 12-57(a) reveals that the gate will be negative with respect to the source. This is the polarity required to enhance a p-channel MOSFET.

Substitution of Eq. 12-44 into Eq. 12-45 yields

$$V_{GS1} = -(I_1 + I_{C3})R_2 = -(I_1R_2 + I_{C3}R_2)$$

$$\simeq -\left[I_1R_2 + \frac{R_2}{R_6 + R_7}V_{BE4}\right] \qquad (12\text{-}46)$$

EXAMPLE 12-21

Determine the range of bias voltage adjustment provided by potentiometer R_6 in Fig. 12-56(a).

SOLUTION If R_6 is adjusted fully clockwise, its resistance will be approximately zero. Equation 12-46 gives

$$V_{GS1} \simeq -\left[I_1R_2 + \frac{R_2}{R_6 + R_7}V_{BE4}\right]$$

$$\simeq -\left[(1.8\text{ mA})(1\text{ k}\Omega) + \frac{1\text{ k}\Omega}{270\ \Omega}(0.7\text{ V})\right] = -4.39\text{ V}$$

If R_6 is adjusted fully counterclockwise, its resistance will be 5 kΩ.

$$V_{GS1} \simeq -\left[I_1R_2 + \frac{R_2}{R_6 + R_7}V_{BE4}\right]$$

$$\simeq -\left[(1.8\text{ mA})(1\text{ k}\Omega) + \frac{1\text{ k}\Omega}{5270\ \Omega}(0.7\text{ V})\right] = -1.93\text{ V}$$

These values seem reasonable when we recall (from Table 12-4) that $V_{GS(\text{th})}$ ranges from -2 to -4 V. ■

With the bias adjust potentiometer fully counterclockwise, V_{GS1} will be -1.93 V [see Fig. 12-57(a)]. Since this is less than the minimum $V_{GS(\text{th})}$ of -2 V, the p-channel MOSFET Q_1 should not conduct any appreciable current. As the trimpot is adjusted clockwise, the additional current drawn by Q_3 through R_2 increases. This causes V_{GS1} to become more negative. At some point, the voltage drop across R_2 will be equal to Q_1's threshold voltage. When this occurs, Q_1 will begin to conduct current.

If the n-channel MOSFET Q_2 is not conducting as much current as Q_1, the output of the amplifier will be forced in a positive direction. The output of the amplifier is connected by R_{10} to the inverting input of the op amp. The feedback of the op amp will cause its output to go in a negative direction. This will cause the current through R_3 to increase. Therefore, the gate-to-source voltage of Q_2 will increase. The conduction of transistor Q_2 will increase until the output of the power amplifier is balanced at zero volts.

In a fashion similar to the V_{BE} multiplier, temperature compensation will be provided. Transistor Q_4 is thermally bonded to transistor Q_1's heatsink. Since Q_4 is biased by a constant-current source, its base–emitter voltage will decrease at a rate of approximately -2 mV/°C.

By taking the derivative of Eq. 12-46, we see that

$$\frac{dV_{GS1}}{dT} = \frac{d}{dT}\left[-I_1R_2 - \frac{R_2}{R_6 + R_7} V_{BE4} \right]$$

$$= -\frac{R_2}{R_6 + R_7} \frac{dV_{BE4}}{dT} \simeq -\frac{R_2}{R_6 + R_7} \frac{\Delta V_{BE4}}{\Delta T}$$

$$\boxed{\frac{dV_{GS1}}{dT} \simeq -\frac{R_2}{R_6 + R_7}(-2 \text{ mV/°C})} \qquad (12\text{-}47)$$

A couple of comments are in order. First, we should emphasize that Eq. 12-47 describes the rate of change in the *bias voltage*, and *not* the MOSFET's temperature coefficient. Second, the bias voltage will have a *positive* temperature coefficient. This is fine when we realize that the $V_{GS(\text{th})}$ of a *p*-channel device also has a positive temperature coefficient.

EXAMPLE 12-22

Determine the range of temperature coefficients offered by the temperature-compensating constant-current source used in Fig. 12-56(a).

SOLUTION When potentiometer R_6 is adjusted fully clockwise, R_6 is approximately zero, and V_{GS1} is -4.39 V. From Eq. 12-47 the temperature coefficient is

$$\frac{dV_{GS1}}{dT} \simeq -\frac{R_2}{R_6 + R_7}(-2 \text{ mV/°C}) = -\frac{1 \text{ k}\Omega}{270 \ \Omega}(-2 \text{ mV/°C})$$

$$\simeq 7.41 \text{ mV/°C}$$

When R_6 is fully counterclockwise, R_6 is 5 kΩ and V_{GS1} is -1.93 V. The corresponding temperature coefficient is

$$\frac{dV_{GS1}}{dT} \simeq -\frac{R_2}{R_6 + R_7}(-2 \text{ mV/°C}) = -\frac{1 \text{ k}\Omega}{5270 \ \Omega}(-2 \text{ mV/°C})$$

$$\simeq 0.380 \text{ mV/°C}$$

This range of temperature coefficients appears to be more than adequate when we recall that the $V_{GS(\text{th})}$ of MOSFETs typically exhibits tempcos from 2.5 to 6 mV/°C. ■

12-21 Common-Source Ac Analysis

To conduct an ac analysis, we need the MOSFET ac parameters. These have been indicated in Table 12-5.

Two negative feedback loops are employed. Local negative feedback in the output stage is provided by resistors R_8 and R_9. Overall system negative feedback is provided by resistors R_{10} and R_{11}, and capacitor C_5. Let us first investigate the effects of R_8 and R_9. A partial schematic diagram has been given in Fig. 12-58(a).

The local negative feedback is voltage-series. Using the rules developed in

TABLE 12-5
Ac Parameters

Parameter	IRF532 (n-channel)	IRF9532 (p-channel)
g_{fs}	4.0 S (min.) @ $I_D = 8$ A 5.5 S (typ.)	2.0 S (min.) @ $I_D = -6.5$ A 3.8 S (typ.)
$r_{DS(ON)}$	0.20 Ω (typ.) and 0.25 Ω (max.)	0.3 Ω (typ.) and 0.4 Ω (max.)
C_{iss}	600 pF (typ.) and 800 pF (max.)	500 pF (typ.) and 700 pF (max.)
C_{rss}	100 pF (typ.) and 150 pF (max.)	100 pF (typ.) and 200 pF (max.)
C_{oss}	300 pF (typ.) and 500 pF (max.)	300 pF (typ.) and 450 pF (max.)

Chapter 10, we arrive at the simplified equivalent circuit shown in Fig. 12-58(b). From Eq. 8-18 we can determine the voltage gain of the Q_A common-emitter stage.

$$A_{vA} = -\frac{g_{mA}r_C}{1 + g_{mA}r_E}$$

In Fig. 12-58(b) we see that

$$r_C \simeq R_2 = 1\ \text{k}\Omega$$

and

$$r_E \simeq R_E + R_8 \,\|\, R_9 = 22\ \Omega + 1\ \text{k}\Omega \,\|\, 470\ \Omega$$
$$= 22\ \Omega + 320\ \Omega = 342\ \Omega$$

Assuming that I_{CA} is approximately 1.8 mA,

$$g_{mA} \simeq \frac{I_{CA}}{26\ \text{mV}} = \frac{1.8\ \text{mA}}{26\ \text{mV}} = 69.2\ \text{mS}$$

and

$$A_{vA} = -\frac{g_{mA}r_C}{1 + g_{mA}r_E} = -\frac{(69.2\ \text{mS})(1\ \text{k}\Omega)}{1 + (69.2\ \text{mS})(342\ \Omega)} = -2.81$$

The voltage gain of the common-source MOSFET stage is

$$A_{v1} = -g_{m1}r_L$$

and from Table 12-5, we see that g_{m1} (g_{fs}) is 2.0 S (minimum). However, note that it is specified for an I_D of -6.5 A. Our average drain current will be considerably less in this application. Therefore, the transconductance will probably be lower than 2.0 S. However, to expedite the analysis, we shall use this value.

$$A_{v1} = -g_{m1}r_L = -(2.0\ \text{S})(8\ \Omega) = -16$$

The overall open-loop output stage voltage gain is given by the product of these individual loaded stage gains.

$$A_{v(\text{ol})} = A_{vA}A_{v1} = (-2.81)(-16) = 45.0$$

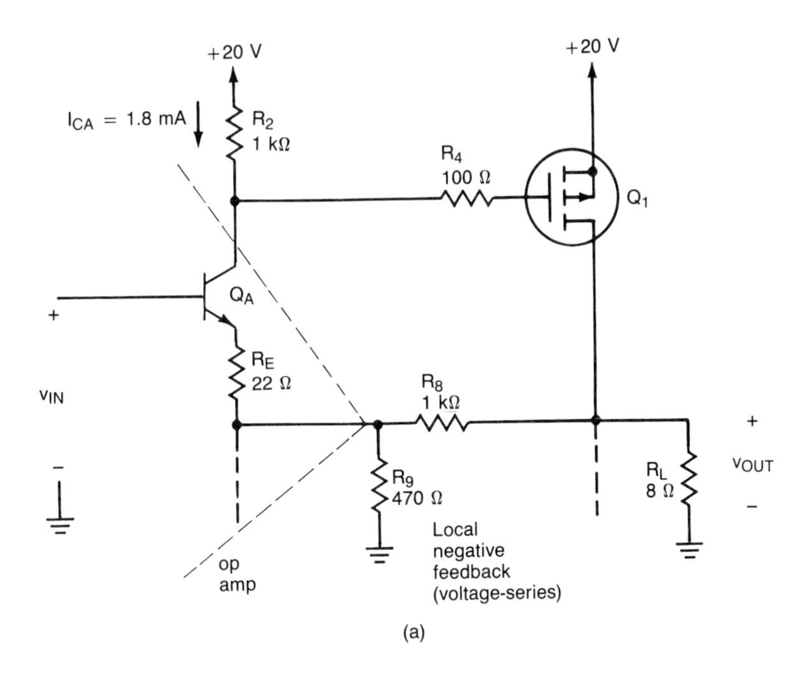

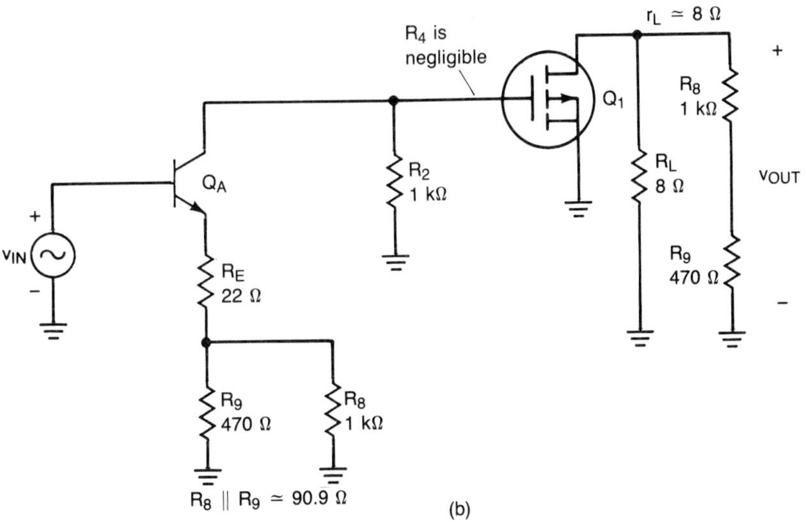

FIGURE 12-58 Finding the voltage gain of the output stage: (a) the output circuitry involved in the positive half-cycle of v_{OUT}; (b) the open-loop output stage ac equivalent circuit.

The local negative feedback factor β is

$$\beta = \frac{R_9}{R_8 + R_9} = \frac{470\ \Omega}{1\ \text{k}\Omega + 470\ \Omega} = 0.320$$

and the closed-loop voltage gain is

$$A_{v(\text{cl})} = \frac{A_{v(\text{ol})}}{1 + \beta A_{v(\text{ol})}} = \frac{45.0}{1 + (0.320)(45.0)} = 2.92$$

Note that since the loop gain is so low, we have used the exact form for $A_{v(cl)}$ rather than approximating it as $1/\beta$ (3.125).

The output stage increases the open-loop voltage gain of the *system* by a factor of 2.92. Therefore, the *loop gain* of the *system* will also be increased with all of the attendant advantages. (Recall that the % THD and the output impedance are inversely proportional to the loop gain, while the bandwidth and the input impedance are directly proportional.)

12-22 Common-Source Slew Rate and Power Bandwidth

Negative feedback does *not* affect the slew rate and power bandwidth. However, adding additonal voltage gain does tend to improve these quantities. We shall not perform a rigorous analysis to support these claims, but rely on our intuition (see Fig. 12-59).

The slew rate of the op amp is limited to a respectable 10 V/μs. However, by adding the additional output stage voltage gain, the op amp's output voltage swing does not have to be as large to produce a large *amplifier system* output voltage swing. Therefore, maximum slew rate of the system will be increased to approximately

FIGURE 12-59 The additional output stage voltage gain improves the system's slew rate and power bandwidth (f_{max}).

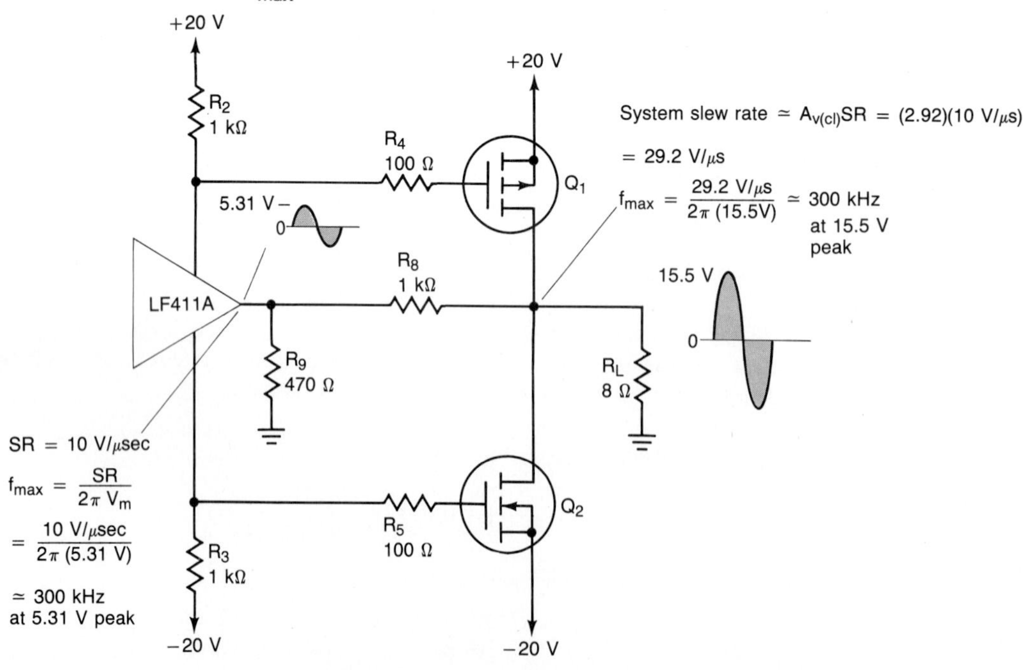

29.2 V/μs as demonstrated in Fig. 12-59. The power bandwidth will also be increased because of its intimate relationship to the slew rate.

In general, the slew rate and power bandwidth improvements will *not* be as great as indicated in Fig. 12-59. This is true because we have not included the speed limitations imposed by the output transistors Q_1 and Q_2. Further, we shall see that additional frequency compensation will be required to ensure freedom from oscillation. The additional frequency compensation will also tend to make the amplifier more "sluggish."

12-23 Common-Source Frequency Compensation

The additional voltage gain provided by the output stage has been shown to offer many advantages. However, it does open up the possibility of oscillation. This is true for two reasons. First, the additional voltage gain *lowers the phase margin* and second, the output stage (Q_1 and Q_2) will add another couple of poles [see Fig. 12-60(a)].

The *system* closed-loop voltage gain is 21 (26.4 dB). If no additional voltage gain were provided (e.g., we had elected to use push-pull source followers), the frequency response of the LF411A op amp would dominate [see Fig. 12-60(b)]. Since the closed-loop gain intersects the open-loop response where it rolls off at −20 dB/decade, the system will be stable.

However, the LF411A is frequency compensated for unity voltage gain using dominate-pole frequency compensation. Therefore, it is quite likely that its second breakpoint occurs at, or near, the unity-gain frequency (4 MHz). This is shown in dashed lines in Fig. 12-60(b).

The common-source MOSFETs and the op amp's output stage provide an additional voltage gain of 2.92 (9.3 dB) [see Fig. 12-60(c)]. This will shift the composite open-loop response upward as depicted in Fig. 12-60(d). Since the additional output stage gain of 9.3 dB is less than the system closed-loop voltage gain of 26.4 dB, the closed-loop response still intersects the open-loop response where it rolls off at a rate of −20 dB/decade. Therefore, the system still appears to be stable. However, the phase margin will be decreased. If we allow ourselves to consider the effects of temperature, component tolerance, and stray capacitances, it becomes increasingly difficult to sleep at night. Consequently, it is highly desirable to increase the phase margin. This can be accomplished by adding capacitor (C_3) as shown in Fig. 12-56(a).

Simply, the purpose of C_3 is to provide additional frequency compensation thereby increasing the phase margin. This becomes apparent in Fig. 12-61(a).

To understand the frequency effects of the R_8-R_9-C_3 network, we must find its voltage gain (V_2/V_1). To simplify matters, we shall first determine the admittance of the parallel combination of R_8 and C_3.

$$Y = \frac{1}{R_8} + j\omega C_3 = \frac{1 + j\omega R_8 C_3}{R_8}$$

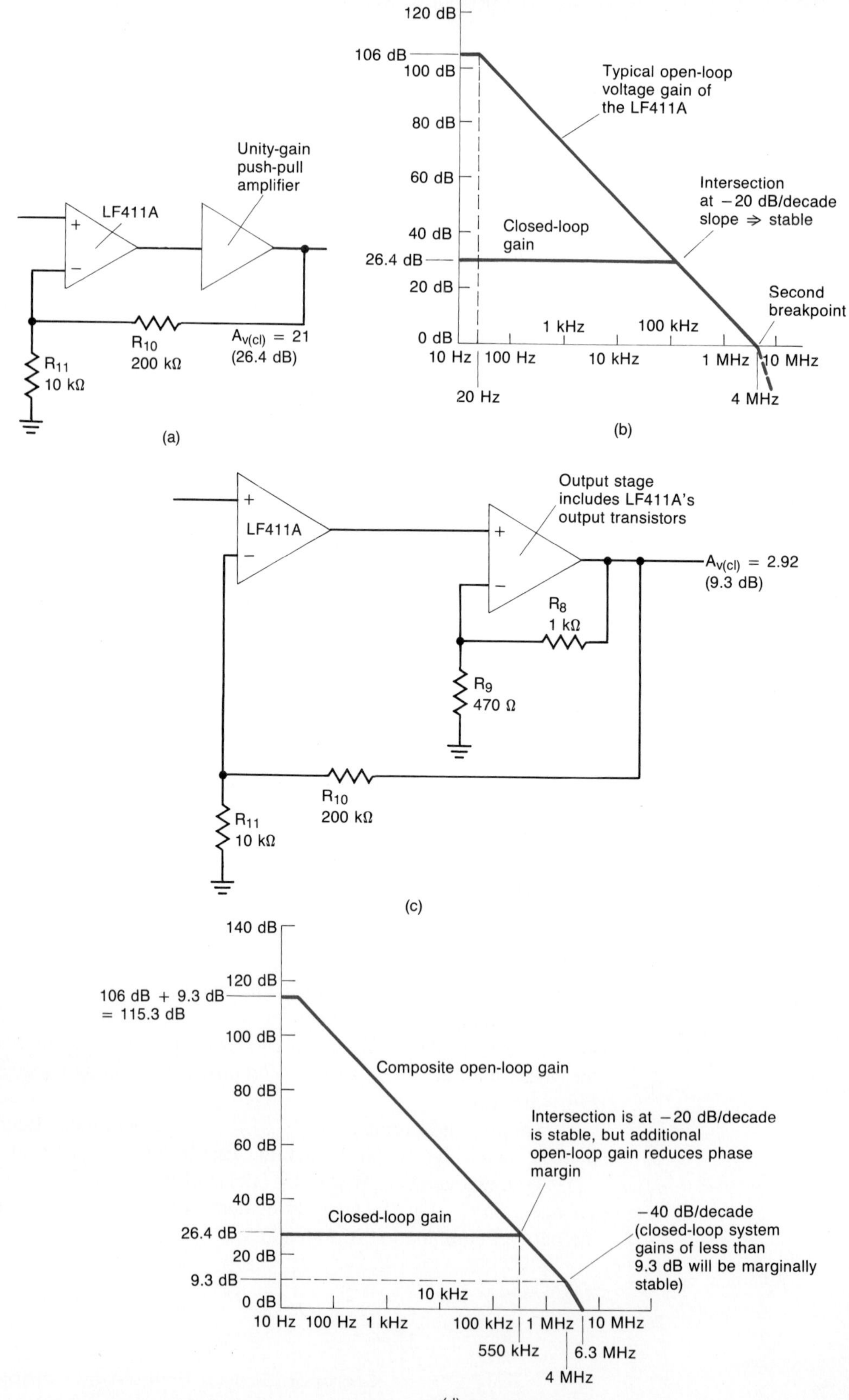

LF411A
Unity-gain push-pull amplifier
R11
10 kΩ
R10
200 kΩ
Av(cl) = 21
(26.4 dB)
(a)
120 dB
106 dB
100 dB
80 dB
60 dB
40 dB
26.4 dB
20 dB
0 dB
Typical open-loop voltage gain of the LF411A
Intersection at −20 dB/decade slope ⇒ stable
Closed-loop gain
Second breakpoint
1 kHz
100 kHz
10 Hz
100 Hz
10 kHz
1 MHz
10 MHz
20 Hz
4 MHz
(b)
LF411A
Output stage includes LF411A's output transistors
Av(cl) = 2.92
(9.3 dB)
R8
1 kΩ
R9
470 Ω
R10
200 kΩ
R11
10 kΩ
(c)
140 dB
120 dB
106 dB + 9.3 dB
= 115.3 dB
100 dB
80 dB
60 dB
40 dB
26.4 dB
20 dB
9.3 dB
0 dB
Composite open-loop gain
Intersection is at −20 dB/decade is stable, but additional open-loop gain reduces phase margin
Closed-loop gain
−40 dB/decade (closed-loop system gains of less than 9.3 dB will be marginally stable)
10 kHz
10 Hz
100 Hz
1 kHz
100 kHz
1 MHz
10 MHz
550 kHz
6.3 MHz
4 MHz
(d)

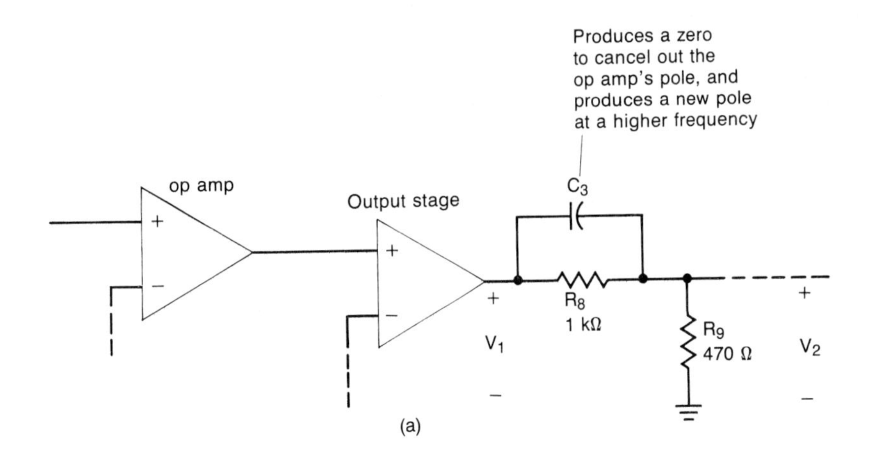

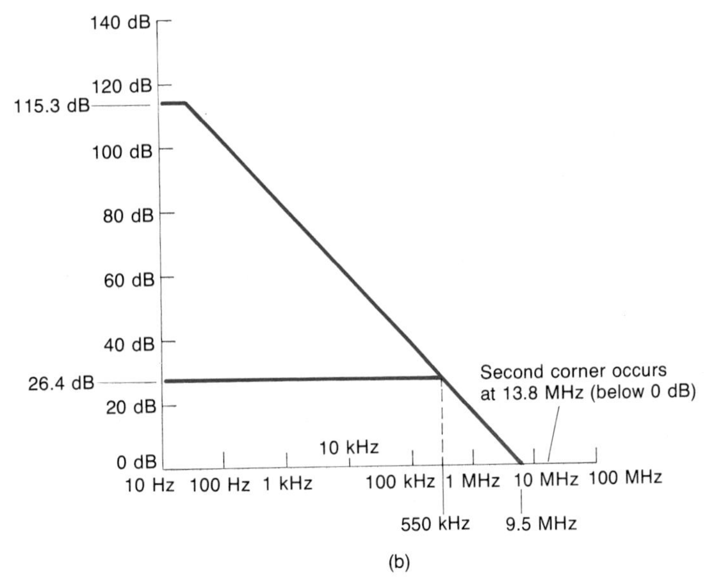

FIGURE 12-61 Providing frequency compensation: (a) compensation capacitor C_3; (b) capacitor C_3 cancels the op amp's pole at 4 MHz to increase the phase margin.

FIGURE 12-60 Effect of the output stage on stability: (a) power amplifier with a push-pull source follower output stage; (b) with a unity-gain output stage the amplifier system will be stable; (c) the common-source output stage increases the open-loop gain by 9.3 dB; (d) additional voltage gain reduces the phase margin.

and

$$Z = \frac{1}{Y} = \frac{R_8}{1 + j\omega R_8 C_3}$$

By employing voltage division between R_9 and Z, we may solve for the voltage gain.

$$A_v = \frac{V_2}{V_1} = \frac{R_9}{R_9 + Z} = \frac{R_9}{R_9 + \dfrac{R_8}{1 + j\omega R_8 C_3}}$$

$$= \frac{R_9}{\dfrac{R_9(1 + j\omega R_8 C_3) + R_8}{1 + j\omega R_8 C_3}} = \frac{R_9(1 + j\omega R_8 C_3)}{R_8 + R_9 + j\omega R_8 R_9 C_3}$$

Factoring out $(R_8 + R_9)$ from the denominator allows us to reach the form given by

$$A_v = \frac{R_9(1 + j\omega R_8 C_3)}{[R_8 + R_9]\left[1 + j\omega \dfrac{R_8 R_9}{R_8 + R_9} C_3\right]}$$

$$A_v = \frac{R_9}{R_8 + R_9} \frac{1 + j\omega R_8 C_3}{\left(1 + j\omega \dfrac{R_8 R_9}{R_8 + R_9} C_3\right)} \tag{12-48}$$

The numerator term gives the gain a zero frequency ω_z and the demoninator term contributes a pole frequency ω_p.

$$\frac{1}{\omega_z} = R_8 C_3 \quad \text{and} \quad \frac{1}{\omega_p} = \frac{R_8 R_9}{R_8 + R_9} C_3$$

We can further simplify Eq. 12-48 if we write the numerator and denominator in terms of their respective zero and pole frequencies. This leads us to

$$A_v = \frac{V_2}{V_1} = \frac{R_9}{R_8 + R_9} \frac{1 + j\omega/\omega_z}{1 + j\omega/\omega_p} = \frac{R_9}{R_8 + R_9} \frac{1 + jf/f_z}{1 + jf/f_p}$$

$$A_v = \frac{R_9}{R_8 + R_9} \frac{1 + jf/f_z}{1 + jf/f_p} \tag{12-49}$$

where $f_z = \dfrac{1}{2\pi R_8 C_3}$ and $f_p = \dfrac{1}{2\pi \dfrac{R_8 R_9}{R_8 + R_9} C_3}$

EXAMPLE 12-23

Determine the frequencies of the zero and pole produced by the R_8-R_9-C_3 feedback network of the power amplifier given in Fig. 12-56(a). Assume that C_3 is a 36-pF disk ceramic capacitor. Sketch the open-loop response.

SOLUTION From Eq. 12-49

$$f_z = \frac{1}{2\pi R_8 C_3} = \frac{1}{2\pi(1\ \text{k}\Omega)(36\ \text{pF})} = 4.42\ \text{MHz}$$

$$f_p = \frac{1}{2\pi \dfrac{R_8 R_9}{R_8 + R_9} C_3} = \frac{1}{2\pi[1\ \text{k}\Omega \parallel 470\ \Omega][6\ \text{pF}]} = 13.8\ \text{MHz}$$

The zero has been placed (as close as standard values will allow) such that the op amp's corner frequency (at approximately 4 MHz) will effectively be canceled. A new pole will be established at 13.8 MHz. The phase margin will increased as shown in Fig. 12-61(b). ■

In the previous example, we have ignored the effects of the poles produced by the MOSFETs. As long as they occur at frequencies that are higher than the op amp's second corner frequency (4 MHz), the above analysis will suffice. However, the capacitances associated with the MOSFETs are quite large (see Table 12-5). Therefore, it is quite probable that they will dominate and the size of C_3 will have to be increased. Let us see if this is the case.

We conduct the analysis by using the parameters for the n-channel MOSFET. This is due to the fact that it has the largest capacitances and will, therefore, produce the lowest pole. We recall that the input corner frequency of an FET is produced by its drain-to-gate capacitance C_{dg} and its gate-to-source capacitance C_{gs} [see Fig. 12-62(a)].

From Table 12-5 we see that

$$C_{dg} = C_{rss} = 150\ \text{pF}$$

and

$$C_{gs} = C_{iss} - C_{rss} = 800\ \text{pF} - 150\ \text{pF} = 650\ \text{pF}$$

The input capacitance (C_{in}) is determined by applying Miller's theorem. Since A_v of the MOSFET stage has been determined previously to be -16,

$$C_{in} = C_{gs} + (1 + A_v)C_{dg}$$
$$= 650\ \text{pF} + (1 + 16)(150\ \text{pF}) = 3200\ \text{pF}$$

The gate pole frequency can be found once the Thévenin's equivalent resistance "seen" by the gate is determined. From Fig. 12-62(a), we see that

$$R_{TH} = R_3 + R_5 = 1\ \text{k}\Omega + 100\ \Omega = 1.1\ \text{k}\Omega$$

The corner frequency f_H is

$$f_H = \frac{1}{2\pi R_{TH} C_{in}} = \frac{1}{2\pi(1100\ \Omega)(3200\ \text{pF})} = 45.2\ \text{kHz}$$

Obviously, the gate pole will dominate since 45.2 kHz is much smaller than the op amp's second corner at approximately 4 MHz [see Fig. 12-62(b)]. Therefore, the zero (f_z) of the R_8-R_9-C_3 network should be set equal to 45.2 kHz.

$$f_z = f_H = 45.2\ \text{kHz}$$

We can use our equation for f_z to solve for the required value for C_3.

$$f_z = \frac{1}{2\pi R_8 C_3}$$

and

$$C_3 = \frac{1}{2\pi R_8 f_z} = \frac{1}{2\pi(1\ \text{k}\Omega)(45.2\ \text{kHz})}$$
$$= 3.52 \times 10^{-9}\ \text{F} = 0.00352\ \mu\text{F}$$

The nearest standard value for C_3 is 0.0033 μF. Using this value, we may find f_z and f_p.

FIGURE 12-62 Including the effects of the FETs: (a) approximate ac equivalent circuit to determine the power MOSFET's pole; (b) open-loop response that includes the MOSFET's pole; (c) frequency compensation has been used to cancel the MOSFET's pole at 45.2 kHz and move it to 151 kHz.

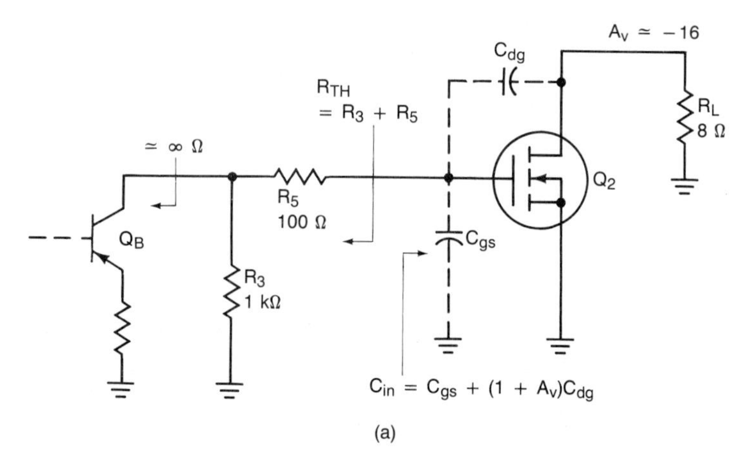

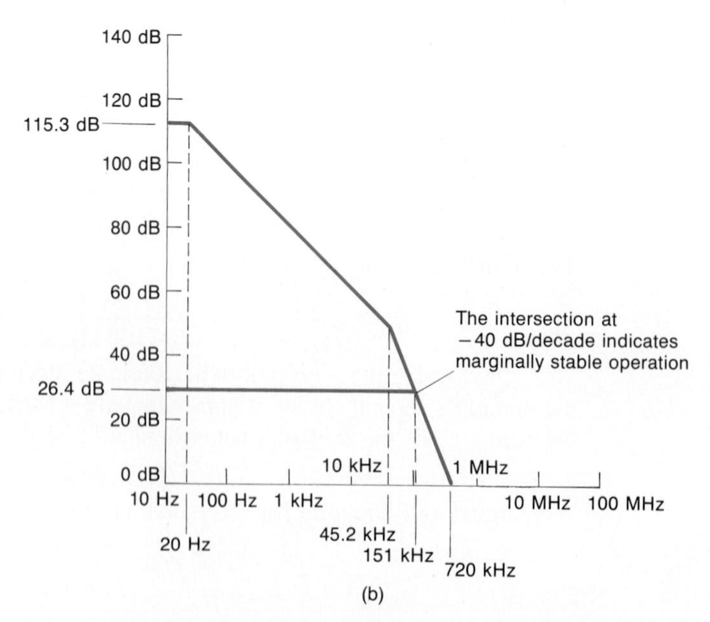

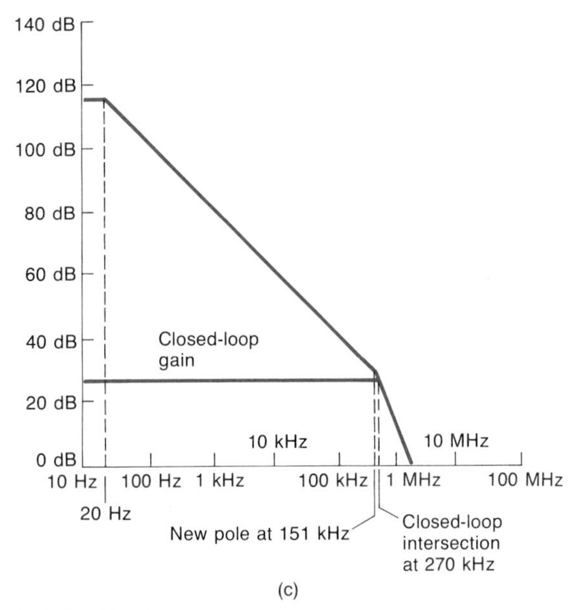

FIGURE 12-62 *(continued)*

$$f_z = \frac{1}{2\pi R_8 C_3} = \frac{1}{2\pi(1\ \text{k}\Omega)(0.0033\ \mu\text{F})} = 48.2\ \text{kHz} \approx 45.2\ \text{kHz}$$

$$f_p = \frac{1}{2\pi(R_8 \parallel R_9)C_3} = \frac{1}{2\pi(1\ \text{k}\Omega \parallel 470\ \Omega)(0.0033\ \mu\text{F})} = 151\ \text{kHz}$$

The new open-loop response has been indicated in Fig. 12-62(c).

A pole exists in the feedback network of virtually every op amp [see Fig. 12-63(a)]. The pole is produced by the stray capacitance (C_s) from the inverting input to ground. In the case of bipolar op amps, the feedback resistors have values on the order of tens of thousands of ohms. In addition, bipolar op amps typically have relatively low bandwidths. Therefore, the additional phase lag produced by the feedback network usually does not cause problems (e.g., oscillation).

However, op amps with JFET inputs (such as our LF411A) have much larger bandwidths, and the feedback resistors are often much larger. The larger feedback resistors will lower the pole frequency. Quite often, the pole will add enough phase shift such that oscillations result.

The pole frequency can be found by applying Thévenin's theorem to arrive at the equivalent circuit shown in Fig. 12-63(b). The resulting corner frequency is given by Eq. 12-50.

$$f_H = \frac{1}{2\pi R_{TH} C_s} \tag{12-50}$$

In Fig. 12-56(a) we have

$$R_{TH} = R_{10} \parallel R_{11} = 200\ \text{k}\Omega \parallel 10\ \text{k}\Omega$$
$$= 9.52\ \text{k}\Omega$$

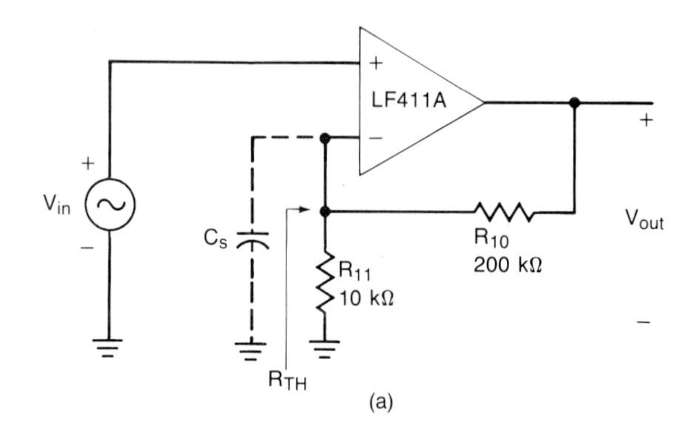

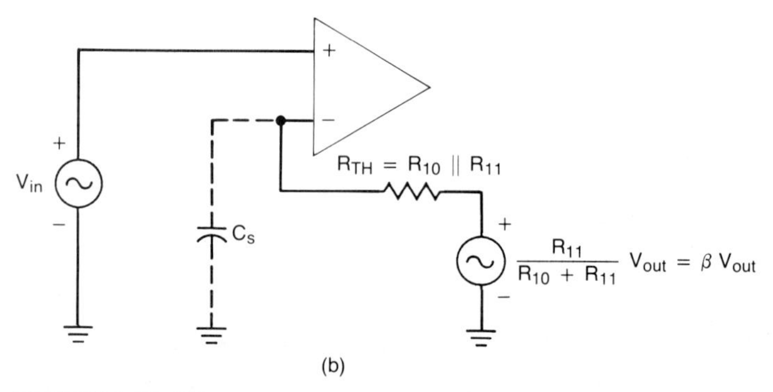

FIGURE 12-63 Analyzing the effects of the feedback pole: (a) stray capacitance (C_s) exists from the inverting input to ground; (b) Thévenin equivalent circuit.

and if we assume that the stray capacitance C_s is 3 pF, we have

$$f_H = \frac{1}{2\pi R_{TH} C_s} = \frac{1}{2\pi(9.52\ \text{k}\Omega)(3\ \text{pF})} = 5.57\ \text{MHz}$$

When we recall that the LF411A has a unity-gain frequency of 4 MHz, it becomes apparent that the feedback pole will add a significant amount of phase shift. Specifically,

$$\theta = -\tan^{-1}\frac{f}{f_H} = -\tan^{-1}\frac{4\ \text{MHz}}{5.57\ \text{MHz}} = -35.7°$$

and this can produce enough additional phase shift to greatly reduce the phase margin (θ_{pm}). Recall that the ideal phase margin is 45° (at unity gain) for an internally frequency-compensated op amp. This requires a loop-gain crossing angle of −135°.

$$\theta_{pm} = 180° + \theta_c = 180° - 135° = 45°$$

However, when the effect of C_s is included, the new crossing angle becomes

$$\theta_c = -135° + (-35.7°) = -170.7°$$

and

$$\theta_{pm} = 180° + \theta_c = 180° - 170.7° = 9.3°$$

Such a small phase margin will invariably result in oscillation.

One solution is to add a small capacitor (e.g., C_4) in parallel with the feedback resistor R_{10} [see Fig. 12-56(a)]. The circuit will be stable if the time constants are selected such that

$$R_{11}C_s \leq R_{10}C_4$$

Typical values for C_4 range from 2 to 100 pF.

In most cases it is recommended to place a small capacitor in parallel with the feedback resistor - particularly in the case of FET-input op amps. Its use tends to limit the bandwidth which minimizes noise and reduces the possibility of stray signal pickup. In Fig. 12-56(a) we see that

$$R_{11}C_s \leq R_{10}C_4$$

$$(10\ \text{k}\Omega)(3\ \text{pF}) \leq (200\ \text{k}\Omega)(2\ \text{pF})$$

$$3 \times 10^{-8}\ s \leq 4 \times 10^{-7}\ s$$

Therefore, the value of C_4 appears to be adequate.

With a little reflection, we recall that the function of capacitor C_3 is to work in concert with resistors R_8 and R_9 to move the MOSFET's pole to 151 kHz. The op amp's pole at 4 MHz occurs below 0 dB [see Fig. 12-62(c)]. However, the calculations above are based on an op amp pole at 4 MHz. Our reasons for doing this were twofold. First, we wanted to illustrate the problem and its solution for frequency-compensated, FET-input op amps. This allows our work to be extended to other applications. Second, the use of C_4 helps cancel out the effects of stray capacitance, and reduces noise in the output of our 15-watt power amplifier. Its presence, or absence, will not have a great effect upon the stability of our power amplifier circuit. However, its use constitutes "good practice," and serves to enhance the overall amplifier performance.

The last components to be discussed are resistors R_4 and R_5. These small (100 Ω) resistors are used to minimize the possibility of parasitic oscillations in the output.

12-24 Common-Source Output Power and Efficiency

In the common-source configuration [Fig. 12-56(a)], only the voltage division between $r_{DS(\text{ON})}$ and R_L serves to limit the maximum output voltage [see Fig. 12-64(a)].

The *p*-channel MOSFET limits the positive peak output voltage.

$$v_{\text{OUT(pos. peak)}} = \frac{R_L}{r_{DS(\text{ON})} + R_L} V_{CC} = \frac{8\ \Omega}{0.40\ \Omega + 8\ \Omega}(20\ \text{V})$$

$$= 19.0\ \text{V}$$

Similarly, the *n*-channel MOSFET limits the negative peak output voltage. Hence

$$v_{\text{OUT(neg. peak)}} = \frac{R_L}{r_{DS(\text{ON})} + R_L}(-V_{CC})$$

$$= \frac{8\ \Omega}{0.25\ \Omega + 8\ \Omega}(-20\ \text{V}) = -19.4\ \text{V}$$

[see Fig. 12-64(b)].

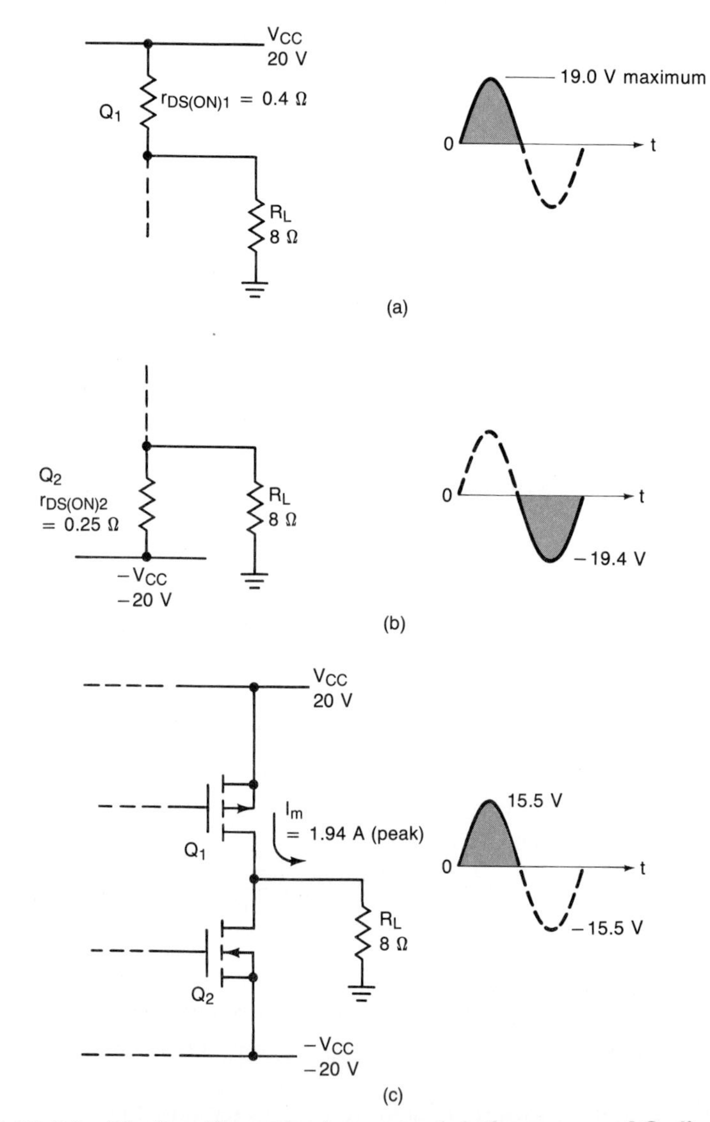

FIGURE 12-64 Finding the output power: (a) the $r_{DS(ON)}$ of Q_1 limits the positive peak; (b) the $r_{DS(ON)}$ of Q_2 limits the negative peak; (c) peak voltages and currents to deliver 15 W to the 8-Ω load.

To calculate the efficiency, the reader is directed to Fig. 12-64(c). A peak output voltage of 15.5 V is required to develop 15 W in an 8-Ω load.

$$I_m = \frac{v_{\text{OUT(peak)}}}{R_L} = \frac{15.5\ \text{V}}{8\ \Omega} = 1.94\ \text{A}$$

$$P_{\text{in(dc)}} = \frac{2V_{CC}I_m}{\pi} = \frac{2(20\ \text{V})(1.94\ \text{A})}{\pi} = 24.7\ \text{W}$$

$$P_{\text{out(ac)}} = \frac{v^2_{\text{OUT(peak)}}}{2R_L} = \frac{(15.5\text{ V})^2}{(2)(8\ \Omega)} = 15\text{ W}$$

$$\%\eta = \frac{P_{\text{out(ac)}}}{P_{\text{in(dc)}}} \times 100\% = \frac{15.0\text{ W}}{24.7\text{ W}} \times 100\% = 60.7\%$$

Admittedly, the power amplifier given in Fig. 12-56(a) is much more complex than any of our previous endeavors. However, providing 15 W into an 8-ohm load at an efficiency of 60.7% is quite an achievement. Laboratory work indicated a power bandwidth in excess of 100 kHz.

Higher power levels at greater efficiencies that exhibit a low % THD, a low TIM, and a wide power bandwidth offer the analog technicians and engineers a formidable challenge. However, the circuit designs and schemes are essentially extensions of the basic principles presented thus far. Performance is limited only by the technology of the power-handling devices, and the cleverness (and tenacity) of the design engineers.

PROBLEMS

Drill, Derivations, and Definitions

Section 12-1

12-1. Define the term ''transducer'' and give three examples.

12-2. Draw the general block diagram of an electronic amplifier system. Identify the small-, medium-, and large-signal voltage amplifiers and the power amplifier stage.

Section 12-2

12-3. Given the common-emitter circuit shown in Fig. 12-2, assume that V_{CC} is reduced to 12 V. Using approximations, determine V_B, V_E, I_C, I_E,V_C, and V_{CE}.

12-4. Repeat Prob. 12-3 if V_{CC} is increased to 20 V.

12-5. Based on the dc values found in Prob. 12-3, determine g_m, r_π, and r_o for the transistor given in Fig. 12-2. (Assume that h_{fe} is 80.)

12-6. Based on the dc values found in Prob. 12-4, determine g_m, r_π, and r_o for the transistor given in Fig. 12-2. (Assume that h_{fe} is 110.)

12-7. Based on the g_m, r_π, and assuming that r_o is negligibly large, use the values obtained in Prob. 12-5 to find $A_{v(\text{oc})}$, A_v, R_{in}, and R_{out} for the amplifier given in Fig. 12-2.

12-8. Based on the g_m, r_π, and assuming that r_o is negligibly large, use the values obtained in Prob. 12-6 to find $A_{v(\text{oc})}$, A_v, R_{in}, and R_{out} for the amplifier given in Fig. 12-2.

12-9. Assume that R_L is disconnected from the amplifier given in Fig. 12-2, and that V_{CC} is 12 V. Draw the dc load line, indicate the Q-point, and draw the ac load line. The dc values found in Prob. 12-3 may assist you. What is the maximum (undistorted) peak-to-peak output voltage swing?

12-10. Assume that R_L is disconnected from the amplifier given in Fig. 12-2 and that V_{CC} is 20 V. Draw the dc load line, indicate the Q-point, and draw the ac load line. The dc values found in Prob. 12-4 may assist you. What is the maximum (undistorted) peak-to-peak output voltage swing?

12-11. Repeat Prob. 12-9 if R_L is connected.

12-12. Repeat Prob. 12-10 if R_L is connected.

Section 12-3

12-13. Given the emitter follower in Fig. 12-9(a), assume that R_L is 2 kΩ and that V_{CC} is increased to 22 V. Perform an approximate dc analysis to determine V_B, V_E, V_C, I_E, I_C, and V_{CE}.

12-14. Repeat Prob. 12-13 if R_L is 1.5 kΩ and V_{CC} is decreased to 18 V.

12-15. Using the dc values found in Prob. 12-13, determine the small-signal parameters for the BJT given in Fig. 12-9(a). Assume that h_{fe} is 120, and find g_m, r_π, and r_o.

12-16. Using the dc values found in Prob. 12-14, determine the small-signal parameters for the BJT given in Fig. 12-9(a). Assume that h_{fe} is 150, and find g_m, r_π, and r_o.

12-17. Based on the small-signal values found in Prob. 12-15, determine $R_{\text{in(base)}}$, R_{in}, A_v, and R_{out}.

12-18. Based on the small-signal values found in Prob. 12-16, determine $R_{\text{in(base)}}$, R_{in}, A_v, and R_{out}.

12-19. Using the data given in Prob. 12-13, sketch the dc load line for the amplifier given in Fig. 12-9(a), and indicate its Q-point. You may wish to use the results of the dc analysis you performed in Prob. 12-13. Also draw the ac load line, and determine the maximum peak-to-peak output voltage.

12-20. Using the data given in Prob. 12-14, sketch the dc load line for the amplifier given in Fig. 12-9(a), and indicate its Q-point. You may wish to use the results of the dc analysis you performed in Prob. 12-14. Also draw the ac load line, and determine the maximum peak-to-peak output voltage.

Section 12-4

12-21. Define the term ''slew rate.'' Why is it an important consideration in power amplifiers?

12-22. The tail current and the frequency-compensation capacitor in an op amp dictate its slew rate. An LM324 op amp has a tail current of 6 μA and a 5-pF frequency-compensation capacitor. Find its slew rate and express it in volts per microsecond.

12-23. Increasing the tail current and/or decreasing the size of C_c will improve the slew rate. What will happen to the unity-gain frequency f_T in an internally compensated op amp? Is this desirable? Explain.

12-24. Will decompensated op amps tend to have a slew rate that is larger, or smaller, than op amps that have been compensated for unity gain? Explain.

12-25. Explain why op amps with a JFET input stage tend to have larger slew rates than op amps with a bipolar input stage.

12-26. What is ''IM''? What is ''TIM''? How does TIM relate to slew rate? What slew rate should a high-performance power amplifier demonstrate if it is to develop a peak output voltage of 10 V?

12-27. What is ''power bandwidth''? How does it relate to the slew rate?

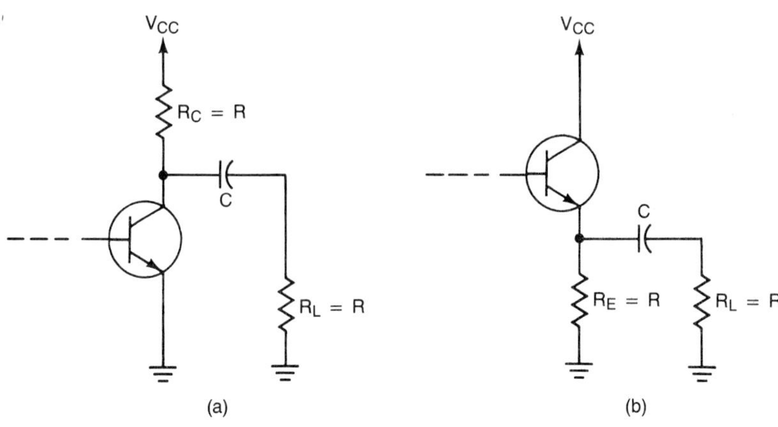

FIGURE 12-65 Circuits for Probs. 12-32 and 12-33.

12-28. A power amplifier has a slew rate of 50 V/μs and provides a peak output voltage of 10 V. Determine its power bandwidth.

12-29. A power amplifier provides 40 V peak across an 8-Ω load. Its power bandwidth is to be at least 25 kHz. What must its slew rate be? Express your answer in volts per microsecond. Also find the power delivered to the load.

Section 12-5

12-30. What is meant by the term "efficiency"? Why is it an important consideration in the case of power amplifiers?

12-31. Prove that the ideal efficiency of a series-fed class A power amplifier [Fig. 12-16(a)] is 25%. Show your work in neat detail.

12-32. The common-emitter power amplifier shown in Fig. 12-56(a) has a load resistor connected across its output. Observe that the collector and load resistors are equal. Also note that the ac load resistance is equal to their parallel combination (r_{ac} is $R/2$). If we analyze the circuit further, we see that

$$I_C = \frac{V_{CC} - V_{CE}}{R_C} = \frac{V_{CC} - V_{CE}}{R}$$

To achieve the maximum possible symmetrical output swing, we must satisfy the condition that

$$V_{CE} = I_C r_L = \frac{I_C R}{2}$$

Using the relationships above, show that

$$I_C = \tfrac{2}{3} I_{C(\mathrm{SAT})}$$

where $I_{C(\mathrm{SAT})}$ is V_{CC}/R. Show your work in neat detail.

12-33. In a fashion similar to Prob. 12-32, we can prove that when R_E is equal to R_L,

$$V_{CE} = \tfrac{1}{3} V_{CE(\mathrm{OFF})} = \tfrac{1}{3} V_{CC}$$

to achieve a maximum symmetrical output swing [see Fig. 12-56(b)]. If the common-collector amplifier is biased to achieve the maximum possible symmetrical output, prove that the theoretical efficiency is only 8.33%.

Section 12-6

12-34. Given the single-supply class A power amplifier in Fig. 12-17(a), assume that V_{CC} has been increased to 15 V. Determine V_{BIAS}, I_E, I_C, and $P_{\text{in(dc)}}$. Draw the ac load line to determine the maximum ac output power $P_{\text{out(ac)}}$ and then find the approximate efficiency.

Section 12-7

12-35. Define class A, class AB, class B, and class C transistor operation.

Section 12-8

12-36. What is a ''complementary pair''?

12-37. Describe the operation of a class B push-pull amplifier that uses a complementary pair.

12-38. What is crossover distortion? How can it be eliminated?

12-39. Prove that the ideal class B push-pull amplifier efficiency is approximately 78.5%. Show your work in step-by-step detail.

12-40. Given the class AB push-pull power amplifier in Fig. 12-21, assume that the collector supply voltages are ± 60 V. If the no-signal bias current is negligible and the peak voltage across the 8-Ω load is 60 V, find $P_{\text{in(dc)}}$, $P_{\text{out(ac)}}$, and the % η.

Section 12-9

12-41. As the magnitude of I_C is increased, what happens to the magnitude of h_{FE}? Explain how this affects the driver stage of a power amplifier.

12-42. Contrast the characteristics of a bipolar power transistor with those of a small-signal bipolar transistor. Specifically, compare their respective values of h_{FE}, $I_{C(\text{max})}$, $BV_{CE(\text{sus})}$, C_{ob}, f_T, and I_{CBO}.

12-43. What is meant by the term ''thermal runaway''? Why is it undesirable?

12-44. What is meant by the term ''second breakdown'' in a bipolar transistor? Describe this effect.

12-45. What is the meaning of the acronym ''SOA''? Make a neat sketch of the SOA for a bipolar power transistor. Label its boundaries.

Section 12-10

12-46. Given the class AB power amplifier in Fig. 12-27(a), assume that R_L is 16 Ω, the supply voltages are ± 12 V, I_C is 10 mA, and the base-to-emitter voltage drops are 0.7 V. Find the maximum output power that can be developed without distortion and the efficiency under these conditions. Restrict the distortion considerations to those imposed by saturation of the output transistors. What will the peak input voltage be under these conditions?

12-47. Continue the analysis of the power amplifier in Fig. 12-27(a) started in Prob. 12-46. Use Eq. 12-26 to determine the peak undistorted output voltage as limited by the diode bias network. Assume that the minimum h_{FE} is 30. Determine corresponding $P_{\text{in(dc)}}$, $P_{\text{out(ac)}}$, and the % η.

12-48. Repeat Prob. 12-47 if the minimum h_{FE} is 60.

12-49. Continue the analysis of the power amplifier given in Fig. 12-27(a) which was started in Prob. 12-46. Find $r_{\text{IN(base)}}$ and the equivalent ac resistance across the output of the op amp. Ignore the diode resistance. Assuming that I_{SC} for the op amp is 10 mA, what is the peak value of the maximum undistorted output voltage? Also find $P_{\text{in(dc)}}$, $P_{\text{out(ac)}}$, and the % η. Use an h_{FE} of 30 to conduct the analysis.

12-50. Repeat Prob. 12-49 if h_{FE} is increased to 60.

12-51. What is meant by the term ''drive requirement'' as it relates to the output stage of a power amplifier? Describe two ways in which it can be reduced.

Section 12-11

12-52. Draw the electrical schematic diagram of an *npn* Darlington pair. Assume that the input transistor has an h_{FE} of 150 and the output transistor has an h_{FE} of 40. Determine the current gain of the Darlington pair [$h_{FE(\text{total})}$].

12-53. Given the power amplifier shown in Fig. 12-33(c), assume that the power supply voltages are increased to ± 12 V. None of the other components are changed. If the peak voltage across R_L is 8 V, find the ac output power, the peak load current, the approximate dc input power, and the efficiency. Ignore the idle current effects.

12-54. Determine the maximum peak-to-peak output voltage that can be developed by the power amplifier given in Fig. 12-36(d). Assume that $h_{FE(\text{total-min})}$ of the Darlington transistors is 3000 and that the collector supply voltages are ± 12 V. Consider the limitations imposed by the emitter resistors, the bias diodes, the op amp's saturation voltage, and I_{SC} limits.

12-55. Repeat Prob. 12-54 if the Darlington power transistors in Fig. 12-36(d) have an $h_{FE(\text{total-min})}$ of 5,000, the collector supply voltages are ± 15 V, R_3 and R_4 are increased to 3.3 kΩ, and both R_5 and R_6 are reduced to 1 Ω.

Section 12-12

12-56. Why are *npn* power transistors preferred over *pnp* power transistors?

12-57. Draw the electrical schematic of a quasi-*pnp* power Darlington transistor. Describe its operation.

12-58. Given the quasi-complementary symmetry power amplifier given in Fig. 12-37(b), assume that the input voltage is a sine wave with a peak value of 300 mV. Sketch the output voltage waveform, the waveform at the emitters of Q_2 and Q_3, the waveforms at the base terminals of Q_1 and Q_3, and the waveform at the op amp's output. Be sure to label both the positive and negative peak values.

Section 12-13

12-59. What is VMOS? DMOS? How do they compare with bipolar power transistors?

12-60. Contrast the SOA of a power MOSFET with a similarly rated bipolar power transistor.

Section 12-14

12-61. Given the V_{BE} multiplier in Fig. 12-42(a), assume that R_1 is 2 kΩ, R_2 is 470 Ω, and R_3 is still a 500-Ω potentiometer. Resistor R_5 is reduced to 47 Ω. All other components are unchanged. Determine the adjustment range of V_{BIAS}. Also determine the minimum and maximum values of the temperature coefficient of V_{BIAS}. Express your answer in mV/°C.

12-62. If the amplifier given in Fig. 12-45 delivers 30 W to the 8-Ω load, what is the peak voltage across the load? The peak load current?

12-63. The balance potentiometer (R_5) in Fig. 12-45 is adjusted such that the dc current through transistor Q_4 is 4.81 mA. The bias adjust potentiometer (R_9) is adjusted such that the collector-to-emitter voltage of Q_3 is 8 V. If the dc voltage across R_L is zero, find the voltage across the bootstrapping capacitor C_4.

Section 12-15

12-64. Given the push-pull output stage in Fig. 12-46(a), assume that the power supply voltages have been reduced to ±20 V and that R_L is 4 Ω. Draw the waveforms for v_{CE1} and i_{C1} and write the equations (Eq. 12-36) that describe them.

12-65. Transfer the SOA graph shown in Fig. 12-46(d) to three-cycle log-log paper. Plot the ac load line on it using the equations found in Prob. 12-64. Does the ac load line always lie within the SOA boundaries?

12-66. Assume that the same peak voltage and peak load current values determined in Prob. 12-64 exist but that the load is 60° lagging. Write the equations (Eq. 12-38) that describe v_{CE1} and i_{C1}. Plot the ac load line on the SOA graph used for Prob. 12-65. Does the ac load line always lie within the SOA boundaries?

Section 12-16

12-67. Determine I_{SC} for the power amplifier given in Fig. 12-49. Why is short-circuit protection necessary? Explain.

12-68. What are the functions of R_{17} and C_8 in Fig. 12-45?

Section 12-17

12-69. What is the primary function of a heat sink? Why are fins sometimes used on heat sinks?

12-70. A transistor has a θ_{ja} of 30°C/W and is used to dissipate 5 W of power. The maximum ambient temperature is 60°C. If the transistor's $T_{j(\max)}$ is 200°C, will the transistor be overheated?

12-71. A 160-W transistor's power rating is to be derated linearly above 25°C at a rate of −914 mW/°C. What is its thermal resistance θ_{jc}? Its $T_{j(\max)}$ is 200°C.

12-72. A silicon power transistor has a heat sink with a θ_{sa} of 1.2°C/W. The transistor is rated at 150 W (25°C) and has a θ_{jc} of 1.17°C/W. A mounting insulator is used that has a θ_{cs} of 0.4°C/W. What is the maximum power that the transistor can dissipate if the maximum ambient temperature is 60°C and T_j is rated at 200°C.

12-73. A TIP120 *npn* power transistor is rated at 65 W (at 25°C). The transistor has a maximum T_j of 150°C and a θ_{jc} of 1.92°C/W. It is to dissipate 15 W in a power amplifier, and the maximum ambient temperature is 40°C. Determine the maximum allowable θ_{sa} of its heat sink. Assume that the transistor is to be mounted to the heat sink with a mica insulator using silicon grease. The transistor has a TO-220 case style.

Section 12-18

12-74. What is the function of thermal shutdown in an IC power amplifier? Explain how the thermal shutdown circuit in Fig. 12-53(c) operates.

12-75. Explain push-pull operations using a single supply and an output coupling capacitor. Should the RC time constant of the output coupling capacitor be long, or small? Under what conditions?

12-76. Assume that the IC power amplifier given in Fig. 12-55(c) is operated from a 12-V supply. What is the dc voltage at pin 1? At pin 2? What will the peak load voltage be across 8 Ω if the amplifier is delivering 1.5 W?

Section 12-20

12-77. Resistor R_7 in Fig. 12-57(a) has been replaced with a 330-Ω unit. Potentiometer R_6 has been changed to 2 kΩ. The constant-current diode D_1 has been replaced with a CR470 which has a nominal current of 4.7 mA. All other components are unchanged. Find V_{GS} when R_6 is fully clockwise and fully counterclockwise.

12-78. Continue Prob. 12-77 by finding the range of temperature coefficients offered by the temperature-compensating constant-current source.

Section 12-21

12-79. Resistor R_8 is changed to 1.2 kΩ and R_9 is 330 Ω in Fig. 12-58(a). All other values are the same. Find the open- and closed-loop voltage gains.

Section 12-22

12-80. Continue the analysis started in Prob. 12-79 by finding the new slew rate and power bandwidth. Use Fig. 12-59 as a guide.

Section 12-23

12-81. Given that R_8 is 1.2 kΩ, R_9 is 330 Ω, and C_3 is 0.0022 μF, find its f_z and f_p [see Fig. 12-61(a)].

12-82. Given that R_{10} is 220 kΩ, R_{11} is 20 kΩ, and the stray capacitance (C_s) is 4 pF, determine the size of the capacitance that should be placed in parallel with R_{10} to cancel the feedback pole [see Fig. 12-63(a)].

Section 12-24

12-83. Assume that Q_1 has an $r_{DS(\text{ON})}$ of 0.3 Ω and Q_2 has an $r_{DS(\text{ON})}$ of 0.2 Ω. Find the maximum positive and negative peak output voltage swings. All other values are the same (see Fig. 12-64).

12-84. Assume that the peak voltage across R_L in Fig. 12-64(c) is increased to 18 V. Find I_m, $P_{\text{in(dc)}}$, $P_{\text{out(ac)}}$, and the % η.

Troubleshooting

The most common failure mode in a power amplifier is for one or both of the output transistors to fail shorted. This usually means that the power supply fuse will open or the supply will enter into its current-limiting mode of operation. With no V_{CC}, troubleshooting is extremely difficult. Therefore, the first step is to remove the suspected power transistor(s). However, before replacing the faulty unit(s) with new ones, it is always advisable to verify that the rest of the circuit is in proper working condition.

12-85. Transistors Q_1 and Q_2 in Fig. 12-27(a) failed. The units have been removed. Since their removal opened the feedback loop, a jumper is installed temporarily between the output terminal of the op amp and the point where R_5 connects to R_6. R_L has been disconnected. Draw the simplified equivalent circuit. Assuming that no input signal is applied, calculate the dc voltages if the circuit is working normally. Specifically, determine the output voltage at pin 10, the voltage at the anode of D_1, the voltage at the cathode of D_2, and the voltage at pin 4 of the op amp.

12-86. Actually measuring the dc voltages in the circuit described in Prob. 12-85 revealed 6 V at the anode of D_1 and 0 V at its cathode. What are the possible causes?

12-87. Actually measuring the dc voltages in the circuit described in Prob. 12-85 revealed -3.6 V at the anode of D_1 and -4.3 V at its cathode. Diode D_2 had a cathode voltage of -5.0 V. What are the possible causes? Explain.

Design

12-88. The power amplifier given in Fig. 12-43(a) is to be redesigned. The basic scheme is sound, but we wish to improve its performance. The 741 op amp used as the driver stage is to be replaced with an LF411A. The LF411A is an op amp with an FET input stage and can operate with supply voltages as large as ± 22 V. The constant-current source consisting of Q_3, R_{10}, R_{11}, D_1, and D_2 is to be replaced by a single Siliconix CR470 (4.70-mA) *constant-current diode* to reduce the parts count. Given that all the other components are the same and the supply voltage is to increased to ± 20 V, draw the new schematic, and find the dc voltage across C_1. Assume that the V_{BE} multipliers have been adjusted to provide a bias of 1.4 V.

12-89. Continue the analysis of Prob. 12-88 by sketching all the waveforms using Fig. 12-43(b) as a guide. Assume that the peak voltage across the load is 15 V.

12-90. Continue the analysis of Probs. 12-88 and 12-89 by determining the distortion limits imposed by (a) saturation of the output transistors, (b) the constant-current source, and (c) output voltage saturation in the op amp. Assume that the current source requires at least 1.4 V across it.

12-91. The emitter resistors [e.g., R_{12} and R_{13} in Fig. 12-43(a)] promote bias stability. However, they also serve as a source of inefficiency. In general,

$$R_E \geq V_{CE}\theta_{ja}|\Delta V_{BE}/°C|$$

where V_{CE} is the collector-to-emitter dc bias, θ_{ja} is the total thermal resistance from junction to ambient, and $\Delta V_{BE}/°C$ is approximately -2.2 mV/°C. Find the minimum values for R_{12} and R_{13} in Fig. 12-43(a) if θ_{ja} is 7°C/W.

Computer

12-92. We wish to minimize our efforts in the development of ac load lines. Write a BASIC program that computes the values of v_{CE} and i_C at 5° intervals (refer to Eq. 12-38).

The program should prompt the user for the phase angle, V_{CC}, and I_m. The program should default to a phase angle of 60°.

12-93. Write a BASIC program that specifies the maximum allowable thermal resistance of a heat sink. It should prompt the user for the pertinent transistor parameters, the method of mounting, the maximum power dissipation, and the maximum ambient temperature.

13

DC Power Supplies: Rectification and Filtering

After Studying Chapter 13, You Should Be Able to:

- Explain the difference between average and rms waveform values.
- Explain and identify the hot, neutral, and ground connections found in a single-phase three-wire power distribution system.
- Apply the ideal transformer concept to analyze transformer circuits.
- Explain and analyze the operation of rectifier circuits: halfwave, fullwave using a center-tapped transformer, fullwave bridge, and dual-complementary.
- Describe the role of the filter capacitor and compute V_m, V_{dcm}, $V_{r(p\text{-}p)}$, $V_{r(rms)}$, V_{DC}, and $V_{dc(min)}$.
- Define ripple factor.
- Describe and compute average and peak repetitive currents.
- Explain and calculate nonrepetitive surge current.
- Determine capacitor ripple current and transformer secondary current.

- Analyze RC pi filter circuits.
- Discuss other filter configurations.
- Explain and use rectifier data sheet specifications.
- Describe the Diode Bridge assembly and its applications.
- Discuss and recognize power supply failure modes.
- Apply the fundamental relationships to the design of a simple power supply.

13-1 Introduction to DC Power Supplies

All electronic circuits and systems require a stable source of dc voltage and current (or dc power) to operate correctly. The source of dc power is used to establish the dc operating points (*Q*-points) for the passive and active electronic devices incorporated in the system.

The dc power supply is typically connected to each and every stage in an electronic system. Consequently, if the dc power supply does not perform correctly, *all* the stages in the system will be affected adversely.

In portable electronic equipment (such as walkie-talkies or electronic games) the dc power supply may simply be a battery. However, most electronic equipment relies on a line-operated dc power supply. The block diagram of a line-operated dc power supply is shown in Fig. 13-1.

FIGURE 13-1 Block diagram of a line-operated dc power supply and its load.

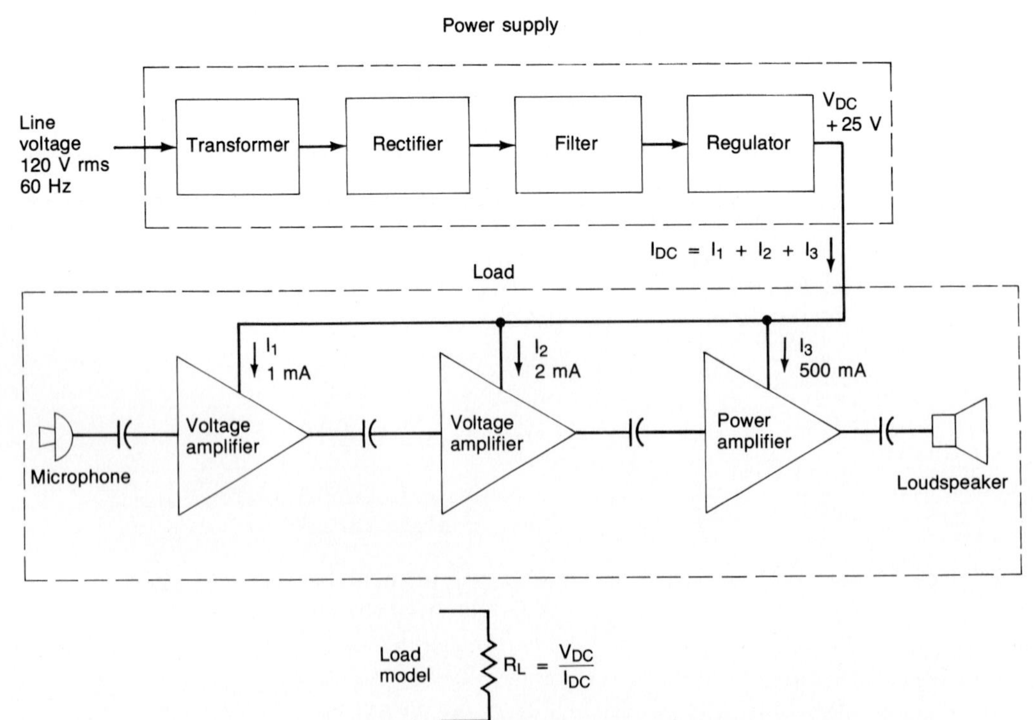

As can be seen in the figure, the basic power supply consists of a *transformer*, *rectifier*, *filter*, and a *regulator*. The output of the dc power supply is used to provide a constant dc voltage to the *load*. To provide an overview of the operation of the dc power supply of Fig. 13-1, let us briefly outline the function of each of the five blocks.

The *transformer* can be used to increase (step up) the amplitude and/or decrease (step down) the amplitude of the ac line voltage. It can provide isolation from the power line ground. It may also include internal shielding to prevent unwanted electrical noise signals on the power line from getting into the power supply and possibly disturbing the load.

The *rectifier* typically requires one, two, or four diodes. Its sole function is to convert the sinusoidal ac voltage into either positive or negative pulsating dc.

The *filter* is a low-pass configuration and is typically constructed from reactive circuit elements such as capacitors and/or inductors and resistors. Its function is to smooth out the pulsating dc produced by the rectifier (by removing its ac ripple content).

The *regulator* may be constructed from a zener diode, and/or discrete transistors, and/or integrated circuits. Its primary function is to maintain a constant dc output voltage. However, it also rejects any ac ripple voltage that is not removed by the filter. The regulator may also include protective functions such as short-circuit protection, current limiting, thermal shutdown, or overvoltage protection.

The *load* can be comprised of several electronic functional blocks (e.g., amplifiers) which require a dc bias across them to operate correctly. *Although the power supply is fundamental to the operation of the load, it is the load that dictates the power supply requirements.*

13-2 The Load

In Fig. 13-1 we note that all the (amplifier) load stages are connected in *parallel* across the output of the power supply. In general, the total dc output current from a dc power supply may be found by adding together the individual dc currents required by each of the stages connected across it. Hence

$$I_{DC} = I_1 + I_2 + I_3 + \cdots + I_{n-1} + I_n \tag{13-1}$$

where I_{DC} is the total power supply dc output current and $I_1, I_2, \ldots, I_n$ are the individual dc currents drawn by the n supply stages connected across the supply.

If we use the *maximum* average dc currents drawn by each of the stages, I_{DC} will be its maximum value. To simplify our work, we can use a static resistance to model the load connected across the power supply:

$$R_L = \frac{V_{DC}}{I_{DC}} \tag{13-2}$$

where R_L = static load resistance
I_{DC} = maximum load current
V_{DC} = voltage across the load

EXAMPLE 13-1

A 25-V dc power supply is used in the three-stage amplifier system shown in Fig. 13-1. If the maximum average dc currents are 1, 2, and 500 mA, respectively, compute I_{DC} and R_L.

SOLUTION By inspection of Fig. 13-1 and by Eq. 13-1,

$$I_{DC} = I_1 + I_2 + I_3 = 1 \text{ mA} + 2 \text{ mA} + 500 \text{ mA} = 503 \text{ mA}$$

The equivalent static resistance of the three stages may be found from Eq. 13-2.

$$R_L = \frac{V_{DC}}{I_{DC}} = \frac{25 \text{ V}}{503 \text{ mA}} = 49.7 \ \Omega$$

Hence the three stages may be modeled by their equivalent static resistance as shown in Fig. 13-1. ■

13-3 Average and RMS Values

In Fig. 13-2 we see the typical waveforms associated with the line-operated dc power supply. The input and output waveforms of the transformer are sinusoidal. However,

FIGURE 13-2 Waveforms associated with a line-operated dc power supply.

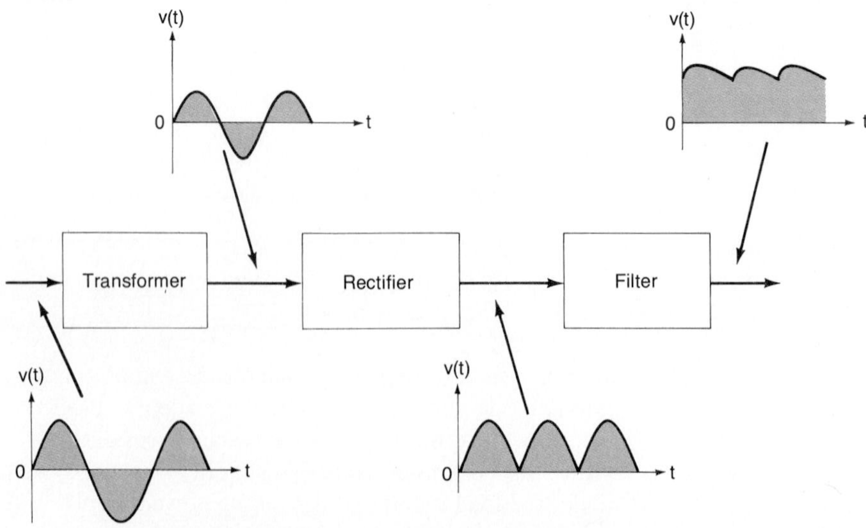

the output of the rectifier and the ac ripple voltage at the output of the filter are *nonsinusoidal*. To describe sinusoidal and nonsinusoidal waveforms, we can refer to their instantaneous values [i.e., $v(t)$] or we can lump their characteristics into *average* and/or *rms* values. In the study of dc power supplies, as well as many other electronic circuits, we find the average and rms descriptions extremely useful.

First, let us define the average, or dc value of a periodic waveform.

The average or dc value of a periodic waveform is given by the algebraic sum of its (positive) area above zero, and its (negative) area below zero, divided by the period of the waveform.

This definition is summarized by

$$\boxed{V_{DC} = \frac{\text{area}}{T}} \tag{13-3}$$

where V_{DC} = dc average value of a periodic voltage waveform
area = algebraic sum of the areas above and below zero for one cycle
T = period

In Fig. 13-2 we indicated the filter's output voltage possesses an ac ripple. Exactly why and how this occurs will be explained later. In Fig. 13-3(a) we have

FIGURE 13-3 Average values: (a) ripple approximation; (b) finding V_{DC} for ac ripple riding on a dc level; (c) relationship between V_{DC} and the ac ripple; (d) the average value of the ac ripple is 0 V; (e) the average value of a sine wave is 0 V.

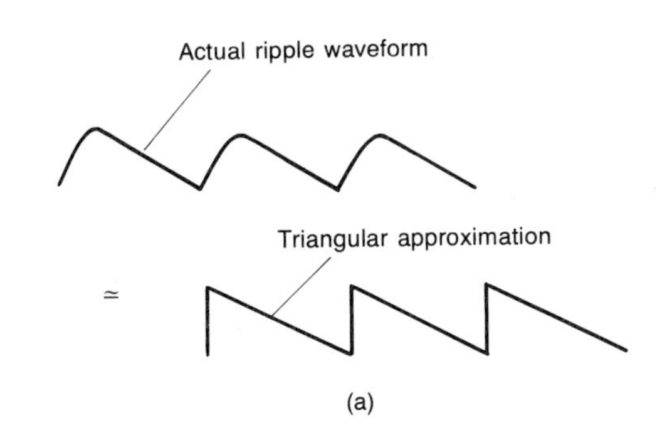

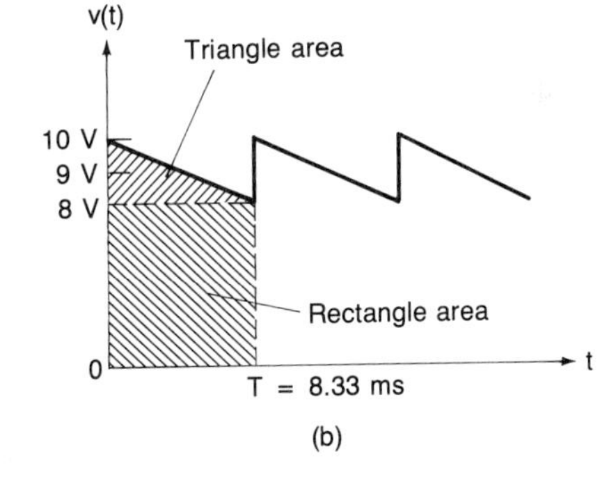

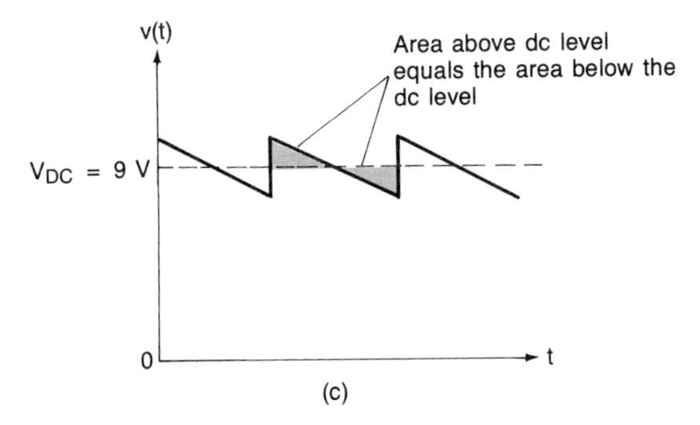

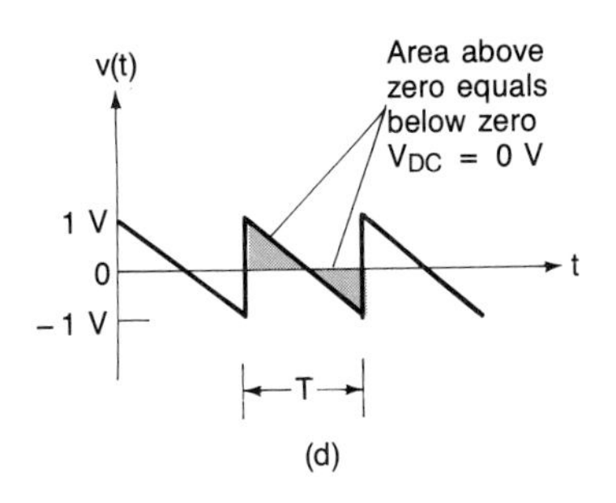

shown a *triangular approximation* of the ripple waveform. Let us now find the dc average value of this triangular waveform when it rides on a dc level [see Fig. 13-3(b)].

EXAMPLE 13-2

Find the dc value of the triangular waveform riding on a dc level as shown in Fig. 13-3(b).

SOLUTION In Fig. 13-3(b) we identify the period T to be 8.33 ms. The area under one cycle can be found easily by finding the area of the triangle and adding it to the area of the rectangle. Hence

$$\begin{aligned}\text{rectangle area} &= \text{base} \times \text{height} = (8\text{ V})(8.33\text{ ms})\\ &= 66.64\text{ mV}\cdot\text{s}\\ \text{triangle area} &= \tfrac{1}{2} \times \text{base} \times \text{height}\\ &= \tfrac{1}{2}(8.33\text{ ms})(2\text{ V}) = 8.33\text{ mV}\cdot\text{s}\\ \text{area} &= \text{rectangle area} + \text{triangle area}\\ &= 66.64\text{ mV}\cdot\text{s} + 8.33\text{ mV}\cdot\text{s} = 74.97\text{ mV}\cdot\text{s}\end{aligned}$$

The dc (average) value is given by applying Eq. 13-3.

$$V_{DC} = \frac{\text{area}}{T} = \frac{74.97\text{ mV}\cdot\text{s}}{8.33\text{ ms}} = 9\text{ V}$$

■

The average value of the total waveform in Fig. 13-3(b) is 9 V. In Fig. 13-3(c) we emphasize that the area of the triangular (ac) waveform *above* the average value is equal to the area *below* the average value. If the *ac* portion of the ripple waveform is redrawn as shown in Fig. 13-3(d), the dc value of this waveform is zero since its area above zero equals its area below zero. In general, this is true for *all* ac waveforms. Consider Fig. 13-2. The dc average value of the sine wave is zero. Consequently, the input and output waveforms of the transformer block shown in Fig. 13-2, have an average value of zero.

The average value of square, rectangular, and triangular waves are all relatively easy to find. This is true because we have simple formulas for finding their areas. However, for finding the areas under many other waveforms (e.g., the sine wave), we must turn to *integral calculus*. Very simply, integral calculus may be used to arrive at the *equations* for finding the areas under waveforms. Just to familiarize the student with calculus notation, consider

$$\text{area} = \int_0^T v(t)\,dt \qquad (13\text{-}4)$$

where $\int$ = mathematical notation used to denote integration
0 and T = limits of the integration
$v(t)$ = mathematical function to be integrated
dt = indicates integration with respect to time (t)

As we have seen, the dc average value of a time-varying waveform is given by Eq. 13-3. If we substitute the general calculus expression for area (Eq. 13-4) into Eq. 13-3, we obtain

$$V_{DC} = \frac{1}{T}\int_0^T v(t)\,dt \tag{13-5}$$

We shall *not* use Eq. 13-5 to any large extent. The important point to remember is that Eq. 13-5 means exactly the same thing as Eq. 13-3. The only difference is that Eq. 13-5 is mathematically more rigorous. Students studying calculus may wish to pursue the mathematics further.

In Fig. 13-4 we see a graph of a dc level. Since this voltage is constant, its dc average value for all time is 12 V. Applying either Eq. 13-3 or 13-5 to the dc level will produce an answer of 12 V. As we have seen, both the triangular and sinusoidal (ac) waveforms in Fig. 13-3 have an average value of 0 V. If these waveforms are allowed to ride on a dc level, such as 12 V, they will contribute *nothing* to the overall average. Consequently, the average value of the total waveform is still 12 V as shown in Fig. 13-4. There are two equally valid ways of describing the total waveform. We may say that we have an ac waveform riding on a 12-V dc level, or we may say that we have a varying dc level.

In Fig. 13-5(a) and (b) we have the pulsating dc waveforms that are produced at the outputs of half-wave and full-wave rectifiers, respectively. If we apply Eq. 13-5 to these waveforms, we will arrive at the results indicated by Eqs. 13-6 and 13-7.

$$V_{DC} = \frac{V_{dcm}}{\pi} = 0.318V_{dcm} \tag{13-6}$$

where V_{DC} is the dc value of a half-wave rectified sine wave and V_{dcm} is the peak value.

$$V_{DC} = 2\frac{V_{dcm}}{\pi} = 0.637V_{dcm} \tag{13-7}$$

where V_{DC} is the dc value of a full-wave rectified sine wave and V_{dcm} is the peak value.

FIGURE 13-4 Effect of an ac waveform on the dc average value.

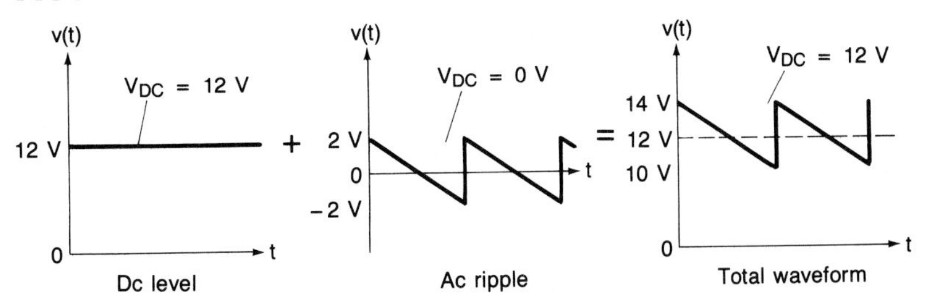

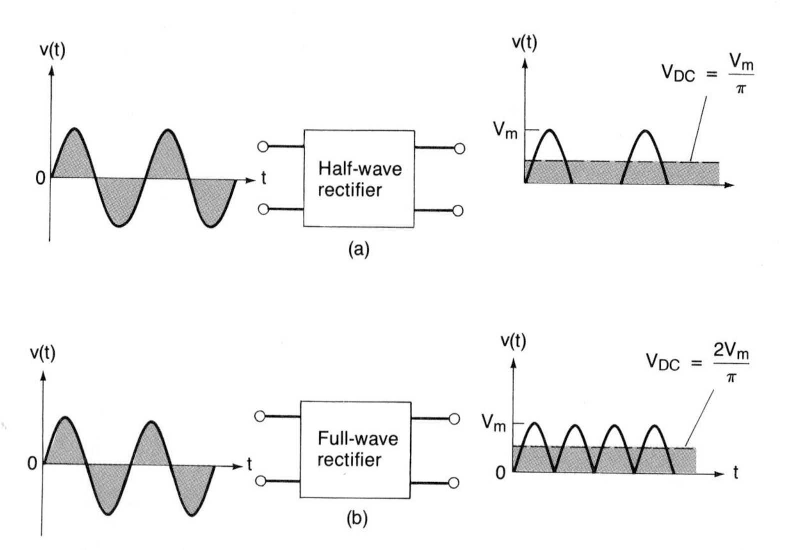

FIGURE 13-5 Rectified waveform average values: (a) half-wave; (b) full-wave.

With a little thought, we can see that the rectifiers convert sine waves with an average value of *zero* into waveforms which have an average value. This is their sole function in a power supply.

As mentioned previously, the *root-mean-square* (rms) or *effective* value is another way to characterize an ac waveform. Let us consider its definition.

The rms or effective value of a waveform is equivalent to dc as far as the power dissipated in a resistor is concerned.

To find the rms value of *any* periodic ac (voltage or current) waveform, we may follow the procedure below.

1. Square the amplitude of the waveform.
2. Find the total area under the squared waveform.
3. Divide by the period.
4. Take the square root of the result.

With a little reflection, the reader should realize that steps 2 and 3 produce an average or *mean* value. Consequently, the procedure provides us with the square *root* of the *mean* of the *squared* waveform or its rms value.

To summarize our procedure, we present Eq. 13-8. Although we have indicated rms *voltage* in this equation, the same general equation applies equally well for finding rms current.

$$V_{\text{rms}} = \sqrt{\frac{1}{T}(\text{area under the waveform}^2)} \tag{13-8}$$

Mathematically, the most difficult part of finding the rms value of a waveform lies in finding the area under the *squared* waveform. This problem is not too difficult in the case of rectangular waves. When the amplitude of rectangular waveforms is squared, rectangles result and we know how to find the area of rectangles.

Unfortunately, most of the waveforms we square do not have simple area formulas. Consequently, we must again invoke integral calculus. Integral calculus may be used to find the area under squared waveforms. The mathematical definition for finding the rms value of a voltage waveform is given by Eq. 13-9.

$$V_{\text{rms}} = \sqrt{\frac{1}{T}\int_0^T v^2(t)\,dt} \qquad (13\text{-}9)$$

where V_{rms} = rms value
$\int$ = mathematical notation for integration
T = period of the waveform
0 and T = limits of the integration
$v^2(t)$ = indicates the waveform is a squared function of time
dt = denotes that we are integrating with respect to time

Admittedly, Eq. 13-9 looks very formidable. However, we will not pursue it to any great length. It is important to understand that Eq. 13-9 directs us to perform exactly the same steps as detailed previously.

A key point to remember is that the rms (or dc) value of any ac waveform depends upon its shape. To emphasize this fact, we have illustrated several ac waveforms in Fig. 13-6 along with their rms and dc relationships.

EXAMPLE 13-3

A triangle wave such as that shown in Fig. 13-6(c) has a peak value of 6 V. Compute its rms value.

SOLUTION From Fig. 13-6(c) we see that

$$V_{\text{rms}} = \frac{V_m}{\sqrt{3}} = \frac{6\text{ V}}{1.732} = 3.46\text{ V rms}$$

■

13-4 Transformers

By inspection of Fig. 13-1, we can see that the transformer forms the "front end" of a dc power supply. Some of the schematic symbols used for the transformers are illustrated in Fig. 13-7. Basically, a transformer such as that shown in Fig. 13-7(a) consists of two windings, or coils of wire, wrapped around a laminated iron core. The winding that connects to the ac source is referred to as the *primary* winding. The winding that connects to the load is called the *secondary* winding. As can be seen in Fig. 13-7(b) and (c), a transformer may have multiple secondaries, and one or more of these windings may be center tapped.

The sinusoidal source voltage produces a sine wave of current in the primary.

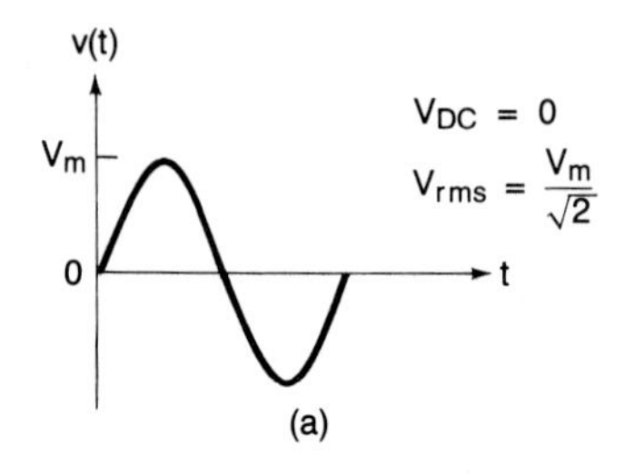

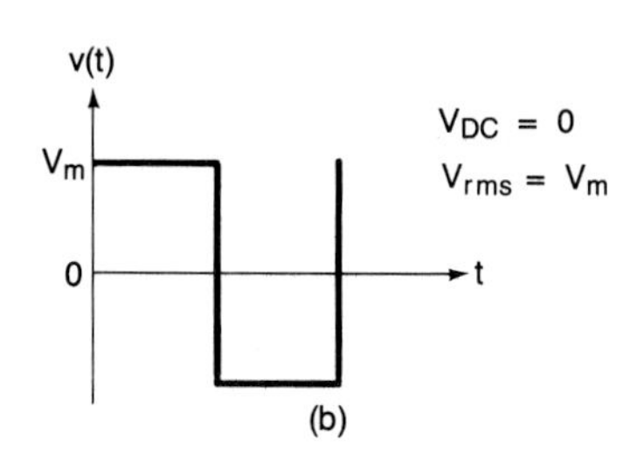

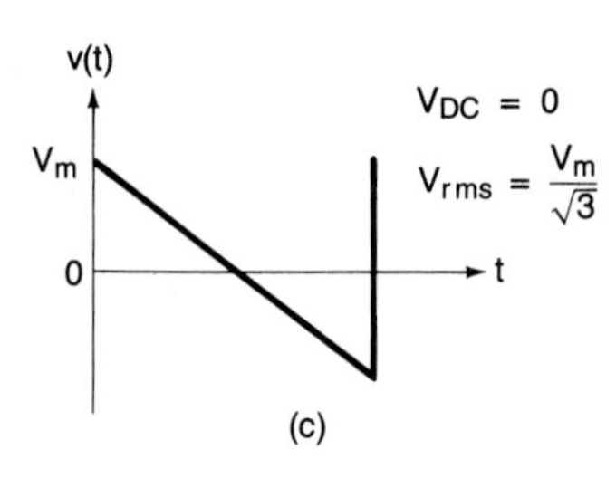

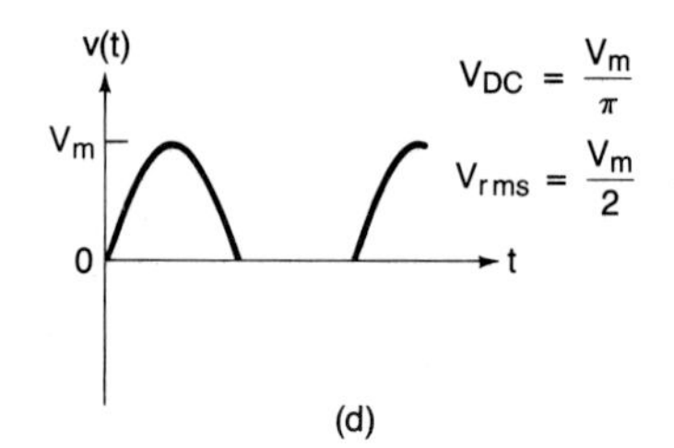

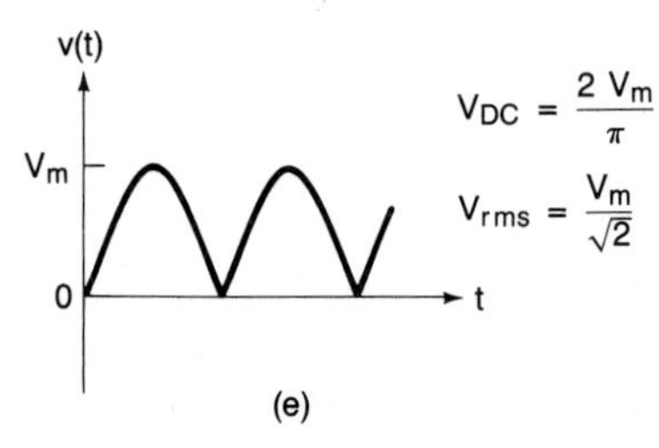

FIGURE 13-6 Dc and rms values: (a) sine wave; (b) square wave; (c) triangle wave; (d) half-wave rectified sine; (e) full-wave rectified sine.

FIGURE 13-7 Transformers: (a) basic symbol; (b) multiple secondaries; (c) center-tapped secondary; (d) transformer packages.

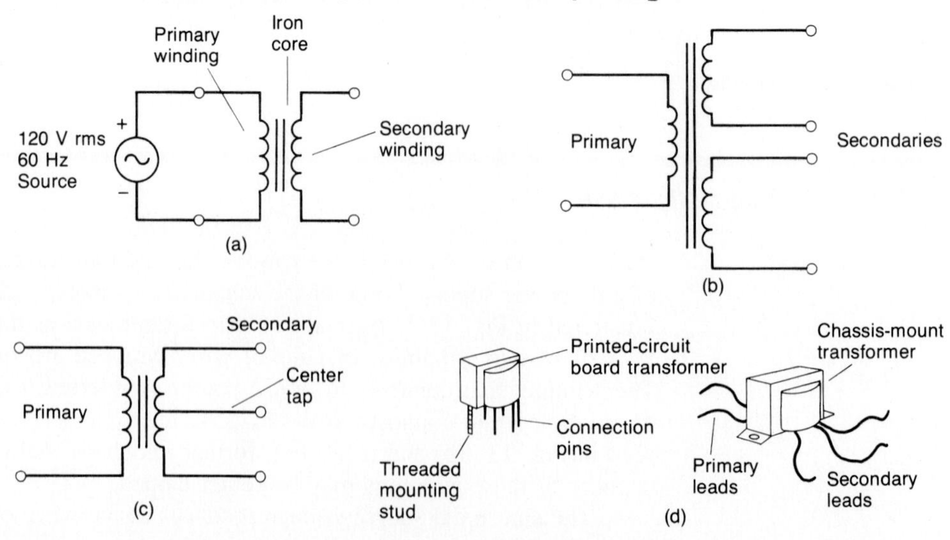

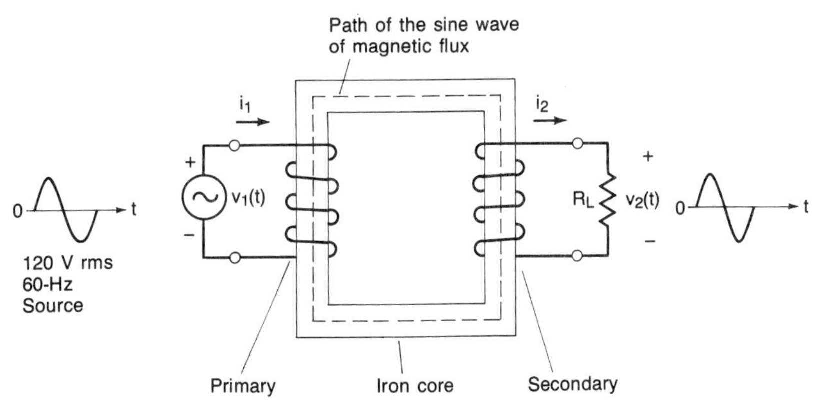

FIGURE 13-8 Basic transformer operation.

The resulting primary current produces a sinusoidal magnetic flux. The *changing* magnetic flux links with the secondary winding to induce a sinusoidal voltage across it (see Fig. 13-8).

The *dot convention* is introduced in Fig. 13-9. The dots are used to indicate the phasing of the transformer. *In most power supply applications, the phasing between the primary and secondary is unimportant.* However, it is quite possible that a phase inversion might be observed between the primary and secondary windings when tracing through a power supply circuit with an oscilloscope. It is for this reason that we have selected to mention it at this time.

The theory underlying transformer operations can become extremely involved. Invariably, such topics as Faraday's law and Lenz's law are drawn upon. However, we can avoid these discussions by utilizing the *ideal transformer* concept.

If we assume that the transformer has perfect magnetic coupling between its primary and secondary winding(s) and it exhibits no losses, we arrive at

$$\boxed{\frac{V_2}{V_1} = \frac{N_2}{N_1}} \tag{13-10}$$

FIGURE 13-9 Transformer dot convention: (a) primary and secondary voltages are in phase; (b) primary and secondary voltages are 180° out of phase.

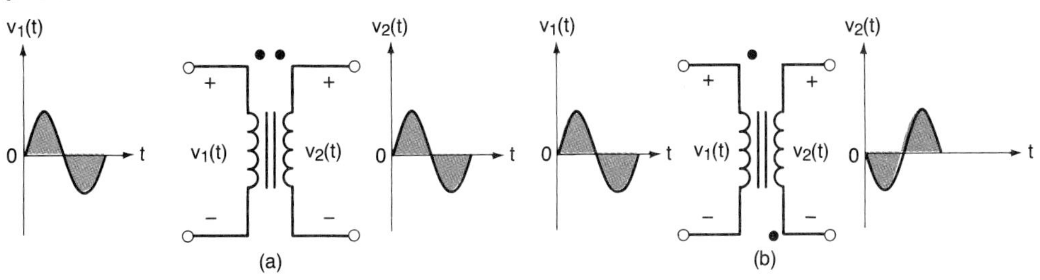

where V_2 = rms value of the secondary voltage
V_1 = rms value of the primary voltage
N_2 = number of secondary turns
N_1 = number of primary turns

EXAMPLE 13-4

The transformer in Fig. 13-9(a) has a primary winding of 100 turns. How many turns must the secondary contain to step a 120 V rms primary voltage down to 12.6 V rms?

SOLUTION From Eq. 13-10

$$\frac{N_2}{N_1} = \frac{V_2}{V_1}$$

$$N_2 = \frac{V_2}{V_1} N_1 = \frac{12.6 \text{ V rms}}{120 \text{ V rms}} (100 \text{ turns}) = 10.5 \text{ turns}$$

■

If we extend the concept of the ideal (lossless) transformer further, we see in Fig. 13-10 that the input power (V_1I_1) is equal to the output power (V_2I_2). From Fig. 13-10 we can state that for the ideal transformer,

$$P_{\text{in}} = P_{\text{out}}$$

$$V_1I_1 = V_2I_2 \tag{13-11}$$

and solving Eq. 13-11 for V_2/V_1 gives us

$$\frac{V_2}{V_1} = \frac{I_1}{I_2} \tag{13-12}$$

From Eqs. 13-10 and 13-12 we can state that

$$\frac{N_2}{N_1} = \frac{V_2}{V_1} = \frac{I_1}{I_2}$$

Hence

$$\boxed{\frac{I_1}{I_2} = \frac{N_2}{N_1}} \tag{13-13}$$

FIGURE 13-10 Ideal transformer.

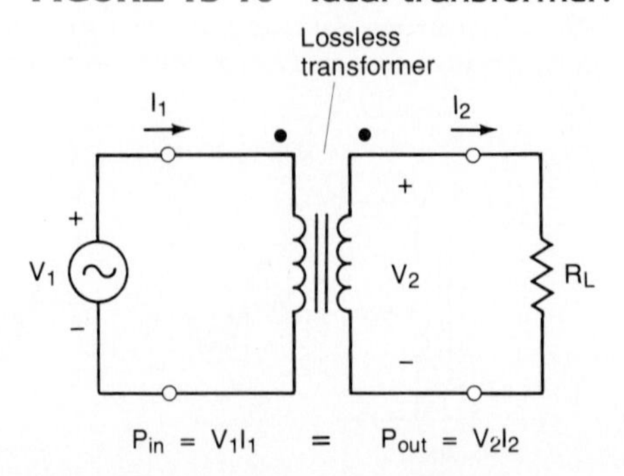

where I_1 = rms value of the primary current
I_2 = rms value of the secondary current
N_2 = number of secondary turns
N_1 = number of primary turns

EXAMPLE 13-5

The transformer in Fig. 13-11(a) has 200 turns on its primary and 40 turns on its secondary. Determine its secondary voltage and current and its primary current.

SOLUTION Solving Eq. 13-10 for V_2, we obtain the secondary voltage,

$$V_2 = \frac{N_2}{N_1} V_1 = \frac{40 \text{ turns}}{200 \text{ turns}} (120 \text{ V rms}) = 24 \text{ V rms}$$

By Ohm's law, the secondary current I_2 is

$$I_2 = \frac{V_2}{R_L} = \frac{24 \text{ V rms}}{100 \ \Omega} = 240 \text{ mA rms}$$

The primary current I_1 may be determined by solving Eq. 13-13 for I_1,

$$I_1 = \frac{N_2}{N_1} I_2 = \frac{40 \text{ turns}}{200 \text{ turns}} (240 \text{ mA rms}) = 48 \text{ mA rms}$$

This solution is illustrated in Fig. 13-11(b). The primary and secondary voltage waveforms have also been indicated. ■

Now that we have explored the ideal transformer concept, let us again cite the reasons why a transformer is used in a dc power supply.

FIGURE 13-11 Transformer circuits for Example 13–5: (a) circuit problem; (b) solution.

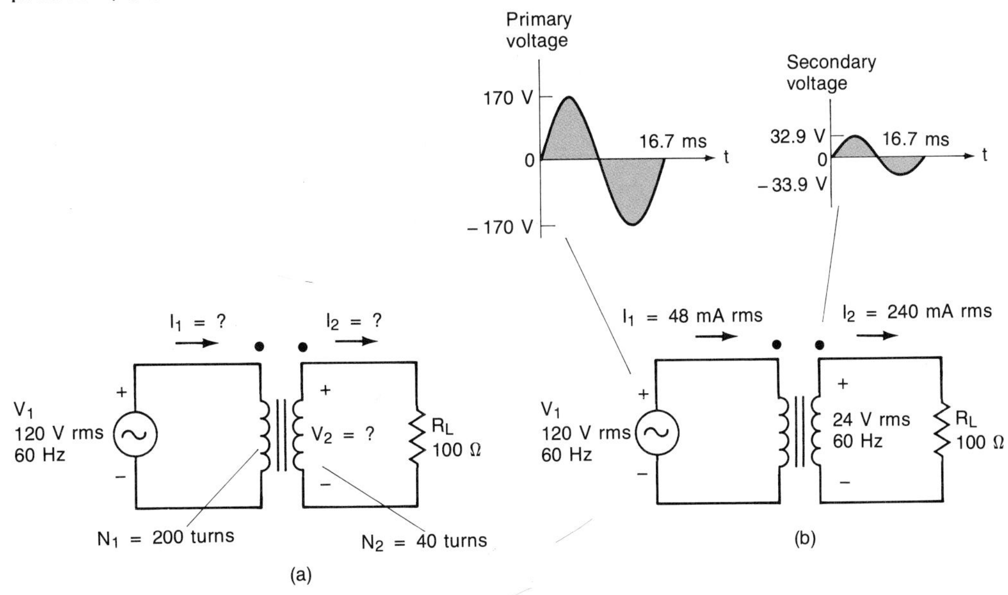

1. The transformer may be used to *step up* the ac line voltage ($N_2 > N_1$).
2. The transformer may be used to *step down* the ac line voltage ($N_2 < N_1$). This is the usual case for solid-state circuits.
3. The transformer is used to provide *isolation* from the ac power-line ground.
4. The transformer can include internal electrostatic shielding to minimize power-line noise coupling into the secondary.

The advantages of using the transformer to step down the 120-V rms line voltage will become more apparent as our studies progress. However, the advantage of electrical isolation from power-line ground is not as obvious [see Fig. 13-12(a)]. Electrical codes require that three wires should be run in 120-V rms, single-phase distribution systems. Specifically, these are the *hot*, *neutral*, and *ground* conductors.

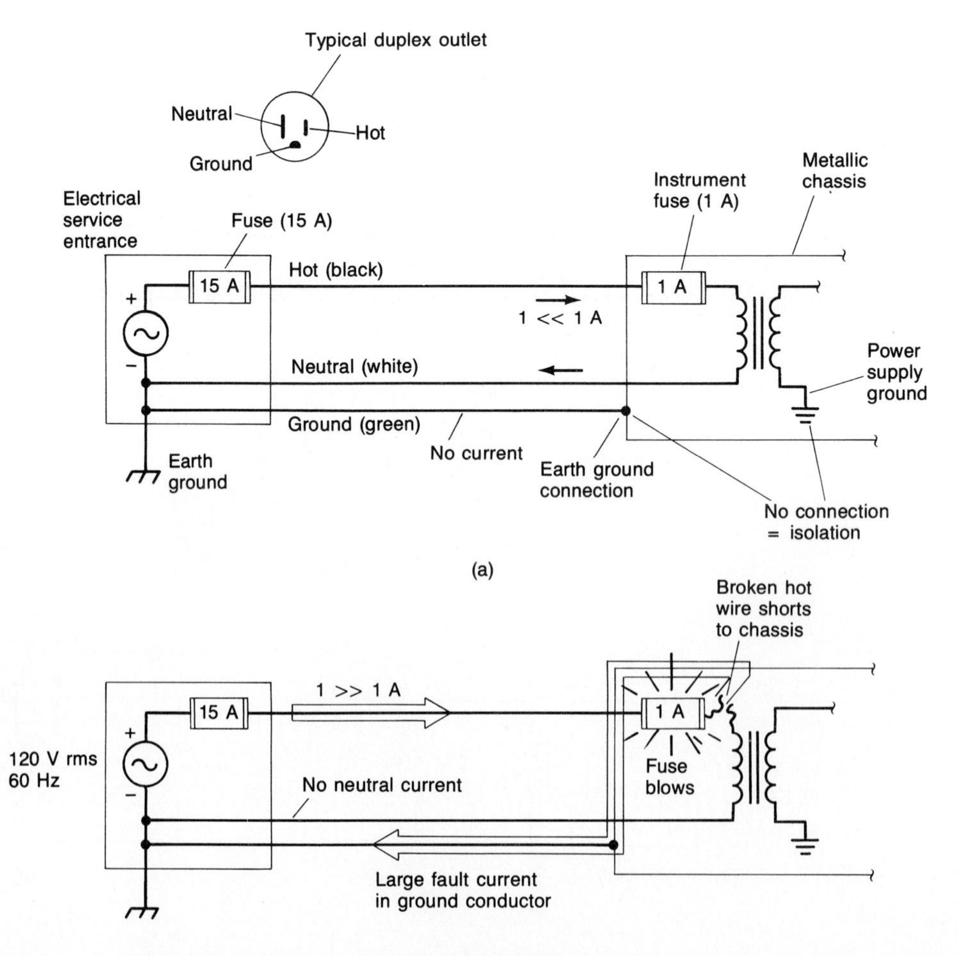

FIGURE 13-12 Three-wire power distribution: (a) normal operation; (b) fault condition.

For safety reasons the code requires that the ground conductor should be tied to the chassis and/or cabinet of electronic equipment. Study Fig. 13-12(a) carefully. Note that the *power supply ground* is *not* tied to the chassis, which is tied to *earth ground*.

The earth ground wire does not carry current unless there is a fault. If the chassis should become "hot" [Fig. 13-12(b)] the earth-ground conductor will short out the 120-V rms line. This serves two purposes. First, the chassis will not stay at 120 V rms (which could pose a serious electrical shock hazard). Second, the large fault current will open the fuse in the electronic equipment, which will disconnect it from the power line.

Since the transformer's secondary is *magnetically* coupled to the power line, there is *no* direct electrical connection between its secondary and earth ground. This is the isolation feature provided by the transformer [refer again to Fig. 13-12(a)].

To minimize electrical noise (unwanted electrical signals) in sensitive electronic systems, they should be tied to earth ground at only *one* point. Since the transformer gives us isolation, the power supply does *not* have to be our earth-ground connection point. This can give us much more flexibility in the design of an electronic *system* that may contain several pieces of electronic equipment, each with its own internal power supply.

13-5 Half-Wave Rectifiers

In Section 3-20 we introduced the basic operation of the half-wave rectifier. Figure 13-13(a) illustrates a half-wave rectifier circuit that incorporates a transformer. Observe that the transformer's primary and secondary waveforms have been illustrated.

In Fig. 13-13(b) we see that during the positive half-cycle, the diode is forward biased. The forward current through the diode develops the positive half-cycle of voltage across R_L. With a little reflection, we recall that the forward voltage drop across the diode is approximately 0.7 V. Therefore, we can model the diode as depicted in Fig. 13-13(c). Applying Kirchhoff's voltage law, we see that the peak voltage across R_L (V_{dcm}) is

$$V_{\text{dcm}} = V_m - 0.7\text{ V} \tag{13-14}$$

Therefore, by Eq. 13-14,

$$V_{\text{dcm}} = V_m - 0.7\text{ V} = 8.91\text{ V} - 0.7\text{ V} = 8.21\text{ V}$$

During the negative half-cycle of the secondary waveform, the diode is reverse biased [Fig. 13-13(d)]. Therefore, no current flows and the voltage across R_L is zero. Since the rectifier acts as an open switch, it drops all of the secondary voltage, and its peak reverse bias (V_{RM}) is

$$V_{\text{RM}} = V_m \tag{13-15}$$

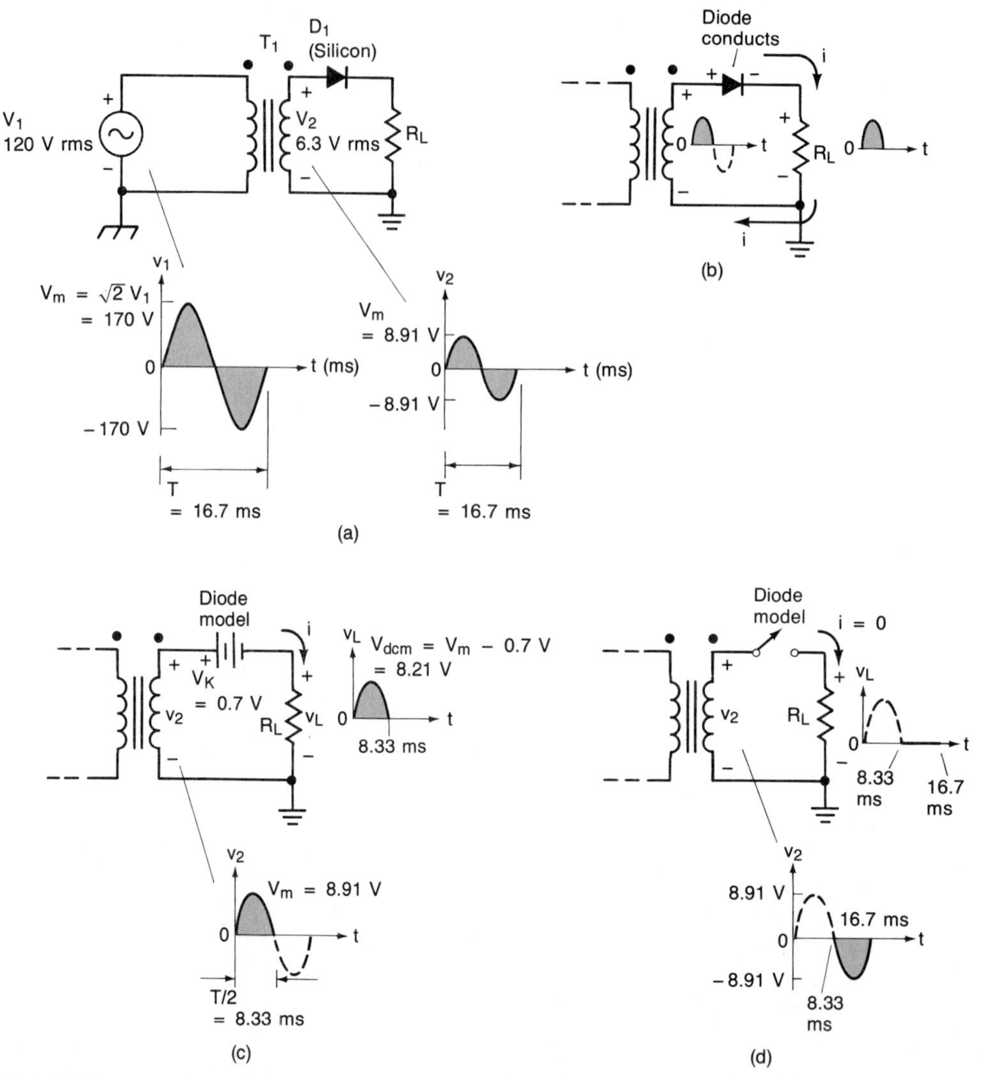

FIGURE 13-13 Operation of a half-wave rectifier: (a) primary and secondary voltages; (b) the diode conducts during the positive half-cycle; (c) the positive half-cycle appears across R_L; (d) the diode blocks the negative half-cycle.

By Eq. 13-15, the diode in Fig. 13-13(d) experiences

$$V_{RM} = 8.91\ \text{V}$$

A summary of the various waveforms associated with the half-wave rectifier is given in Fig. 13-14. Observe that this rectifier is being driven by a transformer with a 12.6-V rms secondary. The reader should study these waveforms carefully. Note that the period of the voltage across R_L is also 16.7 ms. Therefore, it may be regarded as pulsating dc at a frequency of 60 Hz.

The diode voltage [Fig. 13-14(d)] is a constant 0.7 V when it is forward biased.

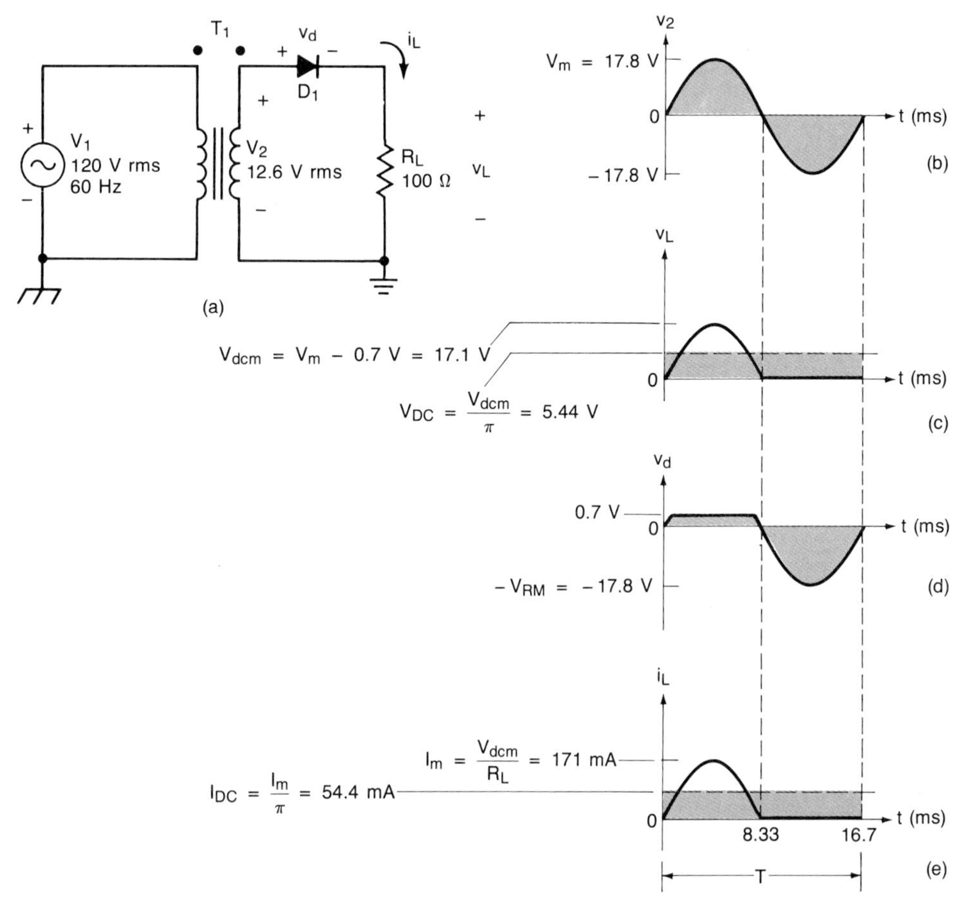

FIGURE 13-14 Summary of the half-wave rectifier waveforms: (a) circuit; (b) secondary voltage; (c) load voltage; (d) diode voltage; (e) load (and diode) current.

However, when it is reverse biased, it drops the negative half-cycle of the secondary voltage. Consequently, it must have a *peak repetitive reverse voltage rating* (V_{RRM}) which is greater than the V_{RM} of 17.8 V.

The simplicity of the half-wave rectifier circuit does *not* compensate for its many disadvantages. For example, it is difficult to filter, and the unidirectional current through the transformer's secondary tends to bias its iron core magnetically. This results in a lower transformer efficiency. Consequently, we find its use somewhat restricted.

EXAMPLE 13-6

Given the half-wave rectifier of Fig. 13-14(a), what value will a dc voltmeter indicate if it is connected across R_L? What value would a milliammeter read if it were placed in series with R_L?

SOLUTION The peak secondary voltage is

$$V_m = \sqrt{2}\, V_2 = (1.414)(12.6 \text{ V rms}) = 17.8 \text{ V}$$

This is illustrated in Fig. 13-14(b). The peak voltage across R_L is given by Eq. 13-14,

$$V_{\text{dcm}} = V_m - 0.7 \text{ V} = 17.8 \text{ V} - 0.7 \text{ V} = 17.1 \text{ V}$$

This is shown in Fig. 13-14(c).

Since both a voltmeter and a milliammeter are *average-responding* instruments, we may predict their respective readings by using Eq. 13-6 and Fig. 13-14(c),

$$V_{\text{DC}} = \frac{V_{\text{dcm}}}{\pi} = 0.318 V_{\text{dcm}} = (0.318)(17.1 \text{ V}) = 5.44 \text{ V}$$

and since the current waveform is also half-wave [Fig. 13-14(c)], we may state that

$$I_{\text{DC}} = \frac{I_m}{\pi} = 0.318 I_m = (0.318)(171 \text{ mA}) = 54.4 \text{ mA}$$ ■

It should be noted in Fig. 13-14(e) that the *peak* current through the diode is 171 mA, and its *average* forward current is 54.4 mA. The average forward current through a diode is usually designated $I_{(\text{AV})}$.

13-6 Full-Wave Rectifiers Using a Center-Tapped Transformer

The operation of the half-wave rectifier may be extended to the full-wave rectifier circuit given in Fig. 13-15. In Fig. 13-16(a) we have a 12.6-V rms, center-tapped transformer with the primary and secondary voltage waveforms indicated. In Fig. 13-16(b) we see that when the upper terminal is positive, diode D_1 is forward biased but diode D_2 is reverse biased. Since D_2 is effectively an open switch, we shall temporarily remove it as shown in Fig. 13-16(c). Current flows down through R_L, and its upper terminal becomes positive with respect to ground.

During the negative half-cycle of the secondary voltage, the lower terminal of the transformer becomes positive with respect to either the center tap or the upper terminal [see Fig. 13-16(d)]. Consequently, D_2 conducts while D_1 is reverse biased.

FIGURE 13-15 Full-wave rectifier using a center-tapped transformer.

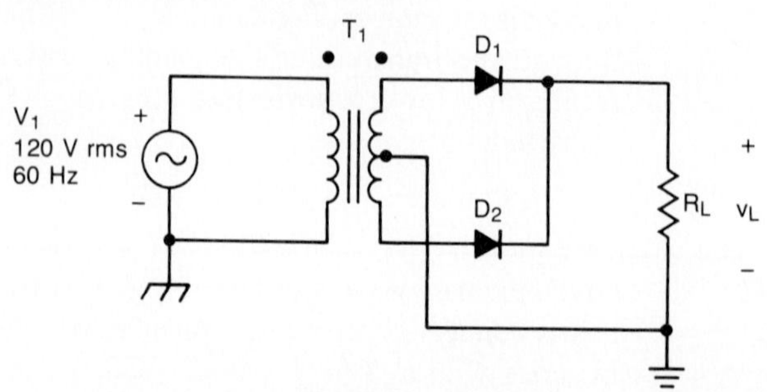

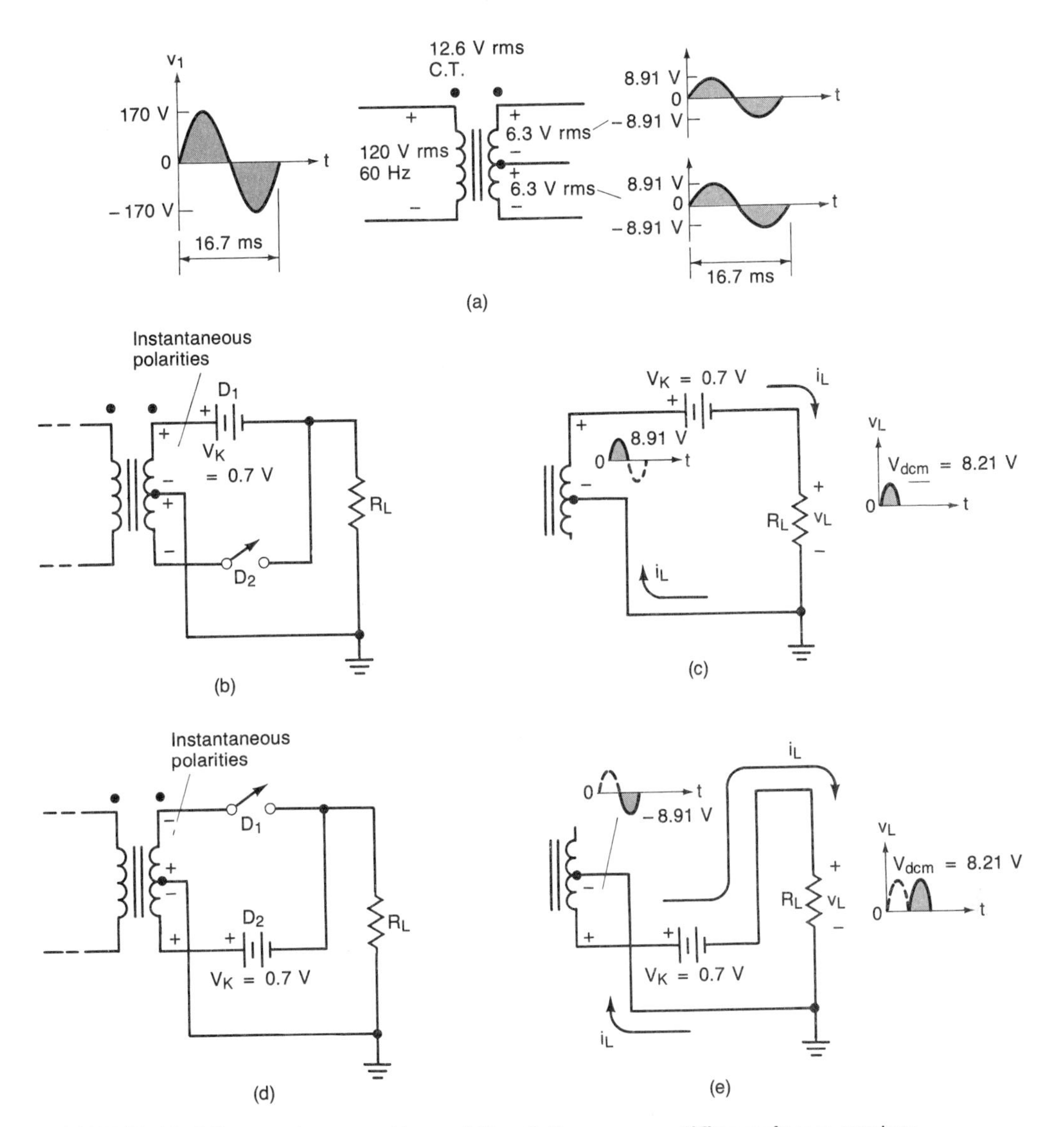

FIGURE 13-16 **Basic operation of the full-wave rectifier using a center-tapped transformer: (a) center-tapped transformer; (b) positive half-cycle (D_1 conducts and D_2 is off); (c) equivalent circuit; (d) negative half-cycle (D_2 conducts and D_1 is off); (e) equivalent circuit.**

In this case, we temporarily remove diode D_1 as shown in Fig. 13-16(e). Note that current will again flow down through R_L. Both half-cycles of the secondary ac voltage will appear across R_L, and we have full-wave rectification.

Observe that the peak voltage (V_{dcm}) across R_L is again reduced by one silicon diode drop (0.7 V) [see Fig. 13-16(c) and (d)]. Therefore, we can state that

$$V_{dcm} = V_m - 0.7 \text{ V} \tag{13-16}$$

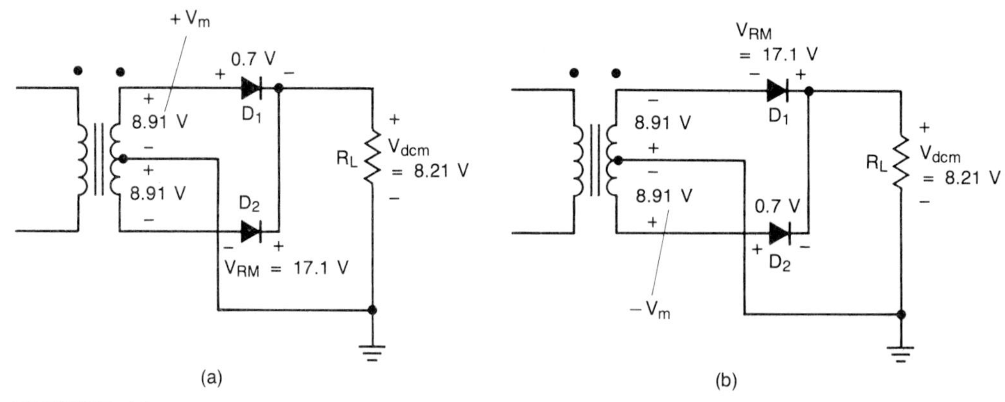

FIGURE 13-17 Finding the diode reverse voltages: (a) peak reverse voltage across D_2; (b) peak reverse voltage across D_1.

where V_{dcm} is the peak rectified voltage of a full-wave rectifier using a center-tapped transformer and V_m is the peak voltage across one half of the secondary. For Fig. 13-16(c) and (e) we can see that

$$V_m = \sqrt{2}\,V_2 = (1.414)(6.3 \text{ V rms}) = 8.91 \text{ V}$$

and by Eq. 13-16,

$$V_{\text{dcm}} = V_m - 0.7 \text{ V} = 8.91 \text{ V} - 0.7 \text{ V} = 8.21 \text{ V}$$

In Fig. 13-17 we can see that when the diodes are reverse biased, they must not only block the peak voltage across one-half of the secondary, but also the peak voltage V_{dcm} across R_L. Therefore, the peak reverse voltage V_{RM} is

$$V_{\text{RM}} = V_{\text{dcm}} + V_m \tag{13-17}$$

Substituting Eq. 13-16 into Eq. 13-17 gives us

$$V_{\text{RM}} = (V_m - 0.7 \text{ V}) + V_m$$

and we arrive at

$$\boxed{V_{\text{RM}} = 2V_m - 0.7 \text{ V}} \tag{13-18}$$

where V_{RM} is the peak reverse voltage across the diodes in a full-wave rectifier using a center-tapped transformer and V_m is the peak voltage across one-half of the secondary. For our example (Fig. 13-18) with a 12.6-V rms, center-tapped transformer, one-half of the secondary is 6.3 V rms (8.91 V peak) produces

$$\begin{aligned} V_{\text{RM}} &= 2V_m - 0.7 \text{ V} \\ &= (2)(8.91 \text{ V}) - 0.7 \text{ V} = 17.1 \text{ V} \end{aligned}$$

Therefore, the diodes utilized in our circuit must have a peak repetitive reverse voltage rating V_{RRM} which exceeds 17.1 V.

To conclude our discussion of this circuit, we have provided the various waveforms associated with it in Fig. 13-18. These should be studied carefully.

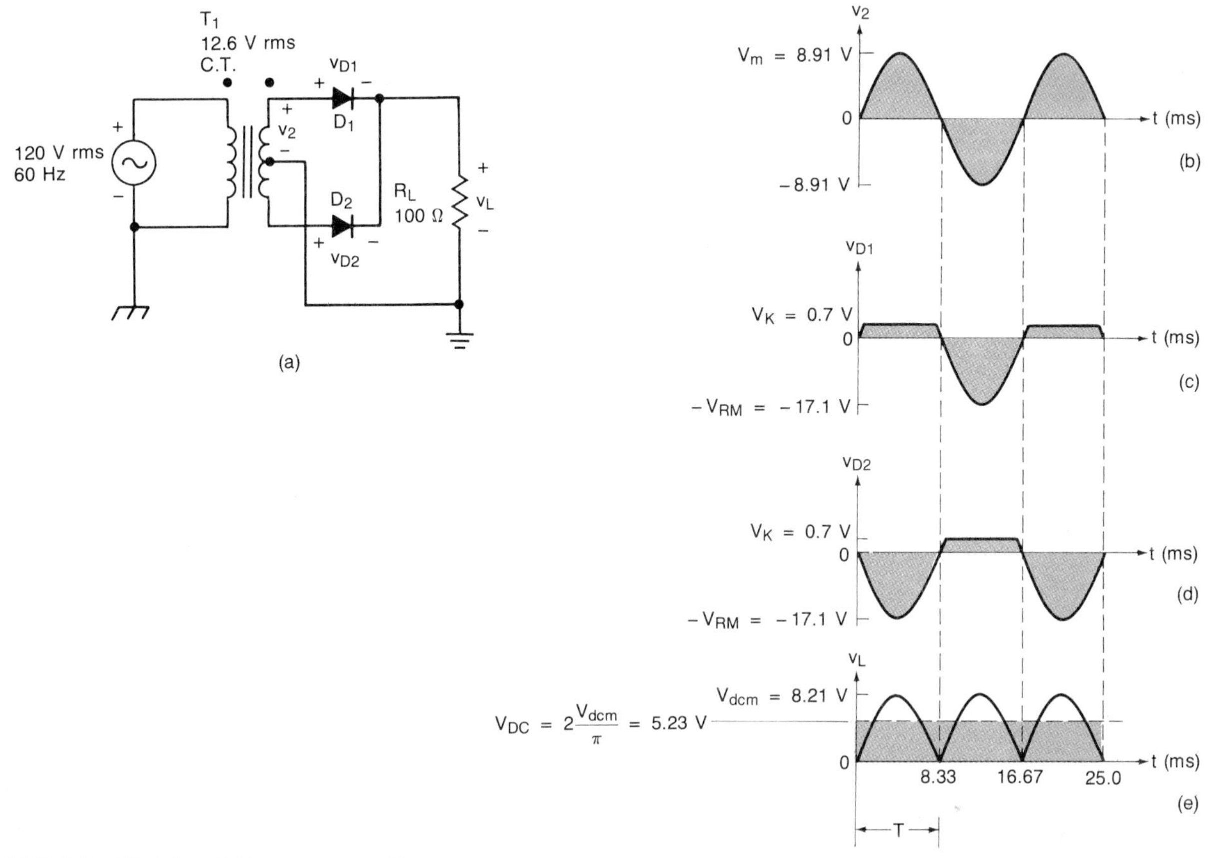

FIGURE 13-18 Full-wave rectifier waveforms: (a) circuit; (b) secondary voltage; (c) voltage across D_1; (d) voltage across D_2; (e) load voltage.

EXAMPLE 13-7

Given the full-wave rectifier of Fig. 13-18(a), what is the average voltage across R_L? If R_L is 100 Ω, what is the average current through it?

SOLUTION Since we have the voltage waveform shown in Fig. 13-18(e) across R_L, we may apply Eq. 13-7.

$$V_{\text{DC}} = 2\frac{V_{\text{dcm}}}{\pi} = 0.637V_{\text{dcm}} = (0.637)(8.21 \text{ V}) = 5.23 \text{ V}$$

Since the current through a resistor is proportional to the voltage across it, it will also be a full-wave rectified waveform. However, we already know the average voltage across R_L; therefore, we may use it to find the average current.

$$I_{\text{DC}} = \frac{V_{\text{DC}}}{R_L} = \frac{5.23 \text{ V}}{100\ \Omega} = 52.3 \text{ mA}$$ ■

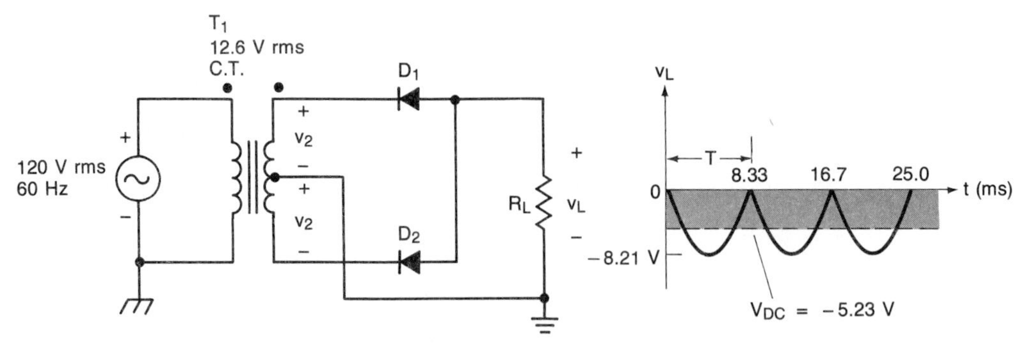

FIGURE 13-19 Full-wave rectifier to produce a negative dc voltage.

If both diodes are turned around, the top of R_L will be negative. Therefore, we will have a negative dc voltage with respect to ground (see Fig. 13-19). The operation of the circuit given in Fig. 13-19 is essentially the same as the circuit given in Fig. 13-18. Consequently, the *magnitudes* of the dc output voltages are both equal to 5.23 V.

In Figs. 13-18(e) and 13-19 we have indicated the full-wave rectified waveforms. Observe that their periods are 8.33 ms. Therefore, the full-wave rectified waveform has a repetition rate of $f = 1/8.33$ ms $= 120$ Hz. *Full-wave rectifiers produce pulsating dc waves with a frequency that is twice the power-line frequency.*

13-7 Full-Wave Bridge Rectifiers

Another extremely popular full-wave rectifier circuit is illustrated in Fig. 13-20. The reader should verify that all the circuits in the figure are electrically equivalent. In our work to come, we shall use the circuit indicated in Fig. 13-20(b). However, for now we shall analyze the full-wave bridge rectifier as it has been drawn in Fig. 13-21(a).

When the upper terminal of the transformer is positive, diodes D_2 and D_4 will conduct while diodes D_1 and D_3 are reverse biased [see Fig. 13-21(b)]. When the lower terminal of the transformer's secondary becomes positive with respect to the upper terminal, diodes D_1 and D_3 will conduct while D_2 and D_4 are reverse biased [see Fig. 13-21(c)].

When diodes D_2 and D_4 are conducting [Fig. 13-21(b)], we may replace them with their knee-voltage models. Diodes D_1 and D_3 may be replaced with open switches since they are reverse biased. This is illustrated in Fig. 13-22(a). Eliminating D_1 and showing D_3 in phantom, we arrive at Fig. 13-22(b). From the resulting equivalent circuit, it is clear that the peak voltage across R_L is the peak secondary voltage minus *two* diode drops [1.4 V]. (Temporarily ignore D_3.)

If we again define the peak voltage across R_L to be V_{dcm}, we can write

$$V_{\mathrm{dcm}} = V_m - 1.4\ \mathrm{V} \tag{13-19}$$

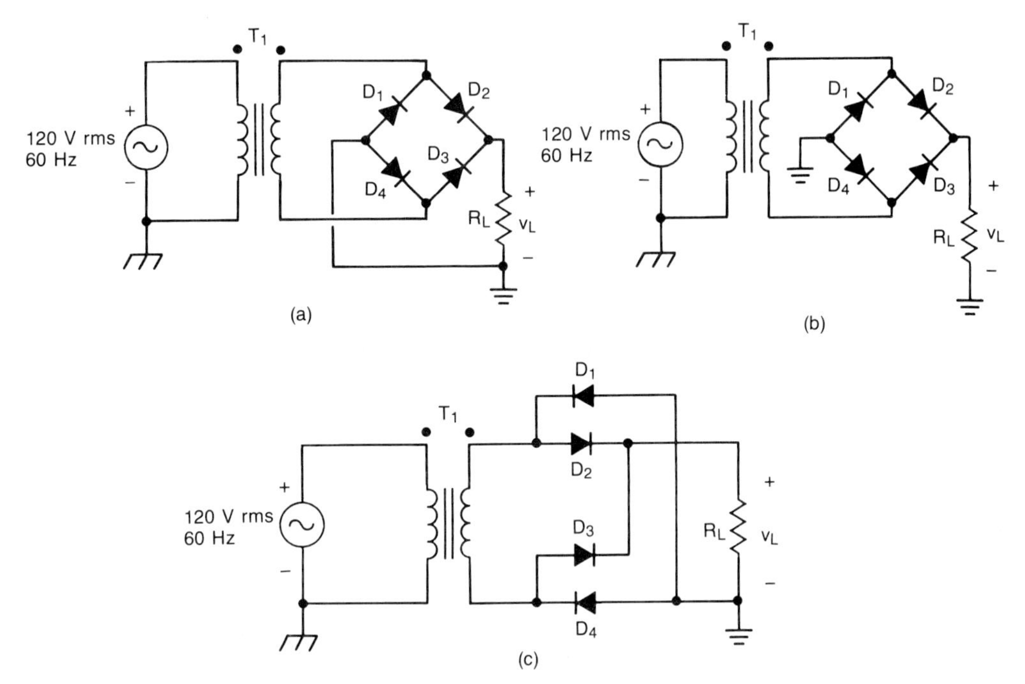

FIGURE 13-20 **Full-wave bridge rectifier.**

where V_{dcm} is the peak voltage across the load of a full-wave bridge rectifier and V_m is the peak secondary voltage.

EXAMPLE 13-8

Given the full-wave bridge rectifier shown in Fig. 13-21(a), if the transformer secondary voltage is 25.2 V rms, compute the average voltage across R_L.

SOLUTION To determine the average voltage across R_L, we must first find the peak secondary voltage, the peak voltage across R_L, and then the average value of the full-wave rectified waveform across R_L. Hence

$$V_m = \sqrt{2}\, V_2 = (1.414)(25.2) = 35.6 \text{ V}$$

as shown in Fig. 13-22(b). From Eq. 13-19,

$$V_{\text{dcm}} = V_m - 1.4 \text{ V}$$

$$= 35.6 \text{ V} - 1.4 \text{ V} = 34.2 \text{ V}$$

The average voltage may be found by employing Eq. 13-7.

$$V_{\text{DC}} = \frac{2V_{\text{dcm}}}{\pi} = 0.637V_{\text{dcm}} = 21.8 \text{ V}$$

■

Diode D_3 has been shown in phantom in Fig. 13-22(b). When diodes D_2 and D_4 are conducting, diodes D_1 and D_3 are reverse biased. If we apply Kirchhoff's voltage law around the loop, including D_3, R_L, and D_4,

$$V_{\text{RM}} - V_K - V_{\text{dcm}} = 0$$

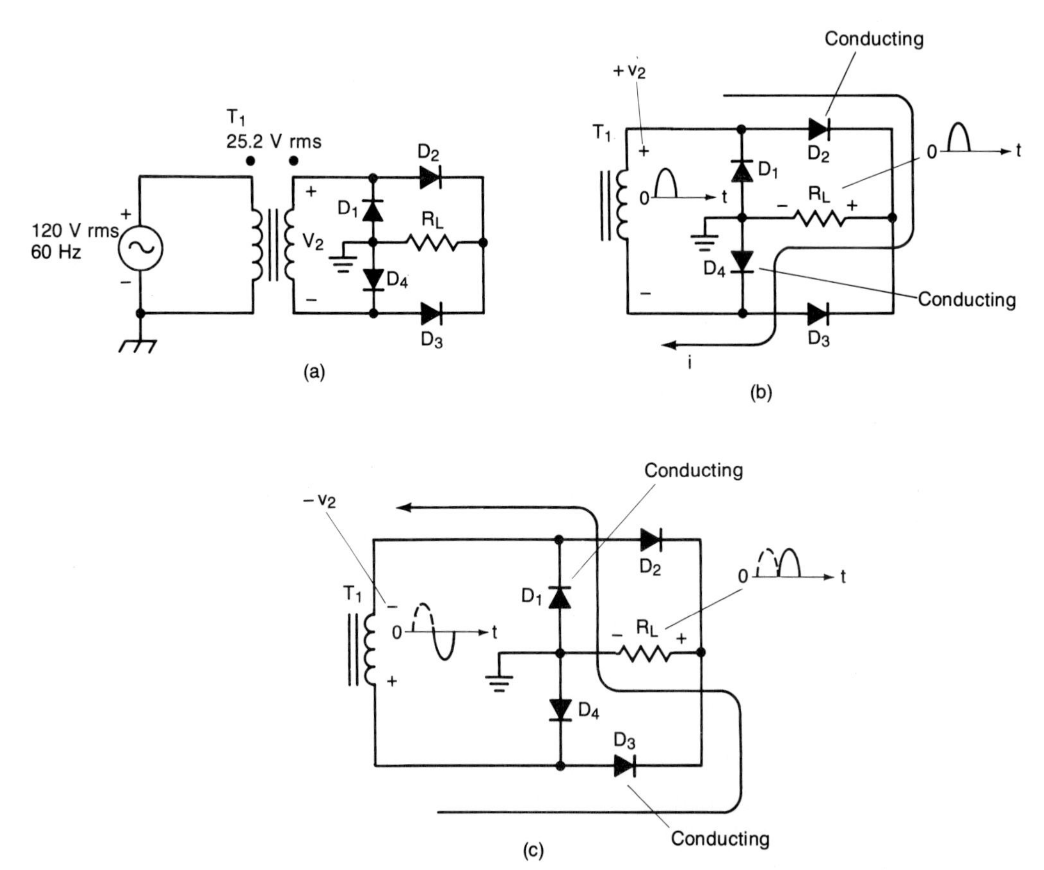

FIGURE 13-21 **Basic operation of the full-wave bridge rectifier: (a) circuit; (b) positive half-cycle of secondary voltage; (c) negative half-cycle of secondary voltage.**

and solving for V_{RM}, we have

$$V_{RM} = V_{dcm} + V_K \tag{13-20}$$

which states that the peak reverse voltage across D_3 is equal to the peak voltage across R_L plus one diode drop. The peak reverse voltage across the diode may also be expressed in terms of the transformer's peak secondary voltage.

Applying Kirchhoff's voltage law around the outside loop of Fig. 13-22(b) consisting of the secondary, D_2, and D_3,

$$-V_m + V_K + V_{RM} = 0$$

Solving for V_{RM}, we arrive at

$$\boxed{V_{RM} = V_m - 0.7\text{ V}} \tag{13-21}$$

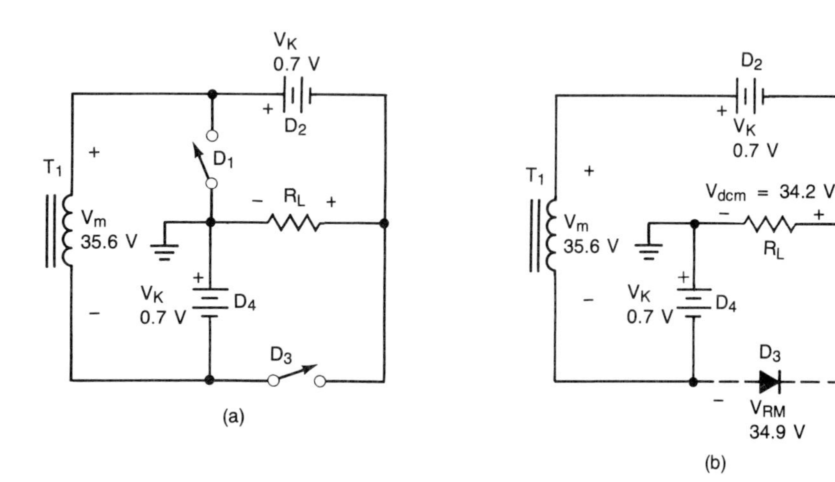

FIGURE 13-22 Full-wave bridge operation using the diode models: (a) the diodes are replaced by their respective models during the positive half-cycle of secondary voltage; (b) simplified circuit.

where V_{RM} is the peak reverse voltage across the diodes in a full-wave bridge rectifier and V_m is the peak secondary voltage.

EXAMPLE 13-9

Compute the peak reverse voltage across the diodes used in the full-wave bridge rectifier of Fig. 13-21(a).

SOLUTION In Example 13-8 we saw that

$$V_m = 35.6\ \mathrm{V}$$

across the secondary, and by Eq. 13-21,

$$\begin{aligned} V_{\mathrm{RM}} &= V_m - 0.7\ \mathrm{V} \\ &= 35.6\ \mathrm{V} - 0.7\ \mathrm{V} = 34.9\ \mathrm{V} \end{aligned}$$

■

To summarize the performance of the full-wave bridge rectifier circuit, we have illustrated the various waveforms associated with it in Fig. 13-23. These should be studied carefully. The waveforms of the voltage across D_2 and D_4 are identical, as are the waveforms across D_1 and D_3 [see Fig. 13-23(c) and (d), respectively]. Since each pair of diodes (D_2, D_4, and D_1, D_3) conduct for only one-half of the secondary voltage waveform on alternate half-cycles, their current waveforms are as shown in Fig. 13-23(e) and (f). The voltage across R_L and the current through it are depicted in Fig. 13-23(g) and (h).

Note that the average current through the diode pairs is only one-half of the average dc load current. This should be verified by the reader.

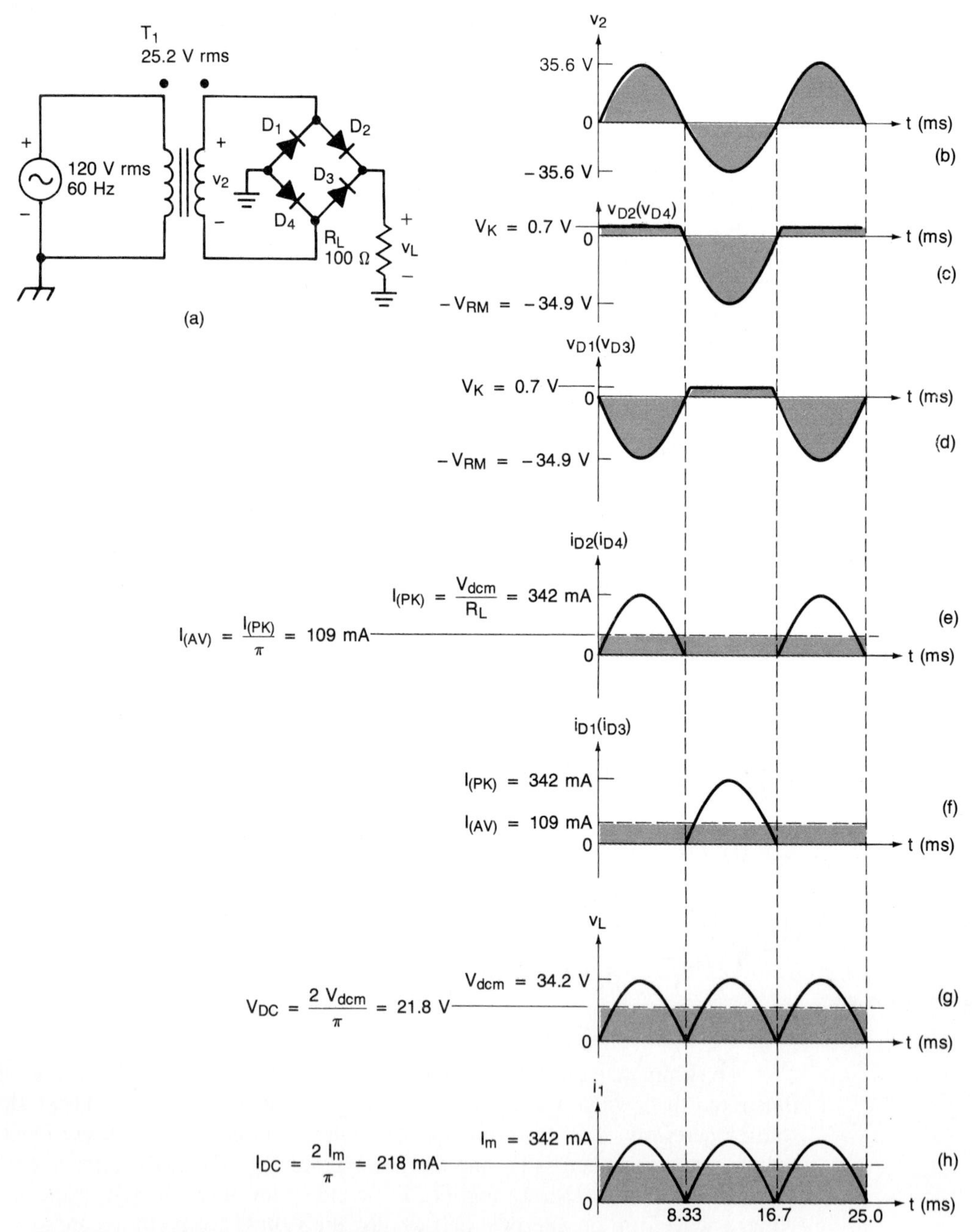

FIGURE 13-23 Full-wave bridge rectifier waveforms: (a) circuit; (b) secondary voltage; (c) D_2 or D_4 voltage; (d) D_1 or D_3 voltage; (e) D_2 or D_4 current; (f) D_1 or D_3 current; (g) load voltage; (h) load current.

13-8 Dual-Complementary Full-Wave Rectifiers

When dual-polarity power supply voltages are required, the *dual-complementary full-wave rectifier* is often used. Essentially, it consists of a positive full-wave rectifier using a center-tapped transformer and a negative full-wave rectifier. Both share the same transformer [refer to Fig. 13-24(a)].

Since the dual-complementary rectifier is often drawn as shown in Fig. 13-24(b), the reader should carefully trace through it to verify that it is equivalent to Fig. 13-24(a). The operation of this rectifier is fundamentally the same as the full-wave rectifier using a center-tapped transformer. Since we examined it thoroughly in Section 13-6, we shall not elaborate further. [Ambitious (and/or confused) students may wish to turn to Example 14-21.]

FIGURE 13-24 **Dual-complementary rectifier: (a) formation; (b) complete circuit.**

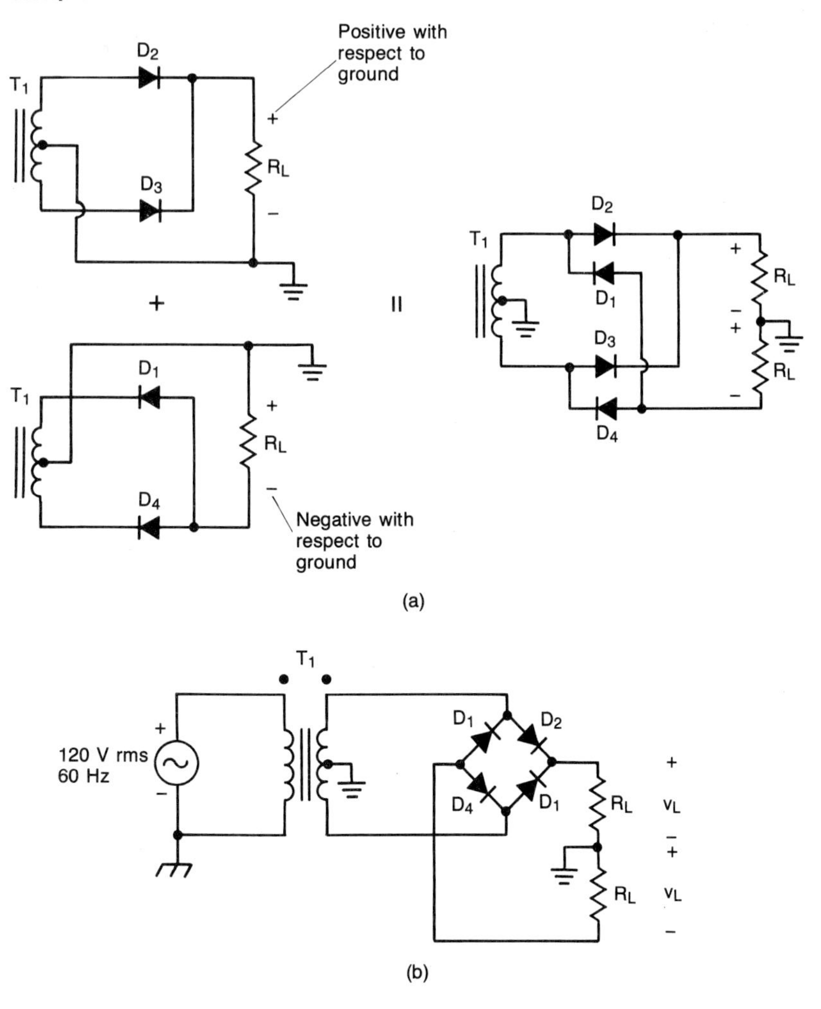

13-9 Filter Capacitor Considerations

Pulsating dc is not suitable for most electronic systems. The dc must be relatively constant. Therefore, the output of the rectifier stage must be filtered (or "smoothed"). The most elementary power supply filter consists of a single capacitor placed across the output of the rectifier in parallel with the load (see Fig. 13-25).

The strategy behind this scheme is based on two basic circuit concepts. First, *capacitors oppose a change in their terminal voltage*, and second, *elements in parallel experience the same voltage*. Therefore, the capacitor will tend to smooth out the dc voltage across the load.

Many filter configurations utilize a capacitor across the output of the rectifier and are termed "capacitive-input." Consequently, we need to make a detailed analysis of its effect. First, let us look at the types of filter capacitors and their characteristics.

Generally, the capacitors employed in power supply filter circuits are *aluminum electrolytics*. Aluminum electrolytic capacitors have the following attributes:

1. They posssess a *large* value of *capacitance* considering their relatively *small physical size* when compared to other types of capacitors.
2. Electrolytic capacitors are *polarized*. As we recall, their positive terminal is indicated on their schematic symbol as shown in Fig. 13-25. If their polarity is inadvertently reversed, or if an ac voltage is placed across them, they may be severely damaged, or *explode*.
3. Electrolytic capacitors tend to pass *more leakage* current than other types of capacitors. However, this is of no great concern in power supply filtering applications.
4. They tend to filter lower frequencies (such as the 120-Hz ripple voltage) much better than the higher frequencies (e.g., above 20 kHz).

Because of their poor high-frequency response, it is not unusual to see a small *disk ceramic capacitor* placed in parallel with a large electrolytic. This tends to improve the high-frequency decoupling action of the power supply filter. As an alternative to the use of aluminum electrolytics, we occasionally see *tantalum electrolytic capacitors* used. Tantalum capacitors offer less leakage, a better temperature range, and a smaller size, but tend to be more expensive. Standard capacitor values are provided in Appendix C.

FIGURE 13-25 Simple capacitor filter.

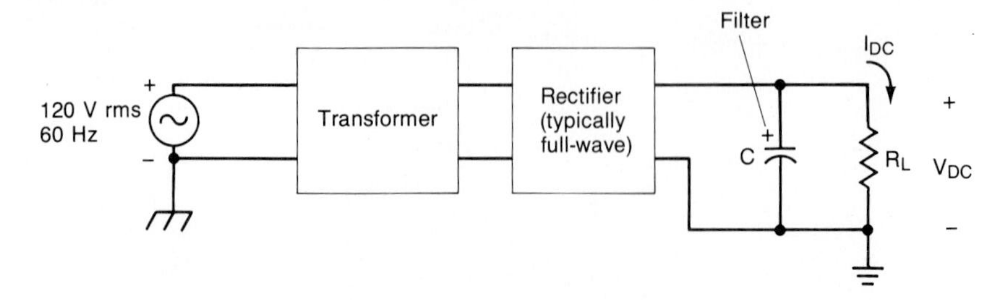

The maximum dc voltage that can be impressed across a capacitor without breaking down its dielectric is called its working voltage rating. This is often denoted WVDC. The WVDC rating of the capacitors used in power supply filters must be greater than the peak dc voltage V_{dcm} impressed across them. To provide a 20% (minimum) safety margin, we offer the condition stated by Eq. 13-22.

$$\boxed{\text{WVDC} \geq 1.2V_{dcm}} \tag{13-22}$$

where WVDC is the filter capacitor working dc working rating and V_{dcm} is the peak rectified voltage ($V_m - 0.7$ V) for full-wave rectifiers using a center-tapped transformer and ($V_m - 1.4$ V) for bridge rectifiers.

As a final comment, we should note that electrolytic capacitors also have a *ripple current rating*. This will be dealt with in Section 13-16.

13-10 Simple Capacitor Filters

To illustrate the effect of the filter capacitor, let us carefully study Fig. 13-26. In Fig. 13-26(a) we have depicted a full-wave rectifier using a center-tapped transformer driving a resistive load R_L. As we saw in Section 13-6, a pulsating dc waveform will develop across R_L, which has a peak voltage V_{dcm}. Recall that

$$V_{dcm} = V_m - 0.7 \text{ V}$$

for this rectifier circuit. The average dc voltage V_{DC} across R_L was shown to be

$$V_{DC} \simeq 0.637V_{dcm}$$

and has been indicated in Fig. 13-26(a).

If we temporarily remove R_L and replace it with a filter capacitor C_1, we have the circuit shown in Fig. 13-26(b). From $t = 0$ to $t = T/4$, the capacitor will quickly charge to V_{dcm}. If the capacitor's leakage current is negligible, the capacitor will not discharge appreciably. Therefore, the voltage across it will remain at V_{dcm}. Note that the rectified voltage has been shown in dashed lines as a reference.

The situation shown in Fig. 13-26(b) is ideal. The average dc voltage has been *increased* from $0.637V_{dcm}$ to V_{dcm}, and the pulsating dc has been converted to a perfectly smooth or constant dc level. However, this occurs only when *no load* is connected across the capacitor.

In Fig. 13-26(c) we have a more realistic situation. Observe that a load has been connected across C_1. As before, the capacitor will initially charge to the peak rectified voltage. Once this occurs, the rectified voltage (again shown in dashed lines) will decrease, and *both rectifier diodes* will be *nonconducting*.

Consequently, the capacitor will start to discharge through R_L. The capacitor will continue to discharge until the rectified voltage becomes large enough to allow one of the diodes to conduct. When this occurs, the capacitor will again quickly charge to the peak value of the pulsating dc.

The key to understanding the waveform in Fig. 13-26(c) is to note that the cathode ends of both diodes D_1 and D_2 are held positive by capacitor C_1. Consequently,

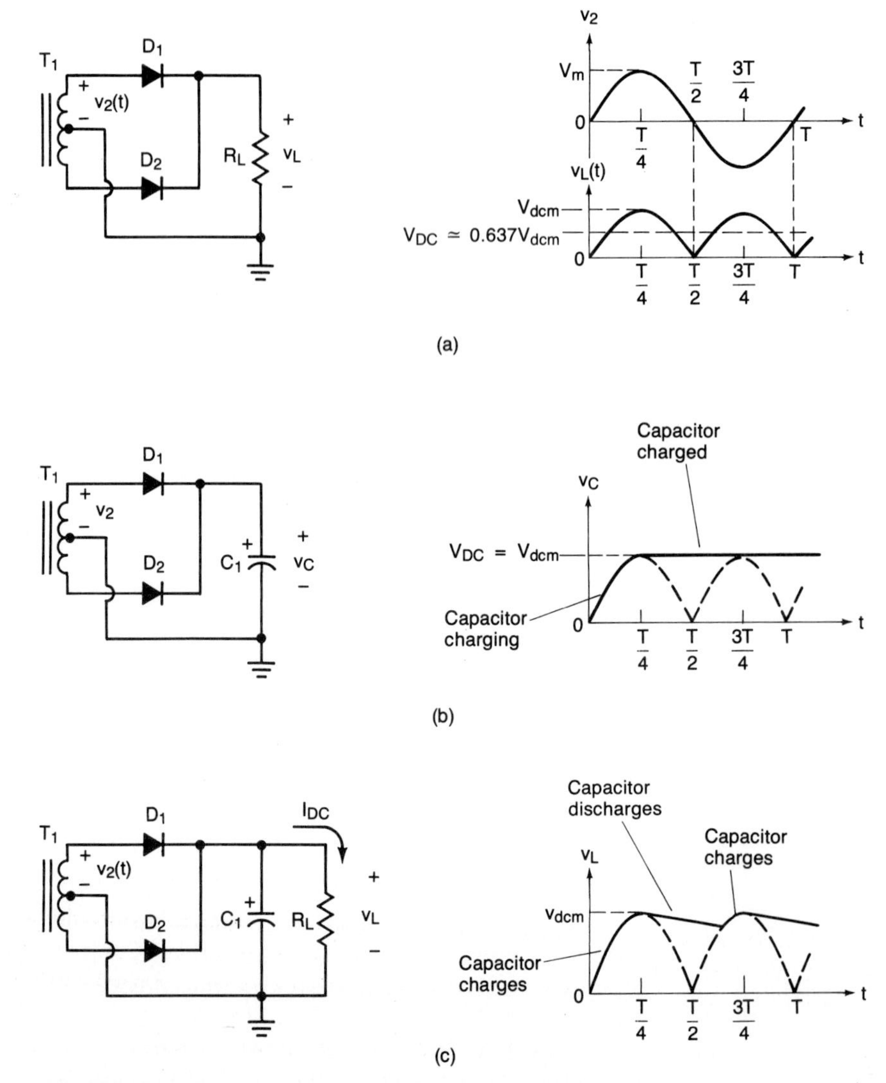

FIGURE 13-26 **Basic action of the filter capacitor: (a) rectifier driving a resistor R_L; (b) rectifier driving a capacitor C_1; (c) rectifier driving a filtered load.**

the only time D_1 or D_2 can conduct is when their *anodes* are *more positive* than their *cathodes*. When either of the diodes conduct, the capacitor will be recharged.

In Fig. 13-27 we see the voltage waveform across the filter capacitor (and R_L). The peak-to-peak ripple voltage shall be designated as $V_{r(\text{p-p})}$. By inspection of Fig. 13-27 it can be seen that the peak dc voltage is V_{dcm}. The instantaneous minimum value of the dc voltage $V_{\text{dc(min)}}$ is given by

$$V_{dc(min)} = V_{dcm} - V_{r(p\text{-}p)} \tag{13-23}$$

where $V_{dc(min)}$ = minimum instantaneous dc voltage
V_{dcm} = peak rectified voltage
$V_{r(p\text{-}p)}$ = peak-to-peak ripple voltage

From our work in Section 13-3 it should be clear that the dc average value of the waveform shown in Fig. 13-27 can be found from

$$V_{DC} = V_{dcm} - \frac{V_{r(p\text{-}p)}}{2} \tag{13-24}$$

where V_{DC} is the average dc voltage across the filter capacitor, in volts.

Before we further quantify our analysis, let us first make some general observations (see Fig. 13-28).

Generally, the *larger* the filter capacitor, the more charge it can hold and the less it will discharge. Hence the peak-to-peak value of the ripple will be *less*, and the average dc level will be *greater*.

[See Fig. 13-28(a).] The converse is also true.

The *smaller* the filter capacitor, the less charge it can hold and the more it will discharge. Therefore, the peak-to-peak value of the ripple will *increase*, and the average dc level will be *less*.

[See Fig. 13-28(b).]

In a similar fashion, for a fixed-value filter capacitance, the ripple will *decrease* and V_{DC} will *increase* as the load current I_{DC} is *decreased*. Conversely, for a fixed value of filter capacitance, the ripple will *increase* and V_{DC} will *decrease* as the load current I_{DC} is *increased*.

Contrast Fig. 13-28(a) and (b). The student should memorize the relationships indicated in the figure.

FIGURE 13-27 Filtered waveform.

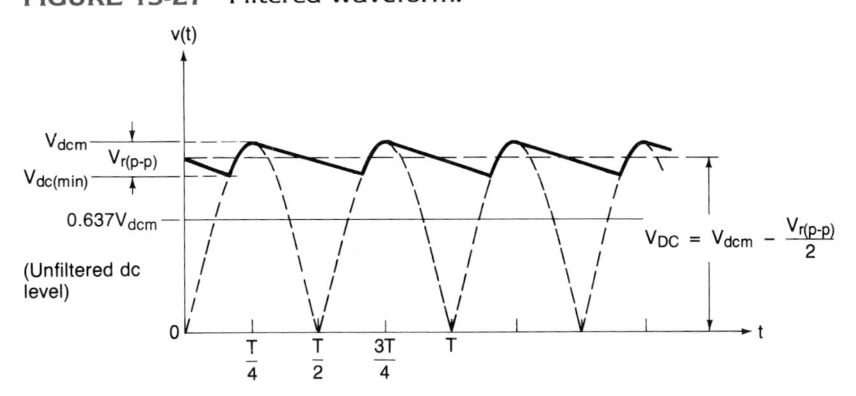

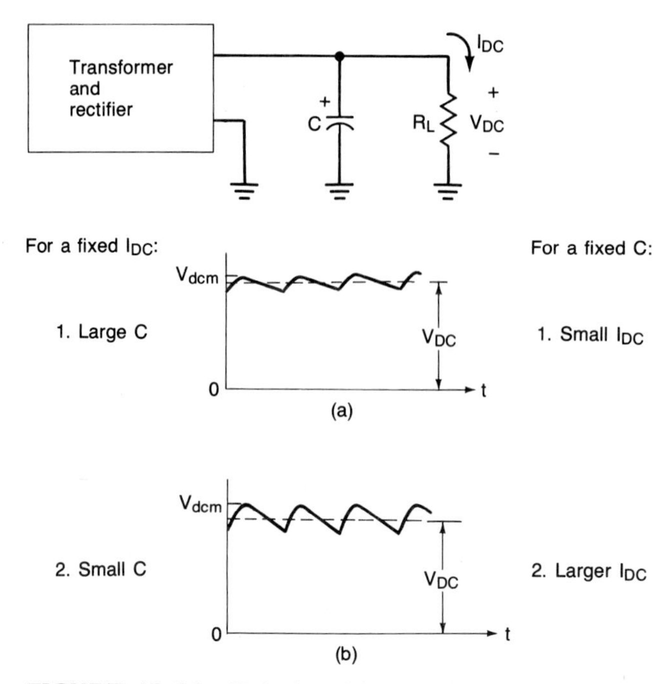

FIGURE 13-28 Relationships between the ripple, V_{DC}, I_{DC}, and C: (a) small ripple and a large V_{DC}; (b) large ripple and a smaller V_{DC}.

13-11 Ripple Factor

To characterize power supply filter circuits, the term *ripple factor* (r) has been defined. Before we provide its definition, let us again take a look at the voltage waveform across R_L. By the principles of superposition, we may regard the load voltage to be the sum of the average dc level and the 120-Hz ac ripple component (see Fig. 13-29).

Since the charge time is very short compared to the discharge time of the capacitor, we may approximate the ripple waveform as indicated in Fig. 13-30. The

FIGURE 13-29 The composite waveform may be decomposed (by superposition) into the sum of the dc level plus an ac ripple component.

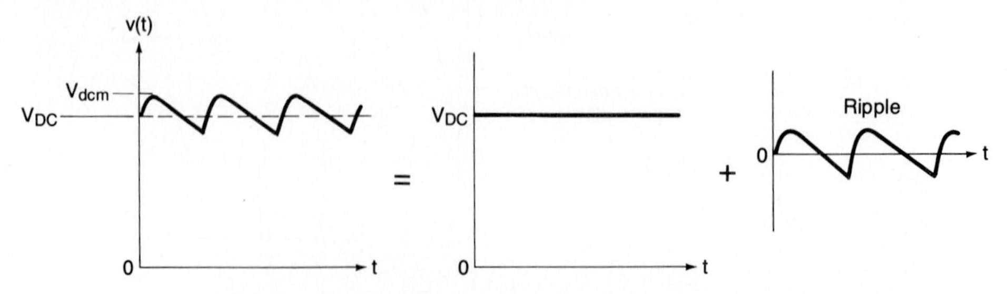

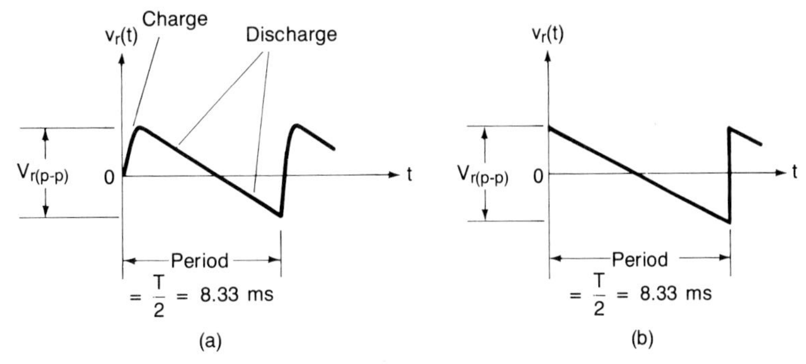

FIGURE 13-30 Ripple waveform approximation: (a) ripple waveform; (b) triangular approximation.

resultant triangular approximation can greatly simplify the mathematical analysis of the filtering action.

From our work in Section 13-3, we may find the rms value of the ripple waveform approximation shown in Fig. 13-30(b). From Fig. 13-6(c), the rms value of a triangular waveform is

$$V_{\text{rms}} = \frac{V_m}{\sqrt{3}}$$

Applying this relationship to Fig. 13-30(b), we arrive at

$$V_{r(\text{rms})} = \frac{V_{r(\text{p-p})}/2}{\sqrt{3}}$$

$$V_{r(\text{rms})} = \frac{V_{r(\text{p-p})}}{2\sqrt{3}} \qquad (13\text{-}25)$$

where $V_{r(\text{rms})}$ is the rms value of the triangular ripple waveform and $V_{r(\text{p-p})}$ is the peak-to-peak ripple voltage.

The ripple factor r is defined by

$$r = \frac{V_{r(\text{rms})}}{V_{\text{DC}}} \qquad (13\text{-}26)$$

where r = ripple factor (dimensionless)
$V_{r(\text{rms})}$ = rms value of the ripple voltage
V_{DC} = average dc level

Very often, the ripple factor is expressed as a percentage:

$$\%\, r = \frac{V_{r(\text{rms})}}{V_{\text{DC}}} \times 100\% \qquad (13\text{-}27)$$

the *ideal* ripple factor is *0%*.

EXAMPLE 13-10

Given the power supply circuit shown in Fig. 13-26(c), assume that its load voltage waveform appears as that indicated in Fig. 13-27 with $V_{dcm} = 15$ V and $V_{r(p\text{-}p)} = 1$ V. Determine the required WVDC rating for the capacitor. Also calculate $V_{dc(min)}$, V_{DC}, and the % r.

SOLUTION Since the peak voltage across the capacitor is (V_{dcm}) 15 V, its working voltage rating should be at *least* that given by Eq. 13-22.

$$\text{WVDC} = 1.2V_{dcm} = (1.2)(15\text{ V}) = 18\text{ V}$$

A standard value of WVDC is 25 V, which should be more than adequate. The minimum dc voltage is given by Eq. 13-23,

$$V_{dc(min)} = V_{dcm} - V_{r(p\text{-}p)}$$
$$= 15\text{ V} - 1\text{ V} = 14\text{ V}$$

The dc average may be found from Eq. 13-24,

$$V_{DC} = V_{dcm} - \frac{V_{r(p\text{-}p)}}{2} = 15\text{ V} - \frac{1\text{ V}}{2} = 14.5\text{ V}$$

To find the % r we must first find $V_{r(rms)}$. From Eq. 13-25,

$$V_{r(rms)} = \frac{V_{r(p\text{-}p)}}{2\sqrt{3}} = \frac{1\text{ V}}{2\sqrt{3}} = 0.289\text{ V rms}$$

and from Eq. 13-27,

$$\%\, r = \frac{V_{r(rms)}}{V_{DC}} \times 100\% = \frac{0.289\text{ V rms}}{14.5\text{ V}} \times 100\% = 1.99\%$$ ■

13-12 Light-Loading Constraint

To analyze (and design) simple capacitor filters and capacitive-input filters, we need to determine V_m, V_{dcm}, $V_{r(p\text{-}p)}$, $V_{r(rms)}$, and V_{DC} in terms of I_{DC} and C. We develop an equation for $V_{r(p\text{-}p)}$ in Section 13-13. To minimize the ripple and keep the mathematics as straightforward as possible, we must ensure that the filter is *lightly loaded*.

If a simple capacitor filter is lightly loaded, the average dc voltage will be very close to V_{dcm}, and the triangular ripple approximation will be valid.

If filter is *not* lightly loaded, ripple noise can cause problems in the load. At the very least, our approximate analysis technique (developed in Section 13-13) is invalid and the mathematics can become extremely involved.

For light loading, the *minimum* value of V_{DC} is given by

$$\text{For light loading:} \quad V_{DC} \geq 0.9V_{dcm} \tag{13-28}$$

13-13 Ripple Voltage Equation

If we develop an equation for the peak-to-peak ripple voltage ($V_{r(\text{p-p})}$), for a lightly loaded filter capacitor, the analysis of power supply filters becomes extremely straightforward. A capacitor opposes a change in its terminal voltage with respect to time. Using calculus notation, we have the mathematical definition provided by

$$i_c = C\frac{dv_c}{dt} \tag{13-29}$$

where i_c = capacitor current
C = capacitance
v_c = voltage across the capacitor
d/dt = calculus notation for the change in one quantity with respect to a change in time

We will not employ Eq. 13-29 directly. However, it is important that we are familiar with its meaning. In differential calculus statements, such as Eq. 13-29, the change in voltage is defined to be infinitesimally small (approaching zero) with respect to an infinitesimally small change in time. If we allow the change in voltage and time to be small but finitely large, we may use the delta (Δ) notation illustrated by

$$i_c \simeq C\frac{\Delta v_C}{\Delta t} \tag{13-30}$$

Simply stated, the current (effectively) through a capacitor is directly proportional to the capacitance, and the rate at which the voltage across the capacitor changes. Based on Eq. 13-30, if the current drawn from a charged capacitor is a *constant*, it will be discharged linearly with respect to time. Therefore, its terminal voltage will also decrease *linearly* with time. Consider Fig. 13-31(a). The voltage across the capacitor decreases from V_0 to V_1 linearly from time t_0 to t_1, assuming that I_C is constant.

We have essentially the same situation in Fig. 13-31(b). The filter capacitor is discharged by I_{DC}. Observe that the magnitude of the capacitor's change in voltage is $V_{r(\text{p-p})}$ during a time interval equal to $T/2$. By drawing on Eq. 13-30, we may derive an equation for $V_{r(\text{p-p})}$.

$$I_{\text{DC}} = C\frac{V_{r(\text{p-p})}}{T/2} = 2C\frac{V_{r(\text{p-p})}}{T}$$

Substituting in $f = 1/T$, we have

$$I_{\text{DC}} = 2fCV_{r(\text{p-p})}$$

and solving for $V_{r(\text{p-p})}$ gives us

$$V_{r(\text{p-p})} = \frac{I_{\text{DC}}}{2fC} \tag{13-31}$$

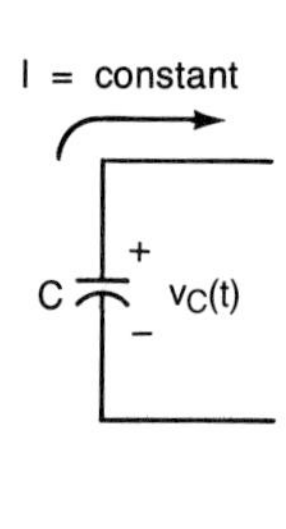

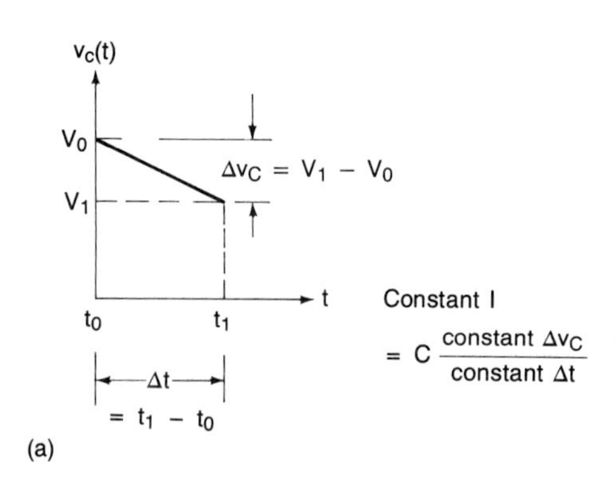

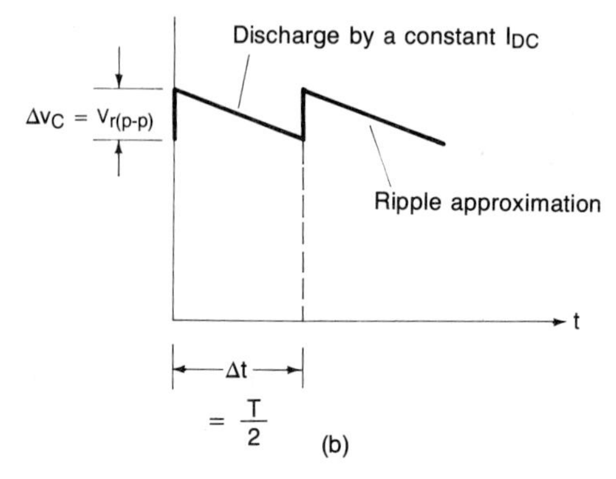

FIGURE 13-31 Linear discharge of a capacitor: (a) capacitor is discharged linearly by a constant current; (b) triangular ripple waveform approximation implies that the filter capacitor is discharged by a constant current.

If we let the ripple frequency be denoted f_r, we arrive at

$$V_{r(\text{p-p})} = \frac{I_{\text{DC}}}{f_r C} \tag{13-32}$$

where $V_{r(\text{p-p})}$ = peak-to-peak ripple voltage across a lightly loaded filter capacitor
I_{DC} = dc load current
f_r = ripple frequency
C = filter capacitance

If the line frequency is 60 Hz and full-wave rectification is used, f_r is 120 Hz. If we are designing a power supply to be used in a foreign country, the line frequency may be 50 Hz. In this case a full-wave rectifier would produce an f_r of 100 Hz.

EXAMPLE 13-11

Given the dc power supply shown in Fig. 13-32(a), calculate the dc voltage across the load and the ripple factor. Make a sketch of the voltage waveform across the load.

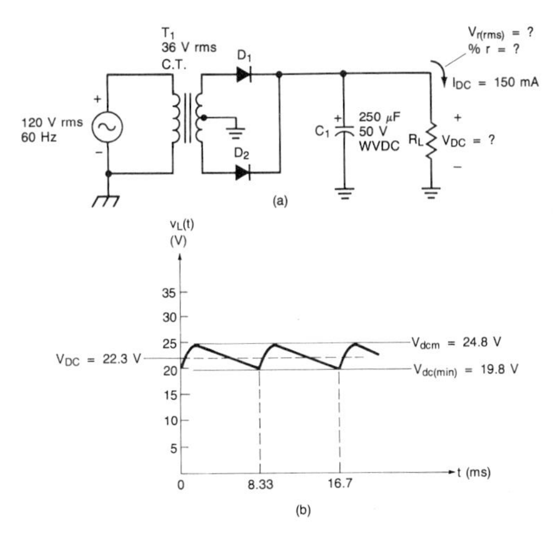

FIGURE 13-32 Circuit and solution for Example 13-11: (a) circuit; (b) solution.

SOLUTION First, we recognize that we have a full-wave rectifier using a center-tapped transformer. Since the transformer secondary voltage is 36 V rms, the voltage across one-half of it is

$$V_2 = 18 \text{ V rms}$$

and from Fig. 13-6(a), we may find the peak secondary voltage.

$$V_m = \sqrt{2}\, V_2 = (1.414)(18 \text{ V rms}) = 25.5 \text{ V}$$

The peak rectified voltage is

$$V_{\text{dcm}} = V_m - 0.7 \text{ V} = 25.5 \text{ V} - 0.7 \text{ V} = 24.8 \text{ V}$$

Since the ripple frequency is 120 Hz, the peak-to-peak ripple voltage across R_L may be found from Eq. 13-32.

$$V_{r(\text{p-p})} = \frac{I_{\text{DC}}}{f_r C_1} = \frac{150 \text{ mA}}{(120 \text{ Hz})(250\ \mu\text{F})} = 5 \text{ V p-p}$$

The rms value of the ripple may be determined by using Eq. 13-25.

$$V_{r(\text{rms})} = \frac{V_{r(\text{p-p})}}{2\sqrt{3}} = \frac{5 \text{ V p-p}}{2\sqrt{3}} = 1.44 \text{ V rms}$$

The dc voltage across the load is given by Eq. 13-24.

$$V_{\text{DC}} = V_{\text{dcm}} - \frac{V_{r(\text{p-p})}}{2} = 24.8 \text{ V} - \frac{5 \text{ V}}{2} = 22.3 \text{ V}$$

The minimum instantaneous dc load voltage is found by using Eq. 13-23.

$$V_{\text{dc(min)}} = V_{\text{dcm}} - V_{r(\text{p-p})} = 24.8 \text{ V} - 5 \text{ V} = 19.8 \text{ V}$$

The ripple factor ($\% \ r$) is arrived at by using Eq. 13-27.

$$\% \ r = \frac{V_{r(\text{rms})}}{V_{\text{DC}}} \times 100\% = \frac{1.44 \text{ V rms}}{22.3 \text{ V}} \times 100\% = 6.46\%$$

As a final check, we employ Eq. 13-28 to verify that we have a lightly loaded filter.

$$V_{DC} \geq 0.9V_{dcm}$$
$$22.3\text{ V} \geq (0.9)(24.8\text{ V})$$
$$22.3\text{ V} \geq 22.3\text{ V}$$

Therefore, our results are valid. The load voltage waveform has been indicated in Fig. 13-32(b). ■

13-14 Rectifier Average and Peak Repetitive Currents

Now that we have investigated the basic operation of the simple capacitor and capacitive-input filters, we need to consider the *average* $I_{(AV)}$ and the *peak* $I_{(PK)}$ currents that occur. This background will help us understand the diode ratings on data sheets (and their derating), the transformer secondary current requirements, and capacitor ripple current ratings.

As we saw previously, each diode in a full-wave rectifier using a center-tapped transformer and each diode *pair* in a full-wave bridge rectifier must pass an average forward current equal to *one-half* of the load current I_{DC}. Therefore, we can state

$$I_{(AV)} = 0.5\, I_{DC} \tag{13-33}$$

where $I_{(AV)}$ is the rectifier diode average forward current and I_{DC} is the dc load current.

Manufacturers typically denote the average forward current *rating* of rectifier diodes as I_o. Obviously, the average forward current *rating* I_o should be greater than the *actual* average forward current $I_{(AV)}$ that the diodes must pass.

To provide a 20% safety margin, we can state the condition given by Eq. 13-34.

$$I_o \geq 1.2\, I_{(AV)} = (1.2)\,(0.5)I_{DC} = 0.6\, I_{DC}$$

$$I_o \geq 1.2\, I_{(AV)} = 0.6\, I_{DC} \tag{13-34}$$

where I_o = average forward current rating
$I_{(AV)}$ = actual forward current
I_{DC} = dc load current

In addition to the average forward current, rectifier diodes must also pass much larger *peak repetitive surge currents* $I_{(PK)}$ which occur each time the filter capacitor is recharged. It is extremely important that we be able to ascertain the magnitude of these peak repetitive surge currents. Very often, manufacturers reduce (derate) the I_o rating on the basis of the $I_{(PK)}/I_{(AV)}$ ratio.

In Fig. 13-33(a) we see that the repetitive surge currents may be monitored with an oscilloscope by using a small, current-sensing resistor. The repetitive surge currents occur each time the filter capacitor is recharged. The *larger* the filter capacitor

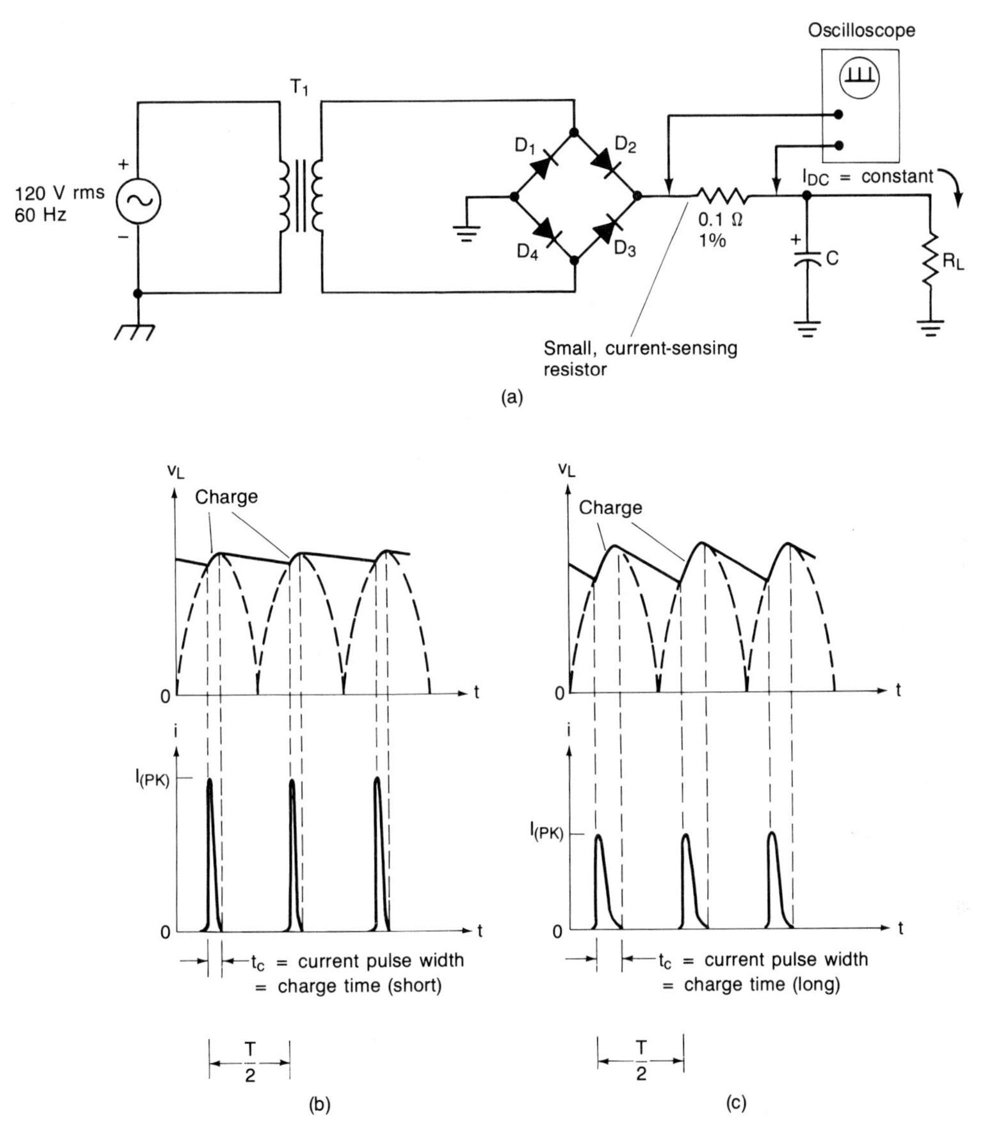

FIGURE 13-33 Repetitive diode surge currents $I_{(PK)}$: (a) measurement of the diode repetitive surge current; (b) large C; (c) small C.

is, the *less* the recharge time interval will be and the *greater* the peak repetitive surge current $I_{(PK)}$ will be [see Fig. 13-33(b)]. The converse is also true, as shown in Fig. 13-33(c). The dc load current is assumed to be the same in both Fig. 13-33(b) and (c).

To determine the peak value of the surge currents, we shall assume that they are approximately *triangular* in shape (see Fig. 13-34). If we assume that the charge into the capacitor replaces the charge removed by the load, the current pulses must

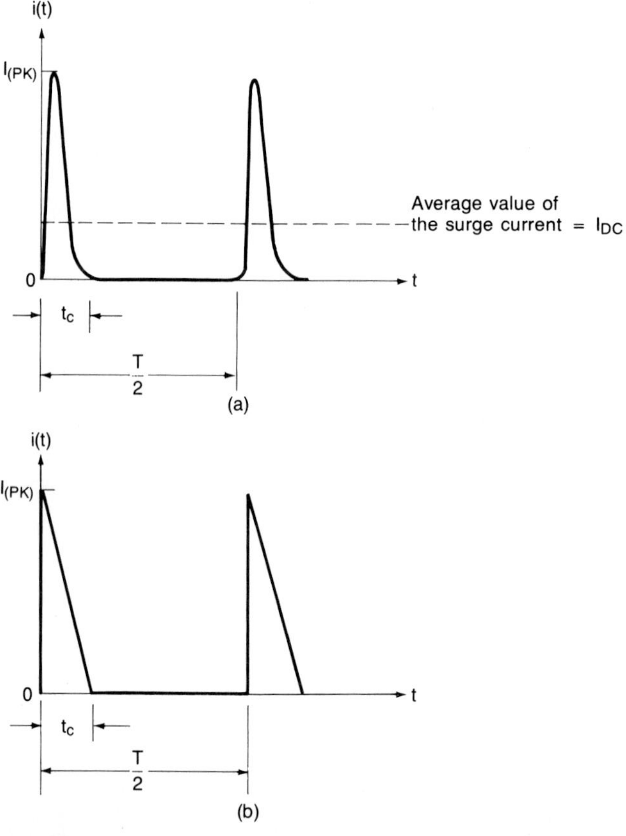

FIGURE 13-34 **Approximation of the repetitive surge current waveform: (a) actual repetitive surge current waveform; (b) triangular approximation.**

have an average value equal to I_{DC}. From our work in Section 13-3, it should be clear that for Fig. 13-34,

$$I_{DC} = \frac{1}{2}\frac{I_{(PK)}\,t_c}{T/2} = \frac{I_{(PK)}\,t_c}{T}$$

and substituting $f = 1/T$, we have

$$I_{DC} = I_{(PK)}\,t_c f$$

Solving for $I_{(PK)}$ gives us

$$\boxed{I_{(PK)} = \frac{I_{DC}}{f\,t_c}} \tag{13-35}$$

The 60-Hz line frequency yields

$$\text{For } f = 60 \text{ Hz:} \qquad I_{(PK)} = \frac{I_{DC}}{60t_c} \tag{13-36}$$

where $I_{(PK)}$ = peak repetitive surge current
I_{DC} = dc load current
t_c = charge time

The only mystery at this point is: How do we determine the charge time t_c? To solve for t_c, we must use Fig. 13-35 and a little trigonometry. The filter capacitor charges from time t_1 to t_2 as illustrated in Fig. 13-35. The charge time is simply the difference between these two points in time.

$$t_c = t_2 - t_1$$

From Fig. 13-35 we can see that $t_2 = T/4$; hence

$$t_c = \frac{T}{4} - t_1 \tag{13-37}$$

The instant in time at which the capacitor starts to charge is t_1, and it can be found by using the equation for a sine wave. Hence

$$V_m \sin \omega t = v$$

and in terms of Fig. 13-35, we can write

$$V_{dcm} \sin \omega t_1 = V_{dc(min)} \tag{13-38}$$

$$\sin \omega t_1 = \frac{V_{dc(min)}}{V_{dcm}}$$

FIGURE 13-35 **Determining the charge time t_c.**

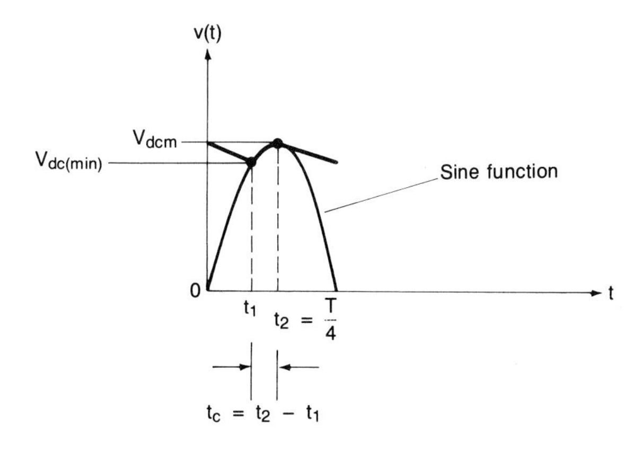

We may solve for t_1 after taking the arcsin or inverse sine of both sides of Eq. 13-38.

$$t_1 = \frac{1}{\omega} \sin^{-1} \frac{V_{dc(min)}}{V_{dcm}}$$

and substituting in $2\pi f = \omega$, we have

$$t_1 = \frac{1}{2\pi f} \sin^{-1} \frac{V_{dc(min)}}{V_{dcm}} \tag{13-39}$$

Assuming that $f = 60$ Hz, and substituting Eq. 13-39 into Eq. 13-37, we arrive at

$$\text{For } f = 60\text{ Hz:} \quad t_c = 4.17\text{ ms} - 2.65 \sin^{-1} \frac{V_{dc(min)}}{V_{dcm}}\text{ ms} \tag{13-40}$$

where t_c = capacitor charge time

$\frac{V_{dc(min)}}{V_{dc}}$ = argument of the arcsin (radians)

EXAMPLE 13-12

For the power supply of Fig. 13-32(a) and Example 13-11, find $I_{(AV)}$, $I_{(PK)}$, and the $I_{(PK)}/I_{(AV)}$ ratio for the rectifier diodes.

SOLUTION From Fig. 13-32(a) we see that I_{DC} is 150 mA, and from Eq. 13-33,

$$I_{(AV)} = 0.5\, I_{DC} = (0.5)\,(150\text{ mA}) = 75\text{ mA}$$

From our work in Example 13-11, we know that V_{dcm} is 24.8 V and $V_{dc(min)}$ is 19.8 V. The capacitor charge time is given by Eq. 13-40.

$$\begin{aligned} t_c &= 4.17\text{ ms} - 2.65 \sin^{-1} \frac{V_{dc(min)}}{V_{dcm}}\text{ ms} \\ &= 4.17\text{ ms} - 2.65 \sin^{-1} \frac{19.8\text{ V}}{24.8\text{ V}}\text{ ms} \\ &= 4.17\text{ ms} - (2.65)\,(0.9246)\text{ ms} = 1.72\text{ ms} \end{aligned}$$

and from Eq. 13-36,

$$I_{(PK)} = \frac{I_{DC}}{60 t_c} = \frac{150\text{ mA}}{(60)\,(1.72\text{ ms})} = 1.46\text{ A}$$

The $I_{(PK)}$ to $I_{(AV)}$ ratio is

$$\frac{I_{(PK)}}{I_{(AV)}} = \frac{1.46\text{ A}}{75\text{ mA}} = 19.5$$

■

From the result in Example 13-12 we can see that the peak currents through the diodes are about *twenty* times their average currents. It should be noted that the surge current has a frequency of 120 Hz. However, *because the diodes alternate in conduction, their individual surge currents have a frequency of* 60 *Hz.*

13-15 Nonrepetitive Diode Surge Current

In addition to the repetitive surge currents through the rectifier diodes, the diodes must also pass a large, *nonrepetitive surge current* when the power supply is first turned on. This nonrepetitive surge current $I_{(SURGE)}$ occurs because the filter capacitor is initially *uncharged*. Since the capacitor has zero volts across it, and a capacitor opposes a change in voltage when power is first applied, it will appear as a *short circuit*.

When power is first applied, the instantaneous voltage across the transformer's secondary could be as large as V_m. Only one diode in a full-wave rectifier using a center-tapped secondary, or one pair of diodes in a full-wave bridge, will be conducting. If we model the transformer secondary as a voltage source, we arrive at the equivalent circuit shown in Fig. 13-36(a). The second diode model shown in the phantom box appears in the equivalent circuit of a full-wave bridge rectifier. Resistance R_2 represents the losses of the secondary winding.

Since the capacitor is initially a short, both it and R_L are not present in the circuit. Note that the resistances of the diodes are significant in this analysis and are essentially equal to the bulk resistances.

The circuit shown in Fig. 13-36(b) is a simplified equivalent. By using it and employing Ohm's law, it becomes a relatively easy matter to find the magnitude of the initial surge current. This is given by

$$I_{(SURGE)} = \frac{V_{dcm}}{R_s} \tag{13-41}$$

where $I_{(SURGE)}$ = initial surge current
V_{dcm} = peak secondary voltage minus the appropriate number of diode drops
R_s = transformer secondary resistance plus the total diode bulk resistance

EXAMPLE 13-13

Compute the nonrepetitive surge currents for the power supply given in Fig. 13-32. (This is the circuit that was used in Examples 13-11 and 13-12.) The transformer secondary resistance is 1 Ω and the rectifiers each have a bulk resistance of 0.149 Ω.

SOLUTION From Example 13-11, V_{dcm} is 24.8 V and

$$R_s = R_2 + r_B$$
$$= 1\ \Omega + 0.149\ \Omega = 1.149\ \Omega$$

Applying Eq. 13-41, we obtain

$$I_{(SURGE)} = \frac{24.8\ \text{V}}{1.149\ \Omega} = 21.6\ \text{A}$$

■

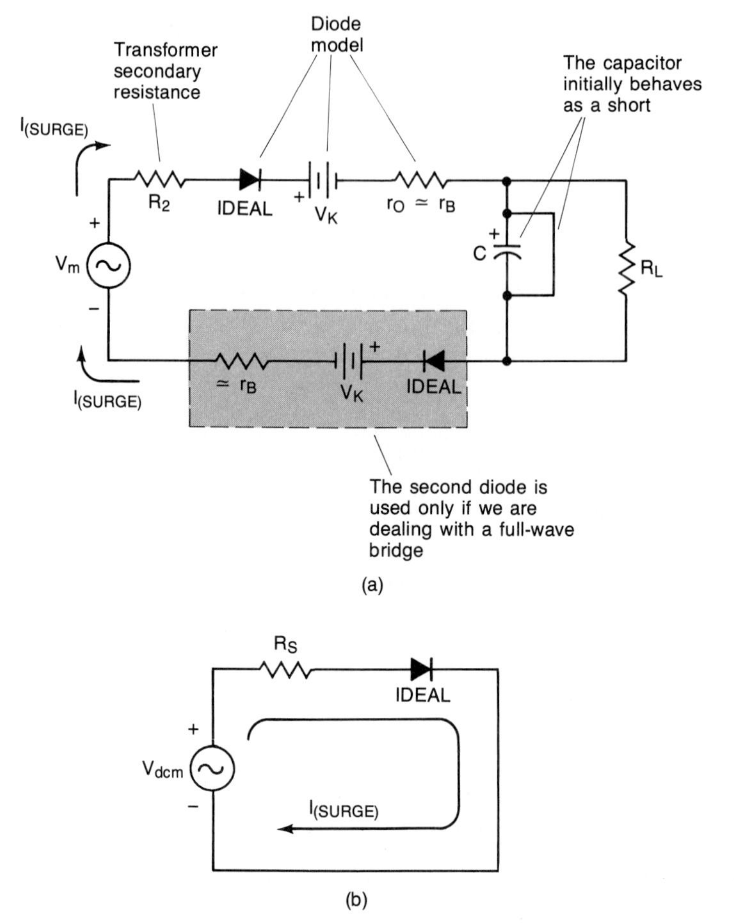

FIGURE 13-36 **Equivalent circuits for the determination of $I_{(SURGE)}$: (a) surge current circuit; (b) simplified equivalent circuit.**

Obviously, the initial surge current can be an extremely large value. Although the surge current has been designated $I_{(SURGE)}$, manufacturers typically denote a rectifier's nonrepetitive surge current *rating* as I_{FSM}.

As a rough but conservative guide, a 20% safety factor produces

$$\boxed{I_{FSM} \geq 1.2 I_{(SURGE)}} \tag{13-42}$$

where I_{FSM} is the diode's peak nonrepetitive surge current rating and $I_{(SURGE)}$ is the actual nonrepetitive surge current.

If the transformer's secondary resistance (R_2) is unknown, the $I_{(SURGE)}$ startup transients may be monitored by using the circuit shown in Fig. 13-33. (A storage oscilloscope should be used.) If the initial surge currents appear to be excessive (Eq. 13-42 is violated), it may be necessary to install a current-limiting resistor permanently in place of the current-sensing resistor. Its value may be computed from

$$R = \frac{V_{\text{dcm}}}{I_{\text{FSM}}} \tag{13-43}$$

13-16 Capacitor Ripple Current

Thus far we have seen that the electrolytic capacitor across the output of the rectifier determines the magnitude of the ripple voltage, $V_{r(\text{rms})}$, and the average dc level, V_{DC}. Further, we have also assumed that the capacitor is discharged by a constant current I_{DC}, and recharged every 8.33 ms by a sharp current pulse $I_{(\text{PK})}$ [see Fig. 13-37(a)]. An approximation of this current waveform is shown in Fig. 13-37(b). This waveform is the *ripple current waveform* $i_r(t)$.

The ripple current waveform in Fig. 13-37 flows "through" the capacitor and can cause internal heating. Specifically, the rms value of this ripple current determines its heating effect.

For lightly loaded capacitors with an I_{DC} below 1 A, capacitor heating is usually not a major concern. However, as I_{DC} becomes larger, the size of the capacitor must be increased to keep the ripple factor low. At higher values of capacitance, the ratio of a capacitor's outside surface area to its volume becomes smaller. For these capacitors, internal heating may become a problem.

Manufacturers typically specify the rms current rating for filter capacitors at a given ambient temperature. To verify that a given capacitor will work, we must first find the rms value of the ripple current and then consult its data sheet.

Using the techniques described in Section 13-3, and from Fig. 13-37(b), we can arrive at an equation for the rms value of the ripple current $I_{r(\text{rms})}$.

For $f = 60$ Hz:

$$I_{r(\text{rms})} = \sqrt{120t_c\left[\frac{I^2_{(\text{PK})}}{3} - I^2_{\text{DC}}\right] + I^2_{\text{DC}}} \tag{13-44}$$

where $I_{r(\text{rms})}$ is the rms value of the capacitor's ripple current. (All of the other terms have been defined previously.)

EXAMPLE 13-14

The power supply circuit of Fig. 13-32(a) has been redrawn in Fig. 13-38(a) together with a summary of our previous work. Determine the rms ripple current rating for the filter capacitor. Also check to see if its WVDC rating is adequate.

SOLUTION From Fig. 13-38(b) and Eq. 13-44, we can find $I_{r(\text{rms})}$.

$$I_{r(\text{rms})} = \sqrt{120t_c\left[\frac{I^2_{(\text{PK})}}{3} - I^2_{\text{DC}}\right] + I^2_{\text{DC}}}$$

$$= \sqrt{120(1.72\text{ ms})\left[\frac{(1.46\text{ A})^2}{3} - (150\text{ mA})^2\right] + (150\text{ mA})^2}$$

$$= 406\text{ mA rms}$$

Therefore, the rms current rating for the filter capacitor should be at least 20% more than 406 mA rms (487 mA rms).

From Fig. 13-38(a) and Eq. 13-22,

$$\text{WVDC} \geq 1.2\,V_{\text{dcm}} = (1.2)\,(24.8\text{V}) = 29.8\text{V}$$

Therefore, the 50-V WVDC rating appears to be much more than adequate. ■

FIGURE 13-37 Ripple current waveform "through" the capacitor: (a) actual current waveform; (b) our approximation.

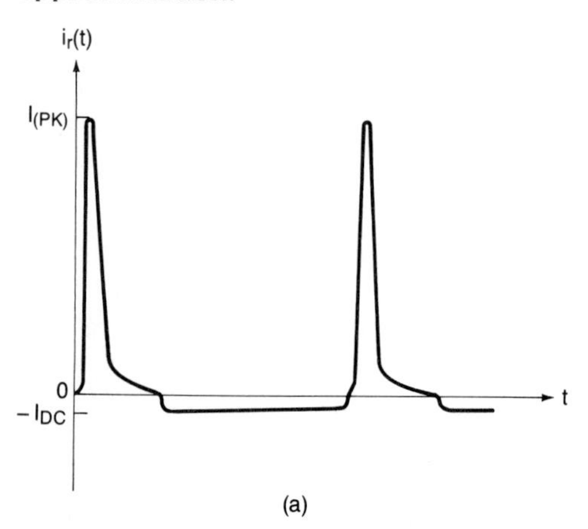

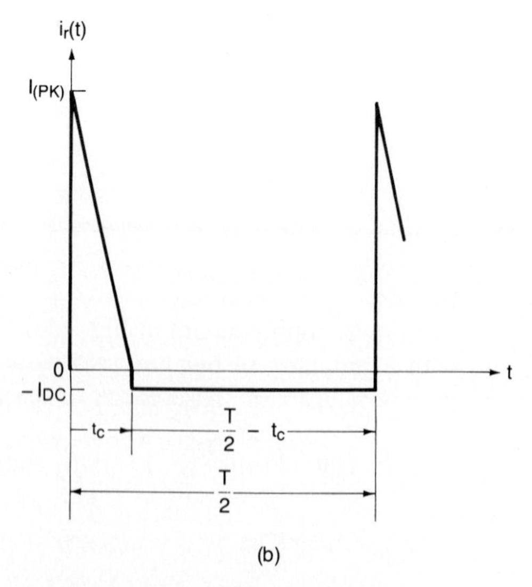

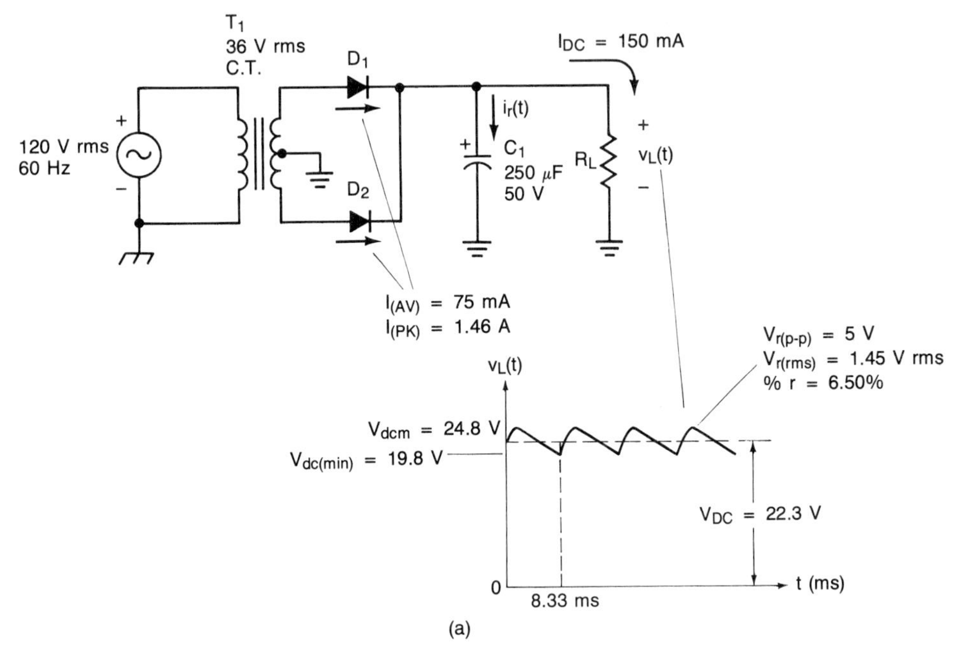

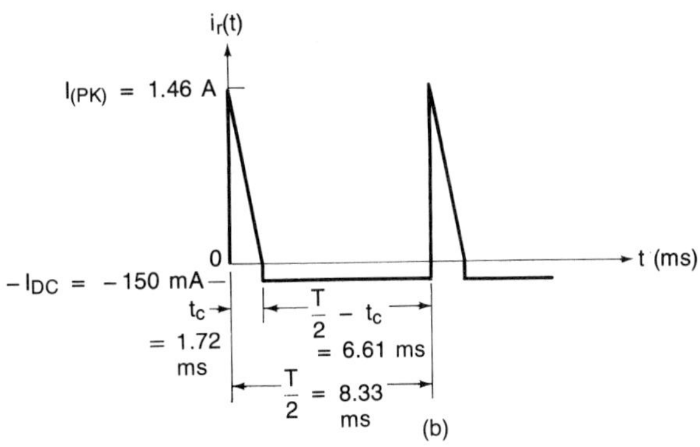

FIGURE 13-38 Circuit and waveform for Example 13-14: (a) power supply circuit and a summary of our analysis; (b) ripple current $i_r(t)$ waveform.

13-17 Transformer Secondary Current

As we have seen, when a capacitor is placed across the output of a rectifier, it is recharged by spikes of current. This current is supplied by the transformer's secondary. In the analysis and design of power supplies, it is often necessary to determine the required transformer secondary rms current rating.

A rigorous mathematical analysis to ascertain the required secondary current rating proves to be rather complex because of the nonsinusoidal secondary current. To circumvent the mathematics, we shall use Table 13-1.

TABLE 13-1
Transformer Secondary RMS Current Ratings for Full-Wave Rectifiers with Capacitive-Input Filters

Rectifier Type	Required Secondary RMS Current Rating
Full-wave using a center-tapped transformer	$1.18 I_{DC}$
Full-wave bridge	$1.65 I_{DC}$
Dual-complementary full-wave	$1.65 I_{DC}$

EXAMPLE 13-15

Determine the secondary current rating required for the transformer in the power supply shown in Fig. 13-38(a).

SOLUTION Since the rectifier is full-wave using a center-tapped transformer, and I_{DC} is 150 mA, we can see from Table 13-1 that the secondary current I_2 is

$$I_2 = 1.18 I_{DC} = (1.18)(150 \text{ mA}) = 177 \text{ mA rms}$$ ■

EXAMPLE 13-16

Determine the required secondary current for the dual-complementary full-wave rectifier given in Fig. 13-39.

SOLUTION From Table 13-1 and Fig. 13-39,

$$I_2 = 1.65 I_{DC} = (1.65)(500 \text{ mA}) = 825 \text{ mA rms}$$ ■

FIGURE 13-39 Circuit for Example 13-16.

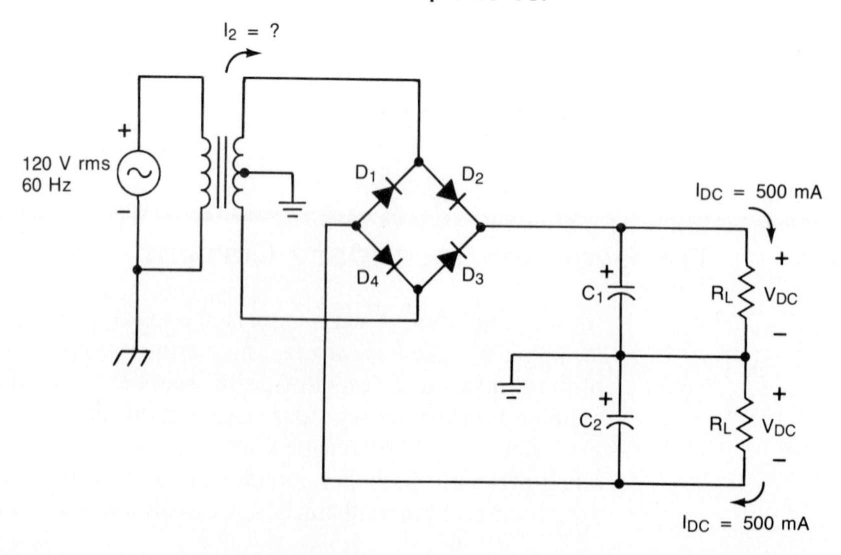

13-18 *RC* Pi Filters

The simple capacitor filter and a three-terminal (integrated circuit) voltage regulator (which is discussed in Chapter 14) will satisfy many power supply requirements. However, to provide improved filtering with its attendant reduction in ripple factor, an additional resistor and capacitor may be added to the simple capacitor filter [see Fig. 13-40(a)]. This capacitive-input filter configuration is referred to as an *RC pi* (π) *filter*. From Fig. 13-40(b) we can see that the term "pi" is derived from the basic shape of the filter.

The analysis of this filter circuit may be simplified by employing five basic assumptions.

1. The input capacitor C_1 provides most of the initial filtering, and the effects of R_1 and C_2 on it are negligible.
2. For light loading, the ripple voltage across C_1 is approximately triangular. Therefore, *all* of our equations for V_{DC}, $V_{dc(min)}$, $V_{r(p\text{-}p)}$, $V_{r(rms)}$, and so on, may be applied to it.
3. Since the ripple voltage across C_1 is nonsinusoidal, it contains high-frequency components. Specifically, it is a combination of sine waves, and their frequencies are even multiples (harmonics) of 60 Hz (i.e., 120 Hz, 240 Hz, 480 Hz, etc.).
4. Because of the filtering action of R_1 and C_2, much of the harmonic content of the ripple voltage across C_2 is bypassed to ground. Consequently, the ac ripple voltage across C_2 is approximately sinusoidal with a frequency of 120 Hz. Therefore, sinusoidal steady-state analysis may be used.
5. By the principles of superposition, separate dc and ac analyses of the effects of R_1 and C_2 may be conducted.

These concepts are illustrated in Figs. 13-41 and 13-42. Be sure to observe the notation that has been defined.

FIGURE 13-40 RC pi filter.

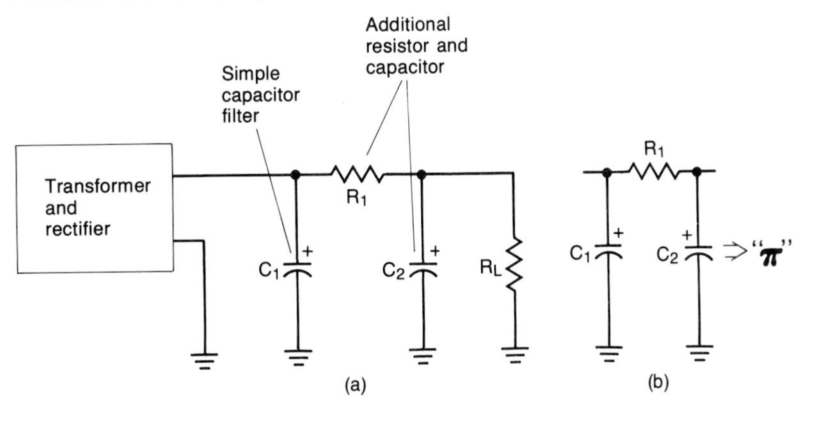

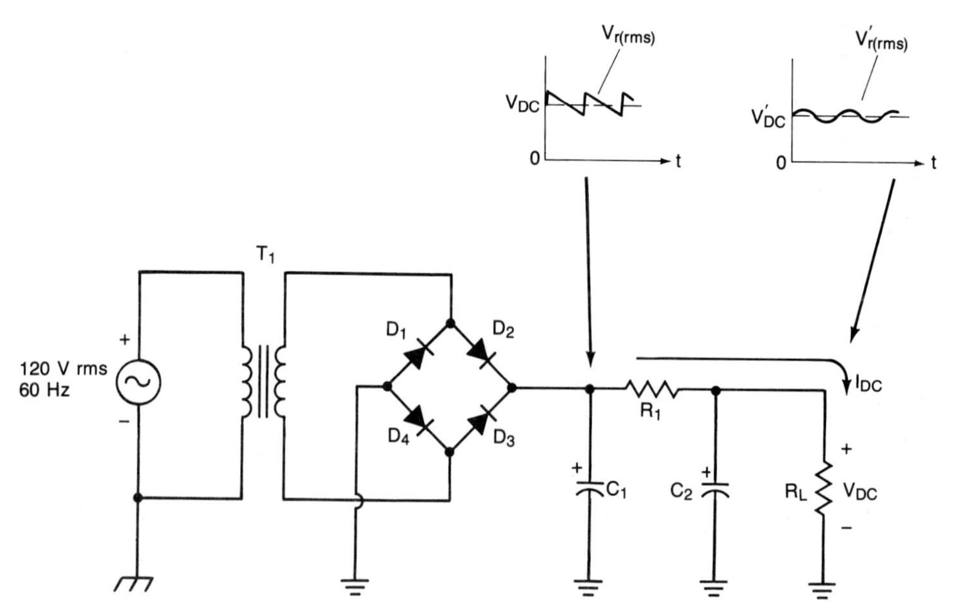

FIGURE 13-41 The ripple voltage across C_1 is triangular, and sinusoidal across C_2.

FIGURE 13-42 RC pi filter equivalent circuits: (a) RC pi filter; (b) equivalent circuit; (c) dc equivalent circuit; (d) ac (ripple) equivalent circuit.

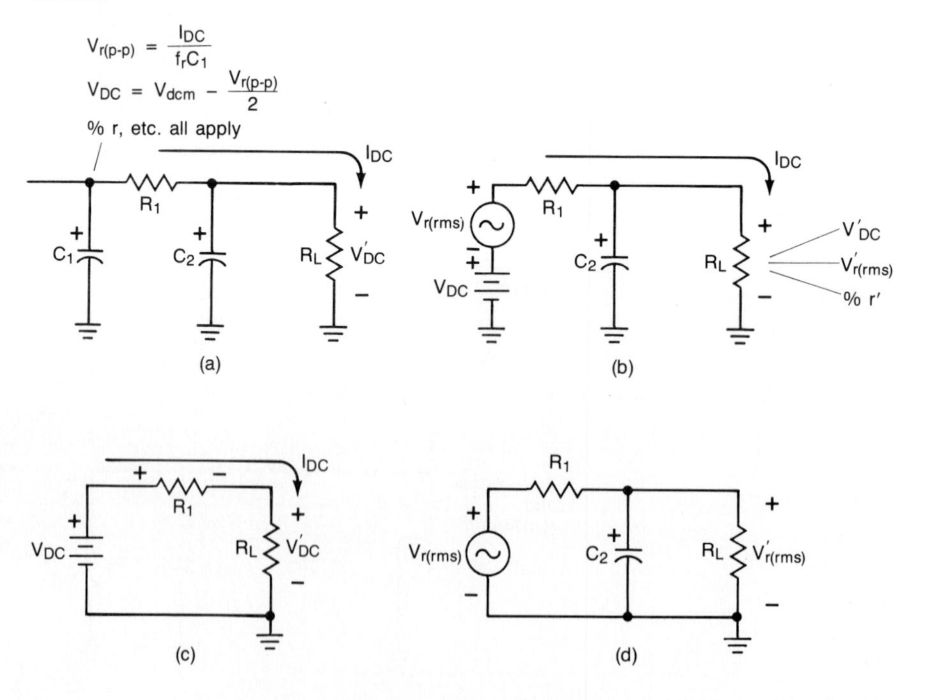

In Fig. 13-42(a) we emphasize that our equations for the simple capacitor filter may be applied to C_1 of the *RC* pi filter. The ac ripple voltage and the dc level across C_1 may be modeled as two voltage sources in series [see Fig. 13-42(b)].

The dc equivalent circuit is shown in Fig. 13-42(c). Capacitor C_2 is not shown since it is an open circuit to dc. We can see that R_1 and R_L are in series as far as dc is concerned. Therefore, by Ohm's law and Kirchhoff's voltage law, we arrive at

$$V'_{DC} = V_{DC} - I_{DC}R_1 \tag{13-45}$$

The ac analysis of Fig. 13-42(d) can be simplified by constraining the capacitive reactance of C_2 to be much smaller than R_L.

$$X_{C2} \leq 0.1R_L \tag{13-46}$$

This is the usual case and poses no serious limitations. Since X_{C2} is so small compared to R_L, the parallel impedance of C_2 and R_L is approximately equal to the capacitive reactance of C_2. Therefore, the ac equivalent circuit may be simplified to that shown in Fig. 13-43.

To find the *magnitude* of the ac ripple voltage across C_2 (and R_L), we may apply ac voltage division.

$$V'_{r(\text{rms})} \simeq \frac{X_{C2}}{\sqrt{R_1^2 + X_{C2}^2}} V_{r(\text{rms})} \tag{13-47}$$

EXAMPLE 13-17

Compute V'_{DC}, $V'_{r(\text{rms})}$, and $\% \ r'$ for the circuit given in Fig. 13-44. Observe that we have merely added R_1 and C_2 to the circuit given in Fig. 13-38(a).

SOLUTION From our previous work, we know that V_{DC} is 22.3 V, and $V_{r(\text{rms})}$ is 1.45 V rms. These voltages are developed across C_1 as indicated in Fig. 13-44. By Eq. 13-45,

$$V'_{DC} = V_{DC} - I_{DC}R_1 = 22.3 \text{ V} - (150 \text{ mA})(10 \ \Omega) = 20.8 \text{ V}$$

The capacitive reactance of C_2 to the 120-Hz ripple is

$$X_{C2} = \frac{1}{2\pi f C_2} = \frac{1}{2\pi(120)(1500 \times 10^{-6})} = 0.884 \ \Omega$$

FIGURE 13-43 Simplified ac equivalent circuit.

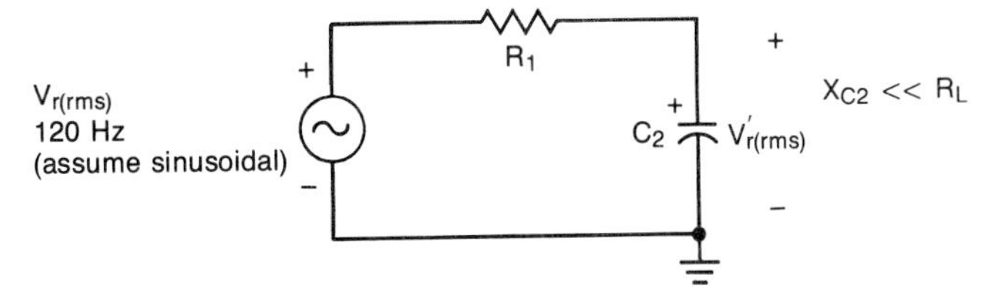

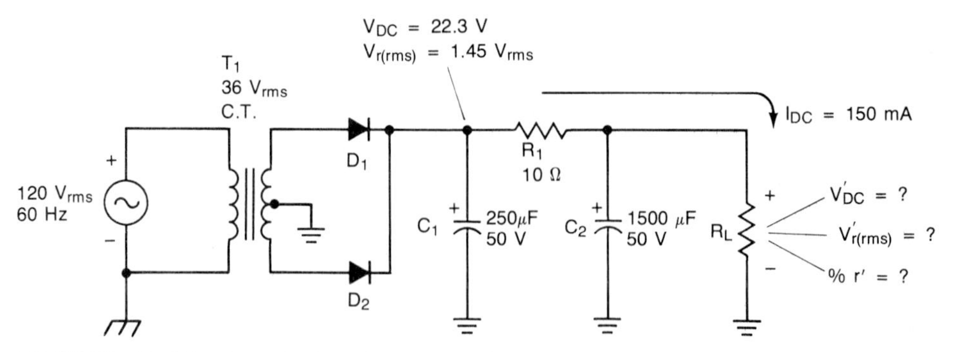

FIGURE 13-44 Circuit for Example 13-17.

and from Eq. 13-47,

$$V'_{r(\mathrm{rms})} \simeq \frac{X_{C2}}{\sqrt{R_1^{\,2} + X_{C2}^{\,2}}} V_{r(\mathrm{rms})}$$

$$= \frac{0.884\ \Omega}{\sqrt{(10\ \Omega)^2 + (0.884\ \Omega)^2}} (1.45\ \text{V rms}) = 0.128\ \text{V rms}$$

The $\% r'$ is found from Eq. 13-27.

$$\% r' = \frac{V'_{r(\mathrm{rms})}}{V'_{\mathrm{DC}}} \times 100\% = \frac{0.128\ \text{V rms}}{20.8\ \text{V}} \times 100\%$$

$$= 0.615\%$$

■

By careful inspection of the results of Example 13-17, we can see that the ripple voltage has been greatly reduced at the expense of a minor reduction in V_{DC}. The selection of R_1 and C_2 merits some discussion at this point.

In practice, we make R_1 as *small* as possible to minimize the dc voltage drop across it. In general, it should drop no more than $0.1V_{\mathrm{DC}}$. Hence

$$\boxed{R_1 \leq \frac{0.1V_{\mathrm{DC}}}{I_{\mathrm{DC}}}} \qquad (13\text{-}48)$$

In Fig. 13-44 the value of R_1 should be

$$R_1 \leq \frac{0.1V_{\mathrm{DC}}}{I_{\mathrm{DC}}} = \frac{(0.1)\,(22.3\ \Omega)}{150\ \text{mA}} = 14.9\ \Omega$$

Therefore, $R_1 = 10\ \Omega$ is a reasonable choice.

Generally, we want C_2 to be as *large* as possible. This keeps its capacitive reactance small to minimize the ac ripple developed across it. However, the larger C_2 is, the more expensive it will be, and its physical size is proportional to its value. Setting a moderate guideline, we have

$$X_{C2} \leq 0.1R_1$$

$$\frac{1}{2\pi f C_2} \leq 0.1R_1$$

Taking the reciprocal of both sides (and reversing the sense of the inequality), we may solve for C_2.

$$C_2 \geq \frac{1}{0.2\pi f R_1} \tag{13-49}$$

For Fig. 13-44 with $R_1 = 10\ \Omega$,

$$C_2 \geq \frac{1}{0.2\pi f R_1} = \frac{1}{0.2\pi(120)(10\ \Omega)} = 1326\ \mu\text{F}$$

Therefore, $C_2 = 1500\ \mu\text{F}$ also appears to be a reasonable choice.

As a final comment, we should note that because R_1 tends to limit current, the initial surge current $I_{(\text{SURGE})}$ and the peak repetitive surge current $I_{(\text{PK})}$ are *essentially* determined by C_1. As a crude approximation, we shall assume that they are determined solely by C_1.

13-19 Other Filter Configurations

The simple capacitor filter and the *RC* pi-filter configurations are the most commonly used. To minimize the dc voltage reduction produced by R_1 (Fig. 13-45) in the *RC* pi filter, the resistor may be replaced by an inductor or *choke*.

When a choke is used instead of a resistor, the filter is called an *LC pi filter* [see Fig. 13-45(a)]. A choke would seem to be the ideal choice. Its opposition to dc

FIGURE 13-45 Other filter configurations: (a) LC pi filter; (b) L-input filter.

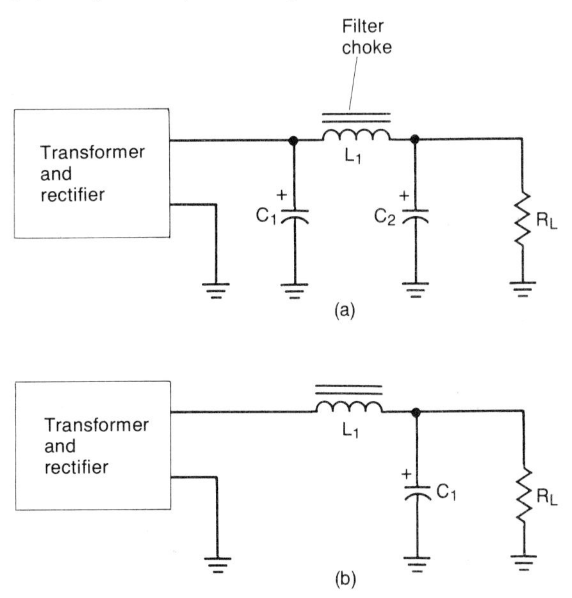

is simply the dc resistance of its coil windings, which can range from several ohms to a fraction of an ohm. Consequently, the dc voltage drop across it can be kept quite small.

However, its opposition to the ac ripple current flow is essentially its inductive reactance X_L. The inductive reactance is given by

$$X_L = 2\pi f L \tag{13-50}$$

where X_L = inductive reactance (ohms)
f = frequency (hertz)
L = inductance (henries)

Since the X_L of the choke can be much larger than its dc resistance, a choke can greatly attenuate the ac ripple while producing a minimal reduction in the dc voltage. Therefore, the *LC* pi filter shown in Fig. 13-45(a) can produce very small ripple factors when compared to the *RC* pi filter. However, the filter choke has several serious disadvantages.

Because the ripple frequency (120 Hz) is so low, a filter choke must have inductance values which range from hundreds of millihenries to several henries in order to possess a sufficiently large inductive reactance. Consequently, the filter choke tends to be physically large, heavy, and expensive for (60-Hz) line-operated power supplies.

The *L*-input filter [Fig. 13-45(b)] is another alternative to filtering the ac ripple. Observe that there is *no* filter capacitor across the output of the rectifier. As we have seen, when a capacitor is placed directly across the output of the rectifier, it charges to the peak value of the rectified voltage. Therefore, for light loading, the average dc voltage approaches the peak value. However, since the *L*-input filter does not incorporate such a capacitor, its average dc output voltage tends to be much *lower* (approximately $0.637V_{dcm}$).

In Section 13-14 we saw that when a capacitor is placed across the output of a rectifier, it is charged by very large, repetitive spikes of current. For power supplies that are required to provide several amperes of current, the peak repetitive surge currents can become excessive. Specifically, it may be difficult and/or expensive to secure rectifiers with adequate current ratings. In this case it may be necessary to use an *L*-input filter, despite the many disadvantages associated with the filter choke.

13-20 Rectifier Specifications

Specifications for diode rectifiers are established by the rectifier JEDEC (Joint Electron Device Engineering Council of the Electronic Industries Association) committee for registered part numbers (e.g., 1NXXXX). Nonregistered parts might *not* adhere to the JEDEC specification requirements. However, most manufacturers do use the JEDEC standards as guidelines.

The JEDEC specifications include a standard set of symbols for diode ratings and electrical characteristics. They also outline the methods and the test circuits to be used by manufacturers for evaluating rectifiers.

In this section we point out some of the more common specifications and the general features offered by most data sheets. In critical applications, the reader would be well advised to search out further information in the many manuals and application notes offered by the manufacturers. If further questions arise, it may be necessary to call the manufacturer and discuss your technical problems with an applications engineer.

In Fig. 13-46 we have one manufacturer's specifications for a series of low-power rectifiers. Specifications sheets generally cover five basic areas: (1) a general description and mechanical characteristics, (2) maximum ratings, (3) electrical characteristics, (4) derating information, and (5) dynamic information. In the discussion that follows, the reader should frequently refer to Fig. 13-46.

General Description and Mechanical Data

Most data sheets will list the device number at the top of the page and provide a short descriptive title. In Fig. 13-46 we see the JEDEC registered numbers 1N5391 through 1N5399. This series of diodes has an average forward current rating of 1.5 A and reverse voltage ratings ranging from 50 to 1000 V.

This series of diodes has a case style designated as 59-04, which conforms to JEDEC requirements. The specific dimensions are also provided. This information can be extremely useful when printed circuit (printed wiring) boards are being designed.

Maximum Ratings

Manufacturers will specify *absolute maximum ratings* for solid-state devices. These are the maximum voltages, currents, and temperatures that a device may experience without resulting in its degradation or failure. The voltage and current ratings are often specified for a *p-n* junction temperature of 25°C. Consequently, these ratings will have to be reduced (derated) at higher temperatures.

It is essential that these ratings are *never* exceeded. For good device reliability, these ratings should not even be approached. A 20% safety margin is generally regarded as a minimum.

V_{RRM} is the peak repetitive reverse voltage *rating*. It should be at least 20% larger than the *actual* peak repetitive voltage (V_{RM}). Similarly, a rectifier's average current rating (I_o) should be at least 20% greater than the actual forward current [$I_{(AV)}$] it must pass. More will be said about I_o when we consider derating.

Electrical Characteristics

Electrical characteristics typically describe such things as a diode's *V-I* characteristics, reverse leakage current, and temperature coefficients. Observe in Fig. 1 of Fig. 13-46 that the manufacturer has provided a *V-I* plot. Note that the current scale is *logarithmic*, rather than linear. For very large forward currents, we see a dramatic increase in the diode's forward voltage drop. This is due, in part, to the diode's lead

1N5391 thru 1N5399

Designers Data Sheet

"SURMETIC" RECTIFIERS

. . . subminiature size, axial lead-mounted rectifiers for general-purpose, low-power applications.

Designers Data for "Worst Case" Conditions

The Designers Data Sheets permit the design of most circuits entirely from the information presented. Limits curves—representing boundaries on device characteristics—are given to facilitate "worst-case" design.

LEAD-MOUNTED SILICON RECTIFIERS

50–1000 VOLTS DIFFUSED JUNCTION

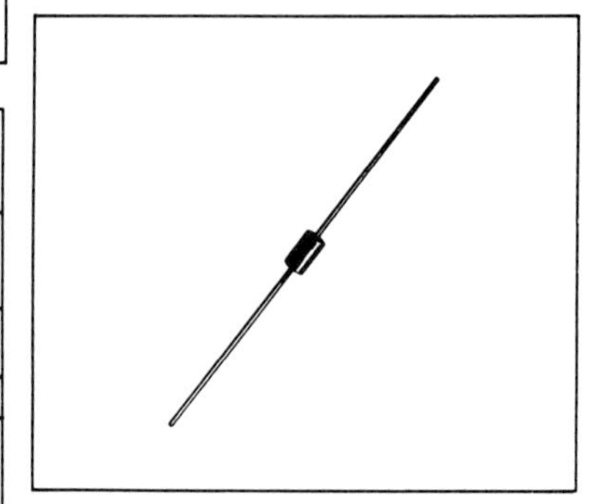

*MAXIMUM RATINGS

Rating	Symbol	1N5391	1N5392	1N5393	1N5395	1N5397	1N5398	1N5399	Unit
Peak Repetitive Reverse Voltage Working Peak Reverse Voltage DC Blocking Voltage	V_{RRM} V_{RWM} V_R	50	100	200	400	600	800	1000	Volts
Nonrepetitive Peak Reverse Voltage (Halfwave, Single Phase, 60 Hz)	V_{RSM}	100	200	300	525	800	1000	1200	Volts
RMS Reverse Voltage	$V_{R(RMS)}$	35	70	140	280	420	560	700	Volts
Average Rectified Forward Current (Single Phase, Resistive Load, 60 Hz, $T_L = 70°C$, 1/2" From Body)	I_O	← 1.5 →							Amp
Nonrepetitive Peak Surge Current (Surge Applied at Rated Load Conditions, See Figure 2)	I_{FSM}	← 50 (for 1 cycle) →							Amp
Storage Temperature Range	T_{stg}	← −65 to +175 →							°C
Operating Temperature Range	T_L	← −65 to +170 →							°C
DC Blocking Voltage Temperature	T_L	← 150 →							°C

*ELECTRICAL CHARACTERISTICS

Characteristic and Conditions	Symbol	Typ	Max	Unit
Maximum Instantaneous Forward Voltage Drop (i_F = 4.7 Amp Peak, $T_L = 170°C$, 1/2 Inch Leads)	v_F	–	1.4	Volts
Maximum Reverse Current (Rated dc Voltage) ($T_L = 150°C$)	I_R	250	300	μA
Maximum Full-Cycle Average Reverse Current (1) (I_O = 1.5 Amp, $T_L = 70°C$, 1/2 Inch Leads)	$I_{R(AV)}$	–	300	μA

*Indicates JEDEC Registered Data.

NOTE 1: Measured in a single-phase, halfwave circuit such as shown in Figure 6.25 of EIA RS-282, November 1963. Operated at rated load conditions I_O = 1.5 A, $V_r = V_{RWM}$, $T_L = 70°C$.

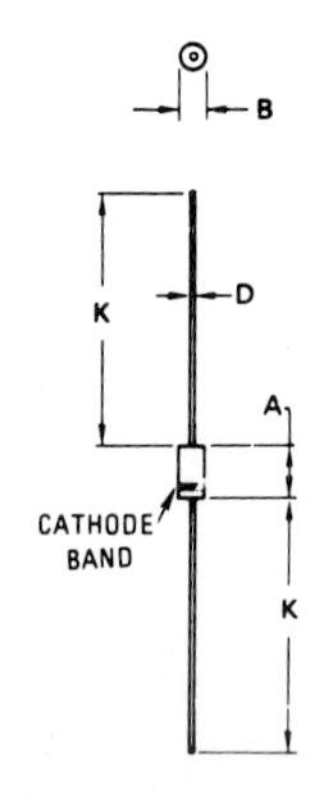

DIM	MILLIMETERS MIN	MILLIMETERS MAX	INCHES MIN	INCHES MAX
A	5.97	6.60	0.235	0.260
B	2.79	3.05	0.110	0.120
D	0.76	0.86	0.030	0.034
K	27.94	–	1.100	–

CASE 59-04

Dimensions Within JEDEC DO-15 Outline.

MECHANICAL CHARACTERISTICS

CASE: Void free, transfer molded

MAXIMUM LEAD TEMPERATURE FOR SOLDERING PURPOSES: 240°C, 1/8" from case for 10 seconds at 5 lbs. tension

FINISH: All external surfaces are corrosion-resistant, leads are readily solderable

POLARITY: Cathode indicated by color band

WEIGHT: 0.40 grams (approximately)

FIGURE 13-46 Specifications for the 1N5390 series of rectifier diodes (reprinted with permission of Motorola, Inc.).

FIGURE 1 – FORWARD VOLTAGE

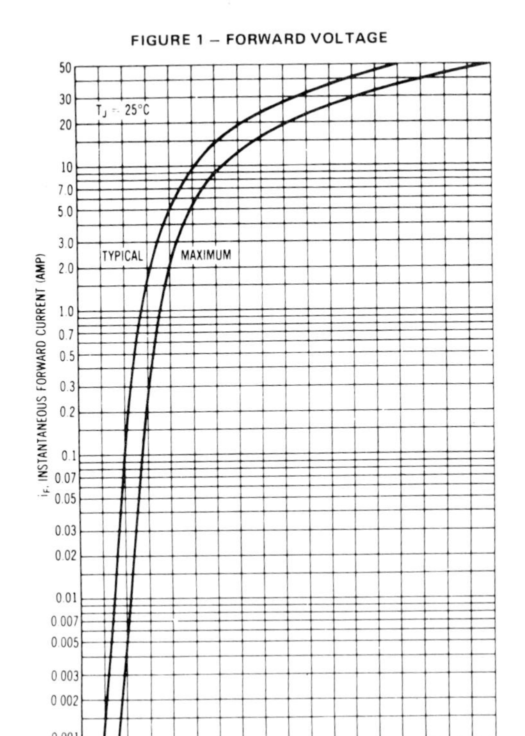

FIGURE 2 – MAXIMUM NONREPETITIVE SURGE CURRENT

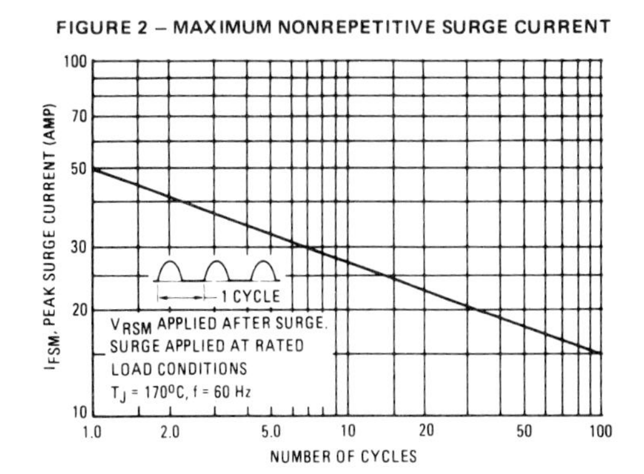

FIGURE 3 – FORWARD VOLTAGE TEMPERATURE COEFFICIENT

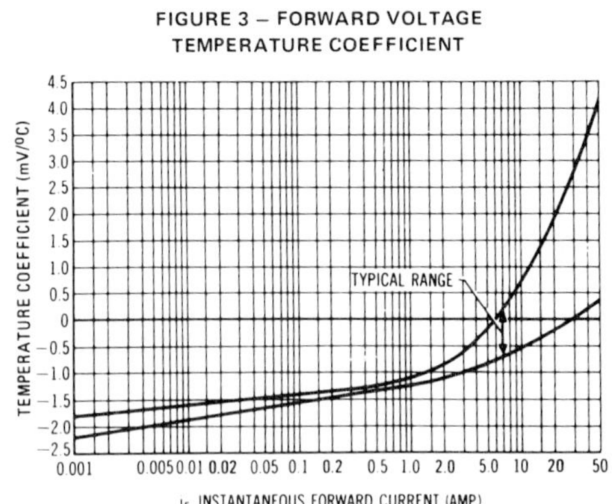
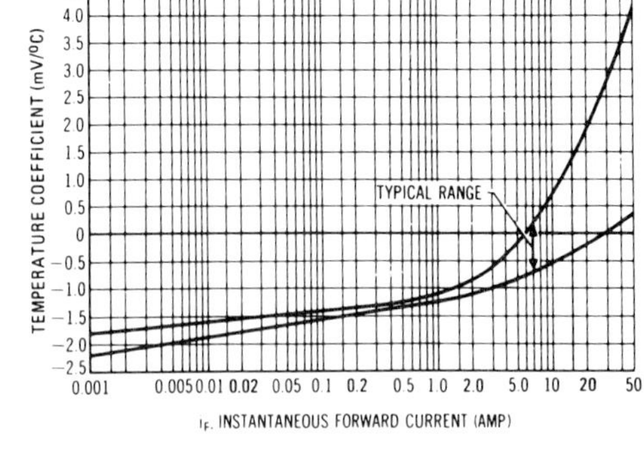

FIGURE 4 – TYPICAL TRANSIENT THERMAL RESISTANCE

DUTY CYCLE, $D = t_p/t_1$
PEAK POWER, P_{pk}, is peak of an equivalent square power pulse

$\Delta T_{JL} = P_{pk} [R_{\theta JL(\infty)} \cdot D + (1-D) \cdot R_{\theta JL(t_1+t_p)} + R_{\theta JL(t_p)} - R_{\theta JL(t_1)}]$
where ΔT_{JL} = increase in junction temperature above the lead temperature.

$R_{\theta JL(t)}$ = value of transient thermal resistance at time t, i.e.
$R_{\theta JL(t_1+t_p)}$ = value of $R_{\theta JL(t)}$ at time t_1+t_p
$R_{\theta JL(t_p)}$ = value of $R_{\theta JL(t)}$ at end of pulse width t_p
$R_{\theta JL(t_1)}$ = value of $R_{\theta JL(t)}$ at time t_1

L = 1″
L = 1/2″
L = 1/32″

$R_{\theta JL}$, JUNCTION-TO-LEAD TRANSIENT THERMAL RESISTANCE (°C/W)

t, TIME (ms)

The temperature of the lead should be measured using a thermocouple placed on the lead as close as possible to the tie point. The thermal mass connected to the tie point is normally large enough so that it will not significantly respond to heat surges generated in the diode as a result of pulsed operation once steady-state conditions are achieved. Using the measured value of T_L, the junction temperature may be determined by:

$$T_J = T_L + \Delta T_{JL}.$$

FIGURE 13-46 ***(continued)***

FIGURE 5 – FORWARD POWER DISSIPATION

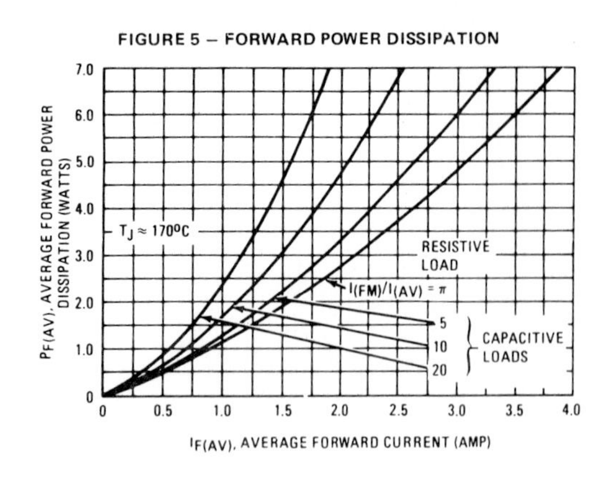

FIGURE 6 – EFFECT OF LEAD LENGTHS, RESISTIVE LOAD

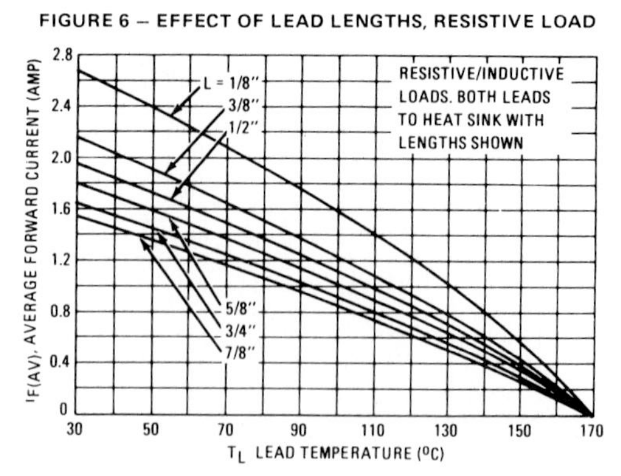

FIGURE 7 – 1/2″ LEAD LENGTH, VARIOUS LOADS

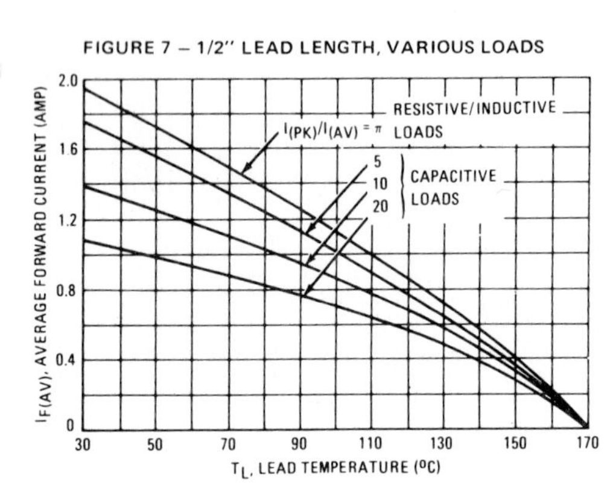

FIGURE 8 – PRINTED CIRCUIT BOARD MOUNTING, VARIOUS LOADS

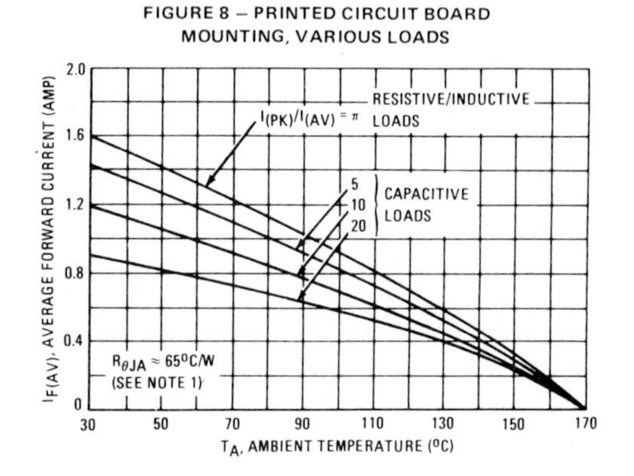

FIGURE 9 – STEADY-STATE THERMAL RESISTANCE

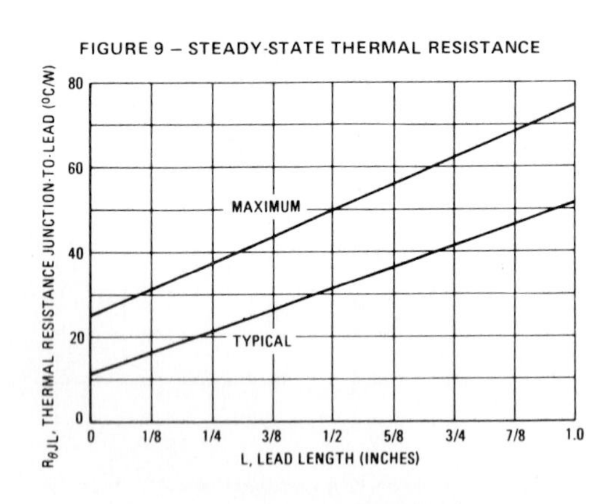

NOTE 1

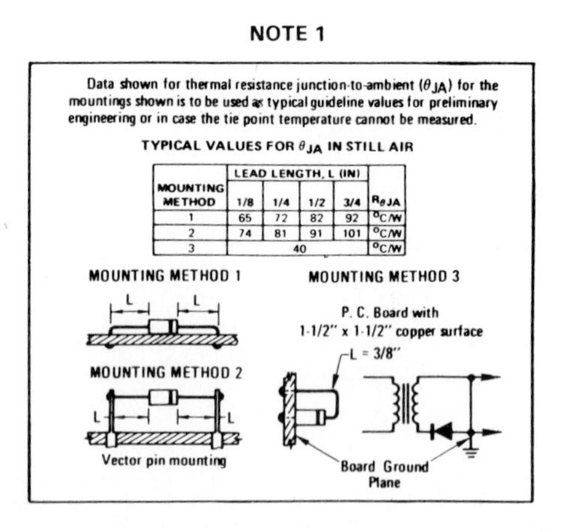

Data shown for thermal resistance junction-to-ambient (θ_{JA}) for the mountings shown is to be used as typical guideline values for preliminary engineering or in case the tie point temperature cannot be measured.

TYPICAL VALUES FOR θ_{JA} IN STILL AIR

MOUNTING METHOD	LEAD LENGTH, L (IN)				$R_{\theta JA}$
	1/8	1/4	1/2	3/4	
1	65	72	82	92	°C/W
2	74	81	91	101	°C/W
3	40				°C/W

FIGURE 13-46 *(continued)*

FIGURE 10 – FORWARD RECOVERY TIME

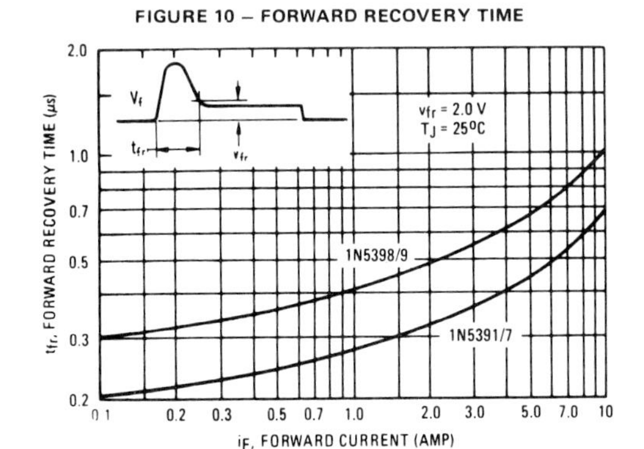

FIGURE 11 – REVERSE RECOVERY TIME

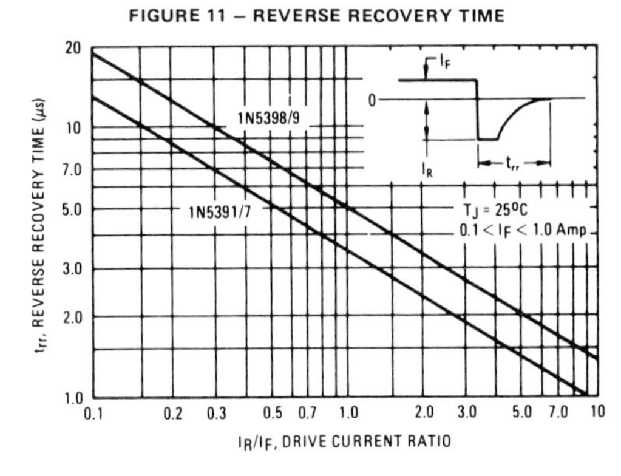

FIGURE 12 – JUNCTION CAPACITANCE

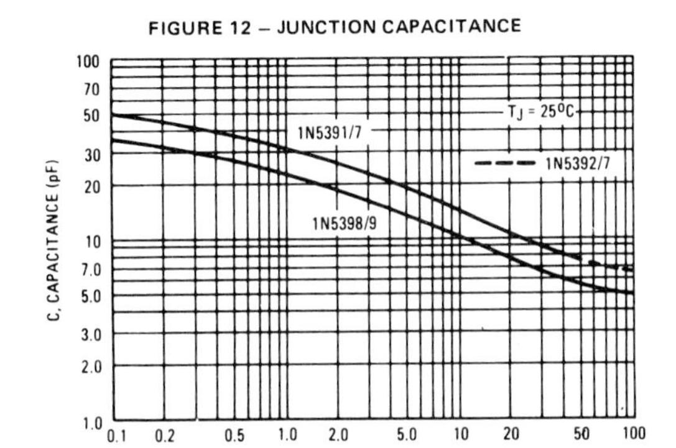

FIGURE 13 – RECTIFICATION WAVEFORM EFFICIENCY FOR SINE WAVE

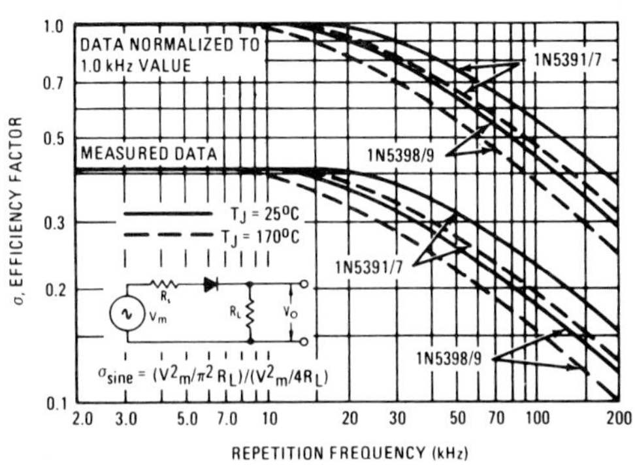

FIGURE 14 – RECTIFICATION WAVEFORM EFFICIENCY FOR SQUARE WAVE

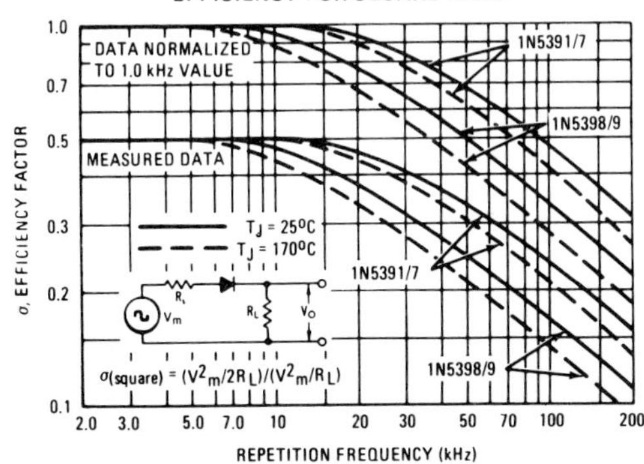

RECTIFIER EFFICIENCY NOTE

The rectification efficiency factor σ shown in Figures 13 and 14 was calculated using the formula:

$$\sigma = \frac{P_{dc}}{P_{rms}} = \frac{\dfrac{V^2{}_{O(dc)}}{R_L}}{\dfrac{V^2{}_{O(rms)}}{R_L}} \bullet 100\% = \frac{V^2{}_{O(dc)}}{V^2{}_{O(ac)} + V^2{}_{O(dc)}} \bullet 100\% \quad (1)$$

For a sine wave input $V_m \sin(\omega t)$ to the diode, assumed lossless, the maximum theoretical efficiency factor becomes 40%; for a square wave input of amplitude V_m, the efficiency factor becomes 50%. (A full wave circuit has twice these efficiencies).

As the frequency of the input signal is increased, the reverse recovery time of the diode (Figure 11) becomes significant, resulting in an increasing ac voltage component across R_L which is opposite in polarity to the forward current thereby reducing the value of the efficiency factor σ, as shown in Figures 13 and 14.

It should be emphasized that Figures 13 and 14 show waveform efficiency only; they do not account for diode losses. Data was obtained by measuring the ac component of V_O with a true rms voltmeter and the dc component with a dc voltmeter. The data was used in Equation 1 to obtain points for the Figures.

FIGURE 13-46 *(continued)*

resistance and a nonlinear interaction between the diode's bulk resistance, and its junction potential.

Derating Information

Figure 2 and Figs. 4 through 9 in Fig. 13-46 illustrate how the specifications are to be derated. In Figs. 5, 7, and 8 we see curves for *capacitive loads*. This is to be interpreted as rectifiers that are driving capacitive-input filters such as the simple capacitor filter or the *RC* pi filter. To find $I_{(PK)}/I_{(AV)}$, we take the ratio of the peak repetitive surge current to the dc average current through the rectifier (refer back to Example 13-12).

Since the heat built up inside the diode is removed via the diode leads, they should be kept as short as possible. Ideally, they should be terminated on relatively large metallic surfaces which can act as heat radiators [see Note 1 on Fig. 13-46]. (Thermal resistance was discussed in Chapter 12.)

Dynamic Information

These characteristics (Figs. 10 through 14) are not of great concern in line-operated (60-Hz) dc power supply rectification. However, in the design of switching regulators and dc-to-dc converters, they are of paramount importance. A rectifier's forward recovery is controlled by its diffusion capacitance (refer to Chapter 3).

EXAMPLE 13-18

Given the power supply shown in Fig. 13-44, select the most appropriate diode from those given in Fig. 13-46.

SOLUTION The peak repetitive reverse voltage may be computed ($V_m = 25.46$ V).

$$V_{RM} = 2V_m - 0.7\text{ V} = 50.9\text{ V} - 0.7\text{ V} = 50.2\text{ V}$$

The V_{RRM} rating should be at least 20% greater,

$$V_{RRM} \geq 1.2V_{RM} = (1.2)\,(50.2\text{ V}) = 60.2\text{ V}$$

The 1N5392 diode has a V_{RRM} rating of 100 V. Therefore, it appears to be the best choice. Our previous work in Example 13-12 showed that $I_{(AV)}$ is 75 mA. Consequently, the 1.5 A I_o rating appears to be more than adequate. However, to ensure that it is sufficient, we must derate it. In Example 13-12 we also determined that $I_{(PK)}/I_{(AV)}$ is 19.2, or approximately 20. If we assume an ambient temperature of 30°C, and the diode is to be mounted with $\frac{1}{8}$-in. leads on a printed circuit board (see Note 1 in Fig. 13-46), we may use Fig. 8 in Fig. 13-46. For $I_{(PK)}/I_{(AV)} = 20$, we see that

$$I_o = 0.9\text{ A}$$

Since this far exceeds our 75-mA requirement, the 1N5392 appears to be the optimum choice. ■

13-21 Diode Assemblies

Because of the widespread use of the full-wave bridge rectifier in both the full-wave bridge rectifier circuits, and in dual-complementary full-wave rectifiers, many manufacturers offer *full-wave bridge diode assemblies*. Although the bridge assemblies do not have JEDEC registered part numbers, their specifications generally follow JEDEC guidelines. Typical bridge assemblies have been illustrated in Fig. 13-47.

13-22 Power Supply Failure Modes

In Section 3-24 we introduced the typical failure modes for rectifier and zener diodes. Obviously, we may apply the results of that discussion to the rectifier and zener diodes used in our power supplies.

By using careful (and conservative) analysis and design techniques, we can minimize the opportunities for a component failure. However, no electronic circuit or system can be 100% reliable. Consequently, from a design and troubleshooting aspect, we need to be able to anticipate probable component failures.

Typically, the most likely components to fail in a power supply are the filter capacitors. The most common failure modes for a capacitor are for it to short out or have excessive leakage current. When the leakage current through a filter capacitor increases, the ripple voltage across it will increase and the dc voltage level will be reduced. If the leakage current becomes excessive, the ripple may become large enough to produce electrical noise in the load. If the load is an audio amplifier system, we may hear a 120-Hz hum.

Although not as likely to fail, the low-voltage transformers found in the majority of the electronic power supplies also exhibit common failure modes. The most probable failures include an open primary, an open secondary, and a primary-to-secondary short circuit. Obviously, the most destructive fault is the primary-to-secondary short circuit.

FIGURE 13-47 Diode bridge assembly: (a) electrical schematic; (b) typical packages.

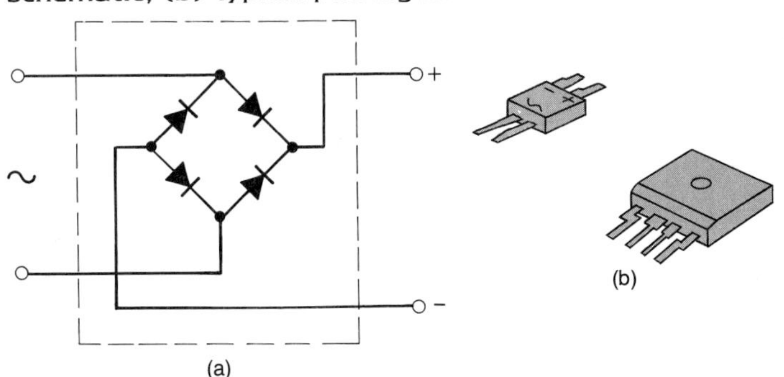

PROBLEMS

Drill, Derivations, and Definitions

Section 13-1

13-1. List the four functional blocks of a line-operated dc power supply.

13-2. List four basic functions that a power supply transformer can serve.

13-3. List two characteristics of a power supply rectifier. What is its primary function?

13-4. List two characteristics of a power supply filter.

13-5. List four attributes of a power supply regulator.

13-6. In your own words, explain the term ''load'' as it relates to dc power supplies.

Section 13-2

13-7. Given an electronic system such as that shown in Fig. 13-1, if I_1 is 2 mA, I_2 is 5 mA, and I_3 is 800 mA, compute the total current drawn from the power supply. Determine the equivalent static load resistance.

13-8. Repeat Prob. 13-7 for an I_1 of 1.5 mA, an I_2 of 10 mA, and an I_3 of 250 mA.

Section 13-3

13-9. What is the definition of the average or dc value of a waveform?

13-10. Given the waveforms shown in Fig. 13-48, calculate their average or dc values. Note that their peak-to-peak values are all 10 V.

13-11. Compute the dc values of the waveforms shown in Fig. 13-49.

13-12. Find the dc values of the waveforms given in Fig. 13-50.

13-13. What is the definition of the rms or effective value of a waveform?

13-14. Calculate the rms values of the waveforms given in Fig. 13-48(a) and (c).

13-15. Calculate the rms values of the waveforms given in Fig. 13-49.

Section 13-4

13-16. Given the circuit shown in Fig. 13-11(a), calculate the secondary voltage, the secondary current, and the primary current. The secondary has 100 turns. All other values are unchanged.

13-17. Repeat Prob. 13-16 if the secondary has 200 turns.

13-18. Explain why the metallic chassis and/or cabinet of a line-operated piece of electronic

FIGURE 13-48 Waveforms for Probs. 13-10 and 13-14.

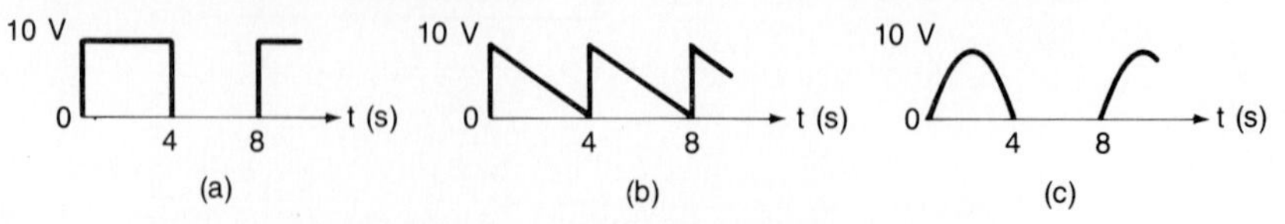

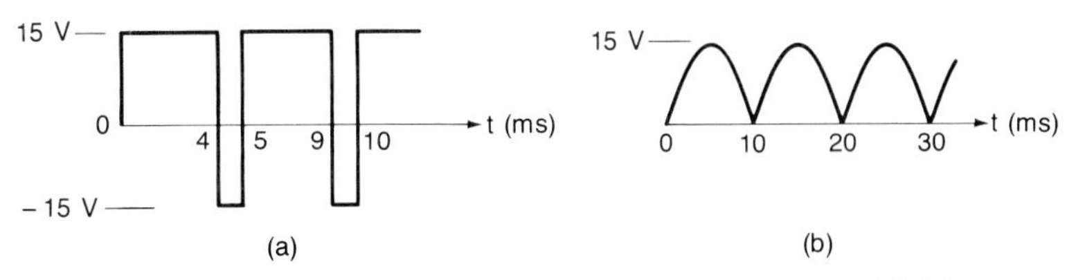

FIGURE 13-49 Waveforms for Probs. 13-11 and 13-15.

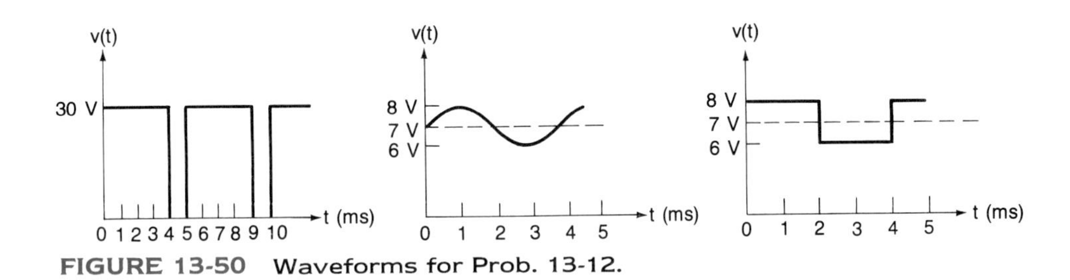

FIGURE 13-50 Waveforms for Prob. 13-12.

equipment should be tied to the earth-ground conductor of a three-wire 120-V rms electrical distribution system.

13-19. What is meant by "isolation" from earth ground? Why is it desirable?

Section 13-5

13-20. Given the half-wave rectifier circuit shown in Fig. 13-14(a), assume that V_2 is 34 V rms. Draw the primary, secondary, and diode voltage waveforms. (Use the standard reference directions for the diode's waveform.) Calculate and label their peak values and the time axes.

13-21. In Prob. 13-20, what are (a) the average voltage across the load, (b) the average diode and load current, and (c) the peak reverse voltage across the diode?

13-22. Given the full-wave rectifier shown in Fig. 13-18(a), assume that the transformer has a 34-V-rms center-tapped secondary. Draw the voltage waveforms for V_2, diodes D_1 and D_2, and across R_L. Also draw the current waveforms for diodes D_1 and D_2 and R_L. (Use the standard reference directions for the diode waveforms.) Calculate and label their peak values and the time axes.

13-23. In Prob. 13-22, what are the peak and average currents through the diodes? What are the peak reverse voltages across the diodes?

13-24. Repeat Prob. 13-22 for the circuit given in Fig. 13-19. Assume that the transformer has a 34-V-rms center-tapped secondary, and that R_L is 200 Ω.

Section 13-7

13-25. Assume that the transformer in Fig. 13-23(a) has a 34-V-rms secondary. Make a sketch of the voltage waveforms across the secondary, across diodes D_1 and D_2, and across the load. Also sketch the current waveforms through the diodes and the load. (Use the standard reference directions for the diode waveforms.) Label the peak values and the time axes.

13-26. For Prob. 13-25, compute the average voltage across the load, and the average current through it. Calculate the average currents through the diodes.

Section 13-8

13-27. Assume that the transformer in Fig. 13-24(b) has a 25.2 V rms center-tapped secondary. Calculate the average voltages across the load resistors with respect to ground. Indicate their polarities. (*Hint*: Remember that each half of the circuit is essentially a full-wave rectifier using a center-tapped transformer.)

Section 13-9

13-28. In your own words, list four basic attributes of aluminum electrolytic capacitors. Why are disk ceramic capacitors occasionally placed in parallel with aluminum electrolytic capacitors?

13-29. List three advantages of tantalum electrolytic capacitors over aluminum electrolytic capacitors. What is their primary disadvantage?

13-30. What is the WVDC rating of a capacitor? What may happen if it is exceeded?

13-31. Given the power supply in Fig. 13-51(a), calculate the minimum WVDC rating of the capacitor. The secondary voltage is 25.2 V rms.

13-32. Repeat Prob. 13-31 for Fig. 13-51(b). The transformer has the same secondary voltage.

FIGURE 13-51 Full-wave rectifier problems.

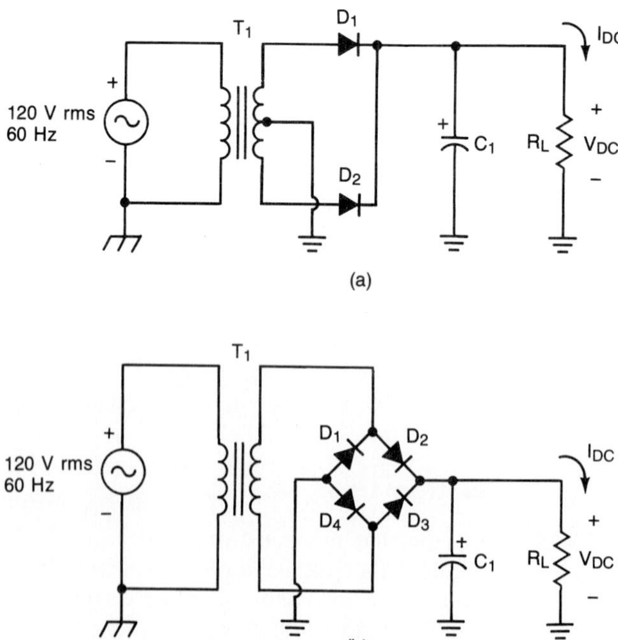

Section 13-11

13-33. Given the dc power supply in Fig. 13-26(c), assume that the load voltage waveform appears as indicated in Fig. 13-27. If V_{dcm} is 25 V and $V_{r(p\text{-}p)}$ is 0.5 V, compute (a) the minimum WVDC rating of the filter capacitor, (b) $V_{dc(min)}$, (c) V_{DC}, and (d) the % r.

13-34. Repeat Prob. 13-33 if V_{dcm} is 27.9 V and $V_{r(p\text{-}p)}$ is 3 V.

Section 13-13

13-35. Given the power supply shown in Fig. 13-51(a), assume that the secondary is 12.6 V rms, I_{DC} is 250 mA, and C_1 is a 2000-μF capacitor. Compute V_{DC}, $V_{dc(min)}$, $V_{r(p\text{-}p)}$, $V_{r(rms)}$, and % r. Make a sketch of the load voltage waveform.

13-36. Repeat Prob. 13-35 if the secondary is 25.2 V rms.

13-37. Repeat Prob. 13-35 for Fig. 13-51(b). The secondary is 12.6 V rms and C_1 is 2200 μF.

13-38. Repeat Prob. 13-35 for Fig 13-51(b). The secondary is 25.2 V rms and C_1 is 5000 μF.

Section 13-14

13-39. Given the circuit shown in Fig. 13-51(a), if the secondary is 36 V rms, I_{DC} is 300 mA, and C_1 is a 2200-μF capacitor, calculate $I_{(AV)}$, $I_{(PK)}$, and the $I_{(PK)}/I_{(AV)}$ ratio.

13-40. Repeat Prob. 13-39 for Fig. 13-51(b). Use the same values.

13-41. Given the circuit shown in Fig. 13-52, if the secondary voltage is 36 V rms, compute V_{dcm}, $V_{dc(min)}$, V_{DC}, $V_{r(rms)}$, $V_{r(p\text{-}p)}$, % r, $I_{(AV)}$, t_c, and $I_{(PK)}$. (*Hint*: Assume symmetry; that is, $|+V_{DC}| = |-V_{DC}|$).

13-42. Repeat Prob. 13-41 if the secondary is 18 V rms.

Section 13-15

13-43. Calculate $I_{(SURGE)}$ for Fig. 13-51(a) if the secondary voltage is 36 V rms, the transformer secondary resistance is 1.5 Ω, and the diodes each have a bulk resistance of 0.3 Ω.

FIGURE 13-52 Dual-complementary rectifier problem.

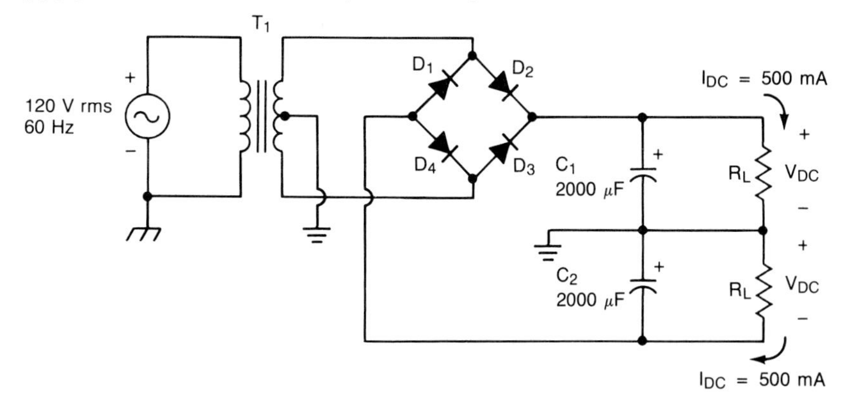

13-44. Repeat Prob. 13-43 for the circuit of Fig. 13-51(b). Use the same values given Prob. 13-41.

Section 13-16

13-45. Calculate the capacitor's rms ripple current $I_{r(\text{rms})}$ in Prob. 13-41. Draw the ripple current waveform.

13-46. Calculate the capacitor's rms ripple current $I_{r(\text{rms})}$ in Prob. 13-42. Draw the ripple current waveform.

Section 13-17

13-47. If the load current I_{DC} is 100 mA for the power supply given in Fig. 13-51(a), calculate the minimum required transformer secondary current rating.

13-48. Given the dc power supplies illustrated in Fig. 13-51(b), calculate the required secondary current if I_{DC} is 400 mA.

Section 13-18

13-49. The *RC* pi filter in Fig. 13-41 has an I_{DC} of 100 mA, a V_{DC} of 14 V, and a $V_{r(\text{rms})}$ of 0.5 V rms applied to it. C_2 is 500 μF and R_1 is 10 Ω. Calculate V'_{DC}, $V'_{r(\text{rms})}$, and $\%\ r'$.

13-50. Repeat Prob. 13-49 if V_{DC} is 20 V and $V_{r(\text{rms})}$ is 1.2 V rms. All other values are the same.

13-51. Calculate the minimum required power rating for R_1 in Fig. 13-41. Assume that R_1 is 10 Ω and I_{DC} is 100 mA. Should the resistor be a $\frac{1}{4}$-, $\frac{1}{2}$-, 1-, or a 2-W unit?

13-52. Repeat Prob. 13-51 if I_{DC} is increased to 200 mA, calculate the minimum required power rating for R_1. Should the resistor be a $\frac{1}{4}$-, $\frac{1}{2}$-, 1-, or a 2-W unit?

13-53. Given the *RC* pi filter shown in Fig. 13-42(a), if V_{DC} is 17 V and I_{DC} is 250 mA, calculate the maximum value for R_1 and its minimum power rating. What is the minimum value for C_2?

13-54. Repeat Prob. 13-53 for a V_{DC} 15 V and an I_{DC} of 750 mA.

Section 13-19

13-55. What is the primary advantage offered by an *LC* pi filter? What disadvantage does it offer?

13-56. When should an *L*-input filter be used? Why? What are its disadvantages?

Troubleshooting and Failure Modes

13-57. Diode D_2 in the full-wave rectifier circuit given in Fig. 13-18(a) has opened. Select the best answers from the following alternatives: (a) the frequency of the pulsating dc will halve, (b) the frequency of the pulsating dc will double, (c) the average dc voltage across the load will halve, and (d) the average dc voltage across the load will double. Explain your answer(s).

13-58. Diode D_1 in the full-wave bridge rectifier circuit given in Fig. 13-23(a) has opened. Using the alternatives given in Prob. 13-57, select the most likely results. Explain your reasoning.

13-59. If diode D_1 in Fig. 13-18(a) shorts, will diode D_2 be damaged? Why, or why not?

13-60. Diode D_1 in Fig. 13-23(a) has shorted. Which of the other diodes (D_2, D_3, and/or D_4) may be damaged? Explain.

13-61. The filter capacitor C_1 in Fig. 13-32 has exploded. After cleaning up the debris, it was determined that diode D_1 had failed shorted, and diode D_2 is open. Why did the capacitor explode?

13-62. The load R_L in Fig. 13-32 is an audio-amplifier system. A loud 120-Hz hum is produced in its loudspeaker. Which of the following alternatives is the most likely: (a) diode D_1 is open, (b) the transformer secondary is shorted, (c) the filter capacitor has become leaky, and (d) the filter capacitance has increased. Explain your answer.

Design

13-63. Design a full-wave rectifier that employs a *center-tapped transformer* [see Fig. 13-18(a)]. The dc load voltage is to be at least 15 V. However, we wish to target our design such that its V_{DC} is as close to 15 V as possible. Select the best transformer secondary rating from the following: (a) 12.6 V, (b) 28 V, (c) 36 V, and (d) 48 V. If the load is 10 Ω, determine the average current through the silicon rectifiers and the peak reverse bias. Should the rectifier *ratings* exceed these values? Explain.

13-64. Design a *full-wave bridge rectifier* that meets the specifications given in Prob. 13-63. Specifically, V_{DC} is to be at least 15 V, and R_L is 10 Ω. Select the transformer secondary rating from the options listed in Prob. 13-63. Also determine the rectifier specifications.

13-65. The transformer input circuit of a dc power supply has been indicated in Fig. 13-12(a). Under full load the secondary current is 5 A rms. The transformer has a secondary voltage of 6.3 V rms when the primary voltage is 120 V rms. Determine the maximum primary current (under full load), and select a standard fuse that has a rating slightly greater than the maximum primary current. The standard (instrument) fuse sizes include: $\frac{1}{8}$, $\frac{1}{4}$, $\frac{1}{2}$, $\frac{3}{4}$, 1, and 5 A.

13-66. A rectifier diode must block a reverse voltage of 50 V, and pass an average forward current of 1 A. The diode is to be mounted on a printed circuit board using $\frac{1}{2}$-in. leads and will operate at a temperature of 30°C. Which one of the 1N5390 series rectifiers (Fig. 13-46) is the most suitable? What is its I_o rating if it experiences an $I_{(PK)}/I_{(AV)}$ ratio of 10?

13-67. Repeat Prob. 13-66 if the required rectifier must block a reverse voltage of 100 V.

A regulated dc power supply has been depicted in Fig. 13-53(a). It is to provide 5 V to a digital logic circuit that incorporates digital integrated circuits which are based on a technology known as "low-power Schottky TTL." The problem below will lead the reader step by step through the design process. The device used across the primary of the transformer is called a *varistor*. It is composed of a metal-oxide material. In essence, it acts like two zener diodes that are back to back. Its *V–I* characteristic curve has been given in Fig. 13-53(b). Its function is to protect the power supply (and its load) from large transient line voltages (e.g., lightning strikes on the power line). If it absorbs too much energy, it fails shorted and causes the fuse to blow. The 7805 is a three-terminal IC voltage regulator. It provides a 5-V dc output voltage. (Three-terminal regulators are discussed in Chapter 14.)

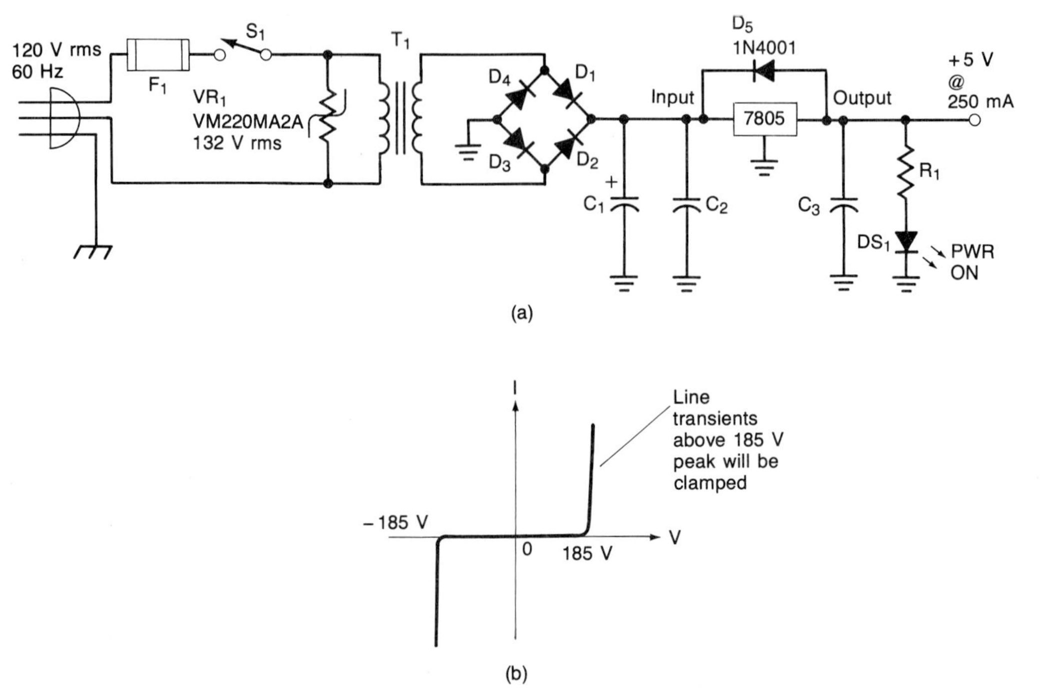

FIGURE 13-53 Circuit for Prob. 13-68: (a) circuit to be designed; (b) V-I characteristics of the metal-oxide varistor (GE VM220MA2A).

13-68. Design the power supply given in Fig. 13-53(a) by answering the following questions.

(a) For the 7805 regulator to maintain its regulated 5-V output, $V_{dc(min)}$ must be at least 2 V greater. Determine its value.

(b) Since a full-wave rectifier is being used, determine the minimum peak secondary voltage rating.

(c) Determine the *minimum* rms value of the transformer secondary voltage V_2. Select the nearest (larger) standard secondary voltage rating from the following choices: 5.0, 6.3, 12.6, and 25.2 V rms.

(d) Using Table 13-1 and the maximum dc load current of 250 mA, compute the minimum required transformer secondary current rating. Based on your result, select the minimum secondary current rating from the following standard values: 0.2, 0.5, 1.0, and 1.5 A rms.

(e) Now that the transformer has been specified, determine V_m and V_{dcm}.

(f) Compute the minimum value of C_1 to ensure that the $V_{dc(min)}$ constraint [part (a)] is met. This can be accomplished by solving Eq. 13-32 for C. Select the nearest *upper* standard value from the values listed in Appendix C. Also compute the minimum WVDC rating given by Eq. 13-22, and use Appendix C to select the nearest upper standard value.

(g) Now that the transformer and filter capacitor have been specified, verify that the light-loading constraint (Eq. 13-28) has been met.

(h) The values of C_2 and C_3 based on the manufacturer's recommendation for the proper operation of the 7805 are 0.22 and 0.01 μF, respectively. Determine their minimum WVDC ratings. Be sure to note that C_3 is across the output of the regulator.

(i) Determine V_R, $I_{(AV)}$, $I_{(PK)}$, and $I_{(PK)}/I_{(AV)}$. With this information we can select the bridge rectifier diodes. Can any of the diodes in Fig. 13-46 be used? Assume that the lead lengths are $\frac{1}{2}$ in. and the maximum lead temperature is 50°C.

(j) Based on the secondary current determined in part (d), compute the maximum primary current. Would a fuse rated at 100 mA rms be adequately large?

(k) Determine the standard value and power rating of resistor R_1 if it is to limit the current through the LED to approximately 15 mA.

Computer

13-69. Write a BASIC computer program that will analyze a power supply that utilizes a simple capacitor filter. The program should prompt the user for the type of rectifier (full-wave using a center-tapped transformer, full-wave bridge, or dual complementary), the secondary voltage, the line frequency, the dc load current, and the size of the filter capacitor. The program should compute V_{DC}, $V_{dc(min)}$, $V_{r(rms)}$, $V_{r(p\text{-}p)}$, $\% r$, $I_{(AV)}$, $I_{(pk)}$, $I_{(PK)}/I_{(AV)}$, and V_R. The program should also caution the use if the light-loading constraint is violated.

13-70. Modify the program described in Prob. 13-69 to also accommodate half-wave rectifiers.

14

Dc Voltage Regulation and Ac Power Control

After Studying Chapter 14, You Should Be Able to:

- Explain and analyze the operation of a simple transistor voltage regulator.
- Explain and analyze the operation of a series voltage regulator with an op amp error amplifier.
- Analyze and describe short-circuit protection, and overvoltage (crowbar) protection using an SCR and an op amp comparator.
- Discuss the dropout voltage limitation of a voltage regulator and techniques for reducing it.
- Analyze and describe foldback current limiting.
- Apply three-terminal integrated circuit voltage regulators.
- Specify IC voltage regulator heat sink requirements.
- Apply adjustable three-terminal IC voltage regulators.
- Discuss and analyze negative voltage regulators and dual-tracking voltage regulators.

- Explain techniques for enhancing the operation of the three-terminal voltage regulators.
- Explain the basic operation of the switching regulators.
- Analyze AC power control circuits using SCR and Triacs.
- Explain the characteristics of trigger devices such as the diac.
- Analyze a solid-state relay circuit.

14-1 Introduction to Voltage Regulation

In Chapter 13 we studied the transformer, rectifier, and filter stages found in a line-operated dc power supply. In this chapter we study the *voltage regulator*. The voltage regulator is the output stage of the line-operated dc power supply (see Fig. 13-1).

Ideally, a voltage regulator should keep the output voltage *constant* for all values of load current. However, in general the output or load voltage (V_L) will tend to *decrease* as the load current (I_L) *increases*. This is often illustrated by means of an *I–V* plot such as that shown in Fig. 14-1.

When the current drawn from the power supply is *zero* (or no load and $I_L = I_{NL}$), the output voltage will be at its *maximum* value ($V_L = V_{NL}$). When the *maximum* or *full-load* current ($I_L = I_{FL}$) is drawn from the power supply, its output voltage will be at its *minimum* value ($V_L = V_{FL}$).

Rather than presenting the load regulation curve for each power supply, its *percent of voltage regulation* (% VR) is given. The % VR is defined by

$$\% \text{ VR} = \frac{V_{NL} - V_{FL}}{V_{FL}} \times 100\% \qquad (14\text{-}1)$$

FIGURE 14-1 Load regulation curve of a dc power supply.

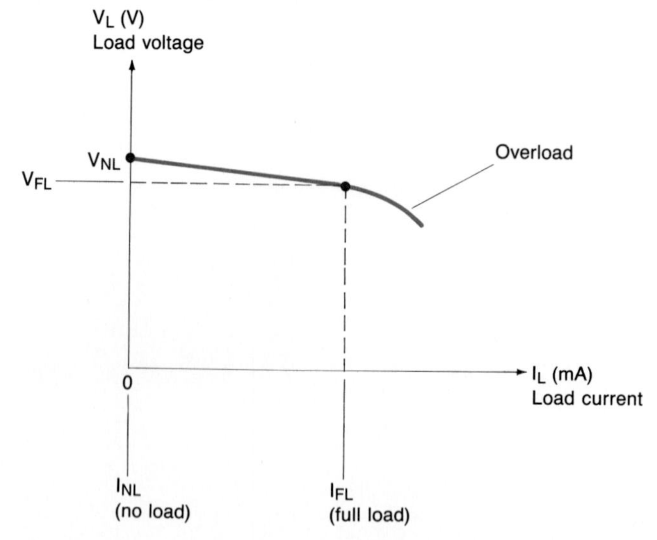

where % VR = percent of voltage regulation
V_{NL} = no-load output voltage
V_{FL} = full-load output voltage

EXAMPLE 14-1

The output voltage of a power supply is 15 V when the load current is zero. When the load current is 250 mA, its output voltage decreases to 14.9 V. Compute its percent of voltage regulation.

SOLUTION From Eq. 14-1,

$$\% \text{ VR} = \frac{V_{NL} - V_{FL}}{V_{FL}} \times 100\% = \frac{15 \text{ V} - 14.9 \text{ V}}{14.9 \text{ V}} \times 100\%$$
$$= 0.671\%$$

■

The *ideal* power supply would have a % VR of 0%. In practice, most electronically regulated power supplies will have a % VR which is very close to zero.

EXAMPLE 14-2

Find the % VR for the power supply given in Fig. 13-44.

SOLUTION When no load current is drawn ($I_{DC} = I_{NL} = 0$), *both* of the filter capacitors C_1 and C_2 will charge to the peak rectified voltage V_{dcm}. From Example 13-11 we know that V_{dcm} is 24.8 V; therefore,

$$V_{NL} = V_{dcm} = 24.8 \text{ V}$$

From Example 13-17 we saw that for an I_{DC} (I_{FL}) of 150 mA, V'_{DC} is 20.8 V; hence

$$V_{FL} = V'_{DC} = 20.8 \text{ V}$$

By Eq. 14-1,

$$\% \text{ VR} = \frac{V_{NL} - V_{FL}}{V_{FL}} \times 100\% = \frac{24.8 \text{ V} - 20.8 \text{ V}}{20.8 \text{ V}} \times 100\%$$
$$= 19.2\%$$

■

Obviously, while the power supply of Fig. 13-44 has a very good ripple factor ($\%r' = 0.615\%$), its voltage regulation (% VR = 19.2%) is rather poor. To improve its voltage regulation, we must add a *voltage regulator*. The simplest voltage regulator consists of a zener diode in parallel with the load. Since elements in parallel experience the same voltage,

$$V_L = V_Z$$

[see Fig. 14-2(a)].

For simple applications, the zener voltage regulator may be quite adequate. However, it suffers from two significant *limitations*:

1. Variations in the load current I_L will cause the zener current I_Z to vary. This, in turn, will cause variations in the zener (load) voltage.
2. The zener's power dissipation will increase as the load current decreases.

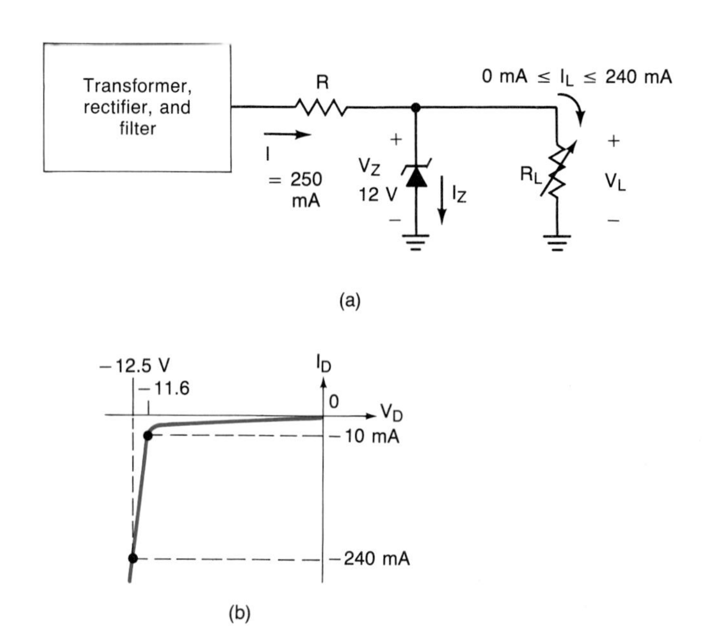

FIGURE 14-2 Simple zener regulator: (a) circuit; (b) zener's V-I characteristic.

Let us examine these limitations more closely. By Kirchhoff's current law,

$$I_Z = I - I_L$$

As long as the circuit is in regulation, the current I will be relatively constant. Therefore, as I_L varies, the zener current also varies. This will cause the voltage across the zener, and therefore the load, to vary [see Fig. 14-2(b)]. Specifically, as I_L increases, I_Z decreases and V_Z decreases. The converse is also true. Consequently, the voltage regulation will not be very "tight."

The zener's power dissipation is *greatest* when the load current is *smallest*. Since

$$I_Z = I - I_L \quad \text{and} \quad P_Z = V_Z I_Z = V_Z (I - I_L)$$

With a little reflection, we can see that the zener's worst-case power dissipation occurs when the load current I_L is zero. Although this may not seem to be a severe limitation, it can produce significant economic impacts.

EXAMPLE 14-3

Given the zener regulator shown in Fig. 14-2(a), assume that V_Z is nominally 12 V, I is a constant 250 mA, and the load current I_L ranges from 0 to 240 mA. Find the variation in V_L and the minimum and maximum zener power dissipation.

SOLUTION The zener current variation can be determined by Kirchhoff's current law. The maximum zener current is

$$I_Z = I - I_L = 250 \text{ mA} - 0 \text{ mA} = 250 \text{ mA}$$

and the minimum zener current is

$$I_Z = I - I_L = 250 \text{ mA} - 240 \text{ mA} = 10 \text{ mA}$$

From Fig. 14-2(b) we can see that when I_Z is 10 mA, V_Z is 11.9 V, and when I_Z is 240 mA, V_Z increases to 12.5 V. Therefore, we can expect a 0.6-V variation in the load voltage V_L. When I_L is 240 mA, I_Z is 10 mA and V_Z is 11.6 V. In this case the zener's power dissipation is

$$P_Z = V_Z I_Z = (11.9 \text{ V})(10 \text{ mA}) = 119 \text{ mW}$$

When I_L is 10 mA, I_Z is 240 mA and V_Z is 12.5 V. Under this condition, the zener's power dissipation becomes

$$P_Z = V_Z I_Z = (12.5 \text{ V})(240 \text{ mA}) = 3 \text{ W}$$

■

If we were actually going to use a zener regulator to supply a nominal voltage of 12 V at a maximum load current of 240 mA, we would have to use a zener diode that can dissipate at least 3 W. A standard 5-W unit would probably be used. Power zener diodes are considerably more expensive than the low-power (e.g., 400 mW) units. In a cost versus performance analysis, the circuit does not hold up very well. Much more cost-effective solutions are available which can dramatically improve performance as well.

The block diagram of a much more elegant voltage regulator has been detailed in Fig. 14-3. Observe that the *fundamental components* include:

1. A voltage reference
2. An error amplifier
3. A series-pass element, and possibly
4. Output protection

In the sections to follow, we introduce and develop the circuitry required to implement the block diagram.

FIGURE 14-3 Voltage regulator block diagram.

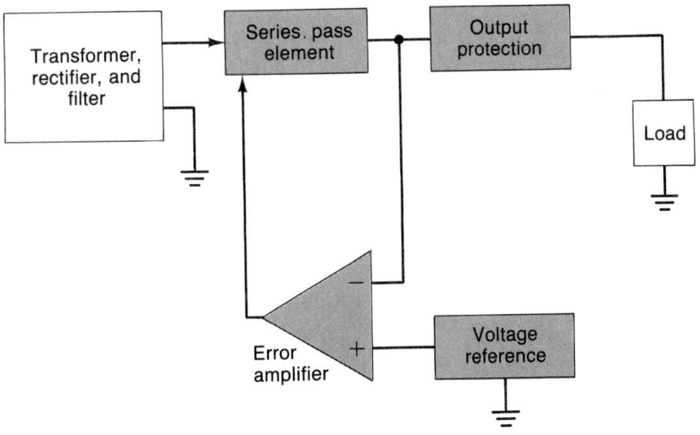

14-2 Simple Transistor Voltage Regulator

A simple transistor voltage regulator that incorporates the blocks indicated in Fig. 14-3 has been depicted in Fig. 14-4(a). Resistor R_1 sets the bias current I_1 through the zener diode D_5. In this case the loading (the current drawn from the zener circuit) is determined by the base current requirements of transistor Q_1.

Transistor Q_1 serves two functions. It acts as the series-pass element and as the error amplifier. By inspection of Fig. 14-4(a), we can see that

$$I_L = I_E$$

and since the emitter and collector currents are approximately equal,

$$I_{B(\max)} = \frac{I_{C(\max)}}{h_{FE(\min)}} \simeq \frac{I_{E(\max)}}{h_{FE(\min)}} = \frac{I_{L(\max)}}{h_{FE(\min)}}$$

The loading on the zener circuit is reduced by the current gain of the transistor. To make a worst-case analysis, we must use the transistor's minimum current gain $h_{FE(\min)}$.

The transistor is being used as an emitter-follower amplifier. Once again, from Fig. 14-4(a), we see that

$$V_L = V_Z - V_{BE}$$

Capacitor C_1 provides most of the initial filtering as discussed in Chapter 13. Additional output filtering is afforded through the use of capacitor C_3. The filtering action produced by capacitor C_2 is rather unique and merits some additional consideration.

To make an ac (ripple) analysis, we shall use the partial ac equivalent circuit given in Fig. 14-4(b). The ac ripple voltage across the output of the regulator has been denoted as v_r.

The base-emitter ripple voltage is typically very small. We shall assume that it is approximately zero. Therefore, the base-to-ground ac voltage v_b will be approximately equal to v_r. Hence

$$v_r \simeq v_b \tag{14-2}$$

and

$$v_b = i_b Z_b \tag{14-3}$$

where Z_b is the equivalent base-to-ground impedance.

The base-to-ground impedance is approximately equal to the capacitive reactance of capacitor C_2.

$$Z_b \simeq \frac{1}{j\omega C_2} \tag{14-4}$$

By substitution of Eqs. 14-3 and 14-4 into Eq. 14-2, we obtain

$$v_r \simeq v_b = i_b Z_b \simeq \frac{i_b}{j\omega C_2} \tag{14-5}$$

Solving for i_b yields

$$i_b \simeq j\omega C_2 v_r \tag{14-6}$$

The ac output ripple current has been denoted as i_r in Fig. 14-4(b). Drawing on our transistor approximation, we see that

$$i_r = \mathrm{i}_e \simeq \mathrm{i}_c \simeq h_{FE} i_b$$

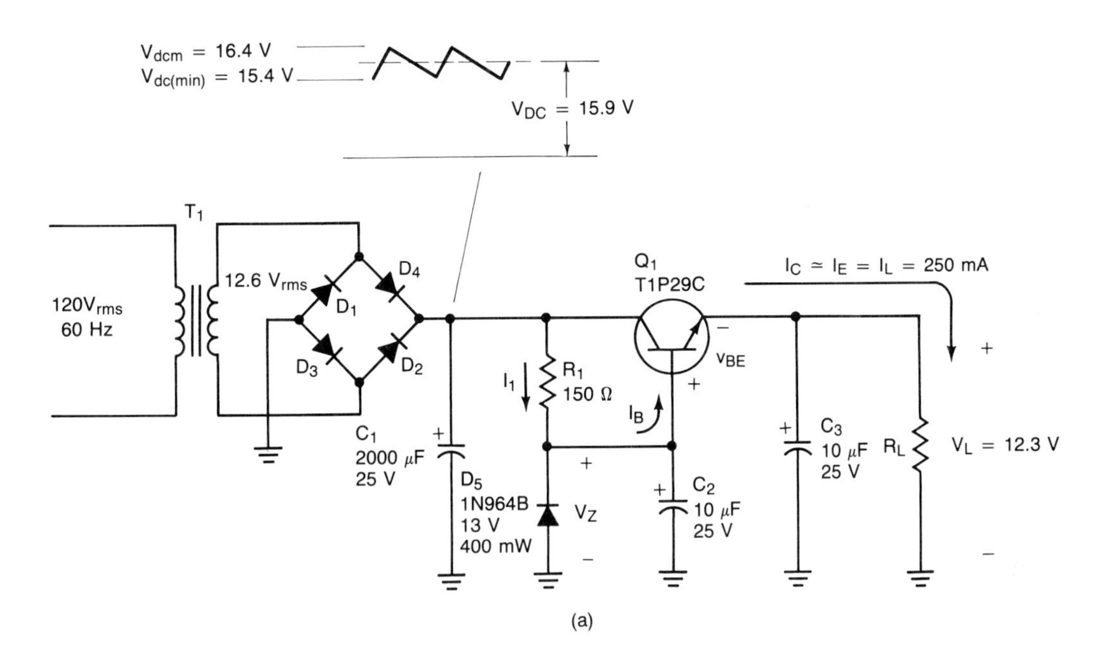

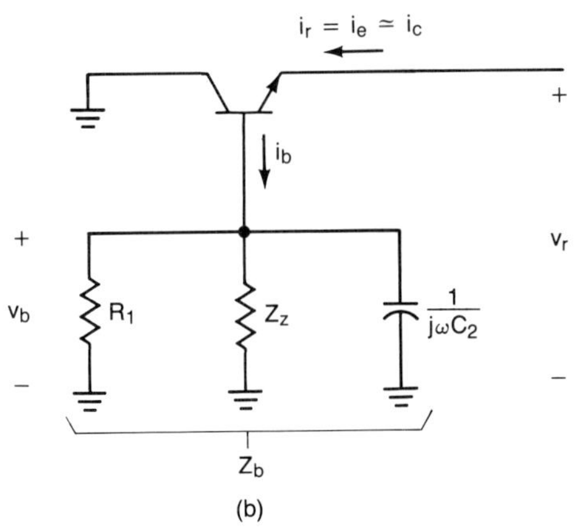

FIGURE 14-4 Simple transistor regulator: (a) regulator analysis; (b) ripple ac equivalent circuit.

Observe that we have approximated h_{fe} (the ac current gain) with h_{FE} (the dc current gain). Once again, by substitution,

$$i_r \simeq h_{FE} i_b = h_{FE}(j\omega C_2 v_r)$$

The output impedance Z_{out} of the regulator is given by

$$Z_{out} = \frac{v_r}{i_r} \simeq \frac{v_r}{j\omega h_{FE} C_2 v_r} = \frac{1}{j\omega h_{FE} C_2} \tag{14-7}$$

This result is extremely significant. The effective capacitance "looking" into the output of the regulator is the capacitance of C_2 *multiplied* by the h_{FE} of the transistor. For this reason, the circuit is often described as a "capacitance multiplier." Therefore, relatively small capacitors (small capacitance means reduced physical size and lower cost) can be used to improve output filtering. The total capacitance across the output is arrived at by recalling that capacitors in parallel add.

$$C_{out} \simeq C_3 + h_{FE}C_2 \tag{14-8}$$

EXAMPLE 14-4

Perform an analysis of the voltage regulator given in Fig. 14-4(a). Specifically, find the peak rectified voltage (V_{dcm}), the minimum instantaneous voltage across C_1 [$V_{dc(min)}$], the average voltage across capacitor C_1 (V_{DC}), the zener bias current, the corresponding zener power dissipation, the load voltage, and the effective regulator output capacitance. Assume that I_L is 250 mA and that $h_{FE(min)}$ is 50.

SOLUTION The transformer's secondary voltage is 12.6 V rms. The peak secondary voltage V_m is

$$V_m = \sqrt{2}\,V_2 = \sqrt{2}\,(12.6 \text{ V rms}) = 17.8 \text{ V peak}$$

Since a full-wave bridge rectifier is used, the peak rectified voltage V_{dcm} is

$$V_{dcm} = V_m - 1.4 \text{ V} = 17.8 \text{ V} - 1.4 \text{ V} = 16.4 \text{ V}$$

The maximum load current is 250 mA. If we neglect the zener bias current, the load current represents the approximate current (I_{DC}) drawn from the filter capacitor C_1. Therefore, the peak-to-peak ripple voltage is

$$V_{r(\text{p-p})} = \frac{I_{DC}}{f_r C_1} = \frac{250 \text{ mA}}{(120 \text{ Hz})\,(2000\ \mu\text{F})} = 1.04 \text{ V p-p}$$

The average dc voltage V_{DC} across C_1 is

$$V_{DC} = V_{dcm} - \frac{V_{r(\text{p-p})}}{2} = 16.4 \text{ V} - \frac{1.04 \text{ V}}{2} = 15.9 \text{ V}$$

The minimum instantaneous voltage across C_1 is given by Eq. 13-23.

$$V_{dc(min)} = V_{dcm} - V_{r(\text{p-p})} = 16.4 \text{ V} - 1.04 \text{ V} = 15.4 \text{ V}$$

The average bias current I_1 is

$$I_1 = \frac{V_{DC} - V_Z}{R_1} = \frac{15.9 \text{ V} - 13 \text{ V}}{150\ \Omega} = 19.3 \text{ mA}$$

At 250 mA, the minimum dc beta of the TIP29C is 50. The maximum base current is

$$I_{B(max)} = \frac{I_{L(max)}}{h_{FE(min)}} = \frac{250 \text{ mA}}{50} = 5 \text{ mA}$$

Under this condition, the zener current is

$$I_Z = I_1 - I_{B(max)} = 19.3 \text{ mA} - 5 \text{ mA} = 14.3 \text{ mA}$$

The zener's power dissipation is easily determined.

$$P_Z = V_Z I_Z = (13 \text{ V})(14.3 \text{ mA}) = 186 \text{ mW}$$

The load voltage may be found by Kirchhoff's voltage law.

$$V_L = V_Z - V_{BE} = 13 \text{ V} - 0.7 \text{ V} = 12.3 \text{ V}$$

The effective output capacitance is given by Eq. 14-8.

$$C_{out} \simeq C_3 + h_{FE}C_2 = 10\ \mu\text{F} + (50)(10\ \mu\text{F}) = 510\ \mu\text{F}$$ ■

With a little more reflection, we can see that the simple transistor voltage regulator is clearly superior to the zener diode regulator. For instance, assume that the load current goes to zero. The average dc voltage across capacitor C_1 will be approximately equal to the peak rectified voltage. Hence

$$V_{DC} \simeq V_{dcm} = 16.4 \text{ V}$$

The base curent will be approximately equal to zero, and the zener's bias current will be equal to I_1.

$$I_1 = \frac{V_{DC} - V_Z}{R_1} = \frac{16.4 \text{ V} - 13 \text{ V}}{150 \ \Omega} = 22.7 \text{ mA}$$

$$I_Z = I_1 - I_B = 22.7 \text{ mA} - 0 \text{ mA} = 22.7 \text{ mA}$$

The zener's power dissipation can now be found.

$$P_Z = V_Z I_Z = (13 \text{ V})(22.7 \text{ mA}) = 295 \text{ mW}$$

As the load current varies from 0 to 250 mA, the zener's bias current changes from 22.7 mA to 14.3 mA. The zener's power dissipation correspondingly decreases from 295 mW to 186 mW. Obviously, the use of the transistor allows us to use a low-power (low-cost) zener diode (e.g., rated at 400 mW) and provides much better voltage regulation since the zener's bias current changes only about 8.4 mA.

14-3 Using an Op Amp Error Amplifier

The performance of the simple transistor voltage regulator can be vastly improved by using an op amp as the error amplifier. In this case, the error amplifier will have much more open-loop voltage gain. This allows us to increase the loop gain. Two major advantages result. First, we obtain much more precise voltage regulation. Second, by adjusting the closed-loop voltage gain, any reasonable reference voltage may be selected. Further, it then becomes easy to provide an *adjustable* regulated dc power supply [consider Fig. 14-5(a)].

The base-emitter junction of Q_1 is enclosed within the negative-feedback loop. Therefore, the base-emitter voltage drop does *not* directly affect the output voltage. Negative feedback is provided by resistors R_2 and R_3. The zener diode serves as the voltage reference. In essence, the circuit given in Fig. 14-5(a) constitutes a noninverting voltage amplifier configuration with the voltage across the zener diode as the input voltage. Consequently, the output (load) voltage is given by

$$V_L = \left(1 + \frac{R_2}{R_3}\right) V_Z \qquad (14\text{-}9)$$

Note that R_2 has been made adjustable. This feature allows us to adjust the output voltage. If R_2 is set to zero, the load voltage will be equal to the zener voltage. In this case the reference sets the *minimum* output voltage.

Observe that the op amp's supply terminal is connected to the unregulated supply voltage across capacitor C_1. This merits a couple of additional comments. *The peak voltage* (V_{dcm}) *must be less than the maximum supply rating of the op amp.* For example, the maximum supply rating of the 741 op amp is ± 18 V, or a maximum of 36 V across the op amp. [The supply terminals are pins 7 and 4 in Fig. 14-5(a).]

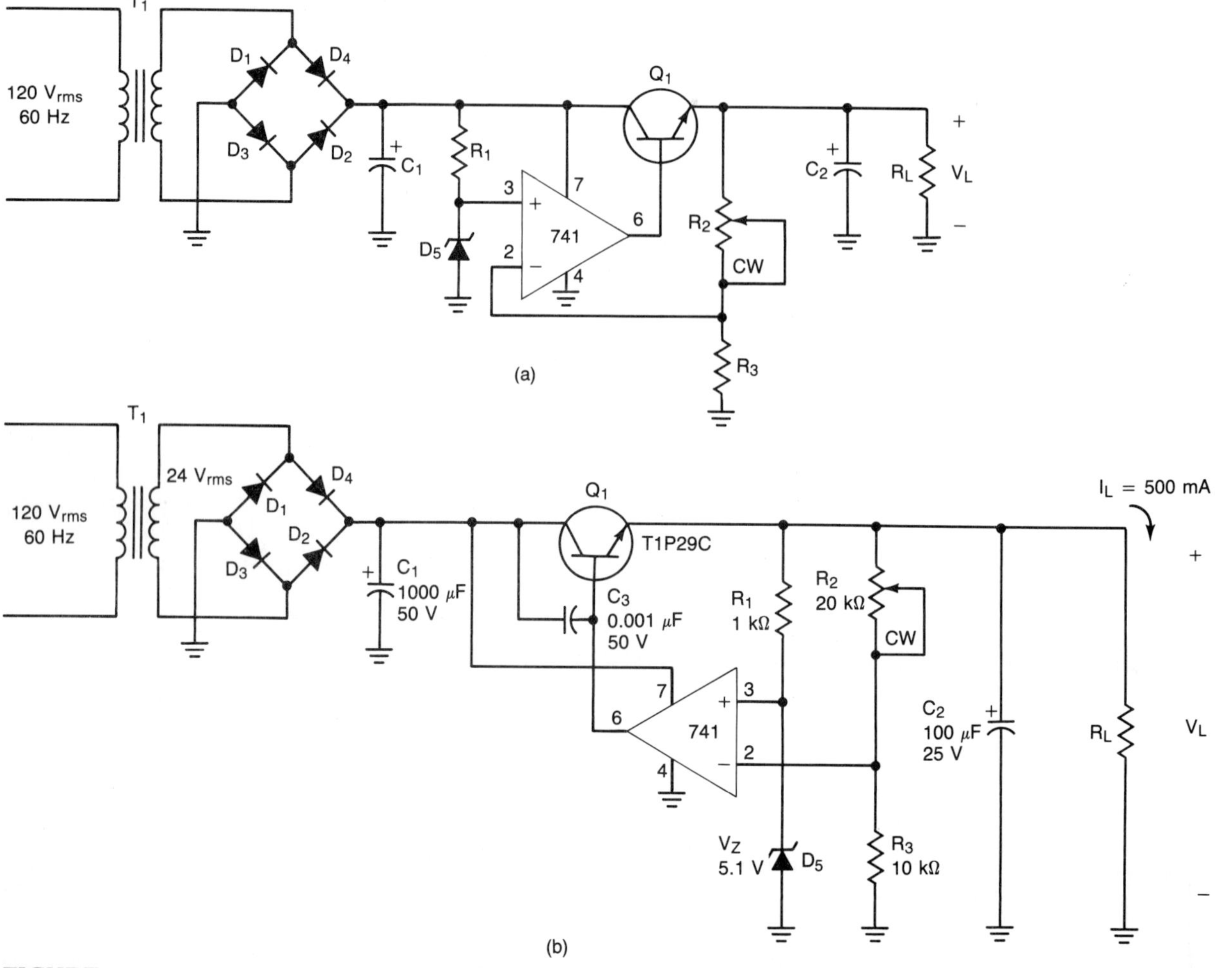

FIGURE 14-5 Using an op amp error amplifier: (a) voltage reference (D_5) on the unregulated side; (b) voltage reference on the regulated side and a start-up capacitor (C_3) is added.

The minimum input voltage [$V_{\text{dc(min)}}$] must be greater than the load voltage V_L. This is necessary to avoid saturation in the op amp, which would cause the circuit to go out of regulation. Specifically, $V_{\text{dc(min)}}$ must be at least 2 V greater than V_L to keep the op amp out of saturation, and another 0.7 V greater to keep transistor Q_1 forward biased.

These constraints are summarized below.

$$V_{\text{dcm}} < V_{\text{op amp supply rating}} \tag{14-10}$$

$$V_{\text{dc(min)}} \geq V_L + 2.7\text{ V} \tag{14-11}$$

The op amp provides the base current drive for transistor Q_1. Since the typical op amp can provide only a maximum output current of 10 mA, a constraint is placed on Q_1's minimum current gain. This drive requirement is expressed in

$$h_{FE(\text{min})} \geq \frac{I_{L(\text{max})}}{I_{\text{op amp output max}}} \tag{14-12}$$

The relationship above is based on the fact that the emitter current is approximately equal to the load current, and the collector current is approximately equal to the emitter current. If the current gain is unrealistically large (e.g., greater than 50), a Darlington power transistor may be used.

The transistor's power dissipation is also an important consideration. The collector power dissipation P_C is given by

$$P_C = V_{CE}I_C \simeq (V_L - V_{\text{DC}})I_L \tag{14-13}$$

Quite often, the series pass transistor will have to have adequate heat sinking in order to dissipate the power given by Eq. 14-13.

A further improvement is possible [refer to Fig. 14-5(b)]. The reference circuitry can be moved to the load side of the series pass transistor. In this case the reference will be much more "solid" since it is being driven from the regulated output voltage as opposed to the unregulated input voltage across capacitor C_1.

Also notice that we have added a small 0.001-μF capacitor between the collector and base terminals of transistor Q_1. When the circuit is first powered up, capacitor C_3 acts as a short circuit to provide initial base drive. This starts Q_1 into conduction and powers up the reference. The op amp will then take over by providing the required base drive. Without capacitor C_3, it is quite possible that the regulator's output will latch at zero volts when the circuit is first powered up. (This is quite likely when a single-supply op amp such as an LM324 is used since its minimum output can be very nearly 0 V.)

All of our previous equations apply to this improved circuit. This will be demonstrated in the following examples.

EXAMPLE 14-5

Analyze the power supply given in Fig. 14-5(b). Determine V_m, V_{dcm}, V_{DC}, and $V_{\text{dc(min)}}$. Assume that I_{DC} is approximately equal to the maximum load current I_L.

SOLUTION First, we must determine the peak voltage V_m across the secondary of the input transformer.

$$V_m = \sqrt{2}\,V_2 = \sqrt{2}\,(24 \text{ V rms}) = 33.9 \text{ V}$$

The peak rectified output voltage developed at the output of the bridge rectifier can now be found.

$$V_{\text{dcm}} = V_m - 1.4 \text{ V} = 33.9 \text{ V} - 1.4 \text{ V} = 32.5 \text{ V}$$

The peak-to-peak ripple voltage is

$$V_{r(\text{p-p})} = \frac{I_{\text{DC}}}{f_r C_1} = \frac{500 \text{ mA}}{(120 \text{ Hz})(1000\ \mu\text{F})} = 4.17 \text{ V p-p}$$

The average dc input voltage developed across C_1 represents the dc input voltage to our regulator.

$$V_{\text{DC}} = V_{\text{dcm}} - \frac{V_{r(\text{p-p})}}{2} = 32.5 \text{ V} - \frac{4.17 \text{ V}}{2} = 30.4 \text{ V}$$

To complete this initial analysis, we shall find the minimum instantaneous voltage across C_1.

$$V_{\text{dc(min)}} = V_{\text{dcm}} - V_{r(\text{p-p})} = 32.5 \text{ V} - 4.17 \text{ V} = 28.3 \text{ V}$$ ■

EXAMPLE 14-6

Determine the minimum and maximum dc output voltage as controlled by the adjustment of R_2 in Fig. 14-5(b). Is the 36-V supply rating of the 741 op amp sufficient? Is the $V_{dc(min)}$ across capacitor C_1 adequate?

SOLUTION The output (load) voltage can be found by using Eq. 14-9.

$$V_{L(max)} = \left[1 + \frac{R_{2(max)}}{R_3}\right]V_Z = \left(1 + \frac{20\ k\Omega}{10\ k\Omega}\right)(5.1\ V) = 15.3\ V$$

$$V_{L(min)} = \left[1 + \frac{R_{2(min)}}{R_3}\right]V_Z = \left(1 + \frac{0}{10\ k\Omega}\right)(5.1\ V) = 5.1\ V$$

The supply rating of the 741 op amp appears to be adequate as indicated by Eq. 14-10.

$$V_{dcm} = 32.5 < 36\ V$$

The minimum instantaneous input voltage across filter capacitor C_1 also appears sufficient. This judgment is based on the constraint offered in Eq. 14-11.

$$V_{dc(min)} \geq V_{L(max)} + 2.7\ V$$
$$28.3\ V \geq 15.3\ V + 2.7\ V = 18.0\ V$$

■

EXAMPLE 14-7

Determine the minimum required current gain of the series pass transistor Q_1 in Fig. 14-5(b) if the maximum load current is 500 mA and the op amp's maximum output current is 10 mA.

SOLUTION The solution to this problem directly follows from Eq. 14-12.

$$h_{FE(min)} \geq \frac{I_{L(max)}}{I_{\text{op amp output max}}} = \frac{500\ mA}{10\ mA} = 50$$

This is the minimum required h_{FE} at a collector current of approximately 500 mA. ■

EXAMPLE 14-8

Determine the worst-case power dissipation of transistor Q_1 in Fig. 14-5(b).

SOLUTION The transistor will dissipate maximum power when the output voltage has been adjusted to its minimum value $V_{L(min)}$ and the regulator is delivering its maximum output current. Hence

$$P_{C(max)} = V_{CE(max)}I_{C(max)} = (V_{DC} - V_{L(min)})I_{L(max)}$$
$$= (30.4\ V - 5.1\ V)(500\ mA) = 12.65\ W$$

■

14-4 Output Protection

In Chapter 12 we saw the need to protect the output of a power amplifier. The output of a dc power supply regulator also requires protection. This is particularly true in the case of general-purpose dc power supplies which are to be used in a laboratory. If the prototype circuit under investigation has a wiring error or a design flaw, we certainly do not want to lose our dc power supply. Two types of basic output protection are often required. The first form of output protection is *overload* or *short-circuit protection*. The second form is output *overvoltage protection*.

14-5 Short-Circuit Protection

If a short circuit occurs within the load, it will draw excessive current. The function of overload or short-circuit protection is automatically to limit the maximum output current. This not only protects the output of the voltage regulator, but tends to minimize damage in the faulty load. Specifically, the electrical conductors and the circuit components between the output of the power supply and the fault within the load circuit will not be extremely overheated. Short-circuit protection circuitry produces a reduction in the output voltage as the output current exceeds a predetermined value (I_{SC}).

Output short-circuit protection circuitry has been added to the power supply circuitry depicted in Fig. 14-5(b) [see Fig. 14-6(a)]. The load current flows through resistor R_{SC} which is used to sense the load current. It will develop a small voltage V_1 in response to the load current passing through it. Thus

$$V_1 = I_L R_{SC}$$

When the load current increases to the point where V_1 ranges from approximately 0.5 to 0.6 V, transistor Q_2 will begin to conduct. The collector current of Q_2 will "steal" some of the base drive current away from transistor Q_1. The reduction in Q_1's base drive will prevent its emitter (and therefore the load) current from exceeding I_{SC}. The value of I_{SC} is given by

$$I_{SC} \simeq \frac{0.6\ \text{V}}{R_{SC}} \tag{14-14}$$

In Fig. 14-6(a) we see that

$$I_{SC} \simeq \frac{0.6\ \text{V}}{R_{SC}} = \frac{0.6\ \text{V}}{1\ \Omega} = 600\ \text{mA}$$

14-6 Overvoltage (Crowbar) Protection

Overvoltage protection circuitry is used to prevent the dc voltage across the load from becoming too large. If the load includes integrated circuits, they can be destroyed if their supply voltage exceeds their maximum rating. Overvoltage protection is designed to protect the load in the event of a power-line transient, a load transient, or even a failure in the power supply. (One of the most common power supply failures is a shorted series pass transistor. If it shorts, the load voltage can reach V_{dcm}.) Overvoltage protection circuitry has been indicated in Fig. 14-6(b).

The circuit illustrated in Fig. 14-6(b) is described as offering *overvoltage crowbar protection*. It incorporates an op amp that is being used as a *voltage comparator* and a three-terminal device called a *silicon-controlled rectifier* (SCR). The voltage comparator and the SCR are important electronic "tools" in their own right. Therefore, it is appropriate that we digress slightly to investigate their operation.

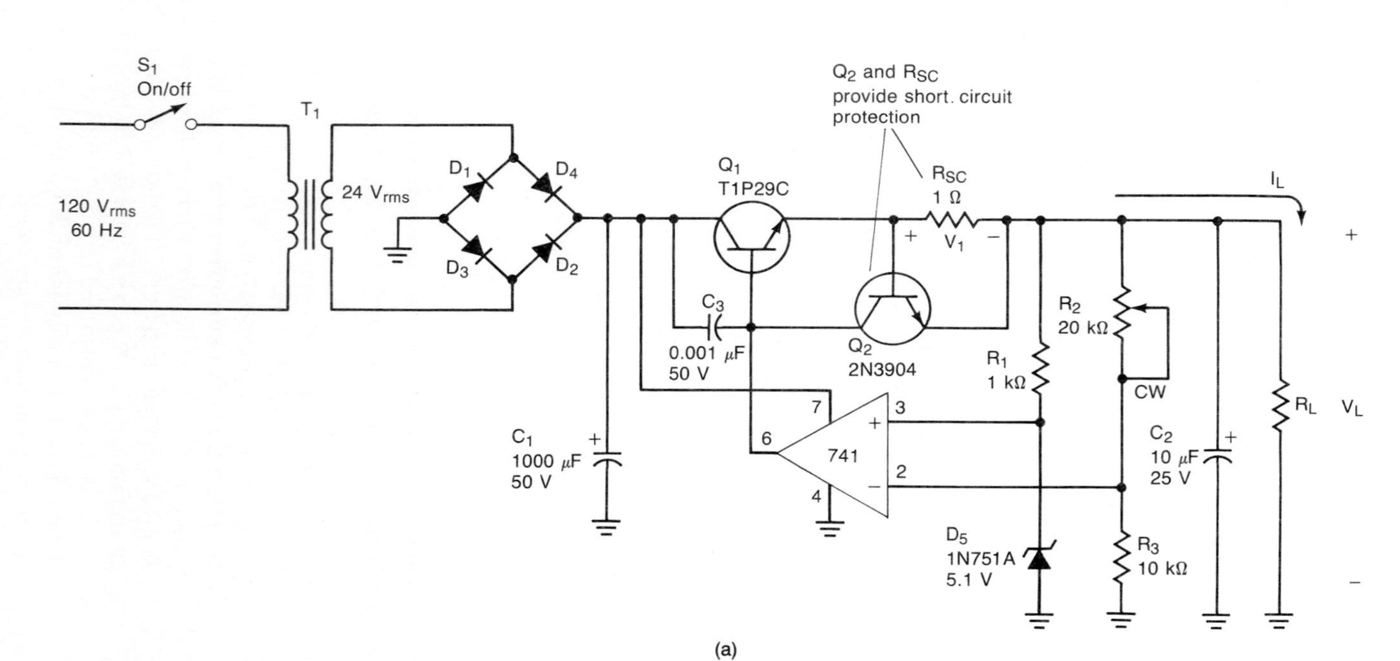

FIGURE 14-6 Adding output protection: (a) adding short-circuit protection; (b) including overvoltage protection.

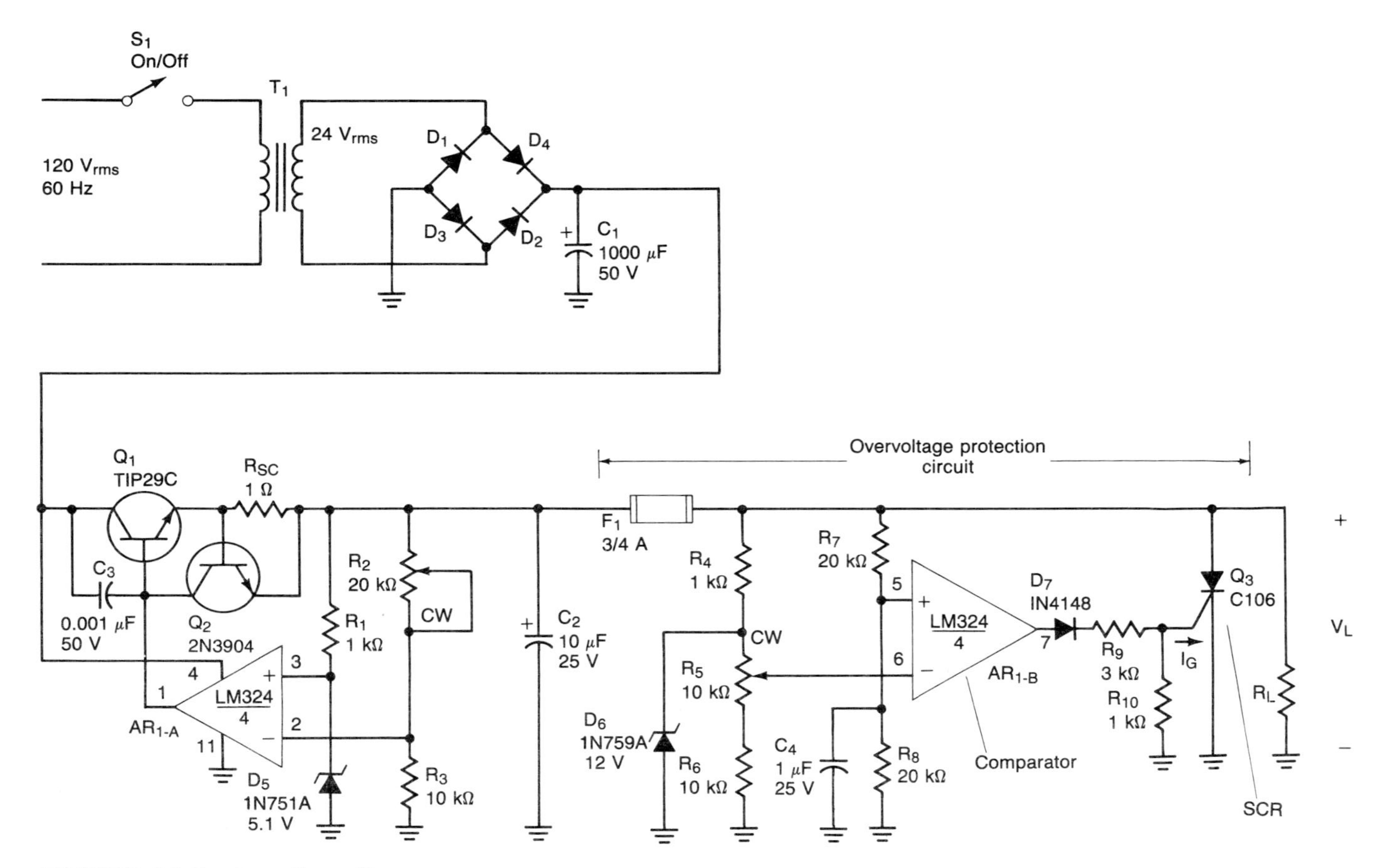

FIGURE 14-6 *(continued)*

14-7 Noninverting Voltage Comparator

The operation of the voltage comparator has been detailed in Fig. 14-7. Notice that the op amp is being used *without* negative feedback. Therefore, the gain of the circuit is equal to the *open-loop* voltage gain (e.g., 50,000 or more) of the op amp. Also

FIGURE 14-7 Noninverting comparator: (a) basic comparator; (b) the comparator's output goes to the positive supply rail when $V_{IN} > V_{REF}$; (c) the comparator's output goes to the negative supply rail when $V_{IN} < V_{REF}$; (d) transfer characteristic curve; (e) large differential input voltages can occur.

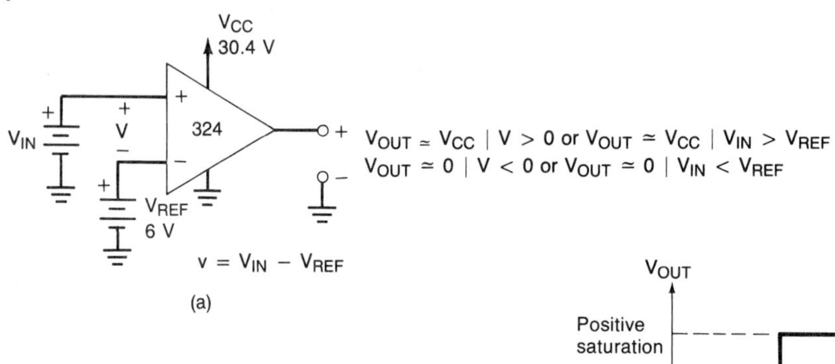

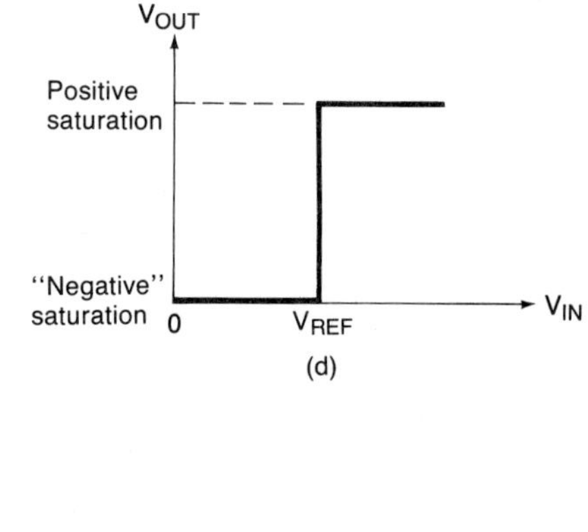

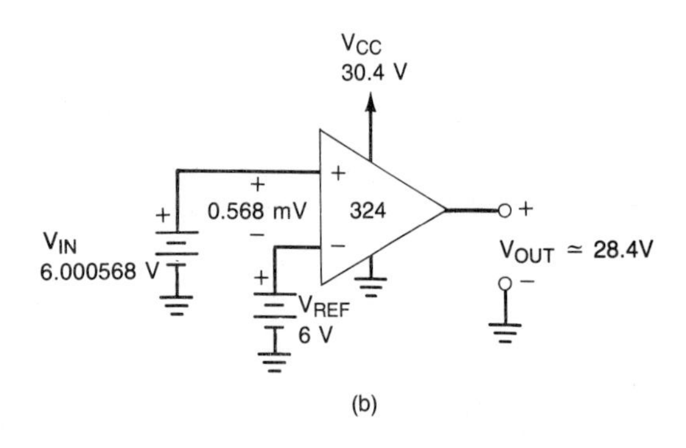

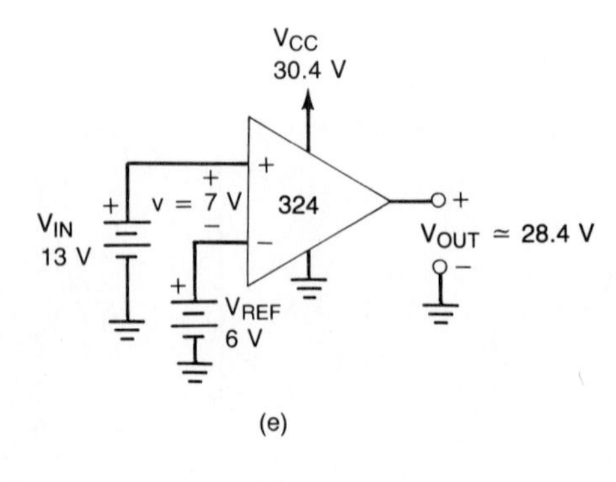

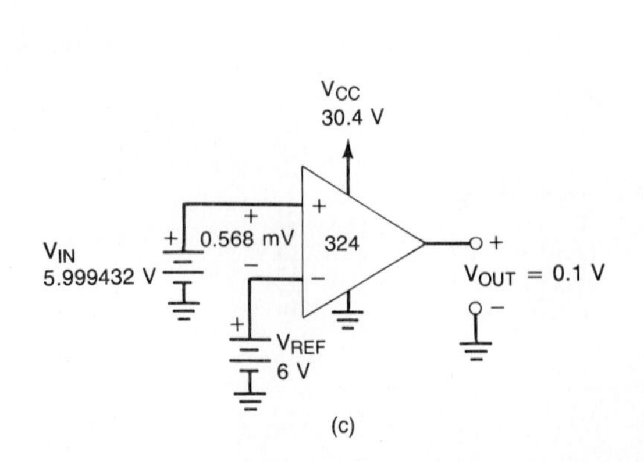

observe that the op amp's inverting input terminal has been tied to a reference voltage (V_{REF}) of 6.0 V.

If the inverting input is more positive than the noninverting input, the op amp's output will be at the negative supply rail. [This happens to be (nearly) ground in Fig. 14-7(a).] If the noninverting input is more positive than the inverting input, the op amp's output will be at the positive supply rail.

The nonlinear response offered by the comparator can best be understood by considering the op amp's open-loop voltage gain, saturation effects, and the op amp's differential input voltage. By inspection of Fig. 14-7(a), we can see that the op amp's differential input voltage v is

$$v = V_{IN} - V_{REF}$$

If V_{IN} is 6.000568 V and V_{REF} is 6.0 V, the differential input voltage is

$$v = V_{IN} - V_{REF} + 6.000568 \text{ V} - 6.0 \text{ V} = 0.568 \text{ mV}$$

The op amp's output voltage will be equal to its differential input voltage times its open-loop voltage gain $A_{v(ol)}$ of 50,000. Hence

$$V_{OUT} = A_{v(ol)}v = (50{,}000)(0.568 \text{ mV}) = 28.4 \text{ V}$$

This value of output voltage represents the op amp's positive output saturation voltage [see Fig. 14-7(b)]. Recall that this will occur when the output reaches the positive supply minus a couple of volts.

If the input voltage is decreased to 5.999432 V, the differential input voltage will become a negative quantity.

$$v = V_{IN} - V_{REF} = 5.999432 \text{ V} - 6.0 \text{ V} = -0.568 \text{ mV}$$

If we take the open-loop voltage gain times this negative differential input voltage, the output voltage becomes negative.

$$V_{OUT} = A_{v(ol)}v = (50{,}000)(-0.568 \text{ mV}) = -28.4 \text{ V}$$

However, since the op amp's negative supply terminal has been taken to ground, the output will saturate at about 0.1 V. (Such a low saturation voltage is made possible through the use of the LM324 single-supply op amp.)

The transfer characteristic of the comparator has been shown in Fig. 14-7(d). Since the output of the comparator goes to the positive rail when V_{IN} is slightly more positive than V_{REF}, the comparator is described as being noninverting. Also note that since the op amp is being used in a nonlinear fashion, the differential input voltage may be several volts [refer to Fig. 14-7(e)].

14-8 Silicon Controlled Rectifiers (SCRs)

The SCR is a four-layer three-terminal device. Its basic construction has been indicated in Fig. 14-8(a). Observe that it possesses *three p-n junctions*.

The SCR's schematic symbol has been given in Fig. 14-8(b). Its three terminals are anode (A), cathode (K), and gate (G). The SCR is being used as a high-current latch in our overvoltage application. However, the SCR is also widely used in high-power (ac) switching applications. (This will be explained later in this chapter.)

The fundamental operation of the device has been explained in Fig. 14-8(c). The SCR is an open switch until a positive gate-to-cathode pulse is applied. The gate

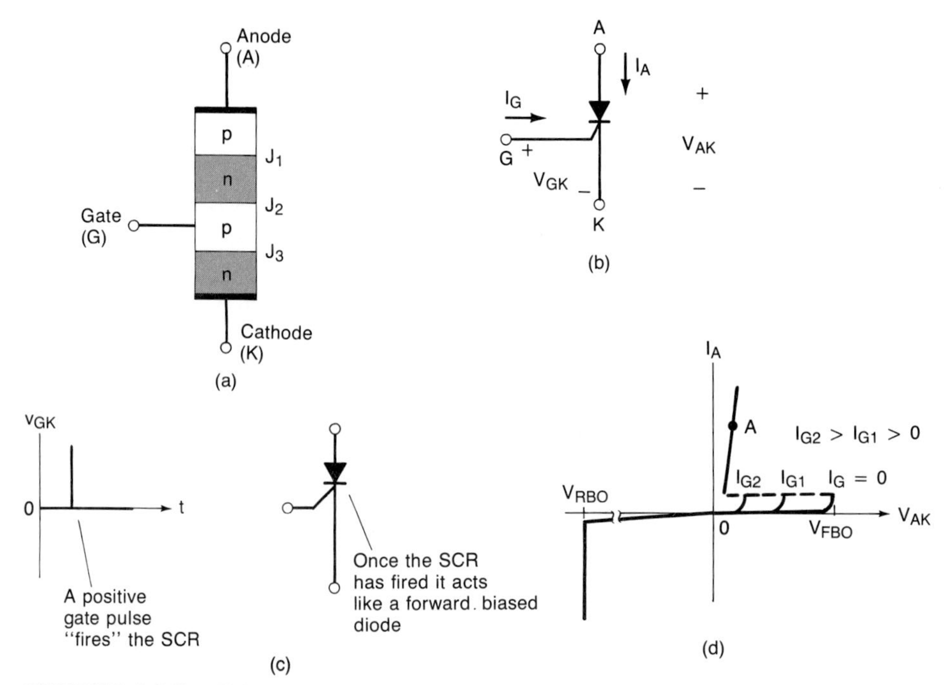

FIGURE 14-8 Silicon-controlled rectifier (SCR): (a) the SCR is a four-layer device; (b) SCR's schematic symbol; (c) SCR's basic operation; (d) V-I characteristic.

pulse will "fire" the SCR. Once the SCR has been triggered, it will act like a forward-biased diode. *The gate will no longer control the device once it has been fired.* The only way the SCR can be turned off again is to reduce the anode current to a value below the SCR's *holding current* (I_H). Once the SCR has been turned off, another positive gate pulse is required to fire it.

The operation of the device can be further detailed by examining its *V–I* characteristic curve [see Fig. 14-8(d)]. Assume that the gate terminal is open. The gate current I_G will be zero. If the anode terminal is negative with respect to the cathode, the operation of the device lies within the third quadrant. The anode current will be approximately zero as only a small reverse leakage current will flow.

If the reverse voltage is increased sufficiently, the reverse breakover voltage V_{RBO} will be reached. When this occurs the SCR will experience avalanche breakdown. As with a normal rectifier, the device can be destroyed if the reverse current is not limited to a safe value. Typically, the SCR should not be allowed to experience avalanche breakdown.

If the gate is still unconnected (I_G is held at zero), and the anode is made positive relative to the cathode, operation lies within the first quadrant. For sufficiently small values of V_{AK}, only a small forward leakage current will flow through the SCR. However, if V_{AK} is increased to the forward breakover voltage V_{FBO}, the SCR will suddenly fire. When this occurs, V_{AK} will decrease and the anode current I_A will increase to a value which is only limited by the circuitry external to the SCR [see point *A* in Fig. 14-8(d)].

If the gate lead is shorted to the cathode, the only apparent effect is a slight increase in V_{FBO} and a reduction in the forward anode-to-cathode leakage current. As the gate current is increased, the SCR's forward characteristic begins to approach that of a standard silicon rectifier diode.

We can gain even further insight into the SCR's operation by applying the two-transistor analogy developed in Fig. 14-9. The four layers of the SCR can be considered to form two transistors, Q_1 (a *pnp* device) and Q_2 (an *npn* device).

If a positive pulse is applied between the gate and the cathode, gate current I_G will flow into the device. The gate current forms base current I_{B2}. The base current I_{B2} will cause collector current I_{C2} to flow. The collector current I_{C2} draws current from the base of transistor Q_1. The base current I_{B1} will produce a collector current I_{C1} that is directed into the base of transistor Q_2. The increase in I_{B2} causes I_{C2} (I_{B1}) to increase, and this causes further increases in I_{C1} and I_{B2}.

The process we have just described is regenerative, or positive, feedback. A small gate current causes transistors Q_1 and Q_2 to latch into conduction. A more rigorous analysis of the SCR produces

$$I_A = \frac{\alpha_{DC2} I_G + (I_{CBO1} + I_{CBO2})}{1 - (\alpha_{DC1} + \alpha_{DC2})} \tag{14-15}$$

Equation 14-15 gives us considerable insight into the SCR. If the gate current is zero, the small anode current that flows is produced by the leakage current terms (I_{CBO1} and I_{CBO2}). As gate current is applied, the collector currents of Q_1 and Q_2 increase. The alphas also tend to increase. Eventually, the alphas will increase to the point where

$$(\alpha_{DC1} + \alpha_{DC2}) \simeq 1$$

When this occurs, the denominator of Eq. 14-15 approaches zero, and the anode current goes toward infinity. The anode current will only be limited by the SCR's load resistance. Once the SCR has fired, the gate current is no longer needed and

FIGURE 14-9 Two-transistor analogy: (a) the SCR can be regarded as forming two transistors; (b) equivalent circuit.

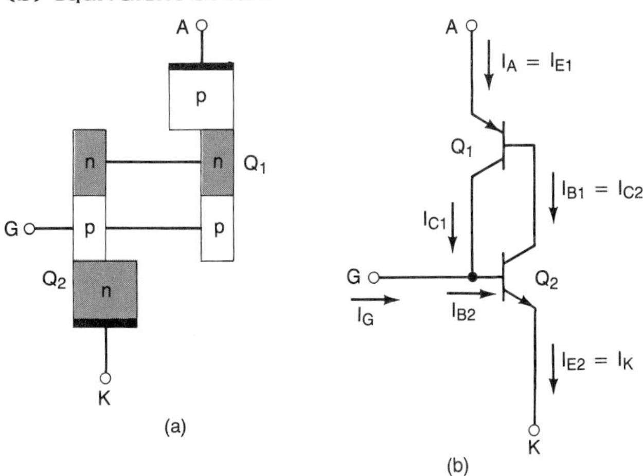

may be removed. To turn the SCR off, the anode current must be reduced to the point where the alpha sum is no longer approximately unity. This occurs when the anode current is reduced below the holding current value I_H.

14-9 Analysis of the Overvoltage Protection Circuit

Now that we understand the comparator and the marvelous SCR, let us return our attention to the overvoltage protection circuit. For convenience, the circuit originally provided in Fig. 14-6(b) has been repeated in Fig. 14-10(a).

The reference voltage (V_{REF}) for the comparator is established by zener diode D_6. Its bias current is provided by resistor R_4. The reference voltage may be adjusted, or trimmed, to the desired value by turning the rotor of potentiometer R_5. In this case, when the rotor is fully clockwise,

$$V_{\text{REF}} = V_{D6} = 12 \text{ V}$$

When the rotor is fully counterclockwise, we see that the voltage division between potentiometer R_5 and resistor R_6 results in

$$V_{\text{REF}} = \frac{R_6}{R_5 + R_6} V_{D6} = \frac{10 \text{ k}\Omega}{10 \text{ k}\Omega + 10 \text{ k}\Omega}(12 \text{ V}) = 6 \text{ V}$$

Therefore, the comparator's reference voltage is adjustable from 6 to 12 V.

The comparator's input voltage is obtained by sensing the load voltage V_L. The actual input voltage is produced by voltage division between R_7 and R_8. Hence

$$V_{\text{IN}} = \frac{R_8}{R_7 + R_8} V_L = \frac{20 \text{ k}\Omega}{20 \text{ k}\Omega + 20 \text{ k}\Omega} V_L = 0.5\, V_L$$

The comparator will fire when the input voltage V_{IN} is approximately equal to the reference voltage V_{REF}. Hence

$$V_{\text{IN}} \simeq V_{\text{REF}} \quad \text{or} \quad 0.5\, V_L \simeq V_{\text{REF}}$$

Multiplying both sides by 2 gives us

$$V_L = 2\, V_{\text{REF}}$$

In summary, we can state that *the output of the comparator will go to the positive supply rail whenever the load voltage becomes approximately equal to twice the reference voltage.*

Since the reference voltage has been shown to range from 6 to 12 V, the comparator will trip whenever the load voltage ranges from 12 to 24 V. The actual trip point is determined by the adjustment of potentiometer R_5.

When the output of the comparator is low (approximately zero volts), no appreciable forward bias is applied to the gate of the SCR (Q_3). To ensure that the low output of the comparator is not large enough to fire the SCR, diode D_7 has been added in series with the op amp's output terminal. This raises the minimum threshold voltage to fire the SCR to two "diode" voltage drops ($V_{D7} + V_{GK}$). Resistor R_{10} is used to pull the gate of the SCR to ground when diode D_7 is off. When the load voltage becomes too large, the output of the comparator will switch to about 28.4 V, and a forward gate bias will be applied through diode D_7 and resistor R_9. If we recall that a base-emitter junction exists between the gate and cathode terminals

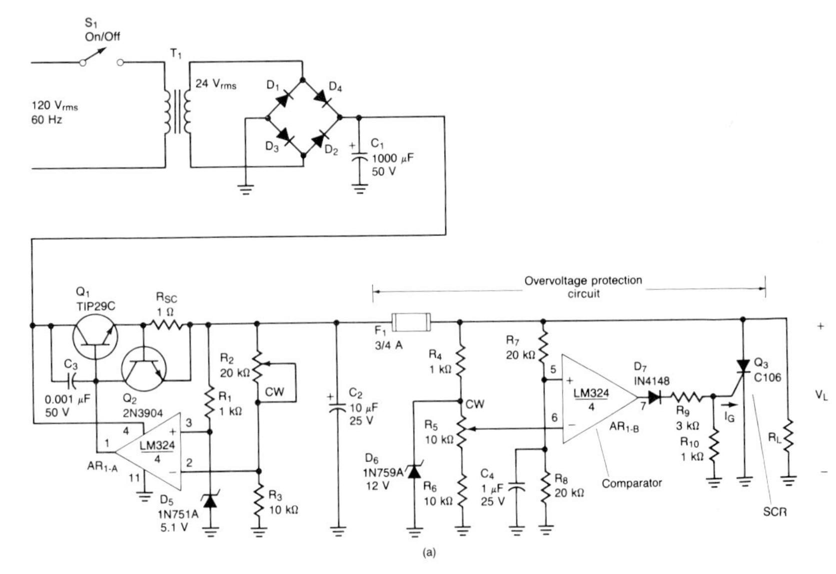

SILICON CONTROLLED RECTIFIERS

C106 Series

4-A Sensitive-Gate Silicon Controlled Rectifiers

For Power-Switching and Control Applications

The RCA-C106 series of sensitive-gate silicon controlled rectifiers are designed for switching ac and dc currents. The types within the series differ in their voltage ratings; the voltage ratings are identified by suffix letters in the type designations.

These SCR's have microampere gate-current requirements which permit operation with low-level logic circuits. They can be used for lighting, power-switching, and motor-speed controls, and for gate-current amplification for driving larger SCR's.

All types in the series utilize the JEDEC-TO-202AB (RCA VERSATAB) plastic package.

Features:

- Microampere gate sensitivity
- 600-V capability
- **3.5-A(rms) on-state current ratings**
- **20-A peak surge capability**
- Glass-passivated chip for stability
- Low thermal resistances
- Surge capability curve
- Package and formed-lead options available

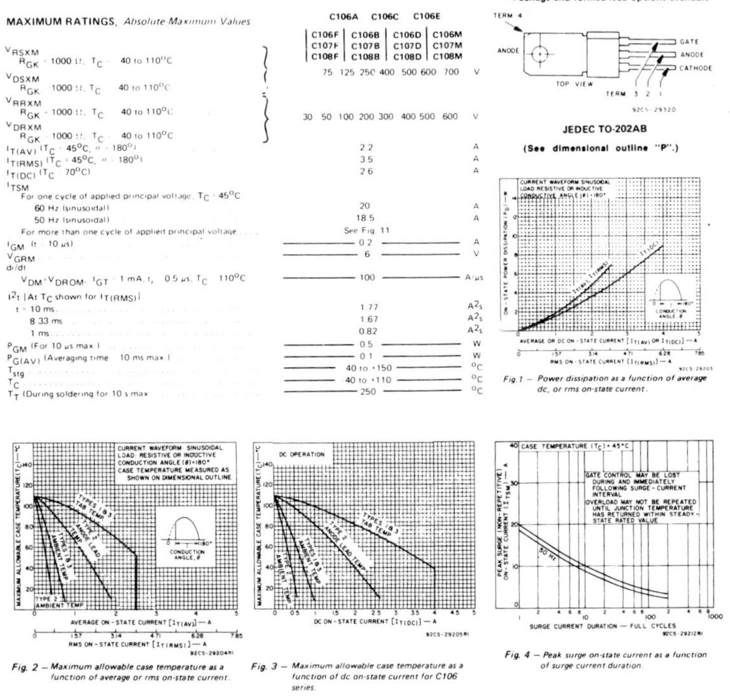

MAXIMUM RATINGS, *Absolute-Maximum Values*

Types: C106A, C106C, C106E; C106F, C107F, C108F; C106B, C107B, C108B; C106D, C107D, C108D; C106M, C107M, C108M

Rating	Value	Unit
V_{RSXM} (R_{GK} = 1000 Ω, T_C = 40 to 110°C); V_{DSXM} (R_{GK} = 1000 Ω, T_C = 40 to 110°C)	75 125 250 400 500 600 700	V
V_{RRXM} (R_{GK} = 1000 Ω, T_C = 40 to 110°C); V_{DRXM} (R_{GK} = 1000 Ω, T_C = 40 to 110°C)	30 50 100 200 300 400 500 600	V
$I_{T(AV)}$ (T_C = 45°C, θ = 180°)	2.2	A
$I_{T(RMS)}$ (T_C = 45°C, θ = 180°)	3.5	A
$I_{T(DC)}$ (T_C = 70°C)	2.6	A
I_{TSM} For one cycle of applied principal voltage, T_C = 45°C: 60 Hz (sinusoidal)	20	A
50 Hz (sinusoidal)	18.5	A
For more than one cycle of applied principal voltage	See Fig. 11	
I_{GM} (t = 10 μs)	0.2	A
V_{GRM}	6	V
di/dt: V_{DM} = V_{DROM}, I_{GT} = 1 mA, t_r = 0.5 μs, T_C = 110°C	100	A/μs
I^2t [At T_C shown for $I_{T(RMS)}$]: t = 10 ms	1.77	A^2s
8.33 ms	1.67	A^2s
1 ms	0.82	A^2s
P_{GM} (For 10 μs max.)	0.5	W
$P_{G(AV)}$ (Averaging time = 10 ms max.)	0.1	W
T_{stg}	40 to +150	°C
T_C	40 to +110	°C
T_T (During soldering for 10 s max	250	°C

Fig. 1 – Power dissipation as a function of average dc, or rms on-state current.

Fig. 2 – Maximum allowable case temperature as a function of average or rms on-state current.

Fig. 3 – Maximum allowable case temperature as a function of dc on-state current for C106 series.

Fig. 4 – Peak surge on-state current as a function of surge current duration.

(b)

FIGURE 14-10 Protected regulator and SCR data: (a) protected regulator; (b) partial data sheet for the C106 SCR (courtesy of RCA Solid State).

of the SCR, it becomes relatively easy to determine the gate current I_G [refer back to Fig. 14-10(a)]. Note that resistor R_{10} diverts 0.7 mA from the gate terminal.

$$I_G = \frac{V_{\text{OP AMP}} - V_{D7} - V_{GK}}{R_9} - \frac{V_{GK}}{R_{10}}$$

$$= \frac{28.4\text{ V} - 0.7\text{ V} - 0.7\text{ V}}{3\text{ k}\Omega} - \frac{0.7\text{ V}}{10\text{ k}\Omega} = 8.30\text{ mA}$$

With 8.30 mA of gate current the SCR (RCA type C106) will certainly latch into conduction. A partial data sheet for the C106 has been given in Fig. 14-10(b).

Once the SCR fires, its anode-to-cathode voltage will reduce to approximately zero. Since the SCR is across the load, the load voltage will also be clamped to zero volts. The SCR acts as if we had suddenly thrown a crowbar across the output of the regulator - shorting it to ground.

If the overvoltage originated in the ac power line or was produced by the load, the short-circuit protection circuit will limit the current through the latched SCR to about 600 mA. If the power switch (S_1) is opened, the current through the SCR will be reduced below its holding value (I_H). This will turn the SCR off. When the power switch is closed again, the circuit will be reset and resume normal operation.

If the overvoltage is produced because of a shorted pass transistor (Q_1), the current-limit circuit will not function. In this case the fuse (F_1) will open to remove power from the SCR and the load. Observe that the fuse has been sized at $\frac{3}{4}$ A. This is above the normal 600-mA current limit.

Capacitor C_4 is used to provide a slight time delay. This allows the comparator to ignore overvoltage transients of short duration. This tends to minimize ''nuisance'' shutdowns that might occur when power is first applied.

14-10 Reducing the Dropout Voltage

The use of an *npn* series pass transistor works quite well. However, to maintain voltage regulation, it is necessary that

$$V_{\text{dc(min)}} \geq V_L + 2.7\text{ V}$$

This was given in Eq. 14-11 [see Fig. 14-11(a)]. If the instantaneous input voltage is *less* than this value, the regulator will drop out of regulation.

With a *pnp* pass transistor, the required input-to-output differential voltage can be reduced to a few tenths of a volt (refer to Fig. 14-11(b)]. In this case the transistor is being used in its common-emitter configuration. As we can see, the op amp's output voltage must be *less* than $V_{\text{dc(min)}}$. Obviously, with this scheme, we do not have to worry about the op amp's output ''bumping its head'' on the positive supply rail. The minimum collector-to-emitter voltage to keep a silicon BJT out of saturation is approximately 0.7 V. Therefore, we have indicated a V_{CE} of -0.7 V for the *pnp* series-pass transistor.

Since we have closed the negative feedback loop around a common-emitter stage, we must concern ourselves with the potential for high-frequency oscillations. For maximum stability, the series-pass transistor should be a single-diffused, wide-base transistor such as the indicated 2N2905A. In higher-current applications a 2N3740

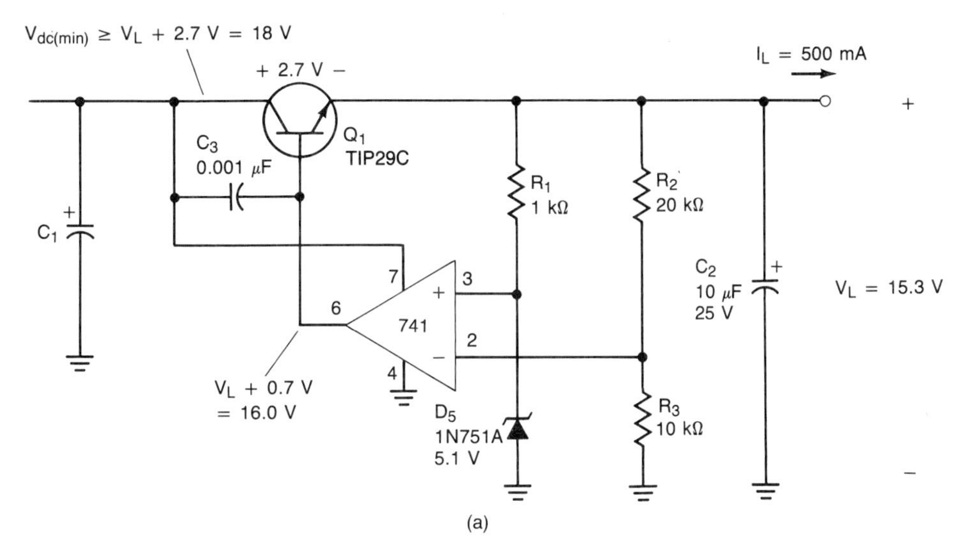

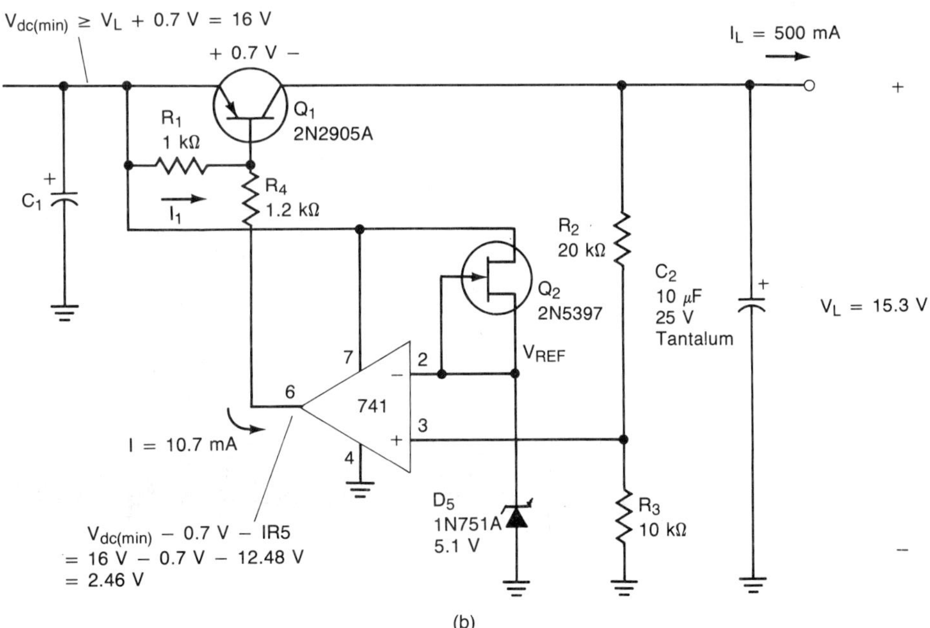

FIGURE 14-11 **Reducing the input-output differential: (a) using an npn series-pass transistor (differential = 2.7 V); (b) using a pnp series-pass transistor (differential = 0.7 V).**

may be employed. Double-diffused planar transistors are more prone to produce oscillations.

Capacitor C_2 is extremely important. It is not merely another 120-Hz ripple filter, but also serves as part of Q_1's collector (ac) load impedance. If high-frequency oscillations are to be avoided, it should offer a low impedance to high frequencies.

Aluminum electrolytic capacitors work well at low frequencies, but at high frequencies their equivalent series impedance increases. Therefore, aluminum electrolytic capacitors are not a good choice in this application (unless they have a small disk ceramic capacitor in parallel). A solid tantalum electrolytic capacitor has a low equivalent series impedance at high frequencies. It is a better choice. Consequently, a 10-μF unit has been specified for C_2.

Observe in Fig. 14-11(b) that *the input connections to the op amp have been reversed*. This is necessary because of the 180° of phase shift produced by the common-emitter stage. Let us see how the circuit operates.

If the output voltage decreases, the voltage at the noninverting input of the op amp will also decrease. Therefore, the output voltage of the op amp will decrease. This action will increase the base current of the series-pass transistor. The increased base drive will cause the collector (load) current to increase. The increase in the load current will produce an increase in the load voltage.

The output of the op amp is required to sink Q_1's base current and the current flow through resistor R_1 (I_1). By Ohm's law, and assuming that the emitter-base voltage drop is 0.7 V,

$$I_1 = \frac{V_{EB}}{R_1} \simeq \frac{0.7 \text{ V}}{R_1}$$

The base current is

$$I_B = \frac{I_C}{h_{FE}} \simeq \frac{I_E}{h_{FE}} = \frac{I_L}{h_{FE}}$$

Now by Kirchhoff's current law, we obtain

$$I = I_1 + I_B = \frac{V_{EB}}{R_1} + \frac{I_L}{h_{FE}} \simeq \frac{0.7 \text{ V}}{R_1} + \frac{I_L}{h_{FE}} \tag{14-16}$$

If we apply Kirchhoff's voltage law, we can determine the instantaneous minimum output voltage of the op amp.

$$V_{\text{op amp}} = V_{\text{dc(min)}} - 0.7 \text{ V} - IR_4 \tag{14-17}$$

EXAMPLE 14-9

Determine the voltage at the output of the op amp in Fig. 14-11(b) if I_L is zero. Assume that the voltage across capacitor C_1 is 16 V.

SOLUTION Since I_L is zero, the emitter and base currents of Q_1 will also be zero. Hence, by Eq. 14-16,

$$I = I_1 + I_B \simeq \frac{0.7 \text{ V}}{R_1} + \frac{I_L}{h_{FE}} = \frac{0.7 \text{ V}}{1 \text{ k}\Omega} + 0 = 0.7 \text{ mA}$$

Now from Eq. 14-17,

$$\begin{aligned} V_{\text{op amp}} &= V_{\text{dc(min)}} - 0.7 \text{ V} - IR_4 \\ &= 16 \text{ V} - 0.7 \text{ V} - (0.7 \text{ mA})(1.2 \text{ k}\Omega) = 14.46 \text{ V} \end{aligned}$$

■

EXAMPLE 14-10

Repeat Example 14-9 if I_L is increased to 500 mA. Assume that the transistor's h_{FE} is 50.

SOLUTION We shall follow the same basic procedure as that outlined in Example 14-9.

$$I = I_1 + I_B \approx \frac{0.7\text{ V}}{R_1} + \frac{I_L}{h_{FE}} = \frac{0.7\text{ V}}{1\text{ k}\Omega} + \frac{500\text{ mA}}{50} = 10.7\text{ mA}$$

$$\begin{aligned} V_{\text{op amp}} &= V_{\text{dc(min)}} - 0.7\text{ V} - IR_4 \\ &= 16\text{ V} - 0.7\text{ V} - (10.7\text{ mA})(1.2\text{ k}\Omega) = 2.46\text{ V} \end{aligned}$$

■

Examples 14-9 and 14-10 demonstrate the operation of the circuit. As the load current is increased, the output voltage of the op amp decreases. The reduction in the op amp's output voltage increases Q_1's base current. The most stringent condition occurs when I_L is zero. The op amp's output voltage is required to go to 14.46 V. This places it only 1.54 V below the supply rail. Although most 741 op amps will work, it might be advisable to use an LM324 instead.

The regulator circuit in Fig. 14-11(b) incorporates voltage-series negative feedback. Its input voltage is V_{REF}. Therefore, the low voltage V_L is easily determined.

$$V_L = \left(1 + \frac{R_2}{R_3}\right)V_{\text{REF}} = \left(1 + \frac{20\text{ k}\Omega}{10\text{ k}\Omega}\right)(5.1\text{ V}) = 15.3\text{ V}$$

By employing Kirchhoff's voltage law, we may ascertain the voltage drop across the transistor.

$$V_{EC} = V_{\text{dc(min)}} - V_L = 16\text{ V} - 15.3\text{ V} = 0.7\text{ V}$$

The collector-to-emitter voltage is

$$V_{CE} = -V_{EC} = -0.7\text{ V}$$

As we mentioned previously, this is the theoretical minimum V_{CE} for linear transistor operation. For example, if V_{CE} were to become only -0.6 V, transistor Q_1 will saturate and the circuit will drop out of regulation.

Another subtle difference exists between the two circuits shown in Fig. 14-11. The *npn* version [Fig. 14-11(a)] has its voltage reference (D_5) located on the regulated side of transistor Q_1. To ensure that the circuit starts properly, a small (0.001 μF) capacitor was installed between the collector and base.

The *pnp* version can also exhibit startup problems, but they are not as easily cured. Therefore, the voltage reference (D_5) is located on the unregulated side. To minimize noise (e.g., ripple) effects, the zener diode is biased by a simple constant-current source [see Fig. 14-11(b)]. The 2N5397 behaves as a constant-current diode and has an I_{DSS} that ranges from 10 to 30 mA and a $V_{GS(\text{OFF})}$ that ranges from -1 to -6 V.

Current limiting has been added to the *pnp* series-pass transistor in Fig. 14-12. Once again, R_{SC} is used to sense the load current. For low values of I_L, the voltage drop across R_{SC} will be insufficient to turn on diodes D_6 and D_7. When the voltage

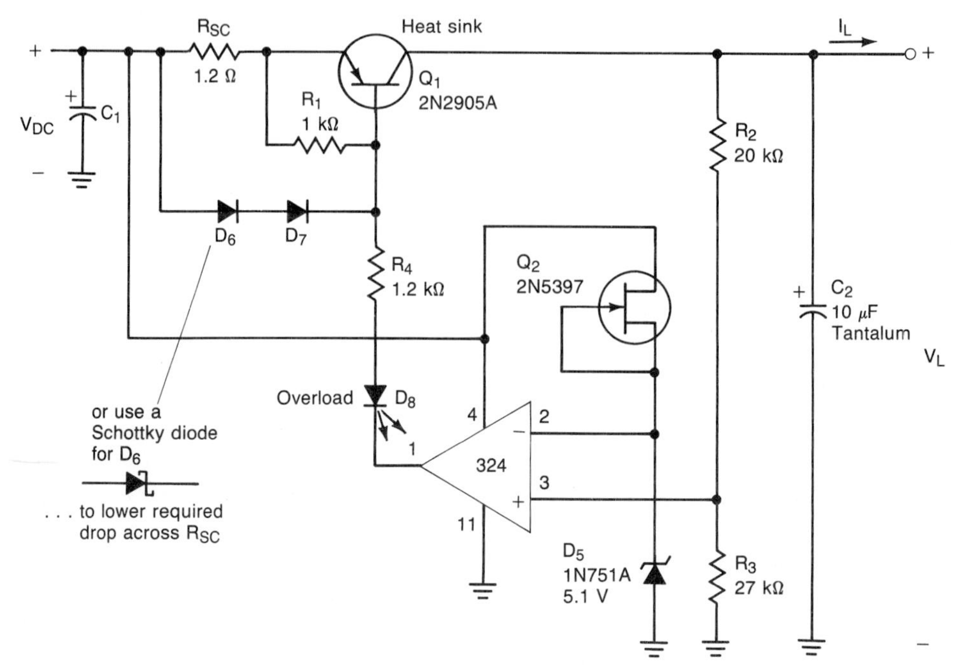

FIGURE 14-12 Adding short-circuit protection and an overload indicator (I_{SC} = 500 mA).

drop across R_{SC} reaches approximately 0.6 V, the voltage across the diode series combination will be approximately 1.3 V. (We are assuming that the base-emitter voltage drop across Q_1 is 0.7 V.) Therefore, the voltage drop across R_{SC} will be clamped to 0.6 V. This locks in the emitter and load currents. Hence

$$I_{SC} \simeq \frac{0.6\text{ V}}{R_{SC}}$$

and I_{SC} will be 500 mA with an R_{SC} of 1.2 Ω.

Since Q_1 will heat up at large values of I_C (e.g., I_{SC}) it is extremely important that either diode D_6 or D_7 be thermally bonded to Q_1's heat sink. If this is *not* done, I_{SC} will increase as Q_1's required V_{BE} decreases with increasing temperature.

If we wish to lower the turn-on voltage across R_{SC}, we can modify the circuit slightly by using a Schottky diode instead of an ordinary silicon rectifier (see Fig. 14-12). Since the knee voltage of the Schottky diode ranges from approximately 0.2 to 0.3 V, the required voltage drop across R_{SC} will be reduced. In this case,

$$I_{SC} \simeq \frac{0.3\text{ V}}{R_{SC}}$$

This scheme offers an even lower dropout voltage. Specifically, the input-to-output differential for Fig. 14-12 with a Schottky diode is approximately 0.3 V plus 0.7 V, or about 1 V. In contrast, an *npn* series-pass transistor with current limiting requires a minimum input-to-output differential voltage on the order of 3.3 V. This is due to the fact that the output voltage of the op amp must be equal to the maximum load voltage plus the base–emitter drop across Q_1, and the 0.6 V drop across the current sense resistor R_{SC}. The positive supply voltage [$V_{dc(min)}$] must be at least 2 V more positive than the op amp's output voltage.

EXAMPLE 14-11

Determine the short-circuit load current (I_{SC}) for the regulator given in Fig. 14-12 if a Shottky diode is used for D_6.

SOLUTION Assuming that the Schottky diode conducts when a forward bias of 0.3 V appears across it, we can easily determine I_{SC}.

$$I_{SC} \simeq \frac{0.3\ \text{V}}{R_{SC}} = \frac{0.3\ \text{V}}{5\ \Omega} = 60\ \text{mA}$$

■

If an LM324 (single-supply) op amp is used, its output can go as low as 0.2 to 0.7 V above ground potential. This would occur when I_L is at its maximum value, or a short circuit occurs across the regulator's output. (The op amp's output will sink a current of approximately 10 mA under this condition.) By adding an LED in series with the op amp's output terminal, we have a simple power supply overload indicator (see D_8 in Fig. 14-12). When the current through the LED reaches around 8 to 10 mA, it will begin to glow. A short across the output of the regulator will cause it to glow brightly.

14-11 Foldback Current Limiting

Simple current limiting works quite well. Only two components are required. See transistor Q_2 and resistor R_{SC} in Fig. 14-13(a). However, let us assume that the output of the regulator is inadvertently shorted to ground. We then have the situation depicted in Fig. 14-13(a).

With 600 mA flowing through it, resistor R_{SC} will develop a voltage drop of 0.6 V. Transistor Q_2 will be fully turned on. The short across the output will essentially remove the input signals to the op amp error amplifier. However, no short circuit is perfect, and we have assumed that a 0.1-V drop exists across the output. Under this condition, the op amp's output will attempt to go to the positive supply rail. This occurs since a *slightly* reverse biased zener will behave as an open circuit. As can be seen in Fig. 14-13(a), the op amp's noninverting input will be more positive than its inverting input. This represents the worst-case condition since the op amp will provide the maximum possible forward bias to transistor Q_1.

The voltage drop across R_{SC} and the base-emitter voltage drop across Q_1 will clamp the op amp's output to approximately 1.4 V. Under this condition, it is a safe assumption that the op amp will enter into its output current limit. [A maximum of 20 mA has been assumed in Fig. 14-13(a).]

If we assume that the beta of Q_1 is 50, 12 mA will be directed into its base terminal. This requires transistor Q_2 to divert 8 mA away from Q_1's base.

In Example 14-8 we saw that the worst-case collector power dissipation of Q_1 is 12.65 W under normal operating conditions. Now let us examine the situation

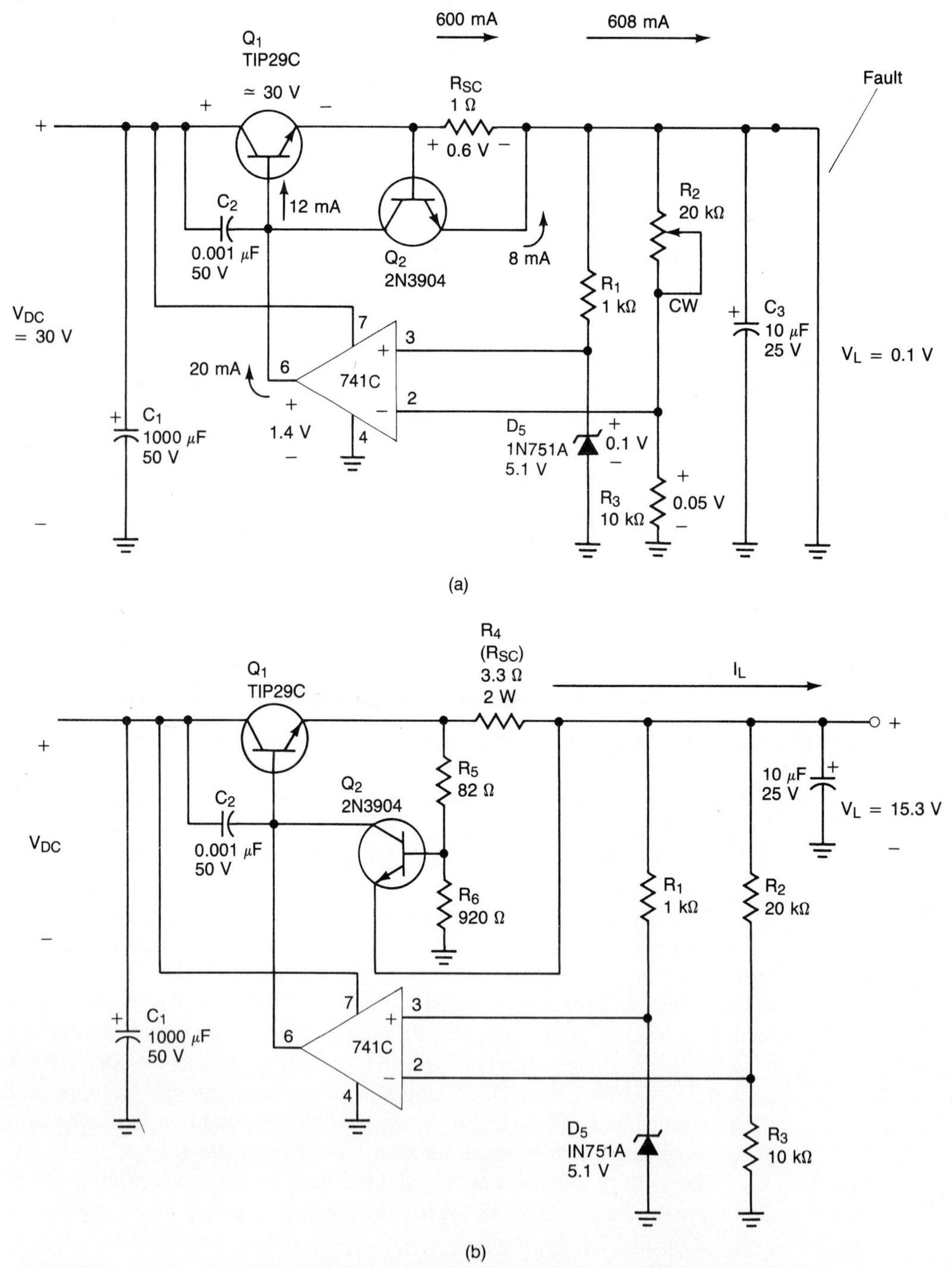

FIGURE 14-13 Foldback current limiting: (a) fault conditions using simple current limiting; (b) foldback current limiting circuit; (c) foldback characteristic; (d) the voltages when $I_{L(max)}$ is reached; (e) as V_L is adjusted to a lower value, $I_{L(max)}$ is also reduced.

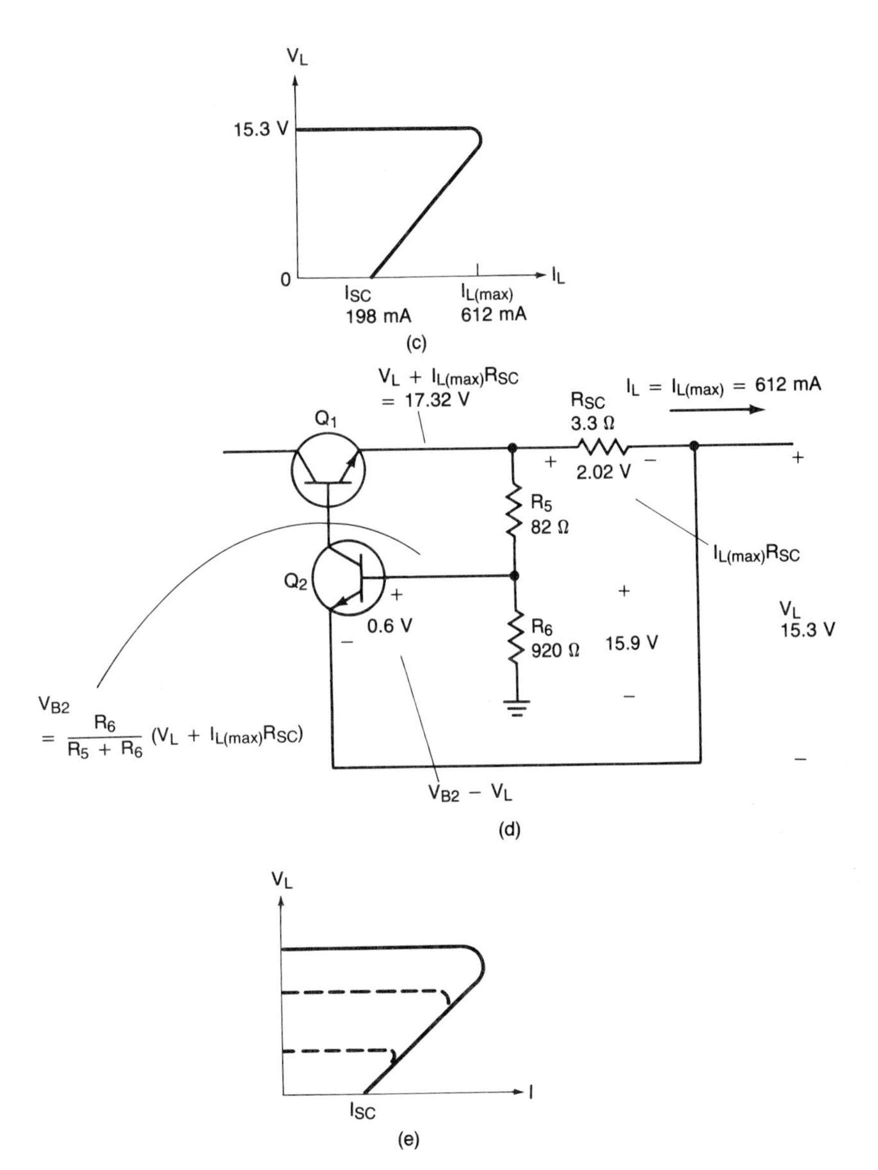

FIGURE 14-13 *(continued)*

when the output is shorted. If 600 mA is being drawn from capacitor C_1, the average voltage across it will be slightly reduced.

$$V_{r\,(\text{p-p})} = \frac{I_{\text{DC}}}{f_r C_1} = \frac{600\text{ mA}}{(120\text{ Hz})(1000\ \mu\text{F})} = 5\text{ V}$$

$$V_{\text{DC}} = V_{\text{dcm}} - \frac{V_{r\,(\text{p-p})}}{2} = 32.5\text{ V} - \frac{5\text{ V}}{2} = 30\text{ V}$$

With the output shorted, approximately all of V_{DC} will be impressed across transistor Q_1. Hence

$$V_{CE1} \simeq V_{DC} = 30 \text{ V}$$

From Fig. 14-13(a) we can see that the collector current through Q_1 will be 600 mA. Therefore, the collector power dissipation is

$$P_{C1} = V_{CE1}I_{C1} = (30 \text{ V})(600 \text{ mA}) = 18 \text{ W}$$

Let us reflect on this result. The normal maximum power dissipation is 12.65 W. However, when the output is shorted, Q_1's power dissipation increases to 18 W. This requires even more heat sinking if the transistor is to withstand this condition indefinitely. This is an expensive proposition. Clearly, we need an alternative.

The ideal situation would be to *reduce* the fault current automatically if the maximum rated output current is exceeded. This is precisely what is done when *foldback current limiting* is incorporated. The cost of foldback current limiting essentially rests in the price of two resistors [see R_5 and R_6 in Fig. 14-13(b)].

Once again, the load current is sensed by R_{SC}. However, in this instance the voltage applied to the base of Q_2 is attenuated by the R_5-R_6 combination. For simplicity, we shall designate the attenuation factor as K.

$$K = \frac{R_6}{R_5 + R_6} \tag{14-18}$$

The emitter voltage of Q_1 may be found by Kirchhoff's voltage law.

$$V_{E1} = I_L R_{SC} + V_L \tag{14-19}$$

The base-to-ground voltage of Q_2 is described below.

$$V_{B2} = \frac{R_6}{R_5 + R_6} V_{E1}$$

Substitution of Eqs. 14-18 and 14-19 into the statement above yields

$$V_{B2} = \frac{R_6}{R_5 + R_6} V_{E1} = K(I_L R_{SC} + V_L) \tag{14-20}$$

The emitter-to-ground voltage of transistor Q_2 is equal to the load voltage. Hence

$$V_{E2} = V_L$$

Q_2's base–emitter bias is given by the difference between its base-to-ground and emitter-to-ground voltage.

$$V_{BE2} = V_{B2} - V_{E2} = V_{B2} - V_L$$

Substitution of Eq. 14-20 leads us to the following result:

$$\begin{aligned} V_{BE2} &= V_{B2} - V_L = K(I_L R_{SC} + V_L) - V_L \\ &= KI_L R_{SC} + KV_L - V_L = KI_L R_{SC} + (K - 1)V_L \end{aligned}$$

Solving for I_L gives us

$$I_L = \frac{V_{BE2} - (K - 1)V_L}{KR_{SC}} = \frac{V_{BE2} + (1 - K)V_L}{KR_{SC}} \tag{14-21}$$

and we have the basic equation for the regulator's output current.

If the output of the regulator is shorted to ground, V_L is zero and I_L will be equal to the short-circuit current I_{SC}. From Eq. 14-21,

$$I_{SC} = \frac{V_{BE2} + (1 - K)V_L}{KR_{SC}} = \frac{V_{BE2} + (1 - K)(0\text{ V})}{KR_{SC}} = \frac{V_{BE2}}{KR_{SC}}$$

and if we assume that the base-emitter drop is 0.6 V, we obtain

$$\boxed{I_{SC} = \frac{0.6\text{ V}}{KR_{SC}}} \qquad (14\text{-}22)$$

where $K = R_6/(R_5 + R_6)$.

The maximum load current $[I_{L(\max)}]$ will be *larger* than I_{SC}. Once again, we employ Eq. 14-21.

$$I_{L(\max)} = \frac{V_{BE2} + (1 - K)V_L}{KR_{SC}} = \frac{V_{BE2}}{KR_{SC}} + \frac{(1 - K)V_L}{KR_{SC}}$$

and if we draw on Eq. 14-22, we have

$$\boxed{I_{L(\max)} = I_{SC} + \frac{(1 - K)V_L}{KR_{SC}}} \qquad (14\text{-}23)$$

where $I_{SC} = 0.6\text{ V}/KR_{SC}$.

A graph of the foldback characteristic is presented in Fig. 14-13(c). As we can see, the load voltage is constant for various values of load current until the knee of the curve $[I_{L(\max)}]$ is reached. Beyond that point, the load current will be reduced, or folded back, as V_L decreases. A V_L of zero represents a short circuit. The current is limited to I_{SC}.

EXAMPLE 14-12

Find the short-circuit current and the maximum load current for the regulator shown in Fig. 14-13(b). What is the collector power dissipation when a short occurs?

First, we must determine K.

$$K = \frac{R_6}{R_5 + R_6} = \frac{920\ \Omega}{82\ \Omega + 920\ \Omega} = 0.918$$

The short-circuit current is given by Eq. 14-22.

$$I_{SC} = \frac{0.6\text{ V}}{KR_{SC}} = \frac{0.6\text{ V}}{(0.918)(3.3\ \Omega)} = 198\text{ mA}$$

The maximum load current before foldback begins may be determined via Eq. 14-23.

$$I_{L(\max)} = I_{SC} + \frac{(1 - K)V_L}{KR_{SC}} = 198\text{ mA} + \frac{(1 - 0.918)(15.3\text{ V})}{(0.918)(3.3\ \Omega)} = 612\text{ mA}$$

The collector power dissipation during a short circuit is

$$P_{C1} \simeq V_{DC}I_{SC} = (30\text{ V})(198\text{ mA}) = 5.94\text{ W}$$

This is quite a reduction from the 18 W obtained when simple current limiting is used. ■

Observe that the maximum load current is approximately three times the value of the short-circuit current. A 3:1 ratio is quite normal for this circuit. The maximum (practical) ratio is on the order of 5:1.

To obtain a better ''feel'' for the operation of the circuit, the voltages, and currents obtained when $I_{L(\text{max})}$ has been reached have been summarized in Fig. 14-13(d). The reader is urged to verify the indicated values.

Admittedly, the finer points of Fig. 14-13(b) may remain slightly obscure at this point. We attempt to remedy this situation by providing a more qualitative description of the circuit's operation.

The load current is sensed by R_{SC}. The current limit transistor Q_2 will stay off until the load current reaches $I_{L(\text{max})}$. This occurs when the emitter voltage of transistor Q_1 is large enough to produce a voltage drop across R_6 that is about 0.6 V more positive than V_L.

When V_{BE2} reaches 0.6 V, transistor Q_2 will begin to conduct and steal some of Q_1's base drive. This, in turn, begins to turn Q_1 off and lowers I_L. The reduction in I_L lowers both the voltage at the emitter of Q_1 and the voltage across the load. These two voltages fall at approximately the same rate since R_{SC} is typically much smaller than R_L.

However, the voltage at the base of transistor Q_2 falls at a much slower rate because of the voltage-divider action produced by resistors R_5 and R_6. Specifically, *both* the emitter and base voltage of transistor Q_2 decrease, but the base voltage decreases at a much *slower rate*. The net result is that V_{BE2} actually *increases* as I_L decreases.

Consequently, this forces Q_2 to conduct harder and further reduces Q_1's base drive. The process continues with Q_2's conduction increasing and Q_1's conduction decreasing until the load voltage (V_L) is approximately zero. The potential at the emitter Q_1 will be just large enough to keep about 0.6 V across Q_2's base-emitter junction. This point is I_{SC}, and it was reached by lowering *both* I_L *and* V_L.

The primary disadvantage of this circuit is its dependence on V_L to determine $I_{L(\text{max})}$. Study Eq. 14-23. As V_L is adjusted to a lower value, $I_{L(\text{max})}$ will also be reduced. This effect has been illustrated in Fig. 14-13(e). Schemes exist for circumventing this problem, but we shall not delve into them at this point.

14-12 Three-Terminal Voltage Regulators

To simplify power supply designs, many engineers and technicians opt to use IC *three-terminal voltage regulators*. The integrated three-terminal voltage regulators typically incorporate many of the functions discussed thus far in a single package [see Fig. 14-14(a)].

The error amplifier is used to maintain a constant output voltage via negative feedback. The internal voltage reference is tightly controlled during the fabrication of the integrated circuit. Consequently, the nominal output voltage of most three-terminal voltage regulators has tolerances that range from $\pm 6\%$ to better than $\pm 2\%$.

The series-pass element is driven by the output of the error amplifier. Its function is to serve as an automatically controlled variable resistor. Its resistance is varied as required to maintain a constant load voltage. The series-pass element is typically a

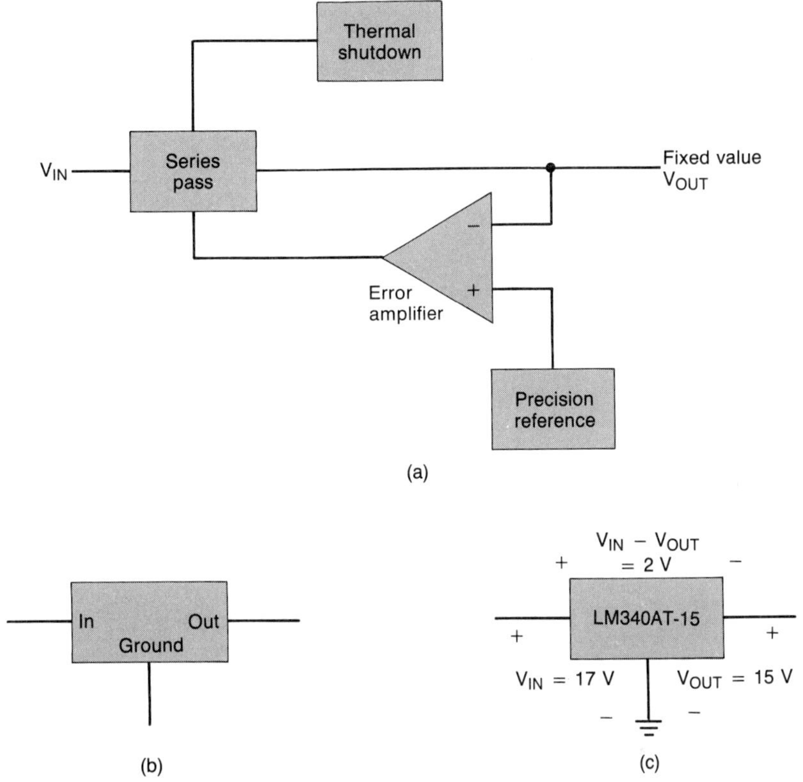

FIGURE 14-14 **Three-terminal voltage regulator: (a) fundamental block diagram of the three-terminal regulator; (b) schematic symbol; (c) the dropout voltage is the minimum input-output differential.**

bipolar transistor that is rated to pass the maximum load current. Three-terminal regulators are available with current ratings that range from 100 mA to as large as 5 A. Generally, adequate heat sinking is required if the regulator is to pass large load currents.

The thermal shutdown circuit is very similar to that discussed in Section 12-18. When the chip becomes too hot (typically, 125 to 150°C) the drive to the series-pass element is reduced and the unit will shut down.

Let us review some of the key characteristics of the three-terminal regulators. The schematic symbol has been given in Fig. 14-14(b).

The output voltage of many three-terminal voltage regulators is fixed. For example, the LM340AT series offers fixed output voltages of 5, 6, 8, 10, 12, 15, 18, and 24 V. Similarly, the LM320T series of negative three-terminal regulators has fixed output voltages of −5, −5.2, −6, −8, −9, −12, −15, −18, and −24 V. The tolerance at 25°C is ±4% of the nominal output voltage for both the LM340AT and the LM320T voltage regulators.

The three-terminal voltage regulators also have a *dropout voltage* specification. This is the minimum permissible input-to-output voltage drop if the regulator is to maintain a constant output voltage. Typical dropout voltage specifications range from

1 to 4 V. As an example, the LM340AT series has a typical dropout voltage of 1.6 to 2 V [see Fig. 14-14(c)].

The regulators will also have a *maximum input voltage* specification. If the maximum input voltage specification is exceeded, the regulator may be damaged. The LM340AT series has a maximum input voltage rating of 35 V.

The output or load current of a three-terminal voltage regulator may vary from zero up to its maximum rating. The LM340AT series is rated at 5 A. However, if adequate heat sinking is not used, the regulator may go into thermal shutdown *before* the maximum current rating is reached. When thermal limiting occurs, the regulator's output voltage will automatically decrease as the series-pass transistor ''chokes'' off the load current. The output current will drop off and remain there until the regulator cools off. This does not constitute true short-circuit protection. However, the regulator can withstand short circuits indefinitely if adequate heat sinking is incorporated.

Voltage regulators will tend to reject any of the remaining ripple voltage that appears across the input filter capacitor. Consequently, *ripple* (R) is another specification often provided on manufacturers' data sheets. It is typically expressed in decibels (dB) and can be used to determine the ripple at the output of a regulator. Let us see exactly how it is developed (refer to Fig. 14-15).

Ripple rejection (R_R) is the ratio of the rms value of the ripple voltage appearing at the input [$V_{r\,(\mathrm{rms})\mathrm{in}}$] to the rms value of the ripple voltage appearing at the output [$V_{r\,(\mathrm{rms})\mathrm{out}}$].

$$R_R = \frac{V_{r\,(\mathrm{rms})\mathrm{in}}}{V_{r\,(\mathrm{rms})\mathrm{out}}} \tag{14-24}$$

The ripple rejection, or attenuation, may be expressed in decibels. Its symbol is $R_{R(\mathrm{dB})}$.

$$R_{R(\mathrm{dB})} = 20 \log R_R = 20 \log \frac{V_{r\,(\mathrm{rms})\mathrm{in}}}{V_{r\,(\mathrm{rms})\mathrm{out}}} \tag{14-25}$$

Since $V_{r\,(\mathrm{rms})\mathrm{in}}$ is always larger than $V_{r\,(\mathrm{rms})\mathrm{out}}$, the ripple rejection [$R_{R(\mathrm{dB})}$] will always be a positive value.

For example, in Fig. 14-15 we see that the input ripple is 0.5 V rms. Thanks to the regulator's ripple rejection, the output ripple is only 0.75 mV rms. The regulator's ripple rejection is given by Eq. 14-25.

$$R_{R(\mathrm{dB})} = 20 \log R_R = 20 \log \frac{V_{r\,(\mathrm{rms})\mathrm{in}}}{V_{r\,(\mathrm{rms})\mathrm{out}}}$$

$$= 20 \log \frac{0.5 \text{ V rms}}{0.75 \text{ mV rms}} = 56.5 \text{ dB}$$

FIGURE 14-15 Regulator ripple rejection.

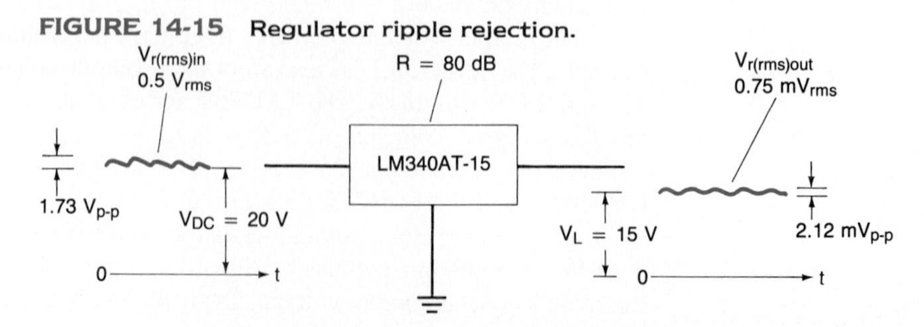

As we stated previously, manufacturers provide a ripple specification called *ripple* (R) in decibels. It is defined by

$$\text{Ripple} = R = 20 \log \frac{V_{r(\text{rms})\text{in}} V_L}{V_{r(\text{rms})\text{out}}} \tag{14-26}$$

By applying basic logarithmic relationships, we see that

$$R = 20 \log \frac{V_{r(\text{rms})\text{in}}}{V_{r(\text{rms})\text{out}}} + 20 \log V_L$$

and substituting in Eq. 14-25 gives us

$$R = R_{R(\text{dB})} + 20 \log V_L$$

If we solve for the ripple rejection, we obtain

$$R_{R(\text{dB})} = R - 20 \log V_L \tag{14-27}$$

The significance of this result is that by knowing the manufacturer's ripple specification (R) and the regulator's output voltage, we may solve for the ripple rejection (R_R). This allows us to predict the ripple voltage at the output of the regulator.

EXAMPLE 14-13

Given the regulator and circuit conditions shown in Fig. 14-15, determine the peak-to-peak ripple voltage across the load.

SOLUTION The ripple specification (R) is 80 dB, and the regulator's dc output voltage (V_L) is 15 V. Hence, by Eq. 14-27,

$$\begin{aligned} R_{R(\text{dB})} &= R - 20 \log V_L \\ &= 80 \text{ dB} - 20 \log 15 = 56.5 \text{ dB} \end{aligned}$$

We shall convert the ripple rejection to a straight ratio.

$$20 \log \frac{V_{r(\text{rms})\text{in}}}{V_{r(\text{rms})\text{out}}} = R_{R(\text{dB})}$$

$$\log \frac{V_{r(\text{rms})\text{in}}}{V_{r(\text{rms})\text{out}}} = \frac{R_{R(\text{dB})}}{20}$$

$$\frac{V_{r(\text{rms})\text{in}}}{V_{r(\text{rms})\text{out}}} = 10^{R_{R(\text{dB})}/20}$$

By taking the inverse of both sides of this result, we arrive at our equation for $V_{r(\text{rms})\text{out}}$.

$$\frac{V_{r(\text{rms})\text{out}}}{V_{r(\text{rms})\text{in}}} = 10^{-R_{R(\text{dB})}/20}$$

$$V_{r(\text{rms})\text{out}} = V_{r(\text{rms})\text{in}} 10^{-R_{R(\text{dB})}/20}$$

Given that the input ripple voltage is 0.5 V rms, we can now solve for the output ripple.

$$\begin{aligned} V_{r(\text{rms})\text{out}} &= V_{r(\text{rms})\text{in}} 10^{-R_{R(\text{dB})}/20} \\ &= (0.5 \text{ V}) 10^{-56.5/20} \\ &= 0.748 \text{ mV rms} \end{aligned}$$

Since the output ripple is approximately sinusoidal, we can easily determine its peak-to-peak value.

$$V_{r(\text{p-p})\text{out}} = 2\sqrt{2}\,(0.748 \text{ mV rms}) = 2.12 \text{ mV p-p}$$

■

This result has been indicated in Fig. 14-15. The 2.12-mV peak-to-peak output ripple can be reduced further by adding an additional output filter capacitor and/or increasing the size of the input filter capacitor.

EXAMPLE 14-14

A regulated dc power supply has been given in Fig. 14-16(a). It provides 15 V at a maximum load current of 750 mA. Is the size of the input filter capacitor C_1 adequate? What is the peak-to-peak output ripple?

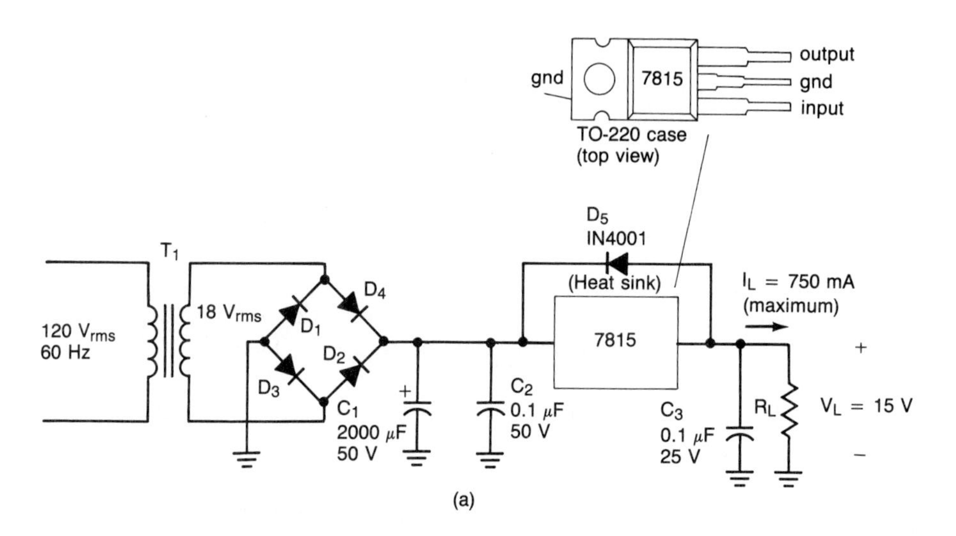

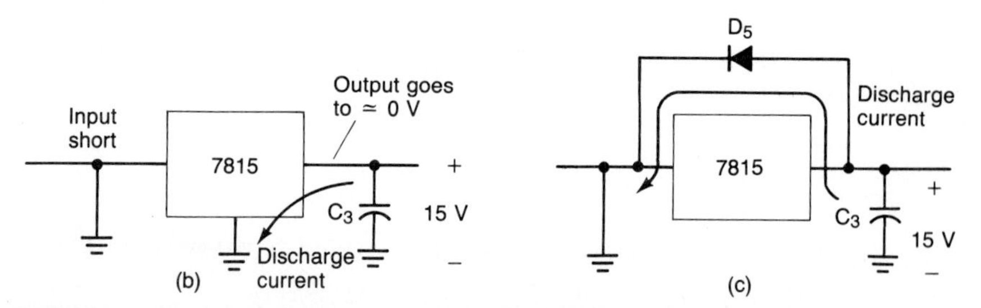

FIGURE 14-16 Regulator circuit for Example 14-14: (a) circuit; (b) the regulator's output can be damaged by C_3's discharge; (c) diode D_5 protects the regulator's output.

SOLUTION We shall analyze the input circuit first to determine V_m, V_{dcm}, $V_{r(\text{p-p})}$, and then $V_{\text{dc(min)}}$.

$$V_m = \sqrt{2}\, V_2 = \sqrt{2}\,(18 \text{ V rms}) = 25.5 \text{ V}$$

$$V_{\text{dcm}} = V_m - 1.4 \text{ V} = 25.5 \text{ V} - 1.4 \text{ V} = 24.1 \text{ V}$$

$$V_{r(\text{p-p})} = \frac{I_{\text{DC}}}{f_r C_1} = \frac{750 \text{ mA}}{(120 \text{ Hz})(2000\ \mu\text{F})} = 3.13 \text{ V}$$

$$V_{\text{dc(min)}} = V_{\text{dcm}} - V_{r(\text{p-p})} = 24.1 \text{ V} - 3.13 \text{ V} = 21 \text{ V}$$

The dropout voltage of the 7815 regulator is from 1.6 to 2 V. Therefore, a $V_{dc(min)}$ of 21 V (which provides an input–output differential of 6 V) is more than adequate. The conclusion to be drawn is that the input filter capacitor C_1 is sufficiently large. The rms value of the peak-to-peak input ripple [$V_{r(rms)in}$] may be found by using our triangular approximation.

$$V_{r(rms)in} = \frac{V_{r(p\text{-}p)}}{2\sqrt{3}} = \frac{3.13\text{ V}}{2\sqrt{3}} = 0.904\text{ V rms}$$

The 7815 voltage regulator is specified as having a ripple (R) of 66 to 80 dB. We take the worst case and use 66 dB in our calculations.

$$\begin{aligned} R_{R(dB)} &= R - 20\log V_L = 66\text{ dB} - 20\log 15 = 42.5\text{ dB} \\ V_{r(rms)out} &= V_{r(rms)in}10^{-R_{R(dB)}/20} \\ &= (0.904\text{ V})10^{-42.5/20} = 6.78\text{ mV rms} \\ V_{r(p\text{-}p)out} &= 2\sqrt{2}\,V_{r(rms)out} = 2\sqrt{2}\,(6.78\text{ mV rms}) \\ &= 19.2\text{ mV p-p} \end{aligned}$$

■

We conclude our introduction to the three-terminal voltage regulators by considering the ancillary components (C_2, C_3, and D_5) in Fig. 14-16(a). These components are used to enhance the operation of the three-terminal regulator. Capacitor C_2 (a 0.1-μF ceramic disk) is used to ensure stability in the regulator. It may not be required if the input of the regulator has less than 2 in. of electrical conductor between it and C_1. Capacitor C_3 (another 0.1-μF ceramic disk) is used to minimize high-frequency noise in the dc output of the regulator and to improve its transient response.

Diode D_5 is used to protect the output of the regulator. If the input capacitor shorts, or merely discharges more rapidly than C_3, capacitor C_3 will discharge into the output of the regulator [see Fig. 14-16(b)]. Diode D_5 will prevent this situation.

Diode D_5 is normally reverse biased since the regulator's input will be more positive than its output. However, if the input becomes less positive than the output (e.g., a short develops across the input), capacitor C_3 will discharge through D_5. This will protect the regulator's output [refer to Fig. 14-16(c)].

14-13 Heat-Sinking Considerations

Any semiconductor device that passes current will dissipate power. Power dissipation will produce an increase in temperature. If the temperature rise exceeds the maximum device rating, the device will fail. This is true of the bipolar and field-effect (VMOS and DMOS) transistors found in audio power amplifiers and the series-pass elements found in voltage regulators. The series-pass transistor included in integrated voltage regulators is the primary source of heat rise in those devices.

In Section 12-17 we introduced the fundamental equation for thermal analysis. It is repeated here:

$$T_j = P_D\theta_{ja} + T_A \tag{14-28}$$

To keep T_j below its maximum rating (typically, 150 to 175°C for IC voltage regulators), we must ensure that the power dissipation P_D, the thermal resistance θ_{ja}, and the ambient temperature T_A are considered carefully. Operating life decreases at high temperatures. Although a regulator may be rated at 150°C, it is not good practice to design for continuous operation at that temperature. A reasonable maximum operating temperature would be 100°C for epoxy-packaged devices [such as the TO-220 plastic package shown in Fig. 14-16(a)] and 125°C for hermetically sealed (TO-3) devices.

Recall that θ_{ja} is the sum of three thermal resistances:

$$\theta_{ja} = \theta_{jc} + \theta_{cs} + \theta_{sa} \tag{14-29}$$

(The reader may wish to refer back to Section 12-17.)

The power dissipation P_D can be controlled by careful circuit design. Some of the options include proper selection of the transformer secondary voltage, the use of foldback current limiting, and relaxing the size of the input filter capacitor.

The transformer's peak secondary voltage must be larger than the maximum required dc output voltage. However, if it is made too large, the rectifier reverse voltage ratings and the filter capacitor WVDC ratings must also be unnecessarily large. This results in increased cost. Further, the average power dissipation in the regulator will also be increased.

As we have seen, the worst-case power dissipation in the regulator occurs when the output is shorted. If this event is quite likely to occur in a particular application, short-circuit protection is required. If foldback current limiting is not employed, we must increase the size of the heat sink. This can add bulk (weight) and additional cost to our design.

If the size of the filter capacitor is reduced, the ripple voltage across it will increase, but the average dc voltage across it will decrease. Lowering the average dc voltage will *lower* the power dissipation in the regulator. The effects of increased ripple can be negated by the ripple rejection provided by the regulator. This approach is quite valid in noncritical, low-cost designs. However, in low-noise (e.g., instrumentation electronics) designs, this approach does not often represent a feasible option.

To ensure that T_j is not exceeded, we must perform a thermal analysis. The basic steps in the analysis are detailed below.

1. Find the device specifications $T_{j(\max)}$, θ_{jc}, and θ_{ja} (if available). The θ_{ja} specification is the thermal resistance of the device if it is operated in free air (e.g., no heat sink).
2. Determine the worst-case power dissipation in the device and the maximum ambient temperature to which it is to be exposed.
3. Determine the maximum allowable θ_{ja} to keep the junction temperature below $T_{j(\max)}$

$$\theta_{ja(\max)} = \frac{T_{j(\max)} - T_{A(\max)}}{P_{D(\max)}}$$

4. If the θ_{ja} specification is *less* than the $\theta_{ja(\max)}$ value calculated above, *no* heat sink is required.

$$\theta_{ja} < \theta_{ja(\max)} \Rightarrow \text{no heat sink required}$$

5. If the θ_{jc} specification of the device is *greater* than the $\theta_{ja(\max)}$ value calculated above, we cannot use this particular device and must select a higher-wattage unit.

$$\theta_{jc} > \theta_{ja(\max)} \Rightarrow \text{use another device}$$

This should be clear since θ_{jc} is merely part of the thermal resistance from junction to ambient when a heat sink is used.

$$\theta_{ja} = \theta_{jc} + \theta_{cs} + \theta_{sa}$$

6. If $\theta_{ja(\text{max})}$ is less than the θ_{ja} of the device and its θ_{jc} is less than $\theta_{ja(\text{max})}$, the device may be used with an appropriate heat sink.

$$\theta_{ja(\text{max})} < \theta_{ja} \quad \text{and}$$

$$\theta_{jc} < \theta_{ja(\text{max})} \Rightarrow \text{we must use the device with a heat sink}$$

7. The thermal resistance of the heat sink (θ_{sa}) is

$$\theta_{sa} < \theta_{ja(\text{max})} - \theta_{jc} - \theta_{cs} \simeq \theta_{ja(\text{max})} - \theta_{jc}$$

Recall that the thermal resistance from the case to the heat sink is controlled by the particular mounting techniques and practices employed. Generally, it is much less than θ_{jc}.

EXAMPLE 14-15

Conduct a thermal analysis of the regulator shown in Fig. 14-16(a). Assume that the highest ambient temperature is 60°C (140°F).

SOLUTION From Example 14-14 we recall that the maximum load current is 750 mA, V_{dcm} is 24.1 V, and $V_{r(\text{p-p})}$ is 3.13 V. Our next step is to find the average voltage across capacitor C_1 (V_{DC}).

$$V_{\text{DC}} = V_{\text{dcm}} - \frac{V_r(\text{p-p})}{2} = 24.1\text{ V} - \frac{3.13\text{ V p-p}}{2} = 22.5\text{ V}$$

The power dissipation in the regulator can now be ascertained.

$$P_{D(\text{max})} = (V_{\text{DC}} - V_L)I_{L(\text{max})} = (22.5\text{ V} - 15\text{ V})(750\text{ mA})$$
$$= 5.625\text{ W}$$

The pertinent data sheet specifications for the 7815 regulator are summarized below.

$$T_{j(\text{max})} = 150°\text{C} \qquad \theta_{jc} = 4°\text{C/W}$$
$$I_{L(\text{max})} = 1.5\text{ A} \qquad \theta_{ja} = 50°\text{C/W}$$
$$P_{D(\text{max})} = 18\text{ W at }25°\text{C}$$

In this application, we see that

$$\theta_{ja(\text{max})} = \frac{T_{j(\text{max})} - T_{A(\text{max})}}{P_{D(\text{max})}} = \frac{150°\text{C} - 60°\text{C}}{5.625\text{ W}} = 16°\text{C/W}$$

This tells us that we must use a heat sink since the 7815 has a free air θ_{ja} of 50°C/W, and 16°C/W is required to keep T_j below 150°C. The 7815 regulator can work in this application since its θ_{jc} is only 4°C/W, which is less than our $\theta_{ja(\text{max})}$ of 16°C/W. Now that we have determined that a heat sink is required, we must find its thermal resistance θ_{sa}.

$$\theta_{sa} \simeq \theta_{ja(\text{max})} - \theta_{jc} = 16°\text{C/W} - 4°\text{C/W} = 12°\text{C/W}$$

The heat sink should be able to accommodate the 7815's TO-220 case style and have a thermal resistance of less than 12°C/W. One choice might be a Thermalloy 6032 heat sink, which has a typical θ_{sa} of 10°C/W in natural convection. ■

14-14 Adjustable Three-Terminal Regulators

Any three-terminal voltage regulator can be made adjustable. This is often accomplished by the addition of a fixed resistor and a potentiometer. Provided that the input-output differential and heat sink requirements are met, the three-terminal regulator will maintain a constant potential difference (V_{REG}) between its output and "ground" terminals [see Fig. 14-17(a)].

If resistor R_1 is a fixed resistor, a constant current will flow through it. If the regulator's quiescent current (I_Q) is negligibly small, a constant current will also flow through R_2. In fact, the three-terminal voltage regulator can be used as a reasonably good *constant-current source*, as indicated in Fig. 14-17(b).

The only problem with a "nonadjustable" three-terminal voltage regulator is that a rather large quiescent current I_Q does flow out of its ground terminal. A typical value for I_Q is 10 mA [refer to Fig. 14-17(a)]. By Kirchhoff's current law,

$$I_2 = I_1 + I_Q$$

and by inspection,

$$I_2 = \frac{V_{REG}}{R_1} + I_Q$$

By Ohm's law and Fig. 14-17(a),

$$V_2 = I_2R_2 = \left(\frac{V_{REG}}{R_1} + I_Q\right)R_2$$

$$= V_{REG}\frac{R_2}{R_1} + I_QR_2$$

The equation for the load voltage may be found via Kirchhoff's voltage law.

$$V_L = V_{REG} + V_2 = V_{REG} + V_{REG}\frac{R_2}{R_1} + I_QR_2$$

$$\boxed{V_L = \left(1 + \frac{R_2}{R_1}\right)V_{REG} + I_QR_2} \qquad (14\text{-}30)$$

FIGURE 14-17 **Using the fixed regulator as an adjustable regulator and a constant-current source: (a) adjustable regulator; (b) constant-current source.**

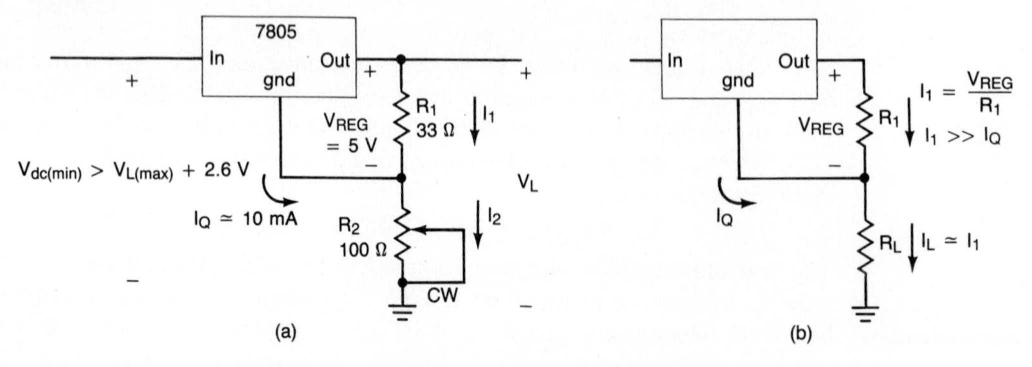

The regulator's quiescent current I_Q plays a significant role in determining the output voltage. This is undesirable because the quiescent current is *not* a tightly controlled parameter. It varies with the regulator's output voltage, the load current, and temperature.

EXAMPLE 14-16

Determine the minimum and maximum load voltage of the regulator given in Fig. 14-17(a). Assume that I_Q is a constant 10 mA.

SOLUTION When potentiometer R_2 is fully counterclockwise it is essentially 0 Ω.

$$V_L = \left(1 + \frac{R_2}{R_1}\right)V_{REG} + I_Q R_2$$

$$= \left(1 + \frac{0\ \Omega}{33\ \Omega}\right)(5\ \text{V}) + (10\ \text{mA})(0\ \Omega) = 5\ \text{V}$$

As we can see, the *minimum* load voltage is equal to V_{REG}. If R_2 is adjusted fully clockwise, its value will be 100 Ω.

$$V_L = \left(1 + \frac{R_2}{R_1}\right)V_{REG} + I_Q R_2$$

$$= \left(1 + \frac{100\ \Omega}{33\ \Omega}\right)(5\ \text{V}) + (10\ \text{mA})(100\ \Omega)$$

$$= 21.15\ \text{V}$$

■

EXAMPLE 14-17

Assume that R_2 in Fig. 14-17(a) has been adjusted to 100 Ω, but I_Q has changed to 7 mA. Recompute V_L.

SOLUTION By Eq. 14-30,

$$V_L = \left(1 + \frac{R_2}{R_1}\right)V_{REG} + I_Q R_2$$

$$= \left(1 + \frac{100\ \Omega}{33\ \Omega}\right)(5\ \text{V}) + (7\ \text{mA})(100\ \Omega)$$

$$= 20.85\ \text{V}$$

■

The variation in I_Q from 7 mA to 10 mA has caused a corresponding change in V_L from 20.85 V to 21.15 V. While a 0.3-V difference is small, the circuit's set-point stability may not be suitable in some applications.

Further, the stability we have achieved has been accomplished by making the current through R_2 much larger than I_Q.

$$I_1 = \frac{V_{REG}}{R_1} = \frac{5\ \text{V}}{33\ \Omega} = 152\ \text{mA} >> I_Q = 10\ \text{mA}$$

Although this approach does tend to make V_L relatively immune to the effects of I_Q, it wastes a great deal of power. The bias current I_1 is drawn from the regulator's output. In the examples above, 152 mA represents 15.2% of the regulator's maximum output current of 1000 mA. Regulators that are designed to be adjustable circumvent this inefficiency.

Adjustable three-terminal voltage regulators offer the same basic features found in the fixed regulators, including thermal overload and short-circuit protection. However, their quiescent current I_Q is much lower (typically, 100 μA or less) and held fairly constant despite changes in the input voltage, load current, and temperature.

The LM317T is an example. It is in a TO-220 case, rated at 1.5 A, has a $T_{j(\max)}$ of 125°C, a $P_{D(\max)}$ of 15 W, a maximum input-to-output differential of 40 V, and a dropout voltage of less than 2.5 V. Its quiescent current is typically 50 μA, with its maximum value being 100 μA. A typical applications circuit has been indicated in Fig. 14-18.

The ''ground'' terminal has been renamed the ''adjustment'' pin. The 100-μA maximum current flowing out of the adjustment terminal produces an error term. To keep it small and essentially constant, the circuitry internal to the LM317T has been designed such that the bulk of the quiescent current actually flows out of the output terminal. Consequently, *the LM317T has a minimum load current requirement of 10 mA*. This requirement is not too severe and is easily accommodated by the proper sizing of R_1.

The LM317T produces an output voltage of 1.25 V. Therefore, to satisfy its output current requirement,

$$R_{1(\max)} = \frac{V_{\text{REG}}}{I_{L(\min)}} = \frac{1.25\text{ V}}{10\text{ mA}} = 125\ \Omega$$

A 75-Ω resistor has been selected.

To ensure that the current which flows out of the adjustment pin (I_Q) introduces negligible error, the value of R_2 should not be too large.

$$R_2 \leq \frac{0.1V_{\text{REG}}}{I_{Q(\max)}} = \frac{(0.1)(1.25\text{ V})}{100\ \mu\text{A}} = 1.25\text{ k}\Omega$$

Observe in Fig. 14-18 that a 1-kΩ potentiometer has been used.

FIGURE 14-18 **A 1.25 to 17.9-V adjustable regulator.**

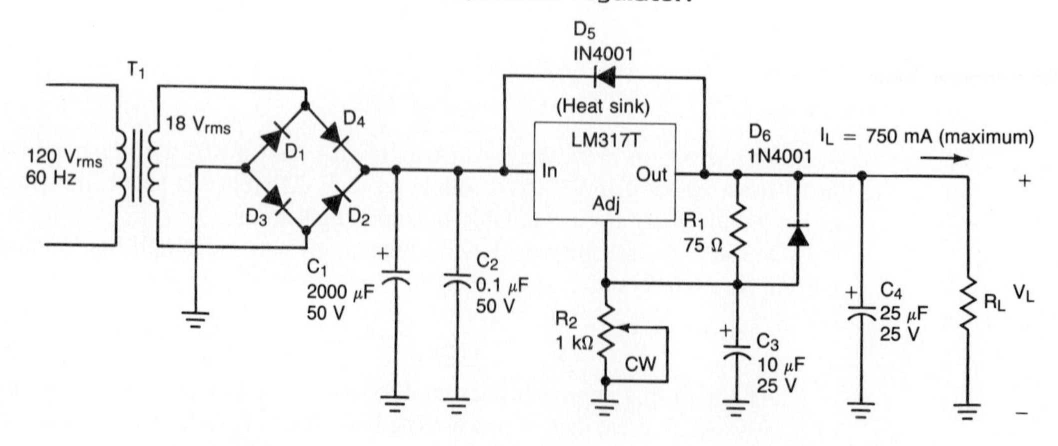

Equation 14-30 also applies to the adjustable regulator circuit given in Fig. 14-18.

$$V_L = \left(1 + \frac{R_2}{R_1}\right)V_{\text{REG}} + I_Q R_2$$

However, the error term $I_Q R_2$ is negligibly small and the load voltage is

$$V_L \simeq \left(1 + \frac{R_2}{R_1}\right)V_{\text{REG}} \tag{14-31}$$

to a close approximation.

The manufacturer recommends a 0.1-μF disk ceramic capacitor (C_2) across the input of the regulator. This tends to minimize the possibility of high-frequency oscillations. The manufacturer also suggests the use of a 10-μF electrolytic capacitor (C_3) from the adjustment pin to ground. This tends to improve the regulator's ripple rejection. The manufacturer claims 80 dB of ripple rejection is obtainable at any output level when C_3 is used. (Without C_3 the regulator's ripple rejection is only 65 dB.) To protect the output of the regulator from C_3's discharge, diode D_6 has also been added. In the event of an input short, diodes D_6 and D_5 serve to provide the discharge path for C_3. If an output short occurs, diode D_6 conducts to protect the adjustment and output pins from C_3's discharge.

Capacitor C_4 is also recommended by the manufacturer. The LM317T is stable even if no output capacitor is used. However, the regulator's output may exhibit ringing in response to transient load changes. Capacitor C_4 minimizes this problem.

EXAMPLE 14-18

Determine the minimum and maximum output voltage for the regulator circuit given in Fig. 14-18.

SOLUTION We may follow the procedure detailed in Example 14-16. However, we shall employ the approximation given in Eq. 14-31. If R_2 is zero ohms, the load voltage will be 1.25 V. If R_2 is adjusted fully clockwise, it will have 1 kΩ of resistance.

$$V_L \simeq \left(1 + \frac{R_2}{R_1}\right)V_{\text{REG}} = \left(1 + \frac{1\ \text{k}\Omega}{75\ \Omega}\right)(1.25\ \text{V}) = 17.9\ \text{V}$$ ■

14-15 Negative Regulators and Dual-Polarity Power Supplies

As we have seen many times previously, positive and *negative* regulated power supply voltages are often required. In response, many manufacturers have developed negative three-terminal regulators. These regulators are available in both "fixed" and adjustable designs. They offer the same types of features and protection as are found in positive regulators.

An example of a dual (positive and negative) regulated supply has been indicated in Fig. 14-19(a). The "front end" consists of a dual-complementary full-wave rectifier. An LM340KC-15 is being used as the positive regulator, and its complement, the LM320KC-15, serves as the negative regulator. The regulators have TO-3 case styles [see Fig. 14-19(b)]. Note that the case is ground for the LM340KC, while the case serves as the input for the LM320KC. This ''detail'' is extremely important and has resulted in many *redesigned* printed circuit boards.

In this circuit a Motorola 3N247 diode bridge assembly (D_1) has been used to reduce the number of discrete components. Diodes D_2 and D_3 are used to protect the outputs of the regulators. Additional output protection diodes are also required in this application [see diodes D_4 and D_5 in Fig. 14-19(a)].

When power is first applied, the two regulators may not turn on simultaneously. If this occurs, the output of the slower regulator will be driven toward the output potential of the faster. As a result, the output circuitry of the ''sluggish'' regulator

FIGURE 14-19 Dual-polarity (±15 V) dc power supply: (a) ±15 V power supply; (b) bottom view of the TO-3 case styles; (c) protecting the regulator outputs during power-up transients.

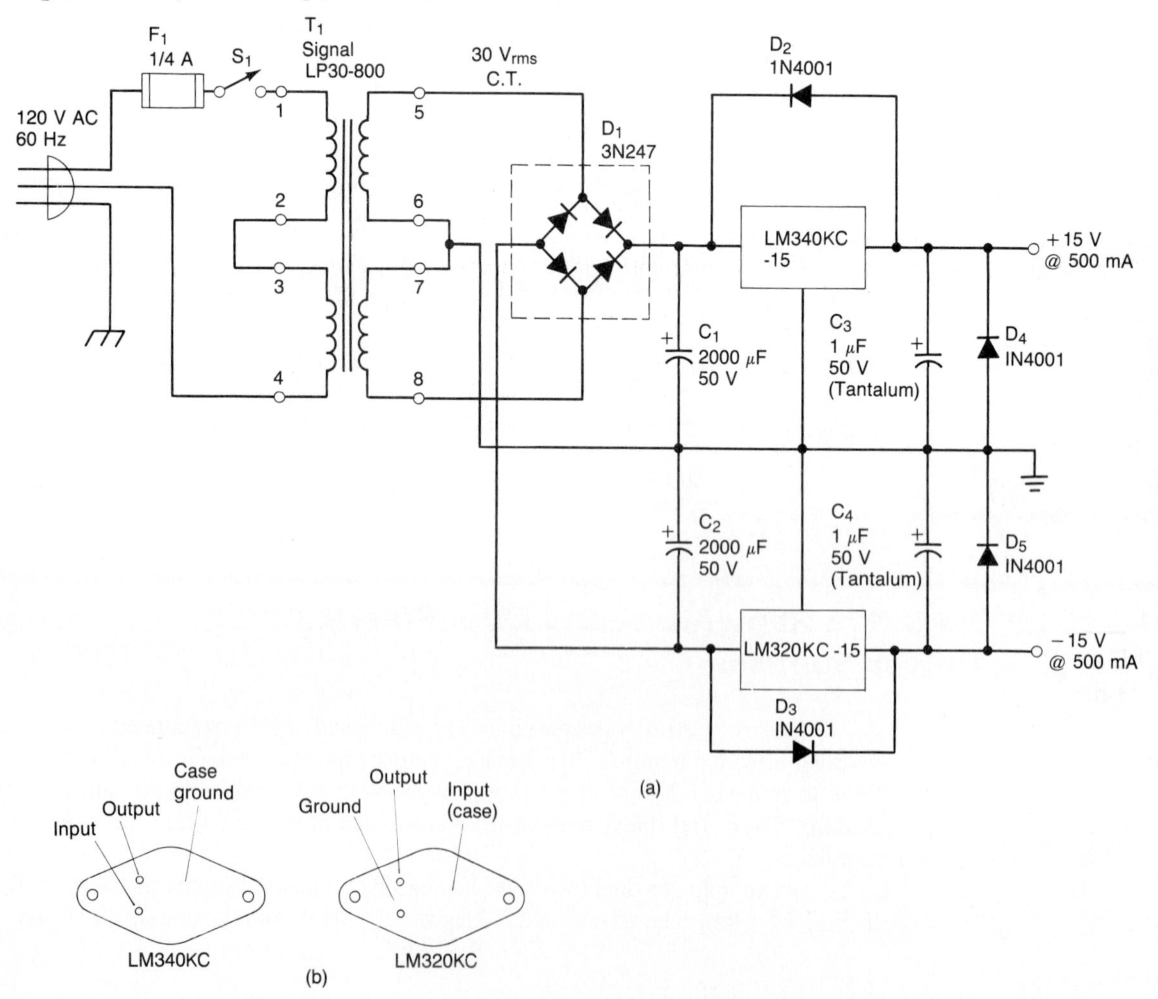

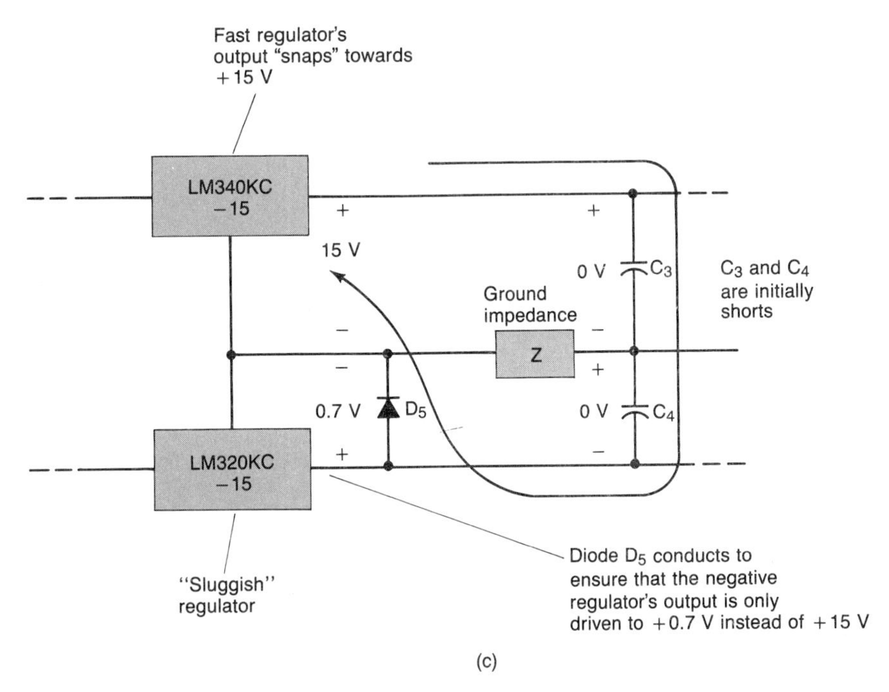

FIGURE 14-19 *(continued)*

may be damaged. Diodes D_4 and D_5 prevent regulator output polarity reversal upon power up. This has been demonstrated in Fig. 14-19(c). Observe that the capacitors are initially shorts. "Z" is used to represent the ground impedance. Ground runs are not perfect, as they possess stray inductance and resistance.

As a final comment, we should point out that the LM320 negative regulator *must* have a capacitor across its output to keep it stable. The manufacturer suggests either a 1-μF tantalum electrolytic or a 25-μF aluminum electrolytic. To help ensure a good output transient response in the LM340, a 1-μF tantalum capacitor has also been connected across its output [see Fig. 14-19(a)].

EXAMPLE 14-19

Determine the dc voltage across capacitors C_1 and C_2 in Fig. 14-19(a) if the dc load current is 500 mA.

SOLUTION From our work in Chapter 13 we recall that the analysis of the dual-complementary full-wave rectifier is similar to that of the full-wave rectifier using a center-tapped transformer.

$$V_m = \sqrt{2}\, V_2 = \sqrt{2}\,(15 \text{ V rms}) = 21.2 \text{ V}$$

$$V_{\text{dcm}} = V_m - 0.7 \text{ V} = 21.2 \text{ V} - 0.7 \text{ V} = 20.5 \text{ V}$$

$$V_{r(\text{p-p})} = \frac{I_{\text{DC}}}{f_r C_1} = \frac{500 \text{ mA}}{(120 \text{ Hz})(2000\ \mu\text{F})} = 2.08 \text{ V p-p}$$

$$V_{DC} = V_{dcm} - \frac{V_{r(p\text{-}p)}}{2} = 20.5\ V - \frac{2.08\ V}{2} = 19.5\ V$$

By symmetry, the dc voltage across C_2 is -19.5 V with respect to ground. ■

EXAMPLE 14-20

Find the power dissipations in the positive and negative voltage regulators used in Fig. 14-19(a).

SOLUTION The power dissipations in each of the regulators will be identical if the positive and negative load current demands are both equal to 500 mA.

$$P_D = (V_{DC} - V_L)I_L = (19.5\ V - 15\ V)(500\ mA)$$
$$= 2.25\ W$$

■

If fixed resistors and potentiometers were added to the positive and negative regulators incorporated in Fig. 14-19(a), we could independently adjust the positive and negative regulated output voltages. However, many applications require *symmetrical* positive and negative supply voltages. To facilitate ease in adjustment, *dual-tracking voltage regulators* are used. In this case a *single* potentiometer is used to adjust *both* the positive and negative supply voltages [see Fig. 14-20(a)].

A "fixed" (LM340K-5.0) 5-V positive regulator and a "fixed" (LM320K-5.0) 5-V negative regulator have been employed. The ground pins of the positive and the negative regulators are driven by a voltage follower and a unity-gain inverting amplifier, respectively. Observe that the two op amps are each one half of an LM1558 dual op amp integrated circuit. This IC package is an eight-pin mini-DIP.

To understand the operation of Fig. 14-20(a), the reader is directed to Fig. 14-20(b). The positive regulator maintains a constant voltage (e.g., 5 V) between its output and ground pins. The voltage follower's output voltage is equal to the voltage which appears at its noninverting input terminal. This implies that the potential difference between its output and noninverting input is zero volts. Therefore, by Kirchhoff's voltage law, V_{REG} must appear across R_1 and R_{2A}. Since the input bias current of the follower is negligibly small, the load voltage is given by the following relationship:

$$V_L = \left(1 + \frac{R_{2B}}{R_1 + R_{2A}}\right)V_{REG}$$

As potentiometer R_2 is varied, the positive reference voltage at the op amp's noninverting input (pin 3) is changed. Consequently, this reference voltage appears at the output of the voltage follower and is used to bias up the positive regulator. Now return to Fig. 14-20(a). The positive reference voltage is made negative by the unity-gain inverting amplifier. Therefore, the adjustment of the potentiometer can be used to control the magnitude of the positive and negative output voltages simultaneously.

In this case the quiescent currents of the regulators may be neglected because the op amp output resistances are very small. However, since the input impedance of the inverting amplifier is equal to R_4, it will load down the R_1-R_2 voltage divider.

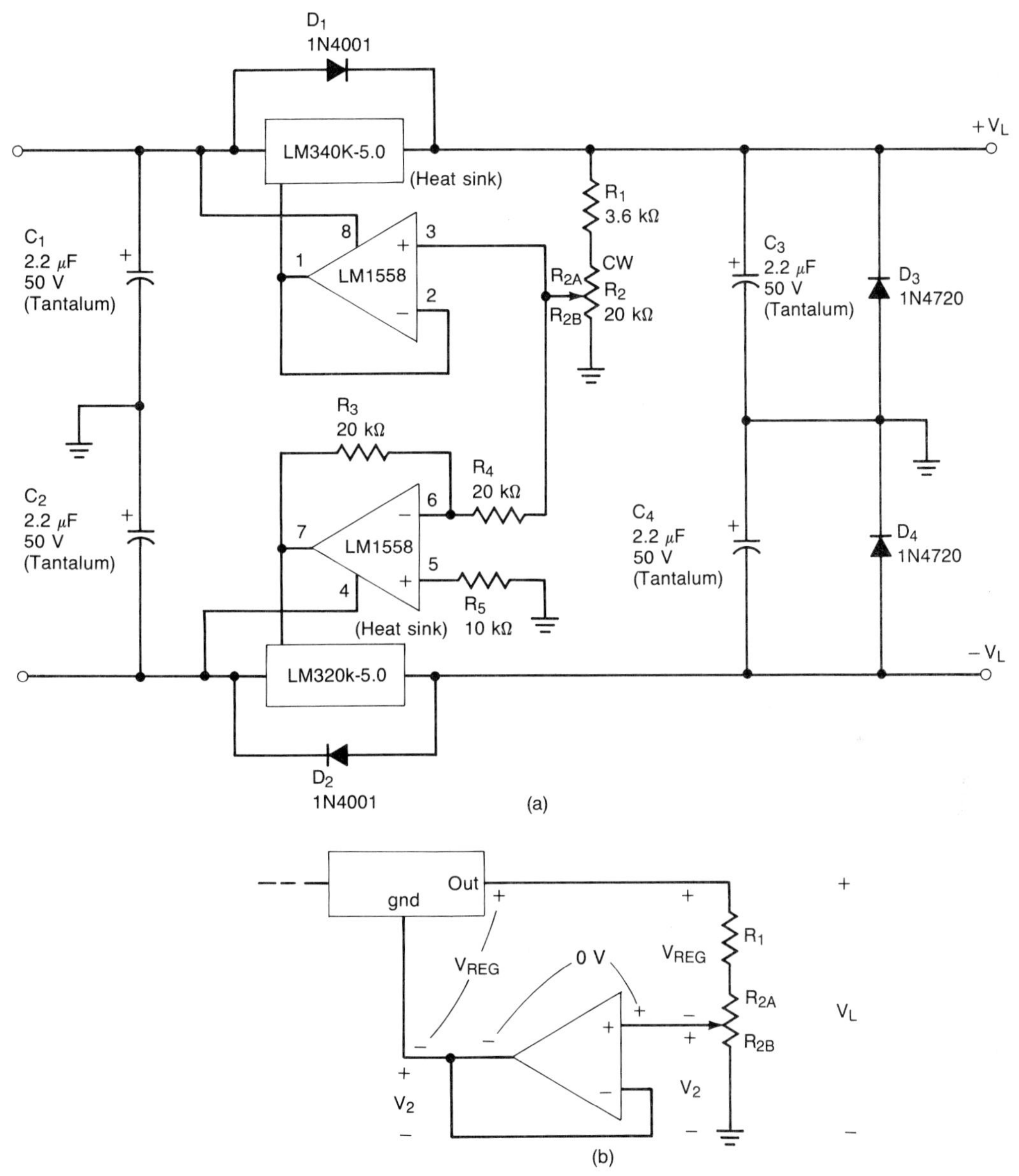

FIGURE 14-20 Dual-tracking voltage regulator with output voltages from ±5 to ±18 V at 1 A: (a) electrical schematic; (b) op amp driver.

In Fig. 14-20(a) the positive load voltage is given by

$$V_L = \left(1 + \frac{R_{2B} \parallel R_4}{R_1 + R_{2A}}\right) V_{REG}$$

The upper portion of R_2 has been denoted as R_{2A} and the lower portion as R_{2B}. This has been noted in Fig. 14-20(a). The parallel combination reflects the loading effect.

EXAMPLE 14-21

Determine the positive and negative load voltages produced by the dual-tracking regulator given in Fig. 14-20(a). Assume that the potentiometer is fully clockwise, adjusted to its center point, and then fully counterclockwise.

SOLUTION When the potentiometer is fully clockwise, R_{2A} is zero and R_{2B} is 20 kΩ.

$$V_L = \left(1 + \frac{R_{2B} \parallel R_4}{R_1 + R_{2A}}\right)V_{REG} = \left(1 + \frac{20\ \text{k}\Omega \parallel 20\ \text{k}\Omega}{3.6\ \text{k}\Omega + 0\ \text{k}\Omega}\right)(5\ \text{V})$$
$$= 18.9\ \text{V}$$

The negative load voltage will be −18.9 V. When the potentiometer is centered, R_{2A} and R_{2B} will both be equal to 10 kΩ.

$$V_L = \left(1 + \frac{R_{2B} \parallel R_4}{R_1 + R_{2A}}\right)V_{REG} = \left(1 + \frac{10\ \text{k}\Omega \parallel 20\ \text{k}\Omega}{3.6\ \text{k}\Omega + 10\ \text{k}\Omega}\right)(5\ \text{V})$$
$$= 7.45\ \text{V}$$

The negative load voltage will be −7.45 V. When the potentiometer is fully counterclockwise R_{2A} will be 20 kΩ and R_{2B} will be equal to zero.

$$V_L = \left(1 + \frac{R_{2B} \parallel R_4}{R_1 + R_{2A}}\right)V_{REG} = \left(1 + \frac{0\ \text{k}\Omega \parallel 20\ \text{k}\Omega}{3.6\ \text{k}\Omega + 20\ \text{k}\Omega}\right)(5\ \text{V})$$
$$= 5\ \text{V}$$

Again, the negative supply voltage will have the same magnitude, and will therefore be −5 V. ■

14-16 Enhancing the Three-Terminal Regulator

Given their inherent simplicity, the three-terminal voltage regulators can be used to solve a wide variety of power supply problems. In this section we gain some insight into how their performance can be augmented by the addition of external circuitry. Two specific examples are given.

In Fig. 14-21 we have a high-voltage adjustable regulator. The circuit is designed around the LM317HVK. This three-terminal adjustable regulator is capable of supplying in excess of 1.5 A over an output voltage range of 1.2 to 57 V. The IC also offers full overload protection.

The problem to be solved is the design of an adjustable, regulated power supply that can deliver 50 V at a maximum load current of 4 A. Since the LM317HVK is rated at only 1.5 A, we might be tempted to dismiss it in this application. However, Fig. 14-21 demonstrates that it can be used by adding a current-boost transistor. If the load requires 4 A, and the regulator is supplying 1.5 A, the *pnp* transistor can supply the balance of the load current (2.5 A).

The regulator satisfies the load needs at low current levels. However, when the load current reaches approximately 300 mA, the voltage drop across resistor R_3 reaches 0.6 V. At this point, transistor Q_1 will begin to conduct and assist the voltage regulator.

Resistor R_4 and diodes D_1 and D_2 normally do not enter the picture. However,

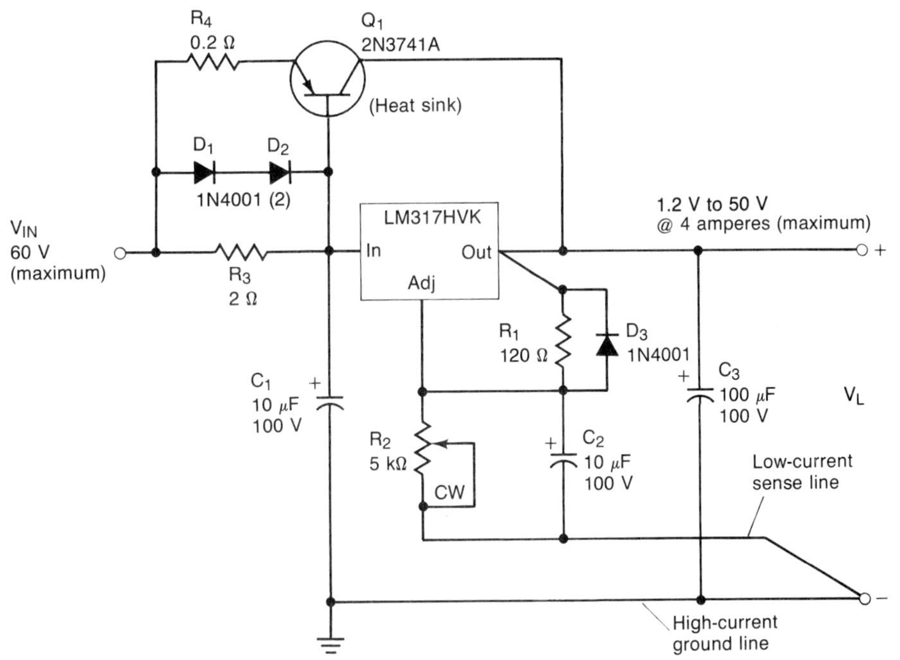

FIGURE 14-21 **High-voltage adjustble regulator using a current-boost transistor.**

if the transistor's collector current reaches approximately 3 A, these components will serve to limit its emitter (and therefore collector) current to this level.

Thus, in the normal operating mode, R_4 and diodes D_1 and D_2 serve to determine the transistor's maximum current. Should an output short circuit occur, the regulator's internal circuitry will serve to limit its power dissipation by folding back its output current. This will reduce the regulator's input current and reduce Q_1's base drive. In essence, the regulator's power limitation also protects the transistor.

To optimize temperature tracking, D_1, D_2, and Q_1 should also be mounted on the regulator's heat sink. It should also be pointed out that at high input-output differentials, the circuit may not provide its full output current capability.

As a final comment, we should mention the circuit's wiring layout. The recommended wiring has been emphasized in the schematic diagram. *To optimize voltage regulation, the load ground return conductor should be run separately.* At high current levels even milliohms of wire resistance can produce significant voltage drops. To preserve the voltage regulation, the high-current path has been separated from the low-current (error amplifier) path. Unfortunately, such documentation efforts are seldom made on conventional electrical schematic diagrams. It then becomes the responsibility of the astute technician to wire the circuit ''correctly.''

In Fig. 14-22(a) a low-voltage adjustable voltage regulator has been shown. In this application a 1.2 V to 25 V dc power supply is required with an *adjustable* current limit. Adjustable current limiting is certainly not a feature offered by three-terminal voltage regulators. Figure 14-22(a) demonstrates one approach to the problem.

The circuit has been designed around an LM338K adjustable voltage regulator.

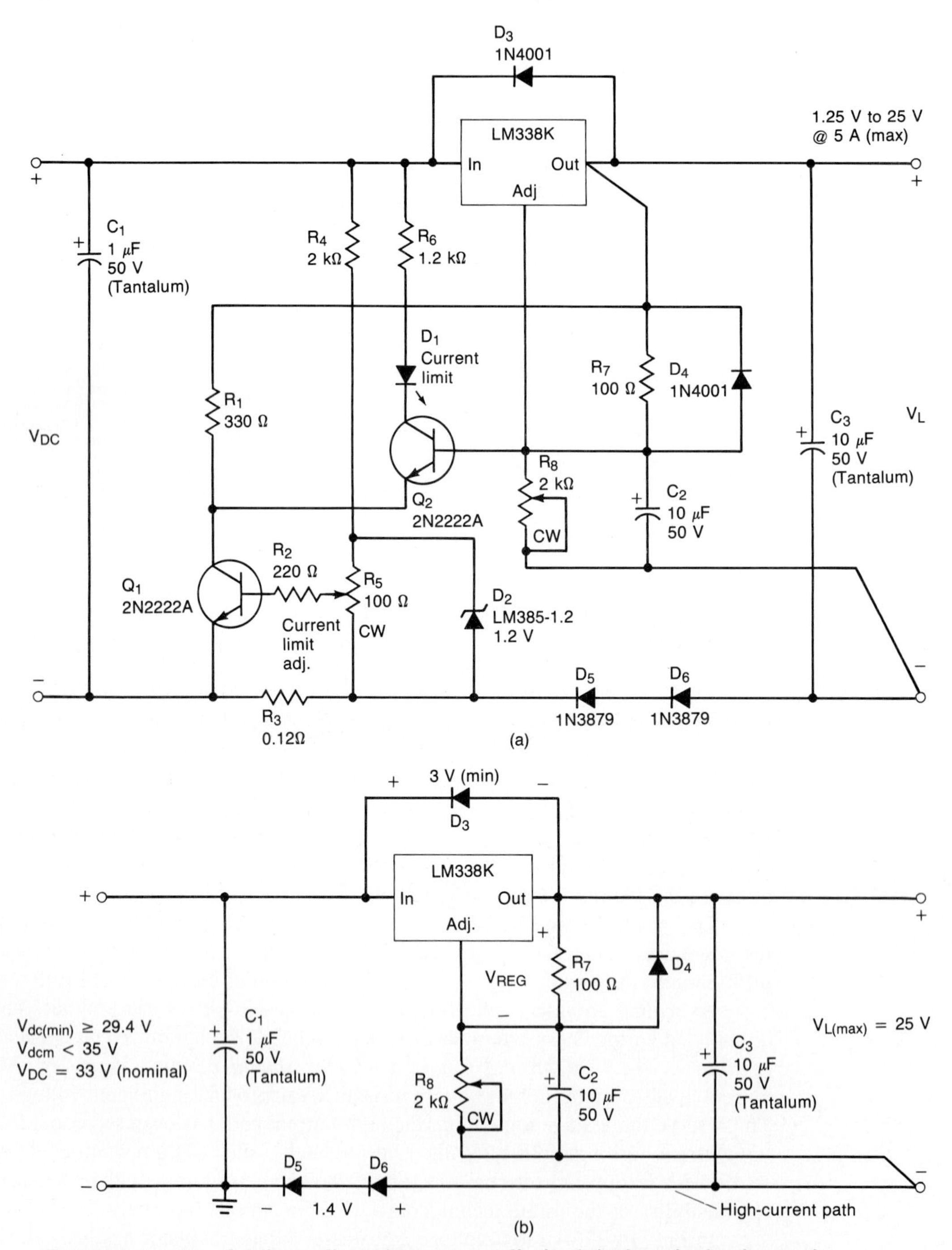

FIGURE 14-22 Adding adjustable current limit: (a) electrical schematic; (b) basic regulator; (c) load voltage when the current limit is set to 5 A; (d) load voltage when the current limit is set to 833 mA.

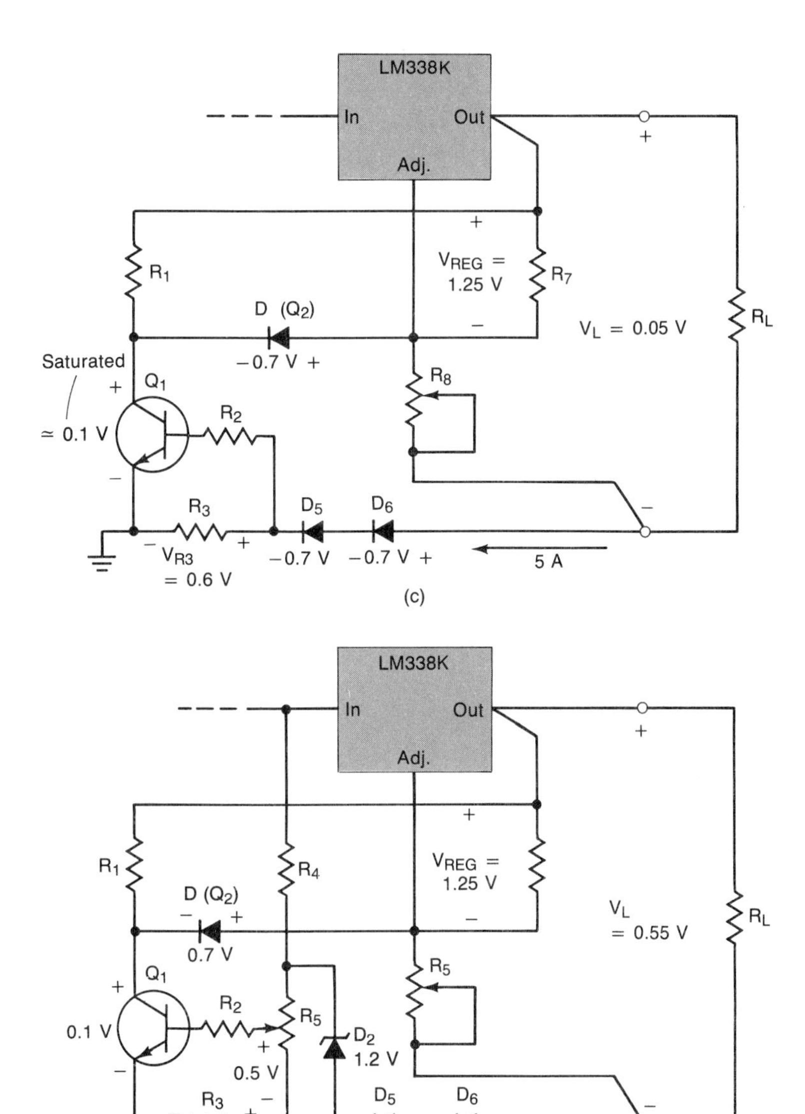

FIGURE 14-22 *(continued)*

The maximum current rating of this device is 5 A. Therefore, a current-boost transistor has not been added. At first glance, the circuit may appear to be slightly overwhelming. However, let us carefully ''pick it apart.'' This is the normal approach to unfamiliar circuits.

The heart of the circuit has been partialed out in Fig. 14-22(b). Capacitor C_1 is a solid tantalum capacitor that has been included to ensure regulator stability. Resistor R_7 and potentiometer R_8 provide the means for adjusting the regulator's output voltage. Capacitor C_2 improves the ripple rejection, and capacitor C_3 improves the regulator's response to transient load changes. Diode D_3 protects the regulator's output in the event of an input short from capacitor C_3's discharge. Diode D_4 is used to protect the output and the adjustment circuitry from capacitor C_2 during either input or output shorts. The role of diodes D_5 and D_6 will become clear as our work progresses.

Despite the presence of diodes D_5 and D_6, the load voltage is given by the following relationship:

$$V_L = \left(1 + \frac{R_8}{R_7}\right)V_{REG}$$

Obviously, if R_8 is adjusted fully counterclockwise, the output voltage will be equal to V_{REG}. The nominal value of V_{REG} is 1.25 V. However, it can range from 1.19 to 1.29 V. If R_8 is fully clockwise, the load voltage for a V_{REG} of 1.19 V is

$$V_L = \left(1 + \frac{R_8}{R_7}\right)V_{REG} = \left(1 + \frac{2\text{ k}\Omega}{100\ \Omega}\right)(1.19\text{ V}) = 24.99\text{ V}$$

and if V_{REG} is 1.29 V,

$$V_L = \left(1 + \frac{R_8}{R_7}\right)V_{REG} = \left(1 + \frac{2\text{ k}\Omega}{100\ \Omega}\right)(1.29\text{ V}) = 27.09\text{ V}$$

Despite this tolerance on the nominal value of V_{REG}, the regulator will regulate the output voltage to within approximately $\pm 0.15\%$ over its full-load current and temperature range.

The presence of diodes D_5 and D_6 and R_3 raises the minimum required input voltage [$V_{dc(min)}$]. The minimum differential of the LM338K is 3 V. If we apply Kirchhoff's voltage law to the circuit given in Fig. 14-22(b) (which ignores R_3), we see that

$$V_{dc(min)} \geq V_{L(max)} + 3\text{ V} + 1.4\text{ V}$$
$$\geq 27.1\text{ V} + 3\text{ V} + 1.4\text{ V} = 31.5\text{ V}$$

The LM338K has a maximum input voltage rating of 35 V. Therefore,

$$V_{dcm} < 35\text{ V}$$

The input voltage constraints have been indicated in Fig. 14-22(b). A good design target for V_{DC} would be 33 V.

The regulator's internal circuitry limits its power dissipation to 50 W. Therefore, when large input-output differentials exist, it may not deliver its rated 5 A of current. For example, assume that the dc input voltage V_{DC} is 33 V. For simplicity, we shall again use Fig. 14-22(b). If the load voltage has been adjusted to 1.25 V, the regulator's maximum load current is

$$I_{L(max)} = \frac{P_{D(max)}}{V_{DC} - V_L - V_{D5} - V_{D6}}$$
$$= \frac{50\text{ W}}{33\text{ V} - 1.25\text{ V} - 0.7\text{ V} - 0.7\text{ V}} = 1.65\text{ A}$$

By the same token, if the output voltage is adjusted to 25 V, the input-to-output differential is reduced and

$$I_{L(\max)} = \frac{P_{D(\max)}}{V_{DC} - V_L - V_{D5} - V_{D6}}$$

$$= \frac{50\text{ W}}{33\text{ V} - 25\text{ V} - 0.7\text{ V} - 0.7\text{ V}} = 7.58\text{ A}$$

The actual current limit is thermally controlled. Consequently, for short periods, the regulator can deliver very large peak currents. (According to the manufacturer, for input-output differentials of 10 V, the regulator can deliver 12 A for 0.2 ms.)

Of course, all of the current-limit discussions above are based on a case temperature of 25°C. Therefore, we must expect considerably less performance when we get rid of the implied ''infinite'' heat sink. Now let us consider the external current limit circuitry.

Resistor R_3 is used to sense the load current and potentiometer R_5 [Fig. 14-22(a)] is used to adjust the trip point. If the potentiometer is adjusted fully clockwise, we have the simplified equivalent circuit shown in Fig. 14-22(c).

When the load current reaches 5 A, 0.6 V will be dropped across R_3, and this will turn on transistor Q_1. When Q_1 is fully turned on, its collector-to-emitter voltage will be approximately zero (e.g., 0.1 V). The output voltage is decreased because the regulator's adjustment pin will be pulled to ground. Normally, this means the output voltage will go to V_{REG} (1.25 V). However, the inclusion of diodes D_5 and D_6 serve to lower the output voltage to approximately zero. This can be understood by applying Kirchhoff's voltage law around the loop indicated in Fig. 14-22(c).

$$\begin{aligned} V_L &= V_{REG} + V_D + V_{CE1} - V_{R3} - V_{D5} - V_{D6} \\ &= 1.25\text{ V} + 0.7\text{ V} + 0.1\text{ V} - 0.6\text{ V} - 0.7\text{ V} - 0.7\text{ V} \\ &= 0.05\text{ V} \end{aligned}$$

The D_2 ''zener'' diode (LM385-1.2) in Fig. 14-22(a) is actually another integrated circuit called a *band-gap reference*. It is a precision voltage reference consisting of several transistors, resistors, and frequency-compensation capacitors. It provides exceptionally low dynamic impedance, long-term stability, low noise, and is temperature compensated. As its name implies, its terminal voltage is referenced to a couple of forward-biased *p-n* junctions.

In this application, the reference voltage is used to (precisely) prebias Q_1's base-emitter. Therefore, less load current is required to turn on transistor Q_1 [refer to Fig. 14-22(d)].

In this case, potentiometer R_5 has been adjusted to apply 0.5 V to Q_1's base–emitter junction. Therefore, a load current of 833 mA is required to develop 0.1 V across R_3 to turn on Q_1 and to place the circuit in its current-limiting mode.

As we can see in Fig. 14-22(d), the load voltage will increase slightly. Once again, by Kirchhoff's voltage law,

$$\begin{aligned} V_L &= V_{REG} + V_D + V_{CE1} - V_{R3} - V_{D5} - V_{D6} \\ &= 1.25\text{ V} + 0.7\text{ V} + 0.1\text{ V} - 0.1\text{ V} - 0.7\text{ V} - 0.7\text{ V} \\ &= 0.55\text{ V} \end{aligned}$$

Diode D in Fig. 14-22(c) and (d) is actually the base-emitter *p-n* junction of transistor Q_2. Transistor Q_2 is used to drive the current limit LED (D_1). When transistor Q_1 turns on, the emitter of transistor Q_2 is taken to ground, and transistor Q_2 also turns on. Transistor Q_2's collector current will flow and the LED will illuminate. If we assume that the LED drops 1.8 V, V_{CE2} is zero, and V_{DC} is 33 V, we can find I_{C2}.

$$I_{C2} = \frac{V_{DC} - V_{D1}}{R_6} = \frac{33\text{ V} - 1.8\text{ V}}{1.2\text{ k}\Omega} = 26\text{ mA}$$

A current of 26 mA should be more than adequate to illuminate the LED. If a high-efficiency (more expensive) LED is incorporated into the design, only about 4 mA of current would be required to illuminate the LED. In this case, a larger value (e.g., 7.5 kΩ) could be used for R_6.

14-17 Introduction to Switching Regulators

All of the voltage regulators we have studied thus far have required the use of a series-pass transistor. This transistor was either a discrete device or contained within an IC regulator. Regardless of the case, the transistor is used in its linear mode of operation. In essence, it is being employed as a class A power amplifier. Consequently, it is not extremely efficient, and a great deal of power is thrown away as heat [see Fig. 14-23(a)]. In fact, as the maximum required load current increases, the power levels handled by the series-pass transistor also increase dramatically. Obviously, the cost of the series-pass transistor increases, and the size of the required heat sink increases. In some cases, a fan may be needed to remove the heat generated by the transistor.

In the quest for improved efficiency, circuit designers have developed a relatively new approach - *switching regulators*. As its name implies, this mode of operation requires the transistor to perform as a *switch*.

When a transistor is saturated, it may be thought of as a closed switch. When a transistor is in cutoff, it behaves as an open switch. In Fig. 14-23(b) we see a transistor that is being driven by a rectangular waveform. When the input waveform is positive, the transistor will be saturated. When the input waveform is zero, the transistor will be in cutoff. Let us analyze Fig. 14-23(b).

If we assume that v_{BE} is 0.7 V and v_{IN} is 10 V, we find that the base current is

$$i_B = \frac{v_{IN} - 0.7\text{ V}}{R_B} = \frac{10\text{ V} - 0.7\text{ V}}{10\text{ k}\Omega} = 0.93\text{ mA}$$

Recall that the maximum collector current is $i_{C(SAT)}$, which occurs when v_{CE} is zero.

$$i_{C(SAT)} = \frac{V_{CC}}{R_C} = \frac{10\text{ V}}{1\text{ k}\Omega} = 10\text{ mA}$$

If the h_{FE} of the transistor is 100 and i_B is 0.93 mA, the collector current will *attempt* to be

$$i_C = h_{FE}i_B = (100)(0.93\text{ mA}) = 93\text{ mA}$$

Obviously, the collector current can never reach this value since R_C is 1 kΩ. The conclusion to be drawn is that when v_{IN} is 10 V, the collector current will be equal to $i_{C(SAT)}$, and v_{CE} will be approximately zero. The transistor is deep into

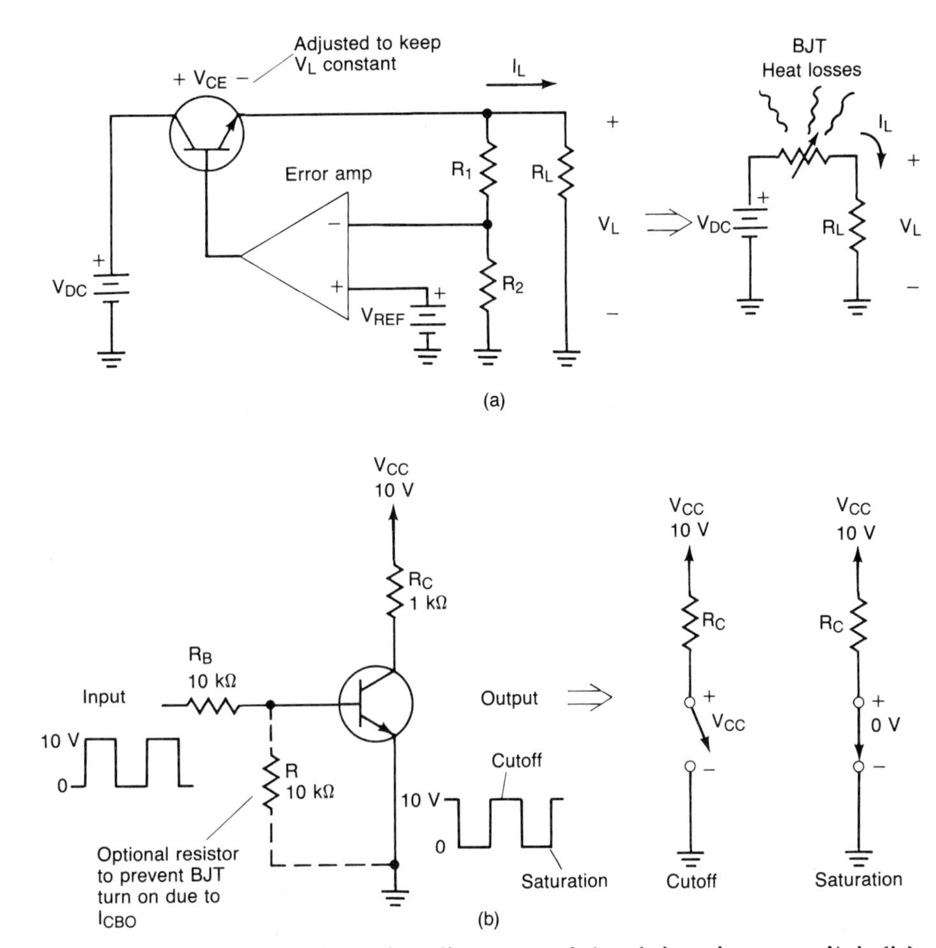

FIGURE 14-23 Transistor in a linear regulator (a) and as a switch (b).

saturation. When a device passes current and has a voltage drop of zero volts, it may be thought of as a closed switch.

When v_{IN} is zero, the base and collector currents will also be zero. Therefore, the voltage drop across R_C will be zero. By Kirchhoff's voltage law, v_{CE} must be equal to V_{CC}, or 10 V. Recall that this is called cutoff. Hence

$$v_{CE(\text{OFF})} = V_{CC} = 10 \text{ V}$$

When a device passes zero current and drops all of the applied voltage, it is essentially an open switch.

When an ideal switch is closed, it passes current, and drops zero volts. Therefore, its power dissipation is zero. Similarly, when a switch is open, it drops voltage, but the current through it is zero. An open switch also dissipates zero power. Therefore, a very good transistor switch will dissipate a minimum amount of power. This is one of the primary motivating factors behind switching regulators. We no longer have to contend with the large losses associated with the series-pass transistor.

A simple switching regulator has been depicted in Fig. 14-24(a). A transistor

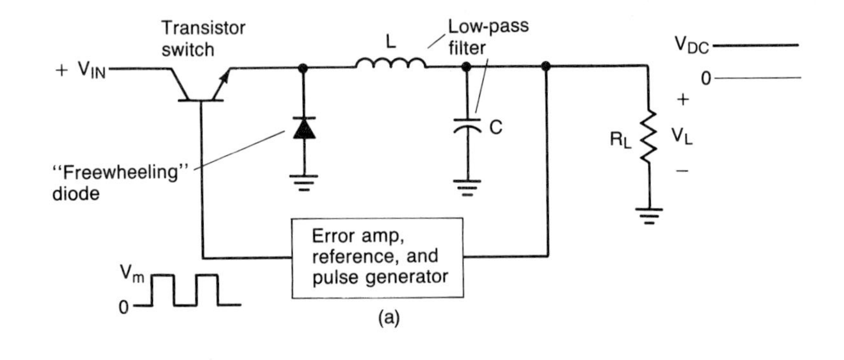

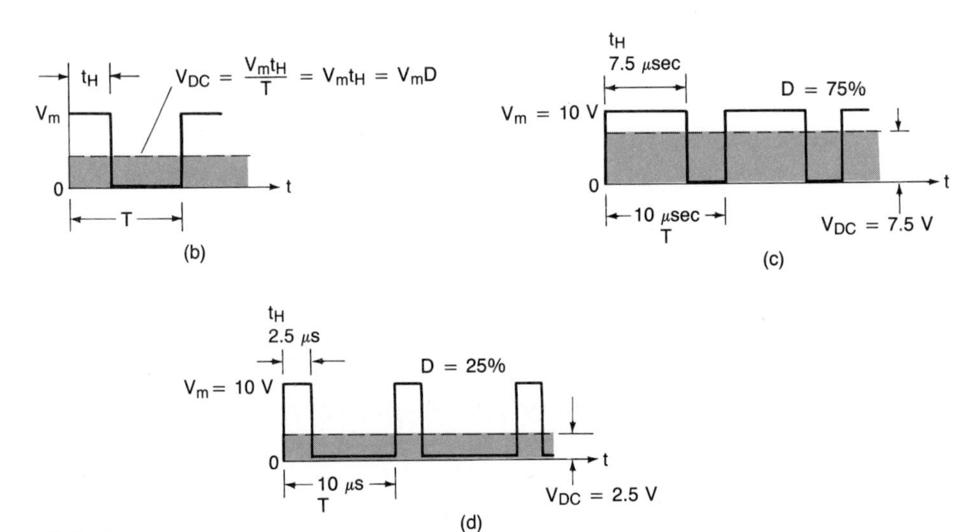

FIGURE 14-24 **Basic switching regulator.**

is being used as a low-loss switch, an inductor and capacitor arrangement is being employed as a low-pass filter, and a pulse generator (with an error amplifier and reference) is used to control the transistor switch. The diode is called a "catch" or "freewheeling" diode. It is used to clamp the negative voltage spikes produced by the inductor as the transistor switches from on to off. Since an inductor opposes a change in current, it will attempt to maintain its current flow by developing a large counter voltage. When the transistor cuts off, the diode continues to provide a path for current flow through the inductor. The diode serves to protect the transistor.

Voltage regulation is achieved by using a technique called *pulse-width modulation*. Very simply, the dc average of a rectangular waveform is developed at the output of a low-pass filter. The longer it is at its maximum (peak) value, the greater its dc average will be. (Waveform average values were discussed in Chapter 13.)

When the transistor switch is closed, V_m is applied to the low-pass filter. When the transistor is open, zero volts is applied to the filter. If V_L is too low, the transistor switch is held closed longer. If V_L is too large, the transistor is not closed as long. The resultant waveform at the emitter of the transistor is characterized by its *duty cycle* (D).

The duty cycle of a rectangular waveform is the ratio of the time its at its maximum value (high) to the period of the waveform [refer to Fig. 14-24(b)]. Its "high" time is denoted t_H, and its period is T. It is typically expressed as a percentage. Hence

$$\% D = \frac{t_H}{T} \times 100\% \tag{14-32}$$

where $\% D$ = duty cycle expressed as a percentage
t_H = time the rectangular waveform is high
T = waveform period

Since the dc average value of a rectangular waveform is its area divided by its period, we see that

$$V_L = \frac{\text{area}}{T} = \frac{V_m t_H}{T} = V_m D \tag{14-33}$$

and the dc average (V_L) is therefore proportional to the duty cycle [refer again to Fig. 14-24(b)].

In other words, the greater the duty cycle, the greater the dc level, and vice versa. Therefore, by raising or lowering the duty cycle, the dc average voltage applied to the low-pass filter in Fig. 14-24(a) can be regulated. In Fig. 14-24(c) we see that a 75% duty cycle produces a dc average of 7.5 V. Figure 14-24(d) illustrates that a 25% duty cycle yields a dc average of 2.5 V.

A more detailed example of a switching regulator has been provided in Fig. 14-25(a). The triangle generator consists of an op amp integrator and a noninverting comparator with hysteresis. The detailed operation of this circuit is covered in Chapter 15.

The output of the triangle generator is fed into the noninverting input of a second comparator. An error amplifier is used to monitor the filtered dc output of the low-pass filter. The output of the error amplifier is directed into the inverting input of the comparator. The comparator develops pulses at its output which are used to drive the transistor switch.

Let us focus our attention on the operation of the comparator and the error amplifier [see Fig. 14-25(b)]. The output of the error amplifier is directly proportional to the dc voltage appearing at the output of the low-pass filter. The output of the error amplifier serves as the reference voltage for the comparator. If the reference voltage is equal to one-half of the triangle's peak voltage, the output of the comparator will be a square wave [which is a rectangular wave with a 50% duty cycle - refer to Fig. 14-25(c)].

If the output voltage increases, the output of the error amplifier will also increase. This raises the comparator's reference and lowers the duty cycle of its output [see Fig. 14-25(d)]. If the output voltage is reduced, the output of the error amplifier will also decrease. This lowers the comparator's voltage reference and raises the duty cycle of its output. This is depicted in Fig. 14-25(e).

In both cases the duty cycle is adjusted to counter the changes in the output voltage in order to maintain a constant output voltage. Now that we have a fundamental understanding of the operation of the switching regulator, let us put it into perspective.

Linear power supplies are relatively easy to design, reliable, inexpensive, and their output voltages can be held to close tolerances. Linear power supplies are still the primary choice for analog systems and low-power digital systems. Switching regulators do not regulate voltage as precisely as linears; they generate appreciable ripple and electrical noise. But "switchers" are smaller, lighter, and more efficient than "linears." This makes them extremely attractive in the more noise-tolerant digital equipment, and in weight-critical avionics applications.

A number of switching regulator categories and topologies exist. For example, isolated switchers employ a transformer, rectifier, and filter at their input. This provides isolation (from earth ground) and serves to simplify the regulator's design. "Off-line" switchers do not use an input transformer. This greatly reduces their weight and size, but it also eliminates their isolation. This can be a serious drawback.

FIGURE 14-25 Pulse-width modulation in a switching regulator.

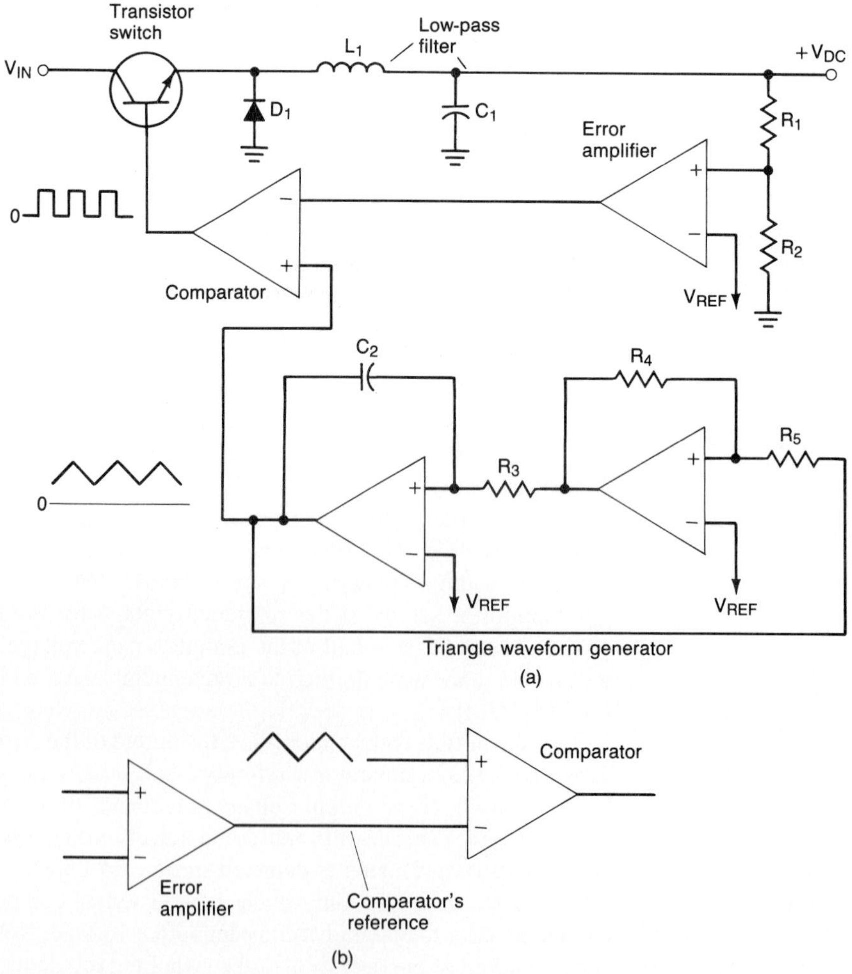

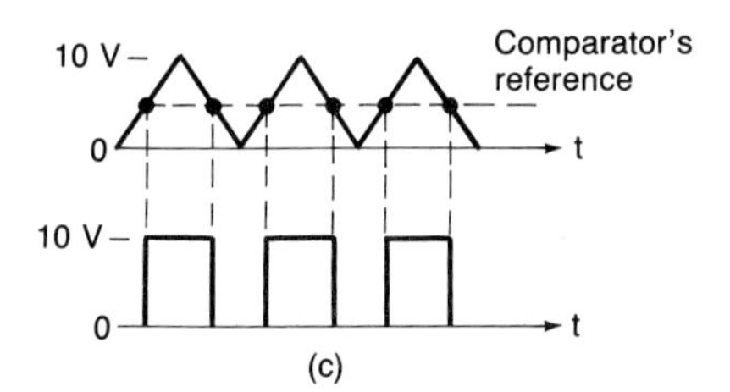

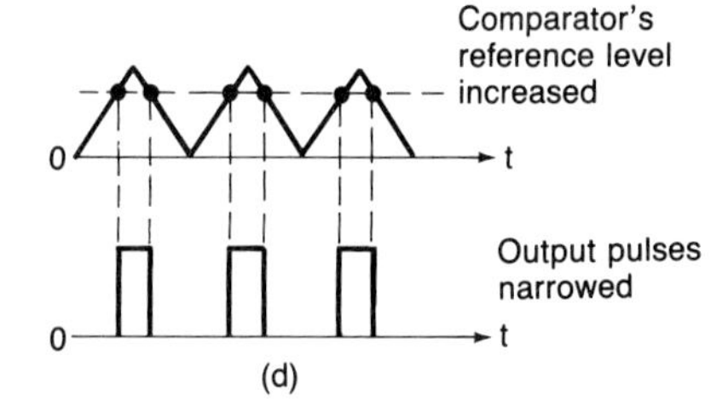

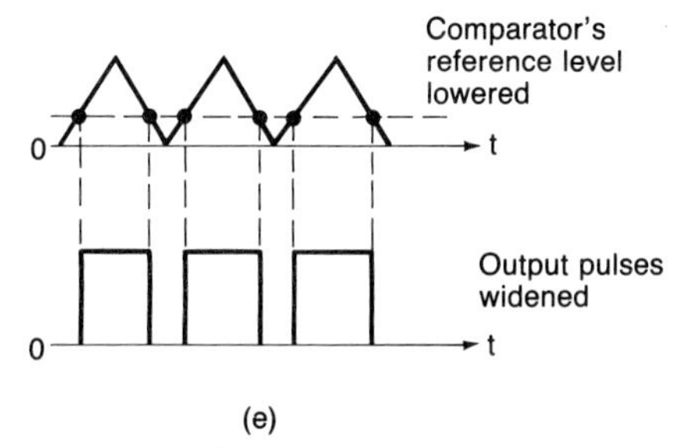

FIGURE 14-25 *(continued)*

Switching power supplies generally fall into one or the other or a combination of two categories - buck or boost. Very simply, buck power supplies deliver pulses of energy directly to the load. Between pulses, the load receives energy from energy storage devices (capacitors and/or inductors). Boost power supply circuits deliver pulses of energy to storage devices which subsequently deliver energy to the load.

The various topologies can be overwhelming to the novice. Examples include the flyback converter (a buck-boost circuit), the forward converter (a buck circuit), and push-pull half-bridge or full-bridge converters (both of which are buck circuits). We shall not delve into the specifics.

The switching frequency [which is set by the triangle generator in Fig. 14-25(a)] is of major concern. The typical switching frequencies range from 10 kHz to 100 kHz. However, National Semiconductor Corp. has developed a 1-MHz switcher that is capable of delivering 15 W to its load, and a 1-MHz switcher has also been developed by Theta-J which is rated at 100 W. Switching speeds up to 10 MHz are on the horizon. Generally, the higher the frequency, the smaller the inductors, and capacitors used in the low-pass filter can be. This can greatly reduce the size and cost of a switcher. There are, of course, other trade-offs.

Getting a transistor to switch several amperes of current at a very high frequency is no easy task. Bipolar transistors are generally used for switchers in the range 10 to 50 kHz. Power MOSFETs are usually selected for switchers that operate above 50 kHz. No clear-cut rules exist, as this is an area of extremely aggressive development.

14-18 Ac Power Control

In addition to regulating dc voltage and current, there are many applications that require regulating *ac power* to loads. In this case the loads may be such things as resistive heaters, incandescent lamps, or ac (e.g., universal) motors as opposed to electronic systems. To investigate this area, we need to answer three fundamental questions.

1. What are the types of solid-state switches used to control ac power?
2. What are the trigger devices used to turn on the ac switches?
3. Why is isolation needed and what are optoisolators?

Ac power control is a fascinating, but often involved area of study. Our intent is to provide an introduction with the hope that the student will pursue this area further in later studies.

14-19 SCR Ac Power Control

In Section 14-8 we studied the operation of the SCR and saw how it can be used as a latch. In Section 14-17 we saw that very efficient voltage regulators can be designed by using the BJT as a switch. Recall that an ideal closed switch passes current without dissipating power since it has zero volts across it. Similarly, an open switch will drop voltage but also dissipates zero watts since no current flows through it. Real switches, of course, will dissipate some power. However, they still offer a very efficient means of controlling dc or ac power.

In Fig. 14-26(a) we see an SCR that is being used as a switch to control the ac power delivered to a resistive load (R_L). In this case, the resistive load is being used as a heater.

The gate circuit is being driven by the ac line voltage (v_s) as shown in Fig. 14-26(a). Remember that between the gate and cathode terminals we essentially have a base-emitter *p-n* junction. Also recall that most base-emitter *p-n* junctions will experience zener breakdown (5 to 6 V) if sufficiently reverse biased. Therefore, to prevent possible damage to the SCR, a 1N4004 silicon rectifier has been placed in series with the gate terminal. (The 1N4004 has a reverse voltage rating of 400 V.)

Resistor R_1 limits the maximum gate current (i_G). The SCR's required *gate trigger current* is denoted as i_{GT}. If Kirchhoff's voltage law is applied to the gate circuit [Fig. 14-26(a)], we can arrive at an equation for i_G.

$$-v_s + i_G R_1 + v_{D1} + v_{GK} = 0$$

$$i_G = \frac{v_s - v_{D1} - v_{GK}}{R_1} \tag{14-34}$$

According to the manufacturer's data, the gate trigger current (i_{GT}) has a typical value of 18 mA and a maximum value of 25 mA. The gate-to-cathode trigger voltage (v_{GT}) ranges from 0.9 to 1.5 V.

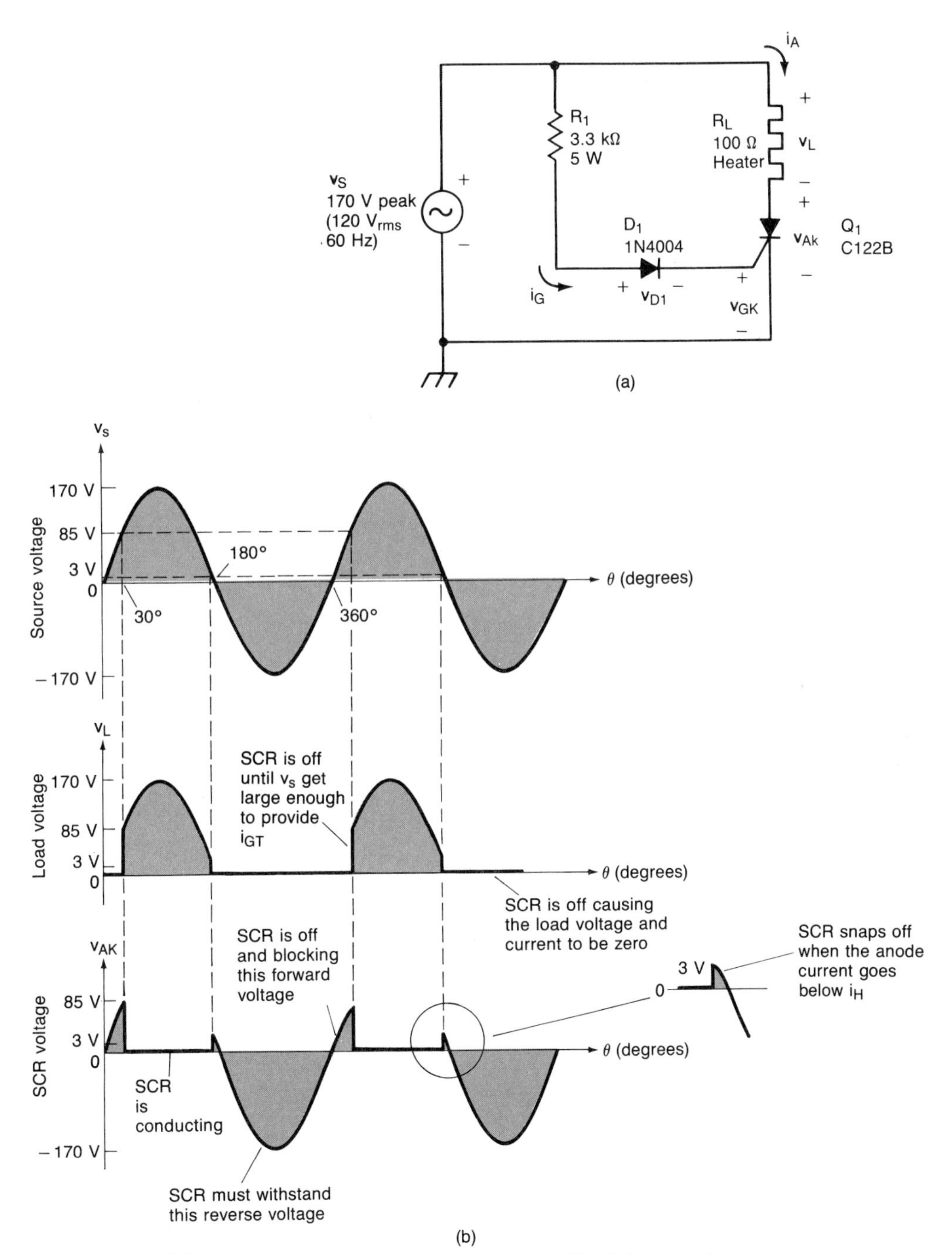

FIGURE 14-26 SCR ac power control: (a) circuit; (b) waveforms.

EXAMPLE 14-22

Determine the required value of R_1 [Fig. 14-26(a)] to trigger a C122B SCR and provide a minimum conduction angle of 150°.

SOLUTION The maximum conduction angle for *any* rectifier is 180°. If the SCR is to conduct for 150°, we must first find the source voltage that occurs at an angle of 30°.

$$v_s(\theta) = V_m \sin\theta$$
$$= v(30°) = 170 \sin 30° = 170(0.5) = 85 \text{ V}$$

Therefore, we want every C122B SCR to fire when v_s reaches 85 V. If we take the worst-case situation, a C122B will require an i_{GT} of 25 mA and have a gate-to-cathode trigger voltage (v_{GT}) of 1.5 V. Using these values and solving Eq. 14-34 for R_1 produces

$$R_1 = \frac{v_s - v_{D1} - v_{GK}}{i_G} = \frac{v_s - v_{D1} - v_{GT}}{i_{GT}}$$
$$= \frac{85 \text{ V} - 0.7 \text{ V} - 1.5 \text{ V}}{25 \text{ mA}} = 3.31 \text{ k}\Omega$$

As can be seen in Fig. 14-26(a), a 3.3-kΩ standard value has been selected. ■

Recall that an SCR will turn off when its anode current is reduced below its holding current. The maximum instantaneous holding current (i_H) for the C122B is 30 mA.

EXAMPLE 14-23

Determine the value of the forward voltage that will shut off the SCR in Fig. 14-26(a). What is the corresponding angle of v_s?

SOLUTION If we apply Kirchhoff's voltage law around the SCR's anode-cathode circuit [Fig. 14-26(a)], we arrive at

$$-v_s + i_A R_L + v_{AK} = 0$$

and solving for v_s yields

$$v_s = i_A R_L + v_{AK}$$

Assuming that v_{AK} is negligibly small (since the SCR is conducting), we obtain

$$v_s = i_A R_L = i_H R_L = (30 \text{ mA})(100\ \Omega) = 3 \text{ V}$$

Now we can find the angle θ.

$$v_s(\theta) = V_m \sin\theta$$
$$\sin\theta = \frac{v_s(\theta)}{V_m}$$
$$\theta = \sin^{-1}\frac{v_s(\theta)}{V_m} = \sin^{-1}\frac{3 \text{ V}}{170 \text{ V}} \simeq 1.01°$$

This corresponds to 178.99°. Obviously, this is not extremely significant. ■

Considerable insight into the operation of the SCR circuit can be gained by observing its associated waveforms [see Fig. 14-26(b)]. Study them carefully.

14-20 Adjustable (0 to 90°) SCR Phase Control

The SCR circuit in Fig. 14-27(a) is capable of adjusting the firing angle of the SCR. Observe that the gate circuit is tied across the SCR rather than across the source [contrast Figs. 14-26(a) and 14-27(a)]. This offers the advantage of reducing the power dissipation in the gate circuit.

Before the SCR fires, it acts like an open circuit and we have the situation shown in Fig. 14-27(b). The SCR will continue to act like an open circuit until the gate current (i_G) exceeds the SCR's required gate trigger current (i_{GT}). After the SCR

FIGURE 14-27 Adjustable SCR circuit: (a) complete schematic; (b) equivalent circuit before SCR fires; (c) equivalent circuit when SCR is in conduction; (d) finding the equivalent gate circuit.

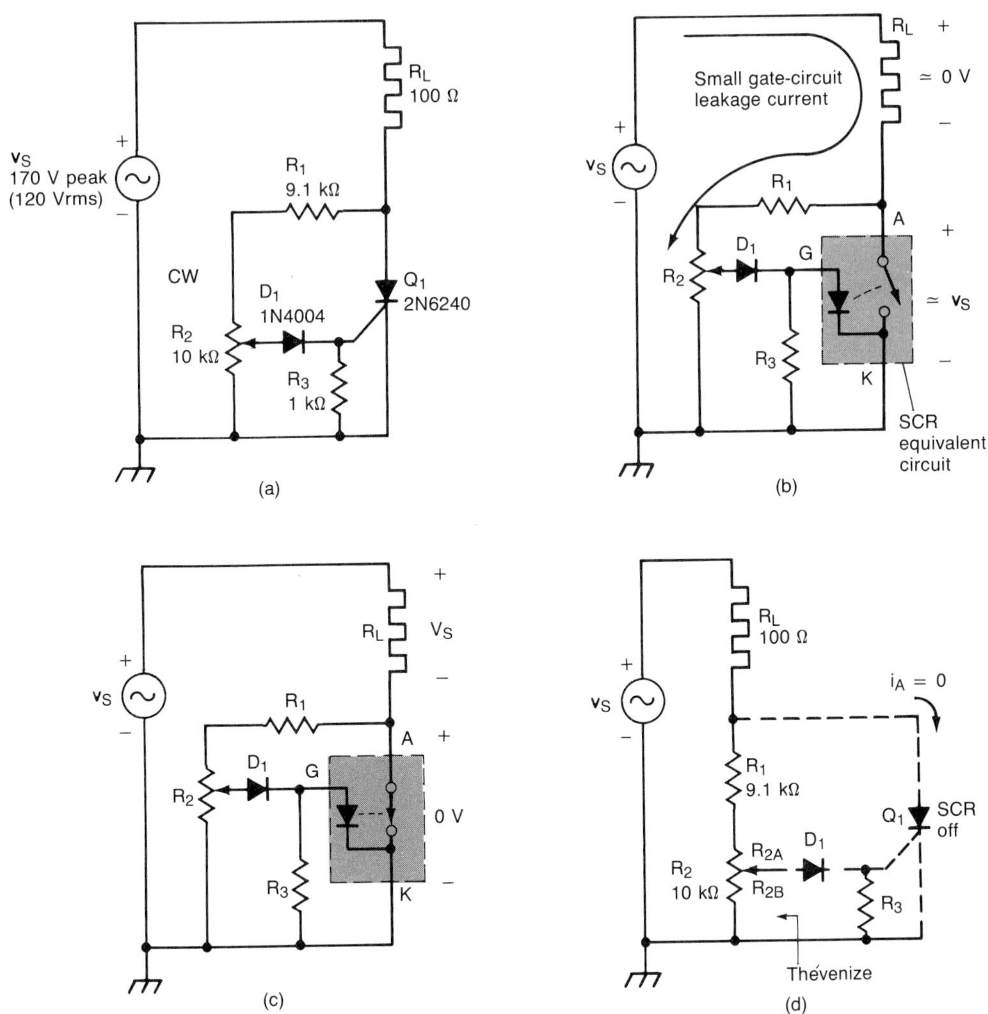

fires, it acts like a closed switch and effectively shorts out the gate circuit [see Fig. 14-27(c)]. Therefore, the power dissipation in the gate circuit goes to zero.

This scheme is quite clever when we remember that once the SCR has fired, it no longer requires gate current. Consequently, this technique is quite popular. The only drawback is the fact that the load current can never go to zero. The gate circuit forms a leakage path [see Fig. 14-27(b)]. This is not a severe problem as long as the gate resistance is much larger than R_L.

To analyze this circuit, we shall first apply Thévenin's theorem to the gate circuit. As can be seen in Fig. 14-27(d), v_{TH} is given by

$$v_{TH} = \frac{R_{2B}}{R_L + R_1 + R_{2A} + R_{2B}} v_s = \frac{R_{2B}}{R_L + R_1 + R_2} v_s \tag{14-35}$$

and R_{TH} is

$$R_{TH} = R_{2B} \parallel (R_{2A} + R_1 + R_L) \tag{14-36}$$

EXAMPLE 14-24

Verify that every 2N6240 SCR used in the circuit shown in Fig. 14-27(a) will fire when the voltage source is at 5° and potentiometer R_2 is adjusted fully clockwise. The 2N6240 has an $i_{GT(\max)}$ of 200 μA and a $v_{GT(\max)}$ of 1 V.

SOLUTION First we find $v_s(\theta)$.

$$v_s(\theta) = V_m \sin\theta = 170 \sin 5° = 14.8 \text{ V}$$

Next we note that if R_2 is fully clockwise, R_{2A} is 0 Ω and R_{2B} is 10 kΩ. Now we can find v_{TH} and R_{TH}.

$$v_{TH} = \frac{R_{2B}}{R_L + R_1 + R_2} v_s = \frac{10 \text{ k}\Omega}{100\ \Omega + 9.1 \text{ k}\Omega + 10 \text{ k}\Omega}(14.8 \text{ V})$$
$$= 7.71 \text{ V}$$
$$R_{TH} = R_{2B} \parallel (R_{2A} + R_1 + R_L)$$
$$= 10 \text{ k}\Omega \parallel (0\ \Omega + 9.1 \text{ k}\Omega + 100\ \Omega) = 4.79 \text{ k}\Omega$$

The Thévenin equivalent circuit is given in Fig. 14-28. The maximum v_{GT} (gate-to-cathode) trigger voltage has been indicated. Since

$$v_{TH} = 7.71 \text{ V} > v_{GT(\max)} = 1 \text{ V}$$

Enough forward-bias gate-to-source voltage exists, but we must also verify that we have enough gate current i_G. If we assume that v_{GK} is 1 V and apply Kirchhoff's voltage law, we may solve for i in Fig. 14-28.

$$-v_{TH} + iR_{TH} + v_D + v_{GK} = 0$$
$$i = \frac{v_{TH} - v_D - v_{GK}}{R_{TH}} = \frac{7.71 \text{ V} - 0.7 \text{ V} - 1 \text{ V}}{4.79 \text{ k}\Omega} = 1.255 \text{ mA}$$

By Kirchhoff's current law, we find that

$$i_G = i - i_3 = i - \frac{v_{GK}}{R_3} = 1.255 \text{ mA} - \frac{1 \text{ V}}{1 \text{ k}\Omega} = 0.255 \text{ mA}$$

and since $i_G = 255\ \mu\text{A} \geq i_{GT(\max)} = 200\ \mu\text{A}$, every 2N6240 SCR will fire at an angle of 5°. ■

The adjustment of R_2 and the corresponding waveforms are summarized in Fig. 14-29. When R_2 is fully counterclockwise, the anode of diode D_1 is taken to ground

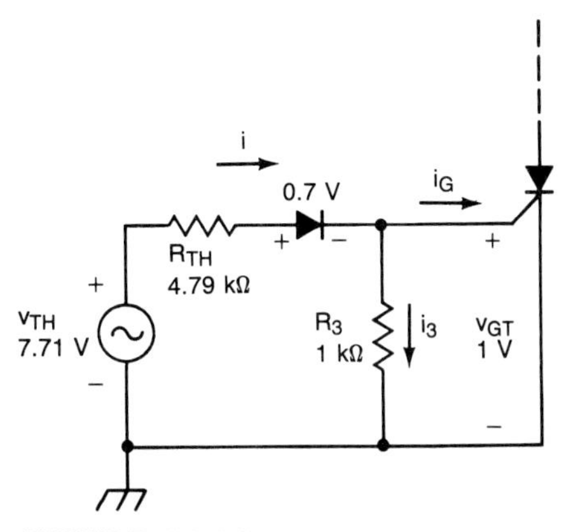

FIGURE 14-28 Thévenin equivalent gate circuit.

FIGURE 14-29 Source (v_s), load (v_L), and SCR(v_{AK}) waveforms for Fig. 14-27(a): (a) R_2 fully clockwise; (b) R_2 adjusted counterclockwise for 90° conduction; (c) R_2 adjusted fully counterclockwise.

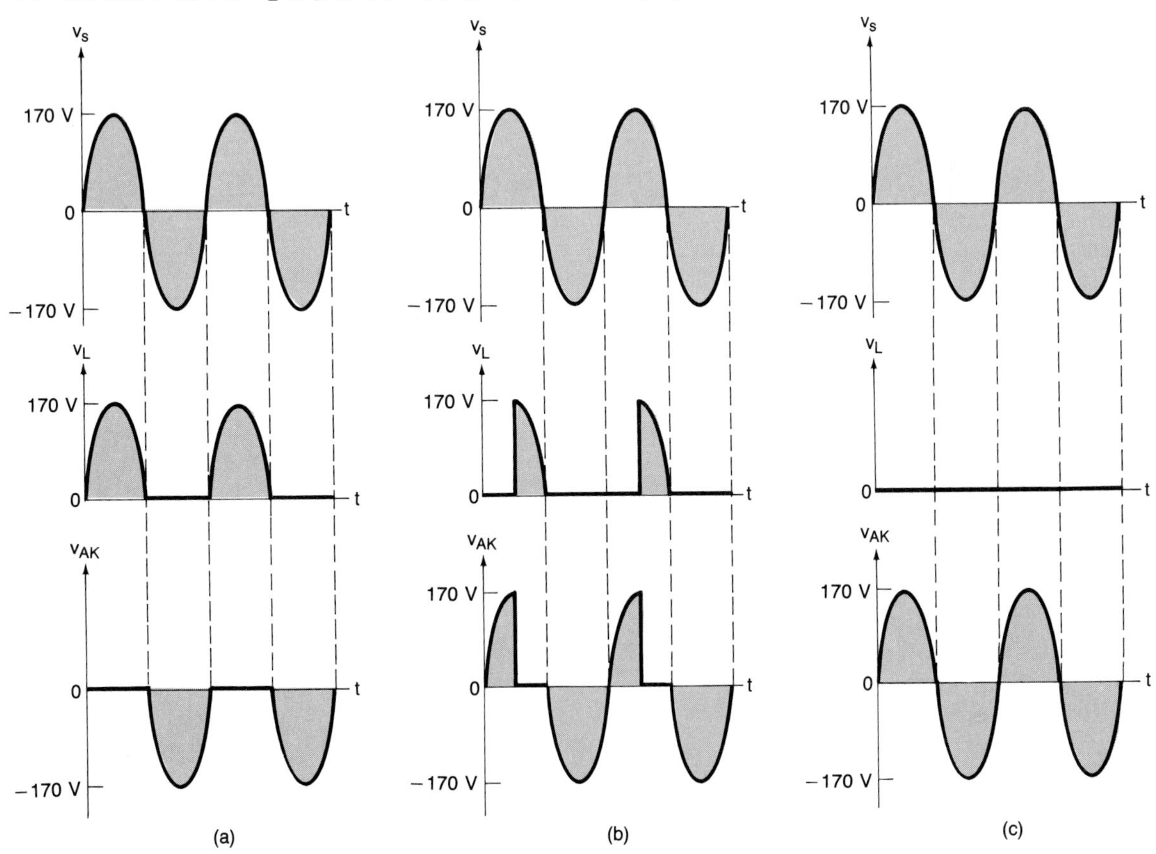

and the SCR will not be fired. Therefore, the SCR can be fired from approximately 5° to 90°. Attempts to adjust the SCR to fire beyond 90° will result in zero volts across R_L.

14-21 Inverse Parallel SCRs

The SCR is a *controlled* rectifier. Its maximum (ideal) conduction angle is 180°. By using a pair of SCRs, it is possible to control *both* the positive and negative half-cycles [see Fig. 14-30(a)].

FIGURE 14-30 Inverse parallel SCRs: (a) circuit; (b) Q_2 triggers on the positive half-cycle when S_1 is closed; (c) Q_1 triggers on the negative half-cycle when S_1 is closed.

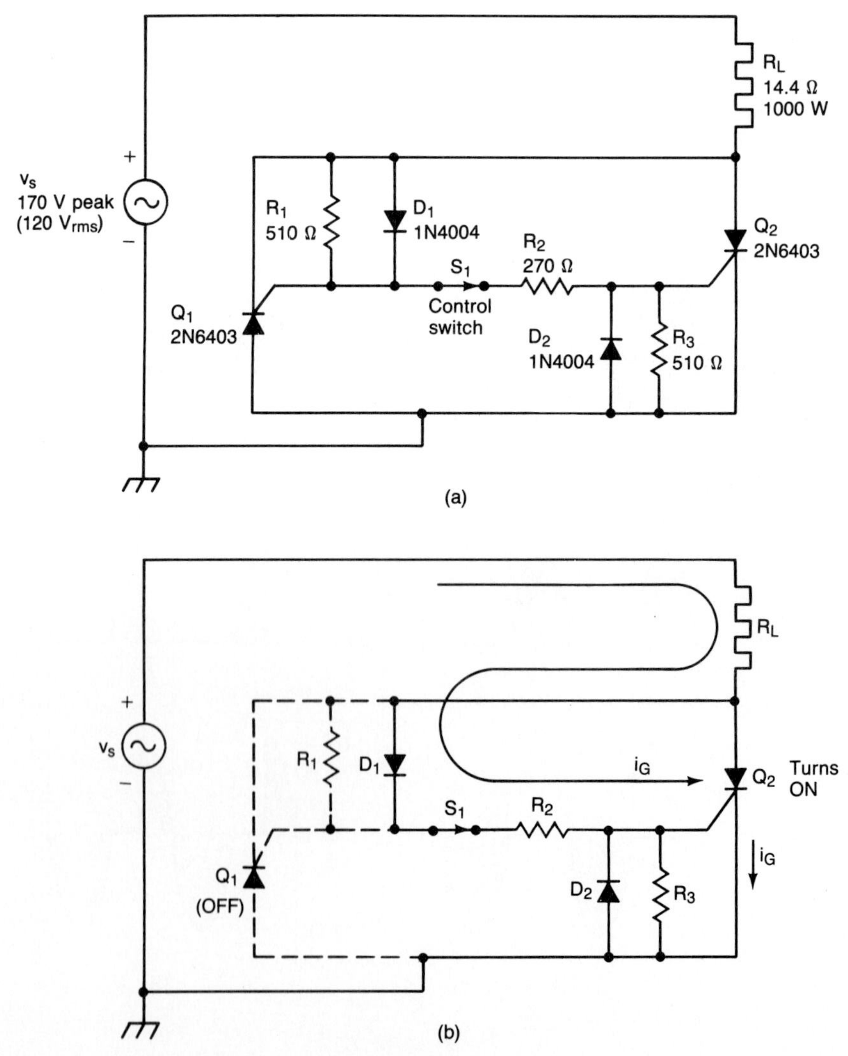

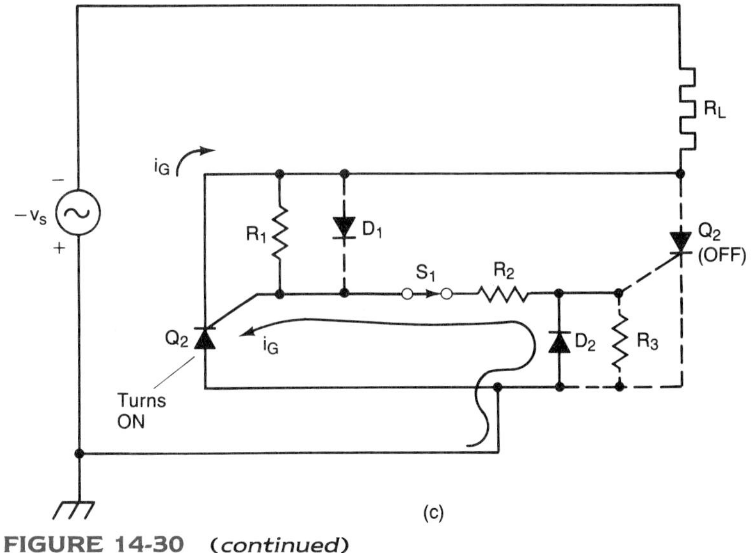

FIGURE 14-30 *(continued)*

The positive half-cycles are controlled by Q_2 and the negative half-cycles are under the control of Q_1. The circuit will fire when the control switch (S_1) is closed. The control switch could be mechanical or electronic. For example, if S_1 is a thermostat, it could be used to control the temperature of the load. If the temperature is too low, the thermostat closes and the SCRs fire to deliver ac power to the heater. When the temperature exceeds the setpoint, the thermostat opens and the SCRs will no longer conduct. The ac power delivered to the load will then go to zero. The switch (e.g., our thermostat) is used to control relatively low-level gate currents as opposed to several amperes of load current. This can increase the life of the switch and reduce its associated cost.

During the positive half-cycle [Fig. 14-30(b)], Q_1's gate current flows through R_L, D_1, S_1, and R_2. During the negative half-cycle, current will flow through D_2, R_2, S_1, into the gate of Q_1, and through R_L to return to the source [see Fig. 14-30(c)].

Diodes D_1 and D_2 serve two purposes. First, they steer the trigger signals into the gates of the SCRs, and second, they protect the gate-to-cathode *p-n* junctions by limiting their reverse biases to one diode drop.

The circuit in Fig. 14-30(a) is *asynchronous* with the power line. This means that it can be fired at virtually any point during either the positive or negative half-cycles [see Fig. 14-31].

As before [Fig. 14-27(a)], once an SCR (either Q_1 or Q_2) fires, the gate signal is removed. Once conduction is initiated on either the positive or negative half-cycle, the particular SCR will continue to conduct until its anode current goes below its holding current value. The 2N6403 SCRs have a maximum holding current of 40 mA.

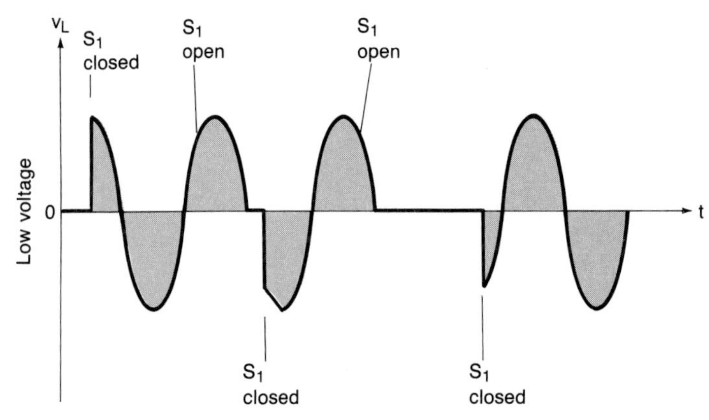

FIGURE 14-31 Control is asychronous with the voltage source.

14-22 Triacs

The triac is another three-terminal ac switch that is triggered into conduction when a low-energy signal is applied to its gate terminal. Unlike the SCR, the triac will conduct in either direction when turned on. The triac also differs from the SCR in that either a positive or a negative gate signal will trigger it into conduction.

In essence, the triac is equivalent to two SCRs connected in inverse parallel [see Fig. 14-32(a)]. It schematic symbol is given in Fig. 14-32(b). Since the triac is a bilateral device, the terms "anode" and "cathode" have no meaning. Therefore, its terminals are designated as main terminal 1 (MT1), main terminal 2 (MT2), and

FIGURE 14-32 Triac: (a) electrical equivalent circuit; (b) schematic symbol; (c) V-I characteristic.

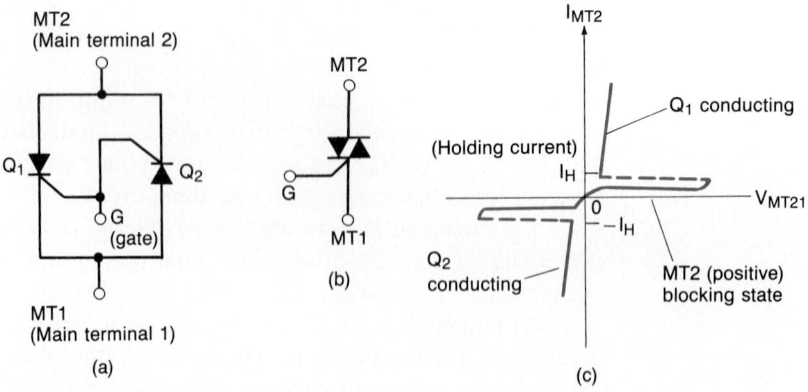

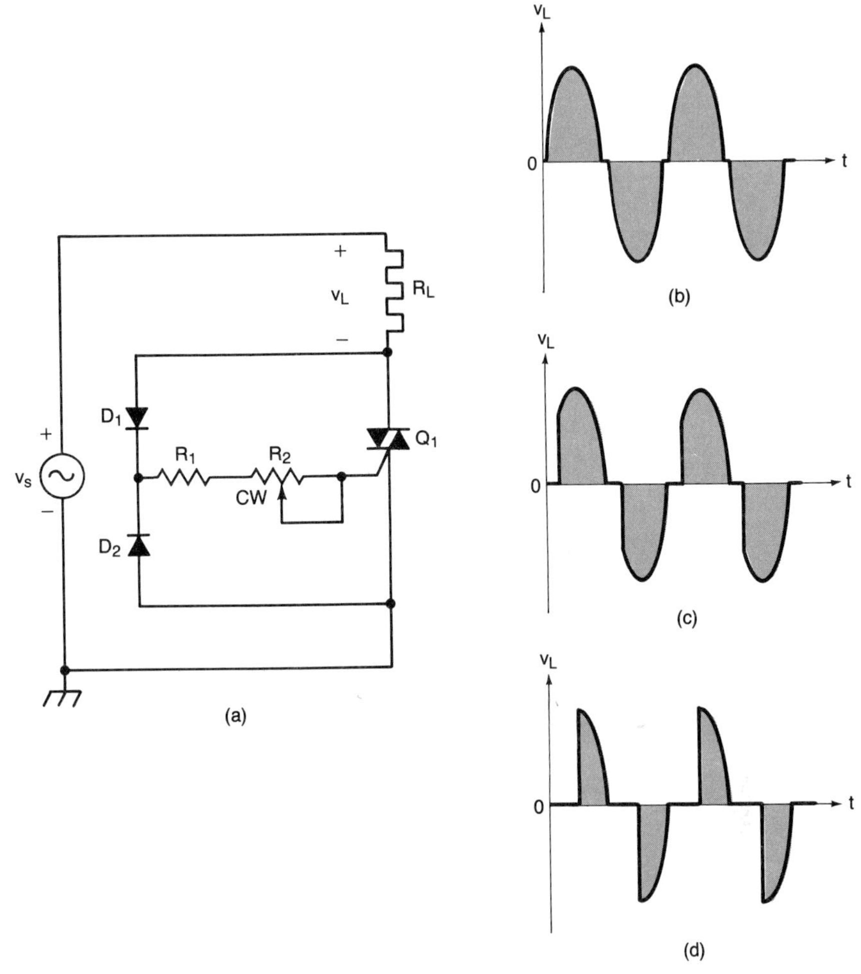

FIGURE 14-33 Simple triac circuit: (a) circuit; (b) load waveform with R_2 fully clockwise; (c) load waveform with R_2 adjusted in a counterclockwise direction; (c) load waveform with R_2 adjusted fully counterclockwise.

gate (G). To minimize confusion, it has become common practice to specify all voltages and currents using MT1 as the reference. The triac's *V-I* characteristic is shown in Fig. 14-32(c).

A simple triac circuit that allows one-half of the device to be turned on during the first 90° of each half-cycle of the supply voltage is shown in Fig. 14-33(a). During the positive half-cycles diode D_1 conducts (while D_2 is reverse biased) to apply gate current through R_1 and R_2 to fire the triac. During the negative half-cycles, diode D_2 conducts (while D_1 is reverse biased) to supply gate current through R_1 and R_2 to fire the triac. Typical waveforms are shown in Fig. 14-33(b) through (d).

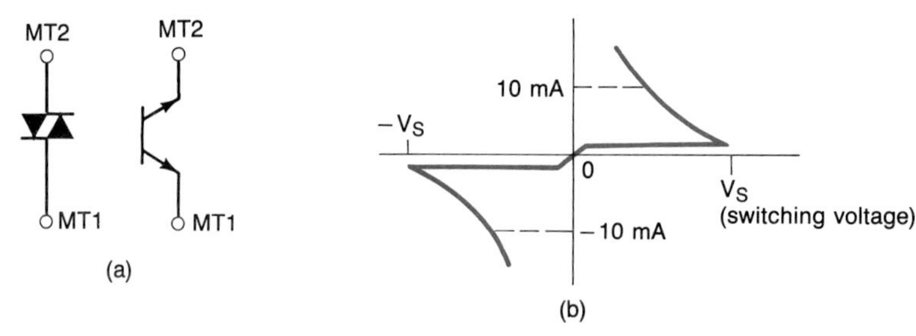

FIGURE 14-34 Diac: (a) schematic symbols; (b) V-I characteristic.

14-23 Trigger Devices: The Diac

Trigger devices are typically placed in series with the gates of SCRs and triacs. The trigger devices act like open circuits until the voltage across them becomes large enough to cause them to go into conduction. Invariably, the trigger devices possess a *negative resistance V-I* characteristic. (Recall that when a device exhibits a negative resistance, its terminal voltage actually *decreases* as the current through it *increases*.)

Examples of trigger devices include (fragile, limited-life) neon lamps, (obsolete) four-layer Shockley diodes that act like an SCR without a gate terminal, a relatively new (high-voltage) device called the Sidac, and the diac. Since the diac is the most popular device, we shall focus our attention on it.

A diac is basically a triac without a gate terminal. Two schematic symbols are given in Fig. 14-34(a). Again, the terminal designations are arbitrary since the diac (like the triac) is also a bilateral device.

The diac acts like an open circuit until its switching (or breakover) voltage is exceeded. At that point the diac will conduct until its current is reduced toward zero. The diac does not switch sharply into a low-voltage condition like the SCR or triac. Instead, once it goes into conduction, the diac maintains an almost continuous negative resistance characteristic [see Fig. 14-34(b)].

In the next section, we shall see how the diac can be employed as a trigger device in an ac power control circuit. The function of any trigger device is to deliver a trigger pulse to the gate of an ac switch such as the SCR or triac.

14-24 Adjustable (0 to 180°) Triac Phase Control Using a Diac

In Fig. 14-35 we see a triac that is being controlled by an RC phase-shift network and a diac. This circuit is an example of a simple lamp dimmer. The adjustment of the potentiometer determines the triac's conduction angle. The longer the triac conducts, the brighter the lamp will be. The 1N5758 diac acts like an open circuit until

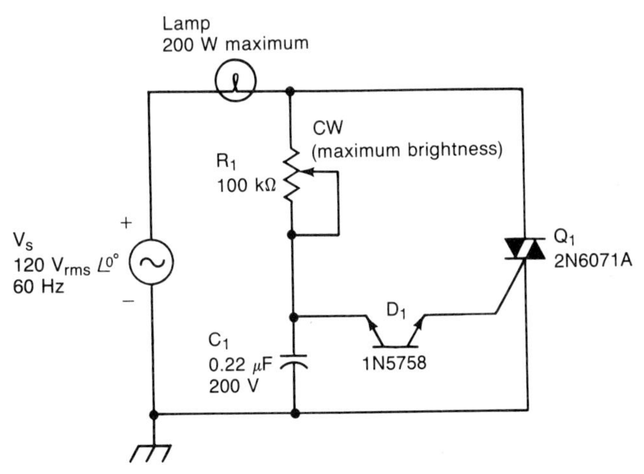

FIGURE 14-35 Triac lamp dimmer circuit.

the voltage across the capacitor exceeds its switch voltage (and the triac's required gate trigger voltage).

If we assume that the triac is an open circuit, and apply voltage division between R_1 and C_1, we can determine the range of phase shifts offered by the control network.

$$V_c = \frac{\dfrac{1}{j\omega C_1}}{R_1 + \dfrac{1}{j\omega C_1}} V_s = \frac{1}{1 + j\omega R_1 C_1} V_s \tag{14-37}$$

EXAMPLE 14-25

Determine the voltage waveform across C_1 in Fig. 14-35 if the diac (and triac) never conduct. Assume that R_1 is adjusted fully counterclockwise.

SOLUTION Using Eq. 14-37 and assuming that R_1 is 100 kΩ, yields

$$V_c = \frac{1}{1 + j\omega R_1 C_1} V_s = \frac{1}{1 + j(60 \text{ Hz})(100 \text{ k}\Omega)(0.22\ \mu\text{F})} 120\ V\angle 0^\circ$$

$$= \frac{1}{1 + j8.294} 120\ V\angle 0^\circ = 14.4\ V\angle -83.1^\circ$$

Expressing the phasor in the time domain produces

$$v_c = 14.4\sqrt{2} \sin(\omega t - 83.1^\circ)\ V$$
$$= 20.3 \sin(\omega t - 83.1^\circ)\ V$$

■

The waveforms when R_1 is adjusted fully counterclockwise (for maximum lamp dimming) are illustrated in Fig. 14-36(a). (Our assumption here is that the diac does not severely distort the voltage waveform across the capacitor. The intent is to show the phase relationship between v_s and v_c.) We can see that v_c lags v_s by 83°. The

1N5758 diac has a minimum switch voltage of 16 V. The typical triac trigger voltage v_{GT} is 1.4 V. Using these values, we can see in Fig. 14-36(a) that v_c reaches the required 17.4 V to fire the triac when v_s has reached a total angle of 142°. (Note that v_c reaches 17.4 V when its angle is 59°, and 83° + 59° is 142°.)

When the voltage across C_1 reaches 17.4 V the diac will conduct and fire the triac. Since the phase-shift network is connected across the triac, the voltage across the gate circuit will be reduced to zero when the triac conducts. If we *assume* that the voltage across the capacitor is not severely distorted, we arrive at the waveforms shown in Fig. 14-36(b).

If the potentiometer is adjusted fully clockwise, the source voltage is impressed across the capacitor. In this case, the diac (and therefore the triac) will fire much sooner and we obtain the waveforms given in Fig. 14-36(c). The lamp will be at maximum brightness.

The only problem with the simple circuit shown in Fig. 14-35 and the idealized waveforms given in Fig. 14-36(b) and (c) is that the capacitor's voltage waveform *will* be distorted when the triac fires. This is particularly true when the potentiometer has been adjusted to its maximum resistance to obtain minimum lamp intensity.

When R_1 is at its maximum resistance the discharge of C_1 through R_1 is negligibly small. Therefore, the discharge of C_1 rests primarily on the characteristics of the (nonlinear) diac. Typically, the diac will not allow the capacitor to discharge com-

FIGURE 14-36 Triac circuit waveforms: (a) phase relationship between the source voltage and the capacitor voltage; (b) potentiometer R_2 adjusted fully counterclockwise for maximum lamp dimming; (c) potentiometer R_2 adjusted fully clockwise for maximum lamp illumination.

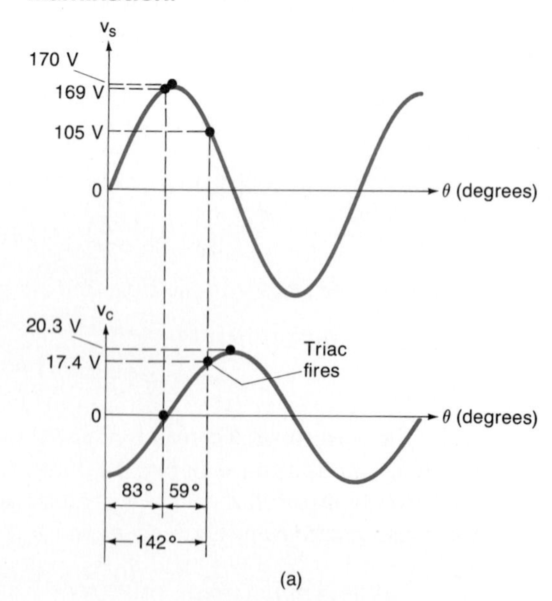

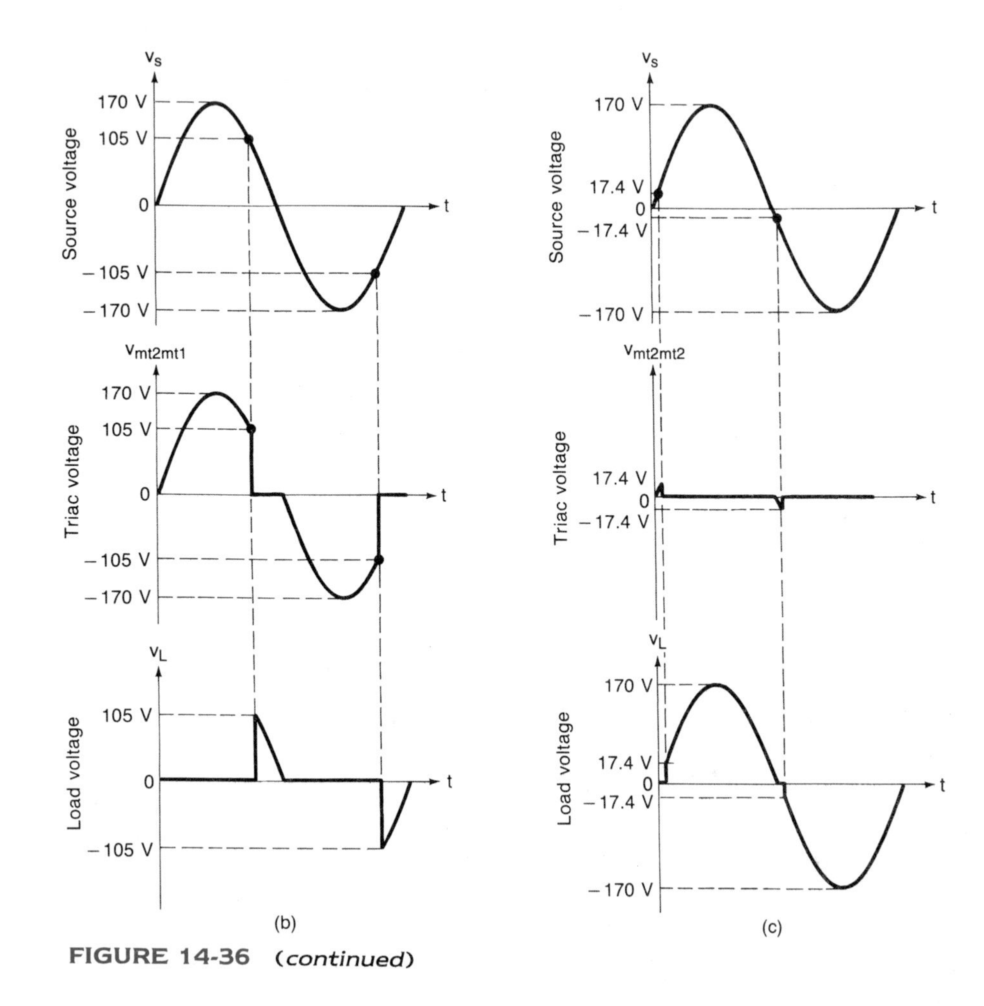

FIGURE 14-36 *(continued)*

pletely. Consequently, more realistic waveforms are shown in Fig. 14-37(a). The end result is that the circuit offers considerably less control. The maximum firing angle will be on the order of 110° as opposed to the 142° predicted previously.

There are modified networks that can be used to augment the control scheme, but we shall not delve into them. Despite its limitations, the network in Fig. 14-35 performs adequately for many applications.

The use of phase control produces very irregular waveforms. The severely distorted sinuosidal waveforms produce a wide spectrum of harmonics. The resulting high frequencies can be *conducted* into the power line or *radiated* into free space. The electrical noise can be severe enough to obliterate the reception of many AM radio stations, or disrupt other electronic systems on the same power line. This is called *electromagnetic interference* (EMI). One solution is to include an EMI filter in the design. A typical example is shown in Fig. 14-38. The use of an EMI filter tends to reduce the conducted and radiated emissions.

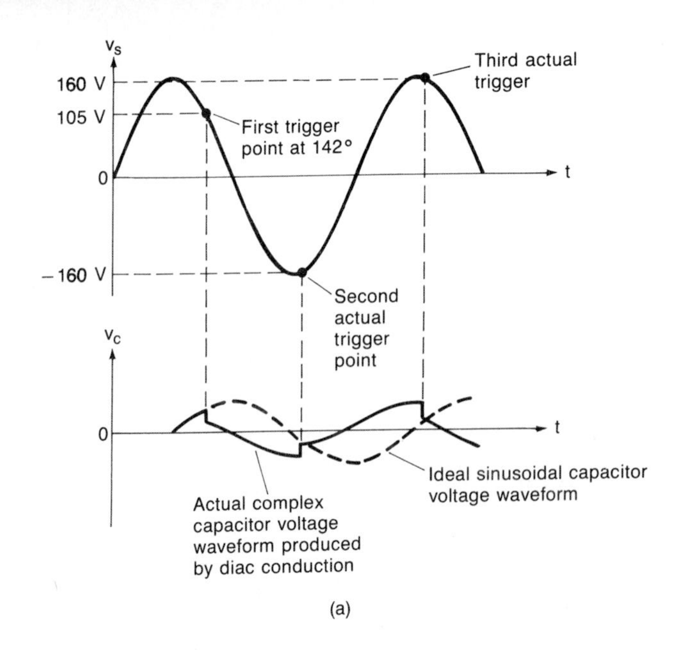

(a)

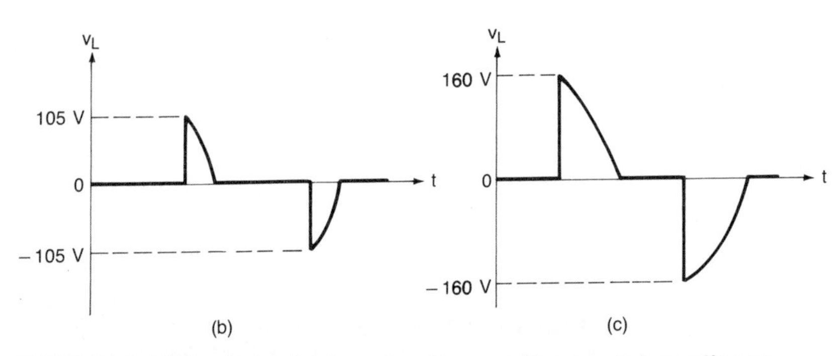

(b) (c)

FIGURE 14-37 Actual triac circuit waveforms: (a) nonlinear capacitor voltage waveform; (b) ideal minimum; (c) actual minimum shows a reduction in the control range.

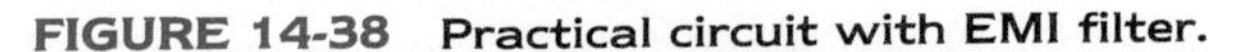

FIGURE 14-38 Practical circuit with EMI filter.

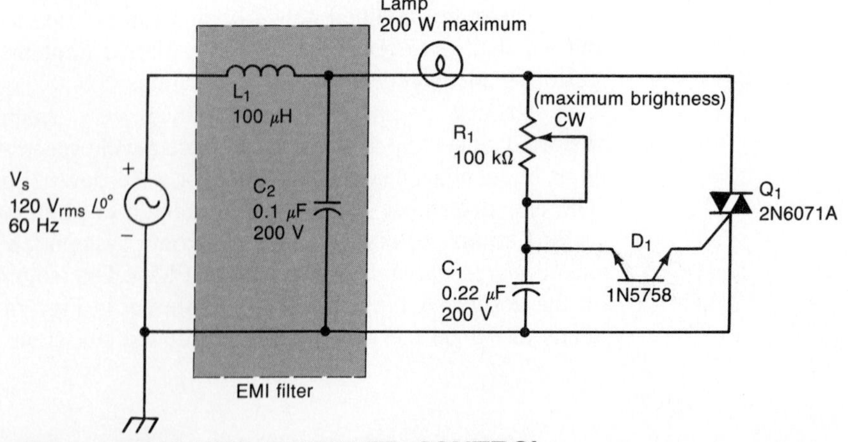

14-25 Isolation Via Optocouplers

In many applications SCR and triac power circuits are under the control of sensitive electronic systems. For example, it is not unusual to have a microprocessor system programmed to turn motors, lights, and heaters on and off. To reduce the possibility of power-line noise being induced into the control electronics, and to protect it in the event of an SCR or triac failure, it is highly desirable to provide *isolation*.

The ideal isolation scheme should only allow signal flow in one direction, should respond to dc levels, and should offer an extremely large resistance between its input and output circuits. These features are available in a class of optoelectronic devices called *optocouplers* or *optoisolators*.

If a *p-n* junction is exposed to light of the proper frequency, the current flow across the junction will tend to increase. If the *p-n* junction is forward biased, the net increase in current will be relatively insignificant. However, if the *p-n* junction is reverse biased, the increase in reverse current will be quite appreciable. The reason why the reverse current tends to increase can be understood by making two simple observations.

1. Reverse current (e.g., I_S or I_{CBO}) through a *p-n* junction is produced by minority carriers.
2. The number of minority carriers is directly related to the level of energy absorbed by the *p-n* junction.

In our previous work, we have been primarily concerned about the effects of thermal energy on reverse current flows. However, we were dealing with devices enclosed in light-tight packages. Their *p-n* junctions were not exposed to ambient light conditions.

In the case of many optoelectronic devices, a *window* is included in their package to promote the exposure of their *p-n* junction(s) to ambient (or directed) photons of light energy. A number of these *light-activated* devices are shown in Fig. 14-39.

In Fig. 14-39(a) we see a *photodiode*. Its dark current is essentially equal to its reverse saturation current (I_S). As the light intensity increases, its reverse current increases. The reverse current component due to light energy has been denoted I_λ. Figure 14-39(b) depicts a *phototransistor*. For high-gain applications a *photo-darlington* transistor can be used [see Fig. 14-39(c)]. The leakage current associated with the SCR (refer to Eq. 14-15) and triac also makes possible the light-activated SCR (LASCR) and phototriac. These are shown in Fig. 14-39(d) and (e), respectively.

The various optoelectronic devices shown in Fig. 14-39 are constructed in opaque packages that are excited by an internal LED. This constitutes the fundamental components of optocouplers/isolators. A number of available arrangements are shown in Fig. 14-40.

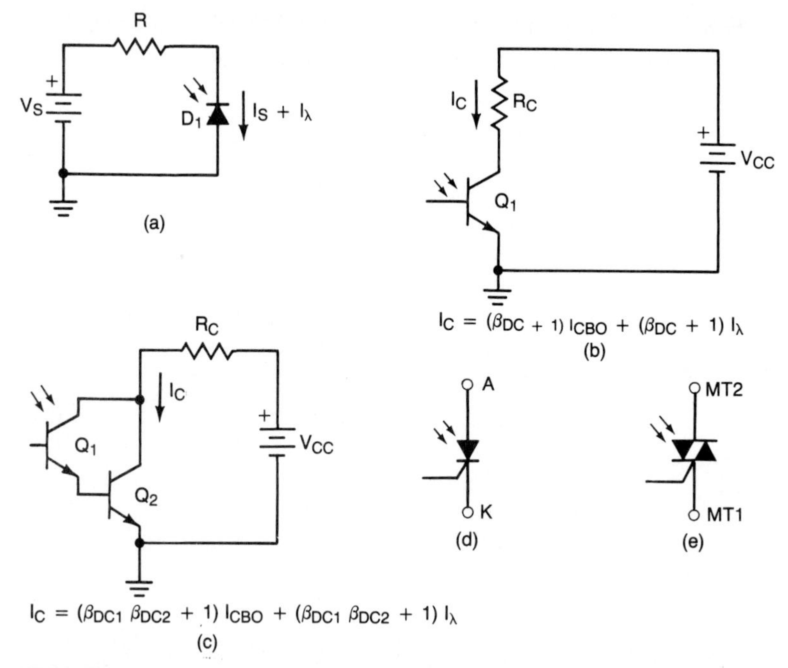

FIGURE 14-39 Optoelectronic devices: (a) photodiode; (b) phototransistor; (c) photodarlington; (d) light-activated SCR (LASCR); (e) phototriac.

FIGURE 14-40 Optocouplers: (a) 4N25 transistor output; (b) 4N29 darlington output; (c) Motorola HC11C2 photo SCR output; (d) Motorola MOC3011 phototriac output.

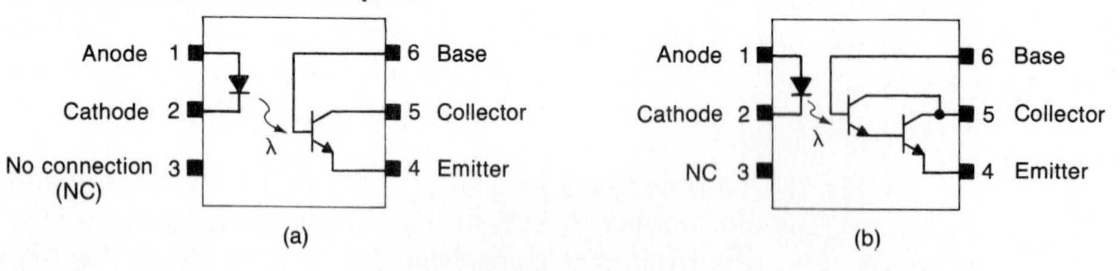

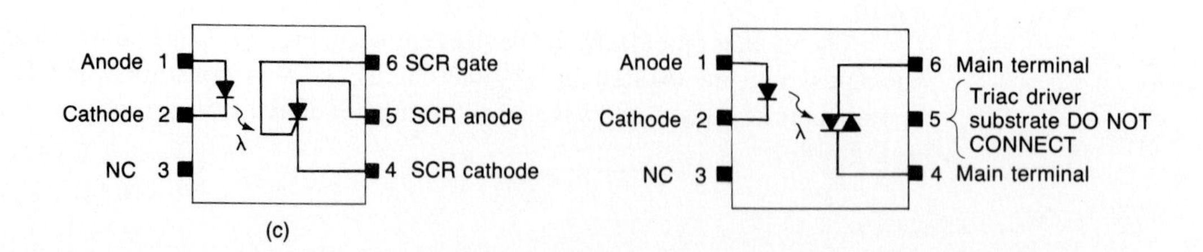

14-26 A Solid-State Relay (SSR)

In Fig. 14-41 we see an example of a *solid-state relay* (SSR). The SSR has been designed around a Motorola MOC3011 (optically-coupled) triac driver. Since SSRs are in such widespread use, it is important that we understand the circuit given in Fig. 14-41. We are *not* suggesting that this circuit is the one employed in all solid-state relays. However, it does possess the same basic features.

Let us examine the input circuit. All LEDs slowly decrease in brightness during their useful life. This characteristic can be accelerated by exposing the LED to high temperatures and high (forward) currents. The LED used in the MOC3011 is guaranteed to trigger the internal phototriac with 10 mA of forward LED current. The manufacturer also states that if the LED current exceeds 50 mA, the useful life of the MOC3011 will be shortened.

The input circuit in Fig. 14-41 (consisting of R_1, D_1, Q_1, and R_2) is designed to allow the LED to be driven by 10 mA for a wide (3 to 30 V) range of input voltages. Consequently, this input circuit is superior to the use of a single current-limiting resistor in series with the LED. Let us see how it works.

FIGURE 14-41 **Solid-state relay (SSR) circuit.**

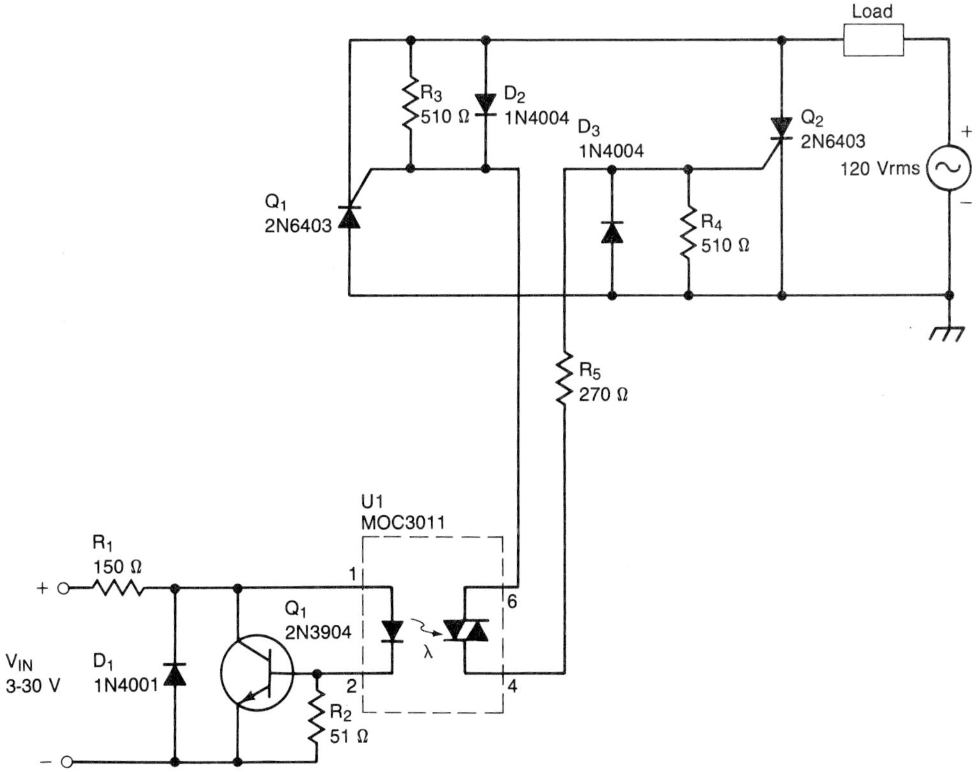

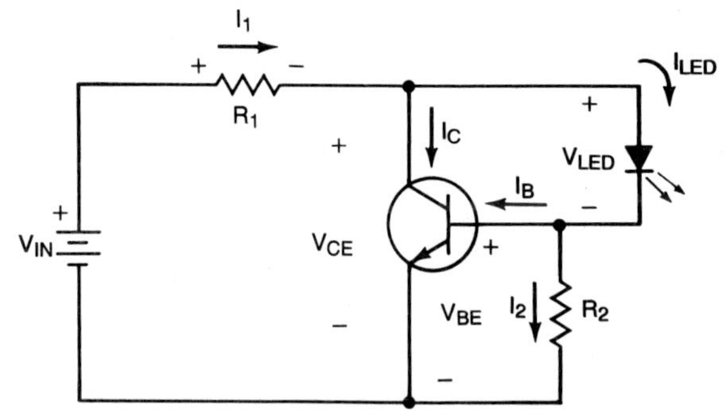

FIGURE 14-42 Simplified LED protection circuit.

In Fig. 14-42 we see that

$$I_2 = \frac{V_{BE}}{R_2} \tag{14-38}$$

and by Kirchhoff's current law,

$$I_{\text{LED}} = I_2 + I_B$$

Assuming the base current is negligibly small leads us to

$$I_{\text{LED}} \simeq \frac{V_{BE}}{R_2} \tag{14-39}$$

The transistor's collector-to-emitter voltage is given by Kirchhoff's voltage law.

$$V_{CE} = V_{BE} + V_{\text{LED}} \tag{14-40}$$

Now we can find I_1 and I_C.

$$I_1 = \frac{V_{\text{IN}} - V_{CE}}{R_1} \tag{14-41}$$

$$I_C = I_1 - I_{\text{LED}} \tag{14-42}$$

Diode D_1 is designed to provide input polarity protection. In normal operation it is reverse biased. If the input voltage becomes (accidentally) reversed, diode D_1 will conduct to protect the transistor and the LED. In this case resistor R_1 limits the current through D_1.

EXAMPLE 14-26

Determine I_2, I_{LED}, V_{CE}, I_1, and I_C in Fig. 14-42 if V_{IN} is 30 V. According to the manufacturer, the nominal voltage drop across the LED is 1.5 V.

SOLUTION The analysis is accomplished by following the procedure detailed in Eqs. 14-38 through 14-42.

$$I_2 = \frac{V_{BE}}{R_2} = \frac{0.7\text{ V}}{51\ \Omega} = 13.7\text{ mA}$$

$$I_{\text{LED}} \simeq I_2 = 13.7\text{ mA}$$

$$V_{CE} = V_{BE} + V_{\text{LED}} = 0.7\text{ V} + 1.5\text{ V} = 2.2\text{ V}$$

$$I_1 = \frac{V_{\text{IN}} - V_{CE}}{R_1} = \frac{30\text{ V} - 2.2\text{ V}}{150\ \Omega} = 185.3\text{ mA}$$

$$I_C = I_1 - I_{\text{LED}} = 185.3\text{ mA} - 13.7\text{ mA} = 171.6\text{ mA}$$

■

The operation of the SSR in Fig. 14-41 is rather simple once we understand the input circuit. When a noncritical (3 to 30 V) dc signal (V_{IN}) is applied to the input, the LED illuminates. The photons of light energy emitted by the LED are used to trigger the triac in the MOC3011. When the triac conducts, the inverse parallel SCRs [explained in Section 14-21] conduct and ac voltage is applied to the load.

PROBLEMS

Drill, Derivations, and Definitions

Section 14-1

14-1. The output voltage of a dc power supply is 12 V when its load current is zero. When its load current is 600 mA, its output voltage drops to 11.5 V. What is its percent of voltage regulation? What should the percent of voltage regulation be if the output voltage is to drop down to only 11.95 V when the load current is 600 mA?

14-2. What are the two primary limitations inherent in a simple zener diode voltage regulator? Explain.

14-3. A zener diode voltage regulator, such as that shown in Fig. 14-2(a), has a V_Z of 15 V, I is a constant 150 mA, and the load current I_L ranges from 0 to 140 mA. Find the minimum and maximum zener power dissipation.

14-4. Repeat Prob. 14-3 if the zener diode is a 5.1-V unit.

14-5. Draw and label the basic block diagram of a voltage regulator that incorporates a series-pass element.

Section 14-2

14-6. Analyze the power supply given in Fig. 14-4(a) if T_1 has a 15-V rms secondary, C_1 is 2200 μF, C_2 and C_3 are both 25 μF, D_5 is a 15-V zener, R_1 is 130 Ω, and Q_1 is a TIP29C with a $h_{FE(min)}$ of 50. The maximum load current is 300 mA. Determine V_m, V_{dcm}, V_{DC}, $V_{dc(min)}$, I_1, $I_{B(max)}$, I_Z, P_Z, V_L, and C_{out} when I_L is 300 mA.

14-7. Repeat Prob. 14-6 if the load current is reduced to zero.

Section 14-3

14-8. What advantage is offered by the regulator shown in Fig. 14-5(b) over the one given in Fig. 14-5(a)? What is the function of capacitor C_3?

14-9. The transformer in Fig. 14-5(b) has been replaced by one with an 18-V rms secondary. All other components and the maximum load current of 500 mA are unchanged. Find V_m, V_{dcm}, V_{DC}, $V_{dc(min)}$, $V_{L(max)}$, and $V_{L(min)}$.

14-10. Repeat Prob. 14-9 if the transformer has a 20-V rms secondary.

14-11. Determine the worst-case power dissipation in transistor Q_1 using the data and the results obtained in Prob. 14-9.

14-12. Determine the worst-case power dissipation in transistor Q_1 using the data and the results obtained in Prob. 14-10.

Section 14-5

14-13. Assume that R_{SC} is 2 Ω in Fig. 14-6(a), and find I_{SC}.

14-14. Assume that I_{SC} is to be 100 mA in Fig. 14-6(a). Determine the required value of R_{SC}.

Section 14-6

14-15. What is the purpose of overvoltage protection in a dc power supply?

Section 14-7

14-16. Assume that the comparator indicated in Fig. 14-7(a) has a V_{REF} of 5 V. Sketch its transfer characteristic curve.

14-17. Assume that the comparator indicated in Fig. 14-7(a) has a V_{REF} of 7 V. Sketch its transfer characteristic curve.

Section 14-8

14-18. Make a sketch of the basic structure of the SCR. Draw its schematic symbol and label its terminals.

14-19. Use the two-transistor analogy to explain the operation of the SCR.

14-20. Describe two ways in which an SCR can be ''fired'' or latched into conduction.

14-21. Make a sketch of the *V-I* characteristic curve of an SCR. Label it completely.

14-22. An SCR circuit has been shown in Fig. 14-43. Initially, the SCR is off and the LED is extinguished. Switch S_1 is a normally open (N.O.) pushbutton, and switch S_2 is a normally closed (N.C.) pushbutton. When switch S_1 is depressed gate current will flow, the SCR will latch on, and the LED will be illuminated. The circuit can only

FIGURE 14-43 Simple SCR latching circuit.

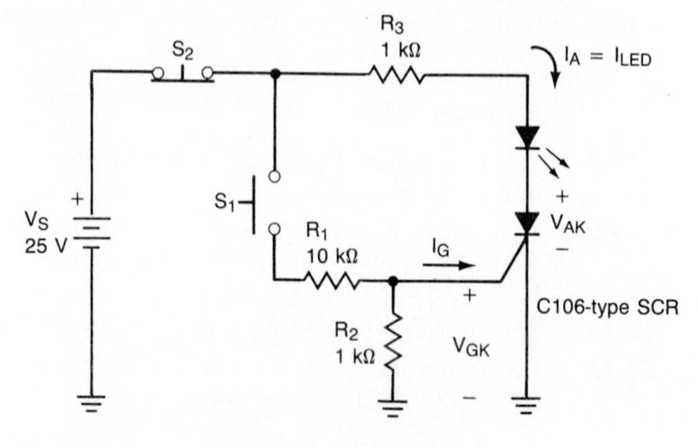

be reset by momentarily depressing switch S_2. Calculate the gate current and the current through R_1 when S_1 is depressed. Assume that V_{GK} is 0.7 V, and be sure to include the presence of R_2 in your calculations. Also calculate the LED's current when the SCR is on. Assume that V_{AK} is 0.1 V and the LED drops 1.8 V.

Section 14-9

14-23. Resistor R_8 in Fig. 14-10(a) has been increased to 22 kΩ. Determine the minimum and maximum trip points as determined by the adjustment of potentiometer R_5.

14-24. Resistor R_7 in Fig. 14-10(a) has been increased to 22 kΩ. Determine the minimum and maximum trip points as determined by the adjustment of potentiometer R_5.

14-25. A discrete overvoltage crowbar circuit has been illustrated in Fig. 14-44. When the load voltage becomes large enough, zener diode D_1 will break down, base current will flow out of Q_1, and its collector current will flow through R_1 and R_2. The collector current will develop a voltage drop across R_2 that will fire the SCR. The load voltage required to break down the zener is determined by the setting of potentiometer R_5. Show that the overvoltage trip point is given by the following relationship:

$$V_{\text{TRIP}} \simeq \left(1 + \frac{R_5 + R_6}{R_4}\right) V_Z$$

14-26. Using the relationship given in Prob. 14-25, determine the minimum and maximum trip points of the circuit given in Fig. 14-44.

Section 14-10

14-27. Explain the purpose of transistor Q_1 in Fig. 14-11. Is it in the common-base, common-collector, or common-emitter amplifier configuration?

FIGURE 14-44 **Discrete crowbar overvoltage protection circuit.**

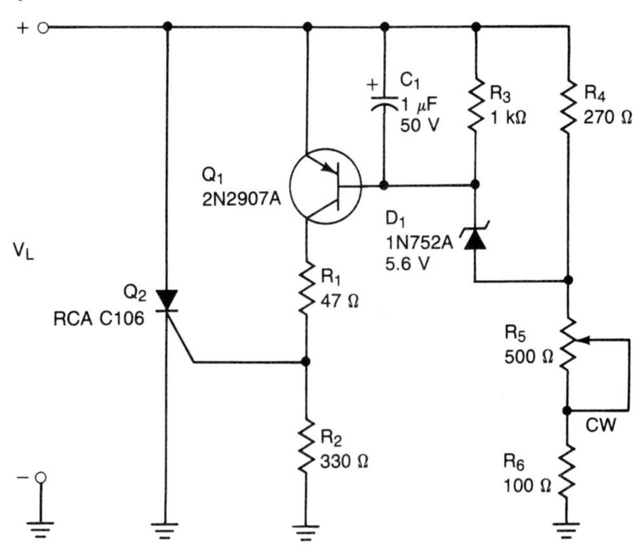

14-28. The zener diode in Fig. 14-12 has been replaced with a 3.3-V unit. Determine the new minimum value of $V_{dc(min)}$. Also find V_L and the op amp's output voltage if I_L is 500 mA, the h_{FE} of the (silicon) transistor is 50, and V_{DC} is 15 V. Diodes D_6 and D_7 are silicon units.

14-29. Determine the required value of R_{SC} if the short-circuit load current in Fig. 14-12 is to be 100 mA. Assume that a Schottky diode has been used for D_6.

Section 14-11

14-30. What is foldback current limiting? What advantage(s) does it offer? What are the disadvantages associated with Fig. 14-13(b)?

14-31. Resistor R_5 in Fig. 14-13(b) has been replaced with a 91-Ω unit. All other component values are the same. Find the short-circuit current, the maximum load current, and the collector power dissipation when a short occurs. Assume that V_{DC} is 20 V.

Section 14-12

14-32. Make a neat sketch of the fundamental block diagram of a three-terminal IC voltage regulator. Explain its operation.

14-33. The peak-to-peak ripple at the input of the regulator in Fig. 14-15 has been increased to 2 V p-p. The LM340AT-15 has been replaced with an LM340AT-5. Given that the regulator's ripple specification (R) is 80 dB and that its load voltage V_L is 5 V, determine (a) the ripple rejection, (b) the rms value of the (triangular) input ripple, (c) the rms value of the ripple appearing across the output, and (d) the peak-to-peak value of the output ripple.

14-34. The input filter capacitor (C_1) in Fig. 14-16(a) has been replaced with a 2500-μF unit. The maximum load current is 850 mA. All other components are the same. Is the size of C_1 adequate if the regulator's dropout voltage is 2 V? Find the peak-to-peak output ripple.

14-35. What is the purpose of D_5 in Fig. 14-16(a)? What roles do capacitors C_2 and C_3 serve?

Section 14-13

14-36. A three-terminal IC voltage regulator experiences a maximum ambient temperature of 70°C and a maximum power dissipation of 4 W. If the regulator's θ_{ja} is 10°C/W, what is its junction temperature? If the regulator has a $T_{j(max)}$ of 125°C, does it require a heat sink?

14-37. Repeat Prob. 14-36 if the maximum power dissipation is increased to 5 W.

14-38. A three-terminal IC voltage regulator has the following specifications: $T_{j(max)} = 150°C$, $\theta_{jc} = 4.5°C/W$, $\theta_{ja} = 60°C/W$, $P_{D(max)} = 20$ W, and $I_{L(max)} = 2.0$ A. If $P_{D(max)}$ is 10 W and $T_{A(max)}$ is 50°C, determine if the regulator can be used in this application. Is a heat sink required? What is the heat sink's required θ_{sa}?

Section 14-14

14-39. A 7812C (with its V_{REG} of 12 V) has a quiescent current of 8.5 mA and is used in the regulator circuit shown in Fig. 14-17(a). Determine the minimum and maximum load voltage (V_L).

14-40. Explain the purpose of capacitors C_2, C_3, and C_4 in Fig. 14-18. Also explain the purpose of diodes D_5 and D_6.

14-41. Potentiometer R_2 has been replaced by a 500-Ω potentiometer in Fig. 14-18. Determine the minimum and maximum load voltage (V_L).

Section 14-15

14-42. The maximum load current in Fig. 14-19(a) has been increased to 600 mA, and the input filter capacitors have also been replaced by 2500-μF units. Determine the dc voltages (V_{DC}) across capacitors C_1 and C_2. Also find the power dissipations in the positive and negative regulators.

14-43. The 20-kΩ potentiometer (R_2) in Fig. 14-20(a) has been replaced with a 10-kΩ potentiometer. Determine the minimum and maximum load voltages. All other components are unchanged.

Section 14-16

14-44. Modify the circuit in Fig. 14-21 to limit the current through Q_1 to 2 A.

14-45. Explain the purpose of diodes D_5 and D_6 in Fig. 14-22(a). Determine the minimum current limit if R_3 is increased to 0.22 Ω.

Section 14-17

14-46. Explain why a transistor switch demonstrates low power dissipations.

14-47. What is the purpose of a ''freewheeling'' or ''catch'' diode? [Refer to Fig. 14-24(a).]

14-48. A 100-kHz waveform with a V_m of 15 V has a duty cycle of 30%. Make a sketch of the waveform, and indicate its dc average value. Repeat the problem if the duty cycle is increased to 80%.

14-49. Explain the use of pulse-width modulation in a switching regulator.

Section 14-19

14-50. The C122B SCR in Fig. 14-26 is to provide a minimum conduction angle of 160°. Determine the required value of R_1 and pick the nearest standard 5%-tolerance value. Assume that v_{GT} is 1.5 V and i_{GT} is 25 mA.

14-51. Repeat Prob. 14-50 if the minimum conduction angle is to be 100°.

Section 14-20

14-52. Resistor R_1 in Fig. 14-27(a) is replaced with an 8.2-kΩ unit. Assume that $i_{GT(\max)} =$ 200 μA and $v_{GT(\max)} = 1$ V. Find the minimum voltage source angle to fire the 2N6240 SCR if R_2 is adjusted fully clockwise.

14-53. Repeat Prob. 14-52 if R_1 is a 10-kΩ unit.

Section 14-21

14-54. Determine the maximum instantaneous gate current in Fig. 14-30(a) if S_1 is closed when v_s is 170 V peak. Assume that v_{GT} for Q_1 and Q_2 is 1 V, and R_2 is 2.2 kΩ.

14-55. Repeat Prob. 14-54 if R_2 is increased to 3.3 kΩ.

Section 14-22

14-56. Name the three terminals of a triac.

14-57. The triac in Fig. 14-33 requires an $i_{GT(\max)}$ of 10 mA to fire. R_L is 10 Ω, R_1 is 1 kΩ, and R_2 is a 20-kΩ potentiometer. If $v_{GT(\max)}$ is 1.5 V, the diodes (D_1 and D_2) are silicon units, and R_2 is adjusted fully clockwise, determine the instantaneous positive value of v_s to fire the triac.

14-58. Repeat Prob. 14-57 if R_2 is adjusted fully counterclockwise. If v_s is a 170-V peak sinewave, will the triac fire?

Section 14-23

14-59. Name three characteristics of a trigger device used to fire SCRs and triacs.

14-60. Name the two terminals of a diac.

Section 14-24

14-61. Capacitor C_1 in Fig. 14-35 is changed to 0.15 μF. Find the phase shift between v_s and v_c if R_1 is adjusted fully counterclockwise. Assume that v_c is sinusoidal. What is the phase shift if R_1 is adjusted fully clockwise?

14-62. Repeat Prob. 14-61 if C_1 is 0.33 μF.

14-63. What is "EMI?"

Section 14-25

14-64. If the light intensity is increased [Fig. 14-39(a)], the reverse current will ______ (increase, decrease) and the voltage across the diode will ______ (increase, decrease).

14-65. If pin 6 is shorted to pin 5 in Fig. 14-40(a), what does the optocoupler become?

Section 14-26

14-66. Determine I_2, I_{LED}, V_{CE}, I_1, and I_C in Fig. 14-41 if V_{IN} is 3 V.

14-67. Repeat Prob. 14-66 if V_{IN} is 5 V.

Troubleshooting

14-68. If the overvoltage protection circuit in Fig. 14-44 produces noise spikes, or triggers erratically, select the most probable cause(s) from the alternatives below. Explain why you feel your answer is correct, and why the other alternatives are not probable causes. The alternatives are: (a) the SCR is shorted, (b) the transistor is open, (c) the capacitor is open, and (d) the SCR is open.

14-69. The overvoltage protection circuit in Fig. 14-44 stays on after an overvoltage condition is corrected. Select the most probable cause(s) from the alternatives below. Explain why you feel your answer is correct and why the other alternatives are not probable causes. The alternatives are: (a) the SCR is shorted, (b) the transistor is shorted, (c) the zener diode is shorted, and (d) the capacitor is shorted.

14-70. The overvoltage protection circuit in Fig. 14-44 refuses to trigger when an overvoltage condition exists. Select the most probable cause(s) from the alternatives below. Explain why you feel that your answer is correct and why the other alternatives are not probable causes. The alternatives are: (a) the SCR is open, (b) the transistor is open, (c) the zener diode is shorted, and (d) the potentiometer is open.

Design

14-71. An LM340KC-12 is a three-terminal voltage regulator in a TO-3 (aluminum) case style. Its electrical specifications have been given in Table 14-1. The LM340KC-12 has a $T_{j(max)}$ of 150°C. It is to be used in a power supply with a V_{DC} of 15 V (the regulator's input voltage) with load current demands ranging from 0 to 500 mA. The maximum ambient temperature is 60°C. Is a heat sink needed? If so, what should its θ_{sa} be? Could an LM7812C with its $T_{j(max)}$ of 125°C be used? Is a heat sink needed? If so, what should its θ_{sa} be? Explain.

TABLE 14-1
Regulator Data

Regulator	Output Current (A)	Output Voltage (V)	V_{IN} max (V)	Ripple (dB)	Dropout Voltage (V)	θ_{jc} (°C/W)	θ_{ja} (°C/W)	P_D (W)
LM340KC-12	1	12	35	66	2.0	4	35	20
LM7812C	1	12	35	66	2.6	4	50	18

14-72. Repeat Prob. 14-71 if the maximum load current is 750 mA.

14-73. Repeat Prob. 14-71 if the maximum load current is 890 mA.

14-74. To achieve high accuracy from a zener diode voltage reference, it should be biased by a constant-current source. The use of a JFET as a constant-current diode in Fig. 14-12(a) minimizes the number of required circuit components. Alternatively, a BJT constant-current source could be used. A circuit has been illustrated in Fig. 14-45. Specify standard ($\pm 5\%$) values of R_1, R_2, and R_3 if the current source is to provide a current (I) of 10 mA.

14-75. Repeat Prob. 14-74 if the current source is to provide 25 mA.

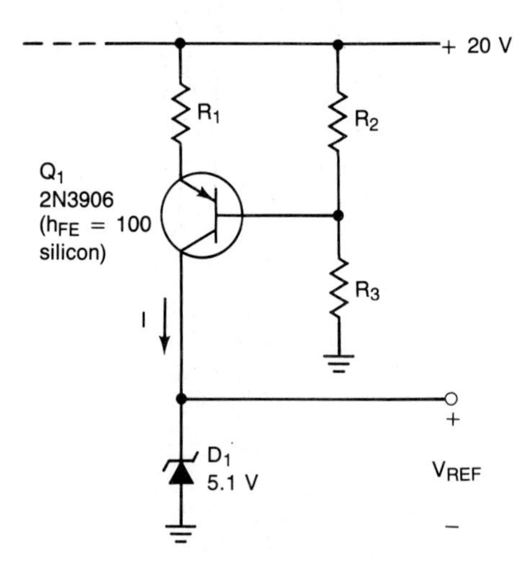

FIGURE 14-45 Using a BJT constant-current source to bias a zener reference.

Computer

14-76. Write a BASIC program that designs heat sinks for IC three-terminal voltage regulators. It should implement the steps detailed in Section 14-13. It should prompt the user for the pertinent data and print out its findings.

14-77. Write a BASIC program that will troubleshoot the circuit given in Fig. 14-44. Specifically, it should ask the technician for the symptoms and then list the most likely problem(s). Use your answers in Probs. 14-68 through 14-70 as a guide.

DISCRETE AND INTEGRATED OSCILLATORS

After Studying Chapter 15, You Should Be Able to:

- Explain the differences between harmonic and relaxation oscillators.
- Analyze and explain the operation of the RC phase-shift and Wien bridge oscillators.
- Analyze and explain the operation of the JFET as a voltage-variable resistor (VVR) and apply it to stabilize the output amplitude of the Wien bridge oscillator.
- Explain and analyze the use of active negative feedback to provide a Wien bridge oscillator with an adjustable output frequency.
- Explain high-frequency harmonic oscillators and the role of series and parallel resonant circuits.
- Analyze and discuss the operation of the Colpitts, Clapp, and Hartley oscillators.
- Discuss the characteristics of the crystal.
- Analyze and explain the operation of the Pierce (and other) crystal oscillator circuits.

- Explain the operation of the unijunction transistor (UJT).
- Explain and analyze the UJT relaxation oscillator.
- Explain and analyze the op amp relaxation oscillator.
- Explain and analyze the op amp triangle waveform generator.

15-1 Role of the Oscillator

In addition to the voltage and power amplifiers, *oscillator* circuits are also found in electronic systems. Very simply, oscillators are used to *generate* signals. Let us put this into perspective. All of our ''serious'' amplifier work thus far has involved carefully avoiding the generation of signals. However, there are many instances in which signal generation is extremely important. For example, oscillators have a wide variety of applications in electronic communication equipment. In AM (amplitude modulation) and FM (frequency modulation) superheterodyne receivers, ''local'' oscillators are used to assist the reduction of the incoming radio frequency (RF) to a lower intermediate frequency (IF). Oscillator circuits are also used in the ''exciter'' section of a transmitter to generate the RF carrier. Other applications include their use as ''clocks'' in digital systems such as microcomputers, in the sweep circuits found in television sets, and oscilloscopes. As we saw in Chapter 14, the triangle generator serves as the ''heart'' of a switching regulator.

In performing its function as a signal generator, the oscillator essentially acts as a converter. It converts power from the dc power supply into ac power. However, we generally think of oscillator circuits as providing an ac *voltage* signal.

There are two broad categories of oscillator circuits: *harmonic* and *relaxation*. Both types can include active devices such as BJTs, FETs, and op amps, and passive components such as resistors, capacitors, and inductors. In harmonic oscillators, the energy always flows in one direction - from the active to the passive components. However, in relaxation oscillators, the energy is exchanged between the active and passive components. In harmonic oscillators, the feedback path determines the frequency of oscillations. However, in relaxation oscillators, the frequency is determined by time constants - specifically, the charge and discharge time constants during the exchange of energy. Harmonic oscillators can produce low-distortion sinusoidal output waveforms, but relaxation oscillators can only produce nonsinusoidal waveforms, such as sawtooth, triangular, or square. Harmonic oscillators are considered first.

15-2 Audio-Frequency Harmonic Oscillators

In Section 10-11 we introduced the problem of closed-loop stability. Specifically, we saw that the potential for oscillation exists in every closed-loop amplifier. By taking the appropriate steps (e.g., frequency compensation, or reducing the amount of negative feedback), oscillation could be avoided. The reader is urged to review Section 10-11 at this point.

The Barkhausen criteria for sustained oscillation set forth two conditions. First, the feedback must be positive. This means that the feedback voltage (or current) must be phased so that it adds to the amplifier's input signal. Second, the loop gain must be larger than unity to allow oscillations to build up, and equal to unity to sustain them. All harmonic oscillators must satisfy the Barkhausen criteria.

Consequently, an audio-frequency range (20 to 20 kHz) harmonic (sinusoidal output) oscillator must contain three basic ingredients:

1. An amplifier
2. A frequency-determining network
3. Positive (regenerative) feedback

The amplifier provides the voltage gain required to sustain oscillation. The frequency-determining network is configured to provide positive feedback at one particular frequency (the desired frequency of oscillation).

In Section 15-4 we shall add a fourth requirement termed *adaptive negative feedback*. Adaptive-negative-feedback circuitry is typically designed to allow the amplifier circuit initially to have a sufficiently large voltage gain to allow the oscillations to build up. This means the loop gain will be greater than unity. Once the desired output signal level is reached, the adaptive negative feedback circuity *automatically* reduces the gain of the amplifier system. This, in turn, reduces the loop gain to unity and ensures a low-distortion sinusoidal output waveform.

15-3 A Phase-Shift Oscillator

To provide positive feedback at one particular frequency, an inverting amplifier may be used with a feedback network that supplies 180° of phase shift at the desired frequency of oscillation [f_o - see Fig. 15-1(a)]. Alternatively, a noninverting amplifier may be used with a frequency-determining feedback network that offers 0° of phase shift at f_o [see Fig. 15-1(b)]. The phase-shift oscillator implements the topology indicated in Fig. 15-1(a) [refer to Fig. 15-1(c)].

The equivalent circuit of the phase-shift oscillator has been indicated in Fig. 15-2(a). When the magnitude of the gain of the amplifier is A_{v1}, the gain (an attenuation) of the feedback network (β) can have a magnitude no less than $1/A_{v1}$ in order to maintain the oscillation. This will provide a loop gain of unity.

When a phase-shift network such as that indicated in Fig. 15-2(a) is employed in a phase-shift oscillator, the R's and C's must be selected so as to produce 180° of phase shift at the desired frequency of oscillation. The output of the voltage amplifier is directed into the input of the phase-shift network. Thus

$$V_1 = V_{out}$$

The output resistance of the amplifier is designed to be very small compared to the input impedance of the phase-shift network. This is not too difficult to achieve when an op amp is employed.

The output voltage of the phase-shift network (V_2) is fed into the input of the

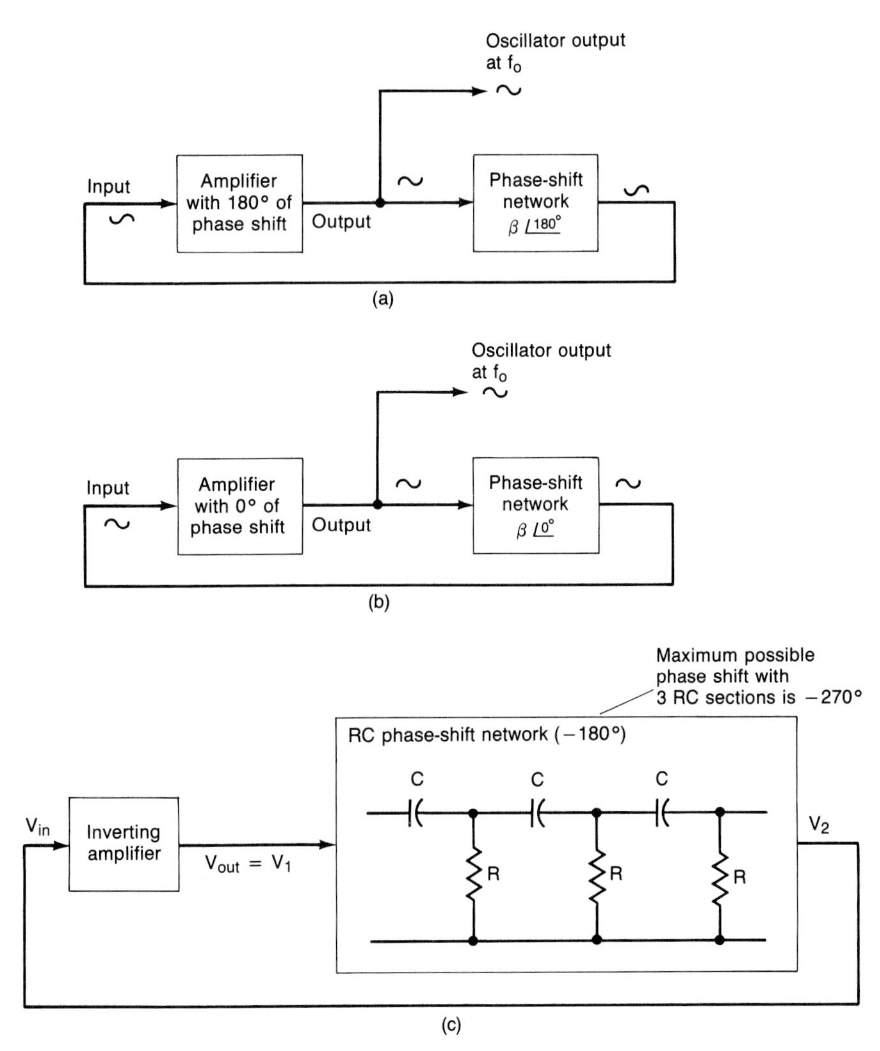

FIGURE 15-1 Basic oscillator configurations and the RC phase-shift oscillator.

amplifier. The amplifier's input resistance must be much larger than the output impedance of the phase-shift network, or its loading effects must be included in the analysis. Initially, we shall assume that the former case exists. From Fig. 15-2(a),

$$V_2 = V_{in}$$

Using the assumptions above, we arrive at the equivalent circuit given in Fig. 15-2(b).

A rigorous analysis of most oscillator circuits can be conducted by performing the following steps:

1. Derive an equation for the voltage gain (β) of the feedback network (e.g., V_2/V_1).

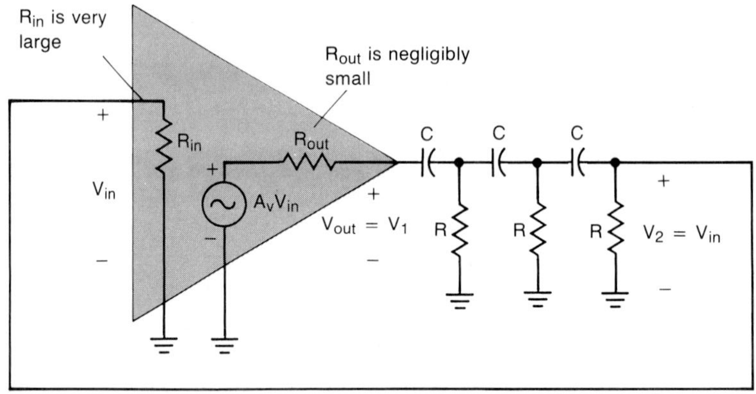

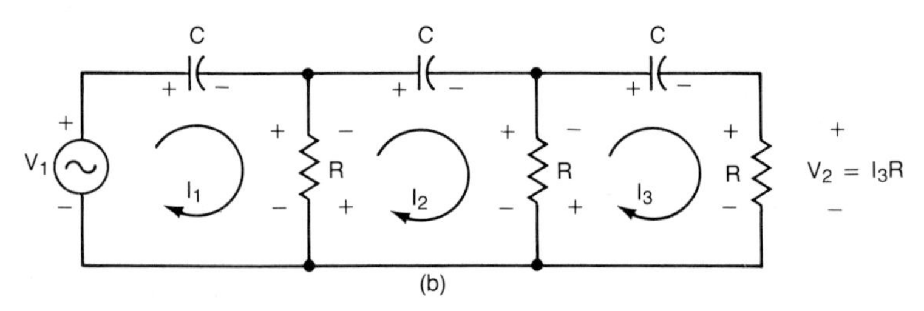

FIGURE 15-2 **Analyzing the RC phase-shift oscillator: (a) basic phase-shift oscillator; (b) phase-shift network.**

2. Place the resulting equation in the form $a + jb$. (If there are no imaginary terms in the numerator, additional algebraic steps can be avoided by stopping the algebraic manipulation when the denominator is in the form above.)
3. Set the imaginary term (j-term) equal to zero, and solve for f_o (the frequency of oscillation).
4. Substitute the result above into the equation for β obtained in step 1, and determine the magnitude of β at f_o.

These steps will produce an equation to determine the frequency of oscillation, and also assist us in the determination of the minimum voltage gain required to ensure oscillation. Consider Fig. 15-2(b).

The applied voltage V_1 produces three mesh currents: I_1, I_2, and I_3. The output voltage of the network is V_2, where

$$V_2 = I_3R$$

To find the voltage gain β of the phase-shift network, we must solve for I_3 since

$$\beta = \frac{\text{output voltage of the feedback network}}{\text{input voltage of the feedback network}}$$

$$= \frac{V_2}{V_1} = \frac{I_3 R}{V_1}$$

The three mesh equations for the network can be written as indicated by

$$-V_1 + \left(-j\frac{1}{\omega C}\right)I_1 + R(I_1 - I_2) = 0 \tag{15-1}$$

$$R(I_2 - I_1) + \left(-j\frac{1}{\omega C}\right)I_2 + R(I_2 - I_3) = 0 \tag{15-2}$$

$$R(I_3 - I_2) + \left(-j\frac{1}{\omega C}\right)I_3 + RI_3 = 0 \tag{15-3}$$

To solve the set of mesh equations for I_3, we shall first place them in standard form and then apply Cramer's rule and determinants. The reader is encouraged to work through the development that follows.

$$\left(R - j\frac{1}{\omega C}\right)I_1 - RI_2 + 0I_3 = V_1 \tag{15-4}$$

$$-RI_1 + \left(2R - j\frac{1}{\omega C}\right)I_2 - RI_3 = 0 \tag{15-5}$$

$$0I_1 - RI_2 + \left(2R - j\frac{1}{\omega C}\right)I_3 = 0 \tag{15-6}$$

Now, by Cramer's rule and determinants, we have

$$I_3 = \frac{\begin{vmatrix} R - j/\omega C & -R & V_1 \\ -R & 2R - j/\omega C & 0 \\ 0 & -R & 0 \end{vmatrix}}{\begin{vmatrix} R - j/\omega C & -R & 0 \\ -R & 2R - j/\omega C & -R \\ 0 & -R & 2R - j/\omega C \end{vmatrix}}$$

The 3 × 3 matrices can be reduced to 2 × 2 matrices by using cofactors. (Do not let this technique hinder you if you are familiar with another.)

$$I_3 = \frac{[R - j/\omega C]\begin{vmatrix} 2R - j/\omega C & 0 \\ -R & 0 \end{vmatrix} - [-R]\begin{vmatrix} -R & 0 \\ 0 & 0 \end{vmatrix} + [V_1]\begin{vmatrix} -R & 2R - j/\omega C \\ 0 & -R \end{vmatrix}}{[R - j/\omega C]\begin{vmatrix} 2R - j/\omega C & -R \\ -R & 2R - j/\omega C \end{vmatrix} - [-R]\begin{vmatrix} -R & -R \\ 0 & 2R - j/\omega C \end{vmatrix} + [0]\begin{vmatrix} -R & 2R - j/\omega C \\ 0 & -R \end{vmatrix}}$$

Tenacity, a bottle of aspirin, and algebraic manipulation lead us to the following result:

$$I_3 = \frac{R^2 V_1}{[R - j/\omega C][(2R - j/\omega C)^2 - R^2] + (R)(-R)(2R - j/\omega C)}$$

$$= \frac{R^2 V_1}{[R - j/\omega C][4R^2 - j4R/\omega C - 1/(\omega C)^2 - R^2] - (R^2)(2R - j/\omega C)}$$

$$= \frac{R^2 V_1}{[R - j/\omega C][3R^2 - j4R/\omega C - 1/(\omega C)^2] - 2R^3 + jR^2/\omega C}$$

Our final equation for I_3 is given by

$$I_3 = \frac{R^2V_1}{R^3 - 5R/\omega^2C^2 + j(1/\omega^3C^3 - 6R^2/\omega C)} \tag{15-7}$$

Now since $V_2 = I_3R$, Eq. 15-7 provides us with the expression

$$V_2 = \frac{R^3}{R^3 - 5R/\omega^2C^2 + j(1/\omega^3C^3 - 6R^2/\omega C)} V_1$$

Dividing both sides by V_1 yields β.

$$\beta = \frac{V_2}{V_1} = \frac{R^3}{R^3 - 5R/\omega^2C^2 + j(1/\omega^3C^3 - 6R^2/\omega C)} \tag{15-8}$$

Dividing the numerator and the denominator by R^3 leads us to

$$\beta = \frac{V_2}{V_1} = \frac{1}{(1 - 5/\omega^2R^2C^2) + j(1/\omega^3R^3C^3 - 6/\omega RC)} \tag{15-9}$$

In order for the feedback network to provide 180° of phase shift, the imaginary term must be equal to zero, and the real term must be negative at the desired frequency of oscillation (f_o). We shall first determine the frequency at which the imaginary term is equal to zero. Note that ω_o is equal to $2\pi f_o$. Hence

$$\frac{1}{\omega_o^3R^3C^3} - \frac{6}{\omega_oRC} = 0$$

$$\frac{\omega_oRC - 6\omega_o^3R^3C^3}{(\omega_o^3R^3C^3)(\omega_oRC)} = 0$$

Multiplying both sides by the denominator yields

$$\omega_oRC - 6\omega_o^3R^3C^3 = 0$$

Dividing both sides by ω_oRC produces

$$1 - 6\omega_o^2R^2C^2 = 0$$

$$6\omega_o^2R^2C^2 = 1$$

Solving for ω_o, we have

$$\omega_o^2 = \frac{1}{6R^2C^2}$$

or

$$\omega_o = \frac{1}{\sqrt{6}\,RC} \quad \text{and} \quad f_o = \frac{1}{2\pi\sqrt{6}\,RC}$$

Therefore, at

$$\omega_o = \frac{1}{\sqrt{6}\,RC}$$

the feedback factor β is

$$\beta = \frac{V_2}{V_1} = \frac{1}{(1 - 5/\omega_o^2R^2C^2) + j(1/\omega_o^3R^3C^3 - 6/\omega_oRC)}$$

$$= \frac{1}{\left\{1 - \left[\frac{5}{\left(\frac{1}{\sqrt{6}\,RC}\right)^2 R^2C^2}\right]\right\} + j0} = \frac{1}{1 - \left[\frac{5}{\left(\frac{1}{6R^2C^2}\right)R^2C^2}\right]}$$

$$= \frac{1}{1 - 30} = -\frac{1}{29}$$

and for oscillation to occur the loop gain (βA_v) must be greater than or equal to unity. Thus for the *RC* phase-shift oscillator, we have the conditions given in

$$f_o = \frac{1}{2\pi\sqrt{6}\,RC} \qquad A_v \geq 29\angle 180^\circ \tag{15-10}$$

where f_o is the *RC* phase-shift oscillator frequency of oscillation in hertz and A_v is the amplifier's required voltage gain.

In Fig. 15-3 we see an op amp phase-shift oscillator. The op amp is being used in its inverting amplifier configuration. As we can see, at f_o the frequency-determining network provides 180° of phase shift. Since we have an inverting amplifier, the total phase shift around the loop is 360° (0°), and the feedback signal will be in phase with the input. Therefore, the feedback will be positive (or regenerative), and the circuit will oscillate.

FIGURE 15-3 Op amp RC phase-shift oscillator.

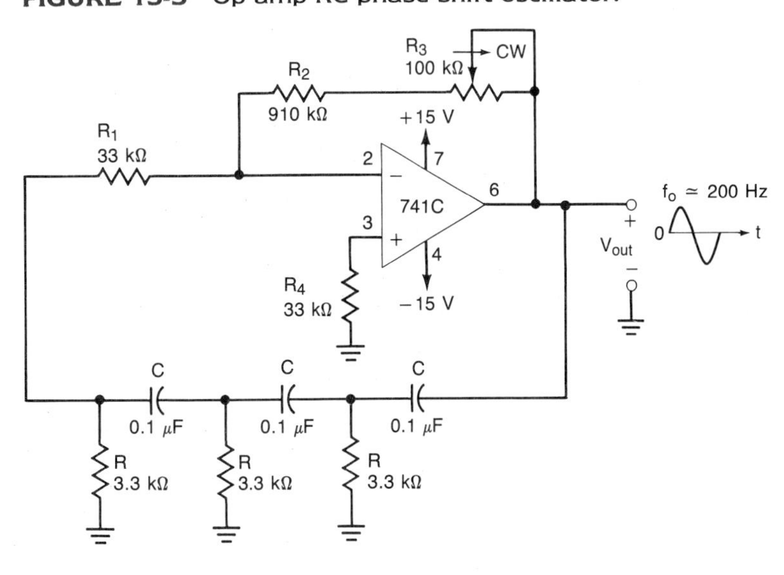

EXAMPLE 15-1

Analyze the phase-shift oscillator given in Fig. 15-3. Determine if the voltage gain and input resistance of the amplifier is adequate. Also find the frequency of oscillation.

SOLUTION First, we determine the range of possible voltage gains. When the potentiometer is fully counterclockwise,

$$A_v = -\frac{R_2 + R_3}{R_1} = -\frac{910\ \text{k}\Omega + 0\ \Omega}{33\ \text{k}\Omega} = -27.6$$

When the potentiometer is fully clockwise, the voltage gain will be at its maximum.

$$A_v = -\frac{R_2 + R_3}{R_1} = -\frac{910\ \text{k}\Omega + 200\ \text{k}\Omega}{33\ \text{k}\Omega} = -33.9$$

Since the maximum voltage gain exceeds -29, it is safe to assume that proper adjustment of the potentiometer will result in oscillation. The output resistance of the inverting amplifier is very small. Therefore, we shall assume that it is negligible. Further inspection of Fig. 15-3 would seem to indicate that the amplifier's input resistance is also adequate.

$$R_{\text{in}} \simeq R_1 = 33\ \text{k}\Omega >> R = 3.3\ \text{k}\Omega$$

Now we may draw on Eq. 15-10 to find the frequency of oscillation.

$$f_o = \frac{1}{2\pi\sqrt{6}\,RC} = \frac{1}{2\pi\sqrt{6}\,(3.3\ \text{k}\Omega)(0.1\ \mu\text{F})} = 197\ \text{Hz}$$

■

The phase-shift oscillator clearly demonstrates the basic principles behind the audio-frequency harmonic oscillator. However, it is not an extremely popular circuit. It is difficult to adjust or trim its frequency of oscillation. A far more popular circuit is the Wien bridge oscillator.

15-4 Wien Bridge Oscillators

The "heart" of all the harmonic oscillators consists of their frequency-determining feedback networks. *The particular feedback circuit arrangement gives rise to the name of a particular oscillator*. Examples include the Pierce, Hartley, Colpitts, Clapp, and Wien bridge oscillators. A Wien bridge oscillator that employs an op amp has been illustrated in Fig. 15-4(a). The others are discussed later.

The Wien bridge oscillator employs a lead–lag network. The phase shift across the network lags with increasing frequency and leads with decreasing frequency. As can be seen in Fig. 15-4, the lead–lag network consists of a parallel RC network in series with a series RC network. *At one particular frequency, the phase shift across the network is* 0°. Therefore, the feedback network is connected to the op amp's noninverting input terminal. The basic topology was shown in Fig. 15-1(b).

The frequency-determining feedback network has been redrawn in Fig. 15-4(b). To analyze it, we shall follow the same procedure applied to the RC phase-shift network. First, we shall find the parallel equivalent impedance of R_1 and C_1.

$$Y = G_1 + jB_{C1} = \frac{1}{R_1} + j\omega C_1 = \frac{1 + j\omega R_1 C_1}{R_1}$$

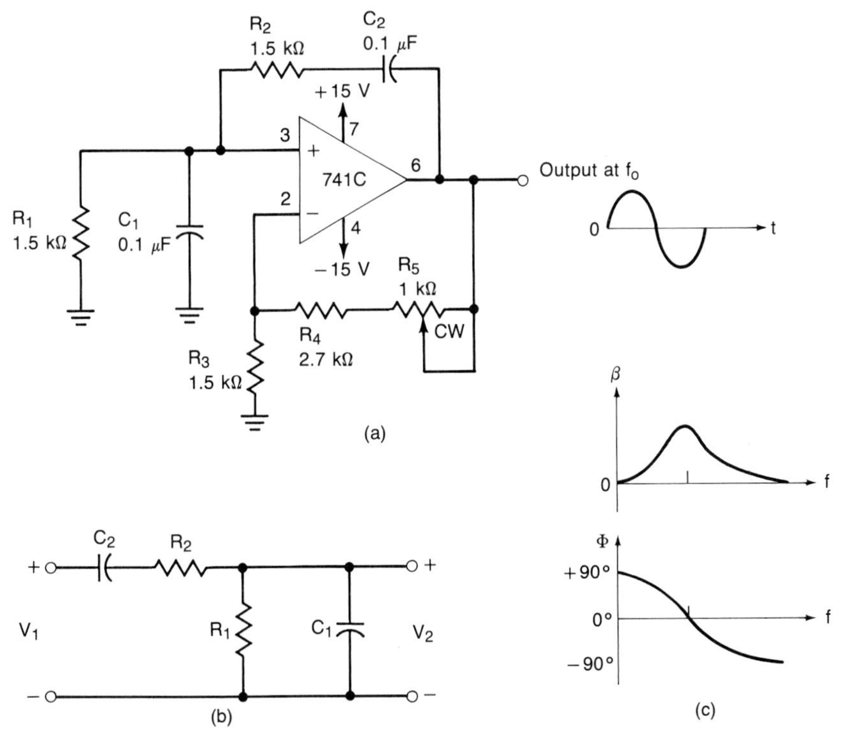

FIGURE 15-4 **Wien bridge oscillator: (a) Wien bridge oscillator that employs an op amp; (b) frequency-determining feedback network; (c) gain (β) and phase of the network.**

and

$$Z = \frac{1}{Y} = \frac{R_1}{1 + j\omega R_1 C_1}$$

By employing voltage division, we shall develop our equation for the network's voltage gain (β). Hence

$$\beta = \frac{V_2}{V_1} = \frac{R_1/(1 + j\omega R_1 C_1)}{R_2 + 1/j\omega C_2 + R_1/(1 + j\omega R_1 C_1)}$$

$$= \frac{\dfrac{\mathrm{R}_1}{(1 + j\omega R_1 C_1)}}{\left[\dfrac{R_2(j\omega C_2)(1 + j\omega R_1 C_1) + 1 + j\omega R_1 C_1 + j\omega R_1 C_2}{j\omega C_2(1 + j\omega R_1 C_1)}\right]}$$

Continuing with algebraic manipulation, we eventually arrive at

$$\beta = \frac{\mathrm{R}_1}{(\cancel{1 + j\omega R_1 C_1})} \frac{j\omega C_2(\cancel{1 + j\omega R_1 C_1})}{j\omega R_2 C_2 - \omega^2 R_1 R_2 C_1 C_2 + 1 + j\omega R_1 C_1 + j\omega R_1 C_2}$$

$$= \frac{j\omega R_1 C_2}{(1 - \omega^2 R_1 R_2 C_1 C_2) + j\omega(R_1 C_1 + R_2 C_2 + R_1 C_2)} \frac{-j}{-j}$$

$$= \frac{\omega R_1 C_2}{\omega(R_1 C_1 + R_2 C_2 + R_1 C_2) + j(\omega^2 R_1 R_2 C_1 C_2 - 1)} \qquad (15\text{-}11)$$

At the frequency of oscillation (f_o or ω_o) the imaginary term (in the denominator) must be equal to zero, and the magnitude of β must be a positive real number. This means that the phase shift across the network will be 0°. Hence

$$\omega_o^2 R_1 R_2 C_1 C_2 - 1 = 0$$

$$\omega_o^2 = \frac{1}{R_1 R_2 C_1 C_2}$$

$$\omega_o = \frac{1}{\sqrt{R_1 R_2 C_1 C_2}} \tag{15-12}$$

The frequency of oscillation in hertz is given by

$$\boxed{f_o = \frac{1}{2\pi \sqrt{R_1 R_2 C_1 C_2}}} \tag{15-13}$$

where f_o is the Wien bridge oscillator frequency of oscillation in hertz.

Generally, to determine the magnitude of β at f_o, we must substitute Eq. 15-12 for ω_o into our equation for β (Eq. 15-11). However, although this approach will work, we can avoid additional algebra by making some simple observations. Specifically, the imaginary term is zero at f_o, and since ω_o appears in the numerator and the denominator, we may cancel them. The development is as follows:

$$\beta = \frac{\omega_o R_1 C_2}{\omega_o(R_1 C_1 + R_2 C_2 + R_1 C_2) + j0}$$

$$= \frac{R_1 C_2}{R_1 C_1 + R_2 C_2 + R_1 C_2}$$

$$= \frac{\dfrac{R_1 C_2}{R_1 C_2}}{\dfrac{R_1 C_1}{R_1 C_2} + \dfrac{R_2 C_2}{R_1 C_2} + \dfrac{R_1 C_2}{R_1 C_2}}$$

$$\boxed{\beta = \frac{1}{1 + R_2/R_1 + C_1/C_2}} \tag{15-14}$$

Quite often the Wien bridge feedback network is designed such that the resistors and capacitors are equal in value. Hence

$$R_1 = R_2 = R \quad \text{and} \quad C_1 = C_2 = C$$

In this case, the equations for f_o and β (at f_o) simplify further.

$$f_o = \frac{1}{2\pi\sqrt{R_1R_2C_1C_2}}$$

$$= \frac{1}{2\pi\sqrt{R^2C^2}}$$

$$f_o = \frac{1}{2\pi RC}\bigg|_{R_1=R_2=R \text{ and } C_1=C_2=C} \tag{15-15}$$

Further,

$$\beta = \frac{1}{1 + R_2/R_1 + C_1/C_2} = \frac{1}{1 + R/R + C/C}$$

$$\beta = \frac{1}{3}\bigg|_{R_1=R_2=R \text{ and } C_1=C_2=C} \tag{15-16}$$

EXAMPLE 15-2

Analyze the Wien bridge oscillator given in Fig. 15-4(a). Specifically, determine its frequency of oscillation, and determine if the noninverting amplifier has enough gain capability to ensure oscillation.

SOLUTION First, we note that the frequency-determining feedback elements are equal in value. Therefore, we may use Eq. 15-15 to determine the frequency of oscillation.

$$f_o = \frac{1}{2\pi RC} = \frac{1}{2\pi(1.5\ \text{k}\Omega)(0.1\ \mu\text{F})} = 1.06\ \text{kHz}$$

From Eq. 15-16 we see that the feedback factor β is $\frac{1}{3}$. Therefore, the noninverting amplifier must have a voltage gain which is slightly greater than 3 to initiate oscillation and a gain of 3 to sustain the oscillation. The gain can be varied by adjusting potentiometer R_5. When R_5 is fully counterclockwise,

$$A_v = 1 + \frac{R_4 + R_5}{R_3} = 1 + \frac{2.7\ \text{k}\Omega + 0\ \text{k}\Omega}{1.5\ \text{k}\Omega} = 2.8$$

which is slightly below the minimum required value. However, when potentiometer R_5 is fully clockwise,

$$A_v = 1 + \frac{R_4 + R_5}{R_3} = 1 + \frac{2.7\ \text{k}\Omega + 1\ \text{k}\Omega}{1.5\ \text{k}\Omega} = 3.47$$

which should be more than adequate. ■

As we saw in Example 15-2 potentiometer R_5 can be adjusted to give our amplifier the required gain. However, if R_5 is adjusted a little low, the voltage gain will be slightly less than 3. Therefore, when power is first applied, the oscillations will die out or fail to start. If potentiometer R_5 is adjusted a little high, the voltage gain will be slightly greater than 3. Under this condition, we can guarantee that the circuit oscillations will build up when power is applied. However, the oscillations will continue to grow until the output waveform is distorted. The oscillator's output waveform will be clipped on the positive and negative peaks as the amplifier enters saturation.

The ideal situation would be an amplifier that has a voltage gain which is initially larger than 3 (a loop gain greater than unity) and reduces to 3 as the output level grows (a loop gain of unity). *Adaptive negative feedback* offers us a solution. One such scheme has been depicted in Fig. 15-5(a). Diodes D_1 and D_2 are signal diodes (such as the 1N914 or the 1N4148). Let us examine the circuit operation.

When power is first applied, electrical noise with a vast spectral content will be produced. The frequency-determining network will ''pick out'' f_o, and the circuit will begin to oscillate at that frequency. Since the output signal is at a low level (i.e., a few millivolts), diodes D_1 and D_2 will not conduct. Therefore, R_5 is in series with an open circuit and has no effect. The feedback network effectively consists of R_4 alone, and the voltage gain is

$$A_v = 1 + \frac{R_4}{R_3} = 1 + \frac{3.3\ \text{k}\Omega}{1.5\ \text{k}\Omega} = 3.2$$

A voltage gain of 3.2 should be more than adequate to allow the oscillations to build up. As the output level builds up, more voltage will be developed across R_4. When the peak voltage drops across R_4 reaches approximately 0.5 V, the diodes will begin to conduct. Diode D_1 conducts when the output is positive, and diode D_2 conducts when the output is negative. The diodes will begin to go into conduction when the output level reaches approximately 0.727 V.

Since the diodes are slightly conducting on the positive and negative peaks, the diodes (and R_5) will be in shunt with feedback resistor R_4. This will lower the amplifier's voltage gain. As the output voltage continues to grow, the positive and negative peaks will drive the diodes further into conduction. Eventually, the diodes will conduct hard enough such that the voltage gain will be reduced to 3.

Predicting the required output voltage to cause adequate diode conduction is a rather difficult nonlinear circuit analysis problem. Specific numerical answers are a strong function of the individual diode characteristics. Therefore, such analysis holds questionable merit. The ''real-world'' solution is to make R_5 adjustable. The potentiometer is used to adjust the output voltage for minimal distortion.

Contrast this. The setting of potentiometer R_5 in Fig. 15-4(a) is critical to ensure oscillation. The circuit given in Fig. 15-5(a) is guaranteed to oscillate. Potentiometer R_5 is merely used to adjust the output for minimum distortion.

The lower the output level, the less the distortion will be. The potentiometer will affect both the output amplitude and the distortion present in the output. Quite often, if potentiometer R_5 in Fig. 15-5(a) is adjusted too low, the output will appear to exhibit crossover distortion. In this case the diodes are producing the distortion as they go in and out of conduction. One remedy is to increase the setting of potentiometer R_5. However, this is occasionally undesirable, as the output level will also be increased.

If we add another op amp and extract the output signal from across the parallel combination of R_1 and C_1 the output will contain less distortion. This occurs because of the filtering action produced by R_1 and C_1. This scheme is illustrated in Fig.

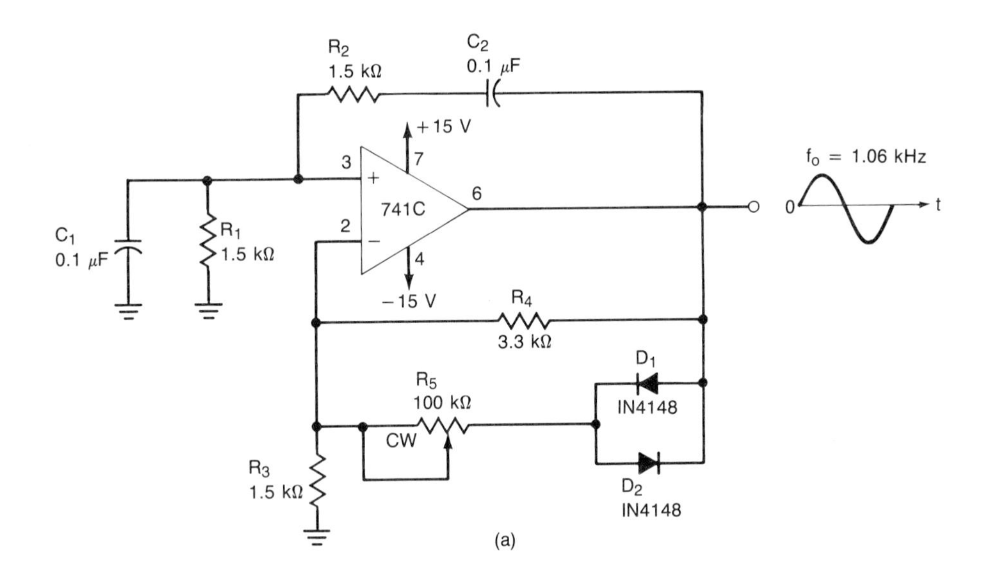

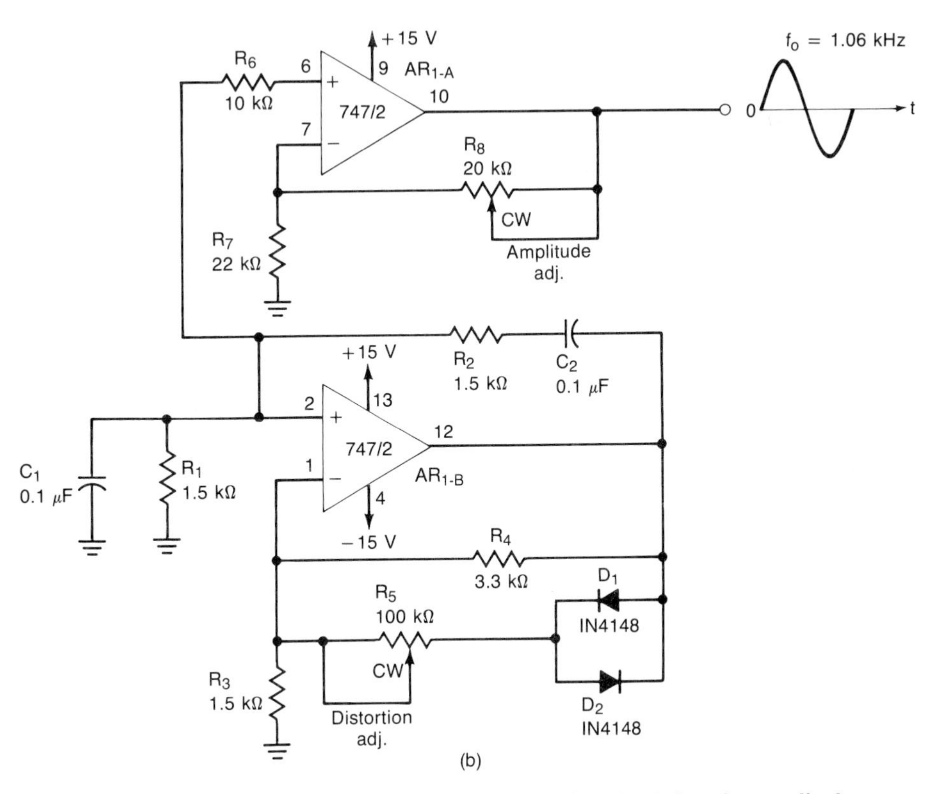

FIGURE 15-5 Adding adaptive negative feedback: (a) using a diode network to provide adaptive negative feedback; (b) providing a low-distortion output with adjustable amplitude.

15-5(b). Observe that the noninverting amplifier is one-half of a 747 dual (741) op amp. The use of a noninverting amplifier produces negligible loading on the feedback network. In this circuit, potentiometer R_5 is adjusted to provide minimal output distortion. Potentiometer R_8 is then used to adjust the output amplitude.

An alternative approach to the use of diodes to provide adaptive negative feedback requires the use of a JFET. The JFET can be used as a *voltage variable resistor* (VVR). Specifically, the JFET's drain-to-source ac resistance can be controlled by the dc bias impressed between its gate and source terminals. It offers the advantage of "softer" gain control and improved linearity. From an oscillator standpoint, this means less output distortion.

15-5 JFETs as Voltage Variable Resistors

Although our immediate concern is providing improved adaptive negative feedback for our Wien bridge oscillator, the use of the JFET as a voltage-variable resistor merits some additional discussion. The VVR is often employed in voltage-controlled attenuator circuits and in automatic gain control (AGC) applications.

In all of our previous work, we have operated our FETs in their pinch-off or constant-current region of operation [refer to Fig. 15-6(a)]. Recall that the locus of points is defined by

$$|V_{DS}| + |V_{GS}| = |V_{GS(\text{OFF})}| \qquad (15\text{-}17)$$

As long as the operating points lie to the right of this boundary line, the JFET is in its pinch-off region. Under this condition, the drain current is given by

$$I_D = I_{DSS}\left[1 - \frac{V_{GS}}{V_{GS(\text{OFF})}}\right]^2 \qquad (15\text{-}18)$$

However, if we operate the JFET to the left of the boundary line given by Eq. 15-17, we are in its *ohmic region* of operation. In this case, the drain current is defined by a new relationship,

$$I_D = \frac{2I_{DSS}}{V^2_{GS(\text{OFF})}}\left[V_{GS} - V_{GS(\text{OFF})} - \frac{V_{DS}}{2}\right]V_{DS} \qquad (15\text{-}19)$$

For very small values of V_{DS}, we can expand the region about the origin as shown in Fig. 15-6(b). As we can see, the *V-I* characteristics are fairly linear, and their slopes are controlled by the magnitude of V_{GS}. The greater the slope, the lower the resistance. This becomes clear when we compare the *V-I* characteristics of various resistors [Fig. 15-6(c)] with Fig. 15-6(b).

The smallest value of drain-to-source resistance occurs when V_{GS} is zero and is designated as $r_{DS(\text{ON})}$. It has been defined graphically in Fig. 15-7. Hence

$$r_{DS(\text{ON})} = \left.\frac{\Delta V_{DS}}{\Delta I_D}\right|_{V_{GS}=0\text{ V}} \qquad (15\text{-}20)$$

The data sheets of FETs which are intended to be used as VVRs or switches often include the parameter $r_{DS(\text{ON})}$. Its value is typically very small, ranging from a few hundred ohms to less than 1 Ω.

As V_{GS} becomes more negative, the value of r_{DS} becomes greater. Note that the

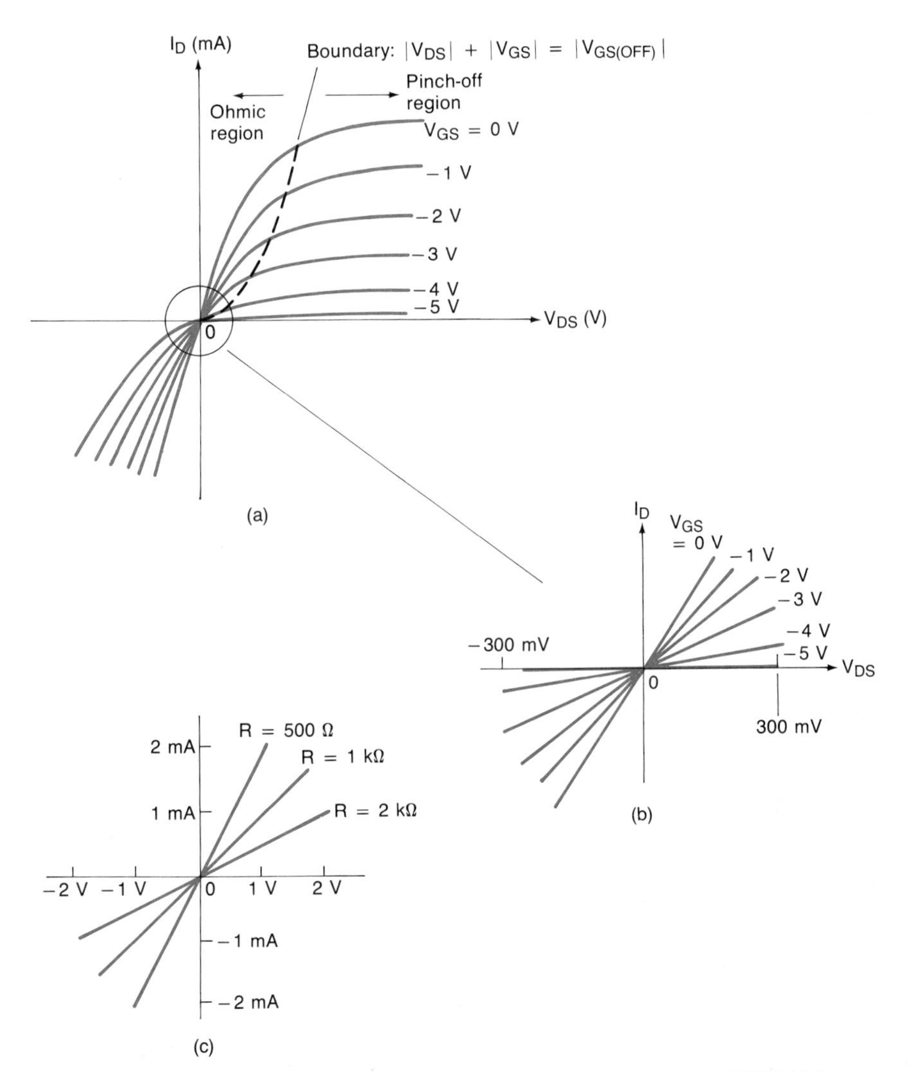

FIGURE 15-6 JFET as a voltage-variable resistor: (a) JFET V-I characteristics; (b) JFET V-I characteristics expanded about the origin; (c) V-I characteristics of three resistors.

parameter r_{DS} should not be confused with r_O. The parameter r_{DS} refers to the (typically) small drain-to-source resistance in the ohmic region, while r_O refers to the much larger drain-to-source resistance in the pinch-off region.

It should be clear that the reciprocal of the drain-to-source resistance (r_{DS}) yields the drain-to-source conductance (g_{DS}). Hence

$$g_{DS} = \frac{1}{r_{DS}} \tag{15-21}$$

and

$$g_{DS} = \frac{\partial I_D}{\partial V_{DS}} \tag{15-22}$$

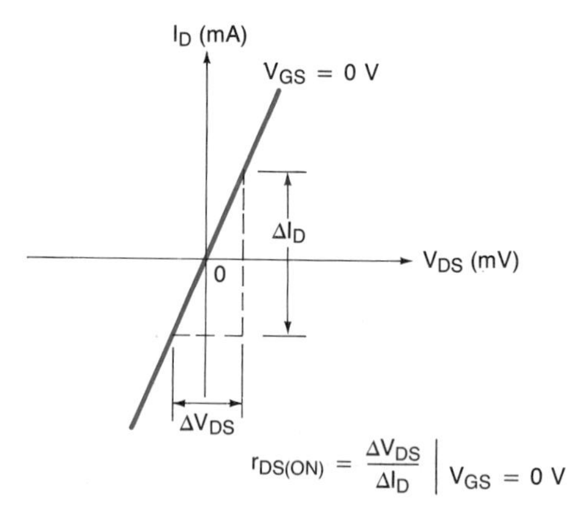

FIGURE 15-7 Graphical definition of $r_{DS(ON)}$.

where $\partial I_D/\partial V_{DS}$ is the partial derivative (rate of change) of the drain current with respect to the drain-to-source voltage.

To obtain a functional relationship for g_{DS}, we must take the partial derivative of Eq. 15-19. Mathematically uninitiated students may ignore the calculus and go directly to Eq. 15-24.

First, we distribute V_{DS}.

$$I_D = \frac{2I_{DSS}}{V^2_{GS(\text{OFF})}}\left[V_{GS} - V_{GS(\text{OFF})} - \frac{V_{DS}}{2}\right]V_{DS}$$

$$= \frac{2I_{DSS}}{V^2_{GS(\text{OFF})}}\left[V_{GS}V_{DS} - V_{GS(\text{OFF})}V_{DS} - \frac{V^2_{DS}}{2}\right]$$

Taking the derivative yields

$$g_{DS} = \frac{\partial I_D}{\partial V_{DS}} = \frac{2I_{DSS}}{V^2_{GS(\text{OFF})}}\frac{\partial}{\partial V_{DS}}\left[V_{GS}V_{DS} - V_{GS(\text{OFF})}V_{DS} - \frac{V^2_{DS}}{2}\right]$$

$$= \frac{2I_{DSS}}{V^2_{GS(\text{OFF})}}[V_{GS} - V_{GS(\text{OFF})} - V_{DS}]$$

In Section 5-13, an equation (Eq. 5-11) was developed for the transconductance (g_{m0}) of a JFET when its V_{GS} is zero. Recalling that g_{fs0} is the symbol used by many manufacturers, we have

$$g_{fs0} = -\frac{2I_{DSS}}{V_{GS(\text{OFF})}}$$

Substituting this relationship into our equation for g_{DS}, we arrive at

$$g_{DS} = \frac{2I_{DSS}}{V^2_{GS(\text{OFF})}}[V_{GS} - V_{GS(\text{OFF})} - V_{DS}]$$

$$= -\frac{2I_{DSS}}{V_{GS(\text{OFF})}}\left[1 - \frac{V_{GS}}{V_{GS(\text{OFF})}} + \frac{V_{DS}}{V_{GS(\text{OFF})}}\right]$$

$$= g_{fs0}\left[1 - \frac{V_{GS}}{V_{GS(\text{OFF})}} + \frac{V_{DS}}{V_{GS(\text{OFF})}}\right] \qquad (15\text{-}23)$$

To ensure linearity in g_{DS}, the FET must *remain in its ohmic region of operation, and V_{DS}* must *be kept small*. From Eq. 15-17,

$$|V_{DS}| << |V_{GS(\text{OFF})}| - |V_{GS}|$$

Hence we obtain the approximation given in

$$g_{DS} \simeq g_{fs0}\left[1 - \frac{V_{GS}}{V_{GS(\text{OFF})}}\right]\Bigg|_{|V_{DS}|<<|V_{GS(\text{OFF})}|-|V_{GS}|} \tag{15-24}$$

$$g_{DS} \simeq \frac{1}{r_{DS(\text{ON})}}\left[1 - \frac{V_{GS}}{V_{GS(\text{OFF})}}\right]\Bigg|_{|V_{DS}|<<|V_{GS(\text{OFF})}|-|V_{GS}|} \tag{15-25}$$

The constraint on V_{DS} is very important. Disregarding it has resulted in VVR circuits that have produced intolerable amounts of distortion and has given the JFET VVR an often undeserved bad reptuation.

EXAMPLE 15-3

A 2N5458 *n*-channel JFET is to be used as a VVR. Its $r_{DS(\text{ON})}$ is not given, but we do know that its minimum $V_{GS(\text{OFF})}$ is -2 V, and its minimum $|y_{fs0}|$ is 1500 μS. If V_{DS} is kept very small, determine r_{DS} if V_{GS} is 0, -0.4, -0.8, -1.2, -1.6, and -1.999 V.

SOLUTION Since we do not know $r_{DS(\text{ON})}$, we shall employ Eq. 15-24. We recall that $|y_{fs0}|$ is approximately equal to g_{fs0}. Hence, for a V_{GS} of 0 V,

$$g_{DS} \simeq g_{fs0}\left[1 - \frac{V_{GS}}{V_{GS(\text{OFF})}}\right]$$

$$\simeq (1.5 \times 10^{-3}\ \text{S})\left(1 - \frac{0\ \text{V}}{-2\ \text{V}}\right) = 1.5\ \text{mS}$$

$$r_{DS} = \frac{1}{g_{DS}} = \frac{1}{1.5\ \text{mS}} = 667\ \Omega$$

and for a V_{GS} of -0.4 V,

$$g_{DS} \simeq g_{fs0}\left[1 - \frac{V_{GS}}{V_{GS(\text{OFF})}}\right]$$

$$\simeq (1.5 \times 10^{-3}\ \text{S})\left(1 - \frac{-0.4\ \text{V}}{-2\ \text{V}}\right) = 1.2\ \text{mS}$$

$$r_{DS} = \frac{1}{g_{DS}} = \frac{1}{1.2\ \text{mS}} = 833\ \Omega$$

The rest of the calculations are very similar and have been summarized in Table 15-1. A graph of the results has been given in Fig. 15-8. ■

TABLE 15-1
r_{DS} Values for Example 15-3

V_{GS} (V)	r_{DS}
0.0	667 Ω
−0.4	833 Ω
−0.8	1.11 kΩ
−1.2	1.67 kΩ
−1.6	3.33 kΩ
−1.999	1.33 MΩ

FIGURE 15-8 Graph of the r_{DS} values for Example 15-3.

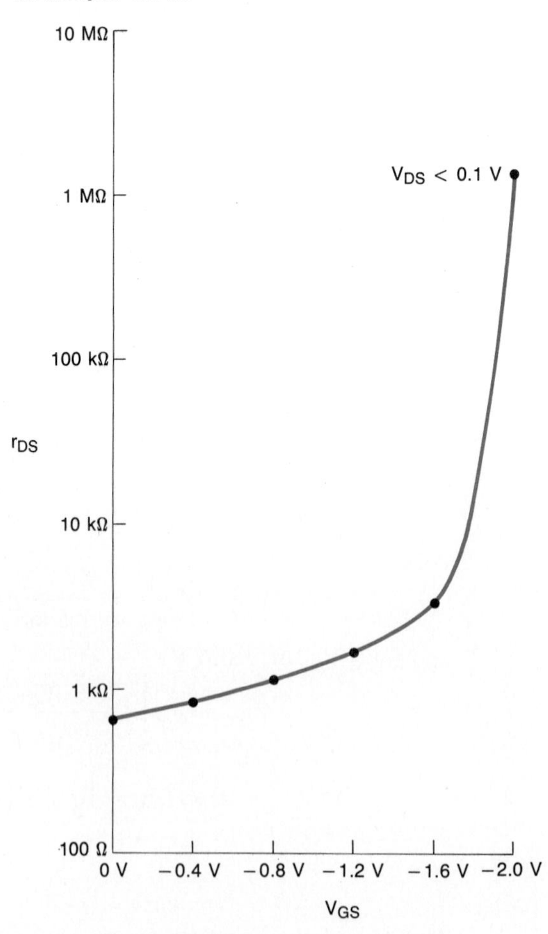

15-6 Amplitude-Stabilized Wien Bridge Oscillator Using a VVR

The Wien bridge oscillator given in Fig. 15-9 employs a JFET as a VVR. The r_{DS} of the JFET is used to control the voltage gain of the noninverting amplifier. Diodes D_1 and D_2 are used to form a half-wave rectifier. When the output voltage becomes sufficiently large, it is converted to negative pulsating dc. Capacitor C_3 is used as a filter to "smooth" the pulsating dc to provide a constant dc bias voltage across the JFET's gate-to-source terminals. Resistor R_5 is used to pull the gate of the JFET to ground. This ensures a V_{GS} of zero when the diodes are not conducting.

When power is first applied, the oscillations will begin to build up. The peak amplitude of the output voltage will be insufficient to cause diodes D_1 and D_2 to go into conduction. Therefore, they will act like open circuits, and V_{GS} will be zero. The JFET's drain-to-source resistance r_{DS} will be its minimum value $r_{DS(ON)}$. This will cause the voltage gain to be at its maximum value. This ensures a loop gain that is greater than unity.

The oscillations will continue to build up. When the (negative) peak output voltage reaches approximately 5.8 V (5.1 V + 0.7 V), the diodes (D_1 and D_2) will

FIGURE 15-9 Amplitude-stabilized Wien bridge oscillator using a VVR.

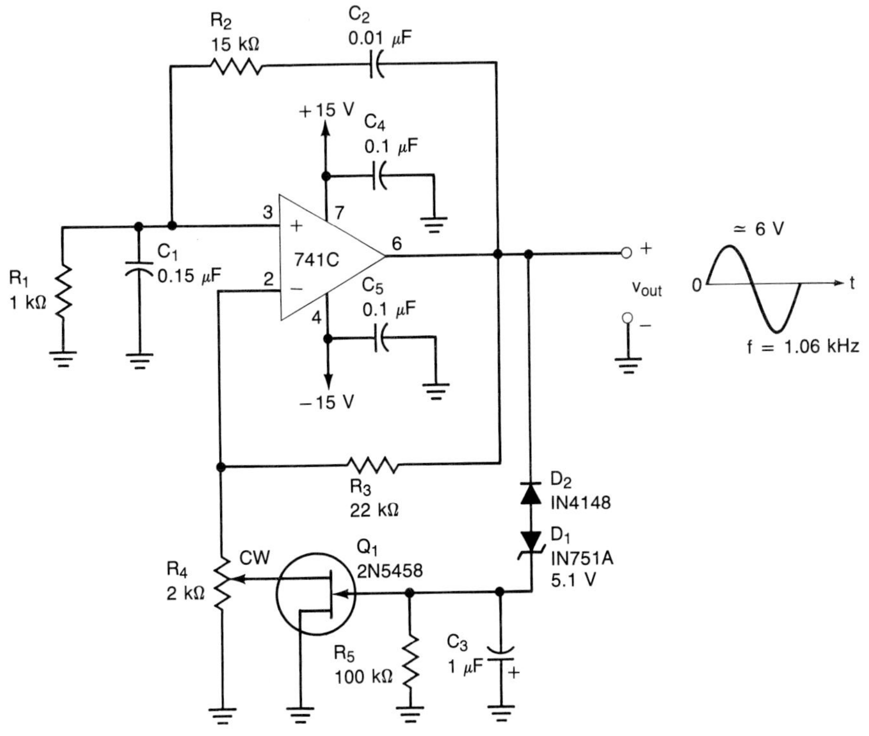

both go into conduction. Any further increases in the peak output voltage will develop a negative voltage V_{GS} across the JFET. This will cause its r_{DS} to increase.

The increase in r_{DS} will increase the total resistance from the op amp's inverting input to ground. This will reduce the voltage gain. The output voltage will continue to grow, and the voltage gain will continue to decrease until the loop gain is reduced to unity. When this condition is reached, the output amplitude will be stabilized.

To further our insight, let us analyze Fig. 15-9 in greater detail. The frequency of oscillation can be determined from Eq. 15-13.

$$f_o = \frac{1}{2\pi\sqrt{R_1 R_2 C_1 C_2}}$$

$$= \frac{1}{2\pi\sqrt{(1\text{ k}\Omega)(15\text{ k}\Omega)(0.15\ \mu\text{F})(0.01\ \mu\text{F})}}$$

$$= 1.06\text{ kHz}$$

Observe that the frequency of oscillation is equal to that of the Wien bridge oscillator given in Fig. 15-5(a). The capacitor and resistor values used in Fig. 15-9 were purposely changed to yield the same f_o and simultaneously *lower* the β of the feedback network. From Eq. 15-14,

$$\beta = \frac{1}{1 + R_2/R_1 + C_1/C_2} = \frac{1}{1 + 15\text{ k}\Omega/1\text{ k}\Omega + 0.15\ \mu\text{F}/0.01\ \mu\text{F}}$$

$$= \frac{1}{31}$$

Obviously, the β of Fig. 15-9 is much smaller than the β of $\frac{1}{3}$ that occurs when the feedback resistors (R_1 and R_2) and capacitors (C_1 and C_2) are equal in value [Fig. 15-5(a)]. The smaller β requires a larger voltage gain (31 in this instance) to provide a loop gain of unity.

Raising the required voltage gain reduces the fraction of the output voltage that appears across R_4 and the drain-to-source of the JFET. This ensures that a small ac voltage will appear across the JFET, and promotes linearity.

If the feedback resistors R_3 and R_4 are selected properly, the voltage gain of 31 should be achieved with the JFET's r_{DS} being close to its $r_{DS(\text{ON})}$ value. This means that V_{GS} will be *close* to zero. In this case, the negative peak output voltage will be

$$v_{\text{out(neg. peak)}} = -(V_{D1} + V_{D2}) + V_{GS}$$

In Fig. 15-9 the negative peak output voltage will be approximately -5.8 V plus the value of V_{GS} required for the circuit to stabilize. Allowing -0.2 V for V_{GS}, this would mean a negative peak output voltage of approximately -6 V. The positive peak will also be approximately 6 V.

The R_5-C_3 network will provide adequate filtering if its time constant is much, much larger than the period of the oscillation frequency. Thus

$$R_5 C_3 >> T$$

We have already determined that f_o is 1.06 kHz. Hence

$$T = \frac{1}{f_o} = \frac{1}{1.06\text{ kHz}} = 0.943\text{ ms}$$

Therefore, in Fig. 15-9 we have

$$R_5C_3 = (100\ \text{k}\Omega)(1\ \mu\text{F}) = 100\ \text{ms} >> 0.943\ \text{ms}$$

Obviously, R_5 and C_3 appear to be adequately sized.

The 2N5458 n-channel JFET has a transconductance (g_{fs0}) which can range from 1500 to 5500 μS. Hence

$$r_{DS(\text{ON-min})} = \frac{1}{g_{fs0(\max)}} = \frac{1}{5500\ \mu\text{S}} = 182\ \Omega$$

$$r_{DS(\text{ON-max})} = \frac{1}{g_{fs0(\min)}} = \frac{1}{1500\ \mu\text{S}} = 667\ \Omega$$

The range of possible $r_{DS(\text{ON})}$ values requires the use of potentiometer R_4 in the design. This allows us to trim in the circuit for optimal performance for any 2N5458 that might happen to be used. This is demonstrated in the following development.

Let us assume that we have a 2N5458 JFET with a g_{fs0} of 1500 μS and the minimum $V_{GS(\text{OFF})}$ of -1 V. In this case we shall allow the circuit to stabilize when V_{GS} reaches 10% of $V_{GS(\text{OFF})}$, or -0.1 V. From Eq. 15-24,

$$g_{DS(\min)} = g_{fs0(\min)}\left[1 - \frac{V_{GS}}{V_{GS(\text{OFF-min})}}\right]$$

$$= (1500\ \mu\text{S})\left(1 - \frac{-0.1\ \text{V}}{-1\ \text{V}}\right) = 1.35\ \text{mS}$$

$$r_{DS(\max)} = \frac{1}{g_{DS(\min)}} = \frac{1}{1.35\ \text{mS}} = 741\ \Omega$$

A 2N5458 with the maximum g_{fs0} of 5500 μS will also have the maximum $V_{GS(\text{OFF})}$ of -7 V. In this case 10% of $V_{GS(\text{OFF})}$ is -0.7 V. Thus

$$g_{DS(\max)} = g_{fs0(\max)}\left[1 - \frac{V_{GS}}{V_{GS(\text{OFF-max})}}\right]$$

$$= (5500\ \mu\text{S})\left(1 - \frac{-0.7\ \text{V}}{-7\ \text{V}}\right) = 4.95\ \text{mS}$$

$$r_{DS(\min)} = \frac{1}{g_{DS(\max)}} = \frac{1}{4.95\ \text{mS}} = 202\ \Omega$$

The peak output voltage in this case will be approximately 6.5 V. The reader should verify this.

EXAMPLE 15-4

Determine the "startup" voltage gain of the Wien bridge oscillator given in Fig. 15-9 if the 2N5458 has a g_{fs0} of 1500 μS and a $V_{GS(\text{OFF})}$ of -1 V. Also determine the "stabilized" voltage gain if V_{GS} is allowed to reach -0.1 V via the setting of potentiometer R_4 [see Fig. 15-10(a)].

SOLUTION From our previous work and Fig. 15-10(a), we see that $r_{DS(\text{ON})}$ is 667 Ω. Notice that the upper portion of the potentiometer has been denoted as R_{4A} and the lower portion as R_{4B}. In this case the equivalent resistance from the inverting input to ground is

$$R_T = R_{4A} + R_{4B} \parallel r_{DS(\text{ON})}$$
$$= 209\ \Omega + 1791\ \Omega \parallel 667\ \Omega = 695\ \Omega$$

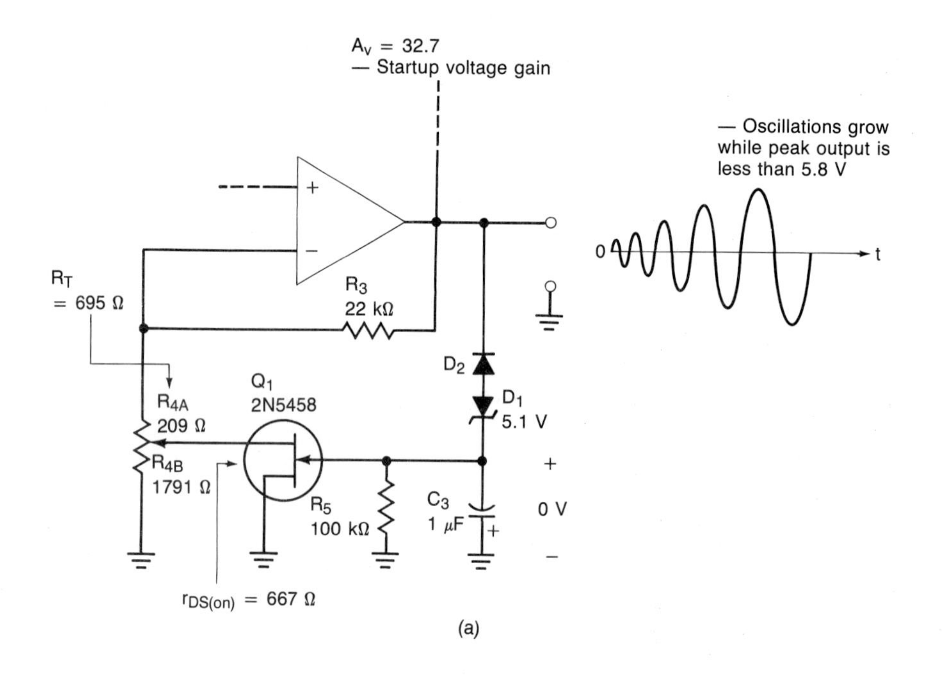

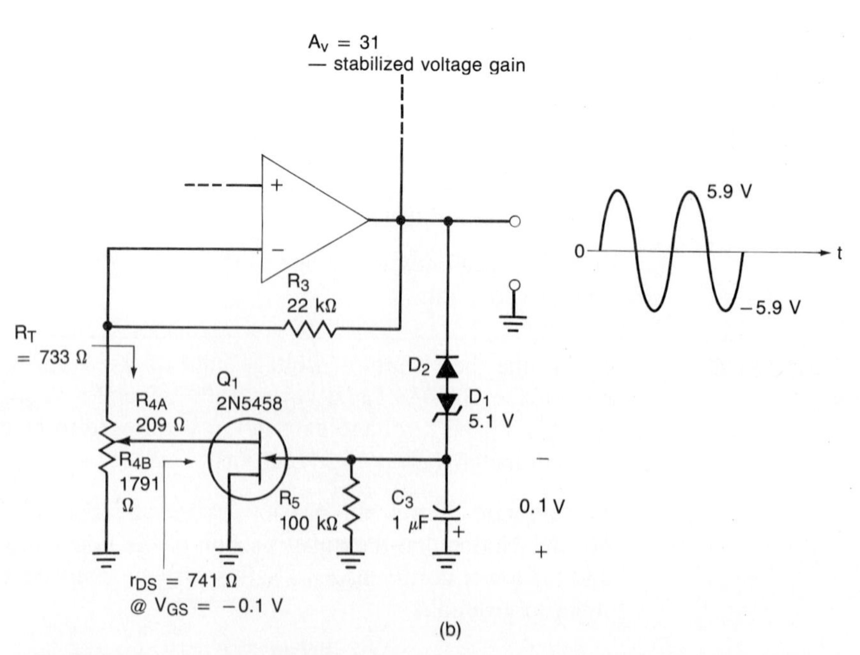

FIGURE 15-10 **Operation of the VVR in the Wien bridge oscillator: (a) oscillator during startup; (b) stabilized oscillator.**

and the corresponding (startup) voltage gain is

$$A_v = 1 + \frac{R_3}{R_T} = 1 + \frac{22 \text{ k}\Omega}{695 \text{ }\Omega} = 32.7$$

which is larger than the required 31 to sustain oscillation. When the peak output voltage reaches 5.9 V, V_{GS} will be equal to -0.1 V. The drain-to-source resistance r_{DS} will be 741 Ω. At this point R_T will be

$$R_T = R_{4A} + R_{4B} \parallel r_{DS(ON)}$$
$$= 209 \text{ }\Omega + 1791 \text{ }\Omega \parallel 741 \text{ }\Omega = 733 \text{ }\Omega$$

and

$$A_v = 1 + \frac{R_3}{R_T} = 1 + \frac{22 \text{ k}\Omega}{733 \text{ }\Omega} = 31$$

A (stabilized) voltage gain of 31 will provide a loop gain of unity, and the oscillations will be sustained [see Fig. 15-10(b)]. ■

15-7 Adjusting the Frequency of the Wien Bridge Oscillator

Quite often it is desirable to have a harmonic oscillator that has an adjustable frequency of oscillation. In the case of the Wien bridge oscillator, this can be accomplished by adjusting one of the elements in the frequency-determining feedback network (e.g., R_1, R_2, C_1, or C_2). However, varying any *one* of these components will change the β of the feedback network. If the β is changed, the loop gain will also change, and the oscillator will either start to distort or will cease to oscillate.

One alternative is to use a *ganged potentiometer* [see Fig. 15-11(a)]. The phantom lines are used to indicate that the rotors of the two potentiometers are mechanically tied together. The two potentiometers are electrically separate but are housed in a single package. This scheme will allow the frequency to be adjusted without changing the β of the feedback network. Therefore, distortion or a cessation of oscillations will be avoided. However, ganged potentiometers are relatively expensive and can be difficult to procure. A rather clever alternative is to utilize active negative feedback [see Fig. 15-11(b)].

The "main" amplifier is AR_1, and amplifier AR_2 provides voltage-gain compensation. The components utilized to achieve an adjustable output frequency are relatively inexpensive and easy to obtain. This makes the circuit given in Fig. 15-11(b) extremely attractive.

Observe that some constraints have been placed on the components. Specifically, resistors R_2, R_3, R_4, and R_5 are all equal in value. Capacitors C_1 and C_2 are also equal in value. These constraints make the operation of the circuit easier to understand. From a hardware standpoint, these constraints also minimize the number of different components required to build the circuit. This is often an advantage.

From Eq. 15-13,

$$f_o = \left. \frac{1}{2\pi\sqrt{R_1 R_2 C_1 C_2}} \right|_{C_1 = C_2 = C,\ R_2 = R} = \frac{1}{2\pi C\sqrt{R_1 R}}$$

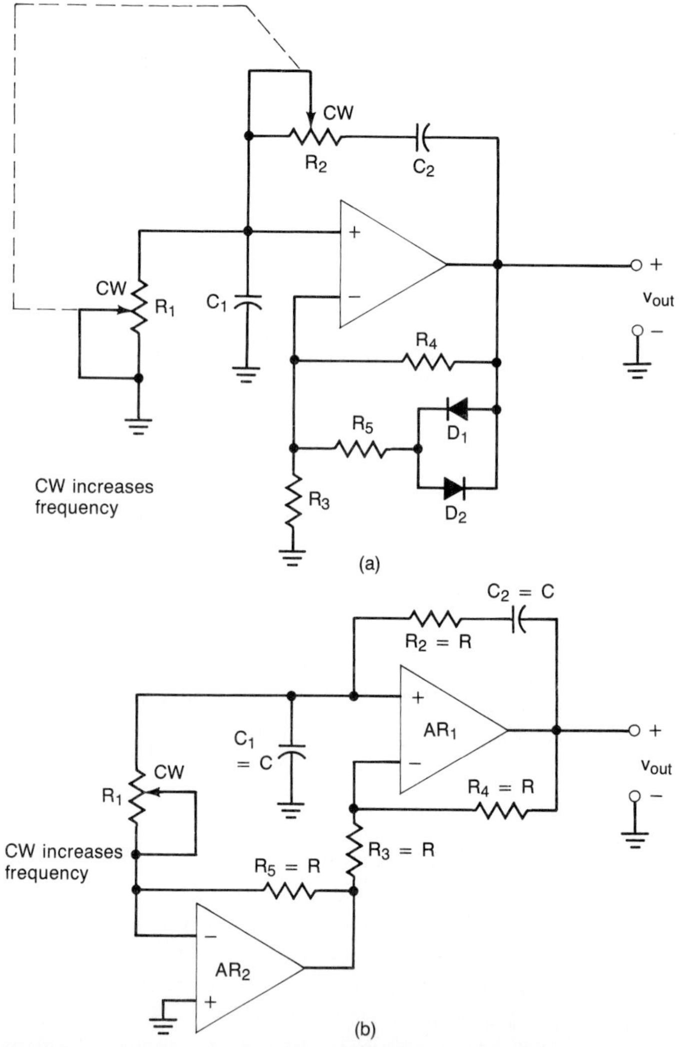

FIGURE 15-11 Adjusting the frequency of the Wien bridge oscillator: (a) using ganged potentiometers; (b) using active negative feedback and a single potentiometer.

and from Eq. 15-14,

$$\beta = \frac{1}{1 + R_2/R_1 + C_1/C_2} = \left.\frac{1}{2 + R/R_1}\right|_{C_1=C_2=C,\ R_2=R}$$

As R_1 is increased, f_o is lowered and β is increased. Conversely, as R_1 is decreased, f_o is increased and β is decreased.

Since the β of the frequency-determining feedback network changes as R_1 is varied to change the frequency, amplifier AR_2 has been added to adjust simultaneously the voltage gain of AR_1. The intent is to keep the loop gain at unity. To see how this works, we must find the equation for the voltage gain of the amplifier system.

The basic amplifier circuit has been given in Fig. 15-12(a). To simplify our analysis, we draw on the substitution and superposition theorems.

The substitution theorem allows us to split v_{in} into equivalent signal sources [see v_{in1} and v_{in2} in Fig. 15-12(b)]. Now we can apply the superposition theorem.

In Fig. 15-12(c) we have set v_{in2} equal to zero. Since the input to AR_2 is zero, its output will also be zero. If we recall that zero volts is ground, we may use the substitution theorem to arrive at the equivalent circuit indicated in Fig. 15-12(c). Obviously, AR_1 will act as a noninverting amplifier, and

$$v'_{out} = \left(1 + \frac{R}{R}\right) v_{in1} = 2v_{in1}$$

In Fig. 15-12(d) we have set v_{in1} to zero. By inspection we can see that AR_1 and AR_2 serve as two cascaded inverting amplifiers. Recalling that the overall voltage gain of a cascaded amplifier system is given by the product of the individual stage gains, we obtain the following result:

$$v''_{out} = \left(-\frac{R}{R_1}\right)\left(-\frac{R}{R}\right) v_{in2} = \frac{R}{R_1} v_{in2}$$

Adding our results produces our equation for v_{out}.

$$v_{out} = v'_{out} + v''_{out} = 2v_{in1} + \frac{R}{R_1} v_{in2}$$

and since

$$v_{in1} = v_{in2} = v_{in}$$

the voltage gain may be found by dividing both sides by v_{in}. Hence

$$v_{out} = 2v_{in} + \frac{R}{R_1} v_{in} = \left(2 + \frac{R}{R_1}\right) v_{in}$$

$$A_v = \frac{v_{out}}{v_{in}} = 2 + \frac{R}{R_1}$$

Now we can see that the loop gain is

$$\beta A_v = \frac{1}{2 + R/R_1}\left(2 + \frac{R}{R_1}\right) = 1$$

Consider this result. If we desire to increase f_o, we must reduce R_1. This will lower the β of the frequency-determining feedback network. However, this will increase the voltage gain of the inverting amplifier AR_2, which serves to raise the voltage gain of the overall amplifier system. The increase in the amplifier gain will counteract the decrease in β. Therefore, the loop gain will remain at unity, and the oscillations will be sustained.

A complete Wien bridge oscillator circuit that is based on this design approach has been given in Fig. 15-13. Although three op amps are used, they are all contained within the same 14-pin package - an LM324 quad op amp. Amplifier AR_1 serves as the main amplifier. Amplifier AR_2 provides the gain compensation. Amplifier AR_3 is used to provide additional voltage gain on the output signal. This allows us to have a relatively wide range of output signal amplitude adjustment.

In operation, potentiometer R_7 is used to adjust the output signal for minimum distortion. Potentiometer R_2 is used to determine the frequency of oscillation. Po-

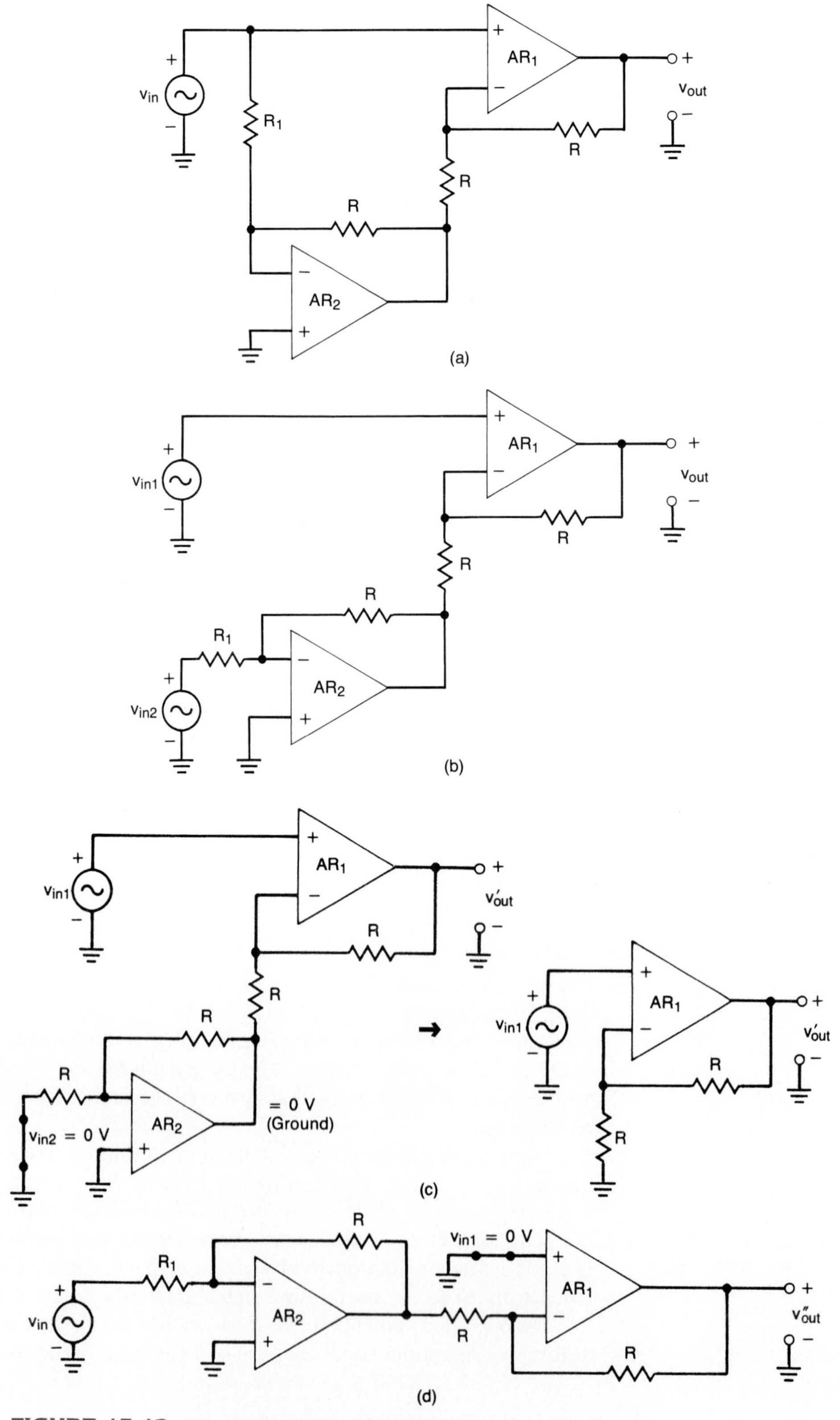

FIGURE 15-12 Analyzing the adaptive negative feedback.

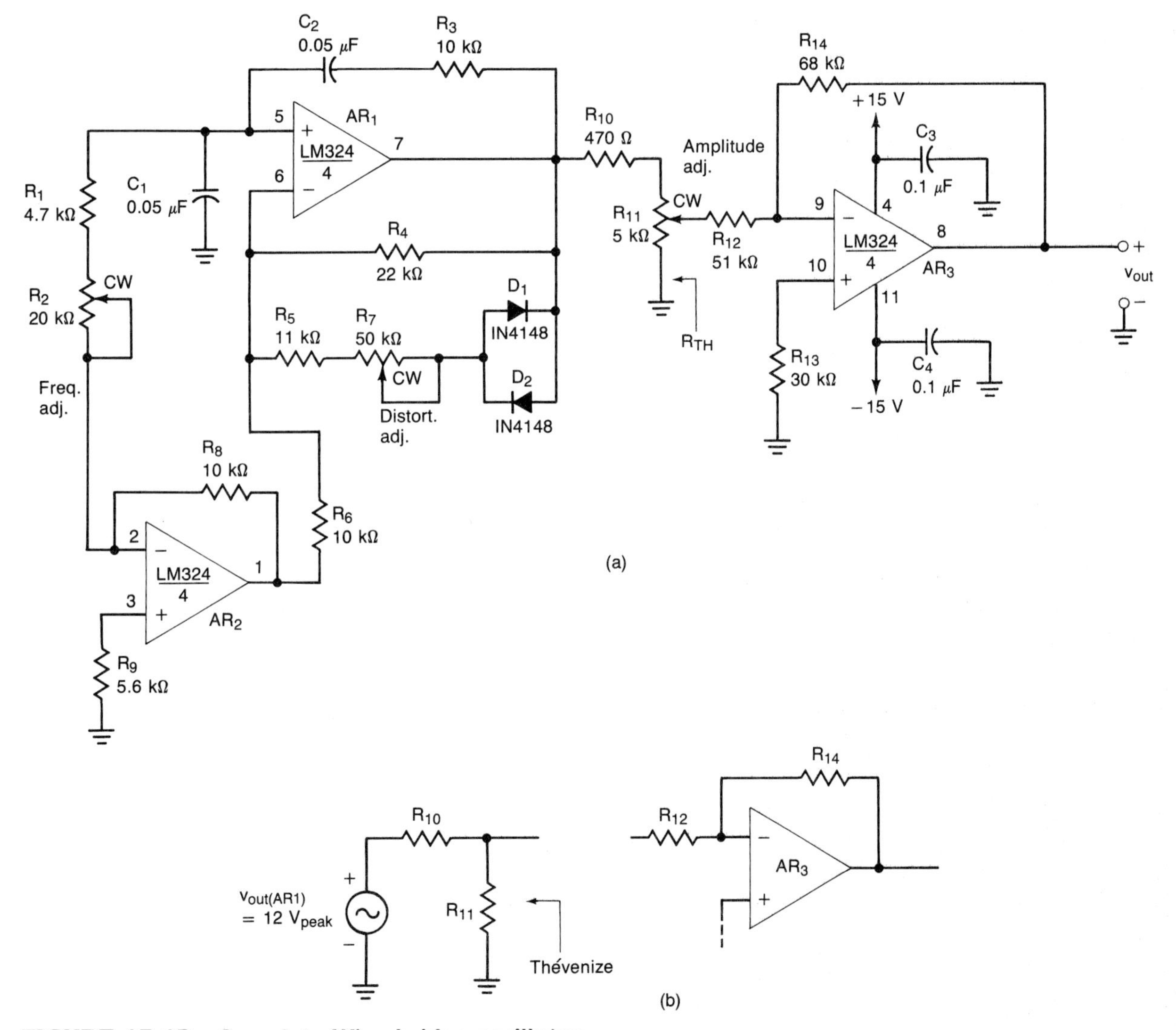

FIGURE 15-13 Complete Wien bridge oscillator.

tentiometer R_{11} may then be used to adjust the output signal amplitude. In practice, the distortion potentiometer (R_7) will be adjusted when the oscillator circuit is initially placed into service. It should not have to be adjusted again.

EXAMPLE 15-5

Determine the minimum and maximum frequency of oscillation of the Wien bridge oscillator given in Fig. 15-13. In the laboratory it was determined that when R_7 was adjusted for miniminal output distortion, a 12-V peak sine wave was produced at AR_1's output. Determine the range of output amplitudes available at AR_3's output.

SOLUTION Using the subscripts employed in Fig. 15-13, we obtain our equation

$$f_o = \frac{1}{2\pi C\sqrt{R_1 R_3}} = \frac{1}{2\pi(0.05\ \mu\text{F})\sqrt{(4.7\ \text{k}\Omega)(10\ \text{k}\Omega)}}$$
$$= 464\ \text{Hz}$$

This frequency will be obtained when R_2 is adjusted fully clockwise. When R_2 is fully counterclockwise, its full 20 kΩ of resistance will be in the circuit. Hence

$$f_o = \frac{1}{2\pi C\sqrt{(R_1 + R_2)R_3}}$$
$$= \frac{1}{2\pi(0.05\ \mu\text{F})\sqrt{(4.7\ \text{k}\Omega + 20\ \text{k}\Omega)(10\ \text{k}\Omega)}}$$
$$= 203\ \text{Hz}$$

Potentiometer R_{11} and resistor R_{10} form a voltage divider. All potentiometers will exhibit a small amount of "end resistance" when they are adjusted to the ends of their travel. The use of R_{10} helps to minimize this effect and allows us to adjust the output voltage to nearly zero. When R_{11} is fully clockwise, we have the situation shown in Fig. 15-13(b). If we apply Thévenin's theorem, we find that

$$R_{\text{TH}} = R_{10} \| R_{11} = 470\ \Omega \| 5\ \text{k}\Omega \simeq 430\ \Omega$$

$$v_{\text{th}} = \frac{R_{11}}{R_{10} + R_{11}} v_{\text{out(AR1)}} = \frac{5\ \text{k}\Omega}{470\ \Omega + 5\ \text{k}\Omega}(12\ \text{V}) \simeq 11\ \text{V}$$

and since R_{TH} forms part of the equivalent resistance in series with AR_3's inverting input, we have

$$v_{\text{out(peak)}} = -\frac{R_{14}}{R_{\text{TH}} + R_{12}} v_{\text{th}} = -\frac{68\ \text{k}\Omega}{430\ \Omega + 51\ \text{k}\Omega}(11\ \text{V})$$
$$= -14.5\ \text{V} \rightarrow \text{a magnitude of 14.5 peak}$$

The maximum peak output voltage is approximately 14.5 V. However, it is quite likely that the op amp's output will saturate before this peak is reached. A conservative estimate would rate our circuit as capable of delivering a peak output voltage of at least 13 V. ■

15-8 High-Frequency Harmonic Oscillators

Quite often, harmonic oscillators are required to oscillate at frequencies considerably above the audio-frequency range. The possibilities encompass 20 kHz to several gigahertz. We restrict our discussion to oscillators which operate in the range 100 kHz to 30 MHz. Higher-frequency oscillators are best left to studies of communications electronics circuitry.

The use of *RC* harmonic oscillator circuits becomes impractical at high frequencies. With a little reflection, we recall that the frequency of oscillation is inversely proportional to the size of the capacitors employed in the frequency-determining feedback network. Consequently, at high frequencies the size of the capacitors must shrink dramatically. When capacitors of only a few picofarads are used, stray capacitances (which are also on the order of a few picofarads) can detune the oscillator or even prohibit oscillation. Therefore, it becomes necessary to use alternative frequency-determining feedback networks. The most common approach is to use series and/or parallel resonant networks.

Resonant circuits require the use of capacitors and inductors (coils). Inductors tend to be impractical at low (e.g., audio) frequencies, but work quite well in RF applications. Further, resonant circuits tend to offer much sharper filtering than do the *RC* networks considered previously. This can make our work easier. Specifically, adaptive negative feedback circuitry may *not* be required. (Typically, some form will be used, but its design is not as critical.) This is true because the narrow frequency response of the resonant circuits tends to filter out the harmonics that may be produced by nonlinearities within the amplifier. However, there are other considerations.

Generally, *discrete* rather than integrated circuits are preferred. High-frequency, high-performance linear integrated circuits are available, but they tend to be relatively expensive. Consequently, discrete BJT and JFET amplifiers are often the most cost-effective approach.

As a result, we have several design details to attend to. For example, to ensure good frequency stability we must provide good bias stability. This can mean that we must include temperature-compensation schemes in our bias circuits. Power supply decoupling also becomes more demanding in RF applications. If we are not careful, the RF produced by the oscillator can ''talk'' to all the other circuits within the system via the power supply connections.

To understand high-frequency oscillators, we must first understand series and parallel resonant circuits. A brief review of their characteristics is provided in the next section. Some students may wish to review their circuit analysis textbook at this point.

15-9 Series and Parallel Resonant Circuits

A series resonant circuit has been illustrated in Fig. 15-14(a). All series resonant circuits contain capacitance, inductance, and resistance. The resistance is often inherent within the coil and need not be a separate component. From basic circuit theory we recall that inductive reactance is directly proportional to the frequency of excitation.

$$jX_L = j2\pi fL$$

Capacitive reactance is inversely proportional to the frequency of the excitation.

$$-jX_C = \frac{1}{j2\pi fC}$$

At *one* particular frequency (the resonant frequency f_o) the magnitude of the inductive reactance and the magnitude of the capacitive reactance will be equal. Using this definition, we may solve for the resonant frequency in terms of the inductance and capacitance. At f_o, $X_L = X_C = X$ and

$$2\pi f_o L = \frac{1}{2\pi f_o C}$$

$$f_o^2 = \frac{1}{(2\pi)^2 LC}$$

$$f_o = \frac{1}{2\pi\sqrt{LC}} \tag{15-26}$$

At resonance the impedance (Z_o) of the series resonant circuit is at its minimum since X_C and X_L cancel one another.

$$Z_o = R + jX_L - jX_C = R + jX - jX = R$$

The net impedance is purely resistive and equal to R. The graph of the impedance of a series resonant circuit versus frequency has been indicated in Fig. 15-14(b).

At frequencies *below* f_o the impedance is primarily *capacitive*. At frequencies

FIGURE 15-14 Series resonant circuit.

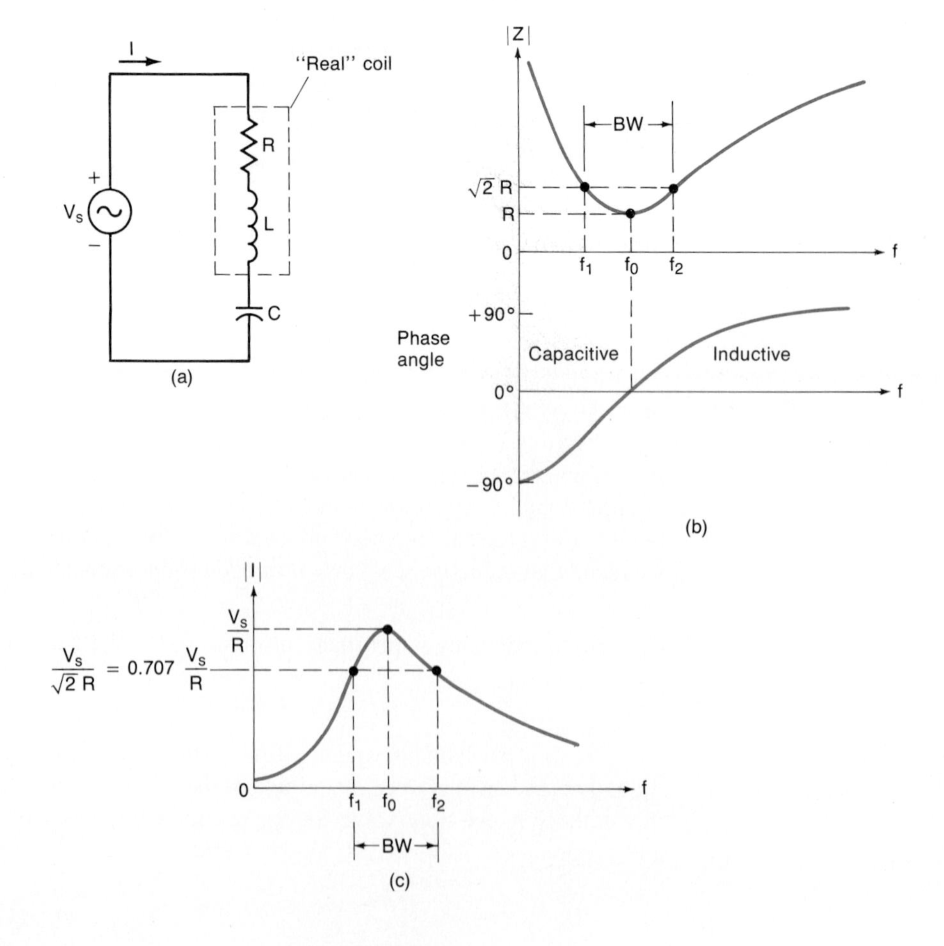

above f_o the impedance is primarily *inductive*. At f_o the impedance is at its *minimum* and resistive. (The phase angle is 0°.)

Below and above f_o at the frequencies f_1 and f_2, respectively, the magnitude of the impedance increases to a value equal to the square root of 2 times R. These two frequencies are used to define the bandwidth (BW) of the series resonant circuit.

$$\text{BW} = f_2 - f_1 \tag{15-27}$$

Since the current through the series resonant circuit is inversely proportional to its impedance, we may also represent the frequency response as indicated in Fig. 15-14(c). When the current is reduced by a factor of approximately 0.707 at f_1 and f_2, it is reduced by 3 dB. Therefore, f_1 and f_2 are also referred to as the "3-dB down points." Consequently, Eq. 15-27 is really no different than our previous definition of bandwidth in Chapter 10.

The frequency selectivity of the series resonant circuit is inversely related to the *quality factor* (Q) of the coil.

$$\text{BW} = \frac{f_o}{Q} \tag{15-28}$$

where BW = bandwidth (hertz)
f_o = resonant frequency (hertz)
Q = coil's quality factor (dimensionless)

The Q of the coil has been defined by

$$Q = \frac{X_L}{R} \tag{15-29}$$

where Q is the coil's quality factor and should be determined at the frequency of interest. Generally, a very narrow bandwidth is highly desirable. This implies that we also desire coils with a very high (10 or more) Q. The resistance R in Eq. 15-29 is associated with the coil. To emphasize this fact, the resistance shown in Fig. 15-14(a) has been enclosed in a dashed-line box with L. The resistance R represents the dc and ac losses associated with all real inductors.

EXAMPLE 15-6

The coil given in Fig. 15-14(a) has an inductance of 10 μH and a resistance of 6 Ω in the frequency range of interest. The capacitor is a 0.0025-μF unit. Determine the resonant frequency f_o, the Q of the coil at f_o, the bandwidth of the circuit, and its impedance at resonance.

SOLUTION From Eq. 15-26,

$$f_o = \frac{1}{2\pi\sqrt{LC}} = \frac{1}{2\pi\sqrt{(10\ \mu\text{H})(0.0025\ \mu\text{F})}} = 1.007\ \text{MHz}$$

To find the Q of the coil, we must first find the magnitude of X_L at f_o.

$$X_L = 2\pi f_o L = 2\pi(1.007 \text{ MHz})(10 \ \mu\text{H}) = 63.3 \ \Omega$$

Now, from Eq. 15-29,

$$Q = \frac{X_L}{R} = \frac{63.3 \ \Omega}{6 \ \Omega} = 10.5$$

This would be classified as a high-Q coil since its Q is greater than 10. The bandwidth may be found from Eq. 15-28.

$$\text{BW} = \frac{f_o}{Q} = \frac{1.007 \text{ MHz}}{10.5} = 95.9 \text{ kHz}$$

The impedance of the series resonant circuit at f_o is equal to the resistance of the coil. Hence

$$Z_o = R = 6 \ \Omega$$

■

A parallel resonant circuit has been given in Fig. 15-15(a). Once again, the effective series resistance of the coil has been indicated. In this case, the series resistance has been denoted as R_s.

If a high-Q coil is used in a parallel resonant circuit, its resonant frequency (f_o) occurs when the magnitude of the capacitive reactance equals the magnitude of the inductive reactance. The resonant frequency is given by Eq. 15-26. A high-Q coil is not an extremely restrictive constraint. Virtually all of the coils employed in our oscillator circuits will be of the high-Q variety. (Again, the Q should be specified at f_o.)

The impedance versus frequency has been depicted in Fig. 15-15(b). For frequencies *below* f_o the inductive susceptance dominates and the circuit has an *inductive*

FIGURE 15-15 **Parallel resonance: (a) parallel resonant circuit; (b) frequency response.**

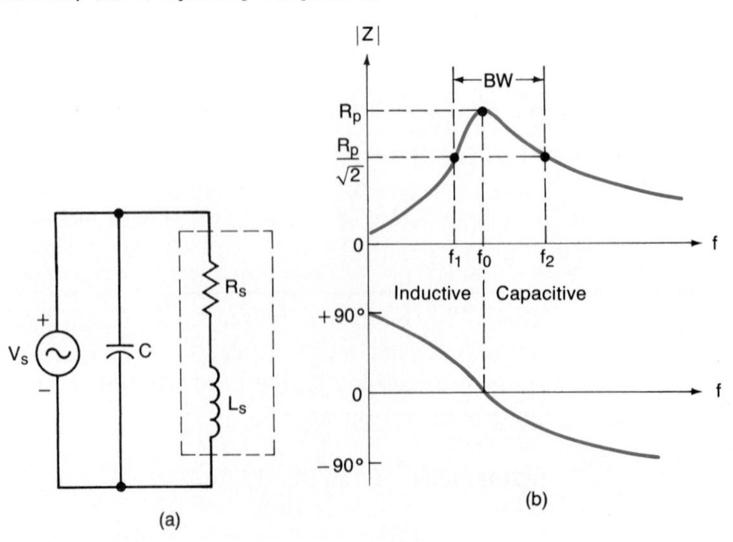

characteristic. For frequencies *above* f_o, the capacitive susceptance dominates and the circuit has a *capacitive* "flavor." The *maximum* impedance of a parallel resonant circuit occurs at f_o and is resistive. The impedance at resonance (with a high-Q coil) is given by

$$R_p = Q^2 R_s \tag{15-30}$$

where R_p = equivalent resistance of a parallel resonant circuit at f_o
Q = coil's quality factor (10 or more)
R_s = coil's internal (dc and ac) resistance

EXAMPLE 15-7

A 10-μH inductor has a Q of 200. (Its R_s is 0.316 Ω.) It is connected in parallel with a 0.0025-μF capacitor [see Fig. 15-15(a)] to form a resonant circuit. Determine its f_o, its bandwidth, and its (resistive) impedance (R_p) at resonance.

SOLUTION Equation 15-26 may be used to determine f_o.

$$f_o = \frac{1}{2\pi\sqrt{LC}} = \frac{1}{2\pi\sqrt{(10\ \mu\text{H})(0.0025\ \mu\text{F})}} = 1.007\ \text{MHz}$$

Its bandwidth is given by Eq. 15-28.

$$\text{BW} = \frac{f_o}{Q} = \frac{1.007\ \text{MHz}}{200} = 5.03\ \text{kHz}$$

The impedance (Z) at resonance is R_p and is given by Eq. 15-30.

$$R_p = Q^2 R_s = (200)^2(0.316\ \Omega) = 12.6\ \text{k}\Omega$$

■

From Example 15-7 we can see that at resonance a high-Q coil offers a high resistance and a very narrow bandwidth. Both of these characteristics are highly desirable in harmonic oscillator circuits. Consequently, the harmonic oscillators we consider in the next few sections will all incorporate high-Q coils.

15-10 General *LC* Oscillator

A number of the high-frequency harmonic oscillators can be represented by the general equivalent circuit given in Fig. 15-16(a). These oscillators are all characterized by the fact that they incorporate resonant circuits in their frequency-determining feedback networks. The three impedances Z_1, Z_2, and Z_3 are comprised of reactive circuit elements. The particular arrangement dictates the name of the oscillator. For example, when the impedances Z_1 and Z_2 are capacitors, and Z_3 is an inductor, we have a Colpitts oscillator. In this case an inverting amplifier is required (see Table 15-2). However, if the feedback elements are arranged such that Z_1 is an inductor while impedances Z_2 and Z_3 are capacitors, a noninverting amplifier may be employed. In some cases a crystal (XTAL) may be incorporated (see the Pierce oscillator in Table 15-2). The crystal is covered in Section 15-14.

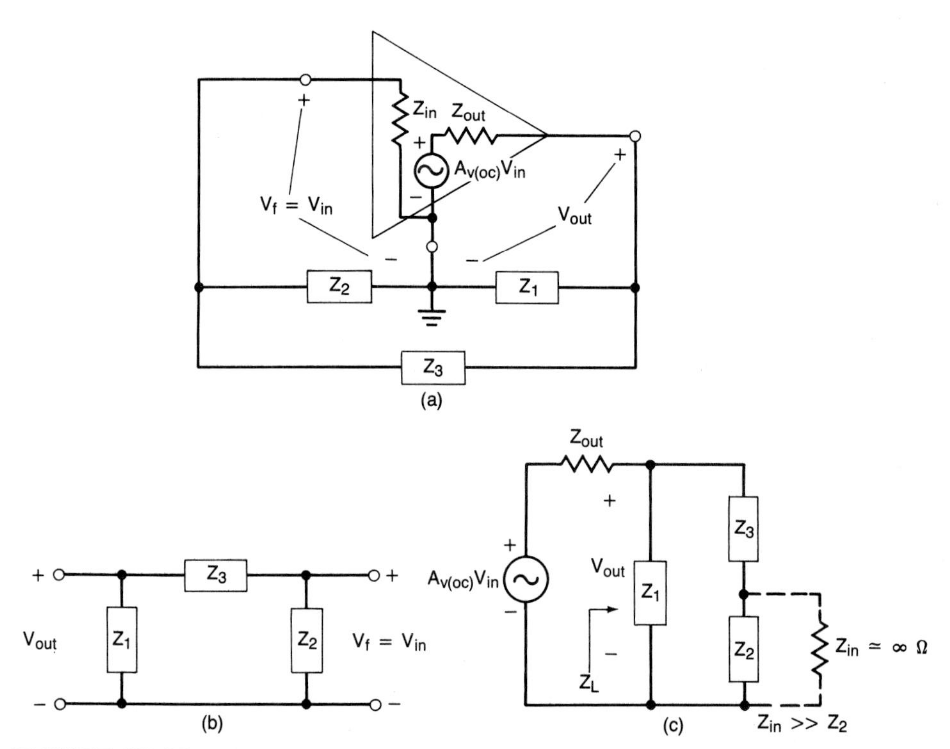

FIGURE 15-16 General LC oscillator: (a) equivalent circuit; (b) determining β; (c) determining A_v.

Each of the oscillators listed in Table 15-2 is investigated in the sections to come. As we progress, the entries in the table will become clear. The table is by no means exhaustive. There are several variations and even fundamental oscillator types which are not mentioned. However, our objective is to provide insight into the basic considerations underlying the various *LC* oscillator types. By using the general *LC* model presented in Fig. 15-16(a), we can develop some fundamental relationships that can readily be extended to a number of specific *LC* oscillators.

TABLE 15-2
Examples of *LC* Oscillators

Oscillator Type	Z_1	Z_2	Z_3	Amplifier
Hartley	*L*	*L*	*C*	Inverting
	L	*C*	*L*	Follower
Colpitts	*C*	*C*	*L*	Inverting
	L	*C*	*C*	Noninverting
Clapp	*C*	*C*	*LC* series	Inverting
Pierce crystal	*C*	*C*	XTAL (*L*)	Inverting

In Fig. 15-16(b) the feedback network has been redrawn. Our first step shall be to develop an equation for the feedback factor β. By inspection of Fig. 15-16(b), we can see that we have simple voltage division between Z_2 and Z_3. Hence

$$V_f = \frac{Z_2}{Z_2 + Z_3} V_{\text{out}}$$

$$\beta = \frac{V_f}{V_{\text{out}}} = \frac{Z_2}{Z_2 + Z_3} \tag{15-31}$$

In Fig. 15-16(c) the feedback network has been redrawn once again. Our intent now is to determine its loading effect on the output of the amplifier. Observe that we have assumed that the input impedance of the amplifier is infinite. Specifically,

$$Z_{\text{in}} >> Z_2$$

This is not merely an assumption but an important design objective in ''real'' oscillator circuits. Our next step will be to determine the equivalent load impedance (Z_L) across the amplifier's output. First we find the admittance.

$$Y_L = \frac{1}{Z_1} + \frac{1}{Z_2 + Z_3} = \frac{Z_1 + Z_2 + Z_3}{Z_1(Z_2 + Z_3)}$$

$$Z_L = \frac{1}{Y_L} = \frac{Z_1(Z_2 + Z_3)}{Z_1 + Z_2 + Z_3}$$

We now work to develop our equation for the amplifier's loaded voltage gain. We substitute in the result above at the appropriate point.

$$V_{\text{out}} = \frac{Z_L}{Z_L + Z_{\text{out}}} A_{v(\text{oc})} V_{\text{in}} = \frac{\dfrac{Z_1(Z_2 + Z_3)}{(Z_1 + Z_2 + Z_3)}}{\left[\dfrac{Z_1(Z_2 + Z_3)}{(Z_1 + Z_2 + Z_3)}\right] + Z_{\text{out}}} A_{v(\text{oc})} V_{\text{in}}$$

$$= \frac{\dfrac{Z_1(Z_2 + Z_3)}{(Z_1 + Z_2 + Z_3)}}{\left[\dfrac{Z_1(Z_2 + Z_3) + Z_{\text{out}}(Z_1 + Z_2 + Z_3)}{(Z_1 + Z_2 + Z_3)}\right]} A_{v(\text{oc})} V_{\text{in}}$$

Dividing both sides by V_{in} provides us with our equation for the loaded voltage gain.

$$A_v = \frac{V_{\text{out}}}{V_{\text{in}}} = \frac{Z_1(Z_2 + Z_3)A_{v(\text{oc})}}{Z_1(Z_2 + Z_3) + Z_{\text{out}}(Z_1 + Z_2 + Z_3)} \tag{15-32}$$

The loop gain (βA_v) can be found by multiplying Eqs. 15-31 and 15-32. This leads us to

$$\beta A_v = \left(\frac{Z_2}{Z_2 + Z_3}\right)\left(\frac{Z_1(Z_2 + Z_3)A_{v(\text{oc})}}{Z_1(Z_2 + Z_3) + Z_{\text{out}}(Z_1 + Z_2 + Z_3)}\right)$$

$$= \frac{Z_1 Z_2 A_{v(\text{oc})}}{Z_1(Z_2 + Z_3) + Z_{\text{out}}(Z_1 + Z_2 + Z_3)} \tag{15-33}$$

At resonance,

$$Z_1 + Z_2 + Z_3 = 0$$

Therefore, we may use this fact to simplify our equation for the loop gain. Hence

$$\beta A_v = \frac{Z_1 Z_2 A_{v(\text{oc})}}{Z_1(Z_2 + Z_3)} = \frac{Z_2}{Z_2 + Z_3} A_{v(\text{oc})}$$

Since

$$Z_1 + Z_2 + Z_3 = 0$$

then

$$Z_2 + Z_3 = -Z_1$$

and we may use this result to further simplify our equation for the loop gain. By substitution, we arrive at

$$\beta A_v = \frac{Z_2}{Z_2 + Z_3} A_{v(\text{oc})} = \frac{Z_2}{-Z_1} A_{v(\text{oc})} = -\frac{Z_2}{Z_1} A_{v(\text{oc})} \qquad (15\text{-}34)$$

We can make some very powerful generalizations at this point. First, the frequency of oscillation can be determined by the following:

$$\text{At } f_o,\ Z_1 + Z_2 + Z_3 = 0$$
$$\text{--- solve for } f_o. \qquad (15\text{-}35)$$

The amplifier gain must be large enough to ensure a loop gain larger than unity. Equation 15-36 states the requirement.

$$\beta A_v \geq 1$$

$$-\frac{Z_2}{Z_1} A_{v(\text{oc})} \geq 1 \rightarrow -A_{v(\text{oc})} > \frac{Z_1}{Z_2}$$

$$A_{v(\text{oc})} < -\frac{Z_1}{Z_2} \qquad (15\text{-}36)$$

where $A_{v(\text{oc})}$ is the required voltage gain to sustain oscillations.

Equations 15-35 and 15-36 will allow us to analyze *LC* oscillators such as the Colpitts, Clapp, Hartley, and Pierce oscillators without ''reinventing the wheel'' each time. This is illustrated in the sections that follow.

15-11 Colpitts Oscillators and Gate Leak Bias

A Colpitts oscillator circuit has been given in Fig. 15-17. The feedback network arrangement and the amplifier have been emphasized in Fig. 15-17(a). As we can see, Z_1 and Z_2 are capacitive reactances and Z_3 is an inductive reactance. In this case, an inverting amplifier is required. Observe that a common-source JFET amplifier has been indicated.

Capacitor C_3 is an input coupling capacitor. It serves to block the dc bias present at the drain of the JFET. Recall that a coil (e.g., L_1) is essentially a short circuit to dc. Therefore, capacitor C_3 is required to keep the large positive drain potential off the gate of the JFET. Resistor R_G is used to reference the gate to ground. As we

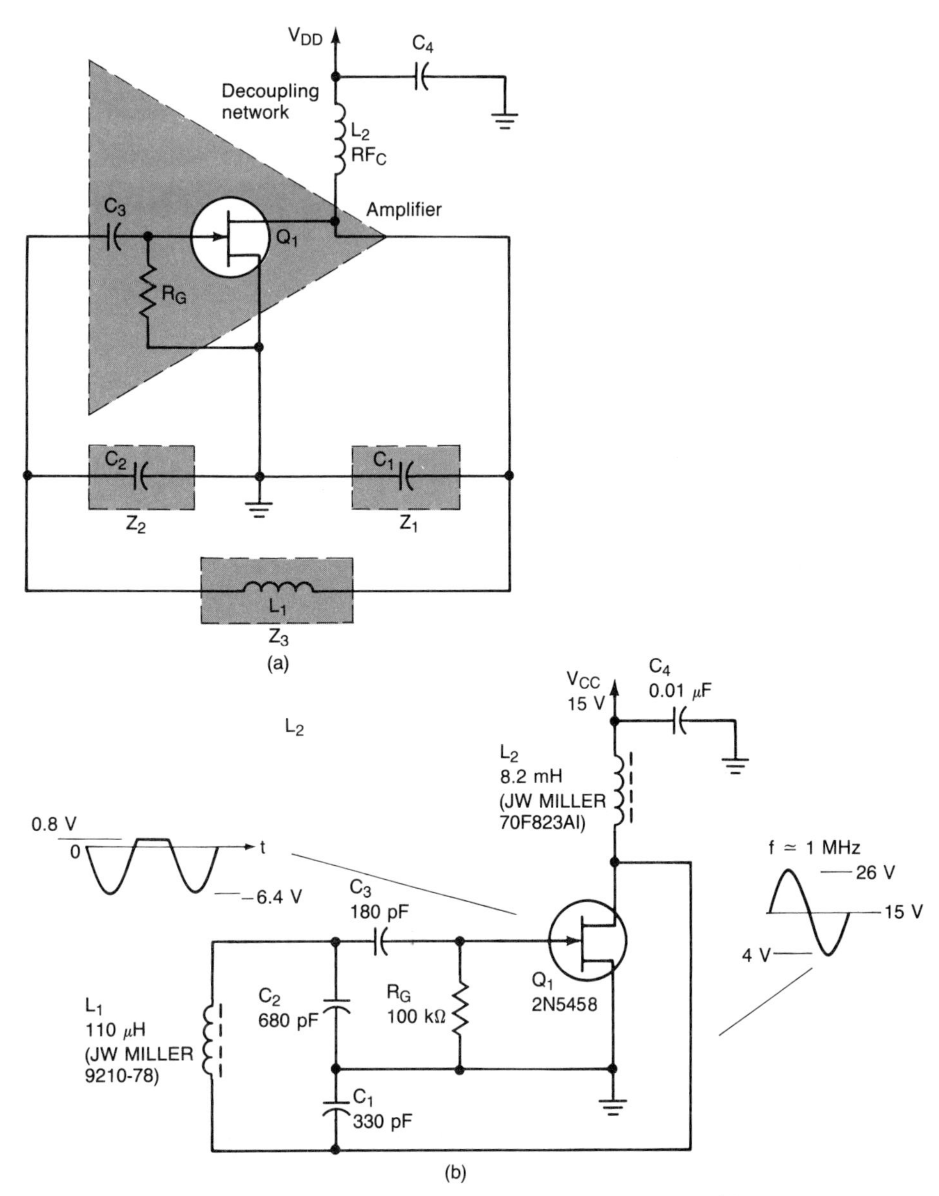

FIGURE 15-17 Colpitts oscillator: (a) basic Colpitts oscillator arrangement; (b) practical oscillator (with actual waveforms).

shall see, resistor R_G and C_3 also form a unique bias scheme referred to as *gate leak bias*. (The gate leak bias provides a form of adaptive negative feedback.)

Inductor L_2 is called a radio-frequency choke (RFC). Its dc resistance is low. Therefore, the drain will have a dc potential of approximately V_{DD} volts. However, the inductive reactance of the RFC is very large at the frequency of oscillation. Its primary function is to keep the RF signal out of the power supply. Before we go any further, let us begin our analysis of the Colpitts oscillator. A more conventional schematic diagram has been indicated in Fig. 15-17(b).

From Eq. 15-35, we recall that at the resonant frequency the reactances cancel one another. Hence

$$Z_1 + Z_2 + Z_3 = 0$$

We may use this relationship to determine the frequency of oscillation (f_o). By substitution,

$$Z_1 + Z_2 + Z_3 = -j\frac{1}{\omega_o C_1} - j\frac{1}{\omega_o C_2} + j\omega_o L_1 = 0$$

$$j\omega_o L_1 = j\left(\frac{1}{\omega_o C_1} + \frac{1}{\omega_o C_2}\right) = j\frac{C_1 + C_2}{\omega_o C_1 C_2}$$

$$\omega_o^2 = \frac{1}{L_1\left[\dfrac{C_1 C_2}{(C_1 + C_2)}\right]}$$

$$f_o = \frac{1}{2\pi\sqrt{L_1 C_T}} \tag{15-37}$$

where f_o = Colpitts oscillator frequency of oscillation (hertz)
L_1 = feedback inductance
C_T = series equivalent capacitance
= $C_1C_2/(C_1 + C_2)$

Now that we have our equation for f_o, we shall use Eq. 15-36 to determine the required amplifier gain to ensure sustained oscillations. Thus

$$A_{v(\text{oc})} < -\frac{Z_1}{Z_2} = -\frac{1/j\omega_o C_1}{1/j\omega_o C_2} = -\frac{1}{j\omega_o C_1}\frac{j\omega_o C_2}{1}$$

$$A_{v(\text{oc})} < -\frac{C_2}{C_1} \tag{15-38}$$

where $A_{v(\text{oc})}$ is the open-circuit voltage gain of the amplifier used in the Colpitts oscillator. This result indicates that the amplifier must be an inverting configuration, and the *magnitude* of its open-circuit voltage gain must be larger than the ratio of C_2 to C_1.

The open-circuit voltage gain may be determined from Fig. 15-18. Observe that the decoupling capacitor [C_4 in Fig. 15-17(b)] is a short circuit to the signal. This effectively places the RFC [L_2 in Fig. 15-18(a)] in parallel with the JFET's output. Consequently, it forms part of the ac load across the JFET.

To find the open-circuit voltage gain, we must disconnect the load [refer to Fig. 15-18(b)]. Next, we note

$$v_{\text{out}} = -g_m v_{gs} r_o$$

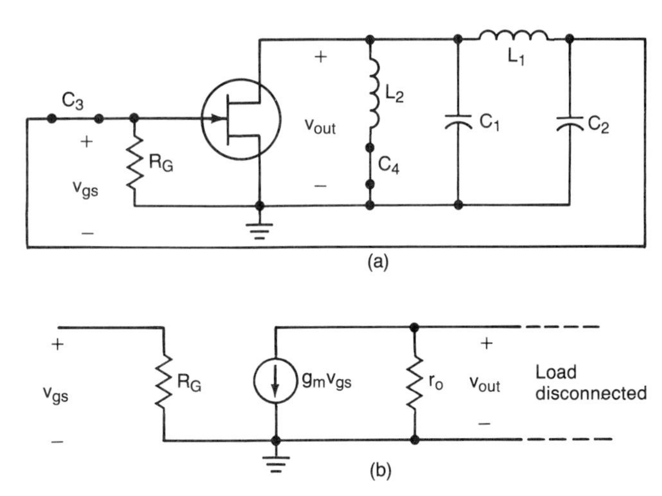

FIGURE 15-18 Ac gain analysis: (a) basic ac equivalent circuit; (b) determining the open-loop voltage gain.

If we divide both sides by v_{gs}, we obtain the open-circuit voltage gain.

$$A_{v(\mathrm{oc})} = -g_m r_o$$

The use of Eqs. 15-37 and 15-38 and the relationship above is demonstrated in Example 15-8. However, before we proceed with the example, let us investigate the effects of the input coupling capacitor C_3 and R_G.

Radio-frequency oscillators are no different than audio-frequency oscillators in that some form of adaptive negative feedback is desirable. The sharp filtering action offered by the *LC* tuning networks greatly reduces the demands on the adaptive negative feedback scheme. Specifically, distortion is much less of a problem.

When the oscillator in Fig. 15-17 is first powered up, V_{GS} is 0 V. (This should not be too surprising since resistor R_G is used to pull the gate to ground.) Recall that when V_{GS} is zero the JFET has its maximum transconductance. Since its voltage gain is directly proportional to its transconductance, it will be at its maximum. Therefore, the loop gain will be large and the oscillations will begin to build up.

As the oscillations build up, so will the magnitude of the feedback voltage [see Fig. 15-19(a)]. When the positive peak of the feedback signal causes v_{GS} to reach approximately 0.5 V, the *p-n* junction between the gate and the source terminals will begin to conduct. A diode analogy will help [see Fig. 15-19(b)].

If we assume that the generator's signal (v_s) reaches a peak value of 3 V, we see that the voltage across the capacitor reaches 2.4 V. This will occur very quickly because the forward resistance of the diode is small. As soon as v_s falls below 2.4 V, the diode will be reverse biased. Consequently, it will be nonconducting, or OFF. The discharge time constant will be much larger than the charge time constant, and the capacitor's voltage will effectively remain at 2.4 V. When the peak value of v_s reaches -3 V, the peak reverse bias across the diode will reach -5.4 V [see Fig. 15-19(c)]. The average value of the reverse bias will be equal to the voltage across the capacitor (-2.4 V), as indicated in Fig. 15-19(d).

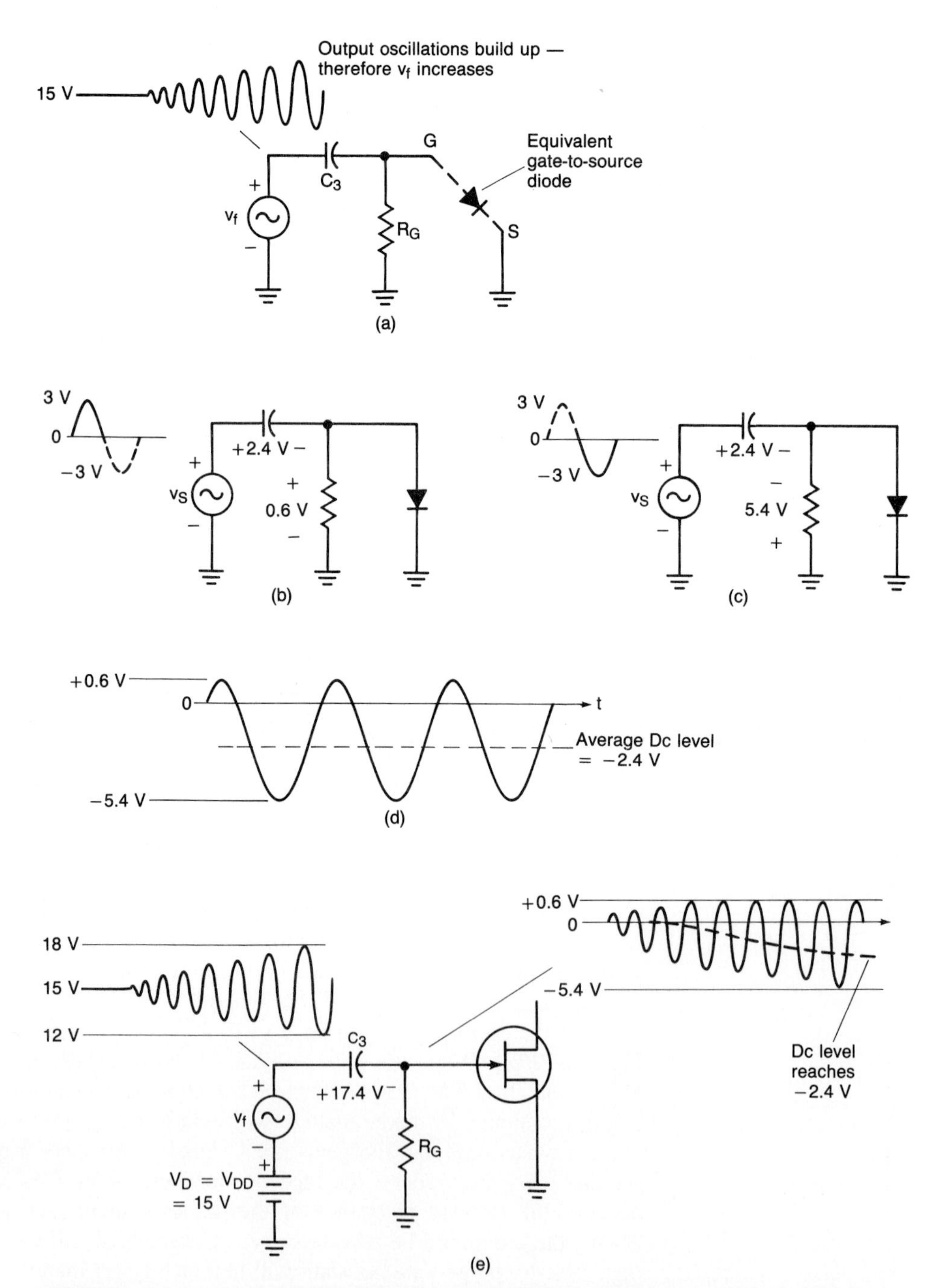

FIGURE 15-19 Gate leak bias: (a) the initially large loop gain allows the oscillations to build up; (b) the capacitor quickly charges to 2.4 V when the diode conducts; (c) the capacitor's discharge is very slow, causing the negative peak reverse bias to reach −5.4 V; (d) total instantaneous voltage across the diode; (e) a negative dc bias builds up across the gate-to-source.

The same basic circuit action occurs when gate leak bias is used. The situation is slightly more involved since the feedback signal rides on a 15-V dc level [see Fig. 15-19(e)]. A peak feedback signal of 3 V causes the total instantaneous feedback signal to go from 12 V up to 18 V. Consequently, the capacitor will charge to 17.4 V. As the oscillations build up, the average gate-to-source bias will reach −2.4 V.

The negative gate-to-source bias will lower the transconductance of the JFET. Consequently, the voltage gain and loop gain will be reduced. The output voltage signal will be very "clean." The forward bias on the gate will cause a drastic amount of distortion in the JFET's drain current. However, the high-Q tuned filter circuit will remove all of the "nasty" harmonics. The reduction in the duration of the flow of the drain current causes the voltage amplifier to leave class A operation and enter class B or even class C operation. Consequently, its efficiency is also increased (refer to Chapter 12).

The RFC (L_2) should have an impedance that is much larger than the capacitive reactance of C_1. A comfortable factor is 100:1. Hence

$$X_{L2} > 100X_{C1}$$

The capacitive reactance of decoupling capacitor C_4 should be small compared to X_{L2}. Again, a ratio of at least 100:1 is desirable. This would place the value of the decoupling capacitor at approximately that of C_1. However, it is often selected to be much larger in practice.

EXAMPLE 15-8

Analyze the JFET Colpitts oscillator given in Fig. 15-17(b). Specifically, determine the frequency of oscillation f_o, the minimum required open-circuit voltage gain, the actual voltage gain when power is first applied, and verify the proper sizing of L_2 and C_4.

SOLUTION From Eq. 15-37,

$$C_T = \frac{C_1C_2}{C_1 + C_2} = \frac{(330 \text{ pF})(680 \text{ pF})}{330 \text{ pF} + 680 \text{ pF}} = 222.2 \text{ pF}$$

$$f_o = \frac{1}{2\pi\sqrt{LC_T}} = \frac{1}{2\pi\sqrt{(110\ \mu\text{H})(222.2 \text{ pF})}} = 1.02 \text{ MHz}$$

The minimum required open-circuit voltage gain may be determined using Eq. 15-38.

$$A_{v(\text{oc})} \leq -\frac{C_2}{C_1} = -\frac{680 \text{ pF}}{330 \text{ pF}} = -2.06$$

If we examine the data sheet for the 2N5458 n-channel JFET, we find that its minimum transconductance $g_{fs0(\text{min})}$ (g_m) is 1500 μS, and the magnitude of its drain-to-source admittance ($|y_{os}|_{(\text{max})}$) is 50 μS (at 1 kHz). This information will allow us to estimate the 2N5458's performance.

$$A_{v(\text{oc})} = -g_{fs0}r_o = -g_{fs0}\frac{1}{|y_{os}|}$$

$$= -(1500\ \mu\text{S})\left(\frac{1}{50\ \mu\text{S}}\right) = -30$$

The minimum voltage gain of the 2N5458 JFET is more than adequate in this circuit. Now let us investigate the gate leak bias circuit. When capacitor C_3 discharges, its discharge path is through R_G. (The gate-to-source *p-n* junction is reverse biased.) For proper operation to occur, the discharge time constant should be at least 10 times longer than the period of the frequency of oscillation. Thus

$$\tau = R_G C_3 = (100\ \text{k}\Omega)(180\ \text{pF}) = 18\ \mu\text{s}$$

and

$$T = \frac{1}{f_o} = \frac{1}{1.02\ \text{MHz}} = 0.980\ \mu\text{s}$$

Since

$$\tau \geq 10T$$

$$18\ \mu\text{s} \geq (10)(0.980\ \mu\text{s}) = 9.80\ \mu\text{s}$$

the time constant appears to be better than required, and C_3 is adequately sized.

Now we shall check the values of the decoupling components L_2 and C_4. At f_o,

$$X_{C1} = \frac{1}{2\pi f_o C_1} = \frac{1}{2\pi(1.02\ \text{MHz})(330\ \text{pF})} = 473\ \Omega$$

The inductive reactance of the RFC is

$$X_{L2} = 2\pi f_o L_2 = 2\pi(1.02\ \text{MHz})(8.2\ \text{mH}) = 52.6\ \text{k}\Omega$$

Hence

$$X_{L2} \geq 100 X_{C1}$$

$$52.6\ \text{k}\Omega \geq (100)(473\ \Omega) = 47.3\ \text{k}\Omega$$

Therefore, the RFC appears to have enough inductance. Since the decoupling capacitor C_4 is much larger than C_1, its capacitive reactance will certainly be small compared to X_{L2}. Therefore, C_4 is adequately sized. ■

As a final comment, we should point out that the Colpitts oscillator given in Fig. 15-17(b) worked very well in the laboratory. The actual drain-to-ground and gate-to-ground waveforms have been illustrated in Fig. 15-17(b). Observe that the positive peak drain voltage is 11 V *above* V_{DD}. This is typical of transistor circuits that have an inductive load. The ac load line indicates that the maximum value is equal to *twice* V_{DD}.

15-12 Clapp Oscillators

The Clapp oscillator is a variation of the Colpitts oscillator. The single inductor found in the Colpitts oscillator is replaced by a series *LC* combination [see L_1 and C_3 in Fig. 15-20(a)]. At the frequency of oscillation the *net* impedance of the *LC* series network will be *inductive*.

The three impedances Z_1, Z_2, and Z_3 have been indicated in Fig. 15-20(a). A more conventional schematic has been depicted in Fig. 15-20(b). Also note that actual measured waveforms have also been indicated. Let us again develop our equations for f_o and the minimum $A_{v(oc)}$.

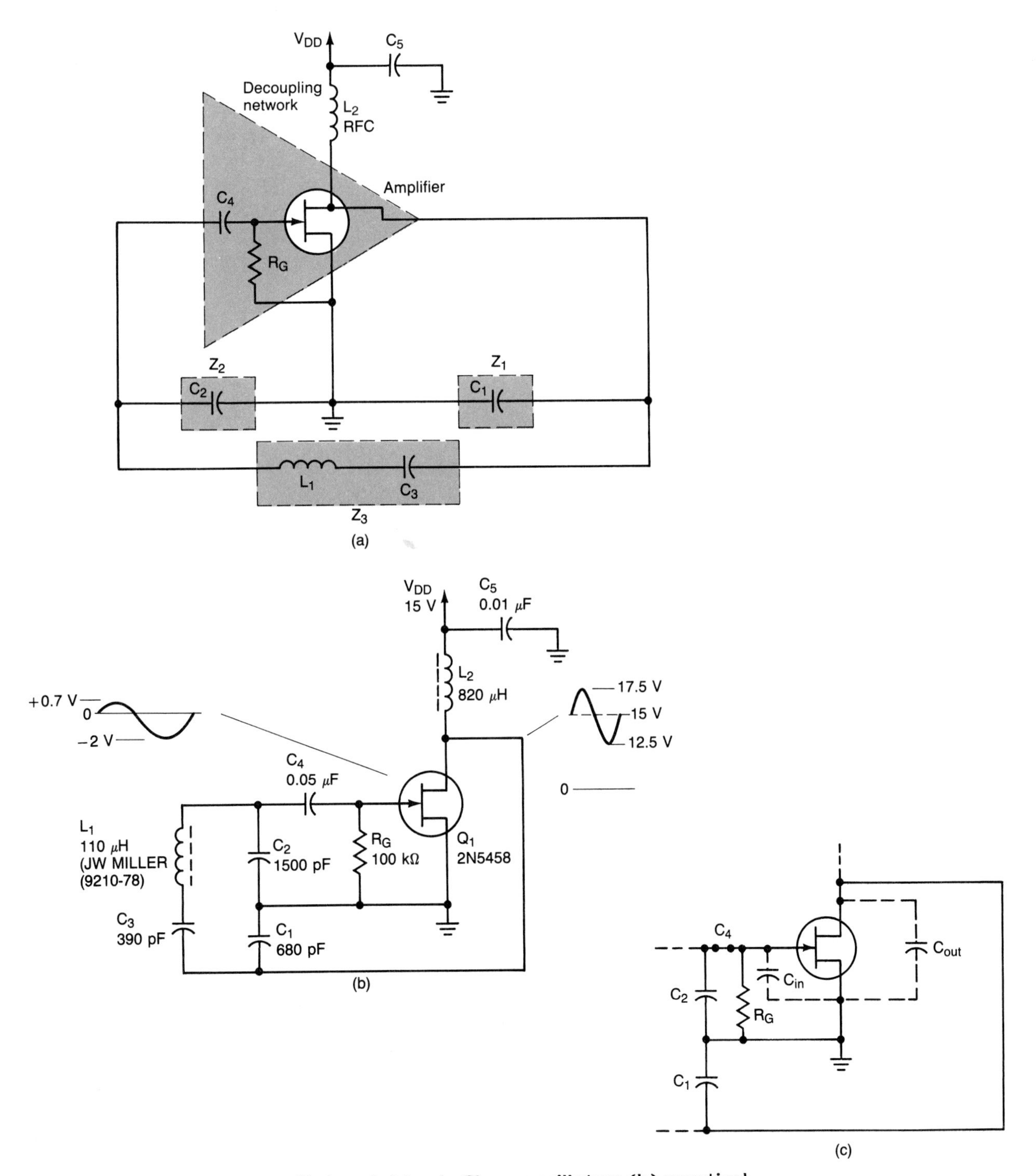

FIGURE 15-20 Clapp oscillator: (a) basic Clapp oscillator; (b) practical oscillator; (c) the JFET's C_{in} is in parallel with C_2, and the JFET's C_{out} is in parallel with C_1.

At resonance we recall that

$$Z_1 + Z_2 + Z_3 = 0$$

Substituting in the specific impedance, we have

$$-j\frac{1}{\omega_o C_1} - j\frac{1}{\omega_o C_2} - j\frac{1}{\omega_o C_3} + j\omega_o L_1 = 0$$

$$j\omega_o L_1 = j\frac{1}{\omega_o}\left(\frac{1}{C_1} + \frac{1}{C_2} + \frac{1}{C_3}\right)$$

The terms on the right side of the equation represent the series combination of C_1, C_2, and C_3. Hence

$$\frac{1}{C_T} = \frac{1}{C_1} + \frac{1}{C_2} + \frac{1}{C_3}$$

Substituting in this standard relationship for capacitors in series leads us to

$$\omega_o^2 = \frac{1}{L_1 C_T}$$

$$f_o = \frac{1}{2\pi\sqrt{L_1 C_T}} \qquad (15\text{-}39)$$

where f_o = Clapp oscillator frequency of oscillation (hertz)
L_1 = feedback inductance
$1/C_T$ = reciprocal of the series equivalent capacitance = $1/C_1 + 1/C_2 + 1/C_3$

The minimum open-circuit voltage gain required to guarantee oscillation is given by Eq. 15-40. It is derived in exactly the same fashion as Eq. 15-38 for the Colpitts oscillator.

$$A_{v(\mathrm{oc})} < -\frac{C_2}{C_1} \qquad (15\text{-}40)$$

where $A_{v(\mathrm{oc})}$ is the open-circuit voltage gain of the amplifier used in the Clapp oscillator.

Let us reflect on these results. The addition of capacitor C_3 *lowers* the total equivalent capacitance C_T. Consequently, the values of capacitors C_1 and C_2 can be *increased* and still produce a relatively small C_T. This advantage becomes more apparent when we look at Fig. 15-20(c).

The total input capacitance of the JFET (C_{in}) is in parallel with C_2, and the JFET's output capacitance (C_{out}) is in parallel with C_1. This increases the net "C_1" capacitance, and the net "C_2" capacitance. Therefore, these small device capacitances tend to lower the frequency of oscillation. To compound the problem, the JFET capacitances are not very tightly controlled parameters and are affected by the dc bias. This introduces a measure of uncertainty in the oscillator design. The situation becomes particularly acute when the oscillator is to be used at very high frequencies (e.g., above 10 MHz) since the values of C_1 and C_2 tend to be small.

Therefore, the fact that we are able to increase the values of C_1 and C_2 (and still obtain a small C_T) tends to negate this effect. The device capacitances can be "swamped out."

Typically, capacitor C_3 is selected to be the smallest capacitor. This makes it dominate in the determination of f_o. However, caution must be exercised. If capacitor C_3 is made too small, the LC branch will not have a net inductive reactance. Under this condition the circuit will refuse to oscillate.

Capacitors C_1 and C_2 [Fig. 15-20(b)] are approximately twice the values given for C_1 and C_2 in Fig. 15-17(b). Capacitor C_3 is the smallest and approximately equal to the series combination of C_1 and C_2 in Fig. 15-20(b). Therefore, the net C_T of the Clapp oscillator is approximately equal to the C_T of the Colpitts oscillator. Since the inductance L_1 is still 110 μH, the frequency of oscillation of the Clapp oscillator will be approximately equal to that of the Colpitts oscillator given in Fig. 15-17(b). The next example will verify this.

EXAMPLE 15-9

Determine f_o, the minimum open-circuit voltage gain, and the net impedance of the L_1-C_3 branch at f_o of the Clapp oscillator given in Fig. 15-20(b).

SOLUTION The frequency of oscillation may be found by using Eq. 15-39.

$$C_T = \frac{1}{1/C_1 + 1/C_2 + 1/C_3} = \frac{1}{1/680\text{ pF} + 1/1500\text{ pF} + 1/390\text{ pF}}$$

$$= 212.7\text{ pF}$$

$$f_o = \frac{1}{2\pi\sqrt{L_1 C_T}} = \frac{1}{2\pi\sqrt{(110\ \mu\text{H})(212.7\text{ pF})}} = 1.04\text{ MHz}$$

The minimum open-circuit voltage gain is given by Eq. 15-40.

$$A_{v(\text{oc})} \leq -\frac{C_2}{C_1} = -\frac{1500\text{ pF}}{680\text{ pF}} = -2.21$$

Now we can determine the net impedance of the L_1-C_3 branch.

$$Z_{\text{net}} = jX_{L1} - jX_{C3} = j2\pi f_o L_1 - j\frac{1}{2\pi f_o C_3}$$

$$= j2\pi(1.04\text{ MHz})(110\ \mu\text{H}) - j\frac{1}{2\pi(1.04\text{ MHz})(390\text{ pF})}$$

$$= j719.1\ \Omega - j392.2\ \Omega = j326.9\ \Omega$$

■

The rest of the analysis of the Clapp oscillator is identical to that conducted on the Colpitts oscillator covered in Example 15-8. Specifically, the JFET's $A_{v(\text{oc})}$ is more than adequate, the gate leak bias time constant is acceptable, and the decoupling action of L_2 and C_5 exceeds the minimum requirements.

15-13 Hartley Oscillators

The Hartley oscillator is almost as popular as the Colpitts oscillator. It may be thought of as the "complement" of the Colpitts oscillator. Specifically, the Colpitt's inductor is replaced by a capacitor, and the two capacitors found in Colpitt's feedback network are replaced by inductors (refer to Table 15-2). The basic Hartley configuration has been depicted in Fig. 15-21(a).

Two separate inductors are seldom employed in "real" Hartley oscillators. Instead, most designs incorporate a single *tapped* inductor [see Fig. 15-21(b)]. A tapped inductor with a single winding will behave as an *autotransformer*. An autotransformer employs a single winding. Consequently, it offers no isolation, but it can be used to either step up or step down a voltage signal [see Fig. 15-21(c) - (e)]. As we shall see, the transformer action will determine the feedback factor (β) of the oscillator. The phasing of the transformer with respect to ground is also configured to provide positive feedback.

FIGURE 15-21 Hartley oscillator using a tapped inductor: (a) two separate inductors; (b) using a single tapped inductor; (c) step-down autotransformer; (d) step-up autotransformer; (e) step-up or step-down with phase inversion.

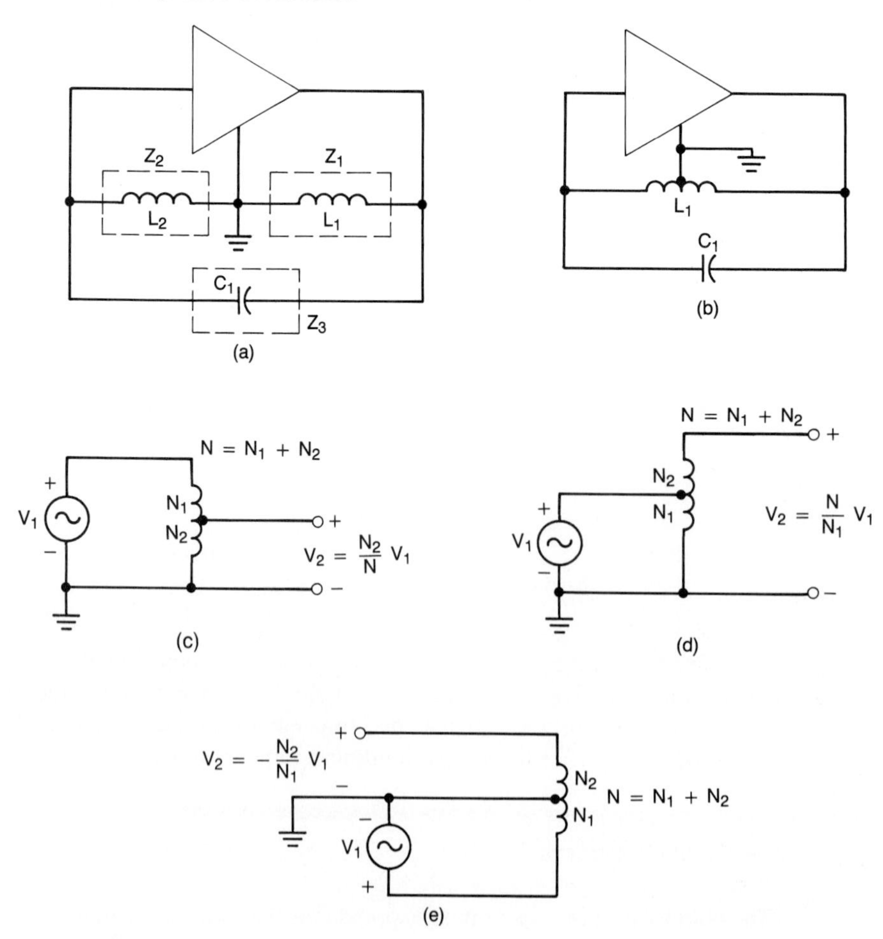

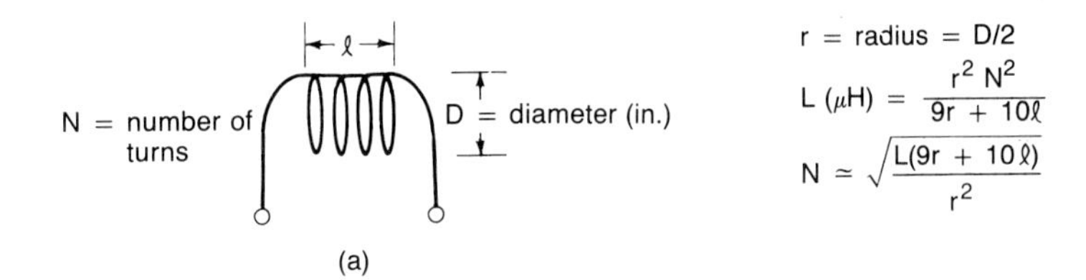

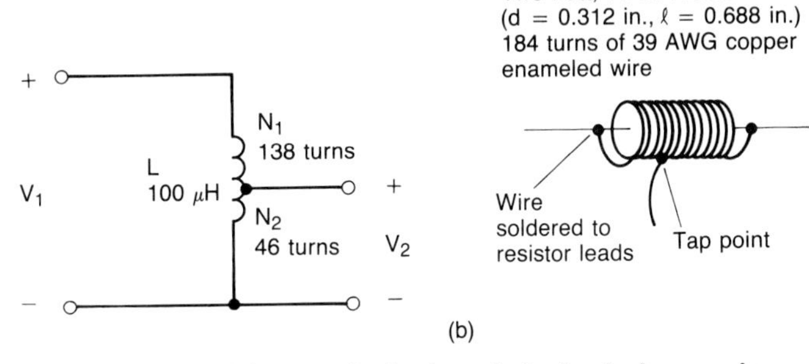

FIGURE 15-22 Air-core inductor: (a) single-layer air-core coil; (b) complete coil specification.

Since the tapped inductor is crucial to proper operation of the Hartley oscillator, it merits additional attention. A single-layer, air-core coil has been illustrated in Fig. 15-22(a). The inductance of the coil is dependent on its geometry and is directly proportional to the square of its turns.

$$L = \frac{r^2N^2}{9r + 10l} \tag{15-41}$$

where L = inductance (microhenries)
r = mean radius of the coil (inches)
N = number of turns
l = length of the coil (inches)

From a design standpoint, we are interested in solving for the required number of turns (N). This is accomplished by solving Eq. 15-41 for N.

$$N = \sqrt{\frac{L(9r + 10l)}{r^2}} \tag{15-42}$$

Equations 15-41 and 15-42 are approximations. They do not take into account variables such as uneven turns and unequal spacing between the turns. Therefore, they serve only to give us a rough estimate.

EXAMPLE 15-10

A single-layer air-core inductor is to be used in a Hartley oscillator. It is to have 100 μH of inductance and be wound on a 2-W carbon resistor. Complete its design.

SOLUTION A 2-W resistor has a length of 0.688 in. and a diameter of 0.312 in. Its radius is one-half of its diameter.

$$r = \frac{d}{2} = \frac{0.312 \text{ in.}}{2} = 0.156 \text{ in.}$$

and since its length (l) is 0.688 in., we can use Eq. 15-42 to determine the required number of turns (N).

$$N = \sqrt{\frac{L(9r + 10l)}{r^2}} = \sqrt{\frac{100[9(0.156) + 10(0.688)]}{(0.156)^2}}$$
$$= 184.5 \simeq 184 \text{ turns}$$

Observe that we have rounded our answer to the nearest turn. Now we can determine the wire size. The wire diameter D is

$$D = \frac{l}{N} = \frac{0.688 \text{ in.}}{184} \simeq 0.00374 \text{ in.}$$

Going to a wire table we find that 39 AWG (American Wire Gauge) enameled copper wire has a nominal diameter of 0.00353 in. The wire has a resistance of 0.832 mΩ/ft. The circumference (c) of the 2-W resistor is

$$c = \pi d = \pi(0.312 \text{ in.}) \simeq 0.980 \text{ in.}$$

Now we can determine the required length of 39 AWG wire.

$$\text{wire length} = Nc = (184)(0.980 \text{ in.}) = 180.3 \text{ in.} = 15.0 \text{ ft}$$

The total dc resistance of the wire used to make the coil is

$$R_{dc} = (0.832 \text{ m}\Omega/\text{ft})(15.0 \text{ ft}) = 0.0125 \ \Omega$$

Such a small dc resistance will promote a high-Q coil. (Remember that ac losses will also come into play, but these are not easily predicted.) ■

EXAMPLE 15-11

The coil designed in Example 15-10 is to be tapped such that

$$\frac{V_2}{V_1} = \frac{N_2}{N_1} = \frac{1}{3}$$

[refer to Fig. 15-21(e)]. At present we ignore the phase inversion.

SOLUTION We shall solve for N_2. First, we apply the relationship above.

$$N_2 = \tfrac{1}{3}N_1$$

By inspection of Fig. 15-21(e),

$$N = N_1 + N_2$$

and solving for N_1 gives us

$$N_1 = N - N_2$$

If we substitute this result into the previous relationship, we can solve for N_2.

$$N_2 = \tfrac{1}{3}N_1 = \tfrac{1}{3}(N - N_2) = \tfrac{1}{3}N - \tfrac{1}{3}N_2$$

$$\tfrac{4}{3}N_2 = \tfrac{1}{3}N$$

$$N_2 = \tfrac{1}{4}N = (\tfrac{1}{4})(184 \text{ turns}) = 46 \text{ turns}$$

The complete inductor has been shown in Fig. 15-22(b). The resistor is 750 kΩ. Its large resistance will have a negligible shunting effect on our coil. We shall use this inductor in our examples to come. ■

A Hartley oscillator that employs a common-source JFET has been indicated in Fig. 15-23(a). Notice that the tapped inductor (L_1) designed in Examples 15-10

FIGURE 15-23 **Operation of the JFET Hartley oscillator: (a) JFET Hartley oscillator; (b) transformer action of the tapped inductor; (c) partial ac equivalent.**

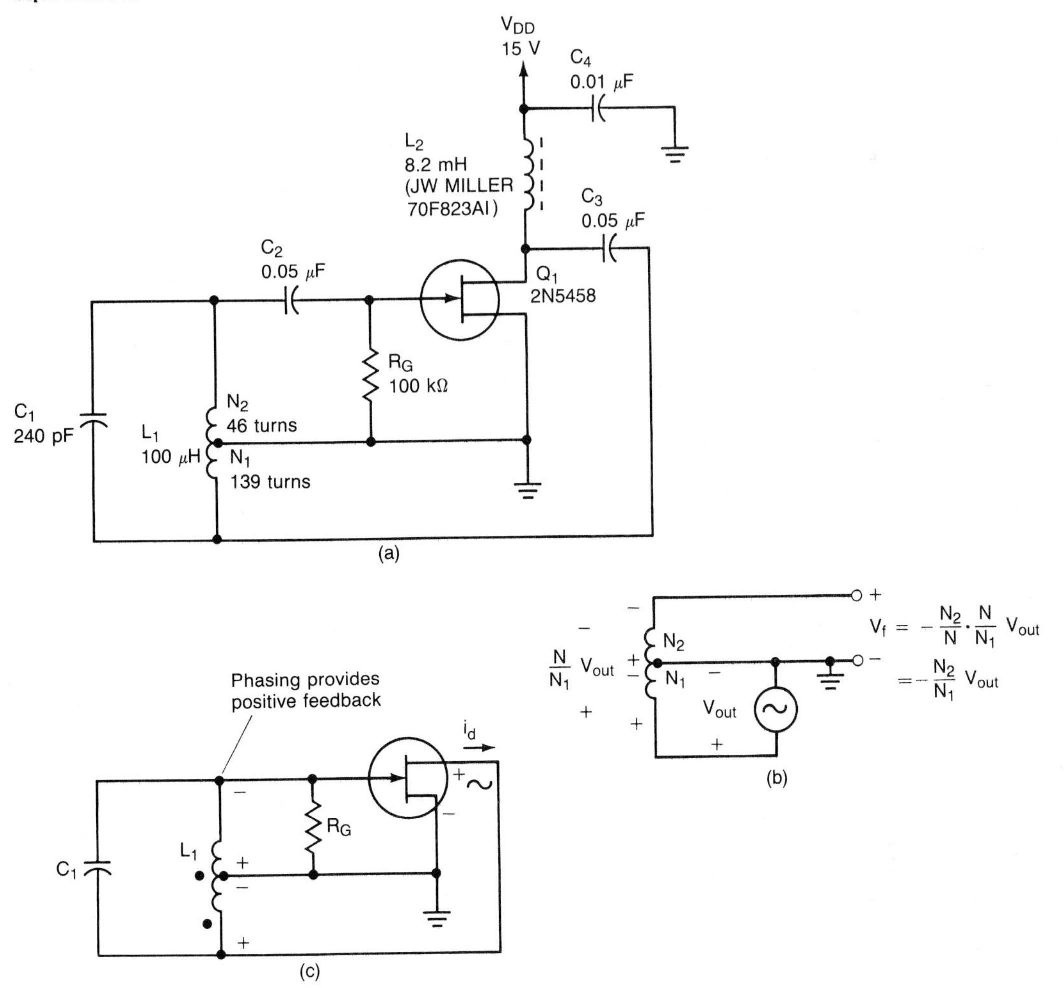

and 15-11 has been incorporated in its resonant circuit. The autotransformer action of L_1 has been emphasized in Fig. 15-23(b). By inspection we see that

$$\beta = \frac{V_f}{V_{out}} = -\frac{N_2}{N_1} = -\frac{46 \text{ turns}}{139 \text{ turns}} = -0.331$$

The step-down action of the inductor provides a β that is less than unity. The negative sign indicates that the transformer provides 180° of phase shift [see Fig. 15-23(c)].

The JFET provides an open-circuit voltage gain of

$$A_{v(oc)} = -g_m r_o = -g_{fs0} r_o$$

From our work in Example 15-8 we recall that the $A_{v(oc)}$ for the 2N5458 JFET is -30. Consequently, the loop gain is (approximately)

$$\beta A_{v(oc)} \simeq (-0.331)(-30) = 9.93 > 1$$

Therefore, the circuit will oscillate. The approximation above has drawn on the *open-circuit* voltage gain. To be rigorous, we should use the *loaded* voltage gain. Some impedance will be reflected back into the primary of our autotransformer, and this reflected impedance will load down the amplifier's output. However, this effect can be ignored except in critical applications.

Now let us determine the frequency of oscillation. It is given by the resonant frequency of the L_1C_1 combination.

$$f_o = \frac{1}{2\pi\sqrt{L_1C_1}} = \frac{1}{2\pi\sqrt{(100\ \mu\text{H})(240\ \text{pF})}} = 1.03\ \text{MHz}$$

The balance of the circuit's operation - the gate leak bias and the decoupling action afforded through the use of L_2, and C_4 - has been examined previously. Now let us consider a common-emitter Hartley oscillator [see Fig. 15-24(a)].

Although more components are involved in this design, its operation will become clear if we do not panic. Capacitor C_4 is a decoupling capacitor. Capacitors C_2 and C_3 are used to block the dc biases. They are shorts to the ac signal. Capacitor C_5 is an emitter bypass capacitor. A partial ac equivalent circuit has been depicted in Fig. 15-24(b).

Once we recognize that L_2 is an RFC, we may redraw the circuit as shown in Fig. 15-24(c). At this point it should be clear that the analysis of the circuit is virtually identical to the JFET Hartley oscillator given in Fig. 15-23(a).

We conclude our discussions of the Hartley oscillator with the circuit given in Fig. 15-25(a). The circuit is novel in that it is designed around an *emitter follower*. The less than unity voltage gain offered by the emitter follower would seem to prohibit oscillations. However, this is *not* the case. The *loop gain* can be made greater than unity because of the step-up transformer action produced by L_1.

The base bias circuit is designed to offer good bias stability, provide power supply decoupling, and simultaneously present a relatively large equivalent resistance to the tank circuit. Let us perform a dc analysis. The oscillator's dc equivalent circuit has been indicated in Fig. 15-25(b).

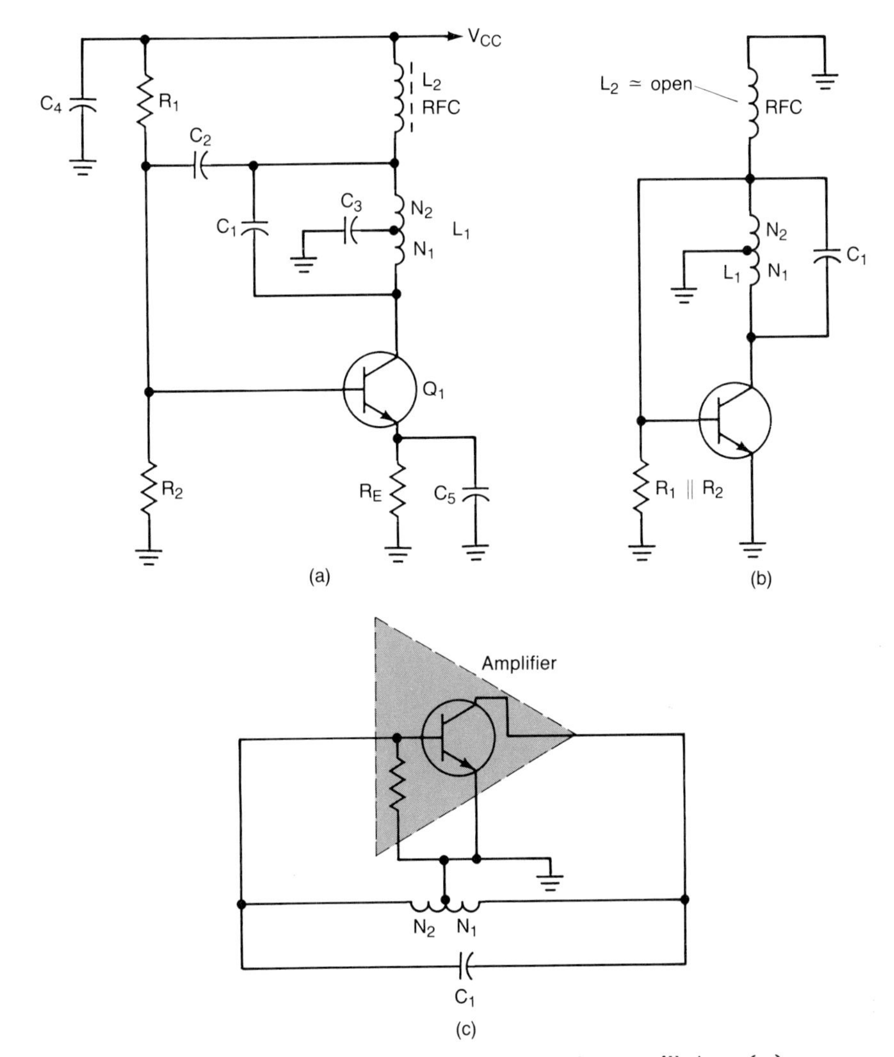

FIGURE 15-24 BJT common-emitter Hartley oscillator: (a) common-emitter BJT Hartley oscillator; (b) partial ac equivalent circuit; (c) basic Hartley oscillator model.

EXAMPLE 15-12

Determine the dc collector current of Q_1 in Fig. 15-25(a). Also determine the voltage drop across R_3.

SOLUTION First, we Thévenize the base circuit as indicated in Fig. 15-25(b).

$$R_{\text{TH}} = R_1 \parallel R_2 + R_3 = 3.3\text{ k}\Omega \parallel 3.3\text{ k}\Omega + 20\text{ k}\Omega = 21.65\text{ k}\Omega$$

$$V_{\text{TH}} = \frac{R_2}{R_1 + R_2} V_{CC} = \frac{3.3\text{ k}\Omega}{3.3\text{ k}\Omega + 3.3\text{ k}\Omega}(15\text{ V}) = 7.5\text{ V}$$

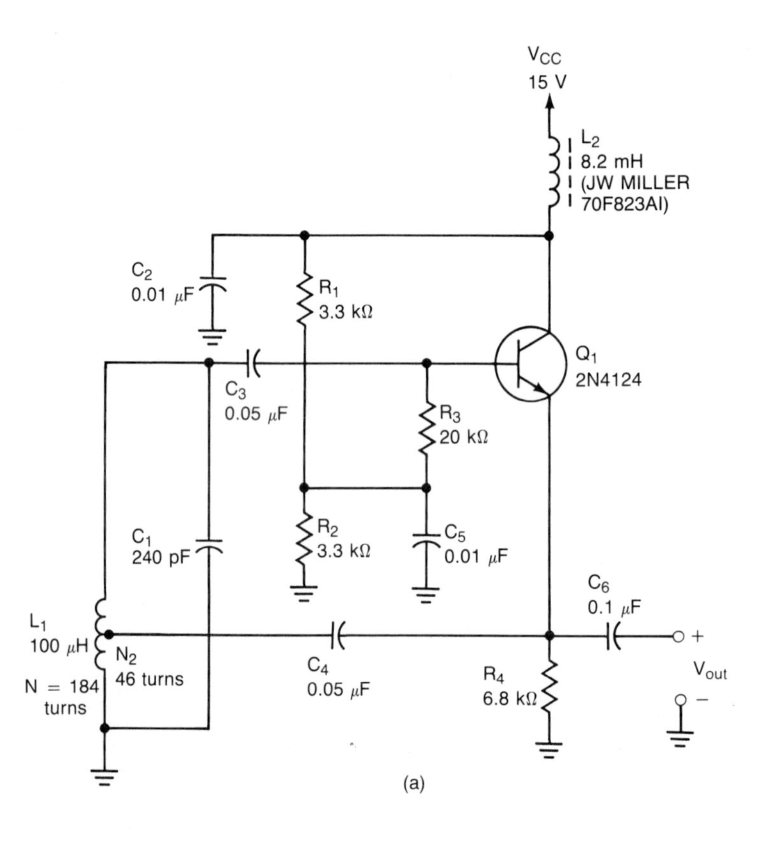

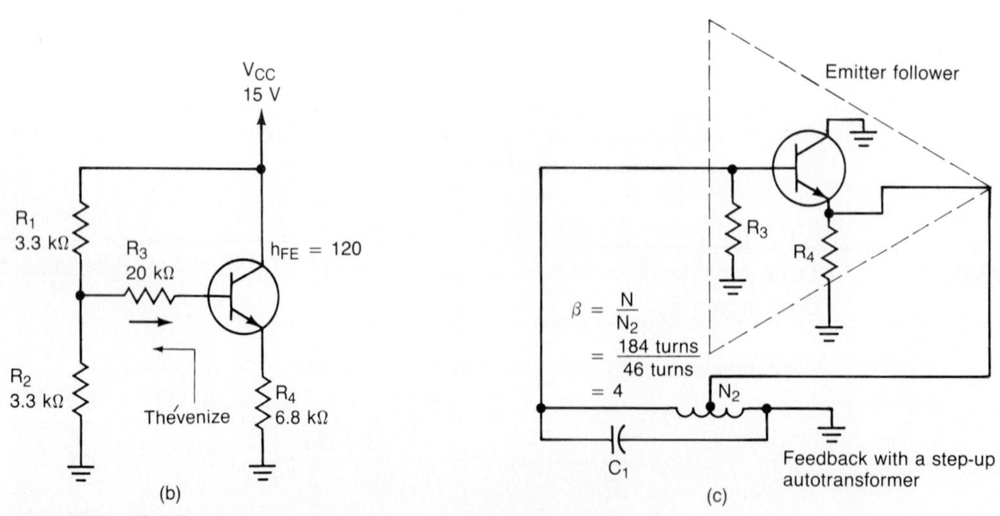

FIGURE 15-25 Hartley oscillator using an emitter follower: (a) complete circuit; (b) dc equivalent; (c) partial ac equivalent.

Now we can determine I_B. The 2N4124 has a minimum h_{FE} of 120.

$$I_B = \frac{V_{\text{TH}} - V_{BE}}{R_{\text{TH}} + (1 + h_{FE})R_4} = \frac{7.5\text{ V} - 0.7\text{ V}}{21.65\text{ k}\Omega + (1 + 120)(6.8\text{ k}\Omega)}$$
$$= 8.05\ \mu\text{A}$$

Now we can find I_C.

$$I_C = h_{FE}I_B = (120)(8.05\ \mu\text{A}) = 0.966\text{ mA}$$

The voltage drop across R_3 is very small.

$$V_{R3} = I_B R_3 = (8.05\ \mu\text{A})(20\text{ k}\Omega) = 0.161\text{ V}$$

■

EXAMPLE 15-13

Determine the open-circuit voltage gain of the emitter follower in Fig. 15-25(a).

SOLUTION First, we must determine the BJT's g_m.

$$g_m = \frac{I_C}{26\text{ mV}} = \frac{0.966\text{ mA}}{26\text{ mV}} = 32.2\text{ mS}$$

Now we can find $A_{v(\text{oc})}$.

$$A_{v(\text{oc})} = \frac{g_m R_4}{1 + g_m R_4} = \frac{(37.2\text{ mS})(6.8\text{ k}\Omega)}{1 + (37.2\text{ mS})(6.8\text{ k}\Omega)} = 0.996$$

■

A partial ac equivalent circuit has been given in Fig. 15-25(c). The output voltage of the emitter follower has been impressed across the "N_2" turns of the tapped inductor. However, the feedback voltage is taken across the entire coil. This will produce step-up (N/N_2) transformer action. Therefore, the β will be greater than unity.

$$\beta = \frac{N}{N_2} = \frac{184\text{ turns}}{46\text{ turns}} = 4$$

Notice that in this scheme there is no phase inversion. Consequently, β is positive, and the loop gain is

$$\beta A_v \simeq \frac{N}{N_2} A_{v(\text{oc})} = (4)(0.996) = 3.98 > 1$$

and the circuit will oscillate. [Notice that we have again used $A_{v(\text{oc})}$ as an approximation of A_v.]

The frequency of oscillation is determined by L_1 and C_1. From our analysis of the common-source JFET circuit we recall that with an L_1 of 100 μH and a C_1 of 240 pF, f_o occurs at 1.03 MHz. The same formula is used.

15-14 The Crystal

Some crystalline materials, such as quartz, tourmaline, and Rochelle salts, demonstrate a property called the *piezoelectric effect*. Very simply, if they are mechanically forced to vibrate, the resulting strain produces an electrical charge on the surface of the crystal and it generates an ac voltage. Conversely, if they are excited by an ac voltage, they will vibrate mechanically. Both modes of operation are commonly used.

In transducer applications the crystal is used to convert mechanical energy into electrical energy. Examples include crystal microphones, dynamic pressure transducers, and accelerometers.

The vibration of a crystal [Fig. 15-26(a) and (b)] that is excited by an ac voltage is capitalized on in RF electrical filters and in the tuned circuits of oscillators. As the frequency of a crystal's excitation voltage is changed, its response to the signal will change. The electrical properties of a piezoelectric crystal are quite dramatic when it enters mechanical resonance. For our purposes, we utilize the crystal's *electrical equivalent circuit* given in Fig. 15-26(c). Capacitance C_m represents the static capacitance produced by the contact wires, the crystal electrodes, and the crystal holder. The series *RLC* combination represents the motional aspects of the crystal. Capacitance C_s models the mechanical elasticity, the inductance L_s is a function of the crystal's mass, and R_s includes all the mechanical and electrical losses in the crystal.

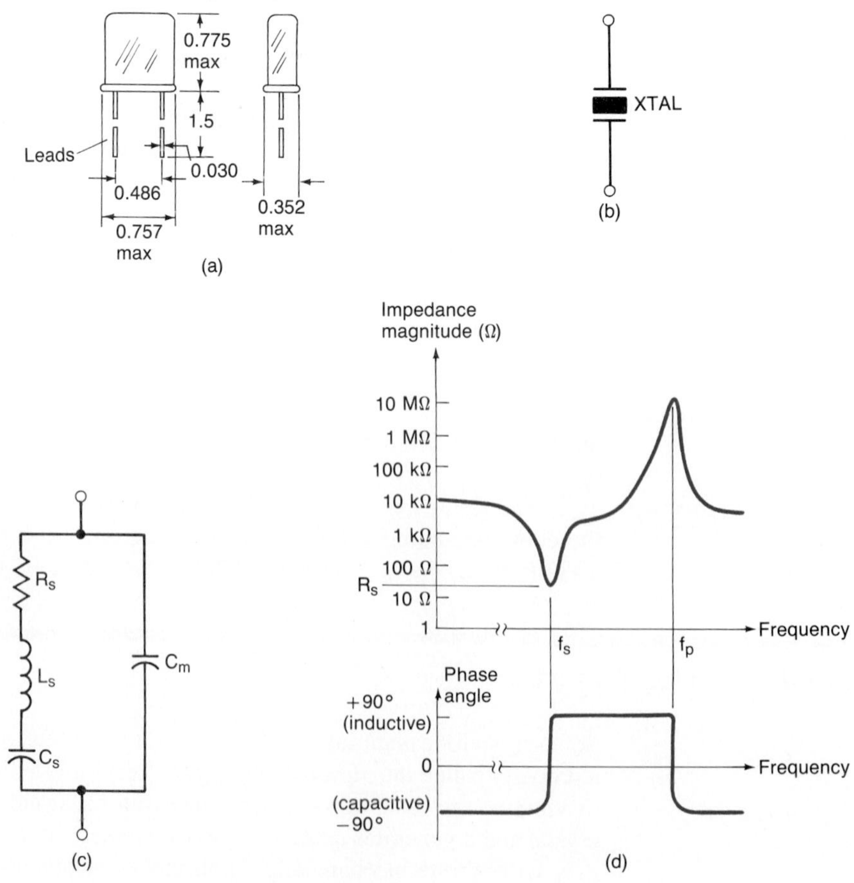

FIGURE 15-26 Crystal: (a) typical crystal package; (b) schematic symbol; (c) electrical equivalent circuit; (d) impedance and phase angle versus frequency.

Quartz is the most often selected material. The fundamental resonant frequencies for quartz typically range from 0.8 kHz to 50 MHz. Higher (up to 250 MHz) resonant frequencies are also available. The higher resonant frequencies are made possible by constructing crystals to operate at multiples (overtones) of their fundamental frequency. The third, fifth, and seventh overtones are the most commonly used.

Approximate values in the equivalent circuit of a particular 2-MHz crystal are a C_s of 0.012 pF, an L_s of 0.5 H, an R_s of 100 Ω, and a C_m of 4 pF. Crystals can offer extremely high Q's - values exceeding 100,000 are not uncommon. The Q of the 2-MHz crystal described above is given as 64,550.

The impedance of a crystal as a function of frequency has been shown in Fig. 15-26(d). At low frequencies the capacitive reactance of C_s dominates. As the frequency is increased to f_s, the crystal will go into series resonance. This occurs when the magnitude of X_{Cs} is equal to the magnitude of X_{Ls}. The net impedance of the crystal is resistive, and equal to R_s.

As the frequency is increased beyond f_s, X_{Ls} will ''swamp out'' X_{Cs}, and the crystal's net impedance will be inductive. If the frequency is increased to f_p, the crystal enters its parallel resonant mode of operation. At f_p the net inductive susceptance of the *RLC* branch equals the capacitive susceptance of C_m. The crystal's impedance will be extremely large. At frequencies beyond f_p, the crystal's impedance will again become capacitive. The capacitive susceptance of C_m will dominate.

EXAMPLE 15-14

Determine the f_s, the impedance at f_s, the Q, the f_p, and the impedance at f_p of a crystal with a C_s of 0.0060 pF, an L_s of 0.165609 H, an R_s of 10 Ω, and a C_m of 13.0 pF.

SOLUTION The series resonant frequency is determined by C_s and L_s. (The magnitude of their respective reactances will be equal.)

$$f_s = \frac{1}{2\pi\sqrt{L_sC_s}} = \frac{1}{2\pi\sqrt{(0.0060\text{ pF})(0.165609\text{ H})}}$$
$$= 5.048967\text{ MHz}$$

The impedance of the crystal at series resonance is equal to R_s (10 Ω). Now we determine the Q.

$$X_{Ls} = 2\pi f_s L_s = 2\pi(5.048967\text{ MHz})(0.165609\text{ H})$$
$$= 5.253713\text{ M}\Omega$$
$$Q = \frac{X_{Ls}}{R_s} = \frac{5.253713\text{ M}\Omega}{10\ \Omega} = 525{,}371$$

The parallel resonant frequency occurs when the equivalent parallel susceptance of the *RLC* branch is equal to the magnitude of the capacitive susceptance of C_m. When the resonant network has a high Q, we may approximate the true condition by stating that the reactances will sum to zero. This requires us first to determine the equivalent series capacitance of C_s and C_m.

$$\frac{1}{C_T} = \frac{1}{C_s} + \frac{1}{C_m} = \frac{1}{0.006\text{ pF}} + \frac{1}{13\text{ pF}} = 1.6674 \times 10^{14}\text{ F}^{-1}$$
$$C_T = 0.005997232\text{ pF}$$

$$f_p = \frac{1}{2\pi\sqrt{L_s C_T}} = \frac{1}{2\pi\sqrt{(0.005997232 \text{ pF})(0.165609 \text{ H})}}$$
$$= 5.050146 \text{ MHz}$$

The impedance of the crystal at parallel resonance may be determined by employing Eq. 15-30. However, we must first determine the *net Q* of the *RLC* branch. At f_p, X_{Ls} is

$$X_{Ls} = 2\pi f_p L_s = 5.254939 \text{ M}\Omega$$

and the capacitive reactance of C_s is

$$X_{Cs} = \frac{1}{2\pi f_p C_s} = 5.252487 \text{ M}\Omega$$

At f_p the net impedance of the *RLC* branch will be inductive.

$$X_{L(\text{net})} = X_{Ls} - X_{Cs} = 5.254939 \text{ M}\Omega - 5.252487 \text{ M}\Omega$$
$$= 2.452 \text{ k}\Omega$$

The net Q can now be found.

$$Q_{(\text{net})} = \frac{X_{L(\text{net})}}{R_s} = \frac{2.452 \text{ k}\Omega}{10 \ \Omega} = 245.2$$

Now from Eq. 15-30,

$$R_p = Q^2_{(\text{net})} R_s = (245.2)^2 (10 \ \Omega) = 601.2 \text{ k}\Omega$$ ■

From Example 15-14 we can see that f_s occurs *before* f_p. The frequency difference between the two is quite small (1.179 kHz). The impedance at series resonance is small (10 Ω) and the parallel resonant impedance is much larger (approximately 601 kΩ).

Quartz crystals are typically specified to operate in either their series or parallel resonant mode. For example, an M-tron MP-2 1-MHz crystal is intended to be used in parallel resonance with an equivalent load capacitance (C_{load}) of 18 pF across it. The crystal has a maximum R_s of 500 Ω with typical values ranging from 100 to 125 Ω.

The load capacitance statement above is very important. *It is possible to "pull," or shift the parallel resonant operating frequency (f_{op}) of a crystal by changing its load capacitance.* The load capacitance is in parallel with the mounting capacitance C_m (see Fig. 15-27).

The equivalent capacitance in Fig. 15-27(a) is the parallel combination of C_m and C_{load} in series with C_s. Therefore, the crystal's operating frequency is given by

$$f_{\text{op}} = \frac{1}{2\pi\sqrt{L_s\{C_s(C_m + C_{\text{load}})/[C_s + (C_m + C_{\text{load}})]\}}} \qquad (15\text{-}43)$$

where f_{op} is the parallel resonant frequency of a crystal with an external load capacitance. The f_{op} of a crystal lies between f_s and f_p. With no load capacitance (a C_{load} of zero) f_{op} will be equal to f_p. As C_{load} becomes larger, f_{op} will approach f_s. In practice,

$$f_s < f_{\text{op}} < f_p \qquad (15\text{-}44)$$

[see Fig. 15-27(b)].

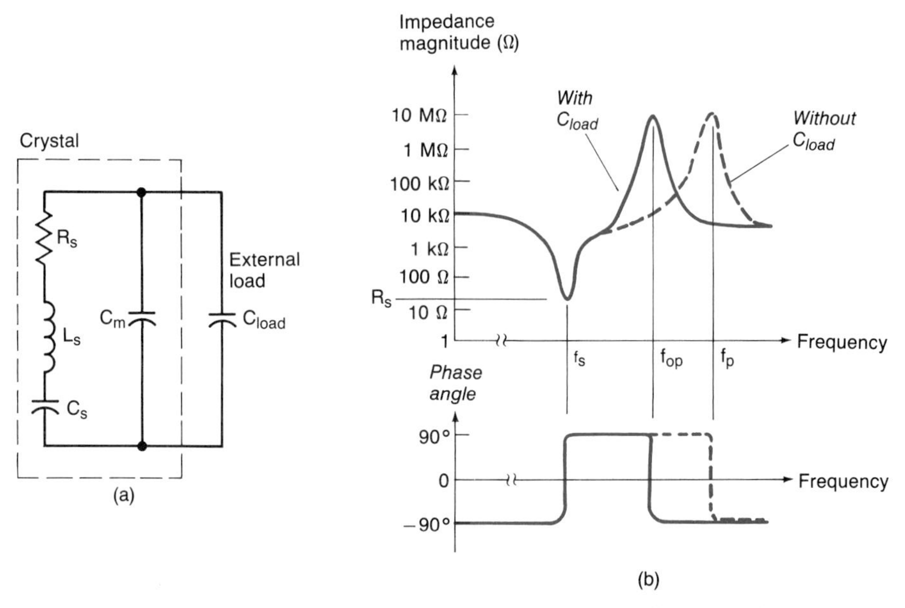

FIGURE 15-27 Crystals designed to operate in their parallel resonant mode have a specified load capacitance: (a) the circuit's parallel resonant (operating) frequency is f_{op}; (b) the C_{load} pulls the parallel resonant frequency toward the series resonant frequency.

EXAMPLE 15-15

Determine the operating frequency (f_{op}) of the crystal analyzed in Example 15-14 if C_{load} is 12 pF.

SOLUTION Using the data given in Example 15-14 and Eq. 15-43, we can determine the crystal's operating frequency.

$$f_{op} = \frac{1}{2\pi\sqrt{L_s\{C_s(C_m + C_{load})/[C_s + (C_m + C_{load})]\}}}$$

$$= \frac{1}{2\pi\sqrt{(0.165609\text{ H})\left\{\dfrac{(0.006\text{ pF})(13\text{ pF} + 12\text{ pF})}{[0.006\text{ pF} + (13\text{ pF} + 12\text{ pF})]}\right\}}}$$

$$= 5.049573\text{ MHz}$$

■

If we compare f_{op} with f_p, we see that C_{load} has lowered the crystal's parallel resonant frequency. Parallel resonance is now only about 606 Hz above the crystal's series resonant frequency. The effect of C_{load} on f_{op} has been illustrated in Fig. 15-28(a). The significance of f_{op} in oscillator circuits will be demonstrated in the next section.

From our previous work it should be clear that the phase shift across the frequency-determining feedback network in oscillator circuits is critical. When an inverting amplifier is used, the ac feedback is intended to provide about 180° of phase

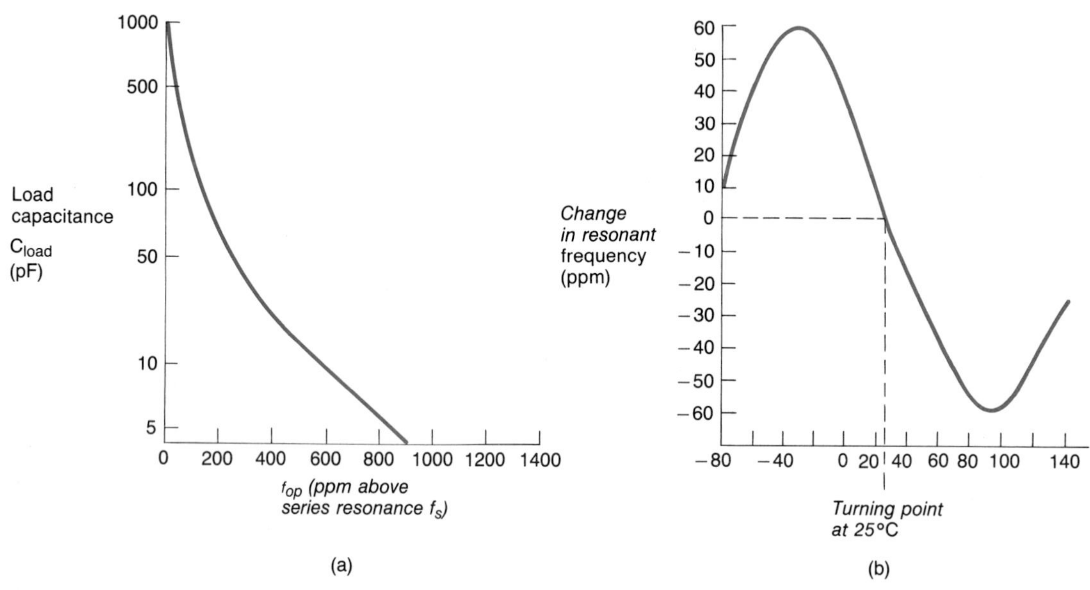

FIGURE 15-28 **Crystal parameters: (a) effect of C_{load} on f_{op}; (b) typical temperature characteristic curve.**

shift at the desired frequency of oscillation (f_{op}). If the Q of the feedback network is low, the change in the phase shift with frequency is gradual. Therefore, the frequency of oscillation is free to ''wander'' or drift over a relatively wide range of values. A stable oscillator demands a high-Q feedback network. A high-Q network maximizes the change in phase shift with frequency.

If we turn back to Fig. 15-26(d), we see that the phase shift offered by a crystal abruptly changes on either side of f_s and f_p. This occurs because of its inherently large Q. With this in mind it should be clear why crystal-based oscillator designs can offer excellent frequency stability.

The resonant frequency of an oscillator's tuned circuit is also critical. In the case of a discrete LC resonant circuit, the temperature coefficient of the capacitors and inductors come into play. Capacitors are available which have positive, negative, and zero temperature coefficients. Coils, however, generally only have positive temperature coefficients. Making a tuned circuit that is insensitive to temperature is a difficult endeavor. To keep the LC product constant, the capacitor's tempco must be negative to compensate for the inductor's positive tempco. The tempcos must be matched, and the components should be thermally bonded together.

The resonant frequency of a crystal also changes with temperature. However, it is much more predictable. A typical temperature curve has been depicted in Fig. 15-28(b). The *turning point* is the temperature at which the crystal's deviation from resonance is zero. For temperatures below the turning point, the crystal's resonant frequency will increase. For temperatures above the turning point, the crystal's resonant frequency will be decreased. We should emphasize that Fig. 15-28(b) shows the general shape of a temperature characteristic. If specific questions arise about a particular crystal, its manufacturer should be consulted. In demanding applications, it is not unusual to find the crystal housed in a small oven. The oven is typically used to regulate the crystal's temperature at its turning-point value.

The unit "ppm" in Fig. 15-28(b) stands for "parts per million." For example, if a 5-MHz crystal's resonant frequency has increased 20 ppm, the change in frequency (Δf) is

$$\Delta f = (5 \times 10^6 \text{ Hz})(20 \times 10^{-6}) = 100 \text{ Hz}$$

The use of ppm units is quite extensive in the electronics industry and *not* just associated with crystals.

15-15 Crystal Oscillators

As we mentioned in Section 15-14, the extremely high Q of a crystal can be used to produce oscillators with excellent frequency stability. In practice, a crystal may be used to replace an entire LC resonant network, or it may be used to replace one or more of the reactances normally found in such a network. We shall see examples of these alternatives in our work to follow.

In Fig. 15-29(a) we see a general form of the Pierce crystal oscillator. In essence, it is a Clapp oscillator in which the inductor and its small series capacitance are replaced by a quartz crystal. The very small C_s of the crystal dominates. *The crystal is typically operated slightly above its series resonant frequency f_s.* Therefore, the net impedance of the crystal is a resistance in series with an inductive reactance. The crystal's equivalent circuit has been emphasized in Fig. 15-29(a). *The Pierce oscillator is one of the most popular crystal oscillator circuits.*

We see a practical Pierce oscillator circuit in Fig. 15-29(b). At first glance, most students will not immediately recognize that the circuit is in the configuration indicated in Fig. 15-29(a). This is because the output (device) capacitance of the JFET is used as the "C_1" of the feedback network, and the JFET's input capacitance is used as the "C_2." This is indicated in Fig. 15-29(c).

The small C_1 and C_2 capacitances produce relatively large capacitive reactances. Therefore, the inductive reactance of the crystal is also large, and it operates considerably above series resonance. Observe that an M-tron MP-2 crystal has been indicated. As we mentioned in Section 15-14, this crystal is intended to operate in the parallel resonant mode. Laboratory work indicated that the circuit in Fig. 15-29(b) operates at a frequency of 999.997 kHz. This is 3 Hz below the crystal's nominal parallel resonant frequency of 1.000 MHz, and 197 Hz above the crystal's series resonant frequency of 999.800 kHz.

Obviously, the crystal oscillator given in Fig. 15-29(b) requires a minimal number of components. However, the addition of a small adjustable capacitor (called a "trimmer") from the gate of the JFET to ground allows the output frequency to be adjusted. As the capacitance of the trimmer is increased, the frequency of oscillation will be lowered. The additional capacitance will pull the crystal's operating frequency toward series resonance.

In Fig. 15-30(a) we see an improved Pierce oscillator circuit. Specifically, capacitors C_1 and C_2 have been added to "swamp out" the JFET's input and output device capacitances. While using the JFET's capacitances to minimize the component

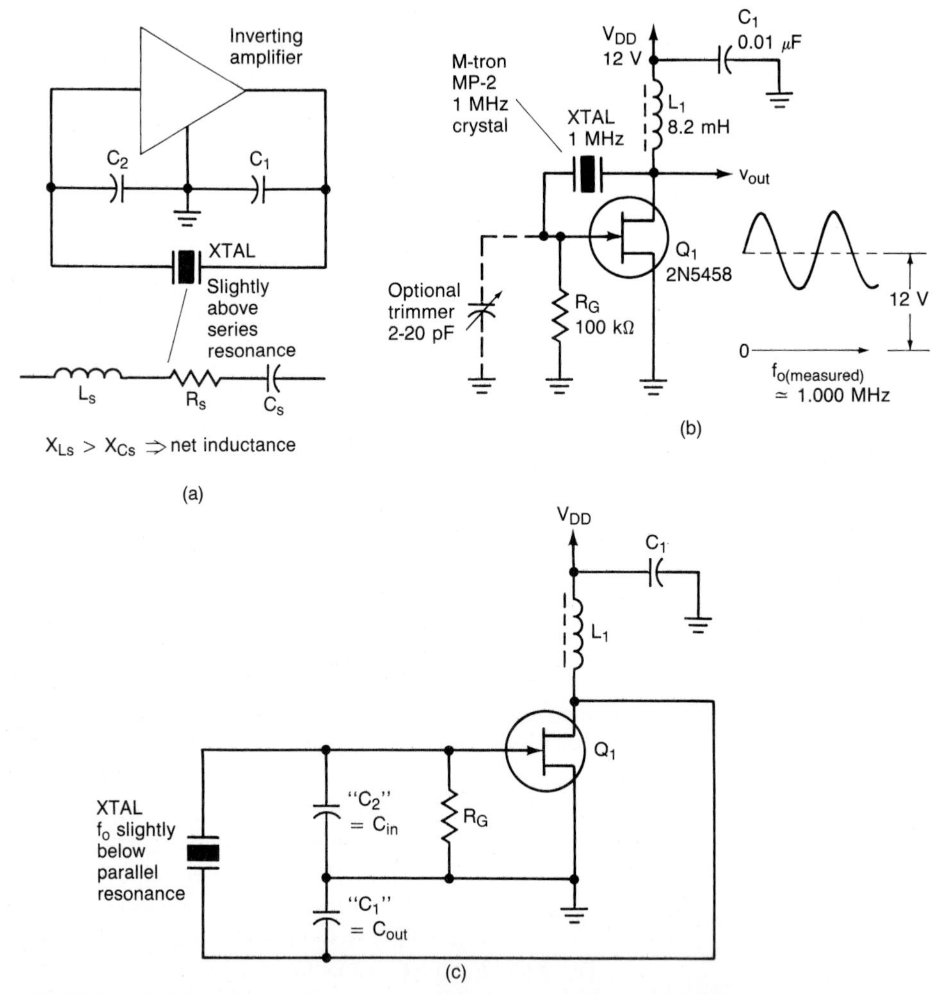

FIGURE 15-29 **Pierce crystal oscillator: (a) the basic Pierce oscillator is similar to the Clapp configuration; (b) simple Pierce oscillator; (c) the oscillator uses device capacitances to complete the circuit.**

count results in a simple circuit, it can also open the door to problems if the circuit enters mass production. Manufacturers will generally limit the maximum values of the device capacitances, but they are not otherwise tightly controlled parameters. Consequently, the mass production of such circuits could lead to a production run of oscillators that refuse to oscillate, or whose frequency is out of tolerance. The use of capacitors C_1 and C_2 tends to minimize this problem.

Capacitor C_3 has been added to provide gate leak bias. This serves to minimize the distortion in the oscillator's output waveform.

The increased capacitance (C_1, C_2, and C_3) across the crystal pulls its operating frequency much closer to series resonance. In a ''good'' Pierce oscillator design, the net impedance of the crystal is only slightly inductive. Typically, the crystal will operate 10 to 50 ppm above series resonance.

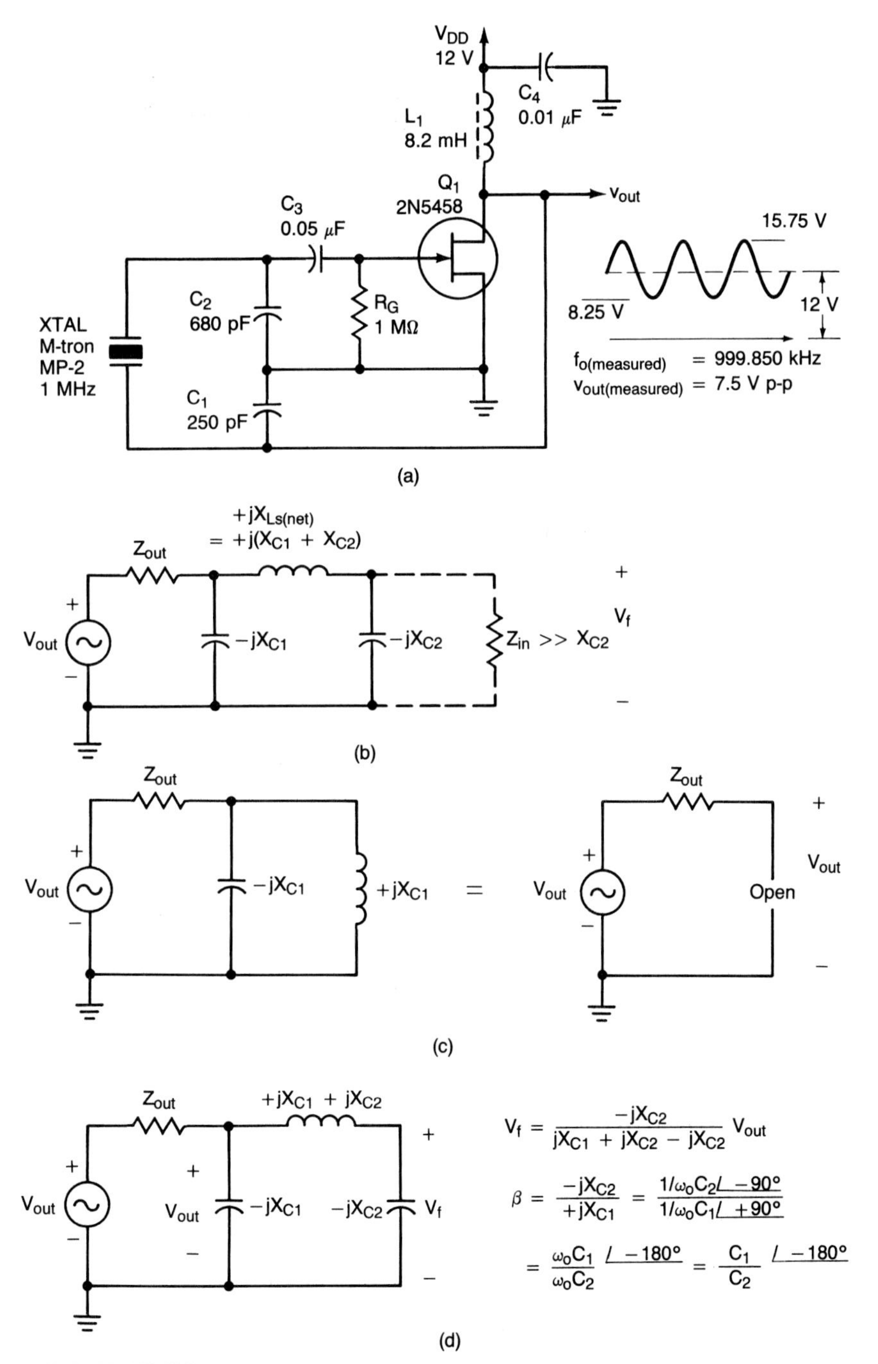

$$V_f = \frac{-jX_{C2}}{jX_{C1} + jX_{C2} - jX_{C2}} V_{out}$$

$$\beta = \frac{-jX_{C2}}{+jX_{C1}} = \frac{1/\omega_o C_2 \angle -90^\circ}{1/\omega_o C_1 \angle +90^\circ}$$

$$= \frac{\omega_o C_1}{\omega_o C_2} \angle -180^\circ = \frac{C_1}{C_2} \angle -180^\circ$$

FIGURE 15-30 Improved Pierce oscillator and its ideal analysis: (a) when capacitances C_1 and C_2 are added the crystal operates much closer to its series resonant frequency; (b) ideal equivalent circuit − $R_s << X_{Ls(net)}$; (c) the amplifier's output drives an ideal parallel resonant circuit; (d) the feedback factor has a magnitude of C_1/C_2 and provides -180° of phase shift.

In Fig. 15-30(a) we see actual laboratory measurements. The output frequency was 999.850 kHz. This was 50 Hz above the crystal's series resonant frequency of 999.800 kHz.

From our previous work, we recall that the main thrust of any oscillator analysis lies in the determination of its feedback factor β. This problem is particularly unwieldy in crystal oscillators. In our previous work we assumed that the inductor's resistance was negligibly small. This greatly simplified our work and resulted in very reasonable approximations. In "real" Pierce oscillators, this is unfortunately *not* the case. The crystal's internal resistance (R_s) is significant.

To understand the effect of R_s, let us first review the ideal analysis from a slightly different perspective [refer to Fig. 15-30(b)]. As we can see, we have again assumed that the amplifier's input impedance (Z_{in}) is large enough to be ignored. Hence

$$Z_{\text{in}} >> X_{C2}$$

We have also assumed that R_s is negligibly small. Observe that the net inductive reactance [$X_{Ls(\text{net})}$] is in series with X_{C2}. Also note that we have assumed that the sum of the three reactances is zero at resonance.

$$+jX_{Ls(\text{net})} - jX_{C1} - jX_{C2} = 0$$

Therefore, we may state that

$$X_{Ls(\text{net})} = X_{C1} + X_{C2}$$

Now we shall solve for the β of this idealized network. First, we find its loading effect on the amplifier's output. The series combination of $X_{Ls(\text{net})}$ and X_{C2} produces

$$+jX_{Ls(\text{net})} - jX_{C2} = (jX_{C1} + jX_{C2}) - jX_{C2} = jX_{C1}$$

The *positive* equivalent impedance jX_{C1} is in parallel with capacitor C_1 ($-jX_{C1}$) [see Fig. 15-30(c)]. The output of the amplifier therefore drives a parallel resonant circuit. Ideally, a parallel resonant circuit acts like an open circuit. Therefore, the output of the amplifier is effectively unloaded. This means that the amplifier's output impedance (Z_{out}) has no effect on the β of the feedback network, and V_{out} appears across capacitor C_1 [see Fig. 15-30(d)].

Voltage division between the inductive reactance and X_{C2} determines the feedback factor β. Hence

$$\beta = \frac{V_f}{V_{\text{out}}} = \frac{-jX_{C2}}{jX_{C1} + jX_{C2} - jX_{C2}} = \frac{-jX_{C2}}{+jX_{C1}}$$

$$= \frac{1/\omega_o C_2 \angle -90^\circ}{1/\omega_o C_1 \angle +90^\circ} = \frac{1}{\omega_o C_2} \frac{\omega_o C_1}{1} \angle -180^\circ$$

$$= \frac{C_1}{C_2} \angle -180^\circ = -\frac{C_1}{C_2}$$

This is the same relationship that we previously determined for both the Colpitts and Clapp oscillators.

The feedback network has given us a gain of C_1/C_2 and provides 180° of phase shift. Therefore,

$$A_{v(\text{oc})} < -\frac{C_2}{C_1}$$

to provide a loop gain with 0° of phase shift and a magnitude greater than unity. Now let us consider the case where R_s is *not* negligibly small [see Fig. 15-31(a)].

The following four points apply to ''real'' Pierce crystal oscillators:

1. The sum of the reactances must be equal to zero to provide 180° of phase shift across the frequency-determining feedback network.

$$-jX_{C1} - jX_{C2} + jX_{Ls} = 0,$$

FIGURE 15-31 More accurate analysis of the Pierce oscillator: (a) the crystal's equivalent series resistance is significant and the sum of the three reactances is zero; (b) approximation of the load across the amplifier's output; (c) determination of the feedback factor β.

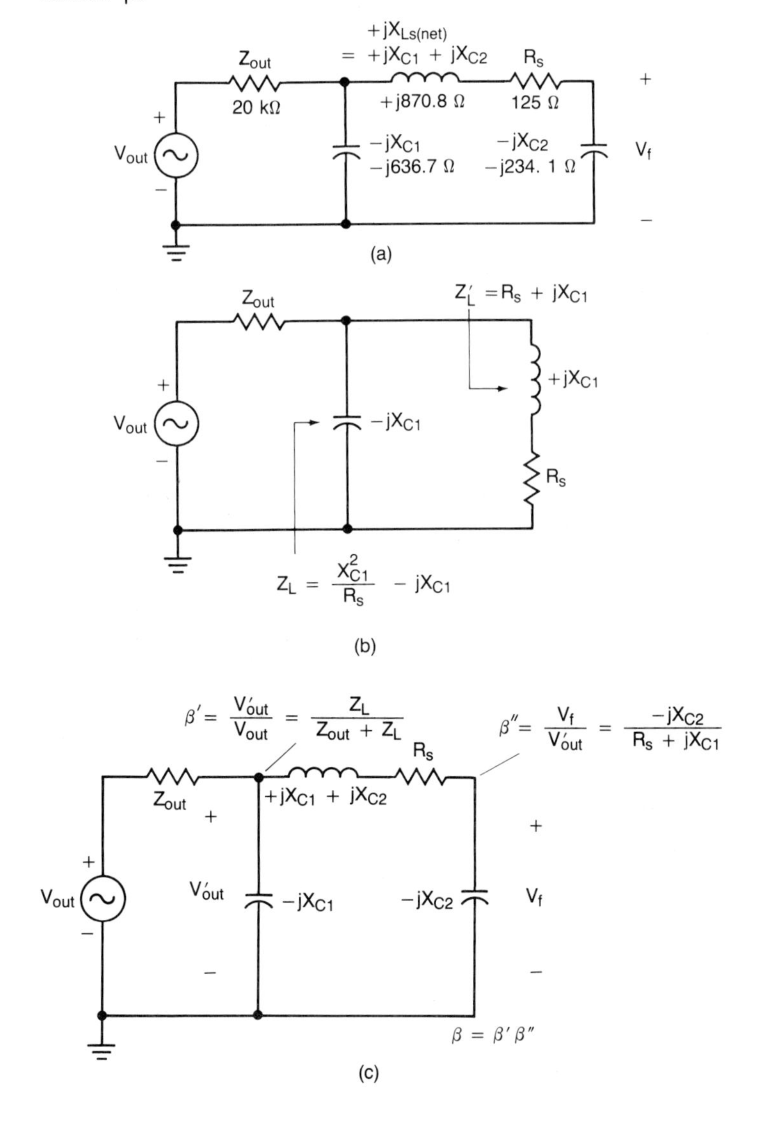

and therefore

$$jX_{Ls} = jX_{C1} + jX_{C2}$$

This statement allows us to determine the net inductive reactance of the crystal as it operates above series resonance.

2. Most designs set the capacitive reactance of C_2 to be much less than the Z_{in} of the amplifier. Therefore, we will assume that

$$X_{C2} << Z_{in}$$

This statement allows us to ignore Z_{in} and greatly simplifies the analysis.

3. The equivalent series resistance of the crystal (R_s) is significant when compared to the other impedance levels. This is an unfortunate fact of life. It is very important to consider R_s in both the analysis and design of crystal oscillators. Fortunately, most crystal manufacturers include R_s on their data sheets.
4. Because R_s is significant, the output of the amplifier does *not* drive an ideal parallel resonant circuit. Therefore, the output impedance of the amplifier must be included in the analysis.

One approach to the analysis of the feedback network is to generate a single equation for the feedback factor β. However, this results in a third-order equation, and the mathematics can quickly become overwhelming. To keep our sanity and to promote a more intuitive understanding, we shall break the analysis down into four relatively simple steps:

1. Find the load (Z_L) across the output of the amplifier.
2. Determine the gain from the amplifier's output to the feedback capacitor C_1 and call it β'.
3. Find the gain from capacitor C_1 to capacitor C_2 and call it β''.
4. Multiply these two loaded gains (β' times β'') to obtain the overall feedback factor β.

The feedback circuit used in Fig. 15-30(a) has been redrawn in Fig. 15-31(a). The indicated reactances were determined at the oscillation frequency of 999.850 kHz. The reader should verify the values.

First, we shall find Z_L [refer to Fig. 15-31(b)]. By inspection we can see that the series combination of $X_{Ls(\text{net})}$, R_s, and X_{C2} reduces to an equivalent inductive reactance (jX_{C1}) in series with R_s. This series combination (Z_L') is in parallel with $-jX_{C1}$. Since admittances in parallel add,

$$Y_L = \frac{1}{-jX_{C1}} + \frac{1}{R_s + jX_{C1}} = \frac{R_s + jX_{C1} - jX_{C1}}{(-jX_{C1})(R_s + jX_{C1})}$$
$$= \frac{R_s}{X_{C1}^2 - jR_sX_{C1}}$$

The reciprocal of the total load admittance produces the load impedance.

$$Z_L = \frac{1}{Y_L} = \frac{X_{C1}^2 - jR_sX_{C1}}{R_s}$$

$$Z_L = \frac{X_{C1}^2}{R_s} - jX_{C1} \qquad (15\text{-}45)$$

where Z_L is the ac load across the output of the amplifier used in the Pierce oscillator.

The gain (β') from the amplifier's output to capacitor C_1 is defined in Fig. 15-31(c). Thus

$$\beta' = \frac{V'_{out}}{V_{out}} = \frac{Z_L}{Z_{out} + Z_L} \qquad (15\text{-}46)$$

where β' is the loaded gain from the amplifier's output to the input of a Pierce oscillator's feedback network. The gain from capacitor C_1 to capacitor C_2 is also defined in Fig. 15-31(c). Hence

$$\beta'' = \frac{V_f}{V'_{out}} = \frac{-jX_{C2}}{R_s + jX_{C1}} \qquad (15\text{-}47)$$

where β'' is the gain of the feedback network used in the Pierce oscillator. The overall feedback factor can now be determined from

$$\beta = \frac{V_f}{V_{out}} = \frac{V'_{out}}{V_{out}}\frac{V_f}{V'_{out}} = \beta'\beta'' \qquad (15\text{-}48)$$

where β is the overall feedback factor of the Pierce oscillator, which includes the loading on the amplifier's output.

EXAMPLE 15-16

Determine the ideal β and the actual β of the Pierce oscillator given in Fig. 15-30(a). Assume that the circuit oscillates at 999.850 kHz. The impedances have been given in Fig. 15-31(a).

SOLUTION Ideally,

$$\beta = -\frac{C_1}{C_2} = -\frac{250\text{ pF}}{680\text{ pF}} = -0.368$$

To determine the actual β we must first find the load across the amplifier's output. From Eq. 15-45,

$$Z_L = \frac{X_{C1}^2}{R_s} - jX_{C1} = \frac{(636.7)^2}{125} - j636.7$$
$$= 3243 - j636.7\ \Omega = 3305\ \Omega \angle -11.1^\circ$$

Now we can determine β' by using Eq. 15-46.

$$\beta' = \frac{Z_L}{Z_L + Z_{out}} = \frac{3305 \angle -11.1^\circ}{(3243 - j636.7) + 20\text{ k}\Omega} = 0.142 \angle -9.5^\circ$$

β'' is given by using Eq. 15-47.

$$\beta'' = \frac{-jX_{C2}}{R_s + jX_{C1}} = \frac{-j234.1}{125 + j636.7} = 0.361 \angle -169^\circ$$

The actual β is determined by Eq. 15-48.

$$\beta = \beta'\beta'' = (0.142 \angle -9.5^\circ)(0.361 \angle -169^\circ)$$
$$= 0.0513 \angle -178.5^\circ \simeq -0.0513$$

Obviously, there is a considerable discrepancy between the ideal and the actual feedback factor. The idealization is simply not a valid approximation. ■

Like all devices, crystals also have maximum ratings. For example, the dc voltage impressed across a crystal must be well below its maximum rating. If such a danger exists, a dc blocking capacitor should be placed in series with the crystal.

Even if the dc levels across the crystal are within acceptable limits, we must avoid overdriving it. Crystals have a maximum (ac) power rating. It is typically referred to as the crystal's maximum *drive level*. The power dissipation within a crystal may be thought of as a power dissipation within the crystal's equivalent series resistance (R_s).

Maximum drive levels range from as low as 0.1 mW to as high as 10 mW. The drive level rating is closely associated with the crystal's geometry, or cut. The most common quartz crystal—the AT cut—is used in the frequency range 500 kHz to 100 MHz, and has a typical maximum drive level of 10 mW.

If a crystal used in an oscillator becomes overdriven, the frequency will be reduced or become unstable. (If the crystal is sufficiently overdriven, it may become fractured, and oscillations will cease.) Therefore, we must take steps to ensure that the crystal is not overdriven.

One technique to limit the drive level is to place an impedance (typically, a resistor) in series with the output of the amplifier. This has been illustrated in Fig. 15-32(a). A practical circuit has been given in Fig. 15-32(b).

The oscillator in Fig. 15-32(b) employs an M-tron MP-2 (AT-cut) crystal. The manufacturer states that its maximum series equivalent resistance is 500 Ω, and its maximum drive level is 10 mW. The manufacturer also recommends drive levels on the order of 100 μW for maximum stability. Let us analyze Fig. 15-32(b).

The dc equivalent circuit of Fig. 15-32(b) has been indicated in Fig. 15-33(a). Diode D_1 has been added to provide a measure of *temperature compensation* for Q_1's dc base bias. The 1N4148 small-signal silicon diode in the base circuit helps the collector current of the transistor remain constant with changes in temperature. (This technique was introduced in Section 12-14. However, in this case a *single* diode is being used.)

EXAMPLE 15-17

Perform a dc analysis of the Pierce oscillator given in Fig. 15-32(b). Determine the approximate values of V_B, V_E, and I_C.

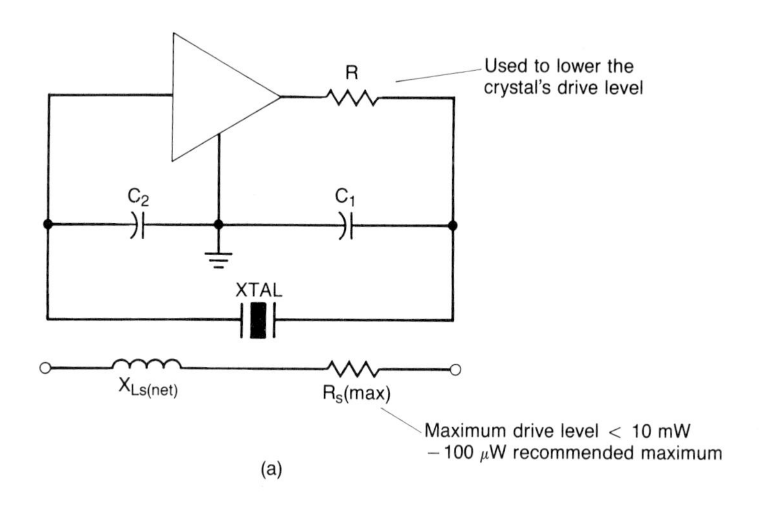

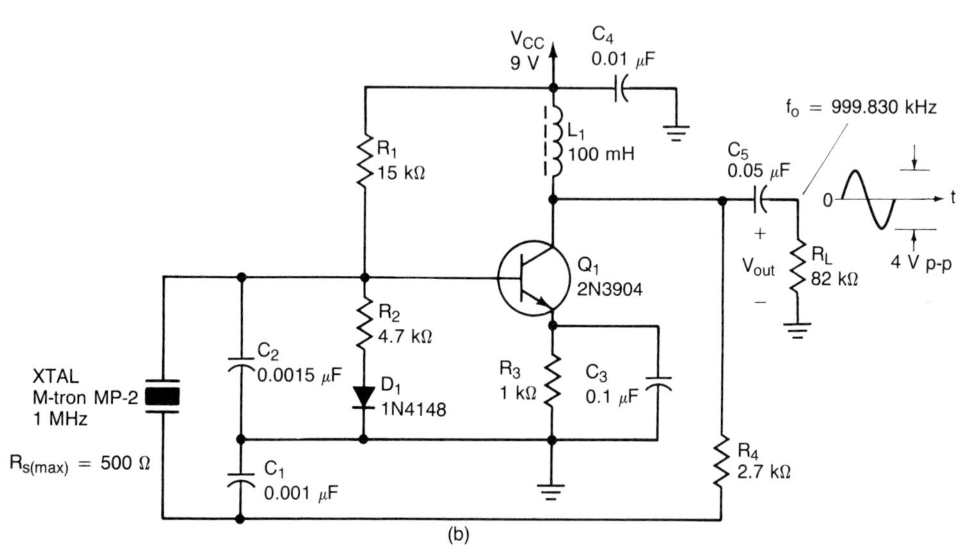

FIGURE 15-32 Improving the short-term stability of the Pierce oscillator: (a) using a resistor to restrict the crystal's drive level; (b) practical circuit.

SOLUTION Using the dc equivalent circuit given in Fig. 15-33(a), we can see that

$$V_B \simeq V_{\text{TH}} = \frac{R_2}{R_1 + R_2}(V_{CC} - 0.7\text{ V}) + 0.7\text{ V}$$

$$\simeq \frac{4.7\text{ k}\Omega}{15\text{ k}\Omega + 4.7\text{ k}\Omega}(9\text{ V} - 0.7\text{ V}) + 0.7\text{ V} = 2.68\text{ V}$$

$$V_E = V_B - V_{BE} = 2.68\text{ V} - 0.7\text{ V} = 1.98\text{ V}$$

$$I_C \simeq I_E = \frac{V_E}{R_3} = \frac{1.98\text{ V}}{1\text{ k}\Omega} = 1.98\text{ mA}$$

■

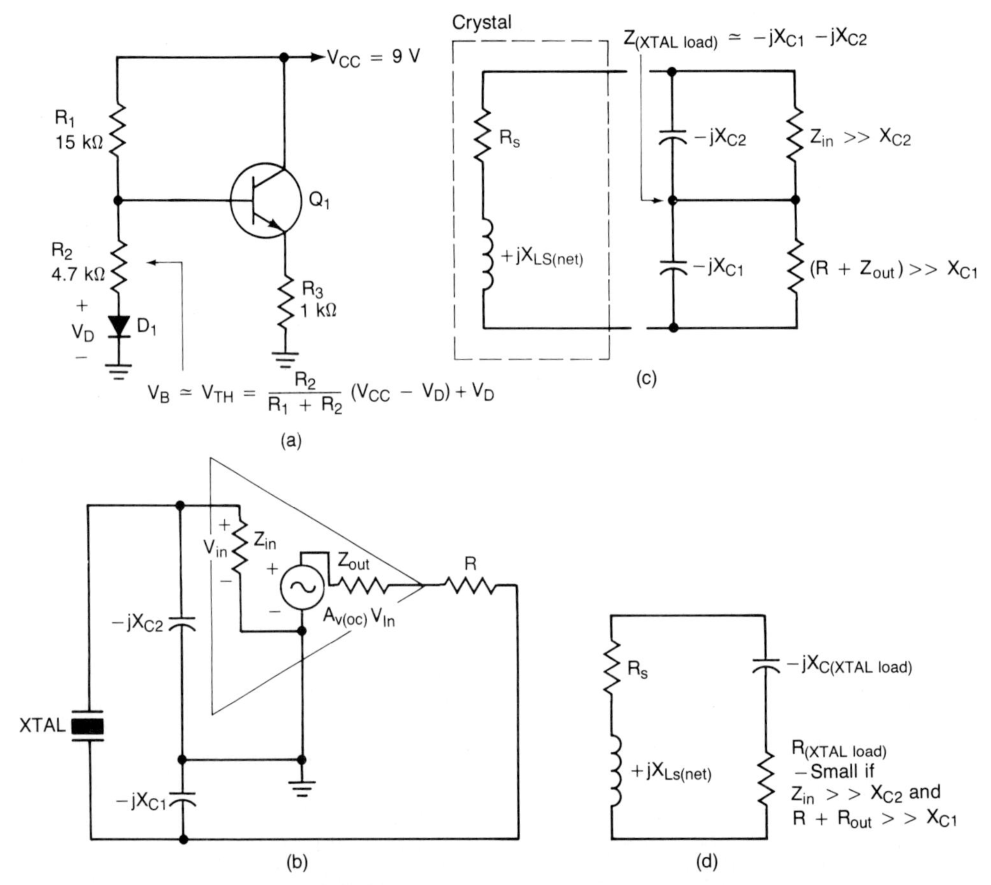

FIGURE 15-33 Analyzing the improved circuit: (a) dc equivalent circuit; (b) design constraints on C_1 and C_2; (c) ac equivalent circuit; (d) the crystal's in-circuit Q will be large.

The ac equivalent circuit of Fig. 15-32(b) has been shown in Fig. 15-33(b). Our next step is to determine the BJT's ac parameters.

EXAMPLE 15-18

Determine g_m, r_o, and r_π for the 2N3904 BJT used in the Pierce oscillator given in Fig. 15-32(b).

SOLUTION First we shall determine the BJT's g_m at its quiescent collector current

$$g_m = \frac{I_C}{26\text{ mV}} = \frac{1.98\text{ mA}}{26\text{ mV}} = 76.2\text{ mS}$$

Now we find r_o.

$$r_o = \frac{200\text{ V}}{I_C} = \frac{200\text{ V}}{1.98\text{ mA}} = 101\text{ k}\Omega$$

The 2N3904 has an h_{fe} of approximately 140 at an I_C of 2 mA. Hence

$$r_\pi = \frac{h_{fe}}{g_m} = \frac{140}{76.2\text{ mS}} = 1.84\text{ k}\Omega$$ ■

EXAMPLE 15-19

Determine Z_{in}, Z_{out}, $A_{v(oc)}$, and the (loaded) voltage gain A_v for the amplifier used in the Pierce oscillator given in Fig. 15-32(b).

SOLUTION To determine Z_{in}, we ignore the diode's dynamic resistance. It will be much, much smaller than R_2.

$$Z_{in} = R_1 \parallel R_2 \parallel r_\pi = 15\text{ k}\Omega \parallel 4.7\text{ k}\Omega \parallel 1.84\text{ k}\Omega$$
$$= 1.21\text{ k}\Omega$$

No collector resistor is used and L_1 is an RFC with an X_L of 628 kΩ at f_o. Therefore, Z_{out} is

$$Z_{out} \simeq r_o = 101\text{ k}\Omega$$

The open-circuit voltage gain is

$$A_{v(oc)} = -g_m r_o = -(76.2\text{ mS})(101\text{ k}\Omega) = -7696$$

To find A_v we must first determine r_C.

$$r_C = r_o \parallel R_L = 101\text{ k}\Omega \parallel 82\text{ k}\Omega = 45.3\text{ k}\Omega$$

and A_v is

$$A_v = -g_m r_C = -(76.2\text{ mS})(45.3\text{ k}\Omega) = 3452$$ ■

For good frequency stability, the crystal's external load [$Z_{(XTAL\ load)}$ in Fig. 15-33(c)] should be as nonresistive as possible. If X_{C1} and X_{C2} are small, the resistive component of the load across the crystal will be negligibly small. As a rough rule of thumb, the reactances X_{C1} and X_{C2} are sized such that their magnitudes are on the same order as R_s. [The specifics are detailed in Fig. 15-33(d).] The symbol $X_{C(XTAL\ load)}$ is the crystal's equivalent load capacitance. The symbol $R_{(XTAL\ load)}$ is the crystal's equivalent load resistance. This will ensure that the crystal's in-circuit Q is large.

$$Q_{(in\text{-}circuit)} = \frac{X_{Ls(net)}}{R_s + R_{(XTAL\ load)}}$$

↑ The smaller this term is, the closer the $Q_{(in\text{-}circuit)}$ will be to the Q of crystal

EXAMPLE 15-20

Given the Pierce oscillator in Fig. 15-32(b), determine if the capacitor values are adequately large, and find the Z_L across the amplifier's output. The frequency of oscillation is 999.830 kHz.

SOLUTION First we determine X_{C1} and X_{C2}.

$$X_{C1} = \frac{1}{2\pi f_o C_1} = \frac{1}{2\pi(999.830\text{ kHz})(0.001\ \mu\text{F})} = 159.2\ \Omega$$

$$X_{C2} = \frac{1}{2\pi f_o C_2} = \frac{1}{2\pi(999.830\text{ kHz})(0.0015\ \mu\text{F})} = 106.1\ \Omega$$

We observe that

$$X_{C1} = 159.2\ \Omega < Z_{\text{out}} + R = 103.7\text{ k}\Omega$$
$$X_{C2} = 106.1\ \Omega < Z_{\text{in}} = 1.21\text{ k}\Omega$$

The capacitive reactances seem reasonable since the MP-2 crystal has a typical R_s of 125 Ω and an $R_{s(\max)}$ of 500 Ω. Since the reactance levels are on the same order as the magnitude of R_s, a high Q will result. However, we must use Eq. 15-45 to find the load across the amplifier's output.

$$Z_L = \frac{X_{C1}^2}{R_s} - jX_{C1} = \frac{(159.2\ \Omega)^2}{500\ \Omega} - j159.2\ \Omega$$
$$= 50.69 - j159.2\ \Omega = 167.1\ \Omega \angle -72.34^\circ$$

■

EXAMPLE 15-21 Determine the loop gain of the Pierce oscillator given in Fig. 15-32(b).

SOLUTION First we find β'. The total Z_{out} includes R_4 *and* R_L. Hence

$$Z_{\text{out}} = R_4 + r_o \| R_L = 2.7\text{ k}\Omega + 101\text{ k}\Omega \| 82\text{ k}\Omega = 48\text{ k}\Omega$$

$$\beta' = \frac{Z_L}{Z_{\text{out}} + Z_L} = \frac{167.1\ \Omega \angle -72.34^\circ}{48\text{ k}\Omega + 50.69\ \Omega - j159.2\ \Omega}$$
$$= 3.478 \times 10^{-3} \angle -72.15^\circ$$

Now we find β''.

$$\beta'' = \frac{-jX_{C2}}{R_s + jX_{C1}} = \frac{-j106.1\ \Omega}{500\ \Omega + j159.2\ \Omega} = \frac{106.1\ \Omega \angle -90^\circ}{524.7\ \Omega \angle 17.66^\circ}$$
$$= 0.2022 \angle -107.66^\circ$$

The total feedback factor β is

$$\beta = \beta'\beta'' = (3.478 \times 10^{-3} \angle -72.15^\circ)(0.2022 \angle -107.66^\circ)$$
$$= 0.000703 \angle -179.81^\circ \simeq -0.000703$$

Now we can find the loop gain.

$$\beta A_v = (-0.000703)(-3452) = 2.43 > 1$$

Therefore, the circuit is definitely going to oscillate. ■

The last step in our analysis of the Pierce oscillator is the determination of the crystal's drive level. When the oscillator is first powered up, we are dealing with small signals. As the oscillations grow in amplitude, we eventually enter the *large-signal* domain. The maximum possible output signal level has been illustrated in Fig. 15-34(a). The resulting large-signal equivalent circuit of the Pierce oscillator [Fig. 15-32(b)] has been given in Fig. 15-34(b).

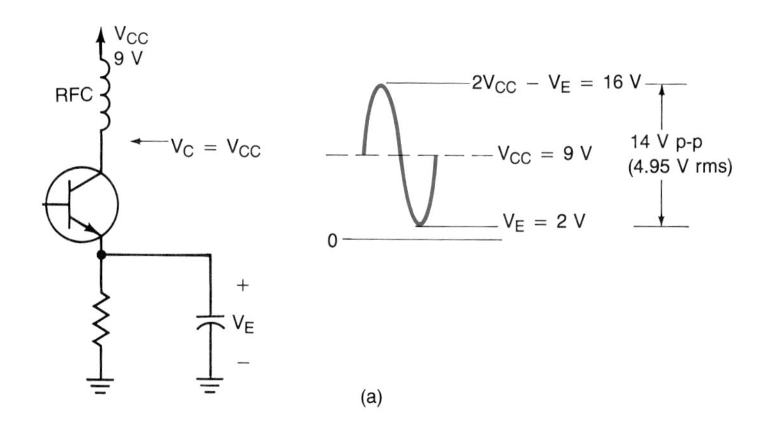

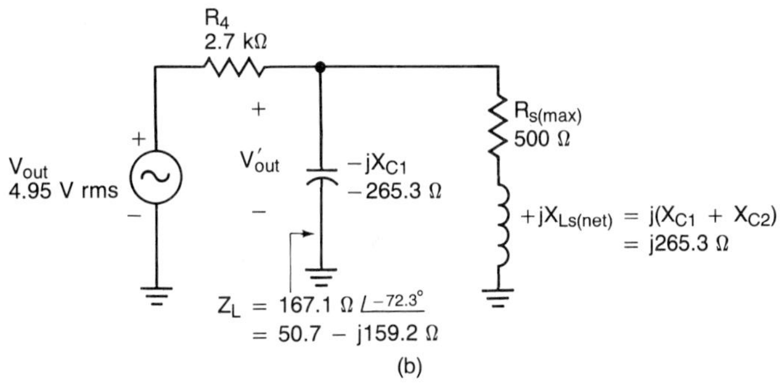

FIGURE 15-34 **Finding the crystal's drive level: (a) determining the maximum possible collector voltage swing; (b) analyzing the large-signal ac equivalent.**

EXAMPLE 15-22

Determine the maximum drive level the crystal will experience in Fig. 15-32(b).

SOLUTION As we can see in Fig. 15-34(b), the output voltage (V_{out}) is 4.95 V rms. By voltage division, we can ascertain V'_{out}.

$$V'_{out} = \frac{Z_L}{Z_L + R_4} V_{out} = \frac{167.1\ \Omega \angle -72.3^\circ}{(50.7\ \Omega - j159.2\ \Omega) + 2.7\ \text{k}\Omega}(4.95\ \text{V})$$
$$= 300\ \text{mV rms} \angle -69^\circ$$

Now we can find the magnitude of the rms current through the crystal.

$$I = \frac{V'_{out}}{R_{s(max)} + jX_{Ls(net)}} = \frac{300\ \text{mV rms} \angle -69^\circ}{500\ \Omega + j265.3\ \Omega}$$
$$= 0.530\ \text{mA rms} \angle -97^\circ$$
$$|I| = 0.530\ \text{mA rms}$$

The crystal's drive level can now be found.

$$P_{XTAL} = |I|^2 R_{s(max)} = (0.530\ \text{mA rms})^2(500\ \Omega)$$
$$= 140\ \mu\text{W}$$

■

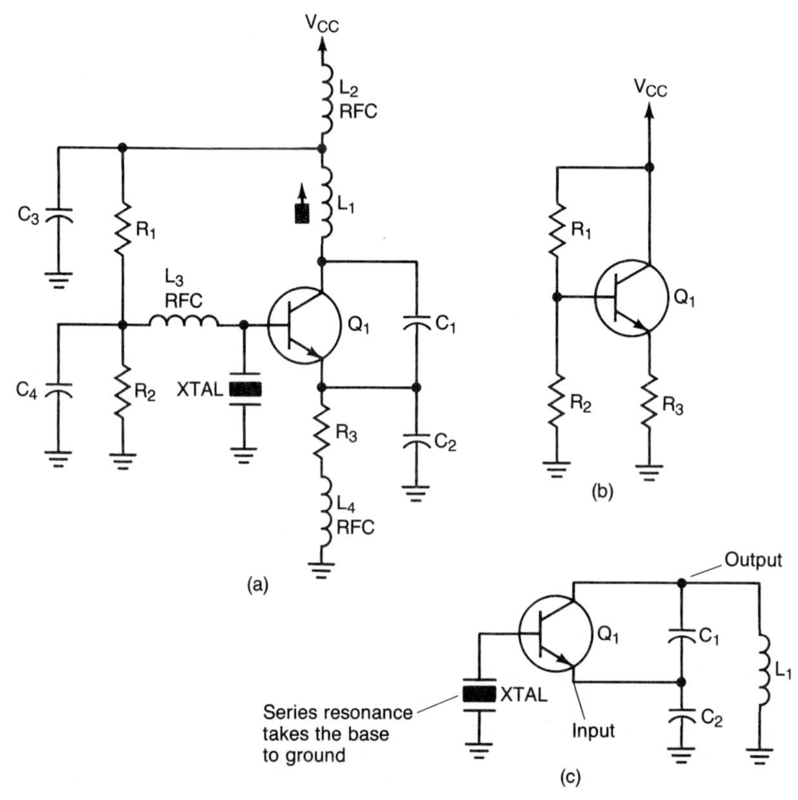

FIGURE 15-35 Crystal-controlled Colpitts oscillator: (a) circuit; (b) dc equivalent; (c) the ac equivalent reveals that we have a common-base BJT amplifier.

The 140 μW of power dissipation is much less than the crystal's maximum rating of 10 mW, and on the order of the manufacturer's recommended nominal drive level of 100 μW. As we can see in Fig. 15-32(b), the actual V_{out} was only 4 V peak to peak. Running through the same calculations, we see that the crystal's drive level is only 13.3 μW.

A rough check of the drive level of a crystal in an oscillator can be made by varying V_{CC}. Normally, increasing V_{CC} will produce a slight increase in the oscillator's output frequency. If the crystal is being overdriven, increasing V_{CC} will result in a decrease in the output frequency, or the output frequency will become unstable.

Using this check, the Pierce oscillator in Fig. 15-32(b) performed very well. V_{CC} was adjusted from 6 to 20 V and its output frequency remained "solid" at 999.830 kHz.

The last crystal oscillator we shall investigate is shown in Fig. 15-35. The circuit may be described as a crystal-controlled Colpitts oscillator. The basic configuration has performed well in applications ranging from 1 MHz to over 30 MHz.

Considerable attention has been paid to decoupling. Three RFCs are used [see L_2, L_3, and L_4 in Fig. 15-35(a)]. Capacitors C_3 and C_4 provide additional decoupling. The dc equivalent circuit has been shown in Fig. 15-35(b).

The ac equivalent circuit has been given in Fig. 15-35(c). The BJT is in its common-base configuration. (The output is at the collector, and the input is at the emitter.) The crystal is operated in its series resonant mode to place the base of the BJT at ac ground. The inductor L_1 is slug tuned. This allows the output frequency to be adjusted over a narrow range.

15-16 UJT Relaxation Oscillators

In Section 15-1 we mentioned that the two basic oscillator categories are harmonic and relaxation. The block diagram of the relaxation oscillator is given in Fig. 15-36.

The circuit oscillates as energy is exchanged between the time constant network and the active device. The output waveforms produced by relaxation oscillators are invariably nonsinusoidal. Typical examples (the "sawtooth," triangular, and rectangular waveforms) have been indicated in Fig. 15-36.

A relatively simple relaxation oscillator can be constructed around a device called the *unijunction transistor* (UJT). In its simplest form, the UJT is made by diffusing a p-type region into a lightly doped n-type (n^-) silicon crystal [see Fig. 15-37(a)]. The resulting p-n junction is located about 70% of the crystal's length away from the base 1 end of the structure. The three terminals of the UJT are the base 1 (B_1), base 2 (B_2), and the emitter (E).

UJTs are packaged in much the same manner as small-signal BJTs. For example, the 2N2646 and 2N2647 UJTs are available in the modified TO-18 case style shown in Fig. 15-37(b). The UJT's schematic symbol is given in Fig. 15-37(c).

The electrically equivalent circuit of a UJT is shown in Fig. 15-38(a). The diode is used to represent the p-n junction between the p region and the n^- region. The resistors R_{B1} and R_{B2} are used to represent the resistance of the lightly doped (n^-) semiconductor.

The resistance between the B_1 and B_2 terminals is called the *interbase resistance*. Its symbol is R_{BB}. Typical values range from 4 to 10 kΩ. The interbase resistance is measured with the emitter open [see the test circuit shown in Fig. 15-38(b)].

$$R_{BB} = R_{B1} + R_{B2} \Big|_{I_E=0} \tag{15-49}$$

where R_{BB} is the interbase resistance parameter of the UJT.

FIGURE 15-36 Block diagram of a relaxation oscillator.

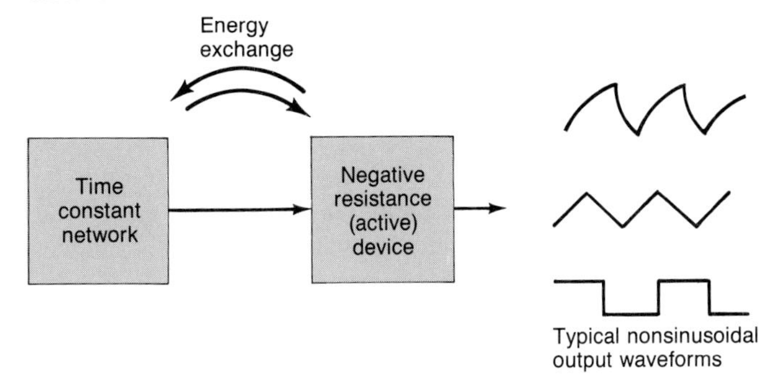

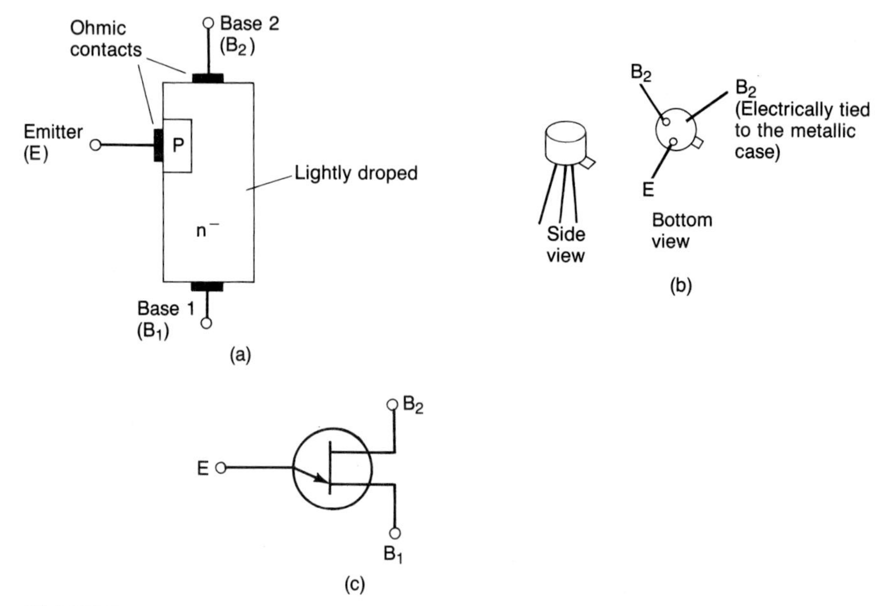

FIGURE 15-37 **Unijunction transistor (UJT): (a) basic structure; (b) modified TO-18 case style; (c) schematic symbol.**

As can be seen in Fig. 15-38(a), base 2 is connected to a positive source (V_{BB}) and base 1 is returned to ground. As a result of the (conventional) current flow from base 2 to base 1, there is a voltage gradient along the length of the n^- silicon structure. Therefore, we have a voltage in the region of the emitter junction which is positive with respect to ground.

The magnitude of this voltage is given by the simple voltage-divider action between R_{B1} and R_{B2}.

$$V = \frac{R_{B1}}{R_{B1} + R_{B2}} V_{B2B1} = \eta V_{BB} \tag{15-50}$$

The Greek letter η (eta) is called the *intrinsic stand-off ratio*. This parameter is also available on most UJT data sheets.

The voltage V defined in Eq. 15-50 is the positive potential that appears on the n-side of the emitter's p-n junction. The emitter junction is reverse biased until the voltage between the emitter terminal and base 1 becomes more positive than the voltage V. Allowing 0.5 V (at 25°C) to be the voltage drop across the p-n junction, the required emitter–base 1 voltage to just cause emitter conduction is

$$V_P = 0.5\text{ V} + \eta V_{B2B1} \tag{15-51}$$

The symbol V_P is used to represent the UJT's *peak-point voltage*.

When V_{EB1} is equal to V_P, the emitter junction becomes forward biased and holes from the emitter are injected into the n^- silicon structure. These holes travel from the emitter toward the base 1 region. The holes in the base 1 region attract electrons from ground.

Since the conductivity of any semiconductor material is a direct function of the number of electrons per unit of volume, the resistance between the emitter and ground

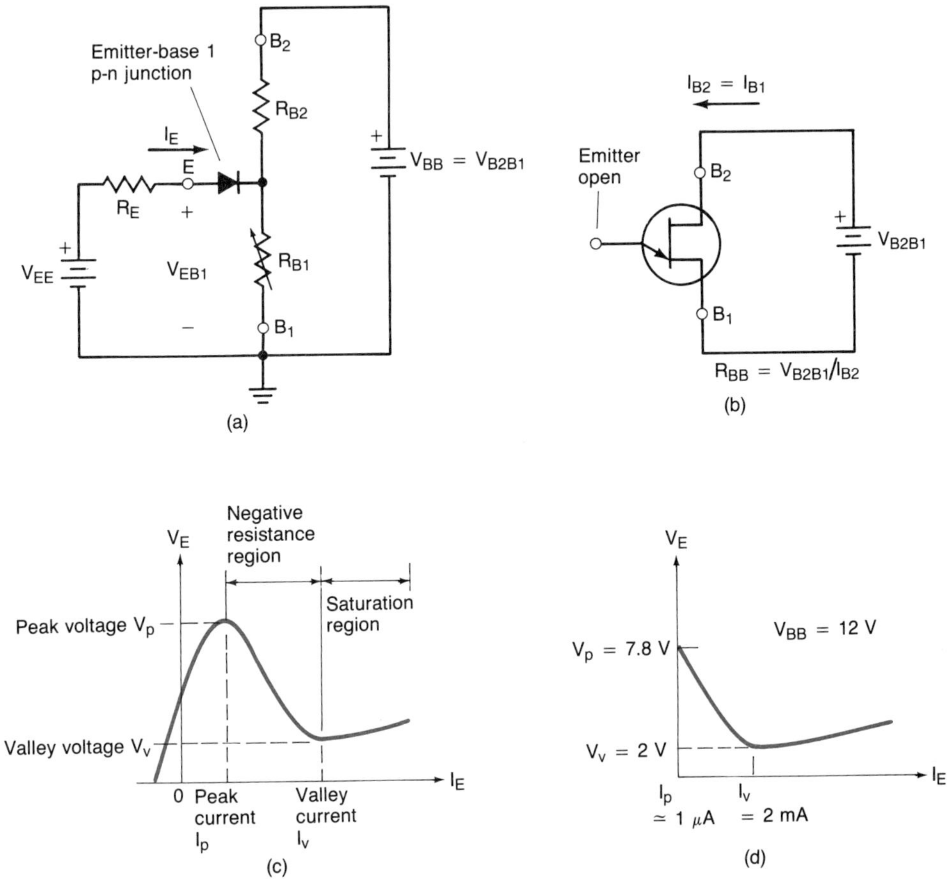

FIGURE 15-38 **UJT's characteristics: (a) UJT's electrical equivalent circuit; (b) determining the interbase resistance R_{BB}; (c) UJT's input characteristic curve; (d) actual measured values for a 2N2646.**

is greatly reduced. As a result, emitter *current rises* while the *voltage* between the emitter and base 1 terminal *decreases*. This gives the emitter–base 1 region a *negative resistance characteristic*.

After a certain point, any further *increases* in the emitter *current* will start to *increase* the *voltage* drop between the emitter and base 1. This is referred to as the UJT's *saturation region*. This will occur when the rate of hole injection is so great as to build up a "positive space charge" in the base 1 region.

The UJT's resulting emitter (or input) *V-I* characteristic curve has been given in Fig. 15-38(c). To emphasize the UJT's peak and valley, it is customary to interchange the voltage and current axes as indicated in the figure. By inspection of Fig. 15-38(c) we can see that I_P, the *peak-point current*, represents the *minimum* amount of emitter current to place the UJT in its negative resistance region. The *valley current* I_V is the *maximum* emitter current within the negative resistance region. Similarly, the *valley voltage* V_V is the *minimum* voltage that can maintain the UJT in its negative resistance region. The peak-point voltage V_P was defined in Eq. 15-51.

The actual emitter characteristic of a 2N2646 UJT was developed in the laboratory. The resulting *V-I* curve has been indicated in Fig. 15-38(d). The indicated values agree with the values found on the manufacturer's data sheet.

For a UJT to function properly in an oscillator, its dc load line must cross the negative resistance region of its emitter characteristic. This has been detailed in Fig. 15-39. Figure 15-39(a) shows the dc equivalent circuit. The dc load line has been constructed in Fig. 15-39(b).

By inspection of Fig. 15-39(c), we can determine the minimum and maximum values of R_E required. Thus

$$R_{E(\max)} = -\frac{\Delta V}{\Delta I} = -\frac{V_{BB} - V_P}{0 - I_P} = \frac{V_{BB} - V_P}{I_P}$$

and

$$R_{E(\min)} = -\frac{\Delta V}{\Delta I} = -\frac{V_{BB} - V_V}{0 - I_V} = \frac{V_{BB} - V_V}{I_V}$$

FIGURE 15-39 UJT's dc load line: (a) dc circuit; (b) dc load line; (c) minimum and maximum values of R_E required for oscillation.

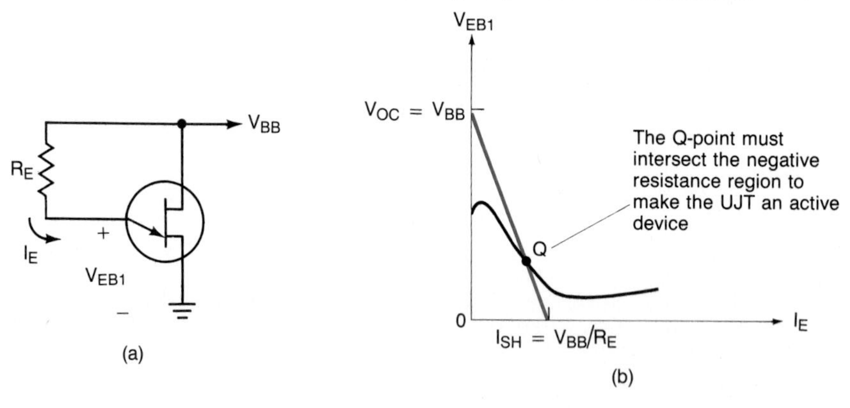

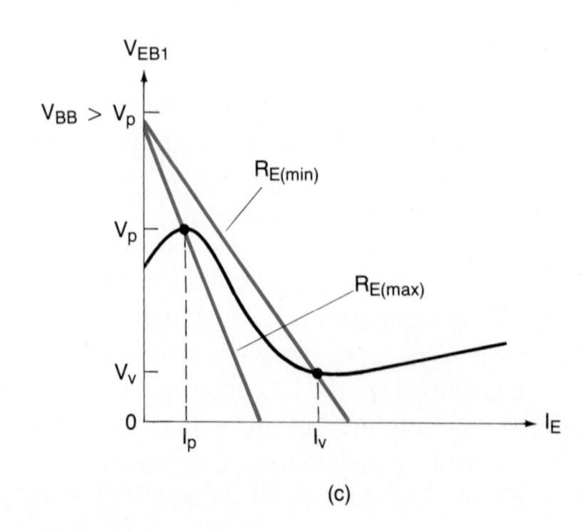

The design rule for R_E is given by

$$\boxed{\frac{V_{BB} - V_V}{I_V} \leq R_E \leq \frac{V_{BB} - V_P}{I_P}} \tag{15-52}$$

where R_E is the emitter resistance required to ensure UJT circuit oscillation.

A basic UJT relaxation oscillator circuit has been given in Fig. 15-40(a). Its operation is relatively simple. When power is first applied, capacitor C_1 will begin to charge through resistor R_E. (The emitter–base 1 diode will be reverse biased. Therefore, the emitter current will be approximately zero.) The voltage across the capacitor will attempt to reach V_{BB}. However, when the voltage across the capacitor

FIGURE 15-40 UJT relaxation oscillator: (a) practical UJT relaxation oscillator; (b) actual waveforms; (c) analyzing the capacitor's charge time (t_1) to find the approximate period (T).

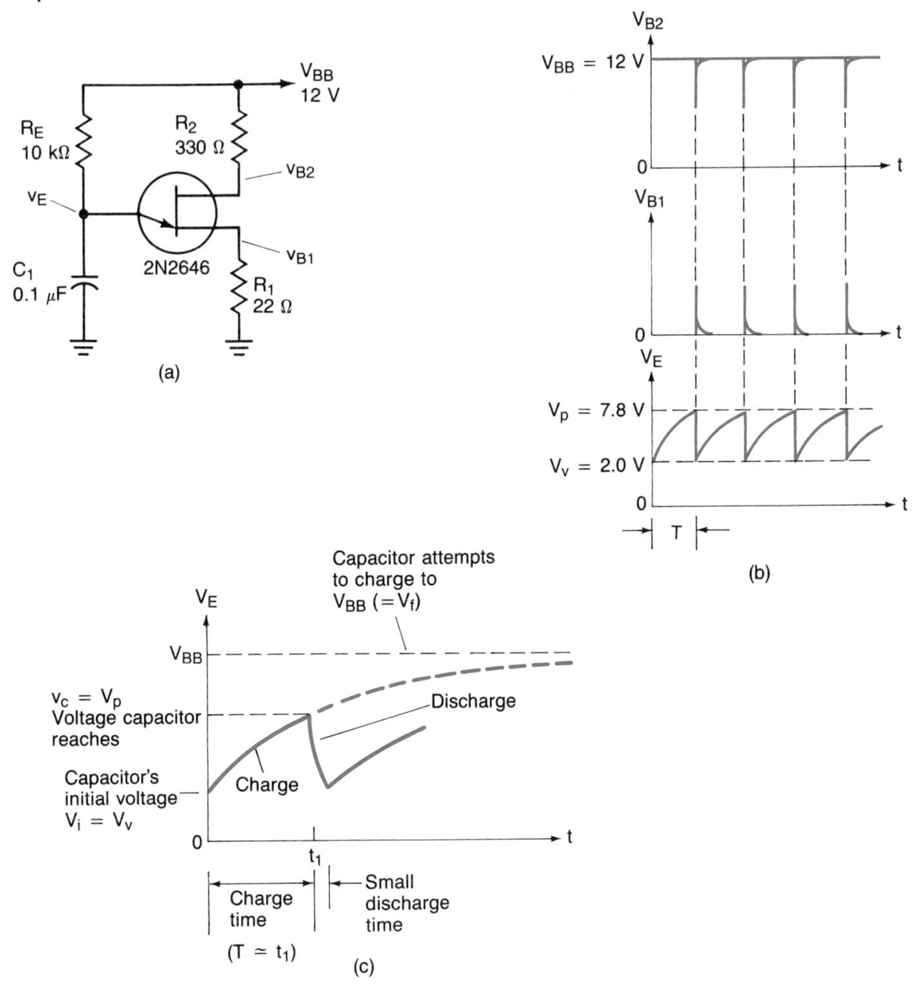

reaches V_P, the resistance between the emitter and base 1 decreases. Simply, the UJT's emitter goes into conduction, or "fires." The sudden increase in the emitter current will serve to discharge the capacitor quickly.

During conduction, the capacitor serves as the driving potential. The capacitor's discharge current is limited by the base 1 resistor R_1, and the dynamic resistance of the emitter diode. When the capacitor discharges to the point where the voltage across it is approximately equal to V_V, the emitter diode stops conducting, and becomes open again. At this point the capacitor will again start to charge toward V_{BB}, and the cycle is repeated.

The charge and discharge of capacitor C_1 produces a sawtooth type of waveform. Voltage "spikes" will be produced at the UJT's base 1 and base 2 terminals. Actual waveforms are indicated in Fig. 15-40(b).

As can be seen in Fig. 15-40(b), the period of the voltage waveform across C_1 is approximately equal to the capacitor's charge time. The equation describing this charging action is

$$v_c = V_I + (V_F - V_I)(1 - e^{-t_1/\tau}) \qquad (15\text{-}53)$$

where v_c = instantaneous capacitor voltage
V_I = initial capacitor voltage
V_F = final voltage to which the capacitor attempts to charge
e = natural number 2.718 · · ·
τ = (tau) the *RC* time constant

For our problem we are interested in finding the length of time required for the capacitor to charge from V_V to V_P. The capacitor is attempting to charge to V_{BB} [see Fig. 15-40(c)]. Hence, from Eq. 15-53,

$$V_P = V_V + (V_{BB} - V_V)(1 - e^{-t_1/\tau})$$

We must solve this relationship for t_1 since it is approximately equal to the period (T) of the output waveform. This will give us an equation for the frequency of oscillation.

$$(1 - e^{-t_1/\tau}) = \frac{V_P - V_V}{V_{BB} - V_V}$$

In a "real" UJT circuit,

$$V_P > V_V \quad \text{and} \quad V_{BB} >> V_V$$

$$1 - e^{-t_1/\tau} = \frac{V_P - V_V}{V_{BB} - V_V} \simeq \frac{V_P}{V_{BB}}$$

To simplify our work further, we can state that

$$V_P = 0.5\text{ V} + \eta V_{B2B1} \simeq \eta V_{B2B1} \simeq \eta V_{BB}$$

The resistors in series with the base 1 and base 2 terminals are very small [refer to Fig. 15-40(a)]. Consequently, V_{B2B1} is approximately equal to V_{BB}. Thus

$$1 - e^{-t_1/\tau} \simeq \frac{V_P}{V_{BB}} \simeq \frac{\eta V_{BB}}{V_{BB}} = \eta$$

$$-e^{-t_1/\tau} \simeq \eta - 1$$

Multiplying both sides by -1 gives us

$$e^{-t_1/\tau} \simeq 1 - \eta$$

Inverting both sides gives us

$$e^{t_1/\tau} \simeq \frac{1}{1 - \eta}$$

To solve for t_1 we must take the natural logarithm of both sides.

$$\frac{t_1}{\tau} \simeq \ln \frac{1}{1 - \eta}$$

$$t_1 \simeq T \simeq \tau \ln \frac{1}{1 - \eta}$$

The time constant τ is given by

$$\tau = R_E C_1$$

and

$$T \simeq R_E C_1 \ln \frac{1}{1 - \eta}$$

$$\boxed{f_o = \frac{1}{T} \simeq \frac{1}{R_E C_1 \ln \left[\dfrac{1}{(1 - \eta)} \right]}} \qquad (15\text{-}54)$$

where f_o is the UJT relaxation oscillator's oscillation frequency.

EXAMPLE 15-23

Analyze the UJT relaxation oscillator given in Fig. 15-40(a). Specifically, verify that R_E is properly sized, and determine f_o.

SOLUTION To check the value of R_E, we must have UJT data. The indicated 2N2646 has the following parameters:

$$V_P = 7.8 \text{ V} \qquad V_V = 2 \text{ V}$$

$$I_P = 1 \ \mu\text{A} \qquad I_V = 2 \text{ mA}$$

From Eq. 15-52,

$$\frac{V_{BB} - V_V}{I_V} \leq R_E \leq \frac{V_{BB} - V_P}{I_P}$$

$$\frac{12 \text{ V} - 2 \text{ V}}{2 \text{ mA}} \leq R_E \leq \frac{12 \text{ V} - 7.8 \text{ V}}{2 \ \mu\text{A}}$$

$$5 \text{ k}\Omega \leq R_E \leq 4.2 \text{ M}\Omega$$

Therefore, an R_E of 10 kΩ appears to be a reasonable value. The circuit will oscillate. The η of the 2N2646 is 0.6. Now we may use Eq. 15-54 to find f_o.

$$f_o \simeq \frac{1}{R_E C_1 \ln \left[\dfrac{1}{(1 - \eta)} \right]}$$

$$\simeq \frac{1}{(10 \text{ k}\Omega)(0.1 \ \mu\text{F}) \ln \left[\dfrac{1}{(1 - 0.6)} \right]}$$

$$\simeq 1091 \text{ Hz}$$

■

The actual measured waveforms have been given in Fig. 15-40(b). Their frequency was 863 Hz. This is about 21% lower than our calculated value. Considering our approximations and the tolerances of R_E and C_1, this is a reasonable result. If the frequency is to be trimmed, "R_E" should be a fixed resistor (e.g., 9.1 kΩ) in series with a potentiometer (e.g., a 2-kΩ unit).

15-17 An Op Amp Relaxation Oscillator

The op amp relaxation oscillator shown in Fig. 15-41(a) is a square-wave generator. In general, square waves are relatively easy to produce. Like the UJT relaxation oscillator, the circuit's frequency of oscillation is dependent on the charge and discharge of a capacitor (C_1) through a resistor (R_1). The "heart" of the oscillator is an inverting op amp comparator [refer to Fig. 15-41(b)]. The comparator uses *positive* feedback.

Positive feedback increases the gain of an amplifier. In a comparator circuit this offers two advantages. First, the high gain causes the op amp's output to switch

FIGURE 15-41 Op amp relaxation oscillator: (a) circuit; (b) the "heart" of the oscillator is an inverting op amp comparator.

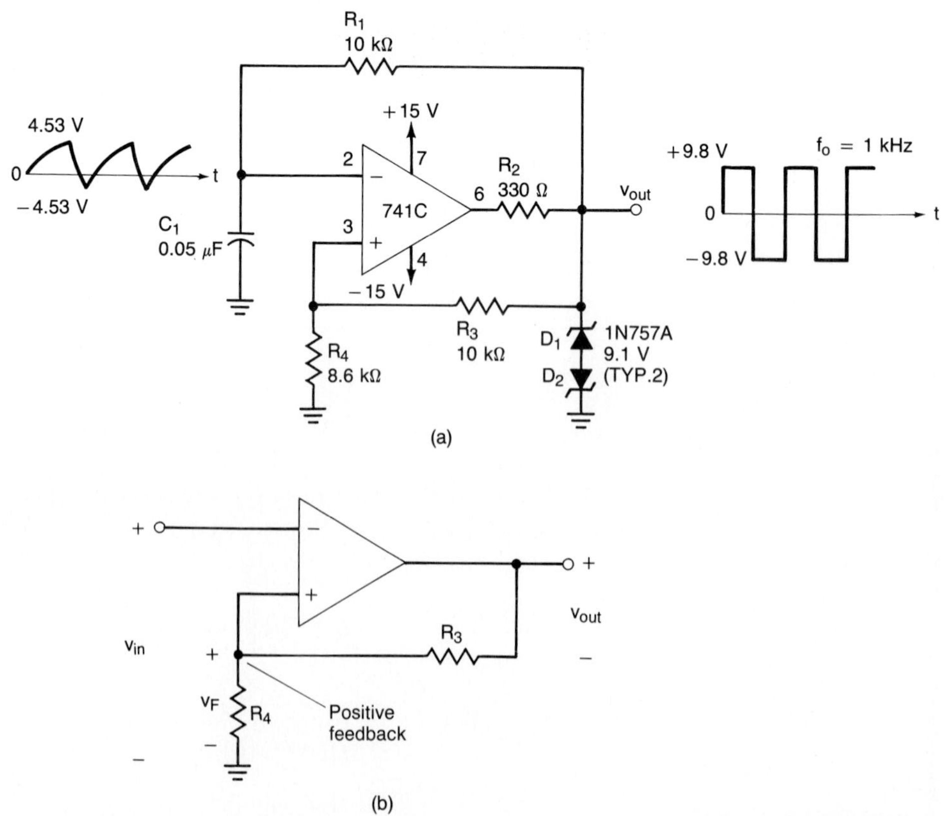

very quickly from rail to rail. Second, the use of positive feedback gives the circuit *hysteresis*. Since the comparator is fundamental to the operation of the square-wave generator, and an important circuit in its own right, it deserves our serious attention. The operation of the inverting comparator has been detailed in Fig. 15-42.

The comparator is a nonlinear circuit. Its output assumes one of two states. It is either at the positive supply rail ($+V_{SAT}$) or at the negative supply rail ($-V_{SAT}$).

FIGURE 15-42 Operation of the inverting comparator and its composite transfer characteristic curve (e).

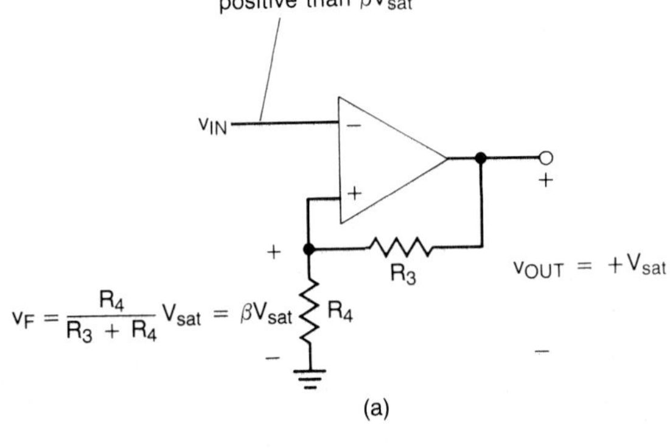

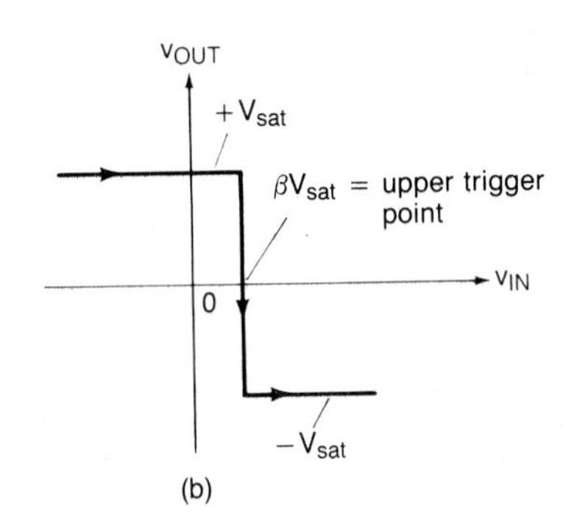

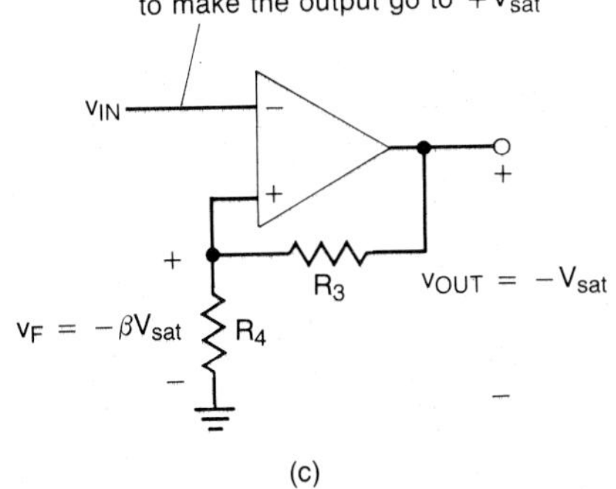

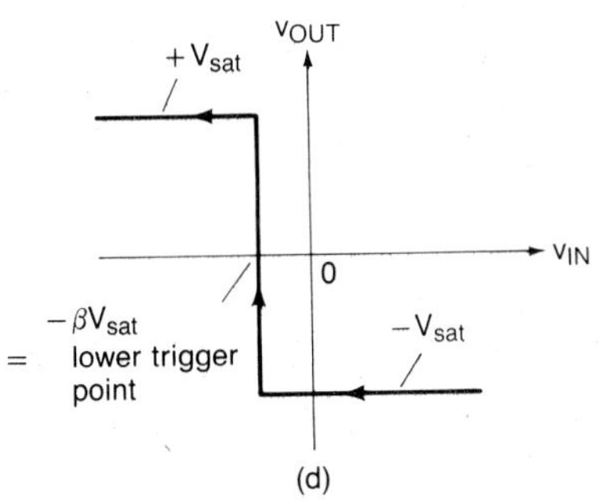

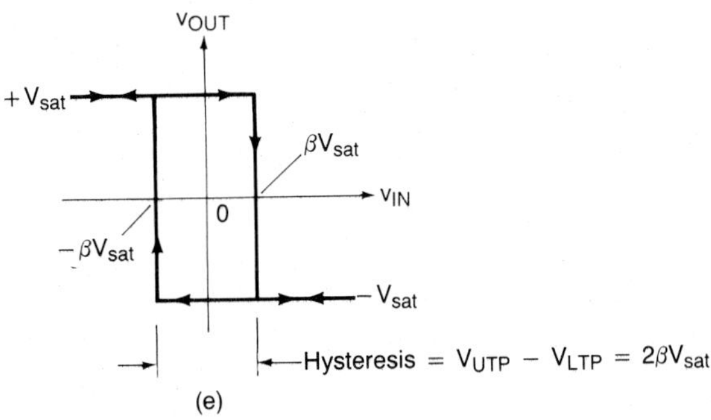

In Fig. 15-42(a) we have assumed that the output is at $+V_{SAT}$. Since the op amp's noninverting input terminal draws negligible current, we may find the feedback voltage v_F by simple voltage division.

$$v_F = \frac{R_4}{R_3 + R_4} v_{OUT} = \frac{R_4}{R_3 + R_4}(+V_{SAT})$$

Let

$$\beta = \frac{R_4}{R_3 + R_4}$$

and we arrive at

$$v_F = \beta(+V_{SAT}) \tag{15-55}$$

As we can see in Fig. 15-42(a), to get the op amp to switch from $+V_{SAT}$ to $-V_{SAT}$, v_{IN} must be slightly more positive than v_F. This particular value of v_{IN} is called the comparator's *upper trigger point* V_{UTP}. A graph of the input-output relationship is given in Fig. 15-42(b).

Once the output of the op amp has switched to $-V_{SAT}$, a negative (feedback) voltage appears at the op amp's noninverting input terminal. Hence

$$v_F = \beta(-V_{SAT}) \tag{15-56}$$

Figure 15-42(c) illustrates that v_{IN} must now be more negative than v_F to get the op amp's output to switch back to $+V_{SAT}$. This particular value of v_{IN} is called the comparator's *lower trigger point* V_{LTP}. A graph of this input-output relationship is provided in Fig. 15-42(d).

The composite transfer characteristic curve is shown in Fig. 15-42(e). If the output is positive, v_{IN} must be increased to a value slightly greater than the upper trigger point V_{UTP}. When this occurs the output will switch to $-V_{SAT}$. To get the output to switch back to $+V_{SAT}$, v_{IN} must be made slightly more negative than the lower trigger point V_{LTP}.

The difference between V_{UTP} and V_{LTP} is called the comparator's *hysteresis* H [see Fig. 15-42(e) and Eq. 15-57].

$$\boxed{H = V_{UTP} - V_{LTP}} \tag{15-57}$$

where H = hysteresis
V_{UTP} = upper trigger point = $\beta(+V_{sat})$
V_{LTP} = lower trigger point = $\beta(-V_{sat})$

An op amp's output saturation voltages ($+V_{SAT}$ and $-V_{SAT}$) are not well defined. We can more precisely determine the output voltage by using back-to-back zener diodes [see Fig. 15-41(a)]. When the op amp's output switches to $+V_{SAT}$, zener diode D_1 enters breakdown and zener diode D_2 becomes forward biased. Similarly, when the op amp's output switches to $-V_{SAT}$, zener diode D_2 enters breakdown and zener diode D_1 becomes forward biased. Therefore, the output voltage becomes

$$v_{OUT} = \pm(V_Z + 0.7\text{ V})$$

and

$$v_F = \beta(V_Z + 0.7\text{ V})$$

Therefore, not only does the output voltage become more well defined, but so do the comparator's upper and lower trigger points.

Resistor R_2 is included to limit the op amp's output current. Since most op amps include output short-circuit protection, its use is optional and falls in the "good practice" category. Now that we understand the operation of the inverting comparator, let us consider the operation of the square-wave generator [Fig. 15-41(a)].

When the output of the comparator is positive, capacitor C_1 will charge through resistor R_1. The capacitor will attempt to charge to v_{OUT}.

$$v_{OUT} = V_Z + 0.7 \text{ V}$$

When the voltage across the capacitor reaches the upper trigger point,

$$V_{UTP} = v_F = \beta v_{OUT} = \beta(V_Z + 0.7 \text{ V})$$

and the comparator's output will immediately switch negative.

$$v_{OUT} = -(V_Z + 0.7 \text{ V})$$

The capacitor will then start to charge from the positive upper trigger point voltage toward the negative output voltage.

When the voltage across the capacitor reaches the lower trigger point,

$$V_{LTP} = v_F = \beta v_{OUT} = -\beta(V_Z + 0.7 \text{ V})$$

the output will again go positive, and the cycle repeats. The waveforms associated with the circuit have been given in Fig. 15-41(a).

To derive an equation for the output frequency, we shall draw on Eq. 15-53. With reference to Fig. 15-41(a), the substitutions below to find the charge time t_1 become clear.

$$\begin{aligned} v_c &= V_I + (V_F - V_I)(1 - e^{-t_1/\tau}) \\ \beta v_{OUT} &= -\beta v_{OUT} + [v_{OUT} - (-\beta v_{OUT})][1 - e^{-t_1/\tau}] \\ &= -\beta v_{OUT} + (v_{OUT} + \beta v_{OUT})(1 - e^{-t_1/\tau}) \end{aligned}$$

Dividing both sides by v_{OUT}, we have

$$\begin{aligned} \beta &= -\beta + (1 + \beta)(1 - e^{-t_1/\tau}) \\ 2\beta &= (1 + \beta)(1 - e^{-t_1/\tau}) \end{aligned}$$

Continuing, we have

$$\begin{aligned} 1 - e^{-t_1/\tau} &= \frac{2\beta}{1 + \beta} \\ -e^{-t_1/\tau} &= \frac{2\beta}{1 + \beta} - 1 = \frac{2\beta - (1 + \beta)}{1 + \beta} = \frac{\beta - 1}{1 + \beta} \end{aligned}$$

Multiplying both sides by -1 yields

$$e^{-t_1/\tau} = \frac{1 - \beta}{1 + \beta}$$

Inverting both sides gives us

$$e^{t_1/\tau} = \frac{1 + \beta}{1 - \beta}$$

We may solve for t_1 by taking the natural log of both sides.

$$\frac{t_1}{\tau} = \ln \frac{1 + \beta}{1 - \beta}$$

$$t_1 = \tau \ln \frac{1+\beta}{1-\beta} = R_1C_1 \ln \frac{1+\beta}{1-\beta}$$

By inspection of Fig. 15-41(a), we can see that the charge time (t_1) from the lower trigger point to the upper trigger point is equal to the charge time from the upper trigger point to the lower trigger point. Alternatively, we can state that t_1 is equal to one-half of the period. Hence

$$T = 2t_1 = 2R_1C_1 \ln \frac{1+\beta}{1-\beta}$$

$$\boxed{f_o = \frac{1}{T} = \frac{1}{2R_1C_1 \ln [(1+\beta)/(1-\beta)]}} \qquad (15\text{-}58)$$

where f_o is the output frequency of the op amp square-wave generator.

Recall that β is determined by simple voltage division between R_3 and R_4. We can simplify Eq. 15-58 if we implement the following constraint:

$$R_4 = 0.859R_3$$

Under this condition β is equal to 0.462, and

$$\ln \frac{1+\beta}{1-\beta} = 1$$

Therefore, Eq. 15-58 simplifies to

$$\boxed{f_o = \left.\frac{1}{2R_1C_1}\right|_{R_4=0.859R_3}} \qquad (15\text{-}59)$$

where f_o is the output frequency of the op amp square-wave generator.

EXAMPLE 15-24

Analyze the op amp square-wave generator given in Fig. 15-41(a). Determine the square-wave peak-to-peak output voltage. Is R_2 adequately sized? (Assume that $+V_{SAT}$ is 13 V and I_{SC} is 10 mA.) Do R_3 and R_4 have the proper relationship? If so, use Eq. 15-59 to determine f_o. Finally, determine the peak-to-peak value of the voltage across capacitor C_1.

SOLUTION Since the 1N757 is a 9.1-V zener diode, the positive peak output voltage will be 9.8 V. Similarly, the negative peak output voltage will be -9.8 V, and the resulting peak-to-peak output voltage will be 19.6 V. Specifically,

$$v_{OUT(peak\text{-}to\text{-}peak)} = 2(V_Z + 0.7\text{ V}) = 2(9.8\text{ V}) = 19.6\text{ V}$$

Now we shall investigate R_2. If $+V_{SAT}$ is 13 V, the op amp's output current I will be

$$I = \frac{+V_{SAT} - (V_Z + 0.7\text{ V})}{R_2} = \frac{13\text{ V} - 9.8\text{ V}}{330\ \Omega}$$

$$= 9.70\text{ mA} < I_{SC} = 10\text{ mA}$$

Therefore, R_2 appears to be adequately sized. Now we shall determine if R_3 and R_4 have the proper relationship to draw on Eq. 15-59.

$$R_4 = 0.859R_3$$

$$\frac{R_4}{R_3} = 0.859$$

$$\frac{R_4}{R_3} = \frac{8.6\ \text{k}\Omega}{10\ \text{k}\Omega} = 0.86 \simeq 0.859$$

Since the constraint on R_3 and R_4 appears to be satisfied, we may use Eq. 15-59 to determine f_o.

$$f_o = \frac{1}{2R_1C_1} = \frac{1}{2(10\ \text{k}\Omega)(0.1\ \mu\text{F})} = 1\ \text{kHz}$$

To determine the peak-to-peak voltage across capacitor C_1, we first recognize that it will be equal to the comparator's hysteresis.

$$\beta = \frac{R_4}{R_3 + R_4} = \frac{8.6\ \text{k}\Omega}{10\ \text{k}\Omega + 8.6\ \text{k}\Omega} = 0.4624$$

$$v_F = \pm\beta v_{OUT} = \pm\beta(V_Z + 0.7\ \text{V})$$
$$= \pm(0.4624)(9.8\ \text{V}) = \pm 4.53\ \text{V} \rightarrow 9.06\ \text{V p-p}$$ ■

We conclude our discussion of the op amp square-wave generator by offering some additional "real-world" considerations. The circuit works reasonably well over the audio-frequency range. To decrease its frequency of oscillation, C_1 and/or R_1 must be increased. However, C_1 cannot be an electrolytic unit. This is true because electrolytics are polarized, and the capacitor must charge to a positive voltage (V_{UTP}) and to a negative voltage (V_{LTP}). Increasing R_1 to lower f_o also reduces the available current to charge C_1. If R_1 is too large, the op amp's input bias current will become significant. This can result in a radical departure from our theoretical calculations.

The maximum oscillation frequencies are limited by the op amp's slew rate. Additional time delays occur as the op amp pulls out of saturation. Both the low- and the high-frequency restrictions can be circumvented by using a reasonably good op amp. Do *not* expect superior performance from the 741C op amp.

15-18 An Op Amp Triangle Generator

The op amp triangle generator is another example of a relaxation oscillator. The oscillator incorporates two stages: a *noninverting comparator with hysteresis* and an *inverting integrator*. As we can see in Fig. 15-43(a), the circuit simultaneously provides two different output waveforms. The comparator's output is a square wave, while the output of the integrator is a triangle wave.

We shall begin our analysis by first examining the noninverting comparator [see Fig. 15-43(b)]. The circuit is nonlinear and incorporates positive feedback. When v_{IN} is positive enough, the noninverting input will be at a positive value. This will send the op amp's output to the positive supply rail ($+V_{SAT}$). The output will remain at $+V_{SAT}$ even when v_{IN} falls, and then goes negative. This is true because of the voltage-divider action between R_1 and R_2. The large positive output voltage of $+V_{SAT}$ will continue to hold the noninverting input terminal positive even for small negative values of v_{IN}.

When v_{IN} goes negative enough, the noninverting input terminal will become negative and send the output to the negative rail $(-V_{SAT})$. The output will remain at $-V_{SAT}$ even for small positive values of v_{IN}. Once again, the voltage divider action offered by R_1 and R_2 tends to hold the noninverting input terminal negative.

Since the circuit is nonlinear, the superposition theorem should *not* be used. However, we can apply Ohm's law and Kirchhoff's voltage law to analyze the circuit. The equivalent circuit is shown in Fig. 15-43(b). We have assumed that the op amp's noninverting input terminal draws a negligibly small input current.

First, we shall find I.

$$I = \frac{+V_{SAT} - V_{IN}}{R_1 + R_2}$$

FIGURE 15-43 Op amp triangle wave generator and its noninverting comparator: (a) basic triangle-wave generator; (b) analyzing its noninverting comparator; (c) the output will be at $+V_{SAT}$ as long as V_{IN} is more positive than $-(R_1/R_2)(+V_{SAT})$; (d) the output will be at $-V_{SAT}$ as long as v_{IN} is more negative than $-(R_1/R_2)(+V_{SAT})$; (e) composite transfer characteristic curve.

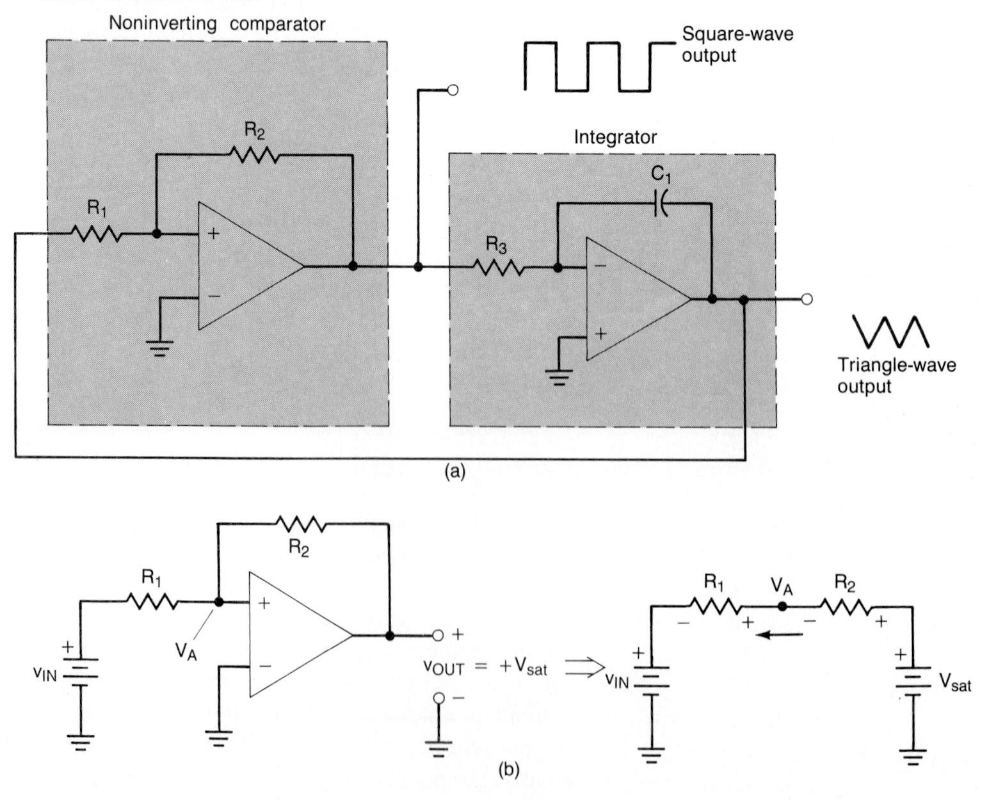

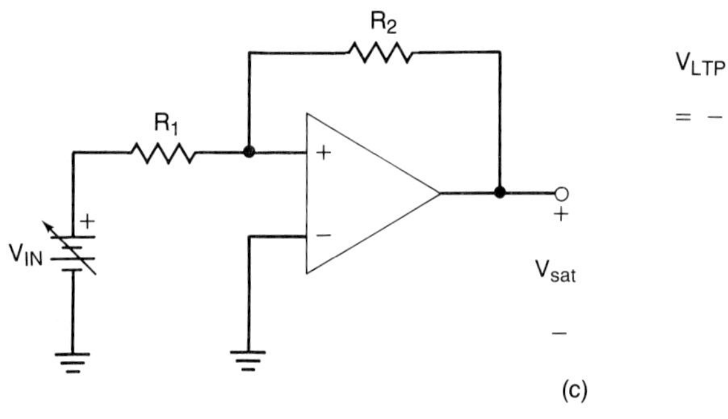

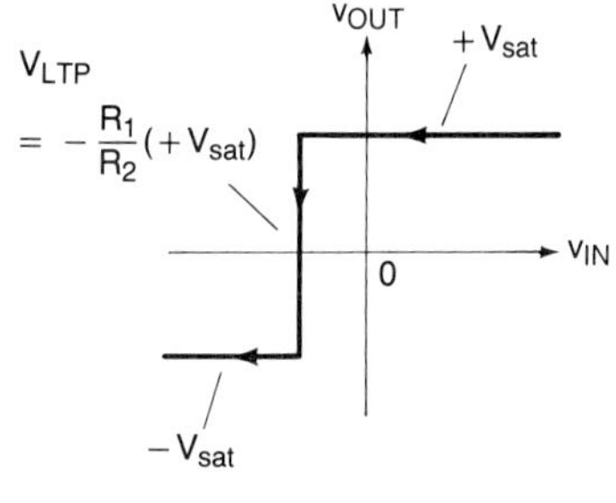

(c)

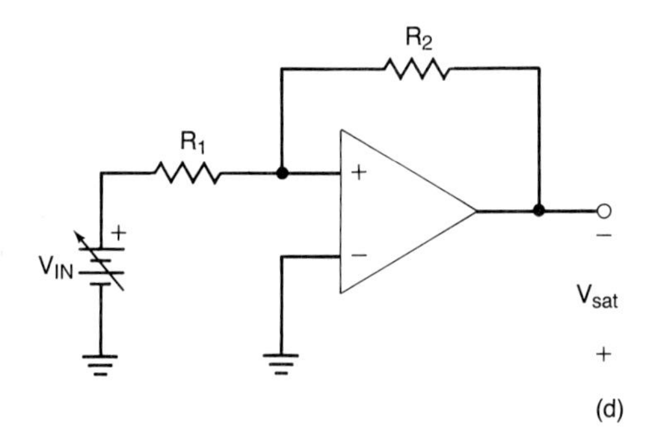

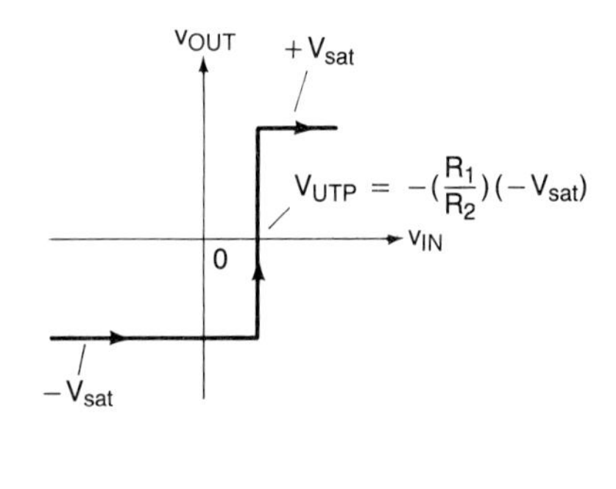

(d)

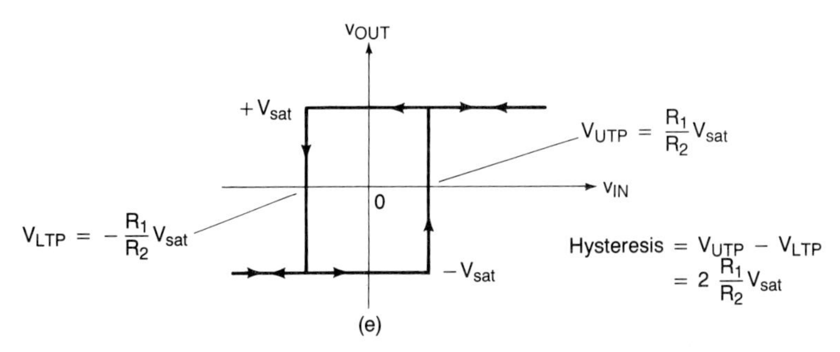

(e)

FIGURE 15-43 ***(continued)***

Now we shall find V_A (the voltage at the noninverting input) by using Kirchhoff's voltage law.

$$V_A = +V_{\text{SAT}} - IR_2$$

By substitution we obtain

$$V_A = +V_{\text{SAT}} - \frac{(+V_{\text{SAT}} - V_{\text{IN}})R_2}{R_1 + R_2}$$

The comparator's output will switch from high ($+V_{\text{SAT}}$) to low ($-V_{\text{SAT}}$) when V_A becomes slightly less than zero. Expanding and setting $V_A < 0$, we can solve for the required value of V_{IN}.

$$+V_{SAT} - \frac{R_2}{R_1 + R_2}(+V_{SAT}) + \frac{R_2}{R_1 + R_2} V_{IN} = V_A < 0$$

$$\frac{R_2}{R_1 + R_2} V_{IN} < \frac{R_2}{R_1 + R_2}(+V_{SAT}) - (+V_{SAT})$$

$$V_{IN} < +V_{SAT} - \frac{R_1 + R_2}{R_2}(+V_{SAT})$$

$$V_{IN} < \frac{R_2 - R_1 - R_2}{R_2}(+V_{SAT}) = -\frac{R_1}{R_2}(+V_{SAT})$$

The result above tells us how negative V_{IN} must be to get the output to switch from $+V_{SAT}$ to $-V_{SAT}$. This is the comparator's lower trigger point [see Fig. 15-43(c)]. By symmetry, we can find the upper trigger point,

$$V_{IN} > -\frac{R_1}{R_2}(-V_{SAT})$$

to get the input voltage V_{IN} positive enough to cause the output to switch from $-V_{SAT}$ to $+V_{SAT}$ [see Fig. 15-43(d)]. The transfer characteristic curve summarizes the operation of the noninverting comparator [see Fig. 15-43(e)].

Since the comparator's trigger points are determined by the op amp's $+V_{SAT}$ and $-V_{SAT}$ output, some degree of precision is lost. However, by once again using zener diodes, it becomes possible to define the trip points more precisely. The use of output limiting, and the resulting transfer characteristic curve, is shown in Fig. 15-44.

FIGURE 15-44 Improving the comparator's performance.

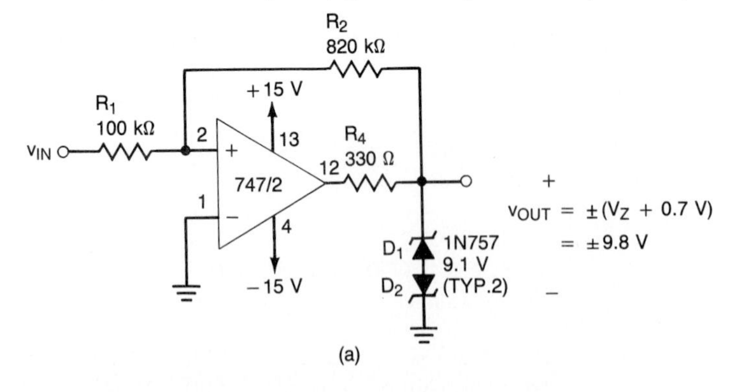

(a)

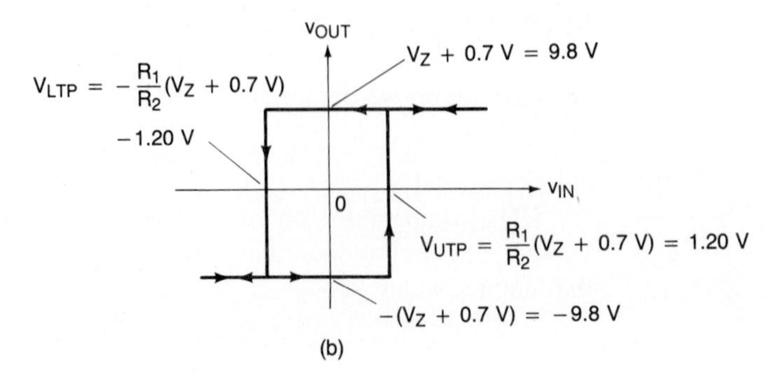

(b)

EXAMPLE 15-25

Sketch the transfer characteristic curve of the noninverting comparator given in Fig. 15-44(a). Also determine the hysteresis.

SOLUTION The general shape of the curve has been given in Fig. 15-44(b). Therefore, our task is to determine the peak output voltages and the upper and lower trigger points. Since the output is clipped by a pair of back-to-back 9.1-V zeners, we recall from Example 15-24 that the output is ± 9.8 V. If v_{OUT} is -9.8 V, v_{IN} must become positive enough to cause the output to switch. Hence

$$v_{IN} = V_{UTP} = \frac{R_1}{R_2}(V_Z + 0.7\text{ V}) = \frac{100\text{ k}\Omega}{820\text{ k}\Omega}(9.8\text{ V}) = 1.20\text{ V}$$

In a similar fashion, the positive output voltage will be 9.8 V, and the lower trigger point is

$$v_{IN} = V_{UTP} = -\frac{R_1}{R_2}(V_Z + 0.7\text{ V}) = -\frac{100\text{ k}\Omega}{820\text{ k}\Omega}(9.8\text{ V}) = -1.20\text{ V}$$

The hysteresis is the difference between the upper and lower trigger points.

$$H = V_{UTP} - V_{LTP} = 1.20\text{ V} - (-1.20\text{ V}) = 2.40\text{ V}$$ ■

Now let us consider the inverting integrator. It has been redrawn in Fig. 15-45(a). Notice that the input voltage has been shown to be a (constant) dc level V_{IN}. The right end of resistor R_3 is at virtual ground. Therefore, all of V_{IN} is dropped across R_3. The resulting current is

$$I = \frac{V_{IN}}{R_3}$$

Since V_{IN} and R_3 are constants, it follows that I will also be constant.

Recall that the op amp's output terminals draw negligible current. Therefore, all of the current I flows into the feedback loop. Consequently, the capacitor (C_1) is being charged by a constant current. To find the voltage developed across the capacitor, we must remind ourselves of two fundamental relationships. First,

$$I = \frac{Q}{t}$$

Therefore,

$$Q = It$$

Let us consider the meaning of this relationship. If the current I is a constant, the total charge is going to increase linearly with time. The second equation we must draw on describes the charge-voltage relationship of a capacitor. Specifically,

$$Q = C_1 v_{C1}$$

Solving for v_{C1} gives us

$$v_{C1} = \frac{Q}{C_1}$$

By substitution,

$$v_{C1} = \frac{Q}{C_1} = \frac{I}{C_1}t$$

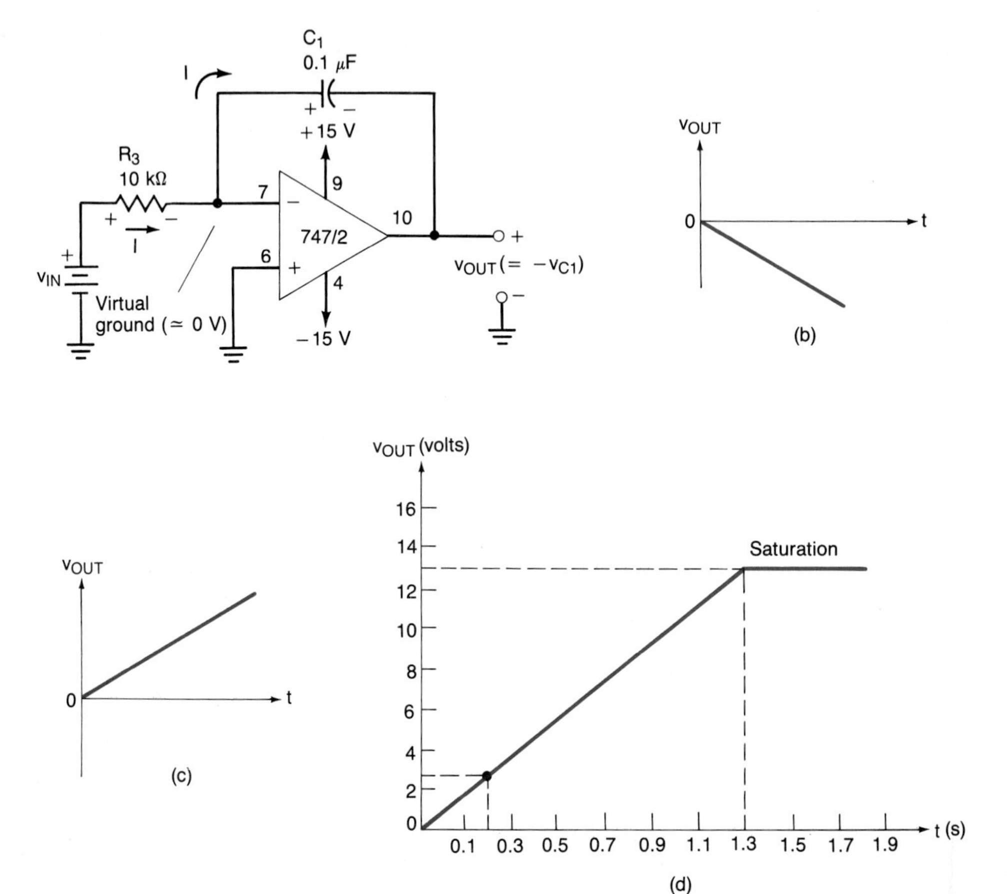

FIGURE 15-45 Op amp integrator: (a) circuit; (b) a positive V_{IN} produces a negative-going ramp; (c) a negative V_{IN} produces a positive-going ramp; (d) integrator's output for a V_{IN} of −10 mV.

and for our op amp integrator,

$$v_{C1} = \frac{I}{C_1}t = \frac{V_{IN}/R_3}{C_1}t = \frac{V_{IN}}{R_3C_1}t$$

Turning again to Fig. 15-45(a), we see that the integrator's output voltage is equal to $-v_{C1}$. This leads us to

$$\boxed{v_{OUT} = -v_{C1} = -\frac{V_{IN}}{R_3C_1}t} \qquad (15\text{-}60)$$

where v_{OUT} is the integrator's output voltage for a constant input voltage (v_{IN}).

Since V_{IN}, R_3, and C_1 are constants, the output voltage will decrease linearly with time. If V_{IN} is a negative input voltage, the output voltage will increase linearly with time. Graphs of the integrator's output voltage for positive and negative input voltages have been given in Fig. 15-45(b) and (c), respectively.

The fact that a constant input voltage produces a linear ramp in the output voltage of an integrator is fundamental to the operation of the triangle generator. However, the op amp integrator is capable of providing an output voltage that is equal to the mathematical integral of the input voltage. This makes the op amp integrator an extremely useful "building block" in many electronic systems. The argument above used to explain the generation of a voltage ramp is actually a *special case*.

In calculus it is learned that mathematical integration allows us to find the area under a curve. If the input voltage is a constant, the area under it increases linearly with time. Therefore, a graph of the integral of a constant is a ramp. Just to familiarize the student with the mathematical notation used to describe the op amp inverting integrator, we present

$$v_{OUT} = -\frac{1}{R_3C_1}\int_0^t v_{IN}\,dt \tag{15-61}$$

We should point out that Eq. 15-61 assumes that the capacitor is uncharged at $t = 0$.

EXAMPLE 15-26

Given the op amp inverting integrator shown in Fig. 15-45(a), find the equation for v_{OUT}. Then assume that v_{IN} is -10 mV, and find v_{OUT} at 0.1 s and at 0.2 s. The integrator's output voltage is initially zero. How long will it take the integrator's output to saturate?

SOLUTION Since R_3 is 10 kΩ, and C_1 is 0.1 μF, we see that

$$v_{OUT} = -\frac{V_{IN}}{R_3C_1}t = -\frac{V_{IN}}{(10\text{ k}\Omega)(0.1\ \mu\text{F})}t = -1000V_{IN}\,t$$

If V_{IN} is -10 mV and t is 0.1 s,

$$\begin{aligned} v_{OUT} &= -1000V_{IN}\,t = -(1000)(-10\text{ mV})(0.1\text{ s}) \\ &= 1\text{ V} \end{aligned}$$

and in 0.2 s,

$$\begin{aligned} v_{OUT} &= -1000V_{IN}\,t = -(1000)(-10\text{ mV})(0.2\text{ s}) \\ &= 2\text{ V} \end{aligned}$$

Assuming that $+V_{SAT}$ is 13 V, we may solve for the time to reach saturation.

$$v_{OUT} = \frac{V_{IN}}{R_3C_1}t = +V_{SAT}$$

$$\begin{aligned} t &= \frac{+V_{SAT}}{-V_{IN}}R_3C_1 = \frac{13\text{ V}}{-(-10\text{ mV})}(10\text{ k}\Omega)(0.1\ \mu\text{F}) \\ &= (1300)(1\text{ ms}) = 1.3\text{ s} \end{aligned}$$

The integrator's output has been sketched in Fig. 15-45(d). ■

Now that we understand the noninverting comparator and the inverting integrator, let us analyze a practical op amp triangle generator. One such circuit is given in Fig. 15-46(a).

Assume that the output of the comparator is negative. Its constant negative output level will cause the integrator to ramp in a positive direction. When the output of the integrator becomes positive enough (equal to V_{UTP}), the comparator's output will switch to its positive value. Its constant positive output level will cause the integrator's output to ramp in a negative direction. As the integrator's output ramps down from V_{UTP}, it will eventually reach V_{LTP}. When this occurs the comparator's output will switch back to its negative output level, and the cycle repeats. The waveforms have been shown in Fig. 15-46(b).

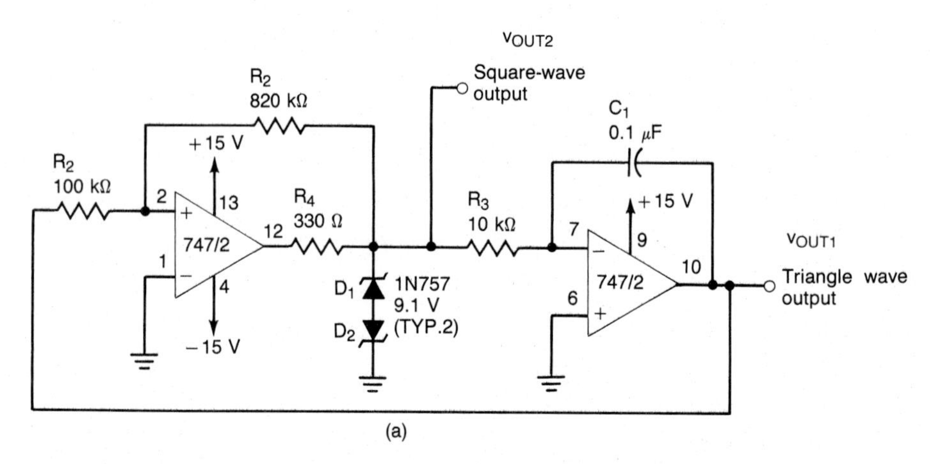

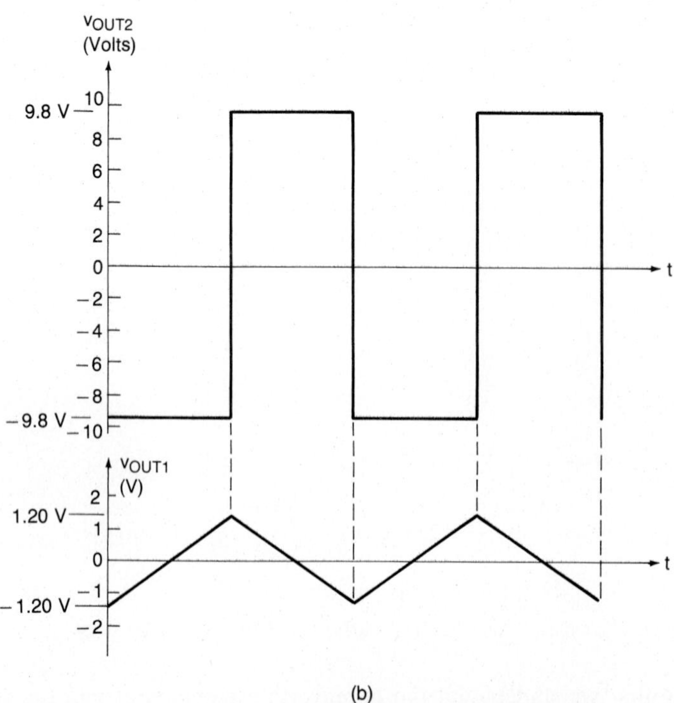

FIGURE 15-46 Practical triangle-waveform generator: (a) circuit; (b) output waveforms (f_o = 2.05 kHz); (c) integrator's output.

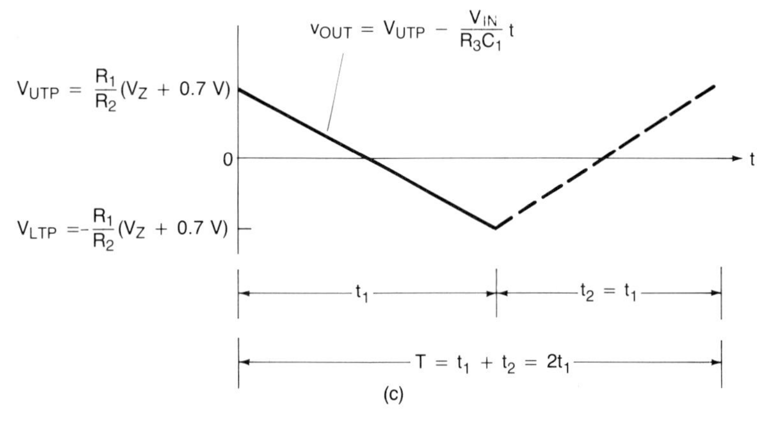

FIGURE 15-46 *(continued)*

Now we determine the formula for the frequency of oscillation of the triangle generator [see Fig. 15-46(c)]. The output of the integrator ramps down from the V_{UTP} to the V_{LTP} and then back to the V_{UTP}. From the figure we see that

$$V_{UTP} = \frac{R_1}{R_2}(V_Z + 0.7\text{ V})$$

and

$$V_{LTP} = -\frac{R_1}{R_2}(V_Z + 0.7\text{ V})$$

The output of the comparator is clamped by a pair of back-to-back zener diodes. The output of the comparator is the "V_{IN}" of the integrator. Hence

$$V_{IN} = V_Z + 0.7\text{ V}$$

and this positive input voltage causes the output of the integrator to ramp in a negative direction from V_{UTP}. The equation that describes the integrator's output voltage is

$$v_{OUT} = V_{UTP} - \frac{V_{IN}}{R_3C_1}t$$

If we substitute in our equations for V_{UTP} and V_{IN}, we arrive at the following result:

$$v_{OUT} = V_{UTP} - \frac{V_{IN}}{R_3C_1}t = \frac{R_1}{R_2}(V_Z + 0.7\text{ V}) - \frac{V_Z + 0.7\text{ V}}{R_3C_1}t$$

The time t_1 is the time required for v_{OUT} to ramp down to the lower trigger point.

$$v_{OUT} = \frac{R_1}{R_2}(V_Z + 0.7\text{ V}) - \frac{V_Z + 0.7\text{ V}}{R_3C_1}t_1 = V_{LTP}$$

After substituting in our relationship for V_{LTP}, we may solve for t_1.

$$\frac{R_1}{R_2}(V_Z + 0.7\text{ V}) - \frac{V_Z + 0.7\text{ V}}{R_3C_1}t_1 = -\frac{R_1}{R_2}(V_Z + 0.7\text{ V})$$

Dividing both sides by $(V_Z + 0.7\text{ V})$, we can then easily solve for t_1.

$$\frac{R_1}{R_2} - \frac{t_1}{R_3C_1} = -\frac{R_1}{R_2}$$

$$-\frac{t_1}{R_3C_1} = -2\frac{R_1}{R_2}$$

$$t_1 = \frac{2R_1R_3C_1}{R_2}$$

By inspection of Fig. 15-46(c), we can see that t_1 is equal to t_2, and that their sum yields the period T. The reciprocal of the period produces the frequency.

$$T = t_1 + t_2 = 2t_1 = \frac{4R_1R_3C_1}{R_2}$$

$$\boxed{f_o = \frac{1}{T} = \frac{R_2}{4R_1R_3C_1}} \qquad (15\text{-}62)$$

where f_o is the frequency of oscillation of the op amp triangle generator.

EXAMPLE 15-27

Find the frequency of oscillation of the op amp triangle generator given in Fig. 15-46(a). Also determine the peak-to-peak voltage of the square wave, and the peak-to-peak voltage of the triangle wave.

SOLUTION From Eq. 15-62,

$$f_o = \frac{R_2}{4R_1R_3C_1} = \frac{820\text{ k}\Omega}{(4)(100\text{ k}\Omega)(10\text{ k}\Omega)(0.1\ \mu\text{F})} = 2.05\text{ kHz}$$

(Both the triangle wave and the square wave will have the same frequency.) The peak-to-peak voltage of the square-wave output is set by the zener diodes at 19.6 V peak to peak. The triangle wave will have a peak-to-peak output voltage that is equal in magnitude to the hysteresis of the comparator. Thus

$$v_{\text{OUT(P-P)}} = H = V_{\text{UTP}} - V_{\text{LTP}} = 2\frac{R_1}{R_2}(V_Z + 0.7\text{ V})$$

$$= 2\frac{100\text{ k}\Omega}{820\text{ k}\Omega}(9.1\text{ V} + 0.7\text{ V}) = 2.4\text{ V}$$

The values have been labeled on Fig. 15-46(b). ■

PROBLEMS

Drill, Derivations, and Definitions

Section 15-1

15-1. In your own words, explain the basic differences between amplifiers and oscillators.

15-2. What are the two basic categories of oscillator circuits? Explain their differences.

Section 15-2

15-3. Explain the Barkhausen criteria for sustained oscillations. Is it applicable to relaxation oscillators?

15-4. What are the four basic requirements on an audio-frequency range harmonic oscillator that is to provide a low-distortion sinusoidal output? Explain each.

Section 15-3

15-5. Given the phase-shift oscillator in Fig. 15-3, assume that the three capacitors (C) have been changed to 0.15 μF. Determine the new f_o.

15-6. Repeat Prob. 15-5 if the three capacitors are changed to 0.05 μF.

Section 15-4

15-7. Given the Wien bridge oscillator in Fig. 15-4(a), assume that C_1 and C_2 are both changed to 0.15 μF. Determine the new f_o.

15-8. Given the Wien bridge oscillator shown in Fig. 15-4(a), assume that C_1 is 0.15 μF and C_2 is 0.1 μF. Determine the f_o and the feedback factor β. Does the amplifier have enough gain to allow the circuit to oscillate? Explain.

15-9. Repeat Prob. 15-8 if R_4 is increased to 3.3 kΩ.

15-10. What is meant by the term "adaptive negative feedback"? In your own words, explain the function of diodes D_1 and D_2 in Fig. 15-5(a).

15-11. Explain the function of $AR_{1\text{-}A}$ in Fig. 15-5(b). Why is its input signal taken across the R_1-C_1 combination?

Section 15-5

15-12. How is the ohmic region of a JFET defined? What is $r_{DS(ON)}$?

15-13. What is a VVR? How does r_{DS} vary with V_{GS}? Explain how the drain-to-source voltage affects linearity.

15-14. Explain the difference between the r_O and the r_{DS} of a JFET.

15-15. An n-channel JFET is to be used as a VVR. Its $r_{DS(ON)}$ is not given, but we do know that its minimum $V_{GS(OFF)}$ is -5 V, and its minimum $|y_{fs0}|$ is 3500 μS. If V_{DS} is kept very small, determine r_{DS} if V_{GS} is 0, -0.5, -1, -1.5, . . . , -4.5, and -4.999 V. Construct a graph of r_{DS} versus V_{GS}. Use five-cycle semilog paper (refer to Fig. 15-8).

15-16. An n-channel JFET has an $r_{DS(ON)}$ of 1 kΩ and a $V_{GS(OFF)}$ of -3.5 V. Determine r_{DS} if V_{GS} is 0, -0.5, -1, . . . , -2.5, and -3.499 V. What is the theoretical r_{DS} if V_{GS} is -3.5 V?

Section 15-6

15-17. An amplitude-stabilized Wien bridge oscillator incorporates a VVR as shown in Fig. 15-9. In your own words, qualitatively explain the operation of the adaptive negative feedback circuit. Explain why the β in the circuit was made 1/31.

Section 15-7

15-18. Given the Wien bridge oscillator in Fig. 15-11(a), assume that C_1 and C_2 are both 0.015-μF units. The ganged potentiometer (R_1-R_2) is adjusted from 500 Ω to 1 kΩ. Determine the minimum and maximum f_o. What is the β?

15-19. What is the primary disadvantage of the circuit given in Fig. 15-11(a)? What will happen if the potentiometer is adjusted fully clockwise?

15-20. Given the Wien bridge oscillator in Fig. 15-13, assume that R_1 is changed to 2 kΩ and R_2 is changed to a 5-kΩ potentiometer. Determine the minimum and maximum f_o and the corresponding β's of the frequency-determining feedback network.

15-21. Repeat Prob. 15-20 if R_1 is 1 kΩ and R_2 is a 10-kΩ potentiometer.

15-22. Assume that the output of AR_1 in Fig. 15-13 is 10 V peak. If R_{10} is changed to 1 kΩ, determine the minimum and maximum peak voltage at the output of AR_3.

Section 15-8

15-23. Explain why *RC* harmonic oscillators are not very practical at very high (e.g., 1 MHz) frequencies.

15-24. Why are *LC* resonant circuits impractical at audio frequencies? Explain.

Section 15-9

15-25. A resonant circuit uses a coil with 10 μH of inductance and 10 Ω of resistance in series with a 270-pF capacitor. Find its series resonant frequency (f_o), the Q of the coil, the impedance at resonance (Z_o), and the bandwidth (BW).

15-26. Repeat Prob. 15-25 if L is increased to 22 μH with all other values unchanged.

15-27. A 12-μH coil with a Q of 15 is in parallel with a 0.005-μF capacitor. Determine the parallel resonant frequency (f_o), the bandwidth (BW), and the impedance at resonance ($Z_o = R_p$).

15-28. Repeat Prob. 15-27 if the Q of the coil is increased to 150.

Section 15-11

15-29. Given that we have a Colpitts oscillator such as that shown in Fig. 15-17(b), determine f_o, the minimum required $A_{v(\mathrm{oc})}$, the actual $A_{v(\mathrm{oc})}$ when power is first applied, and verify the proper sizing of R_G and C_3. The following values are to be used: L_1 is 100 μH, L_2 is 10 mH, C_1 is 820 pF, C_2 is 270 pF, C_3 is 180 pF, C_4 is 0.01 μF, R_G is 150 kΩ, g_{fs0} is 2200 μS, r_O is 30 kΩ, and V_{DD} is 15 V.

15-30. Explain the function of L_2 and C_4 in Fig. 15-17(b).

15-31. In your own words, explain how gate leak bias works. Refer back to the circuit in Fig. 15-17(b). Would it be classified as a form of adaptive negative feedback?

Section 15-12

15-32. Given that we have a Clapp oscillator such as that shown in Fig. 15-20(b), determine f_o, the minimum required $A_{v(\mathrm{oc})}$, the actual $A_{v(\mathrm{oc})}$ when power is first applied, and verify the proper sizing of R_G and C_4. The following values are to be used: L_1 is

100 μH, L_2 is 1 mH, C_1 is 1200 pF, C_2 is 470 pF, C_3 is 330 pF, C_4 is 0.022 μF, C_5 is 0.01 μF, R_G is 150 kΩ, g_{fs0} is 2200 μS, r_O is 200 kΩ, and V_{DD} is 15 V.

Section 15-13

15-33. A single-layer air-core inductor is to have 80 μH of inductance. It is to be wound on a 2-W carbon resistor using 39 AWG enameled copper wire. Determine the required number of turns. How many feet of wire is required? What will the dc resistance of the coil be? Determine the tap point if N_2/N_1 is to be $\frac{1}{3}$.

15-34. Given that we have a Hartley oscillator such as that shown in Fig. 15-23(a), determine f_o, β, the minimum required $A_{v(oc)}$, the actual $A_{v(oc)}$ when power is first applied, and verify the proper sizing of R_G and C_2. The following values are to be used: L_1 is 120 μH (N_1 is 50 turns, and N_2 is 100 turns), L_2 is 10 mH, C_1 is 180 pF, C_2 is 0.05 μF, C_3 is 0.05 μF, C_4 is 0.01 μF, R_G is 150 kΩ, g_{fs0} is 1800 μS, r_O is 220 kΩ, and V_{DD} is 15 V.

15-35. Explain why capacitor C_3 is required in Fig. 15-23(a).

15-36. A common-emitter Hartley oscillator has been indicated in Fig. 15-25(a). None of the component values are changed, but V_{CC} is reduced to 12 V. Will the circuit still oscillate?

Section 15-14

15-37. What is the piezoelectric effect? Explain.

15-38. A quartz crystal has an L_s of 0.15 H, a C_s of 0.003 pF, and R_s of 10 Ω, and a C_m of 15 pF. Determine its f_s, its Q at series resonance, and its impedance at f_s. Also determine its f_p, the net Q of the R_s-L_S-C_s branch at f_p, and the crystal's impedance Z_o at f_p.

15-39. The crystal described in Prob. 15-38 is intended to be operated in its parallel resonant mode with a load capacitance of 18 pF across it. Find its operating frequency (f_{op}).

Section 15-15

15-40. Explain why the Pierce oscillator shown in Fig. 15-30(a) is an improvement over the circuit shown in Fig. 15-29(b). What is its primary disadvantage?

15-41. Capacitors C_1 and C_2 in Fig. 15-30(a) have been exchanged. Specifically, C_1 is 680 pF and C_2 is 250 pF. The circuit's oscillation frequency is to be 999.850 kHz (which is approximately 50 Hz above the crystal's series resonant frequency). Find Z_L, β′, β″, and the overall β of the feedback network. Recalling that the 2N5458 has a g_{fs0} of 1500 μS, and a y_{os} of 50 μS, determine if the circuit will oscillate.

15-42. What is the meaning of the term ''maximum drive level'' as it relates to crystals? Explain.

15-43. What is the general approach to limiting the maximum drive level applied to a crystal used in a Pierce oscillator?

15-44. Qualitatively explain how diode D_1 in Fig. 15-32(b) provides bias temperature compensation.

15-45. The collector supply voltage (V_{CC}) in Fig. 15-32(b) is to be increased to 12 V. Repeat the analyses conducted in Examples 15-17 through 15-22. Assume that f_o remains at 999.830 kHz.

Section 15-16

15-46. Draw the fundamental block diagram of a relaxation oscillator.

15-47. Make a sketch of the basic structure of a UJT. Label it completely.

15-48. In your own words, describe the fundamental operation of the UJT.

15-49. Given the UJT relaxation oscillator in Fig. 15-40(a), assume that the UJT has an η of 0.8, R_E is 100 kΩ, and C_1 is 0.1 μF. Determine f_o.

Section 15-17

15-50. An op amp square-wave generator was presented in Fig. 15-41(a). The following components are used in a similar circuit: R_1 is 22 kΩ, R_2 is 1 kΩ, R_3 is 220 kΩ, R_4 is 180 kΩ, C_1 is 0.1 μF, diodes D_1 and D_2 are 1N751A 5.1-V zeners, and a $\pm$15-V power supply is used. Use Eq. 15-58 to find f_o. Also determine the peak-to-peak voltage across C_1 and the peak-to-peak output voltage.

15-51. Use Eq.15-59 to find the f_o of the oscillator described in Prob. 15-50. Is it a good approximation? Explain.

15-52. Equation 15-59 is a simplification of Eq. 15-58. The simplification was made by setting

$$R_4 = 0.859R_3$$

in order to make

$$\ln \frac{1 + \beta}{1 - \beta} = 1$$

In step-by-step detail, show that this is true.

Section 15-18

15-53. Analyze the op amp triangle generator given in Fig. 15-46(a) to find f_o, the peak-to-peak output voltage, and the peak-to-peak square-wave voltage. The following components are used: R_1 is 150 kΩ, R_2 is 750 kΩ, R_3 is 20 kΩ, R_4 is 1 kΩ, C_1 is 0.05 μF, diodes D_1 and D_2 are 1N751A 5.1-V zeners, and a $\pm$15-V power supply is used.

Troubleshooting and Failure Modes

Oscillators are intriguing when they work and often frustrating when they fail. "Failure" in an oscillator can mean that f_o is too high or too low, the output waveform is distorted, or the circuit refuses to oscillate. In the case of harmonic oscillators, a frequency shift can be produced by a change in the frequency-determining feedback elements. For example, all capacitors can change value over time, but some are more prone to change than others. Ceramic capacitors can change value from 10 to 15% over the first year as their ceramic dielectric relaxes. Inductors can also exhibit a change in value because of core breakage, windings relaxing, or shorts between windings.

In the case of a "dead" oscillator, we must carefully interpret dc bias checks. Consider gate leak bias. If the circuit does not oscillate, the gate-to-ground voltage will not be negative, but zero. If an oscillator's output waveform is severely distorted, we can also observe shifts in the dc biases.

We must also be wary when probing an oscillator with a scope probe. The probe may provide enough capacitance to detune a perfectly good oscillator. Conversely, the scope probe may add enough capacitance to allow a previously "dead" oscillator to generate a signal.

15-54. The Colpitts oscillator in Fig. 15-17(b) is producing 1.1 MHz instead of 1 MHz. Which of the following alternatives is the most likely: (a) capacitor C_3 is open, (b) capacitor C_3 is shorted, (c) the coil L_1 has shorted turns, and/or (d) the JFET has a drain-to-source short. Explain your selection(s) and why the other alternatives are not likely causes.

15-55. The Colpitts oscillator in Fig. 15-17(b) is "dead." The dc voltage at the JFET's drain is 15 V, and the voltage at its gate is also 15 V. Which of the following alternatives are the most likely: (a) capacitor C_3 is shorted, (b) the JFET has a drain-to-gate short, (c) the JFET is open between its gate and source, and/or (d) capacitor C_4 is shorted? Explain your selection(s) and why the other alternatives are not likely causes.

Design

15-56. The phase-shift oscillator in Fig. 15-3 is to oscillate at 100 Hz. Find the required value of C.

15-57. The peak value of the output of the Wien bridge oscillator given in Fig. 15-9 is to be 4 V. Find the required zener diode breakdown voltage for D_1.

15-58. The output frequency of the Wien bridge oscillator in Fig. 15-13(a) is to be adjustable from approximately 100 Hz to 1 kHz. Find the required values for R_1 and R_2.

15-59. The Colpitts oscillator in Fig. 15-17(b) is to oscillate at approximately 560 kHz. Find the required value of L_1. Does C_3 have to be changed? All of the other components are fixed.

15-60. Find the required value of R_1 if the square-wave generator in Fig. 15-41(a) is to have an f_o of approximately 2 kHz.

Computer

15-61. Write a BASIC program that will compute the r_{DS} of a JFET that is being used as a VVR. The program should prompt the user for $V_{GS(OFF)}$, g_{fs0}, V_{DS}, and V_{GS}. The program should use V_{DS} in its calculation. It should also permit a V_{DS} of zero.

15-62. Expand the program developed to satisfy the requirements specified in Prob. 15-61. Specifically, it should include the menu below:

```
          SELECT CHOICE NUMBER

1. RUN AGAIN WITH A NEW JFET
2. RUN AGAIN WITH NEW VDS AND VGS
3. RUN SAME FET WITH NEW VGS
4. END
```

STANDARD 5% TOLERANCE RESISTOR VALUES

The carbon-composition resistor values listed in Table A-1 are available in decade multiples from 0.01 Ω to 100 MΩ. However, the most commonly used values range from 0.1 Ω to 10 MΩ. For example, the table entry 1.2 is to be interpreted as meaning that 1.2-Ω, 12-Ω, 120-Ω, 1.2-kΩ, 12-kΩ, 120-kΩ, and 1.2-MΩ resistors are readily available in a 5% tolerance. The standard power ratings are $\frac{1}{10}$, $\frac{1}{8}$, $\frac{1}{4}$, $\frac{1}{2}$, 1, and 2 W.

TABLE A-1
Standard 5%-Tolerance Resistor Values

1.0	1.1	1.2	1.3	1.5	1.6	1.8	2.0	2.2	2.4
2.7	3.0	3.3	3.6	3.9	4.3	4.7	5.1	5.6	6.2
6.8	7.5	8.2	9.1	—	—	—	—	—	—

TYPICAL POTENTIOMETER VALUES

There are many instances when an adjustable resistor (potentiometer) is required in an electronic circuit. If an adjustment is required for initial circuit calibration, or is infrequent, the potentiometer should probably be located on the printed circuit board. This usually means a small square, or rectangular, trimpot will be used. If the potentiometer is to be routinely adjusted (such as a volume control) it obviously belongs on the front panel. Potentiometers are relatively expensive. They should not be included in a circuit design unless absolutely necessary.

There are two basic types of potentiometer resistive elements: wirewound and conductive plastic. Wirewound potentiometers are available with relatively high power ratings and offer good stability. However, their resolution is limited and they exhibit an inductive characteristic. The conductive plastic types offer infinite resolution and minimal inductance. However, in general, their power ratings and stability are not as good as wirewound units.

Potentiometers may be single-turn or multiple-turn units. Multiple-turn units require several shaft rotations to move the slider across the entire length of their resistive element. This promotes fine adjustment. Standard multiple-turn units include 4, 10, 12, 15, 22, and 25 turns.

In most electronic circuit applications, the required potentiometer power ratings are relatively low. Some of the standard power ratings are: 0.25, 0.5, 1.0, 1.25, and 2 W.

Some of the typical potentiometer resistance values have been given in Table B-1.

TABLE B-1
Nominal Potentiometer Resistance Values

10 Ω	20 Ω	50 Ω	100 Ω	200 Ω	500 Ω
1 kΩ	2 kΩ	5 kΩ	10 kΩ	20 kΩ	50 kΩ
100 kΩ	200 kΩ	500 kΩ	1 MΩ	2 MΩ	5 MΩ
10 MΩ					

C

TYPICAL CAPACITOR VALUES

Next to fixed resistors, capacitors are the most widely used passive circuit components. When selecting or specifying capacitors the most necessary information items are (1) the capacitance, (2) the working voltage, (3) the dielectric type, (4) the capacitance tolerance, and (5) the physical dimensions and mounting. In practice, the cost and lead time are also important considerations.

Tables C-1 through C-3 specify the typical nominal capacitance values. They are not all-inclusive, but readily available.

TABLE C-1
Typical Capacitor Values (pF)[a]

5	10	12	15	18	20	22	24	27	30
33	36	39	43	47	50	51	56	62	68
75	82	100	110	120	150	180	200	220	240
250	270	300	330	360	390	400	470	510	560
600	620	680	750	820	910	1000	1200	1500	1800
2000	2200	2500	2700	3000	3300	3600	4700	5000	5100
5600	6200	6800	7500	8000	8200	9100			

[a] 1000 pF = 0.001 μF.

TABLE C-2
Typical Capacitor Values (μF)

0.01	0.012	0.015	0.018	0.020	0.022	0.033
0.047	0.068	0.1	0.12	0.15	0.18	0.2
0.22	0.25	0.27	0.3	0.33	0.39	0.47
0.5	0.56	0.68	0.82	1.0	1.2	1.5
1.8	2.0	2.2	2.7	3.0	3.3	3.9
4.0	4.7	5.0	5.6	6.0	6.8	8.0
8.2	10	15	18	20	22	25
27	30	33	39	47	50	56
75	80	82				

TABLE C-3
Typical Capacitor Values (μF)[b]

100	150	180	200	240	250	270	300
330	400	470	500	560	680	1000	1200
1300	1500	2200	3300	5100	5600	8200	9600
10,000	12,000	15,000	18,000	20,000	22,000		

[b] These large values usually require the use of an aluminum or tantalum electrolytic capacitor.

Table C-4 describes some of the most commonly used capacitor types. It is not all-inclusive and is intended to list representative values.

TABLE C-4
Typical Characteristics of Commonly Used Capacitors

Type	Capacitance Range	Tolerance (%)	Max. WVDC	Max. Temp. (°C)	Insulation Resistance
Ceramic					
Low *k*	5 pF–0.001 μF	±5–±20	6 kV	125	1000 MΩ
High *k*	100 pF–2.2 μF	+100, −20	100 V	85	100 MΩ
Electrolytic					
Aluminum	1 μF–1 F	+100, −20	700 V	85	<1 MΩ
Tantalum	0.001 μF–1 nF	±5–±20	100 V	125	>1 MΩ
Mica	1 pF–0.1 μF	±0.25–±5	50 kΩ	150	>100 GΩ
Paper	500 pF–50 μF	±10–±20	0.1 MV	125	100 MΩ
Polycarbonate	0.001–5 μF	±1	600 V	140	10 GΩ
Polyester	0.001–15 μF	±10	1 kV	125	10 GΩ
Polystyrene	100 pF–10 μF	±0.5	1 kV	85	10 GΩ
Silvered mica	5 pF–0.1 μF	±1–±20	75 kV	125	1000 MΩ

H-PARAMETER RELATIONSHIPS

While the hybrid-pi model provides an extremely useful description of the bipolar and field-effect transistors (and even the op amp), the *h*-parameter model is preferred by virtually all manufacturers to describe the low-frequency small-signal BJTs. Consequently, it is important that we understand the *h*-parameter model and its relationship to the hybrid-pi model.

The common-emitter BJT and its *h*-parameter model are shown in Fig. D-1(a) and (b), respectively. (Note that the lowercase subscripts mean that the voltages and currents are rms values.)

By applying Kirchhoff's voltage law at the input of the *h*-parameter model and Kirchhoff's current law at its output, we obtain

$$V_{be} = h_{ie}I_b + h_{re}V_{ce} \tag{D-1}$$

$$I_c = h_{fe}I_b + h_{oe}V_{ce} \tag{D-2}$$

FIGURE D-1 **H-parameter model.**

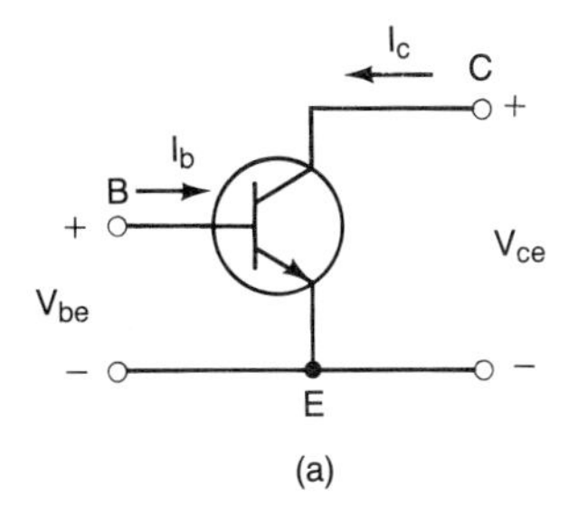

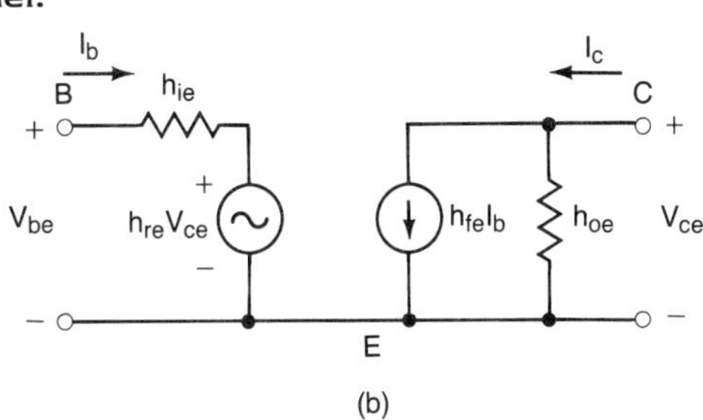

Equations D-1 and D-2 may be used to define the four h-parameters. If we place a short across the output ($V_{ce} = 0$), then Eq. D-1 reduces to

$$V_{be} = h_{ie}I_b \quad \text{and} \quad h_{ie} = \left.\frac{V_{be}}{I_b}\right|_{V_{ce}=0} \tag{D-3}$$

Consequently, h_{ie} (the generic notation is h_{11}) is called the "short-circuit input impedance of the common-emitter BJT."

Similarly, if we open the input ($I_b = 0$), we find

$$V_{be} = h_{re}V_{ce} \quad \text{and} \quad h_{re} = \left.\frac{V_{be}}{V_{ce}}\right|_{I_b=0} \tag{D-4}$$

The parameter h_{re} (generic h_{12}) is described as the "open-circuit reverse voltage gain for the common-emitter BJT."

By imposing the same input and output constraints, we can also define h_{fe} and h_{oe}. Hence, from Eq. D-2,

$$I_c = h_{fe}I_b \quad \text{and} \quad h_{fe} = \left.\frac{I_c}{I_b}\right|_{V_{ce}=0} \tag{D-5}$$

$$I_o = h_{oe}V_{ce} \quad \text{and} \quad h_{oe} = \left.\frac{I_o}{V_{ce}}\right|_{I_b=0} \tag{D-6}$$

The parameter h_{fe} (generic h_{21}) is the "short-circuit forward current gain for the common-emitter BJT," while h_{oe} is the "open-circuit output admittance for the common-emitter BJT."

In practice, the h_{re} parameter is negligibly small (e.g., 1×10^{-4}). Therefore, the h-parameter model is often simplified by assuming an h_{re} of zero [see Fig. D-2(a)].

In Fig. D-2(b) we see the complete, low-frequency hybrid-pi BJT model. The resistance labeled r'_{bb} is called the "base-spreading resistance." It is explained in Section 9-11. By comparing Fig. D-2(a) with Fig. D-2(b), we see that

$$r'_{bb} + r_\pi = h_{ie} \tag{D-7}$$

Typically, $r'_{bb} << r_\pi$ and

$$r_\pi \simeq h_{ie}$$

[see Fig. D-2(c)].

Since [from Fig. D-2(a)]

$$I_b = \frac{V_{be}}{h_{ie}} \quad \text{then} \quad h_{fe}I_b = \frac{h_{fe}}{h_{ie}}V_{be}$$

and by comparison between Fig. D-2(a) and (c),

$$\frac{h_{fe}}{h_{ie}}V_{be} = g_mV_{be}$$

The conclusion to be drawn is

$$g_m = \frac{h_{fe}}{h_{ie}} \tag{D-8}$$

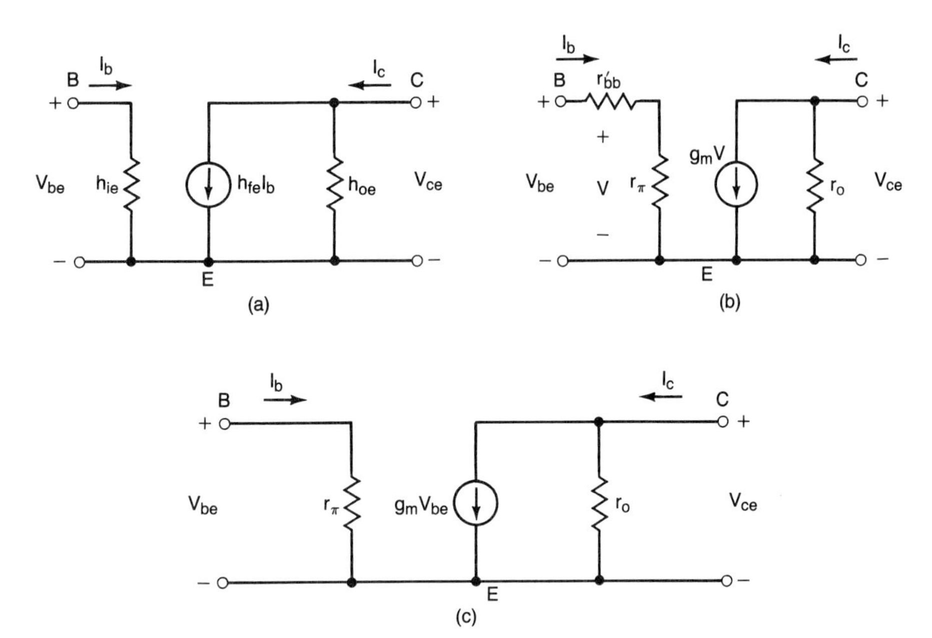

FIGURE D-2 Comparing the h-parameter and hybrid-pi models: (a) simplified h-parameter model; (b) complete low-frequency hybrid-pi model; (c) simplified hybrid-pi model.

Since h_{oe} is an admittance, we can relate its reciprocal to r_o in the hybrid-pi model. Hence

$$r_o = \frac{1}{h_{oe}} \tag{D-9}$$

These conversions have been summarized in Table D-1.

TABLE D-1
Hybrid-Pi to *h*-Parameter Conversions

Hybrid-Pi	*h*-Parameter Equivalent
r_π	h_{ie}
g_m	h_{fe}/h_{ie}
r_o	$1/h_{oe}$

As an example, consider the common-emitter open-circuit voltage gain $A_{v(\text{oc})}$. In Section 8-4 we see that

$$A_{v(\text{oc})} = \frac{V_{ce}}{V_{be}} = -g_m R_C$$

If we use the h-parameter model to derive $A_{v(oc)}$, we find that

$$A_{v(oc)} = \frac{V_{ce}}{V_{be}} = -\frac{h_{fe}}{h_{ie}} R_C$$

This result can be verified by merely drawing on Table D-1.

Even though the h-parameters provided by a manufacturer are *typical* values, they will offer improved accuracy over our normal approximations. Consider the following example.

EXAMPLE D-1

A 2N4124 *npn* BJT is to be operated with an I_C of 5 mA. Determine g_m, r_π, and r_o by using the approximations developed in Chapter 4. Repeat the problem by drawing upon the manufacturer's h-parameters.

SOLUTION From the manufacturer's data we see that for a 2N4124 at $I_C = 5$ mA:

$$h_{ie} = 900\ \Omega \qquad h_{re} = 1.5 \times 10^{-4}$$

$$h_{fe} = 150 \qquad h_{oe} = 30\ \mu\text{S}$$

Now, by drawing on our standard approximations, we find that

$$g_m = \frac{I_C}{26\text{ mV}} = \frac{5\text{ mA}}{26\text{ mV}} = 192\text{ mS}$$

$$r_\pi = \frac{h_{fe}}{g_m} = \frac{150}{192\text{ mS}} = 780\ \Omega$$

$$r_o = \frac{200\text{ V}}{I_C} = \frac{200\text{ V}}{5\text{ mA}} = 40\text{ k}\Omega$$

Repeating the analysis using the h-parameters yields

$$g_m = \frac{h_{fe}}{h_{ie}} = \frac{150}{900\ \Omega} = 167\text{ mS}$$

$$r_\pi = h_{ie} = 900\ \Omega$$

$$r_o = \frac{1}{h_{oe}} = \frac{1}{30\ \mu\text{S}} = 33\text{ k}\Omega$$

■

Our estimation of g_m at 192 mS is 15% higher than the h-parameter computation. Our approximation of r_π at 780 Ω is 13% low. Our approximation of the output resistance r_o at 40 kΩ is 21% high. All things considered - including the fact that h-parameters are not always provided - it appears that our approximations are reasonable.

APPENDIX E

ANSWERS TO SELECTED ODD-NUMBERED PROBLEMS

Chapter 1

3. Figure 3-30 shows a sine wave riding on a dc level; it is a continuous analog waveform **5.** Figure 3-42(d) shows a discontinuous analog waveform **11.** 1.79×10^{-19} J **17.** 5, n^- **19.** 3, p^-

Chapter 2

1. p^-, n^- **5.** lose **7.** narrows **9.** gain **11.** widens **15.** $I_R = I_S + I_{SL}$ **17.** For R = 50 Ω, G = 20 mS; for R = 100 Ω, G = 10 mS; for R = 200 Ω, G = 5 mS; the slopes are in siemens **19.** 30°C, 1.41 nA; 40°C, 2.83 nA; 50°C, 5.66 nA; 60°C, 11.3 nA **21.** 1.2×10^{-5} cm **23.** below, negative **25.** Zener **31.** c, shorted diode most likely; e, an open resistor R_1 is the next most likely cause

Chapter 3

1. For positive V_D, I_D = 0, 4.58 μA, 0.219 mA, 10.3 mA 15.1 mA, 22.1 mA; for negative V_D, I_D = −0.0979 μA, −0.1 μA, −0.1 μA **3.** $I_S(100°C)$ = 18.1 μA, $V_T(100°C)$ = 32.2 mV; for positive V_D, I_D = 0, 388 μA, 9.08 mA, 204 mA, 278 mA, 380 mA; for negative V_D, I_D = −17.3 μA, −18.1 μA, −18.1 μA **7.** I_{SH} = 12 mA, V_{OC} = 12 V, I_{RD} = 4 mA, V_{RD} = 8 V **9.** V_{OC} = 3 V, I_{SH} = 20 mA, I_D = 15.5 mA, V_D = 0.67 V **11.** V_{OC} = 6 V, I_{SH} = 11.8 mA, I_D = 10.5 mA, V_D = 0.65 V **13.** V_{OC} = −5 V, I_{SH} = −500 nA, V_D = −4.6 V, I_D = −40 nA **15.** V_{OC} = −3 V, I_{SH} = −3 mA, $V_Z = -V_D$ = 3 V, $I_Z = -I_D$ = 0 **17.** V_D = 0 V, I_D = 20 mA; the ideal diode model assumes a V_D of 0 V which results in larger values of I_D; the load line method yields the more accurate results **19.** $V_D = V_K$ = 0.7 V; V_O = 0.72 V and r_O = 0.75 Ω **21.** I_Z = 9.9 mA, V_Z = 5.1 V; same results as a dc load line since an idealized V-I characteristic curve was used **23.** I_D = 4 mA **25.** R_{TH} = 750 Ω, V_{TH} = 1 V, I_D = 0, V_L = 1 V **27.** R_{TH} = 6.3 kΩ, V_{TH} = 9 V, I_D = 1.32 mA **29.** I_D = 22.5 mA **31.** V_{TH} = 6.62 V, V_L = 5.1 V **35.** $v_S(t) = -7.5 + 0.05 \sin 20{,}000\pi t$ **39.** r_j = 520 Ω **41.** r_j = 2.60 Ω; same results because I_D is much larger than I_S **43.** r_j = 5.20 Ω, r_{ac} = 6.20 Ω **45.** I_D = 0.808 mA, r_j = 32.2 Ω, r_{ac} = 42.2 Ω, $v_{out}(t)$ = 10.3 mV peak **49.** f_{max} = 10 MHz **53.** 0 V, 50 pF; 0.1 V, 46.8 pF; 1.0 V, 32.1 pF; 10.0 V, 12.8 pF; 20.0 V, 9.2 pF **55.** 13.4 V p-p **57.** V_{DC} = 9.3 V **59.** c, the LED is open **61.** d, resistor R_3 is open; R_{TH} = 4.4 kΩ, V_{TH} = 3 V, I_D = 0.523 mA, V_D = 0.7 V, V_A = 1.75 V **63.** 680 Ω

Chapter 4

1. base, collector, emitter **3.** positive **5.** Holes **7.** collector-base **9.** I_E = 2.04 mA; I_C = 4.0 mA and I_E = 4.08 mA **13.** β_{DC} = 250, α_{DC} = 0.996 **15.** 0.995 **17.** 2.92505 mA **19.** 15.1 μA **21.** 5.00015 mA **25.** positive **29.** $V_{BE}(40°C)$ = 0.617 V; $V_{BE}(10°C)$ = 0.683 V **31.** 2.17 nA **33.** I_E = 0.942 mA, $I_{C(SAT)}$ = −2.35 mA, $V_{CB(OFF)}$ = −12 V, I_C = −0.94 mA, V_{CB} = −7.2 V **35.** I_E = 0.942 mA ≃ I_C, V_{CB} = −7.20 V, $V_{CB(OFF)}$ = −12 V, $I_{C(SAT)}$ = 2.35 mA **37.** r_e' = 23 Ω, I_m = 435 μA peak **39.** g_m = 38.5 mS, r_o = 200 kΩ, r_π = 2.6 kΩ **43.** R_E = 22 kΩ, R_C = 12 kΩ **45.** normally V_{EB} = 0.7 V, if emitter fails open V_E = 12 V **47.** saturated

Chapter 5

1. electrons **7.** 5 V **9.** −0.5 V **13.** Graph the data points: 0 V, 20 mA; −1 V, 15.31 mA; −2 V, 11.25 mA; −3 V, 7.81 mA; −4 V, 5.00 mA; −5 V, 2.81 mA; −6 V, 1.25 mA; −7 V, 0.31 mA; −8 V, 0.00 mA **17.** −8.165 V **19.** −0.87 V **21.** 10.2 μA **23.** increased **29.** Graph the data points: −5 V, 0.00 mA; −4 V, 0.12 mA; −3 V, 0.46 mA; −2 V, 1.04 mA; −1 V, 1.86 mA; 0 V, 2.90 mA; 1 V, 4.18 mA; 2 V, 5.68 mA; 3 V, 7.42 mA; 4 V, 9.40 mA **31.** K = 0.138 mA/V^2; the V_{GS} required to produce $I_{D(MAX)}$ is 16.2 V **33.** V_{GS} = −2.5 V, I_D = 7.56 mA, V_{DS} = 10.9 V; $I_{D(SAT)}$ = 16.7 mA; $V_{DS(OFF)}$ = 20 V **35.** V_{GS} = −0.5 V, I_D = 14.1 mA, V_{DS} = 8.438 V, and A_v = −3.28 **37.** 3771 μS; 1333 μS **39.** 1826 μS; 3651 μS **41.** r_o = 20 kΩ, V_A = 400 V; r_o = 80 kΩ **45.** For V_{DD} = 5 V and $v_{in}(t)$ = 5 V, $v_{gs1}(t)$ = 0 V, $V_{gs2}(t)$ = 5 V and $v_{out}(t)$ = 0 V; if $v_{in}(t)$ = 0 V, $v_{gs1}(t)$ = −5 V and $v_{gs2}(t)$ = 0 V then $v_{out}(t)$ = 5 V **47.** $-V_{GG}$ = −1.68 V, R_D = 1 kΩ **49.** 0.7 V; the gate-to-source p-n junction will be forward biased

Chapter 6

3. 224 **5.** Minimum characteristic: 0 V, 3 mA; −0.2 V, 2.43 mA; −0.4 V, 1.92 mA; −0.6 V, 1.47 mA; −0.8 V, 1.08 mA; −1 V, 0.75 mA; −1.2 V, 0.48 mA; −1.4 V, 0.27 mA; −1.6 V, 0.12 mA; −1.8 V, 0.03 mA; −2 V, 0 mA **9.** 24 V **13.** 9.6 to 13.6 V **17.** h_{FE} = 50: I_B = 14.3 μA, I_C = 0.715 mA, V_{CE} = 9.64 V, $I_{C(SAT)}$ = 2 mA, $V_{CE(OFF)}$ = 15 V; h_{FE} = 150: I_B = 14.3 μA, I_C = 2.14 mA (> $I_{C(SAT)}$), V_{CE} = 0 V (*not* −1.05 V); yes, the transistors with large values of h_{FE} will be saturated **19.** $I_{D(SAT)}$ = 12 mA, $V_{DS(OFF)}$ = 12 V, $I_{D(min)}$ = 1 mA, $V_{DS(max)}$ = 11 V; $I_{D(max)}$ = 12.25 mA >$I_{D(SAT)}$ which means that some 2N5459 JFETs will not work in the circuit **21.** For h_{FE} = 50: I_B = 10.5 μA, I_C = 0.525 mA, V_{CE} = 8.27 V, V_C = 9.06 V, V_E = 0.787 V, V_B = 1.49 V, $I_{C(SAT)}$ = 1.69 mA, $V_{CE(OFF)}$ = $V_{C(OFF)}$ = 12 V; for h_{FE} = 150: I_B = 9.21 μA, I_C = 1.38 mA, V_{CE} = 2.20 V, V_C = 4.27 V, V_E = 2.07 V, V_B = 2.77 V **25.** For $V_{GS(OFF)}$ = −2 V and I_{DSS} = 4 mA: I_D = 0.6 mA, V_S = 1.22 V, V_G = 0 V, V_{GS} = −1.22 V, V_{DS} = 8.40 V, V_D = 9.62 V; for $V_{GS(OFF)}$ = −8 V and I_{DSS} = 16 mA: I_D = 2.4 mA (> $I_{D(SAT)}$ = 2.03 mA) and "maximum" 2N5459 JFETs will be saturated in this circuit **27.** For h_{FE} = 50: I_B = 10.6 μA, I_C = 0.532 mA, V_{CE} = V_C = 8.01 V, $I_{C(SAT)}$ = 1.60 mA, $V_{CE(OFF)}$ = 12 V; for h_{FE} = 150: I_B = 6.23 μA, I_C = 0.935 mA, V_{CE} = V_C = 4.99 V, $I_{C(SAT)}$ = 1.60 mA, $V_{C(OFF)}$ = 12 V **29.** V_{DS} = V_{GS} = 4.5 V; bias line: V_{GS} = V_{DD} − I_DR_D **31.** For h_{FE} = 50: I_B = 18.8 μA, I_C − 0.938 mA, V_E = 1.41 V, V_B = 22.11 V, V_{CE} = 7.21 V, V_C = 8.62 V, $I_{C(SAT)}$ = 1.81 mA, $V_{CE(OFF)}$ = $V_{C(OFF)}$ = 15 V; for h_{FE} = 150: I_B = 6.44 μA, I_C = 0.965 mA, V_E = 1.45 V, V_B = 2.15 V, V_{CE} = 6.99 V, V_C = 8.43 V, $I_{C(SAT)}$ = 1.81 mA, $V_{CE(OFF)}$ = $V_{C(OFF)}$ = 15 V **33.** For $V_{GS(OFF)}$ = −2 V and I_{DSS} = 4 mA: I_D = 0.7 mA, V_G = 3.93 V, V_S = 5.06 V, V_{DS} = 7.26 V, V_D = 12.3 V, $I_{D(SAT)}$ = 1.44 mA, $V_{D(OFF)}$ = 15 V; for $V_{GS(OFF)}$ = −8 V and I_{DSS} = 16 mA: I_D = 1.4 mA, V_G = 3.93 V, V_S = 9.56 V, V_{DS} = 0.386 V, $I_{D(SAT)}$ = 1.44 mA, $V_{D(OFF)}$ = 15 V **35.** For h_{FE} = 50: I_B = 18.6 μA, I_C = 0.931 mA, V_{CE} = 9.06 V, V_C = 8.02 V, V_B = −0.0614 V, V_E = −0.761 V, $I_{C(SAT)}$ = 1.33 mA, $V_{CE(OFF)}$ = 30 V, $V_{C(OFF)}$ = 15 V; for h_{FE} = 150: I_B = 6.30 μA, I_C = 0.946 mA, V_{CE} = 8.72 V, V_C = 7.91 V, V_B = −0.0208 V, V_E = −0.721 V, $I_{C(SAT)}$ = 1.33 mA, $V_{CE(OFF)}$ = 30 V, $V_{C(OFF)}$ = 15 V **37.** The current must be less than the $I_{DSS(min)}$ for a given FET type. **39.** V_B = 0 V, V_G = 0 V, V_E = −0.7 V, I_D = I_C ≃ I_E = 0.953 mA, $V_{S(min)}$ = 1.02 V, $V_{S(max)}$ = 6.05 V, V_D = 10.5 V; the effect on the dc bias is negligible since the base is normally very near ground potential **41.** I_T = 2.10 mA, I_{D1} = I_{S1} = 1.05 mA, V_{D1} = 7.11 V **43.** 1.17 mA **45.** ±4 V or 8 V p-p **47.** V_B = 0.7 V, V_E = 0 V, V_C ≃ 0 V **49.** V_B = 0 V, V_E = −0.7 V,

$V_C = 10.2$ V; the effect on the dc bias is negligible since the base terminal is normally very near ground potential

Chapter 7

1. 3 V peak **3.** 100 μA peak **7.** $A_{v(oc)} = 500$ **9.** $v_{in} = 4.975$ mV peak, $v_{out} = 1.24$ V peak, $A_{vs} = 249$ **11.** $v_{in} = 4.88$ mV peak, $v_{out} = 407$ mV peak, $A_{vs} = 81.3$ **17.** $A_{is} = 90.9$, $i_{out} = 9.09$ mA peak **19.** $A_v = 60.5$, $A_{vs} = 59.5$, $A_i = 72.6$, $A_{is} = 1.19$ **21.** 4392 **23.** 27 dB, 30 dB, 24 dB; doubling A_p increases the power gain by 3 dB; halving A_p decreases the power gain by 3 dB **25.** 46 dB, 43.5 dB, 44.8 dB **27.** −10.4 dB **29.** $A_v = 141$, $A_i = 14.1$ **33.** 8.24 dBm, 6.02 dBV **35.** 40 dB **37.** $A_{vT} = 31{,}063$, 89.8 dB **39.** −0.0435 dB, 43.5 dB, 47.0 dB, −0.677 dB; $|A_{vT}|_{dB} = 89.8$ dB **41.** 4,225; 72.5 dB **47.** 37 dB; no, $70.8 < 100$ **49.** $R_{out} \leq 16.7\ \Omega$

Chapter 8

3. $V_{TH} = 2.17$ V, $R_{TH} = 1{,}882\ \Omega$; for $\beta_{DC} = 50$: $I_B = 18.8\ \mu A$, $I_C = 0.938$ mA, $V_{CE} = 7.22$ V, $V_C = 8.62$ V, $V_E = 1.41$ V, $V_B = 2.11$ V; for $\beta_{DC} = 150$: $I_B = 6.44\ \mu A$, $I_C = 0.966$ mA, $V_{CE} = 6.99$ V, $V_C = 8.43$ V, $V_E = 1.45$ V, $V_B = 2.15$ V; in both cases $I_{C(SAT)} = 1.81$ mA and $V_{CE(OFF)} = 15$ V **7.** $I_C = 0.952$ mA, $r_o = 210$ kΩ, $g_m = 36.6$ mS, $r_\pi = 1366\ \Omega$ **9.** $r_{o(min)} = 277$ kΩ, $g_{m(max)} = 1781\ \mu S$, $g_{m(min)} = 863\ \mu S$, r_π is approximately infinite **11.** $R_{in} = 28.8$ kΩ, $A_{v(oc)} = -5.34$, $R_{out} = 3$ kΩ, $A_v = -4.11$ $A_{vs} \simeq -4.02$, $A_i = -11.8$, $A_{is} = -0.241$, $A_p = 48.5$ **17.** $R_{in(base)} = 75$ kΩ, $R_{in} = 1836\ \Omega$ (minimum), $R_{out} = 6.8$ kΩ, $A_{v(oc)} = -4.45$, $A_v = -2.65$, $A_{vs} = -2.00$, $A_i = -0.486$, $A_{is} = -0.120$, $A_p = 1.29$ **19.** Completely bypassed: $A_{v(oc)} = -2.60$, partially bypassed: $A_{v(oc)} = -1.69$; unbypassed: $A_{v(oc)} = -0.376$ **23.** $R_{in} = 20$ kΩ, $A_{v(oc)} = -7.5$, $R_{out} \simeq 0$, $A_v = -7.5$, $A_{vs} = -7.28$, $A_i = -15$, $A_{is} = -0.437$, $A_p = 112$ **25.** Because the amplifier is direct coupled, the output will be in positive saturation; the solution is to place a dc blocking capacitor in series with the signal source **27.** $R_{in(source)} = 561\ \Omega$, $R_{in} = 519\ \Omega$, $R_{out} = 3$ kΩ, $A_{v(oc)} = 5.34$, $A_v = 4.11$, $A_{vs} = 1.91$, $A_i = 0.213$, $A_{is} = 0.114$, $A_p = 0.875$ **31.** For $V_{IN} = -0.01$ V, $V_{OUT} = -1.01$ V; for $V_{IN} = 0.1$ V, $V_{OUT} = 10.1$ V **33.** $R_{in(base)} = 65.2$ kΩ, $R_{in} = 1.83$ kΩ, $A_{v(oc)} = 0.982$, $R_{TH} = 455\ \Omega$, $R_{out(emitter)} = 35.7\ \Omega$, $R_{out} = 34.9\ \Omega$, $A_v = 0.979$, $A_{vs} = 0.737$, $A_i = 0.179$, $A_{is} = 0.0442$, $A_p = 0.175$ **35.** $R_{in} = 28.8$ kΩ, $R_{out(source)} = 561\ \Omega$, $R_{out} = 519\ \Omega$, $A_{v(oc)} = 0.924$, $A_v = 0.878$, $A_{vs} = 0.860$, $A_i = 2.53$, $A_{is} = 0.0516$, $A_p = 2.22$ **37.** $I_T = 0.942$ mA, $I_{C1} = I_{C2} = 0.471$ mA, $g_m = 18.1$ mS, $r_\pi = 5.52$ kΩ, $A_{vd(oc)} = 217$, $R_{in} = 11.0$ kΩ, $R_{out} = 24$ kΩ **43.** $I_T = 1.09$ mA, $g_{m1} = g_{m2} = 20.9$ mS, $r_{\pi 1} = r_{\pi 2} = 4.77$ kΩ, $g_{m3} = 41.9$ mS, $r_{\pi 3} = 2.39$ kΩ, $r_{o3} = 183$ kΩ, $A_{vd(oc)} = 125$, $r_E = R_{out(col)3} = 13.0$ MΩ, $A_{v(cm)} = -4.61 \times 10^{-4}$, CMRR $= 271 \times 10^3$ or 109 dB **45.** $A_{vd(oc)} = 4.7$, $R_{out} \simeq 0\ \Omega$, $R_{in(+)} = 57$ kΩ, $R_{in(-)} = 10$ kΩ **49.** a, a short circuit exists between the drain and the source which produces simple voltage division between R_D and R_S; c, R_2 is open which forward biases the JFET's gate-to-source p-n junction which saturates the JFET **51.** $R_B = 1$ kΩ, $R_1 = 1$ kΩ, resistor "R_2" consists of a 91-kΩ resistor in series with a 20-kΩ potentiometer; an input coupling capacitor is required

Chapter 9

1. $V_B = 2$, V, $V_E = 1.3$ V, $I_E = 1$ mA $\simeq I_C$, $V_C = 8.2$ V, $V_{CE} = 6.9$ V **3.** $g_m = 38.5$ mS, $r_o = 200$ kΩ, $r_\pi = 2600\ \Omega$, $r_o = 200$ kΩ $>> R_C = 6800\ \Omega$; r_o may be ignored **5.** $R_{in} = 1040\ \Omega$, $A_{v(oc)} = -262$, $R_{out} = 6.8$ kΩ, $A_v = -156$, $A_{vs} = -98.9$ **7.** r_C is the ac equivalent resistance from the collector-to-ground e.g., $R_C \| R_L$; $r_C = 4050\ \Omega$, $A_v = -156$, same A_v as obtained in Prob. 9-5 **9.** $r_D = 1.2$ kΩ, $r_C = 4.05$ kΩ, $R_{in} = 1$ MΩ, $A_{v1} = -2.13$, $A_{v2} = -156$, $A_{v(oc)3} = 0.996$, $A_{vT(oc)} = 331$, $R_{out(emitter)3} = 52.6\ \Omega$, $R_{out} = 51.8\ \Omega$, $A_{vT} = 329$, $A_{vsT} \simeq 329$, $A_{iT} = 3.29 \times 10^3$, $A_{isT} = 19.7$, $A_{pT} = 1.08 \times 10^6$ **11.** $R_{in} = 150$ kΩ, $R_{out} \simeq 0\ \Omega$, $A_{vT(oc)} = A_{vT} = -20$, $A_{iT} = -300$, $A_{isT} = -1.20$, $A_{pT} = 6000$ **13.** $V_{pin1} = 0.60$ V, $V_{pin7} = -6.0$ V, $V_{pin8} = -12.0$ V **17.** The input differential pair and the current mirror it drives determines the op amp's transconductance. **21.** $f_H = 796$ kHz **23.** $f_L = 23.4$ Hz **25.** $R_{TH1} = 1.64$ kΩ, $f_{L1} = 2.06$ Hz, $R_{TH2} = 16.8$ kΩ, $f_{L2} = 0.202$ Hz, $R_{TH3} = 26\ \Omega$, $f_{L3} = 20.4$ Hz **27.** $V_{OUT} = 7.5$ V, $R_{in} = 75$ kΩ, $R_{out} \simeq 0\ \Omega$, $A_{vs} = 50.6$ (34.1 dB), $R_{TH1} = 75.6$ kΩ, $f_{L1} = 1.40$ Hz, $R_{TH2} = 10$ kΩ, $f_{L2} = 15.9$ Hz, $R_{TH3} = 2$ kΩ, $f_{L3} = 7.96$ Hz **29.** increases **31.** small **33.** $r'_{bb} = 400\ \Omega$, $A_v = -135$; lower in magnitude than the A_v of −156 found in Prob.

9-7 **37.** BW = 29.9 kHz; A_{vs} = 27 dB at 30 kHz **41.** b, capacitor C_6 is open (which prohibits an ac signal from appearing across R_L without affecting the dc bias); d, the load (R_L) is shorted to ground—possibly damaging Q_3. **43.** V_{pin3} = 7.5 V, V_{pin6} = 13 V, V_{pin2} = 0.255 V **45.** R_1 = 62 kΩ, R_2 = 10 kΩ, R_C = 3.9 kΩ, R_E = 910 Ω, C_1 = 75 μF, C_2 = 0.82 μF, C_3 = 680 μF

Chapter 10

1. transconductance, voltage; the input resistances should be as large as possible **3.** transresistance, voltage; the output resistances should be as small as possible **5.** output voltage sampling should be used for voltage and transresistance feedback amplifiers; output current sampling should be used with current and transconductance amplifiers **7.** voltage-series negatives feedback is best suited for a voltage amplifier; current-shunt negative feedback is best suited for a current amplifier **9.** $A_{v(cl)}$ = 100.59 minimum, $A_{v(cl)}$ = 100.95 maximum, $A_{v(cl)}$ = 101 ideally **11.** R_{inf} = 124 MΩ **13.** β = 1 for a voltage follower; the loop gain is large since 100% negative feedback is used [loop gain = $\beta A_{v(ol)}$] **15.** $A_{v(cl)}$ = 0.99998, R_{inf} = 25.0005 GΩ, $R_{in(total)}$ = 469.99 kΩ; our approximations are very close with $A_{v(cl)}$ = 1 and $R_{in(total)}$ = 470 kΩ **17.** R_{of} = 50.4 mΩ **19.** f_H = BW = 3.16 Hz **21.** $f_{H(cl)}$ = 1 MHz, f_H = 31.6 Hz **25.** % THD = 8.10% **27.** V_{THD} = 405 mV rms **29.** % $THD_{(cl)}$ = 0.004995% ≃ 0.005% **31.** A factor of two magnitude relationship between two frequencies; 30 kHz, 3.75 kHz **35.** v(t) = 12.732 sin 20,000 πt + 4.244 sin 60,000 πt + 2.546 sin 1 × 10^5 πt **47.** Gain margin = −22 dB, f_c ≃ 2.5 MHz, θ_c = −224.5°, θ_{pm} = −44.5°; no, the amplifier will oscillate **49.** Ideally, $A_{v(cl)}$ = 101 or 40.1 dB, f_c = 2.62 MHz, θ_c = −195.8°, θ_{pm} = −15.8°, the gain margin is −8.4 dB; the circuit will oscillate **51.** −20 dB/decade crossing: absolutely stable; −40 dB/decade crossing: marginally stable; −60 dB/decade and beyond: unstable **57.** 4 MHz **59.** f_p = 1.05 kHz, f_z = 19.9 kHz **61.** f_c = 7.43 MHz, θ_c = −227.6°, θ_{pm} = −47.6°; unstable **63.** f_z = 97.2 kHz ≃ f_{H1} = 100 kHz, f_p = 962 Hz, f_c = 759 kHz, θ_c = −135.8°, θ_{pm} = 44.2°; stable **65.** a, the positive supply connection is open since the positive supply pin has been pulled negative **67.** f_T = 2 MHz (minimum)

Chapter 11

1. For Fig. 11-1(a) $A_{v(cl)}$ = 952.4 and for Fig. 11-1(b) $A_{v(cl)}$ = 996.8; the greater the loop gain, the more stable the closed-loop voltage gain will be **5.** An active device (e.g., an op amp) in the feedback loop constitutes active negative feedback; the feedback amplifier must be faster than the "main" amplifier **7.** $A_{v(cl)1}$ = $A_{v(cl)2}$ = 0.999990, R_{inf1} = 500 GΩ, R'_{in} = 5 GΩ, R_{inT} = 4.95 GΩ **9.** $G_{m(cl)}$ = 4.99875 mS minimum, 4.999375 mS maximum, 5 mS ideal **11.** $G_{m(ol)}$ = 1.67 S, $G_{m(cl)}$ = 99.994 μS, 100 μS ideal **13.** I_E = 10 mA, I_{OUT} = 10 mA, V_1 = 1.70 V, I_1 = 100 μA, V_L = 0.3 V, V_{CE} = 13.7 V **15.** C_1 and C_2 are decoupling capacitors; the purpose of C_1, C_2, and R_1 is to minimize the possibility of parasitic oscillations; resistor R_1 will tend to lower the compliance **17.** For V_{IN} = 3.8 V: V_E = 3.8 V, I_{OUT} = 20 mA, I_B = 200 μA, V_1 = 3.08 V, V_L = 0.6 V, V_{CE} = −3.2 V; for V_{IN} = 12.76 V: V_E = 12.76 V, I_{OUT} = 4.00 mA, I_B = 40 μA, V_1 = 12.06 V, V_L = 0.120 V, V_{CE} = −12.6 V **19.** $R_{m(cl)}$ = −909.1 Ω minimum, −952.4 Ω maximum, −1000 Ω ideal **21.** $R_{m(ol)}$ = −20 MΩ, $R_{m(cl)}$ = −1,999.8 Ω, −2,000 Ω ideal **25.** −4.75 kΩ = −4.75 V/mA **27.** $A_{v(cl)}$ = −49.979, −50 ideally **29.** Voltage follower: $f_{H(cl)}$ = 1 MHz; unity-gain inverting amplifier: $f_{H(cl)}$ = 500 kHz **31.** −11, 20.8 dB **35.** R_2 = 33 kΩ, R_3 = 27 kΩ, R_4 = 10 kΩ potentiometer **37.** c; AR_2 has failed since a large positive input at its noninverting input produces negative saturation; AR_1 is responding to its large negative differential input

Chapter 12

3. V_B = 1.74 V, V_E = 1.04 V, I_E = 0.691 mA ≃ I_C, V_C = 6.33 V, V_{CE} = 5.29 V **5.** g_m = 26.6 mS, r_π = 3010 Ω, r_o = 289 kΩ **7.** $A_{v(oc)}$ = −218, A_v = −120, R_{in} = 1158 Ω, R_{out} = 8.2 kΩ **9.** 10.6 V p-p **11.** 6.23 V p-p **13.** V_B = 11 V, V_E = 10.3 V, I_E = 2.02 mA ≃ I_C, V_{CE} = 11.7 V **15.** g_m = 77.7 mS, r_π = 1545 Ω, r_o = 99 kΩ **17.** $R_{in(base)}$ = 100 kΩ, R_{in} = 2487 Ω, A_v = 0.985, R_{out} = 12.8 Ω **19.** 3.38 V p-p **29.** 6.283 V/μs, 100 W **47.** v_{out} = 4.05 V peak, $P_{in(dc)}$ = 1.93 W, $P_{out(ac)}$ = 512 mW, %η = 26.5% **49.** $r_{IN(base)}$ = 806 Ω, R_{eq} = 208 Ω, v_1 = 2.08 V peak, $P_{out(ac)}$ = 51.1 mW, $P_{in(dc)}$ = 611 mW, %η = 8.36% **53.** $P_{out(ac)}$ = 4 W, i_{out} = 1 A peak, $P_{in(dc)}$ = 7.64 W, %η = 52.4% **55.** Emitter resistor limit: $v_{OUT(peak)}$ = 13.3 V, bias diode limit: $v_{OUT(peak)}$ = 11.3 V, op amp saturation voltage limit: $v_{OUT(peak)}$ = 13 V, $r_{IN(base)}$ = 45 kΩ, R_{eq} = 1.59 kΩ, I_{SC} limit: $v_{OUT(peak)}$ = 15.9 V which cannot be reached; v_{OUT} = 22.6 V p-p **61.** R_3 CW: V_{BIAS} = 3.68 V, $\Delta V_{BIAS}/\Delta T$ = −11.6 mV/°C; R_3 CCW: V_{BIAS} = 2.14 V, $\Delta V_{BIAS}/\Delta T$ = −6.74 mV/°C **63.** 17 V **67.** I_{SC} = 2.77 A; to protect the output circuitry (e.g., Q_4 and Q_6) in the event of an output short circuit to ground **71.** 1.09 °C/W **73.** θ_{sa} = 4.31 °C/W **77.** R_6 fully CW: V_{GS1} = −2.12 V; R_6 fully CCW: V_{GS1} = −300 mV **79.** $A_{v(ol)}$ = 54.1, $A_{v(cl)}$ = 4.27 **81.** f_z = 60.3 kHz, f_p = 280 kHz **83.** max. pos. peak = 19.3 V, max. neg. peak = −19.5 V **85.** Op amp output = 0 V, 0.7 V at the anode of D_1, −0.7 V at the cathode of D_2, and 0 V at the op amp's inverting input terminal **87.** Diodes D_1 and D_2 are forward-biased and apparently good. The op amp has failed and its output is latched to the negative supply rail **91.** R_{12} = $R_{13} \geq 0.231$ Ω

Chapter 13

7. 807 mA, 31.0 Ω **11.** For (a) V_{DC} = 9 V and for (b) V_{DC} = 9.55 V **15.** For (a) V_{rms} = 15 V and for (b) V_{rms} = 10.6 V **17.** V_2 = 60 V rms, I_2 = 600 mA rms, I_1 = 300 mA rms **21.** V_m = 48.1 V, V_{dcm} = 47.4 V, V_{DC} = 15.1 V, I_{DC} = I_D = 151 mA, V_{RM} = 48.1 V **23.** V_{dcm} = 24.0 V, V_{DC} = 15.3 V, I_{DC} = 153 mA, I_{D1} = I_{D2} = 76.5 mA, V_{RM} = 47.4 V **25.** V_m = 48.1 V, V_{dcm} = 46.7 V, i_L = 467 mA peak = i_D, V_{RM} = 47.4 V **27.** V_m = 17.8 V, V_{dcm} = 17.1 V, ± V_{DC} = ± 10.9 V **31.** WVDC ≥ 20.5 V **33.** (a) WVDC ≥ 30 V; (b) $V_{dc(min)}$ = 24.5 V; (c) V_{DC} = 24.75 V; (d) % r = 0.583% **35.** V_{DC} = 7.69 V, $V_{dc(min)}$ = 7.17 V, $V_{r(p\text{-}p)}$ = 1.042 V p-p, $V_{r(rms)}$ = 0.301 V rms, % r = 3.91% **37.** V_m = 17.8 V, V_{dcm} = 16.4 V, $V_{r(p\text{-}p)}$ = 0.947 V p-p, V_{DC} = 15.9 V, $V_{dc(min)}$ = 15.5 V, $V_{r(rms)}$ = 0.273 V rms, % r = 1.72% **39.** V_m = 25.5 V, V_{dcm} = 24.8 V, $V_{r(p\text{-}p)}$ = 1.14 V p-p, $V_{dc(min)}$ = 23.7 V, $I_{(AV)}$ = 150 mA, t_c = 800 μs, $I_{(PK)}$ = 6.25 A, $I_{(PK)}/I_{(AV)}$ = 41.7 **41.** V_m = 25.5 V, V_{dcm} = 24.8 V, $V_{r(p\text{-}p)}$ = 2.08 V p-p, $V_{dc(min)}$ = 22.7 V, ±V_{DC} = ±23.8 V, $V_{r(rms)}$ = 601 mV rms, t_c = 1.11 ms, $I_{(PK)}$ = 7.54 A, $I_{(AV)}$ = 250 mA, % r = 2.53% **43.** $I_{(SURGE)}$ = 13.8 A **45.** $I_{r(rms)}$ = 1.66 A rms **47.** 118 mA rms **49.** V'_{DC} = 13 V, $V'_{r(rms)}$ = 128 mV rms, % r′ = 0.986% **51.** 100 mW, 1/4 W **53.** $R_1 \leq 6.8$ Ω, 1 W; $C_1 \geq 1950$ μF **57.** a, the ripple frequency will halve since the circuit becomes a half-wave rectifier, and b, the dc average voltage will also halve since $V_{DC} = V_{dcm}/\pi$ for half-wave rectifiers **59.** Diode D_2 will also be damaged during the positive half-cycle since D_2 is effectively connected across the secondary without any current limiting **61.** An ac voltage was impressed across the *electrolytic* capacitor **63.** 36 V rms produces a V_{DC} of 15.8 V, $I_{(AV)}$ = 788 ma, V_{RM} = 50.2 V **65.** 262 mA rms, 1/2-A fuse **67.** 1N5393 (200-V rating); I_o = 1.4-A rating

Chapter 14

1. % VR = 4.35%; % VR = 0.418% **3.** 150 mW to 2.25 W **7.** V_m = 17.8 V, V_{dcm} = 16.4 V, V_{DC} = 16.4 V, $V_{dc(min)} \simeq$ 16.4 V, I_1 = 22.7 mA, $I_{B(max)}$ = 0, I_Z = I_1 = 22.7 mA, P_z = 295 mW, V_L = 12.3 V, $C_{out} \simeq$ 510 μF **9.** V_m = 25.5 V, V_{dcm} = 24.1 V, $V_{r(p\text{-}p)}$ = 4.17 V p-p, V_{DC} = 22.0 V, $V_{dc(min)}$ = 19.9 V, $V_{L(max)}$ = 15.3 V, $V_{L(min)}$ = 5.1 V **11.** 8.45 W **13.** I_{SC} = 300 mA **23.** $V_{L(min)}$ = 5.1 V: $V_{REF(min)}$ = 2.43 V, $V_{TRIP(min)}$ = 4.64 V and $V_{REF(max)}$ = 4.86 V, $V_{TRIP(max)}$ = 9.27 V; for $V_{L(max)}$ = 15.3 V: $V_{REF(min)}$ = 6 V, $V_{TRIP(min)}$ = 11.4 V and $V_{REF(max)}$ = 12 V, $V_{TRIP(max)}$ = 22.9 V **29.** R_{SC} = 3.0 Ω **31.** K = 0.910, I_{SC} = 199.8 mA, $I_{L(max)}$ = 799.3 mA, P_C = 3.996 W **33.** $R_{R(dB)}$ = 66 dB, $V_{r(rms)in}$ = 577 mV rms, $V_{r(rms)out}$ = 0.289 mV rms, $V_{r(p\text{-}p)out}$ = 0.818 mV p-p **37.** T_j = 120°C < $T_{j(max)}$ = 125°C; the regulator does not require a heat sink **39.** $V_{L(min)}$ = 12 V, $V_{L(max)}$ = 49.2 V **41.** $V_{L(min)}$ = 1.25 V, $V_{L(max)}$ = 9.58 V **43.** ±$V_{L(min)}$ = ±5 V, ±$V_{L(max)}$ = ±14.26 V **45.** Diodes D_5 and D_6 are used to limit the output voltage to approximately zero when the current limit is reached. I_{CL} = 455 mA(minimum) **51.** R_1 = 6.8 kΩ **53.** v_s = 15.537 V and θ = 5.24° **55.** $i_{G(max)}$ = 48.8 mA **57.** 12.3 V **61.** R_1 fully CCW: −80.0°; R_1 fully CW: 0° **65.** The phototransistor in the optocoupler becomes a (faster) photodiode. **67.** I_2 = 13.7 mA, I_{LED} = 13.7 mA (approximately), V_{CE} = 2.2 V, I_1 = 18.7 mA, I_C = 5.00 mA **69.** Alternatives a, b, and c are possible. Each of these failures would make the circuit appear to be tripped: a shorted SCR acts like a conducting SCR; a shorted transistor will turn on the SCR; a shorted zener diode will turn on the transistor which, in turn, will turn on the SCR. If capacitor C_1 is shorted, the transistor cannot turn on and this will prevent the SCR from firing. **71.** For the LM340KC-12: $P_{D(max)} = 1.5$ W, $\theta_{ja(max)} = 60°C/W$, since $\theta_{ja(max)} = 60°C/W > \theta_{ja} = 35°C/W$, no heat sink is required; for the LM7812C: $P_{D(max)} = 1.5$ W, $\theta_{ja(max)} = 43.3°C/W$, since $\theta_{ja(max)} = 43.3°C/W < \theta_{ja} = 50°C/W$, a heat sink is required with a θ_{sa} of less than 39.3°C/W **73.** For the LM340KC-12: $P_{D(max)} = 2.67$ W, $\theta_{ja(max)} = 33.7°C/W$, since $\theta_{ja(max)} < \theta_{ja} = 35°C/W$, a heat sink is required with a θ_{sa} of less than 29.7°C/W; for the LM7812C: $P_{D(max)} = 2.67$ W, $\theta_{ja(max)} = 24.3°C/W$, since $\theta_{ja(max)} < \theta_{ja} = 50°C/W$ a heat sink is required with a θ_{sa} of less than 20.3°C/W **75.** $R_1 = 82\ \Omega$, $R_2 = 1\ k\Omega$, $R_3 = 6.8\ k\Omega$

Chapter 15

5. 131.3 Hz **7.** 707.4 Hz **9.** 866.3 Hz, $\beta = 1/3.5$, $A_{v(max)} = 3.87$; the circuit has enough (greater than unity) loop gain to oscillate **15.** 0 V, 286 Ω; −0.5 V, 317 Ω; −1 V, 357 Ω; −1.5 V, 408 Ω; −2 V, 476 Ω; −2.5 V, 571 Ω, −3 V, 714 Ω; −3.5 V, 952 Ω; −4 V, 1.43 kΩ; −4.5 V, 2.86 kΩ; −4.999 V, 1.43 MΩ **21.** R_2 CW: $f_o = 1.007$ kHz, $\beta = 1/12$; R_2 CCW: $f_o = 303.5$ Hz, $\beta = 0.344$ **25.** $f_o = 3.06$ MHz, $Q = 19.2$, BW = 159 kHz, $Z_o = R = 10\ \Omega$ **27.** $f_o = 649.7$ kHz, BW = 43.32 kHz, $R_s = 3.266\ \Omega$, $Z_o = 735\ \Omega$ **29.** $C_T = 203.1$ pF, $f_o = 1.117$ MHz, $A_{v(oc)} < -0.329$, start-up $A_{v(oc)} = -66$, $\tau = 27.0\ \mu s \geq 10T = 8.95\ \mu s$, $X_{L2} = 70.1\ k\Omega \geq 100X_{C1} = 17.4\ k\Omega$ **33.** N = 165 turns, 13.5 ft, 0.0112 Ω, tap at 41 turns **39.** $f_{op} = 7.503$ MHz **41.** $R_s = 125\ \Omega$, $Z_L = 438.4 - j234.1\ \Omega = 497\ \Omega$ at $-28.1°$, $Z_{out} = r_o = 20\ k\Omega$, $\beta' = 0.02432$ at $-27.4°$, $\beta'' = 2.399$ at $-151.9°$, $\beta \simeq -0.0583$, the required start-up voltage gain $A_{v(oc)} = -17.1$, the actual start-up $A_{v(oc)} = -30$; the circuit will oscillate **45.** $V_B = 3.40$ V, $V_E = 2.70$ V, $I_C \simeq I_E = 2.70$ mA; $g_m = 104$ mS, $r_o = 74.1\ k\Omega$, $r_\pi = 1.35\ k\Omega$; $Z_{in} = 978\ \Omega$, $Z_{out} \simeq r_o = 74.1\ k\Omega$; $A_{v(oc)} = -7706$; $A_v = -4048$; $X_{C1} = 159.2\ \Omega < (Z_{out} + R) = 76.8\ k\Omega$, $X_{C2} = 106.1\ \Omega < Z_{in} = 978\ \Omega$, $Z_L = 50.7 - j159\ \Omega = 167\ \Omega$ at $-72.3°$; $\beta' = 0.004012$ at $-72.1°$, $\beta'' = 0.2022$ at $-107.7°$, $\beta \simeq -0.000811$, $\beta A_{v(oc)} = 3.28 > 1$; the circuit will continue to oscillate if V_{CC} is increased to 12 V; $V_{out} = 6.58$ V rms, $V'_{out} = 399$ mV rms, $|I| = 0.705$ mA rms, $P_{(crystal)} = 248\ \mu W$ **49.** 62.1 Hz **51.** 227.3 Hz; yes, the ratio of R_4 to R_3 (0.818) should be 0.859 ideally. The further the ratio is from 0.859, the worse the approximation will be. **53.** $f_o = 1.250$ kHz, the square wave is 11.6 V p-p and the triangle wave is 2.32 V p-p **55.** Alternatives a and b would place the gate at 15 V; an open gate-to-source would place the gate at 0 V. If capacitor C_4 is shorted, the power supply would be short circuited. **57.** The zener diode should be a 3.3-V unit. **59.** 364 μH; no, since $\tau = R_G C_3 = 18\ \mu s \geq 10T = 17.9\ \mu s$

INDEX

A

B

C

D

E

F

G

H

I

J-K

L

M

N

O

P

Q

R

S

T

U

V

W

X-Y

Z